THE ART OF ELECTRONICS

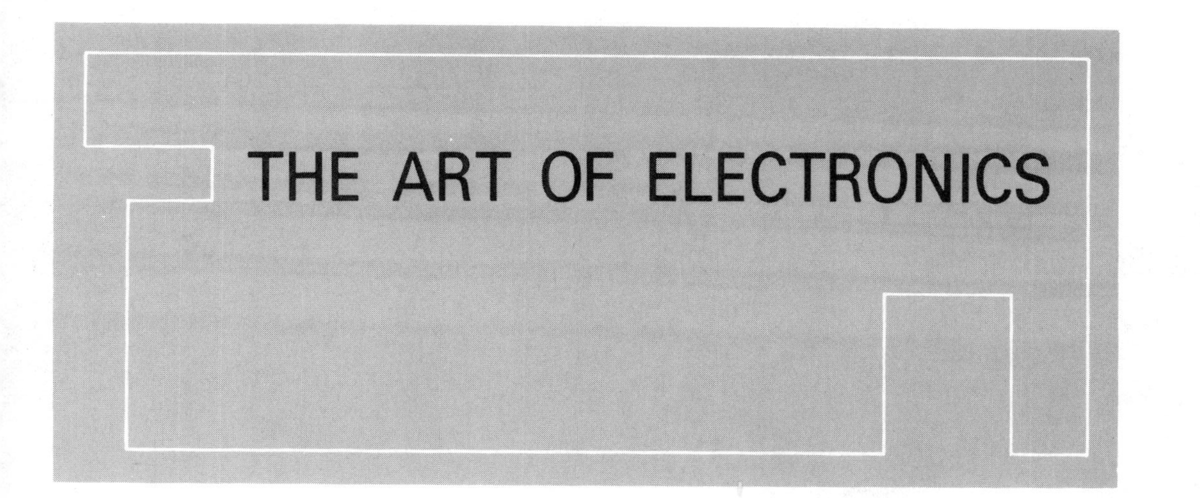

Paul Horowitz HARVARD UNIVERSITY

Winfield Hill SEA DATA CORPORATION, NEWTON, MASSACHUSETTS

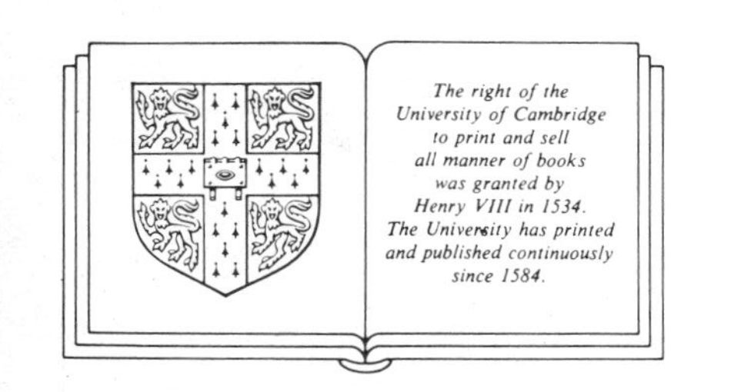

CAMBRIDGE UNIVERSITY PRESS
Cambridge
New York New Rochelle Melbourne Sydney

Published by the Press Syndicate of the University of Cambridge
The Pitt Building, Trumpington Street, Cambridge CB2 1RP
32 East 57th Street, New York, NY 10022, USA
10 Stamford Road, Oakleigh, Melbourne 3166, Australia

First published 1980
Reprinted 1981 (thrice), 1982, 1983 (twice), 1984 (twice), 1985, 1986, 1987 (twice), 1988

Printed in the United States of America

Library of Congress Cataloging in Publication Data

Horowitz, Paul, 1942–

The art of electronics.

1. Electronics. 2. Electronic circuit design.
I. Hill, Winfield, joint author. II. Title.
TK 7815.H67 1980 621.381 79-27170
ISBN 0 521 23151 5 hard covers
ISBN 0 521 29837 7 paperback

TO CAROL, JACOB, AND GINGER

CONTENTS

TABLES

PREFACE

This volume is intended as an electronic circuit design textbook and reference book; it begins at a level suitable for those with no previous exposure to electronics, and carries the reader through to a reasonable degree of proficiency in electronic circuit design. We have used a straightforward approach to the essential ideas of circuit design, coupled with an in-depth selection of topics. We have attempted to combine the pragmatic approach of the practicing physicist with the quantitative approach of the engineer, who wants a thoroughly evaluated circuit design.

This book evolved from a set of notes written to accompany a one-semester course in laboratory electronics at Harvard. That course has a varied enrollment – undergraduates picking up skills for their eventual work in science or industry, graduate students with a field of research clearly in mind, and advanced graduate students and postdoctoral researchers who suddenly find themselves hampered by their inability to "do electronics."

It soon became clear that existing textbooks were inadequate for such a course. Although there are excellent treatments of each electronics specialty, written for the planned sequence of a four-year engineering curriculum or for the practicing engineer, those books that attempt to address the whole field of electronics seem to suffer either from excessive detail (the handbook syndrome), oversimplification (the cookbook syndrome), or poor balance of material. Much of the favorite pedagogy of beginning textbooks is quite unnecessary and, in fact, is not used by practicing engineers, while useful circuitry and methods of analysis in daily use by circuit designers lies hidden in application notes, engineering journals, and hard-to-get data books. In other words, there is a tendency among textbook writers to represent the theory, rather than the art, of electronics.

We collaborated in writing this book with the specific intention of combining the discipline of a circuit design engineer with the perspective of a practicing experimental physicist and teacher of electronics. Thus, the treatment in this book reflects our philosophy that electronics, as currently practiced, is basically a simple art, a combination of some basic laws, rules of thumb, and a large bag of tricks. For these reasons we have omitted entirely the usual discussions of solid-state physics, the *h*-parameter model of transistors, and complicated network theory, and reduced to a bare minimum the mention of load lines and the *s*-plane. The treatment is largely nonmathematical, with strong encouragement of circuit brainstorming, and mental (or, at most, back-of-the-envelope) calculation of circuit values and performance.

In addition to the subjects usually treated in electronics books, we have included the following:

- an easy-to-use transistor model
- extensive discussion of useful subcircuits, such as current sources and current mirrors
- single-supply op-amp design
- easy-to-understand discussions of topics on which practical design information is often difficult to find: op-amp frequency compensation, low-noise circuits, phase-locked loops, and precision linear design.
- simplified design of active filters, with tables and graphs
- a section on noise, shielding, and grounding
- a unique graphical method for streamlined low-noise amplifier analysis
- a chapter on voltage references and regulators, including constant current supplies

- a discussion of monostable multivibrators and their idiosyncrasies
- a collection of digital logic pathology, and what to do about it
- an extensive discussion of interfacing to logic, with emphasis on the new NMOS and PMOS LSI
- a detailed discussion of A/D and D/A conversion techniques
- a section on digital noise generation
- a discussion of minicomputers and interfacing to data buses, with an introduction to assembly language
- a chapter on microprocessors, with actual design examples and discussion – how to design them into instruments, and how to make them do what you want
- a chapter on construction techniques: prototyping, printed circuit boards, instrument design
- a simplified way to evaluate high-speed switching circuits
- a chapter on scientific measurement and data processing: what you can measure and how accurately, and what to do with the data
- bandwidth narrowing methods made clear: signal averaging, multichannel scaling, lock-in amplifiers, and pulse height analysis
- amusing collections of "bad circuits," and collections of "circuit ideas"
- useful appendixes on how to draw schematic diagrams, IC generic types, LC filter design, resistor values, oscilloscopes, mathematics review, and others
- tables of diodes, transistors, FETs, opamps, comparators, regulators, voltage references, microprocessors, and other devices, generally listing the characteristics of both the most popular and the best types.

Throughout we have adopted a philosophy of naming names, often comparing the characteristics of competing devices for use in any circuit, and the advantages of alternate circuit configurations. Example circuits are drawn with real device types, not black boxes. The overall intent is to bring the reader to the point of understanding clearly the choices one makes in designing a circuit – how to choose circuit configurations, device types, and parts values. The use of largely nonmathematical circuit design techniques does not result in circuits that cut corners or compromise performance or reliability. On the contrary, such techniques enhance one's understanding of the real choices and compromises faced in engineering a circuit, and represent the best approach to good circuit design.

This book can be used for a full-year electronic circuit design course at the college level, with only a minimum mathematical prerequisite, namely some acquaintance with trigonometric and exponential functions, and preferably a bit of differential calculus. (A short review of complex numbers and derivatives is included as an appendix.) If the less essential sections are omitted, it can serve as the text for a one-semester course (as it does at Harvard).

A separately available laboratory manual, *Laboratory Manual for the Art of Electronics* (Horowitz and Robinson, 1981) contains twenty-three lab exercises, together with reading and problem assignments keyed to the text.

To assist the reader in navigation we have designated with open boxes in the margin those sections within each chapter that we feel can be safely passed over in an abbreviated reading. For a one-semester course it would probably be wise to omit in addition the materials of Chapter 4 (first half), 7, 12, 13, and possibly 14, as explained in the introductory paragraphs of those chapters.

We would like to thank our colleagues for their thoughtful comments and assistance in the preparation of the manuscript, particularly Mike Aronson, Howard Berg, Dennis Crouse, Carol Davis, David Griesinger, John Hagen, Tom Hayes, Peter Horowitz, Bob Kline, Costas Papaliolios, Jay Sage, and Bill Vetterling. We are indebted to Eric Hieber and Jim Mobley, and to Rhona Johnson and Ken Werner of Cambridge University Press for their imaginative and highly professional work.

Paul Horowitz
Winfield Hill

April 1982

FOUNDATIONS

CHAPTER 1

INTRODUCTION

Developments in the field of electronics have constituted one of the great success stories of this century. Beginning with crude spark-gap transmitters and "cat's-whisker" detectors at the turn of the century, we have passed through a vacuum-tube era of considerable sophistication to a solid-state era in which the flood of stunning advances shows no signs of abating. Calculators, computers, and even talking machines with vocabularies of several hundred words are routinely manufactured on single chips of silicon as part of the technology of large-scale integration (LSI), and current developments in very large scale integration (VLSI) promise even more remarkable devices.

Perhaps as noteworthy is the pleasant trend toward increased performance per dollar. The cost of an electronic microcircuit routinely decreases to a fraction of its initial cost as the manufacturing process is perfected (see Fig. 11.18 for an example). In fact, it is often the case that the panel controls and cabinet hardware of an instrument cost more than the electronics inside.

On reading of these exciting new developments in electronics, you may get the impression that you should be able to construct powerful, elegant, yet inexpensive, little gadgets to do almost any conceivable task – all you need to know is how all these miracle devices work. If you've had that feeling, this book is for you. In it we have attempted to convey the excitement and know-how of the subject of electronics.

In this chapter we begin the study of the laws, rules of thumb, and tricks that constitute the art of electronics as we see it. It is necessary to begin at the beginning – with talk of voltage, current, power, and the components that make up electronic circuits. Since you can't touch, see, smell, or hear electricity, there will be a certain amount of abstraction (particularly in the first chapter), as well as some dependence on such visualizing instruments as oscilloscopes and voltmeters. In many ways the first chapter is also the most mathematical, in spite of our efforts to keep mathematics to a minimum in order to foster a good intuitive understanding of circuit design and behavior.

Once we have considered the foundations of electronics, we will quickly get into the "active" circuits (amplifiers, oscillators, logic circuits, etc.) that make electronics the exciting field it is. The reader with some background in electronics may wish to skip over this chapter, since it assumes no prior knowledge of electronics. Further generalizations at this time would be pointless, so let's just dive right in.

VOLTAGE, CURRENT, AND RESISTANCE

1.01 Voltage and current

There are two quantities that we like to keep track of in electronic circuits: voltage and current. These are usually changing with time; otherwise nothing interesting is happening.

Voltage (symbol: V, or sometimes E). The voltage between two points is the cost in energy (work done) required to move a unit of positive charge from the more negative point (lower potential) to the more positive point (higher potential). Equivalently, it is the energy released when a unit charge moves "downhill" from the higher potential to the lower. Voltage is also called *potential difference* or *electromotive force* (EMF). The unit of measure is the *volt,* with voltages usually expressed in volts (V), kilovolts (1kV = 10^3V), millivolts (1mV = 10^{-3}V), or microvolts (1μV = 10^{-6}V) (see the box on prefixes). A joule of work is needed to move a coulomb of charge through a potential difference of one volt. (The coulomb is the unit of electric charge, and it equals the charge of 6×10^{18} electrons, approximately.) For reasons that will become clear later, the opportunities to talk about nanovolts (1nV = 10^{-9}V) and megavolts (1MV = 10^6V) are rare.

Current (symbol: I). Current is the rate of flow of electric charge past a point. The unit of measure is the ampere, or amp, with currents usually expressed in amperes (A), milliamperes (1mA = 10^{-3}A), microamperes (1μA = 10^{-6}A), nanoamperes (1nA = 10^{-9}A), or occasionally picoamperes (1pA = 10^{-12}A). A current of one ampere equals a flow of one coulomb of charge per second. By convention, current in a circuit is considered to flow from a more positive point to a more negative point, even though the actual electron flow is in the opposite direction.

Important: Always refer to voltage *between* two points or *across* two points in a circuit. Always refer to current *through* a device or connection in a circuit.

To say something like "the voltage through a resistor . . ." is nonsense, or worse. However, we do frequently speak of the voltage *at a point* in a circuit. This is always understood to mean voltage between that point and "ground," a common point in the circuit that everyone seems to know about. Soon you will, too.

We *generate* voltages by doing work on charges in devices such as batteries (electrochemical), generators (magnetic forces), solar cells (photovoltaic conversion of the energy of photons), etc. We *get* currents by placing voltages across things.

At this point you may well wonder how to "see" voltages and currents. The single most useful electronic instrument is the oscilloscope, which allows you to look at voltages (or occasionally currents) in a circuit as a function of time. We will deal with oscilloscopes, and also voltmeters, when we discuss signals shortly; for a preview, see the oscilloscope appendix (Appendix A) and the multimeter box later in this chapter.

In real circuits we connect things together with wires, metallic conductors, each of which has the same voltage on it everywhere (with respect to ground, say). (In the domain of high frequencies or low impedances, that isn't strictly true, and we will have more to say about this later. For now, it's a good approximation.) We mention this now so that you will realize that an actual circuit doesn't have to look like its schematic diagram, since wires can be rearranged.

Here are some simple rules about voltage and current:

1. The sum of the currents into a point in a circuit equals the sum of the currents out (conservation of charge). This is sometimes called Kirchhoff's current law. Engineers like to refer to such a point as a *node*. From this, we get the following: For a series circuit (a bunch of two-terminal things all connected end-to-end) the current is the same everywhere.

2. Things hooked in parallel (Fig. 1.1) have the same voltage across them. Restated, the sum of the "voltage drops" from A to B

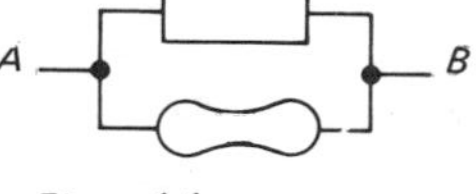

Figure 1.1

via one path through a circuit equals the sum by any other route equals the voltage between A and B. Sometimes this is stated as follows: The sum of the voltage drops around any closed circuit is zero. This is Kirchhoff's voltage law.

3. The power (work per unit time) consumed by a circuit device is

$$P = VI$$

This is simply (work/charge) × (charge/time). For V in volts and I in amps, P comes out in watts. Watts are joules per second (1W = 1J/s).

Power goes into heat (usually), or sometimes mechanical work (motors), radiated energy (lamps, transmitters), or stored energy (batteries, capacitors). Managing the heat load in a complicated system (e.g., a computer, in which many kilowatts of electrical energy are converted to heat, with the energetically insignificant by-product of a few pages of computational results) can be a crucial part of the system design.

Soon, when we deal with periodically varying voltages and currents, we will have to generalize the simple equation $P = VI$, but it's correct as a statement of instantaneous power just as it stands.

Incidentally, don't call current "amperage"; that's strictly bush-league. The same caution will apply to the term "ohmage" when we get to resistance in the next section.

1.02 Relationship between voltage and current: resistors

This is a long and interesting story. It is the heart of electronics. Crudely speaking, the name of the game is to make and use gadgets that have interesting and useful I versus V characteristics. Resistors (I simply proportional to V), capacitors (I proportional to rate of change of V), diodes (I only flows in one direction), thermistors (temperature-dependent resistor), photoresistors (light-dependent resistor), strain gauges (strain-dependent resistor), etc., are examples. We will gradually get into some of these exotic devices; for now, we will start with the most

PREFIXES

These prefixes are universally used to scale units in science and engineering:

Multiple	*Prefix*	*Symbol*
10^{12}	tera	T
10^{9}	giga	G
10^{6}	mega	M
10^{3}	kilo	k
10^{-3}	milli	m
10^{-6}	micro	μ
10^{-9}	nano	n
10^{-12}	pico	p
10^{-15}	femto	f

When abbreviating a unit with a prefix, the symbol for the unit follows the prefix without space. Be careful about upper-case and lower-case letters (especially m and M) in both prefix and unit: 1mW is a milliwatt, or one-thousandth of a watt; 1MHz is 1 million hertz. In general, units are spelled with lower-case letters, even when they are derived from proper names. The unit name is not capitalized when it is spelled out and used with a prefix, only when abbreviated. Thus: hertz and kilohertz, but Hz and kHz; watt, milliwatt, and megawatt, but W, mW, and MW.

Figure 1.2

mundane (and most widely used) circuit element, the resistor (Fig. 1.2).

Resistance and resistors

It is an interesting fact that the current through a metallic conductor (or other partially conducting material) is proportional to the voltage across it. (In the case of wire conductors used in circuits, we usually choose a thick enough gauge of wire so that these "voltage drops" will be negligible.) This is by no means a universal law for all objects. For instance, the current through a neon bulb is a highly nonlinear function of the applied voltage (it is zero up to a critical voltage, at which point it rises dramatically). The same goes for a variety of interesting special devices – diodes, transistors, light bulbs, etc. (If you are interested in understanding why metallic conductors behave this way, read sections 4.3–4.7 in the Berkeley Physics Course, Vol. II, see Bibliography). A resistor is made out of some conducting stuff (carbon, or a thin metal or carbon film, or wire of poor conductivity), with a wire coming out each end. It is characterized by its resistance:

$$R = \frac{V}{I}$$

R is in ohms for V in volts and I in amps. This is known as Ohm's law. Typical resistors of the most frequently used type (carbon composition) come in values from 1 ohm (1Ω) to about 22 megohms ($22\ M\Omega$). Resistors are also characterized by how much power they can safely dissipate (the most commonly used ones are rated at ¼ or ½ watt and by other parameters such as tolerance (accuracy), temperature coefficient, noise, voltage coefficient (the extent to which R depends on applied V), stability with time, inductance, etc. See the box on resistors and Appendixes C and D for further details.

Roughly speaking, resistors are used to convert a voltage to a current, and vice versa. This may sound awfully trite, but you will soon see what we mean.

Resistors in series and parallel

From the definition of R, some simple results follow:

1. The resistance of two resistors in series is $R = R_1 + R_2$ (Fig. 1.3). By putting resistors in series, you always get a *larger* resistor.

Figure 1.3

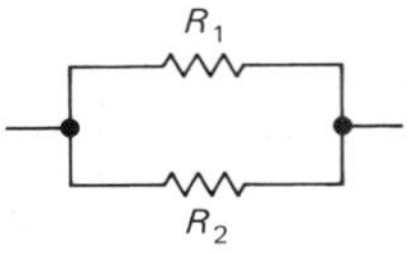

Figure 1.4

2. The resistance of two resistors in parallel (Fig. 1.4) is

$$R = \frac{R_1 R_2}{R_1 + R_2} \quad \text{or} \quad R = \frac{1}{\dfrac{1}{R_1} + \dfrac{1}{R_2}}$$

By putting resistors in parallel, you always get a *smaller* resistor. Resistance is measured in ohms (Ω), but in practice we frequently omit the Ω symbol when referring to

RESISTORS

Resistors are truly ubiquitous. There are almost as many types as there are applications. Resistors are used in amplifiers as loads for active devices, in bias networks, and as feedback elements. In combination with capacitors they establish time constants and act as filters. They are used to set operating currents and signal levels. Resistors are used in power circuits to reduce voltages by dissipating power, to measure currents, and to discharge capacitors after power is removed. They are used in precision circuits to establish currents, to provide accurate voltage ratios, and to set precise gain values. In logic circuits they act as bus and line terminators and as "pull-up" and "pull-down" resistors. In high-voltage circuits they are used to measure voltages and to equalize leakage currents among diodes or capacitors connected in series. In radiofrequency circuits they are even used as coil forms for inductors.

Resistors are available with resistances from 0.01 ohm through 10^{12} ohms, standard power ratings from ⅛ watt through 250 watts, and accuracies from 0.005% through 20%. Resistors can be made from carbon-composition moldings, from metal films, from wire wound on a form, or from semiconductor elements similar to field-effect transistors (FETs). But by far the most familiar resistor is the ¼ or ½ watt carbon-composition resistor. These are available in a standard set of values ranging from 1 ohm to 100 megohms, with twice as many values available for the 5% tolerance as for the 10% types (see Appendix C). We prefer the Allen Bradley type AB (¼ watt, 5%) resistor for general use because of its clear marking, secure lead seating, and stable properties.

Resistors are so easy to use that they're often taken for granted. They're not perfect, though, and it is worthwhile to look at some of their defects. The popular 5% composition type, in particular, although fine for nearly all noncritical circuit applications, is not stable enough for precision applications. You should know about its limitations so that you won't be surprised someday. Its principal defects are variations in resistance with temperature, voltage, time, and humidity. Other defects may relate to inductance (which may be serious at high frequencies), the development of thermal hot spots in power applications, or electrical noise generation in low-noise amplifiers. The following specifications are worst-case values; typically you'll do better, but don't count on it!

SPECIFICATIONS FOR ALLEN BRADLEY AB SERIES TYPE CB

Standard tolerance is ±5% under nominal conditions. Maximum power for 70°C ambient temperature is 0.25 watt, which will raise the internal temperature to 150°C. The maximum applied voltage specification is $(0.25R)^{1/2}$ or 250 volts, whichever is less. A single 5 second overvoltage to 400 volts can cause a permanent change in resistance by 2%.

	Resistance change		*Permanent?*
	(R = 1k)	(R = 10M)	
Soldering (350°C at ⅛ inch)	±2%	±2%	yes
Load cycling (500 ON/OFF cycles in 1000 hours)	+4% − 6%	+4% − 6%	yes
Vibration (20*g*) and shock (100*g*)	±2%	±2%	yes
Humidity (95% relative humidity at 40°C)	+6%	+10%	no
Voltage coefficient (10 volt change)	−0.15%	−0.3%	no
Temperature (25°C to −15°C)	+2.5%	+4.5%	no
Temperature (25°C to 85°C)	+3.3%	+5.9%	no

For applications that require any real accuracy or stability, a 1% metal-film resistor (see Appendix D) should be used. They can be expected to have stability of better than 0.1% under normal conditions and better than 1% under worst-case treatment. Precision wire-wound resistors are available for the most demanding applications.

For power dissipation above about 0.1 watt, a resistor of higher power rating should be used. Carbon-composition resistors are available wtih ratings up to 2 watts, and wire-wound power resistors are available for higher power. For demanding power applications, the conduction-cooled type of power resistor delivers better performance. These carefully designed resistors are available at 1% tolerance and can be operated at core temperatures up to 250°C with dependable long life.

resistors that are more than 1000Ω (1kΩ). Thus a 10kΩ resistor is often referred to as a 10k resistor, and a 1MΩ resistor as a 1M resistor (or 1 meg). On schematic diagrams the symbol Ω is often omitted altogether. If this bores you, please have patience – we'll soon get to numerous amusing applications.

EXERCISE 1.1

You have a 5k resistor and a 10k resistor. What is their combined resistance (a) in series and (b) in parallel?

EXERCISE 1.2

If you place a 1 ohm resistor across a 12 volt car battery, how much power will it dissipate?

EXERCISE 1.3

Prove the formulas for series and parallel resistors.

EXERCISE 1.4

Show that several resistors in parallel have resistance

$$R = \frac{1}{\frac{1}{R_1} + \frac{1}{R_2} + \frac{1}{R_3} + \cdots}$$

A trick for parallel resistors: Beginners tend to get carried away with complicated algebra in designing or trying to understand electronics. Now is the time to begin learning intuition and shortcuts.

Shortcut no. 1. A large resistor in series (parallel) with a small resistor has the resistance of the larger (smaller) one, roughly.

Shortcut no. 2. Suppose you want the resistance of 5k in parallel with 10k. If you think of the 5k as two 10k's in parallel, then the whole circuit is like three 10k's in parallel. Since the resistance of n equal resistors in parallel is $1/n$th the resistance of the individual resistors, the answer in this case is 10k/3, or 3.33k. This trick is handy because it allows you to analyze circuits quickly in your head, without distractions. We want to encourage mental designing, or at least "back of the envelope" designing, for idea brainstorming.

Some more home-grown philosophy: There is a tendency among beginners to want to compute resistor values and other circuit component values to many significant places, and the availability of inexpensive calculators has only made matters worse. There are two reasons you should try to avoid falling into this habit: (a) the components themselves are of finite precision (typical resistors are ±5%; the parameters that characterize transistors, say, are frequently known only to a factor of two); (b) one mark of a good circuit design is insensitivity of the finished circuit to precise values of the components (there are exceptions, of course). You'll also learn circuit intuition more quickly if you get into the habit of doing approximate calculations in your head, rather than watching meaningless numbers pop up on a calculator display.

In trying to develop intuition about resistance, some people find it helpful to think about *conductance*, $G = 1/R$. The current through a device of conductance G bridging a voltage V is then given by $I = GV$ (Ohm's law). A small resistance is a large conductance, with correspondingly large current under the influence of an applied voltage.

Viewed in this light, the formula for parallel resistors is obvious: When several resistors or conducting paths are connected across the same voltage, the total current is the sum of the individual currents. Therefore the net conductance is simply the sum of the individual conductances, $G = G_1 + G_2 + G_3 + \cdots$, which is the same as the formula for parallel resistors derived earlier.

Engineers are fond of defining reciprocal units, and they have designated the unit of conductance the siemens (S = 1/Ω), also known as the mho (that's ohm spelled backward, given the symbol ℧). Although the concept of conductance is helpful in developing intuition, it is not used widely; most people prefer to talk about resistance instead.

Power in resistors

The power dissipated by a resistor (or any other device) is $P = IV$. Using Ohm's law, you can get the alternate forms $P = I^2R$ and $P = V^2/R$.

EXERCISE 1.5

Show that it is not possible to exceed the power rating of a ¼ watt resistor of resistance greater than 1k, no matter how you connect it, in a circuit operating from a 15 volt battery.

EXERCISE 1.6

Optional exercise: New York City requires about 10^{10} watts of electrical power, at 110 volts (this is plausible: 10 million people averaging 1 kilowatt each). A heavy power cable might be an inch in diameter. Let's calculate what will happen if we try to supply the power through a cable 1 foot in diameter made of pure copper. Its resistance is $0.05\mu\Omega$ (5×10^{-8} ohms) per foot. Calculate (a) the power lost per foot from "I^2R losses," (b) the length of cable over which you will lose all 10^{10} watts, and (c) how hot the cable will get, if you know the physics involved ($\sigma = 6 \times 10^{-12}$ W/°K^{4}cm^2).

If you have done your computations correctly, the result should seem preposterous. What is the solution to this puzzle?

1.03 Voltage dividers

This use of resistors is everywhere. Show us a circuit and we'll show you a half dozen voltage dividers. This simple circuit is shown in Figure 1.5. What is V_{out}? Well, the current

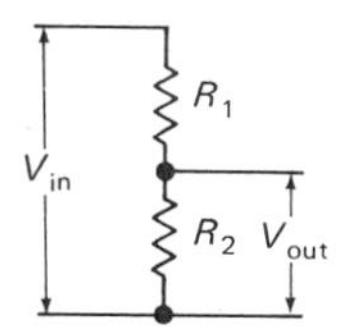

Figure 1.5

(same everywhere, assuming no "load" on the output) is

$$I = \frac{V_{in}}{R_1 + R_2}$$

(We've used the definition of resistance and the series law.) Then, for R_2,

$$V_{out} = IR_2 = \frac{R_2}{R_1 + R_2} V_{in}$$

Note that the output voltage is always less than (or equal to) the input voltage; that's why it's called a divider. You could get amplification (more output than input) if one of the resistances were negative. This isn't as crazy as it sounds; it is possible to make devices with negative "incremental" resistances (e.g., the tunnel diode) or even true negative resistances (e.g., the negative-impedance converter that we will talk about later in the book). However, these applications are rather specialized and need not concern you now.

Voltage dividers are often used in circuits to generate a particular voltage from a larger fixed (or varying) voltage. For instance, if R_2 is an adjustable resistor, you have a "volume control." The humble voltage divider is even more useful, though, as a way of *thinking* about a circuit: the input voltage and upper resistance might represent the output of an amplifier, say, and the lower resistance might represent the input of the following stage. In this case the voltage divider equation tells you how much signal gets to the input of that last stage. This will all become clearer after you know about a remarkable fact (Thévenin's theorem) that will be discussed later. First, though, a short aside on voltage sources and current sources.

1.04 Voltage and current sources

A perfect voltage source is a two-terminal *black box* that maintains a fixed voltage drop across its terminals, regardless of load resistance. (For instance, this means that it must supply a current $I = V/R$ when a resistance R is attached to its terminals. A real voltage source can supply only a finite maximum current, and in addition it generally behaves like a perfect voltage source with a small resistance in series. Obviously, the smaller this series resistance, the better. A voltage source "likes" an open-circuit load and "hates" a short-circuit load, for obvious reasons. The symbols used to indicate a voltage source are shown in Figure 1.6.

A perfect current source is a two-terminal black box that maintains a constant current through the external circuit, regardless of load resistance or applied voltage. In order to

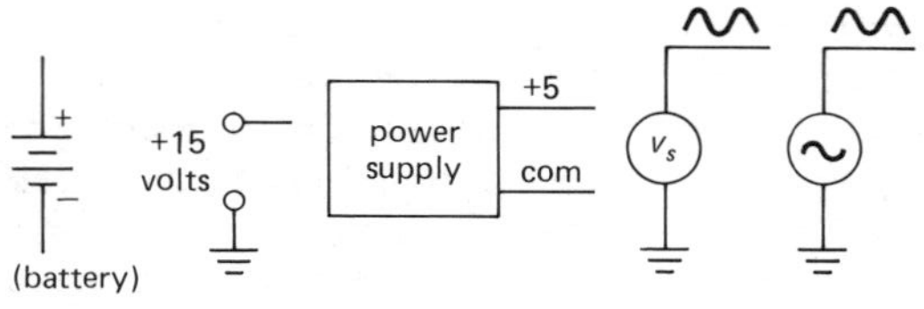

Figure 1.6

do this it must be capable of supplying any necessary voltage across its terminals. Real current sources (a much-neglected subject in most textbooks) have a limit to the voltage they can provide (called the *output voltage compliance,* or just *compliance*), and in addition they do not provide absolutely constant output current. A current source "likes" a short-circuit load and "hates" an open-circuit load. The symbols used to indicate a current source are shown in Figure 1.7.

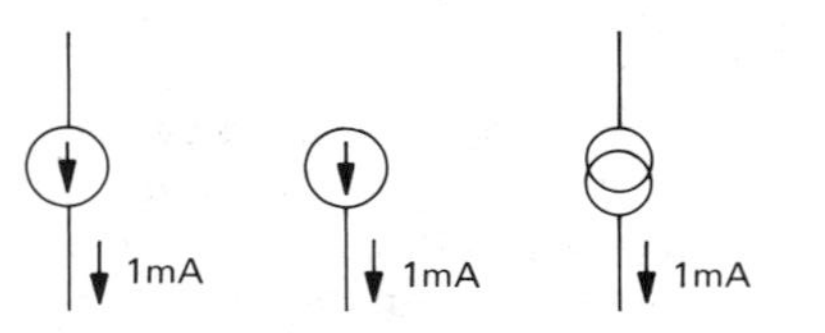

Figure 1.7

A battery is a real-life approximation of a voltage source (there is no analog for a current source). A standard D-size flashlight cell, for instance, has a terminal voltage of 1.5 volts, an equivalent series resistance of about ¼ ohm, and a total energy capacity of about 10,000 watt-seconds (its characteristics gradually deteriorate with use; at the end of its life the voltage may be about 1.0 volt, with an internal series resistance of several ohms). It is easy to construct voltage sources with far better characteristics, as you will learn when we come to the subject of feedback. Except in devices intended for portability, the use of batteries in electronic devices is rare.

1.05 Thévenin's theorem

Thévenin's theorem states that any two-terminal network of resistors and voltage sources is equivalent to a single resistor R in series with a single voltage source V. This is remarkable. Any mess of batteries and resistors can be mimicked with one battery and one resistor (Fig. 1.8). (Incidentally, there's

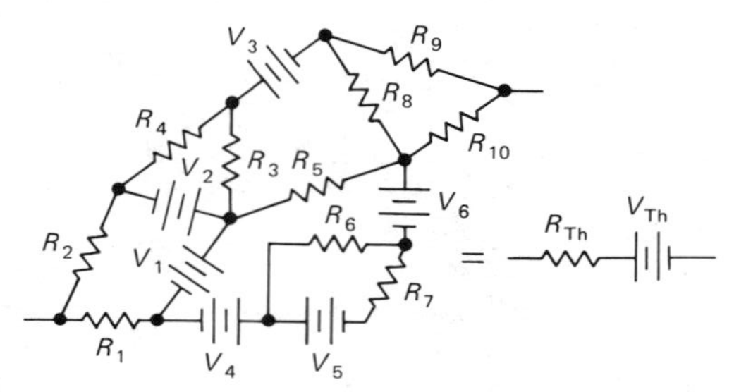

Figure 1.8

another theorem, Norton's theorem, that says you can do the same thing with a current source in parallel with a resistor.)

How do you figure out the Thévenin equivalent R_{Th} and V_{Th} for a given circuit? Easy! V_{Th} is the open-circuit voltage of the Thévenin equivalent circuit; so if the two circuits behave identically, it must also be

MULTIMETERS

There are numerous instruments that let you measure voltages and currents in a circuit. The oscilloscope (see Appendix A) is the most versatile; it lets you "see" voltages versus time at one or more points in a circuit. Logic probes and logic analyzers are special-purpose instruments for troubleshooting digital circuits. The simple multimeter provides a good way to measure voltage, current, and resistance, often with good precision; however, it responds slowly, and thus it cannot replace the oscilloscope where changing voltages of are interest. Multimeters are of two varieties: those that indicate measurements on a conventional moving-pointer meter movement and those of the more recent digital variety.

The standard VOM (volt-ohm-millimmeter) multimeter uses a meter movement that measures current (typically 50μA full scale). (See a less-design-oriented electronics book for pretty pictures of the innards of meter movements; for our purposes, it suffices to say that it uses coils and magnets.) To measure voltage, the VOM puts a resistor in series with the basic movement. For instance, one kind of VOM will generate a 1 volt (full-scale) range by putting a 20k resistor in series with the standard 50μA movement; higher voltage ranges use correspondingly larger resistors. Such a VOM is specified as 20,000 ohms/volt, meaning that it looks like a resistor whose value is 20k multiplied by the full-scale voltage of the particular range selected. Full scale on any voltage range is 1/20,000, or 50μA. It should be clear that one of these voltmeters disturbs a circuit less on a higher range, since it looks like a higher resistance (think of the voltmeter as the lower leg of a voltage divider, with the Thévenin resistance of the circuit you are measuring as the upper resistor). Ideally, a voltmeter should have infinite input resistance.

Nowadays there are various meters with some electronic amplification whose input resistance may be as large as 10^9 ohms. Most digital meters, and even a number of analog-reading meters that use FETs (field-effect transistors, see Chapter 6), are of this type. Warning: Sometimes the input resistance of FET-input meters is very high on the most sensitive ranges, dropping to a lower resistance for the higher ranges. For instance, an input resistance of 10^9 ohms on the 0.2 volt and 2 volt ranges, and 10^7 ohms on all higher ranges, is typical. Read the specifications carefully! For measurements on most transistor circuits, 20,000 ohms/volt is fine, and there will be little loading effect on the circuit by the meter. In any case, it is easy to calculate how serious the effect is by using the voltage divider equation. Typically, multimeters provide voltage ranges from a volt (or less) to a kilovolt (or more), full scale.

A VOM can be used to measure current by simply using the bare meter movement (for our preceding example, the would give a range of 50μA full scale) or by shunting (paralleling) the movement with a small resistor. Since the meter movement itself requires a small voltage drop, typically 0.25 volt, to produce a full-scale deflection, the shunt is chosen by the meter manufacturer (all you do is set the range switch to the range you want) so that the full-scale current will produce that voltage drop through the parallel combination of the meter resistance and the shunt resistance. Ideally, a current-measuring meter should have zero resistance in order not to disturb the circuit under test, since it must be put in series with the circuit. In practice, you tolerate a few tenths of a volt drop with both VOMs and digital multimeters. Typically, multimeters provide current ranges from 50 μA (or less) to an amp (or more), full scale.

Multimeters also have one or more batteries in them to power the resistance measurement. By supplying a small current and measuring the voltage drop, they measure resistance, with several ranges to cover values from an ohm (or less) to 10 megohms (or more).

Important: Don't try to measure "the current of a voltage source," for instance by sticking the meter across the wall plug; the same applies for ohms. This is the leading cause of blown-out meters.

EXERCISE 1.7

What will a 20,000 ohms/volt meter read, on its 1 volt scale, when attached to a 1 volt source with an internal resistance of 10k? What will it read when attached to a 10k-10k voltage divider driven by a "stiff" (zero source resistance) 1 volt source?

EXERCISE 1.8

A 50μA meter movement has an internal resistance of 5k. What shunt resistance is needed to convert it to a 0–1 amp meter? What series resistance will convert it to a 0–10 volt meter?

the open-circuit voltage of the given circuit (which you get by calculation, if you know what the circuit is, or by measurement, if you don't). Then you find R_{Th} by noting that the short-circuit current of the equivalent circuit is V_{Th}/R_{Th}. In other words,

$$V_{Th} = V\text{ (open circuit)}$$

$$R_{Th} = \frac{V\text{ (open circuit)}}{I\text{ (short circuit)}}$$

Let's apply this method to the voltage divider, which must have a Thévenin equivalent:

1. The open-circuit voltage is

$$V = V_{in}\frac{R_2}{R_1 + R_2}$$

2. The short-circuit current is

V_{in}/R_1

So the Thévenin equivalent circuit is a voltage source

$$V_{Th} = V_{in}\frac{R_2}{R_1 + R_2}$$

in series with a resistor

$$R_{Th} = \frac{R_1 R_2}{R_1 + R_2}$$

(It is not a coincidence that this happens to be the parallel resistance of R_1 and R_2. The reason will become clear later.)

From this example it is easy to see that a voltage divider is not a very good battery, in the sense that its output voltage drops severely when a load is attached. As an example, consider Exercise 1.9. You now know everything you need to know to calculate exactly how much the output will drop for a given load resistance: Use the Thévenin equivalent circuit, attach a load, and calculate the new output, noting that the new circuit is nothing but a voltage divider (Fig. 1.9).

EXERCISE 1.9

For the circuit shown in Figure 1.9, with $V_{in} = 30V$ and $R_1 = R_2 = 10k$, find (a) the output voltage with no load attached (the

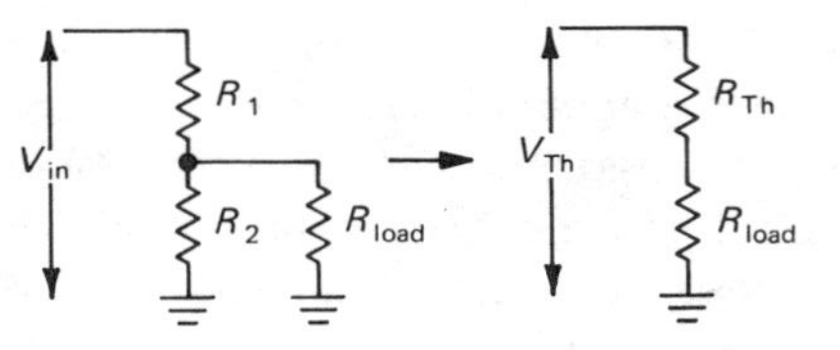

Figure 1.9

open-circuit voltage); (b) the output voltage with a 10k load (treat as voltage divider, with R_2 and R_{load} combined into a single resistor); (c) the Thévenin equivalent circuit; (d) the same as in part b, but using the Thévenin equivalent circuit (again, you wind up with a voltage divider; the answer should agree with the result in part b); (e) the power dissipated in each of the resistors.

Equivalent source resistance and circuit loading

As you have just seen, a voltage divider powered from some fixed voltage is equivalent to some smaller voltage source in series with a resistor. Attaching a load resistor causes the voltage divider's output to drop, owing to the finite *source resistance* (Thévenin equivalent resistance of the voltage divider output, viewed as a source of voltage). This is often undesirable. One solution to the problem of making a stiff voltage source (stiff is used in this context to describe something that doesn't bend under load) might be to use much smaller resistors in a voltage divider. Occasionally this brute-force approach is useful. However, it is usually best to construct a voltage source, or power supply, as it's commonly called, using active components like transistors or operational amplifiers. In this way you can easily make a voltage source with internal (Thévenin equivalent) resistance measured in milliohms (thousandths of an ohm), without the large currents and dissipation of power characteristic of a low-resistance voltage divider delivering the same performance. In addition, with an active power supply it is easy to make the output voltage adjustable.

The concept of equivalent internal resistance applies to all sorts of sources, not just batteries and voltage dividers. Signal sources, (e.g., oscillators, amplifiers, and sensing devices) all have an equivalent internal resistance. Attaching a load whose resistance is less than or even comparable to the internal resistance will reduce the output considerably. This undesirable reduction of the open-circuit voltage (or signal) by the load is called "circuit loading." Therefore you should strive to make $R_{load} \gg R_{internal}$, because a high-resistance load has little attenuating effect on the source. You will see numerous circuit examples in the chapters ahead. This high-resistance condition ideally characterizes measuring instruments such as voltmeters and oscilloscopes. (There are exceptions to this general principle; for example, we will talk about transmission lines and radiofrequency techniques, where you must "match impedances" in order to prevent the reflection and loss of power.)

A word on language: You frequently hear things like "the resistance looking into the voltage divider," or "the output sees a load of so-and-so many ohms," as if circuits had eyes. It's OK (in fact, it's a rather good way to keep straight which resistance you're talking about) to say what part of the circuit is doing the "looking."

Power transfer

Here is an interesting problem: What load resistance will result in maximum power being transferred to the load for a given source resistance? (The terms *source resistance, internal resistance,* and *Thévenin equivalent resistance* all mean the same thing.) It is easy to see that both $R_{load} = 0$ and $R_{load} = \infty$ result in zero power transferred, since $R_{load} = 0$ means that $V_{load} = 0$ and $I_{load} = V_{source}/R_{source}$, so that $P_{load} = V_{load}I_{load} = 0$. But $R_{load} = \infty$ means that $V_{load} = V_{source}$ and $I_{load} = 0$, so that $P_{load} = 0$. There has to be a maximum in between.

EXERCISE 1.10

Show that $R_{load} = R_{source}$ maximizes the power in the load for a given source resistance. Note: Skip this exercise if you don't know calculus, and take it on faith that the answer is true.

Lest this example leave the wrong impression, we would like to emphasize again that circuits are ordinarily designed so that the load resistance is much greater than the source resistance of the signal that drives the load.

1.06 Small-signal resistance

We often deal with electronic devices for which I is not proportional to V; in such cases there's not much point in talking about resistance, since the ratio V/I will depend on V, rather than being a nice constant, independent of V. For these devices it is useful to know the slope of the V–I curve, in other words the ratio of a small change in applied voltage to the resulting change in current through the device, $\Delta V/\Delta I$ (or dV/dI). This quantity has the units of resistance (ohms) and substitutes for resistance in many calculations. It is called the small-signal resistance, incremental resistance, or dynamic resistance.

Zener diodes

As an example, consider the *zener diode,* which has the V–I curve shown in Figure 1.10. Zeners are used to create a constant

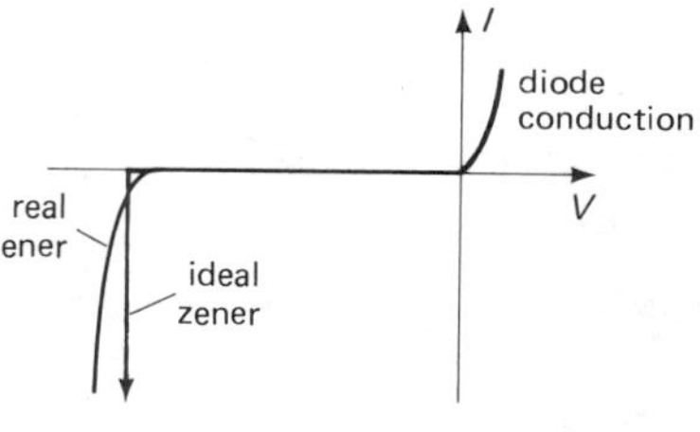

Figure 1.10

voltage inside a circuit somewhere, simply by providing them with a (roughly constant) current derived from a higher voltage within the circuit. It is important to know how the resulting zener voltage will change with applied current; this is a measure of its "regulation" against changes in the driving current provided to it. Included in the specifications of a zener will be its dynamic resistance, given at a certain current. (Useful fact: the dynamic resistance of a zener diode varies roughly in inverse proportion to current.) For example, a zener might have a dynamic resistance of 10 ohms at 10mA, at its zener voltage of 5 volts. Using the definition of dynamic resistance, we find that a 10% change in applied current will therefore result in a change in voltage of

$$\Delta V = R_{dyn}\Delta I = 10 \times 0.1 \times 0.01 = 10\text{m}V$$

or

$$\Delta V/V = 0.002 = 0.2\%$$

thus demonstrating good voltage-regulating ability. In this sort of application you frequently get the zener current through a resistor from a higher voltage available somewhere in the circuit, as in Figure 1.11.

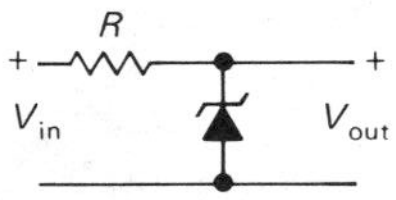

Figure 1.11

Then,

$$I = \frac{V_{in} - V_{out}}{R} \quad \text{and} \quad \Delta I = \frac{\Delta V_{in} - \Delta V_{out}}{R}$$

so

$$\Delta V_{out} = R_{dyn}\Delta I = \frac{R_{dyn}}{R}(\Delta V_{in} - \Delta V_{out})$$

and finally

$$\Delta V_{out} = \frac{R_{dyn}}{R + R_{dyn}}\Delta V_{in}$$

Thus, for *changes* in voltage, the circuit behaves like a voltage divider, with the zener replaced by a resistor equal to its dynamic resistance at the operating current. This is the utility of incremental resistance. For instance, suppose in the preceding circuit we have an input voltage ranging between 15 and 20 volts and use a 1N4733 (5.1V 1W zener diode) in order to generate a stable 5.1 volt power supply. We choose R = 300 ohms, for a maximum zener current of 50mA: (20 − 5.1)/300. We can now estimate the output voltage regulation (variation in output voltage), knowing that this particular zener has a specified maximum dynamic impedance of 7.0 ohms at 50mA. The zener current varies from 50mA to 33mA over the input voltage range; this 17mA change in current then produces a voltage change at the output of $\Delta V = R_{dyn}\Delta I$, or 0.12 volt. You will see more of zeners in Sections 2.04 and 5.13.

In real life, a zener will provide better regulation if driven by a current source, which has, by definition, $R_{incr} = \infty$ (same current regardless of voltage). But current sources are more complex, and therefore in practice we often resort to the humble resistor.

Tunnel Diodes

Another interesting application of incremental resistance is the *tunnel diode,* sometimes called the Esaki diode. Its V–I curve is shown in Figure 1.12. In the region from A to B it has *negative* incremental resistance. This has a remarkable consequence: A voltage *divider* made with a resistor and a tunnel diode can actually be an *amplifier* (Fig. 1.13). For a wiggly voltage v_{sig}, the voltage

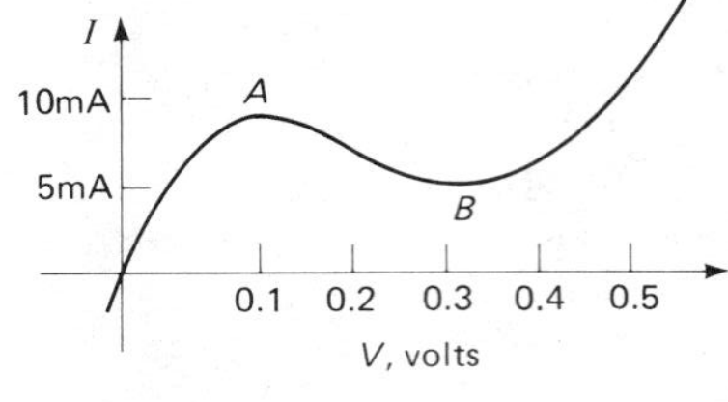

Figure 1.12

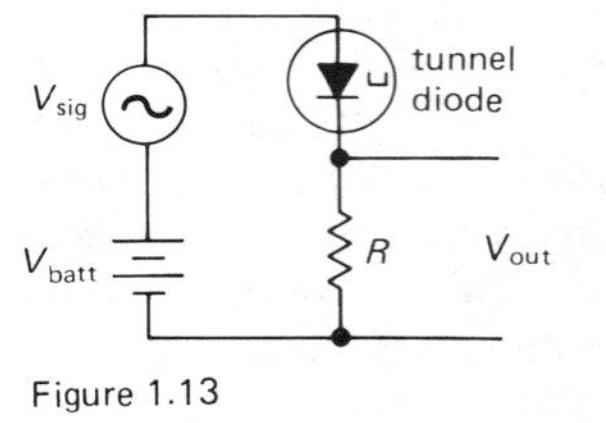

Figure 1.13

divider equation gives us

$$v_{out} = \frac{R}{R + r_t} v_{sig}$$

where r_t is the incremental resistance of the tunnel diode at the operating current and the lower-case symbol v_{sig} stands for a small-signal variation, which we have been calling ΔV_{sig} up to now (we will adopt this widely used convention from now on). The tunnel diode has $r_{t(incr)} < 0$. That is,

$$\Delta V/\Delta I \quad (\text{or } v/i) \quad < 0$$

from A to B on the characteristic curve. If $r_{t(incr)} \approx R$, the denominator is nearly zero, and the circuit amplifies. V_{batt} provides the steady current, or *bias,* to bring the operating point into the region of negative resistance. (Of course, it is always necessary to have a source of power in any device that amplifies.)

A postmortem on these fascinating devices: When tunnel diodes first appeared, late in the 1950s, they were hailed as the solution to a great variety of circuit problems. Because they were fast, they were supposed to revolutionize computers, for instance. Unfortunately, they are difficult devices to use; this fact, combined with stunning improvements in transistors, has made tunnel diodes almost obsolete.

The subject of negative resistance will come up again later, in connection with active filters. There you will see a circuit called a negative-impedance converter that can produce (among other things) a pure negative resistance (not just incremental). It is made with an operational amplifier and has very useful properties.

SIGNALS

A later section in this chapter will deal with capacitors, devices whose properties depend on the way the voltages and currents in a circuit are *changing.* Our analysis of dc circuits so far (Ohm's law, Thévenin equivalent circuits, etc.) still holds, even if the voltages and currents are changing in time. But for a proper understanding of alternating-current (ac) circuits, it is useful to have in mind certain common types of *signals,* voltages that change in time in a particular way.

1.07 Sinusoidal signals

Sinusoidal signals are the most popular signals around; if someone says something like "take a 10 microvolt signal at 1 megahertz," he means a sine wave. Mathematically, what you have is a voltage described by

$$V = A \sin 2\pi ft$$

where A is called the amplitude and f is the frequency in cycles per second, or hertz. A sine wave looks like the wave shown in Figure 1.14. Sometimes it is important to

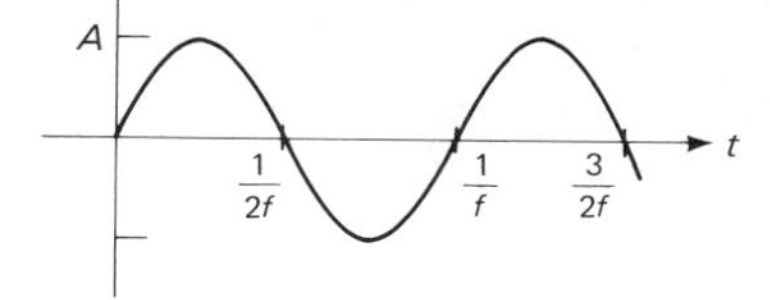

Figure 1.14

know the value of the signal at some arbitrary time $t = 0$, in which case you may see a *phase* ϕ in the expression:

$$V = A \sin (2\pi ft + \phi)$$

The other variation on this simple theme is the use of *angular frequency,* which looks like this:

$$V = A \sin \omega t$$

Here, ω is the angular frequency in radians per second. Just remember the important relation $\omega = 2\pi f$ and you won't go wrong.

The great merit of sine waves (and the cause of their perennial popularity) is the fact that they are the solutions to certain linear differential equations that happen to describe many phenomena in nature as well as the properties of linear circuits. A linear circuit has the property that its output, when driven by the sum of two input signals, equals the sum of its individual outputs when driven by each input signal in turn; i.e., if $O(A)$ represents the output when driven by signal A, then a circuit is linear if $O(A + B) = O(A) + O(B)$. A linear circuit driven by a sine wave responds with a sine wave, although in general the phase and amplitude are changed. No other signal can make this statement. It is standard practice, in fact, to describe the behavior of a circuit by its *frequency response,* the way it alters the amplitude of an applied sine wave as a function of frequency. A high-fidelity amplifier, for instance, should be characterized by a "flat" frequency response over the range 20Hz to 20kHz, at least.

The sine-wave frequencies you will usually deal with range from a few hertz to a few megahertz. Lower frequencies, down to 0.0001Hz or lower, can be generated with carefully built circuits, if needed. Higher frequencies, e.g., up to 2000 MHz, can be generated, but they require special transmission-line techniques. Above that, you're dealing with microwaves, where conventional wired circuits with lumped circuit elements become impractical, and waveguides are used instead.

1.08 Signal amplitudes and decibels

In addition to its amplitude, there are several other ways to characterize the amplitude of a sine wave (or any other signal). You sometimes see it specified by *peak-to-peak amplitude* (pp amplitude), which is just what you would guess, namely twice the amplitude. The other method is to give the *root-mean-square amplitude* (rms amplitude), which is $V_{rms} = (1/\sqrt{2})A = 0.707A$ (this is for sine waves only; the ratio of pp to rms will be different for other waveforms). Odd as it may seem, this is the usual method, because rms voltage is what's used to compute power. The voltage across the terminals of a wall socket (in the United States) is 117 volts rms, 60Hz. The *amplitude* is 165 volts (330 volts pp).

Decibels

How do you compare the relative amplitudes of two signals? You could say, for instance, that signal X is twice as large as signal Y. That's fine, and useful for many purposes. But since we often deal with ratios as large as a million, it is easier to use a logarithmic measure, and for this we present the decibel (it's one-tenth as large as something called a bel, which no one ever uses). By definition, the ratio of two signals, in decibels, is

$$dB = 20 \log_{10} \frac{A_2}{A_1}$$

where A_1 and A_2 are the two signal amplitudes. So, for instance, one signal of twice the amplitude of another is +6dB relative to it, since $\log_{10} 2 = 0.3010$. A signal 10 times as large is +20dB; a signal one-tenth as large is −20dB. It is also useful to express the ratio of two signals in terms of power levels:

$$dB = 10 \log_{10} \frac{P_2}{P_1}$$

where P_1 and P_2 represent the power in the two signals. As long as the two signals have the same kind of waveform, e.g., sine waves, the two definitions give the same result. When comparing unlike waveforms, e.g., a sine wave versus "noise," the definition in terms of power (or the amplitude definition, with rms amplitudes substituted) must be used.

Although decibels are ordinarily used to specify the ratio of two signals, they are sometimes used as an absolute measure of amplitude. What is happening is that you are assuming some reference signal amplitude and expressing any other amplitude in decibels relative to it. There are several standard amplitudes (which are unstated, but understood) that are used in this way; the most common references are (a) dBV; 1 volt rms, (b) dBm; the voltage corresponding to 1mW into 600 ohms load (that's about 0.78 volts rms), and (c) the small noise voltage generated by a resistor at room temperature (this surprising fact is discussed in Section 7.10). In addition to these, there are reference amplitudes used for measurements in other fields. For instance, in acoustics, 0dB SPL is a wave whose rms pressure is 0.0002μbar (a bar is 10^6 dynes per square centimeter, approximately 1 atmosphere); in communications, levels can be stated in dBrnC (relative noise reference weighted in frequency by "curve C"). When stating amplitudes this way, it is best to be specific about the 0dB reference amplitude; say something like "an amplitude of 27 decibels relative to 1 volt rms," or abbreviate "27dB re 1V rms," or define a term like "dBV."

EXERCISE 1.11

Determine the voltage and power ratios for a pair of signals with the following decibel ratios: (a) 3dB, (b) 6dB, (c) 10dB, (d) 20dB.

1.09 Other signals

Ramp

The ramp is a signal that looks like the signal shown in Figure 1.15. It is simply a voltage rising (or falling) at a constant rate. That can't go on forever, of course, even in science fiction movies. It is sometimes approximated by a finite ramp (Fig. 1.16) or by a periodic ramp, or sawtooth (Fig. 1.17).

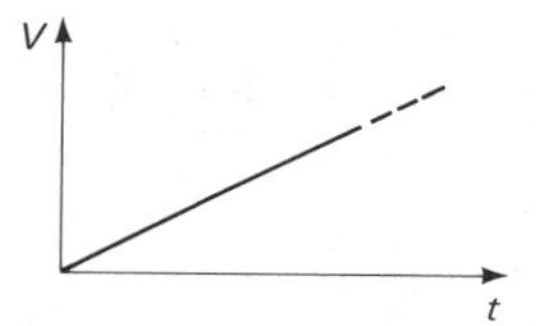

Figure 1.15

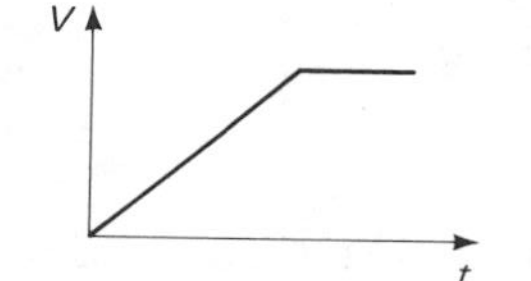

Figure 1.16

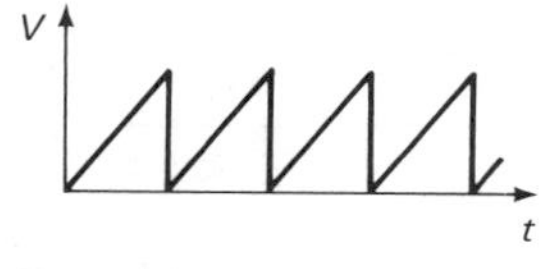

Figure 1.17

Triangle

The triangle wave is a close cousin of the ramp; it is simply a symmetrical ramp (Fig. 1.18).

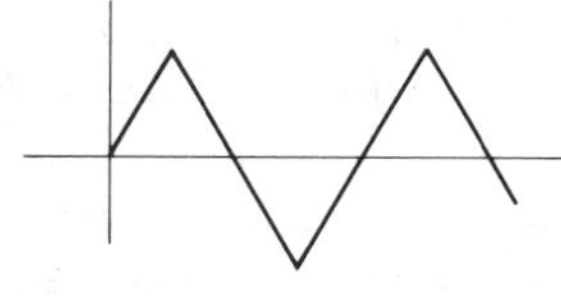

Figure 1.18

Noise

Signals of interest are often mixed with *noise;* this is a catchall phrase that usually applies to random noise of thermal origin. Noise voltages can be specified by their frequency spectrum (power per hertz) or by their amplitude distribution. One of the most common kinds of noise is *band-limited white Gaussian noise,* which means a signal with equal power per hertz in some band of frequencies and a Gaussian (bell-shaped) distribution of amplitudes if large numbers of instantaneous measurements of its amplitude are made. This kind of noise is generated by a resistor (Johnson noise), and it plagues sensitive measurements of all kinds. On an oscilloscope it appears as shown in Figure 1.19. We will study noise and low-noise techniques in some detail in Chapter 7.

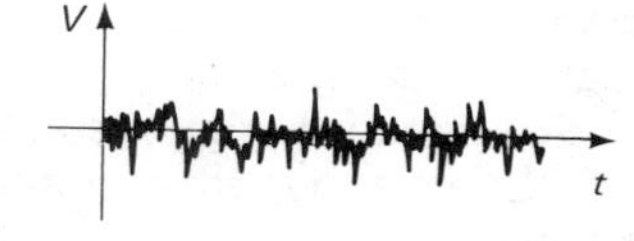

Figure 1.19

Section 9.35 deals with noise-generation techniques.

Square waves

A square wave is a signal that varies in time as shown in Figure 1.20. Like the sine wave,

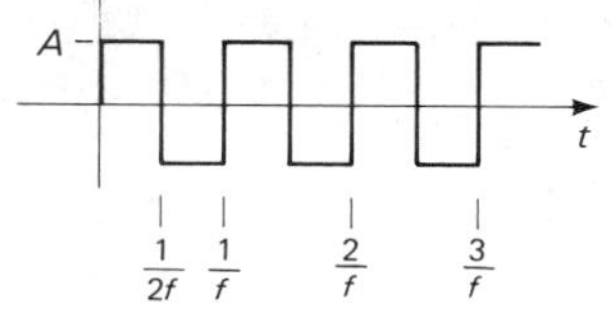

Figure 1.20

it is characterized by amplitude and frequency. A linear circuit driven by a square wave rarely responds with a square wave. For a square wave, the rms amplitude equals the amplitude.

The edges of a square wave are not perfectly square; in typical electronic circuits the rise time t_r ranges from a few nanoseconds to a few microseconds. Figure 1.21

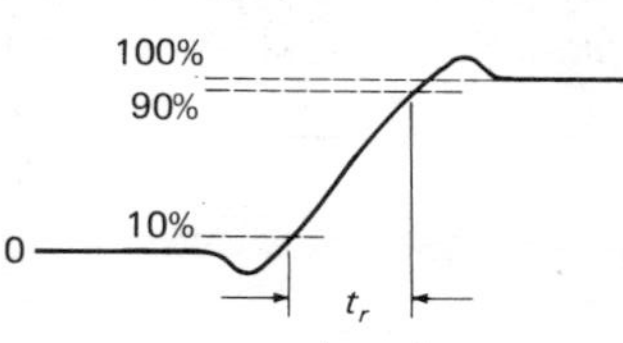

Figure 1.21

shows the sort of thing usually seen. The rise time is defined as the time required for the signal to go from 10% to 90% of its total transition.

Pulses

A pulse is a signal that looks as shown in Figure 1.22. It is defined by amplitude and pulse width. You can generate a train of periodic (equally spaced) pulses, in which case you can talk about the frequency, or pulse repetition rate, and the "duty cycle," the ratio of pulse width to repetition period (duty cycle ranges from zero to 100%).

Pulses can have positive or negative polarity; in addition, they can be "positive-going" or "negative-going." For instance, the second pulse in Figure 1.22 is a negative-going pulse of positive polarity.

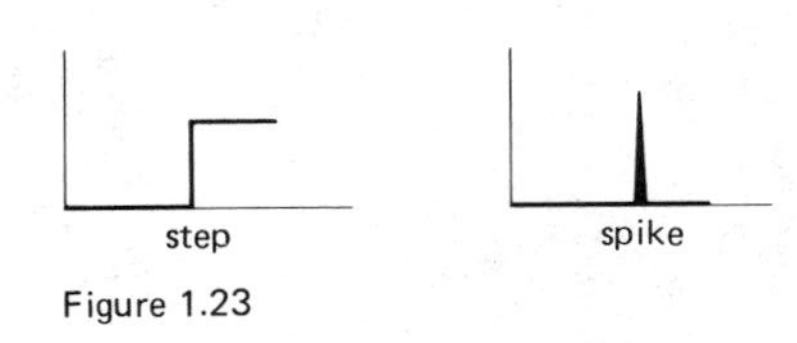

Figure 1.22

Steps and spikes

Steps and spikes are signals that are talked about a lot but are not often used. They provide a nice way of describing what happens in a circuit. If you could draw them,

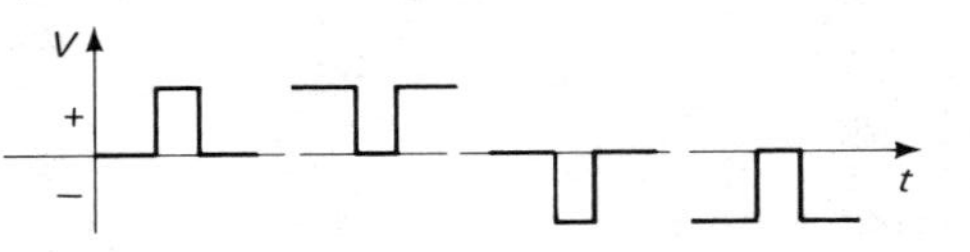

Figure 1.23

they would look something like the example in Figure 1.23. The step function is part of a square wave; the spike is simply a jump of vanishingly short duration.

LOGIC LEVELS

Figure 1.24 shows the ranges of voltages that correspond to the two logic states (HIGH and LOW) for the three most popular families of digital logic. For each logic family it is necessary to specify legal values of both output and input voltages corresponding to the two states HIGH and LOW. The shaded areas above the line show the specified range of output voltages that a logic LOW or HIGH is guaranteed to fall within, with the pair of arrows indicating typical output values (LOW, HIGH) encountered in practice. The shaded areas below the line show the range of input voltages that are guaranteed to be interpreted as LOW or HIGH, with the arrow indicating the typical *logic threshold* voltage, i.e., the dividing line between LOW and HIGH. In all cases a logic HIGH is more positive than a logic LOW.

The meanings of minimum, typical, and maximum in electronic specifications are worth a few words of explanation. Most simply, the manufacturer guarantees that the components

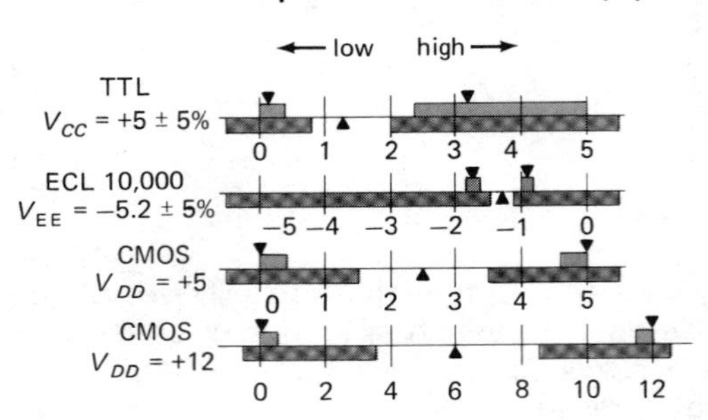

Figure 1.24

will fall in the range minimum–maximum, with many close to "typical." What this means is that typical specifications are what you use when designing circuits; however, those circuits must work properly over the whole range of specifications from minimum to maximum (the extremes of manufacturing variability). In particular, a well-designed circuit must function under the worst possible combination of minimum and maximum values. This is known as *worst-case design,* and it is essential for any instrument produced from off-the-shelf (i.e., not specially selected) components.

1.10 Logic levels

Pulses and square waves are used extensively in digital electronics, where predefined voltage levels represent one of two possible states present at any point in the circuit. These states are called simply HIGH and LOW and correspond to the 0 (false) and 1 (true) states of Boolean logic (the algebra that describes such two-state systems).

Precise voltages are not necessary in digital electronics. You need only to distinguish which of the two possible states is present. Each digital logic family therefore specifies legal HIGH and LOW states. The accompanying box shows logic levels for three popular varieties of digital electronics (CMOS operates with supply voltages from +3 to +18 volts; two popular voltages are listed).

1.11 Signal sources

Often the source of a signal is some part of the circuit you are working on. But for test purposes a flexible signal source is invaluable. They come in three flavors: signal generators, pulse generators, and function generators.

Signal generators

Signal generators are sine wave oscillators, usually equipped to give a wide range of frequency coverage (50kHz to 50MHz is typical), with provision for precise control of amplitude (using a resistive divider network called an *attenuator*). Some units let you *modulate* the output (see Chapter 13). A variation on this theme is the *sweep generator,* a signal generator that can sweep its output frequency repeatedly over some range. These are handy for testing circuits whose properties vary with frequency in a particular way, e.g., "tuned circuits" or filters. Nowadays these devices, as well as many test instruments, are available in configurations that allow you to program the frequency, amplitude, etc., from a computer or other digital instrument.

A variation on the signal generator is the *frequency synthesizer,* a device that generates sine waves whose frequencies can be set precisely. The frequency is set digitally, often to eight significant figures or more, and is internally synthesized from a precise standard (a quartz-crystal oscillator) by digital methods we will discuss later (Sections 9.28–9.33). If your requirement is for no-nonsense accurate frequency generation, you can't beat a synthesizer.

Pulse generators

Pulse generators only make pulses, but what pulses! Pulse width, repetition rate, amplitude, polarity, rise time, etc., may all be adjustable. In addition, many units allow you to generate pulse pairs, with settable spacing and repetition rate, or even coded pulse trains. Most modern pulse generators are provided with logic-level outputs for easy connection to digital circuitry. Like signal generators, these come in the programmable variety.

Function generators

In many ways function generators are the most flexible signal sources of all. You can make sine, triangle, and square waves over an enormous frequency range (0.01Hz to 10 MHz is typical), with control of amplitude and dc offset (a constant dc voltage added to the signal). Many of them have provision for frequency sweeping, often in several modes (linear or logarithmic frequency variation versus time). They are available with pulse outputs (although not with the flexibility you get with a pulse generator), and some of them have provision for modulation.

Like the other signal sources, function generators come in programmable versions and versions with digital readout of frequency (and sometimes amplitude). The most recent addition to the function-generator family is the synthesized function generator, a device that combines all the flexibility of a function generator with the stability and accuracy of a frequency synthesizer. For general use, if you can have only one signal source, the function generator is for you.

CAPACITORS AND AC CIRCUITS

Once we enter the world of changing voltages and currents, or signals, we encounter two very interesting circuit elements that are useless in dc circuits: capacitors and inductors. As you will see, these humble devices, combined with resistors, complete the triad of passive linear circuit elements that form the basis of nearly all circuitry. Capacitors, in particular, are essential in nearly every circuit application. They are used for waveform generation, filtering, and blocking and bypass applications. They are used in integrators and differentiators. In combination with inductors, they make possible sharp filters for separating desired signals from background. You will see some of these applications as we continue with this chapter, and there will be numerous interesting examples in later chapters.

Let's proceed, then, to look at capacitors in detail. Portions of the treatment that follows are necessarily mathematical in nature; the reader with little mathematical preparation may find Appendix B helpful. In any case, an understanding of the details is less important in the long run than an understanding of the results.

1.12 Capacitors

To a first approximation, capacitors (Fig. 1.25) are devices that might be considered

Figure 1.25

simply frequency-dependent resistors. They allow you to make frequency-dependent voltage dividers, for instance. For some applications (bypass, coupling) this is almost all you need to know, but for other applications (filtering, energy storage, resonant circuits) a deeper understanding is needed. For example, capacitors cannot dissipate power, even though current can flow through them, because the voltage and current are 90° out of phase.

A capacitor (the old-fashioned name was *condenser*) is a device that has two wires sticking out of it and has the property

$$Q = CV$$

A capacitor of C farads with V volts across its terminals will contain Q coulombs of stored charge.

Taking the derivative (see Appendix B), you get

$$I = C\frac{dV}{dt}$$

So a capacitor is more complicated than a resistor; the current is not simply proportional to the voltage, but rather to the rate of change of voltage. If you change the voltage across a farad by 1 volt per second, you are supplying an amp. Conversely, if you supply an amp, its voltage changes by 1 volt per second. A farad is very large, and you usually deal in microfarads (μF) or picofarads (pF). (To make matters confusing to the uninitiated, the units are often omitted on capacitor values specified in schematic diagrams. You have to figure it out from the context.) For instance, if you supply a current of 1mA to 1μF, the voltage will rise at 1000 volts per second. A 10 millisecond pulse of this current will increase the voltage across the capacitor by 10 volts (Fig. 1.26).

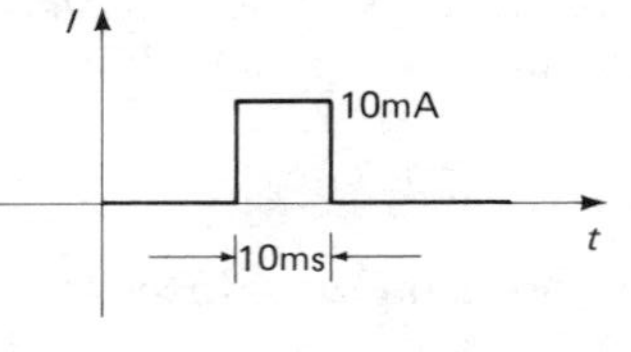

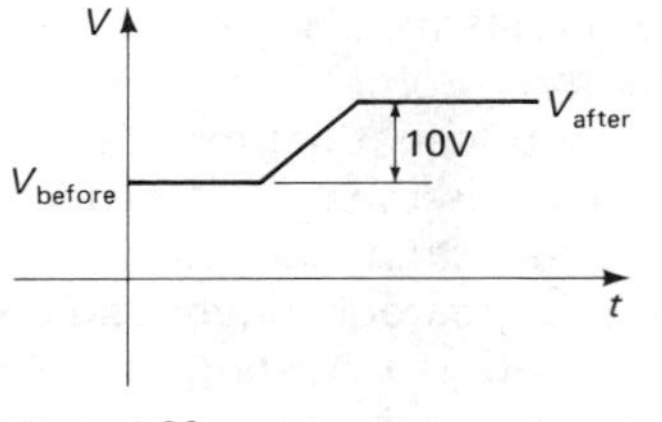

Figure 1.26

Capacitors come in an amazing variety of shapes and sizes; with time, you will come to recognize their more common incarnations. The basic construction is simply two conductors near each other (but not touching); in fact, the simplest capacitors are just that. For greater capacitance, you need more area and closer spacing; the usual approach is to plate some conductor onto a thin insulating material (called a dielectric), for instance, aluminized Mylar film rolled up into a small cylindrical configuration. Other popular types are thin ceramic wafers (disc ceramics), metal foils with oxide insulators (electrolytics), and metallized mica. Each of these types has unique properties; for a brief rundown of their idiosyncrasies, see the box on capacitors. In general, ceramic and Mylar types are used for most noncritical circuit applications; tantalum capacitors are used where greater capacitance is needed, and electrolytics are used for power supply filtering.

Capacitors in parallel and series

The capacitance of several capacitors in parallel is the sum of their individual capacitances. This is easy to see: Put voltage V across the parallel combination; then

$$C_{total}V = Q_{total} = Q_1 + Q_2 + Q_3 + \cdots$$
$$= C_1V + C_2V + C_3V + \cdots$$
$$= (C_1 + C_2 + C_3 + \cdots)V$$

or

$$C_{total} = C_1 + C_2 + C_3 + \cdots$$

For capacitors in series, the formula is like that for resistors in parallel:

$$C_{total} = \frac{1}{\frac{1}{C_1} + \frac{1}{C_2} + \frac{1}{C_3} + \cdots}$$

or (two capacitors only)

$$C_{total} = \frac{C_1C_2}{C_1 + C_2}$$

CAPACITORS

There is wide variety among the capacitor types available. This is a quickie guide to point out their major advantages and disadvantages. Our judgments should be considered somewhat subjective.

Type	Capacitance range	Maximum voltage	Accuracy	Temperature stability	Leakage	Comments
Mica	1pF–0.01μF	100–600	Good		Good	Excellent; good at RF
Tubular ceramic	0.5pF–100pF	100–600		Selectable		Very low values available; several tempcos available (including zero tempco)
Ceramic	10pF–1μF	50–1000		Poor		Small, inexpensive, very popular; can be self-resonant ~ 100kHz
Mylar	0.001μF–10μF	50–600	Good	Poor	Good	Inexpensive, good, very popular
Polystyrene	10pF–0.01μF	100–600	Good		Excellent	High quality, large; good for signal filters
Polycarbonate	100pF–10μF	50–400	Good	Good	Good	High quality; good for integrators
Glass	10pF–1000pF	100–600	Good		Excellent	Long-term stability
Porcelain	100pF–0.1μF	50–400	Good	Good	Good	Good, inexpensive; long-term stability
Tantalum	0.1μF–500μF	6–100	Poor	Poor		High capacitance, with acceptable leakage; polarized, small; low inductance; very popular
Electrolytic	0.1μF–0.2F	3–600	Terrible	Ghastly	Awful	Not recommended except in power supply filters (use tantalum for high-capacitance requirements); polarized; short life
Oil	0.1μF–20μF	200V–10kV			Good	High-voltage filters; large, long life

EXERCISE 1.12

Derive the formula for the capacitance of two capacitors in series. Hint: Since there is no external connection to the point where the two capacitors are connected together, they must have equal stored charges.

1.13 *RC* circuits: *V* and *I* versus time

When dealing with ac circuits (or, in general, any circuits that have changing voltages and currents), there are two possible approaches. You can talk about V and I versus time, or you can talk about amplitude versus signal frequency. Both approaches have their merits, and you find yourself switching back and forth according to which description is most convenient in each situation. We will begin our study of ac circuits in the time domain. Beginning with Section 1.18, we will tackle the frequency domain.

What are some of the features of circuits with capacitors? To answer this question, let's begin with the simple RC circuit (Fig. 1.27). Application of the capacitor rules

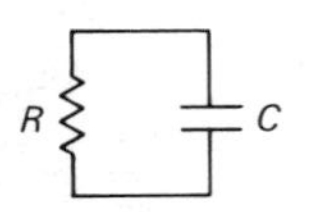

Figure 1.27

gives

$$C\frac{dV}{dt} = I = -\frac{V}{R}$$

This is a differential equation, and its solution is

$$V = Ae^{-t/RC}$$

So a charged capacitor placed across a resistor will discharge as in Figure 1.28.

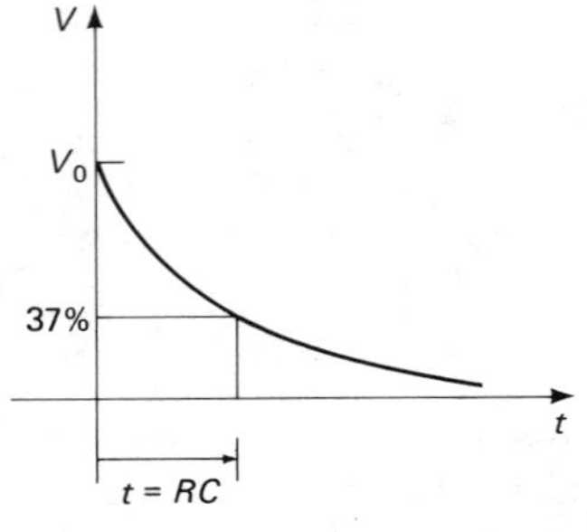

Figure 1.28

Time constant

The product RC is called the *time constant* of the circuit. For R in ohms and C in farads, the product RC is in seconds. A microfarad across 1.0k has a time constant of 1 millisecond; if the capacitor is initially charged to 1.0 volt, the initial current is 1.0mA.

Figure 1.29 shows a slightly different

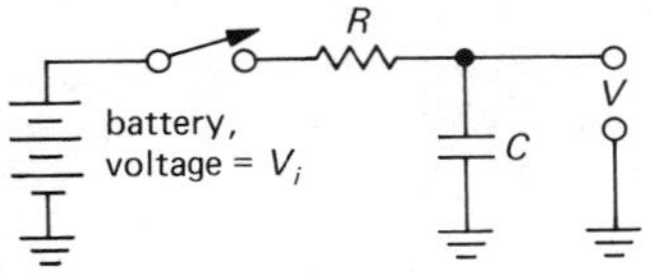

Figure 1.29

circuit. At time $t = 0$, someone connects the battery. The equation for the circuit is then

$$I = C\frac{dV}{dt} = \frac{V_i - V}{R}$$

with the solution

$$V = V_i + Ae^{-t/RC}$$

(Please don't worry if you can't follow the mathematics. What we are doing is getting some important results, which you should remember. Later we will use these results often, with no further need for the mathematics used to derive them.) The constant A is determined by initial conditions (Fig. 1.30): $V = 0$ at $t = 0$; therefore $A = -V_i$,

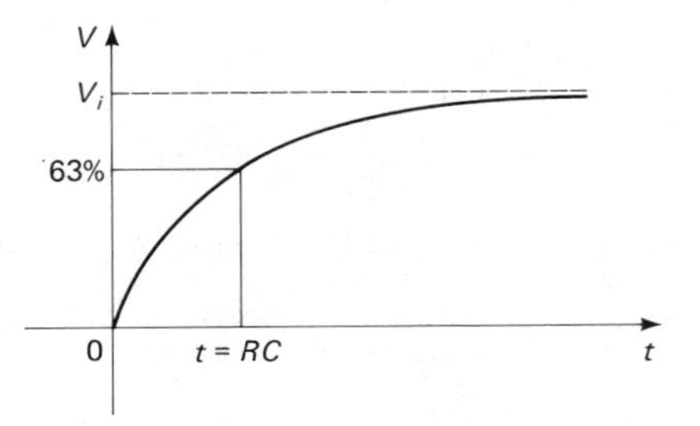

Figure 1.30

and

$$V = V_i(1 - e^{-t/RC})$$

Decay to equilibrium

Eventually (when $t \gg RC$), V reaches V_i. (Presenting the "5RC rule of thumb": a capacitor charges or decays to within 1% of its final value in 5 time constants.) If we then change V_o to some other value (say, 0), V will decay toward that new value with an exponential $e^{-t/RC}$. For example, a square-wave input for V_o will produce the output shown in Figure 1.31.

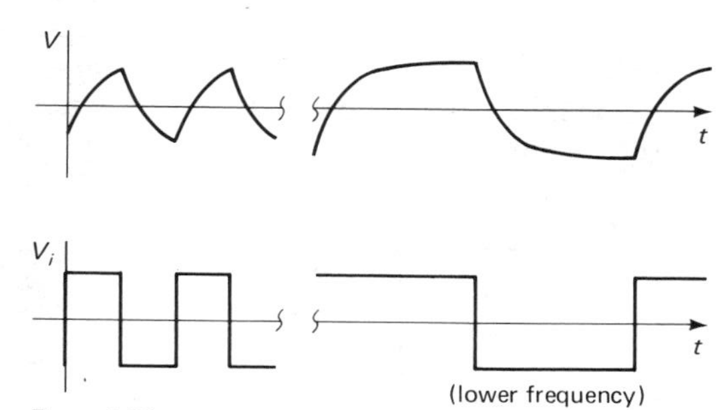

Figure 1.31

EXERCISE 1.13

Show that the rise time (the time required to go from 10% to 90% of its final value) of this signal is $2.2RC$.

You might ask the obvious next question: What about $V(t)$ for arbitrary $V_i(t)$? The solution involves an inhomogeneous differential equation and can be solved by standard methods (which are, however, beyond the scope of this book). You would find

$$V(t) = \frac{1}{RC}\int_{-\infty}^{t} V_i(\tau)e^{-(t-\tau)/RC}\,d\tau$$

That is, the RC circuit averages past history at the input with a weighting factor

$e^{-\Delta t/RC}$

In practice, you seldom ask this question. Instead, you deal in the frequency domain and ask how much of each frequency component present in the input gets through. We will get to this important topic soon. Before we do, though, there are a few other interesting circuits we can analyze simply with this time-domain approach.

Simplification by Thévenin equivalents

We could go ahead and analyze more complicated circuits by similar methods, writing down the differential equations and trying to find solutions. For most purposes it simply isn't worth it. This is as complicated

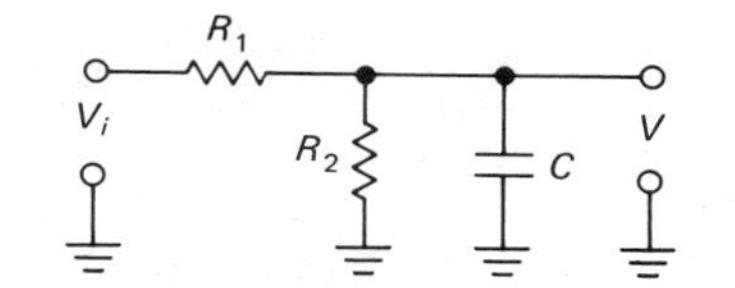

Figure 1.32

an RC circuit as we will need. Many other circuits can be reduced to it (e.g., Fig. 1.32). By just using the Thévenin equivalent of the voltage divider formed by R_1 and R_2, you can find the output $V(t)$ produced by a step input for V_o.

EXERCISE 1.14

$R_1 = R_2 = 10\text{k}$ and $C = 0.1\mu\text{F}$ in the circuit shown in Figure 1.32. Find $V(t)$ and sketch it.

Example: time-delay circuit

We have already mentioned logic levels, the voltages that digital circuits live on. Figure 1.33 shows an application of capacitors to

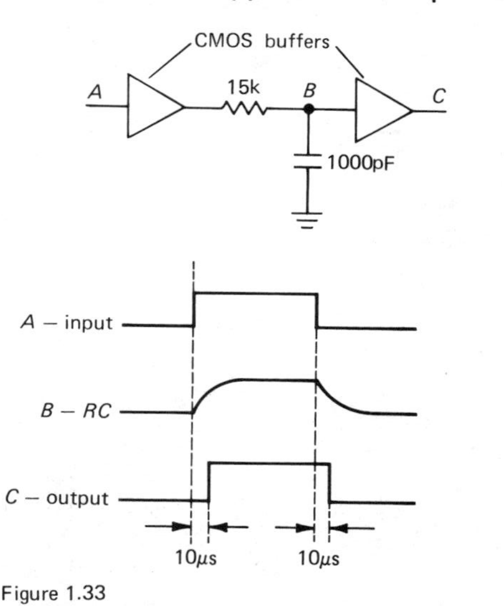

Figure 1.33

produce a delayed pulse. The triangular symbols are "CMOS buffers." They give a HIGH output if the input is HIGH (more than one-half the supply voltage), and vice versa. The first buffer provides a replica of the input signal, but with low source resistance, and prevents input loading by the RC (recall our earlier discussion of circuit loading in Section 1.05). The RC output has the characteristic decays and causes the output buffer to switch 10μs after the input transitions (an *RC* reaches 50% output in 0.7*RC*). In an actual application you would have to consider the effect of the buffer input threshold deviating from one-half the supply voltage, which would alter the delay and change the output pulse width. Such a circuit is sometimes used to delay a pulse so that something else can happen first. In designing circuits you try not to rely on tricks like this, but they're occasionally handy.

1.14 Differentiators

Look at the circuit in Figure 1.34. The voltage across C is $V_{in} - V$, so

$$I = C\frac{d}{dt}(V_{in} - V) = \frac{V}{R}$$

If we choose R and C small enough so that $dV/dt \ll dV_{in}/dt$, then

$$C\frac{dV_{in}}{dt} \approx \frac{V}{R}$$

or

$$V(t) = RC\frac{d}{dt}V_{in}(t)$$

That is, we get an output proportional to the rate of change of the input waveform.

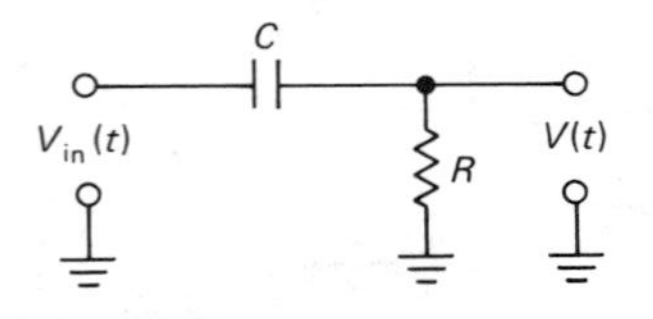

Figure 1.34

To keep $dV/dt \ll dV_{in}/dt$, we make the product *RC* small, taking care not to "load" the input by making R too small (at the transition the change in voltage across the capacitor is zero, so R is the load seen by the input). We will have a better criterion for this when we look at things in the frequency domain. If you drive this circuit with a square wave, the output will be as shown in Figure 1.35.

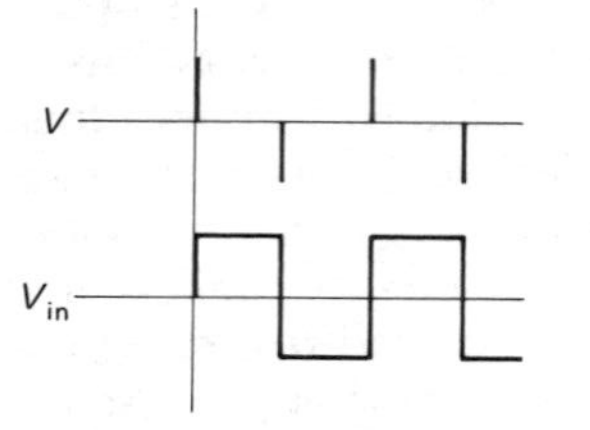

Figure 1.35

Differentiators are handy for detecting *leading edges* and *trailing edges* in pulse signals, and in digital circuitry you sometimes see things like those depicted in Figure 1.36. The *RC* differentiator generates spikes

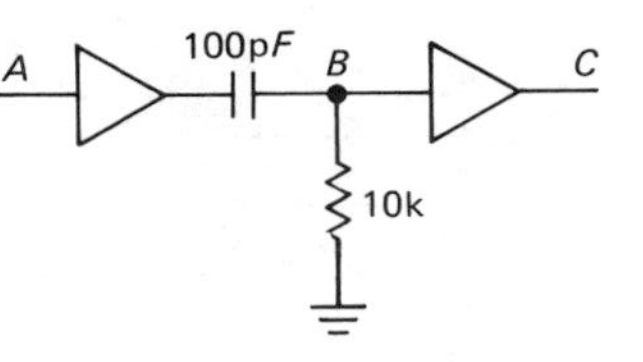

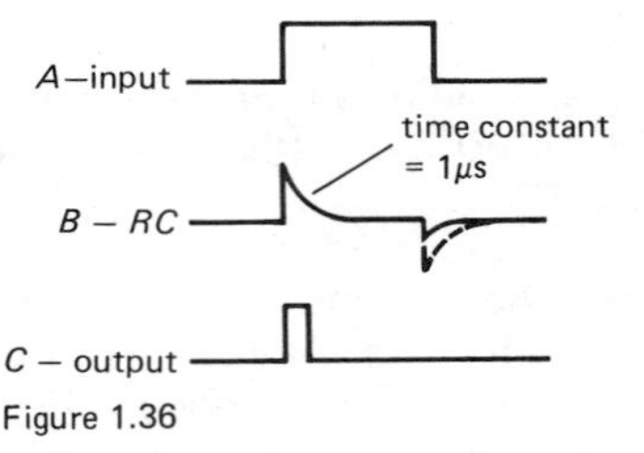

Figure 1.36

at the transitions of the input signal, and the output buffer converts the spikes to short square-topped pulses. In practice, the negative spike will be small because of a diode (a handy device discussed in Section 1.25) built into the buffer.

Unintentional capacitive coupling

Differentiators sometimes crop up unexpectedly, in situations where they're not welcome. You may see signals like those shown in Figure 1.37. The first case is caused by a square wave somewhere in the circuit coupling capacitively to the signal line you're looking at; that might indicate a miss-

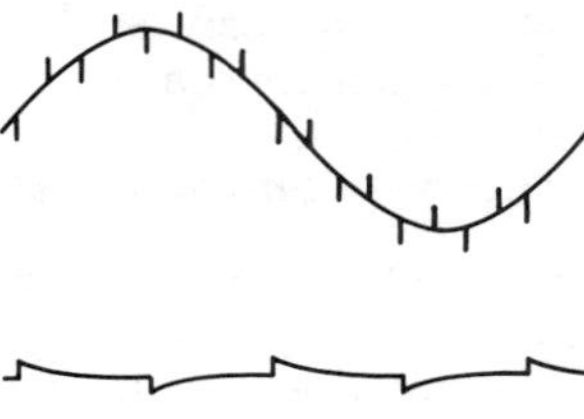
Figure 1.37

ing resistor termination on your signal line. If not, you must either reduce the source resistance of the signal line or find a way to reduce capacitive coupling from the offending square wave. The second case is typical of what you might see when you look at a square wave, but have a broken connection somewhere, usually at the scope probe. The very small capacitance of the broken connection combines with the scope input resistance to form a differentiator. *Knowing that you've got a differentiated "something" can help you find the trouble and eliminate it.*

1.15 Integrators

Take a look at the circuit in Figure 1.38. The voltage across R is $V_{in} - V$, so

$$I = C\frac{dV}{dt} = \frac{V_{in} - V}{R}$$

If we manage to keep $V \ll V_{in}$, by keeping the product RC large, then

$$C\frac{dV}{dt} \approx \frac{V_{in}}{R}$$

or

$$V(t) = \frac{1}{RC}\int^{t} V_{in}(t)\,dt + \text{constant}$$

We have a circuit that performs the integral over time of an input signal! You can see how the approximation works for a square-wave input: $V(t)$ is then the exponential

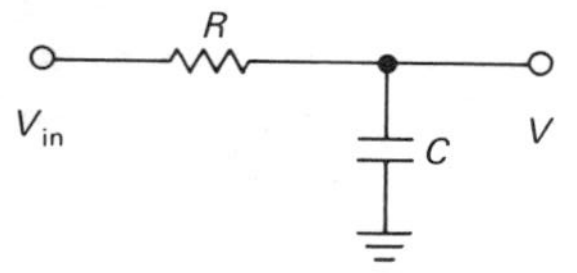

Figure 1.38

charging curve we saw earlier (Fig. 1.39). The first part of the exponential is a ramp, the integral of a constant; as we increase the time constant RC, we pick off a smaller part of the exponential, i.e., a better approximation to a perfect ramp.

Note that the condition $V \ll V_{in}$ is just the same as saying that I is proportional to V_{in}. If we had as input a *current* $I(t)$, rather than a

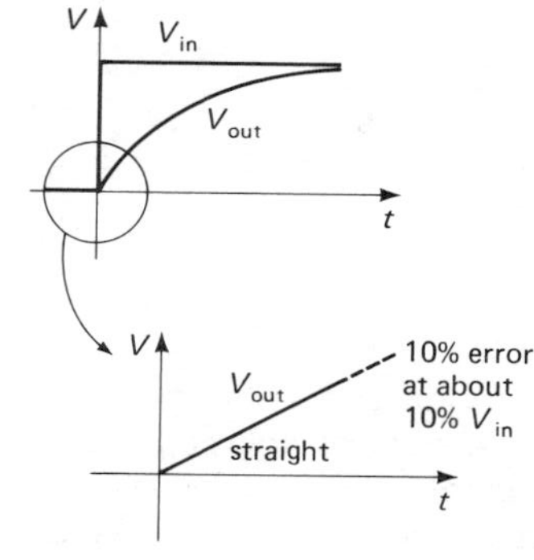

Figure 1.39

voltage, we would have an exact integrator. A large voltage across a large resistance approximates a current source and, in fact, is frequently used as one.

Later, when we get to operational amplifiers and feedback, we will be able to build integrators without the restriction $V_{out} \ll V_{in}$. They will work over large frequency and voltage ranges with negligible error.

The integrator is used extensively in analog computation. It is a useful subcircuit that finds application in control systems, feedback, analog/digital conversion, and waveform generation.

Ramp generators

At this point it is easy to understand how a ramp generator works. This nice circuit is extremely useful, for example in timing circuits, waveform and function generators, oscilloscope sweep circuits, and analog/digital conversion circuitry. The circuit is shown in Figure 1.40. From the capacitor equation

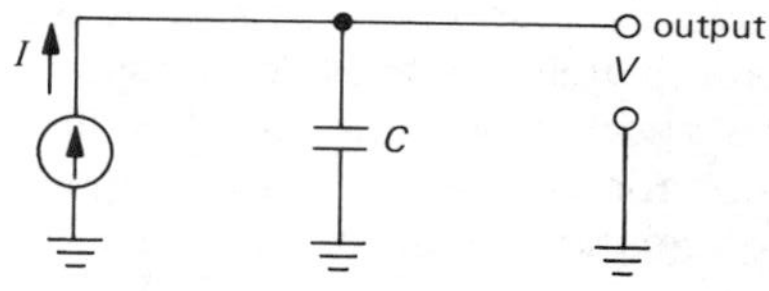

Figure 1.40

$I = C(dV/dt)$, you get $V(t) = (I/C)t$. The output waveform is as shown in Figure 1.41. The ramp stops when the current source "runs out of voltage," i.e., reaches the limit of its compliance. The curve for a simple RC, with the resistor tied to a voltage source equal to the compliance of the current source, and with R chosen so that

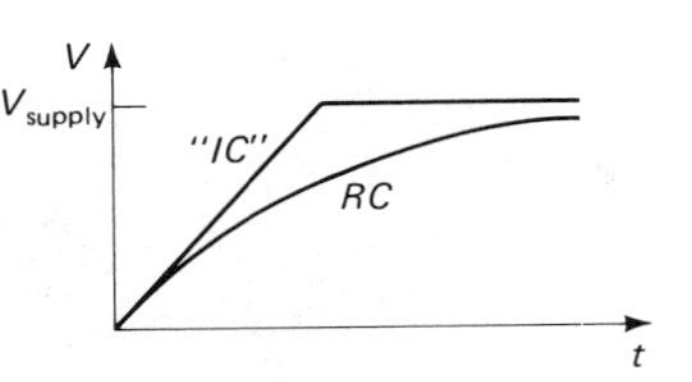

Figure 1.41

the current at zero output voltage is the same as that of the current source, is also drawn for comparison. (Real current sources generally have output compliances limited by the power supply voltages used in making them, so the comparison is realistic.) In the next chapter, which deals with transistors, we will design some current sources, with some refinements to follow in the chapters on operational amplifiers (op-amps) and field-effect transistors (FETs). Exciting things to look forward to!

EXERCISE 1.15

A current of 1mA charges a 1μF capacitor. How long does it take the ramp to reach 10 volts?

INDUCTORS AND TRANSFORMERS

1.16 Inductors

If you understand capacitors, you won't have any trouble with inductors (Fig. 1.42).

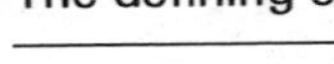

Figure 1.42

They're closely related to capacitors; the rate of current change in an inductor depends on the voltage applied across it, whereas the rate of voltage change in a capacitor depends on the current through it. The defining equation for an inductor is

$$V = L\frac{dI}{dt}$$

where L is called the *inductance* and is measured in henrys (or mH, μH, etc.). Putting a voltage across an inductor causes the current to rise as a ramp (for a capacitor, supplying a constant current causes the voltage to rise as a ramp); 1 volt across 1 henry produces a current that increases at 1 amp per second.

The symbol for an inductor looks like a coil of wire; that's because, in its simplest form, that's all it is. Variations include coils wound on various core materials, the most popular being iron (or iron alloys, laminations, or powder) and ferrite, a black, nonconductive, brittle magnetic material. These are all ploys to multiply the inductance of a given coil by the "permeability" of the core material. The core may be in the shape of a rod, a toroid (doughnut), or even more bizarre shapes, such as a "pot core" (which has to be seen to be understood; the best description we can think of is a doughnut mold split in half, if doughnuts were made in molds).

Inductors find heavy use in radiofrequency (RF) circuits, serving as RF "chokes" and as parts of tuned circuits. A pair of closely coupled inductors form the interesting object known as a transformer. We will talk briefly about them in the next section.

An inductor is, in a real sense, the opposite of a capacitor. You will see how that works out in the next few sections of this chapter, which deal with the important subject of *impedance*.

1.17 Transformers

A transformer is a device consisting of two closely coupled coils (called primary and secondary). An ac voltage applied to the primary appears across the secondary, with a voltage multiplication proportional to the turns ratio of the transformer. The current is, of course, correspondingly reduced. Figure 1.43 shows the circuit symbol for a laminated-core transformer (the kind used for 60Hz ac power conversion).

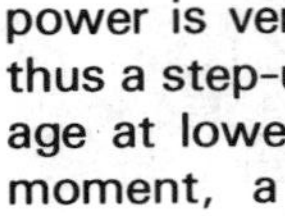

Figure 1.43

Transformers are quite efficient (output power is very nearly equal to input power); thus a step-up transformer gives higher voltage at lower current. Jumping ahead for a moment, a transformer of turns ratio n

increases the impedance by n^2. There is very little primary current if the secondary is unloaded.

Transformers serve two important functions in electronic instruments: They change the ac line voltage to a useful (usually lower) value that can be used by the circuit, and they "isolate" the electronic device from actual connection to the power line, since the windings of a transformer are electrically insulated from each other. "Power transformers" (meant for use from the 110 volt power line) come in an enormous variety of secondary voltages and currents: outputs as low as 1 volt or so up to several thousand volts, current ratings from a few milliamps to hundreds of amps. Typical transformers for use in electronic instruments might have secondary voltages from 10 to 50 volts, with current ratings of 0.1 to 5 amps or so.

Transformers for use at audio frequencies and radiofrequencies are also available. At radiofrequencies you sometimes use tuned transformers, if only a narrow range of frequencies is present. There is also an interesting class of transmission-line transformers that we will discuss briefly in Section 13.10. In general, transformers for use at high frequencies must use special core materials or construction to minimize core losses, whereas low-frequency transformers (e.g., power transformers) are burdened instead by large and heavy cores. The two kinds of transformers are in general not interchangeable.

IMPEDANCE AND REACTANCE

Warning: This section is somewhat mathematical; you may wish to skip over the mathematics, but be sure to pay attention to the results.

Circuits with capacitors and inductors are more complicated than the resistive circuits we talked about earlier, in that they depend on frequency; they "corrupt" inputs such as square waves, as we just saw. Yet it is possible to generalize Ohm's law, replacing the word "resistance" with "impedance," in order to describe any circuit containing these linear passive devices (resistors, capacitors, and inductors). You could think of the subject of impedance and reactance as Ohm's law for circuits that include capacitors and inductors. Some important terminology: Impedance is the "generalized resistance"; inductors and capacitors have *reactance* (they are "reactive"); resistors have *resistance* (they are "resistive"). In other words, impedance = resistance + reactance (more about this later). However, you'll see statements like "the impedance of the capacitor at this frequency is . . ." The reason you don't have to use the word reactance in such a case is that impedance covers everything. In fact, you frequently use the word impedance even when you know it's a resistance you're talking about; you say "the source impedance" or "the output impedance" when you mean the Thévenin equivalent resistance of some source. The same holds for "input impedance."

In all that follows, we will be talking about circuits driven by sine waves at a single frequency. Analysis of circuits driven by complicated waveforms is more elaborate, involving the methods we used earlier (differential equations) or decomposition of the waveform into sine waves (Fourier analysis). Fortunately, these methods are seldom necessary.

1.18 Frequency analysis of reactive circuits

Let's start by looking at a capacitor driven by a sine-wave voltage source (Fig. 1.44). The current is

$$I(t) = C\frac{dV}{dt} = C\omega V_0 \cos \omega t$$

i.e., a current of amplitude I, with the phase leading the input voltage by 90°. If we consider amplitudes only, and disregard phases, the current is

$$I = \frac{V}{1/\omega C}$$

(Recall that $\omega = 2\pi f$.) It behaves like a frequency-dependent resistance $R = 1/\omega C$, but in addition the current is 90° out of phase with the voltage (Fig. 1.45). For example, a 1μF capacitor put across the

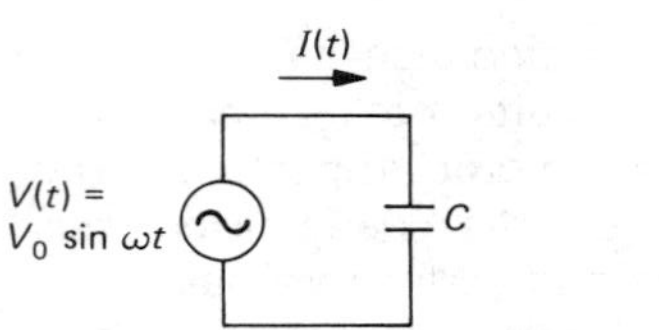

Figure 1.44

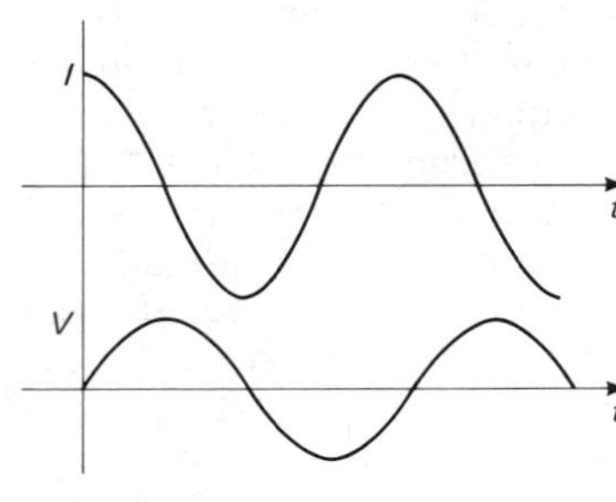

Figure 1.45

110 volt (rms) 60Hz power line draws a current of rms amplitude

$$I = \frac{110}{1/(2\pi \times 60 \times 10^{-6})} = 41.5\text{mA (rms)}$$

Note: At this point it is necessary to get into some complex algebra; you may wish to skip over the math in some of the following sections, taking note of the results as we derive them. A knowledge of the detailed mathematics is not necessary in order to understand the remainder of the book. Very little mathematics will be used in later chapters. The section ahead is easily the most difficult for the reader with little mathematical preparation. Don't be discouraged!

Voltages and currents as complex numbers

As you have just seen, there can be phase shifts between the voltage and current in an ac circuit being driven by a sine wave at some frequency. Nevertheless, as long as the circuit contains only *linear* elements (resistors, capacitors, inductors), the magnitudes of the currents everywhere in the circuit are still proportional to the magnitude of the driving voltage, so we might hope to find some generalization of voltage, current, and resistance in order to rescue Ohm's law. Obviously a single number won't suffice to specify the current, say, at some point in the circuit, since we must somehow have information about both the magnitude and phase shift.

Although we can imagine specifying the magnitudes and phase shifts of voltages and currents at any point in the circuit by writing them out explicitly, e.g., $V(t) = 23.7 \sin(377t + 0.38)$, it turns out that our requirements can be met more simply by using the algebra of complex numbers to *represent* voltages and currents. Then we can simply add or subtract the complex number representations, rather than laboriously having to add or subtract the actual sinusoidal functions of time themselves. Since the actual voltages and currents are real quantities that vary with time, we must develop a rule for converting from actual quantities to their representation, and vice versa. Recalling once again that we are talking about a single sine-wave frequency, ω, we agree to use the following rules:

1. Voltages and currents are *represented* by the complex quantities $\mathbf{V}$ and $\mathbf{I}$. The voltage $V_0 \cos(\omega t + \phi)$ is to be represented by the complex number $V_0 e^{j\phi}$.
2. *Actual* voltages and currents are obtained by multiplying their complex number representations by $e^{j\omega t}$ and then taking the real part: $V(t) = \mathcal{Re}(\mathbf{V}e^{j\omega t})$, $I(t) = \mathcal{Re}(\mathbf{I}e^{j\omega t})$

In other words,

circuit voltage versus time		complex number representation
$V_0 \cos(\omega t + \phi)$	⇄ multiply by $e^{j\omega t}$ and take real part	$V_0 e^{j\phi} = a + jb$

(In electronics, the symbol j is used instead of i in the exponential in order to avoid confusion with the symbol i meaning current.) Thus in the general case the actual voltages and currents are given by:

$$V(t) = \mathcal{Re}(\mathbf{V}e^{j\omega t}) = \mathcal{Re}(\mathbf{V}) \cos \omega t - \mathcal{Im}(\mathbf{V}) \sin \omega t$$

$$I(t) = \mathcal{Re}(\mathbf{I}e^{j\omega t}) = \mathcal{Re}(\mathbf{I}) \cos \omega t - \mathcal{Im}(\mathbf{I}) \sin \omega t$$

For example, a voltage whose complex representation is

$$\mathbf{V} = 5j$$

corresponds to a (real) voltage versus time of

$$V(t) = \mathcal{Re}[5j \cos \omega t + 5j(j) \sin \omega t]$$
$$= -5 \sin \omega t \text{ volts}$$

Reactance of capacitors and inductors

With this convention we can apply complex Ohm's law to circuits containing capacitors and inductors, just as for resistors, once we know the reactance of a capacitor or inductor. Let's find out what these are. We have

$$V(t) = \mathcal{Re}(V_0 e^{j\omega t})$$

For a capacitor, using $I = C(dV/dt)$, we obtain

$$I(t) = -V_0 C\omega \sin \omega t = \mathcal{Re}\left(\frac{V_0 e^{j\omega t}}{-j/\omega C}\right)$$
$$= \mathcal{Re}\left(\frac{V_0 e^{j\omega t}}{X_C}\right)$$

i.e., for a capacitor

$$X_C = -j/\omega C$$

X_C is the reactance of a capacitor at frequency ω. As an example, a 1μF capacitor has a reactance of $-2653j$ ohms at 60Hz and a reactance of $-0.16j$ ohms at 1MHz. Its reactance at dc is infinite.

If we did a similar analysis for an inductor, we would find

$$X_L = j\omega L$$

A circuit containing only capacitors and inductors always has a purely imaginary impedance, meaning that the voltage and current are always 90° out of phase – it is purely reactive. When the circuit contains resistors, there is also a real part to the impedance. The term "reactance" in that case means the imaginary part only.

Ohm's law generalized

With these conventions for representing voltages and currents, Ohm's law takes a simple form. It reads simply

$$\mathbf{I} = \mathbf{V}/\mathbf{Z}$$
$$\mathbf{V} = \mathbf{IZ}$$

where the voltage represented by V is applied across a circuit of impedance $\mathbf{Z}$, giving a current represented by $\mathbf{I}$. The complex impedance of devices in series or parallel obeys the same rules as resistance:

$$\mathbf{Z} = \mathbf{Z}_1 + \mathbf{Z}_2 + \mathbf{Z}_3 + \cdots \qquad \text{(series)}$$

$$\mathbf{Z} = \frac{1}{1/\mathbf{Z}_1 + 1/\mathbf{Z}_2 + 1/\mathbf{Z}_3 + \cdots} \qquad \text{(parallel)}$$

Finally, for completeness we summarize here the formulas for the impedance of resistors, capacitors, and inductors:

$$\mathbf{Z}_R = R \qquad \text{(resistor)}$$
$$\mathbf{Z}_C = -j/\omega C \qquad \text{(capacitor)}$$
$$\mathbf{Z}_L = j\omega L \qquad \text{(inductor)}$$

With these rules we can analyze many ac circuits by the same general methods we used in handling dc circuits, i.e., application of the series and parallel formulas and Ohm's law. Our results for circuits such as voltage dividers will look nearly the same as before. For multiply connected networks we may have to use Kirchhoff's laws, just as with dc circuits, in this case using the complex representations for V and I: The sum of the (complex) voltage drops around a closed loop is zero, and the sum of the (complex) currents into a point is zero. The latter rule implies, as with dc circuits, that the (complex) current in a series circuit is the same everywhere.

EXERCISE 1.16

Use the preceding rules for the impedance of devices in parallel and in series to derive the formulas (Section 1.12) for the capacitance of two capacitors (a) in parallel and (b) in series. Hint: In each case, let the individual capacitors have capacitances C_1 and C_2. Write down the impedance of the parallel or series combination; then equate it to the impedance of a capacitor with capacitance C. Find C.

Let's try out these techniques on the simplest circuit imaginable, an ac voltage applied across a capacitor, which we considered just previously. Then, after a brief look at power in reactive circuits (to finish laying the groundwork), we'll analyze some simple but extremely important and useful *RC* filter circuits.

Imagine putting a 1μF capacitor across a 110 volt (rms) 60Hz power line. What current flows? Using complex Ohm's law, we have

$$\mathbf{Z} = -j/\omega C$$

Therefore the current is given by

$$\mathbf{I} = \mathbf{V}/\mathbf{Z}$$

The phase of the voltage is arbitrary, so let us choose $\mathbf{V} = A$, i.e., $V(t) = A \cos \omega t$, where the amplitude $A = 110\sqrt{2} \approx 156$ volts. Then

$$\mathbf{I} = j\omega CA \approx 0.059 \sin \omega t$$

The resulting current has an amplitude of 59mA (41.5mA rms) and leads the voltage by 90°. This agrees with our previous calculation. Note that if we just wanted to know the magnitude of the current, and didn't care what the relative phase was, we could have avoided doing any complex algebra: If

$$\mathbf{A} = \mathbf{B}/\mathbf{C}$$

then

$$A = B/C$$

where A, B, and C are the magnitudes of the respective complex numbers; this holds for multiplication, also (see Exercise 1.17). Thus in this case

$$I = V/Z = \omega CV$$

This trick is often useful.

Surprisingly, there is no power dissipated by the capacitor in this example. Such activity won't increase your electric bill; you'll see why in the next section. Then we will go on to look at circuits containing resistors and capacitors with our complex Ohm's law.

EXERCISE 1.17

Show that, if $\mathbf{A} = \mathbf{BC}$, then $A = BC$, where A, B, and C are magnitudes. Hint: Represent each complex number in polar form, i.e., $\mathbf{A} = Ae^{i\theta}$.

□ *Power in reactive circuits*

The instantaneous power delivered to any circuit element is always given by the product $P = VI$. However, in reactive circuits where V and I are not simply proportional, you can't just multiply them together. Funny things can happen; for instance, the sign of the product can reverse over one cycle of the ac signal. Figure 1.46 shows an example.

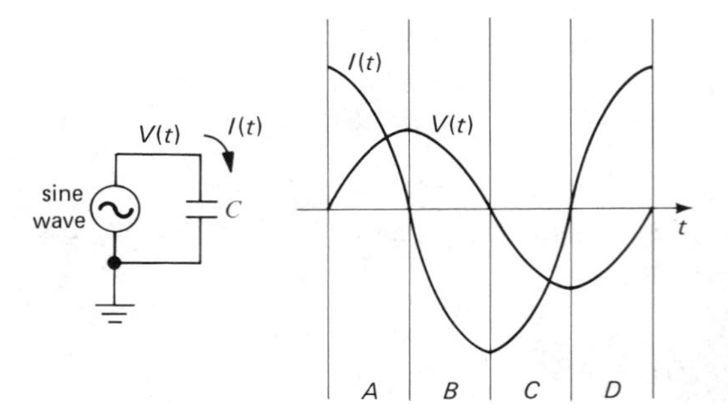

Figure 1.46
Current in a capacitor driven by an applied sine-wave voltage.

During time intervals A and C, power is being delivered to the capacitor (albeit at a variable rate), causing it to charge up; its stored energy is increasing (power is the rate of change of energy). During intervals B and D, the power delivered to the capacitor is negative; it is discharging. The average power over a whole cycle for this example is in fact exactly zero, a statement that is always true for any purely reactive circuit element (inductors, capacitors, or any combination thereof). If you know your trigonometric integrals, the next exercise will show you how to prove this.

EXERCISE 1.18

Optional exercise: Prove that a circuit whose current is 90° out of phase with the driving voltage consumes no power, averaged over an entire cycle.

How do we find the average power consumed by an arbitrary circuit? In general, we can imagine adding up little pieces of VI product, then dividing by the elapsed time.

In other words,

$$P = \frac{1}{T}\int_0^T V(t)\, I(t)\, dt$$

where T is the time for one complete cycle. Luckily, that's almost never necessary. Instead, it is easy to show that the average power is given by

$$P = \mathcal{R}e(\mathbf{VI}^*) = \mathcal{R}e(\mathbf{V}^*\mathbf{I})$$

where **V** and **I** are complex rms amplitudes.

Let's taken an example. Consider the preceding circuit, with a 1 volt (rms) sine wave driving a capacitor. We'll do everything with rms amplitudes, for simplicity. We have

$$\mathbf{V} = 1$$

$$\mathbf{I} = \frac{\mathbf{V}}{j/\omega C} = -j\omega C$$

$$P = \mathcal{R}e(\mathbf{VI}^*) = \mathcal{R}e(j\omega C) = 0$$

That is, the average power is zero, as stated earlier.

As another example, consider the circuit shown in Figure 1.47. Our calculations go like this:

$$\mathbf{Z} = R + \frac{1}{j\omega C} = R - \frac{j}{\omega C}$$

$$\mathbf{V} = V_0$$

$$\mathbf{I} = \frac{\mathbf{V}}{\mathbf{Z}} = \frac{V_0}{R - (j/\omega C)} = \frac{V_0[R + (j/\omega C)]}{R^2 + (1/\omega^2 C^2)}$$

$$P = \mathcal{R}e(\mathbf{VI}^*) = \frac{V_0^2 R}{R^2 + (1/\omega^2 C^2)}$$

(In the third line we multiplied numerator and denominator by the complex conjugate of the denominator, in order to make the denominator real.) This is less than the product of the magnitudes of **V** and **I**. In fact, the ratio is called the *power factor:*

$$|\mathbf{V}||\mathbf{I}| = \frac{V_0^2}{[R^2 + (1/\omega^2 C^2)]^{1/2}}$$

$$\text{power factor} = \frac{\text{power}}{|\mathbf{V}||\mathbf{I}|} = \frac{R}{[R^2 + (1/\omega^2 C^2)]^{1/2}}$$

in this case. The power factor is the cosine of the phase angle between the voltage and the current, and it ranges from 0 (purely reactive circuit) to 1 (purely resistive). A power factor less than 1 indicates some component of reactive current.

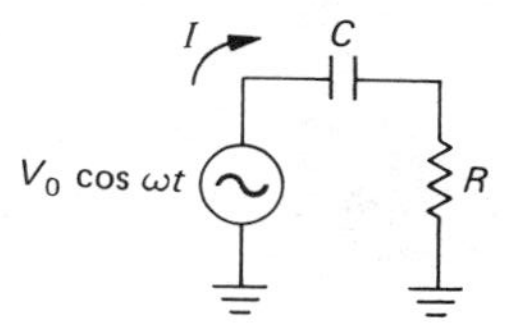

Figure 1.47

EXERCISE 1.19

Show that all the average power delivered to the preceding circuit winds up in the resistor. Do this by computing the value of V^2/R. What is that power, in watts, for a series circuit of a 1μF capacitor and a 1.0k resistor placed across the 110 volt (rms), 60Hz power line?

Power factor is a serious matter in large-scale electrical power distribution, since reactive currents don't result in useful power being delivered to the load, but cost the power company plenty in terms of I^2R heating in the resistance of generators, transformers, and wiring. Although residential users are only billed for "real" power [$\mathcal{R}e(\mathbf{VI}^*)$], the power company charges industrial users according to the power factor. This explains the capacitor yards that you see behind large factories, built to cancel the inductive reactance of industrial machinery (i.e., motors).

EXERCISE 1.20

Show that adding a series capacitor of value $C = 1/\omega^2 L$ makes the power factor equal 1.0 in a series RL circuit. Now do the same thing, but with the word "series" changed to "parallel."

1.19 *RC* filters

By combining resistors with capacitors it is possible to make frequency-dependent voltage dividers, owing to the frequency dependence of a capacitor's impedance $\mathbf{Z}_C = -j/\omega C$. Such circuits can have the desirable property of passing signal frequencies of interest while rejecting undesired signal frequencies. In this section you will see examples of the simplest such *RC* filters, which we will be using frequently throughout the book. Chapter 4 and Appendix H describe filters of greater sophistication.

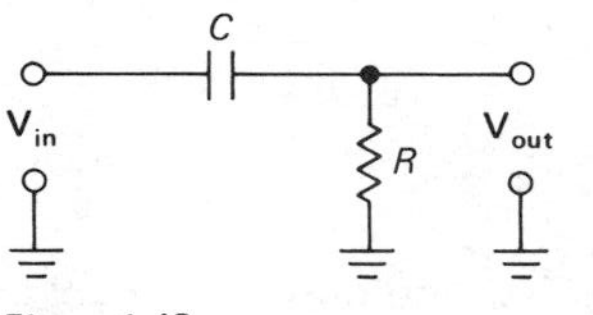

Figure 1.48

High-pass filters

Figure 1.48 shows a voltage divider made from a capacitor and a resistor. Complex Ohm's law gives

$$\mathbf{I} = \frac{\mathbf{V}_{in}}{\mathbf{Z}_{total}} = \frac{\mathbf{V}_{in}}{R - (j/\omega C)} = \frac{\mathbf{V}_{in}[R + (j/\omega C)]}{R^2 + (1/\omega^2 C^2)}$$

(For the last step, multiply top and bottom by the complex conjugate of the denominator.) So the voltage across R is just

$$\mathbf{V}_{out} = \mathbf{I}\mathbf{Z}_R = \mathbf{I}R = \frac{\mathbf{V}_{in}[R + (j/\omega C)]R}{R^2 + (1/\omega^2 C^2)}$$

Most often we don't care about the phase of V_{out}, just its amplitude:

$$V_{out} = (\mathbf{V}_{out}\mathbf{V}^*_{out})^{1/2} = \frac{R}{[R^2 + (1/\omega^2 C^2)]^{1/2}} V_{in}$$

Note the analogy to a resistive divider, where

$$V_{out} = \frac{R_1}{R_1 + R_2} V_{in}$$

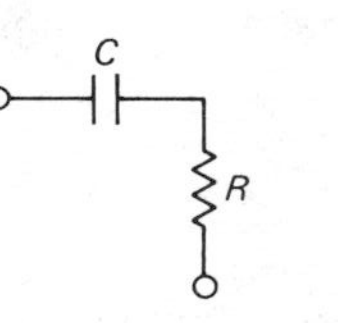

Figure 1.49

Here the impedance of the series *RC* combination (Fig. 1.49) is as shown in Figure 1.50. So the "response" of this circuit, ignoring phase shifts by taking magnitudes of the complex amplitudes, is given by

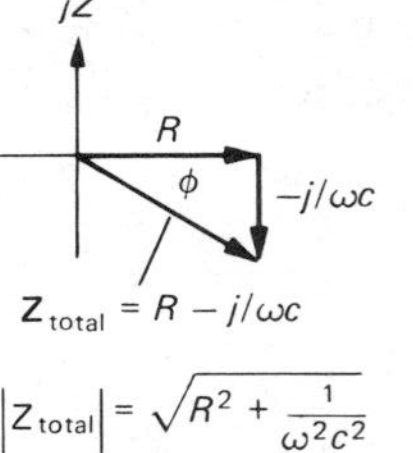

Figure 1.50

$$V_{out} = \frac{R}{[R^2 + (1/\omega^2 C^2)]^{1/2}} V_{in}$$
$$= \frac{2\pi fRC}{[1 + (2\pi fRC)^2]^{1/2}} V_{in}$$

and looks as shown in Figure 1.51. We could have gotten this result immediately by taking the ratio of the *magnitudes* of impedances, as in Exercise 1.17 and the example immediately preceding it; the numerator is the magnitude of the impedance of the lower leg of the divider (R), and the denominator is the magnitude of the impedance of the series combination of R and C.

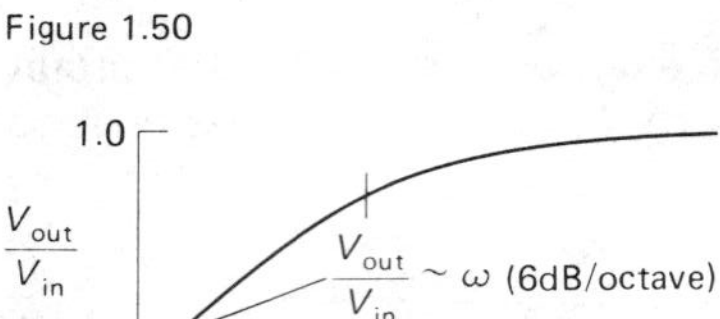

Figure 1.51
High-pass-filter frequency response.

You can see that the output is approximately equal to the input at high frequencies (how high? $\omega \gtrsim 1/RC$) and goes to zero at low frequencies. This is a very important result. Such a circuit is called a high-pass filter, for obvious reasons. It is very common. For instance, the input to the oscilloscope (Appendix A) can be switched to ac coupling. That's just an *RC* high-pass filter with the bend at about 10Hz (you would use ac coupling if you wanted to look at a small signal riding on a large dc voltage).

Note that the capacitor lets no steady current through ($f = 0$). This use as a dc *blocking capacitor* is one of its most frequent applications. Whenever you need to couple a signal from one amplifier to another, you almost invariably use a capacitor. For instance, every hi-fi audio amplifier has all its inputs capacitively coupled, since it doesn't know what dc level its input signals might be riding on. In such a coupling application you always pick R and C so that all frequencies of interest are passed without loss (attenuation).

You often need to know the impedance of a capacitor at a given frequency (e.g., for design of filters). Figure 1.52 provides a very useful graph covering large ranges of capacitance and frequency, giving the value of $|\mathbf{Z}| = 1/2\pi fC$.

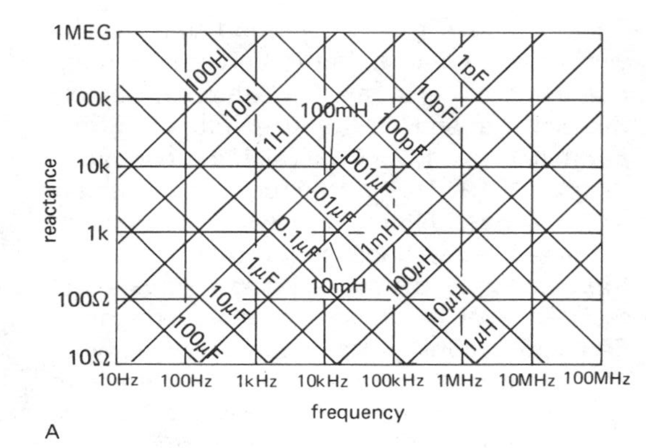

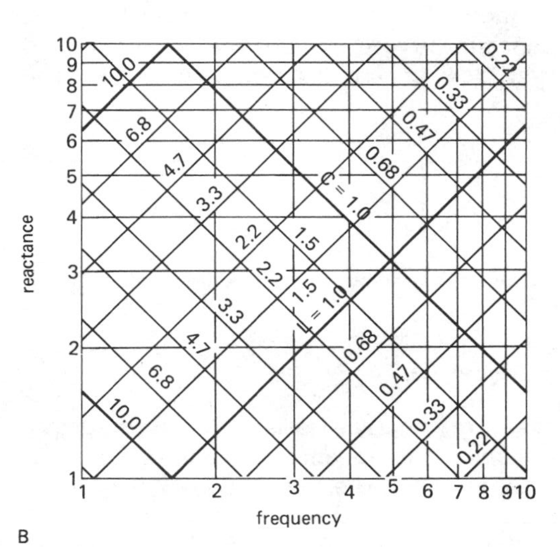

Figure 1.52
A: Reactance of inductors and capacitors versus frequency. Each decade is identical, except for scale, and is expanded in part B. B: A single decade from part A expanded, with standard 20% component values shown.

As an example, consider the filter shown in Figure 1.53. It is a high-pass filter with

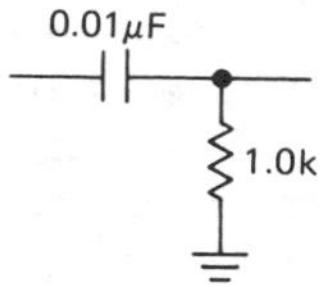

Figure 1.53

0.01μF
1.0k

the 3dB point at 15.9kHz. The impedance of a load driven by it should be much larger than 1.0k in order to prevent circuit loading effects on the filter's output, and the driving source should be able to drive a 1.0k load without significant attenuation (loss of signal amplitude) in order to prevent circuit loading effects by the filter on the signal source.

Low-pass filters

You can get the opposite frequency behavior in a filter by interchanging R and C (Fig. 1.54). You will find

$$V_{out} = \frac{1}{(1 + \omega^2 R^2 C^2)^{1/2}} V_{in}$$

as seen in Figure 1.55. This is called a low-pass filter. The 3dB point is again at a frequency $f = 1/2\pi RC$. Low-pass filters are quite handy in real life. For instance, a low-pass filter can be used to eliminate interference from nearby radio stations, a problem that plagues audio amplifiers and other sensitive electronic equipment.

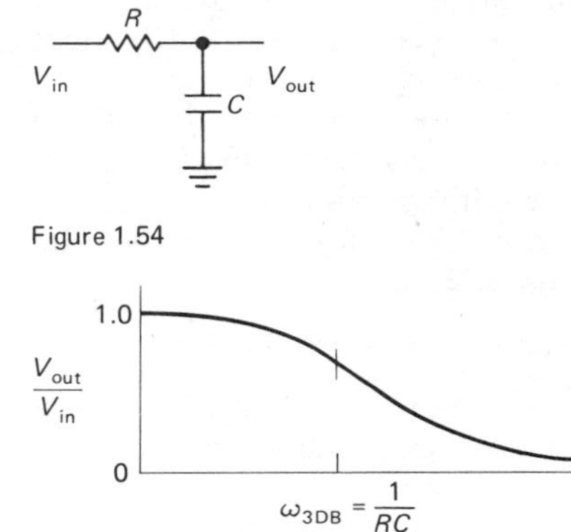

Figure 1.54

Figure 1.55
Low-pass-filter frequency response.

EXERCISE 1.21

Show that the preceding expression for the response of an *RC* low-pass filter is correct.

The low-pass filter's output can be viewed as a signal source in its own right. When driven by a perfect ac voltage source (zero source impedance), the filter's output looks like 1.0k at low frequencies (the perfect signal source can be replaced by a short, i.e., by its small-signal source impedance, for the purpose of impedance calculations). It drops to zero impedance at high frequencies, where the capacitor dominates the output impedance. The signal driving the filter sees a load of 1.0k plus the load resistance at low frequencies, dropping to 1.0k at high frequencies.

RC differentiators and integrators in the frequency domain

The circuit for an *RC* differentiator that we saw in Section 1.14 is the same as a high-pass filter. For the output to be small compared with the input, the signal frequency (or frequencies) must be well below the 3dB point. This is easy to check. Suppose we have an input signal

$$V_{in} = \sin \omega t$$

Then, using the equation we obtained earlier for the differentiator output,

$$V_{out} = RC\frac{d}{dt}\sin \omega t = \omega RC \cos \omega t$$

and so $V_{out} \ll V_{in}$ if $\omega RC \ll 1$, i.e., $RC \ll 1/\omega$. If the input signal contains a range of frequencies, this must hold for the highest frequencies present in the input.

The *RC* integrator (Section 1.15) is the same circuit as the low-pass filter; by similar reasoning, the criterion for a good integrator is that the lowest signal frequencies must be well above the 3dB point.

Inductors versus capacitors

In practice, you rarely see *RL* high- or low-pass filters, even though inductors in combination with resistors can make perfectly good filters. The reason is that inductors tend to be more bulky and expensive and perform less well (i.e., they depart further from the ideal) than capacitors. If you have a choice, use a capacitor. One exception to this general statement is the use of ferrite beads and chokes in high-frequency circuits. You just string a few beads here and there in the circuit; they make the wire interconnections slightly inductive, raising the impedance at very high frequencies and preventing "oscillations," without the added resistance you would get with an *RC* filter. An RF "choke" is an inductor, usually a few turns of wire wound on a ferrite core, used for the same purpose in RF circuits.

□ 1.20 Phasor diagrams

There's a nice graphic method that can be very helpful when trying to understand reactive circuits. Let's take an example, namely the fact that an *RC* filter attenuates 3dB at a frequency $f = 1/2\pi RC$, which we derived in Section 1.19. This is true for both high-pass and low-pass filters. It is easy to get a bit confused here, because at that frequency the reactance of the capacitor equals the resistance of the resistor; so you might at first expect 6dB attenuation. That is what you would get, for example, if you were to replace the capacitor by a resistor of the same impedance (recall that 6dB means half voltage). The confusion arises because the capacitor is reactive, but the matter is clarified by a phasor diagram (Fig. 1.56). The

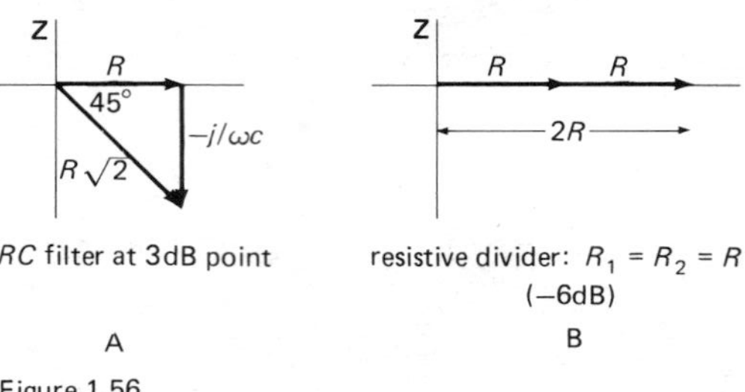

Figure 1.56

axes are the real (resistive) and imaginary (reactive) components of the impedance. In a series circuit like this, the axes also represent the (complex) voltage, since the current is the same everywhere. So for this circuit (think of it as an *R-C* voltage divider) the input voltage (applied across the series *R-C* pair) is proportional to the length of the hypotenuse, and the output voltage (across *R* only) is proportional to the length of the *R* leg of the triangle. The diagram represents the situation at the frequency where the magnitude of the capacitor's reactance equals R, i.e., $f = 1/2\pi RC$, and shows that the ratio of output voltage to input voltage is $1/\sqrt{2}$, i.e., -3dB.

The angle between the vectors gives the phase shift from input to output. At the 3dB point, for instance, the output amplitude equals the input amplitude divided by the square root of 2, and it leads by 45° in phase. This graphic method makes it easy to read off amplitude and phase relationships in RLC circuits. For example, you can use it to get the response of the high-pass filter that we previously derived algebraically.

EXERCISE 1.22

Use a phasor diagram to derive the response of an RC high-pass filter:

$$V_{out} = \frac{R}{[R^2 + (1/\omega^2 C^2)]^{1/2}} V_{in}$$

EXERCISE 1.23

At what frequency does an RC low-pass filter attenuate by 6dB (output voltage equal to half the input voltage)? What is the phase shift at that frequency?

EXERCISE 1.24

Use a phasor diagram to obtain the low-pass filter response previously derived algebraically.

In the next chapter (Section 2.08) you will see a nice example of phasor diagrams in connection with a constant-amplitude phase-shifting circuit.

1.21 "Poles" and decibels per octave

Look again at the response of the RC low-pass filter (Fig. 1.55). Far to the right of the "knee" the output amplitude is dropping proportional to $1/f$. In one octave (as in music, one octave is twice the frequency) the output amplitude will drop to half, or -6dB; so a simple RC filter has a 6dB/octave falloff. You can make filters with several RC sections; then you get 12dB/octave (two RC sections), 18dB/octave (three sections), etc. This is the usual way of describing how a filter behaves beyond the cutoff. Another popular way is to say a "3-pole filter," for instance, meaning a filter with three RC sections (or one that behaves like one). (The word pole derives from a method of analysis that is beyond the scope of this book and that involves complex transfer functions in the complex frequency plane, known by engineers as the "s-plane.")

A caution on multistage filters: You can't simply cascade several identical filter sections in order to get a frequency response that is the concatenation of the individual responses. The reason is that each stage will load the previous one significantly (since they're identical), changing the overall response. Remember that the response function we derived for the simple RC filters was based on a zero-impedance driving source and an infinite-impedance load. One solution is to make each successive filter section have much higher impedance than the preceding one. A better solution involves active circuits like transistor or operational amplifier (op-amp) interstage "buffers," or active filters. These subjects will be treated in Chapters 2 through 4.

1.22 Resonant circuits and active filters

When capacitors are combined with inductors or are used in special circuits called active filters, it is possible to make circuits that have very sharp frequency characteristics (e.g., a large peak in the response at a particular frequency), as compared with the gradual characteristics of the RC filters we've seen so far. These circuits find applications in various audiofrequency and radiofrequency devices. Let's now take a quick look at LC circuits (there will be more on them, and active filters, in Chapter 4 and Appendix H).

First, consider the circuit shown in Figure 1.57. The reactance of the LC combination at frequency f is just

$$\frac{1}{\mathbf{Z}_{LC}} = \frac{1}{\mathbf{Z}_L} + \frac{1}{\mathbf{Z}_C} = \frac{1}{j\omega L} - \frac{\omega C}{j} = j\left(\omega C - \frac{1}{\omega L}\right)$$

i.e.,

$$\mathbf{Z}_{LC} = \frac{j}{(1/\omega L) - \omega C}$$

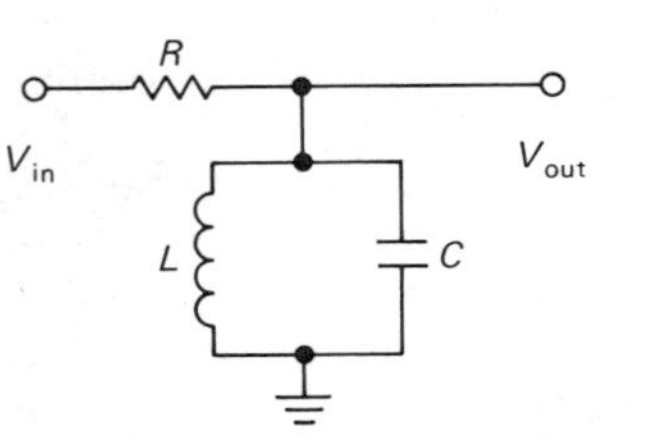

Figure 1.57

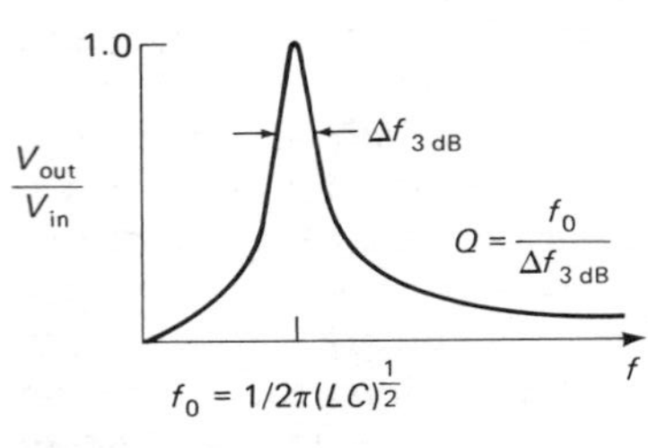

Figure 1.58

In combination with R it forms a voltage divider; because of the opposite behaviors of inductors and capacitors, the impedance of the parallel LC goes to infinity at frequency $f_o = 1/2\pi(LC)^{1/2}$, giving a peak in the response there. The overall response is as shown in Figure 1.58.

In practice, losses in the inductor and capacitor limit the sharpness of the peak, but with good design these losses can be made very small. Alternatively, a Q-spoiling resistor is sometimes added intentionally to reduce the sharpness of the resonant peak. This circuit is known simply as a parallel LC resonant circuit or a tuned circuit and is used extensively in radiofrequency circuits to select a particular frequency for amplification (the L or C can be variable, so you can tune the resonant frequency). The higher the driving impedance, the sharper the peak; it is not uncommon to drive them with something approaching a current source, as you will see later. The *quality factor* Q is a measure of the sharpness of the peak. It equals the resonant frequency divided by the width at the −3dB points.

Another variety of LC circuit is the series LC (Fig. 1.59). By writing down the impedance formulas involved, you can convince yourself that the impedance of the LC goes to zero at resonance [$f_o = 1/2\pi(LC)^{1/2}$]; such a circuit is a "trap" for signals at or near the resonant frequency, shorting them

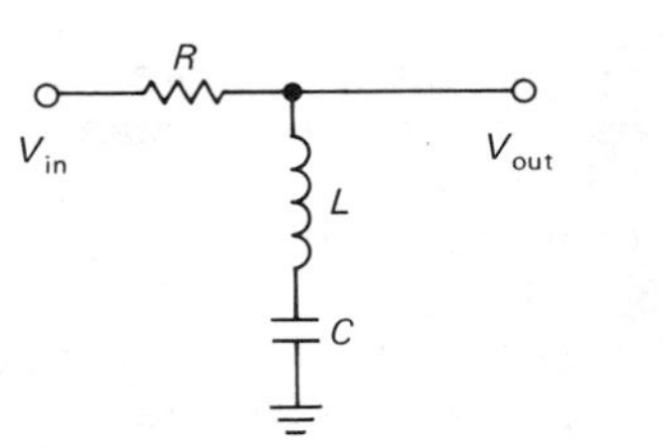

Figure 1.59

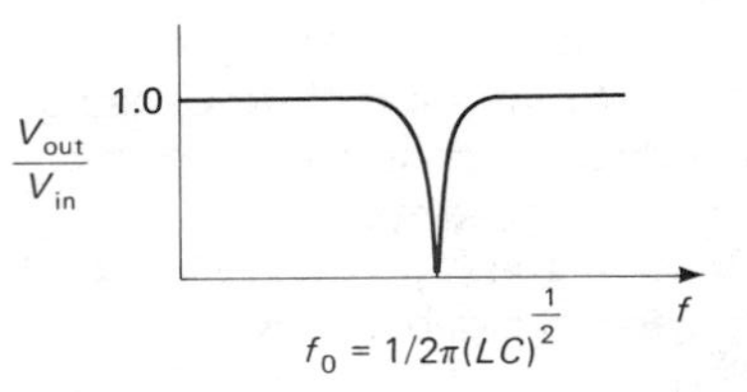

Figure 1.60

to ground. Again, this circuit finds application mainly in radiofrequency circuits. Figure 1.60 shows what the response looks like.

EXERCISE 1.25

Find the response (V_{out}/V_{in} versus frequency) for the series LC trap circuit in Figure 1.59.

1.23 Other capacitor applications

In addition to their uses in filters, resonant circuits, differentiators, and integrators, capacitors are needed for several other important applications. We will treat these in detail later in the book, mentioning them here only as a preview.

Bypassing

The impedance of a capacitor goes down with increasing frequency. This is the basis of another important application: bypassing. There are places in circuits where you want to allow a dc (or slowly varying) voltage, but don't want signals present. Placing a capacitor across that circuit element (usually a resistor) will help to kill any signals there. You choose the capacitor value so that its impedance at signal frequencies is small compared with what it is bypassing. You will see much more of this in later chapters.

Power-supply filtering

Power-supply filtering is really a form of bypassing, although we usually think of it as energy storage. The dc voltages used in electronics are usually generated from the ac line voltage by a process called *rectification* (which will be treated later in this chapter); some residue of the 60Hz input remains, and this can be reduced as much as desired by means of bypassing with suitably large capacitors. These capacitors really are large – they're the big shiny round things you see inside most electronic instruments. You will see how to design power supplies and filters later in this chapter and again in Chapter 5.

Timing and waveform generation

A capacitor supplied with a constant current charges up with a ramp waveform. This is the basis of ramp and sawtooth generators, used in function generators, oscilloscope sweep circuits, analog/digital converters, and timing circuits. *RC* circuits are also used for timing, and they form the basis of digital delay circuits (monostable multivibrators). These timing and waveform applications are important in many areas of electronics and will be covered in Chapters 3, 4, 8, and 9.

1.24 Thévenin's theorem generalized

When capacitors and inductors are included, Thévenin's theorem must be restated: Any two-terminal network of resistors, capacitors, inductors, and signal sources is equivalent to a single complex impedance in series with a single signal source. As before, you find the impedance and the signal source from the open-circuit output voltage and the short-circuit current.

DIODES AND DIODE CIRCUITS

1.25 Diodes

The circuit elements we've discussed so far (resistors, capacitors, and inductors) are all *linear,* meaning that a doubling of the applied signal (a voltage, say) produces a doubling of the response (a current, say). This is true even for the reactive devices (capacitors and inductors). These devices are also *passive,* meaning that they don't have a built-in source of power. And they are all two-terminal devices, which is self-explanatory.

The diode (Fig. 1.61) is a very important and useful two-terminal passive nonlinear device. It has the *V–I* curve shown in Figure 1.62. (In keeping with the general philosophy of this book, we will not attempt to describe the solid-state physics that make such devices possible.)

anode cathode

Figure 1.61

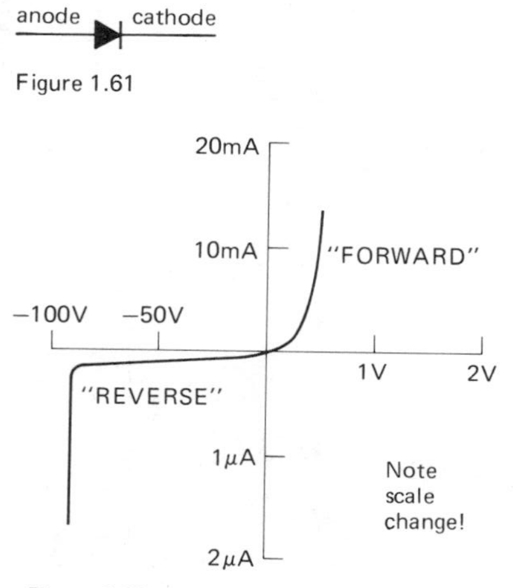

Figure 1.62

The diode's arrow (the anode terminal) points in the direction of forward current flow. The reverse current, which is measured in the nanoamp range for a general-purpose diode (note the different scales in the graph for forward and reverse current), is almost never of any consequence until you reach the reverse breakdown voltage (also called the peak inverse voltage, PIV), typically 75 volts for a general-purpose diode like the 1N914. (Normally you never subject a diode to voltages large enough to cause reverse breakdown; the exception is the zener diode we mentioned earlier.) Frequently, also, the forward voltage drop of about 0.5 to 0.8 volt is of little concern, and the diode can be treated as a good approximation to an ideal one-way conductor. There are other important characteristics that distinghish the thousands of diode types available, e.g., maximum forward current, capacitance, leakage current, and reverse recovery time (see Table 1.1 for characteristics of some typical diodes).

TABLE 1.1. DIODES

Type	$V_{R(max)}$[a] (V)	$I_{R(max)}$[b] (μA)	Continuous V_F (V)	@	Continuous I_F (mA)	Peak V_F (V)	@	Peak I_F (A)	Reverse recovery (ns)	Capacitance (10V) (pF)	Class	Comments
PAD-1	120	1pA @ 20V	0.8		5	—		—	—	0.8	Lowest I_R	Siliconix
FJT1100	30	0.001	—		—	1.1		0.05	—	1.2	Very low I_R	1pA @ 5V, 10pA @ 15V
IN3595	150	3	0.7		10	<1.0		0.2	3000	8.0	Low I_R	1nA @ 125V
IN914 / IN4148	75	5	0.75		10	1.1		0.1	4	1.3	General-purpose signal diode	Industry standard
IN6263	60	10	0.4		1	0.7		0.01	0	1.0	Schottky: low V_F	
IN3062	75	50	<1.0		20	—		—	2	0.6	Low cap, signal	1.0pF @ 0V
IN4305	75	50	0.6		1	—		—	4	1.5	Controlled V_F	
IN4002	100	50	0.9		1000	2.3		25	3500	15	1A rectifier	7-member family
IN4007	1000	50	0.9		1000	2.3		25	5000	10	1A rectifier	7-member family
IN5625	400	50	1.1		5000	2.0		50	2500	45	5A rectifier	Lead-mounted
IN1183A	50	1000	1.1		40000	1.3		100	—	—	High-current rectifier	IN1183RA reverse

(a) $V_{R(max)}$ is repetitive peak reverse voltage, 25°C, 10μA leakage. (b) $I_{R(max)}$ is reverse leakage current at V_R and 100°C ambient temperature.

Before jumping into some circuits with diodes, we should point out two things: (1) A diode doesn't have a resistance (it doesn't obey Ohm's law). (b) If you put some diodes in a circuit, it won't have a Thévenin equivalent.

1.26 Rectification

A rectifier changes ac to dc; this is one of the simplest and most important applications of diodes (diodes are sometimes called rectifiers). The simplest circuit is shown in Figure 1.63. For a sine-wave input that is much larger than the forward drop (about 0.6 volt for silicon diodes, the usual type), the output will look like that in Figure 1.64. If you think of the diode as a one-way conductor, you won't have any trouble understanding how the circuit works. This circuit is called a *half-wave rectifier,* since only half of the input waveform is used.

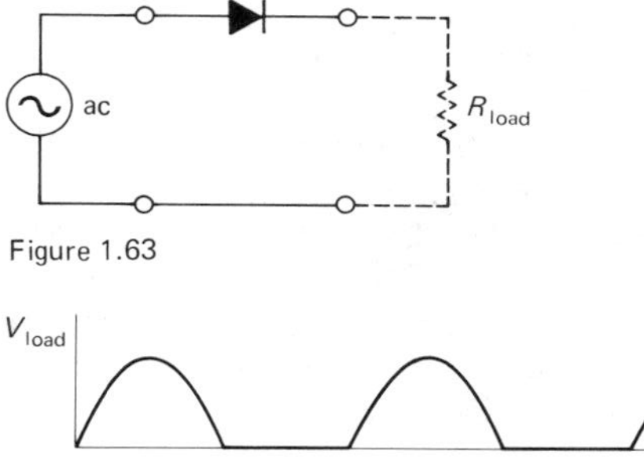

Figure 1.63

Figure 1.64

Figure 1.65 shows another rectifier circuit, a full-wave bridge. Figure 1.66 shows the output, for which the whole input waveform is used. The gaps at zero voltage occur because of the diodes' forward voltage drop. In this circuit, two diodes are always in series with the input; when you design low-

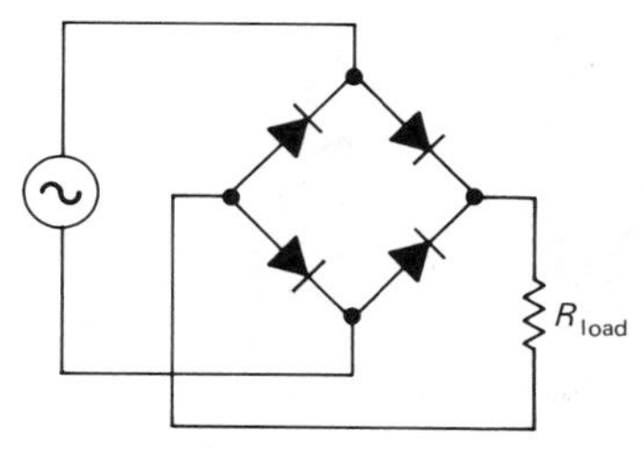

Figure 1.65

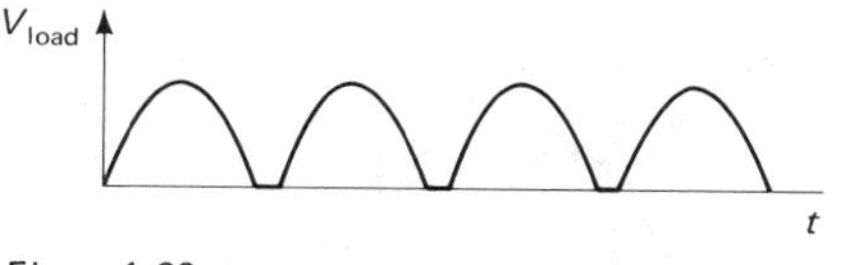

Figure 1.66

voltage power supplies, you have to remember that.

1.27 Power-supply filtering

The preceding rectified waveforms aren't good for much as they stand. They're dc only in the sense that they don't change polarity. But they still have a lot of "ripple" (periodic variations in voltage about the steady value) that has to be smoothed out in order to generate genuine dc. This we do by tacking on a low-pass filter (Fig. 1.67).

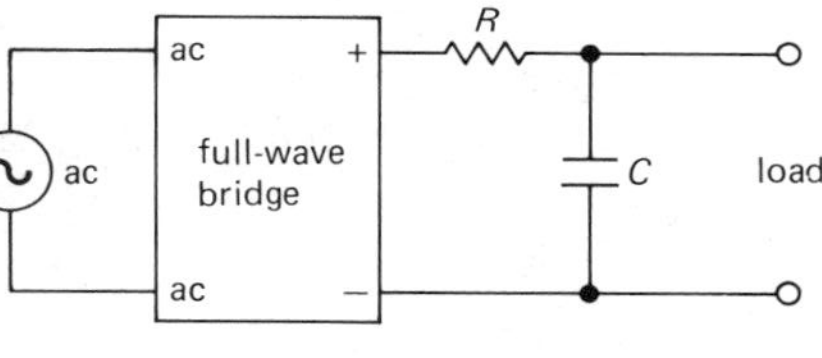

Figure 1.67

Actually, the series resistor is unnecessary and is always omitted (although you sometimes see a very small resistor used to limit the peak rectifier current). The reason is that the diodes prevent flow of current back out of the capacitors, which are really serving more as energy-storage devices than as part of a classic low-pass filter. The energy stored in a capacitor is $U = \frac{1}{2} CV^2$. For C in farads and V in volts, U comes out in joules (watt-seconds).

The capacitor value is chosen so that

$$R_{load}C \gg 1/f$$

(where f is the ripple frequency, here 120Hz) in order to ensure small ripple. We will make this vague statement clearer in the next section.

Calculation of ripple voltage

It is easy to calculate the approximate ripple voltage, particularly if it is small compared with the dc. (see Fig. 1.68). The load causes the capacitor to discharge somewhat

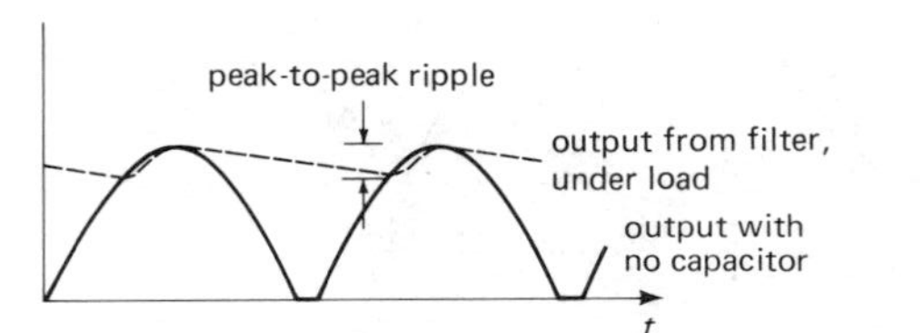

Figure 1.68

between cycles (or half cycles, for full-wave rectification). If you assume that the load current stays constant (it will, for small ripple), you have

$$\Delta V = \frac{I}{C}\Delta t \qquad \left(\text{from } I = C\frac{dV}{dt}\right)$$

Just use $1/f$ (or $1/2f$ for full-wave rectification) for Δt (this estimate is a bit on the safe side, since the capacitor begins charging again in less than a half cycle). You get

$$\Delta V = \frac{I_{load}}{fC} \qquad \text{(half wave)}$$

$$\Delta V = \frac{I_{load}}{2fC} \qquad \text{(full wave)}$$

If you wanted to do the calculation without any approximation, you would use the exact exponential discharge formula. You would be misguided in insisting on that kind of accuracy, though, for two reasons:

1. The discharge is an exponential only if the load is a resistance; many loads are not. In fact, the most common load, a *voltage regulator,* looks like a constant-current load.
2. Power supplies are built with capacitors with typical tolerances of 20% or more. Realizing the manufacturing spread, you design conservatively, allowing for the worst-case combination of component values.

In this case, viewing the initial part of the discharge as a ramp is in fact quite accurate, especially if the ripple is small, and in any case it errs in the direction of conservative design – it overestimates the ripple.

EXERCISE 1.26

Design a full-wave bridge rectifier circuit to deliver 10 volts dc with less than 0.1 volt (pp) ripple into a load drawing up to 10mA. Choose the appropriate ac input voltage, assuming 0.6 volt diode drops. Be sure to use the correct ripple frequency in your calculation.

1.28 Rectifier configurations for power supplies

Full-wave bridge

A dc power supply using the bridge circuit we just discussed looks as shown in Figure 1.69. In practice, you generally buy the

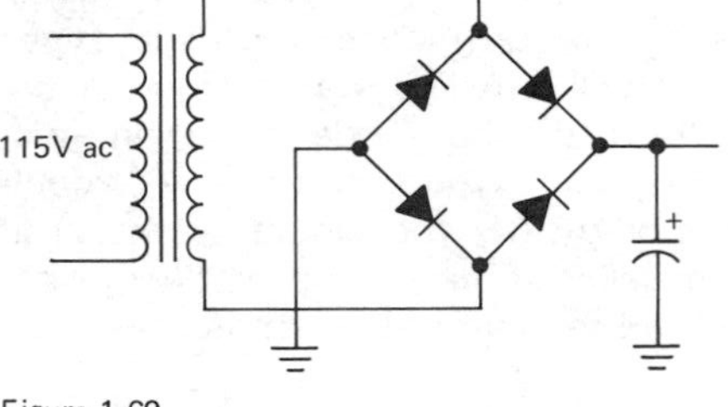

Figure 1.69
Bridge rectifier circuit.

bridge as a prepackaged module. The smallest ones come with ratings of 1 amp, with breakdown voltages going from 100 volts to 600 volts, or even 1000 volts. Giant bridge rectifiers are available with current ratings of 25 amps or more. Take a look at Table 5.4 for a few types.

Center-tapped full-wave rectifier

The circuit in Figure 1.70 is called a center-tapped full-wave rectifier. The output voltage is half what you get if you use a bridge rectifier. It is not the most efficient circuit in terms of transformer design, since each half of the secondary is used only half the time. Thus the current through the winding during

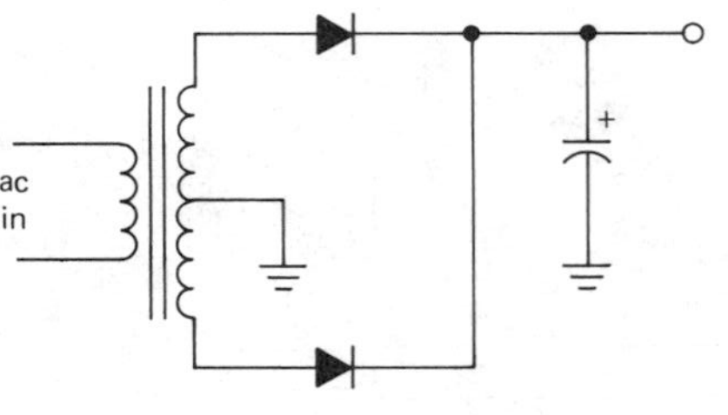

Figure 1.70

that time is twice what it would be for a true full-wave circuit. Heating in the windings, calculated from Ohm's law, is I^2R, so you have four times the heating half the time, or twice the average heating of an equivalent full-wave circuit. You would have to choose a transformer with a current rating 1.4 (square root of 2) times as large, as compared with the (better) bridge circuit; besides costing more, the resulting supply would be bulkier and heavier.

EXERCISE 1.27

This illustration of I^2R heating may help you understand the disadvantage of the center-tapped rectifier circuit. What fuse rating (minimum) is required to pass the current waveform shown in Figure 1.71, which has

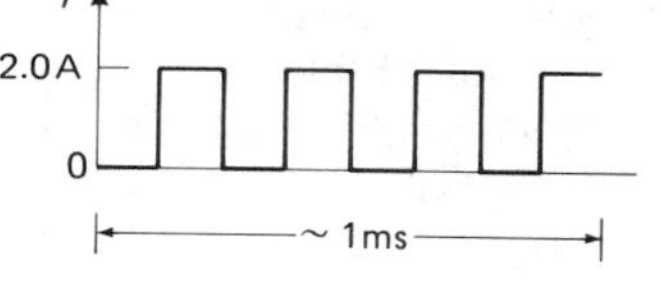

Figure 1.71

1 amp average current? Hint: A fuse "blows out" by melting (I^2R heating) a metallic link, for steady currents larger than its rating. Assume for this problem that the thermal time constant of the fusible link is much longer than the time scale of the square wave, i.e., that the fuse responds to the value of I^2 averaged over many cycles.

Split supply

A popular variation of the center-tapped full-wave circuit is shown in Figure 1.72. It gives you split supplies (equal plus and minus voltages), which many circuits need. It is an efficient circuit, since both halves of the input waveform are used in each winding section.

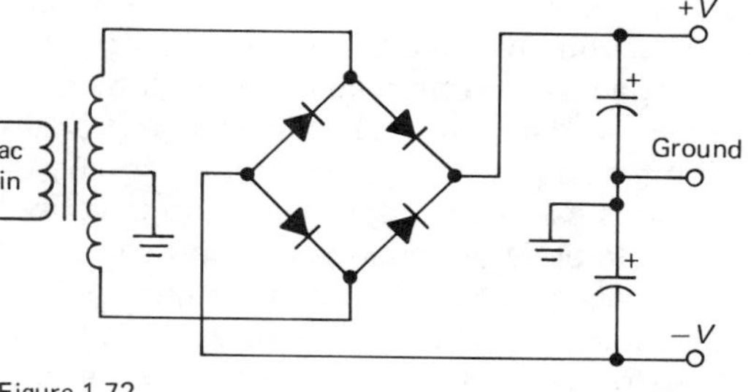

Figure 1.72

□ *Voltage multipliers*

The circuit shown in Figure 1.73 is called a voltage doubler. Think of it as two half-wave

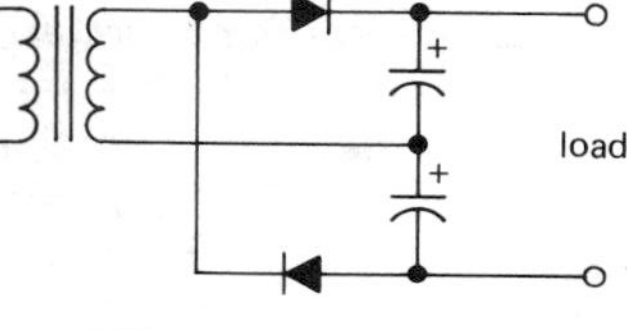

Figure 1.73

rectifier circuits in series. It is officially a full-wave rectifier circuit, since both halves of the input waveform are used – the ripple frequency is twice the ac frequency (120Hz for the 60Hz line voltage in the United States).

Variations of this circuit exist for voltage triplers, quadruplers, etc. Figure 1.74 shows doubler, tripler, and quadrupler circuits that let you ground one side of the transformer.

1.29 Regulators

By choosing capacitors that are sufficiently large, you can reduce the ripple voltage to any desired level. This brute-force approach has two disadvantages:

1. The required capacitors may be prohibitively bulky and expensive.

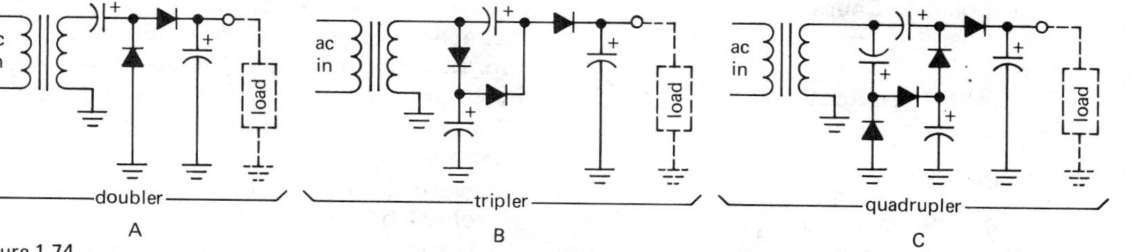

Figure 1.74

2. Even with the ripple reduced to negligible levels, you still have variations of output voltage due to other causes, e.g., the dc output voltage will be roughly proportional to the ac input voltage, giving rise to fluctuations caused by input line voltage variations. In addition, changes in load current will cause the output voltage to change because of the finite internal resistances of the transformer, diode, etc. In other words, the Thévenin equivalent circuit of the dc power supply has $R > 0$.

A better approach to power supply design is to use enough capacitance to reduce ripple to low levels (perhaps 10% of the dc voltage), then use an active *feedback circuit* to eliminate the remaining ripple. Such a feedback circuit "looks at" the output, making changes in a controllable series resistor (a transistor) as necessary to keep the output constant (Fig. 1.75).

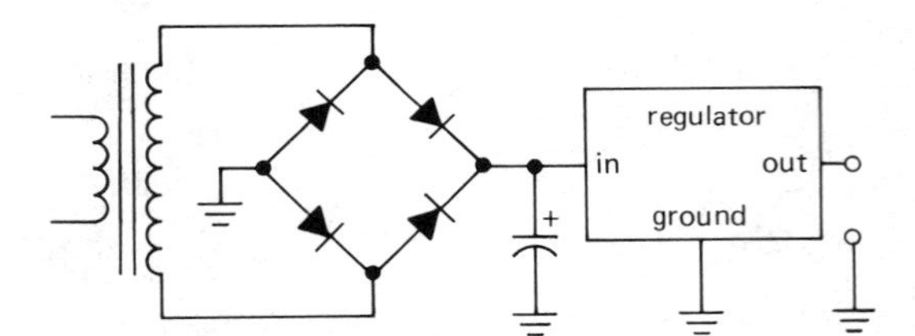

Figure 1.75
Classic dc voltage regulator scheme.

These voltage regulators are used almost universally as power supplies for electronic circuits. Nowadays complete voltage regulators are available as inexpensive integrated circuits (priced under one dollar). A power supply built with a voltage regulator can be made easily adjustable and self-protecting (against short circuits, overheating, etc.), with excellent properties as a voltage source (e.g., internal resistance measured in milliohms). We will deal with regulated dc power supplies in Chapter 5.

1.30 Circuit applications of diodes

Signal rectifier

There are other occasions when you use a diode to make a waveform of one polarity only. If the input waveform isn't a sine wave, you usually don't think of it as a rectification in the sense of a power supply. For instance, you might want a train of pulses corresponding to the rising edge of a square wave. The easiest way is to rectify the differentiated wave (Fig. 1.76). Always

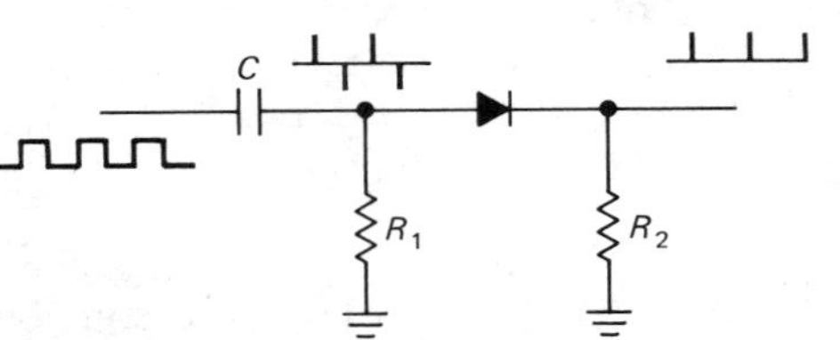

Figure 1.76

keep in mind the 0.6 volt (approximately) forward drop of the diode. This circuit, for instance, gives no output for square waves smaller than 0.6 volt pp. If this is a problem, there are various tricks to circumvent this limitation. One possibility is to use *hot carrier diodes* (Schottky diodes), with a forward drop of about 0.25 volt (another device called a *back diode* has nearly zero forward drop, but its usefulness is limited by very low reverse breakdown voltage).

A possible circuit solution to this problem of finite diode drop is shown in Figure 1.77.

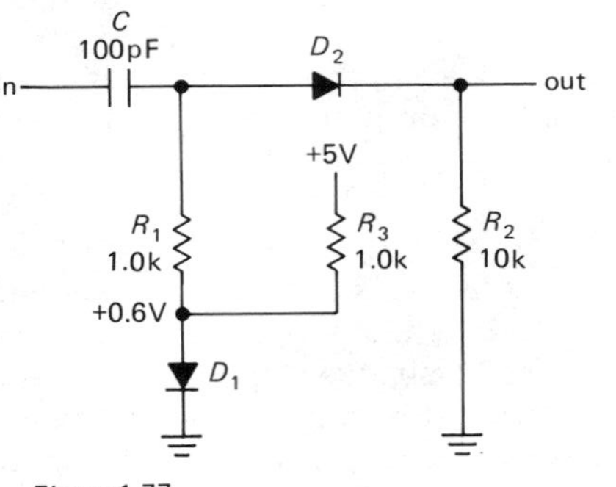

Figure 1.77

Here D_1 compensates D_2's forward drop by providing 0.6 volt of *bias* to hold D_2 at the threshold of conduction. Using a diode (D_1) to provide the bias (rather than, say, a voltage divider) has several advantages: There is nothing to adjust, the compensation will be nearly perfect, and changes of the forward drop (e.g., with changing temperature) will be compensated properly. Later we will see other instances of matched-pair compensation of forward drops in diodes, transistors, and FETs: it is a simple and powerful trick.

Diode gates

Another application of diodes, which we will recognize later under the general heading of *logic,* is to pass the higher of two voltages without affecting the lower. A good example is *battery backup,* a method of keeping something running (e.g., a precision electronic clock) that must not stop when there is a power failure. Figure 1.78 shows a

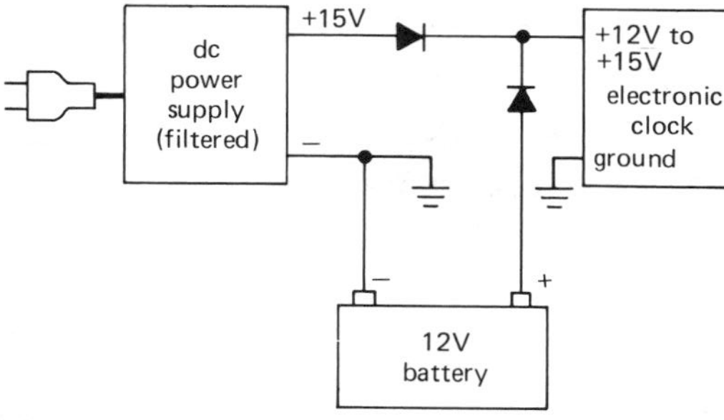

Figure 1.78

circuit that does the job. The battery does nothing until the power fails; then it takes over without interruption.

EXERCISE 1.28

Make a simple modification to the circuit so that the battery is charged by the dc supply (when power is on, of course) at a current of 10mA (such a circuit is necessary to maintain the battery's charge).

Diode clamps

Sometimes it is desirable to limit the range of a signal (i.e., prevent it from exceeding certain voltage limits) somewhere in a circuit. The circuit shown in Figure 1.79 will accomplish this. The diode prevents the output from exceeding about +5.6 volts, with no effect on voltages less than that

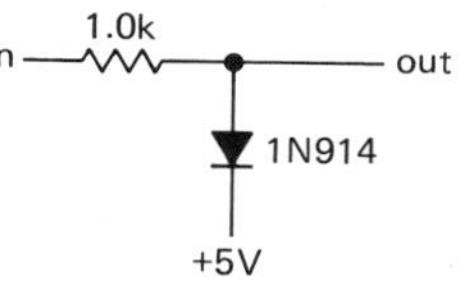

Figure 1.79

(including negative voltages); the only limitation is that the input must not go so negative that the reverse breakdown voltage of the diode is exceeded (e.g., −70 volts for a 1N914). Diode clamps are standard equipment on all inputs in the CMOS family of digital logic. Without them, the delicate input circuits are easily destroyed by static electricity discharges during handling.

EXERCISE 1.29

Design a symmetrical clamp, i.e., one that confines a signal to the range −5.6 volts to +5.6 volts.

A voltage divider can provide the reference voltage for a clamp (Fig. 1.80). In this case you must ensure that the impedance looking into the voltage divider (R_{vd}) is small compared with R, since what you have looks as shown in Figure 1.81, when the voltage divider is replaced by its Thévenin equivalent circuit. When the diode conducts (input voltage exceeds clamp voltage), the output is really just the output of a voltage divider, with the Thévenin equivalent resistance of the voltage reference as the lower resistor (Fig. 1.82). So, for the values shown, the output of the clamp for a triangle-wave input would look as shown in Figure 1.83. The problem is that the voltage divider doesn't provide a stiff reference, in the language of electronics. A stiff voltage source is one that doesn't bend easily, i.e., it has low internal (Thévenin) impedance.

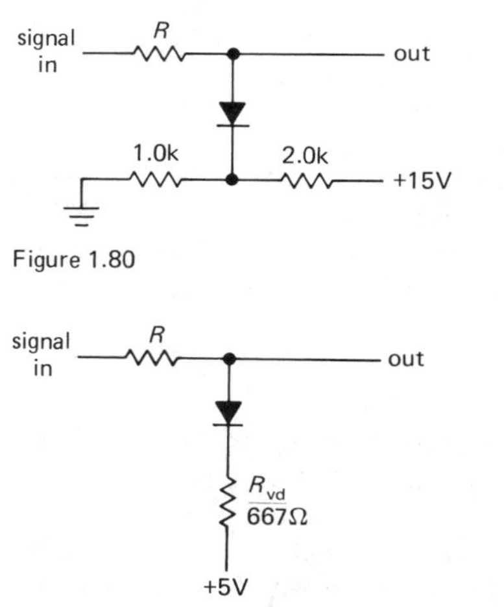

Figure 1.80

Figure 1.81

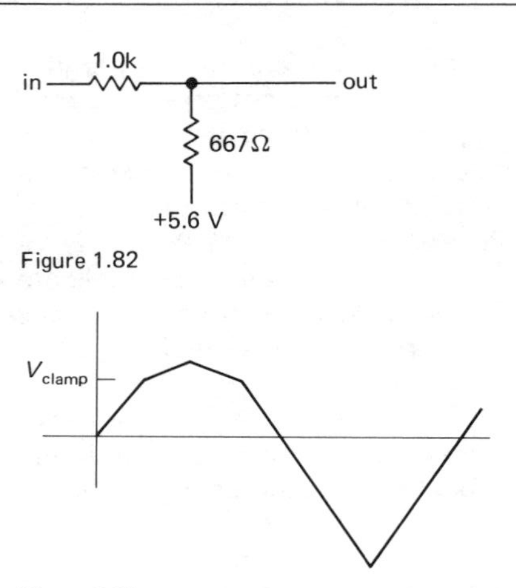

Figure 1.82

Figure 1.83

In practice, the problem of finite impedance of the voltage divider reference can be easily solved using a transistor or operational amplifier (op-amp). This is usually a better solution than using very small resistor values, because it doesn't consume large currents, but it provides impedances of a few ohms or less. Furthermore, there are other ways to construct a clamp, using an op-amp as part of the clamp circuit. You will see these methods in Chapter 3.

One interesting clamp application is "dc restoration" of a signal that has been ac-coupled (capacitively coupled). Figure 1.84

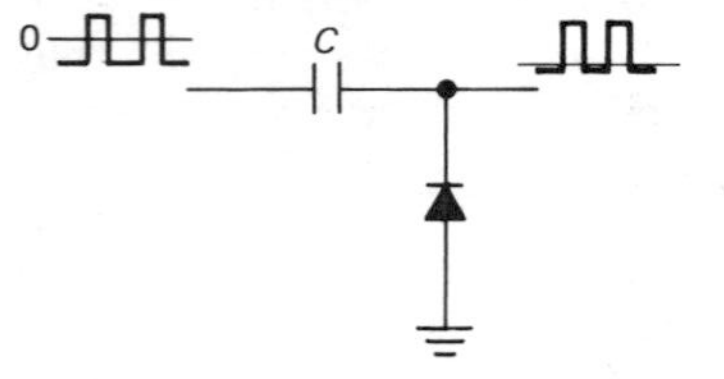

Figure 1.84

shows the idea. This is particularly important for circuits whose inputs look like diodes (e.g., a transistor with grounded emitter); otherwise on ac-coupled signal will just fade away.

Limiter

One last clamp circuit is shown in Figure 1.85. This circuit limits the output "swing" (again, a common electronics term) to one diode drop, roughly 0.6 volt. That might seem awfully small, but if the next stage is an amplifier with large voltage amplification, its input will always be near zero volts;

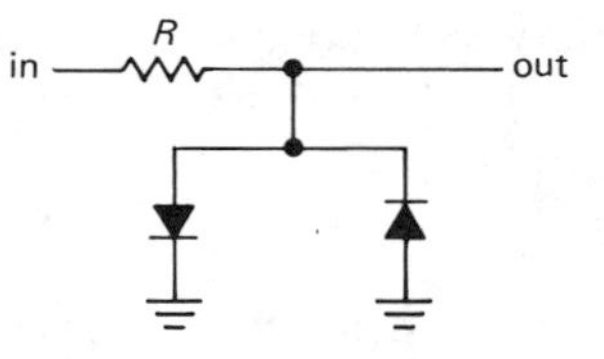

Figure 1.85

otherwise the output is in "saturation" (e.g., if the next stage has a gain of 1000 and operates from ±15 volt supplies, its input must stay in the range ±15mV in order for its output not to saturate). This clamp circuit is often used as input protection for a high-gain amplifier.

Diodes as nonlinear elements

To a good approximation the forward current through a diode is proportional to an exponential function of the voltage across it at a given temperature (for a discussion of the exact law see Section 2.10). So you can use a diode to generate an output voltage proportional to the logarithm of a current (Fig. 1.86). Since V hovers in the region of 0.6 volt, with only small voltage changes that reflect input current variations, you can generate the input current with a resistor if the input voltage is much larger than a diode drop (Fig. 1.87).

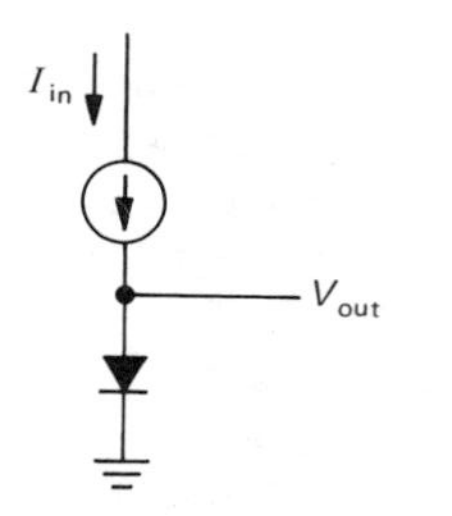

Figure 1.86

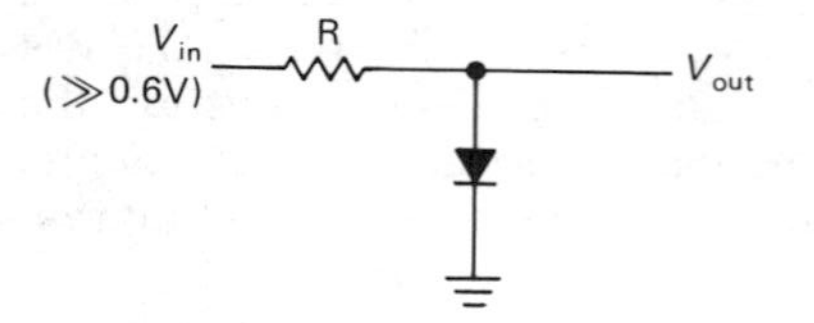

Figure 1.87

$$I = \frac{V_{in} - V_{out}}{R} \approx \frac{V_{in} - 0.6 \text{ volts}}{R}$$

$$\approx \frac{V_{in}}{R} \qquad (\text{if } V_{in} \gg 0.6 \text{ volts})$$

In practice you may want an output voltage that isn't offset by the 0.6 volt diode drop. In addition, it would be nice to have a circuit that is insensitive to changes in temperature. The method of diode drop compensation is helpful here (Fig. 1.88). R_1

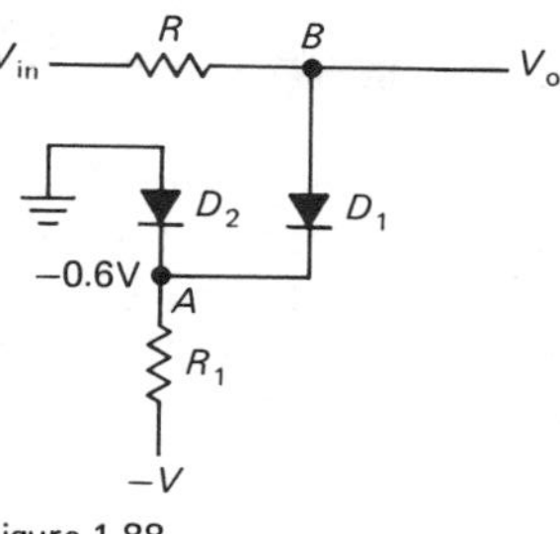

Figure 1.88

makes D_2 conduct, holding point A at about −0.6 volt. Point B is the near ground (making I_{in} accurately proportional to V_{in}, incidentally). As long as the two (identical) diodes are at the same temperature, there is good cancellation of the forward drops, except, of course, for the difference owing to input current through D_1, which produces the desired output. In this circuit R_1 should be chosen so that the current through D_2 is much larger than the maximum input current, in order to keep D_2 in conduction.

In the chapter on op-amps we will examine better ways of constructing logarithmic converter circuits, along with careful methods of temperature compensation. With such methods it is possible to construct logarithmic converters accurate to a few percent over six decades or more of input current. A better understanding of diode and transistor characteristics, along with an understanding of op-amps, is necessary first. This section is meant to serve only as an introduction for things to come.

1.31 Inductive loads and diode protection

What happens if you open a switch that is providing current to an inductor? Since inductors have the property

$$V = L\frac{dI}{dt}$$

it is not possible to turn off the current suddenly, since that would imply an infinite voltage across the inductor's terminals. What happens instead is that the voltage across the inductor suddenly rises and keeps rising until it forces current to flow. Electronic devices controlling inductive loads can be easily damaged, especially the component that "breaks down" in order to satisfy the inductor's craving for continuity of current. Consider the circuit in Figure 1.89.

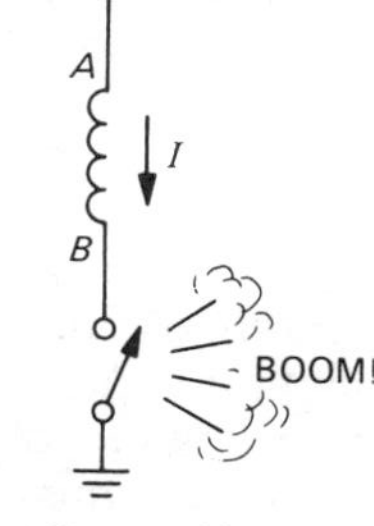

Figure 1.89

The switch is initially closed, and current is flowing through the inductor (which might be a relay, as will be described later). When the switch is opened, the inductor "tries" to keep current flowing from A to B, as it had been. That means that terminal B goes positive relative to terminal A. In a case like this it may go 1000 volts positive before the switch contact "blows over." This shortens the life of the switch and also generates impulsive interference that may affect other circuits nearby. If the switch happens to be a transistor, it would be an understatement to say that its life is shortened; its life is ended!

The best solution is to put a diode across the inductor, as in Figure 1.90. When the switch is on, the diode is back-biased (from the dc drop across the inductor's winding

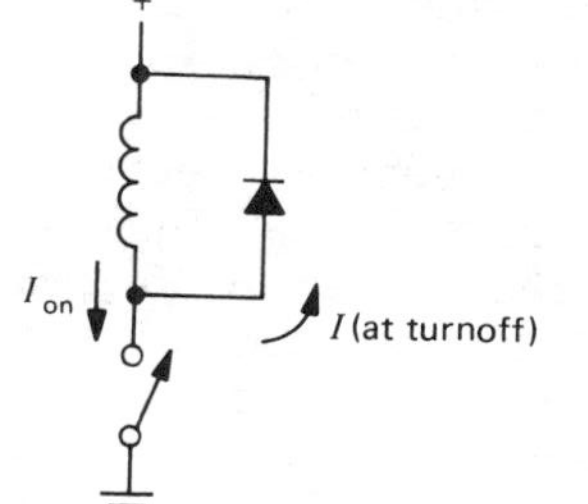

Figure 1.90

resistance). At turn-off the diode goes into conduction, putting the switch terminal a diode drop above the positive supply voltage. The diode must be able to handle the initial diode current, which equals the steady current that had been flowing through the inductor; something like a 1N4004 is fine for nearly all cases.

The only disadvantage of this protection circuit is that it lengthens the decay of current through the inductor, since the rate of change of inductor current is proportional to the voltage across it. For applications where the current must decay quickly (high-speed impact printers, high-speed relays, etc.), it may be better to put a resistor across the inductor, choosing its value so that $V_{supply} + IR$ is less than the maximum allowed voltage across the switch. (For fastest decay with a given maximum voltage, a zener could be used instead, giving a ramp-down of current rather than an exponential decay.)

For inductors driven from ac (transformers, ac relays), the diode protection just described will not work, since the diode will conduct on alternate half cycles when the switch is closed. In that case a good solution is an *RC* "snubber" network (Fig. 1.91). The

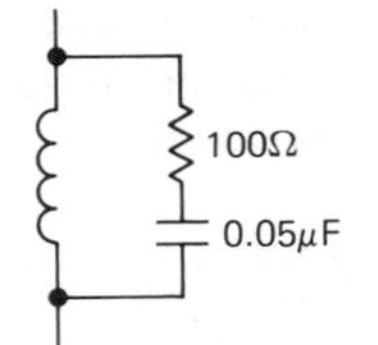

Figure 1.91

values shown are typical for small inductive loads driven from the ac power line. Such a snubber should be included in all instruments that run from the ac power line, since a transformer is inductive. An alternative protection device is a metal-oxide varistor, or transient suppressor, an inexpensive device that looks something like a disc ceramic capacitor and behaves electrically like a bidirectional zener diode. They are available at voltage ratings from 10 to 1000 volts and can handle transient currents up to thousands of amperes (see Section 5.10 and Table 5.2). Putting a transient suppressor across the ac power-line terminals makes good sense in any piece of electronic equipment, not only to prevent inductive spike interference to other nearby instruments but also to prevent occasional large power-line spikes from damaging the instrument itself.

OTHER PASSIVE COMPONENTS

In the following sections we would like to introduce briefly an assortment of miscellaneous but essential components. If you are experienced in electronic construction, you may wish to proceed to the next chapter.

1.32 Electromechanical devices

Switches

These mundane but important devices seem to wind up in most electronic equipment. It is worth spending a few paragraphs on the subject. Figure 1.92 shows some common switch types.

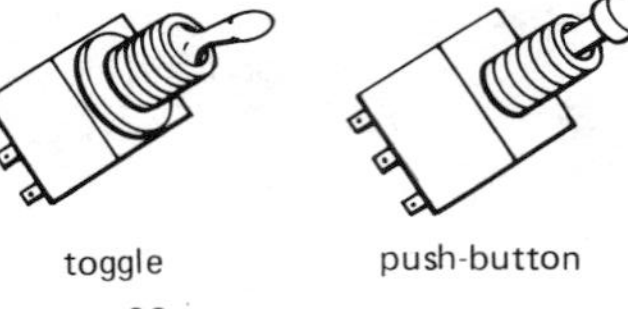
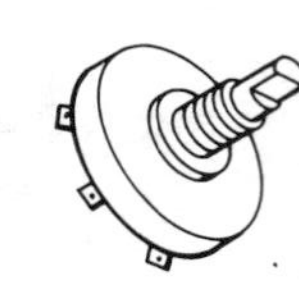

Figure 1.92

Toggle switches. The simple toggle switch is available in various configurations, depending on the number of poles; Figure 1.93 shows the usual ones (SPDT indicates

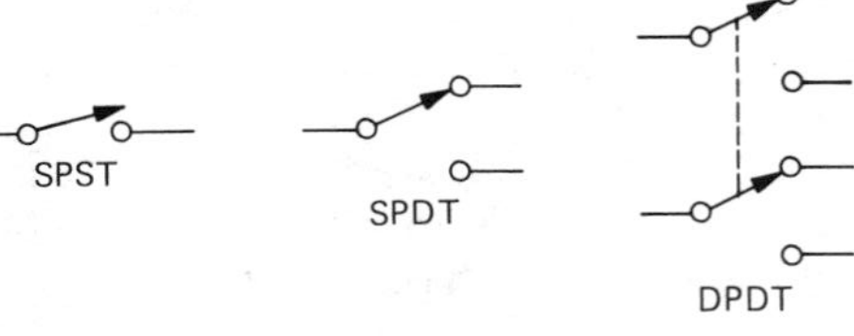

Figure 1.93

a single-pole double-throw switch, etc.). Toggle switches are also available with "center OFF" positions and with up to 4 poles switched simultaneously. Toggle switches are always "break before make," e.g., the moving contact never connects to both terminals in an SPDT switch.

Push-button switches. Push-button

switches are useful for momentary-contact applications; they are drawn schematically as shown in Figure 1.94 (NO and NC mean

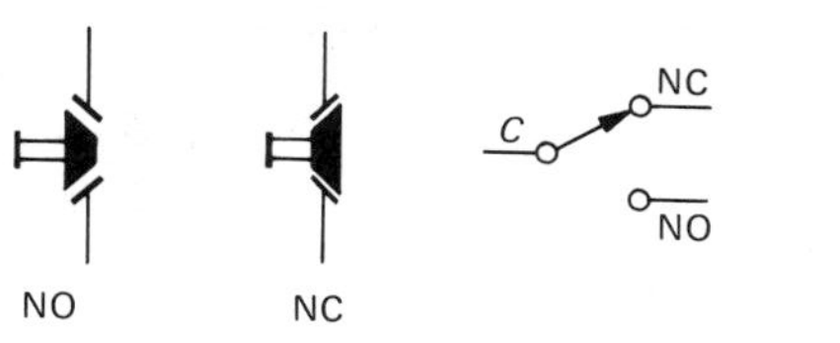

Figure 1.94

normally open and normally closed). For SPDT momentary-contact switches, the terminals must be labeled NO and NC, whereas for SPST types the symbol is self-explanatory. Momentary-contact switches are always "break before make." In the electrical (as opposed to electronic) industry, the terms form A, form B, and form C are used to mean SPST (NO), SPST (NC) and SPDT, respectively.

Rotary switches. Rotary switches are available with many poles and many positions, often as kits with individual wafers and shaft hardware. Both shorting (make before break) and nonshorting (break before make) types are available, and they can be mixed on the same switch. In many applications the shorting type is useful to prevent an open circuit between switch positions, because circuits can go amok with unconnected inputs. Nonshorting types are necessary if the separate lines being switched to one common line must not ever be connected to each other.

Other switch types. In addition to these basic switch types, there are available various exotic switches such as Hall-effect switches, reed switches, proximity switches, etc. All switches carry maximum current and voltage ratings; a small toggle switch might be rated at 150 volts and 5 amps. Operation with inductive loads drastically reduces switch life because of arcing during turn-off.

Switch examples. As an example of what can be done with simple switches, let's consider the following problem: Suppose you want to sound a warning buzzer if the driver of a car is seated and one of the car doors is open. Both doors and the driver's seat have switches, all normally open. Figure 1.95 shows a circuit that does what you

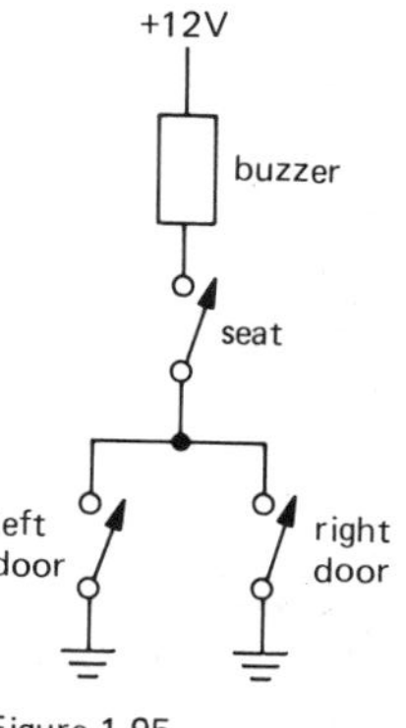

Figure 1.95

want. If one OR the other door is open (switch closed) AND the seat switch is closed, the buzzer sounds. The words OR and AND are used in a logic sense here, and we will see this example again in Chapters 2 and 8 when we talk about transistors and digital logic.

Figure 1.96 shows a classic switch circuit

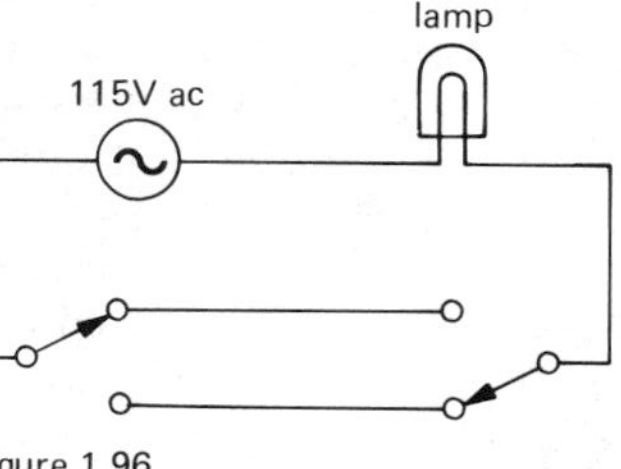

Figure 1.96

used to turn a ceiling lamp on or off from a switch at either of two entrances to a room.

Relays

Relays are electrically controlled switches. In the usual type, a coil pulls in an armature when sufficient coil current flows. Many varieties are available, including "latching" and "stepping" relays; the latter provided the cornerstone for telephone switching stations, and they're still popular in pinball machines. Relays are available for dc or ac excitation, and coil voltages from 5 volts up to 110 volts are common. "Mercury-wetted" and "reed" relays are intended for high-speed (~1ms) applications, and giant relays intended to switch thousands of amps are used by power companies. Many previous relay applications are now handled with transistor or FET switches, and devices known as solid-state relays are now available to handle ac switching applications. The primary uses of relays are in remote switching and high-voltage (or high-current) switching. Since it is important to keep electronic circuits electrically isolated from the ac power line, relays are useful to switch ac power while keeping the control signals electrically isolated.

Connectors

Bringing signals in and out of an instrument, routing signal and dc power around between the various parts of an instrument, providing flexibility by permitting circuit boards and larger modules of the instrument to be unplugged (and replaced) – these are the functions of the connector, an essential ingredient (and usually the most unreliable part) of any piece of electronic equipment. Connectors come in a bewildering variety of sizes and shapes.

Single-wire connectors. The simplest kind of connector is the simple pin jack or banana jack used on multimeters, power supplies, etc. It is handy and inexpensive, but not as useful as the shielded-cable or multiwire connectors you often need. The humble binding post is another form of single-wire connector, notable for the clumsiness it inspires in those who try to use it.

Shielded-cable connectors. In order to prevent capacitive pickup, and for other reasons we'll go into in Chapter 13, it is usually desirable to pipe signals around from one instrument to another in shielded coaxial cable. The most popular connector is the BNC type that adorns most instrument front panels. It connects with a quarter-turn twist and completes both the shield (ground) circuit and inner conductor (signal) circuit simultaneously. Like all connectors used to mate a cable to an instrument, it comes in both panel-mounting and cable-terminating varieties (Fig. 1.97).

Among the other connectors for use with coaxial cable are the nearly obsolete UHF type, the high-performance type N, the miniature SMA, the subminiature LEMO connectors, and the MHV, a high-voltage version of the standard BNC connector. The so-called phono jack used in audio equipment is a nice lesson in bad design, since the inner conductor mates before the shield when the jack is plugged in.

Multipin connectors. Very frequently electronic instruments demand multiwire cables and connectors. There are literally dozens of different kinds. The simplest example is a 3-wire line cord connector. Among the more popular are the excellent type D subminiature, the Winchester MRA series, the venerable MS type, and the flat ribbon-cable mass-termination connectors (Fig. 1.98). Beware of connectors that can't tolerate being dropped on the floor (the miniature

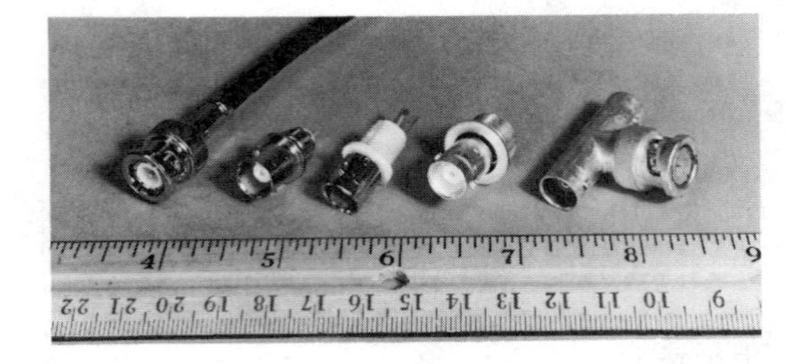

Figure 1.37
BNC connectors are the most popular type for use with shielded (coaxial) cable. From left to right: a male connector on a length of cable, a standard panel-mounting female connector, two varieties of insulated panel-mounting female connectors, and a BNC T, a handy device to have in the laboratory.

Figure 1.98
A selection of popular multipin connectors. From left to right: type-D subminiature, available in panel- and cable-mounting versions, with 9, 15, 25, 37, or 50 pins; the venerable MS/AN connector, available in many (too many!) pin and mounting configurations, including types suitable for shielded cables; a miniature rectangular connector (Winchester MRA type) with integral securing jackscrews, available in several sizes; a circuit-board-mounting mass-termination connector with its mating female ribbon connector.

hexagon connectors are classic) or that don't provide a secure locking mechanism (e.g., the Jones 300 series).

Card-edge connectors. The most common method used to make connection to printed-circuit cards is the card-edge connector, which mates to a row of gold-plated contacts at the edge of the card. Card-edge connectors may have from 15 to 100 pins, and they come with different lug styles according to the method of connection. You can solder them to a "motherboard" or "backplane," which is itself just another printed-circuit board containing the interconnecting wiring between the individual circuit cards. Alternatively, you may want to use edge connectors with standard solder-lug terminations, particularly in a system with only a few cards (see Chapter 12 for some photographs).

1.33 Indicators

Meters

To read out the value of some voltage or current, you have a choice between the time-honored moving-pointer type of meter and digital-readout meters. The latter are more expensive and more accurate. Both types are available in a variety of voltage and current ranges. There are, in addition, exotic panel meters that read out such things as VU (volume units, an audio dB scale), expanded-scale ac volts (e.g., 105 to 130 volts), temperature (from a thermocouple), percentage motor load, frequency, etc. Digital panel meters often provide the option of logic-level outputs, in addition to the visible display, for internal use by the instrument.

Lamps and LEDs

Flashing lights, screens full of numbers and letters, eerie sounds – these are the stuff of science fiction movies, and except for the latter, they form the subject of lamps and displays. Small incandescent lamps have been standard for front-panel indicators, but they are being replaced by light-emitting diodes (LEDs). The latter behave electrically like ordinary diodes, but with a forward voltage drop in the range of 1.5 to 2.5 volts. When current flows in the forward direction, they light up. Typically, 5mA to 20mA produces adequate brightness. LEDs are cheaper than incandescent lamps, they last forever, and they are even available in three colors (red, yellow, and green). They come in convenient panel-mounting packages.

LEDs are also used for digital displays, most often the familiar 7-segment numeric display you see in calculators. For displaying letters as well as numbers (alphanumeric display), you can get 16-segment displays or dot-matrix displays. For low power or outdoor use, liquid-crystal displays are superior.

1.34 Variable components

Resistors

Variable resistors (also called volume controls, potentiometers, pots, or trimmers) are useful as panel controls or internal adjustments in circuits. The most common panel type is known as a 2 watt type AB potentiometer; it uses the same basic material as the fixed carbon-composition resistor, with a rotatable "wiper" contact. Other panel types are available with ceramic or plastic resistance elements, with improved characteristics. Multiturn types (3, 5, or 10 turns) are available, with counting dials, for improved resolution and linearity. "Ganged" pots (several independent sections on one shaft) are also manufactured, although in limited variety, for applications that demand them.

For use inside an instrument, rather than on the front panel, *trimmer pots* come in single-turn and multiturn styles, most intended for printed-circuit mounting. These are handy for calibration adjustments of the "set-and-forget" type. Good advice: Resist the temptation to use lots of trimmers in your circuits. Use good design instead.

The symbol for a variable resistor, or pot, is shown in Figure 1.99. Sometimes the

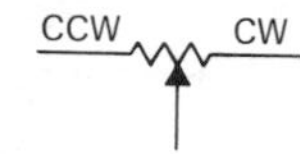

Figure 1.99

symbols CW and CCW are used to indicate clockwise and counterclockwise.

One important point about variable resistors: Don't attempt to use a potentiometer as a substitute for a precise resistor value somewhere within a circuit. This is tempting, since you can trim the resistance to the value you want. The trouble is that potentiometers are not as stable as good (1%) resistors, and in addition they may not have good resolution (i.e., they can't be set to a precise value). If you must have a precise and settable resistor value somewhere, use a combination of a 1% (or better) precision resistor and a potentiometer, with the fixed resistor contributing most of the resistance. For example, if you need a 23.4k resistor, use a 22.6k (a 1% value) 1% fixed resistor in series with a 2k trimmer pot. Another possibility is to use a series combination of several precision resistors, choosing the last (and smallest) resistor to give the desired series resistance.

As you will see later, it is possible to use FETs as voltage-controlled variable resistors in some applications. Transistors can be used as variable-gain amplifiers, again controlled by a voltage. Keep an open mind when design brainstorming.

Capacitors

Variable capacitors are primarily confined to the smaller capacitance values (up to about 1000pF) and are commonly used in radiofrequency (RF) circuits. Trimmers are available for in-circuit adjustments, in addition to the panel type for user tuning. Figure 1.100 shows the symbol for a variable capacitor.

Figure 1.100

Diodes operated with applied reverse voltage can be used as voltage-variable capacitors; in this application they're called varactors, or sometimes varicaps or epicaps. They're very important in RF applications, especially automatic frequency control (AFC), modulators, and parametric amplifiers.

Inductors

Variable inductors are usually made by arranging to move a piece of core material in a fixed coil. In this form they're available with inductances ranging from microhenrys to henrys, typically with a 2:1 tuning range for any given inductor. Also available are rotary inductors (coreless coils with a rolling contact).

Transformers

Variable transformers are very handy devices, especially the ones operated from the 115 volt ac line. They're usually "autoformers," which means that they have only one winding, with a sliding contact. They're also commonly called Variacs, the name used by the General Radio Company (GenRad). Typically they provide 0 to 135 volts ac output when operated from 115 volts, and they come in current ratings from 1 amp to 20 amps or more. They're good for testing instruments that seem to be affected by power-line variations, and in any case to verify worst-case performance. Warning: Don't forget that the output is not electrically isolated from the power line, as it would be with a transformer!

ADDITIONAL EXERCISES

(1) Find the Norton equivalent circuit (a current source in parallel with a resistor) for the voltage divider in Figure 1.101. Show

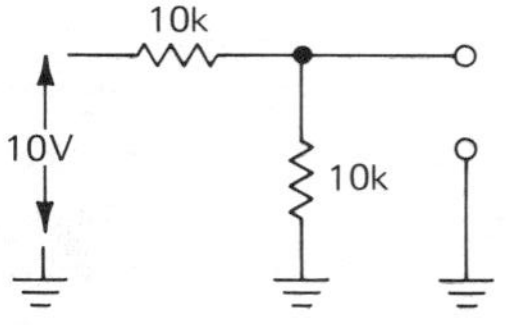

Figure 1.101

that the Norton equivalent gives the same output voltage as the actual circuit when loaded by a 5k resistor.

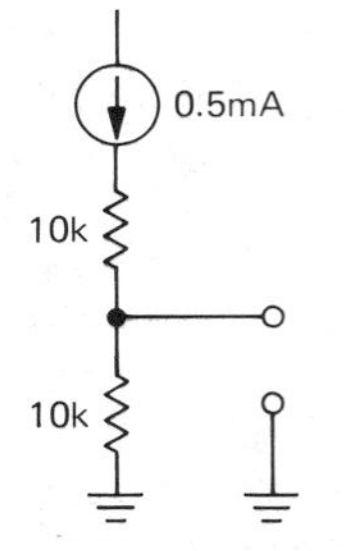

Figure 1.102

(2) Find the Thévenin equivalent for the circuit shown in Figure 1.102. Is it the same as the Thévenin equivalent for exercise 1?
(3) Design a "rumble filter" for audio. It should pass frequencies greater than 20Hz (set the −3dB point at 10Hz). Assume zero source impedance (perfect voltage source) and 10k (minimum) load impedance (that's important so that you can choose R and C such that the load doesn't affect the filter operation significantly).
(4) Design a "scratch filter" for audio signals (3dB down at 10kHz). Use the same source and load impedances as in exercise 3.
(5) How would you make a filter with Rs and Cs to give the response shown in Figure 1.103?

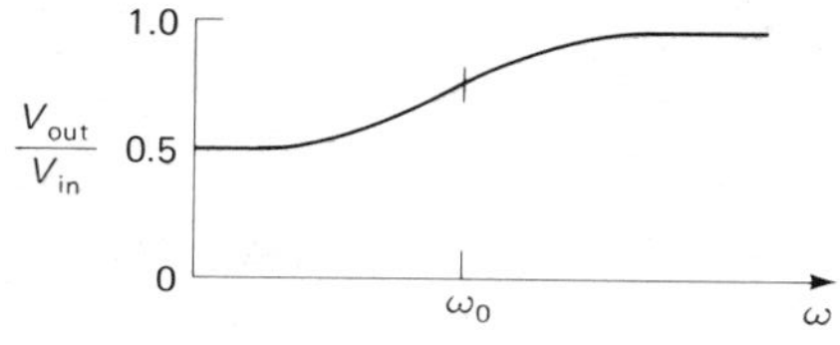

Figure 1.103

(6) Design a bandpass RC filter (Fig. 1.104); f_1 and f_2 are the 3dB points. Choose impedances so that the first stage isn't much affected by the loading of the second stage.

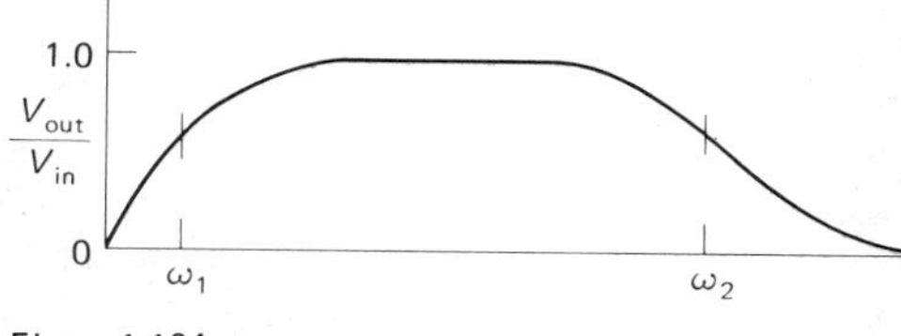

Figure 1.104

(7) Sketch the output for the circuit shown in Figure 1.105.

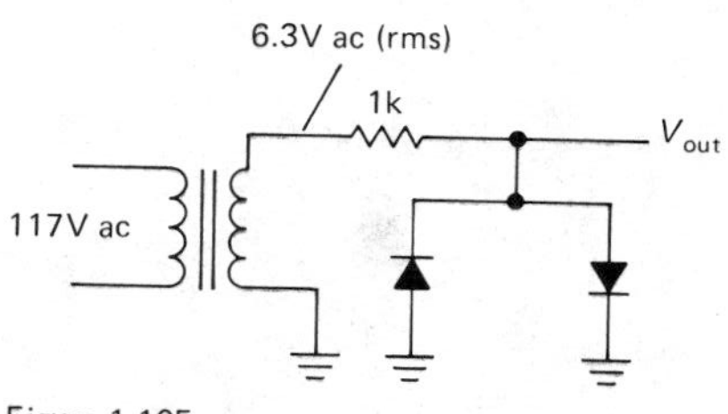

Figure 1.105

(8) Design an oscilloscope "×10 probe" (see Appendix A) to use with a scope whose input impedance is 1MΩ in parallel with 20pF. Assume that the probe cable adds an additional 100pF and that the probe components are placed at the tip end (rather than at the scope end) of the cable (Fig. 1.106). The resultant network should have 20dB (×10) attenuation at all frequencies, including dc. The reason for using a ×10 probe is to increase the load impedance seen by the circuit under test, which reduces loading effects. What input impedance (R in parallel with C) does your ×10 probe present to the circuit under test, when used with the scope?

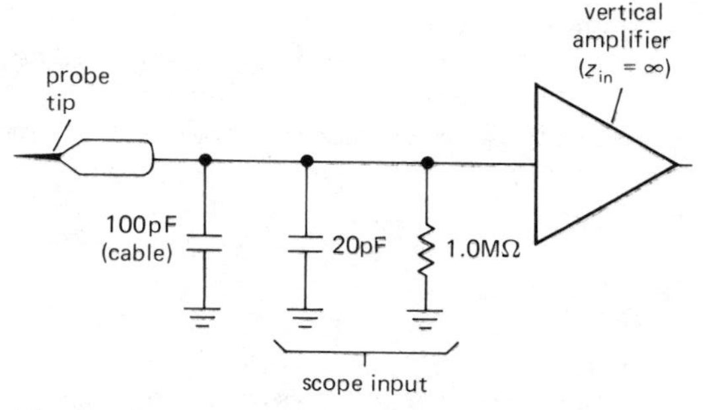

Figure 1.106

CHAPTER 2 TRANSISTORS

INTRODUCTION

The transistor is our most important example of an "active" component, a device that can amplify, producing an output signal with more power in it than the input signal. The additional power comes from an external source of power (the power supply, to be exact). Note that voltage amplification isn't what matters, since, for example, a step-up transformer, a "passive" component just like a resistor or capacitor, has voltage gain but no power gain. Devices with power gain are distinguishable by their ability to make oscillators, by feeding some output signal back into the input.

It is interesting to note that the property of power amplification seemed very important to the inventors of the transistor. Almost the first thing they did to convince themselves that they had really invented something was to power a loudspeaker from a transistor, observing that the output signal sounded louder than the input signal.

The transistor is the essential ingredient of every electronic circuit, from the simplest amplifier or oscillator to the most elaborate digital computer. Integrated circuits (ICs), which are rapidly replacing circuits constructed from discrete transistors, are themselves merely arrays of transistors and other components built from a single chip of semiconductor material.

A good understanding of transistors is very important, even if most of your circuits are made from ICs, because you need to understand the input and output properties of the IC in order to connect it to the rest of your circuit and to the outside world. In addition, the transistor is the single most powerful resource for interfacing, whether between ICs and other circuitry or between one subcircuit and another. Finally, there are frequent (some might say too frequent) situations where the right IC just doesn't exist, and you have to rely on discrete transistor circuitry to do the job. As you will see, transistors have an excitement all their own. Learning how they work can be great fun.

Our treatment of transistors is going to be quite different from that of many other books. It is common practice to use the h-parameter model and equivalent circuit. In our opinion that is unnecessarily complicated and unintuitive. Not only does circuit behavior tend to be revealed to you as something that drops out of elaborate equations, rather than deriving from a clear understanding in your own mind as to how the circuit functions; you also have the tendency to lose sight of which parameters of transistor behavior you can count on and, more important, which ones can vary over large ranges.

In this chapter we will build up instead a very simple introductory transistor model and immediately work out some circuits with it. Soon its limitations will become apparent; then we will expand the model to include the respected Ebers-Moll conventions. With the Ebers-Moll equations and a simple 3-terminal model, you will have a good understanding of transistors; you won't need to do a lot of calculations, and your designs will be first-rate. In particular, they will be largely independent of the poorly controlled transistor parameters such as current gain.

Some important engineering notation should be mentioned. Voltage at a transistor terminal (relative to ground) is indicated by a single subscript (*C*, *B*, or *E*): V_C is the collector voltage, for instance. Voltage between two terminals is indicated by a double subscript: V_{BE} is the base-to-emitter voltage drop, for instance. If the same letter is repeated, that means a power-supply voltage: V_{CC} is the (positive) power-supply voltage associated with the collector, and V_{EE} is the (negative) supply voltage associated with the emitter.

2.01 First model: current amplifier

Let's begin. A transistor is a 3-terminal device (Fig. 2.1) available in two flavors (*npn* and *pnp*), with properties that meet the following rules for *npn* transistors (for *pnp* simply reverse all polarities):

1. The collector must be more positive than the emitter.
2. The base-emitter and base-collector circuits behave like diodes (Fig. 2.2). Normally the base-emitter diode is conducting and the base-collector diode is reverse-biased.

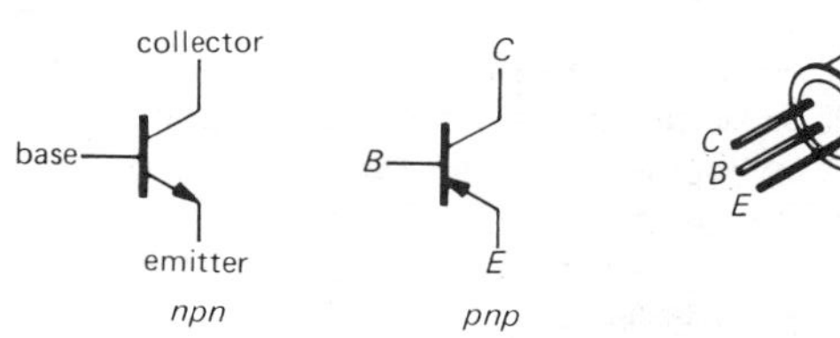

Figure 2.1

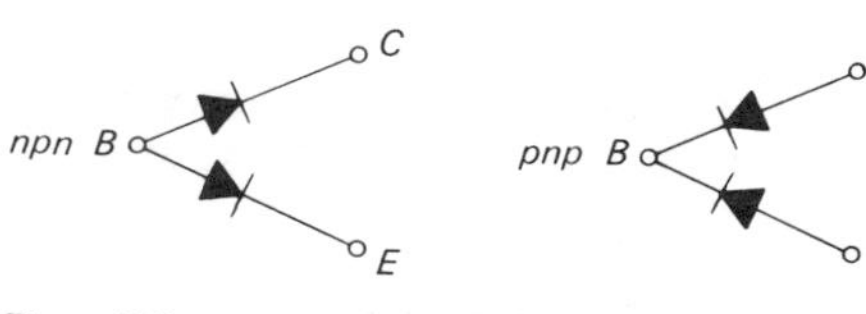

Figure 2.2

3. Any given transistor has maximum values of I_C, I_B, and V_{CE} that cannot be exceeded without costing the exceeder the price of a new transistor (for typical values, see Table 2.1). There are also other limits, such as power dissipation ($I_C V_{CE}$), temperature, V_{BE}, etc., that you must keep in mind.
4. When rules 1–3 are obeyed, I_C is roughly proportional to I_B and can be written as

$$I_C = h_{FE} I_B = \beta I_b$$

where h_{FE}, the current gain (also called beta), is typically about 100. Note: Don't confuse this with forward conduction of the base-collector diode; that diode is reverse-biased. Just think of it as "transistor action."

Property 4 gives the transistor its usefulness: A small current flowing into the base controls a much larger current flowing into the collector.

Warning: h_{FE} is not a "good" transistor parameter; for instance, its value can vary from 50 to 250 for different specimens of a given transistor type. It also depends on collector current, collector-to-emitter voltage, and temperature. *A circuit that depends on a particular value for* h_{FE} *is a bad circuit.*

Note particularly the effect of property 2. This means you can't go sticking a voltage across the base-emitter terminals, because an enormous current will flow if the base is more positive than the emitter by more than about 0.6 to 0.8 volt (forward diode drop). This rule also implies that an operating transistor has $V_B \approx V_E + 0.6$ volt ($V_B = V_E + V_{BE}$). Again, polarities are normally given for *npn* transistors; reverse them for *pnp*.

Let us emphasize again that you shouldn't try to think of the collector current as diode conduction. It isn't, since the collector-base diode normally has voltages applied across it in the reverse direction. Furthermore, collector current varies very little with collector voltage (it behaves like a not-too-great current source), unlike forward diode conduction, where the current rises very rapidly with applied voltage

SOME BASIC TRANSISTOR CIRCUITS

2.02 Transistor switch

Look at the circuit in Figure 2.3. This application is called a transistor switch. From the

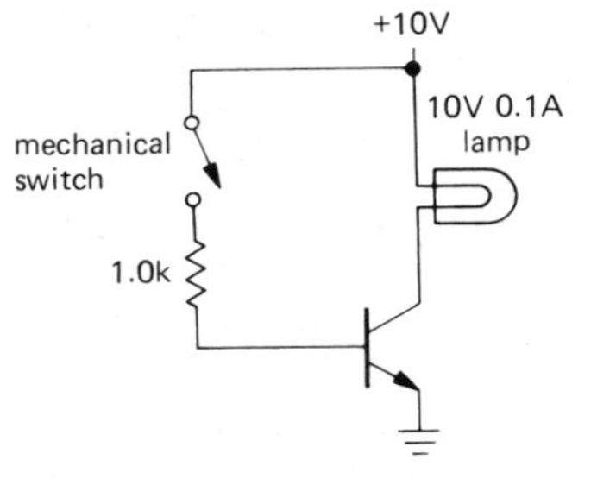

Figure 2.3

preceding rules it is easy to understand. When the mechanical switch is open, there is no base current. So, from rule 4, there is no collector current. The lamp is off.

When the switch is closed, the base rises to 0.6 volt (base-emitter diode is in forward conduction). The drop across the base resistor is 9.4 volts, so the base current is 9.4mA. Blind application of rule 4 gives I_C = 940mA (for a typical beta of 100). That is wrong. Why? Because rule 4 holds only if rule 1 is obeyed; at a collector current of 100mA the lamp has 10 volts across it. To get higher current you would have to pull the collector below ground. A transistor can't do this, and the result is what's called saturation – the collector goes as close to ground as it can (typical saturation voltages are about 0.05–0.2V, see Appendix G) and stays there. In this case the lamp goes on, with its rated 10 volts across it.

Overdriving the base (we used 9.4mA when 1.0mA would have barely sufficed) makes the circuit conservative; in this particular case it is a good idea, since a lamp draws more current when cold (the resistance of a lamp when cold is 5 to 10 times lower than its resistance at operating current). Also, transistor beta drops at low collector-to-base voltages, so some extra base current is necessary to bring a transistor into full saturation (see Appendix G). Incidentally, in a real circuit you would probably put a resistor from base to ground (perhaps 10k in this case) to make sure the base is at ground with the switch open. It wouldn't affect the "on" operation, since it would sink only 0.06mA from the base circuit.

There are certain cautions to be observed when designing transistor switches:

1. Choose the base resistor conservatively to get plenty of excess base current, especially when driving lamps, because of the reduced beta at low V_{CE}. This is also a good idea for high-speed switching, because of capacitive effects and reduced beta at very high frequencies (many megahertz). A small "speedup" capacitor is often connected across the base resistor to improve high-speed performance.
2. If the load swings below ground for some reason (e.g., it is driven from ac, or it is inductive), use a diode in series with the collector (or a diode in the reverse direction to ground) to prevent collector-base conduction on negative swings.
3. For inductive loads, protect the transistor with a diode across the load, as shown in Figure 2.4. Without the diode the inductor

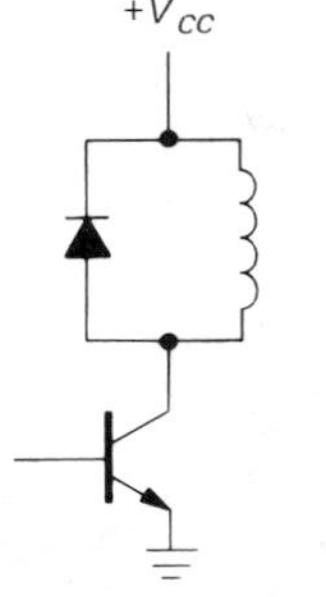

Figure 2.4

will swing the collector to a large positive voltage when the switch is opened, most likely exceeding the collector-emitter breakdown voltage, as the inductor tries to maintain its "on" current from V_{CC} to the collector (see the discussion of inductors in Section 1.31).

Transistor switches enable you to switch very rapidly, typically in a small fraction of a microsecond. Also, you can switch many different circuits with a single control signal. One further advantage is the possibility of remote *cold switching,* in which only dc control voltages snake around through cables to reach front-panel switches, rather than the electrically inferior approach of having the signals themselves traveling through cables and switches (if you run lots of signals through cables, you're likely to get capacitive pickup as well as some signal degradation).

"Transistor Man"

Figure 2.5 presents a cartoon that will help you understand some limits of transistor

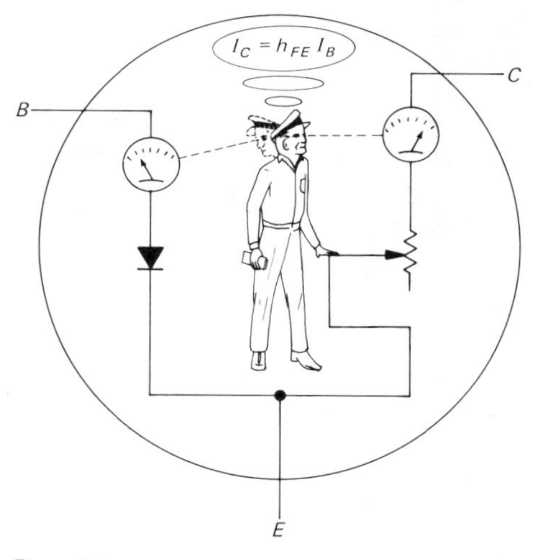

Figure 2.5

behavior. The little man's perpetual task in life is to try to keep $I_C = h_{FE}I_B$; however, he is allowed to turn only the knob on the variable resistor. Thus he can go from a short circuit (saturation) to an open circuit (transistor in the "off" state), or anything in between, but he isn't allowed to use batteries, current sources, etc. One warning is in order here: Don't think that the collector of a transistor looks like a resistor. It doesn't. Rather, it looks approximately like a poor-quality constant-current sink (the value of current depending on the signal applied to the base), primarily because of this little man's efforts.

Another thing to keep in mind is that, at any given time, a transistor may be (a) cut off (no collector current), (b) in the active region (some collector current, and collector voltage more than a few tenths of a volt above the emitter), or (c) in saturation (collector within a few tenths of a volt of the emitter). See Appendix G on transistor saturation for more details.

2.03 Emitter follower

Figure 2.6 shows an *emitter follower.* It is called that because the output terminal is the

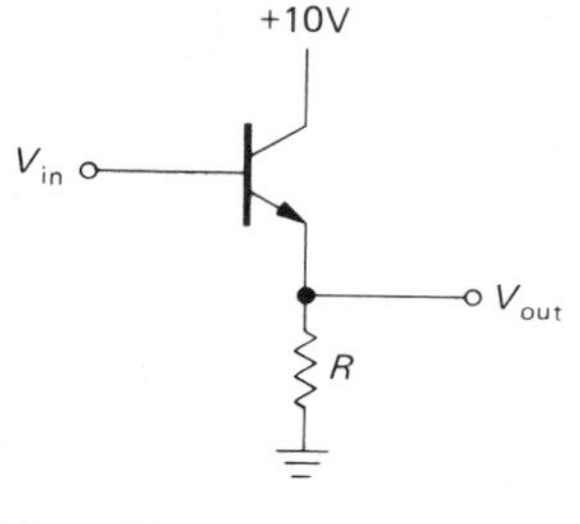

Figure 2.6

emitter, which follows the input (the base), less one diode drop:

$$V_E \approx V_B - 0.6 \text{ volt}$$

The output is a replica of the input, but 0.6 to 0.7 volt less positive. For this circuit, V_{in} must stay at +0.6 volt or more, or else the output will sit at ground. By returning the emitter resistor to a negative supply voltage, you can permit negative voltage swings as well. Note that there is no collector resistor in an emitter follower.

At first glance this circuit may appear useless, until you realize that the input impedance is much larger than the output impedance, as will be demonstrated shortly. This means that the circuit requires less power from the signal source to drive a given load than would be the case if the signal source were to drive the load directly. Or a signal of some internal impedance (in the Thévenin sense) can now drive a load of comparable or even lower impedance without loss of amplitude (from the usual voltage-divider effect). In other words, an emitter follower has current gain, even though it has no voltage gain. It has power gain. Voltage gain isn't everything!

Input and output impedances of emitter followers

As you have just seen, the emitter follower is useful for changing impedances of signals or loads. To put it bluntly, that's the whole point of an emitter follower.

Let's calculate the input and output impedances of the emitter follower. In the preceding circuit we will consider R to be the load (in practice it sometimes is the load; otherwise the load is in parallel with R, but with R dominating the parallel resistance anyway). Make a voltage change ΔV_B at the base; the corresponding change at the emitter is $\Delta V_E = \Delta V_B$. Then the change in emitter current is

$$\Delta I_E = \Delta V_B / R$$

so

$$\Delta I_B = \frac{1}{h_{fe} + 1} \Delta I_E = \frac{\Delta V_B}{R(h_{fe} + 1)}$$

(using $I_E = I_C + I_B$). The input resistance is $\Delta V_B / \Delta I_B$. Therefore

$$r_{in} = (h_{fe} + 1)R$$

The transistor beta (h_{fe}) is typically about 100, so a low-impedance load looks like a much higher impedance at the base; it is easier to drive.

In the preceding calculation, as in Chapter 1, we have used lower-case symbols such as h_{fe} to signify small-signal (incremental) quantities. The distinction between dc current gain (h_{FE}) and small-signal current gain (h_{fe}) isn't always made clear, and the term beta is used for both. That's alright, since $h_{fe} \approx h_{FE}$ (except at very high frequencies), and you never assume you know them accurately, anyway.

Although we used resistances in the preceding derivation, we could generalize to complex impedances by allowing ΔV_B, ΔI_B, etc., to become complex numbers. We would find that the same transformation rule applies for impedances: $\mathbf{Z}_{in} = (h_{fe} + 1)\mathbf{Z}_{load}$.

We could do a similar calculation to find that the output impedance $\mathbf{Z}_{out}$ of an emitter follower (the impedance looking into the emitter) driven from a source of internal impedance $\mathbf{Z}_{source}$ is given by

$$\mathbf{Z}_{out} = \frac{\mathbf{Z}_{source}}{h_{fe} + 1}$$

Strictly speaking, the output impedance of the circuit should also include the parallel resistance of R, but in practice $\mathbf{Z}_{out}$ (the impedance looking into the emitter) dominates.

EXERCISE 2.1

Show that the preceding relationship is correct. Hint: Hold the source voltage fixed, and find the change in output current for a given change in output voltage. Remember that the source voltage is connected to the base through a series resistor.

Because of these nice properties, emitter followers find application in many situations, e.g., making low-impedance signal sources within a circuit (or at outputs), making stiff voltage references from higher-impedance references (formed from voltage dividers, say), and generally isolating signal sources from the loading effects of subsequent stages.

EXERCISE 2.2

Use a follower with base driven from a voltage divider to provide a stiff source of +5 volts from an available regulated +15 volt supply. Load current (max) = 25mA. Choose your resistor values so that the output voltage doesn't drop more than 5% under full load.

Important points about followers

1. Remember that in an emitter follower the *npn* transistor can only "source" current. For instance, in the loaded circuit shown in Figure 2.7 the output can swing to within a transistor saturation voltage drop of V_{CC} (about +9.9V), but it cannot go more negative than −5 volts. That is because on the extreme negative swing the transistor can do no more than turn off, which it does at −4.4 volts input (−5V output). Further negative swing at the input results in back-biasing of the base-emitter junction, but no further change in output. The output, for a

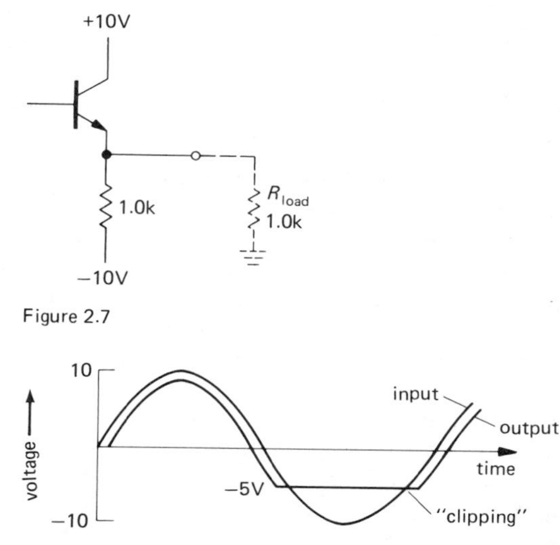

Figure 2.7

Figure 2.8

10-volt-amplitude sine-wave input, looks as shown in Figure 2.8.

Another way to view the problem is to say that the emitter follower has low small-signal output impedance. Its large-signal output impedance is much larger (as large as R_E). The output impedance changes over from its small-signal value to its large-signal value at the point where the transistor goes out of the active region (in this case at an output voltage of -5V). To put this very important point another way, a low value of small-signal output impedance doesn't necessarily mean that the circuit can generate large signal swings into a low-resistance load. Low small-signal output impedance doesn't imply large output current capability.

Possible solutions to this problem involve either decreasing the value of the emitter resistor (with greater power dissipation in resistor and transistor), using a *pnp* transistor (if all signals are negative only), or using a "push-pull" configuration, in which two complementary transistors (one *npn,* one *pnp*) are used (more on this later). This sort of problem can also come up when the load of an emitter follower contains voltage or current sources of its own. This happens most often with regulated power supplies (the output is usually an emitter follower) driving a circuit that has other power supplies.

2. Always remember that the base-emitter breakdown voltage for silicon transistors is small, quite often as little as 6 volts. Input swings large enough to take the transistor out of conduction can easily result in breakdown (with consequent degradation of h_{FE}) unless a protective diode is added (Fig. 2.9).

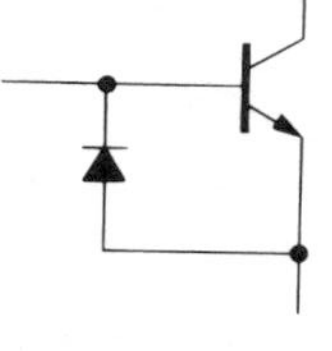

Figure 2.9

3. The voltage gain of an emitter follower is actually slightly less than 1.0, because the base-emitter voltage drop is not really constant, but depends slightly on collector current. You will see how to handle that later in the chapter, when we have the Ebers-Moll equation.

2.04 Emitter followers as voltage regulators

The simplest regulated supply of voltage is simply a zener (Fig. 2.10). Some current

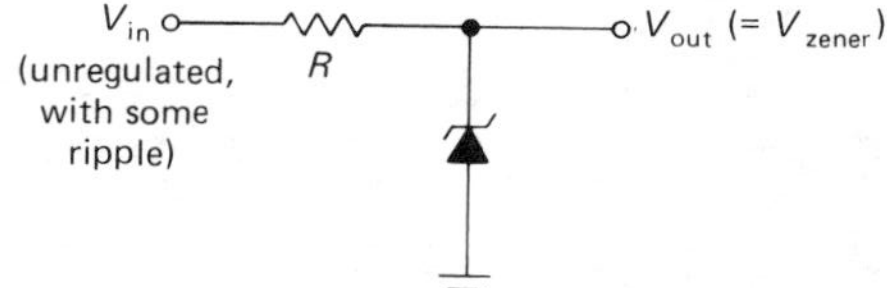

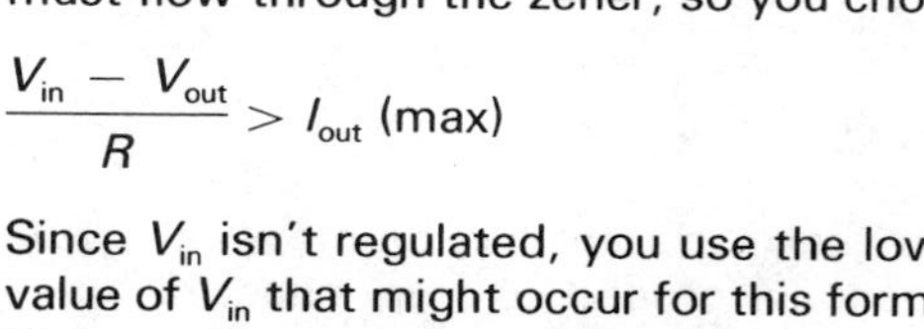

Figure 2.10

must flow through the zener, so you choose

$$\frac{V_{in} - V_{out}}{R} > I_{out}\,(\text{max})$$

Since V_{in} isn't regulated, you use the lowest value of V_{in} that might occur for this formula. This is called worst-case design. In practice you would also worry about component tolerances, line voltage limits, etc., designing to accommodate the worst possible combination that could ever occur.

The zener must be able to dissipate

$$P_{zener} = \left(\frac{V_{in} - V_{out}}{R} - I_{out}\right) V_{zener}$$

Again, for worst-case design you would use V_{in} (max), R_{min}, and I_{out} (min).

EXERCISE 2.3

Design a +10 volt regulated supply for load currents from 0 to 100mA; the input voltage is +20 to +25 volts. Allow at least 10mA zener current under all (worst-case) conditions. What power rating must the zener have?

This simple zener-regulated supply is sometimes used for noncritical circuits, or circuits using little supply current. However, it has limited usefulness, for several reasons:

1. V_{out} isn't adjustable, or settable to a precise value.
2. Zener diodes give only moderate ripple rejection and regulation against changes of input or load, owing to their finite dynamic impedance.
3. For widely varying load currents a high-power zener is often necessary to handle the dissipation at low load current.

By using an emitter follower to isolate the zener, you get the improved circuit shown in Figure 2.11. Now the situation is much

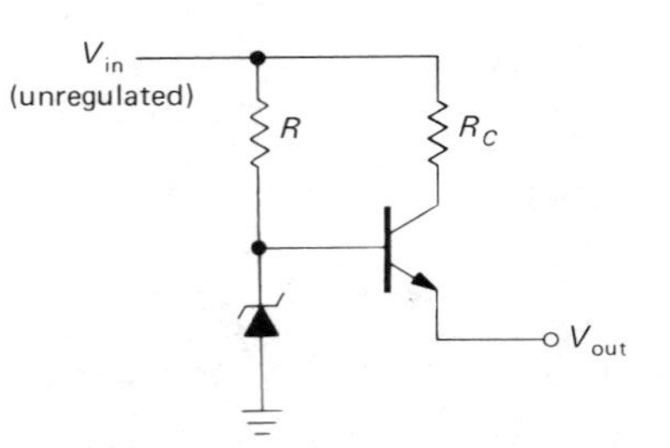

Figure 2.11

better. Zener current can be made relatively independent of load current, since the transistor base current is small, and far lower zener power dissipation is possible (reduced by as much as $1/h_{FE}$). The resistor R_C is important to protect the transistor from momentary output short circuits by limiting the current, even though it could be omitted, since an emitter follower normally has no collector resistor. Choose R_C so that the voltage drop across it is less than the drop across R for the highest normal load current.

EXERCISE 2.4

Design a +10 volt supply with the same specifications as in Exercise 2.3. Use a zener and emitter follower. Calculate worst-case dissipation in transistor and zener. What is the percentage change in zener current from the no-load condition to full load? Compare with your previous circuit.

A nice variation of this circuit aims to eliminate the effect of ripple current (through R) on the zener voltage by supplying the zener current from a current source, which is the subject of Section 2.06. An alternative method uses a low-pass filter in the zener bias circuit (Fig. 2.12). R is chosen to

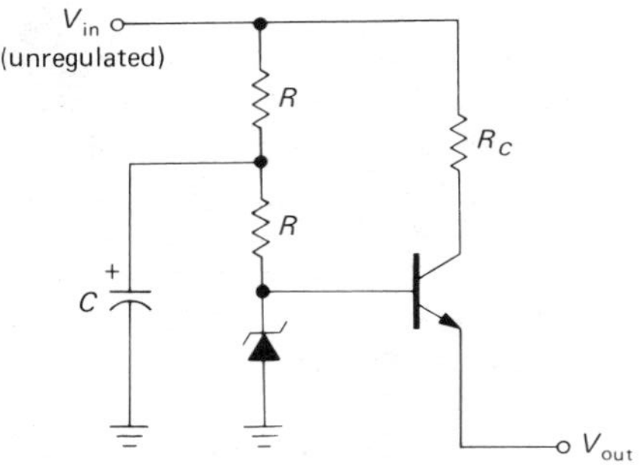

Figure 2.12

provide sufficient zener current. Then C is chosen large enough so that $RC \gg 1/f_{ripple}$. (In a variation of this circuit, the upper resistor is replaced by a diode.)

Later you will see better voltage regulators, ones in which you can vary the output easily and continuously, using feedback. They are also better voltage sources, with output impedances measured in milliohms, temperature coefficients of a few parts per million per degree centigrade, etc.

2.05 Emitter follower biasing

When an emitter follower is driven from a preceding stage in a circuit, it is usually OK to connect its base directly to the previous stage's output, as shown in Figure 2.13.

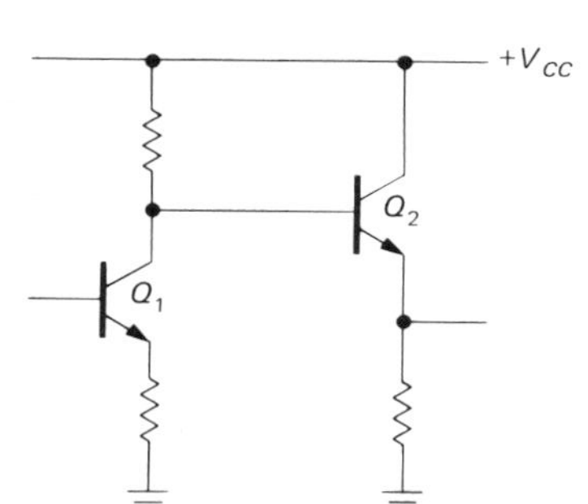

Figure 2.13

Since the signal on Q_1's collector is always within the range of the power supplies, Q_2's base will be between V_{CC} and ground, and therefore Q_2 is in the active region (neither cut off nor saturated), with its base-emitter diode in conduction and its collector at least a few tenths of a volt more positive than its emitter. Sometimes, though, the input to a follower may not be so conveniently situated with respect to the supply voltages. A typical example is a capacitively coupled (or ac-coupled) signal from some external source (e.g., an audio signal input to a high-fidelity amplifier). In that case the signal's average voltage is zero, and direct coupling to an emitter follower will give an output like that in Figure 2.14.

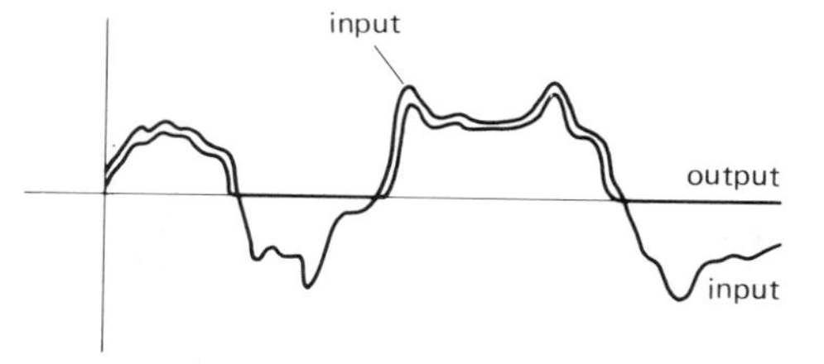

Figure 2.14

It is necessary to *bias* the follower (in fact, any transistor amplifier) so that collector current flows during the entire signal swing. In this case a voltage divider is the simplest way (Fig. 2.15). R_1 and R_2 are chosen to put the base halfway between ground and V_{CC} with no input signal, i.e., R_1 and R_2 are equal. The process of selecting the operating voltages in a circuit, in the absence of applied signals, is known as setting the *quiescent point*. In this case, as

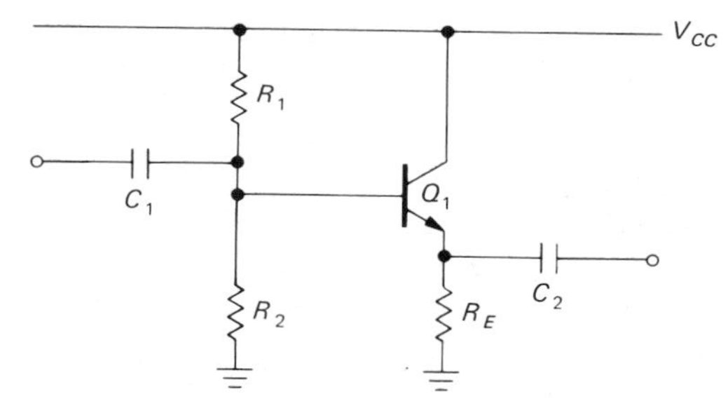

Figure 2.15

in most cases, the quiescent point is chosen to allow maximum symmetrical signal swing of the output waveform without *clipping* (flattening of the top or bottom of the waveform). What values should R_1 and R_2 have? Applying our general principle (Section 1.05), we make the impedance of the dc bias source (the impedance looking into the voltage divider) small compared with the load it drives (the dc impedance looking into the base of the follower). In this case,

$$R_1 \| R_2 \ll h_{FE} R_E$$

This is approximately equivalent to saying that the current flowing in the voltage divider should be large compared with the current drawn by the base.

Emitter follower design example

As an actual design example, let's make an emitter follower for audio signals (20Hz to 20kHz). V_{CC} is $+15$ volts, and quiescent current is to be 1mA.

Step 1. Choose V_E. For the largest possible symmetrical swing without clipping, $V_E = 0.5\ V_{CC}$, or $+7.5$ volts.
Step 2. Choose R_E. For a quiescent current of 1mA, $R_E = 7.5\text{k}$.
Step 3. Choose R_1 and R_2. V_B is $V_E + 0.6$, or 8.1 volts. This determines the ratio of R_1 to R_2 as 1:1.17. The preceding loading criterion requires that the parallel resistance of R_1 and R_2 be about 75k or less (one-tenth of 7.5k times h_{FE}). Suitable standard values are $R_1 = 130\text{k}$, $R_2 = 150\text{k}$.
Step 4. Choose C_1. C_1 forms a high-pass filter with the impedance it sees as a load, namely the impedance looking into the base in parallel with the impedance looking into

the base voltage divider. If we assume that the load this circuit will drive is large compared with the emitter resistor, then the impedance looking into the base is $h_{FE}R_E$, about 750k. The divider looks like 70k. So the capacitor sees a load of about 63k, and it should have a value of at least 0.15μF so that the 3dB point will be below the lowest frequency of interest, 20Hz.

Step 5. Choose C_2. C_2 forms a high-pass filter in combination with the load impedance, which is unknown. However, it is safe to assume that the load impedance won't be smaller than R_E, which gives a value for C_2 of at least 1.0μF, to put the 3dB point below 20Hz. Since there are now two cascaded high-pass-filter sections, the capacitor values should be increased somewhat to prevent large attenuation (reduction of signal amplitude, in this case 6dB) at the lowest frequency of interest. $C_1 = 0.5\mu$F and $C_2 = 3.3\mu$F might be good choices.

Followers with split supplies

Since signals are often "near ground," it is convenient to use symmetrical positive and negative supplies. This simplifies biasing and eliminates coupling capacitors (Fig. 2.16).

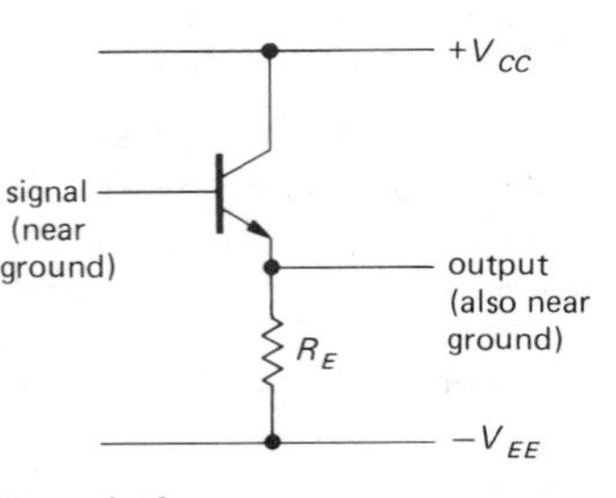

Figure 2.16

Warning: You must always provide a dc path for base bias current, even if it goes only to ground. In the preceding circuit it is assumed that the signal source has a dc path to ground. If not (e.g., if the signal is capacitively coupled), you must provide a resistor to ground (Fig. 2.17). R_B could be about one-tenth of $h_{FE}R_E$, as before.

EXERCISE 2.5

Design an emitter follower with ±15 volt supplies to operate over the audio range

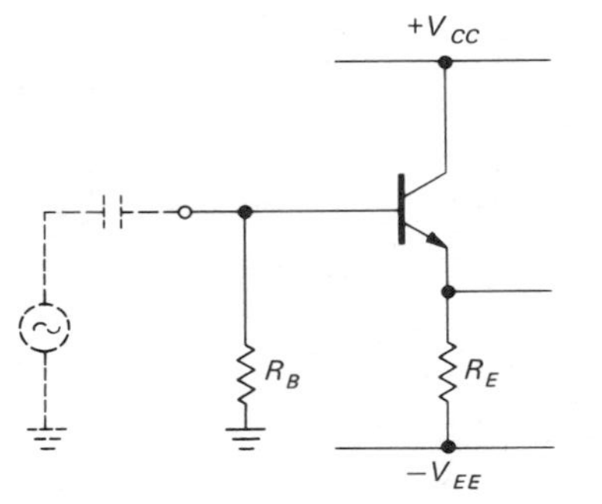

Figure 2.17

(20Hz–20kHz). Use 5mA quiescent current and capacitive input coupling.

Bad biasing

Unfortunately, you sometimes see circuits like the disaster shown in Figure 2.18. R_B

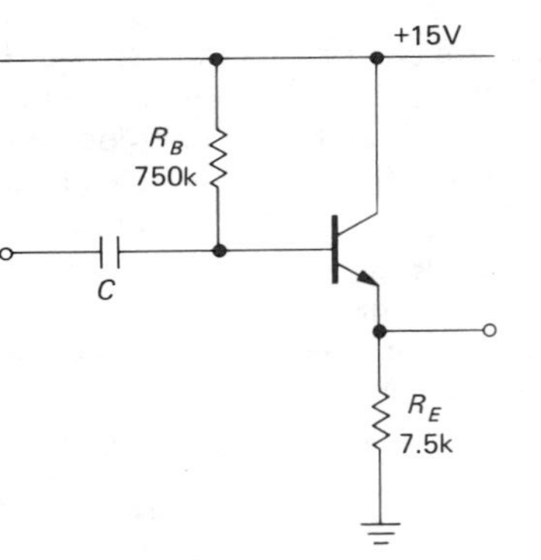

Figure 2.18

was chosen by assuming a particular value for h_{FE} (100), estimating the base current, and then hoping for a 7 volt drop across R_B. This is bad design; h_{FE} is not a good parameter and will vary considerably. By using voltage biasing with a stiff voltage divider, as in the detailed example presented earlier, the quiescent point is insensitive to variations in transistor beta. For instance, in the previous design example the emitter voltage will increase by only 0.35 volt (5%) for a transistor with $h_{FE} = 200$ instead of the nominal $h_{FE} = 100$. As with this emitter follower example, it is just as easy to fall into this trap and design bad transistor circuits in the other transistor configurations (e.g., the common-emitter amplifier, which we will treat later in this chapter).

2.06 Transistor current source

Current sources, although often neglected, are as important and as useful as voltage sources. They often provide an excellent way to bias transistors, and they are unequaled as "active loads" for super-gain amplifier stages and as emitter sources for differential amplifiers. Integrators, sawtooth generators, and ramp generators need current sources. They provide wide-voltage-range pull-ups within amplifier and regulator circuits. And, finally, there are applications in the outside world that require constant-current sources, e.g., electrophoresis or electrochemistry.

Resistor plus voltage source

The simplest approximation to a current source is shown in Figure 2.19. As long as

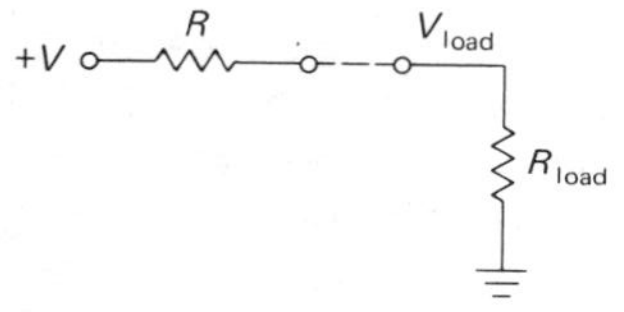

Figure 2.19

$R_{load} \ll R$ (in other words, $V_{load} \ll V$), the current is nearly constant and is approximately

$I = V/R$

The load doesn't have to be resistive. A capacitor will charge at a constant rate, as long as $V_{capacitor} \ll V$; this is just the first part of the exponential charging curve of an *RC*.

There are several drawbacks to a simple resistor current source. In order to make a good approximation to a current source, you must use large voltages, with lots of power dissipation in the resistor. In addition, the current isn't easily *programmable,* i.e., controllable over a large range via a voltage somewhere else in the circuit.

EXERCISE 2.6

If you want a current source constant to 1% over a load voltage range of 0 to +10 volts, how large a voltage source must you use in series with a single resistor?

EXERCISE 2.7

Suppose you want a 10mA current in the preceding problem. How much power is dissipated in the series resistor? How much gets to the load?

Transistor current source

Fortunately, it is possible to make a very good current source with a transistor (Fig. 2.20). It works like this: Applying V_B to the

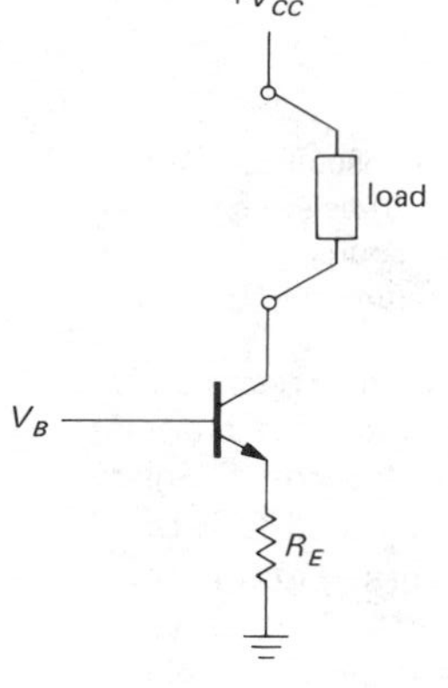

Figure 2.20

base, with $V_B > 0.6$ volt, ensures that the emitter is always conducting:

$V_E = V_B - 0.6$ volt

So

$I_E = V_E/R_E = (V_B - 0.6 \text{ volt})/R_E$

But, since $I_E \approx I_C$ for large h_{FE},

$I_C \approx (V_B - 0.6 \text{ volt})/R_E$

independent of V_C, as long as the transistor is not saturated ($V_C > V_E + 0.2$ volt).

Current source biasing

The base voltage can be provided in a number of ways. A voltage divider is OK, as long as it is stiff enough. As before, the criterion is that its impedance should be much less than the dc impedance looking into the base ($h_{FE}R_E$). Or you can use a zener diode, biased from V_{CC}, or even a few forward-biased diodes in series from base to the corresponding emitter supply. Figure 2.21 shows some examples. In the last example (Fig. 2.21C), a *pnp* transistor

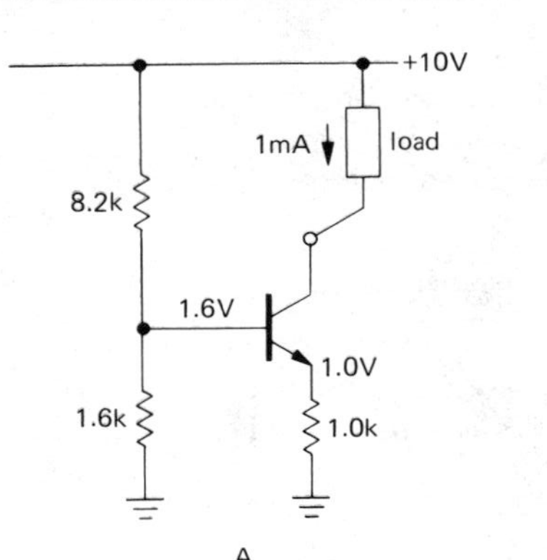

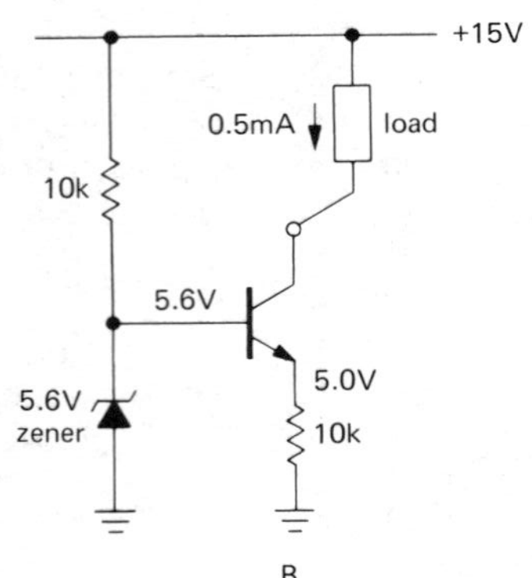

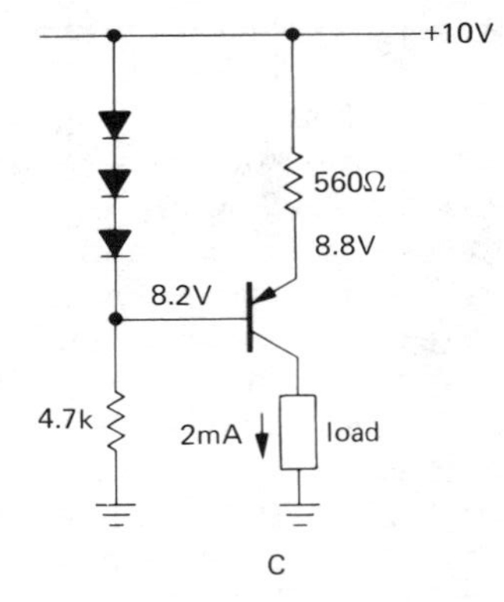

Figure 2.21
Transistor current source circuit illustrating three methods of base biasing; *npn* transistors sink current, whereas *pnp* transistors source current. The circuit in C illustrates a load returned to ground.

sources current to a load returned to ground. The other examples (using *npn* transistors) should properly be called current sinks, but the usual practice is to call all of them current sources. In the first circuit, the voltage divider impedance of ~1.3k is very stiff compared with the impedance looking into the base of about 100k (for $h_{FE} = 100$), so any changes in beta with collector voltage will not much affect the output current by causing the base voltage to change. In the other two circuits the biasing resistors are chosen to provide several milliamps to bring the diodes into conduction.

Compliance

A current source can provide constant current to the load only over some finite range of load voltage. To do otherwise would be equivalent to providing infinite power. The output voltage range over which a current source behaves well is called its output *compliance*. For the preceding transistor current sources, the compliance is set by the requirement that the transistors stay in the active region. Thus in the first circuit the voltage at the collector can go down until the transistor is almost in saturation, perhaps +1.2 volts at the collector. The second circuit, with its higher emitter voltage, can sink current only down to a collector voltage of about +5.2 volts.

In all cases the collector voltage can range from a value near saturation all the way up to the supply voltage. For example, the last circuit can source current to the load for any voltage between zero and about +8.6 volts across the load. In fact, the load might even contain batteries or power supplies of its own, carrying the collector beyond the supply voltage. That's OK, but you must watch out for transistor breakdown (V_{CE} must not exceed BV_{CEO}, the specified breakdown voltage of the collector-emitter junction) and also for excessive power dissipation (set by $I_C V_{CE}$). As you will see in Section 5.07. there is an additional safe-operating-area constraint on power transistors.

EXERCISE 2.8

You have +5 and +15 volt regulated supplies available in a circuit. Design a 5mA *npn* current source (sink) using the +5 volts on the base. What is the output compliance?

A current source doesn't have to have a fixed voltage at the base. By varying V_B you get a voltage-programmable current source. The input signal swing v_{in} (remember, lower-case symbols mean *variations*) must stay small enough so that the emitter voltage never drops to zero, if the output current is to reflect input voltage variations smoothly. The result will be a current source with variations in output current proportional to the variations in input voltage, $i_{out} = v_{in}/R_E$.

□ *Deficiencies of current sources*

To what extent does this kind of current source depart from the ideal? In other words, does the load current vary with voltage, say, and if so why? There are two kinds of effects:

1. Both V_{BE} (Early effect) and h_{FE} vary slightly with collector-to-emitter voltage at a given collector current. The changes in V_{BE} produced by voltage swings across the load cause the output current to change, because the emitter voltage (and therefore the emitter current) changes, even with a fixed applied base voltage. Changes in h_{FE} produce small changes in output (collector) current for fixed emitter current, since $I_C = I_E - I_B$; in addition, there are small changes in applied base voltage produced by the variable loading of the nonzero bias source impedance as h_{FE} (and therefore the base current) changes. These effects are small. For instance, the current from the circuit in Figure 2.21A varied about 0.5% in actual measurements with a 2N3565 transistor. In particular, for load voltages varying from zero to 8 volts, the Early effect contributed 0.5%, and transistor heating effects contributed 0.2%. In addition, variations in h_{FE} contributed 0.05% (note the stiff divider). Thus these variations result in a less-than-perfect current source: The output current depends slightly on voltage and therefore has less than infinite impedance. Later you will see methods that get around this difficulty.

2. V_{BE} and also h_{FE} depend on temperature. This causes drifts in output current with changes in ambient temperature; in addition, the transistor junction temperature varies as the load voltage is changed (because of variation in transistor dissipation), resulting in departure from ideal current source behavior. The change of V_{BE} with ambient temperature can be compensated with a circuit like that shown in Figure 2.22, in which Q_2's base-emitter drop is compensated by the drop in emitter follower Q_1, with similar temperature dependence. R_3, incidentally, is a pull-up resistor for Q_1, since Q_2's base sinks current, which Q_1 cannot source.

□ *Improving current source performance*

In general, the effects of variability in V_{BE}, whether caused by temperature dependence

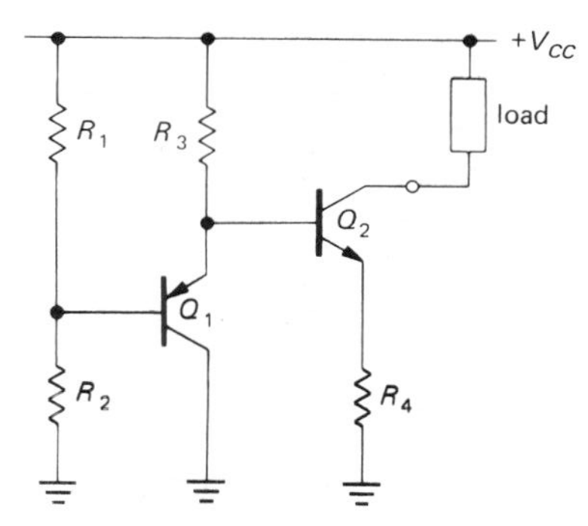

Figure 2.22
One method of temperature-compensating a current source.

(approximately -2mV/°C) or by dependence on V_{CE} (the Early effect, given roughly by $\Delta V_{BE} \approx -0.001\, \Delta V_{CE}$), can be minimized by choosing the emitter voltage to be large enough (at least 1V, say) so that changes in V_{BE} of tens of millivolts will not result in large fractional changes in the voltage across the emitter resistor (remember that the *base* voltage is what is held constant by your circuit). For instance, choosing $V_E = 0.1$ volt (i.e., applying about 0.7V to the base) would cause 10% variations in output current for 10mV changes in V_{BE}, whereas the choice $V_E = 1.0$ volt would result in 1% current variations for the same V_{BE} changes. Don't get carried away, though. Remember that the lower limit of output compliance is set by the emitter voltage. Using a 5 volt emitter voltage for a current source running from a +10 volt supply limits the output compliance to slightly less than 5 volts (the collector can go from about $V_E + 0.2$V to V_{CC}, i.e., from 5.2V to 10V).

Figure 2.23 shows a circuit modification that improves current source performance significantly. Current source Q_1 functions as before, but with collector voltage held fixed by Q_2's emitter. The load sees the same current as before, since Q_2's collector and emitter currents are nearly equal (large h_{FE}). But with this circuit the V_{CE} of Q_1 doesn't change with load voltage, thus eliminating the small changes in V_{BE} from Early effect and dissipation-induced temperature changes. Measurements with 2N3565s gave 0.1% current variation for load voltages from 0 to 8 volts; to obtain performance of this accuracy it is important to use

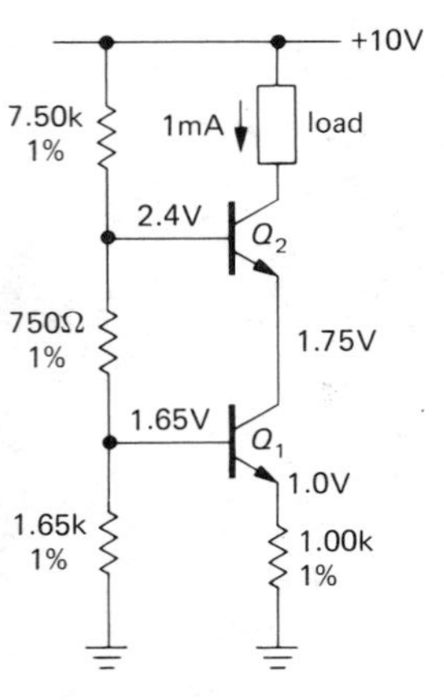

Figure 2.23
Cascode current source for improved current stability with load voltage variations.

stable 1% resistors, as shown. (Incidentally, this circuit connection also finds use in high-frequency amplifiers, where it is known as the "cascode.") Later you will see current source techniques using op-amps and feedback that circumvent the problem of V_{BE} variation altogether.

The effects of variability of h_{FE} can be minimized by choosing transistors with large h_{FE}, so that the base current contribution to the emitter current is relatively small.

Figure 2.24 shows one last current source, whose output current doesn't depend on supply voltage. In this circuit, Q_1's V_{BE} across R_2 sets the output current, independent of V_{CC}:

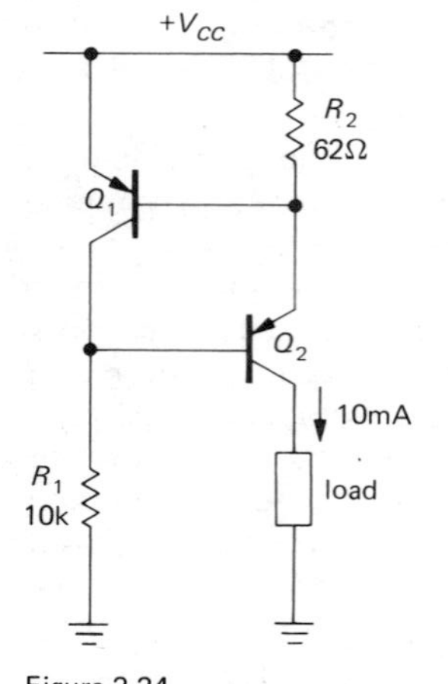

Figure 2.24
Transistor V_{BE}-referenced current source.

$$I_{out} = V_{BE}/R_2$$

R_1 biases Q_2 and holds Q_1's collector at two diode drops below V_{CC}, eliminating Early effect as in the previous circuit. This circuit is not temperature-compensated; the voltage across R_2 decreases approximately 2.1mV/°C, causing the output current to decrease approximately 0.3%/°C.

2.07 Common-emitter amplifier

Consider a current source with a resistor as load (Fig. 2.25). The collector voltage is

$$V_C = V_{CC} - I_C R_C$$

We could capacitively couple a signal to the base to cause the collector voltage to vary. Consider the example in Figure 2.26. C is chosen so that all frequencies of interest are passed by the high-pass filter it forms in combination with the parallel resistance of the base biasing resistors (the impedance looking into the base itself will usually be much larger because of the way the base

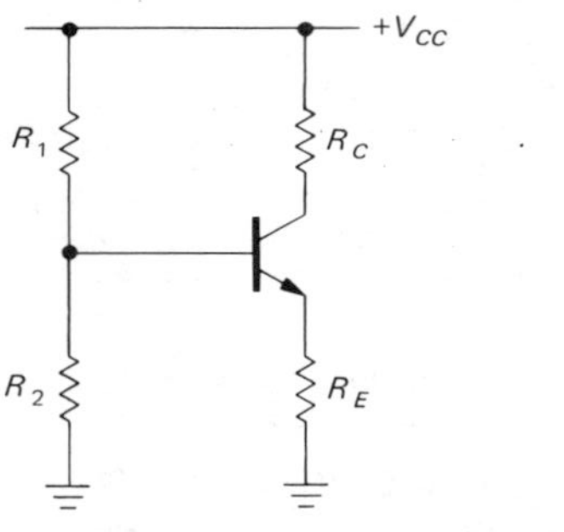

Figure 2.25

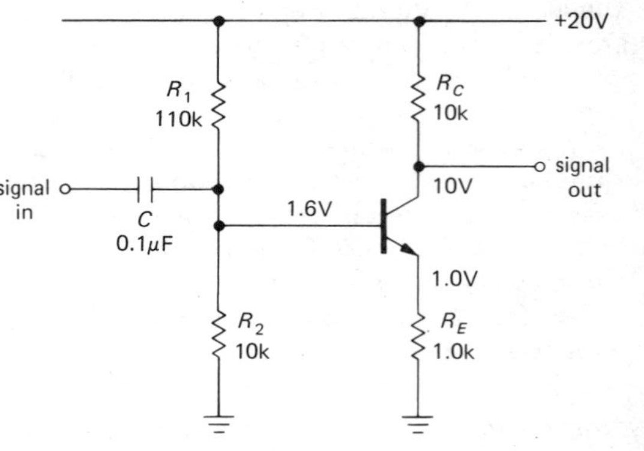

Figure 2.26
An ac common-emitter amplifier.

resistors are chosen, and it can be ignored); that is,

$$C \geq \frac{1}{2\pi f(R_1 \| R_2)}$$

The quiescent collector current is 1.0mA because of the applied base bias and the 1.0k emitter resistor. That current puts the collector at +10 volts (+20V, minus 1.0mA through 10k). Now imagine an applied wiggle in base voltage v_B. The emitter follows with $v_E = v_B$, which causes a wiggle in emitter current

$$i_E = v_E/R_E = v_B/R_E$$

and nearly the same change in collector current (h_{fe} is large). So the initial wiggle in base voltage finally causes a collector voltage wiggle

$$v_C = -i_C R_C = -v_B(R_C/R_E)$$

Aha! It's a *voltage amplifier*, with a voltage amplification (or "gain") given by

$$\text{gain} = v_{out}/v_{in} = -R_C/R_E$$

In this case the gain is −10,000/1000, or −10. The minus sign means that a positive wiggle at the input gets turned into a negative wiggle (10 times as large) at the output. This is called a *common-emitter amplifier* with emitter degeneration.

Input and output impedance of the common-emitter amplifier

We can easily determine the input and output impedances of the amplifier. The input signal sees, in parallel, 110k, 10k, and the impedance looking into the base. The latter is about 100k (h_{fe} times R_E), so the input impedance (dominated by the 10k) is about 8k. The input coupling capacitor thus forms a high-pass filter, with the 3dB point at 200Hz. The signal driving the amplifier sees 0.1μF in series with 8k, which to signals of normal frequencies (well above the 3dB point) just looks like 8k.

The output impedance is 10k in parallel with the impedance looking into the collector. What is that? Well, remember that if you snip off the collector resistor, you're simply looking into a current source. The collector impedance is very large (measured in megohms), and so the output impedance is just the value of the collector resistor, 10k. It is worth remembering that the impedance looking into a transistor's collector is high, whereas the impedance looking into the emitter is low (as in the emitter follower). Although the output impedance of a common-emitter amplifier will be dominated by the collector load resistor, the output impedance of an emitter follower will not be dominated by the emitter load resistor, but rather by the impedance looking into the emitter.

2.08 Unity-gain phase splitter

Sometimes it is useful to generate a signal and its inverse, i.e., two signals 180° out of phase. That's easy to do – just use an emitter-degenerated amplifier with gain of −1 (Fig. 2.27). The quiescent collector voltage is set to 0.75V_{CC}, rather than the usual

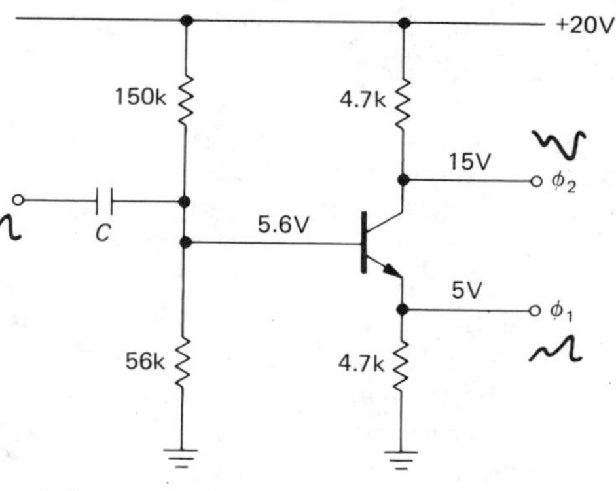

Figure 2.27

0.5V_{CC}, in order to achieve the same result – maximum symmetrical output swing without clipping at either output. The collector can swing from 0.5V_{CC} to V_{CC}, whereas the emitter can swing from ground to 0.5V_{CC}.

Note that the phase splitter outputs must be loaded with equal (or very high) impedances at the two outputs in order to maintain gain symmetry.

Phase shifter

A nice use of the phase splitter is shown in Figure 2.28. This circuit gives (for a sinewave input) an output sine wave of adjustable phase (from zero to 180°), but with

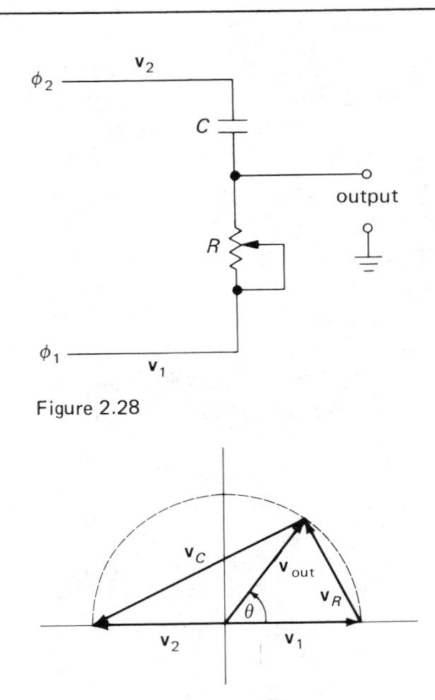

Figure 2.28

Figure 2.29

constant amplitude. It can be best understood with a phasor diagram of voltages (see Chapter 1); representing the input signal by a unit vector along the real axis, the signals look as shown in Figure 2.29. Signal vectors $\mathbf{V}_R$ and $\mathbf{V}_C$ must be at right angles, and they must add to form a vector of constant length along the real axis. There is a theorem from geometry that says that the locus of such points is a circle. So the resultant vector (the output voltage) always has unit length, i.e., the same amplitude as the input, and its phase can vary from nearly zero to nearly 180° relative to the input wave as R is varied from nearly zero to a value much larger than Z_C at the operating frequency. However, note that the phase shift also depends on the frequency of the input signal for a given setting of the potentiometer R. It is worth noting that a simple RC high-pass (or low-pass) network could also be used as an adjustable phaseshifter. However, its output amplitude would vary over an enormous range as the phase shift was adjusted.

An additional concern here is the ability of the phase splitter circuit to drive the RC phase shifter as a load. Ideally, the load should present an impedance that is large compared with the collector and emitter resistors. As a result, this circuit is of limited utility where a wide range of phase shifts is required. You will see improved phase splitter techniques in the next chapter.

2.09 Transconductance

In the preceding section we figured out the operation of the emitter-degenerated amplifier by (a) imagining an applied base voltage swing and seeing that the emitter voltage had the same swing, then (b) calculating the emitter current swing; then, ignoring the small base current contribution, we got the collector current swing and thus (c) the collector voltage swing. The voltage gain was then simply the ratio of collector (output) voltage swing to base (input) voltage swing.

There's another way to think about this kind of amplifier. Imagine breaking it apart, as in Figure 2.30. The first part is a voltage-

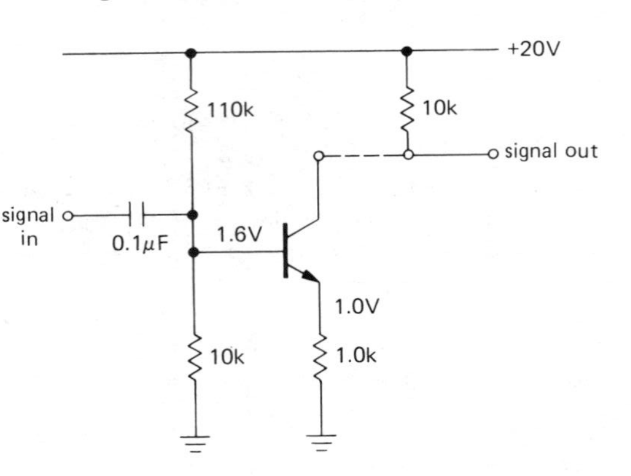

Figure 2.30

controlled current source, with quiescent current of 1.0mA and gain of −1mA/V. Gain means the ratio output/input; in this case the gain has units of current/voltage, or 1/resistance. The inverse of resistance is called *conductance* (the inverse of reactance is *susceptance*, and the inverse of impedance is *admittance*) and has a special unit, the *mho* (ohm spelled backward) which has recently been renamed the *siemens*. An amplifier whose gain has units of conductance is called a *transconductance* amplifier; the ratio I_{out}/V_{in} is called the transconductance, g_m.

Think of the first part of the circuit as a transconductance amplifier, i.e., a voltage-to-current amplifier with transconductance g_m (gain) of 1mA/V (1000 μmho, or 1mS, which is just $1/R_E$). The second part of the circuit is the load resistor, an "amplifier" that converts current to voltage. This resistor could be called a *transresistance* amplifier, and its gain (r_m) has units of voltage/current, or resistance. In this case its quiescent voltage is V_{CC}, and its gain (transresistance) is 10kV/A (10kΩ), which is just R_C. Connecting the two parts together gives you a voltage amplifier. You get the overall gain by multiplying the two gains. In this case $G = g_m R_C = R_C/R_E$, or -10, a unitless number equal to the ratio (output voltage)/(input voltage).

This is a useful way to think about an amplifier, because you can analyze performance of the sections independently. For example, you can analyze the transconductance part of the amplifier by evaluating g_m for different circuit configurations or even different devices, such as field-effect transistors (FETs). Then you can analyze the transresistance (or load) part by considering gain versus voltage swing tradeoffs. If you are interested in the overall voltage gain, it is given by $G_V = g_m r_m$, where r_m is the transresistance of the load. Ultimately the substitution of an active load (current source), with its extremely high transresistance, can yield one-stage voltage gains of 10,000 or more. The *cascode* configuration, which we will discuss later, is another example easily understood with this approach.

In Chapter 3, which deals with operational amplifiers, you will see further examples of amplifiers with voltages or currents as inputs or outputs: voltage amplifiers (voltage to voltage), current amplifiers (current to current), transconductance amplifiers (voltage to current), and transresistance amplifiers (current to voltage).

Turning up the gain: limitations of the simple model

The voltage gain of the emitter-degenerated amplifier is $-R_C/R_E$, according to our model. What happens as R_E is reduced toward zero? The equation predicts that the gain will rise without limit. But if we made actual measurements of the preceding circuit, keeping the quiescent current constant at 1mA, we would find that the gain would level off at about 400 when R_E is zero, i.e., with the emitter grounded. We would also find that the amplifier would become significantly nonlinear (the output would not be a faithful replica of the input), the input impedance would become small and nonlinear, and the biasing would be critical and unstable with temperature. Clearly our transistor model is incomplete and needs to be modified in order to handle this circuit situation, as well as others we will talk about shortly. Our fixed-up model, which we will call the transconductance model, will be accurate enough for the remainder of the book.

EBERS-MOLL MODEL APPLIED TO BASIC TRANSISTOR CIRCUITS

2.10 Improved transistor model: transconductance amplifier

The important change is in property 4 (Section 2.01), where we said earlier that $I_C = h_{FE} I_B$. We thought of the transistor as a current amplifier whose input circuit behaved like a diode. That's roughly correct, and for some applications it's good enough. But to understand differential amplifiers, logarithmic converters, temperature compensation, and other important applications, you must think of the transistor as a *transconductance* device – collector current is determined by base-to-emitter *voltage*.

Here's the modified property 4:

4. When rules 1–3 (Section 2.01) are obeyed, I_C is related to V_{BE} by

$$I_C = I_S\left[\exp\left(\frac{V_{BE}}{V_T}\right) - 1\right]$$

where $V_T = kT/q = 25.3$mV at room temperature (68°F, 20°C), q is the electron charge (1.60×10^{-19} coulombs), k is Boltzmann's constant (1.38×10^{-23} joules/°K), T is the absolute temperature in degrees Kelvin (°K = °C + 273.16), and I_S is the saturation current of the particular transistor (depends on T). Then the base current, which also depends on V_{BE}, can be approximated by

$$I_B = I_C/h_{FE}$$

where the "constant" h_{FE} is typically in the range 20 to 1000, but depends on transistor type, I_C, V_{CE}, and temperature. I_S represents the reverse leakage current. In the active region, $I_C \gg I_S$, and therefore the -1 term can be neglected in comparison with the exponential.

The equation for I_C is known as the Ebers-Moll equation. It also describes the current versus voltage for a diode. For transistors it is important to realize that the collector current is accurately determined by the base-emitter voltage, rather than by the base current (the base current is then roughly determined by h_{FE}), and that this exponential law is accurate over an enormous range of currents, typically from nanoamps to milliamps. Figure 2.31 makes the point graphically. If you measure the base current at various collector currents, you will get a graph of h_{FE} versus I_C like that in Figure 2.32.

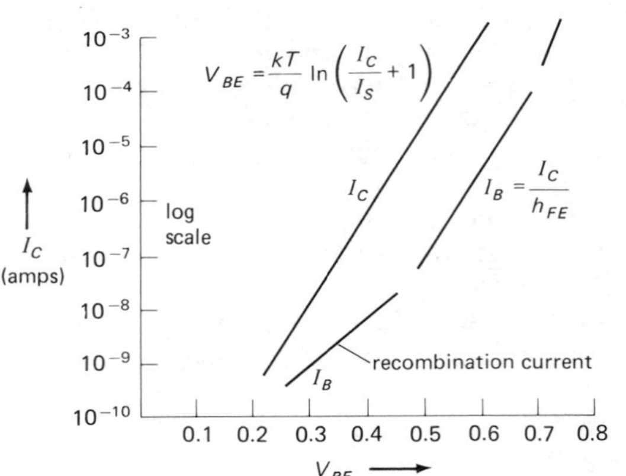

Figure 2.31
Transistor base and collector currents as functions of base-to-emitter voltage V_{BE}.

Although the Ebers-Moll equation tells us that the base-emitter voltage "programs" the collector current, this property may not be directly usable in practice (biasing a transistor by applying a base voltage) because of the large temperature coefficient of base-emitter voltage. You will see later how the

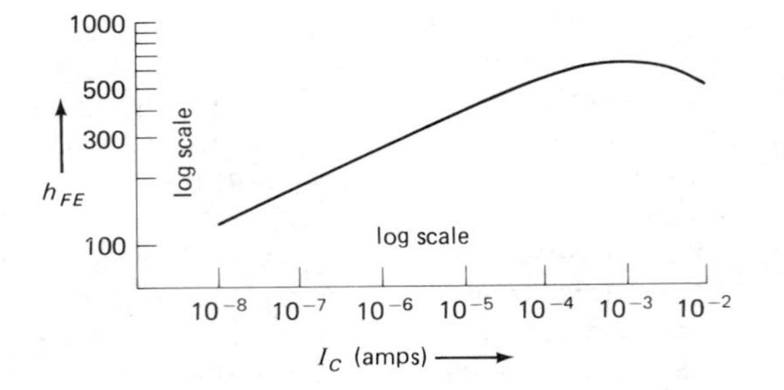

Figure 2.32

Ebers-Moll equation provides insight and solutions to this problem.

Rules of thumb for transistor design

From the Ebers-Moll equation we can get several important quantities we will be using often in circuit design:

1. The steepness of the diode curve. How much do we need to increase V_{BE} in order to increase I_C by a factor of 10? From the Ebers-Moll equation, that's just $V_T \log_e 10$, or 60mV at room temperature. *Base voltage increases 60mV per decade of collector current.*

2. The small-signal impedance looking into the emitter, for the base held at a fixed voltage. Taking the derivative of V_{BE} with respect to I_C, you get

$$r_e = V_T/I_C = 25/I_C \text{ ohms}$$

where I_C is in milliamps. The numerical value $25/I_C$ ohms is for room temperature. This *intrinsic* emitter resistance, r_e, acts as if it is in series with the emitter in all transistor circuits. It limits the gain of a grounded emitter amplifier, causes an emitter follower to have a voltage gain slightly less than unity, and prevents the output impedance of an emitter follower from reaching zero. It is a small-signal parameter. Note that the transconductance of a grounded emitter amplifier is $g_m = 1/r_e$.

3. The temperature dependence of V_{BE}. Because of the temperature dependence of I_S, V_{BE} *decreases* about 2.1mV/°C. It is roughly proportional to $1/T_{abs}$, where T_{abs} is the absolute temperature.

There is one additional quantity we will need on occasion, although it is not derivable from the Ebers-Moll equation. It is the Early effect we described in Section 2.06, and it sets important limits on current source performance. For example:

4. (Early effect) V_{BE} varies slightly with changing V_{CE} at constant I_C. This effect is caused by changing effective base width, and it is given, approximately, by

$$\Delta V_{BE} \approx -0.001 \Delta V_{CE}$$

These are the essential quantities we need. With them we will be able to handle most problems of transistor circuit design, and we will have little need to refer again to the Ebers-Moll equation itself.

2.11 The common-emitter amplifier revisited

Previously we got wrong answers for the voltage gain of the common-emitter amplifier with emitter resistor (sometimes called emitter degeneration) when we set the emitter resistor equal to zero. The problem is that the transistor has $25/I_C$(mA) ohms of built-in (intrinsic) emitter resistance r_e that must be added to the actual external emitter resistor. This extra resistance is significant only when small emitter resistors (or none at all) are used. So, for instance, the amplifier we considered previously will have a voltage gain of $-10\text{k}/r_e$, or -400, when the external emitter resistor is zero. The input impedance is not zero, as we would have predicted earlier ($h_{FE}R_E$); it is approximately $h_{fe}r_e$, or in this case (1mA quiescent current) about 2.5k.

The terms "grounded emitter" and "common emitter" are sometimes used interchangeably, and they can be confusing. We will use the phrase grounded emitter amplifier to mean a common-emitter amplifier with $R_E = 0$. A common-emitter amplifier stage may have an emitter resistor; what matters is that the emitter circuit is common to the input circuit and the output circuit.

Shortcomings of the single-stage grounded emitter amplifier

The extra voltage gain you get by using $R_E = 0$ comes at the expense of other properties of the amplifier. In fact, the grounded emitter amplifier, in spite of its popularity in textbooks, should be avoided except in circuits with overall negative feedback. In order to see why, consider Figure 2.33.

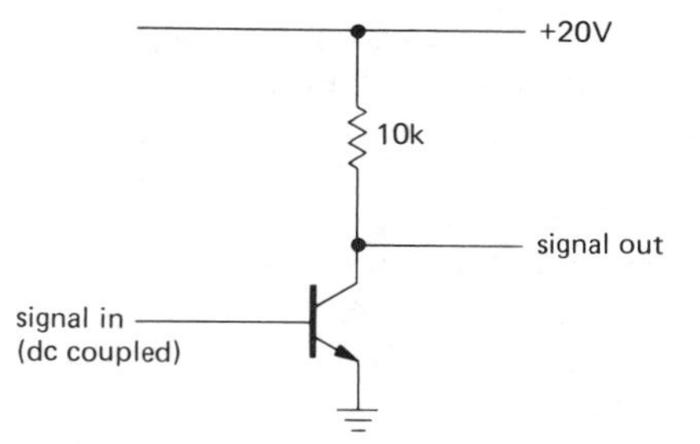

Figure 2.33

1. Nonlinearity. The gain is $G = -g_m R_C = -R_C/r_e = -R_C I_C(\text{mA})/25$, so for a quiescent current of 1mA, the gain is -400. But I_C varies as the output signal varies. For this example, the gain will vary from -800 ($V_{out} = 0$, $I_C = 2\text{mA}$) down to zero ($V_{out} = V_{CC}$, $I_C = 0$). For a triangle-wave input, the output will look like that in Figure 2.34. The amplifier has high distortion, or poor linearity. The grounded emitter amplifier without feedback is useful only for small signal swings about the quiescent point. By contrast, the emitter-degenerated amplifier has gain almost entirely independent of collector current, as long as $R_E \gg r_e$, and can be used for undistorted amplification even with large signal swings.

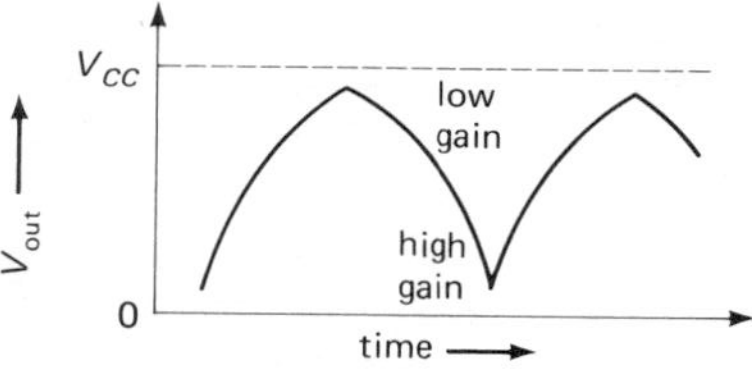

Figure 2.34

2. Input impedance. The input impedance is roughly $Z_{in} = h_{fe}r_e = 25h_{fe}/I_C(\text{mA})$ ohms. Once again, I_C varies over the signal swing, giving a varying input impedance. Unless the signal source driving the base has low impedance, you will wind up with nonlinearity due to the nonlinear variable voltage divider formed from the signal source and the amplifier's input impedance. By contrast, the input impedance of an emitter-degenerated amplifier is constant and high.

3. ***Biasing.*** The grounded emitter amplifier is difficult to bias. It might be tempting just to apply a voltage (from a voltage divider) that gives the right quiescent current according to the Ebers-Moll equation. That won't work, because of the temperature dependence of V_{BE} (at fixed I_C), which varies about $-2.1\text{mV}/°\text{C}$ (it actually decreases with increasing T because of the variation of I_S with T; as a result, V_{BE} is roughly proportional to $1/T$, the absolute temperature). This means that the collector current (for fixed V_{BE}) will increase by a factor of 10 for a 30°C rise in temperature. Such unstable biasing is useless, because even rather small changes in temperature will cause the amplifier to saturate. For example, a grounded emitter stage biased with the collector at half the supply voltage will go into saturation if the temperature rises by 8°C.

EXERCISE 2.9

Verify that an 8°C rise in ambient temperature will cause a base-voltage-biased grounded emitter stage to saturate, assuming that it was initially biased for $V_C = 0.5 V_{CC}$.

Some solutions to the biasing problem will be discussed in the following sections. By contrast, the emitter-degenerated amplifier achieves stable biasing by applying a voltage to the base, most of which appears across the emitter resistor, thus determining the quiescent current.

Emitter resistor as feedback

Adding an external series resistor to the intrinsic emitter resistance r_e (emitter degeneration) improves many properties of the common-emitter amplifier, at the expense of gain. You will see the same thing happening in the next two chapters, when we discuss *negative feedback*, an important technique for improving amplifier characteristics by feeding back some of the output signal to reduce the effective input signal. The similarity here is no coincidence; the emitter-degenerated amplifier itself uses a form of negative feedback. Think of the transistor as a transconductance device, determining collector current (and therefore output voltage) according to the voltage applied between base and emitter; but the input to the amplifier is the voltage from base to ground. So the voltage from base to emitter is the input voltage, *minus a sample of the output* ($I_E R_E$). That's negative feedback, and that's why emitter degeneration improves most properties of the amplifier (improved linearity and stability and increased input impedance; also, the output impedance would be reduced if the feedback were taken directly from the collector). Great things to look forward to in Chapters 3 and 4!

2.12 Biasing the common-emitter amplifier

If you must have the highest possible gain (or if the amplifier stage is inside a feedback loop), it is possible to arrange successful biasing of a common-emitter amplifier. There are three solutions that can be applied, separately or in combination: bypassed emitter resistor, matched biasing transistor, and dc feedback.

Bypassed emitter resistor

Use a bypassed emitter resistor, biasing as for the degenerated amplifier, as shown in Figure 2.35. In this case R_E has been chosen about $0.1R_C$, for ease of biasing; if R_E is too small, the emitter voltage will be much smaller than the base-emitter drop, leading to temperature instability of the quiescent point as V_{BE} varies with temperature. The emitter bypass capacitor is chosen by

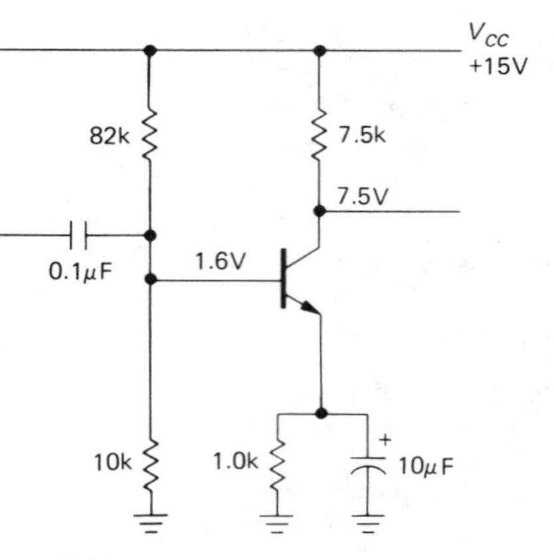

Figure 2.35

making its impedance small compared with r_e (*not* R_E) at the lowest frequency of interest. In this case its impedance is 25 ohms at 650Hz. At signal frequencies the input coupling capacitor sees an impedance of 10k in parallel with the base impedance, in this case h_{fe} times 25 ohms, or roughly 2.5k. At dc the impedance looking into the base is much larger (h_{FE} times the emitter resistor, or about 100k), which is why stable biasing is possible.

A variation on this circuit consists of using two emitter resistors in series, one of them bypassed. For instance, suppose you want an amplifier with a voltage gain of 50, quiescent current of 1mA, and V_{CC} of +20 volts, for signals from 20Hz to 20kHz. If you try to use the emitter-degenerated circuit, you will have the circuit shown in Figure 2.36. The collector resistor is chosen to put

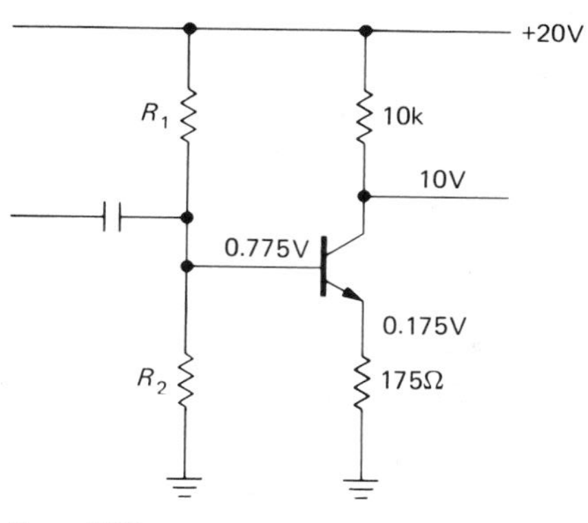

Figure 2.36

the quiescent collector voltage at $0.5V_{CC}$. Then the emitter resistor is chosen for the required gain, including the effects of the r_e of $25/I_C$(mA). The problem is that the emitter voltage of only 0.175 volt will vary significantly as the ~0.6 volt of base-emitter drop varies with temperature (−2.1mV/°C, approximately), since the base is held at constant voltage by R_1 and R_2; for instance, you can verify that an increase of 20°C will cause the collector current to increase by nearly 25%.

The solution here is to add some bypassed emitter resistance for stable biasing, with no change in gain at signal frequencies (Fig. 2.37). As before, the collector resistor is chosen to put the collector at 10

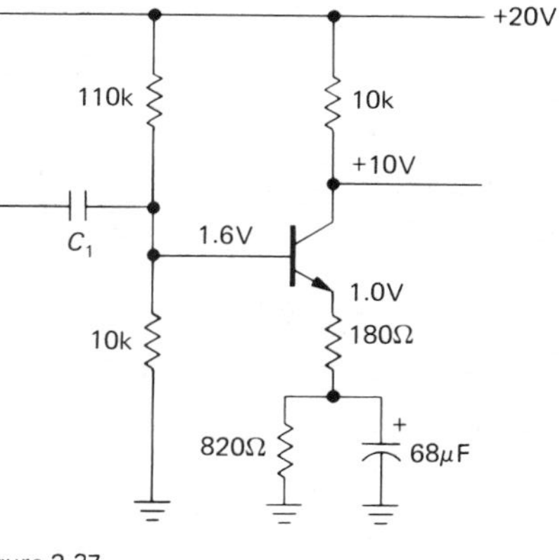

Figure 2.37

volts ($0.5V_{CC}$). Then the unbypassed emitter resistor is chosen to give a gain of 50, including the intrinsic emitter resistance $r_e = 25/I_C$(mA). Enough bypassed emitter resistance is added to make stable biasing possible (one-tenth of the collector resistance is a good rule). The base voltage is chosen to give 1mA of emitter current, with impedance about one-tenth the dc impedance looking into the base (in this case about 100k). The emitter bypass capacitor is chosen to have low impedance compared with 180 ohms at the lowest signal frequencies. Finally, the input coupling capacitor is chosen to have low impedance compared with the *signal-frequency* input impedance of the amplifier, which is equal to the voltage divider impedance in parallel with $(180 + 25)h_{fe}$ ohms (the 820Ω is bypassed, and looks like a short at signal frequencies).

An alternative circuit splits the signal and dc paths (Fig. 2.38). This lets you vary the

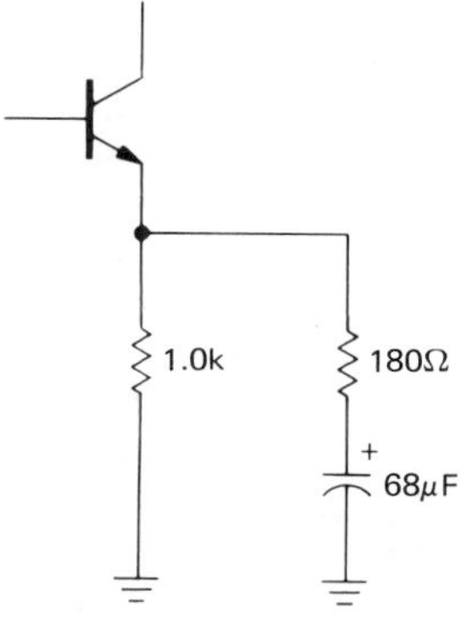

Figure 2.38

gain (by changing the 180Ω resistor) without bias change.

□ *Matched biasing transistor*

Use a matched transistor to generate the correct base voltage for the required collector current; this ensures automatic temperature compensation (Fig. 2.39). Q_1's collector

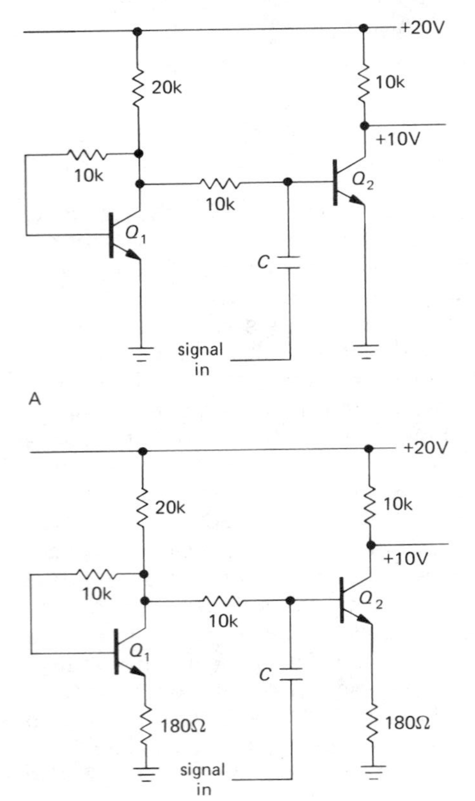

Figure 2.39

is drawing 1mA, since it is guaranteed to be near ground (about one V_{BE} drop above ground, to be exact); if Q_1 and Q_2 are a matched pair (available as a single device, with the two transistors on one piece of silicon), then Q_2 will also be biased to draw 1mA, putting its collector at +10 volts and allowing a full ±10 volt symmetrical swing on its collector. Changes in temperature are of no importance, as long as both transistors are at the same temperature. This is a good reason for using a "monolithic" dual transistor.

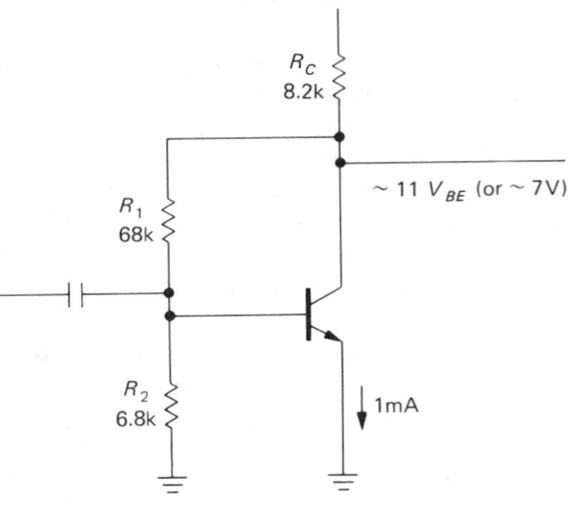

Figure 2.40

Feedback at dc

Use dc feedback to stabilize the quiescent point. Figure 2.40 shows one method. By taking the bias voltage from the collector, rather than from V_{CC}, you get some measure of bias stability. The base sits one diode drop above ground; since its bias comes from a 10:1 divider, the collector is at 11 diode drops above ground, or about 7 volts. Any tendency for the transistor to saturate (e.g., if it happens to have unusually high beta) is stabilized, since the dropping collector voltage will reduce the base bias. This scheme is acceptable if great stability is not required. The quiescent point is liable to drift a volt or so as the ambient (surrounding) temperature changes, since the base-emitter voltage has a significant temperature coefficient. Better stability is possible if several stages of amplification are included within the feedback loop. You will see examples later in connection with feedback.

A better understanding of feedback is really necessary to understand this circuit. For instance, feedback acts to reduce the input and output impedances. The input signal sees R_1's resistance effectively reduced by the voltage gain of the stage. In this case it is equivalent to a resistor of about 300 ohms to ground. In the next chapter we will treat feedback in enough detail so that you will be able to figure the voltage gain and terminal impedances of this circuit.

Note that the base bias resistor values could be increased in order to raise the input impedance, but you should then take into account the non-negligible base current. Suitable values might be $R_1 = 220k$ and $R_2 = 33k$. An alternative approach might be to bypass the feedback resistance in order to eliminate feedback (and therefore lowered input impedance) at signal frequencies (Fig. 2.41).

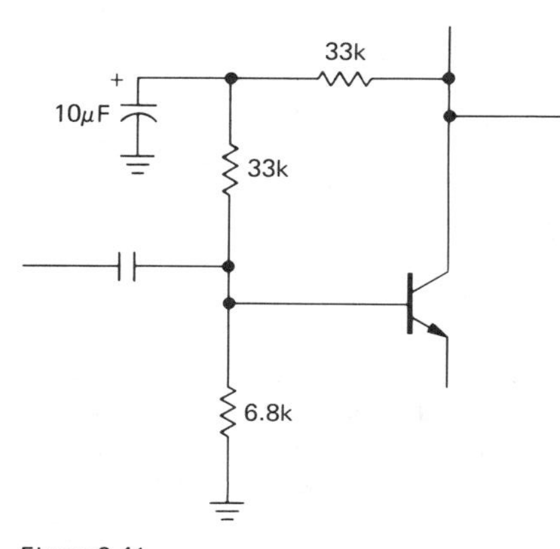

Figure 2.41

Comments on biasing and gain

One important point about grounded emitter amplifier stages: You might think that the voltage gain can be raised by increasing the quiescent current, since the intrinsic emitter resistance r_e drops with rising current. Although r_e does go down with increasing collector current, the smaller collector resistor you need to obtain the same quiescent collector voltage just cancels the advantage. In fact, you can show that the small-signal voltage gain of a grounded emitter amplifier biased to $0.5V_{CC}$ is given by $G = 20V_{CC}$, independent of quiescent current.

EXERCISE 2.10

Show that the preceding statement is true.

If you need more voltage gain in one stage, one approach is to use a current source as an *active load.* Since its impedance is very high, single-stage voltage gains of 1000 or more are possible. Such an arrangement cannot be used with the biasing schemes we have discussed, but must be part of an overall dc feedback loop, a subject we will discuss in the next chapter. You should be sure such an amplifier looks into a high-impedance load; otherwise the gain obtained by high collector load impedance will be lost. Something like an emitter follower, a field-effect transistor (FET), or an op-amp presents a good load.

In radiofrequency amplifiers intended for use only over a narrow frequency range, it is common to use a parallel *LC* circuit as a collector load; in that case very high voltage gain is possible, since the *LC* circuit has high impedance (like a current source) at the signal frequency, with low impedance at dc. Since the *LC* is "tuned," out-of-band interfering signals (and distortion) are effectively rejected. Additional bonuses are the possibility of peak-to-peak output swings of $2V_{CC}$ and the use of transformer coupling from the inductor.

EXERCISE 2.11

Design a tuned common-emitter amplifier stage to operate at 100kHz. Use a bypassed emitter resistor, and set the quiescent current at 1.0mA. Assume $V_{CC} = +15$ volts and $L = 1.0$mH, and put a 6.2k resistor across the *LC* to set $Q = 10$ (to get a 10% bandpass; see Section 1.22). Use capacitive input coupling.

2.13 Current mirrors

The technique of matched base-emitter biasing can be used to make what is called a *current mirror* (Fig. 2.42). You "program"

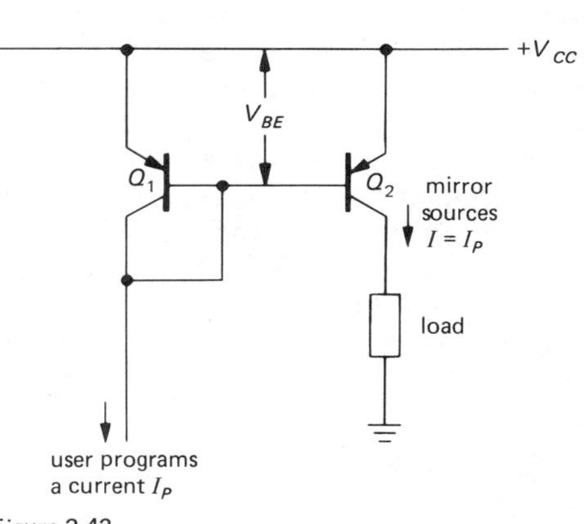

Figure 2.42
Classic bipolar transistor matched-pair current mirror.

the mirror by sinking a current from Q_1's collector. That causes a V_{BE} for Q_1 appropriate to that current at the circuit temperature and for that transistor type. Q_2, matched to Q_1 (a monolithic dual transistor is ideal), is thereby programmed to source the same current to the load. The small base currents are unimportant.

One nice feature of this circuit is voltage compliance of the output transistor source to within a few tenths of a volt of V_{CC}, since there is no emitter resistor drop to contend with. Also, in many applications it is handy to be able to program a current with a current. An easy way to generate the control current I_P is with a resistor (Fig. 2.43). Since

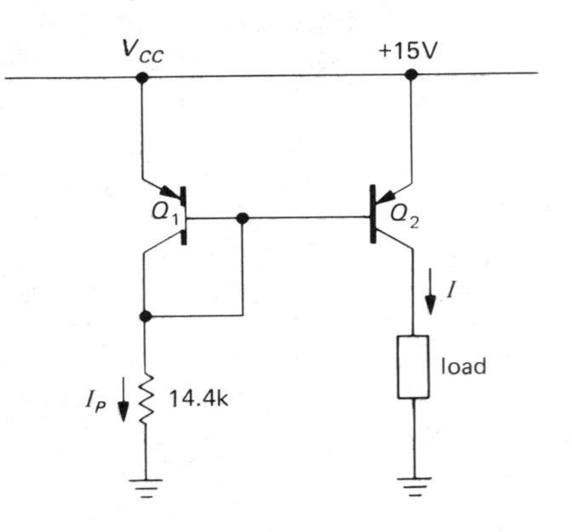

Figure 2.43

the bases are a diode drop below V_{CC}, the 14.4k resistor produces a control current, and therefore an output current, of 1mA. Current mirrors can be used in transistor circuits whenever a current source is needed. They're very popular in integrated circuits, where (a) matched transistors abound and (b) the designer tries to make circuits that will work over a large range of supply voltages. There are even resistorless integrated circuit op-amps in which the operating current of the whole amplifier is set by one external resistor, with all the quiescent currents of the individual amplifier stages inside being determined by current mirrors.

Current mirror limitations due to Early effect

One problem with the simple current mirror is that the output current varies a bit with changes in output voltage, i.e., the output impedance is not infinite. This is because of the slight variation of V_{BE} with collector voltage at a given current in Q_2 (due to Early effect); in other words, the curve of collector current versus collector-emitter voltage at a fixed base-emitter voltage is not flat (Fig. 2.44). In practice, the current might vary

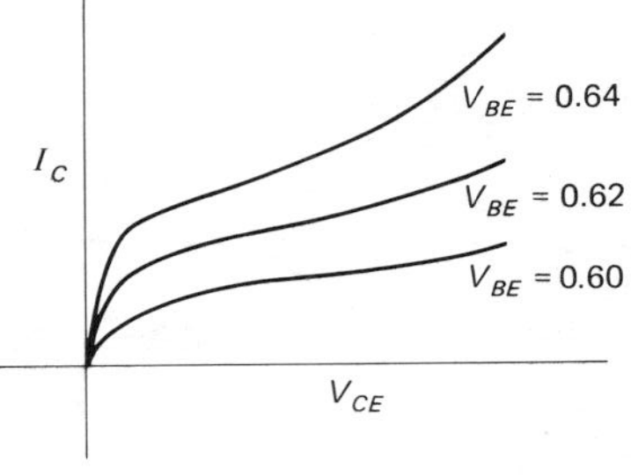

Figure 2.44

25% or so over the output compliance range – much poorer performance than the current source with emitter resistor discussed earlier.

One solution, if a better current source is needed (it often isn't), is the circuit shown in Figure 2.45. The emitter resistors are

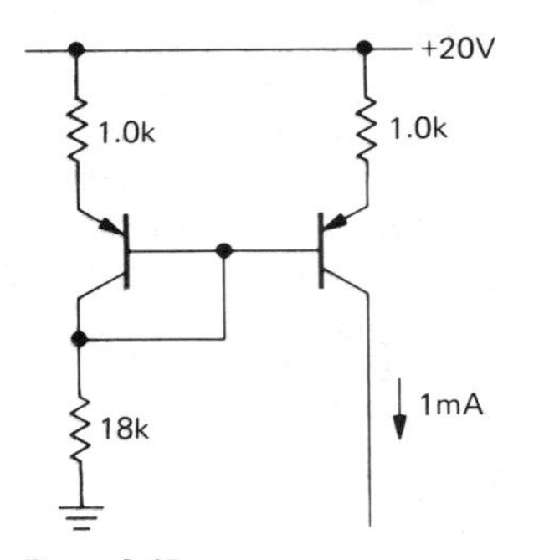

Figure 2.45

chosen to have at least a few tenths of a volt drop; this makes the circuit a far better current source, since the small variations of V_{BE} with V_{CE} are now negligible in determining the output current. Again, matched transistors should be used.

Wilson mirror

Another current mirror with very constant current is shown in the clever circuit of Figure 2.46. Q_1 and Q_2 are in the usual mirror configuration, but Q_3 now keeps Q_1's

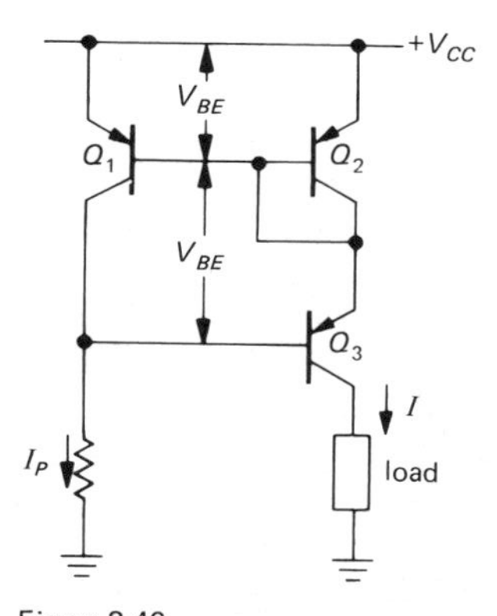

Figure 2.46
Wilson current mirror. Good current stability with load voltage variations is achieved through cascode transistor Q_3, which reduces voltage variations across Q_1.

collector fixed at two diode drops below V_{CC}. That circumvents the Early effect in Q_1, whose collector is now the programming terminal, with Q_2 now sourcing the output current. Q_3 does not affect the balance of currents, since its base current is negligible; its only function is to pin Q_1's collector. The result is that both current-determining transistors (Q_1 and Q_2) have fixed collector-emitter drops; you can think of Q_3 as simply passing the output current through to a variable-voltage load (a similar trick is used in the cascode connection, which you will see later in the chapter). Q_3, by the way, does not have to be matched to Q_1 and Q_2.

Multiple outputs and current ratios

Current mirrors can be expanded to source (or sink, with *npn* transistors) current to several loads. Figure 2.47 shows the idea. Note that if one of the current source transistors saturates (e.g., if its load is disconnected), its base robs current from the

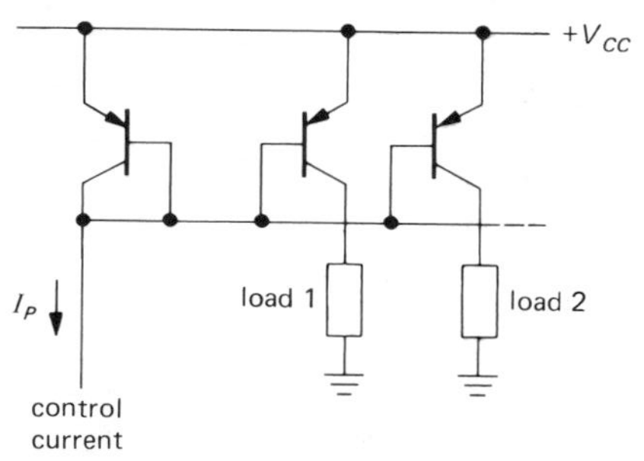

Figure 2.47

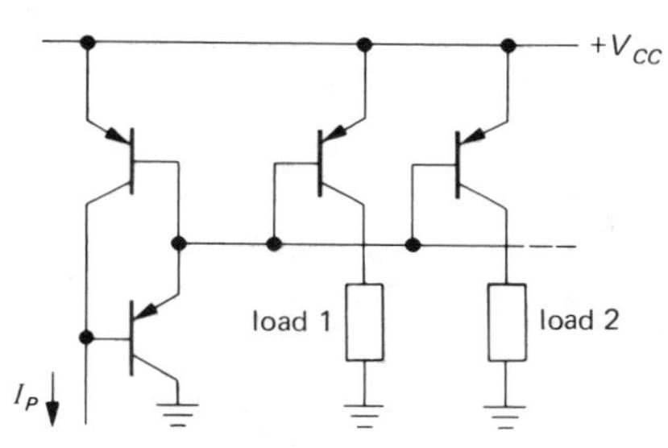

Figure 2.48

shared base reference line, reducing the other output currents. The situation is rescued by adding another transistor (Fig. 2.48).

Figure 2.49 shows two variations on the multiple-mirror idea. These circuits mirror

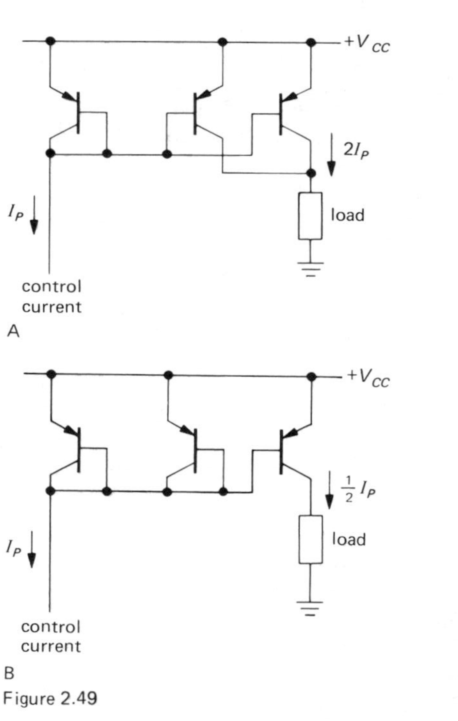

Figure 2.49

twice (or half) the control current. In the design of integrated circuits, current mirrors with any desired current ratio can be made by adjusting the size of the emitter junctions appropriately.

Another way to generate an output current that is a fraction of the programming current is to add a resistor in the emitter circuit of the output transistor (Fig. 2.50). In any circuit where the transistors are operat-

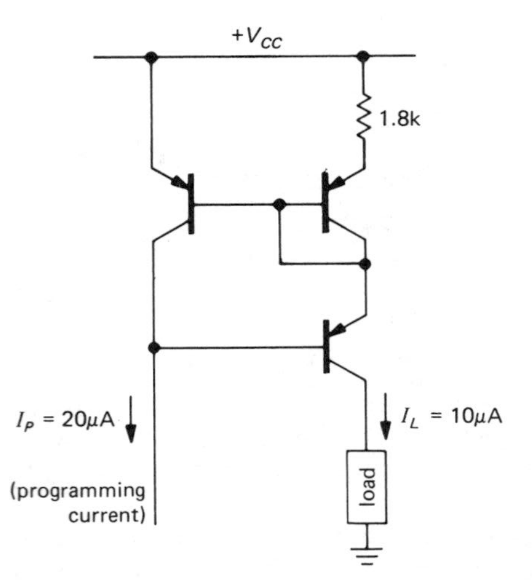

Figure 2.50
Altering current source output with an emitter resistor.

ing at different current densities, the Ebers-Moll equation predicts that the difference in V_{BE} depends only on the ratio of the current densities. For matched transistors, the ratio of collector currents equals the ratio of current densities. The graph in Figure 2.51 is handy for determining the difference in base-emitter drops in such a situation. This makes it easy to design a "ratio mirror."

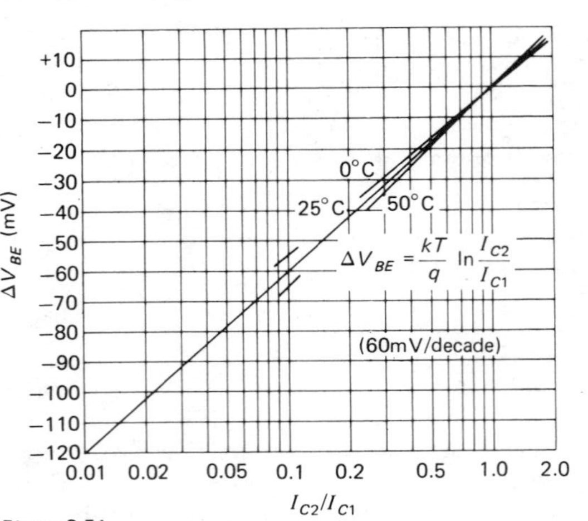

Figure 2.51
Collector current ratios for matched transistors as determined by the difference in applied base-emitter voltages.

EXERCISE 2.12

Show that the ratio mirror in Figure 2.50 works as advertised.

SOME AMPLIFIER BUILDING BLOCKS

2.14 Push-pull output stages

As we mentioned earlier in the chapter, an *npn* emitter follower cannot sink current, and a *pnp* follower cannot source current. The result is that a single-ended follower operating between split supplies can drive a ground-returned load only if a high quiescent current is used (this is sometimes called a class A amplifier). The quiescent current must be at least as large as the maximum output current during peaks of the waveform, resulting in high quiescent power dissipation. For instance, Figure 2.52 shows

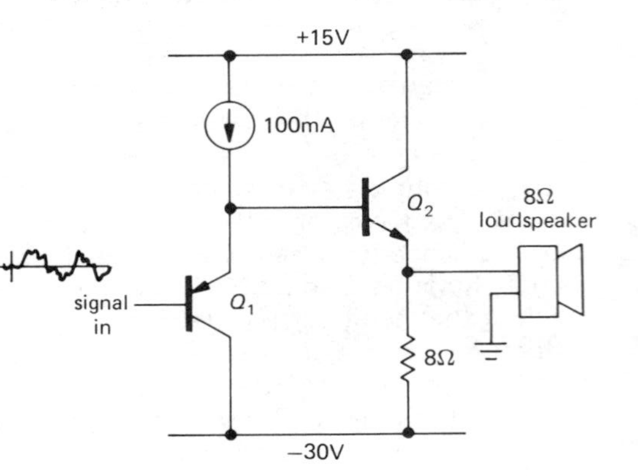

Figure 2.52

a follower circuit to drive an 8 ohm load with up to 10 watts of audio. The *pnp* follower Q_1 is included to reduce drive requirements and to cancel Q_2's V_{BE} offset (zero volts input gives zero volts output). Q_1 could, of course, be omitted for simplicity. The hefty current source in Q_1's emitter load is used to ensure that there is sufficient base drive to Q_2 at the top of the signal swing. A resistor as emitter load would be inferior because it would have to be a rather low value (50Ω or less) in order to guarantee at least 50mA of base drive to Q_2 at the peak of the swing, when load current would be maximum and the drop across the resistor would be minimum; the resultant quiescent current in Q_1 would be excessive.

The output of this example circuit can swing to nearly ±15 volts (peak) in both directions, giving the desired output power (9v rms across 8Ω). However, the output transistor dissipates 55 watts with no signal, and the emitter resistor dissipates another 110 watts. Quiescent power dissipation many times greater than the maximum output power is characteristic of this kind of class A circuit (transistor always in conduction); this obviously leaves a lot to be desired in applications where any significant amount of power is involved.

Figure 2.53 shows a push-pull follower to do the same job. Q_1 conducts on positive

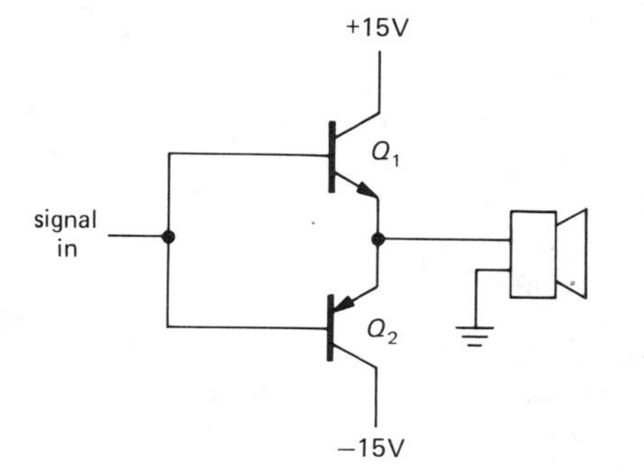

Figure 2.53

swings, Q_2 on negative swings. With zero input voltage there is no collector current and no power dissipation. At 10 watts output power there is less than 10 watts dissipation in each transistor.

Crossover distortion in push-pull stages

There is a problem with the preceding circuit as drawn. The output trails the input by a V_{BE} drop; on positive swings the output is about 0.6 volt less positive than the input, and the reverse for negative swings. For an input sine wave the output would look as shown in Figure 2.54. In the language of the audio

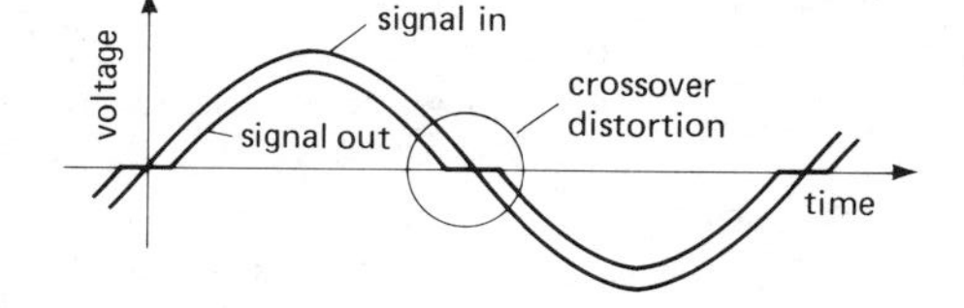

Figure 2.54

business, this is called crossover distortion. The best cure (feedback offers another method, although it is not entirely satisfactory) is to bias the push-pull stage into slight conduction, as in Figure 2.55.

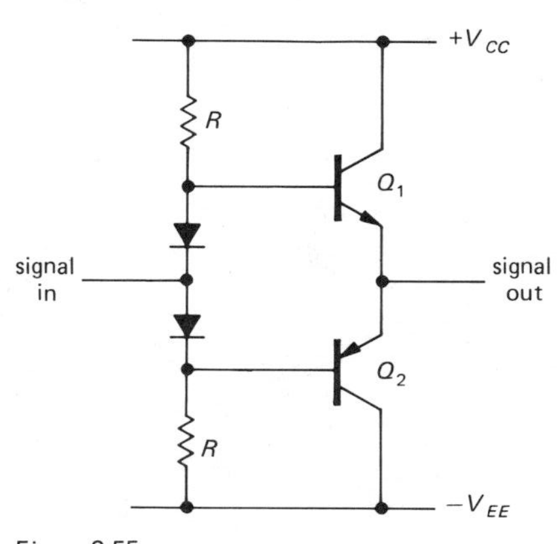

Figure 2.55

The bias resistors R bring the diodes into forward conduction, holding Q_1's base a diode drop above the input signal and Q_2's base a diode drop below the input signal. Now, as the input signal crosses through zero, conduction passes from Q_2 to Q_1; one of the output transistors is always on. R is chosen to provide enough base current for the output transistors at the peak output swing. For instance, with ±20 volt supplies and an 8 ohm load running up to 10 watts sine-wave power, the peak base voltage is about 13.5 volts, and the peak load current is 1.6 amps. Assuming a transistor beta of 50 (power transistors generally have lower current gain than small-signal transistors), the 32mA of necessary base current will require base resistors of about 220 ohms (6.5V from V_{CC} to base at peak swing).

Thermal stability in class B push-pull amplifiers

The preceding amplifier (sometimes called a class B amplifier, meaning that each transistor conducts over half the cycle) has one bad feature: It is not thermally stable. As the output transistors warm up (and they will get hot, because they are dissipating power when signal is applied), their V_{BE} drops, and quiescent collector current begins to flow. The added heat this produces causes the situation to get worse, with the strong possibility of what is called *thermal runaway* (whether it runs away or not depends on a number of factors, including how large a "heat sink" is used, how well the diode temperature tracks the transistor tempera-

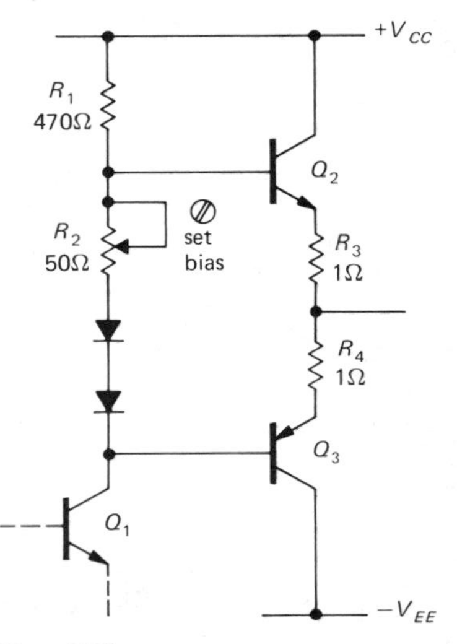

Figure 2.56

ture, etc.). Even without runaway, better control over the circuit is needed, usually with the sort of arrangement shown in Figure 2.56.

For variety, the input is shown coming from the collector of the previous stage; R_1 now serves the dual purpose of being Q_1's collector resistor and providing current to bias the diodes and bias-setting resistor in the push-pull base circuit. Here R_3 and R_4, typically a few ohms or less, provide a "cushion" for the critical quiescent current biasing: The voltage between the bases of the output transistors must now be a bit greater than two diode drops, and you provide the extra with adjustable biasing resistor R_2 (often replaced by a third series diode). With a few tenths of a volt across R_3 and R_4, the temperature variation of V_{BE} doesn't cause the current to rise very rapidly (the larger the drop across R_3 and R_4, the less sensitive it is), and the circuit will be stable. Stability is improved by mounting the diodes in physical contact with the output transistors (or their heat sinks).

You can estimate the thermal stability of such a circuit by remembering that the base-emitter drop decreases by about 2.1mV for each 1°C rise and that the collector current increases by a factor of 10 for every 60mV increase in base-emitter voltage. For example, if R_2 were replaced by a diode, you would have three diode drops between the bases of Q_2 and Q_3, leaving about one diode drop across the series combination of R_3 and R_4. (The latter would then be chosen to give an appropriate quiescent current, perhaps 50mA for an audio power amplifier.) The worst case for thermal stability occurs if the biasing diodes are not thermally coupled to the output transistors.

Let us assume the worst and calculate the increase in output-stage quiescent current corresponding to a 30°C temperature rise in output transistor temperature. That's not a lot for a power amplifier, by the way. For that temperature rise, the V_{BE} of the output transistors will decrease by about 63mV at constant current, raising the voltage across R_3 and R_4 by about 20% (i.e., the quiescent current will rise about 20%). The corresponding figure for the preceding amplifier circuit without emitter resistors (Fig. 2.55) will be a factor of 10 rise in quiescent current (recall that I_C increases a decade per 60mV increase in V_{BE}), i.e., 1000%. The improved thermal stability of this biasing arrangement is evident.

This circuit has the additional advantage that by adjusting the quiescent current, you have some control over the amount of residual crossover distortion. A push-pull amplifier biased in this way to obtain substantial quiescent current at the crossover point is sometimes referred to as a class AB amplifier, meaning that both transistors conduct simultaneously during a portion of the cycle. In practice, you choose a quiescent current that is a good compromise between low distortion and excessive quiescent dissipation. Feedback, the subject of the next chapter, is almost always used to reduce distortion still further.

An alternative method for biasing a push-pull follower is shown in Figure 2.57. Q_4 acts as an adjustable diode: The base resistors are a divider, and therefore Q_4's collector-emitter voltage will stabilize at a value that puts 1 diode drop from base to emitter, since any greater V_{CE} will bring it into heavy conduction. For instance, if both resistors were 1k, the transistor would turn on at 2 diode drops, collector to emitter. In this case the bias adjustment lets you set the push-pull interbase voltage anywhere from 1 to 3.5 diode drops. The 10μF capacitor

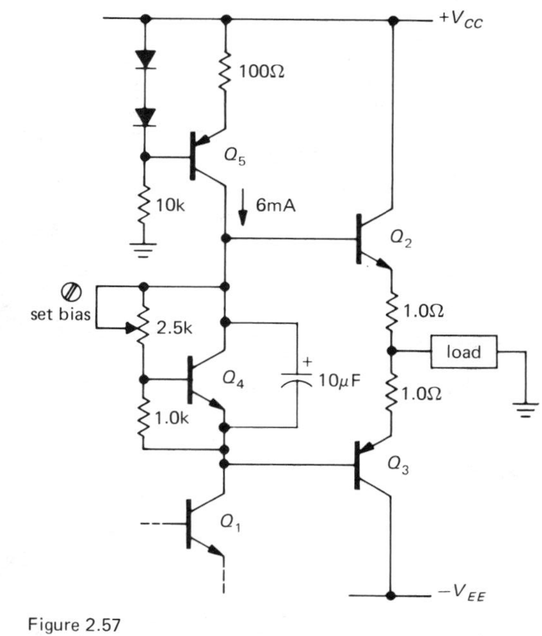

Figure 2.57
Biasing a push-pull output stage for low crossover distortion.

ensures that both output transistor bases see the same signal; such a bypass capacitor is a good idea for any biasing scheme you use. In this circuit, Q_1's collector resistor has been replaced by current source Q_5. That's a useful circuit variation, because with a resistor it is sometimes difficult to get enough base current to drive Q_2 near the top of the swing. A resistor small enough to drive Q_2 sufficiently results in high quiescent collector current in Q_1 (with high dissipation), and also reduced voltage gain (remember that $G = -R_{\text{collector}}/R_{\text{emitter}}$). Another solution to the problem of Q_2's base drive is the use of *bootstrapping,* a technique that will be discussed shortly.

2.15 Darlington connection

If you hook two transistors together as in Figure 2.58, the result behaves like a single transistor with beta equal to the product of the two transistor betas. This can be very handy where high currents are involved (e.g., voltage regulators or power amplifier output stages), or for input stages of amplifiers where very high input impedance is necessary.

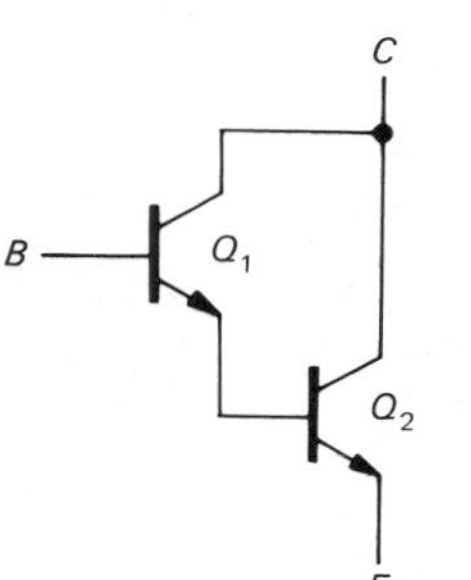

Figure 2.58
Darlington transistor configuration.

For a Darlington transistor the base-emitter drop is twice normal, and the saturation voltage is at least one diode drop (since Q_1's emitter must be a diode drop above Q_2's emitter). Also, the combination tends to act like a rather slow transistor because Q_1 cannot turn off Q_2 quickly. This problem is usually taken care of by including a resistor from base to emitter of Q_2 (Fig. 2.59). R also

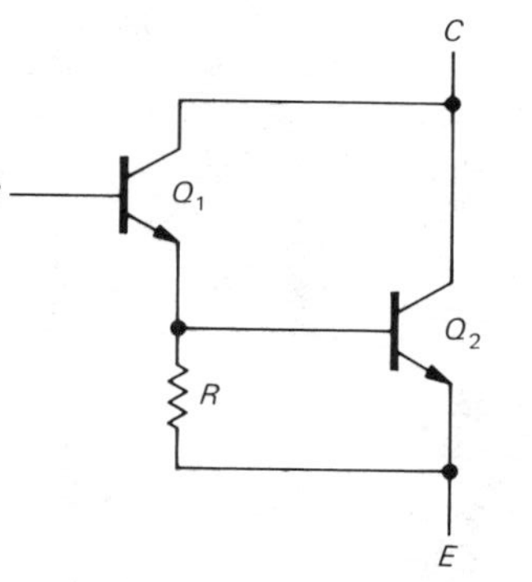

Figure 2.59

prevents leakage current through Q_1 from biasing Q_2 into conduction; its value is chosen so that Q_1's leakage current (nanoamps for small-signal transistors, as much as hundreds of microamps for power transistors) produces less than a diode drop across R and so that R doesn't sink a large proportion of Q_2's base current when it has a diode drop across it. Typically R might be a few hundred ohms in a power transistor Darlington, or a few thousand ohms for a small-signal Darlington.

Darlington transistors are available as single packages, usually with the base-emitter resistor included. A typical example is the *npn* power Darlington 2N6285, with current gain of 4000 (typically) at a collector current of 10 amps.

Sziklai connection

A similar beta-boosting configuration is the Sziklai connection, sometimes referred to as a complementary Darlington (Fig. 2.60).

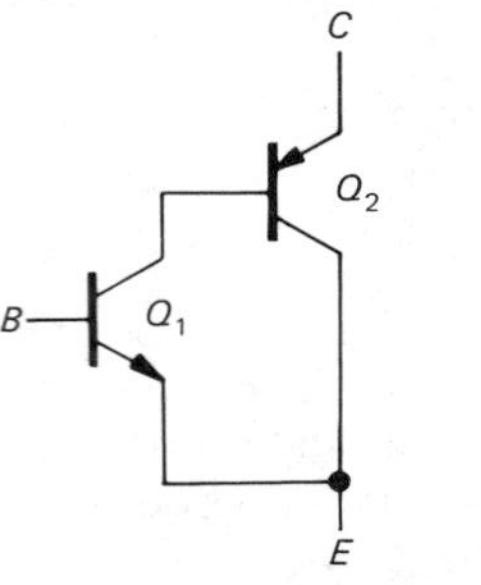

Figure 2.60

This combination behaves like an *npn* transistor, again with large beta. It has only a single base-emitter drop, but it also cannot saturate to less than a diode drop. A small resistor from base to emitter of Q_2 is advisable. This connection is common in push-pull power output stages where the designer wishes to use one polarity of output transistor only. Such a circuit is shown in Figure 2.61. As before, R_1 is Q_1's collector resistor. Darlington Q_2Q_3 behaves like a single *npn* transistor with high current gain. The Sziklai connected pair Q_4Q_5 behaves like a single high-gain *pnp* power transistor. As before, R_3 and R_4 are small. This circuit is sometimes called a pseudocomplementary push-pull follower. A true complementary stage would use a Darlington-connected *pnp* pair for Q_4Q_5.

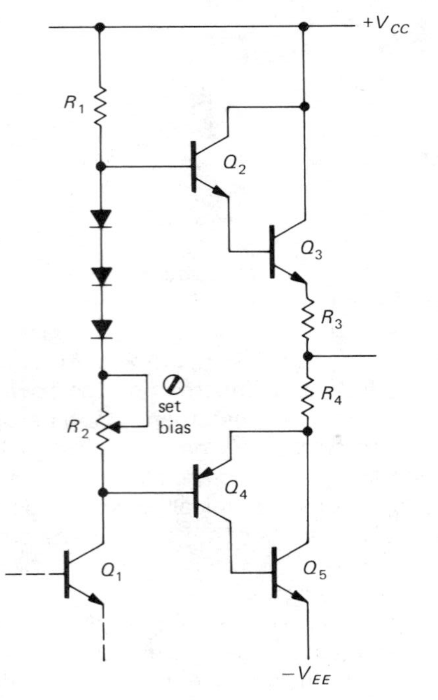

Figure 2.61

Superbeta transistor

The Darlington connection and its near relatives should not be confused with the so-called superbeta transistor, a device with very high h_{FE} achieved through the manufacturing process. A typical superbeta transistor is the 2N5963, with a guaranteed minimum current gain of 900 at collector currents from 10μA to 10mA; it belongs to the 2N5961-2N5963 series, with a range of maximum V_{CE}s of 30 to 60 volts (if you need higher collector voltage, you have to settle for lower beta). Superbeta matched pairs are available for use in low-level amplifiers that require matched characteristics, a topic we will discuss in Section 2.17. Examples are the LM114, LM394, MAT-01, and CA3095; these provide high-gain transistor pairs whose V_{BE}s are matched to a fraction of a millivolt (as little as 50μV in the best versions) and whose h_{FE}s are matched to about 1%.

It is possible to combine superbeta transistors in a Darlington connection. Some commercial devices (e.g., the LM11 and LM316 op-amps) achieve base bias currents as low as 50 *pico*amps this way.

□ 2.16 Bootstrapping

When biasing an emitter follower, for instance, you choose the base voltage divider resistors so that the divider presents a stiff voltage source to the base, i.e., their parallel impedance is much less than the impedance looking into the base. For this reason the resulting circuit has an input impedance dominated by the voltage divider – the driving signal sees a much lower impedance than would otherwise be necessary. Figure 2.62 shows an example.

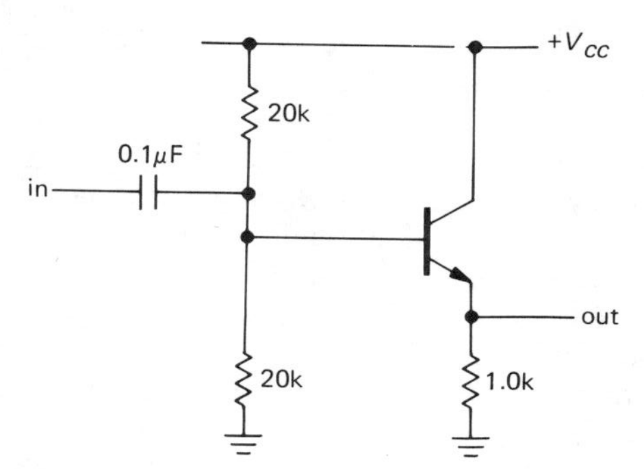

Figure 2.62

The input resistance of about 9k is mostly due to the voltage divider impedance of 10k. It is always desirable to keep input impedances high, and anyway it's a shame to load the input with the divider, which, after all, is only there to bias the transistor. Bootstrapping is the colorful name given to a technique that circumvents this problem (Fig. 2.63). The transistor is biased by the divider

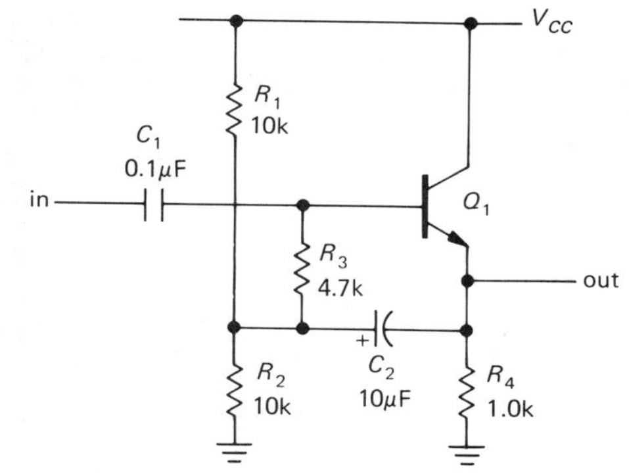

Figure 2.63

R_1R_2 through series resistor R_3. C_2 is chosen to have low impedance at signal frequencies compared with the bias resistors. As always, bias is stable if the dc impedance seen from the base (in this case 9.7k) is much less than the dc impedance looking into the base (in this case approximately 100k). But now the signal-frequency input impedance is no longer the same as the dc impedance. Look at it this way: An input wiggle v_{in} results in an emitter wiggle $v_E \approx v_{in}$. So the change in current through bias resistor R_3 is $i = (v_{in} - v_E)/R_3 \approx 0$, i.e., Z_{in} (due to bias string) $= v_{in}/i_{in} \approx$ infinity. We've made the loading (shunt) impedance of the bias network very large *at signal frequencies*.

Another way of seeing this is to notice that R_3 always has the same voltage across it at signal frequencies (since both ends of the resistor have the same voltage changes), i.e., it's a current source. But a current source has infinite impedance. Actually, the effective impedance is less than infinity because the gain of a follower is slightly less than 1. That is so because the base-emitter drop depends on collector current, which changes with the signal level. You could have predicted the same result from the voltage-dividing effect of the impedance looking into the emitter [$r_e = 25/I_C$(mA) ohms] combined with the emitter resistor. If the follower has voltage gain A ($A \approx 1$), the effective value of R_3 at signal frequencies is

$$R_3/(1 - A)$$

In practice the value of R_3 is effectively increased by a hundred or so, and the input impedance is then dominated by the transistor's base impedance. The emitter-degenerated amplifier can be bootstrapped in the same way, since the signal on the emitter follows the base. Note that the bias divider circuit is driven by the low-impedance emitter output at signal frequencies, thus isolating the input signal from this usual task.

□ *Bootstrapping collector load resistors*

The bootstrap principle can be used to increase the effective value of a transistor's collector load resistor, if that stage drives a follower. That can increase the voltage gain of the stage substantially [recall that $G_V = -g_m R_C$, with $g_m = 1/(R_E + r_e)$]. Figure 2.64 shows an example of a bootstrapped push-pull output stage similar to the push-pull follower circuit we saw earlier. Since the output follows Q_2's base signal, C bootstraps Q_1's collector load, keeping a constant voltage across R_2 as the signal varies (C must be chosen to have low impedance compared with R_1 and R_2 at all signal frequencies). That makes R_2 look like a current source, raising Q_1's voltage gain and maintaining good base drive even at the peaks of the signal swing. When the signal gets near V_{CC}, the junction of R_1 and R_2

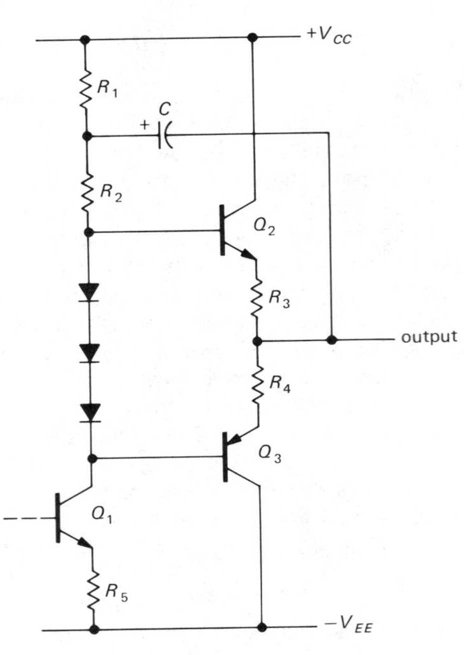

Figure 2.64

actually rises above V_{CC} because of the stored charge in C. In this case, if $R_1 = R_2$ (not a bad choice) the junction between them rises to 1.5 times V_{CC} when the output reaches V_{CC}. This circuit has enjoyed considerable popularity in commercial audio amplifier design, although a simple current source in place of the bootstrap is superior, since it maintains the improvement at low frequencies and eliminates the undesirable electrolytic capacitor.

2.17 Differential amplifiers

The differential amplifier is a very common configuration used to amplify the difference voltage between two input signals. In the ideal case the output is entirely independent of the individual signal levels – only the difference matters. Some commonly used terms: When both inputs change levels together, that's a *common-mode* input change. A differential change is called *normal mode*. A good differential amplifier has a high *common-mode rejection ratio* (CMRR), the ratio of response for a normal-mode signal to the response for a common-mode signal of the same amplitude. CMRR is usually specified in decibels. The common-mode input range is the voltage level over which the inputs may safely vary.

Differential amplifiers are important in applications where weak signals are contaminated by "pickup" and other miscellaneous noise. Examples include digital signals transferred over long cables (usually twisted pairs of wires), audio signals (the term "balanced" means differential, usually 600Ω impedance, in the audio business), radiofrequency signals (twin-lead cable is differential), electrocardiogram voltages, magnetic-core memory readout signals, and numerous other applications. A differential amplifier at the receiving end restores the original signal if the common-mode signals are not too large. Differential amplifiers are universally used in operational amplifiers, which we will come to soon. They're very important in dc amplifier design (amplifiers that amplify clear down to dc, i.e., have no coupling capacitors) because their symmetrical design is inherently compensated against thermal drifts.

Figure 2.65 shows the basic circuit. The output is taken off one collector with respect

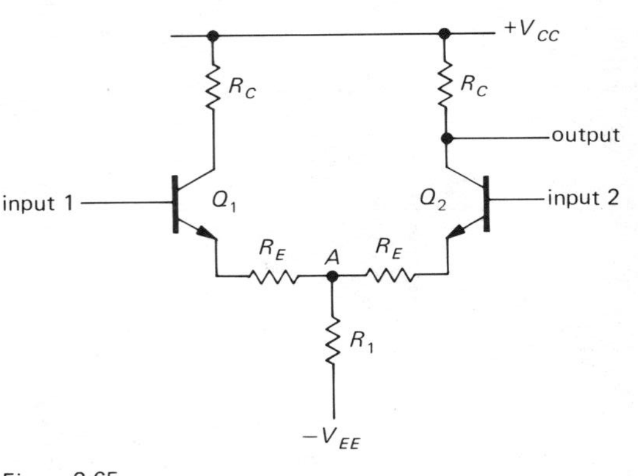

Figure 2.65
Classic transistor differential amplifier.

to ground; that is called a *single-ended output* and is the most common configuration. You can think of this amplifier as a device that amplifies a difference signal and converts it to a single-ended signal so that ordinary subcircuits (followers, current sources, etc.) can make use of the output. (If, instead, a differential output is desired, it is taken between the collectors.)

What is the gain? That's easy enough to calculate: Imagine a symmetrical input signal

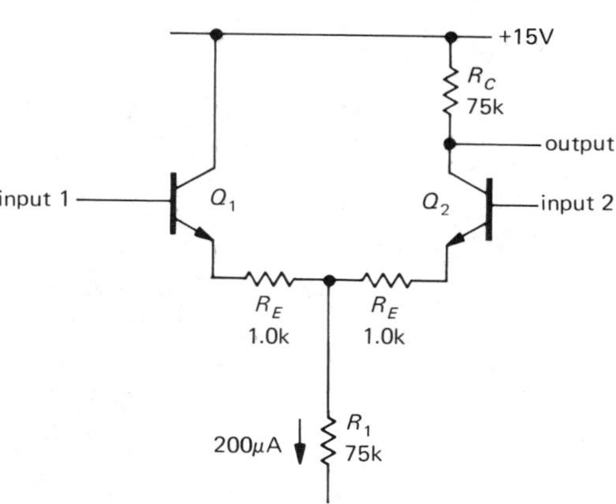

Figure 2.66

$$G_{diff} = \frac{v_{out}}{v_1 - v_2} = \frac{R_C}{2(R_E + r_e)}$$

$$G_{CM} = -\frac{R_C}{2R_1 + R_E + r_e}$$

$$\text{CMRR} \approx \frac{R_1}{R_E + r_e}$$

wiggle, in which input 1 rises by v_{in} (a small-signal variation) and input 2 drops by the same amount. As long as both transistors stay in the active region, point A remains fixed. The gain is then determined as before, remembering that the input change is actually twice the wiggle on either base: $G_{diff} = R_C/2(r_e + R_E)$. Typically R_E is small, 100 ohms or less, or it may be omitted entirely. Differential voltage gains of a few hundred are typical.

The common-mode gain can be determined by putting identical signals v_{in} on both inputs. If you think about it correctly (remembering that R_1 carries both emitter currents), you'll find $G_{CM} = -R_C/(2R_1 + R_E)$. Here we've ignored the small r_e, because R_1 is typically large, at least a few thousand ohms. We really could have ignored R_E as well. The CMRR is roughly $R_1/(r_e + R_E)$. Let's look at a typical example (Fig. 2.66) to get some familiarity with differential amplifiers.

R_C is chosen for a quiescent current of 100μA. As usual, we put the collector at $0.5V_{CC}$ for large dynamic range. Q_1's collector resistor can be omitted, since no output is taken there. R_1 is chosen to give total emitter current of 200μA, split equally between the two sides when the (differential) input is zero. This amplifier has a differential gain of 30 and a common-mode gain of 0.5. Omitting the 1.0k resistors raises the differential gain to 150, but drops the (differential) input impedance from about 250k to about 50k (you can substitute Darlington transistors in the input stage to raise the impedance into the megohm range, if necessary).

Remember that the maximum gain of a single-ended grounded emitter amplifier biased to $0.5V_{CC}$ is $20V_{CC}$. In the case of a differential amplifier the maximum differential gain ($R_E = 0$) is half that figure, or (for arbitrary quiescent point) 20 times the voltage across the collector resistor. The corresponding maximum CMRR (again with $R_E = 0$) is equal to 20 times the voltage across R_1.

EXERCISE 2.13

Verify that these expressions are correct. Then design a differential amplifier to your own specifications.

The differential amplifier is sometimes called a "long-tailed pair," because if the length of a resistor symbol indicated its magnitude, the circuit would look like Figure

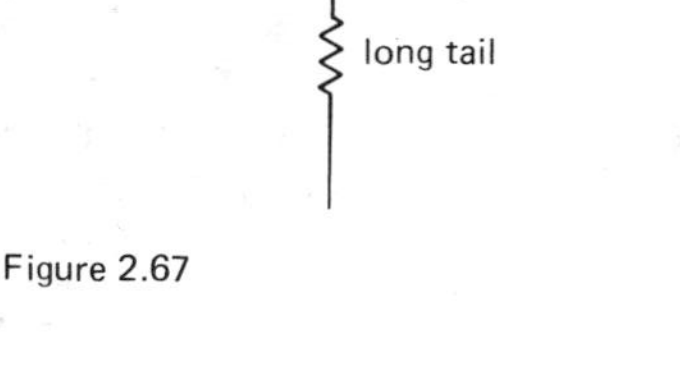

Figure 2.67

2.67. The long tail determines the common-mode gain, and the small inter-emitter resistance (including intrinsic emitter resistance r_e) determines the differential gain.

Current source biasing

The common-mode gain of the differential amplifier can be reduced enormously by substituting a current source for R_1. Then R_1 effectively becomes very large, and the common-mode gain is nearly zero. If you prefer, just imagine a common-mode input swing; the emitter current source maintains a constant total emitter current, shared equally by the two collector circuits, by symmetry. The output is therefore unchanged. Figure 2.68 shows an example.

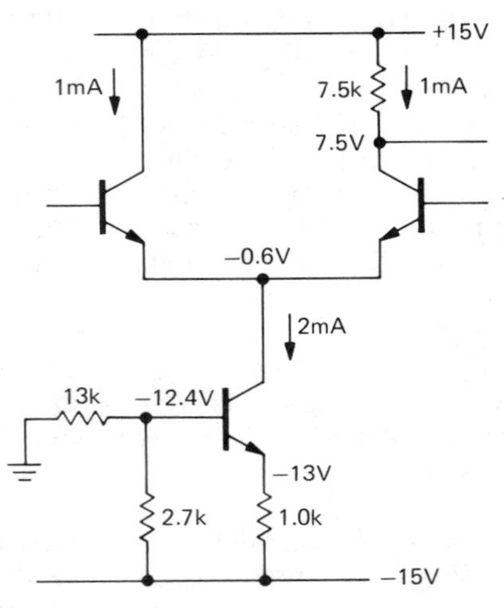

Figure 2.68

The CMRR of this circuit, using an LM394 monolithic transistor pair for Q_1 and Q_2 and a 2N5963 current source, is 100,000:1 (100dB). The common-mode input range for this circuit goes from −12 volts to +7 volts; it is limited at the low end by the compliance of the emitter current source and at the high end by the collector's quiescent voltage.

Be sure to remember that this amplifier, like all transistor amplifiers, must have a dc bias path to the bases. If the input is capacitively coupled, for instance, you must have base resistors to ground.

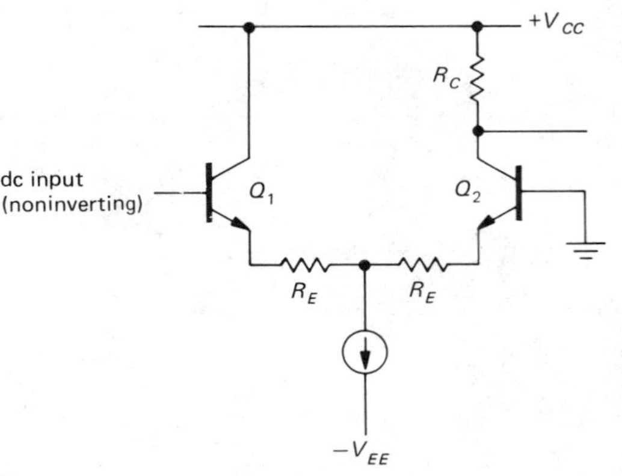

Figure 2.69

Use in single-ended dc amplifiers

A differential amplifier makes an excellent dc amplifier, even for single-ended inputs. You just ground one of the inputs and connect the signal to the other (Fig. 2.69). You might think that the "unused" transistor could be eliminated. Not so! The differential configuration is inherently compensated for temperature drifts, and even when one input is at ground that transistor is still doing something: A temperature change causes both V_{BE}s to change the same amount, with no change in balance or output. That is, changes in V_{BE} are not amplified by G_{diff} (only by G_{CM}, which can be made essentially zero). Furthermore, the cancellation of V_{BE}s means that there are no 0.6 volt drops at the input to worry about. The quality of a dc amplifier constructed this way is limited only by mismatching of input V_{BE}s or their temperature coefficients. Commercial monolithic transistor pairs and commercial differential amplifier ICs are available with extremely good matching (e.g., the MAT-01 *npn* monolithic matched pair has a typical drift of V_{BE} between the two transistors of $0.15\mu V/°C$ and $0.2\mu V$ per month).

Either input could have been grounded in the preceding circuit example. The choice depends on whether or not the amplifier is supposed to invert the signal. The connection shown is noninverting, and so the inverting input has been grounded. This terminology carries over to op-amps, which are simply high-gain differential amplifiers.

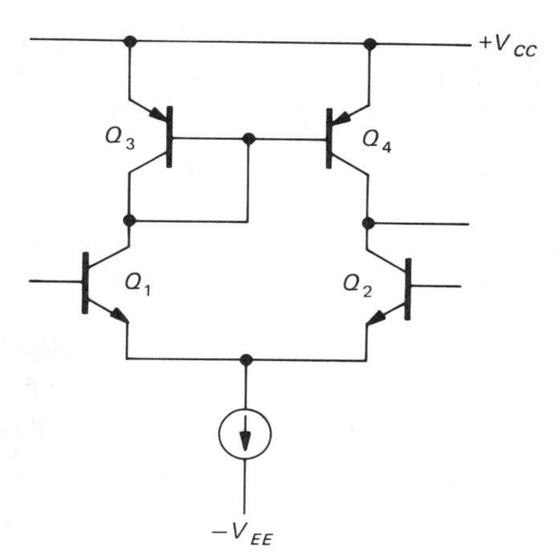

Figure 2.70
Differential amplifier with active current mirror load.

Current mirror active load

As with the simple grounded emitter amplifier, it is sometimes desirable to have a single-stage differential amplifier with very high gain. An elegant solution is a current mirror active load (Fig. 2.70). Q_1Q_2 is the differential pair with emitter current source. Q_3 and Q_4, a current mirror, form the collector load. The high effective collector load impedance provided by the mirror yields voltage gains of 5000 or more, assuming no load at the amplifier's output. Such an amplifier is usually used only within a feedback loop, or as a comparator (discussed in the next section). Be sure to load such an amplifier with a high impedance, or the gain will drop enormously.

Differential amplifiers as phase splitters

The collectors of a symmetrical differential amplifier generate equal signal swings of opposite phase. By taking outputs from both collectors, you've got a phase splitter. Of course, you could also use a differential amplifier with both differential inputs and differential outputs. This differential output signal could then be used to drive an additional differential amplifier stage, with greatly improved overall common-mode rejection.

Differential amplifiers as comparators

Because of its high gain and stable characteristics, the differential amplifier is the main building block of the *comparator,* a circuit that tells which of two inputs is larger. They are used for all sorts of applications: switching on lights and heaters, generating square waves from triangles, detecting when a level in a circuit exceeds some particular threshold, class D amplifiers and pulse code modulation, switching power supplies, etc. The basic idea is to connect a differential amplifier so that it turns a transistor switch on or off, depending on the relative levels of the input signals. The linear region of amplification is ignored, with one or the other of the two input transistors cut off at any time. A typical hookup is illustrated in the next section by a temperature-controlling circuit that uses a resistive temperature sensor (thermistor).

2.18 Capacitance and Miller effect

In our discussion so far we have used what amounts to a dc, or low-frequency, model of the transistor. Our simple current amplifier model and the more sophisticated Ebers-Moll transconductance model both deal with voltages, currents, and resistances seen at the various terminals. With these models alone we have managed to go quite far, and in fact these simple models contain very nearly everything you will ever need to know to design transistor circuits. However, one important aspect that has serious impact on high-speed and high-frequency circuits has been neglected: the existence of capacitance in the external circuit and in the transistor junctions themselves. Indeed, at high frequencies the effects of capacitance often dominate circuit behavior; at 100 MHz a typical junction capacitance of 5pF has an impedance of 320 ohms!

We will deal with this important subject in detail in Chapter 13. At this point we would merely like to state the problem, illustrate some of its circuit incarnations, and suggest some methods of circumventing the problem. It would be a mistake to leave this chapter without realizing the nature of this problem. In the course of this brief discussion we will encounter the famous *Miller effect* and the use of configurations such as the cascode to overcome it.

Junction and circuit capacitance

Capacitance limits the speed at which the voltages within a circuit can swing, owing to finite driving impedance or current. When a capacitance is driven by a finite source resistance, you see *RC* exponential charging behavior, whereas a capacitance driven by a current source leads to slew-rate-limited waveforms (ramps). As general guidance, reducing the source impedances and load capacitances and increasing the drive currents within a circuit will speed things up. However, there are some subleties connected with feedback capacitance and input capacitance. Let's take a brief look.

The circuit in Figure 2.71 illustrates most of the problems of junction capacitance. The

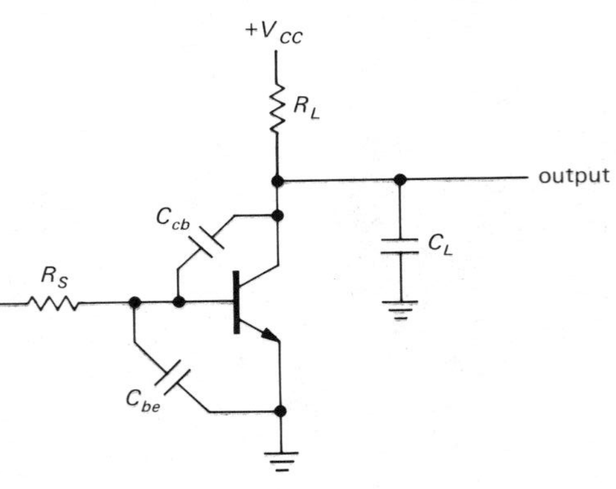

Figure 2.71

output capacitance forms a time constant with the output resistance R_L (R_L includes both the collector and load resistances, and C_L includes both junction and load capacitances), giving a rolloff starting at some frequency $f = 1/2\pi R_L C_L$. The same is true for the input capacitance in combination with the source impedance R_S.

Miller effect

C_{cb} is another matter. The amplifier has some overall voltage gain G_V, so a small voltage wiggle at the input results in a wiggle G_V times larger (and inverted) at the collector. This means that the signal source sees a current through C_{cb} that is $G_V + 1$ times as large as if C_{cb} were connected from base to ground; i.e., for the purpose of input rolloff frequency calculations, the feedback capacitance behaves like a capacitor of value $C_{cb}(G_V + 1)$ from input to ground. This effective increase of C_{cb} is known as the Miller effect. It often dominates the rolloff characteristics of amplifiers, since a typical feedback capacitance of 4pF can look like several hundred picofarads to ground.

There are several methods available to beat the Miller effect. It is absent altogether in a grounded base stage. You can decrease the source impedance driving a grounded emitter stage by using an emitter follower. Figure 2.72 shows two other possibilities.

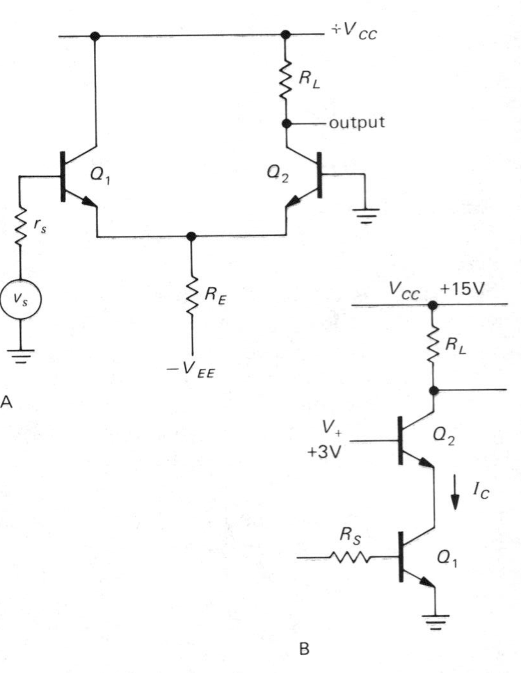

Figure 2.72

The differential amplifier circuit (with no collector resistor in Q_1) has no Miller effect; you can think of it as an emitter follower driving a grounded base amplifier. The second circuit is the famous cascode configuration. Q_1 is a grounded emitter amplifier, with R_L as its collector resistor. Q_2 is interposed in the collector path to prevent Q_1's collector from swinging (thereby eliminating the Miller effect) while passing the collector current through to the load resistor unchanged. V_+ is a fixed bias voltage, usually set a few volts above Q_1's emitter voltage to pin Q_1's collector and keep it in the active region. This fragment is incomplete as shown; you could either include a bypassed emitter resistor and base divider for biasing (as we did earlier in the chapter) or include it within an overall loop with feedback at dc. V_+ might be provided from a divider or zener, with bypassing to keep it stiff at signal frequencies.

EXERCISE 2.14

Explain in detail why there is no Miller effect in either transistor in the preceding differential amplifier and cascode circuits.

Capacitive effects can be somewhat more complicated than this brief introduction might indicate. In particular: (a) The rolloffs due to feedback and output capacitances are not entirely independent; in the terminology of the trade there is pole splitting, an effect we will explain in the next chapter. (b) The input capacitance still has an effect, even with a stiff input signal source. In particular, current that flows through C_{be} is not amplified by the transistor. This base current "robbing" by the input capacitance causes the transistor's small-signal current gain h_{fe} to drop at high frequencies, eventually reaching unity at a frequency known as f_T. (c) To complicate matters, the junction capacitances depend on voltage. C_{be} changes so rapidly with base current that it is not even specified on transistor data sheets; f_T is given instead. (d) When a transistor is operated as a switch, effects associated with charge stored in the base region of a saturated transistor cause an additional loss of speed. We will take up these and other topics having to do with high-speed circuits in Chapter 13.

2.19 Field-effect transistors

In the next few chapters we will explore some of the applications of transistors in operational amplifier and feedback circuits. However, it would be a mistake to leave this chapter without a few words of explanation about the other kind of transistor, the field-effect transistor (FET), which we will take up in detail in Chapter 6. The FET behaves in many ways like an ordinary bipolar transistor. It is a 3-terminal amplifying device, available in both polarities, with a terminal (the gate) that controls the current flow between the other two terminals (source and drain). It has a unique property, though: The gate draws no current, except for leakage. This means that extremely high input impedances are possible, limited only by capacitance and leakage effects. Input currents measured in picoamperes are commonplace. Yet the FET is a rugged and capable device, with voltage and current ratings comparable to those of bipolar transistors (except for the highest currents).

Most of the available devices fabricated with transistors (matched pairs, differential and operational amplifiers, comparators, high-current switches and amplifiers, radiofrequency amplifiers, and digital logic) are also available with FET construction. It would only be confusing to introduce these fascinating and useful devices now, but keep them in mind as you read through the subsequent chapters.

SOME TYPICAL TRANSISTOR CIRCUITS

To illustrate some of the ideas of this chapter, let's look at a few examples of circuits with transistors. The range of circuits we can cover is necessarily limited, since real-world circuits often use negative feedback, a subject we will cover in the next chapter.

2.20 Regulated power supply

Figure 2.73 shows a very common configuration. R_1 normally holds Q_1 on; when the output reaches 10 volts, Q_2 goes into conduction (base at 5V), preventing further rise of output voltage by shunting base current from Q_1's base. The supply can be

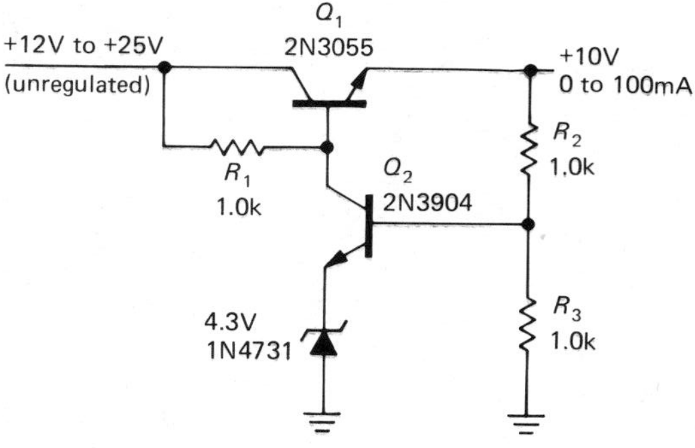

Figure 2.73

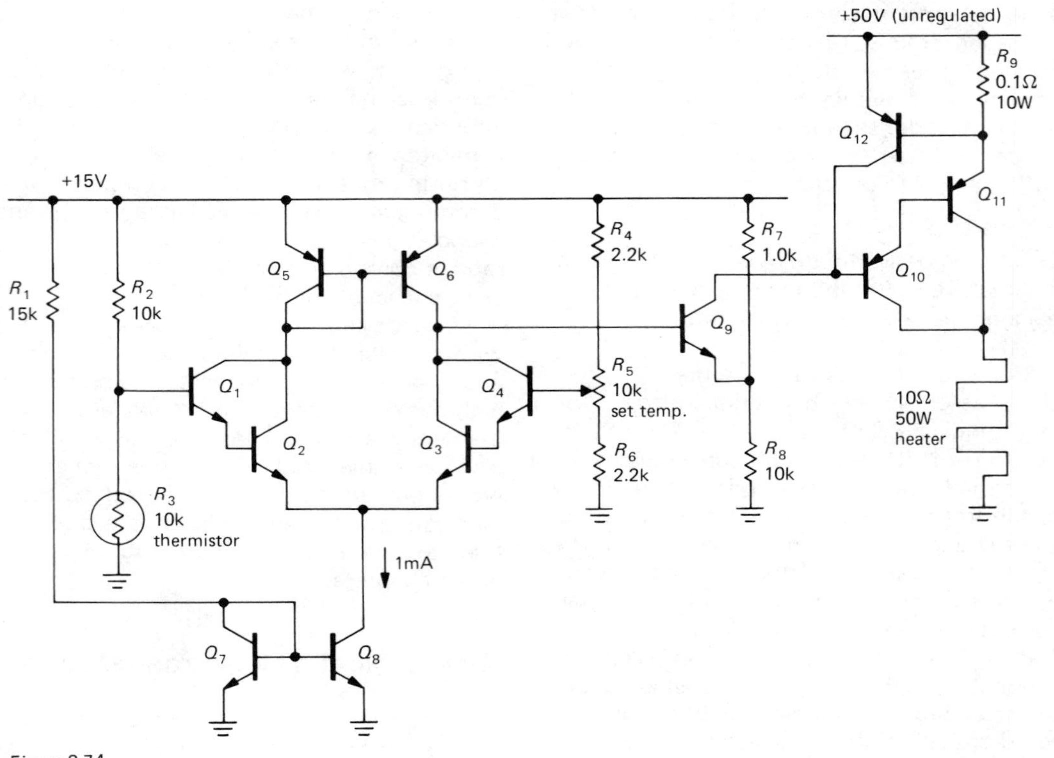

Figure 2.74
Temperature controller for 50 watt heater.

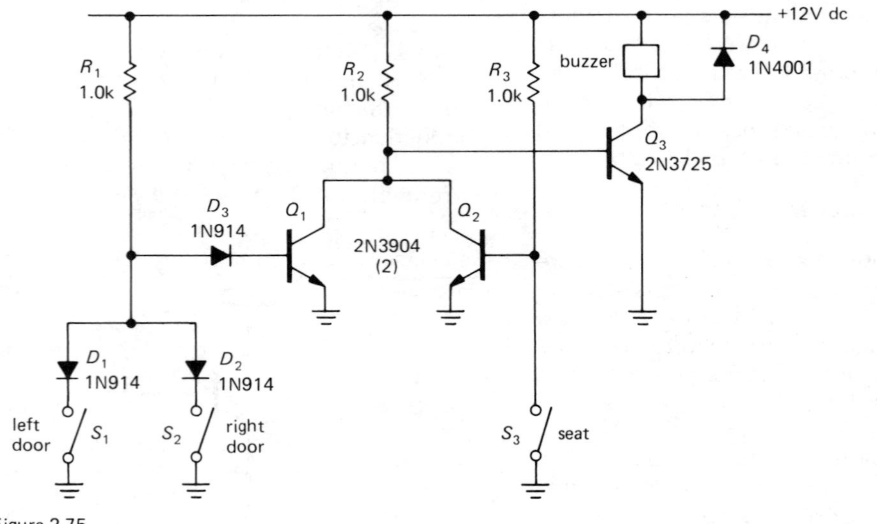

Figure 2.75

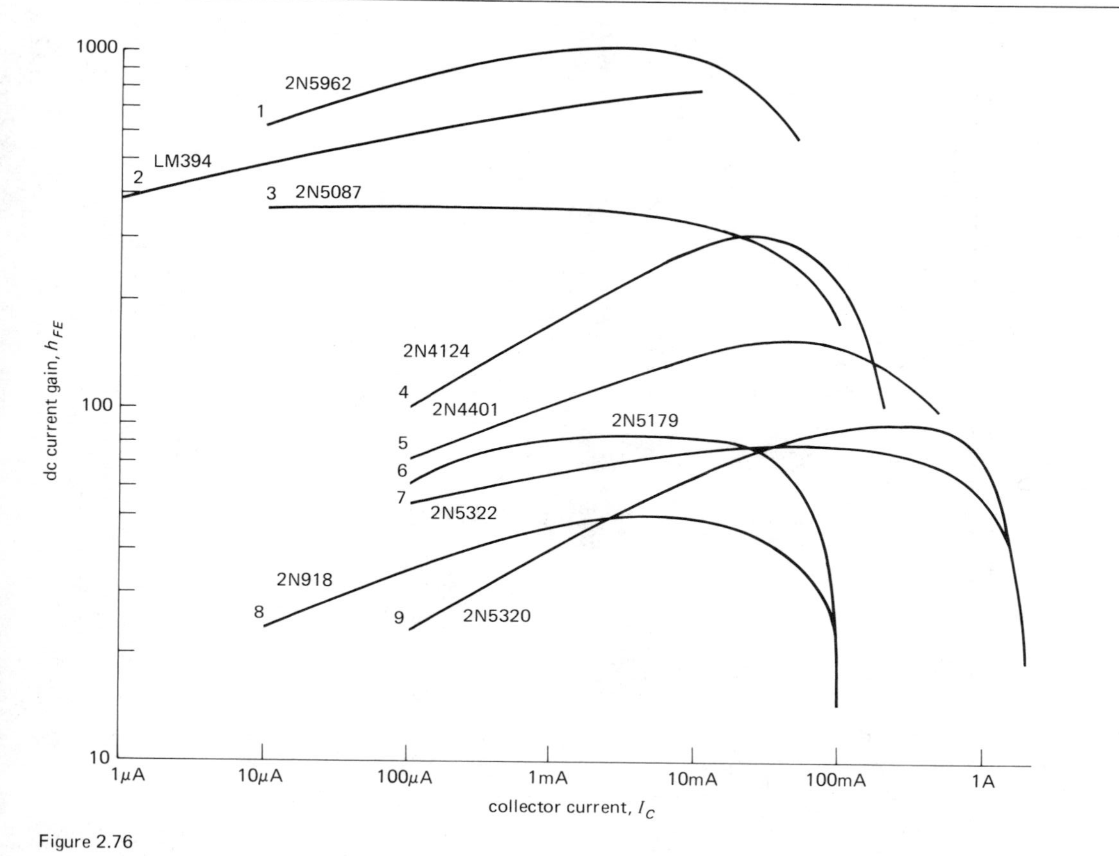

Figure 2.76
Curves of typical transistor current gain, h_{FE}, for a selection of transistors from Table 2.1. These curves are taken from manufacturers' literature.

made adjustable by replacing R_2 and R_3 by a potentiometer. This is actually an example of negative feedback: Q_2 "looks at" the output and does something about it if the output isn't at the right voltage.

2.21 Temperature controller

The schematic diagram in Figure 2.74 shows a temperature controller based on a *thermistor* sensing element, a device that changes resistance with temperature. Differential Darlington Q_1–Q_4 compares the voltage of the adjustable reference divider R_4–R_6 with the divider formed from the thermistor and R_2. (By comparing *ratios* from the same supply, the comparison becomes insensitive to supply variations; this particular configuration is called a Wheatstone bridge.) Current mirror Q_5Q_6 provides an active load to raise the gain, and mirror Q_7Q_8 provides emitter current. Q_9 compares the differential amplifier output with a fixed voltage, saturating Darlington $Q_{10}Q_{11}$, which supplies power to the heater, if the thermistor is too cold. R_9 is a current-sensing resistor that turns on protection transistor Q_{12} if the output current exceeds about 6 amps; that removes base drive from $Q_{10}Q_{11}$, preventing damage.

2.22 Simple logic with transistors and diodes

Figure 2.75 shows a circuit that performs a task we illustrated in Chapter 1: sounding a buzzer if either car door is open and the driver is seated. In this circuit the transistors all operate as switches (either off or saturated). Diodes D_1 and D_2 form what is called

TABLE 2.1. SELECTED SMALL-SIGNAL TRANSISTORS[a]

	V_{CEO} (V)	I_C max (mA)	h_{FE} typ	@ I_C (mA)	C_{cb} typ[b] (pF)	f_T typ (MHz)	Gain curve[c]	Metal: TO-5[d] npn	Metal: TO-5[d] pnp	Metal: TO-18[e] npn	Metal: TO-18[e] pnp	Plastic: TO-105 npn	Plastic: TO-105 pnp	Plastic: TO-106 npn	Plastic: TO-106 pnp	Plastic: TO-92[f] npn	Plastic: TO-92[f] pnp
General purpose	20	500	100	150	16	200		—	—	—	—	5136	5142	5137	5143	—	—
	25	200	200	2	1.8–2.8	300	4	—	—	—	—	—	—	—	—	4124	4126
—	40	200	200	10	1.8–2.8	300		—	—	3947	3251	—	4122	—	—	3904	3906
High gain, low noise	25	50	300	10	2–7	150		—	—	—	—	—	—	3565	—	(3707[f] 3391A[f])	4058[f]
	25	300	250	50	4	300		—	—	—	—	—	—	—	—	6008[f]	6009[f]
	25	50	500	5	1.5–4	500		—	—	—	—	—	—	—	4250	5089	
	40	20	700	1	14	200	2	LM394	—	—	—	—	—	—	—	—	—
	45	50	1000	10	1.5	300	1	—	—	—	—	—	—	—	—	5962	
	50	50	350	5	1.8	400	3	—	—	2848	3965	—	—	—	—	(4967 5210)	(4965 5087)
High current	30–60	600	150	150	5	300	5	2219	2905	2222	(2907 3251)	3643	3645	3566	3638	4401	4403
	50	1000	100	200	7	450		3725	5022	4014	—	—	—	—	—	—	—
	60	500	180	50	15	200		—	—	—	—	3568	4355	—	—	—	—
—	60	1000	70	80	15	100		(3107 2102)	4036	—	—	—	—	—	—	—	—
	75	2000	70	500	20	60	7,9	5320	5322	—	—	—	—	—	—	—	—
—																	
High voltage	150	600	100	10	3–6	250		—	4929	—	—	—	—	5965	—	5550	5401
	300	1000	50	50	10	50		3439	5416	—	—	—	—	—	—	—	—
High speed	12	50	80	3	0.7	1500	6	—	—	5179	—	—	—	—	—	3662[f]	—
	12	100	50	8	1.5	900	8	—	—	918	4208	—	—	3563	—	5770	—
	12	200	75	25	3	500		—	—	2369	2894	—	—	—	4313	5769	5771

[a] All transistors are 2Nxxxx numbers, except for the LM394 dual transistor. Devices listed on a single row are similar in characteristics and in some cases are electrically identical. [b] At V_{CB} = 10 volts.
[c] See Figure 2.76. [d] Or TO-39. [e] Or TO-72, TO-46. [f] TO-92 and its variants have two basic pinouts: EBC and ECB. Transistors with superscript *f* are ECB; all others are EBC.

an OR gate, turning off Q_1 if either door is open (switch closed). However, the collector of Q_1 stays near ground, preventing the buzzer from sounding unless switch S_3 is also closed (driver seated); in that case R_2 turns Q_3 on, putting 12 volts across the buzzer. D_3 provides a diode drop so that Q_1 is off with S_1 or S_2 closed, and D_4 protects Q_3 from the buzzer's inductive turn-off transient. In Chapter 8 we will discuss logic circuitry in detail.

Table 2.1 presents a selection of useful and popular small-signal transistors, and Figure 2.76 shows corresponding curves of current gain.

SELF-EXPLANATORY CIRCUITS

2.23 Bad circuits

A lot can be learned from your own mistakes or someone else's mistakes. In this section we present a gallery of blunders (Fig. 2.77). You can amuse yourself by thinking of variations on these bad circuits, and then avoiding them!

ADDITIONAL EXERCISES

(1) Design a transistor switch circuit that allows you to switch two loads to ground via saturated *npn* transistors. Closing switch *A* should cause both loads to be powered, whereas closing switch *B* should power only one load. Hint: Use diodes.

(2) Consider the current source in Figure 2.78. ***(a)*** What is I_{load}? What is the output compliance? Assume V_{BE} is 0.6 volt. ***(b)*** If h_{FE} varies from 50 to 100 for collector voltages within the output compliance range, how much will the output current vary? (There are two effects here.) ***(c)*** If V_{BE} varies according to $\Delta V_{BE} = -0.001\Delta V_{CE}$ (Early effect), how much will the load current vary over the compliance range? ***(d)*** What is the temperature coefficient of output current assuming that h_{FE} does not vary with temperature? What is the temperature coefficient of output current assuming that h_{FE} increases from its nominal value of 100 by 0.4%/°C?

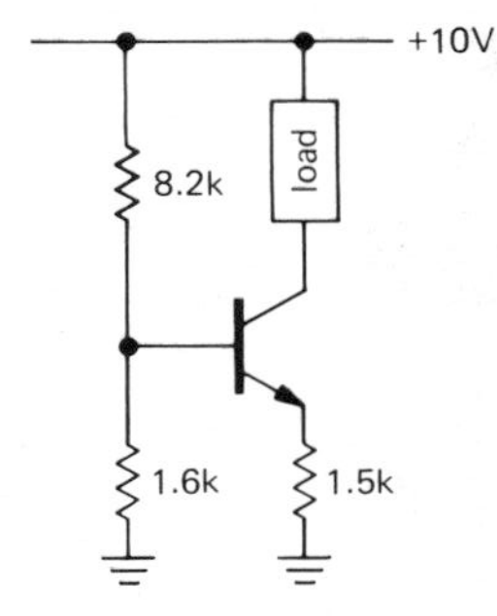

Figure 2.78

(3) Design a common-emitter *npn* amplifier with voltage gain of 15, V_{CC} of +15 volts, and I_C of 0.5mA. Bias the collector at $0.5V_{CC}$, and put the low-frequency 3dB point at 100Hz.

(4) Bootstrap the circuit in the preceding problem in order to raise the input impedance. Choose the rolloff of the bootstrap appropriately.

(5) Design a dc-coupled differential amplifier with voltage gain of 50 (to a single-ended output) for input signals near ground, supply voltages of ±15 volts, and quiescent

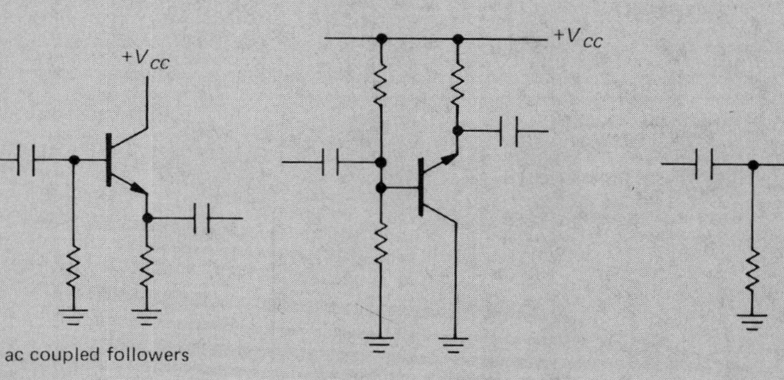

A ac coupled followers

Bad circuits

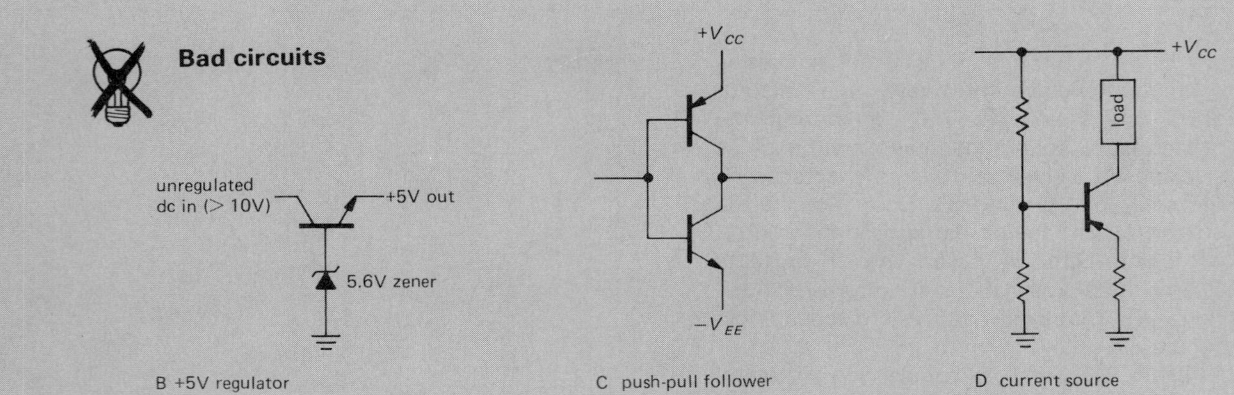

B +5V regulator

C push-pull follower

D current source

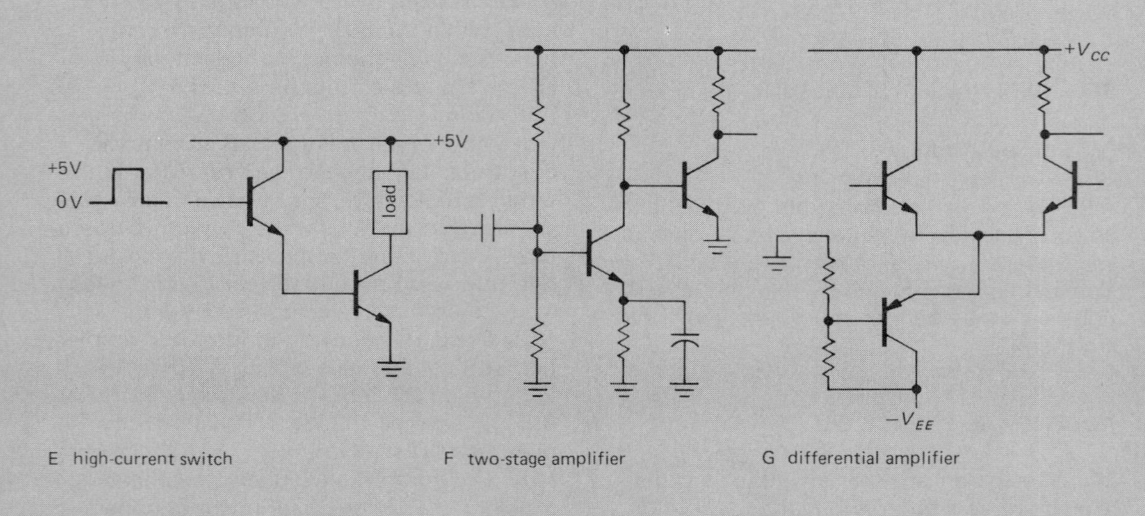

E high-current switch

F two-stage amplifier

G differential amplifier

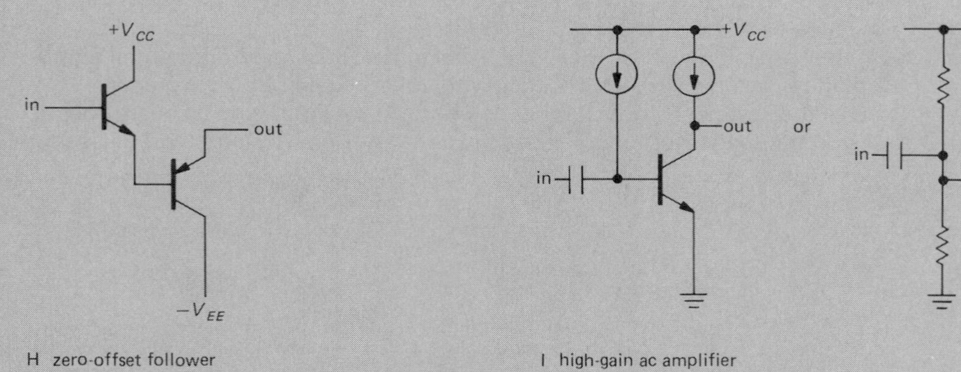

H zero-offset follower

I high-gain ac amplifier

Figure 2.77

currents of 0.1mA in each transistor. Use a current source in the emitter and an emitter follower output stage.

(6) In this problem you will ultimately design an amplifier whose gain is controlled by an externally applied voltage (in Chapter 6 you will see how to do the same thing with FETs). ***(a)*** Begin by designing a long-tailed-pair differential amplifier with emitter current source and no emitter resistors (undegenerated). Use ± 15 volt supplies. Set I_C (each transistor) at 1mA, and use R_C = 1.0k. Calculate the voltage gain from a single-ended input (other input grounded) to a single-ended output. ***(b)*** Now modify the circuit so that an externally applied voltage controls the emitter current source. Give an approximate formula for the gain as a function of controlling voltage. (In a real circuit you might arrange a second set of voltage-controlled current sources to cancel the quiescent point shift that gain changes produce in this circuit, or a differential-input second stage could be added to your circuit.)

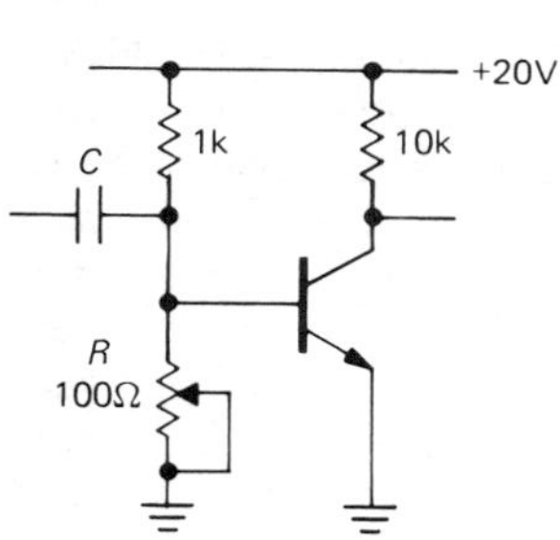

Figure 2.79

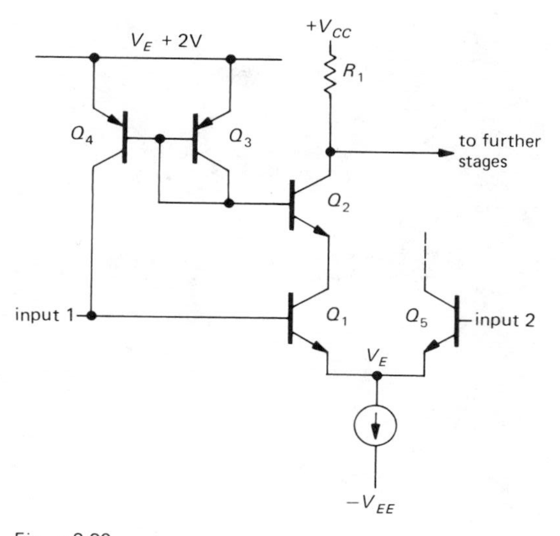

Figure 2.80

(7) Disregarding the lessons of this chapter, a disgruntled student builds the amplifier shown in Figure 2.79. He adjusts R until the quiescent point is $0.5V_{CC}$. ***(a)*** What is Z_{in} (at high frequencies where $Z_C \approx 0$)? ***(b)*** What is the small-signal voltage gain? ***(c)*** What rise in ambient temperature (roughly) will cause the transistor to saturate?

(8) Several commercially available precision op-amps (e.g., the OP-07) use the circuit in Figure 2.80 to cancel input bias current (only half of the symmetrical-input differential amplifier is shown in detail; the other half works the same way). Explain how the circuit works. Note: Q_1 and Q_2 are a beta-matched pair. Hint: It's all done with mirrors.

FEEDBACK AND OPERATIONAL AMPLIFIERS

CHAPTER 3

INTRODUCTION TO FEEDBACK AND OPERATIONAL AMPLIFIERS

Feedback has become such a well-known concept that the word has entered the general vocabulary. In control systems, feedback consists in comparing the output of the system with the desired output and making a correction accordingly. The ''system'' can be almost anything: for instance, the process of driving a car down the road, in which the output (the position and velocity of the car) is sensed by the driver, who compares it with expectations and makes corrections to the input (steering wheel, throttle, brake). In amplifier circuits the output should be a multiple of the input, so in a feedback amplifier the input is compared with an attenuated version of the output.

3.01 Introduction to feedback

Negative feedback is the process of coupling the output back in such a way as to cancel some of the input. You might think that this would only have the effect of reducing the amplifier's gain and would be a pretty stupid thing to do. Harold S. Black, who attempted to patent negative feedback in 1928, was greeted with the same response. In his words, ''Our patent application was treated in the same manner as one for a perpetual-motion machine.'' (See the fascinating article in *IEEE Spectrum,* December 1977.) True, it does lower the gain, but in exchange it also improves other characteristics, most notably freedom from distortion and nonlinearity, flatness of response (or conformity to some desired frequency response), and predictability. In fact, as more negative feedback is used, the resultant amplifier characteristics become less dependent on the characteristics of the open-loop (no-feedback) amplifier and finally depend only on the properties of the feedback network itself. Operational amplifiers are typically used in this *high-loop-gain* limit, with *open-loop* voltage gain (no feedback) of a million or so.

A feedback network can be frequency-dependent, to produce an equalization amplifier (with specific gain-versus-frequency characteristics, an example being the famous RIAA phono amplifier characteristic), or it can be amplitude-dependent, producing a nonlinear amplifier (a popular example is a logarithmic amplifier, built with feedback that exploits the logarithmic V_{BE} versus I_C of a diode or transistor). It can be arranged to produce a current source (near-infinite output impedance) or a voltage source (near-zero output impedance), and it can be connected to generate very high or very low input impedance. Speaking in general terms, the property that is sampled to produce feedback is the property that is improved. Thus, if you feed back a signal proportional to the output current, you will generate a good current source.

Feedback can also be *positive*; that's how you make an oscillator, for instance. As much fun as that may sound, it simply isn't as important as negative feedback. More often it's a nuisance, since a negative-feedback circuit may have large enough phase shifts at some high frequency to produce positive feedback and oscillations. It is surprisingly easy to have this happen, and the prevention of unwanted oscillations is the object of what is called *compensation*, a subject we will treat briefly at the end of the chapter.

Having made these general comments, we will now look at a few feedback examples with operational amplifiers.

3.02 Operational amplifiers

Most of our work with feedback will involve operational amplifiers, very high gain dc-coupled differential amplifiers with single-ended outputs. You can think of the classic long-tailed pair (Section 2.17) with its two inputs and single output as a prototype, although real op-amps have much higher gain (typically 10^5 to 10^6) and lower output impedance and allow the output to swing through most of the supply range (you usually use a split supply, most often ± 15V). Operational amplifiers are now available in literally hundreds of types, with the universal symbol shown in Figure 3.1, where the (+) and (−) inputs do as expected: The output goes positive when the noninverting input (+) goes more positive than the inverting input (−), and vice versa. The (+) and (−) symbols don't mean that you have to keep one positive with respect to the other, or anything like that; they just tell you the relative phase of the output (which is important to keep negative feedback negative). Using the words "noninverting" and "inverting," rather than "plus" and "minus," will help avoid confusion. Power-supply connections are frequently not displayed, and there is no ground terminal. Operational amplifiers have enormous voltage gain, and they are *never* used without feedback. Think of an op-amp as fodder for feedback. The open-loop gain is so high that, for any reasonable closed-loop gain, the characteristics depend only on the feedback network. Of course, at some level of scrutiny this generalization must fail. We will start with a naive view of op-amp behavior and fill in some of the finer points later, when we need to.

The most common op-amp is the μA741C, or just 741 for short. It is a wee beastie packaged in the so-called mini-DIP (dual in-line package), and it looks as shown in Figure 3.2. It is very popular, being a good performer, easy to use, inexpensive (about 25 cents), and readily available (at least a dozen companies make them). Inside is a piece of silicon containing 20 transistors and

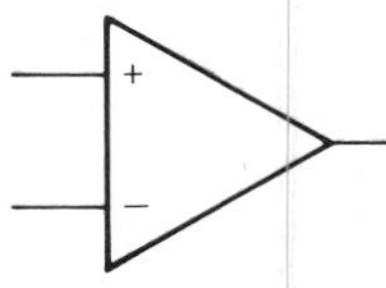

Figure 3.1

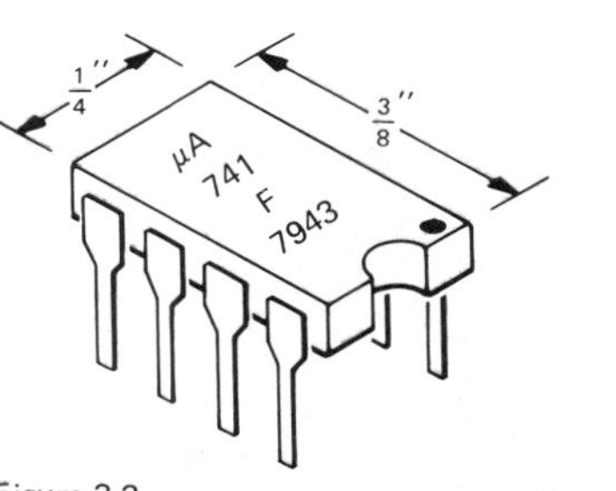

Figure 3.2
Mini-DIP integrated circuit.

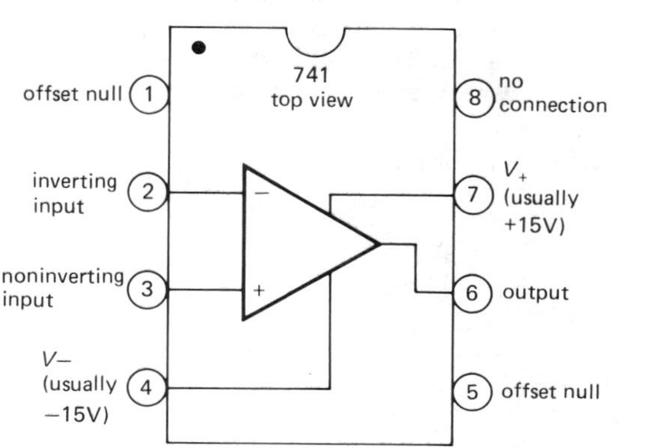

Figure 3.3

11 resistors. The pin connections are shown in Figure 3.3. The dot in the corner, or notch at the end of the package, identifies the end from which to begin counting the pin numbers. As with most electronic packages, you count pins counterclockwise, viewing from the top. The "offset null" terminals have to do with correcting (externally) the small asymmetries that are unavoidable when making the op-amp. You will learn about this later in the chapter.

3.03 The golden rules

Here are the simple rules for working out op-amp behavior with external feedback. They're good enough for almost everything you'll ever do.

First, the op-amp voltage gain is so high that a fraction of a millivolt between the input terminals will swing the output over its full range, so we ignore that small voltage and state golden rule I:

I. The output attempts to do whatever is necessary to make the voltage difference between the inputs zero.

Second, op-amps draw very little input current (0.08μA for the 741; picoamps for FET-input types); we round this off, stating golden rule II:

II. The inputs draw no current.

One important note of explanation: Golden rule I doesn't mean that the op-amp actually changes the voltage at its *inputs*. It can't do that. (How could it, and be consistent with golden rule II?) What it does is "look" at its input terminals and swing its output terminal around so that the external feedback network brings the input differential to zero (if possible).

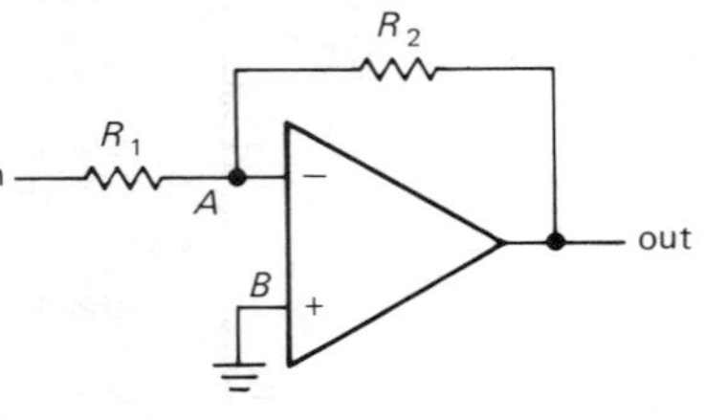

Figure 3.4

These two rules get you quite far. We will illustrate with some basic and important op-amp circuits, and these will prompt a few cautions listed in Section 3.08.

BASIC OP-AMP CIRCUITS

3.04 Inverting amplifier

Let's begin with the circuit shown in Figure 3.4. The analysis is simple, if you remember your golden rules:

1. Point *B* is at ground, so rule I implies that point *A* is also.
2. This means that (a) the voltage across R_2 is V_{out} and (b) The voltage across R_1 is V_{in}.
3. So, using rule II, we have

$$V_{out}/R_2 = -V_{in}/R_1$$

in other words,

$$\text{voltage gain} = V_{out}/V_{in} = -R_2/R_1.$$

Later you will see that it's often better not to ground *B* directly, but through a resistor. However, don't worry about that now.

Our analysis seems almost too easy! In some ways it obscures what is actually happening. To understand how feedback works, just imagine some input level, say +1 volt. For concreteness, imagine that R_1 is 10k and R_2 is 100k. Now, suppose the output decides to be uncooperative, and sits at zero volts. What happens? R_1 and R_2 form a voltage divider, holding the inverting input at +0.91 volt. The op-amp sees an enormous input unbalance, forcing the output to go negative. This action continues until the output is at the required −10.0 volts, at which point both op-amp inputs are at the same voltage, namely ground. Similarly, any tendency for the output to go more negative than −10.0 volts will pull the inverting input below ground, forcing the output voltage to rise.

What is the input impedance? Simple. Point *A* is always at zero volts (it's called a *virtual ground*). So $Z_{in} = R_1$. At this point you don't yet know how to figure the output impedance; for this circuit, it's a fraction of an ohm.

Note that this analysis is true even for dc – it's a dc amplifier. So if you have a signal source offset from ground (collector of a previous stage, for instance), you may want to use a coupling capacitor (sometimes called a blocking capacitor, since it blocks dc but couples the signal). For reasons you will see later (having to do with departures of op-amp behavior from the ideal), it is usually a good idea to use a blocking capacitor if you're only interested in ac signals anyway.

This circuit is known as an *inverting amplifier.* Its one undesirable feature is the low input impedance, particularly for amplifiers with large (closed-loop) voltage gain, where R_1 tends to be rather small. That is remedied in the next circuit (Fig. 3.5).

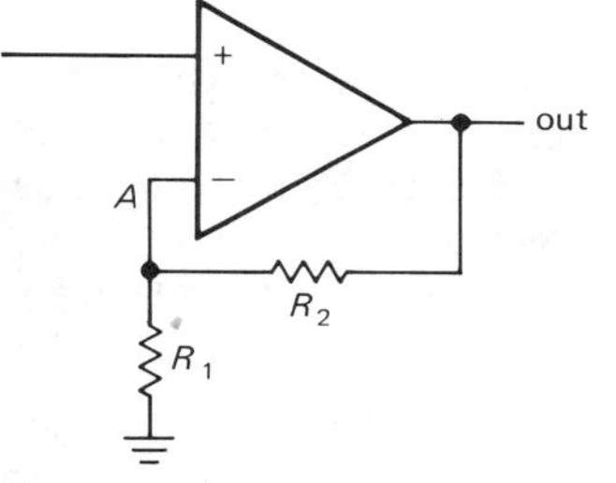

Figure 3.5

3.05 Noninverting amplifier

Consider Figure 3.5. Again, the analysis is simplicity itself:

$V_A = V_{in}$

But V_A comes from a voltage divider:

$V_A = V_{out}R_1/(R_1 + R_2)$

Set $V_A = V_{in}$, and you get

gain $= V_{out}/V_{in} = 1 + R_2/R_1$

This is a *noninverting amplifier.* In the approximation we are using, the input impedance is infinite (with a 741 it would be hundreds of megohms, at least; a FET op-amp might have $Z_{in} \approx 10^{12}\Omega$ or more). The output impedance is still a fraction of an ohm. As with the inverting amplifier, a detailed look at the voltages at the inputs will persuade you that it works as advertized.

Once again we have a dc amplifier. If the signal source is ac-coupled, you must provide a return to ground for the (very small) input current, as in Figure 3.6. The component values shown give a voltage gain of 10 and a low-frequency 3dB point of 16Hz.

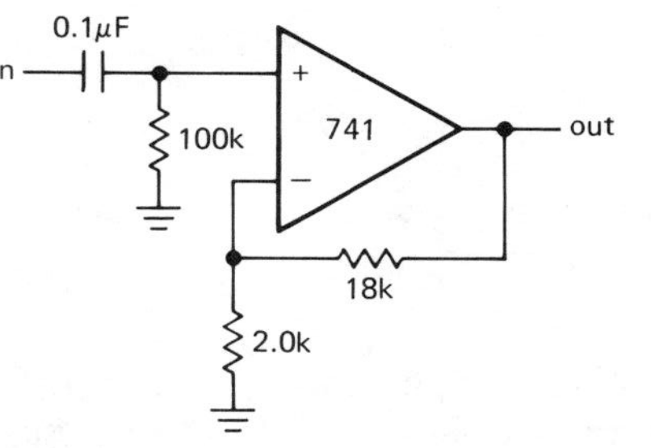

Figure 3.6

ac Amplifier

Again, if only ac signals are being amplified, it is often a good idea to "roll off" the gain to unity at dc, especially if the amplifier has large voltage gain, in order to reduce the effects of finite "input offset voltage." The circuit in Figure 3.7 has a low-frequency 3dB point of 17Hz, the frequency at which the impedance of the capacitor equals 2.0k. Note the large capacitor value required. For noninverting amplifiers with high gain, the capacitor in this ac amplifier configuration may be undesirably large. In that case it may be preferable to omit the capacitor and trim the offset voltage to zero, as we will discuss later (Section 3.12). An alternative is to raise R_1 and R_2, perhaps using a T network for the latter (Section 3.17).

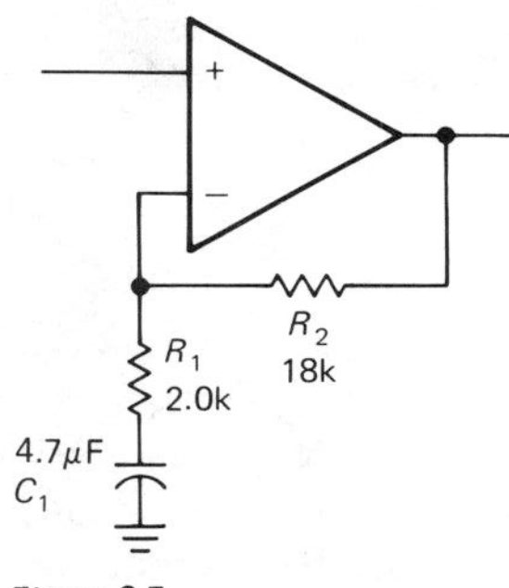

Figure 3.7

3.06 Follower

Figure 3.8 shows the op-amp version of an emitter follower. It is simply a noninverting

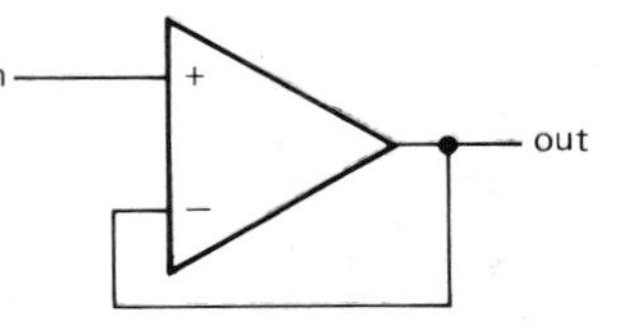

Figure 3.8

amplifier with R_1 infinite and R_2 zero (gain = 1). There are special op-amps, usable only as followers, with improved characteristics (mainly higher speed), e.g., the LM310.

An amplifier of unity gain is sometimes called a *buffer* because of its isolating properties (high input impedance, low output impedance).

3.07 Current sources

The circuit in Figure 3.9 approximates an ideal current source, without the V_{BE} offset

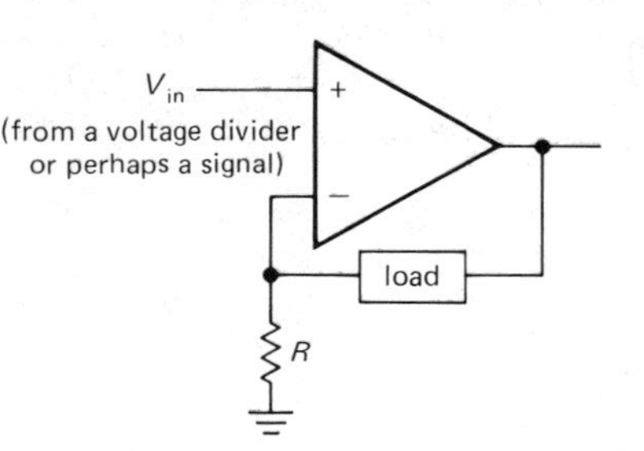

Figure 3.9

of a transistor current source. Negative feedback results in V_{in} at the inverting input, producing a current $I = V_{in}/R$ through the load. The major disadvantage of this circuit is the ''floating'' load (neither side grounded). You couldn't generate a usable sawtooth wave with respect to ground with this current source, for instance. One solution is to float the whole circuit (power supplies and all) so that you can ground one side of the load (Fig. 3.10). The circuit in the box is the previous current source, with its power supplies shown explicitly. R_1 and R_2 form a voltage divider to set the current. If this circuit seems confusing, it may help to remind yourself that ''ground'' is a relative concept. Any one point in a circuit could be called ground. This circuit is useful for generating currents into a load that is returned to ground, but it has the disadvantage that the

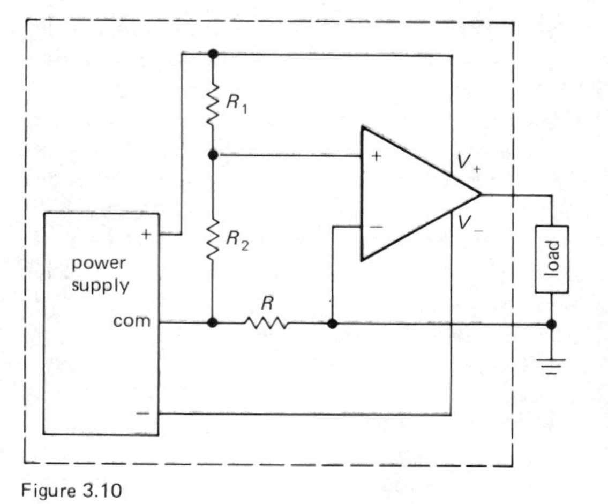

Figure 3.10
Current source with grounded load and floating power supply.

control input is now floating, so you cannot program the output current with an input voltage referenced to ground. Some solutions to this problem are presented in Chapter 5 in the discussion of constant-current power supplies.

Current sources for loads returned to ground

With an op-amp and external transistor it is possible to make a simple high-quality current source for a load returned to ground; a little additional circuitry makes it possible to use a programming input referenced to ground (Fig. 3.11). In the first circuit, feedback forces a voltage $V_{CC} - V_{in}$ across R, giving an emitter current (and therefore an output current) $I_E = (V_{CC} - V_{in})/R$. There are no V_{BE} offsets, or their variations with temperature, I_C, V_{CE}, etc., to worry about. The current source is imperfect only insofar as the small base current may vary somewhat with V_{CE} (assuming the op-amp draws no input current), not too high a price to pay for the convenience of a grounded load; a Darlington for Q_1 would reduce this error considerably. This error comes about, of course, because the op-amp stabilizes the *emitter* current, whereas the load sees the *collector* current. A variation of this circuit, using a FET instead of a bipolar transistor, avoids this problem altogether, since FETs draw no gate current (the FET's gate is the analog of the transistor's base).

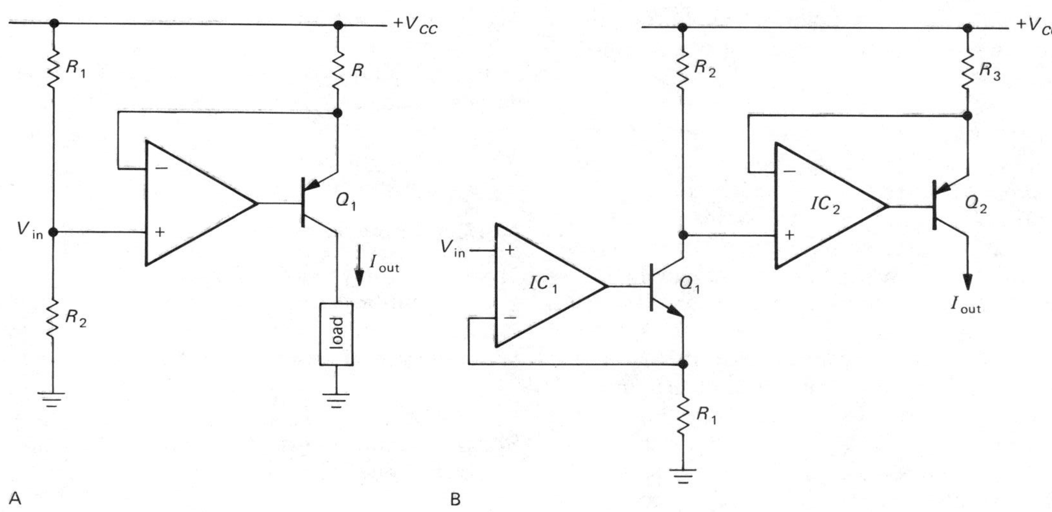

Figure 3.11
Current sources for grounded loads that don't require a floating power supply.

With this circuit the output current is proportional to the voltage drop below V_{CC} applied to the op-amp's noninverting input; in other words, the programming voltage is referenced to V_{CC}, which is fine if V_{in} is a fixed voltage generated by a voltage divider, but an awkward situation if an external input is to be used. This is remedied in the second circuit, in which a similar current source with *npn* transistor is used to convert an input voltage (referenced to ground) to a V_{CC}-referenced input to the final current source. Op-amps and transistors are inexpensive. Don't hesitate to use a few extra components to improve performance or convenience in circuit design.

One important note about the last circuit: The op-amp must be able to operate with its inputs near or at the positive supply voltage. An op-amp like the 307 or 355 is good here. Alternatively, the op-amp could be powered from a separate V_+ voltage higher than V_{CC}.

EXERCISE 3.1

What is the output current in the last circuit for a given input voltage V_{in}?

Howland current source

Figure 3.12 shows a nice "textbook" current source. If the resistors are chosen so that $R_3/R_2 = R_4/R_1$, then it can be shown that $I_{load} = -V_{in}/R_2$.

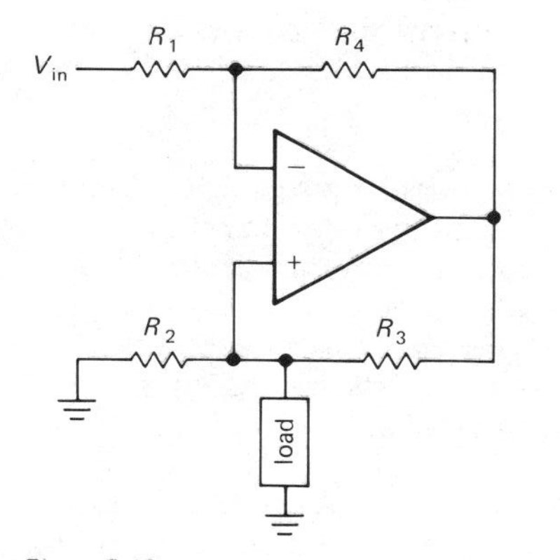

Figure 3.12

EXERCISE 3.2

Show that the preceding result is correct.

This sounds great, but there's a hitch: The resistors must be matched exactly; otherwise it isn't a perfect current source. Even so, its performance is limited by the CMRR of the op-amp. For large output currents, the resistors must be small, and the compliance is limited. As clever as it looks, it's rarely used.

3.08 Basic cautions for op-amp circuits

1. In all op-amp circuits golden rules I and II (Section 3.03) will be obeyed only if the op-amp is in the active region, i.e., inputs and outputs not saturated at one of the supply voltages.

For instance, overdriving one of the amplifier configurations will cause output clipping at output swings near V_{CC} or V_{EE}. During clipping the inputs will no longer be maintained at the same voltage. The op-amp output cannot swing beyond the supply voltages (typically it can swing only to within 2V of the supplies). Likewise, the output compliance of an op-amp current source is set by the same limitation. The current source with floating load, for instance, can put a maximum of $V_{CC} - V_{in}$ across the load in the "normal" direction (current in the same direction as applied voltage) and $V_{in} - V_{EE}$ in the reverse direction (the load could be rather strange, e.g., it might contain batteries, requiring the reverse sense of voltage to get a forward current; the same thing might happen with an inductive load driven by changing currents).

2. The feedback must be arranged so that it is negative. This means (among other things) that you must not mix up the inverting and noninverting inputs.
3. There must always be feedback at dc in an op-amp circuit. Otherwise the op-amp is guaranteed to go into saturation.

For instance, we were able to put a capacitor from the feedback network to ground in the noninverting amplifier (to reduce gain at dc to 1), but we could not similarly put a capacitor in series between the output and the inverting input.

4. Many op-amps have a relatively small maximum differential input voltage limit. The maximum voltage difference between the inverting and noninverting inputs might be limited to as little as 5 volts in either polarity. Breaking this rule will cause large input currents to flow, with degradation or destruction of the op-amp.

We will take up some more issues of this type in Section 3.11, and again in Section 7.06 in connection with precision circuit design.

AN OP-AMP SMORGASBORD

In the following examples we will skip the detailed analysis, leaving that fun for you, the reader.

3.09 Linear circuits

Optional inverter

The circuits in Figure 3.13 let you invert, or amplify without inversion, by flipping a switch. The voltage gain is either +1 or −1, depending on the switch position.

EXERCISE 3.3

Show that the circuits in Figure 3.13 work as advertised.

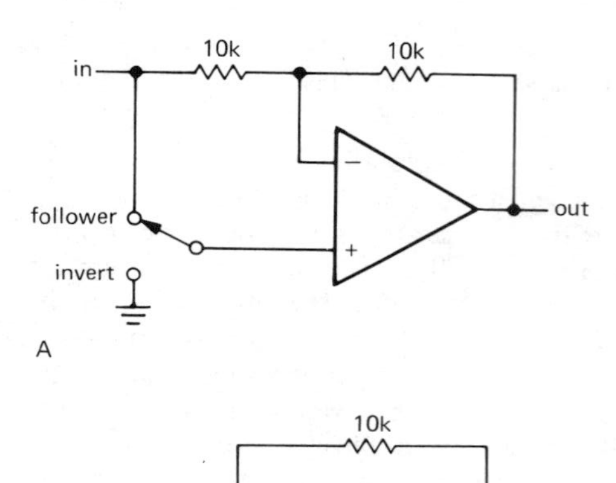

Figure 3.13

Follower with bootstrap

As with transistor amplifiers, the bias path can compromise the high input impedance you would otherwise get with an op-amp, particularly with ac-coupled inputs, where a resistor to ground is mandatory. If that is a problem, the bootstrap circuit shown in Figure 3.14 is a possible solution. As in the transistor bootstrap circuit (Section 2.16), the 1μF capacitor makes the upper 100k

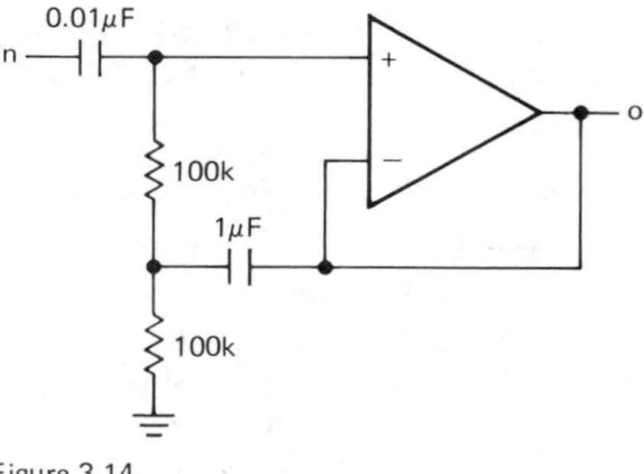

Figure 3.14

resistor look like a high-impedance current source to input signals. The low-frequency rolloff for this circuit will begin at about 10Hz, and the response will drop at 12dB per octave for frequencies somewhat below this, where both capacitors contribute to the rolloff.

Ideal current-to-voltage converter

Remember that the humble resistor is the simplest *I*-to-*V* converter. However, it has the disadvantage of presenting a nonzero impedance to the source of input current; this can be fatal if the device providing the input current has very little compliance or does not produce a constant current as the output voltage changes. A good example is a *photovoltaic cell,* a fancy name for a sun battery. Even the garden-variety signal diodes you use in circuits have a small photovoltaic effect (there are amusing stories of bizarre circuit behavior finally traced to this effect). Figure 3.15 shows the good way to convert current to voltage while holding the input strictly at ground. The inverting input is a virtual ground; this is fortunate, since a photovoltaic diode can generate only a few tenths of a volt. This particular circuit has an output of 1 volt per microamp of input current. The resistor from the noninverting input to ground could be omitted, but it is highly desirable; its function will be explained shortly in connection with op-amp shortcomings.

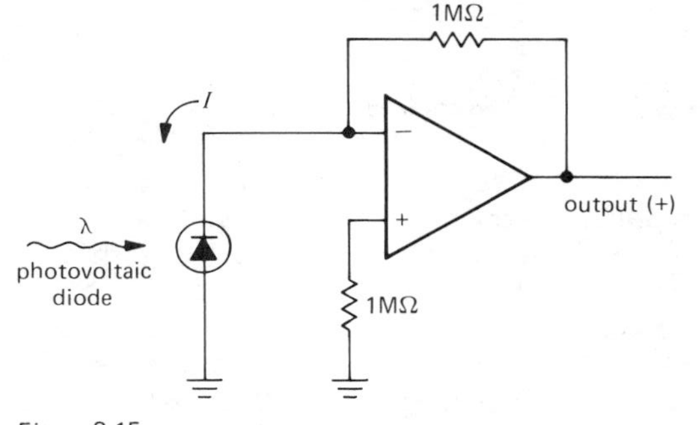

Figure 3.15

Of course, this *transresistance* configuration can be used equally well for devices that source their current via some positive excitation voltage, such as V_{CC}. Photomultiplier tubes and phototransistors (both devices that source current from a positive supply when exposed to light) are often used this way (Fig. 3.16).

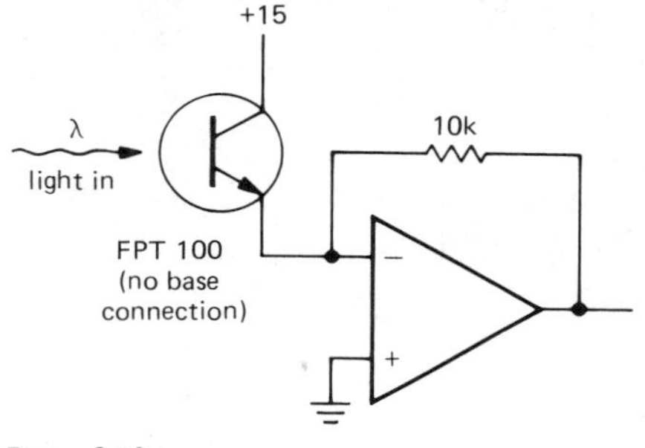

Figure 3.16

EXERCISE 3.4

Use a 741 and a 1mA (full scale) meter to construct a "perfect" current meter (i.e., one with zero input impedance) with 5mA full scale. Design the circuit so that the meter will never be driven more than ±150% full scale. Assume that the 741 output can swing to ±13 volts (±15V supplies) and that the meter has 500 ohms internal resistance.

Differential amplifier

The circuit in Figure 3.17 is a differential amplifier with gain R_2/R_1. As with the

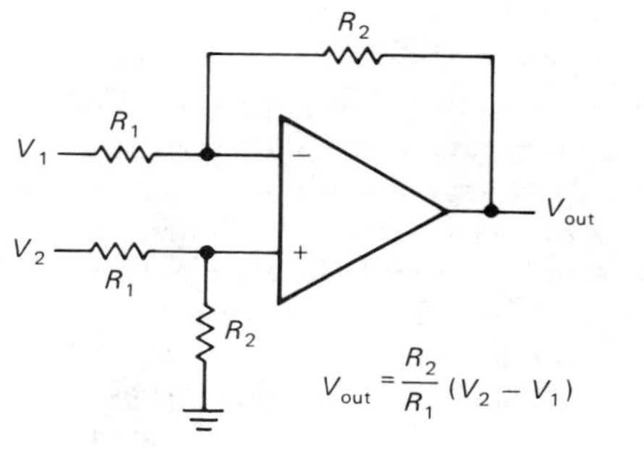

Figure 3.17

current source that used matched resistor ratios, this circuit requires precise resistor matching to achieve high common-mode rejection ratios. The best procedure is to stock up on a bunch of 100k 0.01% resistors next time you have a chance. All your differential amplifiers will have unity gain, but that's easily remedied with further (single-ended) stages of gain. We will treat differential amplifiers in more detail in Chapter 7.

Summing amplifier

The circuit shown in Figure 3.18 is just a variation of the inverting amplifier. Point X is

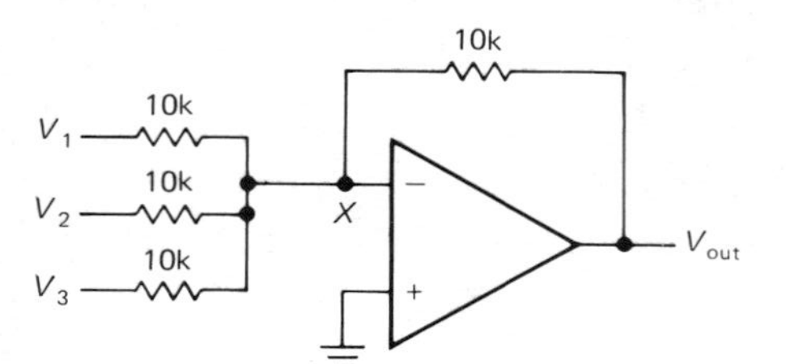

Figure 3.18

a virtual ground, so the input current is $V_1/R + V_2/R + V_3/R$. That gives $V_{out} = -(V_1 + V_2 + V_3)$. Note that the inputs can be positive or negative. Also, the input resistors need not be equal; if they're unequal, you get a weighted sum. For instance, you could have four inputs, each of which is +1 volt or zero, representing binary values 1, 2, 4, and 8. By using input resistors of 10k, 5k, 2.5k, and 1.25k you will get an output in volts equal to the binary count input. This scheme can be easily expanded to several digits. It is the basis of digital-to-analog conversion, although a different input circuit (an R-$2R$ ladder) is usually used.

EXERCISE 3.5

Show how to make a two-digit digital-to-analog converter by appropriately scaling the input resistors in a summing amplifier. The digital input represents two digits, each consisting of four lines that represent the values 1, 2, 4, and 8 for the respective digit. An input line is either at +1 volt or at ground, i.e., the eight input lines represent 1, 2, 4, 8, 10, 20, 40, and 80. Since op-amp outputs generally cannot swing beyond ±13V, you will have to settle for an output in volts equal to one-tenth the value of the input number.

RIAA preamp

The RIAA preamp is an example of an amplifier with a specifically tailored frequency response. Phonograph records are cut with approximately flat amplitude characteristics; magnetic pickups, on the other hand, respond to velocity, so a playback amplifier with rising bass response is required. The circuit shown in Figure 3.19 produces the

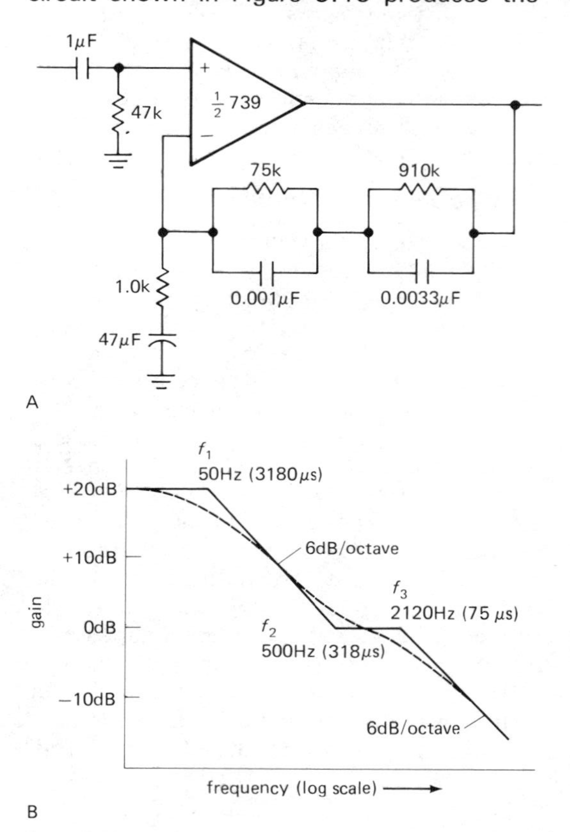

Figure 3.19
Op-amp RIAA phono playback amplifier.

required response. The RIAA playback amplifier frequency response (relative to 0dB at 1kHz) is shown in the graph, with the breakpoints given in terms of time constants. The 47μF capacitor to ground rolls off the gain to unity at dc, where it would otherwise be about 1000; as we have hinted earlier, the reason is to avoid amplification of dc input "offsets." The 739 is a low-noise dual op-amp intended for audio applications.

Power booster

For high output current, a power transistor follower can be hung on an op-amp output (Fig. 3.20). In this case a noninverting

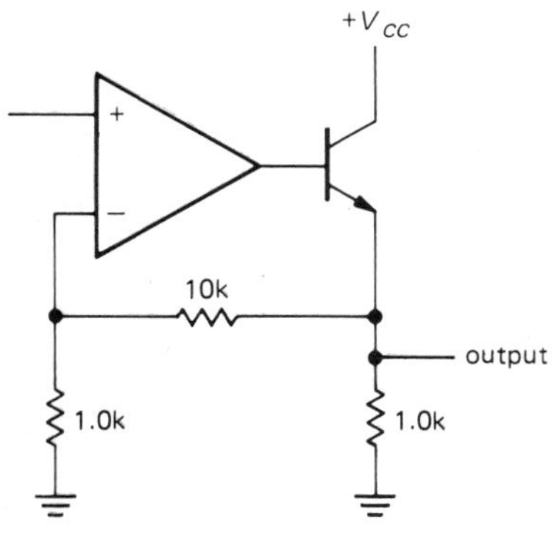

Figure 3.20

amplifier has been drawn; the follower can be added to any op-amp configuration. Notice that feedback is taken from the emitter; thus feedback enforces the desired output voltage in spite of the V_{BE} drop. This circuit has the usual problem that the follower output can only source current. As with transistor circuits, the remedy is a push-pull booster (Fig. 3.21). You will see later that the limited speed with which the op-amp can move its output (slew rate) seriously limits the speed of this booster in the crossover region, creating distortion. For slow-speed applications you don't need to bias the push-pull pair into quiescent conduction, because feedback will take care

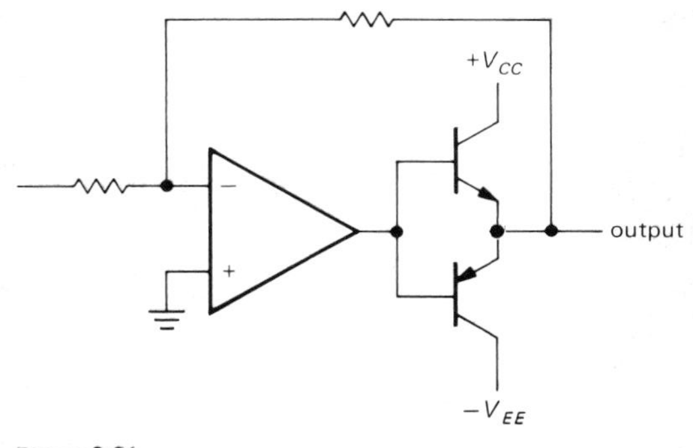

Figure 3.21

of most of the crossover distortion. Commercial op-amp power boosters are available, e.g., the MC1438, LH0063, and 3553. These are unity-gain push-pull amplifiers capable of 200mA of output current and operation to 100 MHz and above. You can include them inside the feedback loop without any worries (see Table 7.3).

Power supply

An op-amp can provide the gain for a feedback voltage regulator (Fig. 3.22). The op-

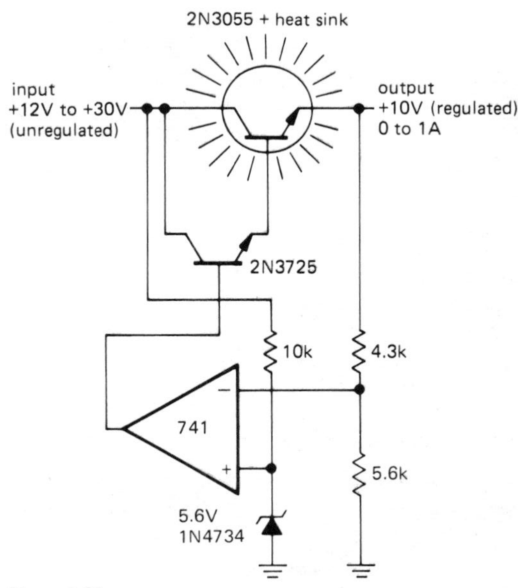

Figure 3.22

amp compares a sample of the output with the zener reference, changing the drive to the Darlington "pass transistor" as needed. This circuit supplies 10 volts regulated, at up to 1 amp load current. Some notes about this circuit:

1. The voltage divider that samples the output could be a potentiometer, for adjustable output voltage.
2. For reduced ripple at the zener, the 10k resistor should be replaced by a current source. Another approach is to bias the zener from the output; that way you take advantage of the regulator you have built. Caution: When using this trick, you must analyze the circuit carefully to be sure it will start up when power is first applied.

3. The circuit as drawn could be damaged by a temporary short circuit across the output, because the op-amp would attempt to drive the Darlington pair into heavy conduction. Regulated power supplies should always have circuitry to limit "fault" current (see Chapter 5 for more details).
4. Integrated circuit voltage regulators are available in tremendous variety, from the time-honored 723 to the recent 3-terminal adjustable regulators with internal current limit and thermal shutdown (see Tables 5.7–5.9). These devices, complete with temperature-compensated internal zener reference and pass transistor, are so easy to use that you will almost never use a general-purpose op-amp as a regulator. The exception might be to generate a stable voltage within a circuit that already has a stable power-supply voltage available.

In Chapter 5 we will discuss voltage regulators and power supplies in detail, including special ICs intended for use as voltage regulators.

□ 3.10 Nonlinear circuits

□ *Power-switching driver*

For loads that are either on or off, a switching transistor can be driven from an op-amp. Figure 3.23 shows how. Note the diode to

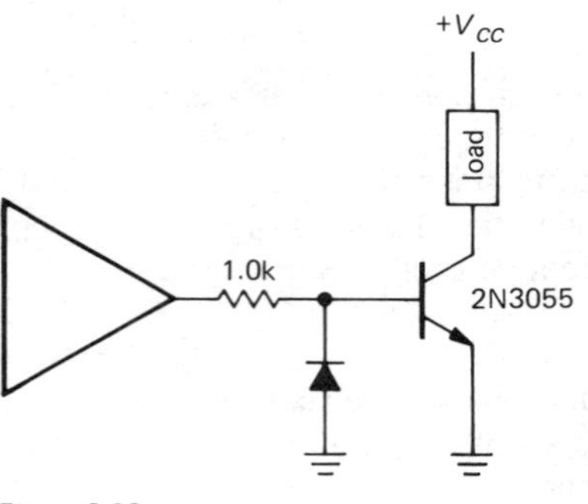

Figure 3.23

prevent reverse base-emitter breakdown (op-amps easily swing more than −5V). The 2N3055 is everyone's power transistor for noncritical high-current applications. A Darlington can be used if currents greater than about 1 amp need to be driven.

□ *Active Rectifier*

Rectification of signals smaller than a diode drop cannot be done with a simple diode-resistor combination. As usual, op-amps come to the rescue, in this case by putting a diode in the feedback loop (Fig. 3.24). For

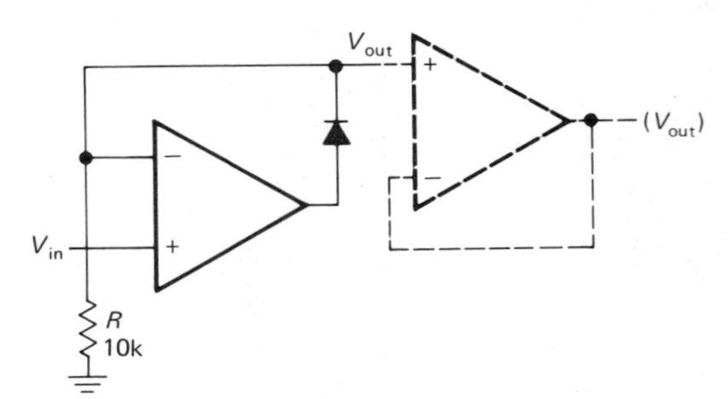

Figure 3.24

V_{in} positive, the diode provides negative feedback; the output follows the input, coupled by the diode, but without a V_{BE} drop. For V_{in} negative, the op-amp goes into negative saturation, and V_{out} is at ground. R could be chosen smaller for lower output impedance, with the tradeoff of higher op-amp output current. A better solution is to use an op-amp follower at the output, as shown, to produce very low output impedance regardless of the resistor value.

There is a problem with this circuit that becomes serious with high-speed signals. Because an op-amp cannot swing its output infinitely fast, the recovery from negative saturation (as the input waveform passes through zero from below) takes some time, during which the output is incorrect. It looks something like the curve shown in Figure 3.25. The output (heavy line) is an accurate rectified version of the input (light line),

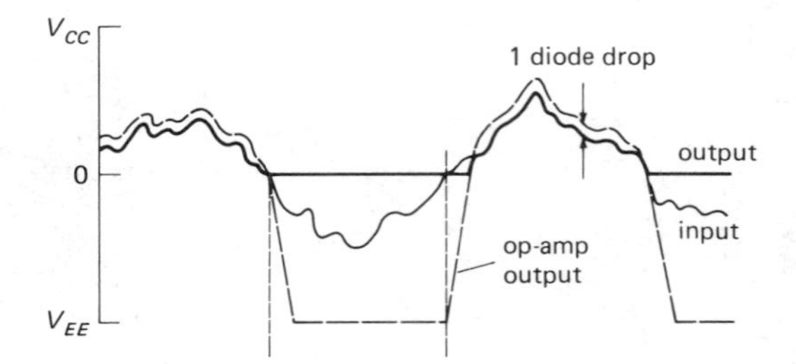

Figure 3.25

except for a short time interval after the input rises through zero volts. During that interval the op-amp output is racing up from saturation near $-V_{EE}$, so the circuit's output is still at ground. A general-purpose op-amp like the 741 has a *slew rate* of 0.5 volt per microsecond; recovery from negative saturation therefore takes about 30μs, which is a long time on the scale of signals in the kilohertz range and is hopeless for signals of microsecond speed. A circuit modification improves the situation considerably (Fig. 3.26).

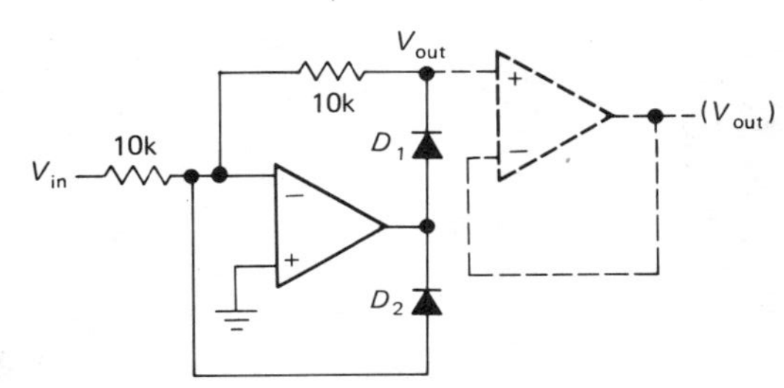

Figure 3.26

D_1 makes the circuit a unity-gain inverter for negative input signals. D_2 clamps the op-amp's output at one diode drop below ground for positive inputs, and since D_1 is then back-biased, V_{out} sits at ground. The improvement comes because the op-amp's output swings only two diode drops as the input signal passes through zero. Since the op-amp output has to slew only about 1.2 volts instead of V_{EE} volts, the "glitch" at zero crossings is reduced more than tenfold. This rectifier is inverting, incidentally. If you require a noninverted output, attach a unity-gain inverter to the output.

The performance of these circuits is improved if you choose an op-amp with a high slew rate. Slew rate also influences the performance of the other op-amp applications we've discussed, for instance the simple voltage amplifier circuits. At this point it is worth pausing for a while to see in what ways real op-amps depart from the ideal, since that influences circuit design, as we have hinted on several occasions. A good understanding of op-amp limitations and their influence on circuit design and performance will help you choose your op-amps wisely and design with them effectively.

A DETAILED LOOK AT OP-AMP BEHAVIOR

Figure 3.27 shows the schematic of the 741, a very popular op-amp. Its circuit is relatively straightforward, in terms of the kind of transistor circuits we discussed in the last chapter. It has a differential input stage with current mirror load, followed by a common-emitter *npn* stage (again with active load) that provides most of the voltage gain. A *pnp* emitter follower drives the push-pull emitter follower output stage, which includes current-limiting circuitry. This circuit is typical of many op-amps now available. For many applications the properties of these amplifiers approach ideal op-amp performance characteristics. We will now take a look at the extent to which real op-amps depart from the ideal, what the consequences are for circuit design, and what to do about it.

3.11 Departures from ideal op-amp performance

The ideal op-amp has these characteristics:

1. Input impedance (differential or common mode) = infinity
2. Output impedance (open loop) = 0
3. Voltage gain = infinity
4. Common-mode voltage gain = 0
5. $V_{out} = 0$ when both inputs are at the same voltage (zero "offset voltage")
6. Output can change instantaneously (infinite slew rate)

All of these characteristics are independent of temperature and supply voltage changes.

Real op-amps depart from these characteristics in the following ways (see Table 3.1 for some typical values).

Input current

The input terminals sink (or source, depending on the op-amp type) a small current called the input bias current, I_B, which is defined as half the sum of the input currents with the inputs tied together (the two input currents are approximately equal and are simply the base currents of the input transistors). For the 741 the bias current is typi-

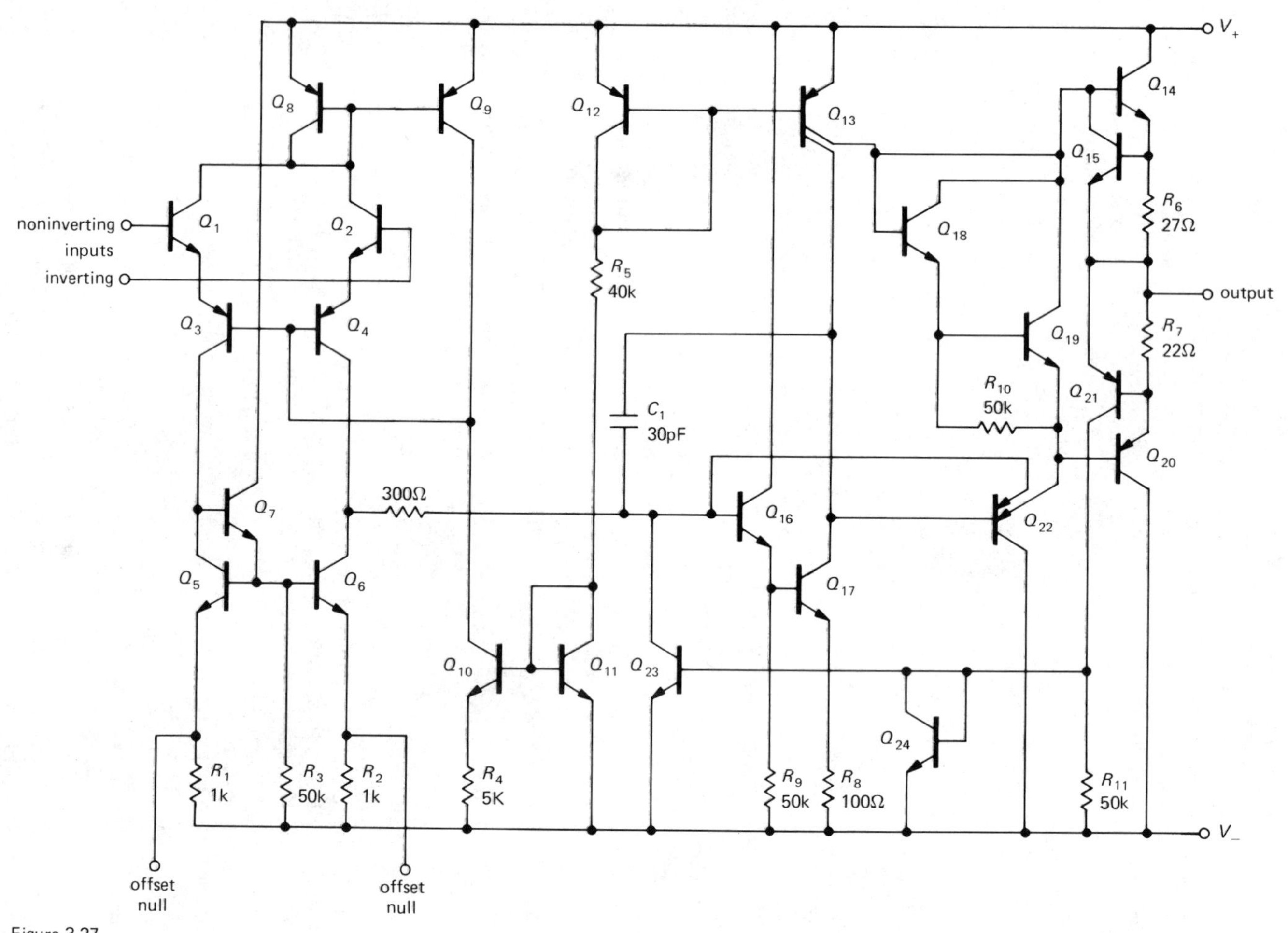

Figure 3.27

Schematic of the popular 741 op-amp. (Fairchild Camera and Instrument Corp.)

cally 80nA (0.08μA). The significance of input bias current is that it causes a voltage drop across the resistors of the feedback network, bias network, or source impedance. How small a resistor this restricts you to depends on the dc gain of your circuit and how much output variation you can tolerate. You will see how this works later.

Op-amps are available with input bias currents down to a nanoamp or less for (bipolar) transistor-input circuit types or down to a few picoamps (10^{-6}μA) for FET-input circuit types. The very lowest bias currents are typified by the superbeta Darlington LM11, with an input current of 25pA, and the MOSFET ICH8500, with an input current of 0.01pA. In general, transistor op-amps intended for high-speed operation have higher bias currents.

Input offset current

Input offset current is a fancy name for the difference in input currents between the two inputs. Unlike input bias current, the offset current, I_{os}, is a result of manufacturing variations, since an op-amp's symmetrical input circuit would otherwise result in identical bias currents at the two inputs. The significance is that even when it is driven by identical source impedances the op-amp will see unequal voltage drops and hence a difference voltage between its inputs. You will see shortly how this influences design.

Typically, the offset current is about one-tenth the bias current. For the 741, I_{offset} = 10nA, typical.

□ *Input impedance*

Input impedance refers to the differential input resistance (impedance looking into one input, with the other input grounded), which is usually much less than the common-mode resistance (a typical input stage looks like a long-tailed pair with current source). For the 741 it is about 2MΩ. FET-input op-amps typically have $R_{in} \approx 10^{12}$ ohms or more. Because of the input bootstrapping effect of negative feedback (it attempts to keep both inputs at the same voltage, thus eliminating most of the differential input signal), Z_{in} in practice is raised to very high values and usually is not as important a parameter as input bias current.

Common-mode input range

The inputs to an op-amp must stay within a certain voltage range, typically less than the full supply range, for proper operation. If the inputs go beyond this range, the gain of the op-amp may change drastically, even reversing sign! For a 741 operating from ±15 volt supplies, the common-mode input range is ±12 volts. There are op-amps available with common-mode input ranges down to the negative supply (e.g., the LM358, a dual op-amp in mini-DIP package, or the 3130/3140 series) and up to the positive supply (the 301/307, or the 355-357 series, for instance). In addition, there are maximum allowable input voltages beyond which damage will result. For the 741, they are ±15 volts or the supply voltages, whichever are less.

Differential input range

Some op-amps allow only a limited voltage between the inputs, sometimes as small as ±0.5 volt, although most are more forgiving, permitting differential inputs nearly as large as the supply voltages. Exceeding the specified maximum can degrade or destroy the op-amp.

□ *Output impedance; output swing versus load resistance*

Output impedance R_o means the op-amp's intrinsic output impedance *without feedback*. For the 741 it is about 75 ohms, but with some low-power op-amps it can be as high as several thousand ohms (see Fig. 7.9). Feedback lowers the output impedance into insignificance (or raises it, for a current source); so what usually matters more is the maximum output current, with typical values of 20mA or so. This is frequently given as a graph of output voltage swing V_{om} as a function of load resistance, or sometimes just a few values for typical load resistances. For the ubiquitous 741, output swings to within about 2 volts of V_{CC} and V_{EE} are possible into load resistances greater than about 2k. Load resistances significantly less than that will permit only a small swing. Some op-amps can produce output swings all the way down to the negative supply (e.g., the LM358), a particularly useful feature for circuits operated from a single positive supply, since output swings all the way to ground are then possible. Finally, op-amps with MOS transistor outputs (e.g., the CA3130 and CA3160) can swing all the way to both supply rails.

□ *Voltage gain and phase shift*

Typically the voltage gain A_{vo} at dc is 100,000 to 1,000,000 (often specified in decibels), dropping to unity gain at a frequency (called f_T) of 1MHz to 10MHz. This is usually given as a graph of open-loop voltage gain as a function of frequency. For *internally compensated* op-amps this graph is simply a 6dB/octave rolloff beginning at some fairly low frequency (for the 741 it begins at about 100Hz), an intentional characteristic necessary for stability, as you will see in Section 3.31. This rolloff (the same as a simple *RC* low-pass filter) results in a constant 90° lagging phase shift from input to output (open-loop) at all frequencies above the beginning of the rolloff, increasing to 120° to 160° as the open-loop gain approaches unity. Since a 180° phase shift at a frequency where the voltage gain equals 1 will result in positive feedback (oscillations), the term "phase margin" is used to specify the difference between the phase shift at f_T and 180°.

Input offset voltage

Op-amps don't have perfectly balanced input stages, owing to manufacturing variations. If you connect the two inputs together for zero input signal, the output will usually saturate at either V_{CC} or V_{EE} (you can't predict which). The difference in input voltages necessary to bring the output to zero is called the input offset voltage V_{os} (it's as if there were a battery of that voltage in series with one of the inputs). Usually op-amps make provision for trimming the input offset voltage to zero. For a 741 you use a 10k pot between pins 1 and 5, with the wiper connected to V_{EE}.

Of greater importance for precision applications is the drift of the input offset voltage with temperature and time, since any initial offset can be trimmed to zero. A 741 has a typical offset voltage of 2mV (6mV maximum) and unspecified coefficients of offset drift with temperature and time. The OP-07, a precision op-amp, is laser-trimmed for a typical offset of 30 *micro*volts, with temperature coefficient TCV_{os} of 0.2μV/°C and long-term drift of 0.2μV/month.

Slew rate

The op-amp "compensation" capacitance and small internal drive currents act together to limit the rate at which the output can change, even when a large input unbalance occurs. This limiting speed is usually specified as *slew rate* or slewing rate (SR). For the 741 it is 0.5V/μs; a high-speed op-amp might slew at 100V/μs, and the LH0063C "damn fast buffer" slews at 6000V/μs. The slew rate limits the amplitude of a sine-wave output swing above some critical frequency (the frequency at which the full supply swing requires the maximum slew rate of the op-amp), thus explaining the "output voltage swing as a function of frequency" graph. A sine wave of frequency f hertz and amplitude A volts requires a minimum slew rate of $2\pi Af$ volts per second.

For externally compensated op-amps the slew rate depends on the compensation network used. In general, it will be lowest for "unity gain compensation," increasing to perhaps 30 times faster for ×100 gain compensation. This is discussed further in Section 3.31.

Temperature dependence

All these parameters have some temperature dependence. However, this usually doesn't make any difference, since small variations in gain, for example, are almost entirely compensated by feedback. Furthermore, the variations of these parameters with temperature are typically small compared with the variations from unit to unit.

The exceptions are input offset voltage and input offset current. This will matter, particularly if you've trimmed the offsets approximately to zero, and will appear as drifts in the output. When high precision is important, a low-drift "instrumentation" op-amp should be used, with external loads kept above 10k to minimize the horrendous effects on input-stage performance caused by temperature gradients. We will have much more to say about this subject in Chapter 7.

For completeness, we should mention here that op-amps are also limited in common-mode rejection ratio (CMRR), power-supply rejection ratio (PSRR), input noise voltage and current (e_n, i_n), and output crossover distortion. These become significant limitations only in connection with precision circuits and low-noise amplifiers, and they will be treated in Chapter 7.

3.12 Effects of op-amp limitations on circuit behavior

Let's go back and look at the inverting amplifier with these limitations in mind. You will see how they affect performance, and you will learn how to design effectively in spite of them. With the understanding you will get from this example, you should be able to handle other op-amp circuits. Figure 3.28 shows the circuit again.

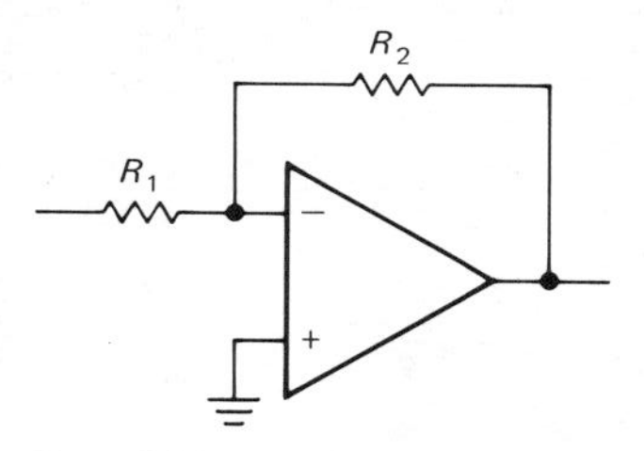

Figure 3.28

Open-loop gain

Because of finite open-loop gain, the voltage gain of the amplifier with feedback (closed-loop gain) will begin dropping at a frequency where the open-loop gain approaches R_2/R_1 (Fig. 3.29). For garden-variety op-amps like the 741, this means that you're dealing with a relatively low frequency amplifier; the open-loop gain is down to 1000 at 1kHz, and f_T is 1MHz. Note that the closed-loop gain is always less than the open-loop gain; this means, for instance, that a ×1000 amplifier built with a 741 will show a noticeable falloff of gain for frequencies approaching 1kHz. Later in the chapter (Section 3.24), when we deal with transistor feedback circuits with finite open-loop gains, we

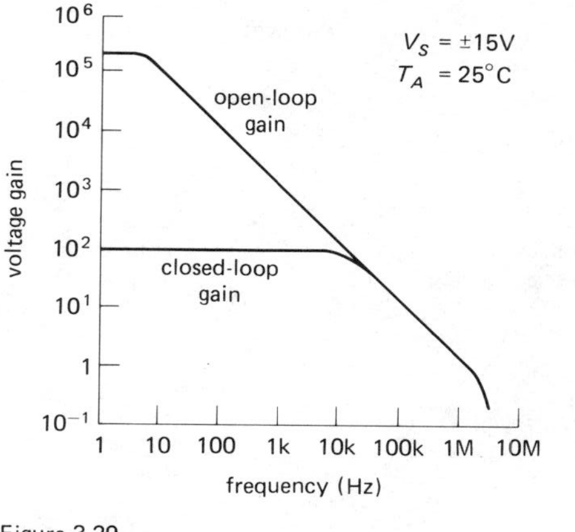

Figure 3.29

will have a more accurate statement of this behavior.

Slew rate

Because of limited slew rate, the maximum sine-wave output swing drops above a certain frequency. Figure 3.30 shows the

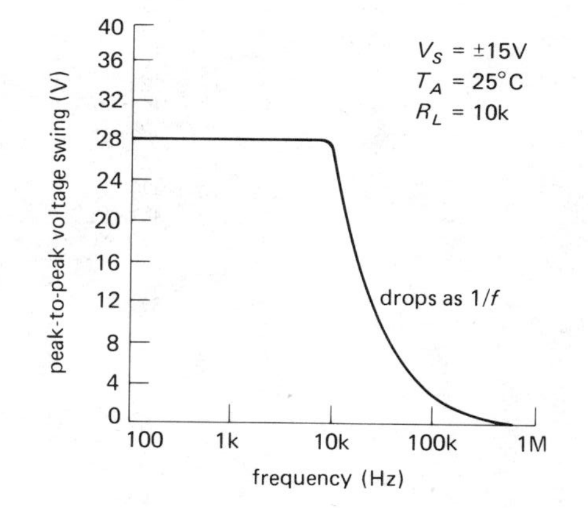

Figure 3.30

curve for a 741, with its 0.5V/μs slew rate. For slew rate S, the output amplitude is limited to A (pp) $\leq S/\pi f$ for a sine wave of frequency f, thus explaining the $1/f$ dropoff of the curve. The flat portion of the curve reflects the power-supply limits of output voltage swing.

Output current

Because of limited output current capability, an op-amp's output swing is reduced for

TABLE 3.1. OPERATIONAL AMPLIFIERS

Type	Single[a]	Dual[a]	Quad[a]	FET	Offset adj.	Ext. comp.[b]	Min. gain[c]	Total supply voltage min (V)	Total supply voltage max (V)	Supply curr.[d] (mA)	Input Voltage Offset typ (mV)	Input Voltage Offset max (mV)	Input Voltage Drift typ (μV/°C)	Input Voltage Drift max (μV/°C)	Input Current Offset typ (nA)	Input Current Offset max (nA)	Input Current Bias typ (nA)	Input Current Bias max (nA)
741 type																		
741C[1]	✓	*	*	—	✓	—	1	10	36	2.8	2	6	—	—	20	200	80	500
OP-01E	✓			—	✓	—	1	10	44	3	1	2	3	10	1	5	20	50
OP-02E[2]	✓	*		—	✓	—	1	10	44	2	0.3	0.5	2	8	0.5	2	18	30
OP-11E[3]			✓	—	—	—	1	10	44	6	0.3	0.5	2	10	8	20	180	300
349			✓	—	—	—	5	10	36	4.5	1	6	—	—	4	50	30	200
AD542L	✓			✓	✓	—	1	10	36	1.5	—	0.5	—	5	0.002	—	—	0.025
AD741L	✓			—	✓	—	1	10	44	2.8	0.2	0.5	2	5	2	5	30	50
748C	✓			—	✓	✓	U	10	36	3.3	2	6	—	—	20	200	80	500
μA777	✓			—	✓	✓	U	10	44	2.8	0.7	5	4	30	0.7	20	25	100
1458S		✓		—	✓	—	1	—	36	12	—	6	—	—	30	200	200	500
1741S	✓			—	✓	—	1	8	44	3.5	1	5	3	—	30	200	200	500
ULN2171	✓			—	✓	—	1	6	40	3.1	0.7	5	—	—	8	20	30	50
4131	✓			—	✓	—	1	7	36	1.9	1.5	5	5	20	3	20	70	150
HA4741			✓	—	—	—	1	4	40	7	1	5	5	—	30	50	60	300
LF13741	✓			✓	✓	—	1	10	36	4	5	15	10	—	0.01	0.05	0.05	0.2
4136 type																		
4136C			✓	—	—	—	1		36	11	0.5	6	—	—	5	200	40	500
1456	✓			—	✓	—	1	10	36	3	5	10	—	—	5	10	15	30
RC4156			✓	—	—	—	1	6	40	7	1	5	5	—	30	50	60	300
RC4157			✓	—	—	—	5	6	40	7	1	5	5	—	30	50	60	300
4558C		✓		—	—	—	1		36	5.6	2	6	—	—	20	200	80	500
HA4605			✓	—	—	—	1	10	40	6.5	0.5	3.5	2	—	30	100	130	300
HA4625			✓	—	—	—	10	10	40	6.5	0.5	3.5	2	—	30	100	130	300
301 type																		
301A[4]	✓	*		—	✓	✓	U	10	44	2.5	2	7.5	6	30	3	50	70	250
AD301AL	✓			—	✓	✓	U	10	44	3	0.3	0.5	2	5	3	5	15	30
307	✓			—	—	—	1	10	44	2.5	2	7.5	6	30	3	50	70	250
NE5534[5]	✓	*		—	✓	✓	3	6	44	8	0.5	4	—	—	20	300	500	1500
324 type																		
324[6]	*	*	✓	—	—	—	1	3	32	3	2	7	7	—	5	50	45	250
324A[6]	*	*	✓	—	—	—	1	3	32	3	2	3	7	30	5	30	45	100
μA799[6]	✓	*		—	✓	—	1	3	36	3	2	5	10	—	10	25	50	100
739 type																		
μA739		✓		—	—	✓	U	8	36	14	1	6	—	—	50	10000	300	2000
μA749		✓		—	—	✓	U	8	36	10	1	3	3	—	50	400	300	750
MC1303		✓		—	—	✓	U	10	30	15	1.5	10	—	—	200	400	1000	10k
RC4739		✓		—	—	—	1	8	36	5.2	2	6	—	—	5	200	40	500
725 type																		
μA725	✓			—	✓	✓	U	6	44	3	0.5	1	2	5	2	20	40	100
OP-05E[7]	✓	*		—	✓	—	1	6	44	4	0.2	0.5	0.7	2	1.2	3.8	1.2	4
OP-07A[8]	✓			—	✓	—	1	6	44	4	0.01	0.025	0.2	0.6	0.3	2	0.7	2
OP-07C[9]	✓			—	✓	—	1	6	44	5	0.06	0.15	0.5	1.8	0.8	6	1.8	7
OP-07E[10]	✓			—	✓	—	1	6	44	4	0.03	0.08	0.3	1.3	0.5	3.8	1.2	4
AD504L	✓			—	✓	✓	U	10	36	3	0.2	0.5	—	2	—	10	—	80
AD510L	✓			—	✓	—	1	10	36	3	—	0.025	—	—	—	2.5	—	10
AD517J	✓			—	✓	—	1	10	36	4	—	0.15	—	3	—	1	—	5
AD517L	✓			—	✓	—	1	10	36	3	—	0.025	—	0.5	—	0.25	—	1
3510CM	✓			—	✓	✓	10	6	40	3.5	—	0.06	—	0.5	—	10	—	15
308 type																		
LM308[11]	✓	*		—	—	✓	U	10	36	0.8	2	7.5	6	30	0.2	1	1.5	7
OP-08E	✓			—	—	✓	U	10	40	0.5	0.07	0.15	0.5	2.5	0.05	0.2	0.8	2
LM11	✓			—	✓	✓	1	5	40	0.6	0.1	0.3	1	3	0.5pA	10pA	25pA	50pA
OP-12E	✓			—	—	—	1	10	40	0.5	0.07	0.15	0.5	2.5	0.05	0.2	0.8	2
LH0044A	✓			—	—	✓	10	4	40	3	0.008	0.025	0.1	0.5	1.5	2.5	8.5	15
LM308A-1	✓			—	—	✓	U	10	36	0.8	0.3	0.5	0.6	1	0.2	1	1.5	7
LM312	✓			—	✓	✓	1	10	40	0.8	2	7.5	6	30	0.2	1	1.5	7
LM316A	✓			—	✓	✓	1	6	40	0.5	—	3	—	—	—	0.05	—	0.15

Slew rate[e] (typ) (V/μs)	f_T[e] (typ) (MHz)	CMRR min (dB)	CMRR typ (dB)	PSRR min (dB)	PSRR typ (dB)	Gain min (×1000)	Gain typ (×1000)	Max. output current (mA)	Max. diff'l. input[f] (V)	Swing to supplies?[g] In V_+	In V_-	Out V_+	Out V_-	Comments
0.5	1.2	70	90	76	90	20	200	20	30	—	—	—	—	Gen. purp., industry st'd.
18	2.5	80	100	80	100	50	100	6	30	—	—	—	—	Fast (feedforward), precision
0.5	1.3	90	110	90	110	100	250	6	30	—	—	—	—	Precision, low current
1.0	2	110	120	90	110	100	650	6	30	—	—	—	—	Precision quad
2[h]	4[h]	70	90	77	96	25	160	15	36	—	—	—	—	Decomp. 348 (quad 741)
3	1	80	—	80	—	300	—	10	20	—	—	—	—	Precision FET
0.5	1	90	110	96	106	50	200	15	30	—	—	—	—	Precision 741
0.5	1.2	70	90	76	90	50	200	15	30	—	—	—	—	Uncomp.741 (301 pinout)
0.5	1	70	95	76	96	25	250	20	30	—	—	—	—	Low-bias 741
20	1	70	90	76	100	20	100	10	30	—	—	—	—	Fast-slew 1458 (dual 741)
12	1	70	90	76	100	50	200	10	30	—	—	—	—	Fast-slew 741
1.5	—	80	100	80	100	25	100	—	30	—	—	—	—	Low bias
2	4	70	100	70	100	25	160	10	30	—	—	—	—	Fast 741
1.6	3.5	74	—	80	—	25	50	10	30	—	—	—	—	Quad fast 741
0.5	1	70	90	77	96	25	100	15	30	✓	—	—	—	FET follower + 741
1.0	3	70	90	76	90	20	300	20	30	—	—	—	—	Gen. purp. med. speed quad
2.5	1	70	110	74	84	70	100	5	40	—	—	—	—	
1.6	3.5	80	—	80	—	25	100	20	30	—	—	—	—	Fast 348 (quad 741)
8[h]	19[h]	80	—	80	—	25	100	20	30	—	—	—	—	Decomp. 4156
1.0	2.5	70	90	74	84	20	200	15	30	—	—	—	—	Fast 1458 (4559 = low noise)
4	8	80	—	80	—	75	250	10	7	—	—	—	—	Faster 348
20[i]	70[i]	80	—	80	—	75	250	10	7	—	—	—	—	Decomp. 4605
0.5	1	70	90	70	96	25	160	10	30	✓	—	—	—	Gen. purp. uncomp.
0.5	1	—	100	—	100	—	300	10	30	✓	—	—	—	Precision, low bias
0.5	1	70	90	70	96	15	—	10	30	✓	—	—	—	Compensated 301
6	10	70	100	80	100	25	100	20	0.5	—	—	—	—	Low noise, fast, good for audio
0.5	1	65	70	65	100	25	100	20	30	—	✓	—	✓	Gen. purp. quad. sing. supp.
0.5	1	65	85	65	100	25	100	20	30	—	✓	—	✓	Improved 324
0.6	1	70	90	76	90	50	200	20	30	—	✓	—	✓	Sim. to 324, but low distortion
1	6	70	90	—	85	6.5	20	1.5	5	—	—	—	✓	Low-noise audio op-amp
2	6	70	90	74	86	20	50	1.5	5	—	—	—	✓	Improved 739
—	—	—	—	—	—	6	10	10	40	—	—	—	—	Similar to 739
1	3	70	100	76	100	15	200	7	30	—	—	—	—	739 with push-pull output
0.005	0.08	110	120	100	116	250	3000	15	5	—	—	—	—	Orig. precision op-amp
0.17	0.6	110	123	94	107	200	500	10	30	—	—	—	—	
0.17	0.6	110	126	100	110	300	500	10	30	—	—	—	—	Precision premium op-amp
0.17	0.6	100	120	90	104	120	400	10	30	—	—	—	—	
0.17	0.6	106	123	94	107	200	500	10	30	—	—	—	—	
0.12	0.3	110	120	—	100	1000	8000	15	30	—	—	—	—	
0.1	0.3	110	—	—	100	1000	—	10	30	—	—	—	—	Low offset
0.1	0.25	94	—	88	—	1000	—	10	30	—	—	—	—	
0.1	0.25	110	—	96	—	1000	—	10	30	—	—	—	—	Precision premium op-amp
0.8[i]	0.4	110	—	110	130	1000	10M	10	40	—	—	—	—	Precision
0.15	0.3	80	100	80	96	25	300	5	0.5	—	—	—	—	Orig. low curr. (superbeta)
0.12	0.8	104	120	104	120	80	300	5	0.5	—	—	—	—	Precision 308
0.3	0.5	110	130	100	118	100	300	2	0.5	—	—	—	—	Lowest I_B bipolar op-amp; precision
0.12	0.8	104	120	104	120	80	300	5	0.5	—	—	—	—	Precision 312
0.06	0.4	120	145	120	145	1000	20M	4	1	—	—	—	—	Precision
0.15	0.3	96	110	96	110	80	300	5	0.5	—	—	—	—	Precision 308
0.15	0.3	80	110	96	96	25	300	5	0.5	—	—	—	—	Compensated 308
0.15	0.3	80	—	80	—	30	—	5	0.4	—	—	—	—	Superbeta Darlington

TABLE 3.1 (*cont.*)

Type	Single[a]	Dual[a]	Quad[a]	FET	Offset adj.	Ext. comp.[b]	Min. gain[c]	Total supply voltage min (V)	Total supply voltage max (V)	Supply curr.[d] (mA)	Input Voltage Offset typ (mV)	Input Voltage Offset max (mV)	Input Voltage Drift typ ($\mu V/°C$)	Input Voltage Drift max ($\mu V/°C$)	Input Current Offset typ (nA)	Input Current Offset max (nA)	Input Current Bias typ (nA)	Input Current Bias max (nA)
355 type																		
355	✓			✓	✓	—	1	10	36	4	3	10	5	—	0.003	0.02	0.03	0.2
OP-15E	✓			✓	✓	—	1	10	44	4	0.2	0.5	2	5	0.003	0.01	0.015	0.05
OP-16E	✓			✓	✓	—	1	10	44	7	0.2	0.5	2	5	0.003	0.01	0.015	0.05
355A	✓			✓	✓	—	1	10	36	4	1	2	3	5	0.003	0.01	0.03	0.15
LFT355	✓			✓	✓	—	1	10	44	4	—	0.5	3	5	0.003	0.01	0.03	0.05
356	✓			✓	✓	—	1	10	36	10	3	10	5	—	0.007	0.04	0.07	0.2
356A	✓			✓	✓	—	1	10	36	10	1	2	3	5	0.007	0.02	0.07	0.1
LFT356	✓			✓	✓	—	1	10	44	7	—	0.5	3	5	0.003	0.02	0.07	0.1
357	✓			✓	✓	—	5	10	36	10	3	10	5	—	0.007	0.04	0.07	0.2
357A	✓			✓	✓	—	5	10	36	10	1	2	3	5	0.007	0.02	0.07	0.1
TL081 type																		
TL081C[12]	✓	•	•	✓	✓	–	1		36	2.8	5	15	10	—	0.005	0.02	0.03	0.4
TL081B[12]	✓	•	•	✓	✓	–	1		36	2.8	2	3	10	—	0.005	0.01	0.03	0.2
TL061C[13]	✓	•	•	✓	✓	–	1		36	0.25	3	15	10	—	0.005	0.2	0.03	0.4
TL071C[14]	✓	•	•	✓	✓	–	1		36	2.5	3	10	10	—	0.005	0.05	0.03	0.2
LF351[15]	✓	•	•	✓	✓	–	1	10	36	3.4	5	10	10	—	0.025	0.1	0.05	0.2
LF351A[15]	✓	•	•	✓	✓	–	1	10	36	2.8	1	2	10	—	0.025	0.05	0.05	0.1
AD544L	✓			✓	✓	–	1	10	36	2.5	—	0.5	—	5	0.005	—	—	0.05
μA771A[16]	✓	•	•	✓	✓	–	1		36	3	—	2	10	—	—	0.05	0.05	0.1
MOSFET type																		
CA3160A[17]	✓			✓	✓	✓	1	5	16	15	2	5	10	—	0.0005	0.02	0.005	0.03
CA3140A[18]	✓	•		✓	✓	–	1	4	44	6	2	5	6	—	0.0005	0.02	0.01	0.04
ICL7600	✓			✓	–	–	1	4	18	5	0.002	0.005	0.01	0.1	—	—	0.3	3
ICL7612B	✓			✓	–	–	1	3	18	2.5	—	5	5	—	0.0005	0.03	0.001	0.05
ICL7641B[19]	•	•	✓	✓	–	–	1	1	18	2.5	—	5	5	—	0.0005	0.03	0.001	0.05
ICH8500A	✓			✓	✓	–	1	16	36	2.5	—	50	—	—	—	0.01pA	—	0.01pA[j]
MC14573-1			✓	✓	–	–	1	3	18	1.5[k]	—	10	20	—	—	0.02	—	1
FET: low bias																		
LH0022	✓			✓	✓	–	1	11	40	2.8	2	4	5	10	0.2pA	2pA	5pA	10pA
LH0052	✓			✓	✓	–	1	11	40	3.8	0.1	0.5	2	5	0.01pA	0.5pA	0.5pA	2.5pA
LH0062C	✓			✓	✓	✓	1	10	40	8	2	5	5	25	0.2pA	2pA	5pA	10pA
AD506L	✓			✓	✓	–	1	10	44	7	0.4	1	5	10	—	2pA	—	2pA
AD515J	✓			✓	✓	–	1	10	36	1.5	0.4	3	—	50	—	0.3pA	—	0.3pA
AD515L	✓			✓	✓	–	1	10	36	1.5	0.4	1	—	25	—	0.08pA	—	0.08pA
AD545L	✓			✓	✓	–	1	10	36	1.5	—	0.5	—	5	—	—	—	1pA
3521L	✓			✓	✓	–	1	10	40	4	—	0.25	—	1	2pA	—	—	10pA
3527CM	✓			✓	✓	–	1	10	40	4	0.1	0.25	1	2	0.3pA	—	2pA	5pA
3528BM	✓			✓	✓	–	1	10	40	1.5	0.1	0.25	2	5	0.04pA	—	—	0.15pA
High voltage																		
LM343	✓			–	✓	–	1	10	68	5	2	8	—	—	1	10	8	40
LM344	✓			–	✓	✓	U	10	68	5	2	8	—	—	1	10	8	40
1436	✓			–	✓	–	1	10	80	5	5	10	—	—	5	10	15	40
HA2645	✓			–	✓	✓	1	20	80	4.5	2	6	15	—	12	30	15	30
3583	✓			✓	✓	–	1	100	300	8.5	—	3	—	25	—	0.1	—	0.1
3584	✓			✓	✓	✓	U	140	300	6.5	—	3	—	25	—	0.1	—	0.1
High speed																		
LM318	✓			–	✓	✓	1	10	40	10	4	10	—	—	30	200	150	500
LH0024C	✓			–	✓	✓	20		36	15	5	8	25	—	4μA	15μA	18μA	40μA
LH0032C	✓			✓	✓	✓	50	10	36	22	5	15	25	—	0.01	0.05	0.025	0.2
AD518J	✓			–	✓	✓	1	10	40	10	4	10	10	—	30	200	120	500
AD528J	✓			✓	✓	✓	1	10	40	7	1	3	25	50	—	0.005	0.01	0.03
NE530[20]	✓	•		–	✓	–	1		36	3	2	5	6	—	15	40	65	150
NE538[21]	✓	•		–	✓	–	5	10	36	2.8	2	5	6	—	15	40	65	150
NE531	✓			–	✓	✓	U	12	44	10	2	6	—	—	50	200	400	1500
NE535[22]	✓	•		–	✓	–	1	10	36	2.8	2	5	6	—	15	40	65	150
1435	✓			–	✓	✓	10	24	32	30	2	5	5	25	—	—	10μA	20μA
HA2505	✓			–	✓	✓	1	20	40	6	4	8	20	—	20	50	125	250

Slew rate[e] (typ) (V/μs)	f_T[e] (typ) (MHz)	CMRR min (dB)	CMRR typ (dB)	PSRR min (dB)	PSRR typ (dB)	Gain min (×1000)	Gain typ (×1000)	Max. output current (mA)	Max. diff'l. input[f] (V)	Swing to supplies?[g] In V_+	In V_-	Out V_+	Out V_-	Comments
5	2.5	80	100	80	100	25	100	20	30	✓	—	—	—	Popular gen. purp. BiFET
7	6	86	100	86	100	100	240	15	40	—	—	—	—	Precision fast 355
5	8	86	100	86	100	100	240	20	40	—	—	—	—	Precision fast 356 (OP-17 = decomp.)
5	2.5	85	100	85	100	50	100	20	30	✓	—	—	—	—
5	2.5	95	—	85	100	50	200	20	30	✓	—	—	—	Precision 355
2	4.5	80	100	80	100	25	100	20	30	✓	—	—	—	Fast 355
2	4.5	85	100	85	100	50	100	20	30	✓	—	—	—	
2	4.5	95	—	85	100	50	200	20	30	✓	—	—	—	Precision 356
0[h]	20[h]	80	100	80	100	25	100	20	30	✓	—	—	—	Decompensated 356
0[h]	20[h]	85	100	85	100	50	100	20	30	✓	—	—	—	
3	3	70	76	70	76	25	200	10	30	—	—	—	—	Inexpensive gen. purp. BiFET
3	3	80	86	80	86	50	200	10	30	—	—	—	—	
3.5	1	70	76	70	95	3	10	5	30	—	—	—	—	Low power
3	3	70	76	70	76	25	200	10	30	—	—	—	—	Lower noise
3	4	70	100	70	100	25	100	10	30	✓	—	—	—	Similar to TL071
3	4	80	100	80	100	50	100	10	30	✓	—	—	—	—
3	2	80	—	80	—	50	—	15	20	—	—	—	—	Precision, low noise
3	3	80	—	80	—	50	100	10	30	—	—	—	—	Similar to TL071
0	4	80	90	76	94	50	320	12	8	—	✓	✓	✓	MOS in/out (3130 = uncomp.)
7	3.7	70	90	76	80	20	100	+10, −1	8	—	✓	—	✓	Gen. purp. MOSFET op-amp
0.5	0.3	—	88	—	110	30	160	+5, −10	18	—	—	✓	✓	CAZ (lowest offset, low freq.)
1.6	1.4	60	87	70	77	10	100	5[m]	18	✓	✓	✓	✓	Programmable
1.6	1.4	60	87	70	77	10	100	5[m]	18	✓	—	✓	✓	CMOS, gen. purp., low voltage
0.5	0.5	60	75	—	80	—	100	10	0.5	—	—	—	—	Lowest input current
8	—	—	80	—	70	—	50	1.3[ℓ]	18	—	✓	✓	✓	CMOS
3	1	80	90	80	90	100	200	10	30	—	—	—	—	LH0042 = low-cost version
3	1	74	90	74	90	100	200	10	30	—	—	—	—	Precision
0	15	80	90	80	90	50	200	10	30	—	—	—	—	Fast
6	1	80	90	80	86	50	120	15	4	—	—	—	—	
1	0.4	66	94	68	86	40	—	10	20	—	—	—	—	
1	0.4	70	94	74	86	50	—	10	20	—	—	—	—	Lowest current precision
1	0.7	76	80	—	74	40	—	10	20	—	—	—	—	Precision
0.6	1.5	—	90	—	92	50	—	10	40	—	—	—	—	Precision
0.9	1	—	76	—	80	100	300	20	40	—	—	—	—	Precision
0.7	0.7	80	86	80	92	100	—	10	40	—	—	—	—	Precision, low current
2.5	1	70	90	74	100	70	180	10	68	—	—	—	—	
0[i]	10[i]	70	90	74	100	70	180	10	68	—	—	—	—	Uncompensated 343
2	1	70	110	80	96	70	500	10	80	—	—	—	—	
5	4	74	100	74	90	100	200	10	37	—	—	—	—	
0	5	—	110	—	84	50	900	75	300	—	—	—	—	Fast high-voltage FET
0[n]	7	—	110	—	84	100	1000	15	300	—	—	—	—	High-voltage uncomp FET
0	15	70	100	65	80	20	—	10	0.5	—	—	—	—	Popular
0	70	—	60	—	60	3	4	10	5	—	—	—	—	
0	70	50	60	50	60	1	2.5	15	30	—	—	—	—	FET
0	12	70	100	65	80	25	100	15	—	—	—	—	—	
0	10	70	90	70	90	25	100	15	20	—	—	—	—	FET, compensated
5	3	70	90	76	90	50	200	10	30	✓	—	—	—	Fast dual (5530)
0[h]	5[h]	70	90	76	90	50	200	10	30	✓	—	—	—	Fast dual (5538)
5	1	70	100	76	100	20	60	—	15	✓	—	—	—	
5	1	70	90	76	90	50	200	10	30	✓	—	—	—	Fast dual (5535)
0	1000	80	90	—	75	10	20	10	2	—	—	—	—	Fast settling (2035 buffer)
0	12[i]	74	90	74	90	15	25	10	15	✓	—	—	—	2507 = mini-DIP

TABLE 3.1 (*cont.*)

								Total supply voltage			Input							
											Voltage				Current			
											Offset		Drift		Offset		Bias	
Type	Single[a]	Dual[a]	Quad[a]	FET	Offset adj.	Ext. comp.[b]	Min. gain[c]	min (V)	max (V)	Supply curr.[d] (mA)	typ (mV)	max (mV)	typ (μV/°C)	max (μV/°C)	typ (nA)	max (nA)	typ (nA)	max (nA)
High speed (*cont.*)																		
HA2515	✓			—	✓	✓	1	20	40	6	5	10	30	—	20	50	125	250
HA2525[23]	✓			—	✓	✓	3	20	40	6	5	10	30	—	20	50	125	250
HA2535	✓			—	—	✓	10	20	40	6	0.8	5	5	—	5	20	15	200
HA2605	✓			—	✓	✓	1	10	45	4	3	5	10	—	5	25	5	25
HA2625[24]	✓			—	✓	✓	5	10	45	4	3	5	10	—	5	25	5	25
HA2655		✓		—	✓	—	1	4	40	4	2	5	8	—	2	60	50	200
CA3100	✓			—	✓	✓	10	13	36	11	1	5	—	—	50	400	700	2000
HA5105	✓			✓	✓	✓	1	20	40	8	0.5	1.5	15	—	0.005	0.05	0.05	0.1
HA5115	✓			✓	✓	✓	10	20	40	8	0.5	1.5	15	—	0.005	0.05	0.05	0.1
HA5155	✓			✓	✓	✓	1	20	40	7	—	1	10	—	—	—	—	0.0
HA5165	✓			✓	✓	✓	10	20	40	7	—	1	10	—	—	—	—	0.0
HA5195	✓			—	—	—	5	20	35	25	3	6	20	—	1μA	4μA	5μA	15μA
NE5539	✓			—	—	✓	7	6	24	15	2.5	5	5	10	—	—	10μA	20μA
8017C	✓			—	—	✓	U		36	8	2	7	10	—	—	—	50	200
Miscellaneous																		
LM10	✓			—	✓	—	1	1	45	0.4	0.3	2	2	—	0.3	0.7	10	20
1439[25]	✓			—	✓	✓	U	20	36	6	2	7.5	3	—	20	100	250	1000
HA2705	✓			—	✓	—	1	11	40	0.2	1	5	5	—	2.5	15	5	40
HA2905	✓			—	—	—	1	20	42	5	0.02	0.08	0.2	—	0.05	0.5	0.15	1
72088	✓			—	—	—	1		36	10	0.07	0.15	—	—	0.2	0.6	0.6	10

[a] A check (✓) indicates the number of op-amps per package for the listed type; an asterisk (*) indicates other packages that are available, as described in the numbe footnotes. [b] A check means that pins are provided for external compensation. [c] A number gives the minimum closed-loop gain without instability. Op-an with pins for external compensation can generally be operated at lower gain, if an appropriate external compensation network is used. The letter U means that op-amp is uncompensated – external capacitance is necessary for any small value of closed-loop gain. [d] Maximum for $V_{supply} = \pm 15V$. [e] For unity ç compensation, unless otherwise noted. [f] The maximum value without damage to the chip; not to exceed the total supply voltage used, if that is less. a check in an In column means that the input operating common-mode range includes that supply rail; a check in an Out column means that the op-amp can swing a output all the way to the supply voltage. [h] At $G \geq 5$. [i] At $G \geq 10$. [j] Over full temperature range. [k] At $V_{DD} = +15V$. [l] At V_{DD} + 10V. [m] For $V_{\pm} = \pm 7.5V$. [n] At $G \geq 100$. [o] At $G = 2$.

small load resistances. Figure 3.31 shows the graph for a 741. For precision applications it is a good idea to avoid large output currents in order to prevent on-chip thermal gradients produced by excessive power dissipation in the output stage.

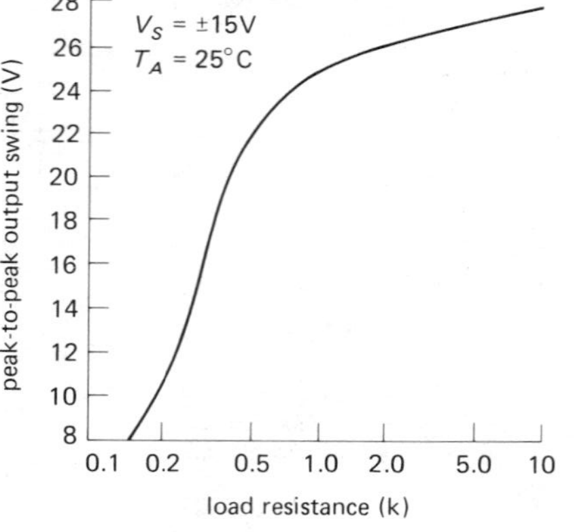

Figure 3.31

Offset voltage

Because of input offset voltage, a zero input produces an output of $V_{out} = G_{dc} V_{os}$. For an inverting amplifier with voltage gain of 100 built with a 741, the output could be as large as ±0.6 volt when the input is grounded (V_{os} = 6mV max). Solutions: (a) If you don't need gain at dc, use a capacitor to drop the gain to unity at dc, as in the RIAA amplifier circuit earlier. In this case you could do that by capacitively coupling the input signal. (b) Trim the voltage offset to zero using the manufacturer's recommended trimming network. (c) Use an op-amp with smaller V_{os}.

Slew rate[e] (typ) V/μs)	f_T[e] (typ) (MHz)	CMRR min (dB)	CMRR typ (dB)	PSRR min (dB)	PSRR typ (dB)	Gain min (×1000)	Gain typ (×1000)	Max. output current (mA)	Max. diff'l. input[f] (V)	Swing to supplies?[g] In V_+	In V_-	Out V_+	Out V_-	Comments
0	12[i]	74	90	74	90	7.5	15	10	15	✓	—	—	—	2517 = mini-DIP
0[h]	20[i]	74	90	74	90	7.5	15	10	15	✓	—	—	—	Popular (2527 = mini-DIP)
0[i]	70[i]	80	100	80	100	100	2000	35	0.5	—	—	—	—	Fast
7	12	74	100	74	90	80	150	10	12	—	—	—	—	2607 = mini-DIP
5[h]	100[n]	74	100	74	90	80	150	10	12	—	—	—	—	Popular (2627 = mini-DIP)
5	8	74	100	74	100	20	40	10	30	—	—	—	—	
5	30	76	90	60	70	0.8	1.1	15	12	—	—	—	—	
8	18[i]	80	86	80	94	50	100	10	40	—	—	—	—	
0[i]	50[i]	80	86	80	94	50	100	10	40	—	—	—	—	
0	50	86	—	86	—	50	—	10	40	—	—	—	—	Fast FET monolithic, precision
0	100	86	—	86	—	50	—	10	40	—	—	—	—	Decomp 5155
0[h]	150[i]	74	—	70	90	10	30	25	6	—	—	—	—	Fast settling
0[o]	1200[i]	70	85	66	74	0.2	0.25	40	10	—	—	—	✓	Video, small output swing
0	10	—	—	70	—	25	1000	15	30	—	—	—	—	Popular, fast
0.12	0.1	93	102	90	96	120	400	20	40	—	✓	✓	✓	"1 volt op-amp," precision, volt. ref.
4.2	1	80	100	75	90	20	100	10	36	—	—	—	—	
0	1	80	106	80	100	200	300	10	18	—	—	—	—	Fast, low power
2.5	3[i]	120	160	120	160	1k	50k	7	15	—	—	—	—	Chopper, noisy
5	3	—	80	—	70	1k	10k	15	15	—	—	—	—	Chopper, noisy, poor CMRR and PSRR

ual: 747, 1458 (mini-DIP). Quad: MC4741, 348. [2]Dual: OP-04, OP-14 (mini-DIP). [3]OP-09 has same specs, different pinout. [4]Dual: 01. [5]Dual: 5532, 5533. [6]Similar to 534, 2902, 3403. Dual: 358, 798, 2904, 532. Single: 799 (with offset). [7]Dual: OP-10E. [8]Also 714A. [9]Also μA714C. [10]Also μA714E. [11]Dual: 2308. [12]Dual: TL082B,C, TL083B,C. Quad: TL084A,C. Uncompensated: 080. [13]Dual: TL062C, TL063C. Quad: TL064C. [14]Dual: TL072C, TL073C. Quad: TL074C, TL075C. Similar to μA771-4 and LF347-54 ies. [15]Dual: LF353,4. Quad: LF347. Similar to TL071-4 and μA771-4 series. [16]Dual: μA772A. Quad: μA774A. Similar to TL071-4 and LF347-54 ies. [17]3130 is uncompensated version. [18]Dual: CA3240A. [19]Similar to ICL7611, 7614 (single), and ICL7621, 7622 (dual). [20]Dual: 5530. [21]Dual: NE5538. [22]Dual: NE5535. [23]Also 1322, 3507J, AD509J. [24]Also 1321, 3508J, AD507J. [25]Also 2139.

Input bias current

Because of finite input current, zero input voltage still won't give zero output, even if the input offset voltage is trimmed to zero. The inverting input terminal sees a driving impedance of $R_1 \parallel R_2$, so the bias current produces a voltage $V_{in} = I_B(R_1 \parallel R_2)$, which is then amplified by the gain at dc, $-R_2/R_1$. For a 741 connected with R_1 = 100k and R_2 = 1MΩ, the output (for grounded input) could be as large as 10 × 200nA × 91k, or 0.18 volt ($G_{dc}I_BR_{unbalance}$; I_B = 200nA max).

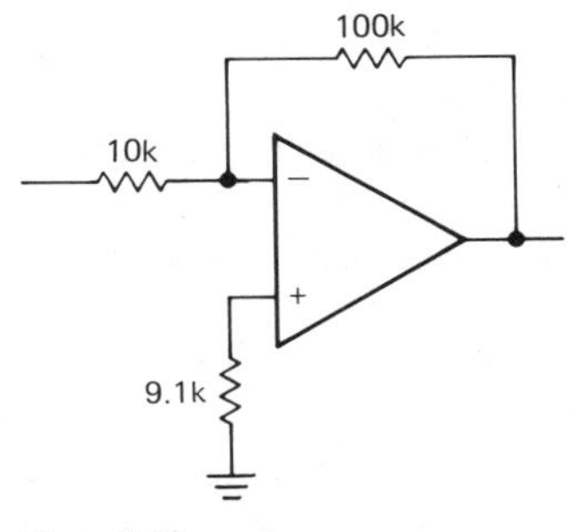

Figure 3.32

To minimize the effects of input bias current, it is a good idea to ensure that both inputs see the same dc driving resistance, as in Figure 3.32. In this case, 9.1k is chosen as the parallel resistance of 10k and 100k. In addition, it is best to keep the resistance of the feedback network small enough so that bias current doesn't produce large offsets; typical values for the resistance seen from the op-amp inputs are 1k to 100k or so. A third cure involves dropping the gain to unity at dc, as before. If you're still in trouble with errors produced by bias current, use an op-amp with lower bias current. The inexpensive LM308 has I_B = 1.5nA (typ), its improved cousin (the LM11) has I_B = 25pA (typ), the popular LF355–LF357 series of FET-input op-amps has I_B = 30pA, and the

TABLE 3.2. HIGH-VOLTAGE OP-AMPS

Type	Total supply min (V)	Total supply max (V)	Max diff'l. input[a] (V)	FET input	Ext. compen.	Offset adj.	f_T (typ) (MHz)	Slew rate (typ) (V/μs)	Max. output curr. (±mA)	Max P_{diss} ($T_C = 50°C$) (W)	Thermal limit	Case	Comments
MC1436	10	80	80	—	—	✓	1	2	10	0.6	—	TO-99	Original, still good
LM343	10	68	68	—	—	✓	1	2	20	0.6	—	TO-99	Superbeta
LM344	10	68	68	—	✓	—	1	2.5[b]	20	0.6	—	TO-99	Superbeta
HA2645	20	80	74	—	✓	✓	4	5	10	0.6	✓	TO-99	Same as Philbrick 1332
3583	100	300	300	✓	—	✓	5	30	75	10	✓	TO-3	Power jobs
3584	140	300	300	✓	✓	✓	7[c]	150[c]	15	4.5	✓	TO-3	Power jobs

[a] Not to exceed total supply voltage. [b] Slew rate = 30V/μs when compensated for $G \geq 10$. [c] Compensated for $G \geq 100$.

recent CA3140 and CA080 series, priced at under a dollar, have $I_B = 10\text{pA}$ (typ).

Input offset current

Because of input offset current, even a circuit with balanced input impedances at dc will suffer some output offset. This differential effect is usually quite a bit smaller than the output offset produced by input bias current with unbalanced input resistances, so it is worth balancing the input resistances. If the remaining output offset is too large for a given application, the only solution (besides lower resistances in the feedback network) is a lower input current op-amp.

Limitations imply tradeoffs

The limitations of op-amp performance we have talked about will have an influence on component values in nearly all circuits. For instance, the feedback resistors must be large enough so that they don't load the output significantly, but they must not be so large that input bias current produces sizable offsets. High impedances in the feedback network also increase susceptibility to capacitive pickup of interfering signals and

POPULAR OP-AMPS

Sometimes a new op-amp comes along at just the right time, filling a vacuum with its combination of performance, convenience, and price. Several companies begin to manufacture it (it becomes "second-sourced"), designers become familiar with it, and you have a hit. Here is a list of the popular favorites of today:

301 First easy-to-use op-amp; first use of "lateral *pnp*." External compensation. National.

741 The industry standard (see accompanying box). Internal compensation. Fairchild.

1458 Motorola's answer to the 741; two 741s in a mini-DIP, with no offset pins.

308 National's precision op-amp. Low power, superbeta, guaranteed drift specifications.

324 Popular quad op-amp (358 = dual, mini-DIP). Single-supply operation. National.

355 All-purpose bi-FET op-amp (356, 357 faster). Practically as precise as bipolar, but faster and lower input current. National. (Fairchild tried to get the FET ball rolling with their 740, which flopped because of poor performance. Would you believe 0.1V input offset?)

TL081 Texas Instruments' answer to the 355 series. Low-cost comprehensive series of singles, duals, quads; low power, low noise, many package styles.

increase the loading effects of stray capacitance. These tradeoffs typically dictate resistor values of 2k to 100k with general-purpose op-amps like the 741.

Similar sorts of tradeoffs are involved in almost all electronic design, including the simplest circuits constructed with transistors. For instance, the choice of quiescent current in a transistor amplifier is limited at the high end by device dissipation, increased input current, excessive supply current, and reduced current gain, whereas the lower limit of operating current is limited by leakage current, reduced current gain, and reduced speed (from stray capacitance in combination with the high resistance values). For these reasons you typically wind up with collector currents in the range of a few tens of microamps to a few tens of milliamps (higher for power circuits, sometimes a bit lower in "micropower" applications), as mentioned in Chapter 2. In the next three chapters we will look more carefully at some of these problems in order to give you a good understanding of the tradeoffs involved.

THE 741 AND ITS FRIENDS

Bob Widlar designed the first really successful monolithic op-amp back in 1965, the Fairchild μA709. It achieved great popularity, but it had some problems, in particular the tendency to go into a latch-up mode when the input was overdriven and its lack of output short-circuit protection. It also required external frequency compensation (two capacitors and one resistor) and had a clumsy offset trimming circuit (again requiring three external components). Finally, its differential input voltage was limited to 5 volts.

Widlar moved from Fairchild to National, where he went on to design the LM301, an improved op-amp with short-circuit protection, freedom from latch-up, and a 30 volt differential input range. Widlar didn't provide internal frequency compensation, however, because he liked the flexibility of user compensation. The 301 could be compensated with a single capacitor, but because there was only one unused pin remaining, it still required three external components for offset trimming.

Meanwhile, over at Fairchild the answer to the 301 (the now-famous 741) was taking shape. It had the advantages of the 301, but Fairchild engineers opted for internal frequency compensation, freeing two pins to allow simplified offset trimming with a single external trimmer. Since most circuit applications don't require offset trimming (Widlar was right), the 741 in normal use requires no components other than the feedback network itself. The rest is history – the 741 caught on like wildfire and has become firmly entrenched as the industry standard.

There are now numerous 741-type op-amps, essentially similar in design and performance, but with various features such as FET inputs, dual or quad units, versions with improved specifications, decompensated and uncompensated versions, etc. We list some of them here for reference and as a demonstration of man's instinct to clutch onto the coattails of the famous (see Table 3.1 for a more complete listing).

Single units		*Dual units*		*Quad units*	
741S	fast (10V/μs)	747	dual 741	MC4741	quad 741 (alias 348)
MC741N	low noise	OP-04	precision	OP-11	precision
OP-02	precision	1458	mini-DIP package	4136	fast (3MHz)
4132	low power (35μA)	4558	fast (15V/μs)	HA4605	fast (4V/μs)
LF13741	FET low input current	TL082	FET, fast (similar to LF353)	TL084	FET, fast (similar to LF347)
748	uncompensated				
NE530	fast (25V/μs)				
TL081	FET, fast (similar to LF351)				

EXERCISE 3.6

Draw a dc-coupled inverting amplifier with gain of 100 and $Z_{in} = 10k$. Include compensation for input bias current, and show offset voltage trimming network (10k pot between pins 1 and 5, wiper tied to V_-). Now add circuitry so that $Z_{in} \gtrsim 10^8$ ohms.

□ 3.13 Low-power and programmable op-amps

For battery-powered applications there is a popular group of op-amps known as "programmable op-amps," because all of the internal operating currents are set by an externally applied current at a bias programming pin. The internal quiescent currents are all related to this bias current by current mirrors, rather than by internal resistor-programmed current sources. As a consequence, such amplifiers can be programmed to operate over a wide range of supply currents, typically from a few microamps to a few milliamps. The slew rate, gain-bandwidth product f_T, and input bias current are all roughly proportional to the programmed operating current. When programmed to operate at a few microamps, programmable op-amps are extremely useful in battery-powered circuits.

The 4250 was the original programmable op-amp, and it is still a good unit for many applications. Developed by Union Carbide, this classic is now "second-sourced" by many manufacturers, and it even comes in duals and triples (the 8022 and 8023, respectively). As an example of the sort of performance you can expect for operation at low supply currents, let's look at the 4250 running at 10μA. To get that operating current, we have to supply a bias current of 1.5μA with an external resistor. When it is operated at that current, f_T is 75kHz, the slew rate is 0.05V/μs, and the input bias current I_B is 3nA. At low operating currents the output drive capability is reduced considerably, and the open-loop output impedance rises to astounding levels, in this case about 3.5k. At low operating currents the input noise voltage rises, while the input noise current drops (see Chapter 7). The 4250 specifications claim that it can run from as little as 1 volt total supply voltage, but the claimed minimum supply voltages of op-amps may not be terribly relevant in an actual circuit, particularly where any significant output swing or drive capability is needed.

The 776 (or 3476) is an upgraded 4250, with better output-stage performance at lower currents. The 346 is a nice quad programmable op-amp, with three sections programmed by one of the programming inputs, and the fourth programmed by the other. Some other programmable op-amps constructed with ordinary bipolar transistors are the L144, HA4725, HA4735, CA3078, and L144. Programmable op-amps constructed with FETs (see Chapter 6) are also available: The XR094 and TL061 use JFET inputs, for low bias current (30pA); they have ordinary transistor output stages, with the same drive capabilities as other op-amps. These op-amps have the disadvantage of a relatively large minimum supply voltage, a drawback that is remedied in the ICL7612 and MC14573 CMOS programmable op-amps. The 7612 family, in particular, claims a minimum supply voltage of just 1 volt, and the CMOS outputs of these op-amps can swing all the way to both supply voltages. The 7612 family includes every imaginable package style. These op-amps suffer from different problems, namely larger input offset voltage, higher noise, and lower output drive currents when operated at low supply voltages.

In addition to these op-amps, there are several nonprogrammable op-amps that have been designed for low supply currents and low-voltage operation, and therefore they should be considered for low-power applications. Notable among these is the outstanding LM10, an op-amp that is fully specified at 1 volt total supply voltage ($\pm$0.5V, for example). This is extraordinary, considering that V_{BE} increases with decreasing temperature and is close to 1 volt at $-55°$C, the lower limit of the LM10's operating range. The HA2705 offers extraordinary speed for a low-power op-amp, with a slew rate of 20V/μs and an f_T of 1MHz at its operating current of 75μA. The OP-20 is a precision low-power op-amp, with a typical offset voltage of only 0.1mV. Other low-power op-amps worth looking at are the classic 308, the single-supply 358 (and 324), and the low-current RA4132. The LM11, an improved version of the 308, offers an unbeatable combination of low input offset voltage and low bias current.

A DETAILED LOOK AT SELECTED OP-AMP CIRCUITS

The performance of the next few circuits is affected significantly by the limitations of op-amps; we will go into a bit more detail in their description.

3.14 Logarithmic amplifier

The circuit shown in Figure 3.33 exploits the logarithmic dependence of V_{BE} on I_C to produce an output proportional to the logarithm of a positive input voltage. R_1 converts V_{in} to a current, owing to the virtual ground at the inverting input. That current flows through Q_1, putting its emitter one V_{BE} drop below ground, according to the Ebers-Moll equation. Q_2 is essential for temperature compensation. The current source (which can be a resistor, since point B is always within a few tenths of a volt of ground) sets the input current at which the output voltage is zero. The second op-amp is a noninverting amplifier with a voltage gain of 16, in order to give an output voltage of -1.0 volt per decade of input current (recall that V_{BE} increases 60mV per decade of collector current).

Some further details: Q_1's base could have been connected to its collector, but the base current would then have caused an error (remember that V_{BE} is an accurate exponential function of I_C). In this circuit the base is at the same voltage as the collector because of the virtual ground, but there is no base current error. Q_1 and Q_2 should be a matched pair, thermally coupled (a matched monolithic pair like the LM394 or MAT-01 is ideal). This circuit will give accurate logarithmic output over seven decades of current or more (1nA to 10mA, approximately), providing that low-leakage transistors and a low-bias-current input op-amp are used. An op-amp like the 741 with 80nA of bias current is unsuitable, and a FET-input op-amp is usually required to achieve the full seven decades of linearity. Furthermore, in order to give good performance at low input currents, the input op-amp must be accurately trimmed for zero offset voltage, since V_{in} may be as small as a few tens of microvolts at the lower limit of current. If possible, it is better to use a current input to this circuit, omitting R_1 altogether.

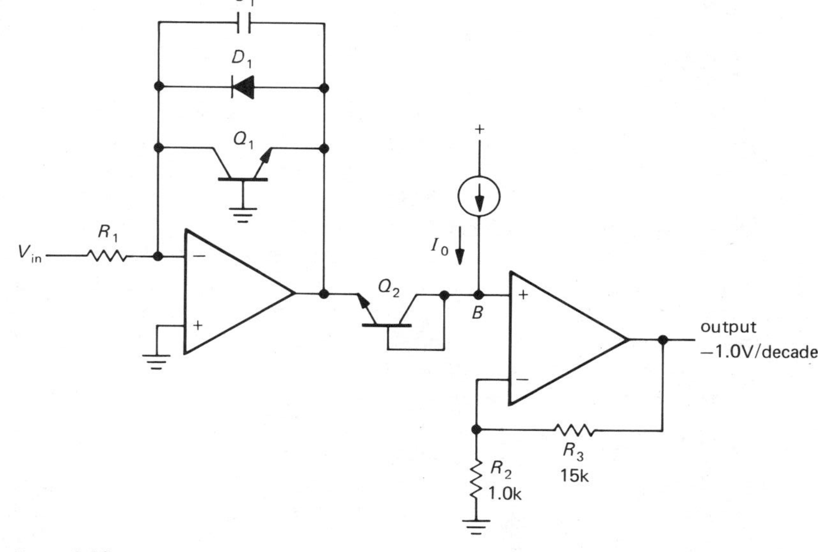

Figure 3.33

The capacitor C_1 is necessary to stabilize the feedback loop, since Q_1 contributes voltage gain inside the loop. Diode D_1 is necessary to prevent base-emitter breakdown (and destruction) of Q_1 in the event the input voltage goes negative, since Q_1 provides no feedback path for positive op-amp output voltage. Both these minor problems are avoided if Q_1 is wired as a diode, i.e., with its base tied to its collector.

□ *Temperature compensation of gain*

Q_2 compensates changes in Q_1's V_{BE} drop as the ambient temperature changes, but the changes in the slope of the curve of V_{BE} versus I_C are not compensated. In Section 2.10 we saw that the "60mV/decade" is proportional to absolute temperature. The output voltage of this circuit will look as shown in Figure 3.34. Compensation is

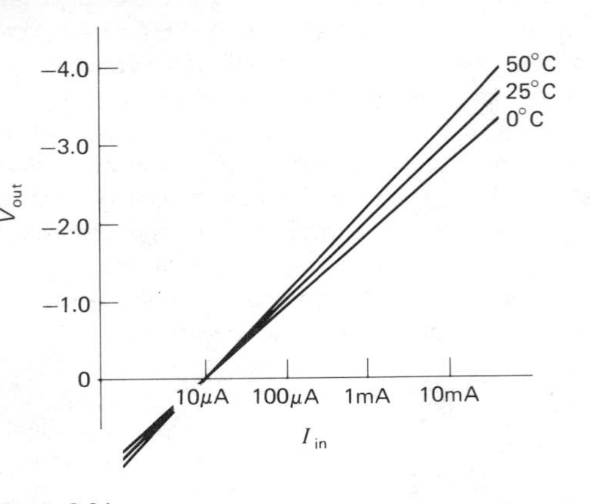

Figure 3.34

perfect at an input current equal to I_0, Q_2's collector current. A change in temperature of 30°C causes a 10% change in slope, with corresponding error in output voltage. The usual solution to this problem is to replace R_2 with a series combination of an ordinary resistor and a resistor of positive temperature coefficient. Knowing the temperature coefficient of the resistor (e.g., the TG 1/8 type manufactured by Texas Instruments has a coefficient of +0.67%/°C) allows you to calculate the value of the ordinary resistor to put in series in order to effect perfect compensation. For instance, with the 2.7k TG 1/8 type "sensistor" just mentioned, a 2.4k series resistor should be used.

There are several logarithmic converter modules available as complete integrated circuits. These offer very good performance, including internal temperature compensation. Some manufacturers are Analog Devices, Burr Brown, Philbrick, Fairchild, Intersil, and National Semiconductor.

EXERCISE 3.7

Finish up the log converter circuit by (a) drawing the current source explicitly and (b) using a TG 1/8 resistor (+0.67%/°C tempco) for thermal slope compensation. Choose values so that $V_{out} = +1$ volt per decade, and provide an output offset control so that V_{out} can be set to zero for any desired input current (do this with an inverting amplifier offset circuit, not by adjusting I_0).

□ 3.15 Active peak detector

There are numerous applications in which it is necessary to determine the peak value of some input waveform. The simplest method is a diode and capacitor (Fig. 3.35). The

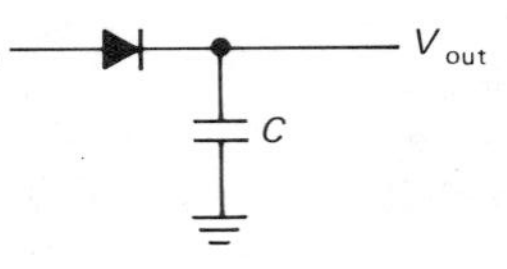

Figure 3.35

highest point of the input waveform charges up C, which holds that value while the diode is back-biased.

This method has some serious problems. The input impedance is variable and is very low during peaks of the input waveform. Also, the diode drop makes the circuit insensitive to peaks less than about 0.6 volt and inaccurate (by one diode drop) for larger peak voltages. Furthermore, since the diode drop depends on temperature and current, the circuit's inaccuracies depend on the ambient temperature and on the rate of change of output; recall that $I = C(dV/dT)$. An input emitter follower would improve the first problem only.

Figure 3.36 shows a better circuit, using feedback. By taking feedback from the voltage at the capacitor, the diode drop doesn't cause any problems. The sort of output

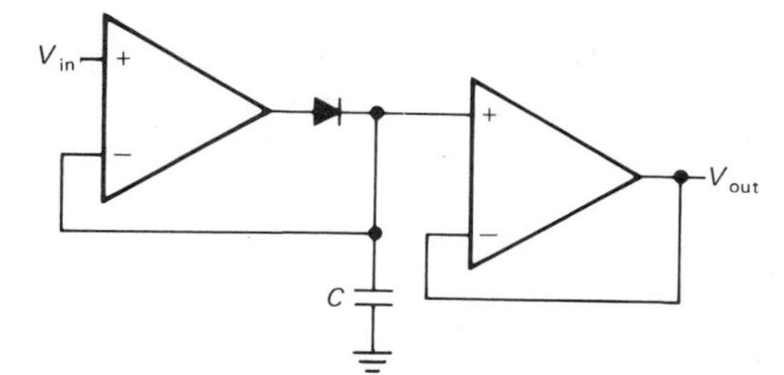

Figure 3.36
Op-amp peak detector.

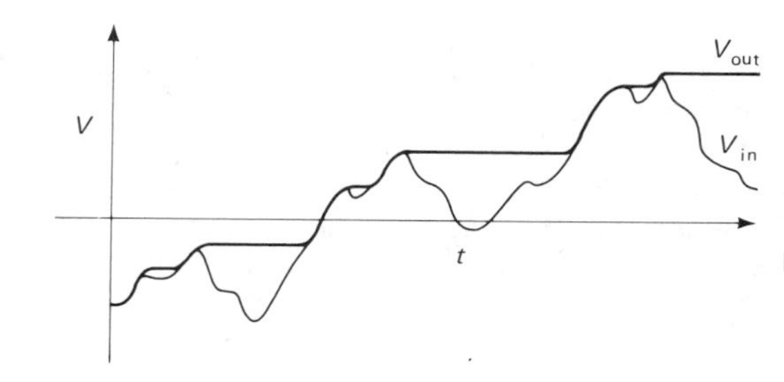

Figure 3.37

waveform you might get is shown in Figure 3.37.

Op-amp limitations affect this circuit in two ways: (a) Input bias current causes a slow discharge (or charge, depending on the sign of the bias current) of the capacitor. This is sometimes called "droop," and it is best avoided by using op-amps with very low bias current. For the same reason, the diode must be a low-leakage type (e.g., the FJT1100, with less than 1pA reverse current at 20V, or a "FET diode" such as the PAD-1 from Siliconix or the ID101 from Intersil), and the following stage must also present high impedance (ideally it should also be a FET or FET-input op-amp). (b) The maximum op-amp output current limits the rate of change of voltage across the capacitor, i.e., the rate at which the output can follow a rising input. Thus the choice of capacitor value is a compromise between low droop and high output slew rate.

For instance, a 1μF capacitor used in this circuit with a 741 (which would be a poor choice because of its high bias current) would droop at $dV/dt = I_B/C = 0.08$V/s and would follow input changes only up to $dV/dt = I_{output}/C = 0.02$V/$\mu$s. This maximum follow rate is much less than the op-amp's slew rate of 0.5V/μs, being limited by the maximum output current of 20mA driving 1μF. By decreasing C you could achieve greater output slewing rate at the expense of greater droop. A more realistic choice of components would be the popular LF355 FET-input op-amp as driver and output follower (30pA bias current, 20mA output current) and a value of $C = 0.01\mu$F. With this combination you would get a droop of only 0.006V/s and an overall circuit slew rate of 2V/μs. For better performance, use a FET op-amp like the ICL8007, AD515, or AD545 with input currents of 1pA or less. Capacitor leakage may then limit performance even if unusually good capacitors are used, e.g., polystyrene or polycarbonate (see Section 7.05).

□ *A circuit cure for diode leakage*

Quite often a clever circuit configuration can provide a solution to problems caused by nonideal behavior of circuit components. Such solutions are aesthetically pleasing as well as economical. At this point we yield to the temptation to take a closer look at such a high-performance design, rather than delaying until Chapter 7, where we treat such subjects under the heading of precision design.

Suppose we want the best possible performance in a peak detector. i.e., highest ratio of output slew rate to droop. If the lowest-input-current op-amps are used in a peak-detector circuit (some are available with bias currents as low as 0.01pA), the droop will be dominated by diode leakage; i.e., the best available diodes have higher leakage currents than the op-amps' bias currents. Figure 3.38 shows a clever circuit

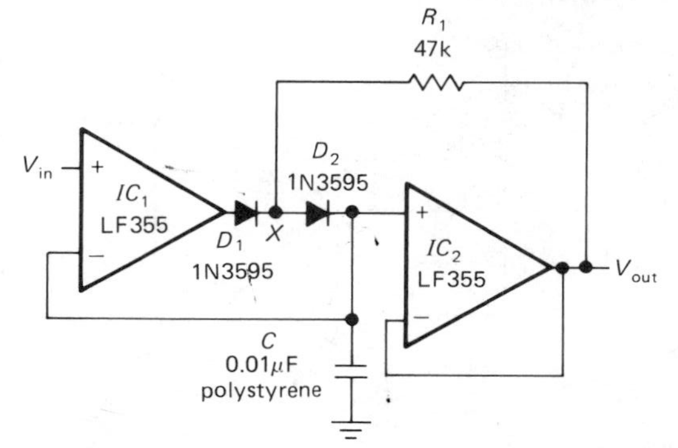

Figure 3.38

solution. As before, the voltage on the capacitor follows a rising input waveform: IC_1 charges the capacitor through both diodes and is unaffected by IC_2's output. When the input drops below the peak value, IC_1 goes into negative saturation, but IC_2 holds point X at the capacitor voltage, eliminating leakage altogether in D_2. D_1's small leakage current flows through R_1, with negligible drop across the resistor. Of course, both op-amps must have low bias current. This circuit is analogous to the so-called guard circuits used for high-impedance or small-signal measurements.

□ *Resetting a peak detector*

In practice it is usually desirable to reset the output of a peak detector in some way. One possibility is to put a resistor across the output so that the circuit's output decays with a time constant RC. In this way it holds only the most recent peak values. A better method is to put a transistor switch across C; a short pulse to the base then zeros the output. A FET switch is often used instead; they will be treated in detail in Chapter 6.

□ 3.16 Active clamp

Figure 3.39 shows a circuit that is an active version of the clamp function we discussed in Chapter 1. For the values shown, $V_{in} < +10$ volts puts the op-amp output at positive saturation, and $V_{out} = V_{in}$. When V_{in} exceeds $+10$ volts the diode closes the feedback loop, clamping the output at 10 volts. In this circuit, op-amp slew rate limitations allow small glitches as the input reaches the clamp voltage from below (Fig. 3.40).

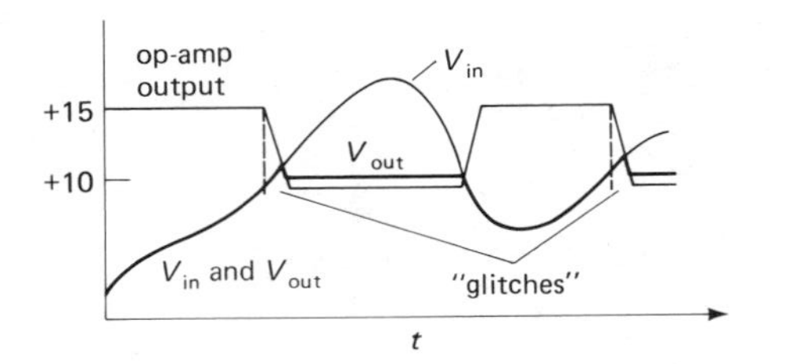

Figure 3.39

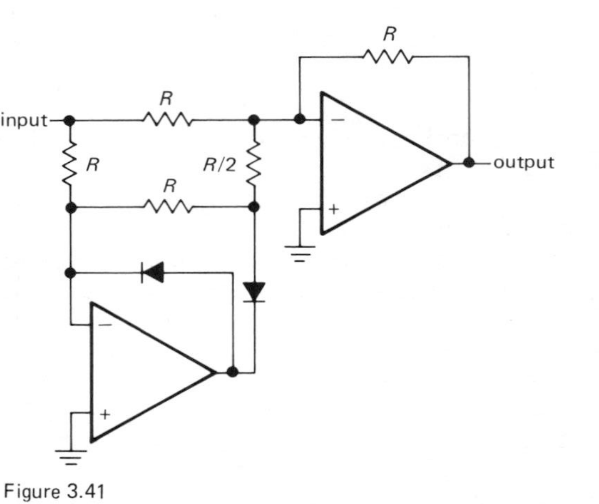

Figure 3.40

□ 3.17 Absolute-value circuit

The circuit shown in Figure 3.41 gives a positive output equal to the magnitude of the input signal; it is a full-wave rectifier. As usual, the use of op-amps and feedback eliminates the diode drops of a passive full-wave rectifier.

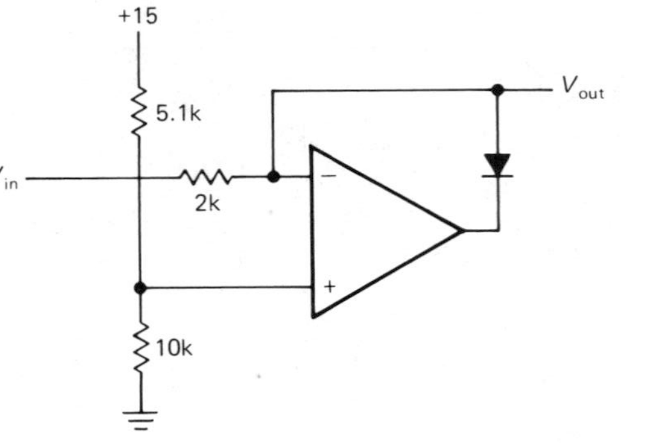

Figure 3.41

EXERCISE 3.8

Figure out how the circuit in Figure 3.41 works.

Figure 3.42 shows another absolute-value circuit. It is readily understandable as a simple combination of an optional inverter (IC_1) and an active clamp (IC_2). For positive input levels the clamp is out of the circuit, with its output at negative saturation, making IC_1 a follower. For negative input levels the clamp holds point X at ground, making IC_1 a unity-gain inverter. Thus the output is equal to the absolute value of the input voltage. By running IC_2 from a single positive supply, you avoid problems of slew rate limitations in the clamp, since its output

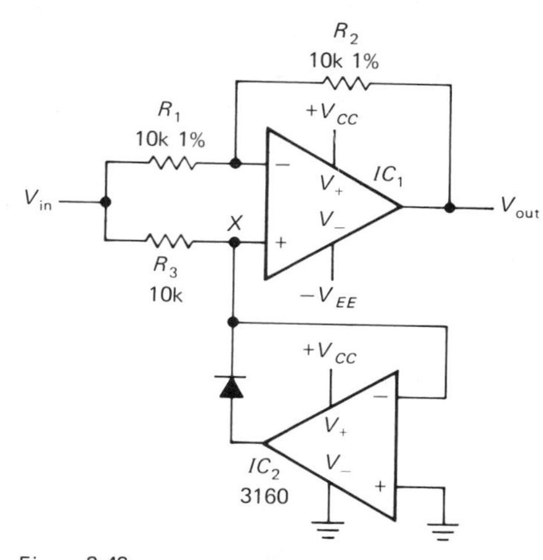

Figure 3.42

moves over only one diode drop. Note that no great accuracy is required of R_3.

3.18 Integrators

Op-amps allow you to make nearly perfect integrators, without the restriction that $V_{out} \ll V_{in}$. Figure 3.43 shows how it's done.

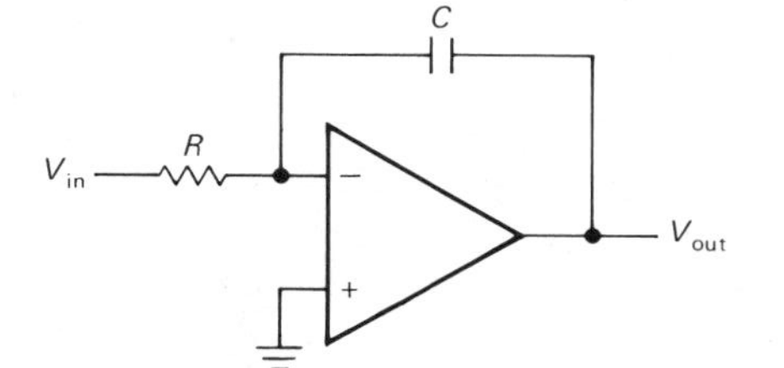

Figure 3.43

Input current V_{in}/R flows through C. Since the inverting input is a virtual ground, the output voltage is given by

$$V_{in}/R = -C(dV_{out}/dt)$$

or

$$V_{out} = -\frac{1}{RC}\int V_{in}\, dt + \text{constant}$$

The input can, of course, be a current, in which case R will be omitted. One problem with this circuit as drawn is that the output tends to wander off, even with the input grounded, due to op-amp offsets and bias current. This problem can be minimized by using an FET op-amp for low input current and offset, trimming the op-amp input offset voltage, and using large R and C values. In addition, in many applications the integrator is zeroed periodically by closing a switch placed across the capacitor (usually an FET), so only the drift over short time scales matters. As an example, an inexpensive FET op-amp like the LF355 (30pA bias current) trimmed to a voltage offset of 0.2mV and used in an integrator with $R = 10M\Omega$ and C $= 10\mu F$ will produce an output drift of less than 0.003 volt in 1000 seconds.

If the residual drift of the integrator is still too large for a given application, it may be necessary to put a large resistor R_2 across C to provide dc feedback for stable biasing. The effect is to roll off the integrator action at very low frequencies, $f < 1/R_2C$. Figure 3.44 shows integrators with FET zeroing switch and with resistor bias stabilization.

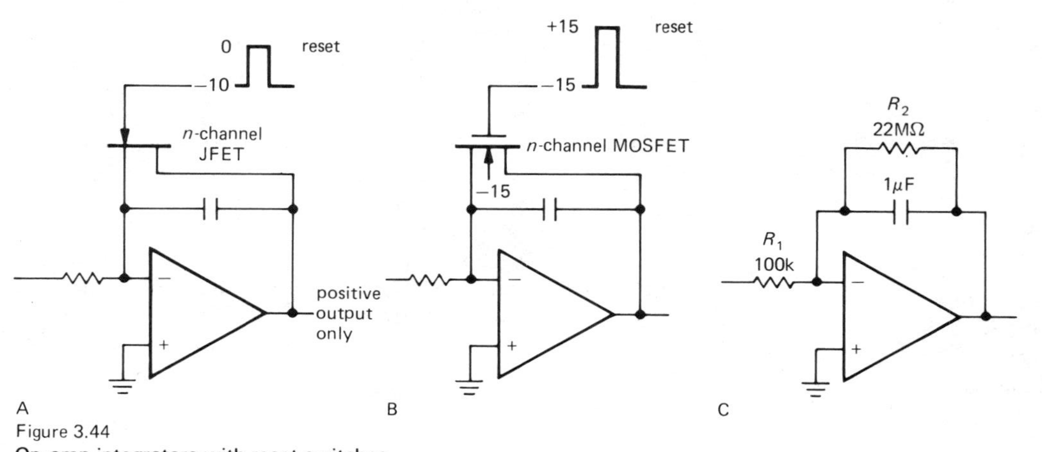

Figure 3.44
Op-amp integrators with reset switches.

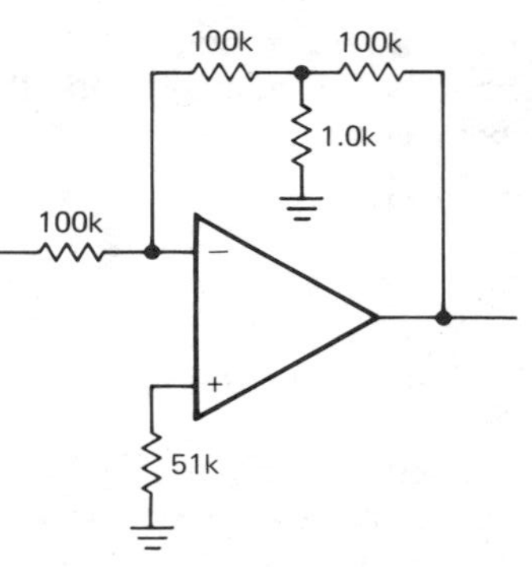

Figure 3.45

The feedback resistor may become rather large in this sort of application. Figure 3.45 shows a trick for producing the effect of a large feedback resistor using smaller values. In this case the feedback network behaves like a single 10MΩ resistor in the standard inverting amplifier circuit giving a voltage gain of −100. This technique has the advantage of using resistors of convenient values without the problems of stray capacitance, etc., that occur with very large resistor values.

□ 3.19 Differentiators

Differentiators are similar to integrators, but with R and C reversed (Fig. 3.46). Since the

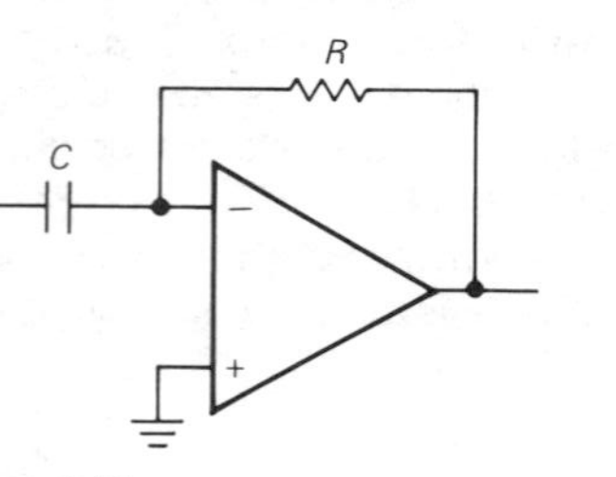

Figure 3.46

inverting input is at ground, the rate of change of input voltage produces a current $I = C(dV_{in}/dt)$ and hence an output voltage

$$V_{out} = -RC\frac{dV_{in}}{dt}$$

Differentiators are bias-stable, but they generally have problems with noise and instabilities at high frequencies because of the op-amp's high gain and internal phase shifts. For this reason it is necessary to roll off the differentiator action at some maximum frequency. The usual method is shown in Figure 3.47. The choice of the rolloff components R_1 and C_2 depends on the noise level of the signal and the bandwidth of the op-amp. At high frequencies this circuit becomes an integrator, due to R_1 and C_2.

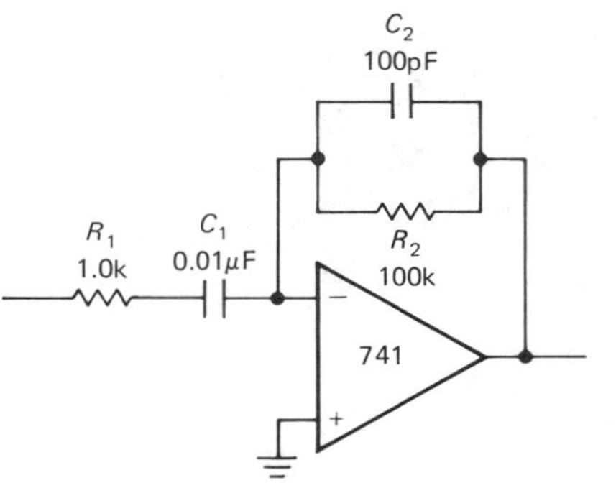

Figure 3.47

□ OP-AMP OPERATION WITH A SINGLE POWER SUPPLY

Op-amps don't require ±15 volt regulated supplies. They can be operated from split supplies of lower voltages, or from unsymmetrical supply voltages (e.g., +12 and −3), as long as the total supply voltage ($V_+ - V_-$) is within specifications (see Table 3.1). Unregulated supply voltages are often adequate because of the high "power supply rejection ratio" you get from negative feedback (for the 741, it's 90 dB typ). But there are many occasions when it would be nice to operate an op-amp from a single supply, say +12 volts. This can be done with ordinary op-amps by generating a "reference" voltage above ground, if you are careful about minimum supply voltages, output swing limitations, and maximum common-mode input range. With some of the more recent op-amps whose input and output ranges include the negative supply (i.e., ground, when run from a single positive supply), single-supply operation is attractive because of its simplicity. Keep in mind, though, that operation with symmetrical split supplies remains the usual technique for nearly all applications.

□ 3.20 Biasing single-supply ac amplifiers

For a general-purpose op-amp like the 741, the inputs and output can typically swing to

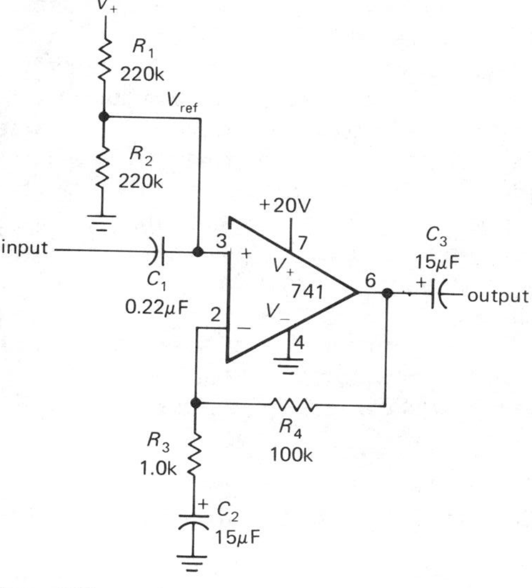

Figure 3.48

within about 1.5 volts of either supply. With V_- connected to ground, you can't have either of the inputs or the output at ground. Instead, by generating a reference voltage (e.g., $0.5V_+$) you can bias the op-amp for successful operation (Figure 3.48). This circuit is an audio amplifier with 40dB gain. $V_{ref} = 0.5V_+$ gives an output swing of about 17 volts pp before onset of clipping. Capacitive coupling is used at the input and output to block the dc level, which equals V_{ref}.

□ 3.21 Single-supply op-amps

There are two kinds of op-amps now available that permit simplified operation with a single positive supply voltage:

1. The LM324(quad)/LM358(dual) or μA798(dual)/799(single with offset trim) type. These have input common-mode ranges all the way down to 0.3 volt *below* V_-, and the output can swing down to V_-. Both inputs and output can go to within 1.5 volts of V_+. (If you need an input range up to V_+, use something like an LM301/307 or a 355; an example is illustrated in Section 5.24 in the discussion of constant-current supplies.) In order to understand some of the subtleties of this sort of op-amp, it is helpful to look at the schematic (Fig. 3.49). It is a reasonably straightforward differential amplifier, with current mirror active load on the input stage and push-pull complementary output stage with current limiting. The special things to remember are these (calling V_- ground):

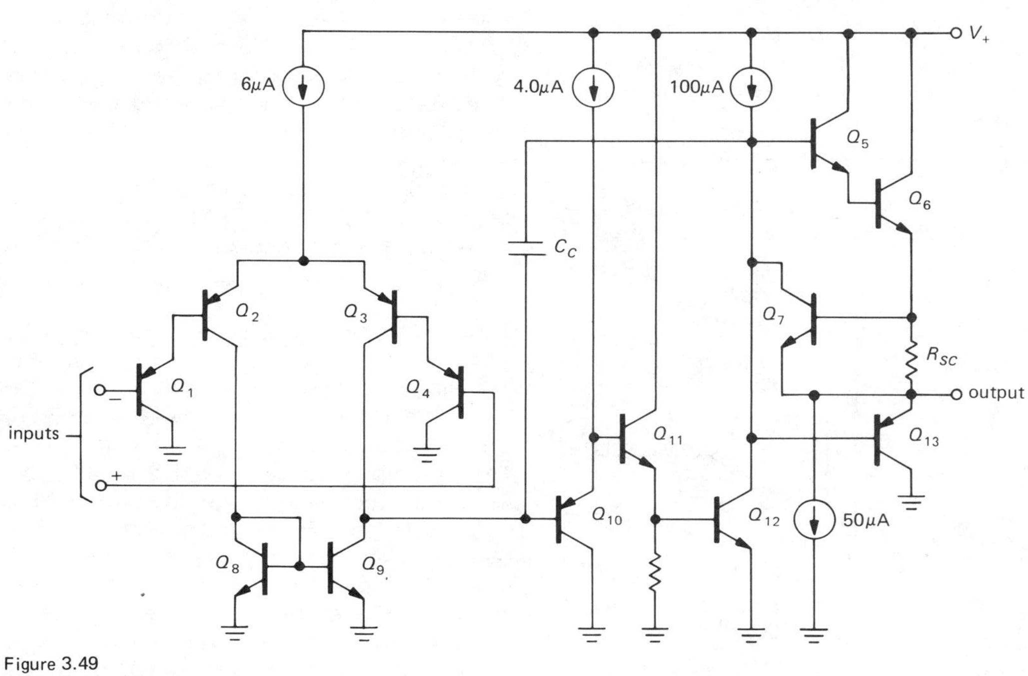

Figure 3.49
Schematic of the popular 324 and 358 op-amps. (National Semiconductor Corp.)

Inputs: The *pnp* input structure allows swings of 0.3 volts below ground; if that is exceeded by either input, weird things happen at the output (it may go negative, for instance).

Output: Q_{13} pulls the output down and can sink plenty of current, but it goes only to within a diode drop of ground. Outputs below that are provided by the 50μA current sink, which means you can't drive a load that sources more than 50μA and get closer than a diode drop above ground. Even for "nice" loads (an open circuit, say), the current source won't bring the output lower than a saturation voltage (0.1V) above ground. If you want the output to go clear down to ground, the load should sink a small current to ground; it could be a resistor to ground, for instance.

We will illustrate the use of these op-amps with some circuits, after mentioning the other kind of op-amp that lends itself well to single-supply operation.

2. The CA3130/3160-type FET op-amp. This kind of op-amp uses complementary FETs in the output stage. When saturated, they look like a small resistance from the output to the supply (V_+ or V_-). Thus the output can swing all the way to either supply. In addition, the inputs can go 0.5 volt below V_-. Unfortunately, the CA3130 and 3160 are limited to 16 volts (max) total supply voltage and ±8 volts differential input voltage.

□ ***Example: single-supply photometer***

Figure 3.50 shows a typical example of a circuit for which single-supply operation is convenient. We discussed a similar circuit earlier under the heading of current-to-voltage converters. Since a photocell circuit might well be used in a portable light-measuring instrument, and since the output is known to be positive only, this is a good candidate for a battery-operated single-supply circuit. R_1 cancels the effect of input bias current, R_2 sets the gain at 1 volt output per microamp, and R_3 trims the input offset voltage.

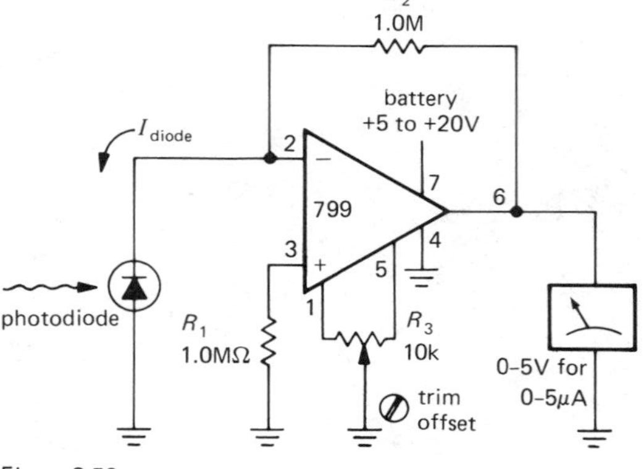

Figure 3.50

Better performance at low light levels results if the photodiode is connected as in the circuit shown in Figure 3.82B.

COMPARATORS AND SCHMITT TRIGGER

It is quite common to want to know which of two signals is larger, or to know when a given signal exceeds a predetermined value. For instance, the usual method of generating triangle waves is to supply positive or negative currents into a capacitor, reversing the polarity of the current when the amplitude reaches a preset peak value. Another example is a digital voltmeter. In order to convert a voltage to a number, the unknown voltage is applied to one input of a comparator, with a linear ramp (capacitor + current source) applied to the other. A digital counter counts cycles of an oscillator while the ramp is less than the unknown voltage and displays the result when equality of amplitudes is reached. The resultant count is proportional to the input voltage. This is called single-slope integration; in most sophisticated instruments a dual-slope integration is used (see Chapter 9).

3.22 Comparators

The simplest form of comparator is a high-gain differential amplifier, made either with transistors or with an op-amp (Fig. 3.51). The op-amp goes into positive or negative saturation according to the difference of the input voltages. Because the voltage gain typically exceeds 100,000, the inputs will have to be equal to a fraction of a millivolt in order for the output not to be saturated. Although an ordinary op-amp can be used as a comparator (and frequently is), there are

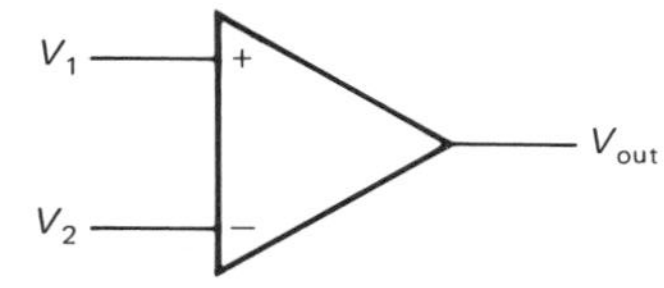

Figure 3.51

special integrated circuits intended for use as comparators. Some examples are the LM306, LM311, LM393, and NE529. These chips are designed for very fast response and aren't even in the same league as op-amps. For example, the high speed NE521 slews at several thousand volts per microsecond. With comparators, the term "slew rate" isn't usually used; you talk instead about "propagation delay versus input overdrive."

Comparators generally have more flexible output circuits than op-amps. Whereas an ordinary op-amp uses a push-pull output stage to swing between the supply voltages (±13V, say, for a 741 running from ±15V supplies), a comparator chip usually has an "open-collector" output with grounded emitter. By supplying an external "pull-up" resistor (that's accepted terminology, believe it or not) connected to a voltage of your choice, you can have an output swing from +5 volts to ground, say. You will see later that logic circuits have well-defined voltages they like to operate between; the preceding example would be ideal for driving a TTL circuit, a popular type of digital logic. Figure 3.52 shows the circuit. The output switches

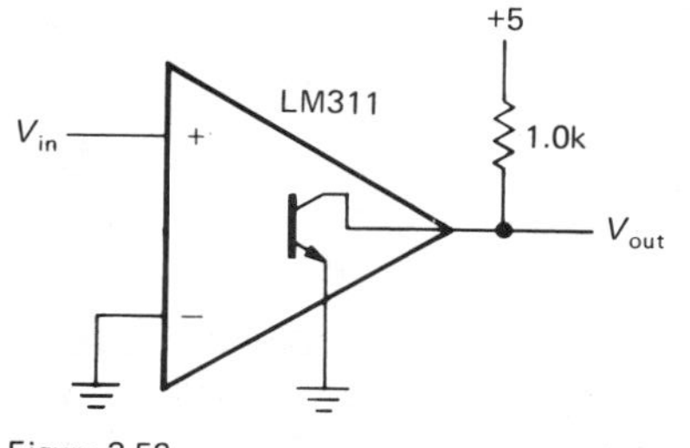

Figure 3.52

from +5 volts to ground when the input signal goes negative. This use of a comparator is really an example of analog-to-digital conversion.

This is the first example we have presented of an *open-collector* output; this is a common configuration in logic circuits, as you will see throughout Chapters 8–11. If you like, you can think of the external pull-up resistor as completing the comparator's internal circuit by providing a collector load resistor for an *npn* output transistor. Since the output transistor operates as a saturated switch, the resistor value is not at all critical, with values typically between a few hundred ohms and a few thousand ohms; smaller values yield improved switching speed and noise immunity at the expense of increased power dissipation. Incidentally, in spite of their superficial resemblance to op-amps, comparators are never used with negative feedback because they would not be stable (see Sections 3.31–3.33). However, some *positive* feedback is often used, as you will see in the next section.

Comments on comparators

Some points to remember (a) Since there is no negative feedback, golden rule I is not obeyed. The inputs are not at the same voltage. (b) The absence of negative feedback means that the (differential) input impedance isn't bootstrapped to the high values characteristic of op-amp circuits. As a result, the input signal sees a changing load and changing (small) input current as the comparator switches; if the driving impedance is too high, strange things may happen. (c) Some comparators permit only limited differential input swings, as little as ±5 volts in some cases. Check the specs! See Table 9.2 and the discussion in Section 9.07 for the properties of some popular comparators.

3.23 Schmitt trigger

The simple comparator circuit in Figure 3.52 has two disadvantages. For a very slowly varying input, the output swing can be rather slow. Worse still, if the input is noisy, the output may make several transitions as the input passes through the trigger point (Fig. 3.53). Both these problems can be remedied by the use of *positive* feedback (Fig. 3.54). The effect of R_3 is to make the circuit have two thresholds, depending on the output state. In the example shown, the threshold when the output is at ground (input high) is 4.76 volts, whereas the threshold with the

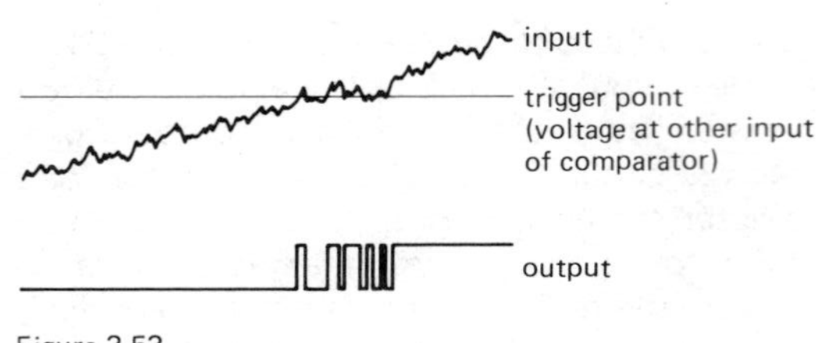

Figure 3.53

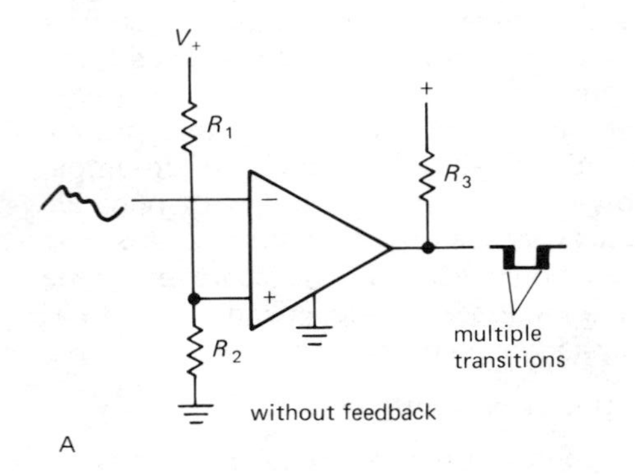

A

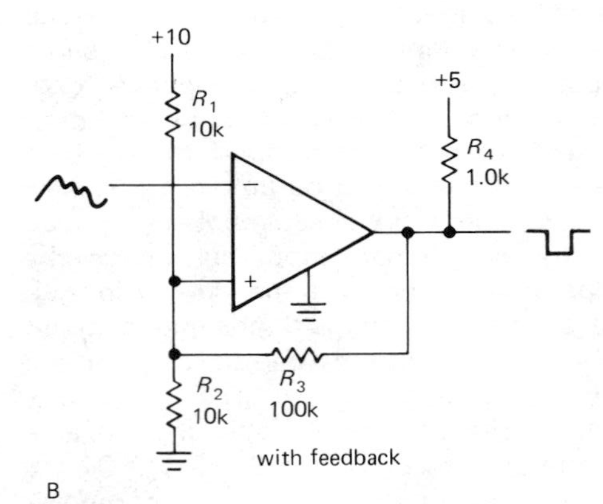

B

Figure 3.54

output at +5 volts is 5.0 volts. A noisy input is less likely to produce multiple triggering (Fig. 3.55). Furthermore, the positive feedback ensures a rapid output transition, regardless of the speed of the input waveform. (A small ''speedup'' capacitor of 10–100pF is often connected across R_3 to enhance switching speed still further.) This configuration is known as a Schmitt trigger. (If an op-amp were used, the pull-up would be omitted.) The output depends both on the input voltage and on its recent history, an effect called *hysteresis*. This can be illustrated with a diagram of output versus input, as in Figure 3.56.

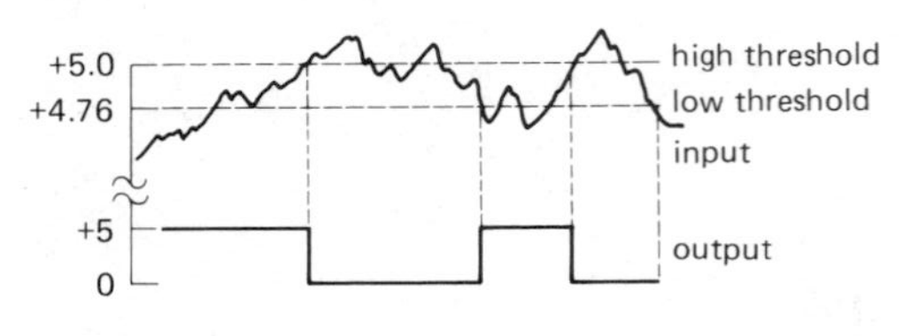

Figure 3.55

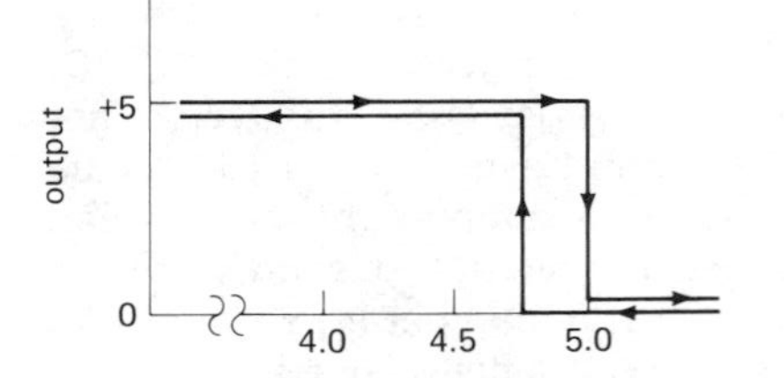

Figure 3.56

Discrete-transistor Schmitt trigger

A Schmitt trigger can also be made simply with transistors (Fig. 3.57). Q_1 and Q_2 share an emitter resistor. It is essential that Q_1's collector resistor be larger than Q_2's. In that way the threshold to turn on Q_1, which is one diode drop above the emitter voltage,

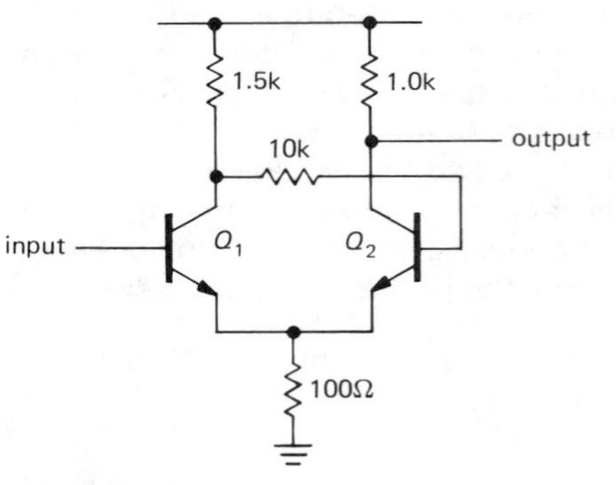

Figure 3.57

rises when Q_1 is turned off, since the emitter current is higher with Q_2 conducting. This produces hysteresis in the trigger threshold, just as in the preceding integrated-circuit Schmitt trigger.

EXERCISE 3.9

Design a Schmitt trigger using a 311 comparator (open-collector output) with thresholds at +1.0 volt and +1.5 volts. Use a

1.0k pull-up resistor to +5 volts, and assume that the 311 is powered from ±15 volt supplies.

FEEDBACK WITH FINITE-GAIN AMPLIFIERS

We mentioned in Section 3.12 that the finite open-loop gain of an op-amp limits its performance in a feedback circuit. Specifically, the closed-loop gain can never exceed the open-loop gain, and as the open-loop gain approaches the closed loop gain, the amplifier begins to depart from the ideal behavior we have come to expect. In this section we will quantify these statements so that you will be able to predict the performance of a feedback amplifier constructed with real (less than ideal) components. This is important also for feedback amplifiers constructed entirely with discrete components (transistors), where the open-loop gain is usually much less than with op-amps. In these cases the output impedance, for instance, will not be zero. Nonetheless, with a good understanding of feedback principles you will be able to achieve the performance required in any given circuit.

3.24 Gain equation

Let's begin by considering an amplifier of finite voltage gain, connected with feedback (Fig. 3.58). The amplifier has open-loop

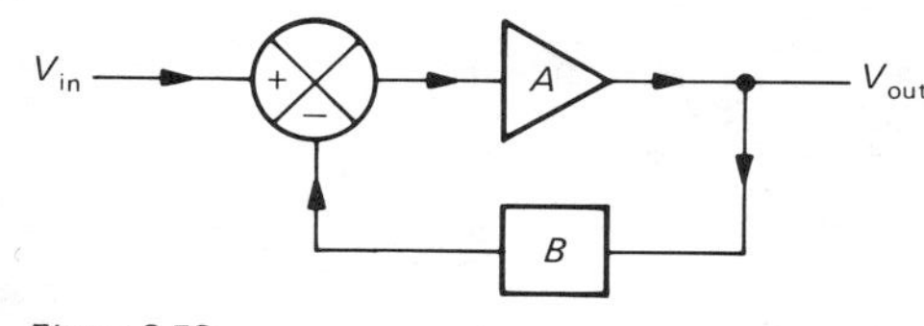

Figure 3.58

voltage gain A, and the feedback network subtracts a fraction B of the output voltage from the input. (Later we will generalize things so that inputs and outputs can be currents or voltages.) The input to the gain block is then $V_{in} - BV_{out}$. But the output is just the input times A:

$$A(V_{in} - BV_{out}) = V_{out}$$

In other words,

$$V_{out} = \frac{A}{1 + AB} V_{in}$$

and the closed-loop voltage gain, V_{out}/V_{in}, is just

$$G = \frac{A}{1 + AB}$$

Some terminology: The standard designations for these quantities are as follows: G = closed-loop gain, A = open-loop gain, AB = loop gain, $1 + AB$ = return difference, or desensitivity. The feedback network is sometimes called the beta network (no relation to transistor beta, h_{fe}).

3.25 Effects of feedback on amplifier circuits

Let's look at the important effects of feedback. The most significant are predictability of gain (and reduction of distortion), changed input impedance, and changed output impedance.

Predictability of gain

The voltage gain is $A/(1 + AB)$. In the limit of infinite open-loop gain A, $G = 1/B$. We saw this result in the noninverting amplifier configuration, where a voltage divider on the output provided the signal to the inverting input. The closed-loop voltage gain was just the inverse of the division ratio of the voltage divider. For finite gain A, feedback still acts to reduce the effects of variations of A (with frequency, temperature, amplitude, etc.). For instance, suppose A depends on frequency as in Figure 3.59. This will surely satisfy anyone's definition of a poor amplifier (the gain varies over a factor of 10 with frequency). Now imagine we introduce feed-

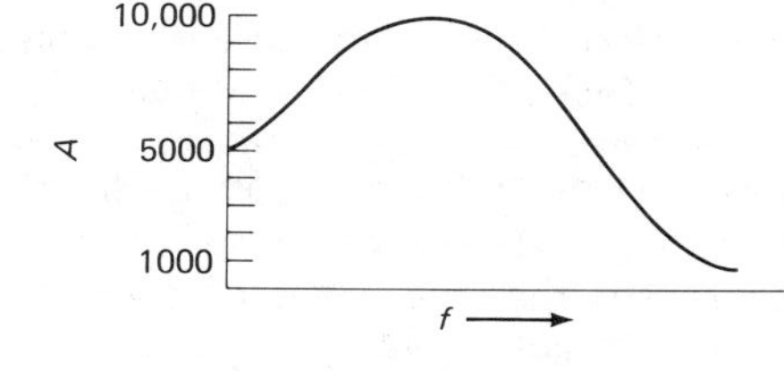

Figure 3.59

back, with $B = 0.1$ (a simple voltage divider will do). The closed-loop voltage gain now varies from $1000/[1 + (1000 \times 0.1)]$, or 9.90, to $10{,}000/[1 + (10{,}000 \times 0.1)]$, or 9.99, a variation of just 1% over the same range of frequency! To put it in audio terms, the original amplifier is flat to ±10dB, whereas the feedback amplifier is flat to ±0.04dB. We can now recover the original gain of 1000 with nearly this linearity by just cascading three such stages. It was for just this reason (namely, the need for extremely flat telephone repeater amplifiers) that negative feedback was invented. As the inventor, Harold Black, described it in his first open publication on the invention (*Electrical Engineering,* 53:114, 1934), "by building an amplifier whose gain is made deliberately, say 40 decibels higher than necessary (10,000-fold excess on energy basis) and then feeding the output back to the input in such a way as to throw away the excess gain, it has been found possible to effect extraordinary improvement in constancy of amplification and freedom from nonlinearity."

It is easy to show that relative variations in the open-loop gain are reduced by the desensitivity:

$$\frac{\Delta G}{G} = \frac{1}{1 + AB}\,\frac{\Delta A}{A}$$

Thus for good performance the loop gain AB should be much larger than 1. That's equivalent to saying that the open-loop gain should be much larger than the closed-loop gain.

A very important consequence of this is that nonlinearities, which are simply gain variations that depend on signal level, are reduced in exactly the same way.

Input impedance

Feedback can be arranged to subtract a voltage or a current from the input (these are sometimes called *series feedback* and *shunt feedback,* respectively). The noninverting op-amp configuration, for instance, subtracts a sample of the output voltage from the differential voltage appearing at the input, whereas in the inverting configuration a current is subtracted from the input. The effects on input impedance are opposite in the two cases: Voltage feedback multiplies the open-loop input impedance by $1 + AB$, whereas current feedback reduces it by the same factor. In the limit of infinite loop gain the input impedance (at the amplifier's input terminal) goes to infinity or zero, respectively. This is easy to understand, since voltage feedback tends to subtract signal from the input, resulting in a smaller change (by the factor AB) across the amplifier's input resistance; it's a form of bootstrapping. Current feedback reduces the input signal by bucking it with an equal current.

Let's see explicitly how the effective input impedance is changed by feedback. We will illustrate the case of voltage feedback only, since the derivations are similar for the two cases. We begin with an op-amp model with (finite) input resistance as shown in Figure 3.60. An input V_{in} is reduced by BV_{out},

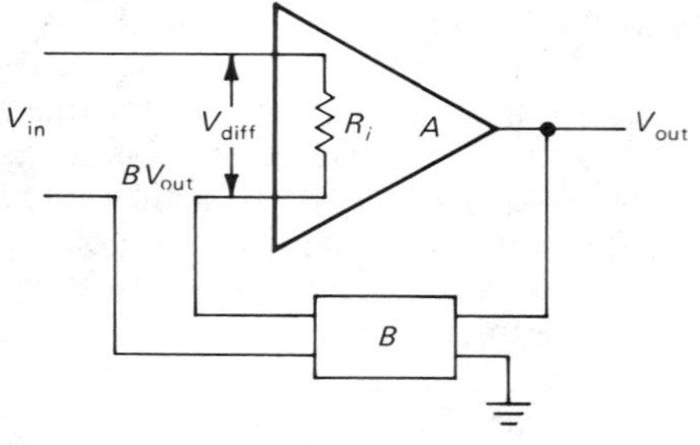

Figure 3.60

putting a voltage $V_{diff} = V_{in} - BV_{out}$ across the inputs of the amplifier. The input current is therefore

$$I_{in} = \frac{V_{in} - BV_{out}}{R_i} = \frac{V_{in}\left(1 - B\dfrac{A}{1 + AB}\right)}{R_i} = \frac{V_{in}}{(1 + AB)R_i}$$

giving an effective input resistance

$$R_i' = V_{in}/I_{in} = (1 + AB)R_i$$

The classic op-amp noninverting amplifier is exactly this feedback configuration, as shown in Figure 3.61. In this circuit, $B = R_1/(R_1 + R_2)$, giving the usual voltage gain expression $G_v = 1 + R_2/R_1$ and an infinite input impedance for the ideal case of infinite

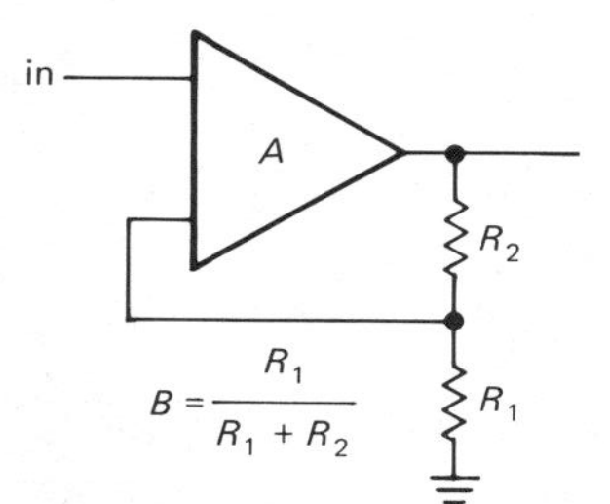

Figure 3.61

open-loop voltage gain A. For finite loop gain the equations as previously derived apply.

In the case of current (shunt) feedback, the input to the amplifier is at the "summing junction" (the inverting input of the amplifier), where the currents from feedback and input signals are combined (this amplifier connection is really a "transresistance" configuration; it converts a current input to a voltage output). Feedback reduces the impedance looking into the summing junction (including the resistance of the feedback resistor) by the factor $1 + AB$. In cases of very high loop gain (e.g., an op-amp) the input impedance is reduced to a fraction of an ohm, a good characteristic for a current-input amplifier. Some good examples are the photometer amplifier in Section 3.21 and the logarithmic converter in Section 3.14.

The classic op-amp inverting amplifier connection is a combination of a shunt feedback amplifier and a series input resistor, as shown in Figure 3.62. As a result, the input

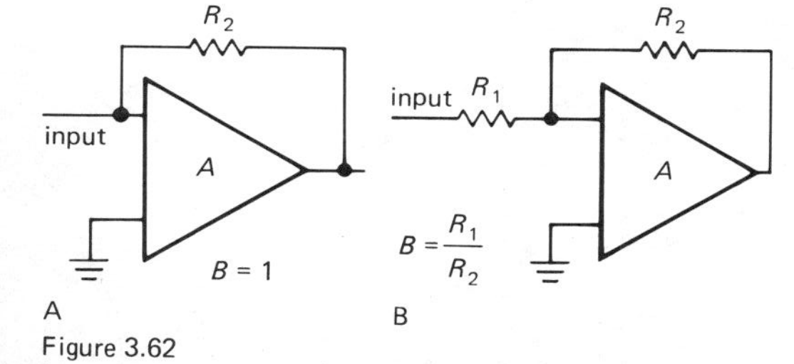

Figure 3.62
Input impedance of the inverting amplifier.

impedance equals the sum of R_1 and the impedance looking into the summing junction. For the ideal case of infinite open-loop voltage gain, the input impedance of the entire amplifier circuit is just R_1, and the gain is $-R_2/R_1$.

Output impedance

Again, feedback can extract a sample of the output voltage or the output current. In the first case the open-loop output impedance will be reduced by the factor $1 + AB$, whereas in the second case it will be increased by the same factor. We will illustrate this effect for the case of voltage sampling. We begin with the model shown in Figure 3.63. This time we have shown the

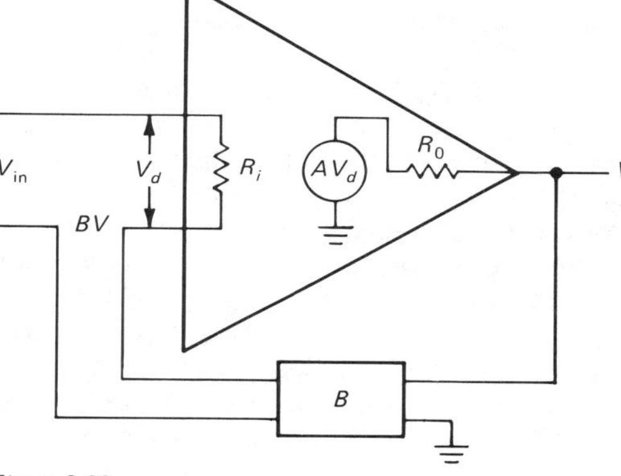

Figure 3.63

output impedance explicitly. The calculation is simplified by a trick: Short the input, and apply a voltage V to the output; by calculating the output current I, we get the output impedance $R_o' = V/I$. Voltage V at the output puts a voltage $-BV$ across the amplifier's input, producing a voltage $-ABV$ in the amplifier's internal generator. The output current is therefore

$$I = \frac{V - (-ABV)}{R_o} = \frac{V(1 + AB)}{R_o}$$

giving an effective output impedance

$$R_o' = V/I = R_o \;/(1 + AB)$$

If feedback is connected instead to sample the output current, the expression becomes

$$R_o' = R_o(1 + AB)$$

It is possible to have multiple feedback paths, sampling both voltage and current. In the general case the output impedance is given by Blackman's impedance relation

$$R'_o = R_o \frac{1 + (AB)_{SC}}{1 + (AB)_{OC}}$$

where $(AB)_{SC}$ is the loop gain with the output shorted to ground and $(AB)_{OC}$ is the loop gain with no load attached. Thus feedback can be used to generate a desired output impedance. This equation reduces to the previous results for the usual situation in which feedback is derived from either the output voltage or the output current.

Loading by the feedback network

In feedback computations, you usually assume that the beta network doesn't load the amplifier's output. If it does, that must be taken into account in computing the open-loop gain. Likewise, if the connection of the beta network at the amplifier's input affects the open-loop gain (feedback removed, but network still connected), you must use the modified open-loop gain. Finally, the preceding expressions assume that the beta network is unidirectional, i.e., it does not couple signal from the input to the output.

3.26 Two examples of transistor amplifiers with feedback

Figure 3.64 shows a transistor amplifier with negative feedback.

Circuit description

It may look complicated, but it is extremely straightforward in design and is relatively easy to analyze. Q_1 and Q_2 form a differential pair, with common-emitter amplifier Q_3 amplifying its output. R_6 is Q_3's collector load resistor, and push-pull pair Q_4 and Q_5 form the output emitter follower. The output voltage is sampled by the feedback network consisting of voltage divider R_4 and R_5, with C_2 included to reduce the gain to unity at dc for stable biasing. R_3 sets the quiescent current in the differential pair, and since overall feedback guarantees that the quiescent output voltage is at ground, Q_3's quiescent current is easily seen to be 10mA (V_{EE} across R_6, approximately). As we have discussed earlier (Section 2.14), the diodes bias the push-pull pair into conduction, leaving one diode drop across the series pair R_7 and R_8, i.e., 60mA quiescent current. That's

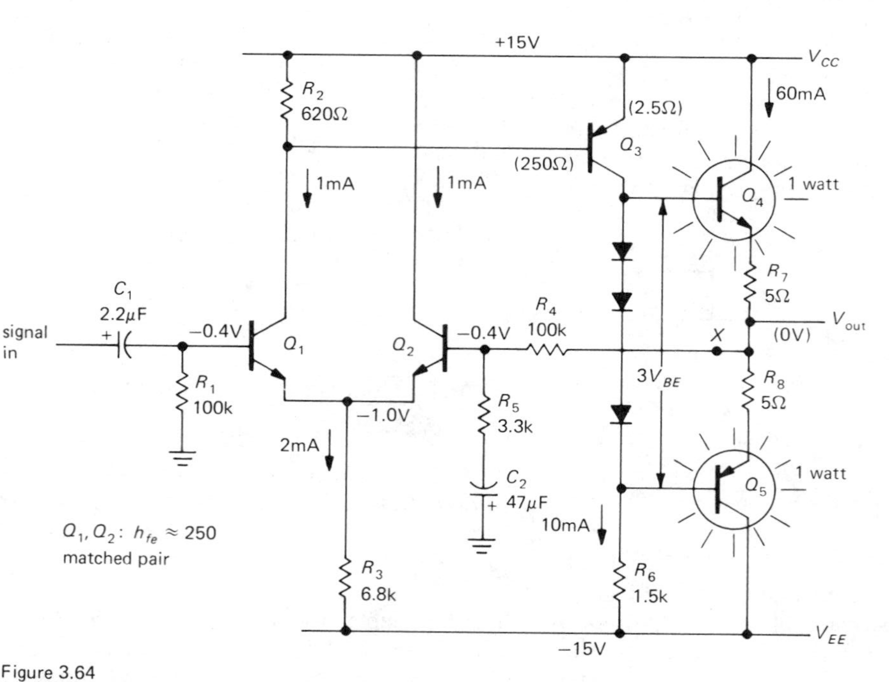

Figure 3.64
Transistor power amplifier with negative feedback.

class AB operation, good for minimizing crossover distortion, at the cost of 1 watt standby dissipation in each output transistor.

From the point of view of our earlier circuits, the only unusual feature is Q_1's quiescent collector voltage, one diode drop below V_{CC}. That is where it must sit in order to hold Q_3 in conduction, and the feedback path ensures that it will. (For instance, if Q_1 were to pull its collector closer to ground, Q_3 would conduct heavily, raising the output voltage, which in turn would force Q_2 to conduct more heavily, reducing Q_1's collector current and hence restoring the status quo.) R_2 was chosen to give a diode drop at Q_1's quiescent current in order to keep the collector currents in the differential pair approximately equal at the quiescent point. In this transistor circuit the input bias current is not negligible ($4\mu A$), resulting in a 0.4 volt drop across the 100k input resistors. In transistor amplifier circuits like this, in which the input currents are considerably larger than in op-amps, it is particularly important to make sure that the dc resistances seen from the inputs are equal, as shown (a Darlington input stage would probably be better here).

□ *Analysis*

Let's analyze this circuit in detail, determining the gain, input and output impedances, and distortion. To illustrate the utility of feedback, we will find these parameters for both the open-loop and closed-loop situations (recognizing that biasing would be hopeless in the open-loop case). To get a feeling for the linearizing effect of the feedback, the gain will be calculated at +10 volts and −10 volts output, as well as the quiescent point (zero volts).

□ *Open loop.* Input impedance: We cut the feedback at point X and ground the right side of R_4. The input signal sees 100k in parallel with the impedance looking into the base. The latter is h_{fe} times twice the intrinsic emitter resistance plus the impedance seen at Q_2's emitter due to the feedback network at Q_2's base. For $h_{fe} \approx 250$, $Z_{in} \approx 250 \times [(2 \times 25) + (3.3k/250)]$; i.e., $Z_{in} \approx$ 16k.

Output impedance: Since the impedance looking back into Q_3's collector is high, the output transistors are driven by a 1.5k source (R_6). The output impedance is about 15 ohms ($h_{fe} \approx 100$) plus the 5 ohm emitter resistance, or 20 ohms. The intrinsic emitter resistance of 0.4 ohm is negligible.

Gain: The differential input stage sees a load of R_2 paralleled by Q_3's base resistance. Since Q_3 is running 10mA quiescent current, its intrinsic emitter resistance is 2.5 ohms, giving a base impedance of about 250 ohms (again, $h_{fe} \approx 100$). The differential pair thus has a gain of

$$\frac{250 \parallel 620}{2 \times 25\Omega} \quad \text{or} \quad 3.5$$

The second stage, Q_3, has a voltage gain of 1.5k/2.5 ohms, or 600. The overall voltage gain at the quiescent point is 3.5 × 600, or 2100. Since Q_3's gain depends on its collector current, there is substantial change of gain with signal swing, i.e., nonlinearity. The gain is tabulated in the following section for three values of output voltage.

□ *Closed loop.* Input impedance: This circuit uses series feedback, so the input impedance is raised by (1 + loop gain). The feedback network is a voltage divider with $B = 1/30$ at signal frequencies, so the loop gain AB is 70. The input impedance is therefore 70 × 16k, still paralleled by the 100k bias resistor, i.e., about 92k. The bias resistor now dominates the input impedance.

Output impedance: Since the output *voltage* is sampled, the output impedance is reduced by (1 + loop gain). The output impedance is therefore 0.3 ohm. Note that this is a small-signal impedance and does not mean that a 1 ohm load could be driven to nearly full swing, for instance. The 5 ohm emitter resistors in the output stage limit the large signal swing. For instance, a 4 ohm load could be driven only to 10 volts pp, approximately.

Gain: The gain is $A/(1 + AB)$. At the quiescent point that equals 30.84, using the exact value for B. In order to illustrate the gain stability achieved with negative feedback, the overall voltage gain of the circuit with and without feedback is tabulated at three values of output level at the end of this paragraph. It should be obvious that negative feedback has brought about considerable improvement in the amplifier's characteristics, although in fairness it should be pointed out that the amplifier could have been designed for better open-loop performance, e.g., by using a current source for Q_3's collector load and degenerating its emitter, by using a current source for the differential-pair emitter circuit, etc. Even so, feedback would still make a large improvement.

	Open Loop			*Closed Loop*		
V_{out}	−10	0	+10	−10	0	+10
Z_{in}	16k	16k	16k	92k	92k	92k
Z_{out}	20Ω	20Ω	20Ω	0.3Ω	0.3Ω	0.3Ω
Gain	1360	2100	2400	30.60	30.84	30.90

□ *Series feedback pair*

Figure 3.65 shows another transistor amplifier with feedback. Thinking of Q_1 as an

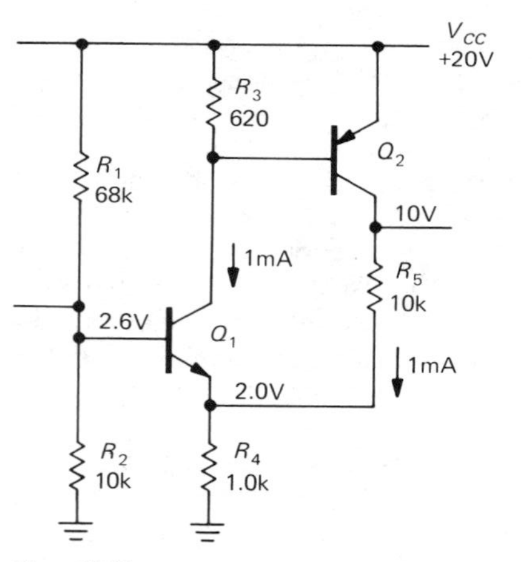

Figure 3.65

amplifier of its base-emitter voltage drop (thinking in the Ebers-Moll sense), the feedback samples the output voltage and subtracts a fraction of it from the input signal. This circuit is a bit tricky because Q2's collector resistor doubles as the feedback network. Applying the techniques we used earlier, you should be able to show that G (open loop) ≈ 200, loop gain ≈ 20, Z_{out} (open loop) ≈ 10k, Z_{out} (closed loop) ≈ 500 ohms, and G (closed loop) ≈ 9.5.

SOME TYPICAL OP-AMP CIRCUITS

3.27 General-purpose lab amplifier

Figure 3.66 shows a dc-coupled "decade amplifier" with settable gain, bandwidth, and wide-range dc output offset. IC_1 is a FET-input op-amp with noninverting gain from unity (0dB) to ×100 (40dB) in accurately calibrated 10dB steps; a vernier is provided for variable gain. IC_2 is an inverting amplifier; it allows offsetting the output over a range of ±10 volts, accurately calibrated via R_{14}, by injecting current into the summing junction. C_2–C_4 set the high-frequency rolloff, since it is often a nuisance to have excessive bandwidth (and noise). IC_5 is a power booster for driving low-impedance loads or cables; it can provide ±300mA output current.

Some interesting details: a 10MΩ input resistor is small enough, since the bias current of the 356 is 30pA (0.3mV error with open input). R_2, in combination with D_1 and D_2, limits the input voltage at the op-amp to the range V_- to $V_+ + 0.7$. D_3 is used to generate a clamp voltage at $V_- + 0.7$, since the input common-mode range extends only to V_- (exceeding V_- causes the output to reverse phase). With the protection components shown, the input can go to ±150 volts without damage.

EXERCISE 3.10

Check that the gain is as advertised. How does the variable offset circuitry work?

3.28 Voltage-controlled oscillator

Figure 3.67 shows a clever circuit, borrowed from the application notes of several manufacturers. IC_1 is an integrator, rigged up so that the input current (V_{in}/200k) changes sign, but not magnitude, when Q_1 conducts. IC_2 is connected as a Schmitt trigger, with thresholds at one-third and two-thirds of V_+. Q_1 is an *n*-channel MOSFET, used here as a switch; it is simpler to use than bipolar transistors in this sort of application, but an alternative circuit using *npn* transistors is shown in addition. In either case the bottom side of R_4 is pulled to ground when the

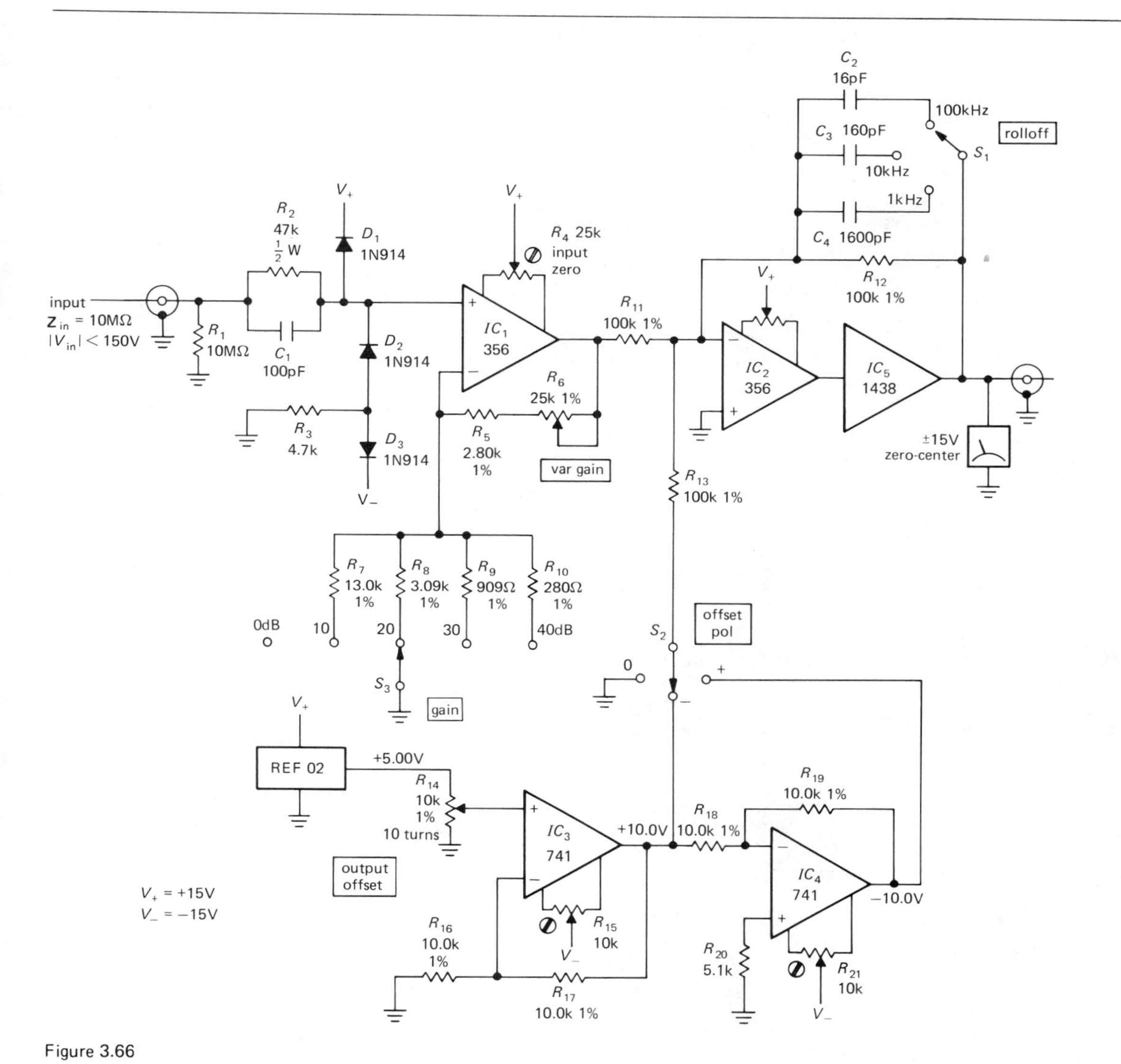

Figure 3.66
Laboratory dc amplifier with output offset.

output is HIGH and open-circuited when the output is LOW.

An unusual feature of this circuit is its operation from a single positive supply. The 3160 (internally compensated version of the 3130) has FETs as output transistors, guaranteeing a full swing between V_+ and ground at the output; this ensures that the thresholds of the Schmitt don't drift, as they would with an op-amp of conventional output-stage design, with its ill-defined limits of output swing. In this case this means that the frequency and amplitude of the triangle wave will be stable. Note that the frequency depends on the ratio V_{in}/V_+; this means that if V_{in} is generated from V_+ by a resistive divider (made from some sort of resistive transducer, say), the output frequency won't vary with V_+, only with changes in resistance.

EXERCISE 3.11

Show that the output frequency is given by $f\,(\text{Hz}) = 150V_{in}/V_+$. Along the way, verify

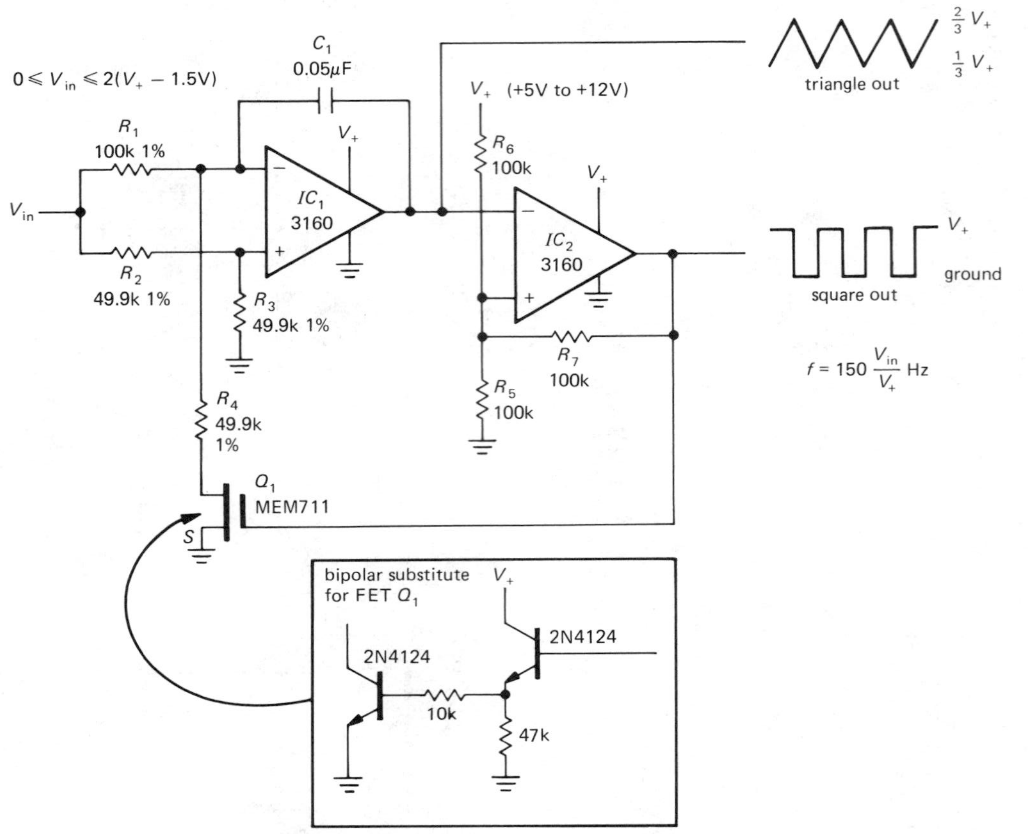

Figure 3.67
Voltage-controlled waveform generator.

that the Schmitt thresholds and integrator currents are as advertised.

□ 3.29 TTL zero-crossing detector

The circuit shown in Figure 3.68 generates an output square wave for use with TTL logic (zero to +5V range) from an input wave of any amplitude up to 100 volts. R_1, combined with D_1 and D_2, limits the input swing to −0.6 volts to +5.6 volts, approximately. Resistive divider R_2R_3 is necessary to limit negative swing to less than 0.3 volt, the limit for a 393 comparator. R_5 and R_6 provide hysteresis, with R_4 setting the trigger points symmetrically about ground. The input impedance is nearly constant, because of the large R_1 value relative to the other resistors in the input attenuator. A 393 is used because its inputs can go all the way to ground, making single-supply operation simple.

EXERCISE 3.12

Verify that the trigger points are at ±25mV at the input signal.

□ 3.30 Load-current-sensing circuit

The circuit shown in Figure 3.69 provides a voltage output proportional to load current, for use with a current regulator, metering circuit, or whatever. The voltage across the 4-terminal resistor R_S goes from zero to 0.1 volt, with probable common-mode offset due to the effects of resistance in the ground lead (note that the power supply is grounded at the output). For that reason the op-amp is wired as a differential amplifier, with gain of 100. Voltage offset is trimmed externally

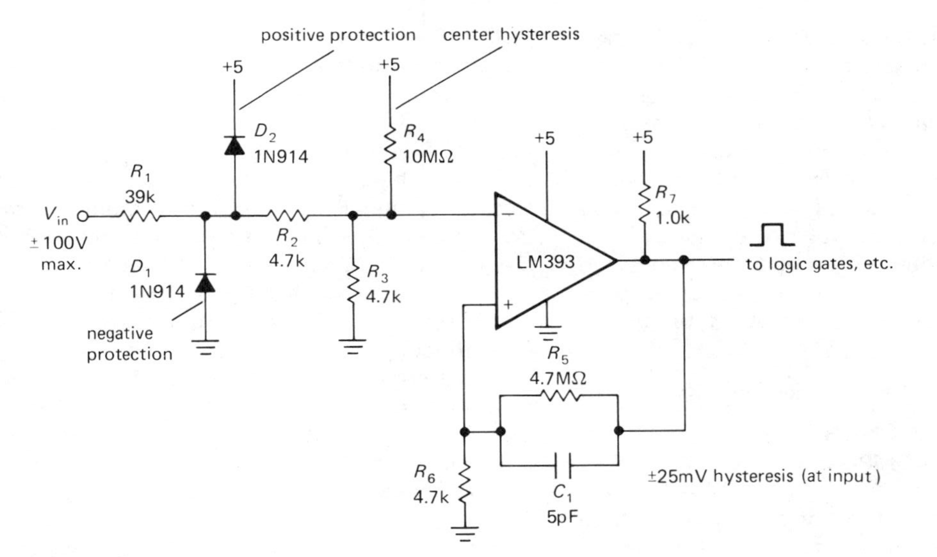

Figure 3.68
Zero-crossing level detector with input protection.

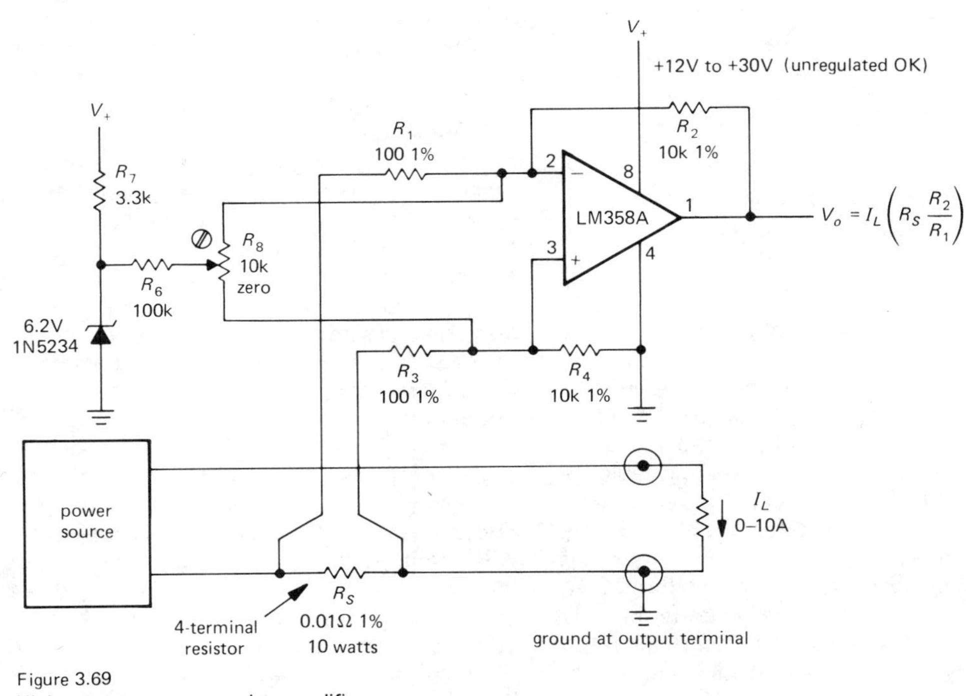

Figure 3.69
High-power current-sensing amplifier.

with R_8, since the 358 doesn't have internal trimming circuitry (the similar 799 does, however). A zener reference with a few percent stability is adequate for trimming, since the trimming is itself a small correction (you hope!). The versatile 358 has been chosen for single-supply operation because both inputs and output go all the way to ground. V_+ could be unregulated, since the power-supply rejection of the op-amp is more than adequate, 100dB (typ) in this case.

FEEDBACK AMPLIFIER FREQUENCY COMPENSATION

If you look at a graph of open-loop voltage gain versus frequency for several op-amps, you'll see something like the curves in Figure 3.70. From a superficial look at such a *Bode*

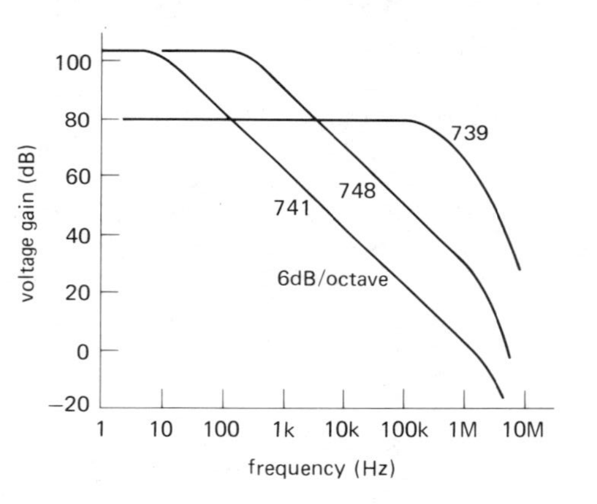

Figure 3.70

plot (a log-log plot of gain and phase versus frequency) you might conclude that the 741 is an inferior op-amp, since its open-loop gain drops off so rapidly with increasing frequency. In fact, that rolloff is built into the op-amp intentionally and is recognizable as the same −6dB/octave curve characteristic of an *RC* low-pass filter. The 748, by comparison, is identical to the 741 except that it is *uncompensated* (as is the 739). Op-amps are generally available in internally compensated varieties and uncompensated varieties; let's take a look at this business of frequency compensation.

3.31 Gain and phase shift versus frequency

An op-amp (or, in general, any multistage amplifier) will begin to roll off at some frequency because of the low-pass filters formed by signals of finite source impedance driving capacitive loads within the amplifier stages. For instance, it is common to have an input stage consisting of a differential amplifier, perhaps with current mirror load (see the LM358 schematic in Fig. 3.49), driving a common-emitter second stage. For now, imagine that the capacitor labeled C_C in that circuit is removed. The high output impedance of the input stage, in combination with junction capacitance C_{ie} and feedback capacitance C_{cb} (Miller effect, see Sections 2.18 and 13.04) of the following stage, forms a low-pass filter whose 3dB point might fall somewhere in the range of 100Hz to 10kHz.

The decreasing reactance of the capacitor with increasing frequency gives rise to the characteristic 6dB/octave rolloff: At sufficiently high frequencies (which may be below 1kHz), the capacitive loading dominates the collector load impedance, resulting in a voltage gain $G_V = g_m X_C$, i.e., the gain drops off as $1/f$. It also produces a 90° lagging phase shift at the output relative to the input signal. (You can think of this as the tail of an *RC* low-pass filter characteristic, where *R* represents the equivalent source impedance driving the capacitive load. However, it is not necessary to have any actual resistors in the circuit.)

In a multistage amplifier there will be additional rolloffs at higher frequencies, caused by low-pass filter characteristics in the other amplifier stages, and the overall open-loop gain will look something like that shown in Figure. 3.71. The open-loop gain begins dropping at 6dB/octave at some low frequency f_1, due to capacitive loading of the first-stage output. It continues dropping off with that slope until an internal *RC* of another stage rears its ugly head at frequency f_2, beyond which the rolloff goes at 12dB/octave, and so on.

What is the significance of all this? Remember that an *RC* low-pass filter has a

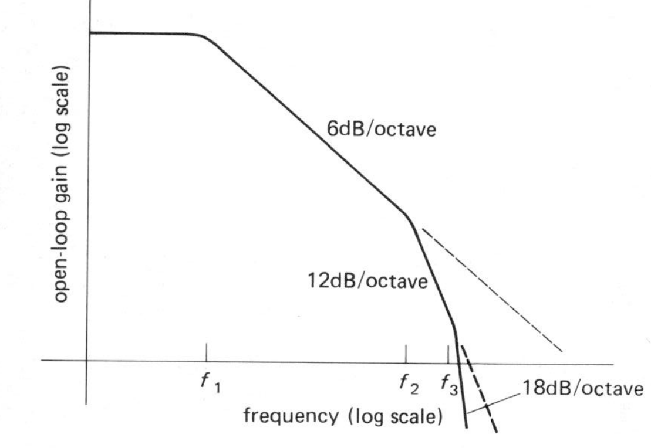

Figure 3.71

phase shift that looks as shown in Figure 3.72. Each low-pass filter within the amplifier has a similar phase shift characteristic, so the overall phase shift of the hypothetical amplifier will be as shown in Figure 3.73.

Now here's the problem: If you were to connect this amplifier as an op-amp follower, for instance, it would oscillate. That's because the open-loop phase shift reaches 180° at some frequency at which the gain is still greater than one (negative feedback becomes positive feedback at that frequency). That's all you need to generate an oscillation, since any signal whatsoever at that frequency builds up each time around

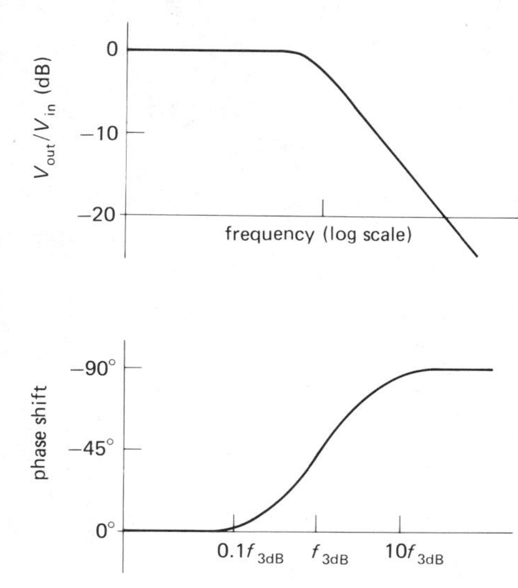

Figure 3.72

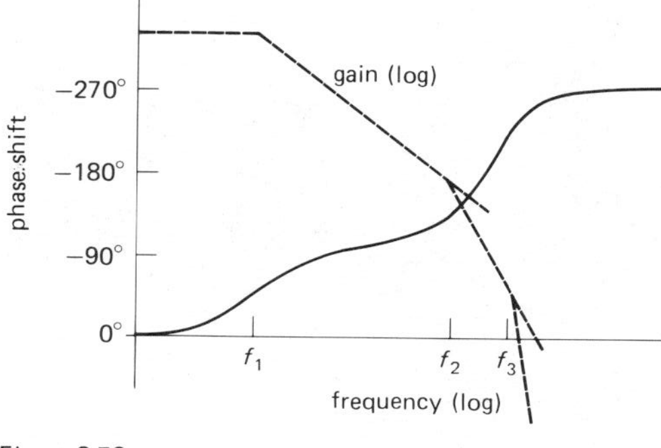

Figure 3.73

the feedback loop, just like a public address system with the gain turned up too far.

Stability criterion

The criterion for stability against oscillation for a feedback amplifier is that its open-loop phase shift must be less than 180° at the frequency at which the loop gain (in the feedback configuration) is unity. This criterion is hardest to satisfy when the amplifier is connected as a follower, since the loop gain then equals the open-loop gain, the highest it can be. Internally compensated op-amps are designed to satisfy the stability criterion even when connected as followers; thus they are stable when connected for any closed-loop gain with a simple resistive feedback network. As we hinted earlier, this is accomplished by deliberately modifying an existing internal rolloff in order to put the 3dB point at some low frequency, typically 1Hz to 20Hz. Let's see how that works.

3.32 Amplifier compensation methods

Dominant-pole compensation

The goal is to keep the open-loop phase shift much less than 180° at all frequencies for which the loop gain is greater than 1. Assuming that the op-amp may be used as a follower, the words "loop gain" in the last sentence can be replaced by "open-loop gain." The easiest way to do this is to add enough capacitance at the point in the circuit that produces the initial 6dB/octave rolloff, so that the open-loop gain drops to unity at about the 3dB frequency of the next "natu-

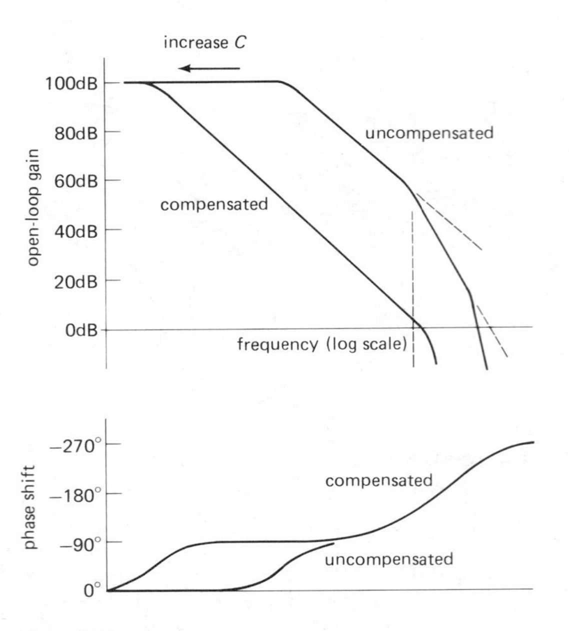

Figure 3.74

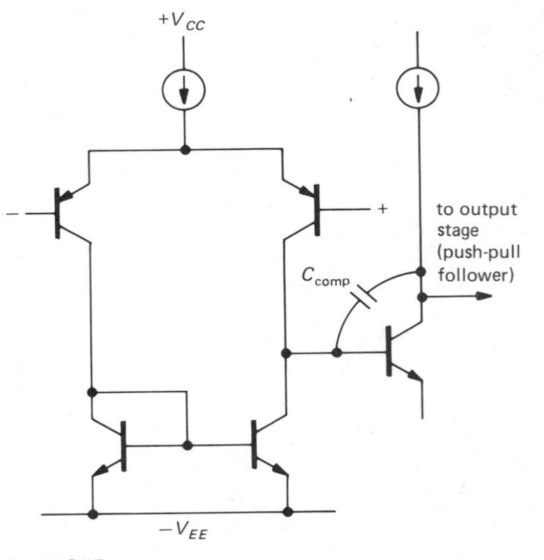

Figure 3.75
Classic op-amp input stage with compensation.

ral'' RC filter. In this way the open-loop phase shift is held at a constant 90° over most of the passband, increasing toward 180° only as the gain approaches unity. Figure 3.74 shows the idea. Without compensation the open-loop gain drops toward 1, first at 6dB/octave, then at 12dB/octave, etc., resulting in phase shifts of 180° or more before the gain has reached 1. By moving the first rolloff down in frequency (forming a "dominant pole"), the rolloff is controlled so that the phase shift begins to rise above 90° only as the open-loop gain approaches unity. Thus, by sacrificing open-loop gain, you buy stability. Since the natural rolloff of lowest frequency is usually caused by Miller effect in the stage driven by the input differential amplifier, the usual method of dominant-pole compensation consists simply of adding additional feedback capacitance around the second-stage transistor, so that the combined voltage gain of the two stages is $g_m X_C$ or $g_m/2\pi f C_{comp}$ over the compensated region of the amplifier's frequency response (Fig. 3.75). In practice, Darlington-connected transistors would probably be used for both stages.

By putting the dominant-pole unity-gain crossing at the 3dB point of the next rolloff, you get a phase margin of about 45° in the worst case (follower), since a single RC filter has a 45° lagging phase shift at its 3dB frequency, i.e., the phase margin equals 180° − (90° + 45°), with the 90° coming from the dominant pole.

An additional advantage of using a Miller-effect pole for compensation is that the compensation is inherently insensitive to changes in voltage gain with temperature, or manufacturing spread of gain: Higher gain causes the feedback capacitance to look larger, moving the pole downward in frequency in exactly the right way to keep the unity-gain crossing frequency unchanged. In fact, the actual 3dB frequency of the compensation pole is quite irrelevant; what matters is the point at which it intersects the unity-gain axis (Fig. 3.76).

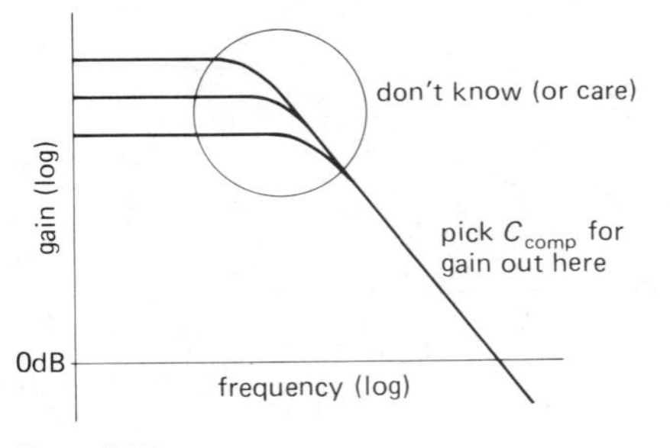

Figure 3.76

Uncompensated op-amps

If an op-amp is used in a circuit with closed-loop gain greater than 1 (i.e., not a follower), it is not necessary to put the pole (the term for the "corner frequency" of a low-pass filter) at such a low frequency, since the stability criterion is relaxed because of the lower loop gain. Figure 3.77 shows the situation graphically.

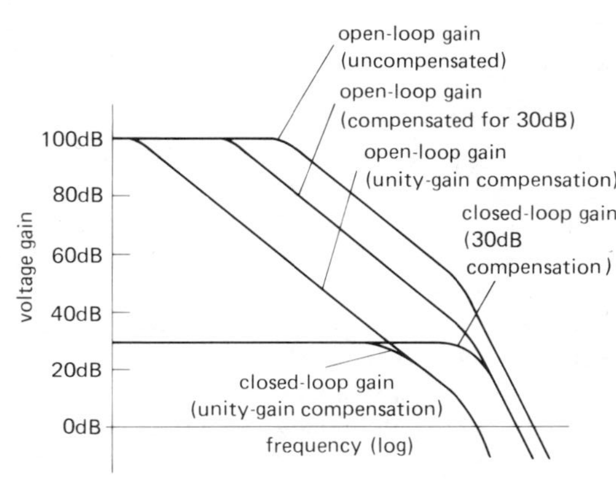

Figure 3.77

For a closed-loop gain of 30dB, the loop gain (which is the ratio of the open-loop gain to the closed-loop gain) is less than for a follower, so the dominant pole can be placed at a higher frequency. It is chosen so that the open-loop gain reaches 30dB (rather than 0dB) at the frequency of the next natural pole of the op-amp. As the graph shows, this means that the open-loop gain is higher over most of the frequency range, and the resultant amplifier will work at higher frequencies. Most op-amps are available in uncompensated versions [e.g., the 748 is an uncompensated 741: the same is true for the 308 (312), 3130 (3160), etc.], with recommended external capacitance values for a selection of minimum closed-loop gains. They are worth using if you need the added bandwidth and your circuit operates at high gain. An alternative is to use "decompensated" (a better word might be "undercompensated") op-amps such as the 349 or 357, which are internally compensated for closed-loop gains greater than some minimum ($A_V > 5$ in the case of the 349 and 357).

□ *Pole-zero compensation*

It is possible to do a bit better than with dominant-pole compensation by using a compensation network that begins dropping (6dB/octave, a "pole") at some low frequency, then flattens out again (it has a "zero") at the frequency of the second natural pole of the op-amp. In this way the amplifier's second pole is "canceled," giving a smooth 6dB/octave rolloff up to the amplifier's third pole. Figure 3.78 shows a

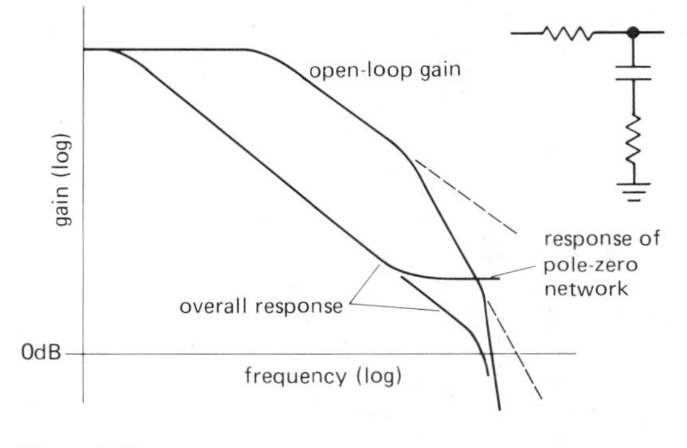

Figure 3.78

frequency response plot. In practice, the zero is chosen to cancel the amplifier's second pole; then the position of the first pole is adjusted so that the overall response reaches unity gain at the frequency of the amplifier's third pole. A good set of data sheets will often give suggested component values (an R and a C) for pole-zero compensation, as well as the usual capacitor values for dominant-pole compensation.

As you will see in Section 13.06, moving the dominant pole downward in frequency actually causes the second pole of the amplifier to move upward somewhat in frequency, an effect known as "pole splitting." The frequency of the canceling zero will be chosen accordingly.

□ 3.33 Frequency response of the feedback network

In all of the discussion thus far we have assumed that the feedback network has a flat frequency response; this is usually the case, with the standard resistive voltage divider as a feedback network. However, there are occasions when some sort of equalization amplifier is desired (integrators and differentiators are in this category) or when the frequency response of the feedback network is modified to improve amplifier stability. In such cases it is important to remember that the Bode plot of loop gain versus frequency is what matters, rather than the curve of open-loop gain. To make a long story short, the curve of ideal closed-loop gain versus frequency should intersect the curve of open-loop gain with a difference in slopes of 6dB/octave. As an example, it is common practice to put a small capacitor (a few picofarads) across the feedback resistor in the usual inverting or noninverting amplifier. Figure 3.79 shows the circuit and Bode plot.

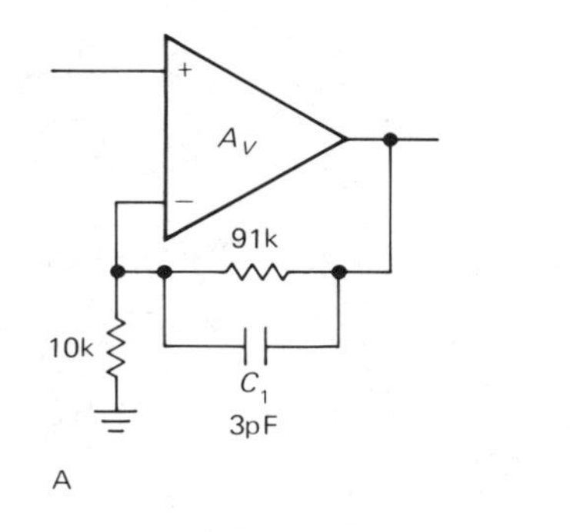

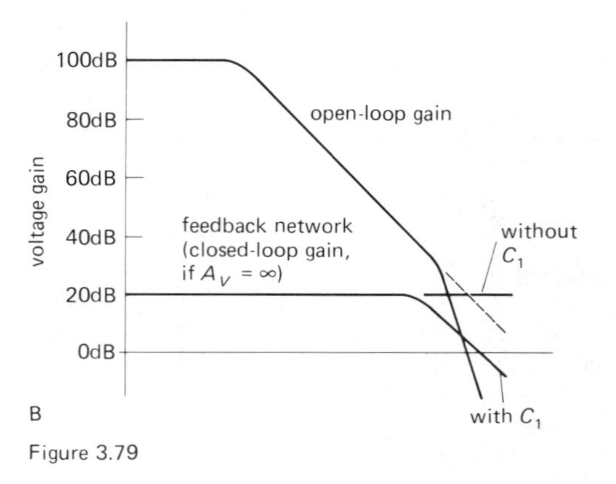

Figure 3.79

The amplifier would have been close to instability with a flat feedback network, since the loop gain would have been dropping at nearly 12dB/octave where the curves meet. The capacitor causes the loop gain to drop at 6dB/octave near the crossing, guaranteeing stability. This sort of consideration is very important when designing differentiators, since an ideal differentiator has a closed-loop gain that rises at 6dB/octave; it is necessary to roll off the differentiator action at some moderate frequency, preferably going over to a 6dB/octave rolloff at high frequencies. Integrators, by comparison, are very friendly in this respect, owing to their 6dB/octave closed-loop rolloff. It takes real talent to make a low-frequency integrator oscillate!

□ *What to do*

In summary, you are generally faced with the choice of internally compensated or uncompensated op-amps. It is simplest to use the compensated variety, and that's the usual choice. At the present time the internally compensated 741 is by far the most popular op-amp. If you need greater bandwidth or slew rate, and the closed-loop gain is greater than unity (as it usually is), you can either use an uncompensated op-amp with an external capacitor (and possibly a resistor), as specified by the manufacturer for the gain you are using, or simply choose a faster compensated op-amp. Several amplifiers, including the popular 356 op-amp, offer another choice: a decompensated version (the 357) with internal compensation good for a closed-loop gain of 5 or more.

□ *Example: 60Hz power source*

Uncompensated op-amps also give you the flexibility of overcompensating, a simple solution to the problem of additional phase shifts introduced by other stuff in the feedback loop. Figure 3.80 shows a nice example. This is a low-frequency amplifier designed to generate a 115 volt ac power output from a variable 60Hz low-level sinewave input (it goes with the 60Hz synthesizer circuit described in Section 8.30). The op-amp, together with R_2 and R_3, forms a $\times 100$ gain block; this is then used as the relatively low "open-loop gain" for overall feedback. The op-amp output drives the push-pull output stage, which in turn drives the transformer primary. Low-frequency feedback is taken from the transformer output via R_{10}, in order to generate low distortion and a stable output voltage under

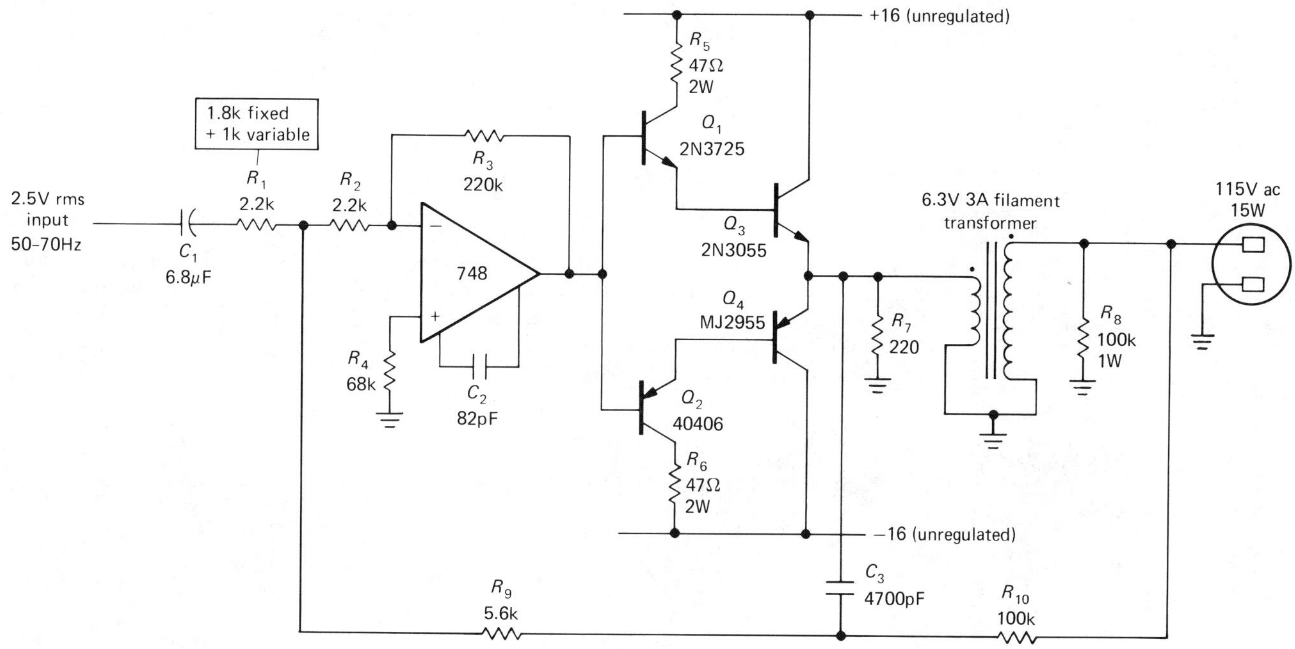

Figure 3.80
Output amplifier for 60Hz-power-source.

load variations. Because of the unacceptably large phase shifts of such a transformer at high frequencies, the circuit is rigged up so that at higher frequencies the feedback comes from the low-voltage input to the transformer, via C_3. The relative sizes of R_9 and R_{10} are chosen to keep the amount of feedback constant at all frequencies. Even though high-frequency feedback is taken directly from the push-pull output, there are still phase shifts associated with the reactive load (the transformer primary) seen by the transistors. In order to ensure good stability, even with reactive loads at the 115 volt output, the op-amp has been overcompensated with an 82pF capacitor (30pF is the normal value for unity gain compensation). The loss of bandwidth that results is unimportant in a low-frequency application like this.

An application such as this represents a compromise, since ideally you would like to have plenty of loop gain to stabilize the output voltage against variations in load current. But a large loop gain increases the tendency of the amplifier to oscillate, especially if a reactive load is attached. This is because the reactive load, in combination with the transformer's finite output impedance, causes additional phase shifts within the low-frequency feedback loop. Since this circuit was built to drive a telescope's synchronous driving motors (highly inductive loads), the loop gain was intentionally kept low. Figure 3.81 shows a graph of the ac output voltage versus load, which illustrates good (but not great) regulation.

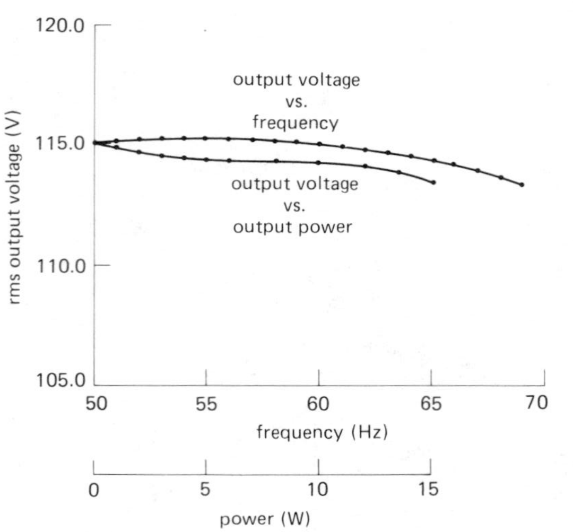

Figure 3.81

□ *Motorboating*

In ac-coupled feedback amplifiers, stability problems can also crop up at very low frequencies, due to the accumulated *leading* phase shifts caused by several capacitively coupled stages. Each blocking capacitor, in combination with the input resistance due to bias strings and the like, causes a leading phase shift that equals 45° at the low-frequency 3dB point and approaches 90° at lower frequencies. If there is enough loop gain, the system can go into a low-frequency oscillation picturesquely known as "motorboating." With the widespread use of dc-coupled amplifiers, motorboating is almost extinct. However, old-timers can tell you some good stories about it.

SELF-EXPLANATORY CIRCUITS

3.34 Circuit ideas

Some interesting circuit ideas, mostly lifted from manufacturers' data sheets, are shown in Figure 3.82.

3.35 Bad circuits

Figure 3.83 presents a zoo of intentional (mostly) blunders to amuse, amaze, and educate you. There are a few real howlers here this time. These circuits are guaranteed not to work. Figure out why. All op-amps run from ±15 volts unless shown otherwise.

ADDITIONAL EXERCISES

(1) Design a "sensitive voltmeter" to have $Z_{in} = 1M\Omega$ and full-scale sensitivities of 10mV to 10V in four ranges. Use a 1mA meter movement and an op-amp. Trim voltage offsets if necessary, and calculate what the meter will read with input open, assuming (a) I_B = 30pA (typical for a 355) and (b) I_B = 80nA (typical for a 741). Use some form of meter protection (e.g., keep its current less than 200% of full scale), and protect the amplifier inputs from voltages outside the supply voltages. What do you conclude about the suitability of the 741 for low-level high-impedance measurements?

(2) Design an audio amplifier, using a 5534 op-amp (low noise, good for audio), with the following characteristics: gain = 20dB, Z_{in} = 10k, −3dB point = 20Hz. Use the noninverting configuration, and roll off the gain at low frequencies in such a way as to reduce the effects of input offset voltage. Use proper design to minimize the effects of input bias current on output offset. Assume that the signal source is capacitively coupled.

(3) Design a unity-gain phase splitter (see Chapter 2) using 741s. Strive for high input impedance and low output impedances. The circuit should be dc-coupled. At roughly what maximum frequency can you obtain full swing (27V pp, with ±15V supplies), owing to slew rate limitations?

(4) El Cheapo brand loudspeakers are found to have a treble boost, beginning at 2kHz (+3dB point) and rising 6dB/octave. Design a simple *RC* filter, buffered with AD544 op–amps (another good audio chip) as necessary, to be placed between preamp and amplifier to compensate this rise. Assume that the preamp has Z_{out} = 50k and that the amplifier has Z_{in} = 10k, approximately.

(5) A 741 is used as a simple comparator, with one input grounded; i.e., it is a zero-crossing detector. A 1-volt-amplitude sine wave is fed into the other input (frequency = 1kHz). What voltage(s) will the input be when the output passes through zero volts? Assume that the slew rate is 0.5V/μs and that the op-amp's saturated output is ±13 volts.

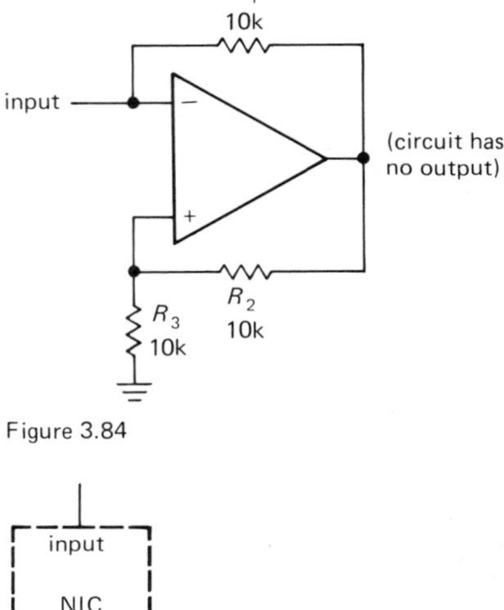

Figure 3.84

Figure 3.85

(6) The circuit in Figure 3.84 is an example of a "negative-impedance converter." (a) What is its input impedance? (b) If the op-amp's output range goes from V_+ to V_-, what range of input voltages will this circuit accommodate without saturation?

(7) Consider the circuit in the preceding problem as the 2-terminal black box (Fig. 3.85). Show how to make a dc amplifier with a gain of −10. Why can't you make a dc amplifier with a gain of +10? (Hint: The circuit is susceptible to a latchup condition for a certain range of source resistances. What is that range? Can you think of a remedy?)

Circuit ideas

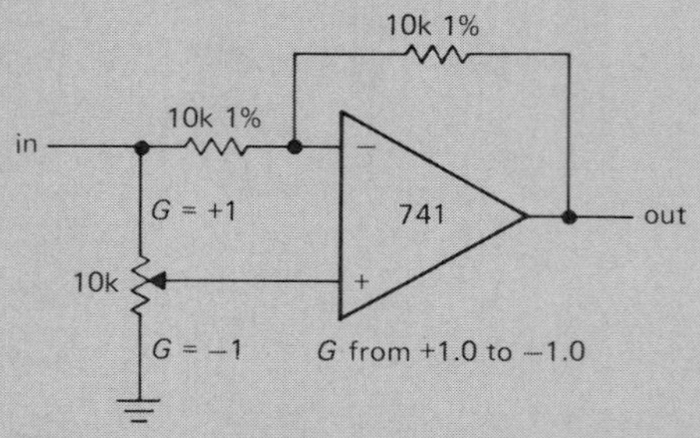

A. continuously variable gain

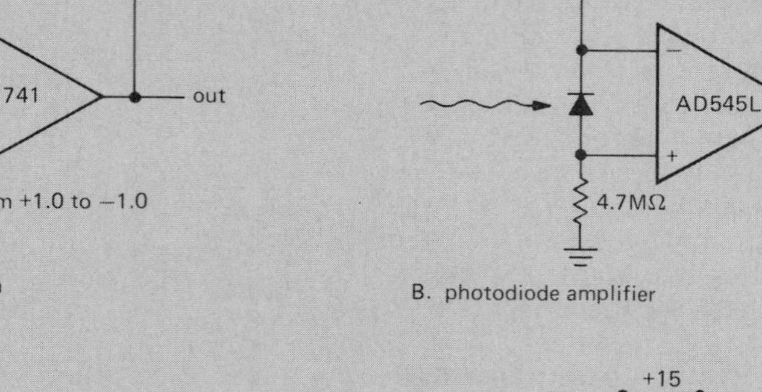

B. photodiode amplifier

+15
10k
ICL 8069
(1.23V)
1.21MΩ
1%
LF355
I_{out}
≈1μA

C. current source

D. differential amplifier biasing for zero tempco of gain

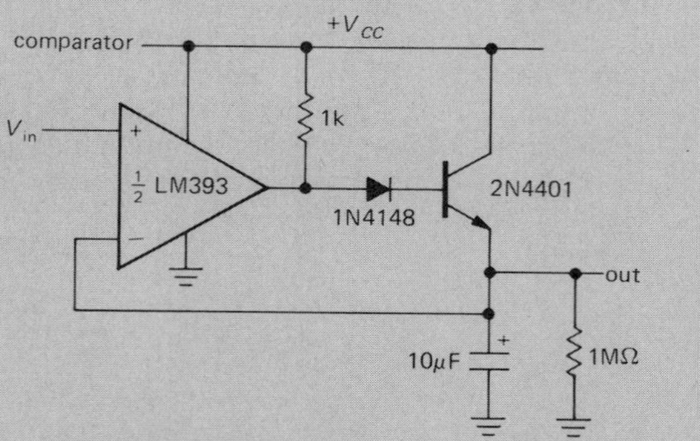

E. positive peak detector

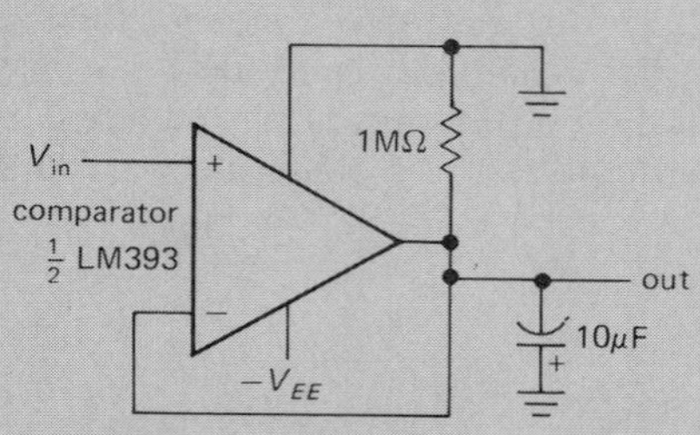

F. negative peak detector

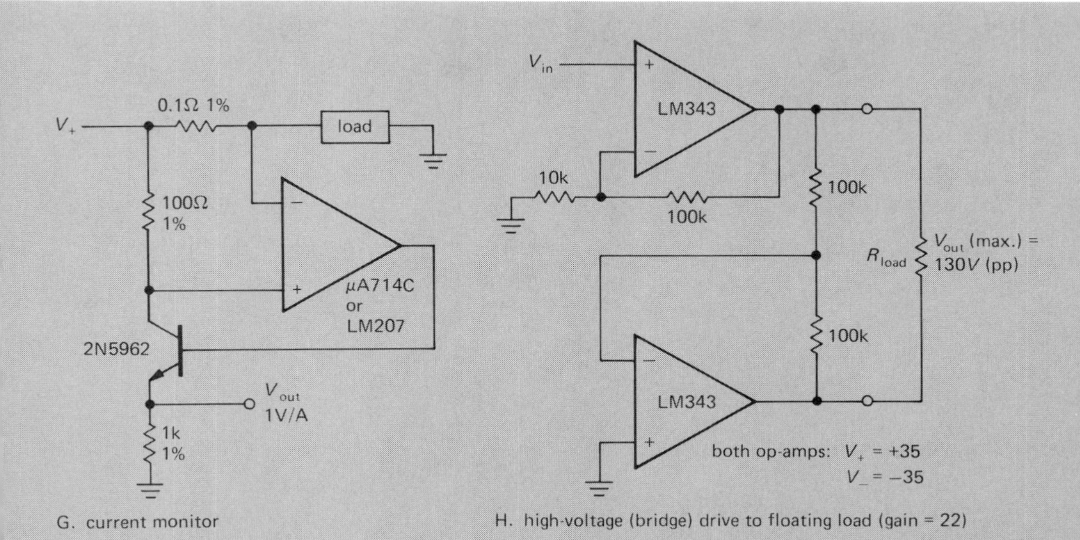

G. current monitor

H. high-voltage (bridge) drive to floating load (gain = 22)

I. fast logarithmic converter

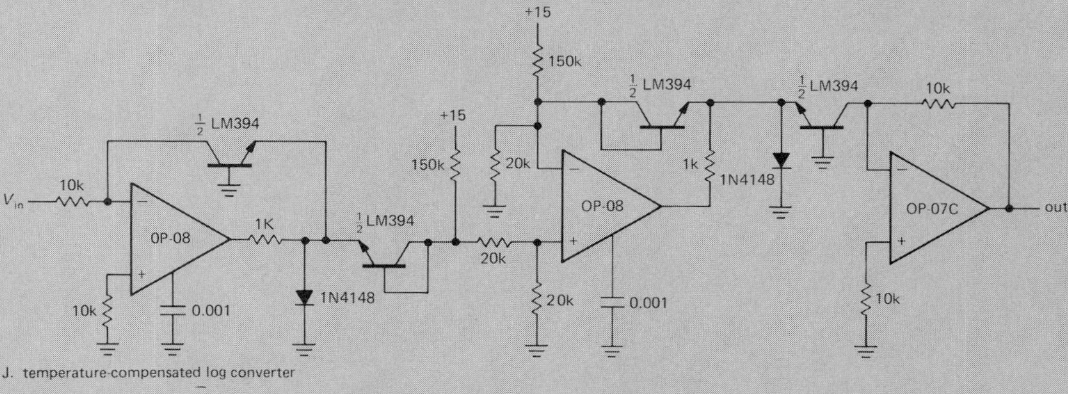

J. temperature-compensated log converter

Figure 3.82

Bad circuits

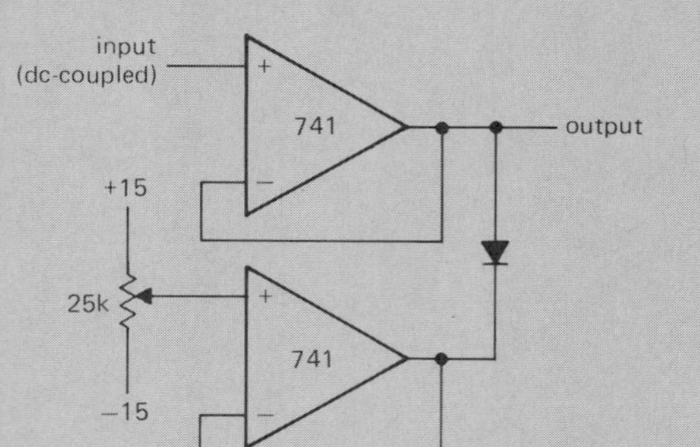

A. adjustable clamp

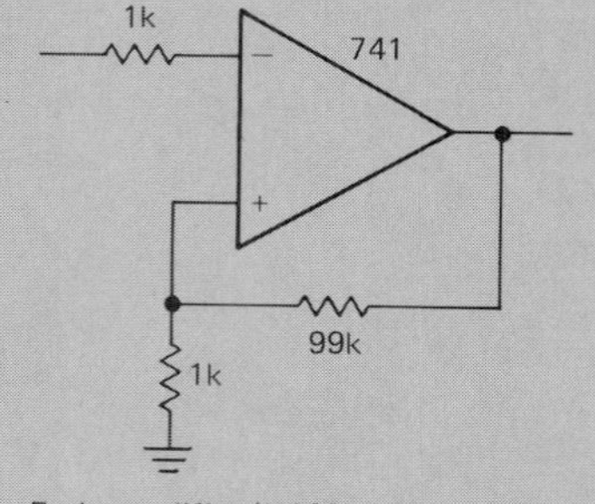

E. dc amplifier (×100)

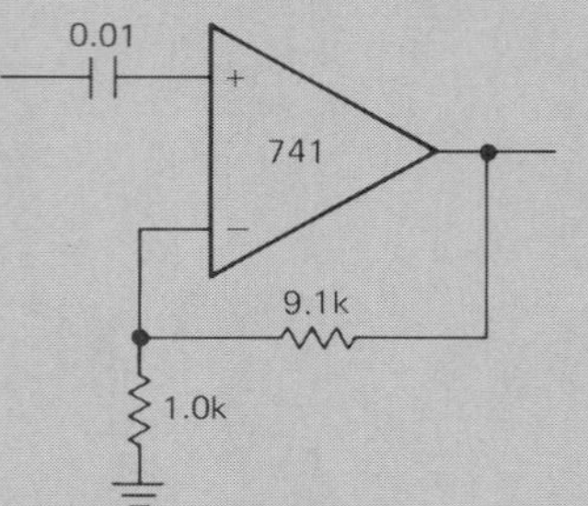

B. ac-coupled ×10 amplifier

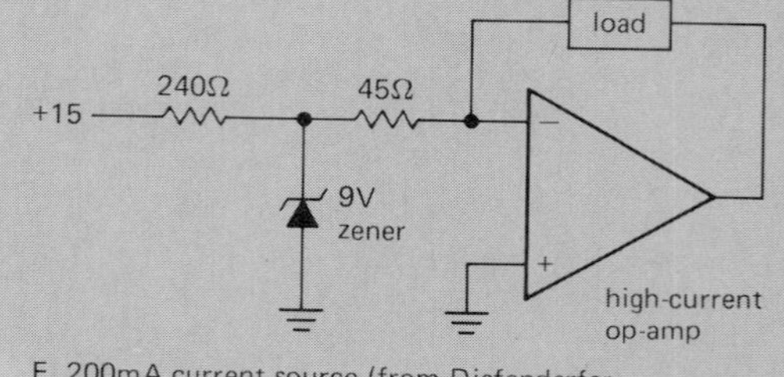

F. 200mA current source (from Diefenderfer, not intended as a "bad circuit")

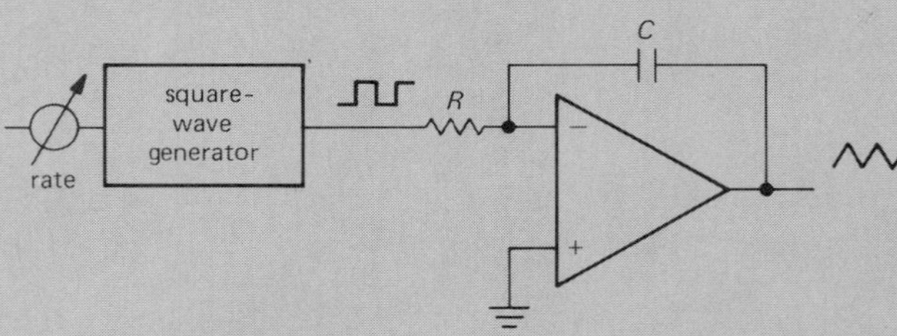

C. triangle-wave generator

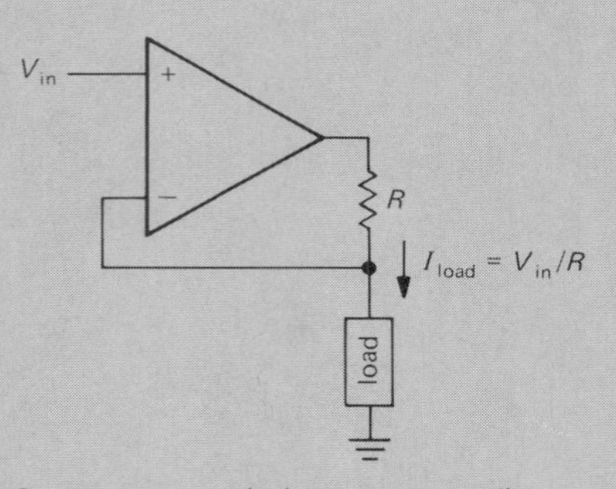

D. current source (voltage-programmed)

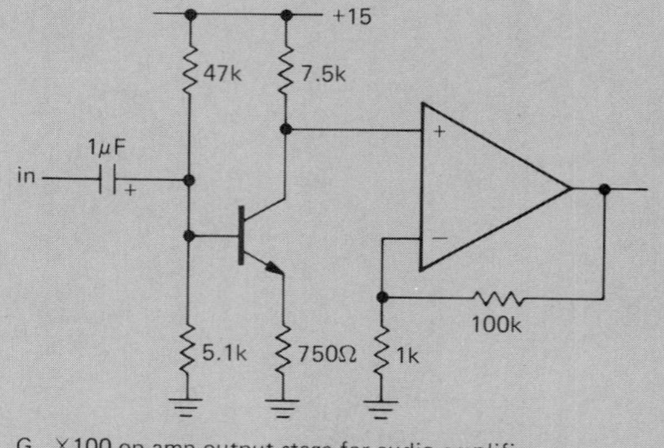

G. ×100 op-amp output stage for audio amplifier

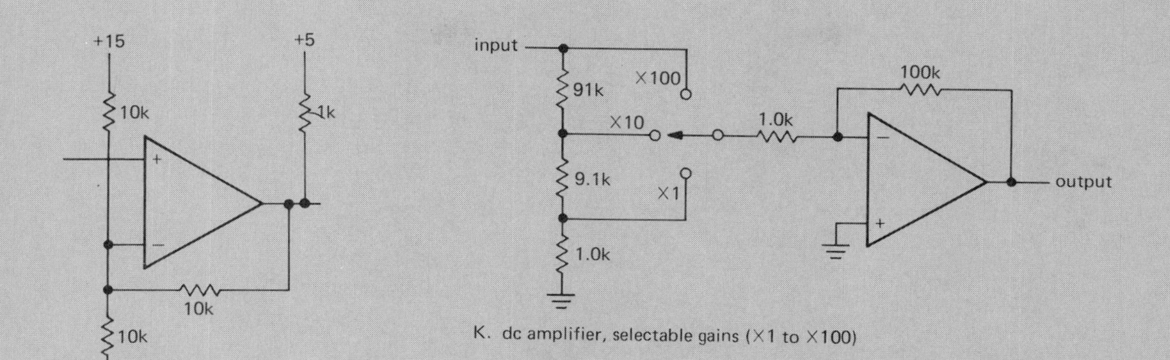

H. Schmitt trigger

K. dc amplifier, selectable gains (×1 to ×100)

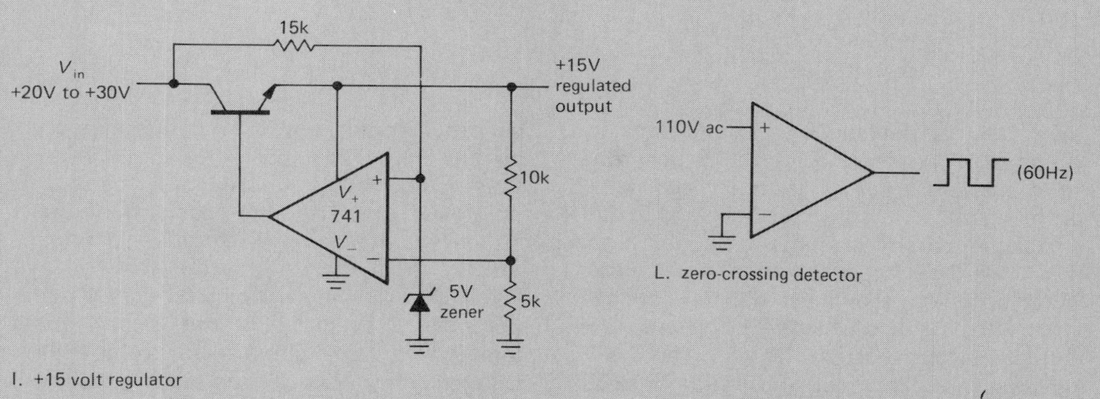

I. +15 volt regulator

L. zero-crossing detector

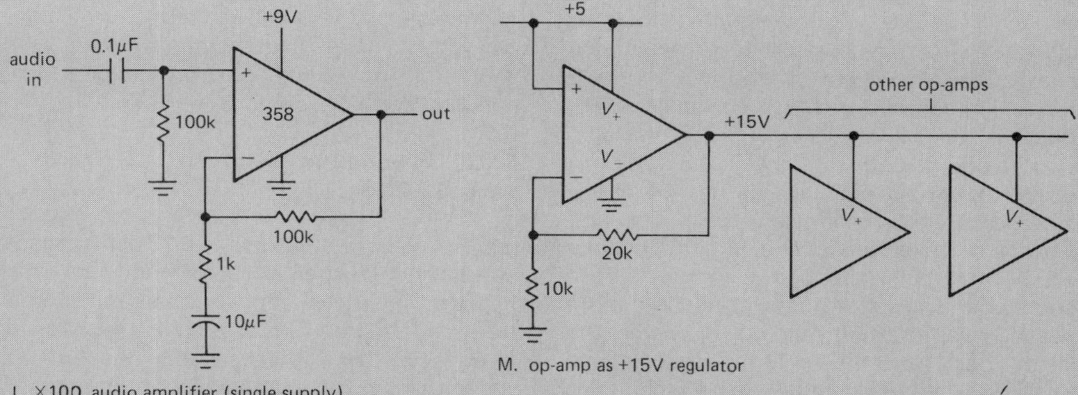

J. ×100 audio amplifier (single supply)

M. op-amp as +15V regulator

Figure 3.83

ACTIVE FILTERS AND OSCILLATORS

CHAPTER 4

With only the techniques of transistors and op-amps it is possible to delve into a number of interesting areas of linear (as contrasted with digital) circuitry. We believe that it is important to spend some time doing this now, in order to strengthen your understanding of some of these difficult concepts (transistor behavior, feedback, op-amp limitations, etc.) before introducing more new devices and techniques and getting into the large area of digital electronics. In this chapter, therefore, we will treat briefly the areas of active filters and oscillators. Additional analog techniques are treated in Chapter 5 (voltage regulators and high-current design), Chapter 6 (FETs), Chapter 7 (precision circuits and low noise), Chapter 13 (radiofrequency techniques), and Chapter 14 (measurements and signal processing). The first part of this chapter (active filters, Sections 4.01 to 4.10) describes techniques of a somewhat specialized nature, and it can be passed over in a first reading. However, the latter part of this chapter (oscillators, Sections 4.11 to 4.16) describes techniques of broad utility and should not be omitted.

ACTIVE FILTERS

In Chapter 1 we began a discussion of filters made from resistors and capacitors. Those simple *RC* filters produced gentle high-pass or low-pass gain characteristics, with a 6dB/octave falloff well beyond the −3dB point. By cascading high-pass and low-pass filters, we showed how to obtain bandpass filters, again with gentle 6dB/octave "skirts." Such filters are sufficient for many purposes, especially if the signal being rejected by the filter is far removed in frequency from the desired signal passband. Some examples are bypassing of radiofrequency signals in audio circuits, "blocking" capacitors for elimination of dc levels, and separation of modulation from a communications "carrier" (see Chapter 13).

4.01 Frequency response with *RC* filters

Often, however, filters with flatter passbands and steeper skirts are needed. This happens whenever signals must be filtered from other interfering signals nearby in frequency. The obvious next question is whether or not (by cascading a number of identical low-pass filters, say) we can generate an approximation to the ideal "brick-wall" low-pass frequency response, as in Figure 4.1.

We know already that simple cascading won't work, since each section's input impedance will load the previous section

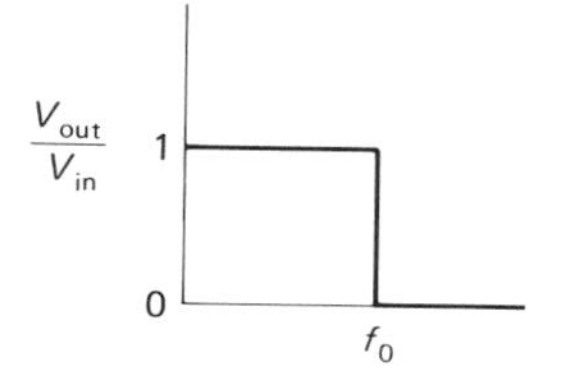

Figure 4.1

seriously, degrading the response. But with buffers between each section (or by arranging to have each section of much higher impedance than the one preceding), it would seem possible. Nonetheless, the answer is "No." Cascaded *RC* filters do produce a steep *ultimate* falloff, but the "knee" of the curve of response versus frequency is not sharpened. We might restate this as "many soft knees do not a hard knee make." To make the point graphically, we have plotted some graphs of gain response (i.e., V_{out}/V_{in}) versus frequency for low-pass filters constructed from 1, 2, 4, 8, 16, and 32 identical *RC* sections, perfectly buffered (Fig. 4.2).

The first graph shows the effect of cascading several *RC* sections, each with its 3dB point at unit frequency. As more sections are added, the overall 3dB point is pushed downward in frequency, as you could easily have predicted. To compare filter characteristics fairly, the rolloff frequencies of the individual sections should be adjusted so that the overall 3dB point is always at the same frequency. The other graphs in Figure 4.2, as well as the next few graphs in this chapter, are all "normalized" in frequency, meaning that the −3dB point (or breakpoint, however defined) is at a frequency of 1 radian per second. To determine the response of a filter whose breakpoint is set at some other frequency, simply multiply the values on the frequency axis by the actual breakpoint frequency. In general, we will also stick to the log-log graph of frequency response when talking about filters, because it tells the most about the frequency response. It lets you see the approach to the ultimate rolloff slope, and it permits you to read off accurate values of attenuation. In this case (cascaded *RC* sections) the normalized graphs in Figure 4.2B and 4.2C demonstrate the soft knee characteristic of passive *RC* filters.

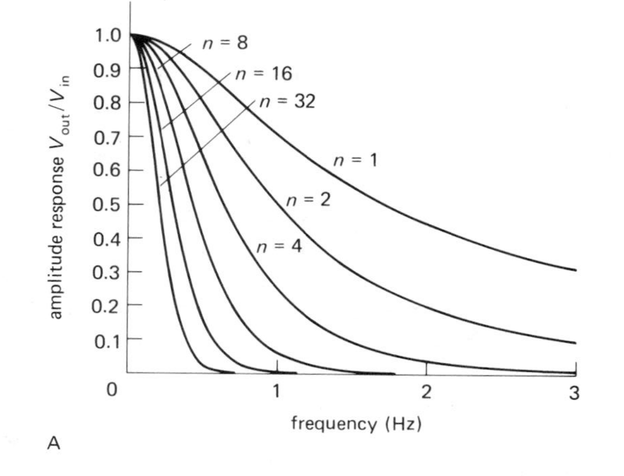

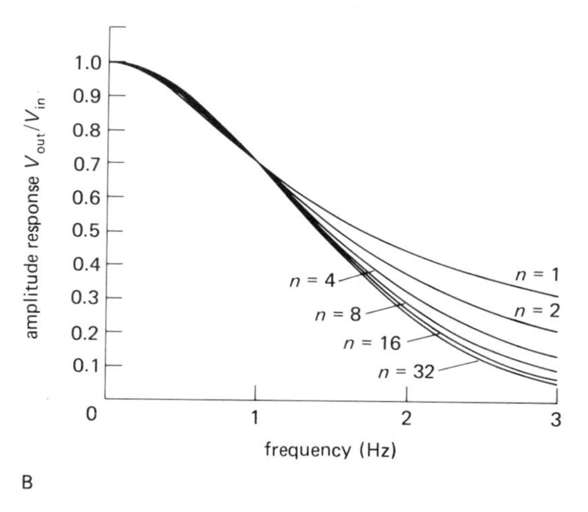

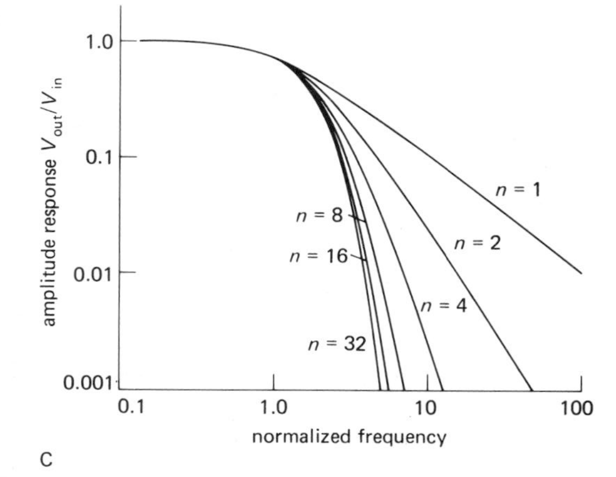

Figure 4.2
Frequency responses of multisection *RC* filters. Graphs A and B are linear plots, whereas C is logarithmic. The filter responses in B and C have been normalized (or scaled) for 3dB attenuation at unit frequency.

4.02 Ideal performance with *LC* filters

As we pointed out in Chapter 1, filters made with inductors and capacitors can have very sharp responses. The parallel *LC* resonant circuit is an example. By including inductors in the design, it is possible to create filters with any desired flatness of passband combined with sharpness of transition and steepness of falloff outside the band. Figure 4.3 shows an example of a telephone filter and its characteristics.

Obviously the inclusion of inductors into the design brings about some magic that cannot be performed without them. In the terminology of network analysis, that magic consists in the use of "off-axis poles." Even so, the complexity of the filter increases according to the required flatness of passband and steepness of falloff outside the band, accounting for the large number of components used in the preceding filter. The transient response and phase shift characteristics are also generally degraded as the amplitude response is improved to approach the ideal brick-wall characteristic.

The synthesis of filters from passive components (*R, L, C*) is a highly developed subject, as typified by the authoritative handbook by Zverev (see chapter references at end of book). The only problem is that inductors as circuit elements frequently leave much to be desired. They are often bulky and expensive, and they depart from the ideal by being "lossy," i.e., by having significant series resistance, as well as other "pathologies" such as nonlinearity, distributed winding capacitance, and susceptibility to magnetic pickup of interference. What is needed is a way to make inductorless filters with the characteristics of ideal *RLC* filters.

4.03 Enter active filters: an overview

By using op-amps as part of the filter design, it is possible to synthesize any *RLC* filter characteristic without using inductors. Such

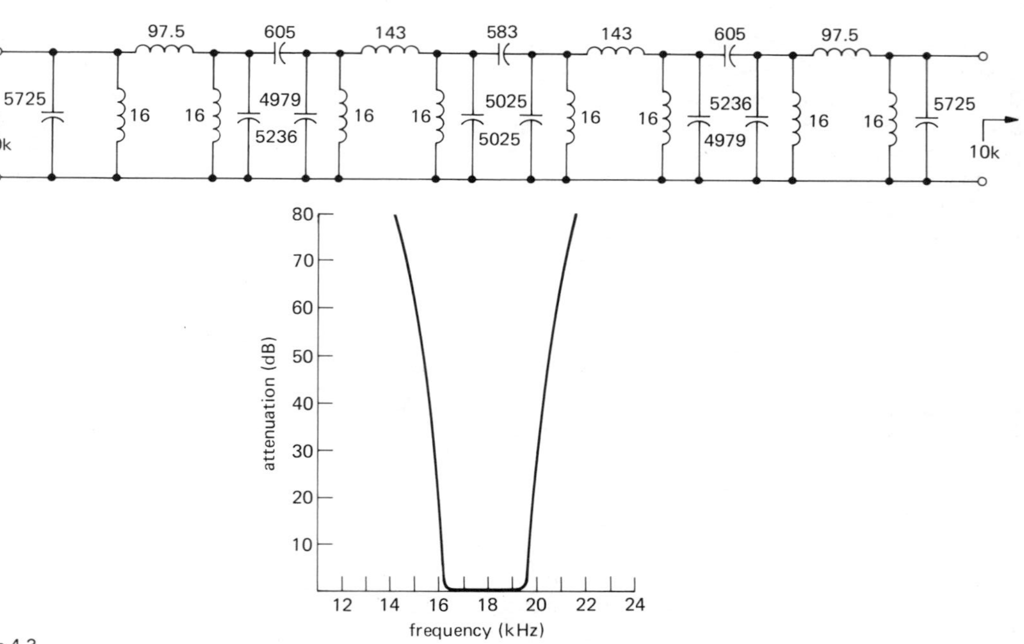

Figure 4.3
An unusually good passive bandpass filter implemented from inductors and capacitors; inductances are in mH, capacitances in pF. Bottom: Measured response of the filter circuit. (Based on Figures 11 and 12 from Orchard, H.J., and Sheahan, D.F., IEEE Journal of Solid-State Circuits, Vol. SC-5, No. 3 (1970).)

inductorless filters are known as active filters because of the inclusion of an active element (the amplifier).

Active filters can be used to make low-pass, high-pass, bandpass, and band-reject filters, with a choice of filter types according to the important features of the response, e.g., maximal flatness of passband, steepness of skirts, or uniformity of time delay versus frequency (more on this shortly). In addition, "all-pass filters" with flat amplitude response but tailored phase versus frequency can be made (they're also known as "delay equalizers"), as well as the opposite – a filter with constant phase shift but tailored amplitude response.

□ *Negative-impedance converters and gyrators*

Two interesting circuit elements that should be mentioned in any overview are the negative-impedance converter (NIC) and the gyrator. These devices can mimic the properties of inductors, while using only resistors and capacitors in addition to op-amps. Once you can do that, you can build inductorless filters with the ideal properties of any *RLC* filter, thus providing at least one way to make active filters.

The NIC converts an impedance to its *negative,* whereas the gyrator converts an impedance to its *inverse.* The following exercises will help you discover for yourself how that works out.

EXERCISE 4.1

Show that the circuit in Figure 4.4 is a negative-impedance converter, in particular

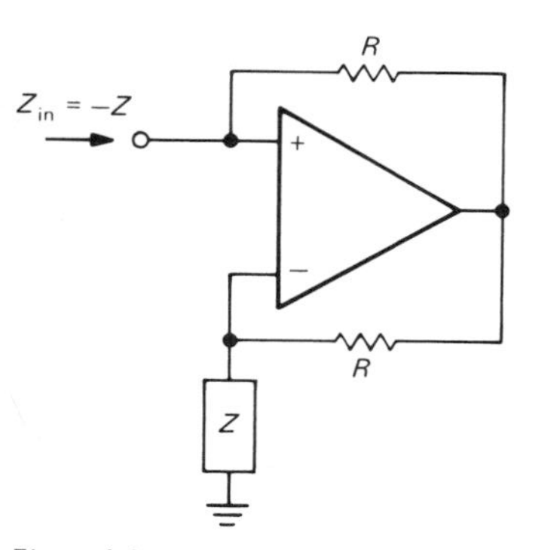

Figure 4.4
Negative-impedance converter.

that $Z_{in} = -Z$. Hint: Apply some input voltage V, and compute the input current I. Then take the ratio to find $Z_{in} = V/I$.

EXERCISE 4.2

Show that the circuit in Figure 4.5 is a gyrator, in particular that $Z_{in} = R^2/Z$. Hint: You

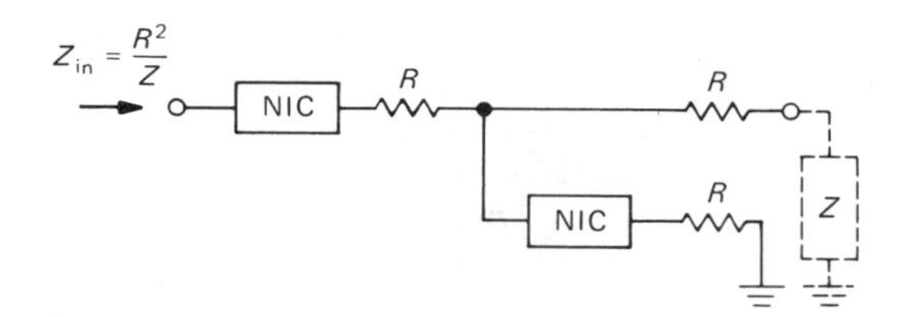

Figure 4.5
Gyrator circuit made from NICs.

can analyze it as a set of voltage dividers, beginning at the right.

The NIC therefore converts a capacitor to a "backward" inductor:

$$Z_C = 1/j\omega C \longrightarrow Z_{in} = j/\omega C$$

i.e., it is inductive in the sense of generating a current that lags the applied voltage, but its impedance has the wrong frequency dependence (it goes down, instead of up, with increasing frequency). The gyrator, on the other hand, converts a capacitor to a true inductor:

$$Z_C = 1/j\omega C \longrightarrow Z_{in} = j\omega CR^2$$

i.e., an inductor with inductance $L = CR^2$.

The existence of the gyrator makes it intuitively reasonable that inductorless filters can be built to mimic any filter using inductors: Simply replace each inductor by a gyrated capacitor. The use of gyrators in just that manner is perfectly OK, and in fact the telephone filter illustrated previously was built that way. In addition to simple gyrator substitution into preexisting *RLC* designs, it is possible to synthesize many other filter configurations. The field of inductorless filter design is extremely active, with new designs appearing in the journals every month.

Sallen and Key filter

Figure 4.6 shows an example of a simple and even partly intuitive filter. It is known as a Sallen and Key filter, after its inventors. The unity-gain amplifier can be an op-amp

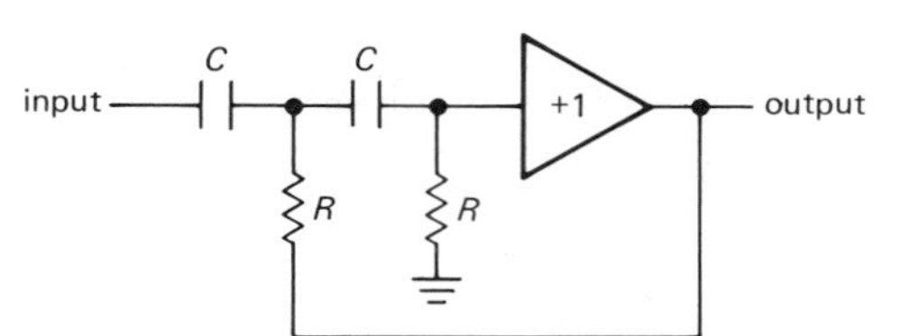

Figure 4.6

connected as a follower, or just an emitter follower. This particular filter is a 2-pole high-pass filter. Note that it would be simply two cascaded *RC* high-pass filters except for the fact that the bottom of the first resistor is bootstrapped by the output. It is easy to see that at very low frequencies it falls off just like a cascaded *RC,* since the output is essentially zero. As the output rises at increasing frequency, however, the bootstrap action tends to reduce the attenuation, giving a sharper knee. Of course, such hand-waving cannot substitute for honest analysis, which luckily has already been done for a prodigious variety of nice filters. We will come back to active filter circuits in Section 4.06.

4.04 Key filter performance criteria

There are some standard terms that keep appearing when we talk about filters and try to specify their performance. It is worth getting it all straight at the beginning.

Frequency domain

The most obvious characteristic of a filter is its gain versus frequency, typified by the sort of low-pass characteristic shown in Figure 4.7.

The *passband* is the region of frequencies that are relatively unattenuated by the filter. Most often the passband is considered to extend to the −3dB point, but with certain filters (most notably the "equiripple" types) the end of the passband may be defined somewhat differently. Within the passband the response may show variations, or *ripples,* defining a *ripple band,* as shown. The *cutoff frequency,* f_c, is the end of the passband. The response of the filter then drops off through a *transition region* (also colorfully known as the *skirt* of the filter's response) to a *stopband,* the region of significant attenuation. The stopband may be defined by some minimum attenuation, e.g., 40dB.

Along with the gain response, the other parameter of importance in the frequency domain is the *phase shift* of the output signal relative to the input signal. In other words, we are interested in the *complex* response of the filter, which usually goes by the name of **H(s)**, where **s** = $j\omega$, where **H, s,** and ω all are complex. Phase is important because a signal entirely within the passband of a filter will emerge with its waveform distorted if the time delay of different frequencies in going through the filter is not constant. Constant time delay corresponds to a phase shift increasing linearly with frequency; hence the term *linear-phase filter* applied to a filter ideal in this respect. Figure 4.8 shows a typical graph of phase shift and time delay for a low-pass filter that is definitely not a linear-phase filter. Graphs of phase shift versus frequency are best plotted on a linear-frequency axis.

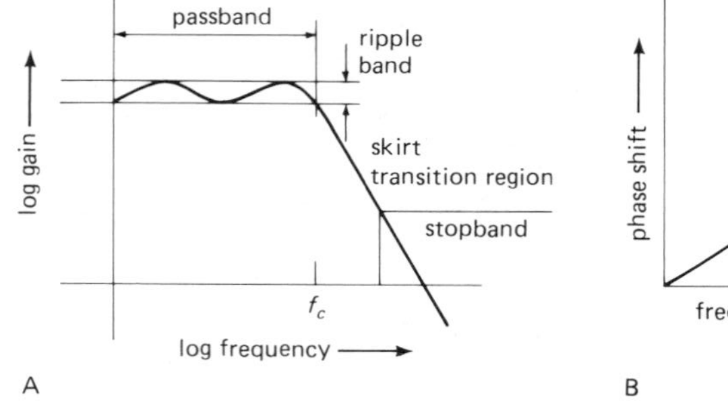

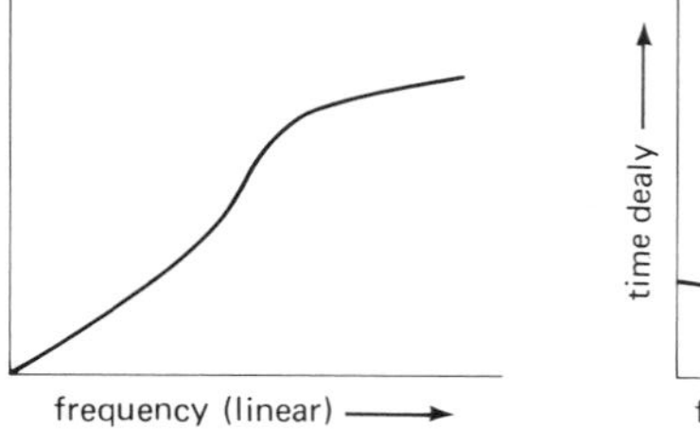

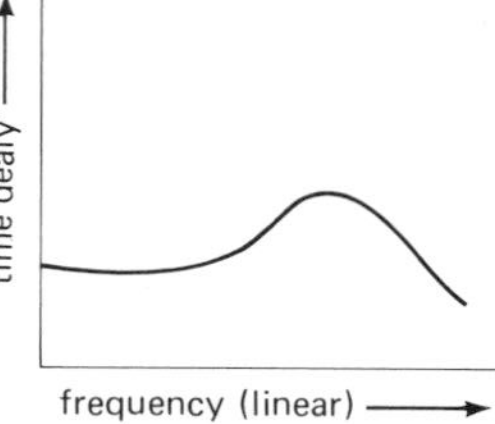

Figure 4.7
Filter characteristics versus frequency.

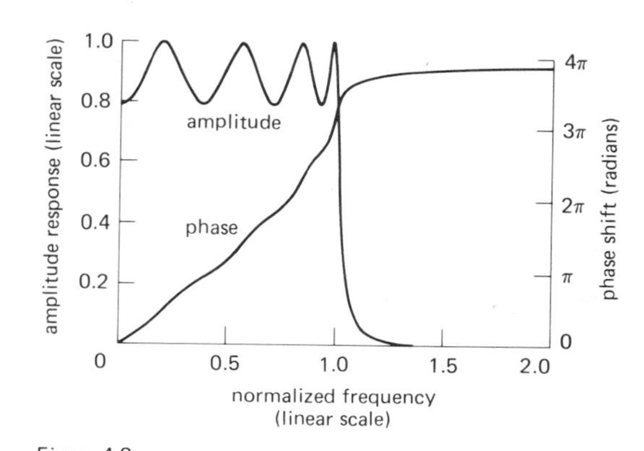

Figure 4.8
Phase and amplitude response for an 8-pole Chebyshev low-pass filter (2dB passband ripple).

Time domain

As with any ac circuit, filters can be described in terms of their *time-domain* properties: rise time, overshoot, ringing, and settling time. This is of particular importance where steps or pulses may be used. Figure 4.9 shows a typical low-pass-filter step response. *Rise time* is the time required to reach 90% of the final value, whereas *settling time* is the time required to get within some specified amount of the final value and stay there. *Overshoot* and *ringing* are self-explanatory terms for some undesirable properties of filters.

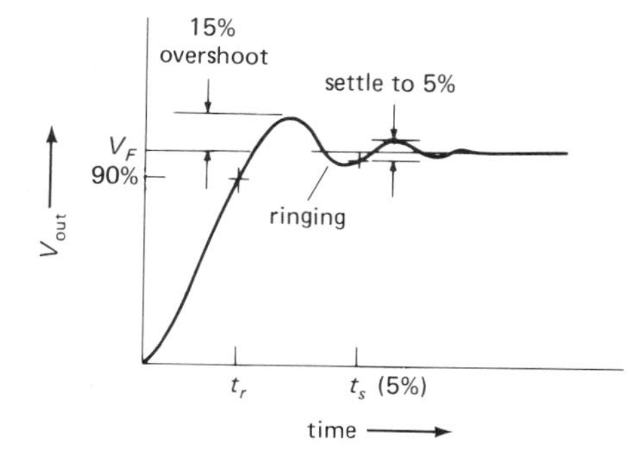

Figure 4.9

4.05 Filter types

Suppose you want a low-pass filter with flat passband and sharp transition to the stopband. The ultimate rate of falloff, well into the stopband, will always be 6ndB/octave, where n is the number of "poles." You need one capacitor (or inductor) for each pole, so the required ultimate rate of falloff of filter response determines, roughly, the complexity of the filter.

Now, assume that you have decided to use a 6-pole low-pass filter. You are guaranteed an ultimate rolloff of 36dB/octave at high frequencies. It turns out that the filter design can now be optimized for maximum flatness of passband response, at the expense of a slow transition from passband to stopband. Alternatively, by allowing some ripple in the passband characteristic, the transition from passband to stopband can be steepened considerably. A third criterion that may be important is the ability of the filter to pass signals within the passband without distortion of their waveforms caused by phase shifts. You may also care about rise time, overshoot, and settling time.

There are filter designs available to optimize each of these characteristics, or combinations of them. In fact, rational filter selection will not be carried out as just described; rather, it normally begins with a set of requirements on passband flatness, attenuation at some frequency outside the passband, and whatever else matters. You will then choose the best design for the job, using the number of poles necessary to meet the requirements. In the next few sections we will introduce the three popular favorites, the Butterworth filter (maximally flat passband), the Chebyshev filter (steepest transition from passband to stopband), and the Bessel filter (maximally flat time delay). Each of these filter responses can be produced with a variety of different filter circuits, some of which we will discuss later. They are all available in low-pass, high-pass, and bandpass versions.

Butterworth and Chebyshev filters

The Butterworth filter produces the flattest passband response, at the expense of steepness in the transition region from passband to stopband. As you will see later, it also has poor phase characteristics. The amplitude response is given by

$$\frac{V_{out}}{V_{in}} = \frac{1}{[1 + (f/f_c)^{2n}]^{1/2}}$$

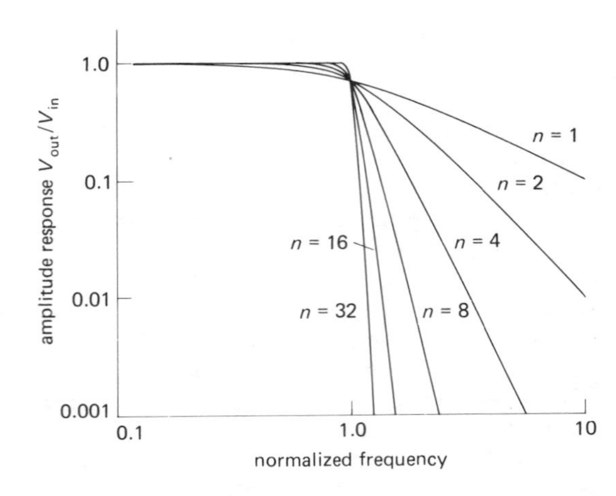

Figure 4.10
Normalized low-pass Butterworth filter response curves. Note the improved attenuation characteristics for the higher-order filters.

where n is the order of the filter (number of poles). Increasing the number of poles flattens the passband response and steepens the stopband falloff, as shown in Figure 4.10.

The Butterworth filter trades off everything else for maximum flatness of response. It starts out extremely flat at zero frequency and bends over near the cutoff frequency f_c (f_c is usually the -3dB point).

In most applications, all that really matters is that the wiggles in the passband response be kept less than some amount, say 1dB. The Chebyshev filter responds to this reality by allowing ripples throughout the passband, with greatly improved sharpness of the knee. A Chebyshev filter is specified in terms of its number of poles and passband ripple. By allowing greater passband ripple, you get a sharper knee. The amplitude response is given by

$$\frac{V_{out}}{V_{in}} = \frac{1}{[1 + \epsilon^2 C_n^2 (f/f_c)]^{1/2}}$$

where C_n is the Chebyshev polynomial of the first kind of degree n, and ϵ is a constant that sets the passband ripple. Like the Butterworth, the Chebyshev has phase characteristics that are less than ideal.

Figure 4.11 presents graphs comparing the responses of Chebyshev and Butterworth 6-pole low-pass filters. As you can see, they're both tremendous improvements over a 6-pole *RC* filter.

Actually the Butterworth, with its maximally flat passband, is not as attractive as it might appear, since you are always accepting some variation in passband response anyway (with the Butterworth it is a gradual rolloff near f_c, whereas with the Chebyshev it is a set of ripples spread throughout the passband). Furthermore, active filters constructed with components of finite tolerance will deviate from the predicted response, which means that a real Butterworth filter will exhibit some passband ripple anyway. The graph in Figure 4.12 illustrates the effects of worst-case variations in resistor and capacitor values on filter response.

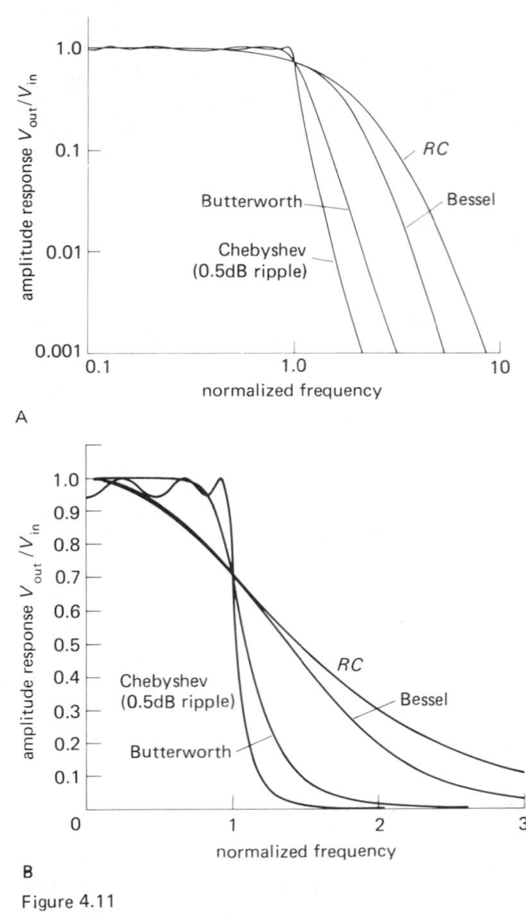

Figure 4.11
Comparison of some common 6-pole low-pass filters. The same filters are plotted on both linear and logarithmic scales.

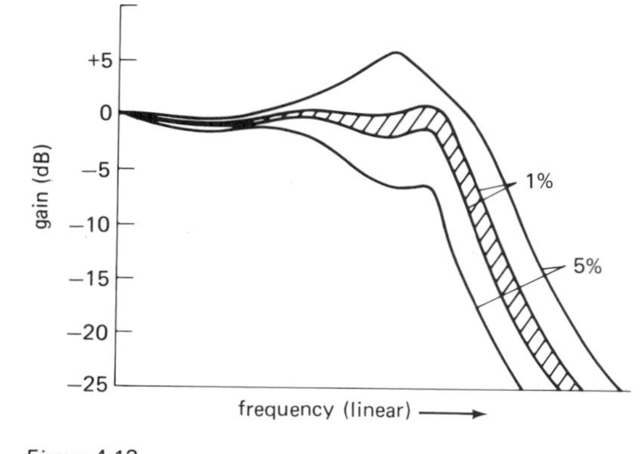

Figure 4.12
The effect of component tolerance on active filter performance.

Viewed in this light, the Chebyshev is a very rational filter design. It is sometimes called an equiripple filter: It manages to improve the situation in the transition region by spreading equal-size ripples throughout the passband, the number of ripples increasing with the order of the filter. Even with rather small ripples (as little as 0.1dB) the Chebyshev filter offers considerably improved sharpness of the knee as compared with the Butterworth. To make the improvement quantitative, suppose that you need a filter with flatness to 0.1dB within the passband and 20dB attenuation at a frequency 25% beyond the top of the passband. By actual calculation that will require a 19-pole Butterworth, but only an 8-pole Chebyshev.

The idea of accepting some passband ripple in exchange for improved steepness in the transition region, as in the equiripple Chebyshev filter, is carried to its logical limit in the so-called elliptic (or Cauer) filter by trading ripple in both passband and stopband for an even steeper transition region than that of the Chebyshev filter. With computer-aided design techniques, the design of elliptic filters is as straightforward as for the classic Butterworth and Chebyshev filters.

Bessel filter

As we hinted earlier, the amplitude response of a filter does not tell the whole story. A filter characterized by a flat amplitude response may have large phase shifts. The result is that a signal in the passband will suffer distortion of its waveform. In situations where the shape of the waveform is paramount, a linear-phase filter (or constant-time-delay filter) is desirable. A filter whose phase shift varies linearly with frequency is equivalent to a constant time delay for signals within the passband, i.e., the waveform is not distorted. The Bessel filter (also called the Thomson filter) has maximally flat time delay within its passband, in analogy with the Butterworth, which has maximally flat amplitude response. To see the kind of improvement in time-domain performance you get with the Bessel filter, look at Figure 4.13 for a comparison of time delay versus

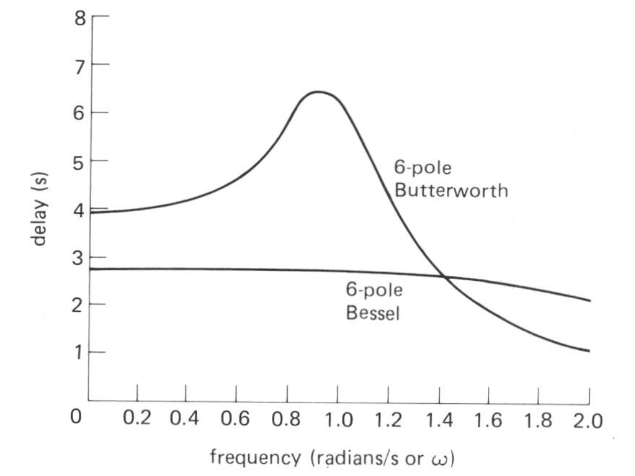

Figure 4.13
Comparison of time delays for 6-pole Bessel and Butterworth low-pass filters. The excellent time-domain performance of the Bessel filter minimizes waveform distortion.

normalized frequency for 6-pole Bessel and Butterworth low-pass filters. The poor time-delay performance of the Butterworth gives rise to effects such as overshoot when driven with pulse signals. On the other hand, the price you pay for the Bessel's constancy of time delay is an amplitude response with even less steepness than that of the Butterworth in the transition region between passband and stopband.

There are numerous filter designs that attempt to improve on the Bessel's good time-domain performance by compromising some of the constancy of time delay for improved rise time and amplitude-versus-frequency characteristics. The Gaussian filter has phase characteristics nearly as good as those of the Bessel, with improved step response. In another class there are interesting filters that allow uniform ripples in the passband time delay (in analogy with the Chebyshev's ripples in its amplitude response) and yield approximately constant time delays even for signals well into the stopband. Another approach to the problem of getting filters with uniform time delays is to use all-pass filters, also known as delay equalizers. These have constant amplitude response with frequency, with a phase shift that can be tailored to individual requirements. Thus they can be used to improve the time-delay constancy of any filter, including Butterworth and Chebyshev types.

Filter comparison

In spite of the preceding comments about the Bessel filter's transient response, it still has vastly superior properties in the time domain, as compared with the Butterworth and Chebyshev. The Chebyshev, with its highly desirable amplitude-versus-frequency characteristics, actually has the poorest time-domain performance of the three. The Butterworth is in between in both frequency and time-domain properties. Table 4.1 and Figure 4.14 give more information about time-domain performance for these three kinds of filters to complement the frequency-domain graphs presented earlier. They make it clear that the Bessel is a very desirable filter where performance in the time domain is important.

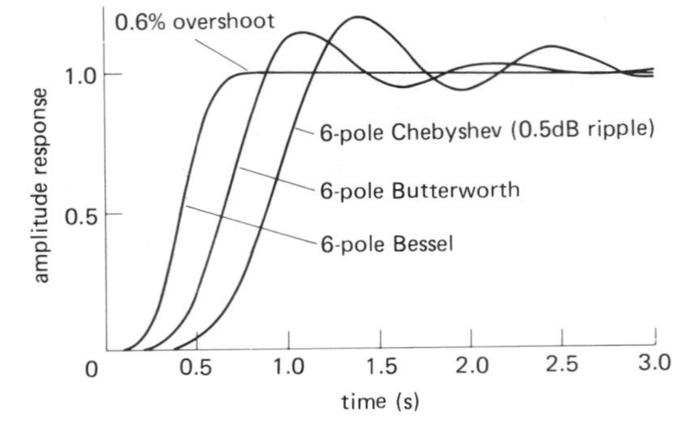

Figure 4.14
Step-response comparison for 6-pole low-pass filters normalized for 3dB attenuation at 1Hz.

ACTIVE FILTER CIRCUITS

A lot of ingenuity has been used in inventing clever active filter circuits, each of which can

TABLE 4.1. TIME-DOMAIN PERFORMANCE COMPARISON FOR LOW-PASS FILTERS[a]

	f_{3dB} (Hz)	Poles	Step rise time (0 to 90%) (s)	Overshoot (%)	Settling time to 1% (s)	Settling time to 0.1% (s)	Stopband $f = 2f_c$ (dB)	Attenuation $f = 10f_c$ (dB)
Bessel (−3.0dB at	1.0	2	0.4	0.4	0.6	1.1	10	36
f_c = 1.0Hz)	1.0	4	0.5	0.8	0.7	1.2	13	66
	1.0	6	0.6	0.6	0.7	1.2	14	92
	1.0	8	0.7	0.3	0.8	1.2	14	114
Butterworth	1.0	2	0.4	4	0.8	1.7	12	40
(−3.0dB at	1.0	4	0.6	11	1.0	2.8	24	80
f_c = 1.0Hz)	1.0	6	0.9	14	1.3	3.9	36	120
	1.0	8	1.1	16	1.6	5.1	48	160
Chebyshev 0.5dB	1.39	2	0.4	11	1.1	1.6	8	37
ripple (−0.5dB	1.09	4	0.7	18	3.0	5.4	31	89
at f_c = 1.0Hz)	1.04	6	1.1	21	5.9	10.4	54	141
	1.02	8	1.4	23	8.4	16.4	76	193
Chebshev 2.0dB	1.07	2	0.4	21	1.6	2.7	15	44
ripple (−2.0dB	1.02	4	0.7	28	4.8	8.4	37	96
at f_c = 1.0Hz)	1.01	6	1.1	32	8.2	16.3	60	148
	1.01	8	1.4	34	11.6	24.8	83	200

[a] A design procedure for these filters is presented in Section 4.07.

be used to generate response functions such as the Butterworth, Chebyshev, etc. You might wonder why the world needs more than one active filter circuit. The reason is that various circuit realizations excel in one or another desirable property, so there is no all-around best circuit.

Some of the features to look for in active filters are (a) small numbers of parts, both active and passive, (b) ease of adjustability, (c) small spread of parts values, especially the capacitor values, (d) undemanding use of the op-amp, especially requirements on slew rate, bandwidth, and output impedance, (e) the ability to make high-Q filters, and (f) sensitivity of filter characteristics to component values and op-amp gain (in particular, the gain-bandwidth product, f_T). In many ways the last feature is one of the most important. A filter that requires parts of high precision is difficult to adjust, and it will drift as the components age; in addition, there is the nuisance that it requires components of good initial accuracy. The VCVS circuit probably owes most of its popularity to its simplicity and its low parts count, but it suffers from high sensitivity to component variations. By comparison, recent interest in the more complicated gyrator-like realizations stems from their desirable properties of insensitivity to small component variability.

In this section we will present several circuits for low-pass, high-pass, and bandpass active filters. We will begin with the popular VCVS, or controlled-source type, then show the state-variable designs available as integrated circuits from several manufacturers, and finally mention the twin-T sharp rejection filter and some interesting new directions in gyrator realizations.

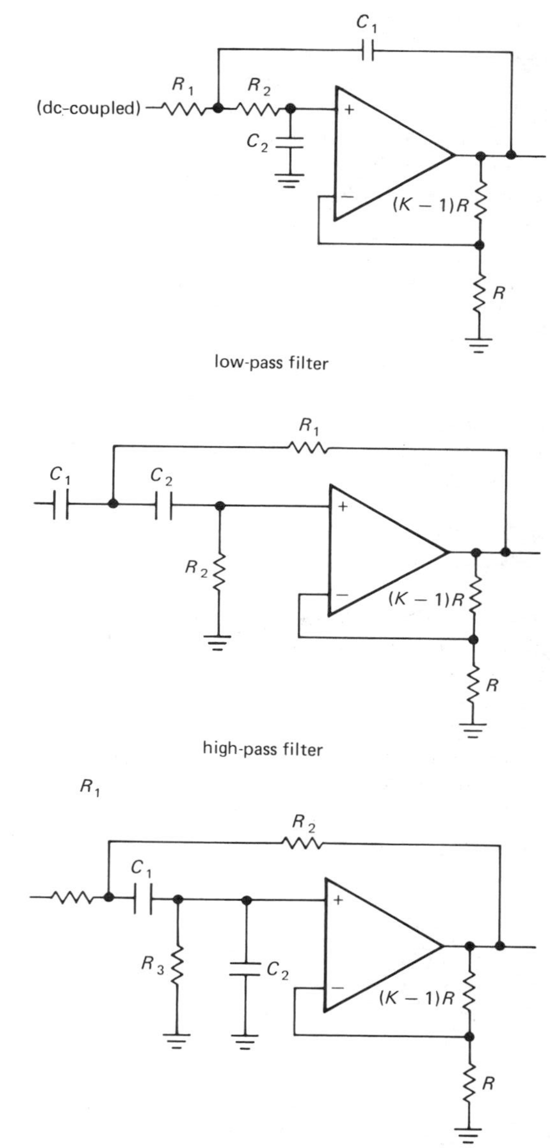

Figure 4.15
VCVS active filter circuits.

4.06 VCVS circuits

The voltage-controlled voltage-source (VCVS) filter, also known simply as a controlled-source filter, is a variation of the Sallen and Key circuit shown earlier. It replaces the unity-gain follower with a noninverting amplifier of gain greater than 1. Figure 4.15 shows the circuits for low-pass, high-pass, and bandpass realizations. The resistors at the outputs of the op-amps create a noninverting voltage amplifier of voltage gain K, with the remaining Rs and Cs contributing the frequency response properties of the filter. These are 2-pole filters, and they can be Butterworth, Bessel, etc., by suitable choice of component values, as we will show later. Any number of VCVS 2-pole sections may be cascaded to generate higher-order filters. When that is done, the individual filter sections are, in general, not identical. In fact, each section represents a quadratic polynomial factor of the *n*th-order polynomial describing the overall filter.

There are design equations and tables in most standard filter handbooks for all the standard filter responses, usually including separate tables for each of a number of ripple amplitudes for Chebyshev filters. In the next section we will present an easy-to-use design table for VCVS filters of Butterworth, Bessel, and Chebyshev responses (0.5dB and 2dB passband ripple for Chebyshev filters) for use as low-pass or high-pass filters. Bandpass and band-reject filters can be easily made from combinations of these.

4.07 VCVS filter design using our simplified table

To use Table 4.2, begin by deciding which filter response you need. As we mentioned earlier, the Butterworth may be attractive if maximum flatness of passband is desired, the Chebyshev gives the fastest rolloff from passband to stopband (at the expense of some ripple in the passband), and the Bessel provides the best phase characteristics, i.e., constant signal delay in the passband, with correspondingly good step response. The frequency responses for all types are shown in the accompanying graphs (Fig. 4.16).

To construct an n-pole filter (n is an even number), you will need to cascade $n/2$ VCVS sections. Only even-order filters are shown, since an odd-order filter requires as many op-amps as the next higher-order filter. Within each section, $R_1 = R_2 = R$, and $C_1 = C_2 = C$. As is usual in op-amp circuits, R will typically be chosen in the range 10k to 100k. (It is best to avoid small resistor values, because the rising open-loop output impedance of the op-amp at high frequencies adds to the resistor values and upsets calculations.) Then all you need to do is set the gain, K, of each stage according to the table entries. For an n-pole filter there are $n/2$ entries, one for each section.

Butterworth low-pass filters

If the filter is a Butterworth, all sections have the same values of R and C, given simply by $RC = 1/2\pi f_c$, where f_c is the desired −3dB frequency of the entire filter. To make a 6-pole low-pass Butterworth filter, for example, you cascade three of the low-pass sections shown previously, with gains of 1.07, 1.59, and 2.48 (preferably in that order, to avoid dynamic range problems), and with identical Rs and Cs to set the 3dB point. The telescope drive circuit in Section 8.30 shows such an example, with f_c = 88.4Hz (R = 180k, $C = 0.01\mu$F).

Bessel and Chebyshev low-pass filters

To make a Bessel or Chebyshev filter with the VCVS, the situation is only slightly more complicated. Again we cascade several 2-pole VCVS filters, with prescribed gains for each section. Within each section we again use $R_1 = R_2 = R$, and $C_1 = C_2 = C$. However, unlike the situation with the Butterworth, the RC products for the different sections are different and must be scaled by the normalizing factor f_n (given for each section in Table 4.2) according to $RC = 1/2\pi f_n f_c$. Here f_c is again the −3dB point for the Bessel filter, whereas for the Chebyshev filter it defines the end of the passband, i.e., it is the frequency at which the amplitude response falls out of the ripple band on its way into the stopband. For example, the response of a Chebyshev low-pass filter with 0.5dB ripple and f_c = 100Hz will be flat within +0dB to −0.5dB from dc to 100Hz, with 0.5dB attenuation at 100Hz and a rapid falloff for frequencies greater than

TABLE 4.2. VCVS LOW-PASS FILTERS

Poles	Butterworth	Bessel		Chebyshev (0.5dB)		Chebyshev (2.0dB)	
	K	f_n	K	f_n	K	f_n	K
2	1.586	1.274	1.268	1.231	1.842	0.907	2.114
4	1.152	1.432	1.084	0.597	1.582	0.471	1.924
	2.235	1.606	1.759	1.031	2.660	0.964	2.782
6	1.068	1.607	1.040	0.396	1.537	0.316	1.891
	1.586	1.692	1.364	0.768	2.448	0.730	2.648
	2.483	1.908	2.023	1.011	2.846	0.983	2.904
8	1.038	1.781	1.024	0.297	1.522	0.238	1.879
	1.337	1.835	1.213	0.599	2.379	0.572	2.605
	1.889	1.956	1.593	0.861	2.711	0.842	2.821
	2.610	2.192	2.184	1.006	2.913	0.990	2.946

100Hz. Values are given for Chebyshev filters with 0.5dB and 2.0dB passband ripple; the latter have a somewhat steeper transition into the stopband (Fig. 4.16).

High-pass filters

To make a high-pass filter, use the high-pass configuration shown previously, i.e., with the *R*s and *C*s interchanged. For Butterworth filters, everything else remains unchanged (use the same values for *R*, *C*, and *K*). For the Bessel and Chebyshev filters, the *K* values remain the same, but the normalizing factors f_n must be inverted, i.e., for each section the new f_n equals $1/(f_n$ listed in Table 4.2).

A bandpass filter can be made by cascading overlapping low-pass and high-pass filters. A band-reject filter can be made by summing the outputs of nonoverlapping low-pass and high-pass filters. However, such cascaded filters won't work well for high-*Q* filters (extremely sharp bandpass filters) because there is great sensitivity to the component values in the individual (uncoupled) filter sections. In such cases a high-*Q* single-stage bandpass circuit (e.g., the VCVS bandpass circuit illustrated previously) should be used instead. Even a

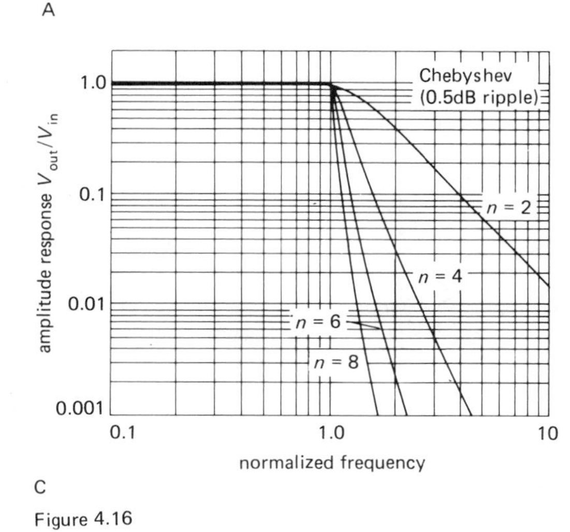

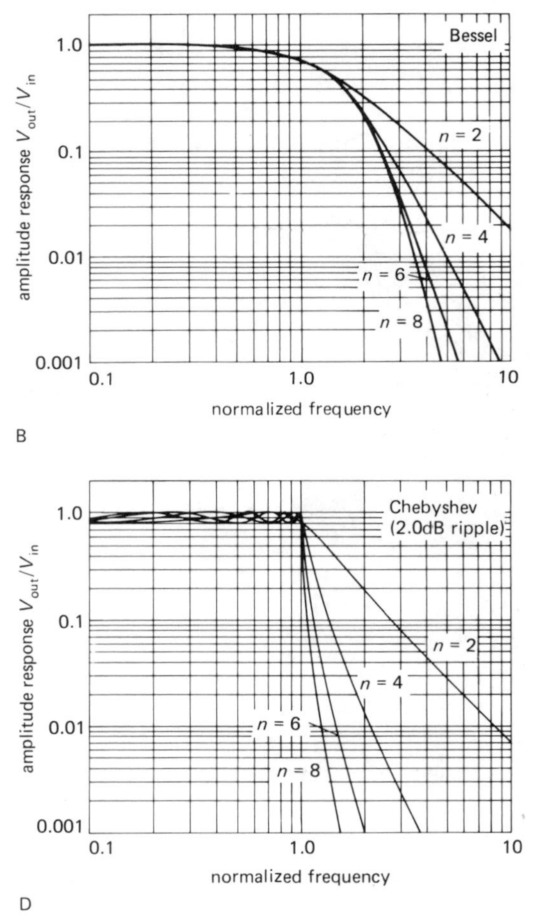

Figure 4.16
Normalized frequency response graphs for the 2-, 4-, 6-, and 8-pole filters in Table 4.2. The Butterworth and Bessel filters are normalized to 3dB attenuation at unit frequency, whereas the Chebyshev filters are normalized to 0.5dB and 2dB attenuations.

single-stage 2-pole filter can produce a response with an extremely sharp peak. Information on such filter design is available in the standard references.

VCVS filters minimize the number of components needed (2 poles/op-amp) and offer the additional advantages of noninverting gain, low output impedance, small spread of component values, easy adjustability of gain, and the ability to operate at high gain or high Q. They suffer from high sensitivity to component values and amplifier gain, and they don't lend themselves well to applications where a tunable filter of stable characteristics is needed.

EXERCISE 4.3

Design a 6-pole Chebyshev low-pass VCVS filter with a 0.5dB passband ripple and 100Hz cutoff frequency f_c. What is the attenuation at 1.5 f_c?

4.08 State-variable filters

The 2-pole filter shown in Figure 4.17 is far more complex than the VCVS circuits, but it is popular because of its improved stability and ease of adjustment. It is called a state-variable filter, or "biquad" filter, and is available as an IC from National (the AF100), Burr-Brown (the UAF series), and others. Because it is a manufactured module, all components except R_{in}, R_Q, R_{F1}, and R_{F2} are built in. Among its nice properties is the availability of high-pass, low-pass, and bandpass outputs from the same circuit; also, its frequency can be tuned while maintaining constant Q (or, alternatively, constant bandwidth) in the bandpass characteristic. As with the VCVS realizations, multiple stages can be cascaded to generate higher-order filters.

Extensive design formulas and tables are provided by the manufacturers for the use of these convenient ICs. They show how to choose the external resistor values to make Butterworth, Bessel, and Chebyshev filters for a wide range of filter orders, for low-pass, high-pass, bandpass, and band-reject responses. Among the nice features of these hybrid ICs is integration of the capacitors into the module, so that only external resistors need be added.

□ 4.09 Twin-T notch filters

The passive RC network shown in Figure 4.18 has infinite attenuation at a frequency $f_c = 1/2\pi RC$. Infinite attenuation is uncharacteristic of RC filters in general; this one works by effectively adding two signals that have been shifted 180° out of phase at the cutoff frequency. It requires good matching of components in order to obtain a good null at f_c. It is called a twin-T, and it can be used to remove an interfering signal, such as

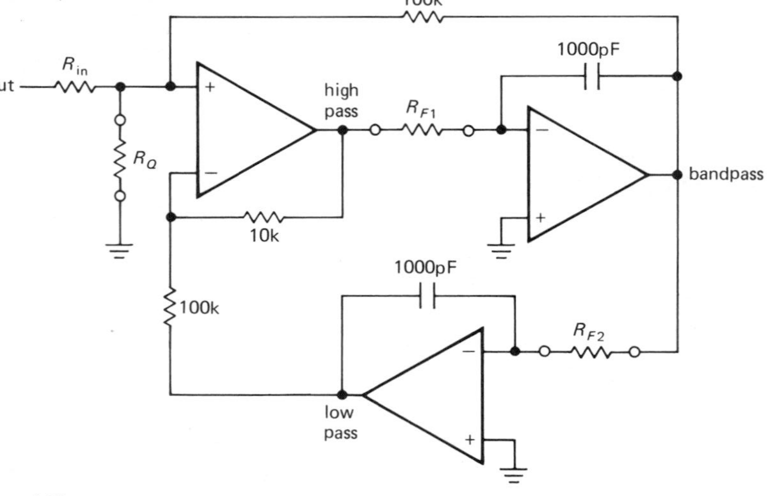

Figure 4.17
State-variable active filter.

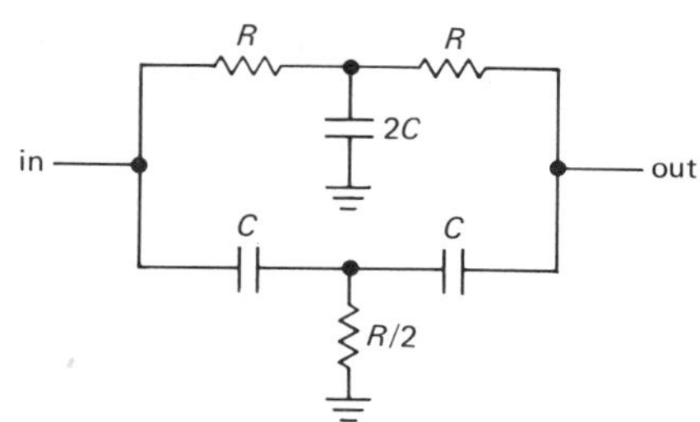

Figure 4.18

60Hz power-line pickup. The problem is that it has the same "soft" cutoff characteristics as all passive *RC* networks, except, of course, near f_c, where its response drops like a rock. For example, a twin-T driven by a perfect voltage source is down 10dB at twice (or half) the notch frequency and 3dB at four times (or one-fourth) the notch frequency. One trick to improve its notch characteristics is to "activate" it in the manner of a Sallen and Key filter (Fig. 4.19). This technique looks good in principle, but it is generally disappointing in practice, owing to the impossibility of maintaining a good filter null. As the filter notch becomes sharper (more gain in the bootstrap), its null becomes less deep.

Twin-T filters are available as prefab modules, going from 1Hz to 50kHz, with notch depths of about 60dB (with some deterioration at high and low temperatures). They are easy to make from components, but resistors and capacitors of good stability and low temperature coefficient should be used to get a deep and stable notch. One of the components should be made trimmable.

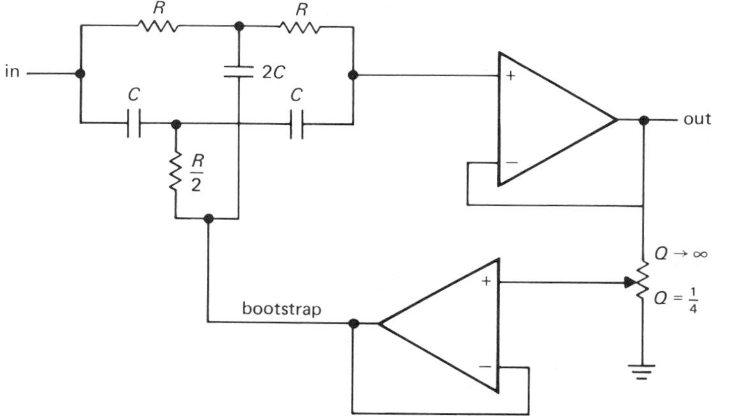

Figure 4.19

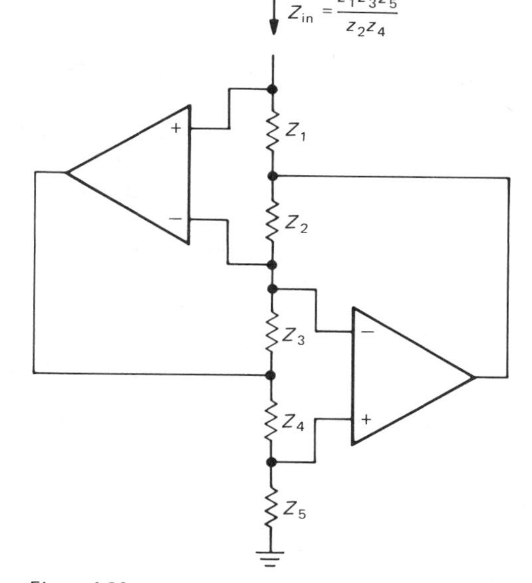

Figure 4.20

□ 4.10 Gyrator filter realizations

There has recently been a reawakening of interest in active filters made with gyrators; basically they are used to substitute for inductors in traditional filter designs. The gyrator circuit shown in Figure 4.20 is popular. Z_4 will ordinarily be a capacitor, with the other impedances being replaced by resistors, creating an inductor $L = kC$, where $k = R_1R_3R_5/R_2$. It is claimed that these gyrator-substituted filters have the lowest sensitivity to component variations, exactly analogous to their passive *RLC* prototypes.

□ *Double capacitors*

One problem with gyrator circuits has been the difficulty of making "floating" inductors (inductors in which both terminals are accessible) for use in low-pass-filter designs where there are networks of series inductors with capacitors to ground. Although it is possible to make a floating inductor with two gyrators, this requires many op-amps (usually four). A clever alternative solution is to go ahead with the passive filter design on paper, using *R*s, *L*s, and *C*s as required, and then divide all terms in the equation that gives the transfer function (V_{out}/V_{in}, as a complex number) by $j\omega$. This converts inductors to resistors (of value L), resistors to capacitors (value $1/R$), and capacitors to frequency-dependent negative resistors (FDNR) with resistance equal to $-1/\omega^2 C$. An FDNR behaves like a "double capacitor" (i.e., it has a 180° phase shift and a 12dB/octave decrease in impedance) and is represented by the suggestive symbol shown in Figure 4.21. It can be easily made

Figure 4.21

from the gyrator circuit in Figure 4.20 by substituting capacitors for Z_1 and Z_3 and substituting resistors for Z_2, Z_4, and Z_5. With the availability of good gyrator circuits and clever tricks like this, active filters made with gyrators are becoming quite popular.

OSCILLATORS

4.11 Introduction to oscillators

Within nearly every electronic instrument it is essential to have an oscillator or waveform generator of some sort. Apart from the obvious cases of signal generators, function generators, and pulse generators themselves, a source of regular oscillations is necessary in any cyclical measuring instrument, in any instrument that initiates measurements or processes, and in any instrument whose function involves periodic states or periodic waveforms. That includes just about everything. For example, oscillators or waveform generators are used in digital multimeters, oscilloscopes, radiofrequency receivers, computers, every computer peripheral (tape, disc, printer, alphanumeric terminal), nearly every digital instrument (counters, timers, calculators, and anything with a "multiplexed display"), and a host of other devices too numerous to mention. A device without an oscillator either doesn't do anything or expects to be driven by something else (which probably contains an oscillator). It is not an exaggeration to say that an oscillator of some sort is as essential an ingredient in electronics as a regulated supply of dc power.

Depending on the application, an oscillator may be used simply as a source of regularly spaced pulses (e.g., a "clock" for a digital system), or demands may be made on its stability and accuracy (e.g., the time base for a frequency counter), its adjustability (e.g., the local oscillator in a transmitter or receiver), or its ability to produce accurate waveforms (e.g., the horizontal-sweep ramp generator in an oscilloscope).

In the following sections we will treat briefly the most popular oscillators, from the simple *RC* relaxation oscillators to the stable quartz-crystal oscillators. Our aim is not to survey everything in exhaustive detail, but simply to make you acquainted with what is available and what sorts of oscillators are suitable in various situations.

4.12 Relaxation oscillators

A very simple kind of oscillator can be made by charging a capacitor through a resistor (or a current source), then discharging it rapidly when the voltage reaches some threshold, beginning the cycle anew. Alternatively, the external circuit may be arranged to reverse the polarity of the charging current when the threshold is reached, thus generating a triangle wave rather than a sawtooth. Oscillators based on this principle are known as relaxation oscillators. They are inexpensive and simple, and with careful design they can be made quite stable in frequency.

In the past, negative-resistance devices such as unijunction transistors and neon bulbs were used to make relaxation oscillators, but current practice favors op-amps or special timer ICs. Figure 4.22 shows a classic *RC* relaxation oscillator. The operation is

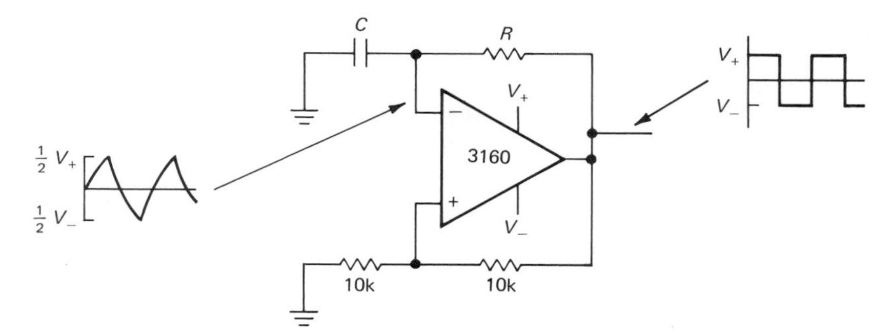

Figure 4.22
Op-amp relaxation oscillator.

simple: Assume that when power is first applied the op-amp output goes to positive saturation (it's actually a toss-up which way it will go, but it doesn't matter). The capacitor begins charging up toward V_+, with time constant RC. When it reaches one-half the supply voltage, the op-amp switches into negative saturation (it's a Schmitt trigger), and the capacitor begins discharging toward V_- with the same time constant. The cycle repeats indefinitely, with period $2.2RC$, independent of supply voltage. A CMOS output-stage op-amp (see Section 6.19) was chosen because its outputs saturate cleanly at the supply voltages.

EXERCISE 4.4

Show that the period is as stated.

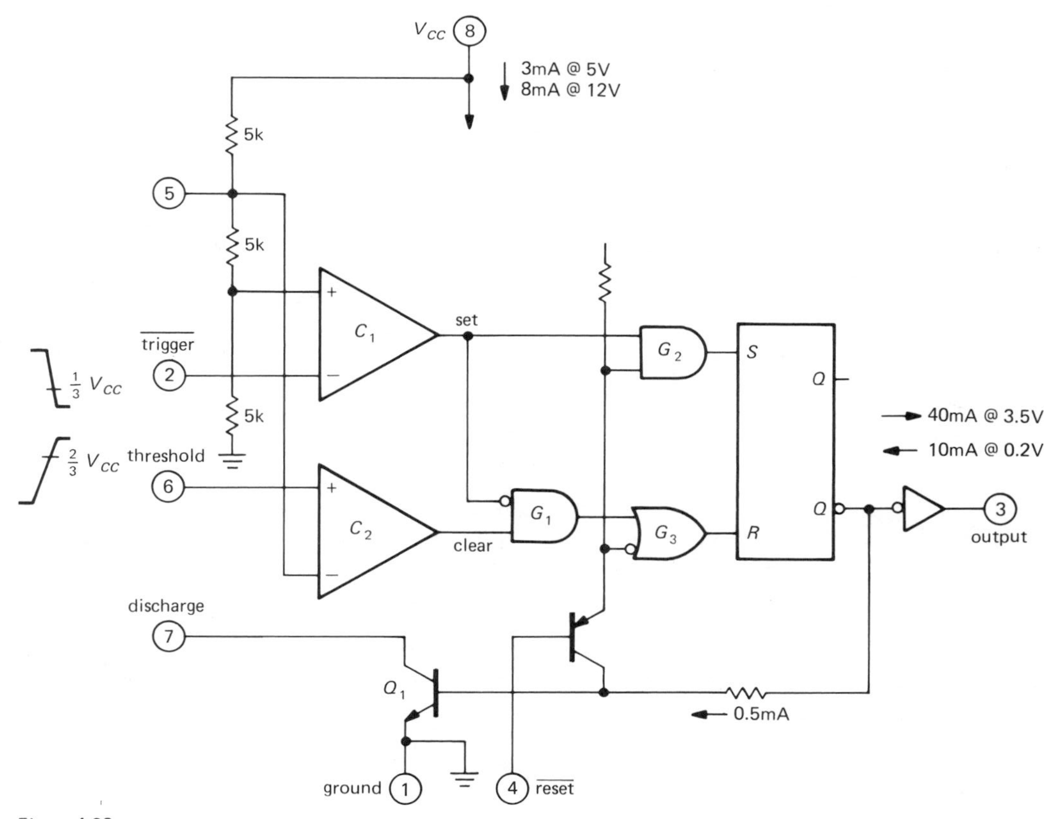

Figure 4.23
Simplified 555 schematic.

By using current sources to charge the capacitor, a good triangle wave can be generated. A clever circuit using that principle was shown in Section 3.28.

4.13 The classic timer chip: the 555

The next level of sophistication involves the use of timer or waveform-generator ICs as relaxation oscillators. The most popular chip around is the 555. It is also a misunderstood chip, and we intend to set the record straight with the equivalent circuit shown in Figure 4.23. Some of the symbols belong to the digital world (Chapter 8 and following), so you won't become a 555 expert for a while yet. But the operation is simple enough: The output goes HIGH (near V_{CC}) when the 555 receives a TRIGGER′ input, and it stays there until the THRESHOLD input is driven, at which time the output goes LOW (near ground) and the DISCHARGE transistor is turned on. The TRIGGER′ input is activated by an input level below $\frac{1}{3}V_{CC}$, and the THRESHOLD is activated by an input level above $\frac{2}{3}V_{CC}$.

The easiest way to understand the workings of the 555 is to look at an example (Fig. 4.24). When power is applied, the capacitor is discharged; so the 555 is triggered, causing the output to go HIGH, the discharge transistor Q_1 to turn off, and the capacitor to begin charging toward 10 volts through $R_A + R_B$. When it has reached $\frac{2}{3}V_{CC}$, the THRESHOLD input is triggered, causing the output to go LOW and Q_1 to turn on, discharging C toward ground through R_B. Operation is now cyclic, with C's voltage going between $\frac{1}{3}V_{CC}$ and $\frac{2}{3}V_{CC}$, with period $T = 0.693(R_A + 2R_B)C$. The output you generally use is the square wave at the output.

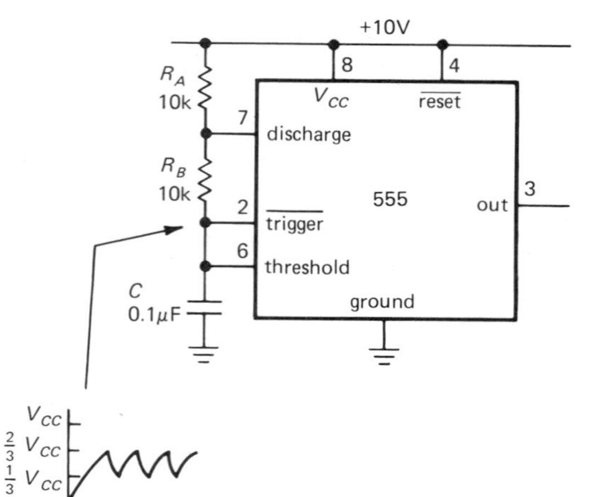

Figure 4.24
The 555 connected as an oscillator.

EXERCISE 4.5

Show that the period is as advertised, independent of supply voltage.

The 555 makes a respectable oscillator, with stability approaching 1%. It can run from a single positive supply of 4.5 to 16 volts, maintaining good frequency stability with supply voltage variations because the thresholds track the supply fluctuations. The 555 can also be used to generate single pulses of arbitrary width, as well as a bunch of other things. It is really a small kit, containing comparators, gates, and flip-flops. It has become a game in the electronics industry to try to think of new uses for the 555. Suffice it to say that many succeed at this new form of entertainment.

There are some other excellent IC timers available. The 556 is a dual 555 in a 14-pin DIP. The Intersil 7555 is a low-power 555 constructed with CMOS; it requires only 80μA supply current and will operate from supply voltages of 2 to 18 volts. The 7556 is a dual 7555. The 322 timer from National includes its own internal voltage reference for determining the threshold.

Voltage-controlled oscillators

Other IC oscillators are available as voltage-controlled oscillators (VCOs), with the output rate variable over some range according to an input control voltage. The best of these have frequency ranges exceeding 1000:1. Examples are the original 566 and the newer LM331, 8038, 2206, 74LS124, 74LS325-327, 74S124, and MC4024.

The 74S124 can operate at frequencies as high as 80MHz. The 74*xxx* types, as well as the 4024, use external *RC*s to set the nominal frequency and generate logic-level outputs. The LM331 is actually a linear voltage-to-frequency (V/F) converter, a technique we will discuss in Section 9.22.

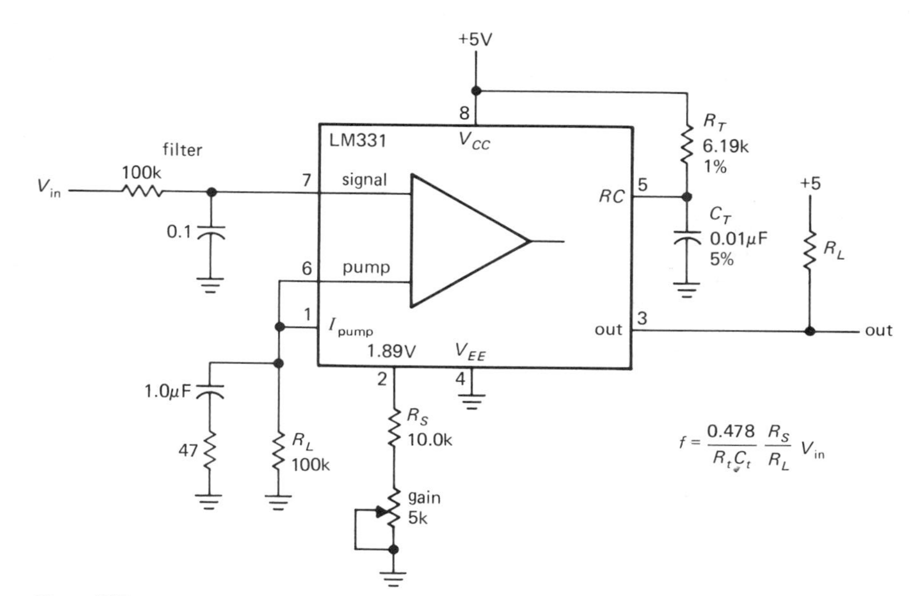

Figure 4.25
Typical V/F converter IC (zero to 10kHz VCO).

The others use internal current sources to generate triangle-wave outputs, and the 8038 and 2206 even include a set of "soft" clamps to convert the triangle wave to a not-too-great sine wave. VCO chips sometimes have an awkward reference for the control voltage (e.g., the positive supply) and complicated symmetrizing schemes for the sine-wave output. It is our opinion that the ideal VCO has yet to be developed. Many of these chips can be used with an external quartz crystal, as we will discuss shortly, for much higher accuracy and stability; in such cases the crystal simply replaces the capacitor. Figure 4.25 shows a VCO circuit with an output frequency range of 10Hz to 10kHz built with the LM331.

□ 4.14 Wien bridge and *LC* oscillators

When a low-distortion sine wave is required, none of the preceding methods is generally adequate. Although wide-range function generators do use the technique of "corrupting" a triangle wave with diode clamps, the resulting distortion can rarely be reduced below 1%. By comparison, most hi-fi audiophiles insist on distortion levels below 0.1% for their amplifiers. To test such low-distortion audio components, pure sine-wave signal sources with residual distortion less than 0.05% or so are required.

At low to moderate frequencies the Wien bridge oscillator (Fig. 4.26) is a good source of low-distortion sinusoidal signals. The idea is to make a feedback amplifier with 180° phase shift at the desired output frequency, then adjust the loop gain so that a self-sustaining oscillation just barely takes place. For equal-value *R*s and *C*s as shown, the voltage gain from the noninverting input to op-amp output should be exactly +3.00. With less gain the oscillation will cease, and with more gain the output will saturate. The distortion is low if the amplitude of oscillation remains within the linear region of the amplifier, i.e., it must not be allowed to go into a full-swing oscillation. Without some trick to control the gain, that is exactly what will happen, with the amplifier's output increasing until the effective gain is reduced to 3.0 because of saturation. The tricks involve some sort of long-time-constant gain-setting feedback, as you will see.

In the first circuit an incandescent lamp is used as a variable-resistance feedback element. As the output level rises, the lamp heats slightly, reducing the noninverting gain. In the second circuit an amplitude discriminator consisting of the diodes and

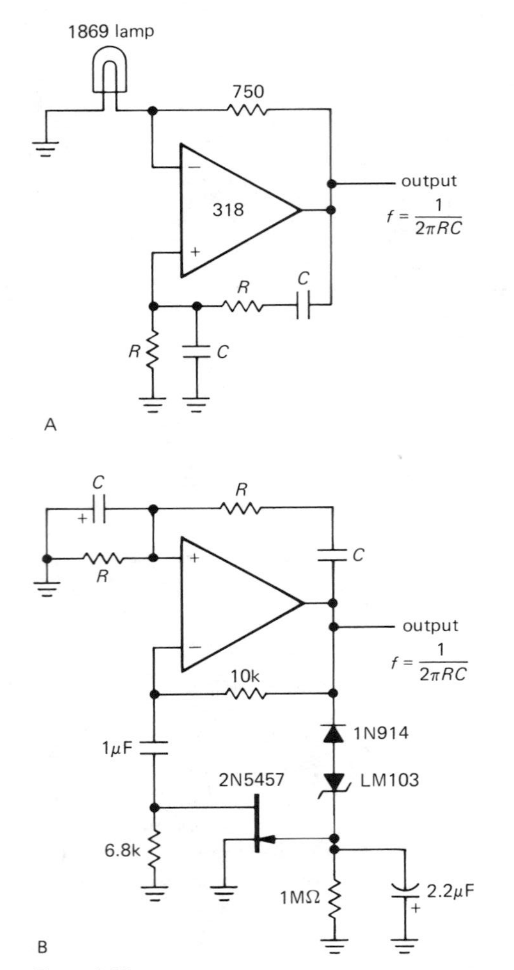

Figure 4.26
Wien-bridge low-distortion oscillators.

RC adjusts the ac gain by varying the resistance of the FET, a form of transistor that behaves like a voltage-variable resistance for small applied voltages. Note the long time constant used (2 seconds); this is essential to avoid distortion, since fast feedback will distort the wave by attempting to control the amplitude within the time of one cycle.

□ 4.15 *LC* oscillators

At high frequencies the favorite method of sine-wave generation is an *LC*-controlled oscillator, in which a tuned *LC* is connected in an amplifier-like circuit to provide gain at its resonant frequency. Overall positive feedback is then used to cause a sustained oscillation to build up at the *LC*'s resonant frequency; such circuits are self-starting.

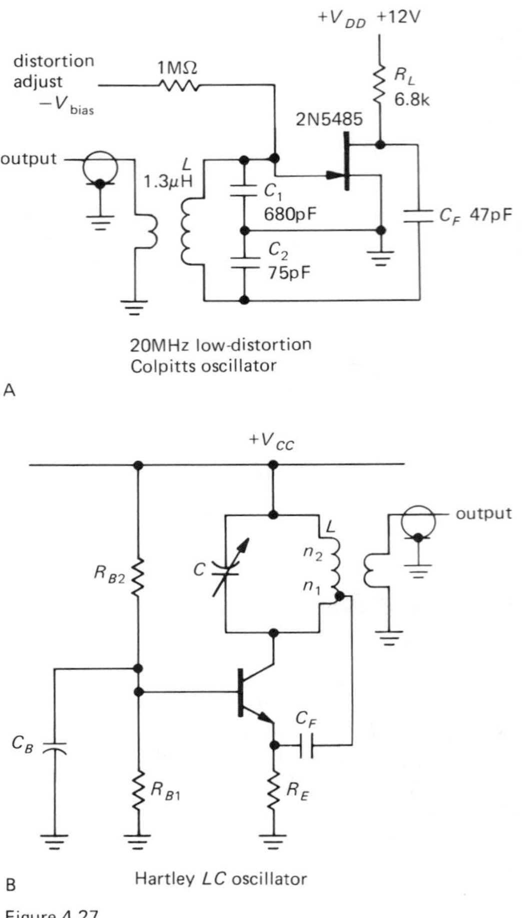

Figure 4.27

Figure 4.27 shows two popular configurations. The first circuit is the trusty Colpitts oscillator, a parallel tuned *LC* at the input, with positive feedback from the output. It uses a JFET, a device we will discuss in Chapter 6, and it is claimed that its distortion is less than −60dB. The second circuit is a Hartley oscillator, built with an *npn* transistor. The variable capacitor is for frequency adjustment. Both circuits use *link coupling,* just a few turns of wire acting as a step-down transformer.

For historical reasons we should mention a close cousin of the *LC* oscillator, namely the tuning-fork oscillator. It used the high-*Q* oscillations of a tuning fork as the frequency-determining element of an oscillator, and it found use in low-frequency standards (stability of a few parts per million, if run in a constant-temperature oven) as well as wristwatches. These objects have been superseded by quartz oscillators, which are discussed in the next section.

☐ *Parasitic oscillations*

Suppose you have just made a nice amplifier and are testing it out with a sine-wave input. You switch the input function generator to a square wave, but the output remains a sine wave! You don't have an amplifier; you've got trouble.

Parasitic oscillations aren't normally as blatant as this. They are normally observed as fuzziness on part of a waveform, erratic current-source operation, unexplained op-amp offsets, or circuits that behave normally with the oscilloscope probe applied, but go wild when the scope isn't looking. These are bizarre manifestations of untamed high-frequency parasitic oscillations caused by unintended Hartley or Colpitts oscillators employing lead inductance and interelectrode capacitances.

The circuit in Figure 4.28 shows an oscillating current source born in an electronics lab course where a VOM was used to measure the output compliance of a standard transistor current source. The current seemed to vary excessively (5% to 10%)

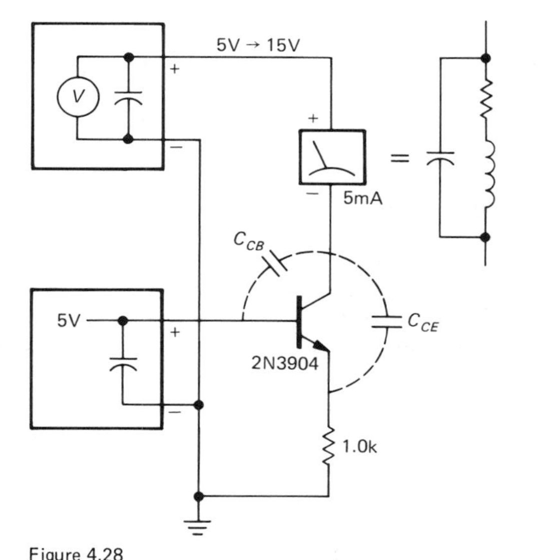

Figure 4.28
Parasitic oscillation example.

with load voltage variations within its expected compliance range, a symptom that could be "cured" by sticking a finger on the collector lead! The collector-base capacitance of the transistor and the meter capacitance resonated with the meter inductance in a classic Hartley oscillator circuit, with feedback provided by collector-emitter capacitance. Adding a small base resistor suppressed the oscillation by reducing the high-frequency common-base gain. This is one trick that often helps.

4.16 Quartz-crystal oscillators

RC oscillators can easily attain stabilities approaching 0.1%, with initial predictability of 5% to 10%. That's good enough for many applications, such as the *multiplexed display* in a pocket calculator, in which a multidigit numerical display is driven by lighting one digit after another in rapid succession (a 1kHz rate is typical). Only one digit is lit at any time, but your eye sees the whole display. In such an application the precise rate is quite irrelevant – you just want something in the ballpark. As stable sources of frequency, *LC* oscillators can do a bit better, with stabilities of 0.01% over reasonable periods of time. That's good enough for oscillators in radiofrequency receivers and television sets.

For real stability there's no substitute for a crystal oscillator. This uses a piece of quartz (same chemical as glass, silicon dioxide) that is cut and polished to vibrate at a certain frequency. Quartz is *piezoelectric* – (a strain generates a voltage, and vice versa), so acoustic waves in the crystal can be driven by an applied electric field and in turn can generate a voltage at the surface of the crystal. By plating some contacts on the surface, you wind up with an honest circuit element that can be modeled by an *RLC* circuit, pretuned to some frequency. In fact, its equivalent circuit contains two capacitors, giving a pair of closely spaced (within 1%) series and parallel resonant frequencies (Fig. 4.29). The effect is to produce a rapidly changing reactance with frequency (Fig. 4.30). The quartz crystal's high *Q* (typically around 10,000) and good stability make it a natural for oscillator control, as well as for

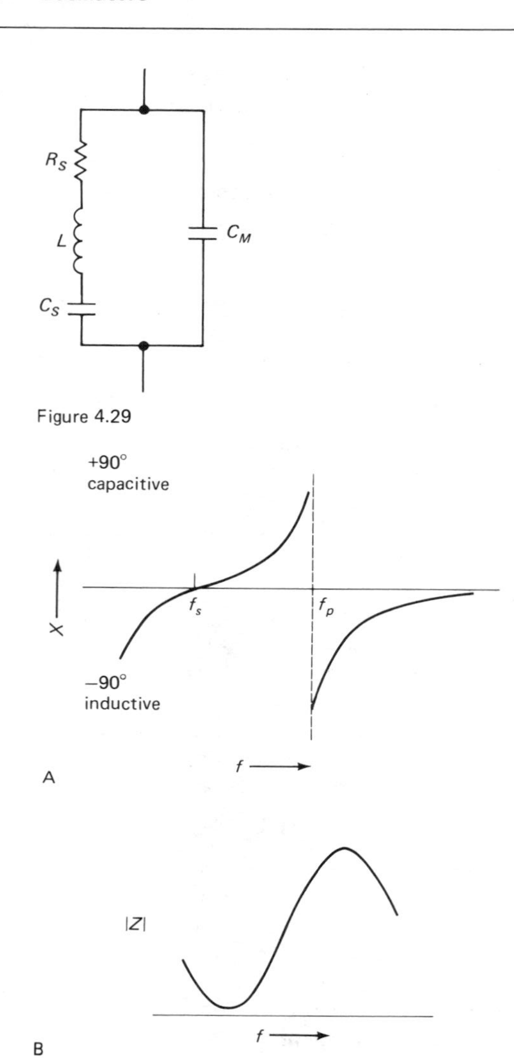

Figure 4.29

Figure 4.30

high-performance filters (see Section 13.12). As with *LC* oscillators, the crystal's equivalent circuit provides positive feedback and gain at the resonant frequency, leading to sustained oscillations.

Figure 4.31 shows some crystal oscillator circuits. In A the classic Pierce oscillator is shown, using the versatile FET (see Chapter 6). The Colpitts oscillator, with a crystal instead of an *LC,* is shown in B. An *npn* bipolar transistor with the crystal as feedback element is used in C. The remaining circuits generate logic-level outputs using CMOS digital logic functions (D) and the TTL VCOs mentioned earlier (E,F).

A convenient alternative to these crystal oscillator circuits, particularly when a logic-compatible output isn't needed, is provided by the SL680/SL1680 series of "crystal oscillator maintaining circuits" from Plessey Semiconductors. These chips are intended as oscillator circuits for crystals in the range 100kHz to 100MHz and are designed to give excellent frequency stability (effectively limited only by the properties of the crystal itself) by carefully limiting the amplitude of oscillation via internal amplitude discrimination and limiting circuitry. Typical frequency stability (assuming a perfect crystal) is 0.001ppm/°C and 0.1ppm/V supply variation. The chip also includes an on-chip voltage regulator and, for the SL680 version, current output as well as voltage output at the oscillation frequency.

Quartz crystals are available from about 10kHz to about 10MHz, with overtone-mode crystals going to about 250MHz. Although crystals have to be ordered for a given frequency, most of the commonly used frequencies are available off the shelf. Frequencies such as 100kHz, 1.0MHz, 2.0MHz, 4.0MHz, 5.0MHz, and 10.0MHz are always easy to get. A 3.579545MHz crystal (available for less than a dollar) is used in TV color-burst oscillators. Digital wristwatches use 32.768kHz (divide by 2^{15} to get 1Hz), and other powers of 2 are also common. A crystal oscillator can be adjusted slightly by varying a series or parallel capacitor, as shown in Figure 4.31D. Given the low cost of crystals (typically about 5 dollars), it is worth considering a crystal oscillator in any application where you would have to strain the capabilities of *RC* relaxation oscillators.

Without great care you can obtain frequency stabilities of a few parts per million over normal temperature ranges with crystal oscillators. By using temperature-compensation schemes you can make a TCXO (temperature-compensated crystal oscillator) with somewhat better performance. Both TCXOs and uncompensated oscillators are available as complete modules from many manufacturers, e.g., Bliley, CTS Knights, Motorola, Reeves Hoffman, Statek, and Vectron. They come in various sizes, ranging down to DIP packages and TO-5

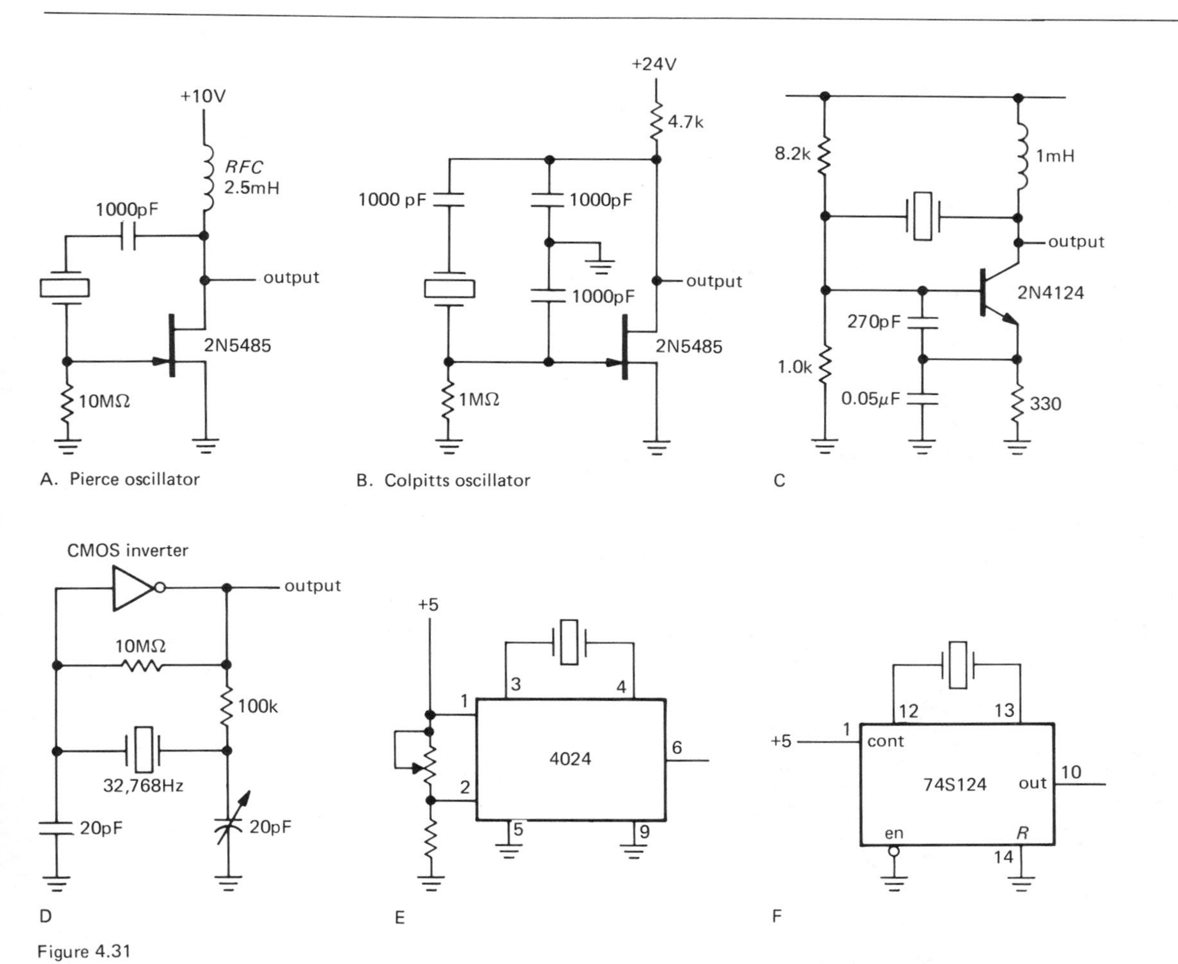

Figure 4.31
Various crystal oscillators.

standard transistor cans. TCXOs deliver stabilities of 1ppm over the range 0°C to 50°C (inexpensive) down to 0.1ppm over the same range (expensive).

Temperature-stabilized oscillators

For the utmost in stability, you may need a crystal oscillator in a constant-temperature oven. A crystal with a zero temperature coefficient at some elevated temperature (80°C to 90°C) is used, with the thermostat set to maintain that temperature. Such oscillators are available as small modules for inclusion into an instrument or as complete frequency standards ready for rack mounting. The 10544 from Hewlett-Packard is typical of high-performance modular oscillators, delivering 10MHz with stabilities of a few parts in 10^{11} over periods of seconds to hours.

When thermal instabilities have been reduced to this level, the dominant effects become crystal "aging" (the frequency tends to decrease continuously with time), power-supply variations, and environmental influences such as shock and vibration (the latter are the most serious problems in quartz wristwatch design). To give an idea of the aging problem, the oscillator mentioned previously has a specified aging rate at delivery of 5 parts in 10^{10} per day, maximum. Aging effects are due in part to the gradual relief of strains, and they tend to settle down after a few months, particularly in a well-manufactured crystal. Our specimen of the 10544 oscillator ages about 1 part in 10^{11} per day.

Atomic frequency standards are used where the stability of ovenized-crystal standards is insufficient. These use a microwave absorption line in a rubidium gas cell, or atomic transitions in an atomic cesium beam, as the reference to which a quartz crystal is stabilized. Accuracy and stability of a few parts in 10^{12} can be obtained. Cesium-beam standards are the official timekeepers in this country, with timing transmissions from the National Bureau of Standards and the Naval Observatory. Atomic

Circuit ideas

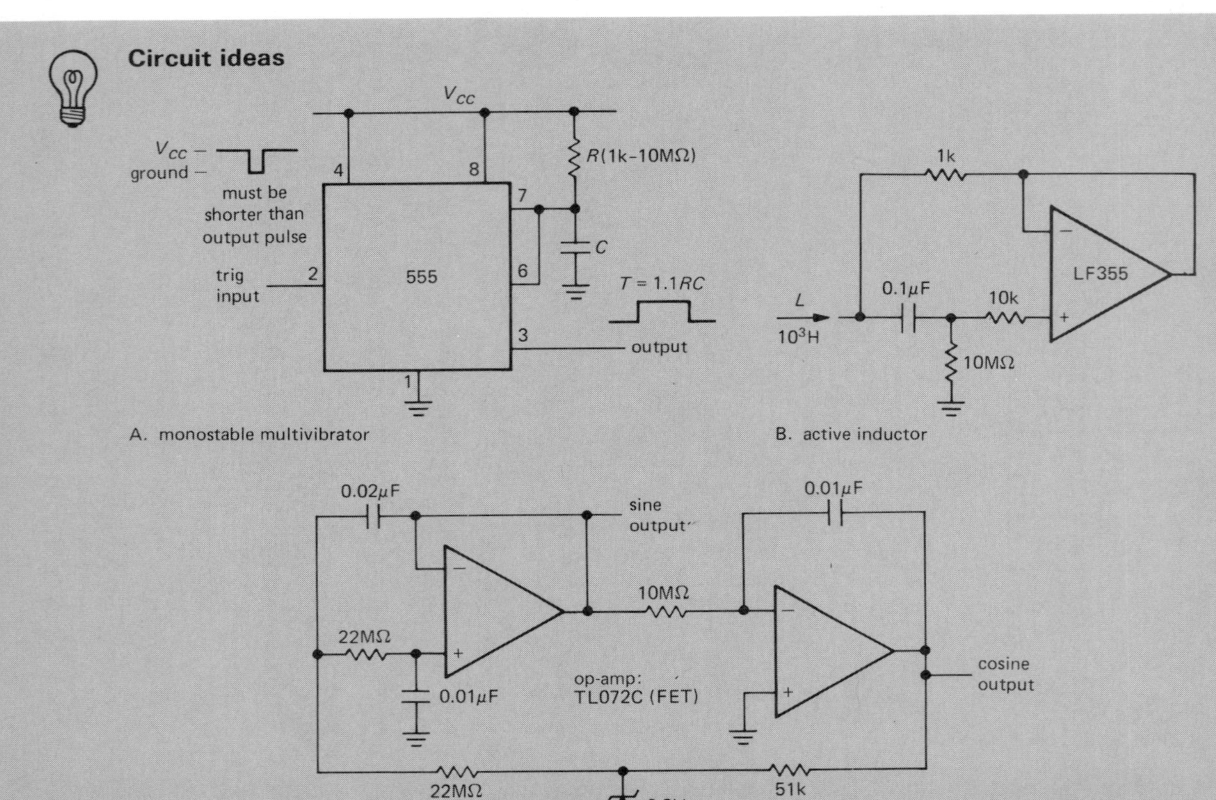

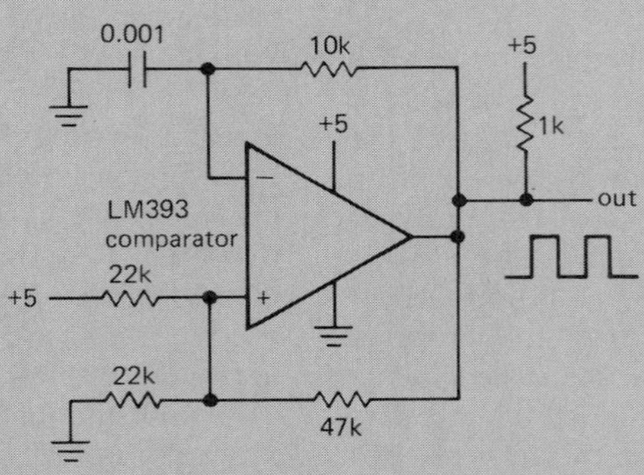

Figure 4.32

hydrogen masers have been suggested as the ultimate in stable clocks, with claimed stabilities approaching a few parts in 10^{14}.

SELF-EXPLANATORY CIRCUITS

4.17 Circuit ideas

Figure 4.32 presents a variety of circuit ideas, mostly taken from manufacturers' data sheets and applications literature.

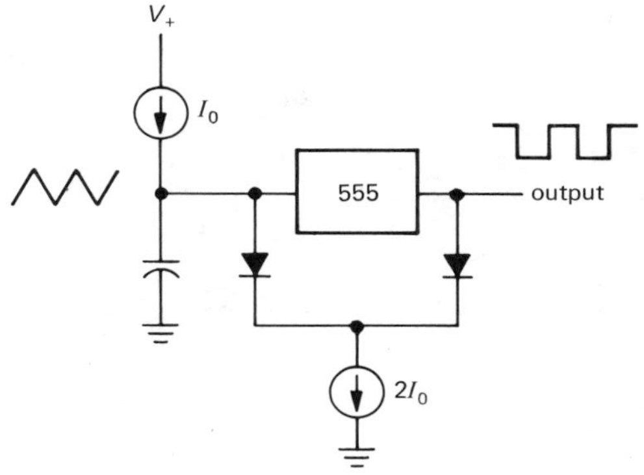

ADDITIONAL EXERCISES

(1) Design a 6-pole high-pass Bessel filter with cutoff frequency of 1kHz.
(2) Design a 60Hz twin-T notch filter with op-amp input and output buffers.
(3) Design a sawtooth wave oscillator, to deliver 1kHz, by replacing the charging resistor in the 555 oscillator circuit with a transistor current source. Be sure to provide enough current-source compliance. What value should R_B (Fig. 4.24) have?
(4) Make a triangle-wave oscillator with a 555. Use a pair of current sources I_0 (sourcing) and $2I_0$ (sinking). Use the 555's output to switch the $2I_0$ current sink on and off appropriately. Figure 4.33 shows one possibility.

Figure 4.33

VOLTAGE REGULATORS AND POWER CIRCUITS

CHAPTER 5

Nearly all electronic circuits, from simple transistor and op-amp circuits up to elaborate digital and microprocessor systems, require one or more sources of stable dc voltage. The simple transformer-bridge-capacitor unregulated power supplies we discussed in Chapter 1 are not generally adequate because their output voltages change with load current and line voltage and because they have significant amounts of 120Hz ripple. Fortunately, it is easy to construct stable power supplies using negative feedback to compare the dc output voltage with a stable voltage reference. Such regulated supplies are in universal use and can be simply constructed with integrated-circuit voltage regulator chips, requiring only a source of unregulated dc input (from a transformer-rectifier-capacitor combination, a battery, or some other source of dc input) and a few other components.

In this chapter you will see how to construct voltage regulators using special-purpose integrated circuits. The same circuit techniques can be used to make regulators with discrete components (transistors, resistors, etc.), but because of the availability of inexpensive high-performance regulator chips, there is usually no advantage to using discrete components in new designs. Voltage regulators get us into the domain of high power dissipation, so we will be talking about heat sinking and techniques like "foldback limiting" to limit transistor operating temperatures and prevent circuit damage. These techniques can be used for all sorts of power circuits, including power amplifiers. With the knowledge of regulators you will have at that point, we will be able go back and discuss the design of the unregulated supply in some detail. In this chapter we will also look at voltage references and voltage-reference ICs, devices with uses outside of power-supply design.

BASIC REGULATOR CIRCUITS WITH THE CLASSIC 723

5.01 The 723 regulator

The μA723 voltage regulator is a classic. Designed by Bob Widlar and first introduced in 1967, it is a flexible, easy-to-use regulator with excellent performance. Although you might not choose it for a new design nowadays, it is worth looking at in some detail, since more recent regulators work on the same principles. Its circuit is shown in Figures 5.1 and 5.2. As you can see, it is really a power-supply kit, containing a temperature-compensated voltage refer-

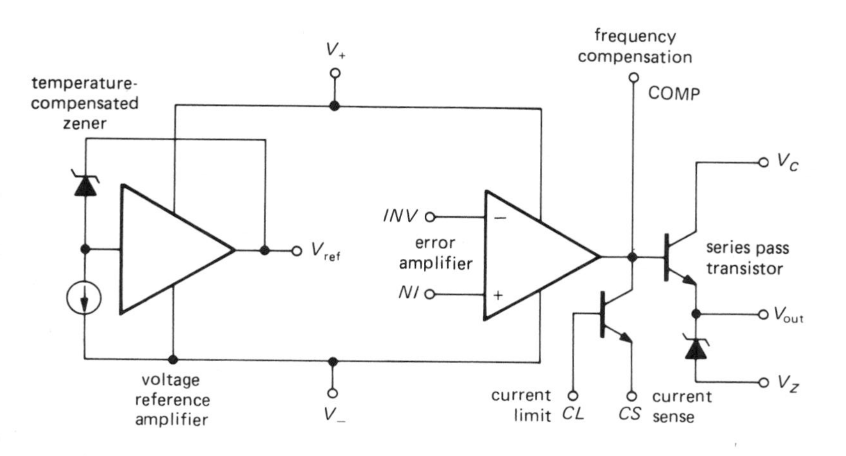

Figure 5.1
Simplified circuit of the 723 regulator. (Fairchild Camera and Instrument Corp.)

ence, differential amplifier, series pass transistor, and current-limiting protective circuit. As it comes, the 723 doesn't regulate anything. You have to hook up an external circuit to make it do what you want. Before going on to design regulators with it, let's look briefly at its internal circuit. It is straightforward and easy to understand (the innards of many ICs aren't).

The heart of the regulator is the temperature-compensated zener reference. Zener D_2 has a positive temperature coefficient, so its voltage is added to Q_6's base-emitter drop (remember, V_{BE} has a negative temperature coefficient of roughly -2mV/°C) to form a voltage reference (nominally 7.15V) of nearly zero temperature coefficient (typically 0.003%/°C). Q_4 through Q_6 are arranged to

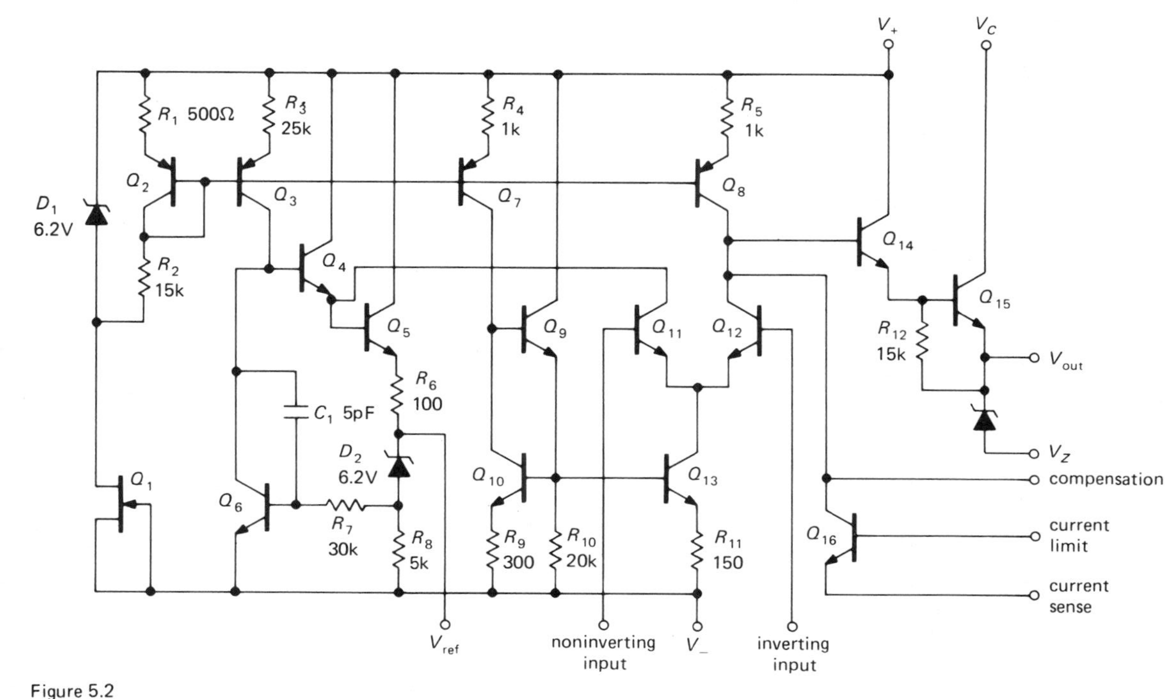

Figure 5.2
Schematic of the 723 regulator. (Fairchild Camera and Instrument Corp.)

bias D_2 at $I = V_{BE}/R_8$ via negative feedback at dc, as indicated on the block diagram. Q_2 and Q_3 form an unsymmetrical current mirror to bias the reference; current to the mirror is set by D_1 and R_2 (their junction is fixed at 6.2V below V_+), which in turn is biased by Q_1, a junction FET (we will talk about them in Chapter 6; in this configuration the FET behaves roughly like a current source).

Q_{11} and Q_{12} form the differential amplifier (sometimes called the "error amplifier," thinking of the whole thing as an exercise in negative feedback), a classic long-tailed pair with emitter current source Q_{13}. The latter is half of a current mirror (Q_9, Q_{10}, and Q_{13}), driven in turn from current mirror Q_7 (Q_3, Q_7, and Q_8 all mirror the current generated by the D_1 reference, as we mentioned in Section 2.13). Q_{11}'s collector is tied to the fixed positive voltage at Q_4's emitter, and the error amplifier's output is taken from Q_{12}'s collector. Current mirror Q_8 supplies the latter's collector load. Q_{14} drives the pass transistor Q_{15}, in a not-quite-Darlington connection. Note that Q_{15}'s collector is brought out separately, to allow for separate positive supplies. By turning on Q_{16} you cut off drive to the pass transistors; this is used to limit output currents to nondestructive levels. Unlike many of the newer regulators, the 723 does not incorporate internal shutdown circuitry to protect against excessive load current or chip dissipation. The SG3532 and LAS1000 are improved 723-type regulators, with low-voltage bandgap reference (Section 5.14), internal current limiting, and thermal-overload shutdown circuitry.

5.02 Positive regulator

Figure 5.3 shows how to make a positive voltage regulator with the 723. All the components except the four resistors and the two capacitors are contained on the 723. Voltage divider R_1R_2 compares a fraction of the output with the voltage reference, and the 723 components do the rest; this circuit is identical to the op-amp noninverting amplifier with emitter follower, with V_{ref} as the "input." R_4 is chosen for about 0.5 volt drop at maximum desired output current, since a V_{BE} drop applied across the CL-CS inputs will turn on the current-limiting transistor (Q_{16} in Fig. 5.2), shutting off base drive to the output pass transistor. The 100pF capacitor stabilizes the loop. R_3 (sometimes omitted) is chosen so that the

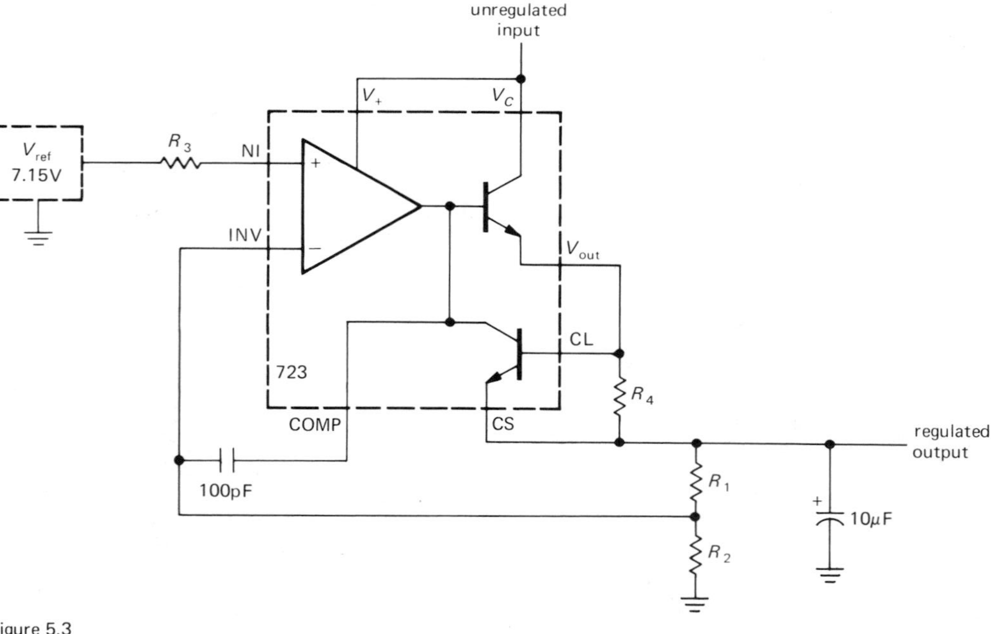

Figure 5.3

differential amplifier sees equal impedances at its inputs. This makes the output insensitive to changes in bias current (with changes in temperature, say), in the same way as we saw with op-amps (Section 3.12).

With this circuit a regulated supply with output voltage ranging from V_{ref} to the maximum allowable output voltage (37V) can be made. Of course, the input voltage must stay a few volts more positive than the output at all times, including the effects of ripple on the unregulated supply. The "dropout voltage" (the amount by which the input voltage must exceed the regulated output voltage) is specified as 3 volts (minimum) for the 723, a value typical of most regulators. R_1 or R_2 is usually made adjustable, or trimmable, so the output voltage can be set precisely. The production spread in V_{ref} is 6.8 to 7.5 volts.

It is usually a good idea to put a capacitor of a few microfarads across the output, as shown. This keeps the output impedance low even at high frequencies, where the feedback becomes less effective. It is best to use the output capacitor value recommended on the specification sheet, since oscillations can occur otherwise. In general, it is a good idea to bypass power-supply leads to ground liberally throughout a circuit, using a combination of ceramic types (0.01–0.1μF) and electrolytic or tantalum types (1–10μF).

For output voltages less than V_{ref}, you just put the voltage divider on the reference (Fig. 5.4). Now the full output voltage is compared with a fraction of the reference. The values shown are for +5 volts 50mA max. With this circuit configuration, output voltages from +2 volts to V_{ref} can be produced. The output cannot be adjusted down to zero volts because the differential amplifier will not operate below 2 volts input. This is given as a manufacturer's specification (see Table 5.8). With this circuit the unregulated input voltage must never drop below +9.5 volts, the voltage necessary to power the reference.

A third variation of this circuit is necessary if you want a regulator that is continuously adjustable through a range of output voltages around V_{ref}. In such cases just compare a divided fraction of the output with a fraction of V_{ref} chosen to be less than the minimum output voltage desired.

EXERCISE 5.1

Design a regulator to deliver up to 50mA load current over an output voltage range of +5 to +10 volts, using a 723. Hint: Compare a fraction of the output voltage with $0.5V_{ref}$.

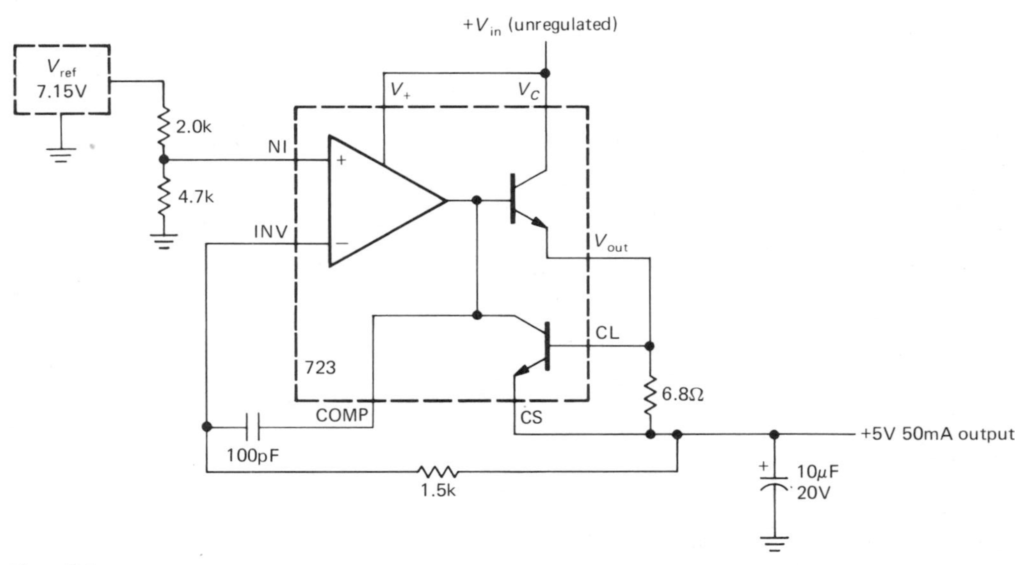

Figure 5.4

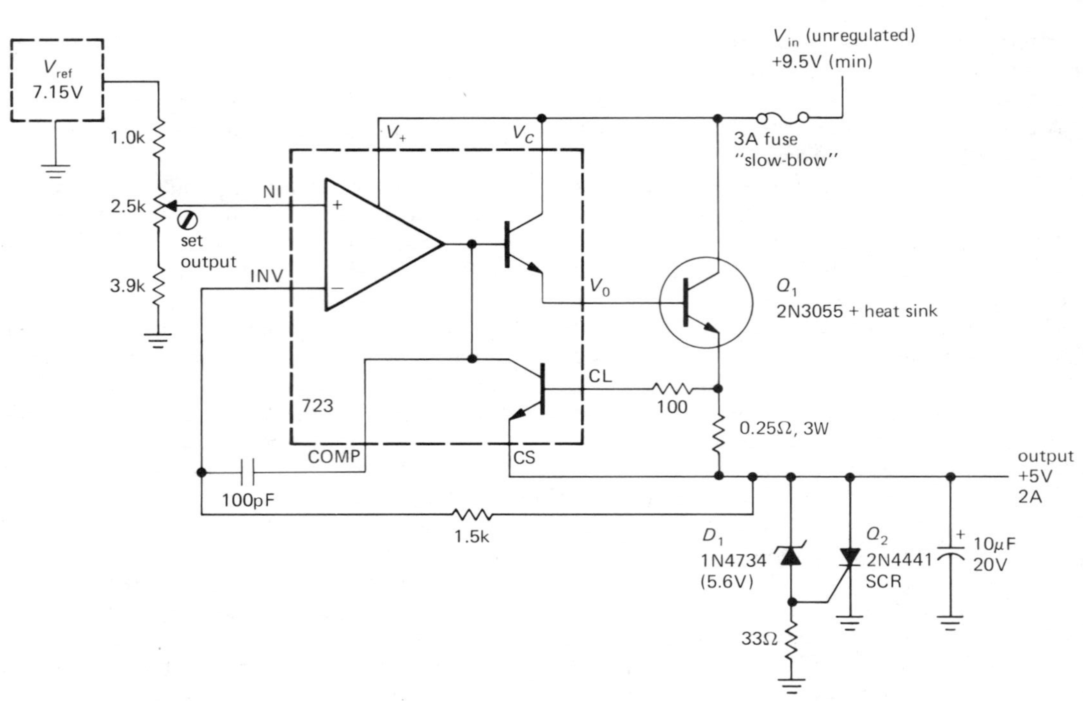

Figure 5.5

5.03 High-current regulator

The internal pass transistor in the 723 is rated at 150mA maximum; in addition, the power dissipation must not exceed 1 watt at 25°C (less at higher ambient temperatures; the 723 must be "derated" at 8.3mW/°C above 25°C in order to keep the junction temperature within safe limits). Thus, for instance, a 5 volt regulator with +15 volts input cannot deliver more than about 80mA to the load. To construct a higher-current supply, an external pass transistor must be used. It is easy to add one as a Darlington pair with the internal transistor (Fig. 5.5). Q_1 is the external pass transistor; it must be mounted on a *heat sink*, most often a finned metal plate designed to carry off heat (alternatively, the transistor can be mounted to one wall of the metal chassis housing the power supply). We will deal with thermal problems like these in the next section. A trimmer potentiometer has been used so that the output can be set accurately to +5 volts; its range of adjustment should be sufficient to allow for resistor tolerances as well as the maximum specified spread in V_{ref} (this is an example of worst-case design), and in this case it allows about ±1 volt adjustment from the nominal output voltage. Note the low-resistance high-power current-limiting resistor necessary for a 2 amp supply.

Pass transistor dropout voltage

One problem with this circuit is the high power dissipation in the pass transistor (at least 10 W at full load current). This is unavoidable if the regulator chip is powered by the unregulated input, since it needs a few volts of "headroom" to operate (specified by the dropout voltage). With the use of a separate low-current supply for the 723 (e.g., +12V), the minimum unregulated input to the external pass transistor can be only a volt or so above the regulated output voltage (although you will always have to allow at least a few volts, since worst-case design dictates proper operation even at 105V ac line input).

Overvoltage protection

Also shown in this circuit is an *overvoltage crowbar* protection circuit consisting of D_1,

Q_2, and the 33 ohm resistor. Its function is to short the output if some circuit fault causes the output voltage to exceed about 6.2 volts (this could happen if one of the resistors in the divider were to open up, for instance, or if some component in the 723 were to fail). Q_2 is an SCR (silicon-controlled rectifier), a device that is normally nonconducting but that goes into saturation when the gate-cathode junction is forward-biased. Once turned on, it will not turn off again until anode current is removed externally. In this case, gate current flows when the output exceeds D_1's zener voltage plus a diode drop. When that happens, the regulator will go into a current-limiting condition, with the output held near ground by the SCR. If the failure that produces the abnormally high output also disables the current-limiting circuit (e.g., a collector-to-emitter short in Q_1), then the crowbar will sink a very large current. For this reason it is a good idea to include a fuse somewhere in the power supply, as shown. We will treat overvoltage crowbar circuits in more detail in Section 5.06.

HEAT AND POWER DESIGN

5.04 Power transistors and heat sinking

As in the preceding circuit, it is often necessary to use power transistors or other high-current devices like SCRs or power rectifiers that can dissipate many watts. The 2N3055, an inexpensive power transistor of great popularity, can dissipate as much as 115 watts if properly mounted. All power devices are packaged in cases that permit contact between a metal surface and an external heat sink. In most cases the metal surface of the device is electrically connected to one terminal (e.g., for power transistors the case is always connected to the collector).

The whole point of heat sinking is to keep the transistor junction (or the junction of some other device) below some maximum specified operating temperature. For silicon transistors in metal packages the maximum junction temperature is usually 200°C, whereas for transistors in plastic packages it is usually 150°C. Table 5.1 lists some useful power transistors, along with their thermal properties. Heat sink design is then simple: Knowing the maximum power the device will dissipate in a given circuit, you calculate the junction temperature, allowing for the effects of heat conductivity in the transistor, heat sink, etc., and the maximum ambient temperature in which the circuit is expected to operate. You then choose a heat sink large enough to keep the junction temperature well below the maximum specified by the manufacturer. It is wise to be conservative in heat sink design, since transistor life drops rapidly at operating temperatures near or above maximum.

Thermal resistance

To carry out heat sink calculations, you use *thermal resistance*, θ, defined as heat rise (in degrees) divided by power transferred. For heat transferred entirely by conduction, the thermal resistance is a constant, independent of temperature, that depends only on the mechanical properties of the joint. For a succession of thermal joints in "series," the total thermal resistance is the sum of the thermal resistances of the individual joints. Thus for a transistor mounted on a heat sink, the total thermal resistance from transistor junction to the outside (ambient) world is the sum of the thermal resistance from junction to case, θ_{JC}, the thermal resistance from case to heat sink, θ_{CS}, and the thermal resistance from heat sink to ambient, θ_{SA}. The temperature of the junction is therefore

$$T_J = T_A + (\theta_{JC} + \theta_{CS} + \theta_{SA})P$$

where P is the power being dissipated.

Let's take an example. The preceding power-supply circuit, with external pass transistor, has a maximum transistor dissipation of 20 watts for an unregulated input of +15 volts (10V drop, 2A). Let's assume that the power supply is to operate at ambient temperatures up to 50°C, not unreasonable for electronic equipment packaged together in close quarters. And let's try to keep the junction temperature below 150°C, well below its specified maximum of 200°C. The thermal resistance from junction to case is 1.5°C per watt. A TO-3

TABLE 5.1. POWER TRANSISTORS

npn	*pnp*	Pkg.[a]	V_{CEO} max (V)	I_C max (A)	h_{FE} typ	@ I_C (A)	f_T min (MHz)	C_{cb}[b] typ (pF)	P_D $T_C = 25°C$ (W)	θ_{JC} (°C/W)	T_J max (°C)	Comments
Regular power: V_{CE} (sat) = 0.4V (typ); V_{BE} (on) = 0.8V (typ)												
2N5191	2N5194	A	60	4	100	0.2	2	80	40	3.1	150	Low cost, gen. purpose
2N5979	2N5976	B	80	5	50	0.5	2	60	70	1.8	150	
2N3055	MJ2955	TO-3	60	15	50	2	2.5	125	115	1.5	200	Metal, industry standard
MJE3055	MJE2955	B	60	10					90	1.4	150	Plastic, industry standard
2N5886	2N5884	TO-3	80	25	50	10	4	400	200	0.9	200	
2N5686	2N5684	TO-3	80	50	30	25	2	700	300	0.6	200	For real power jobs
2N6338	2N6437	TO-3	100	25	50	8	40	200	200	0.9	200	Premium audio output stages
2N6275	2N6379	TO-3	120	50	50	20	30	400	250	0.7	200	Premium audio output stages
Darlington power: V_{CE} (sat) = 0.8V (typ); V_{BE} (on) = 1.4V (typ)												
2N6038	2N6035	A	60	4	2000	2	—	30	40	3.1	150	Low cost
2N6044	2N6041	B	80	8	2500	4	4	80	75	1.7	150	
2N6059	2N6052	TO-3	100	12	3500	5	4	100	150	1.2	200	
2N6284	2N6287	TO-3	100	20	3000	10	4	150	160	1.1	200	High current

(a) A: small plastic power package (TO-126). B: large plastic power package (TO-127). (b) C_{cb} (*npn*) at V_{CB} = 10 volts; C_{cb} (*pnp*) ≈ $2C_{cb}$ (*npn*).

power transistor package mounted with an insulating washer and heat-conducting compound has a thermal resistance from case to heat sink of about 0.3 °C per watt. Finally, a Wakefield model 641 heat sink (Fig. 5.6) has a thermal resistance from sink to ambient of about 2.3 °C per watt. So the total thermal resistance from junction to ambient is about 4.1 °C per watt. At 20 watts dissipation the junction will be 84 °C above ambient, or 134 °C (at maximum ambient temperature) in this example. The chosen heat sink will be adequate; in fact, a smaller one could be used if necessary to save space.

Comments on heat sinks

1. Where very high power dissipation (several hundred watts, say) is involved, forced air cooling may be necessary. Large heat sinks designed to be used with a blower are available with thermal resistances (sink to

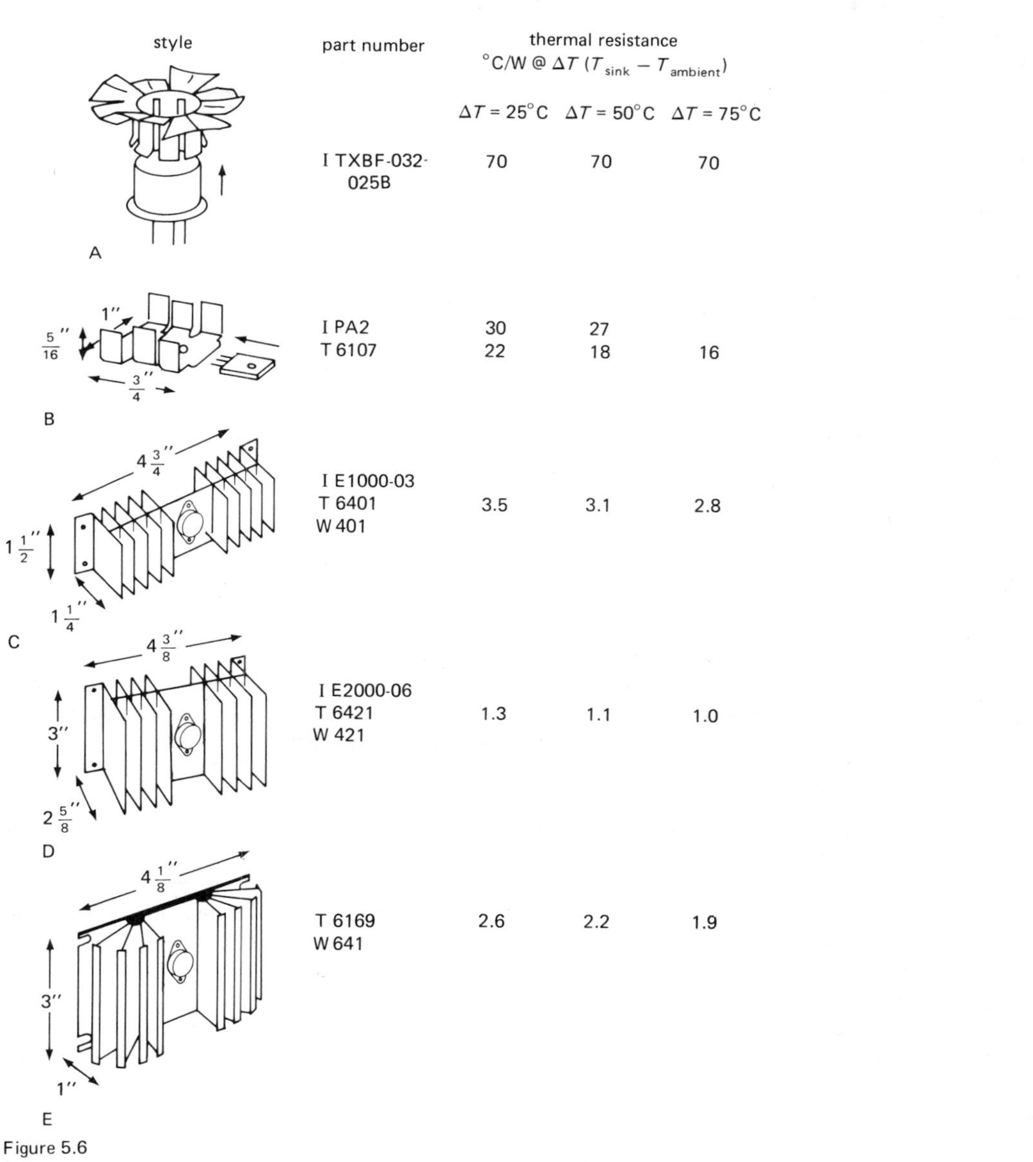

style	part number	thermal resistance °C/W @ ΔT ($T_{sink} - T_{ambient}$) $\Delta T = 25°C$	$\Delta T = 50°C$	$\Delta T = 75°C$
A	I TXBF-032-025B	70	70	70
B	I PA2	30	27	
	T 6107	22	18	16
C	I E1000-03 T 6401 W 401	3.5	3.1	2.8
D	I E2000-06 T 6421 W 421	1.3	1.1	1.0
E	T 6169 W 641	2.6	2.2	1.9

Figure 5.6
Power transistor heat sinks. I-IERC; T-Thermalloy; W-Wakefield.

ambient) as small as 0.05 °C to 0.2 °C per watt.

2. When the transistor must be insulated from the heat sink, as is usually necessary (especially if several transistors are mounted on the same sink), a thin insulating washer is used between the transistor and sink, and insulating bushings are used around the mounting screws. Washers are available in standard transistor-shape cutouts made from mica, insulated aluminum, or beryllia (BeO_2). Used with heat-conducting grease, these add from 0.14 °C per watt (beryllia) to about 0.5 °C per watt.

3. Small heat sinks are available that simply clip over the small transistor packages (like the standard TO-5). In situations of relatively low power dissipation (a watt or two) this often suffices, avoiding the nuisance of mounting the transistor remotely on a heat sink with its leads brought back to the circuit. An example is shown in Figure 5.6. In addition, there are various small heat sinks intended for use with the plastic power packages (many regulators, as well as power transistors, come in this package) that mount right on a printed-circuit board underneath the package. These are very handy in situations of a few watts dissipation; a typical unit is illustrated in Figure 5.6.

4. Sometimes it may be convenient to mount power transistors directly to the chassis or case of the instrument. In such cases it is wise to use conservative design (keep it cool), especially since a hot case will subject the other circuit components to high temperatures and shorten component life.

5. If a transistor is mounted to a heat sink without insulating hardware, the heat sink must be insulated from the chassis. The use of insulating washers (e.g., Wakefield model 103) is recommended (unless, of course, the transistor case happens to be at ground). When the transistor is insulated from the sink, the heat sink may be attached directly to the chassis. But if the transistor is accessible from outside the instrument (e.g., if the heat sink is mounted externally on the rear wall of the box), it is a good idea to use an insulating cover over the transistor (e.g., Thermalloy 8903N) to prevent someone from accidentally coming in contact with it, or shorting it to ground.

6. The thermal resistance from heat sink to ambient is usually specified for the sink mounted with the fins vertical and with unobstructed flow of air. If the sink is mounted differently, or if the air flow is obstructed, the efficiency will be reduced (higher thermal resistance); usually it is best to mount it on the rear of the instrument with fins vertical.

7. There are special heat sinks designed for forced air cooling. When used with a small instrument blower, these sinks can carry off prodigious amounts of heat (many hundreds of watts). We will go into more detail on the subject of forced air cooling in Chapter 12.

EXERCISE 5.2

A 2N5320, with a thermal resistance from junction to case of 17.5 °C per watt, is fitted with an IERC TXBF slip-on heat sink of the type shown in Figure 5.6. The maximum permissible junction temperature is 200 °C. How much power can you dissipate with this combination at 25 °C ambient temperature? How much must the dissipation be decreased per degree rise in ambient temperature?

5.05 Foldback current limiting

For a regulator with simple current limiting, transistor dissipation is maximum when the output is shorted to ground (either accidentally or through some circuit malfunction), and it usually exceeds the maximum value of dissipation that would otherwise occur under normal load conditions. For instance, the pass transistor in the preceding +5 volt 2 amp regulator circuit will dissipate 30 watts with the output shorted (+15V input, current limit at 2A), whereas the worst-case dissipation under normal load conditions is 20 watts (10V drop at 2A). The situation is even worse in circuits in which the voltage normally dropped by the pass transistor is a smaller fraction of the output voltage. For instance, in a +15 volt 2 amp regulated supply with +25 volt unregulated input, the transistor dissipation rises from 20 watts (full load) to 50 watts (short circuit).

You get into a similar problem with push-pull power amplifiers. Under normal conditions you have maximum load current when the voltage across the transistors is minimum (near the extremes of output swing), and you have maximum voltage across the transistors when the current is nearly zero (zero output voltage). With a short-circuit load, on the other hand, you have maximum load current at the worst possible time, namely, with full supply voltage across the transistor. This results in much higher transistor dissipation than normal.

The brute-force solution to this problem is to use massive heat sinks and transistors of higher power rating (and safe operating area, see Section 5.07) than necessary. Even so, it isn't a good idea to have large currents flowing into the powered circuit under fault conditions, since other components in the circuit may then be damaged. The best solution is to use *foldback* current limiting, a circuit technique that reduces the output current under short-circuit or overload conditions. Figure 5.7 shows the basic configuration, again illustrated with a 723 with external pass transistor.

The divider at the base of the current-limiting transistor Q_L provides the foldback. At +15 volts output (the normal value) the circuit will limit at about 2 amps, since Q_L's base is then at +15.5 volts while its emitter is at +15 (V_{BE} is about 0.5V at the elevated temperatures at which regulator chips are normally run). But the short-circuit current is less; with the output shorted to ground, the output current is about 0.5 amp, holding Q_1's dissipation down to less than in the full-load case. This is highly desirable, since excessive heat sinking is not now required, and the thermal design need only satisfy the full-load requirements. The choice of the three resistors in the current-limiting circuit sets the short-circuit current, for a given full-load current limit. Warning: Use care in choosing the short-circuit current, since it is possible to be overzealous and design a supply that will not "start up" into a normal load. The short-circuit current should not be too small; as a rough guide, the short-circuit current limit should be set at about one-third the maximum load current at full output voltage.

EXERCISE 5.3

Design a 723 regulator with outboard pass transistor and foldback current limiting to

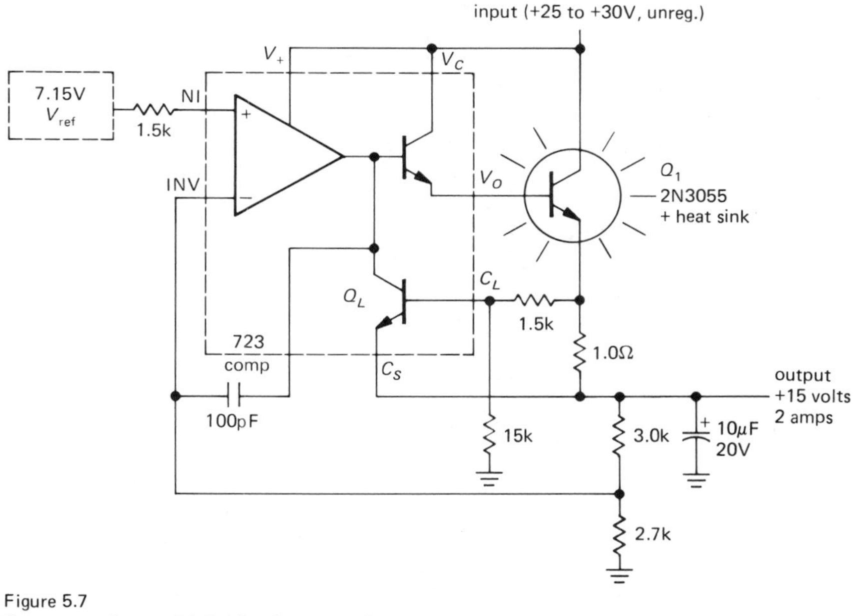

Figure 5.7
Power regulator with foldback current limiting.

provide up to 1.0 amp when the output is at its regulated value of +5.0 volts, but only 0.4 amp into a short-circuit load.

5.06 Overvoltage crowbars

As we remarked in Section 5.03, it is often a good idea to include some sort of overvoltage protection at the output of a regulated supply. Take, for instance, a +5 volt supply used to power a large digital system (you'll see lots of examples beginning in Chapter 8). The input to the regulator is probably in the range of +10 to +15 volts. If the series pass transistor fails by shorting its collector to emitter (a common failure mode), the full unregulated voltage will be applied to the circuit, with devastating results. Although a fuse probably will blow, what's involved is a race between the fuse and the "silicon fuse" that is constituted by the rest of the circuit; the rest of the circuit will probably respond first! This problem is most serious with TTL logic, which operates from a +5 volt supply, but cannot tolerate more than +7 volts without damage. Another situation with considerable disaster potential arises when you operate something from a wide-range "bench" supply, where the unregulated input may be 40 volts or more, regardless of the output voltage.

Zener sensing

Figure 5.8 shows a popular crowbar circuit. You hook it between the regulated output terminal and ground. If the voltage exceeds the zener voltage plus a diode drop (about 6.2V for the zener shown), the SCR is turned on, and it remains in a conducting state until its anode current drops below a few milliamps. An inexpensive SCR like the 2N4441 can sink 5 amps continuously and withstand 80 amp surge currents; its voltage drop in the conducting state is typically 1.0 volt at 5 amps. The 68 ohm resistor is provided to generate a reasonable zener current (10mA) at SCR turn-on, and the capacitor is added to prevent crowbar triggering on harmless short spikes.

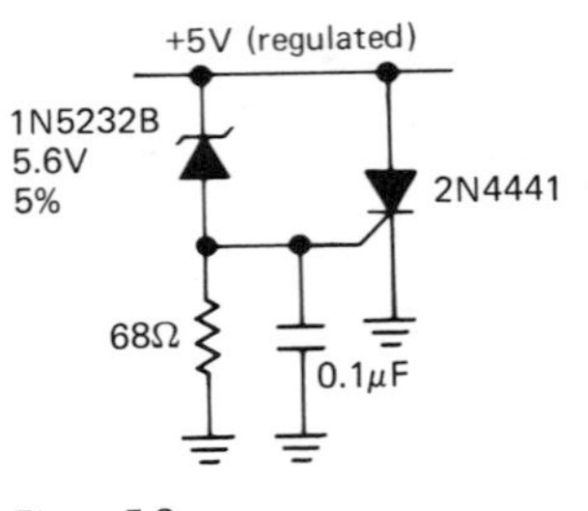

Figure 5.8

The preceding circuit, like all crowbars, puts an unrelenting 1 volt "short circuit" across the supply when triggered by an overvoltage condition, and it can be reset only by turning off the supply. Since the SCR maintains a low voltage while conducting, there isn't much problem with the crowbar itself failing from overheating. As a result, it is a reliable crowbar circuit. It is essential that the regulated supply have some sort of current limiting, or at least fusing, to handle the short. There may be overheating problems with the supply when the crowbar fires. In particular, if the supply includes internal current limiting, the fuse won't blow, and the supply will sit in the "crowbarred" state, with the output at low voltage, until someone notices. Foldback current limiting of the regulated supply would be a good solution here.

There are several problems with this simple crowbar circuit, mostly involving the choice of zener voltage. Zeners are available in discrete values only, with generally poor tolerances and (often) soft knees in the *VI* characteristic. The desired crowbar trigger voltage may involve rather tight tolerances. Consider a 5 volt supply used to power digital logic. There is typically a 5% or 10% tolerance on the supply voltage, meaning that the crowbar cannot be set less than 5.5 volts. The minimum permissible crowbar voltage is raised by the problem of transient response of a regulated supply: When the load current is changed quickly, the voltage can jump, creating a spike followed by some "ringing." This problem is exacerbated by remote sensing via long (inductive) sense leads. The resultant ringing puts glitches on the supply that should not trigger the crowbar. The result is that the crowbar voltage should not be set less than about 6.0 volts, but it cannot exceed 7.0 volts without risk of damage to the logic circuits. When you fold in zener tolerance, the discrete voltages actually available, and SCR trigger voltage tolerances, you've got a tricky problem. In the example shown earlier, the crowbar threshold could lie between 5.9 volts and 6.6 volts, even using the relatively expensive 5% zener indicated.

□ *IC sensing*

A nice solution to the problems of predictability and lack of adjustability in the simple zener/SCR crowbar circuit is to use a special crowbar trigger IC such as the MC3423. This is an inexpensive chip in a mini-DIP with adjustable threshold and response time; it even has an indicator output to signal the crowbar condition. The chip includes an internal reference and several comparators and drivers, and it requires only three external resistors, an optional capacitor, and an SCR to form a complete crowbar.

□ *Clamps*

Another possible solution to overvoltage protection is to put a power zener, or its equivalent, across the supply terminals. This avoids the problems of false triggering on spikes, since the zener will stop drawing current when the overvoltage condition disappears (unlike an SCR, which has the memory of an elephant). Figure 5.9 shows the circuit of an "active zener." Unfortunately, a crowbar constructed from a power zener clamp has its own problems. If the regulator fails, the crowbar has to contend with high power dissipation ($V_{zener}I_{limit}$) and may itself fail. We witnessed just such a failure in a commercial 15 volt 4 amp magnetic disc supply. When the pass transistor failed, the 16 volt 50 watt zener found itself dissipating more than rated power, and it proceeded to fail too.

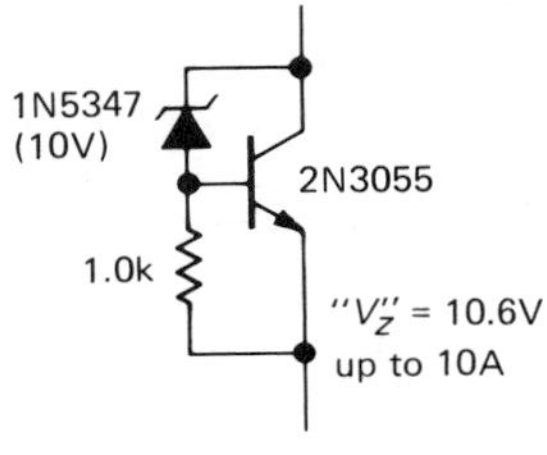

Figure 5.9

□ 5.07 Further considerations in high-current power-supply design

□ *Separate high-current unregulated supply*

As we mentioned in Section 5.03, it is usually a good idea to use a separate supply to power the regulator in very high current supplies. In that way the dissipation in the pass transistors can be minimized, since the unregulated input to the pass transistor can then be chosen just high enough to allow sufficient "headroom" (regulators like the 723 have separate V_+ terminals for this purpose). For instance, a +5 volt 10 amp regulator might use a 10 volt unregulated input with a volt or two of ripple, with a separate low-current +15 volt supply for the regulator components (reference, error amplifier, etc.). As mentioned earlier, the unregulated input voltages must be chosen large enough to allow for worst-case ac power-line voltage (105V) as well as transformer and capacitor tolerances.

□ *Connection paths*

With high-current supplies, or supplies of highly precise output voltage, careful thought must be given to the connection paths, both within the regulator and between the regulator and its load. If several loads are run from the same supply, they should connect to the supply at the place where the output voltage is sensed; otherwise, fluctuations in the current of one load will affect the voltage seen by the other loads (Fig. 5.10).

In fact, it is a good idea to have one common ground point (a "mecca"), as shown, to which the unregulated supply, reference, etc., are all returned. The problem of unregulated voltage drops in the connecting leads from power supply to high-current load is sometimes solved by remote sensing: The connections back to the error amplifier and reference are brought out to the rear of the supply separately and may either be connected to the output terminals right there (the normal method) or brought out and connected to the load at a remote location along with the output voltage leads (this

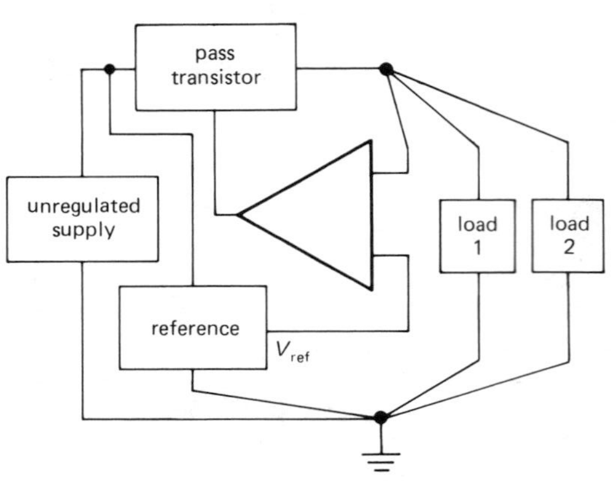

Figure 5.10

requires four wires, two of which must be able to handle the high load currents). Most commercially available power supplies come with jumpers at the rear that connect the sensing circuitry to the output and that may be removed for remote sensing. Four-wire resistors are used in an analogous manner to sense load currents accurately when constructing precision constant-current supplies. This will be discussed in greater detail in Section 5.24.

□ *Parallel pass transistors*

When very high output currents are needed, it may be necessary to use several pass transistors in parallel. Since there will be a spread of V_{BE}s, it is necessary to add a small resistance in series with each emitter, as in Figure 5.11. The *R*s ensure that the current is shared approximately equally among the pass transistors. *R* should be chosen for about 0.2 volt drop at maximum output

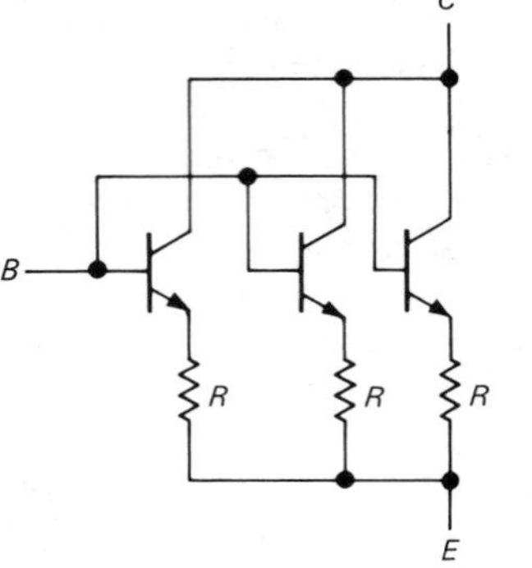

Figure 5.11

current. Power FETs can be connected in parallel without any external components, owing to their negative temperature coefficient of drain current. They will be discussed in Section 6.20.

□ *Safe operating area*

One last point about power transistors: A phenomenon known as "second breakdown" restricts the simultaneous voltage and current that may be applied for any given transistor, and it is specified on the data sheet as the safe operating area (SOA) (it's a family of safe voltage-versus-current regions, as a function of time duration). Second breakdown involves the formation of "hot spots" in the transistor junctions, with consequent uneven sharing of the total load. Except at low collector-to-emitter voltages, it sets a limit that is more restrictive than the maximum power dissipation specification. As an example, Figure 5.12 shows the SOA

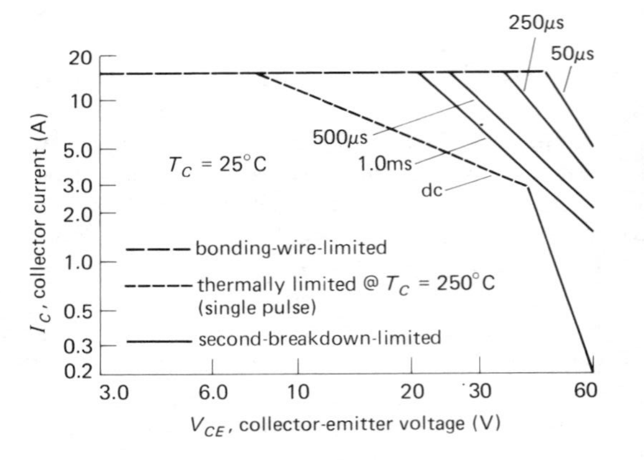

Figure 5.12
Power transistor SOA. (Courtesy Motorola Semiconductor Products Inc.)

for the ever-popular 2N3055. For $V_{CE} > 40$ volts, second breakdown limits the dc collector current to values corresponding to less than the maximum allowable dissipation of 115 watts. Power MOSFETs (Section 6.20), with their negative coefficient of drain current with increasing temperature, are immune from thermal runaway and second breakdown; their safe operating area is limited only by maximum power dissipation, within the bounds of maximum voltage and current.

□ 5.08 Programmable supplies

There is frequently the need for power supplies that can be adjusted right down to zero volts, especially in bench applications where a flexible source of power is essential. In addition, it is often desirable to be able to "program" the output voltage with another voltage or with a digital input (via digital thumbwheel switches, for instance). Figure 5.13 shows the classic scheme for a supply

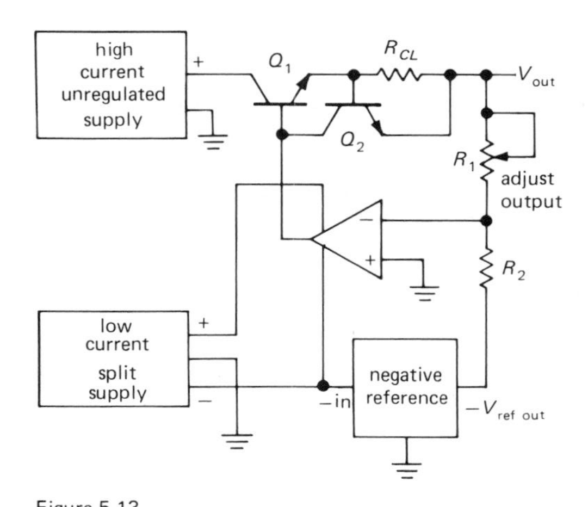

Figure 5.13
Regulator adjustable down to zero volts.

that is adjustable down to zero output voltage (as our 723 circuits so far are not). A separate split supply provides power for the regulator and also generates an accurate negative reference voltage (more on references in Sections 5.13 and 5.14). R_1 sets the output voltage (since the inverting input will be at ground), which can be adjusted all the way down to zero (at zero resistance). When the regulator circuitry (which can be an integrated circuit or discrete components) is run from a split supply, no problems are encountered at low output voltages.

To make the supply programmable with an external voltage, just replace V_{ref} with an externally controlled voltage (Fig. 5.14). The rest of the circuit is unchanged. R_1 now sets the scale of $V_{control}$.

Digital programmability can be added by replacing V_{ref} with a device called a DAC (digital-to-analog converter) with current-sinking output. These devices, which we will discuss later, convert a binary input code to

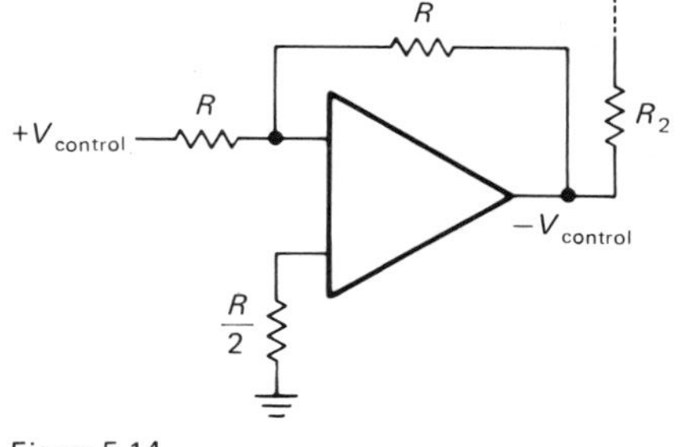

Figure 5.14

a proportional current (or voltage) output. A popular unit is the MC1408, a monolithic 8-bit DAC with current-sinking output and a price tag of about 5 dollars. By replacing V_{ref} and R_2 with the DAC, you get a digitally programmed supply. Since the inverting input is a virtual ground, the DAC doesn't even have to have any output compliance. In practice, R_1 will be adjusted to set a convenient scale for the output, say 1mV per input digit.

□ 5.09 Power-supply circuit example

The "laboratory" bench supply shown in Figure 5.15 should help pull all these design ideas together. It is important to be able to adjust the regulated output voltage right down to zero volts in a general-purpose bench supply, so an additional split supply is used to power the regulator. IC_1 is a high-voltage op-amp of the type intended for regulators, e.g., the MC1436, which can operate with 68 volts total supply voltage. Paralleled pass transistors provide plenty of dissipation and safe operating area, necessary even for moderate output current when such a wide range of output voltage is provided. This is because the unregulated input voltage has to be high enough for the maximum regulated output voltage, resulting in a large voltage drop across the pass transistors when the regulated output voltage is low. Some supplies solve this problem by having several ranges of output voltage, switching the unregulated input voltage accordingly. There are even supplies with the unregulated supply driven from a variable-voltage transformer ganged to the same control as the output voltage. In both cases you lose the capability of remote programmability.

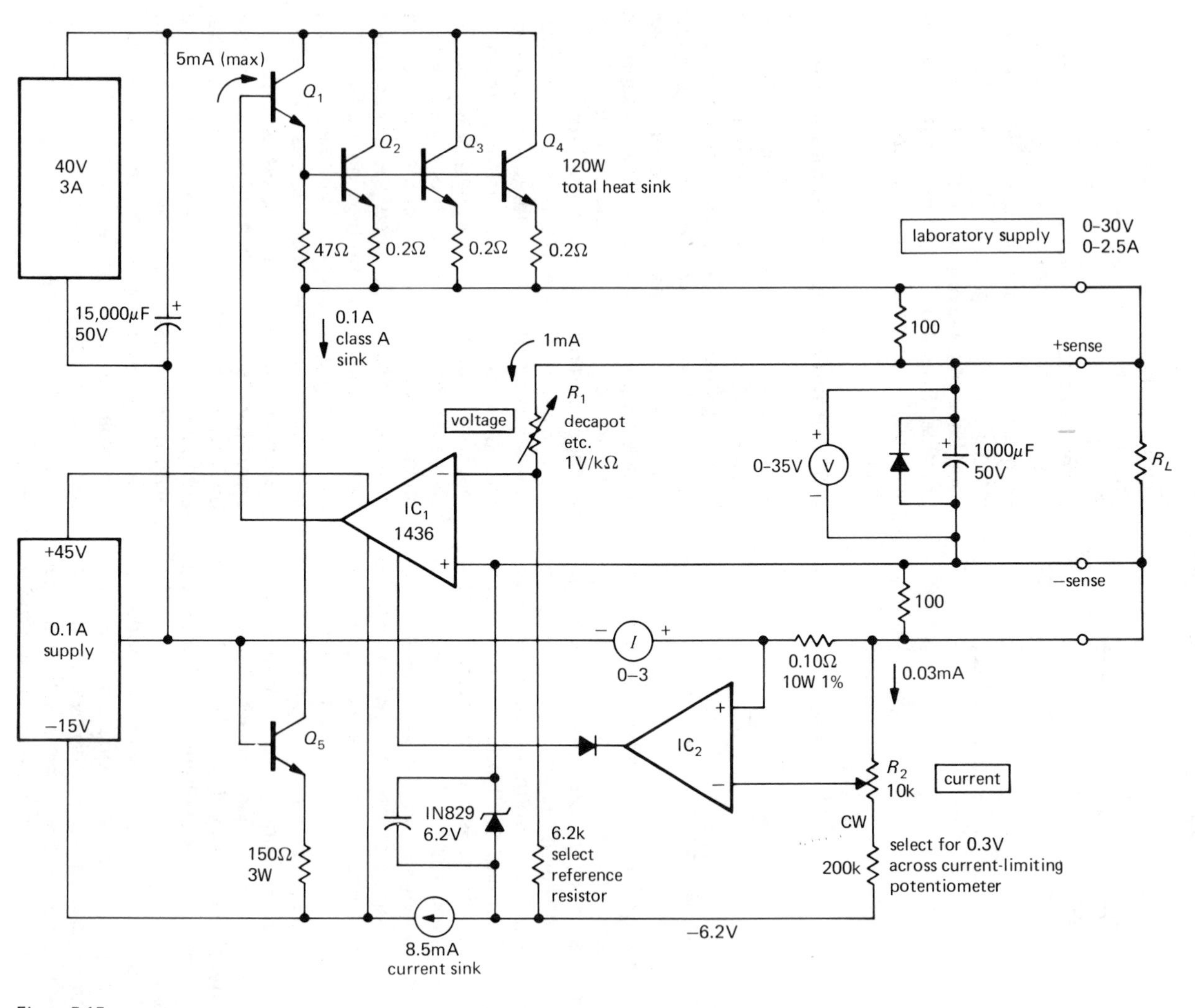

Figure 5.15
Laboratory bench supply.

R_1 is a precision multidecade potentiometer for precise and linear adjustment of the output voltage. The output voltage is referenced to the 1N829 precision zener (5ppm/°C tempco at 7.5mA zener current). The current-limiting circuitry is considerably better than the simple protective current limiters we have been discussing, since it is sometimes desirable to be able to set a precise and stable current limit when using a bench supply. With that capability, the supply becomes a flexible constant-current source. Q_5 provides a constant 100mA load, maintaining good performance near zero output voltage (or current) by keeping the pass transistors well into the active region. This current sink also allows the load to source some current into the supply without its output voltage rising. This is useful with the bizarre loads you sometimes encounter, e.g., an instrument that contains some additional supplies of its own capable of sourcing some current into the power-supply output terminal.

Note the external sense leads, with default connection to the power-supply output terminals. For precise regulation of output voltage at the load, you would bring external sense leads to the load itself, eliminating (through feedback) voltage drops in the connecting leads.

THE UNREGULATED SUPPLY

All regulated supplies require a source of "unregulated" dc, a subject we introduced in Section 1.27 in connection with rectifiers and ripple calculations. Let's look at this subject in more detail, beginning with the circuit shown in Figure 5.16. This is an unregulated +13 volt (nominal) supply for use with a +5 volt 2 amp regulator. Let's go through it from left to right, pointing out some of the things to keep in mind when you do this sort of design.

5.10 ac Line components

Three-wire connection

Always use a 3-wire line cord with neutral (green) connected to the instrument case. Instruments with ungrounded cases can become lethal devices in the event of transformer insulation failure or accidental connection of one side of the power line to the case. With a grounded case, such a failure simply blows a fuse.

Line filter and transient suppressor

In this supply we have used a simple *LC* line filter. Although they are often omitted, such filters are a good idea, since they serve the purpose of preventing possible radiation of radiofrequency interference (RFI) from the instrument via the power line as well as filtering out incoming interference that may be present on the power line. Power-line filters with excellent performance characteristics are available from several manufacturers, e.g., Corcom, Cornell-Dubilier, and Sprague. Studies have shown that spikes as large as 1kV to 5kV are occasionally present on the power lines at most locations, and

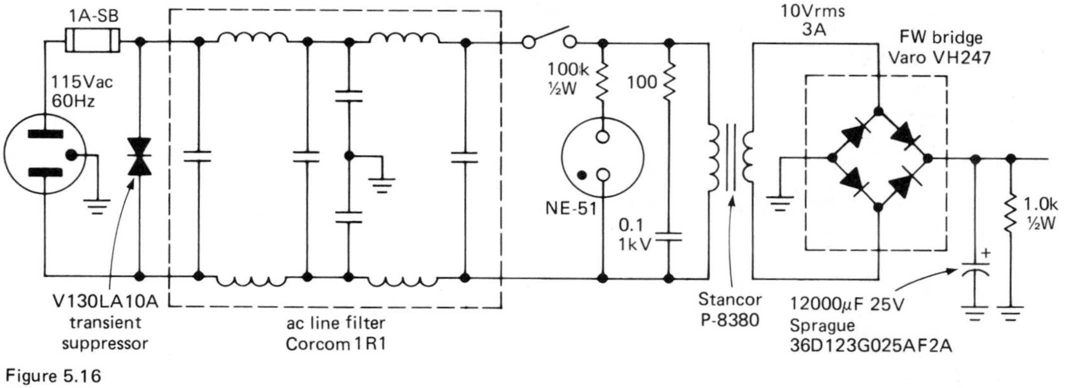

Figure 5.16
Unregulated supply with ac line connections.

TABLE 5.2. 130 VOLT AC TRANSIENT SUPPRESSORS

Manufact.	Part no.	Approx. Dia. (disc) (in)	Energy (W-s)	Peak current (A)	Capacitance (pF)
GE	V130LA1	0.34	4	500	180
Siemens	S07K130	0.35	6	500	130
GE	V130LA10A	0.65	30	4000	1000
Siemens	S14K130	0.67	22	2000	1000
GE	V130LA20B	0.89	50	6000	1900
Siemens	S20K130	0.91	44	4000	2300

smaller spikes occur quite frequently. Line filters are reasonably effective in reducing such interference.

In many situations it is desirable to use a "transient suppressor," as shown, a device that conducts when its terminal voltage exceeds certain limits (it's like a bidirectional high-power zener). These are inexpensive and small and can short out hundreds of amperes of potentially harmful current in the form of spikes. Transient suppressors are made by a number of companies, e.g., GE and Siemens. Tables 5.2 and 5.3 list some useful RFI filters and transient suppressors.

Fuse

A fuse is essential in every piece of electronic equipment. The large wall fuses (typically 15–20A) in house or lab won't protect electronic equipment, since they are chosen to blow only when the current rating of the wiring in the wall is exceeded. For instance, a house wired with 14 gauge wire will have 15 amp fuses. Now, if the filter capacitor in the preceding supply becomes short-circuited someday (a typical failure mode), the transformer might then draw 5 amps primary current (instead of its usual 0.25A). The house fuse won't blow, but your instrument becomes an incendiary device, with its transformer dissipating over 500 watts!

Some notes on fuses: It is best to use a "slow-blow" type in the power-line circuit, because there is a large current transient at turn-on (the filter capacitors have to be charged, for instance). Use a fuse at least 50% larger than the nominal load current, for two reasons: Fuses blow out eventually

TABLE 5.3. 115 VOLT AC POWER-LINE FILTERS

Manufact.	Part no.	Circuit	Current (A)	Typical atten. (50Ω/50Ω) 150kHz (dB)	300kHz (dB)	1MHz (dB)	Terminals
CDE	APF-110L	L	1	10	20	42	2-wire, solder terminals (for IEC connector use suffix -CEE; for 3-wire line cord receptacle use -CL)
	APF-140L	L	1	40	50	60	
	APF-510L	L	5	10	20	42	
	APF-540L	L	5	40	50	60	
Corcom	1EF1	pi	1	12	20	42	2-wire, IEC female connector
	1EF6	pi	6	12	20	42	
	1R1	dual-T	1	45	60	70	2-wire, solder terminals
	1R3	dual-T	3	45	60	70	
Erie	9011-100-1010	pi	0.5	75	80	80	1-wire, feedthrough (threaded bushing)
	9001-100-1021	pi	1	20	40	70	
	9011-100-1012	pi	3	48	66	80	
Sprague	1JX5201A	pi	1	20	30	50	2-wire, solder terminals (line cord receptacle use suffix -D)
	3JX5303A	dual-T	3	45	60	70	2-wire, solder terminals
	3JX5421A	pi	3	20	30	40	2-wire, IEC female connector

from "fatigue" if run near their rated current. Also, worst-case design implies some margin of conservatism, since load current may rise when the line voltage is high, etc. When wiring cartridge-type fuse holders (used with the popular 3AG fuse, which is almost universal in electronic equipment), connect the leads so that anyone changing the fuse cannot come in contact with the power line. This means connecting the "hot" lead to the rear terminal of the fuse holder (one of the authors learned this the hard way!).

Shock hazard

Incidentally, it is a good idea to insulate all exposed 110 volt power connections inside any instrument, using Teflon heat-shrink tubing, for instance (the use of "friction tape" or electrical tape inside electronic instruments is strictly bush-league). Since most transistorized circuits operate on relatively low dc voltages ($\pm 15V$ to $\pm 30V$ or so), from which it is not possible to receive a shock, the power-line wiring is the only place where any shock hazard exists in most electronic devices (there are exceptions, of course). The front-panel ON/OFF switch is particularly insidious in this respect, since it is close to other low-voltage wiring. Your test instruments (or, worse, your fingers) can easily come in contact with it when you go to pick up the instrument while testing it.

Miscellany

A pilot light is shown in the 110 volt circuit, using a neon lamp and dropping resistor. Nowadays it is probably better to use a light-emitting diode (LED) run from the regulated voltage. It has a longer life and a more pleasing appearance; it is also less expensive.

The series combination of 100 ohms and $0.1\mu F$ capacitor across the transformer primary prevents the large inductive transient that would otherwise occur at turn-off. This is often omitted, but it is highly desirable, particularly in equipment intended for use near computers or other digital devices. Sometimes this *RC* "snubber" network is wired across the switch, which is equivalent.

5.11 Transformer

Now for the transformer. Never build an instrument to run off the power line without a transformer! To do so is to flirt with disaster. Transformerless power supplies, which are popular in some consumer electronics (radios and televisions, particularly) because they're cheap, put the circuit at high voltage with respect to external ground (water pipes, etc.). This has no place in instruments intended to interconnect with any other equipment and should always be avoided. And use extreme caution when servicing any such equipment; just connecting your oscilloscope probe to the chassis can be a shocking experience.

The choice of transformer is more involved than you might at first expect. One problem is that manufacturers have been slow to introduce transformers with voltages and currents appropriate for transistorized circuitry (the catalogs are still cluttered with transformers designed for vacuum tubes), and you wind up making compromises you'd rather avoid. We have found the Signal Transformer Company unusual, with their nice selection of transformers and quick delivery. Don't overlook the possibility of having transformers custom-made if your application requires more than a few transformers.

Even assuming that you can get the transformer you want, you still have to decide what voltage and current are best. The lower the input voltage to the regulator, the lower the dissipation in the pass transistors. But you must be absolutely certain the input to the regulator will never drop below the minimum necessary for regulation, typically 2 to 3 volts above the regulated output voltage, or you may encounter 120Hz dips in the regulated output. The amount of ripple in the unregulated output is involved here, since it is the minimum input to the regulator that must stay above some critical voltage, but it is the average input to the regulator that determines the transistor dissipation.

As an example, for a +5 volt regulator you might use an unregulated input of +10 volts at the minimum of the ripple, which itself might be a volt or two. From the secondary voltage rating you can make a pretty good guess of the dc output from the bridge, since the peak voltage (at the top of the ripple) is approximately 1.4 times the rms secondary voltage, less two diode drops. But it is essential to make actual measurements if you are designing a power supply with near-minimum drop across the regulator, because the actual output voltage of the unregulated supply depends on poorly specified parameters of the transformer, such as winding resistance and voltage under load. Be sure to make measurements under worst-case conditions: full load and low power-line voltage (105V). Remember that large filter capacitors typically have loose tolerances: −30% to +100% about the nominal value is not unusual. It is a good idea to use transformers with multiple taps on the primary, when available, for final adjustment of output voltage. The Triad F-90X series and the Stancor TP series are very flexible this way.

One further note on transformers: Current ratings are sometimes given as rms secondary current, particularly for transformers intended for use into a resistive load (filament transformers, for instance). Since a rectifier circuit draws current only over a small part of the cycle (during the time the capacitor is actually charging), the rms current, and therefore the I^2R heating, is likely to exceed specifications for a load current approaching the rated rms current of the transformer. The situation gets worse as you increase capacitor size to reduce preregulator ripple; this simply requires a transformer of larger rating. Full-wave rectification is better in this respect, since a greater portion of the transformer waveform is used.

5.12 dc Components

Filter capacitor

The filter capacitor is chosen large enough to provide acceptably low ripple voltage, with voltage rating sufficient to handle the worst-case combination of no load and high line voltage (125–130V rms). For the circuit shown in Figure 5.16, the ripple is about 1.5 volts pp at full load. Good design practice calls for the use of computer-type electrolytics (they come in a cylindrical package with screw terminals at one end), e.g., the Sprague 36D type. In smaller capacitance values most manufacturers provide capacitors of equivalent quality in an axial-lead package (one wire sticking out each end), e.g., the Sprague 39D type. Watch out for the loose capacitance tolerance!

At this point it may be helpful to look back at Section 1.27, where we first discussed the subject of ripple. With the exception of switching regulators (see Section 5.21 and following), you can always calculate ripple voltage by assuming a constant-current load equal to the maximum output load current. In fact, the input to a series regulator looks just like a constant-current sink. This simplifies your arithmetic, since the capacitor discharges with a ramp, and you don't have to worry about time constants or exponentials (Fig. 5.17).

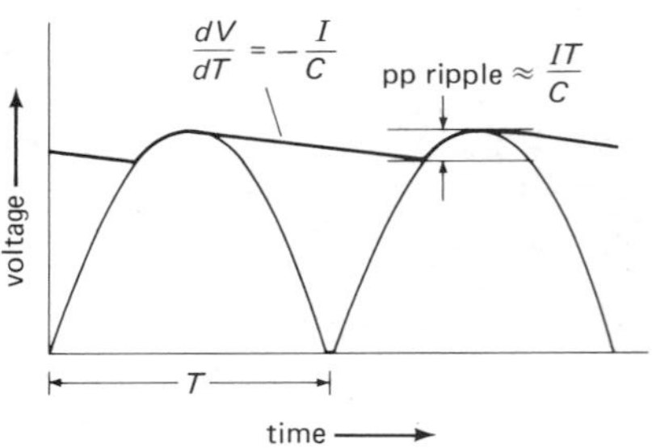

Figure 5.17

For example, suppose you want to choose a filter capacitor for the unregulated portion of a +5 volt 1 amp regulated supply, and suppose you have already chosen a transformer with a 10 volt rms secondary, to give an unregulated dc output of 12 volts (at the peak of the ripple) at full load current. With a typical regulator dropout voltage of 2 volts, the input to the regulator should never dip below +7 volts (the 723 will require +9.5V, but the convenient 3-terminal regulators discussed in Section 5.15 are more friendly). Since you have to contend with a ±10% worst-case line-voltage variation, you should keep ripple to less than 2 volts pp. Therefore,

$$2 = T(dV/dT) = TI/C = 0.008 \times 1.0/C$$

from which $C = 4000\mu F$. A $5000\mu F$ 25 volt electrolytic would be a conservative choice, with allowance for a 20% tolerance in capacitor value.

The "bleeder" resistor shown across the output in Figure 5.16 discharges the capacitor in a few seconds under no-load conditions. This is a good feature, because power supplies that stay charged after things have been shut off can easily lead you to damage some circuit components if you mistakenly think that no voltage is present.

TABLE 5.4. RECTIFIERS

Type	Breakdown voltage V_{br} (V)	Forward drop V_F typ (V)	@ Average current I_0 (A)	Package	Comments
General purpose					
1N4001-7	50–1000	0.9	1	Lead-mounted	Popular
1N5059-62	200–800	1.0	2	Lead-mounted	
1N5624-7	200–800	1.0	5	Lead-mounted	
1N1183A-90A	50–600	0.9	40	Stud-mounted	Popular, suffix R for rev. pol.
Fast recovery ($t_{rr} = 0.1\mu s$ typ)					
1N4933-7	50–600	1.0	1	Lead-mounted	
1N5415-9	50–500	1.0	3	Lead-mounted	
1N3879-83	50–400	1.2	6	Stud-mounted	Suffix R for rev. pol.
1N3899-3903	50–400	1.0	20	Stud-mounted	Suffix R for rev. pol.
Schottky (low V_F, very fast)					
1N5817-19	20–40	0.6 max	1	Lead-mounted	
1N5820-22	20–40	0.5 max	3	Lead-mounted	
1N5826-28	20–40	0.5 max	15	Stud-mounted	
1N5832-34	20–40	0.6 max	40	Stud-mounted	
Full-wave bridge					
Motorola 3N246-252	50–1000	0.9	1	Plastic SIP	
3N253-259	50–1000		2	Plastic SIP	
MDA980-1 to -6	50–600	0.85	12	Plastic chassis-mounting	
MDA3500-10	50–1000		35	Plastic chassis-mounting	
Semtech SCAJ2-6	200–600	1.2	5	Metal chassis-mounting	Suffix F for fast recov.
SCBH2-6	200–600	1.0	10	Metal chassis-mounting	Suffix F for fast recov.
SCBA2-6	200–600	1.0	15	Metal chassis-mounting	
Varo VM08-108	50–1000		1	Mini-DIP	Convenient PC package
VE27-67	200–600	1.2 max	1	TO-105	Controlled avalanche
VH247-647	200–600	1.2	6	Plastic chassis-mounting	Controlled avalanche
VK048-1048	50–1000		30	Plastic chassis-mounting	Controlled avalanche
VS148X-448X	100–400	1.5 max	2	Plastic chassis-mounting	Fast recov. ($0.2\mu s$)
Exotic					
GE A570A-A640L	100–2000	1.0 max	1500	Giant button	High current!
Semtech SCH5000–25000	5000–25,000	7–33 max	0.5	Lead-mounted	High voltage, curr.; fast ($0.2\mu s$)
Varo VF25-5 to -40	5000–40,000	12–50 max	0.025	Lead-mounted	High voltage
Semtech SCKV100K3-200K3	100kV–200kV	150–300	0.1	Plastic rod	Very high voltage!

Rectifier

The first point to be made is that the diodes used in power supplies are quite different from the small 1N914-type signal diodes used in circuitry. Signal diodes are generally designed for high speed (a few nanoseconds), low leakage (a few nanoamps), and low capacitance (a few picofarads), and they can generally handle currents up to about 100mA, with breakdown voltages rarely exceeding 100 volts. By contrast, rectifier diodes and bridges for use in power supplies are hefty objects with current ratings going from 1 amp to 25 amps or more and breakdown voltages going from 100 volts to 1000 volts. They have relatively high leakage currents (in the range of microamps to milliamps) and plenty of junction capacitance. They are not intended for high speed. Table 5.4 lists a selection of popular types.

Typical of rectifiers is the popular 1N4001–1N4007 series, rated at 1 amp, with reverse-breakdown voltages ranging from 50 to 1000 volts. The 1N5625 series is rated at 3 amps, which is about the highest current available in a lead-mounted (cooled by conduction through the leads) package. The popular 1N1183A series typifies high-current stud-mounted rectifiers, with a current rating of 40 amps and breakdown voltages to 600 volts. Plastic-encapsulated bridge rectifiers are quite popular also, with lead-mounted 1 and 2 amp types and chassis-mounting packages in ratings up to 25 amps or more. For rectifier applications where high speed is important (e.g., dc-to-dc converters, see Section 5.22), fast-recovery diodes are available, e.g., the 1N4933 series of 1 amp diodes. For low-voltage applications it may be desirable to use Schottky barrier rectifiers, e.g., the 1N5823 series, with forward drops of less than 0.4 volt at 5 amps.

□ VOLTAGE REFERENCES

There is frequently the need for good voltage references within a circuit. For instance, you might wish to construct a precision regulated supply with characteristics better than those you can obtain using complete regulators like the 723 (since integrated voltage regulator chips usually dissipate considerable power because of the built-in pass transistor, they tend to heat up, with consequent drift). Or you might want to construct a precision constant-current supply. Another application that requires a precision reference, but not a precision power supply, is design of an accurate voltmeter, ohmmeter, or ammeter.

□ 5.13 Zener diodes

The simplest form of voltage reference is the zener diode, a device we discussed in Section 1.06. Basically, it is a diode operated in the reverse-bias region, where current begins to flow at some voltage and increases dramatically with further increases in voltage. To use it as a reference, you simply provide a roughly constant current; this is often done with a resistor from a higher supply voltage, forming the most primitive kind of regulated supply.

Zeners are available in selected voltages from 2 to 200 volts (they come in the same series of values as standard 5% resistors), with power ratings from a fraction of a watt to 50 watts and tolerances of 1% to 20%. As attractive as they might seem for use as general-purpose voltage references, zeners are actually somewhat difficult to use, for a variety of reasons. It is necessary to stock a selection of values, the voltage tolerance is poor except in high-priced precision zeners, they are noisy, and the zener voltage depends on current and temperature. As an example of the last two effects, a 27 volt zener in the popular 1N5221 series of 500mW zeners has a temperature coefficient of +0.1%/°C, and it will change voltage by 1% when its current varies from 10% to 50% of maximum.

There is an exception to this generally poor performance of zeners. It turns out that in the neighborhood of 6 volts, zener diodes become very stiff against changes in current and simultaneously achieve a nearly zero temperature coefficient. The graphs in Figure 5.18, plotted from measurements on zeners with different voltages, illustrate the effects. This peculiar behavior comes about because "zener" diodes actually employ two different mechanisms: zener breakdown (low voltage) and avalanche breakdown (high

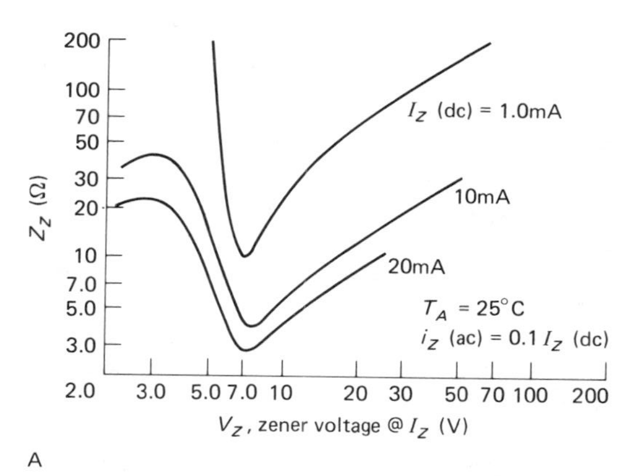

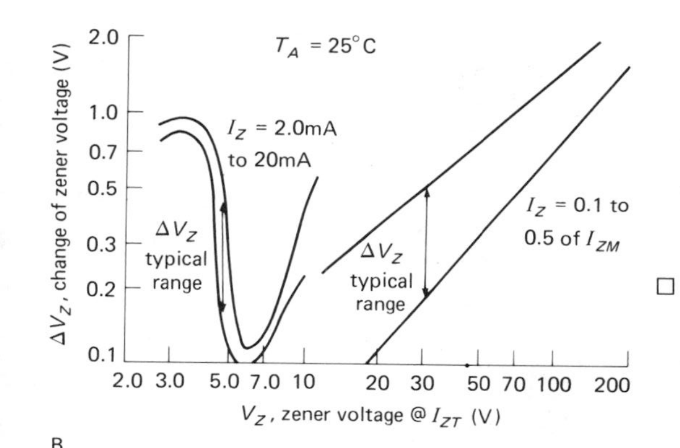

Figure 5.18
Zener diode impedance and regulation for zener diodes of various voltages. (Courtesy Motorola Semiconductor Products Inc.)

voltage). If you need a zener for use as a stable voltage reference only, and you don't care what voltage it is, the best thing to use is one of the compensated zener references constructed from a 5.6 volt zener (approximately) in series with a forward-biased diode. The zener voltage is chosen to give a positive coefficient to cancel the diode's temperature coefficient of −2.1mV/°C.

As you can see from the graph in Figure 5.19, the temperature coefficient depends on operating current and also on the zener voltage. Therefore, by choosing the zener current properly, you can "tune" the temperature coefficient somewhat. Such zeners with built-in series diodes make particularly good references. As an example, the 1N821 series of inexpensive 6.2 volt references offers temperature coefficients

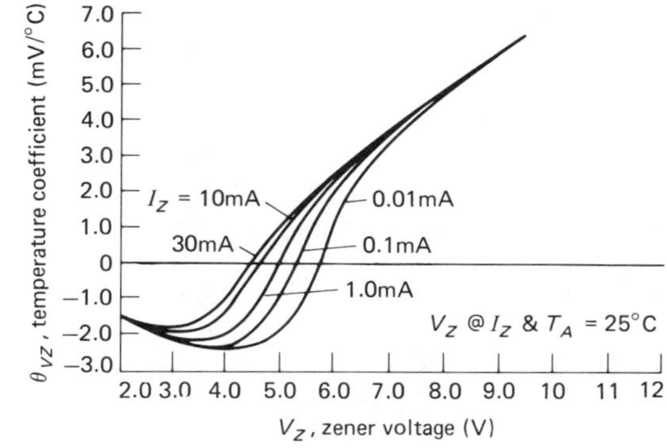

Figure 5.19
Temperature coefficient of zener diode breakdown voltage versus the voltage of the zener diode. (Courtesy Motorola Semiconductor Products Inc.)

going from 100ppm/°C (1N821) down to 5ppm/°C (1N829); the 1N940 and 1N946 are 9 volt and 11.7 volt references with tempcos of 2ppm/°C.

□ *Providing operating current*

These compensated zeners can be used as stable voltage references within a circuit, but they must be provided with constant current. The 1N821 series is specified as 6.2 volts ±5% at 7.5mA, with an incremental resistance of about 15 ohms; thus a change in current of 1mA changes the reference voltage three times as much as a change in temperature from −55°C to +100°C for the 1N829. Figure 5.20 shows a simple way to provide constant bias current for a precision zener. The op-amp is wired as a noninverting amplifier in order to generate an output of exactly +10.0 volts.

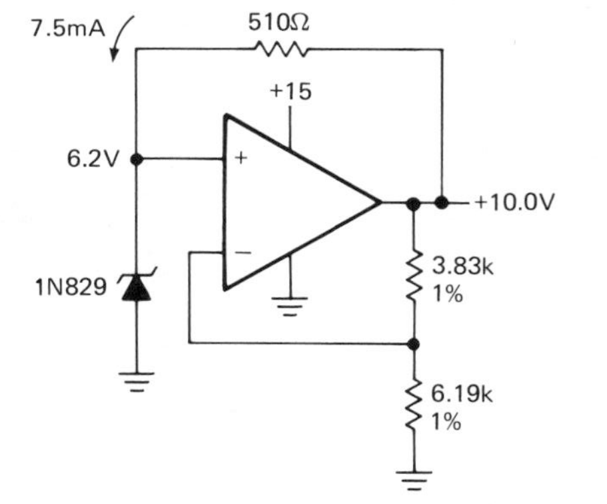

Figure 5.20

That stable output is itself used to provide a precision 7.5mA bias current. This circuit is self-starting, but it can turn on with either polarity of output! For the "wrong" polarity, the zener operates as an ordinary forward-biased diode. Running the op-amp from a single supply, as shown, overcomes this bizarre problem.

There are special compensated zeners available with guaranteed stability of zener voltage with *time,* a specification that normally tends to get left out. Examples are the 1N3501 and 1N4890 series. Zeners of this type are available with guaranteed stability of better than 5ppm/1000h. They're not cheap. Table 5.5 lists the characteristics of some useful zeners and reference diodes.

☐ *IC zeners*

The 723 regulator uses a compensated zener reference to achieve its excellent performance (30ppm/°C stability of V_{ref}). The 723, in fact, is quite respectable as a voltage reference all by itself, and you can use the other components of the IC to generate a stable reference output at any desired voltage.

The 723 used as a voltage reference is an example of a 3-terminal reference, meaning that it requires a power supply to operate. Precision temperature-compensated zener ICs are available as 2-terminal devices also; electrically they look just like zeners, although they actually include a number of active devices to give improved performance (most notably, constancy of "zener" voltage with applied current). An example is the LM329, with a zener voltage of about 6.9 volts. Its best version has a temperature coefficient of 6ppm/°C (typ), 10ppm/°C (max), when provided with a constant current of 1mA.

Unfortunately, like their discrete counterparts, IC zeners are noisy. The noise is substantially worse for the avalanche-mode

TABLE 5.5 ZENER AND REFERENCE DIODES

Type	Zener voltage V_Z (V)	@	Test current I_{ZT} (mA)	Tolerance (±%)	Tempco max (ppm/°C)	Regulation ΔV for ±10% I_{ZT} max (mV)	Maximum power (W)	Comments
Reference zeners								
1N821A-	6.2		7.5	5	±100	7.5	0.4	5 member family, graded by tempco; best and worst shown
1N829A	6.2		7.5	5	±5	7.5	0.4	
1N4890-	6.35		7.5	5	±20		0.4	Long-term stability < 100ppm/1000h
1N4895	6.35		7.5	5	±5		0.4	Long-term stability < 10ppm/1000h
LM129A	6.9		1.0	5	±10	0.1	0.2	IC 2-terminal reference
LM399	6.9		1.0	5	±1	0.1	0.15	Thermostated 2-terminal IC reference
Regulator zeners								
1N5221A	2.4		20	10	−850	60	0.5	60 member family, 2.4V to 200V in "5% resistor values" plus some extras; ±20%, ±10%, and ±5% avail.; popular
1N5231A	5.1		20	10	±300	34	0.5	
1N5281A	200		0.65	10	+1100	160	0.5	
1N4728A	3.3		76	10	−750	76	1.0	37 member family, 3.3V to 100V in "5% resistor values"; ±20%, ±10%, and ±5% avail.; popular
1N4735A	6.2		41	10	+500	8	1.0	
1N4764A	100		2.5	10	+1100	88	1.0	
TL431	2.75		10	2	±50	3	0.8	IC 2-terminal regulator; V_Z settable from 2.75V to 40V

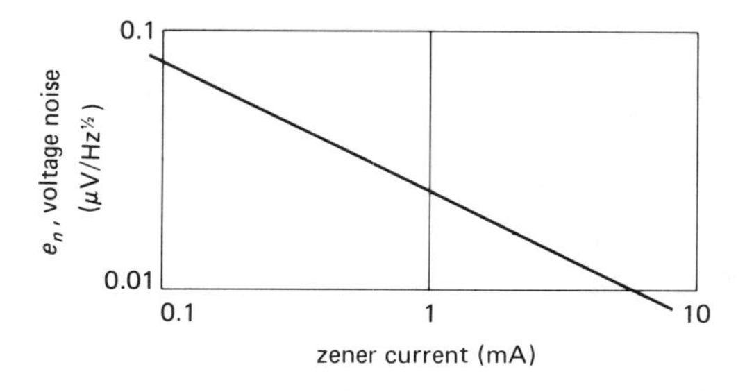

Figure 5.21
Voltage noise for a low-noise zener reference diode similar to the type used in the 723 regulator.

regulators, i.e., zener voltages higher than about 6 volts. The graph in Figure 5.21 illustrates the noise spectrum for a 723-type zener reference.

The use of a so-called buried (or subsurface) zener structure can improve zener stability and reduce zener noise substantially. This is used in one of the most stable monolithic references (see the later section entitled Temperature-stabilized references).

□ 5.14 Bandgap (V_{BE}) reference

More recently, a circuit known as a "bandgap" reference has become popular. It should properly be called a V_{BE} reference, and it is easily understandable using the Ebers-Moll diode equation. Basically, it involves the generation of a voltage with a positive temperature coefficient the same as V_{BE}'s negative coefficient; when added to a V_{BE}, the resultant voltage has zero tempco.

We start with a current mirror with two transistors operating at different emitter current densities (typically a ratio of 10:1) (see Fig. 5.22). Using the Ebers-Moll equation, it is easy to show that I_{out} has a positive temperature coefficient, since the difference

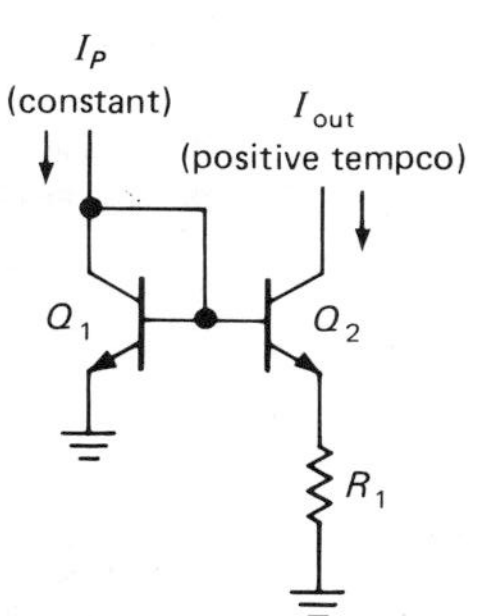

Figure 5.22

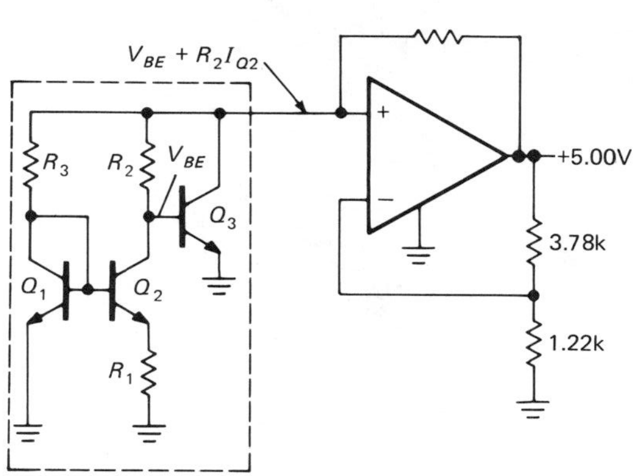

Figure 5.23
Classic V_{BE} bandgap voltage reference.

in V_{BE}s is just $(kT/q)\log_e r$, where r is the ratio of current densities (see the graph in Fig. 2.51). You may wonder where we get the constant programming current I_P. Don't worry; you'll see the clever method at the end. Now all you do is convert that current to a voltage with a resistor and add a normal V_{BE}. Figure 5.23 shows the circuit. R_2 sets the amount of positive-coefficient voltage you have added to V_{BE}, and by choosing it appropriately, you get zero overall temperature coefficient. It turns out that zero temperature coefficient occurs when the

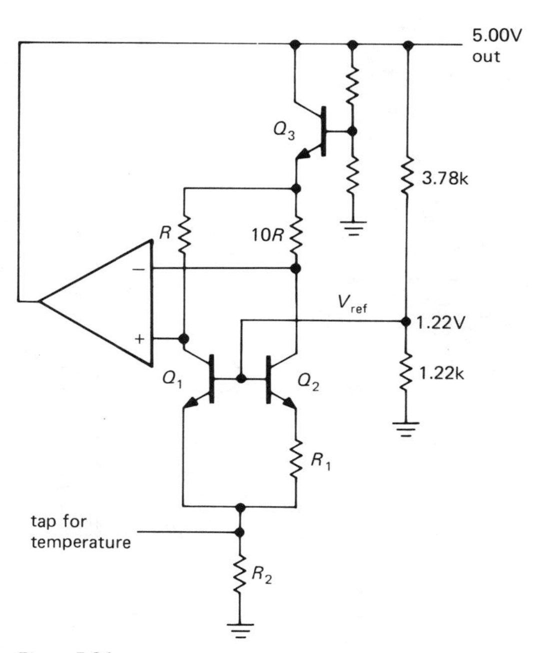

Figure 5.24

TABLE 5.6. IC VOLTAGE REFERENCES

Type	Bandgap	Zener	Terminals	Trim	Voltage (V)	Accuracy (%)	Tempco typ (ppm/°C)	Min. supply voltage (V)	Supply current (mA)	Max. output current (mA)	Noise voltage 0.1–10Hz typ (μV pp)	Long-term stability typ (ppm/1000h)	Regulation: Line typ (%/V)	Regulation: Load 0–10mA typ (%)
Regulator type														
LM10C	✓	—	8	✓	0.20	5	30	1.1	0.3	20			0.001	0.01[a]
μA723C	—	✓	14	✓	7.15	3	20	9.5	2.3	65		1000	0.003	0.03
SG3532J	✓	—	10	✓	2.50	4	50	4.5	1.6	150		300	0.005	0.02
Two-terminal (zener) type														
LM113-1	✓	—	2	—	1.22	1	100		1	20[b]				0.2
LM129A	—	✓	2	—	6.9	5	6		1	15[b]		20		0.1
LM329C	—	✓	2	—	6.9	5	30		1	15[b]		20		0.1
LM299A[c]	—	✓	4	—	6.95	2	0.2	9.0	17	10[b]		20		0.1
LM399[c]	—	✓	4	—	6.95	5	0.3	9.0	17	10[b]		20		0.1
LM3999[c]	—	✓	3	—	6.95	5	2.0	9.0	17	10[b]		20		0.1
LM336	✓	—	3	✓	2.50	4	10		1	10[b]		20		0.1
TL430	✓	—	3	✓	2.75	5	120		10	100[b]	50			0.5
TL431	✓	—	3	✓	2.75	2	10		10	100[b]	50			0.5
ICL8069A	✓	—	2	—	1.23	2	10		0.5	10[b]				0.2[a]
LM385A	✓	—	2	—	1.23	1	20		0.1[d]	20[b]	25			0.02[a]

Three-terminal type														
REF-01A	✓	—	8	✓	10.0	0.3	3	12	1	15	20		0.006	0.005
REF-01C	✓	—	8	✓	10.0	1	20	12	1	15	25		0.009	0.006
REF-02A	✓	—	8	✓	5.0	0.3	3	7	1	10	10		0.006	0.005
REF-02C	✓	—	8	✓	5.0	1	20	7	1	10	12		0.009	0.006
LH0070-1	—	✓	3	—	10.0	0.01	4	12.5	3	10	20		0.001	0.01
AD580K	✓	—	3	—	2.5	2	40	4.5	1	10	60	25	0.04	0.4
AD580M	✓	—	3	—	2.5	1	10	4.5	1	10	60	25	0.04	0.4
AD581L	✓	—	3	—	10.0	0.05	5	12	0.75	10	50	25	0.005	0.002
AD584JH	✓	—	8	✓	2.5	0.3	30	5	0.75	18	50	25	0.005	0.002
					5.0	↓	↓	7.5	↓	15	↓	↓	↓	↓
					7.5	↓	↓	10	↓	13	↓	↓	↓	↓
					10.0	↓	↓	12.5	↓	10	↓	↓	↓	↓
AD584LH	✓	—	8	✓	2.5	0.05	10	5	0.75	18	50	25	0.005	0.002
					5.0	0.06	5	7.5	↓	15	↓	↓	↓	↓
					7.5	0.06	5	10	↓	13	↓	↓	↓	↓
					10.0	0.1	5	12.5	↓	10	↓	↓	↓	↓
MC1403A	✓	—	8	—	2.5	1	10	4.5	1.2	10			0.002	0.06
MC1404AU5	✓	—	8	✓	5.0	1	10	7.5	1.2	10	12	25	0.001	0.06
MC1404AU6	✓	—	8	✓	6.25	↓	↓	8.8	↓	↓	↓	↓	0.001	↓
MC1404AU10	✓	—	8	✓	10.0	↓	↓	12.5	↓	↓	↓	↓	0.0006	↓
HA1600-5[c]	—	✓	14	—	10.0	0.02	1.0	12	25	2			0.001	0.001
HA1610	—	✓	14	—	10.0	0.01	3	12	1.9	10			0.004	0.04
ICL8212	✓	—	8	✓	1.15	3	200	1.8	0.035	20			0.2	

[a]Zero to 1mA. [b]Maximum zener current. [c]On-chip heater/thermostat. [d]Micropower: specified down to 10 μA operating current.

total voltage equals the silicon bandgap voltage (extrapolated to absolute zero), about 1.22 volts. The circuit in the box is the reference. Its own output is used (via R_3) to create the constant current we initially assumed.

Figure 5.24 shows another very popular bandgap reference circuit (it replaces the components in the box in Figure 5.23). Q_1 and Q_2 are a matched pair, forced to operate at a ratio of emitter currents of 10:1 by feedback from the collector voltages. The difference in V_{BE}s is $(kT/q)\log_e 10$, making Q_2's emitter current proportional to T (the preceding voltage applied across R_1). But since Q_1's collector current is larger by a factor of 10, it also is proportional to T. Thus the total emitter current is proportional to T, and therefore it generates a positive tempco voltage across R_2. That voltage can be used as a thermometer output, by the way, as will be discussed shortly. R_2's voltage is added to Q_1's V_{BE} to generate a stable reference of zero tempco at the base. Bandgap references appear in many variations, but they all feature the summation of V_{BE} with a voltage generated from a pair of transistors operated with some ratio of current densities.

□ *IC bandgap references*

An example of a bandgap reference is the LM113, with an output voltage of 1.22 volts. One of the best references now available (1980) is the REF-02; its best version has a temperature coefficient of 3ppm/°C (typ), 8.5ppm/°C (max). Internal circuitry is used to generate an output voltage of +5.0 volts. The REF-02 has provision for trimming the output voltage to precisely +5.0 volts. Another reference, the AD584, generates an output of 2.5, 5.0, or 10.0 volts, selectable through external connections. These references are 3-terminal types and require a dc supply. Some of their characteristics, along with those of other IC voltage references, are listed in Table 5.6.

One other interesting voltage reference is the TL431C. It is an inexpensive "programmable zener" reference, and it is used as shown in Figure 5.25. The "zener" (made from a V_{BE} circuit) turns on when the control voltage reaches 2.75 volts; the device

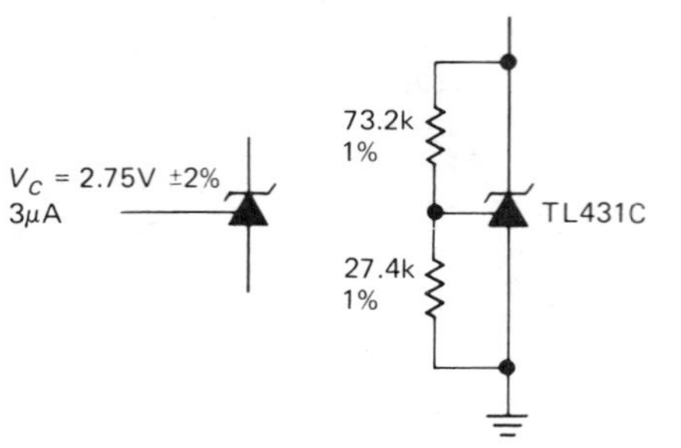

Figure 5.25

draws only a few microamps from the control terminal and gives a typical tempco of output voltage of 10ppm/°C. The circuit values shown give a zener voltage of 10.0 volts, for example. This device comes in a mini-DIP package and can handle currents to 100mA.

Voltage references can be made in other interesting ways. For example, Section 6.17 discusses a FET low-current "pinch-off reference."

□ *Bandgap temperature sensors*

The predictable V_{BE} variation with temperature can be exploited to make a temperature-measuring IC. The REF-02, for instance, generates an additional output voltage that varies linearly with temperature (see preceding discussion). With some simple external circuitry you can generate an output voltage that tells you the chip temperature, accurate to 1% over the full "military" temperature range (−55°C to +125°C). The AD590, intended for temperature measurement only, generates an accurate current of 1μA/°K. It's a 2-terminal device; you just put a voltage across it (4–30V) and measure the current.

□ *Temperature-stabilized references*

Another approach to achieving excellent temperature stability in a voltage reference circuit (or any other circuit, for that matter) is to hold the reference, and perhaps its associated electronics, at a constant elevated temperature. You will see simple techniques for doing this in Chapter 14 (one obvious method is to use a bandgap temperature sensor to control a heater). In this way the circuit can deliver equivalent performance with a greatly relaxed temperature coefficient, since the actual circuit components are isolated from external temperature fluctuations. Of greater interest for precision circuitry is the ability to deliver significantly improved performance by putting an already well-compensated reference circuit into a constant-temperature environment.

This technique of temperature-stabilized or "ovenized" circuits has been used for many years, particularly for ultrastable oscillator circuits. There are commercially available power supplies and precision voltage references that use ovenized reference circuits. This method works well, but it has the drawbacks of bulkiness, relatively large heater power consumption, and sluggish warm-up (typically 10 min or more). These problems are effectively eliminated if the thermal stabilization is done at the chip level by integrating a heater circuit (with sensor) onto the integrated circuit itself. This approach was pioneered in the 1960s by Fairchild with the μA726 and μA727 temperature-stabilized differential pair and preamp, respectively. More recently, temperature-stabilized voltage references such as the Harris HA1600-5 and the National LM199 series have appeared. The latter offers a temperature coefficient of 0.00002%/°C (typ), which is a mere 0.2ppm/°C. These references are packaged in standard metal transistor cans (TO-46); they consume about 0.25 watt of heater power and come up to temperature in 3 seconds. Users should be aware that subsequent op-amp circuitry, and even precision wire-wound resistors with their ±2.5ppm/°C tempco, may degrade performance considerably, unless extreme care is used in design. In particular, low-drift precision op-amps such as the OP-07, with 0.2μV/°C (typ) input-stage drift, are essential. These aspects of precision circuit design are discussed in Sections 7.01 to 7.06.

THREE-TERMINAL AND FOUR-TERMINAL REGULATORS

5.15 Three-terminal regulators

For most noncritical applications the best choice for a voltage regulator is the simple

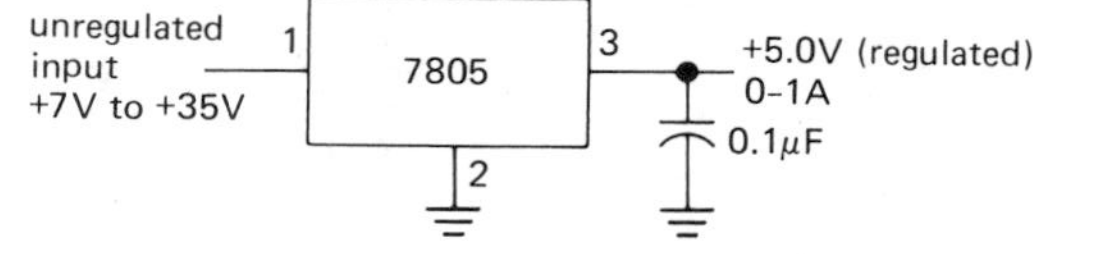

Figure 5.26

3-terminal type. It has only three connections (input, output, and ground) and is factory-trimmed to provide a fixed output. Typical of this type is the 78*xx*. The voltage is specified by the last two digits of the part number and can be any of the following: 05, 06, 08, 10, 12, 15, 18, or 24. Figure 5.26 shows how easy it is to make a +5 volt regulator, for instance, with one of these regulators. The capacitor across the output improves transient response and keeps the impedance low at high frequencies (an input capacitor of at least 0.33μF should be used in addition if the regulator is located a considerable distance from the filter capacitors). The 7800 series is available in plastic or metal power packages (same as power transistors). A low-power version, the 78L*xx*, comes in the same plastic and metal packages as small-signal transistors (see Table 5.7). The 7900 series of negative regulators works the same way (with negative input voltage, of course). Some other examples of this type of regulator are the LM320 and LM340 series. The 7800 series can provide up to 1 amp load current and has on-chip circuitry to prevent damage in the event of overheating or excessive load current; the chip simply shuts down, rather than blowing out. In addition, on-chip circuitry prevents operation outside the transistor safe operating area (see Section 5.07) by reducing available output current for large input–output voltage differential. These regulators are inexpensive and easy to use, and they make it practical to design a system with many printed-circuit boards in which the unregulated dc is brought to each board and regulation is done locally on each circuit card. Table 5.7 lists the characteristics of a representative selection of 3-terminal fixed voltage regulators.

5.16 Four-terminal regulators

For applications where you would like to be able to adjust the output voltage, there are

TABLE 5.7. FIXED VOLTAGE REGULATORS

Type	Pkg.	V_{out} (V)	Output current max[a] @ 75°C case I_{out} (A)	No sink[b] I_{out} (A)	No sink[b] P_{diss} (W)	Regulation typ Load[c] (mV)	Regulation typ Line[d] (mV)	θ_{JC} (°C/W)	Input voltage min (V)	Input voltage max (V)	120Hz ripple reject typ (dB)	Temp. stab.[e] typ (mV)	Long-term stab.[f] max (%)	Output impedance 10Hz (Ω)	Output impedance 10KHz (Ω)	Comments
Positive[g]																
μA78L05AWC / LM240LAZ-5.0	TO-92	5	0.1	0.1	0.6	5	50	35	7	35	50	—	0.25	0.2	0.2	Small package
μA78M05HC	TO-39	5	0.5	0.2	0.7	10	3	18	7	35	80	40	0.4	0.01	0.05	Small package
LM309K	TO-3	5	1.0	0.6	2.2	20	4	3.0	7	35	80	50	0.4	0.04	0.05	Original +5V regulator
μA7805UC / LM340T-5	TO-220	5	1.0	0.45	1.7	10	3	3.0	7	35	80	30	0.4	0.01	0.03	Popular series
μA7815UC / LM340T-15	TO-220	15	1.0	0.15	1.7	12	4	3.0	17	35	70	100	0.4	0.02	0.05	Popular series
μA7805KC / LM340K-5	TO-3	5	1.0	0.6	2.2	10	3	3.5	7	35	80	30	0.4	0.01	0.03	
LAS1605	TO-3	5	2.0	0.75	2.8	30[h]	100[h]	2.5	7.6	30	75	—	—	0.002	0.02	Lambda; 1.5A, 3A, 5A, 8A avail.
LM323K	TO-3	5	3.0	0.6	2	25	5	2.0	7	20	70	30	0.7	0.01	0.02	
μA78H05KC	TO-3	5	5.0	0.8	3	10	10	2.0	8	25	60	50	—	0.002	0.003	Hybrid; 12V and 15V also
LAS3905	TO-3	5	8.0	0.8	3	20[h]	100[h]	0.7	7.6	25	60[h]	100	—	0.004	0.01	Monolithic
μA78P05	TO-3	5	10.0	0.8	3	5	10	1.6	7.5	40	60[h]	75	—	—	—	Hybrid
Negative[g]																
LM79L15ACZ / LM320LZ-15	TO-92	−15	0.1	0.05	0.6	75[h]	45[h]	35	−17	−35	40	—	0.4[i]	0.05	0.05	Small package
μA79M15AHC / LM320H-15	TO-39	−15	0.4	0.6	0.7	45	7	18	−16.5	−35	60	60	—	0.06	0.07	Small package
μA7915UC / LM320T-15	TO-220	−15	1.0	0.15	1.7	4	3	3.0	−16.5	−35	60	60	0.4	0.06	0.07	Popular series
μA7915KC / LM320KC-15	TO-3	−15	1.0	0.2	2.2	4	3	3.5	−16.5	−35	60	60	0.4	0.06	0.07	
LM345K-5.0	TO-3	−5	3.0	0.2	2.1	10	5	2.0	−7.5	−20	65	25	1.0	0.02	0.04	

[a] $V_{in} = 1.75\ V_{out}$. [b] For 50°C ambient. [c] 10% to 50% I_{max}. [d] For $\Delta V_{in} \approx 15$ volts. [e] ΔV_{out} for 0°C to 100°C junction temperature. [f] 1000 hours. [g] All have ±4% voltage tolerance (max). All include internal thermal shutdown and current-limiting circuitry. Most are available in ± 5, 6, 8, 10, 12, 15, 18, and 24 volt units. A few are also available in −2, −3, −4, −5.2, −9, +2.6, +6.2, +9, and 17 volt units. [h] Maximum. [i] Typical.

TABLE 5.8. ADJUSTABLE VOLTAGE REGULATORS

Type	Pol.	Pkg.	Output voltage min (V)	Output voltage max (V)	I_{max} (A)	Regulation typ Load[a] (%)	Regulation typ Line[b] (%)	θ_{JC} (°C/W)	Input voltage min (V)	Input voltage max (V)	Dropout voltage max (V)	120Hz ripple reject typ (dB)	Temp. stab.[c] typ (%)	Long-term Stab.[d] max (%)	Output impedance 10Hz (Ω)	Output impedance 10kHz (Ω)	Thermal Shutdown	Int. curr. limit	Ext. curr. limit	Comments
Three-terminal																				
LM317H	+	TO-39	1.2	37	0.5	0.1	0.2	12	—	40[e]	2[f]	80	0.6	0.3	0.01	0.03	√	√	—	Small package
LM337H	−	TO-39	−1.2	−37	0.5	0.3	0.2	12	—	−40[e]	2[f]	75	0.5	0.3	0.02	0.02	√	√	—	Negative 317H
LM317T	+	TO-220	1.2	37	1.5	0.1	0.2	4	—	40[e]	2.5[f]	80	0.6	0.3	0.01	0.03	√	√	—	Popular
LM317HVK	+	TO-3	1.2	57	1.5	0.1	0.2	2.3	—	60[e]	2.5[f]	80	0.6	0.3	0.01	0.03	√	√	—	High voltage 317
LM337T	−	TO-220	−1.2	−37	1.5	0.3	0.2	4	—	−40[e]	2.5[f]	75	0.5	0.3	0.02	0.02	√	√	—	Negative 317T
LM337HVK	−	TO-3	−1.2	−47	1.5	0.3	0.2	2.3	—	−50[e]	2.5[f]	75	0.5	0.3	0.02	0.02	√	√	—	High-voltage 337
LM350K	+	TO-3	1.2	32	3.0	0.1	0.1	2	—	35[e]	2.5[f]	80	0.6	0.3	0.005	0.02	√	√	—	
LM338K	+	TO-3	1.2	32	5.0	0.1	0.1	2	—	35[e]	2.5[f]	80	0.6	0.3	—	—	√	√	—	
Four-terminal																				
μA78MGHC	+	TO-39	5	30	0.5	1[g]	1[g]	18	7.5	40	2.5	80	3[g]	—	—	—	√	√	—	Small package
μA79MGHC	−	TO-39	−2.5	−30	0.5	1[g]	1[g]	18	−7	−40	2[f]	65	3[g]	—	—	—	√	√	—	Small package
μA78GU1C	+	TO-220	5	30	1.0	1[g]	1[g]	7.5	7.5	40	2.5	80	3[g]	—	—	—	√	√	—	
μA79GU1C	−	TO-220	−2.5	−30	1.0	1[g]	1[g]	7.5	−7	−40	2[f]	60	3[g]	—	—	—	√	√	—	
LAS15U	+	TO-3	4	30	1.5	0.6[g]	2[g]	3	6.5	40	2.4	70	3[g]	—	0.003	0.02	√	√	—	Lambda
LAS18U	−	TO-3	−2.6	−30	1.5	0.6[g]	2[g]	3	−5	−40	2.1	60	3[g]	—	0.02	0.04	√	√	—	Lambda
LAS16U	+	TO-3	4	30	2.0	0.6[g]	2[g]	2.5	6.5	35	2.6	70	2[g]	—	0.002	0.02	√	√	—	Lambda
LAS14AU	+	TO-3	4	35	3.0	0.6[g]	1[g]	1.5	6.5	40	2.3	70	2[g]	—	0.001	0.01	√	√	—	Lambda
LAS19U	+	TO-3	4	30	5.0	0.6[g]	1[g]	0.9	6.5	35	2.6	65	2[g]	—	0.0005	0.004	√	√	—	Lambda
Multiterminal																				
LM376N	+	mini-DIP	5	37	0.025	0.2[g]	0.6[g]	190[h]	9	40	3	60[g]	1[g]	—	—	—	—	—	√	
LM305AH	+	TO-5	4.5	40	0.045	0.03	0.3	45	8.5	50	3	80	0.3	0.1	—	—	—	—	√	
LM304H	−	TO-5	0	−40	0.025	1mV	0.2	45	−8	−40	2	65	0.3	0.01	—	—	—	—	√	Original neg. reg.
μA723PC	+	DIP	2	37	0.15	0.03	0.1	150[h]	9.5	40	3	75	0.3	0.1	0.05	0.1	—	—	√	Classic
SG3532J	+	DIP	2	38	0.17	0.1	0.1	125[h]	4.7	40	2	66	0.5	0.3	—	—	√	√	√	Improved 723
NE550N	+	DIP	2	40	0.15	0.03	0.08	150[h]	8.5	40	3	90	0.2	0.1[f]	0.1	0.1	—	—	√	
LAS1000	+	TO-5	3	38	0.15	0.1[g]	0.2	150[h]	5	40	2	60[g]	1.5[g]	—	0.004	0.05	√	√	√	Lambda, improved 723
LAS1100	+	TO-5	3	48	0.15	0.1[g]	0.2	150[h]	5	50	2	69[g]	1.5[g]	—	0.004	0.05	√	√	√	High voltage
LH0075C	+	TO-8	0	27	0.2	0.04	0.1	100[h]	8	32	3.5	80	0.3	—	0.03	0.1	—	—	√	Several fixed voltages; pin prog. (±0.1%)
LH0076C	−	TO-8	0	−27	0.2	0.02	0.05	100[h]	−8	−32	3.5	70	0.3	—	0.01	0.05	—	—	√	(see above)
MC1469R	+	TO-66	2.5	32	0.6	0.005	0.05	7	9	35	3	100	0.2	—	0.05	0.1	—	—	√	Precision; may oscillate
MC1463R	−	TO-66	−3.8	−32	0.6	0.005	0.05	17	−9	−35	3	90	0.2	—	0.02	0.03	—	—	√	Neg. MC1469
MC1466L	+	DIP	0	1000	—	0.02	0.05	170[h]	—	—	2[f]	70	0.4	—	—	—	—	—	√	Lab supply; good curr. limit; floats with aux. pwr. supply

[a] 10% to 50% I_{max}. [b] For $\Delta V_{in} \approx$ 15 volts. [c] ΔV_{out} for 0°C to 100°C junction temperature. [d] 1000 hours. [e] Maximum $V_{in} - V_{out}$. [f] Typical. [g] Minimum or maximum (worst case). [h] θ_{JA}.

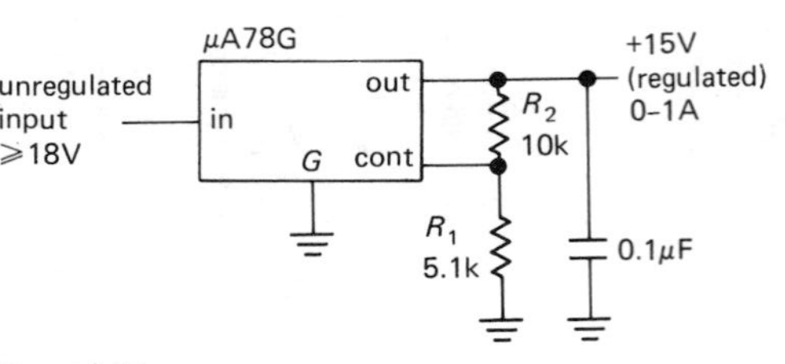

Figure 5.27

4-terminal regulators that are almost as easy to use as 3-terminal regulators. An example is the μA78G (the μA79G is the same thing, for negative output voltages). The fourth terminal is the "control" terminal, which you connect to a voltage divider across the output (Fig. 5.27). The voltage divider is chosen to put +5 volts (−2.23V for the μA79G) on the control terminal. In practice, you will usually use an adjustable trimmer potentiometer for R_1 to set the output voltage precisely. This kind of regulator is good if you want an unusual output voltage or precise settability. Like the 7800 series, it is not a precision regulator. The output-voltage temperature coefficient and the regulation of output voltage against changes in input-voltage ("line") or load current are mediocre (Table 5.8). However, most well-designed circuits are not critical in their power-supply requirements, and for them these regulators are entirely adequate. As with 3-terminal regulators, there is internal thermal protection, short-circuit protection, and safe operating area protection.

5.17 Three-terminal adjustable regulators

A regulator with improved characteristics as well as simplicity of application is the LM317, a 3-terminal adjustable positive regulator. This regulator has no ground terminal; instead, it adjusts V_{out} to maintain a constant 1.25 volts (bandgap) from the output terminal to the adjustment terminal. Figure 5.28 shows the easiest way to use it. The regulator puts 1.25 volts across R_1, so 5mA flows through it. The adjustment terminal draws very little current (50–100μA), so the output voltage is just

$$V_{out} = 1.25(1 + R_2/R_1) \text{ volts}$$

In this case the output voltage is adjustable

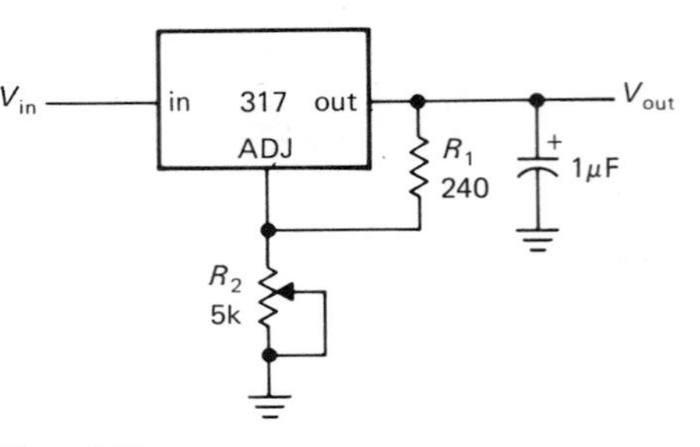

Figure 5.28

from 1.25 volts to 25 volts. For a fixed-output-voltage application, R_2 will normally be adjustable only over a narrow range, to improve settability (use a fixed resistor in series with a trimmer). The 317 is available in several packages, including the plastic power package, the metal power package (TO-3), and the small transistor package. In the power package it can deliver up to 1.5 amps, with proper heat sinking. Because it doesn't "see" ground, it can be used for high-voltage regulators, as long as the input–output voltage differential doesn't exceed the rated maximum of 40 volts.

EXERCISE 5.4

Design a +5 volt regulator with the 317. Provide ±20% voltage adjustment range with a trimmer pot.

Three-terminal adjustable regulators are available in higher current ratings, e.g., the LM350 at 3 amps and the LM338 at 5 amps. A negative version is also available, the LM337, with properties otherwise identical to those of the LM317 (see Table 5.8 for more information).

5.18 Additional comments about 3-terminal regulators

General characteristics of 3- and 4-terminal regulators

The following specifications are typical for most 3- and 4-terminal regulators, and they may be useful as a rough guide to the performance you can expect:

Output voltage tolerance	2%
Dropout voltage	2 volts
Maximum input voltage	35 volts

Ripple rejection	0.05 to 0.1%
Spike rejection	0.1 to 0.3%
Load regulation	0.1 to 0.5%, over max. load change
dc input rejection	0.2%
Temperature stability	2% over full temp. range

Outboard pass transistors

Three-terminal fixed regulators are available with 5 amps or more of output current, e.g., the 78H00 series and the LM323 and LM345. However, such high-current operation may be undesirable, since the maximum chip operating temperature for these regulators is lower than for power transistors, mandating oversize heat sinks. Also, they are expensive. An alternative solution is the use of external pass transistors, which can be added to the 3- and 4-terminal regulators just as with the classic 723. Figure 5.29 shows the basic circuit.

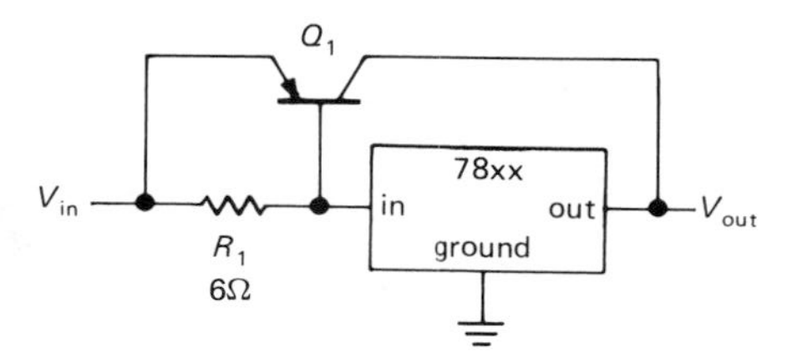

Figure 5.29
Three-terminal regulator with current-boosting outboard transistor.

The circuit works as before, for load currents less than 100mA. For greater load currents, the drop across R_1 turns on Q_1, limiting the actual current through the 3-terminal regulator to about 100mA. The 3-terminal regulator maintains the output at the correct voltage, as usual, by reducing input current and hence drive to Q_1 if the output voltage rises, and vice versa. It never even realizes the load is drawing more than 100mA! With this circuit the input voltage must exceed the output voltage by the dropout voltage of the 7800 (2V) plus a V_{BE} drop.

In practice, the circuit must be modified to provide current limiting for Q_1, which could otherwise supply an output current equal to h_{FE} times the regulator's internal current limit, i.e., 20 amps or more! That's

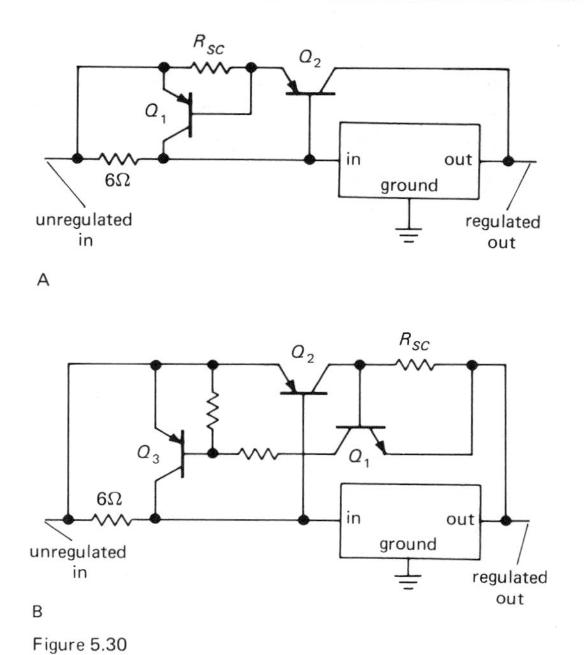

Figure 5.30
Current limit circuits for outboard transistor booster.

enough to destroy Q_1, as well as the unfortunate load that happens to be connected at the time. Figure 5.30 shows two methods of current limiting.

In both circuits Q_2 is the high-current pass transistor, and its emitter-to-base resistor has been chosen to turn it on at 100mA load current. In the first circuit Q_1 senses the load current via the drop across R_{SC}, cutting off Q_2's drive when the drop exceeds a diode drop. There are a couple of drawbacks to this circuit: The input voltage must now exceed the regulated output voltage by the dropout voltage of the 3-terminal regulator plus two diode drops for load currents near the current limit. Also, Q_1 must be capable of handling high currents (equal to the current limit of the regulator), and it is difficult to add foldback limiting because of the small resistor values required in Q_1's base. The second circuit helps solve these problems, at the expense of some additional complexity. With high-current regulators a low dropout voltage is often important to reduce power dissipation to acceptable values. To add foldback limiting to the latter circuit, just tie Q_1's base to a divider from Q_2's collector to ground, rather than directly to Q_2's collector.

External pass transistors can be added to the adjustable 3- and 4-terminal regulators in exactly the same way. See the manufacturer's data sheets for further details.

Current source

The 317 makes a handy constant-current source. Figure 5.31 shows one to source 1

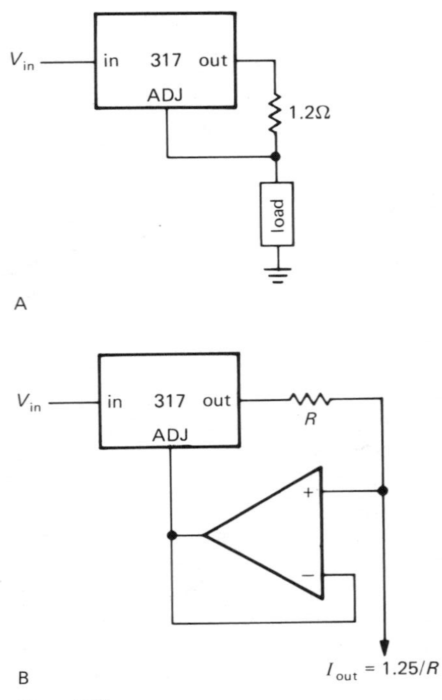

Figure 5.31

amp. The addition of an op-amp follower, as in the second circuit, is necessary if the circuit is used to source small currents, since the ADJ (adjust) input contributes a current error of about 50μA. As with the previous regulators, there is on-chip current limit, thermal overload protection, and safe operating area protection.

EXERCISE 5.5

Design an adjustable current source for output currents from 10μA to 1mA using a 317. If $V_{in} = +15$ volts, what is the output compliance? Assume a dropout voltage of 2 volts.

□ SPECIAL-PURPOSE POWER-SUPPLY CIRCUITS

□ 5.19 Dual-tracking regulators

For circuits that require equal positive and negative supply voltages (most op-amp circuits, and many other circuits that operate with signals near ground), the best power-supply circuit is often provided by the so-called dual-tracking regulator. In its classic form it looks as shown in Figure 5.32. Q_1 is

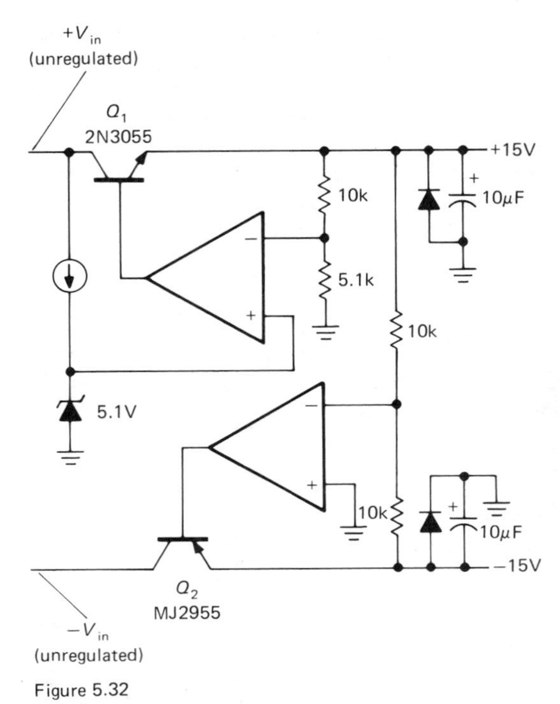

Figure 5.32
Dual-tracking regulator.

the pass transistor for a conventional positive regulated supply. The positive regulated output is then simply used as the reference for a negative supply. The lower error amplifier controls the negative output by comparing the average of the two output voltages with ground, thereby giving equal 15 volt positive and negative regulated outputs. The positive supply can be any of the configurations we have already talked about; if it is an adjustable regulator, the negative output follows any changes in the positive regulated

TABLE 5.9. DUAL-TRACKING REGULATORS

Type	Pkg.	V_{out} (V)	Adj. output	Balance trim	Adj. curr. limit	Thermal limit	Max. $V_+ - V_-$ Input (V)	Maximum output current[a] (each supply): @ 75°C case I_{out} (mA)	No sink[b] I_{out} (mA)	No sink[b] P_{diss} (W)	Regulation typ: Load[c] (mV)	Regulation typ: Line[d] (mV)	θ_{JC} (°C/W)	120Hz ripple reject typ (dB)	Temp. stab.[e] typ (mV)	Noise[f] (μV rms)	Comments
Motorola																	
MC1468L	DIP	±15	✓	✓	✓	—	60	55	30	0.5	10[g]	10[g]	50	75	45	100	
MC1468R	TO-66	±15	✓	✓	✓	—	60	100	65	1.2	10[g]	10[g]	17	75	45	100	
National																	
LM325S	P-DIP	±15	—	—	✓	✓	60	100	80	1.4	6	2	12	75	45	150	Intended for use with a pair of external pass transistors (LM325S–LM327S)
LM325N	DIP	±15	—	—	✓	✓	60	—	30	0.5	6	2	150[h]	75	45	150	
LM326S	P-DIP	±12	—	—	✓	✓	60	100	100	1.4	6	2	12	75	35	100	
LM326N	DIP	±12	—	—	✓	✓	60	—	35	0.5	6	2	150[h]	75	35	100	
LM327S	P-DIP	+5; −12	—	—	✓	✓	60	100	100	1.4	6	2	12	75	15; 35	40; 100	
Raytheon																	
RC4194DB	DIP	adj.	✓	✓	—	✓	70	30[i]	25[i]	0.5	0.1%	0.2%	160[h]	70	0.2%	250[j]	
RC4194TK	TO-66	adj.	✓	✓	—	✓	70	250[i]	90[i]	1.8	0.2%	0.2%	7	70	0.2%	250[j]	Our favorites
RC4195NB	mini-DIP	±15	—	✓	—	✓	60	—	20	0.35	2	2	210[h]	75	75	60	
RC4195TK	TO-66	±15	—	✓	—	✓	60	150	70	1.2	3	2	11	75	75	60	Our favorites
Signetics																	
NE5553N	DIP	±12	✓	✓	—	✓	64	100	55	0.8	10	100	95[h]	50	100	55	
NE5553U1	TO-220	±12	✓	✓	—	✓	64	300	90	1.25	10	100	12	50	100	55	
NE5554N	DIP	±15	✓	✓	—	✓	64	80	45	0.8	10	100	95[h]	50	100	55	
NE5554U1	TO-220	±15	✓	✓	—	✓	64	230	70	1.25	10	100	12	50	100	55	
NE5555U1	TO-220	+5; −12	✓	✓	—	✓	64	300	125	1.25	10	100	12	50	100	55	
Silicon general																	
SG3501AJ	DIP	±15	—	✓	—	✓	60	60	30	0.6	30	20	125[h]	75	150	50	
SG3502	DIP	adj.	✓	✓	✓	✓	50	50[i]	30	0.6	0.3%	0.2%	125[h]	75	1%	50	

(a) $V_{in} = 1.6\ V_{out}$ (each supply). (b) For 50°C ambient. (c) 10% to 50% I_{max} (d) For $\Delta V_{in} \approx 15$ volts. (e) ΔV_{out} for 0°C to 100°C T_J. (f) 100Hz to 10kHz. (g) Maximum. (h) θ_{JA}. (i) 10 volt drop (each supply). (j) 10Hz to 100kHz.

output. In practice, it is wise to include current-limiting circuitry.

□ *Reverse-polarity protection*

An additional caution with dual supplies: Almost any electronic circuit will be damaged extensively if the supply voltage is reversed. The only way that can happen with a single supply is if you connect the wires backward; sometimes you see a high-current rectifier connected across the circuit in the reverse direction to protect against this error. With circuits that use several supply voltages (a split supply, for instance), extensive damage can result if there is a component failure that shorts the two supplies together; a common situation is a collector-to-emitter short in one transistor of a push-pull pair operating between the supplies. In that case the two supplies find themselves tied together, and one of the regulators will win out. The opposite supply voltage is then reversed in polarity, and the circuit starts to smoke. For this reason it is wise to connect a power rectifier (e.g., a 1N4004) in the reverse direction from each regulated output to ground, as shown.

□ *IC dual-tracking regulators*

As with single-polarity regulators, dual-tracking regulators are available as complete integrated circuits in both fixed and adjustable versions. Table 5.9 lists the characteristics of most types now available. Typical are the 4194 and 4195 regulators, which are used as shown in Figure 5.33. The 4195 is factory-trimmed for ±15 volt outputs, whereas the 4194 outputs are adjustable via R_1. Both regulators are available in power packages as well as the small transistor packages, and both have internal thermal shutdown and current limiting. As with the preceding regulators, outboard pass transistors can be added for greater output current.

Many of the preceding regulators (e.g., the 4-terminal adjustable regulators) can be connected as dual-tracking regulators. The manufacturer's data sheets often given suggested circuit configurations. It is worth keeping in mind that the idea of referencing one supply's output to another supply can be used even if the two supplies are not of equal and opposite voltages. For instance,

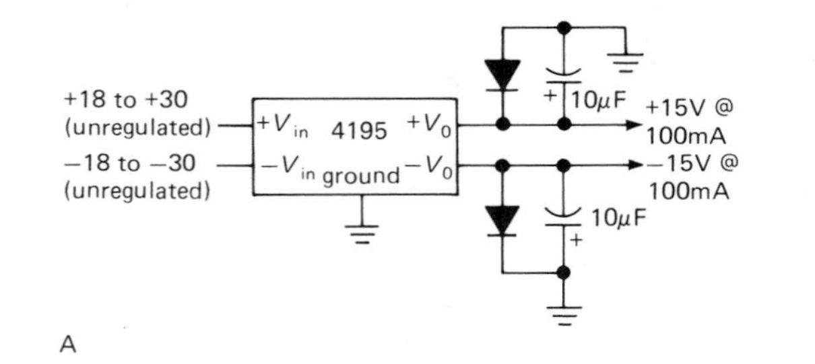

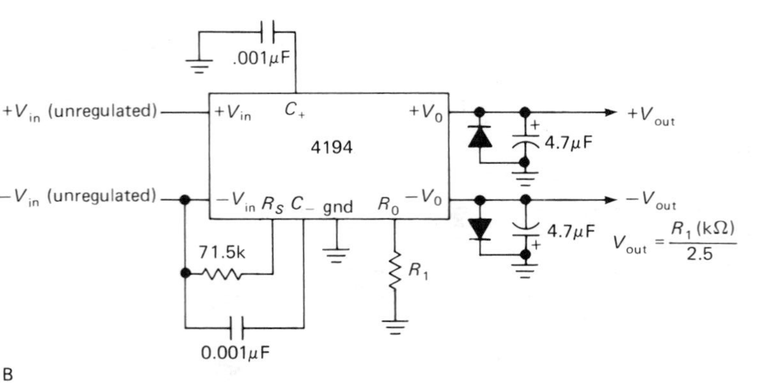

Figure 5.33

once you have a stable +15 volt supply, you can use it to generate a regulated +5 volt output, or even a regulated −12 volt output.

EXERCISE 5.6

Design a ±12 volt regulator using the 4194.

□ 5.20 High-voltage regulators

Some special problems arise when you design regulators to deliver high voltages. Since ordinary transistors typically have breakdown voltages of less than 100 volts, supplies to deliver voltages higher than that require some clever circuit trickery. This section will present a collection of such techniques.

□ *Brute force: high-voltage components*

Power transistors are available with breakdown voltages higher than 1000 volts, and they're not even very expensive. The DTS704 from Delco is typical of this kind of transistor, with a breakdown voltage of 1400 volts and a maximum collector current of 3 amps (not at the same time, of course). With the error amplifier near ground (the output-voltage-sensing divider gives a low-voltage sample of the output), only the pass transistor and its driver need to have high breakdown voltage. Figure 5.34 shows the idea. Q_1 and Q_2 are high-voltage types, with breakdown voltages of 600 volts or more. The diodes are used to prevent breakdown; D_1 protects Q_2's base-emitter junction during turn-off, when the stored charge on C_2 could hold Q_2's emitter more positive than its base. Similarly, D_2 protects the base-collector and emitter-collector junctions. Zener diode D_3 protects the input stage of the error amplifier. R_5, R_6, and C_3 are included for frequency compensation, i.e., stability. This circuit is an exception to the general rule that transistor circuits do not present a shock hazard!

If a high-voltage regulator is designed to provide a fixed output voltage only, the pass transistor does not need to have a large breakdown voltage. In the preceding circuit, omitting R_3 results in a fixed +500 volt regulator. A 100 volt pass transistor will

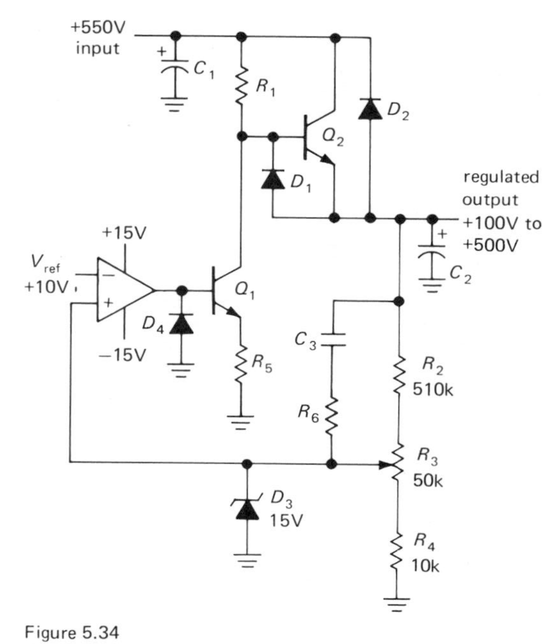

Figure 5.34
Conventional high-voltage regulator.

then be fine, provided that the circuit ensures that the voltage across it never exceeds 100 volts, even during turn-on, turn-off, and output short-circuit conditions. As drawn, the pass transistor sees the full input voltage at turn-on, because C_2 holds the output at ground; replacing D_2 by a zener solves that problem. If the zener can handle high current, it can also protect the pass transistor against short-circuit loads, if suitable fusing is provided ahead of the regulator. The active zener circuit mentioned in Section 5.06 would be a good choice here.

□ *Regulating the ground return*

Figure 5.35 shows another way to regulate high voltages with low-voltage components. Q_1 is a series pass transistor, but it is connected in the low side of the supply; its "output" goes to ground. It has only a fraction of the output voltage across it, and it sits near ground, simplifying the driver circuitry. As before, protection must be provided during on/off transients and overloads. The simple zener protection shown is adequate, but remember that the zener must be able to handle the full short-circuit current.

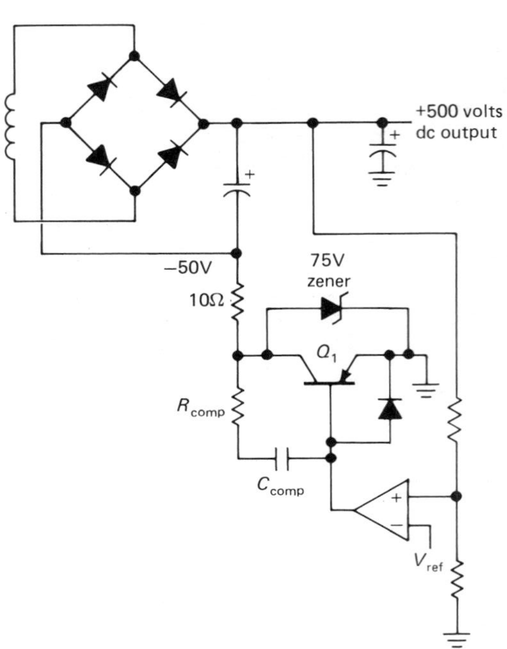

Figure 5.35

□ *Lifting the regulator above ground*

Another method sometimes used to extend the voltage range of regulators, including the simple 3-terminal type, is to raise the common terminal off ground with a zener (Fig. 5.36). D_1 raises the common terminal off ground, adding its voltage to the normal output of the regulator. D_2 sets the drop across the regulator via follower Q_1 and provides protection during short circuit because of D_3.

□ *Optically coupled transistor*

There is another way to handle the problem of transistor breakdown ratings in high-voltage supplies, especially if the pass transistor can be a relatively low voltage unit because of fixed (known) output voltage. In such cases only the driver transistor has to

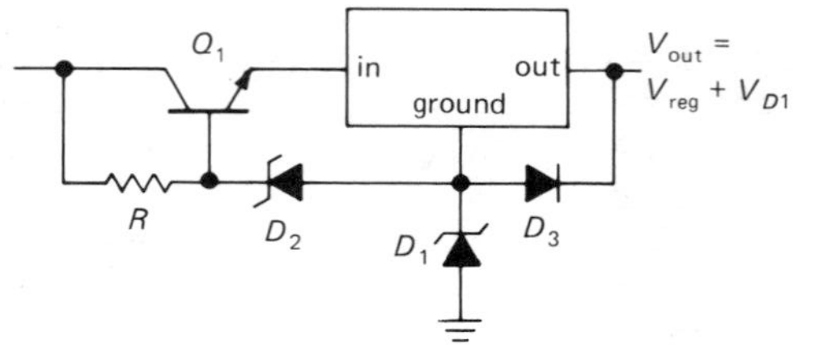

Figure 5.36

withstand high voltage, and even that can be avoided by using optically coupled transistors. These devices, which we will talk about further in connection with digital interfacing in Chapter 9, actually consist of two units electrically insulated from each other: a light-emitting diode (LED), which lights up when current flows through it in the forward direction, and a phototransistor (or photo-Darlington) mounted in close proximity in an opaque package. Running current through the diode causes the transistor to conduct, just as if there were base current. As with an ordinary transistor, you apply collector voltage to put the phototransistor in the active region. In many cases no separate base lead is actually brought out. Optically coupled devices are typically insulated to withstand several thousand volts between input and output.

Figure 5.37 shows a couple of ways to use an optically coupled transistor in a high-voltage supply. In the first circuit, phototransistor Q_2 shuts off pass transistor Q_3 when the output rises too high. In the second version, for which only the pass transistor circuitry is shown, phototransistor Q_2 increases the output voltage when driven, so the error amplifier inputs should be reversed. Both circuits generate some output current through the pass transistor biasing circuit, so some load from output to ground is needed to keep the output voltage from rising under no-load conditions. The output divider resistors can do the job, or a separate bleeder resistor can be connected across the output, which is always a good idea in a high-voltage supply.

□ *Floating regulator*

Another way to avoid applying large voltages to the control components of a high-voltage power supply is to "float" the control circuitry at the pass transistor potential, comparing the drop across its own voltage reference with the drop down to ground. The excellent MC1466 regulator chip is intended for this kind of application, which normally requires an auxiliary low-current floating power supply to provide dc (20–30V) for the chip itself. The output voltage is limited only by the pass transistors and the insulation of the auxiliary

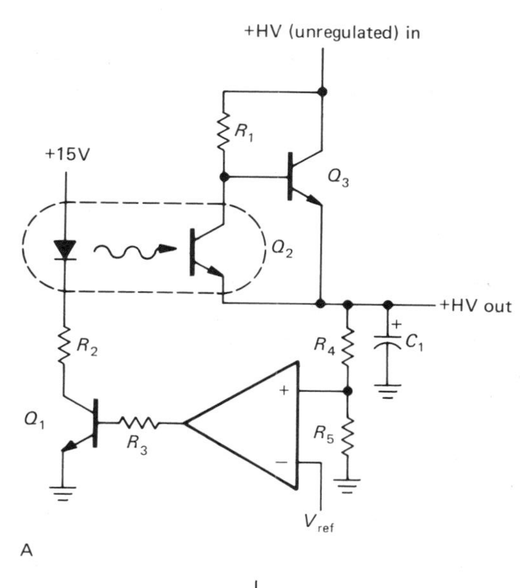

A

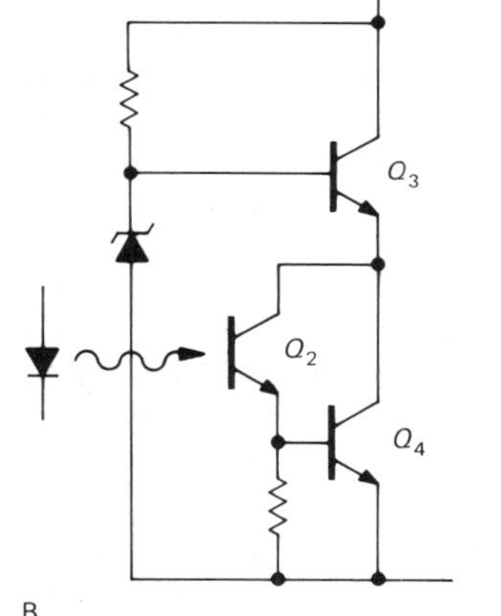

B

Figure 5.37
Optoisolated high-voltage regulator.

supply. The MC1466 features very good regulation and precise current-limiting circuitry and is well suited for accurate "laboratory" power supplies.

An elegant way to rig up a floating regulator is provided by the recently introduced LM-10 op-amp/voltage reference combination, a remarkable breakthrough in chip technology from the legendary Widlar (see Section 5.01) that will operate from a single 1.2 volt supply. Such a chip can be powered from the base-emitter drop of a Darlington pass transistor! Figure 5.38 shows an example. If you enjoy analogies, think of a giraffe who measures his height by looking at the distance to the ground, then stabilizes it by craning his neck accordingly.

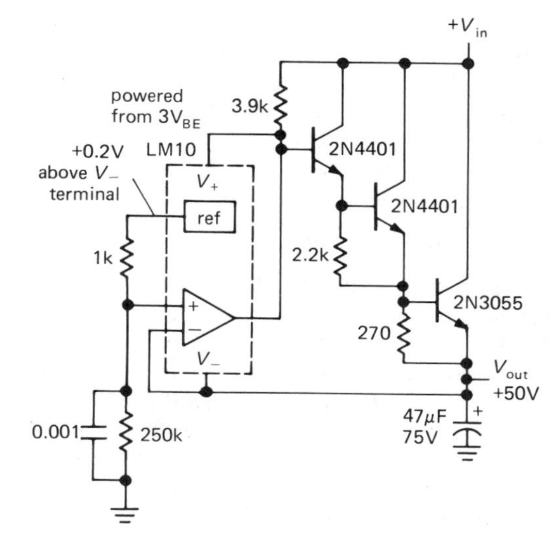

Figure 5.38

□ *Transistors in series*

Figure 5.39 shows a trick for connecting transistors in series to increase the breakdown voltage. Driver Q_1 drives series-connected transistors Q_2–Q_4, which share the large voltage from Q_2's collector to the output. The base resistors are chosen small enough to drive the transistors to full output

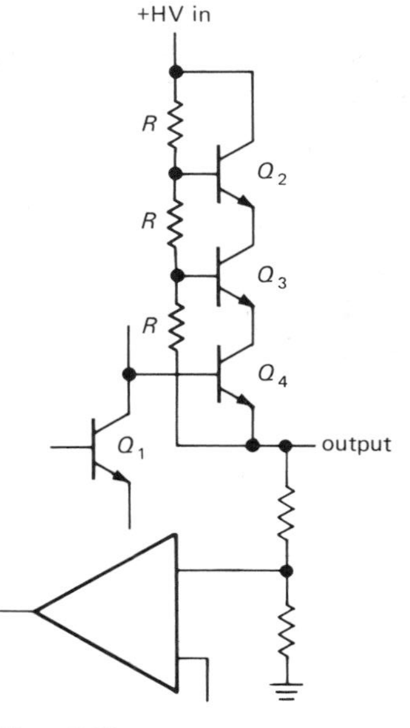

Figure 5.39

current. As in the preceding circuits, the bias resistors give some output current even when the transistors are cut off, so there must be some load to ground to prevent the output from rising above its regulated voltage. Watch out for resistor voltage and power ratings with this sort of circuit.

□ *Regulating the input*

Another technique sometimes used in high-voltage supplies is to regulate the input rather than the output. This is usually done with high-frequency dc-to-dc switching supplies, since attempting to regulate the 60Hz ac input will result in poor regulation and plenty of residual ripple. Figure 5.40 shows the general idea. T_1 and associated circuitry generate unregulated dc at some manageable voltage, say 40 volts. A high-frequency square-wave power oscillator runs from this, with its output full-wave-rectified and filtered. This filtered dc is the output, which is sampled and fed back to control the oscillator's duty cycle or amplitude in response to the output voltage. Since the oscillator runs at high frequency, the response is rapid, and its rectified waveform is easy to filter, especially since it is a full-wave-rectified square wave. T_2 must be designed for high-frequency operation, since ordinary laminated-core power transformers will have excessive core losses. Suitable transformers are built with iron powder, ferrite, or "tape-wound" toroidal cores and are much smaller and lighter than conventional power transformers of the same power rating. No high-voltage components are used, except, of course, for the output bridge rectifier and capacitor.

□ 5.21 Switching regulators

Conventional series pass regulators are the best choice for most power-supply requirements, but for high-current supplies the large pass transistor power dissipation makes the supply large (and hot!). For applications where this is undesirable, the solution is the so-called switching regulator (sometimes called a periodic energy-transfer regulator), in which the pass transistor is switched rapidly (5–50kHz typ) from saturation to open circuit. The output (a nasty square wave going from ground to the unregulated input voltage) is smoothed with a low-pass filter made from a series inductor and a capacitor to ground (actually, it is more accurate to think of the inductor as an energy-storage device rather than as part of a low-pass filter). The use of high-frequency switching makes it relatively easy to filter. As with conventional regulators, you compare the dc output with a stable reference, but now you modify the "duty cycle" of the switching waveform (the ratio of ON time to OFF time), instead of the base drive current, to change the output voltage in response to the error signal.

Switching regulators have very low power dissipation, because the pass transistor

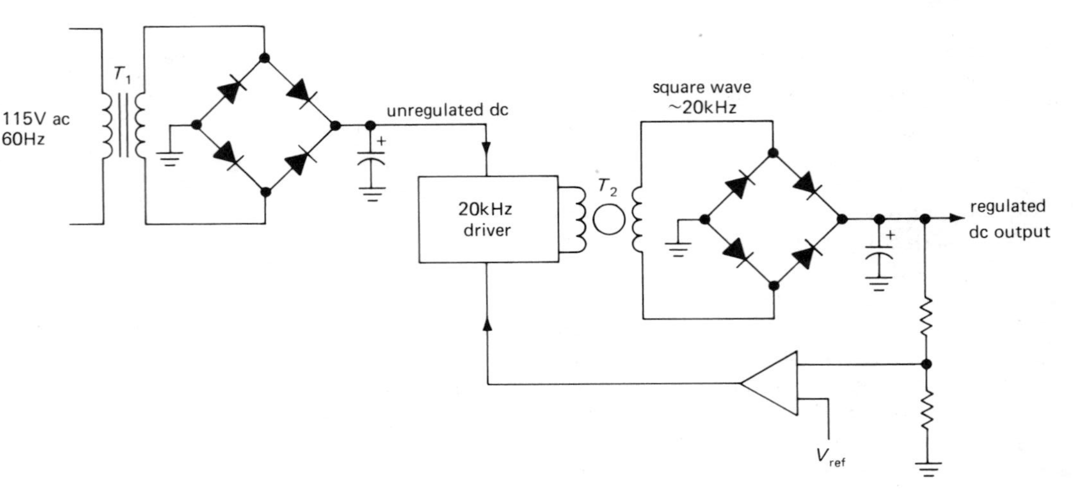

Figure 5.40

either is in saturation or is nonconducting, and inductors do not dissipate any power except for losses due to their winding resistance and core losses, both of which can be made small. The result is a high-current regulated supply with very high efficiency, with a curious property: In a switching regulator whose output is at a lower voltage than the unregulated input, the current drawn from the unregulated supply is less than the output current delivered to the load. In fact, the unregulated supply voltage can be much larger than the regulated output. For instance, the Nova 3/4 minicomputer uses a switching supply that generates all its regulated voltages, including +5 volts at 20 amps, from a single +35 volt high-current unregulated supply; with a conventional series pass regulator such a high unregulated input voltage would be unimaginable, since it would correspond to 600 watts of transistor dissipation at full load. With switching regulators you can generate output voltages greater than the unregulated input voltage, and it is even possible to generate outputs of polarity opposite to that of the unregulated input, a trick that's also used in the minicomputer supply mentioned earlier.

The details of designing switching regulators are beyond the scope of this book. In practice, their design can be tricky, and they have some serious disadvantages, most notably the generation of electrical interference from the high-current switching transients, problems of no-load performance, and, often, the generation of an objectionable high-level audible screech. We suggest that if you need a switching regulator, buy it, paying particular attention to its noise filtering, its short-circuit protection, and its fault-condition protection.

□ 5.22 dc-to-dc Converters

On some occasions it is convenient or essential to be able to convert one dc voltage to another. For instance, in digital systems it is common to have +5 volts available at high current nearly everywhere, but you may wish to operate a single op-amp that needs only a few milliamps of ±15 volt dc. Or you may wish to operate some high-power equipment from a 12 volt automobile storage battery.

□ *Oscillator plus transformer*

One of the most popular methods of dc-to-dc conversion, especially where considerable output power is needed, is to use a high-power square-wave oscillator and transformer. The two can be combined in a self-excited power oscillator, as shown in Figure 5.41. Q_1 and Q_2 form a push-pull oscillator that drives transformer T_1 into magnetic saturation each half cycle. It is common to use a "square-loop" toroidal core for the transformer, although conventional laminated-core materials will work, at lower efficiency. The resistors and diode in the base are to ensure oscillation at start-up. The dots next to the windings are the universal way to indicate relative winding polarites. Conventional series pass regulators are usually used, once the filtered but unregulated dc has been generated from the transformer secondary. Filtering is easy because the output is a full-wave-rectified square wave, already almost perfect dc. The oscillator is usually operated at frequencies of at least a few kilohertz in order to reduce the transformer core size. At these frequencies special core materials or fabrication techniques are used to reduce core losses. Complete dc-to-dc converters are available from many of the regular power-supply manufacturers, particularly for popular voltages like +5 volts dc input, ±15 volts output.

Warning: In selecting a commercial dc-to-dc converter, it is important to watch out for some nasty problems that can crop up. Check to see that the converter (a) will start up into your worst-case load, (b) won't blow up when the output is shorted, and (c) doesn't put spikes or other undesirable stuff on the input or output dc lines.

□ *Transformerless conversion*

For moderate current applications it is possible to make transformerless dc-to-dc converters, using just a square-wave oscillator and rectifier (or voltage multiplier). Figure 5.42 shows an example. This circuit generates +5 volt and −5 volt regulated outputs from a single +12 volt unregulated input (a storage battery, for instance). The 555 generates 1kHz square waves going from ground to +12 volts. This is rectified by D_1, D_2, and the two capacitors to give an unregulated source of about −10 volts dc. The dual-tracking regulator uses the pair of unregulated voltages to give ±5 volts regulated output.

An example of a ready-made IC for transformerless dc voltage conversion is the Intersil ICL7660. It uses MOS switches (see Chapter 6) in a "flying capacitor" technique to generate a negative dc output voltage equal in magnitude to the positive dc voltage used to power it, e.g., —5 volts output from a positive 5 volt supply. It operates from 1.5V to 10V, it has a conversion efficiency approaching 100%, and it provides up to 40mA of output current. This convenient chip comes in a mini-DIP package and requires just two external components (capacitors). By stacking several ICL7660s you can generate a negative output voltage equal to a multiple of the supply voltage.

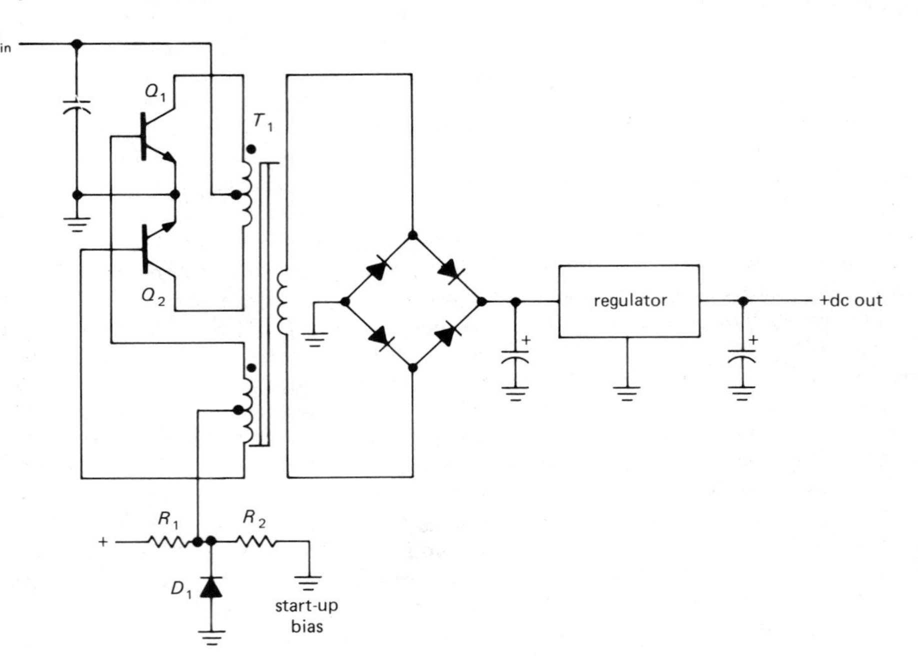

Figure 5.41

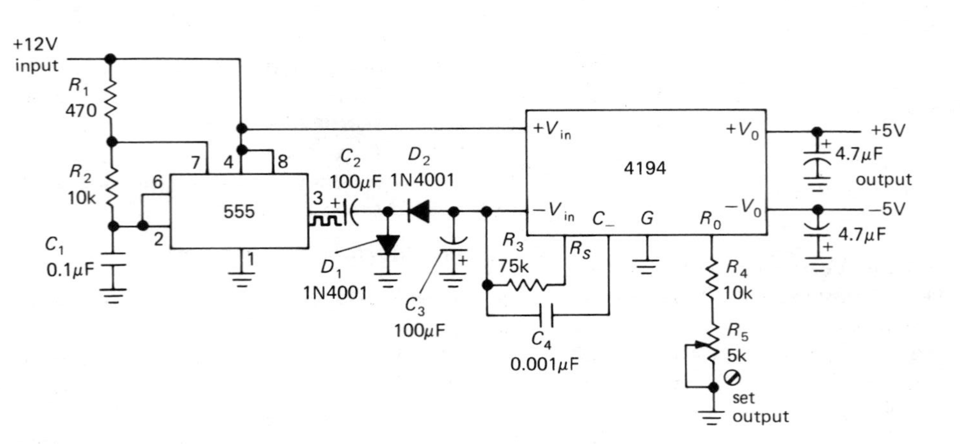

Figure 5.42

□ 5.23 Energy-storage inductor

The techniques illustrated below are examples of switching regulators, in one incarnation or another. They all share the property of storing energy in an inductor, then releasing it into a filter capacitor and load (see the general comments in Section 5.21).

□ *Step-up regulator*

A bizarre technique of limited application is the use of a "flyback" inductor, in which the $\frac{1}{2}LI^2$ of energy stored in the inductor's magnetic field is converted to dc. Figure 5.43 shows an example. This circuit generates regulated +12 volts from a +5 volt supply, commonly available in logic circuits. This particular circuit takes advantage of the convenient TL497 switching regulator IC, although a similar circuit can be constructed from standard components like comparators and oscillators. The internal oscillator periodically generates 40μs pulses, turning Q_1 on and pulling L_1 to ground. The current through L_1 is a ramp during the time of this pulse [$V = L(dI/dT)$], and when Q_1 is turned off, the inductor's voltage rises to maintain the current.

With such a scheme the no-load output voltage will rise continuously, since the inductor just keeps dumping charge into C_2, whatever its voltage (eventually, of course, D_1, Q_1, or C_2 will break down). In this circuit the output voltage is sampled and compared with a 1.2 volt (bandgap) reference. The output of the comparator inhibits the pulse generator if the output voltage is high enough, thus regulating the output without a pass transistor. In this circuit, D_1 blocks reverse current during the 40μs pulses, and R_1 provides the ever-important current limiting. C_1 sets the pulse width, and C_2 filters the output to maintain low ripple. Since regulation occurs only in spurts, a relatively large output capacitor is necessary.

□ *Flyback supply*

For high-voltage supplies, flyback circuits are sometimes built with a transformer. The primary is driven as described previously, but the large turns ratio gives a high-voltage-rectified output from the secondary. Figure 5.44 shows the idea. Q_1 is driven by pulses, pulling the primary to ground. It may be self-excited, as in the first dc-to-dc converter we mentioned earlier. D_1 is a "damper" diode that prevents Q_1's collector from rising too high during the flyback. D_2, connected to the high-voltage secondary winding, rectifies the output. Flyback circuits like this are used in television and cathode-ray tube (CRT) video displays to generate the high anode voltage, typically 10kV at a few microamps. The circuit is operated at frequencies of 15kHz or so, which means

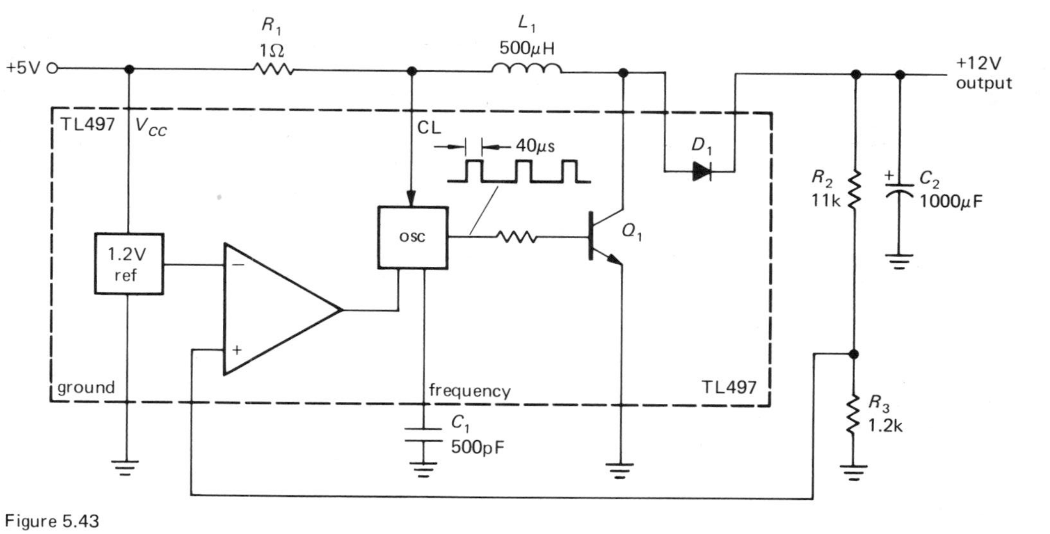

Figure 5.43
Step-up converter with integrated circuit.

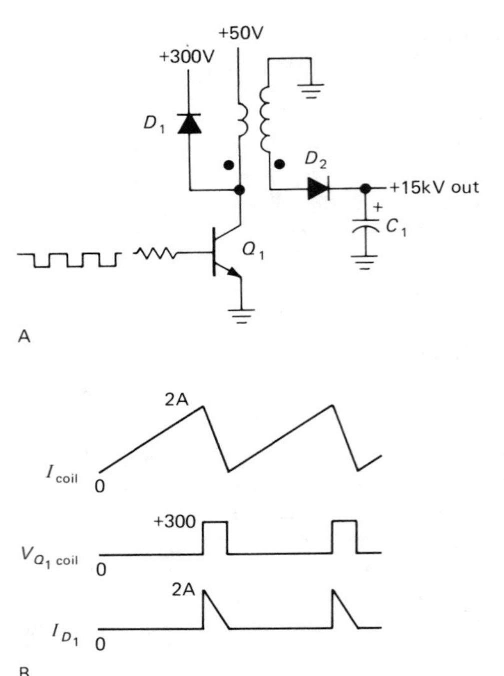

Figure 5.44

that filter capacitor C_1 can be as small as a few hundred picofarads. The rising current ramp during the time Q_1 is on (as much as several amps) is often used to generate the magnetic field to scan the CRT beam: in such cases the flyback frequency sets the horizontal scan rate. A related circuit is the so-called blocking oscillator, which generates its own excitation pulses.

□ *Micropower regulators*

Another application for flyback converters is in micropower circuits such as those in electronic watches, calculators, and other devices that must operate for extended periods from small batteries. For instance, Figure 5.45 shows a micropower circuit for a heart pacemaker regulated supply that converts an input voltage in the range +5 volts down to +3 volts (as the battery ages) to a regulated +5.5 volt supply. The power supply has a quiescent current of 1μA and provides line and load regulation of about 5% with 85% conversion efficiency under full load for all battery voltages. A conventional circuit using an oscillator, doubler, and series pass regulator would be far less efficient because of regulator losses following the higher unregulated dc voltage. The flyback technique is effectively like a variable-ratio voltage multiplier, which yields extremely high efficiency, making it an attractive technique for micropower applications.

The 2N6028 programmable unijunction transistor (PUJT) is a versatile relaxation

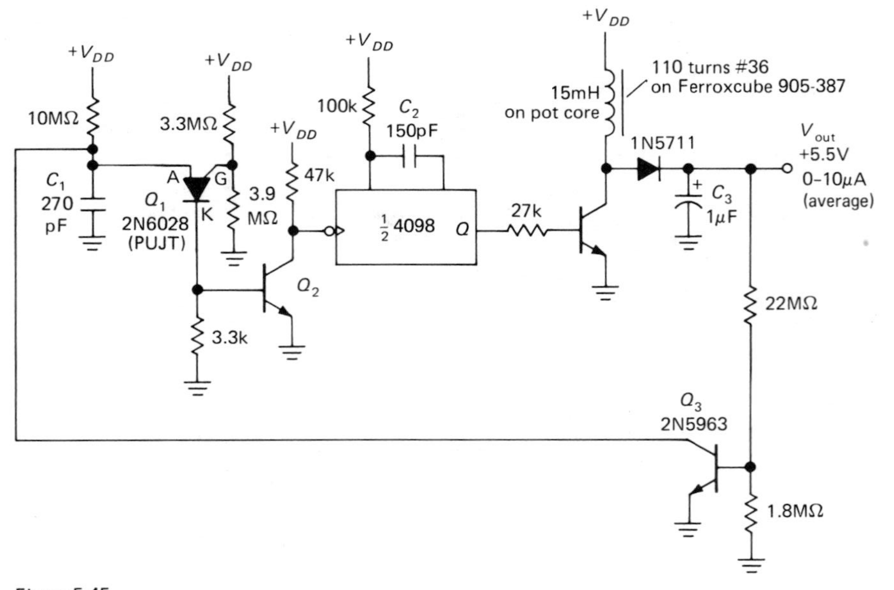

Figure 5.45
Micropower switching regulator.

oscillator component. Its sense terminal (the anode) draws no current until its voltage exceeds the gate programming voltage by a diode drop, at which point it goes into heavy conduction from anode to cathode, discharging the capacitor.

In this circuit Q_3 senses the output voltage and robs charging current from C_1, reducing the energy-transfer pulsing rate of the inductor as necessary to maintain the desired output voltage. Note the large resistor values throughout the circuit. Temperature compensation is not necessary here because the circuit operates in a stable 98.6°F oven.

□ *Step-down switching regulator*

Switching supplies can, of course, be built in the familiar step-down regulator configuration we have been discussing throughout most of the chapter. The advantages of a switching regulator in this kind of circuit are small size (low power dissipation) and the ability to generate a low output voltage from a relatively high unregulated input voltage without the enormous power dissipation of a conventional linear series pass regulator. Figure 5.46 shows an example. This is a regulator to provide +5 volts at up to 5 amps, beginning with an unregulated source of 24 to 28 volts. A conventional linear regulator will dissipate 100 watts and will require 5 amps input current, whereas a switching regulator requires less than 2 amps input current. To put it in other terms, the linear regulator is 20% efficient, whereas the switching regulator is about 70% efficient.

The 78S40 is a "universal switching regulator subsystem," including an internal 1.25 volt (bandgap) reference, comparator, op-amp, oscillator, switching transistor, and diode. In the configuration shown, the oscillator is set by C_T to generate an output pulse train with 200μs ON time and 20μs OFF time, i.e., at about a 5kHz rate. Its output is inhibited by the comparator's output, according to the final output voltage. An external *pnp* switching transistor Q_3 is connected as a Sziklai pair (complementary Darlington) with internal Darlington Q_1Q_2. Both Q_3 and D_2 are relatively fast, to maintain high efficiency at these switching speeds. The particular values chosen for the inductor, timing capacitor, and filter capacitor were determined by design tables

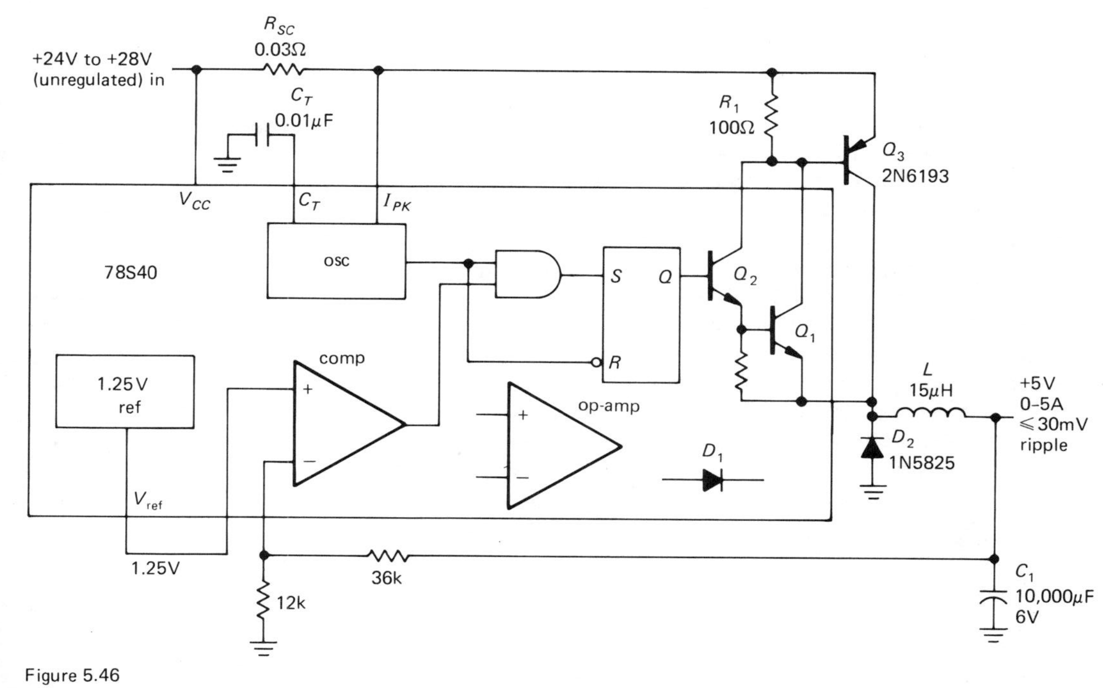

Figure 5.46
Switching converter with IC regulator.

provided on the 78S40 data sheet. Note that the internal op-amp and diode are unused, the latter because its peak current rating of 1.5 amps is too low in this application.

□ 5.24 Constant-current supplies

In Sections 2.06 and 2.13 we described some methods for generating constant currents within a circuit, including voltage-programmed currents with floating or grounded loads and various forms of current mirrors. Later, in Section 6.06, we will discuss the use of FETs to construct some simple current-source circuits, including "current-regulator diodes" (a FET with gate tied to source) such as the 1N5283 series. There is often a need for a flexible constant-current supply as a complete instrument, and we will look at some of the more successful circuit techniques.

□ *Three-terminal regulator*

Perhaps the easiest method is to use a 3-terminal regulator with the common terminal floating: Connect a resistor from output to the common (ground) terminal; then connect the load from the common terminal to ground. The regulator maintains a fixed voltage across the resistor, and thus a constant current; that current then flows through the load, in addition to the (small) operating current of the regulator itself. This type of current source has some disadvantages: At low output currents the regulator's operating current, typically a few milliamps, contributes a large error, and at high output currents the drop across the resistor (5V minimum for available regulators) results in unnecessary power dissipation. The best solution is to use a regulator like the LM317, which maintains a mere 1.25 volts from output to common terminal and operates with 50μA current at the common terminal. This circuit was discussed in Section 5.18.

□ *Supply-line current sensing*

A simple technique that yields good performance involves constructing a conventional series pass regulator, with current sensing at the input to the pass transistor (Fig. 5.47). R_2 is the current-sensing resis-

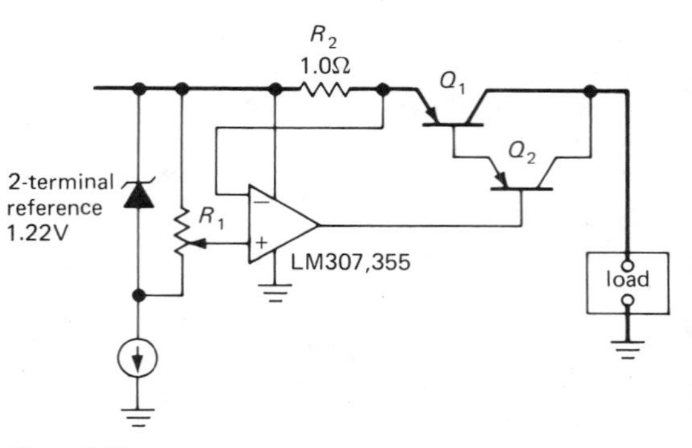

Figure 5.47

tor, and preferably it should be a low-temperature-coefficient type. For very high current or high-precision applications, you should use a 4-wire resistor, intended for current-sensing applications, in which the sensing leads are connected internally; the sensing voltage does not depend on the connection resistance of the joints to the current-carrying leads, which, for clarity, are drawn with heavy lines in this schematic.

For this circuit you must use an op-amp that has an input common-mode range all the way to the positive supply. The 301 and 307 have that virtue (as do the 355–357 FET op-amps). Note that *npn* output transistors (connected as followers) can be used instead of *pnp* if the input connections to the op-amp are reversed. However, the current source will then have a low output impedance at frequencies approaching f_T of the op-amp loop, since the output is actually an emitter follower. This is a common error in current-source design, since the dc analysis shows correct performance.

Since the output current includes the pass transistor base current, a Darlington should be used to minimize that error. This error term can be eliminated completely by using a new class of devices known as vertical MOSFETs, or simply VFETs, which are now available commercially (see Section 6.20). They behave like MOSFETs, which we will discuss in the next chapter, but they can handle up to 10 amps or more of current. The important point is that FETs draw no gate current, and so they can be used as pass transistors with zero current into the control terminal (the gate). They are attractive for other areas of design, also, including audio and high-frequency applications.

□ *Return-line current sensing*

A good way to make a precise current source is to sense the voltage across a precision resistor directly in series with the load, since this makes it easier to meet the simple criterion for eliminating current-source errors due to base drive currents; the base drive current must either pass through both the load and sense amplifier or pass through neither. However, to meet this criterion it is necessary to "float" either the load or the power supply by at least the voltage drop across the current-sensing resistor. Figure 5.48 shows a couple of circuits that use floating loads.

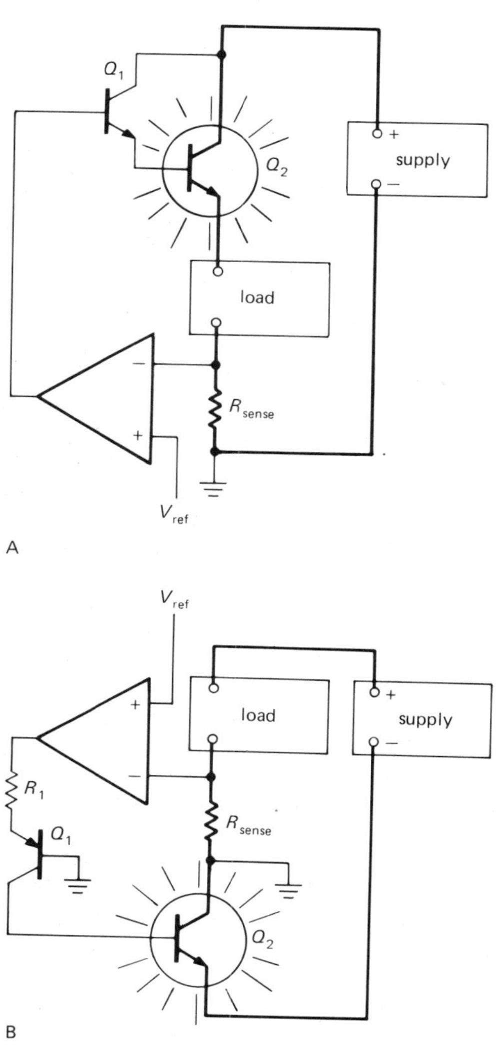

Figure 5.48

The first circuit is a conventional series pass circuit, with the error signal derived from the drop across the small resistor in the load's return path to ground. The high-current path is again drawn with bold lines. The Darlington connection is used here not

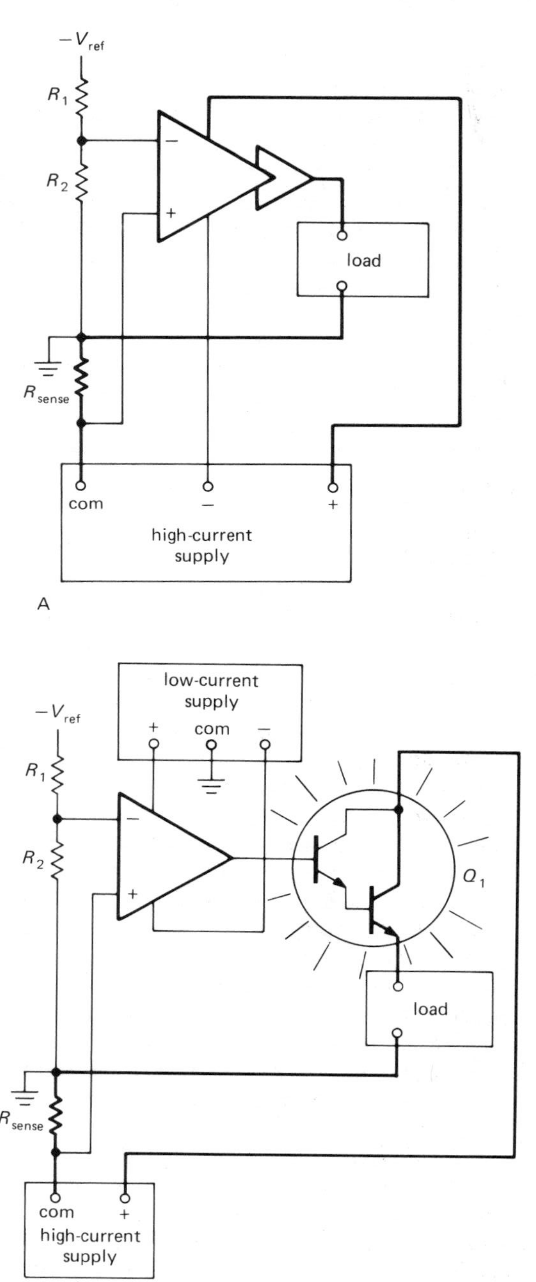

Figure 5.49

to avoid base-current error, since the actual current through the load is sensed, but rather to keep the drive current down to a few milliamps so that ordinary op-amps can be used for the error amplifier. The sensing resistor should be a precision power resistor of low temperature coefficient, preferably a 4-wire resistor. In the second circuit the regulating transistor Q_2 is in the ground return of the high-current supply. The advantage here is that its collector is at ground, so you don't have to worry about insulating the transistor case from the heat sink.

In both circuits R_{sense} will normally be chosen to drop a volt or so at typical operating currents; its value is a compromise between op-amp input offset errors at one extreme and a combination of reduced current-source compliance and increased resistor dissipation at the other. If the circuit is meant to operate over large ranges of output current, R_{sense} should probably be a set of precision power resistors, with the appropriate resistor selected by a range switch.

□ *Grounded load*

If it is important for the load to be returned to circuit ground, a circuit with floating supply can be used. Figure 5.49 shows two examples. In the first circuit, the funny-looking op-amp represents an error amplifier with a high-current buffer output, run from a single split supply (it can be a 723, for instance, for currents up to 150mA). The high-current supply has a common terminal that floats relative to circuit ground, and it is important that the error amplifier (or at least its buffer output) be powered from the floating supply so that base drive currents return through R_{sense}. An additional low-current supply with grounded common would be needed if other op-amps, etc., were in the same instrument. A negative reference (relative to circuit ground) programs the output current. Note the polarity at the error amplifier inputs.

The second circuit illustrates the use of a second low-power supply when an ordinary op-amp is used as error amplifier. Q_1 is the outboard pass transistor, which must be a Darlington, since the base current returns through the load but not through the sense resistor. The error amplifier is now powered from the same split supply with grounded common that powers the rest of the instrument. This circuit is well suited as a simple bench-instrument current source, with the low-current split supply built in and the high-current supply connected externally. The latter's voltage and current capability will be chosen to fit each application.

SELF-EXPLANATORY CIRCUITS

5.25 Circuit ideas

Figure 5.50 presents a variety of current ideas, mostly taken from manufacturers' data sheets.

5.26 Bad circuits

Figure 5.51 presents some circuits that are guaranteed not to work. Figure them out, and you will avoid these pitfalls.

ADDITIONAL EXERCISES

(1) Design a regulated supply to deliver exactly +10.0 volts at currents up to 10mA using a 723. You have available a 15 volt (rms) 100mA transformer, diodes by the bucketful, various capacitors, a 723, resistors, and a 1k trimmer pot. Choose your resistors so that they are standard (5%) values and so that the range of adjustment of the trimmer will be sufficient to accommodate the production spread of internal reference voltages (6.80V to 7.50V).

(2) Design +5 volt 50mA voltage regulators, assuming +10 volt unregulated input, using the following: **(a)** zener diode plus emitter follower; **(b)** 7805 3-terminal regulator; **(c)** 723 regulator; **(d)** 723 plus outboard *npn* pass transistor; use foldback current limiting with 100mA onset (full-voltage current limit) and 25mA short-circuit current limit; **(e)** a 317 3-terminal adjustable positive regulator; (f) discrete components, with zener reference and feedback. Be sure to show component values; provide 100mA current limiting for (a), (c), and (f).

(3) Design a complete +5 volt 500mA power supply for use with digital logic. Begin at the beginning (the 115V ac wall socket), specifying such things as transformer voltage and current ratings, capacitor values, etc. To make your job easy, use a 7805 3-terminal regulator. Don't squander excess capacitance, but make your design conservative by allowing for ±10% variation in all parameters (power-line voltage, transformer and capacitor tolerances, etc.). When you're finished, calculate worst-case dissipation in the regulator.

Next, modify the circuit for 2 amp load capability by incorporating an outboard pass transistor. Include a 3 amp current limit.

Circuit ideas

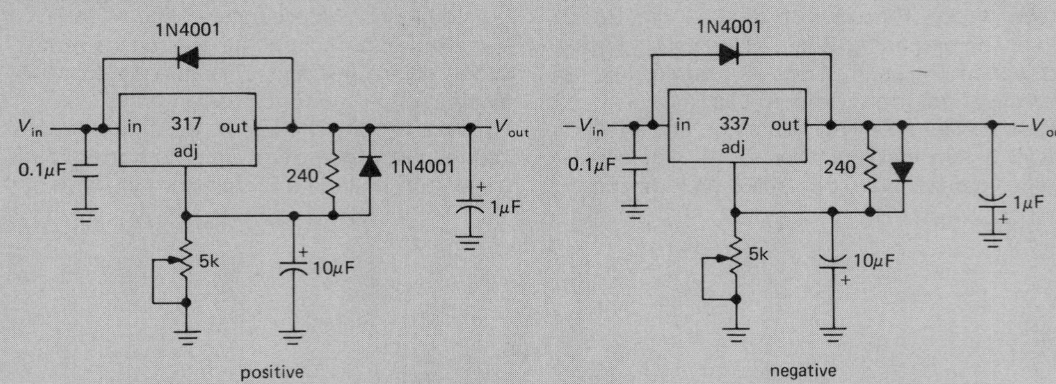

A. 3-terminal regulators with improved ripple rejection (diodes protect against input/output shorts)

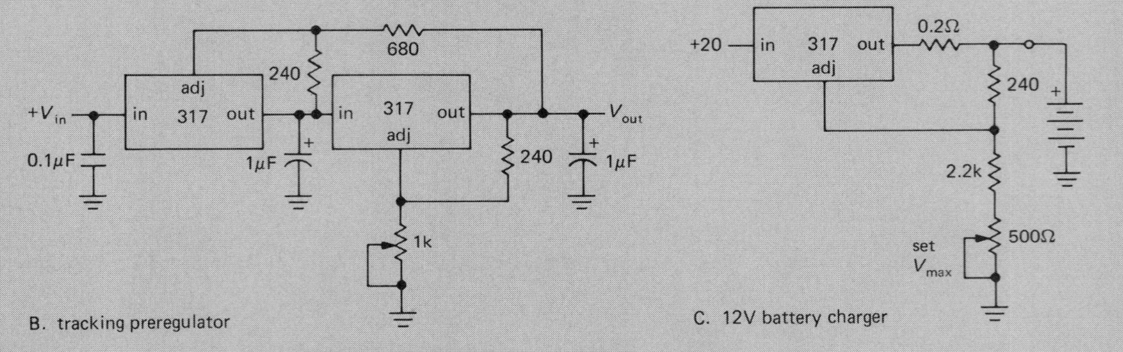

B. tracking preregulator

C. 12V battery charger

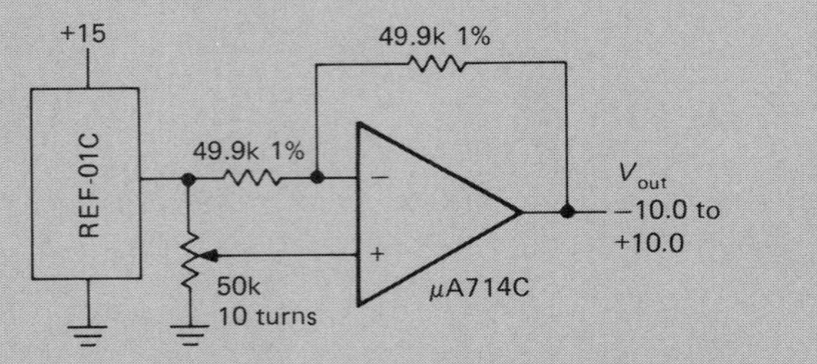

D. adjustable stable bipolar voltage reference

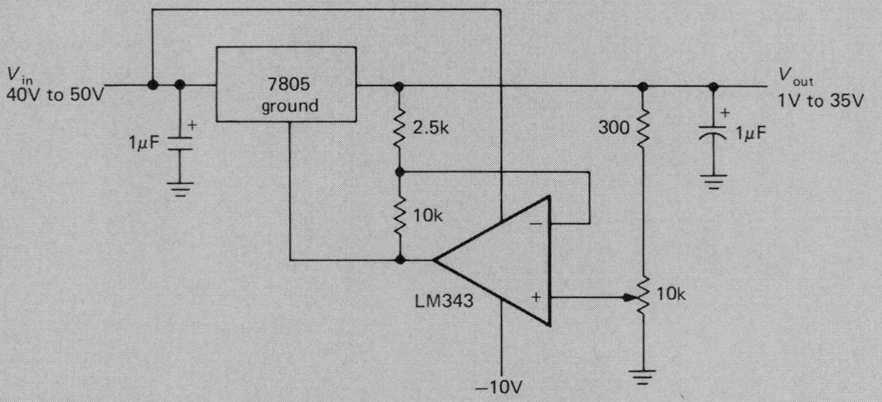

E. wide-range regulator (1V to 35V)

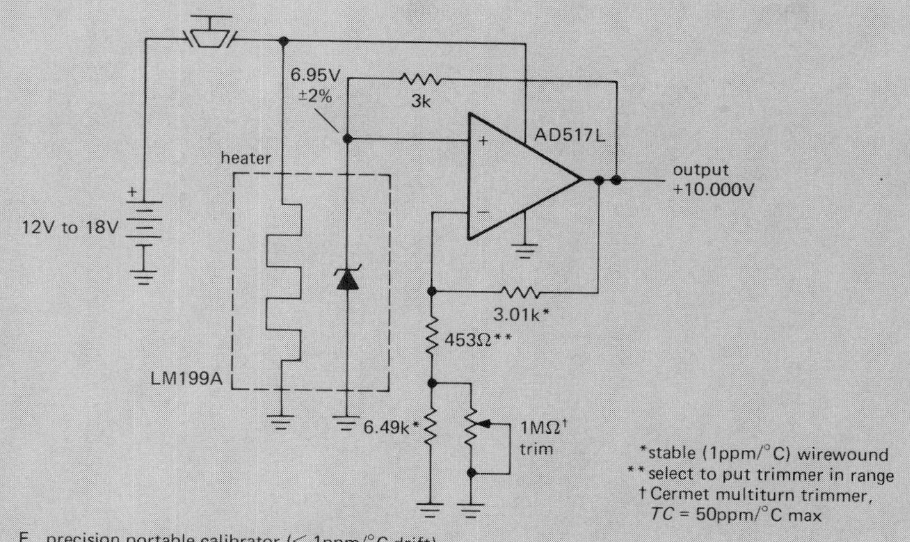

F. precision portable calibrator (< 1ppm/°C drift)

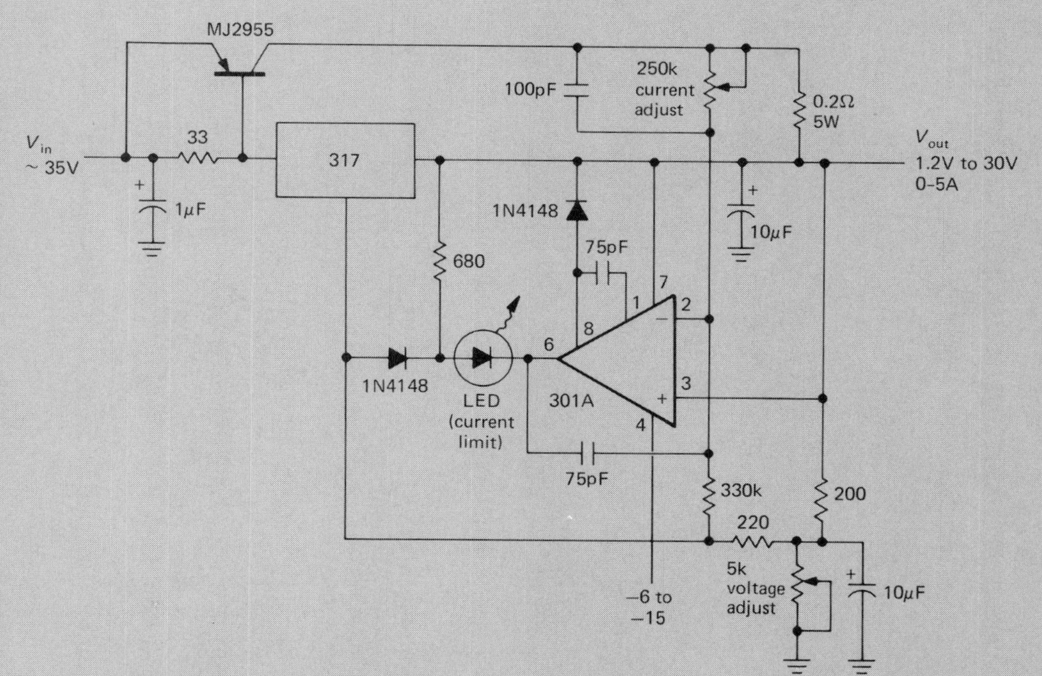

G. constant voltage/constant current supply

Figure 5.50

Bad circuits

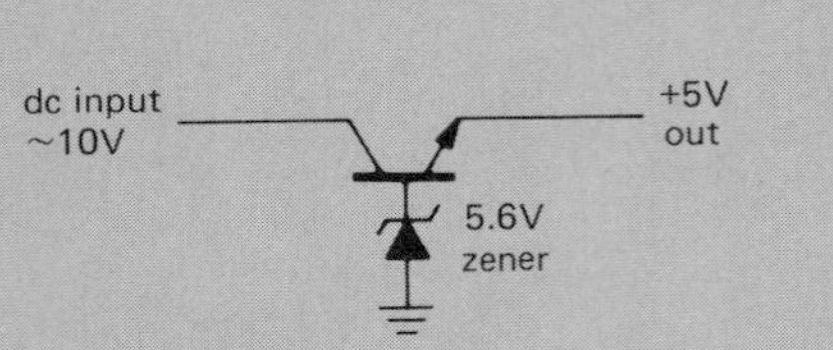

A. simple regulated supply

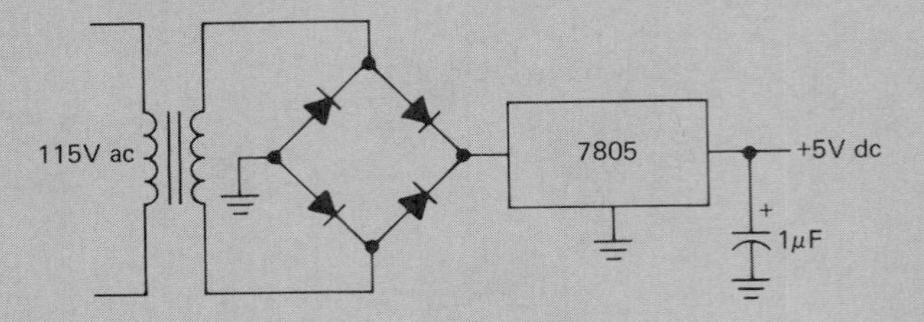

B. +5V supply

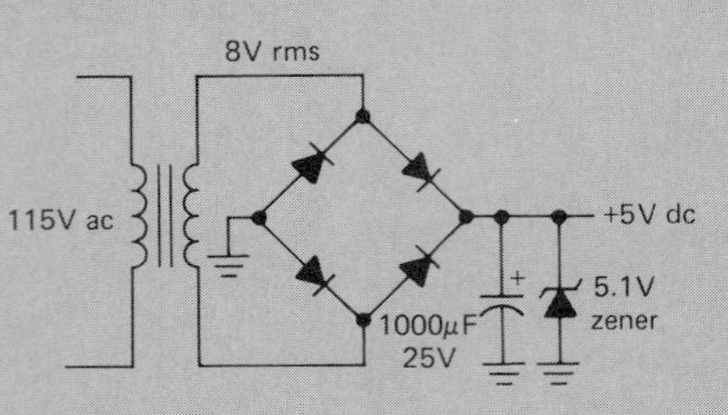

C. +5V supply

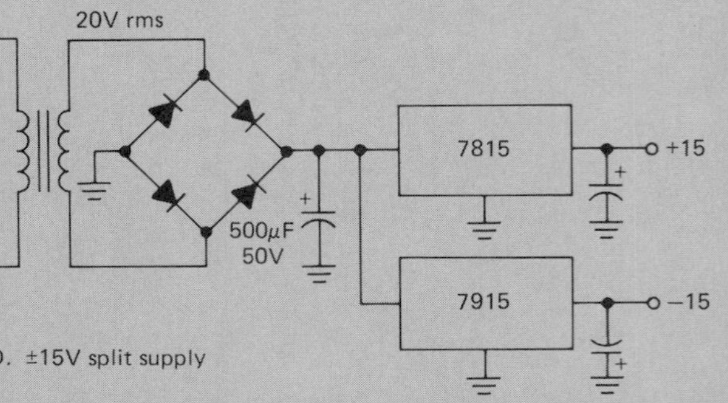

D. ±15V split supply

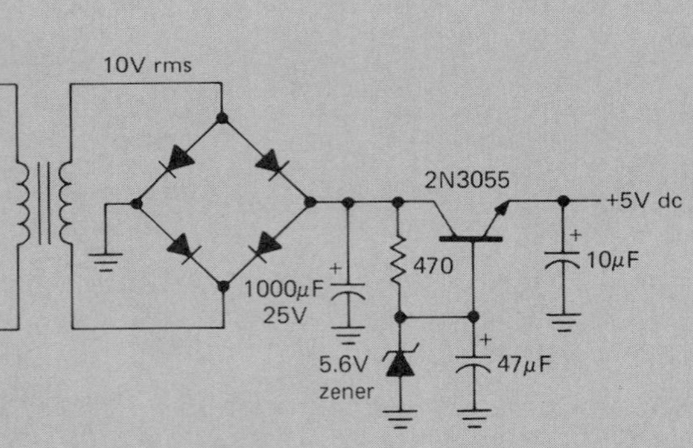

E. +5V supply

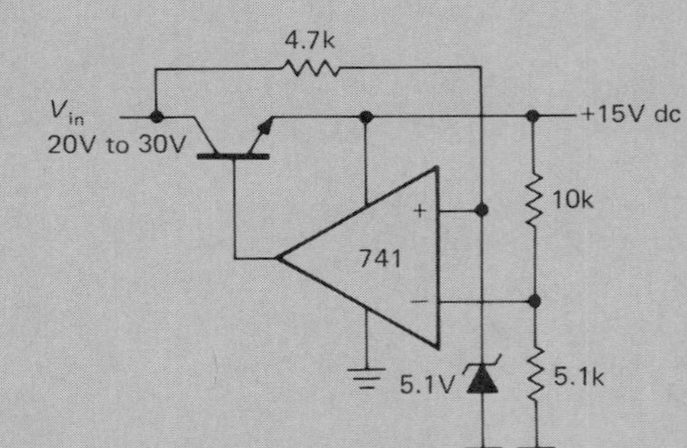

F. +15V regulator

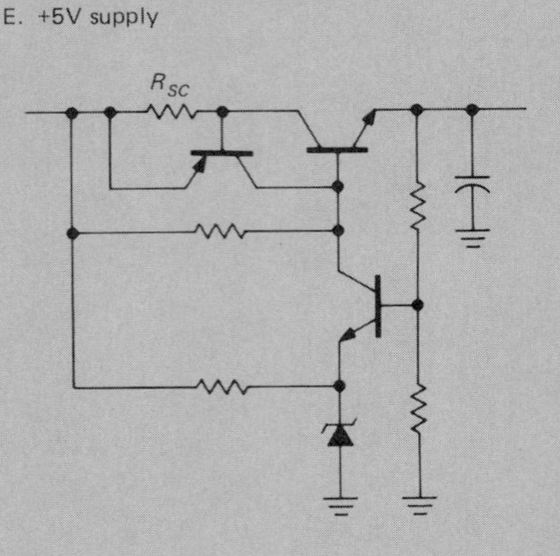

G. regulator with upstream current limiter

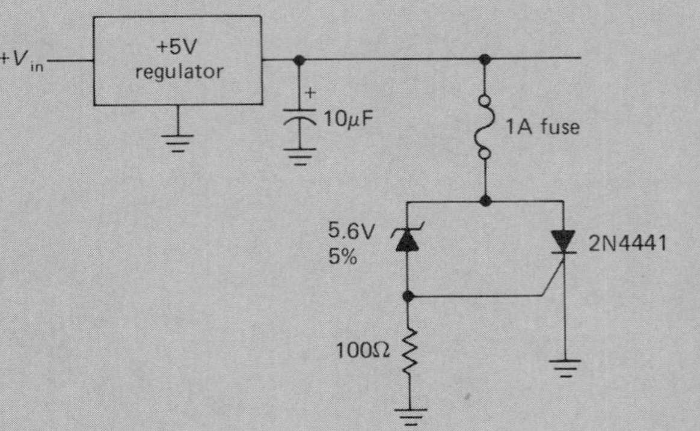

H. crowbar with SCR protection

Figure 5.51

FIELD-EFFECT TRANSISTORS

CHAPTER 6

Field-effect transistors (FETs) are transistors with properties quite different from those of ordinary transistors, which are sometimes called bipolar transistors to distinguish them from FETs. As the name suggests, FETs work by controlling a current with an *electric field* produced by an applied voltage, rather than with a base current. The result is that the control electrode (the *gate*) draws virtually no current, except for leakage. The resulting high input impedance (which can be greater than $10^{14}\Omega$) is essential in many applications, and in any case it makes circuit design simple and fun. For applications like analog switches and amplifiers of ultrahigh input impedance, FETs have no equal. FETs are very useful as voltage-controlled resistors and as current sources. Since many FETs can be constructed in a small area, they are especially useful for large-scale-integration (LSI) digital circuits such as calculator chips, microprocessors, and memories. In addition, the recent availability of high-current FETs (10A or more) means that FETs can replace bipolar transistors in many applications, often providing simpler circuits with improved performance.

You might use a FET instead of a bipolar transistor in a circuit in order to get superior performance. However, you often have to take extra care to be sure you get that improved performance. FETs have some subtleties, and we are going to treat them rather carefully and thoroughly here. Stick with us through the first few sections of this chapter, and you will be rewarded with some pleasing circuits and powerful techniques.

FET CHARACTERISTICS

There are two fundamental varieties of FETs, each of which is available in two polarities (*n*-channel, which is like *npn,* and *p*-channel, which is like *pnp*), namely, junction FETs or JFETs and metal-oxide-semiconductor FETs or MOSFETs.

6.01 JFETs

In a JFET you begin with a conducting semiconductor bar, with the ends labeled drain (*D*) and source (*S*), in the middle of which you diffuse a gate (*G*). A voltage applied to the gate controls the conductivity of the bar, or "channel." The symbols for JFETs are shown in Figure 6.1. As the figure suggests, drain, source, and gate are analogous to collector, emitter, and base. An *n*-channel FET is normally operated with its drain more positive than its source, for

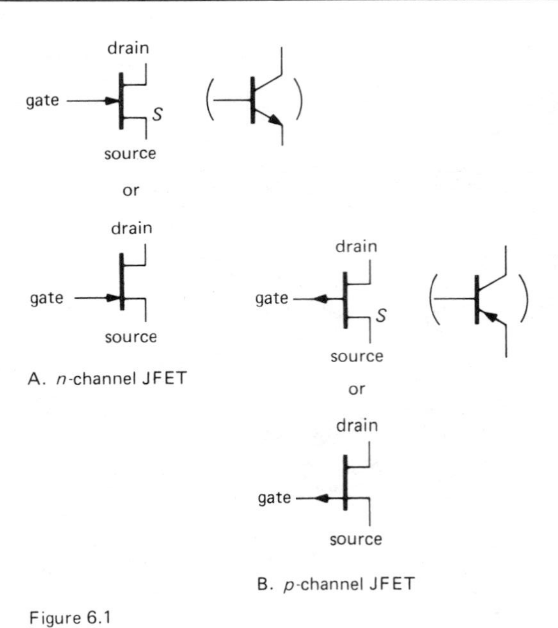

Figure 6.1

instance. However, unlike the situation with an *npn* transistor, current will flow from drain to source with the gate grounded, and the gate junction must be reverse-biased a few volts to cut off the drain current. A JFET is never operated with its gate forward-biased, so no current (except leakage) flows in the gate circuit. The circuit symbol for FETs doesn't distinguish source and drain, which is unfortunate. FETs are nearly symmetrical, but the gate-drain capacitance is designed to be less than the gate-source capacitance, for instance, thus making the drain the preferred output terminal. Sometimes you write an *S* and a *D* on the schematic, to clarify things, or use the alternate symbol in which the gate arrow is drawn opposite the source; otherwise you figure it out from the circuit context. The same ambiguity exists for MOSFETs.

6.02 MOSFETs

In the JFET the gate forms a diode junction with the drain-source channel. This limits performance because there is leakage current (as much as a nanoamp or so at ordinary temperatures, and considerably more at large drain voltages), and if the gate becomes forward-biased with respect to either source or drain, regular diode conduction occurs, with drastic reduction of the input impedance. In the *insulated-gate* FET, or MOSFET (metal-oxide-semiconductor FET, sometimes called IGFET), the gate region is separated from the conducting channel by a thin layer of SiO_2 (glass) grown onto the channel. This gate is truly insulated from the source-drain circuit (input resistance $\sim 10^{14}\Omega$), and it affects the source-drain current only by its electric field.

The symbols for MOSFETs are shown in Figure 6.2. The extra terminal is the "body,"

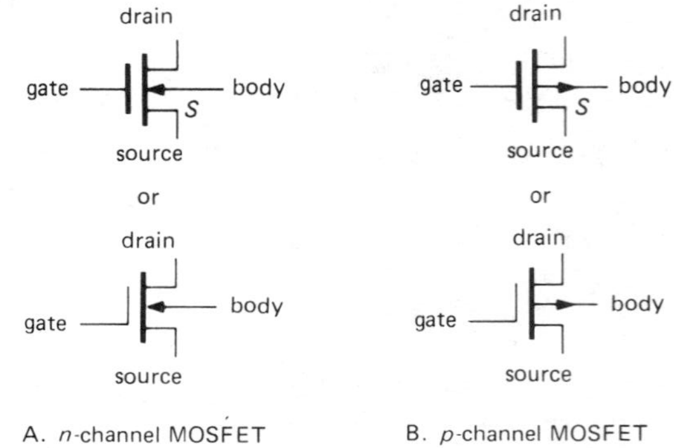

Figure 6.2

or "substrate." Since it forms a diode junction with the channel, it must be held at a nonconducting voltage. It can be tied to the source or to a point in the circuit more negative (positive) than the source for *n*-channel (*p*-channel) MOSFETs.

With MOSFETs the gate can go either polarity relative to the source without gate current flowing, since it is electrically insulated from the source-drain circuit. This often simplifies circuit design. It also allows the construction of two varieties of MOSFETs, the *depletion* and *enhancement* types. Depletion-mode MOSFETs conduct with the gate tied to the source (or forward-biased) and are cut off with the gate back-biased a few volts (just like JFETs). Enhancement-mode MOSFETs are cut off with the gate tied to the source (or back-biased) and conduct only when the gate is forward-biased.

A graph of drain current versus gate-source voltage, at a fixed value of drain voltage, may help clarify this distinction (Fig. 6.3). The enhancement-mode device draws no drain current until the gate is brought positive (these are *n*-channel FETs) with respect to the source, whereas the deple-

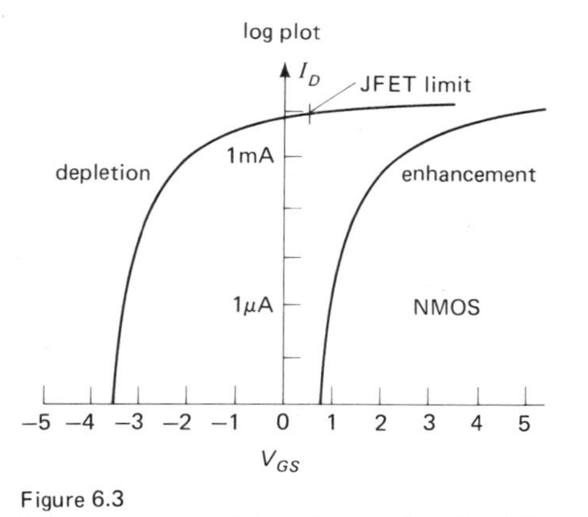

Figure 6.3
Enhancement- and depletion-mode FETs differ only by a gate-source voltage shift.

tion-mode device is operating at nearly its maximum value of drain current when the gate is at the same voltage as the source. In some sense the two categories are artificial, since the two curves are identical except for a shift along the V_{GS} axis. In fact, it is possible to manufacture "in-between" MOSFETs. Nevertheless, commercially available MOSFETs are labeled as enhancement or depletion, and the distinction is an important one when it comes to circuit design.

Note that JFETs are always depletion-mode devices and that the gate cannot be brought more than about 0.5 volt more positive (for *n*-channel) than the source, since the gate-channel diode junction will conduct.

6.03 Universal FET characteristics

A family tree (Fig. 6.4) and a map of input/output voltage (source grounded) (Fig. 6.5) may help simplify things. The different

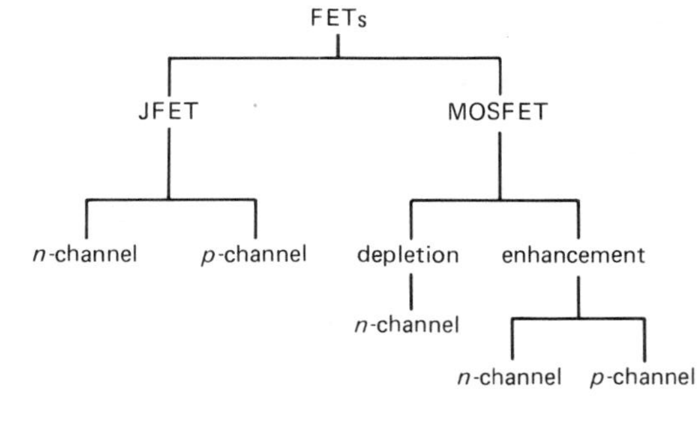

Figure 6.4

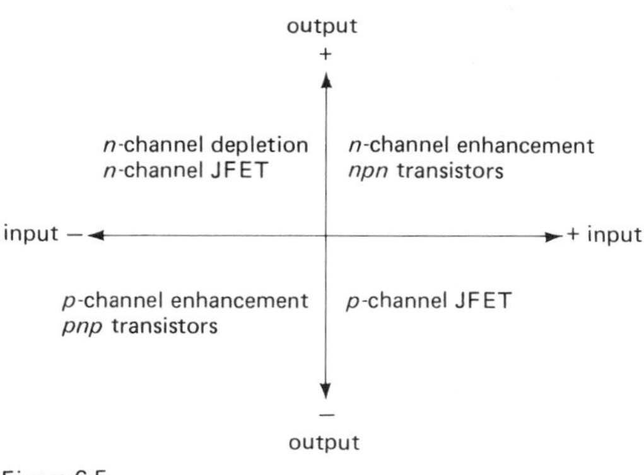

Figure 6.5

devices (including garden-variety bipolar *npn* and *pnp* transistors) are drawn in the quadrant that characterizes their input and output voltages when they are in the active region with source (or emitter) grounded. As you can see, there are five kinds of FETs normally used. However, you don't have to remember the properties of each, because they're all basically the same.

First, with the source grounded, a FET is turned on (brought into conduction) by bringing the gate voltage "toward" the active drain supply voltage. This is true for all five types. For example, an *n*-channel depletion-mode MOSFET uses a positive drain supply, as do all *n*-type devices. Thus a positive-going gate voltage tends to turn on the FET. The subtlety for depletion-mode devices is that the gate must be back-biased for zero drain current, whereas for enhancement-mode devices zero gate voltage is sufficient to give zero drain current.

Second, because of the near symmetry of source and drain, either terminal can act as the effective source. When thinking of FET action, and for purposes of calculation, the effective source terminal is always the one most "away" from the active drain supply. For example, suppose a FET is used to switch a line to ground, and both positive and negative signals are present on the switched line, which is usually selected to be the FET drain. If the switch is an *n*-channel enhancement-mode MOSFET, and a negative voltage happens to be present on the (turned-off) drain terminal, then that terminal is actually the "source" for purposes of gate turn-on voltage calculations. As we will discuss in Section 6.12, this means that a

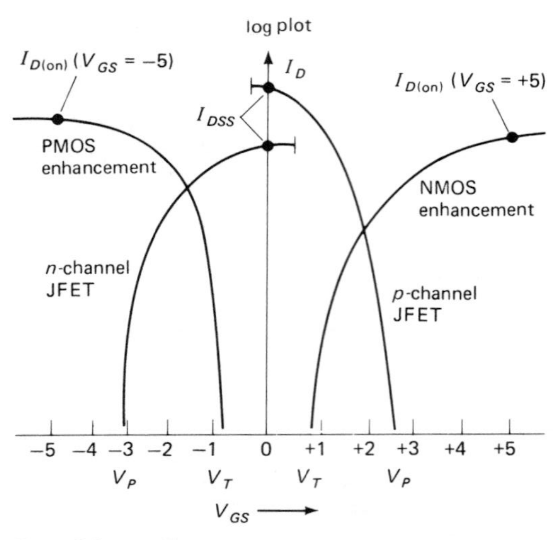

Figure 6.6
Relationships between enhancement and depletion FETs of both polarities.

negative gate voltage larger than the most negative signal, rather than ground, is needed to ensure turn-off.

The graph in Figure 6.6 may help you sort out all these confusing ideas. Again, the difference between enhancement and depletion is merely a question of displacement along the V_{GS} axis – whether there is a lot of drain current or no drain current when the gate is at the same voltage as the source. The *n*-channel and *p*-channel FETs are complementary in the same way as *npn* and *pnp* bipolar transistors, although you will see some subtle differences later.

With FETs it is easy to get confused about polarities. For example, *n*-channel devices, which usually have the drain positive with respect to the source, can have positive or negative gate voltage and positive (enhancement) or negative (depletion) threshold voltages. To make matters worse, the drain can be (and often is) operated negative with respect to the source. Of course, all these statements go in reverse for *p*-channel devices. In order to minimize confusion, we will always assume *n*-channel devices unless explicitly stated otherwise.

6.04 FET drain characteristics

FETs behave like pretty good transconductance devices, i.e., a constant gate-source voltage gives a nearly constant drain current (except at small drain-source voltages). Figure 6.7 shows typical curves of I_D versus V_{DS} for a set of values of V_{GS}. The first is an *n*-channel JFET, and the second is an *n*-channel enhancement-mode MOSFET. For depletion-mode devices (all JFETs, and depletion-mode MOSFETs), the value of drain current with gate shorted to source is specified on the data sheets as I_{DSS} and is nearly the maximum drain current possible. For enhancement-mode MOSFETs the equivalent specification is $I_{D(ON)}$, given at some forward gate voltage ("I_{DSS}" would be zero for any enhancement-mode device).

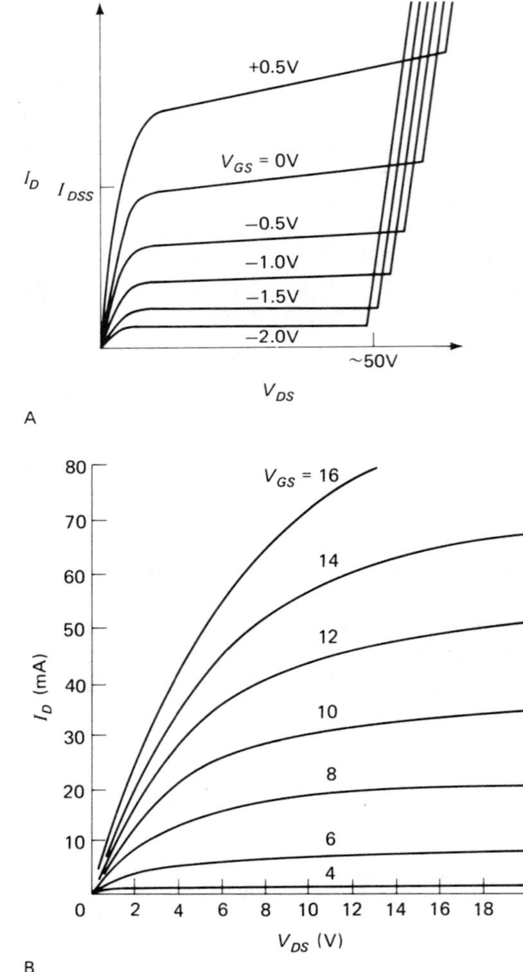

Figure 6.7
Drain characteristics. A: JFET. B: Enhancement-mode MOSFET.

For depletion-mode devices (all JFETs, and depletion-mode MOSFETs), the gate-source voltage at which drain current approaches zero is called the "gate-source cutoff voltage," $V_{GS(OFF)}$, and is typically in the range of -3 to -10 volts (positive for *p*-channel, of course). In JFET terminology the alternative term "pinch-off voltage," V_P, is widely used, but this is confusing because the same term describes something completely different when applied to MOSFETs. For enhancement-mode MOSFETs the "threshold voltage," V_T (or $V_{GS(th)}$), is the gate-source voltage at which drain current begins to flow. V_T is typically in the range of 0.5 to 5 volts, in the "forward" direction, of course. Incidentally, don't confuse the MOSFET V_T with the V_T in the Ebers-Moll equation that describes bipolar transistor collector current. They have nothing to do with each other.

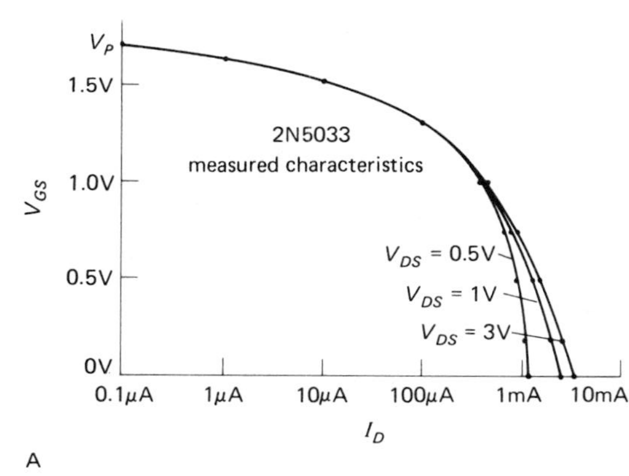

A

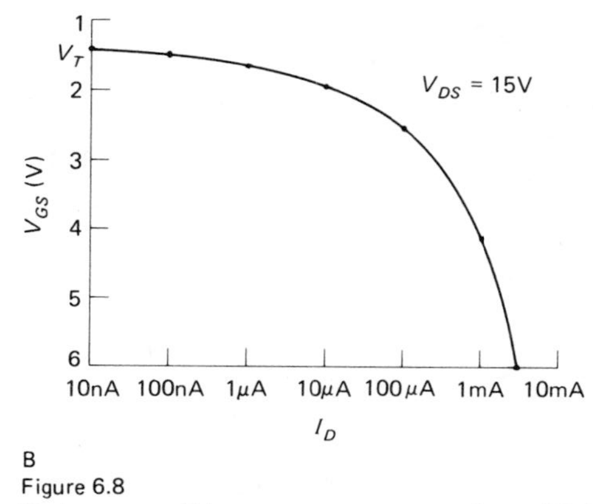

B

Figure 6.8

Drain current (I_D) versus gate-source voltage (V_{GS}), graphed over a wide range of I_D. A: P-channel JFET. B: N-channel enhancement-mode MOSFET.

The graphs in Figure 6.8 show FET drain characteristics down to very small currents (log scale). From these graphs you can see that a FET could be used as a current source simply by applying a constant V_{GS}. In fact, they are often used this way.

Pinch-off and threshold voltages

The business of threshold voltages deserves further explanation. From the graphs just preceding you can see that drain current does not have a sharp cutoff. It just keeps on dropping. To find out what is going on, it is best to plot the square root of I_D versus V_{GS} for a fixed value of V_{DS} (Fig. 6.9). The

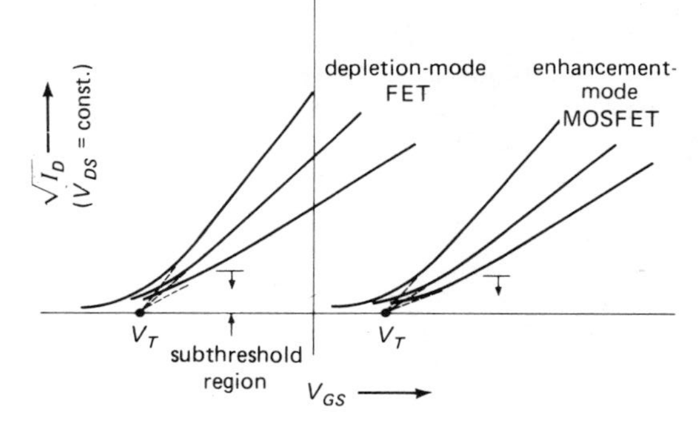

Figure 6.9

drain current follows an accurate square-root law over most of the drain current range, but it bends away at very low drain currents. The threshold voltage (or pinch-off voltage, for JFETs) is the point where the extrapolated straight-line portion of the curve would intersect the V_{GS} axis. To put it another way, the threshold voltage is chosen so that the FET drain current is given (approximately) by

$$I_D = k(V_{GS} - V_T)^2$$

with significant deviations only in the subthreshold region of very low drain currents (typically below 100μA). In FET data sheets the threshold voltage is sometimes given as the gate-source voltage corresponding to a drain current of 10μA; since V_T has an enormous production spread (see Section 6.05), it hardly matters what definition they use.

Note again that enhancement- and depletion-mode devices differ only in their positions along the V_{GS} axis. In other words, what matters is $V_{GS} - V_T$, the amount by which the gate-source voltage is above threshold.

These curves are given for a few values of k, which is proportional to the width/length ratio of the channel and to a "conduction factor" for the manufacturing process. These variables (k and V_T) are the essential manufacturing parameters for FETs. Once they have been set, the curve of I_D versus V_{GS} is fully determined. In some sense they are analogous to the parameter I_S in the Ebers-Moll equation.

I_D versus V_{GS}

Once we have defined the threshold voltage (or JFET pinch-off voltage, or gate-source cutoff voltage) as just described, I_D is simply related to V_{GS}, or, more accurately, to $V_{GS} - V_T$. Figure 6.10 explains the situation sche-

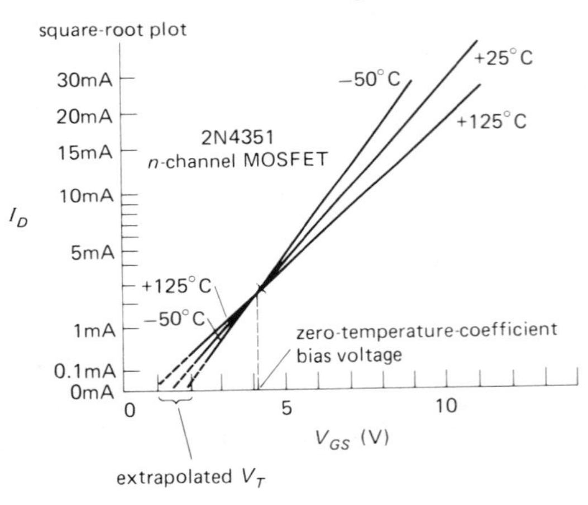

Figure 6.10

matically. The drain current increases linearly with V_{DS} up to a voltage $V_{DS(sat)}$, at which it bends over and becomes a current source. The slope in the linear region, I_D/V_{DS}, is proportional to $V_{GS} - V_T$. Furthermore, the drain voltage at which the curves enter the "saturation region," $V_{DS(sat)}$, equals $V_{GS} - V_T$, making the saturation drain current, $I_{D(sat)}$, proportional to $(V_{GS} - V_T)^2$, as given in the preceding formula. To summarize:

$$I_D = 2k[(V_{GS} - V_T)V_{DS} - 0.5V_{DS}^2] \quad \text{(linear region)}$$

$$I_D = k(V_{GS} - V_T)^2 \quad \text{(saturation region)}$$

These equations assume that the body is connected to the source. The constant k is proportional to the width/length ratio of the channel, and some other parameters (oxide capacitance, carrier mobility). What is important is its negative temperature coefficient:

$$k \propto T^{-3/2}$$

This effect alone would tend to decrease I_D with increasing temperature. However, it turns out that V_T also depends slightly on temperature, with a temperature coefficient of 2–5mV/°C in the direction that increases the magnitude of $V_{GS} - V_T$ with increasing temperature; i.e., the temperature dependence of V_T alone will tend to increase I_D with increasing temperature. As a result, the curve of drain current versus temperature in a FET looks as shown in Figure 6.11.

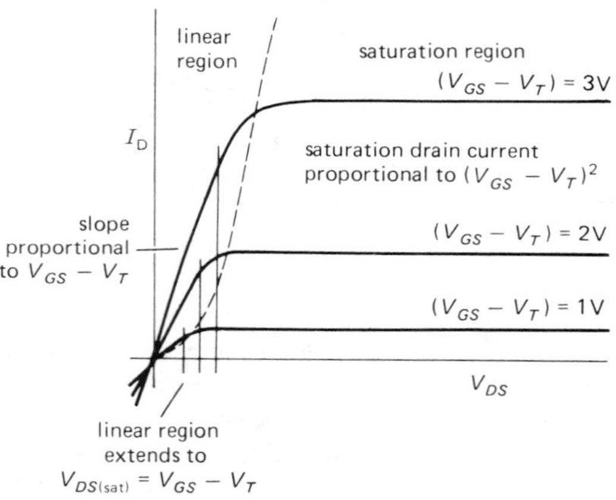

Figure 6.11

At large drain currents the negative temperature coefficient of k causes the drain current to decrease with increasing temperature – goodbye thermal runaway! As a consequence, FETs of a given type can be paralleled without external current-equalizing resistors. At low drain currents (where the temperature coefficient of V_T dominates), I_D has a positive temperature coefficient, with a point of zero temperature coefficient at some drain current in between. This effect is exploited in BiFET op-amps, as we will discuss in Section 6.19.

Incidentally, don't confuse the saturation drain current region with bipolar transistor saturation; they're almost exact opposites, since transistor saturation is the state of small V_{CE}. It might help to think of the FET saturation region as the region where the FET is giving you all the drain current it has to offer, for a given gate voltage V_{GS}.

The preceding equation for I_D in the linear region has a nonlinear term in V_{DS}^2. This is less important when the gate voltage is well above threshold. In Section 6.10 you will see an elegant way to cancel the nonlinearity, using the FET as a variable-resistance gain control.

For depletion-mode FETs, the preceding equations can be written in terms of I_{DSS}, the drain current when $V_{GS} = 0$. In particular,

$$I_{D(sat)} = I_{DSS}(V_{GS} - V_T)^2/V_T^2$$
$$= I_{DSS}(1 - V_{GS}/V_T)^2$$

For enhancement-mode MOSFETs there isn't such a convenient benchmark. Often a value for $I_{D(ON)}$ (namely, a saturation drain current at some reasonably large value of V_{GS}) is given. From that you can get an expression for I_D versus V_{GS}, noting that

$$k = I_{D(ON)}/(V_{GS(ON)} - V_T)^2$$

We will use these forms shortly to get an expression for g_m in terms of drain current.

EXERCISE 6.1

Finish the preceding work, getting expressions for I_D versus V_{GS} and V_{DS} in both the linear and saturation regions for both enhancement-mode and depletion-mode FETs. The formulas should be in terms of $I_{D(ON)}$ and $V_{GS(ON)}$ (enhancement mode) and I_{DSS} (depletion mode).

Effect of MOSFET body voltage

The body terminal of a MOSFET is either internally connected to the source or brought out as a separate terminal. When it is brought out, you usually want to connect it to the source, but for reasons of leakage, capacitance, or signal polarities, it may be desirable to connect it to some other voltage, as we will discuss later in the chapter. When the body is at some other potential, it has the effect of shifting the threshold voltage by an amount

$$\Delta V_T = 0.5 V_{BS}^{1/2}$$

in a direction that tends to decrease the drain current for a given V_{GS}.

6.05 Manufacturing spread of FET characteristics

Before we look at some circuits, let's take a look at the range of FET parameters (such as I_{DSS}), as well as their spread from one device to another, in order to get a better idea of the FET. By "spread" we mean the manufacturing variation from one device to the next you might find among FETs of the same type. Unfortunately, many of the characteristics of FETs show much greater process spread than the corresponding characteristics of bipolar transistors, a fact that the designer must keep in mind.

Characteristic	*Available range*	*Spread*
$I_{D(ON)}$	1mA to 1A	×5
$R_{DS(ON)}$	1Ω to 10k	×5
g_m @ 1mA	500–3000μmho	×5
V_p (JFETs)	0.5–10V	5V
V_T (MOSFETs)	0.5–5V	2V
$BV_{DS(OFF)}$	6–50V	
$BV_{GS(OFF)}$	6–50V	

$R_{DS(ON)}$ is the drain-source resistance when the FET is conducting fully, e.g., with the gate grounded in the case of JFETs and depletion-mode MOSFETs or with a large applied gate-source voltage (10V) for enhancement-mode MOSFETs. V_P is the pinch-off voltage (JFETs), V_T is the turn-on gate threshold voltage (enhancement MOSFETs), and the BVs are breakdown voltages. As you can see, a FET with grounded source may be a good current source, but you can't predict very well what the current will be. Likewise, V_{GS} for some value of drain current (corresponding to V_{BE} for bipolar transistors) can vary considerably. For this reason it is difficult to manufacture FET pairs with small offset voltage, and FET op-amps tend to have large input offset voltages (10–25mV is not uncommon).

Matching of characteristics

As you can see, FETs are inferior to bipolar transistors in V_{GS} predictability, i.e., they have a large spread in V_{GS} for a given I_D. Devices with a large process spread will, in general, have larger offsets when used as differential pairs. For instance, typical run-of-the-mill bipolar transistors might show a spread in V_{BE} of 25mV or so, at some collector current, for a selection of off-the-shelf transistors. The comparable figure for

TABLE 6.1 JFETs

Type	BV_{GSS} (V)	I_{DSS} min (mA)	I_{DSS} max (mA)	$V_{GS(OFF)}$ V_P min (V)	$V_{GS(OFF)}$ V_P max (V)	C_{iss} max (pF)	C_{rss} max (pF)	Comments
***n*-channel**								
2N4117A-	40	0.03	0.09	0.6	1.8	3	1.5	Low leakage: 1pA (max)
2N4119A	40	0.24	0.6	2	6	4	1.5	
2N4416	30	5	15	2.5	6	4	0.8	VHF/UHF low noise: 2dB (max) @ 100MHz
2N4867A-	40	0.4	1.2	0.7	2	25	5	Low freq., low noise: $10nV/Hz^{1/2}$ (max) @ 10Hz
2N4869A	40	2.5	7.5	1.8	5	25	5	
2N5265-	60	0.5	1	—	3	7	2	Series of 6, tight I_{DSS} spec.; 2N5358-64 *p*-channel compl.
2N5270	60	7	14	—	8	7	2	
2N5432	25	150	—	4	10	30	15	Switch: R_{ON} = 5Ω (max)
2N5457-	25	1	5	0.5	6	7	3	Gen. purpose; 2N5460-62 *p*-channel comp.
2N5459	25	4	16	2	8	7	3	
2N5484-	25	1	5	0.3	3	5	1	Low noise RF; inexpensive
2N5486	25	8	20	2	6	5	1	
***p*-channel**								
2N5114	30	30	90	5	10	25	7	Switch: R_{ON} = 75Ω (max)
2N5358-	40	0.5	1	0.5	3	6	2	Series of 7, tight I_{DSS} spec.; 2N5265-70 *n*-channel compl.
2N5364	40	9	18	2.5	8	6	2	
2N5460-	40	1	5	0.75	6	7	2	Gen. purpose; 2N5457-59 *n*-channel compl.
2N5462	40	4	16	1.8	9	7	2	

enhancement-mode MOSFETs is more like 1 volt! Since FETs have desirable characteristics otherwise, it is worthwhile to put in some extra effort to reduce these offsets in specially manufactured matched pairs. IC designers use techniques like interdigitation (two devices sharing the same general piece of IC real estate) and thermal-gradient cancellation schemes to improve performance (Fig. 6.12).

TABLE 6.2. MOSFETs

Enchancement mode type	Gate protect.	$R_{DS(ON)}$ max (Ω)	@ V_{GS} (V)	$V_{GS(th)}$ min (V)	$V_{GS(th)}$ max (V)	$I_{D(ON)}$ (V_{DS} = 10V) min (mA)	C_{rss} max (pF)	BV_{DS} (V)	BV_{GS} (V)	I_{GSS} (nA)	Comments
***n*-channel**											
SD210	—	45	10	0.5	2	—	0.5	30	40	0.1	Low r_{ON}
SD211	✓	45	10	0.5	2	—	0.5	30	15	10	Low r_{ON}
MEM563C	—	150	10	0.5	4	15	0.8	20	30	0.01	
MEM571C	✓	—	—	0.3	4	—	0.5	20	6	10	VHF amp.
MEM711	✓	100	10	0.5	1.5	10	0.8	25	30	1	Popular
MEM712A	✓	50	10	0.5	2	10	0.8	30	30	1	Low r_{ON}
2N4351	—	300	10	1.5	5	3	2.5	25	35	0.01	Popular
***p*-channel**											
3N163	—	250	20	2	5	5	0.7	40	40	0.01	
MEM517C	✓	100	10	2.5	5	20	15	25	25	1	
MEM560	✓	150	15	1.5	3	15	3.5	35	35	0.5	
MEM817	—	350	15	2.5	5.5	3	2.5	45	200	0.001	High BV_{GS}, low I_{GSS}
IT1700	—	400	10	2	5	2	1.2	40	40	0.01	
2N4065	—	1500	15	3	6	3	0.7	30	25	0.003	Oldie, now obsolete
2N4352	—	600	10	1.5	6	2	2.5	25	35	0.01	Popular
2N4353	✓	300	20	3	5	30	4	30	30	1	Popular

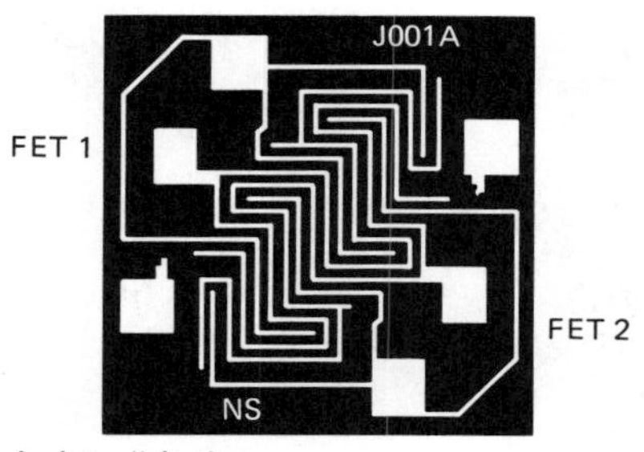

A. interdigitation

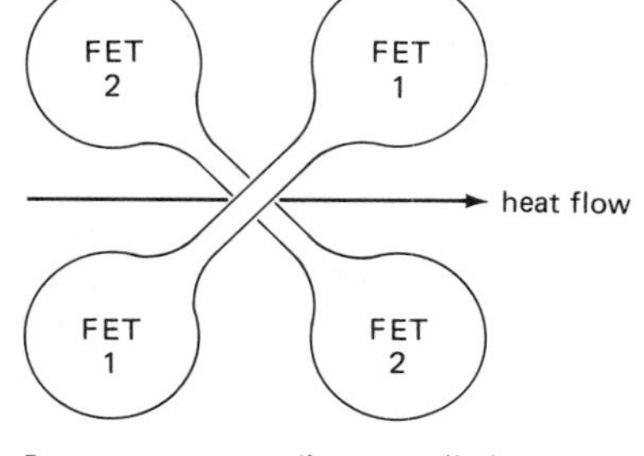

B. temperature-gradient cancellation

Figure 6.12

The results are impressive. Although FET devices still cannot equal bipolar transistors in V_{GS} matching, their performance is adequate for most applications. For example, the best available matched FET has a voltage offset of 0.5mV and tempco of 5μV/°C (max), whereas the best bipolar pair has values of 25μV and 0.6μV/°C (max), roughly 10 times better. One variety of FET op-amp is available with V_{offset} = 3mV (typ) for less than a dollar. These have input bias currents measured in picoamps, and they are generally faster than their bipolar transistor counterparts.

BASIC FET CIRCUITS

6.06 JFET current sources

The simplest FET current source is shown in Figure 6.13. From the preceding graph of

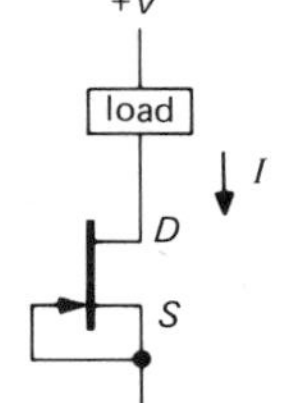

Figure 6.13

drain characteristics, you can see that the current will be reasonably constant for V_{DS} larger than a couple of volts. However, because of I_{DSS} spread, the current is unpredictable. For example, the 2N5484 (a typical *n*-channel JFET) has a specified I_{DSS} of 1mA to 5mA. Still, the circuit is attractive because of the simplicity of a two-terminal constant-current device. If that appeals to you, you're in luck. There are commercially available "current-regulator diodes" that are nothing more than JFETs with gate tied to source, sorted according to current. Here are

TABLE 6.3. DUAL MATCHED *n*-CHANNEL JFETS

Type	V_{os} max (mV)	Drift max (μV/°C)	I_{GSS} (V_{DG} = 20V) max (pA)	CMRR min (dB)	$V_{GS(OFF)}$ / V_P min (V)	$V_{GS(OFF)}$ / V_P max (V)	e_n (10Hz) max (nV/Hz$^{1/2}$)	C_{rss} (V_{DG} = 10V) max (pF)	Comments
U421	10	10	0.2	90	0.4	2	50	1.5	Siliconix
2N3954A	5	5	100	—	1	3	150[a]	1.2	Gen. purpose, low drift
2N3955	5	25	100	—	1	4.5	150[a]	1.2	Popular
2N3958	25	—	100	—	1	4.5	150[a]	1.2	
2N5196	5	5	15	—	0.7	4	20[b]	2	
2N5520	5	5	100	100	0.7	4	15	5	
2N5906	5	5	2	90[c]	0.6	4.5	70[c]	1.5	Low gate leakage
2N5911	10	20	100	—	1	5	20[d]	1.2	Low noise at high freq.
2N6483	5	5	100	100	0.7	4	10	3.5	Low noise at low freq.
NDF9406	5	5	5	120	0.5	4	30	0.1	National "cascode FET"; low C_{rss}, high CMRR

[a] At 100 Hz. [b] At 1 kHz. [c] Typical. [d] At 10kHz.

the characteristics of the CR022–CR470 series manufactured by Siliconix (also available as the 1N5283–1N5313):

Currents available	0.22mA to 4.7mA
Tolerance	10%
Temperature coefficient	0.15%/°C
Voltage compliance	1–100V for ±5% current variation
Impedance	0.8M (worst), 4M (typ)

Source self-biasing

You can make an adjustable current source by adding a source self-biasing resistor, as in Figure 6.14. You choose R by looking at the

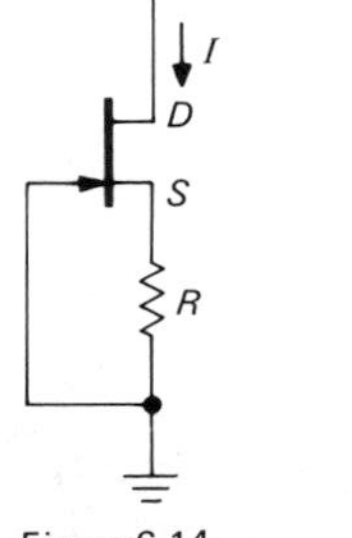

Figure 6.14

drain characteristic curves or at the curve of drain current versus V_{GS}. The resistor allows you to set the current, as well as making it more predictable. Furthermore, the circuit is a better current source (higher impedance) because (a) the source resistor provides feedback to increase the output impedance and (b) FETs are better current sources anyway when the gate is back-biased, as can be seen from the flatness of the family of curves of I_D versus V_{DS} at low currents (Fig. 6.7) or from the convergence of the separate V_{DS} lines in the log I_D versus V_{GS} plot (Fig. 6.8). Remember, though, that actual values of I_D for some value of V_{GS} obtained with a real FET may differ from the values read from a set of characteristic curves by a large factor, owing to manufacturing spread. You may therefore want to use an adjustable source resistor, if it is important to have a specific current. It is also important to realize that a good bipolar transistor or op-amp current source will give far better predictability and stability. A FET current source might vary 5% over a typical temperature range and load voltage variation, even after being set to the desired current by trimming the source resistor, whereas an op-amp/transistor current source is predictable and stable to 0.5% without great effort.

EXERCISE 6.2

Choose R to provide 100μA, using a p-channel version of the circuit of Figure 6.14 with a 2N5033, for which the (logarithmic) curve of drain current versus V_{GS} is given in Figure 6.8.

□ 6.07 FET amplifiers

Source followers and common-source FET amplifiers are analogous to emitter followers and common-emitter amplifiers made with bipolar transistors. With depletion-mode FETs it is convenient to use the same self-biasing scheme described earlier, with a single gate-biasing resistor to ground, whereas for enhancement-mode devices the gate must be biased with a divider from the drain supply, as with bipolar transistors (Fig. 6.15). The gate-biasing resistors can be quite large (a megohm or so), since the gate leakage current is measured in nanoamps.

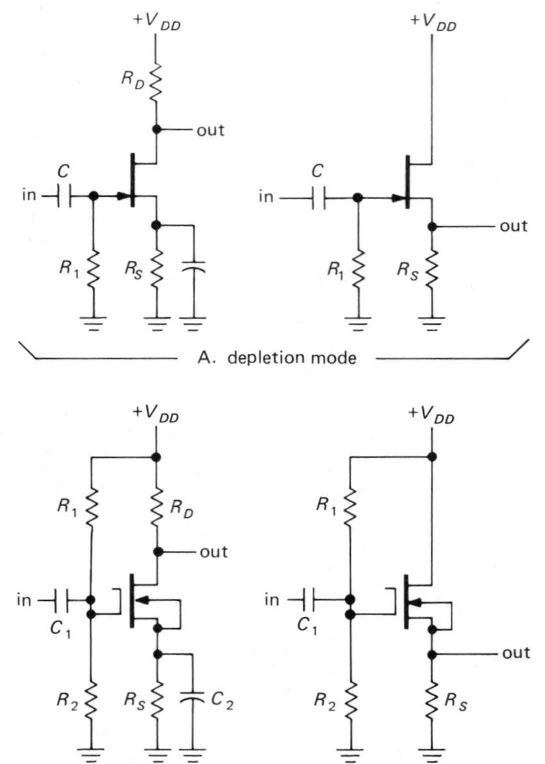

Figure 6.15

□ *Transconductance*

The gain can be estimated from the characteristic curves or by using the transconductance, g_m (also called g_{fs} or y_{fs}):

$$g_m(I_D) = i_d/v_{gs}$$

(Remember that lower-case letters indicate quantities that are small-signal variations.) From this you get

$$G_{voltage} = v_d/v_{gs} = -R_D i_d/v_{gs} = -g_m R_D$$

Typically, FETs have transconductances of a few thousand micromhos at a few milliamps. The value of g_m depends on drain current, as you will see later, so you've got to use the transconductance at the operating point. There will be some variation of gain (nonlinearity) over the waveform as the drain current varies, just as we have with grounded emitter amplifiers, where g_m is proportional to collector current. A curve of g_m versus V_{GS} is usually given. If not, you can use the approximate relation we presented earlier for the (saturation) drain current,

$$I_D = k(V_{GS} - V_T)^2$$

to find

$$g_m = 2k(V_{GS} - V_T) = 2(kI_D)^{1/2}$$

If g_m is known at a specific value of drain current I_{D0}, you can use instead

$$g_m = g_{m0}\,(I_D/I_{D0})^{1/2}$$

where g_{m0} is the transconductance measured at that reference drain current I_{D0}.

EXERCISE 6.3

Derive this last result for g_m.

In the case of depletion-mode FETs, g_m is always specified at $V_{GS} = 0$, i.e., at I_{DSS}. In that case g_m at some other current I_D is simply

$$g_m(\text{at } I_D) = (I_D/I_{DSS})^{1/2} g_m(\text{at } I_{DSS})$$

An expression for g_m can also be written in terms of gate voltage:

$$g_m(\text{at } V_{GS}) = \frac{V_{GS} - V_T}{V_{G0} - V_T}\, g_m(\text{at } V_{G0})$$

giving the transconductance at arbitrary gate voltage in terms of its value measured at a gate voltage of V_{G0}.

The formula for saturation drain current versus V_{GS} can also be used to figure the operating point with source self-bias (for a depletion-mode FET amplifier) if you know I_{DSS} and V_P. Exercise 6.4 suggests how.

EXERCISE 6.4

Use the fact that $V_S = I_D R_S$ to show that in order to have drain current I_D, the source resistor in a current source or amplifier should have value

$$R_S = \frac{V_T}{I_D}[1 - (I_D/I_{DSS})^{1/2}]$$

□ *Transconductance of bipolar transistors versus FETs*

A major drawback of the FET amplifier is that its transconductance is much lower than that of a bipolar transistor at the same current (except possibly at extremely low currents). For instance, an amplifier running at 1mA quiescent current from a +15 volt supply, biased to half the supply voltage for maximum output swing, will have a voltage gain of only −7.5 for a typical transconductance of 1000μmho, whereas a bipolar transistor will have a transconductance of $1/r_e$, or 40,000μmho, at the same current, giving a voltage gain of −300.

The gain situation can be improved by using a current-source load, but once again the bipolar transistor will be better in the same circuit. For this reason you seldom see FETs used as simple amplifiers, unless it's important to take advantage of their unique input properties (extremely high input impedance, low input current, and low feedback capacitance).

EXERCISE 6.5

Show that the gain of a self-biased common-source amplifier with unbypassed source resistor is $G = R_D/(1/g_m + R_S)$.

The comparison of FET transconductance with that of bipolar transistors reveals some interesting features, as shown in Figure 6.16. In the subthreshold region of drain current, both FETs and bipolar transistors display transconductance that increases linearly with drain (collector) current. However, the FET's transconductance is somewhat lower, about I/40mV for PMOS and

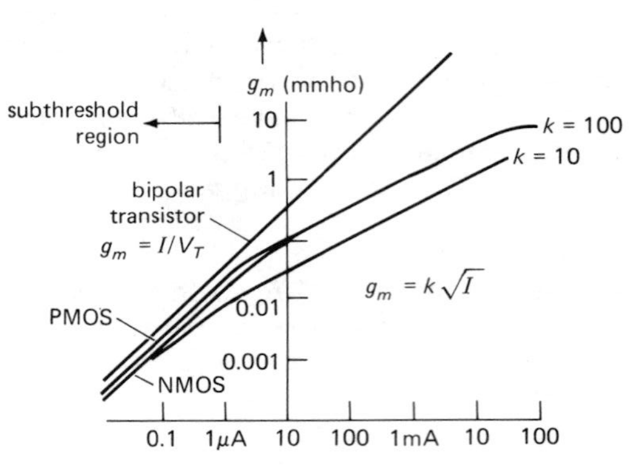

Figure 6.16
Comparison of g_m for bipolar transistors and FETs.

I/60mV for NMOS, as compared with the bipolar transistor's $g_m = 1/r_e = I_C/25\text{mV}$. As the drain current is increased into the region where I_D is proportional to V_{GS}^2, the FET transconductance varies as the square root of I_D, as in the preceding formulas, and is well below the transconductance of a bipolar transistor at the same operating current. Increasing the constant k in our preceding equations by raising the channel width/length ratio increases the transconductance (and the drain current, for a given V_{GS}) in the region above threshold, but the transconductance still remains less than that of a bipolar transistor at the same current.

□ *Differential amplifiers*

Matched FETs can be used to construct high-input-impedance front-end stages for comparators, op-amps, and discrete transistor amplifiers. As we mentioned earlier, the substantial FET V_{GS} offsets will generally result in larger input-voltage offsets and offset drifts than with a comparable amplifier constructed entirely with bipolar transistors, but of course the input impedance will be raised enormously. We will discuss such circuits, along with matched FET followers, in Section 6.09.

□ *Oscillators*

In general, FETs have characteristics that make them suitable substitutes for bipolar transistors in almost any circuit that can benefit from their uniquely high input impedance and low bias current. A particular instance is their use in high-stability *LC* and crystal oscillators, with examples in Sections 4.15, 4.16, and 13.11.

□ *Active load*

Just as with transistor amplifiers, it is possible to replace the drain-load resistor in a FET amplifier with an active load, i.e., a current source. The voltage gain you get that way can be very large:

$G_V = -g_m R_D$ (with a drain resistor as load)

$G_V = -g_m R_o$ (with a current source as load)

where R_o is the impedance looking into the drain (g_{oss}), typically in the range of 100k to 1M.

One possibility for an active load is a current mirror as the drain load for a differential FET pair, just as is often done in bipolar transistor op-amps (e.g., see Fig. 3.27). You will see another nice example of the active load technique in Section 6.16 when we discuss the CMOS linear amplifier.

6.08 Source followers

FETs are often used as source followers. This is a good way to achieve high input impedance, and a FET follower is commonly used as the input stage in oscilloscopes as well as other measuring instruments. There are many applications in which the signal source is intrinsically high impedance, e.g., capacitor microphones, pH probes, the voltage from a microelectrode used to probe muscle or nerve potentials; in these cases a FET input stage is a good solution. Within circuits there are situations where the following stage must draw little or no current. Common examples are analog "sample-and-hold" and "peak-detector" circuits, in which the level is stored on a capacitor and will "droop" if the next amplifier draws too much input current. Unlike the situation described in the last section, in which bipolar transistors usually provide

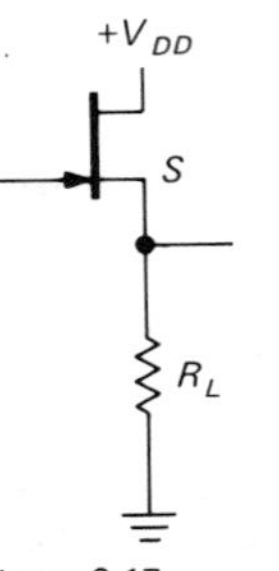

Figure 6.17

superior performance as common-emitter amplifiers, the source follower is often a very attractive alternative to the bipolar emitter follower.

Figure 6.17 shows the simplest source follower. As before, we can figure out the amplitude of the output signal by using the transconductance. We have

$$v_s = R_L i_d$$

since i_g is negligible, but

$$i_d = g_m v_{gs} = g_m(v_g - v_s)$$

so

$$v_s = \frac{R_L g_m}{1 + R_L g_m} v_g$$

For $R_L >> 1/g_m$ it is a good follower ($v_s \approx v_g$), with gain approaching, but always less than, unity.

Output impedance

The preceding equation for v_s/v_g is exactly what you would predict if the FET follower's output impedance were equal to $1/g_m$ (try the calculation, assuming a source voltage of v_g in series with $1/g_m$ driving a load of R_L). This is exactly analogous to the emitter follower situation, where the output impedance was $r_e = 25/I_C$, or $1/g_m$. It can be easily shown explicitly that a source follower has output impedance $1/g_m$ by figuring the

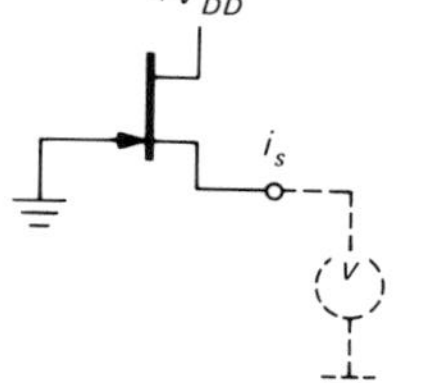

Figure 6.18

source current for a signal applied to the output with grounded gate (Fig. 6.18). The drain current is

$$i_d = g_m v_{gs} = g_m v$$

so

$$r_{out} = v/i_d = 1/g_m$$

typically a few hundred ohms at currents of a few milliamps. As you can see, FET source followers aren't nearly as stiff as emitter followers. There are two drawbacks to this circuit:

1. The relatively high output impedance means that the output swing may be significantly less than the input swing, even with high load impedance, because R_L alone forms a divider with the source's output impedance. Furthermore, since the drain current is changing over the signal waveform, g_m and therefore the output impedance will vary, producing some nonlinearity (distortion) at the output. The situation is improved if FETs of high transconductance are used, of course, but a combination FET–bipolar follower is often a better solution.

2. Since V_{GS} is a poorly controlled parameter in FET manufacture, a source follower has an unpredictable dc offset, a serious drawback for dc-coupled circuits.

Active load

The addition of a few components improves the source follower enormously. Let's take it in stages:

First, replace R_L with a (pull-down) current source (Fig. 6.19). The constant

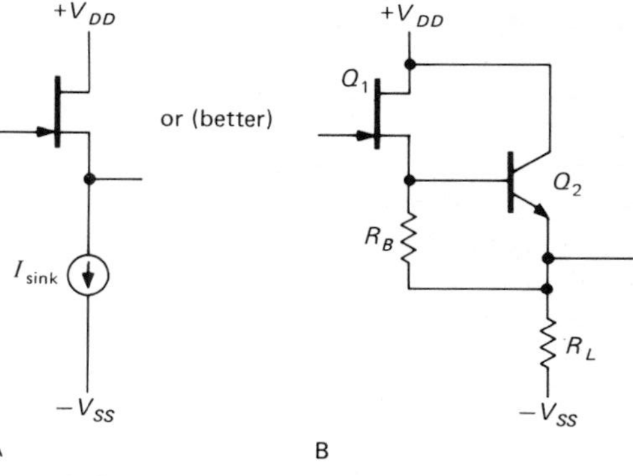

Figure 6.19

source current makes V_{GS} approximately constant, thus reducing nonlinearities. You can think of this as the previous case, with R_L infinite, which is what a current source is. The circuit on the right has the advantage of providing low output impedance, while still providing a (roughly) constant source current of V_{BE}/R_B. We still have the problem of unpredictable (and therefore nonzero) offset voltage (from input to output) of V_{GS} ($V_{GS} + V_{BE}$ for the circuit on the right). Of course, we could simply adjust I_{sink} to the particular value of I_{DSS} for the given FET (in the first circuit) or adjust R_B (in the second). This is a poor solution, for two reasons: (a) It requires individual adjustment for each FET. (b) Even so, I_D may vary by a factor of two over the normal operating temperature range for a given V_{GS}.

A better circuit uses a matched FET pair to achieve zero offset (Fig. 6.20). Q_1 and Q_2

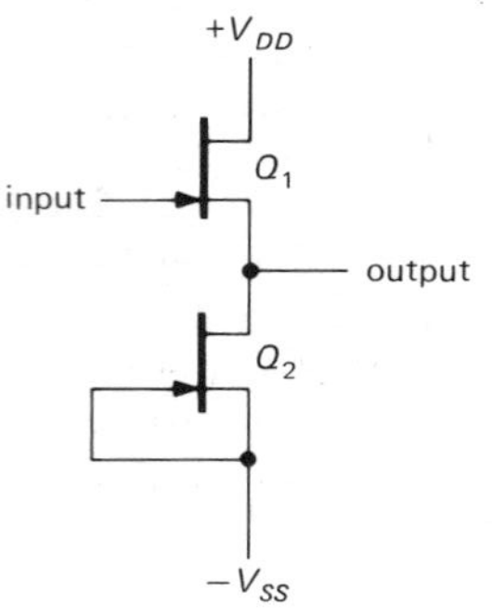

Figure 6.20

are a matched pair, on a single chip of silicon. Q_2 sinks a current exactly appropriate to the condition V_{GS}= 0. So, for both FETs V_{GS} = 0, and Q_1 is therefore a follower with zero offset. Since Q_2 tracks Q_1 in temperature, the offset remains near zero independent of temperature.

You usually see the preceding circuit with source resistors added (Fig. 6.21). A little thought should convince you that R_1 is necessary and that $R_1 = R_2$ guarantees that $V_{out} = V_{in}$ if Q_1 and Q_2 are matched. This circuit modification gives better I_D predictability, allows you to set the drain current to

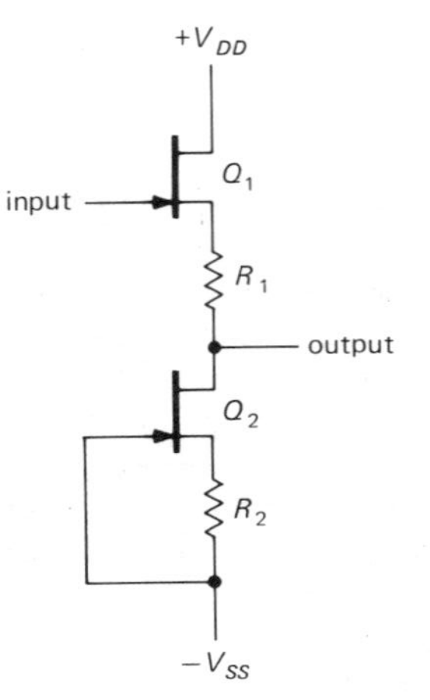

Figure 6.21

some value other than I_{DSS}, and gives improved linearity, since FETs are better current sources when operated at currents below I_{DSS}. This follower circuit is popular as the input stage for oscilloscope vertical amplifiers.

□ ***Example: a no-holds-barred follower***

By adding a bipolar output transistor, you can reduce the output impedance of a FET follower. Figure 6.22 shows such a circuit,

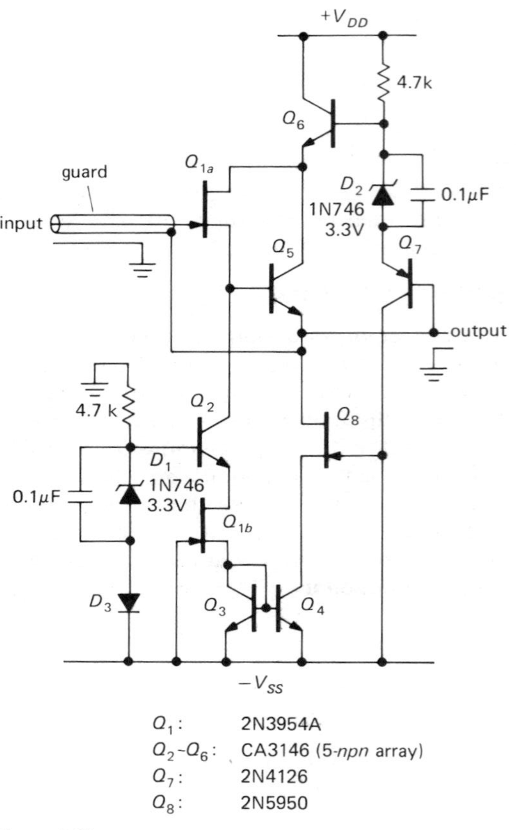

Figure 6.22
A FET-buffered source follower with bootstrap and zero voltage offset.

also featuring a bootstrap to lower input capacitance (a "guard" electrode) and a bootstrapped active drain load to reduce input capacitance and ensure matched V_{DS} conditions in the two FETs.

This circuit has extremely high Z_{in}, low C_{in}, zero dc offset from input to output, low output impedance, and good frequency response. Q_{1a} is the input follower, with companion Q_{1b} providing a constant-current sink to generate constant and identical V_{GS}s.

The current is set by Q_{1b}, whose drain current is whatever value corresponds to a gate-source reverse bias of one diode drop (the voltage across Q_3). Q_5 arranges things so that the amplifier has no dc offset at the output relative to the input: Both halves of Q_1 are running the same current; since Q_{1b}'s gate is one V_{BE} below its source, the same must be true of Q_{1a}. But Q_5 makes the amplifier's output one V_{BE} drop below Q_{1a}'s source and therefore at the same voltage as the input. The circuit is even arranged so that the corresponding V_{BE} drops are matched: Since Q_3 and Q_5 are matched (Q_2 through Q_6 are all part of a monolithic array with well-matched V_{BE}s), and current mirror Q_3Q_4 sets Q_5's collector current at the same value as Q_3's, the V_{BE} drops of Q_3 and Q_5 are nearly identical.

Q_6 bootstraps Q_{1a}'s drain to eliminate the effects of drain-gate capacitance. Q_2 plays the same trick on Q_{1b} to keep the drain-source voltages of the matched FET pair, and therefore the gate-source voltages, identical. Cascode transistor Q_8 is included to pin Q_4's collector at a constant voltage, improving the current source properties of Q_4 by defeating the Early effect (as discussed in Section 2.06). Q_8's I_{DSS} should be higher than the operating current of the circuit (i.e., Q_1's drain current) for it to work in this cascode arrangement. (Q_8 could be replaced by an *npn* transistor, with its base connected to Q_2's base.) Finally, the output signal is used to drive the input guard in order to reduce the effects of cable capacitance, which would otherwise be devastating for the high source impedances that you might see with this sort of amplifier.

6.09 JFET input impedance and gate leakage

Matched FETs as input buffers

In some applications an unadorned FET source follower will work fine as an input buffer. For instance, you might want to use a particular comparator, because of short delay time or good output properties, but its input current might be too high for the application. The sort of input buffer shown in Figure 6.23 will solve the problem. The monolithic dual FET provides reasonably low offset (5mV) and very high input impedance. The fact that such a follower may have gain

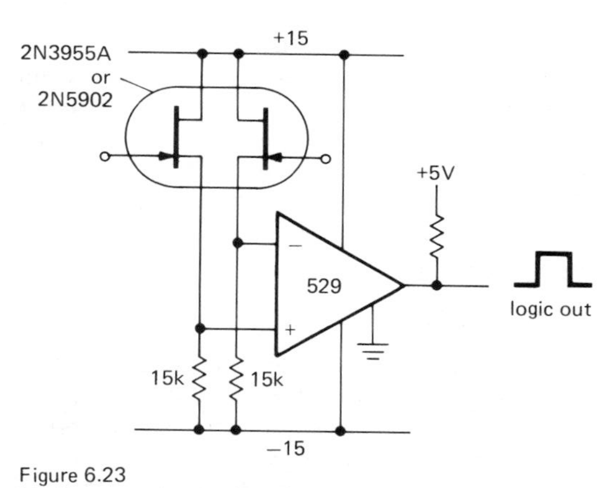

Figure 6.23
Matched FET follower pair with op-amp.

significantly less than 1 (because of low g_m) is irrevelant in a comparator circuit. The 529 is a fast comparator with input current of 5μA.

Although a matched FET source follower is the most common form of input buffer, it is possible to use a matched FET pair as a common-source differential amplifier. This can have the advantage of improved common-mode rejection ratio, if one of the monolithic dual cascode FETs is used. Figure 6.24 shows an example. The particular matched cascode FET shown has a CMRR of 125dB in this circuit. With an input-stage gain of 35dB, the op-amp need have only a

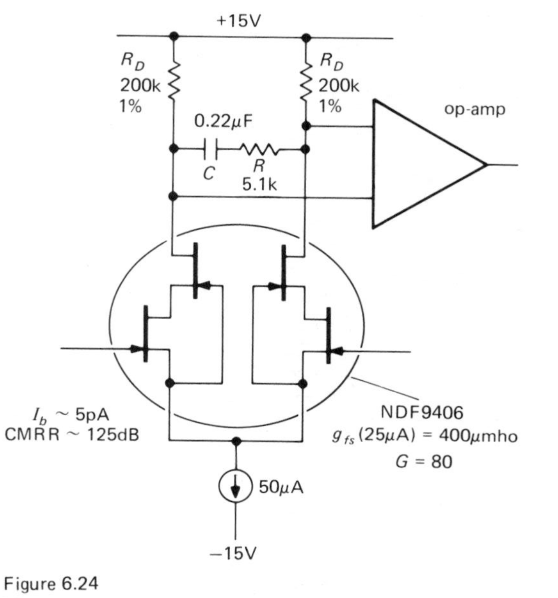

Figure 6.24
Matched FET differential amplifier with op-amp, for improved CMRR.

modest 90dB of CMRR in order for the circuit to deliver 125dB of CMRR, a value difficult to achieve in an ordinary op-amp. The compensation components R and C are necessary because of the added loop gain provided by the input stage. R will be chosen $\sim 1/g_m$, with C then selected for a 1ms RC time constant.

Gate leakage

The low-frequency input impedance of a JFET buffer, whether follower or amplifier, is limited by gate leakage. FET data sheets usually specify a breakdown voltage, BV_{GSS}, defined as the voltage from gate to channel (source and drain connected together) at which the gate current reaches 1μA. For smaller applied gate-channel voltages, the gate leakage current, I_{GSS}, again measured with the drain and source connected together, is considerably smaller, dropping quickly to the picoamp range for gate-drain voltages well below breakdown.

These low gate leakage currents are, of course, what make FETs so useful for buffering high-impedance signals. However, there is a serious snag. It turns out that n-channel JFETs can exhibit rather large gate leakage currents once you disconnect drain and source (for data sheet purposes) and start running some drain current (for real-life purposes). The graph in Figure 6.25 shows what happens. The gate leakage current remains near the I_{GSS} value until you reach a critical drain-gate voltage, at which point it rises precipitously. This extra "impact-ionization" current is proportional to drain current, and it rises exponentially with voltage and temperature. The onset of this current occurs at drain-gate voltages of about 25% of BV_{GSS}, and it can reach gate currents of a microamp or more. Obviously a "high-impedance buffer" with a microamp of input current is worthless. That's what you would get if you used a 2N4868A as a follower, running 1mA of drain current from a 40 volt supply.

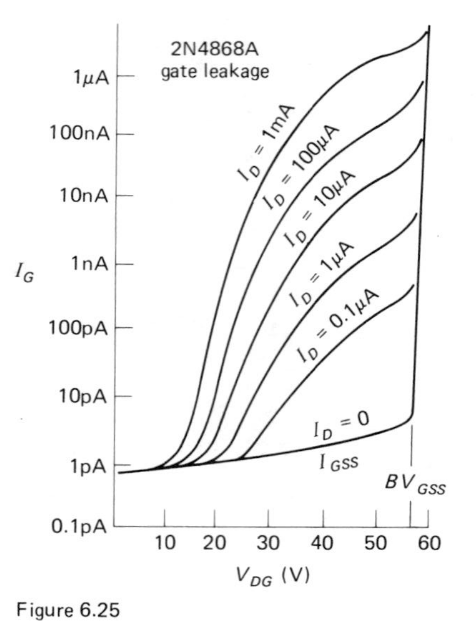

Figure 6.25
JFET gate leakage increases disastrously at higher drain-gate voltages and is proportional to drain current.

The graph in Figure 6.26, plotting gate

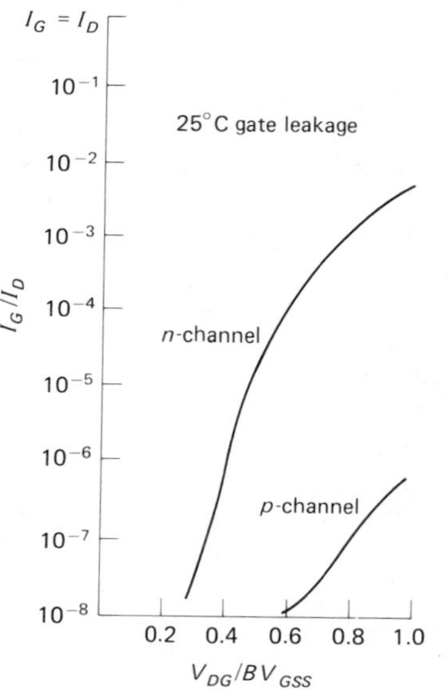

Figure 6.26

leakage (as a fraction of drain current) versus gate-drain voltage (as a fraction of breakdown voltage), illustrates the fact that this gate leakage is primarily an n-channel JFET disease and that it is proportional to drain current.

This gate leakage problem can seriously limit the usefulness of JFETs as high-impedance buffers and analog switches. There are several possible solutions:

1. Operate at low gate-drain voltages. This may mean *very* low source-drain voltages.
2. Use a cascode connection, thereby holding V_{DG} to a small value in the input FET.

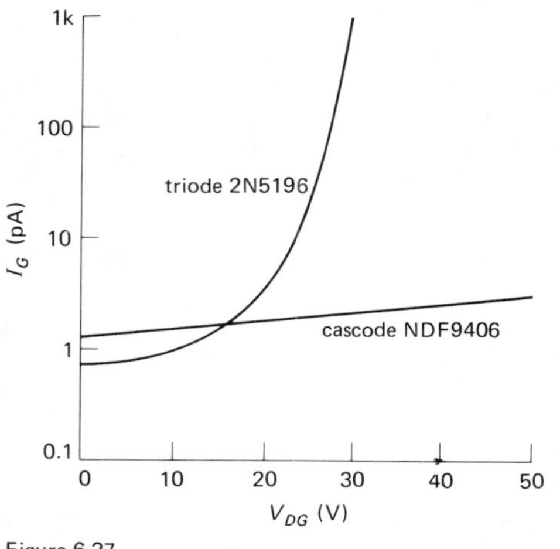

Figure 6.27
Gate current versus drain-gate voltage at I_D = 200μA.

JFET cascodes are available as single devices, e.g., the NDF9401 series from National Semiconductor. Figure 6.27 shows a graph comparing gate leakage current for one of these devices with normal JFET gate leakage.

3. Use *p*-channel JFETs instead of *n*-channel JFETs. Their lower carrier mobility reduces this effect enormously, as shown in Figure 6.26.

4. Use a MOSFET.

Note that this effect is peculiarly an *n*-channel JFET problem and that it is more serious for high-*k*-value FETs (large channel width/length ratio) designed for high g_m, low $R_{DS(ON)}$, and low noise. Figure 6.28 compares gate leakage currents for a few popular *n*-channel JFETs.

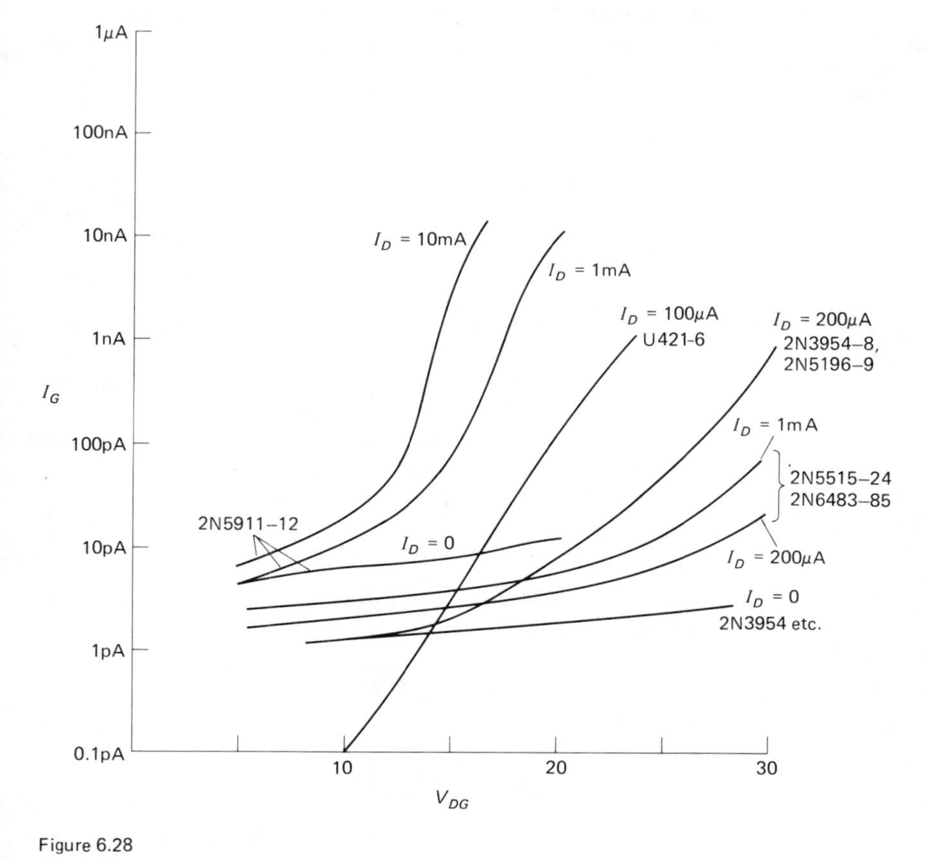

Figure 6.28
Gate leakage current for some typical n-channel JFETs.

□ 6.10 FETs as variable resistors

Figure 6.29 shows the region of the FET characteristic curves (drain current versus V_{DS} for various values of V_{GS}) for small drain-source voltages. The I_D versus V_{DS} curves are approximately straight lines for V_{DS} smaller than $V_{GS} - V_T$, and they extend in both directions, i.e., the device can be used as a voltage-controlled resistor for small signals of either polarity. From the equation for I_D versus V_{GS} in the "linear region" (Section 6.04), we find for the variation of R with V_{GS}

$$1/R_{DS} = 2k[(V_{GS} - V_T) - \tfrac{1}{2} V_{DS}]$$

i.e., for drain voltages substantially less than the amount by which the gate is above threshold ($V_{DS} \rightarrow 0$), the FET behaves like a linear resistance:

$$R_{DS} = 1/2k(V_{GS} - V_T)$$

Note that $R_{DS} = 1/g_m$, so that the channel resistance in the linear (unsaturated) region is the inverse of the transconductance in the saturated region. This is a handy thing to know when only one parameter is given on the data sheet.

If the resistance is known at some value of gate voltage, you can use this formula to find the resistance at any other gate voltage:

$$R_{DS} = \frac{V_{G0} - V_T}{V_G - V_T} R_0$$

where R_0 is the resistance measured at a gate voltage of V_{G0}.

EXERCISE 6.6

Derive the preceding "scaling" formula.

Typically, the values of resistance you can produce with FETs vary from a few tens of ohms (as low as 0.1Ω for power FETs) all the way up to an open circuit. A typical application might be an automatic-gain-control (AGC) circuit in which the gain of an amplifier is adjusted (via feedback) to keep the output within the linear range. In such an AGC circuit you must be careful to put the variable-resistance FET at a place in the circuit where the signal swing is small, preferably less than 200mV or so.

The range of V_{DS} over which the FET behaves like a good resistor depends on the particular FET and is roughly proportional to the amount by which the gate voltage exceeds V_P (or V_T). Typically, you might have nonlinearities of about 2% for $V_{DS} < 0.1\,(V_G - V_P)$, and perhaps 10% nonlinearity for $V_{DS} \approx 0.25(V_G - V_P)$. MOSFETs, with their larger V_{GS} range, permit linear operation at higher V_{DS}. Matched FETs make it easy to design a ganged variable resistor to control several signals at once. FETs intended for use as variable resistors are available (Siliconix VCR series) with resistance tolerances of 30%, specified at some V_{GS}.

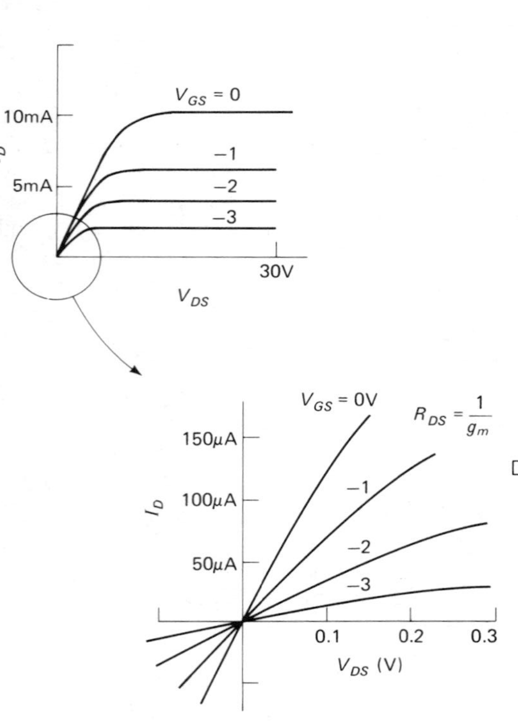

Figure 6.29

□ *Electronic gain control*

By looking at the preceding equation for R_{DS}, you can see that the linearity will be nearly perfect if you can add to the gate voltage a voltage equal to one-half the drain-source voltage. Figure 6.30 shows a nice variable-gain amplifier circuit that does exactly that. The JFET forms half of the feedback network in a conventional noninverting amplifier configuration. The overall gain is

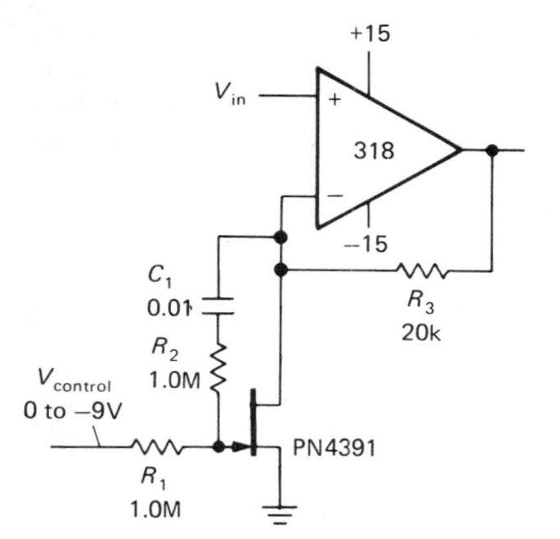

Figure 6.30
Variable gain ac amplifier with low distortion.

adjustable from 1 to 1000 via $V_{control}$. R_1, R_2, and blocking capacitor C_1 improve the linearity by adding a voltage of $0.5V_{DS}$ to V_{GS}, as just discussed. With that trick the circuit claims low distortion for output signals up to 8.5 volts.

When considering FETs for an application requiring a gain control, e.g., an AGC or "modulator" (in which the amplitude of a high-frequency signal is varied at an audio rate, say), it is worthwhile to look also at "analog multiplier" ICs. These are high-accuracy devices with good dynamic range that are normally used to form the product of two voltages. One of the voltages can be a dc control signal, setting the multiplication factor of the device for the other input signal, i.e., the gain. Analog multipliers exploit the g_m versus I_C characteristic of bipolar transistors ($g_m = I_C$(mA)/25 mho), using matched arrays to circumvent problems of offsets and bias shifts. At very high frequencies (100MHz and above) passive "balanced mixers" (Section 13.12) are often the best devices to accomplish the same task.

It is important to remember that a FET in conduction at low V_{DS} behaves like a good resistance (not like the current source that characterizes a bipolar transistor's collector) and that it will continue to act like a resistor all the way down to zero volts from drain to source (there are no diode drops or the like to worry about). There are op-amps (e.g., the 3130/3160) and logic families (CMOS) that take advantage of this nice property.

6.11 Op-amp controlled current sources

The total absence of gate current makes FETs ideal for current sources when used in conjunction with op-amps. Figure 6.31

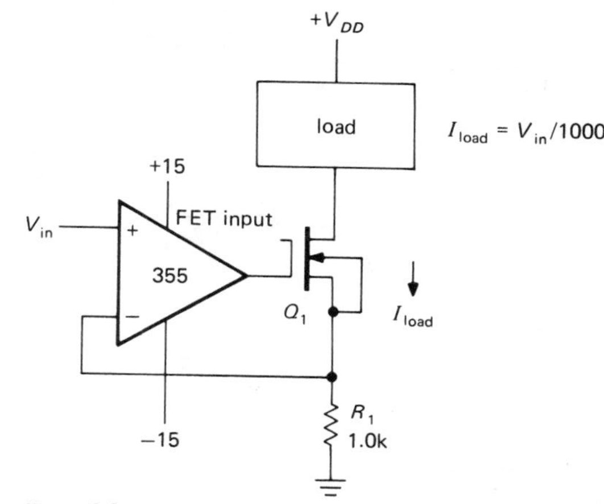

Figure 6.31

shows an example. The n-channel MOSFET Q_1 sinks current from the load; the current is sampled across R_1 and compared with the voltage at the noninverting input to the op-amp. Since there is no gate current, the output current is sampled at the source resistor without error, eliminating the base current error of the equivalent bipolar transistor circuit. Any departures from ideal current-source behavior are due only to nonlinearities in the current-sampling resistor, and to errors in the op-amp input circuit, such as offsets and drift. For best performance at low output currents it is wise to use a FET-input op-amp to eliminate errors caused by input bias current. With this sort of circuit it is easy to make a current source of 0.1% accuracy or better.

If Q_1 is an enhancement-mode MOSFET, the op-amp needs to swing only to ground to cut off output current, so the circuit can be operated from a single positive supply. In this case it is important to use an op-amp whose input common-mode range goes down to ground and that can swing its output to ground; something like the 358, 799, or 3160 would be good.

Bipolar transistor booster

For output currents greater than $I_{DS(ON)}$, just add a bipolar transistor to the FET current source (Fig. 6.32). Q_2 begins to conduct

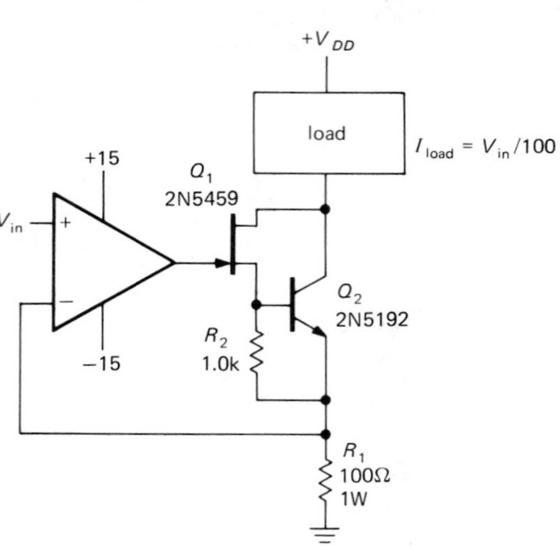

Figure 6.32

when Q_1 is drawing about 0.6mA drain current. With Q_1's minimum I_{DSS} of 4mA and a reasonable value for Q_2's beta, load currents of 100mA or more can be generated (Q_2 can be replaced by a Darlington for much higher currents, and in that case R_1 should be reduced accordingly). As before, this current source is "perfect" in having no base current error. For variety, a JFET has been used in this circuit, although a MOSFET would be fine. With a JFET, which is a depletion-mode device, the op-amp must be run from split supplies to ensure a gate voltage range sufficient for pinch-off.

FET SWITCHES

6.12 FET linear switches

One of the most frequent uses of FETs, particularly MOSFETs, is as analog switches. Their combination of low ON resistance, extremely high OFF resistance, low leakage currents, and low capacitance makes them ideal as voltage-controlled switch elements for analog signals. An ideal linear, or analog, switch behaves like a perfect mechanical switch: It passes a signal through to a load without attenuation or nonlinearity. Note that this is not the same as the transistor "switch" discussed in Chapter 2. That was an example of a logic switch, unsuitable for linear signals.

Let's look at an example (Fig. 6.33). Q_1 is

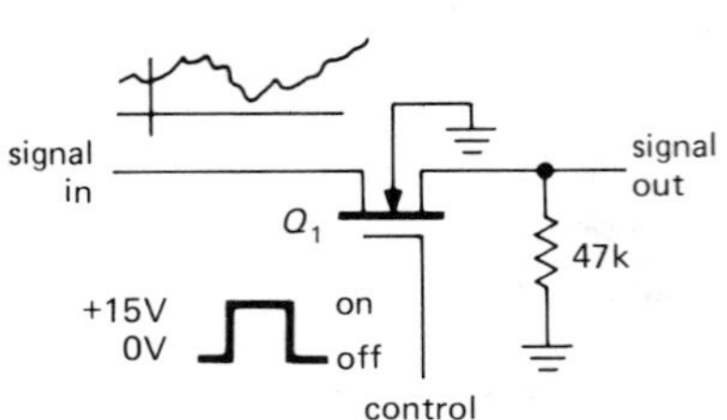

Figure 6.33

an *n*-channel enhancement-mode FET, and it is nonconducting when the gate is grounded or negative. In that state the drain-source resistance (R_{OFF}) is typically more than 10,000M, and no signal gets through. Bringing the gate to +15 volts puts the drain-source channel into conduction, typically 25 to 100 ohms (R_{ON}) in FETs intended for use as analog switches. The gate signal level is not at all critical, as long as it is sufficiently more positive than the largest signal (to maintain R_{ON} low), and it could be provided from logic circuitry (CMOS logic levels would be fine; TTL levels could be used to generate a full-supply swing with an external transistor; see the box on logic levels in Chapter 1) or even from an op-amp: ±13 volts from a 741 output would do nicely, since gate breakdown voltages in MOSFETs are frequently 20 volts or more. The reverse bias you would have on the gate during negative swings of the op-amp would have the added advantage of allowing the switching of analog signals of either polarity, as will be described later. Note that the FET switch is a bidirectional device; signals can go either way through it. Ordinary mechanical switches work that way, too, so it should be easy to understand.

The circuit as shown will work for positive signals up to about 10 volts; for larger signals the gate drive is insufficient to hold the FET in conduction (R_{ON} begins to rise), and negative signals will cause the FET to turn on with the gate grounded (it will also forward-bias the channel-body junction). If you want to switch signals that are of both polarities (e.g., signals in the range $-10V$ to $+10V$), you can use the same circuit, but with the gate driven from -15 volts (OFF) to $+15$ volts (ON); the body should then be tied to -15 volts.

With any FET switch it is important to provide a load resistance in the range of 10k to 100k in order to prevent capacitive feedthrough of the input signal during the OFF state that would otherwise result. The value of load resistance is a compromise. Low values reduce feedthrough, but they begin to attenuate the input signal because of the voltage divider formed by R_{ON} and the load. Since R_{ON} varies over the input signal swing, this attenuation produces some undesirable nonlinearity. Excessively low load resistance appears at the switch input, of course, loading the signal source as well. Several possible solutions to this problem (multiple-stage switches, R_{ON} cancellation) are shown in Section 6.18. An attractive alternative is to use a second FET switch section to connect the output to ground when the FET that switches the signal is in the OFF state, thus effectively forming an SPDT switch (more on this in Section 6.18).

CMOS linear switches

Frequently it is necessary to switch signals that may go nearly to the supply voltages. In that case the simple switch circuit just described won't work, since the gate is not forward-biased at the peak of the signal swing. The solution is to use paralleled complementary MOSFET (CMOS) switches (Fig. 6.34). The triangular symbol is a digital inverter, which we will discuss shortly; it inverts a HIGH input to a LOW output, and

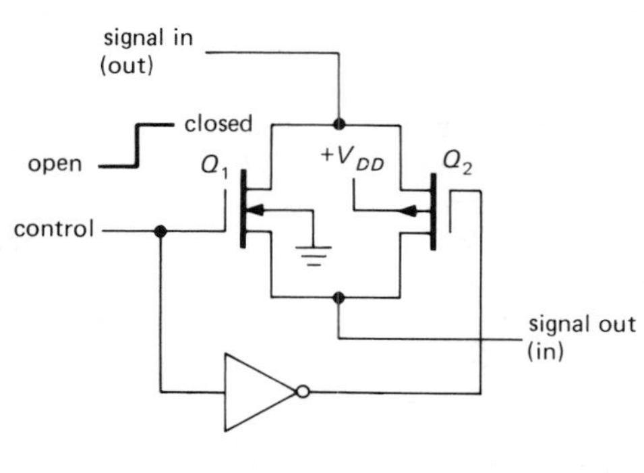

Figure 6.34

vice versa. When the control input is high, Q_1 is held on for signals from ground to within a few volts of V_{DD}. Q_2 is likewise held on (by its grounded gate) for signals from V_{DD} to within a few volts of ground. Thus signals anywhere between V_{DD} and ground are passed through with low series resistance. Bringing the control signal to ground turns off both FETs, providing an open circuit. The result is an analog switch for signals between ground and V_{DD}. This is the basic construction of the 4066 CMOS "transmission gate." It is bidirectional, like the switches described earlier; either terminal can be the input.

There is a variety of CMOS switches available, with various switch configurations (e.g., several independent sections with several poles each). The 4066 is the classic 4000 series CMOS "analog transmission gate," just another name for an analog switch for signals between ground and a single positive supply. The IH5040 and IH5140 series from Intersil and the DG305 series from Siliconix are very convenient to use; they accept TTL control voltages, they will handle analog signals to ± 15 volts, they come in a variety of configurations, and they have relatively low ON resistance (35Ω for some members of these families). Analog Devices and Harris also manufacture nice analog switches.

JFET linear switches

JFETs can also be used conveniently as linear switches, but you have to be more careful about gate signals so that gate conduction doesn't occur. Figure 6.35 shows a typical arrangement. The gate is held well below ground to keep the JFET pinched off. This means that if the input signals go negative, the gate must be held at least V_P below the most negative input

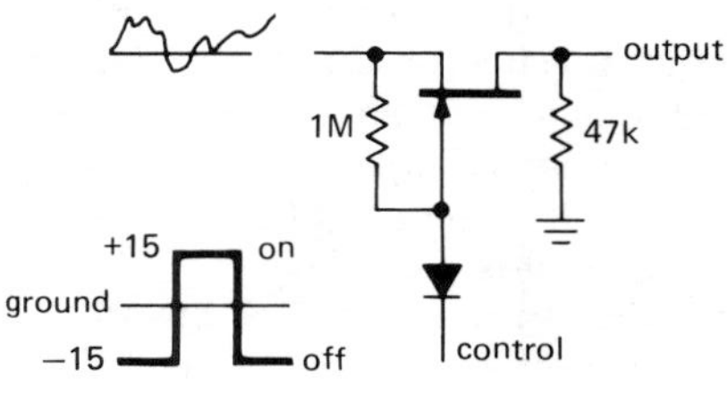

Figure 6.35

swing. To bring the FET into conduction, the control input is brought more positive than the most positive input excursion. The diode is then reverse-biased, and the gate rides at source voltage via the 1M resistor.

The awkwardness of this circuit probably accounts for much of the popularity of MOSFETs in linear switch applications. However, it is possible to devise an elegant JFET linear switch circuit if you use an op-amp, since you can tie the JFET source to the virtual ground at the summing junction of an inverting amplifier. Then you simply bring the gate to ground potential to turn the JFET on. This arrangement has the added advantage of providing a method of canceling precisely the errors caused by finite R_{ON} and its nonlinearity. Figure 6.36 shows the circuit.

There are two noteworthy features of the circuit in Figure 6.36: (a) When Q_1 is ON

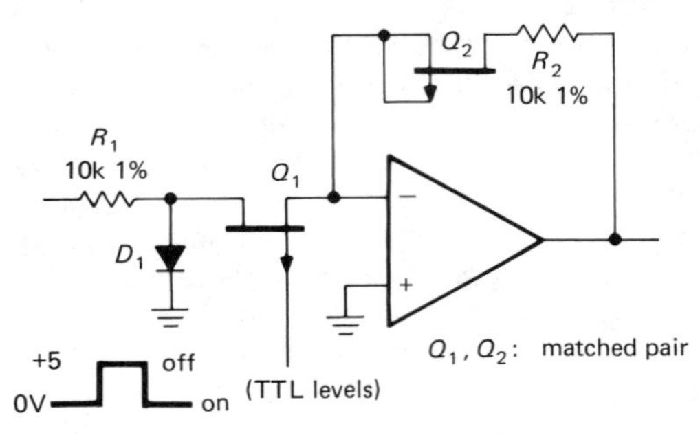

Figure 6.36
JFET-switched amplifier with R_{ON} cancellation.

(gate grounded), the overall circuit is an inverter with identical impedances in the input and feedback circuits. That results in the cancellation of any effects of finite or nonlinear ON resistance, assuming the FETs are matched in R_{ON}. (b) Because of the low pinch-off voltage of JFETs, the circuit will work well with a control signal of zero to +5 volts, which is convenient with TTL logic. The inverting configuration, with Q_1's source connected to a virtual ground (the summing junction), simplifies circuit operation, since there are no signal swings on Q_1's source in the ON state; D_1 prevents FET turn-on for positive input swings when Q_1 is OFF, and it has no effect when the switch is closed.

There are *p*-channel JFETs with low pinch-off voltages available in useful configurations at low prices. For example, the IH5009–IH5024 family includes devices with four input FETs and one cancellation FET in a single DIP package, with R_{ON} of 100 ohms and a price less than two dollars. Add an op-amp and a few resistors and you've got a 4-input multiplexer (see Section 6.14).

The same R_{ON} cancellation trick can be used with MOSFET switches.

Let's make a short digression at this point to discuss the application of FETs to "logic switching," the control of ON/OFF logic levels rather than the continuous analog signals we've been discussing. You will see much more of this in Chapter 8, but it is important to understand logic inverters now, in order to go further with analog switching applications.

6.13 FETs as logic switches

FETs can be used to switch a load to ground, just as with bipolar transistors (Fig. 6.37).

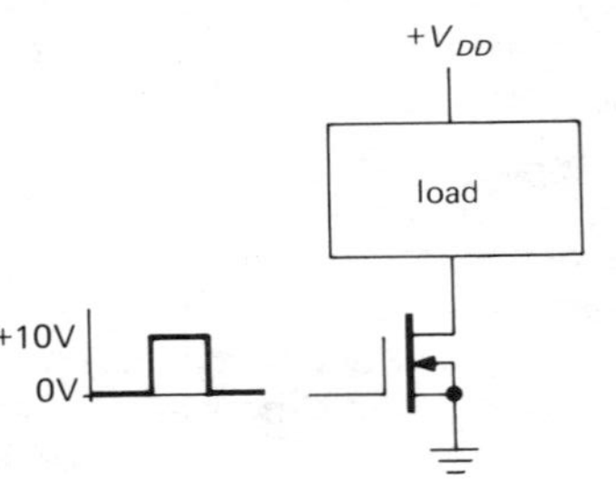

Figure 6.37

This is really a logic switch; it's either ON or OFF (output LOW or HIGH, respectively), when driven from logic levels. A FET used this way is a particularly convenient interface from logic levels to high-current or high-voltage loads, since the gate draws no current from the driving logic. With VMOS power FETs you can switch 10 amps or more with a single FET driven from CMOS levels. A simple variation is the FET inverter (Fig. 6.38). Both circuits are inverters. The *n*-channel version pulls the output to ground when the gate goes high, whereas the *p*-channel version pulls the resistor high for grounded input. We will talk about these NMOS and PMOS logic switches in greater detail in Chapter 9.

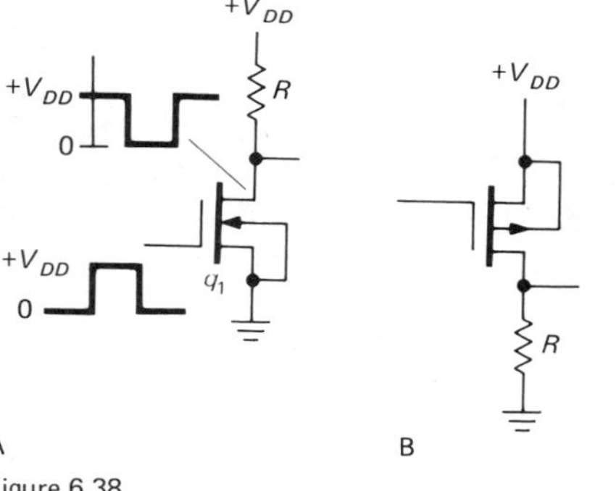

Figure 6.38

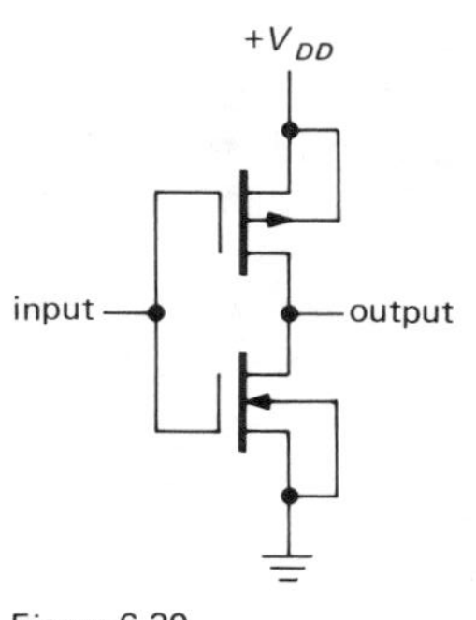

Figure 6.39
CMOS logic inverter.

CMOS inverter

Both of the preceding circuits have the disadvantage of drawing current in the ON state and having relatively high output impedance in the OFF state. Figure 6.39 shows a better circuit. You might think of it as a "push-pull switch": Input grounded cuts off the bottom transistor and turns on the top transistor, pulling the output high. High input ($+V_{DD}$) does the reverse, pulling the output to ground. It's an inverter with low output impedance in either state and no quiescent current whatsoever. It's called a CMOS (complementary MOS) inverter, and it is the basic structure of the CMOS logic family. You will see much more of this later; for now, it should be evident that the CMOS family is a low-power digital family with high-impedance inputs, and outputs that swing the full supply range. It is important to note that CMOS inverters have one unpleasant property: As the input jumps between the supply voltage and ground, there is a region where both FETs are conducting, resulting in large current spikes from V_{DD} to ground. You will see some consequences of this in Chapters 8 and 9 in connection with CMOS logic.

6.14 FET linear switch applications

Integrators

Figure 6.40 shows two op-amp integrator circuits that have FET switches to "zero" the capacitor. In the first circuit, a JFET is used to short the integration capacitor. Since it is an *n*-channel depletion-mode device, the gate must be held negative to keep the switch "open." The diode and resistor in the gate circuit ensure that the gate is brought only to zero volts during the discharge pulse. Since the source is also at ground (the inverting input is a virtual ground), zero gate voltage puts the FET into conduction. This circuit works only for positive output polarity.

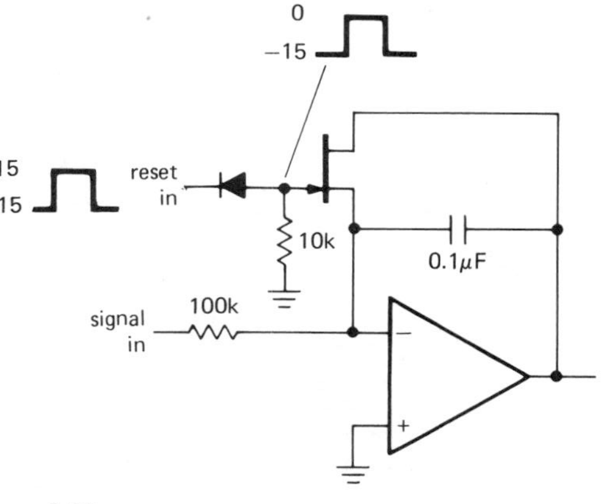

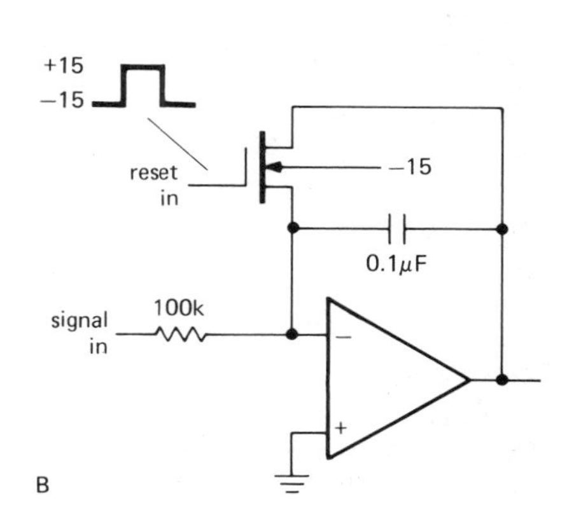

Figure 6.40

In the second circuit, an enhancement-mode MOSFET does the switching. Here the gate circuit is simplified, and the circuit works with signals of either polarity. If the output is known to be of positive polarity only, the gate signal can go from ground to some positive voltage, with the body terminal grounded.

See Section 6.18 for an interesting circuit modification that eliminates the effects of leakage currents in the FET.

Multiplexers

A nice application of FET switches is the "multiplexer" (or MUX), a circuit that allows you to select any of several inputs, as specified by a digital control signal. Since a FET that is ON looks like a small resistor (R_{ON}), such a circuit is an analog (or linear) multiplexer, and it will faithfully pass through to the output the actual voltage present on the selected input. Figure 6.41 shows the basic

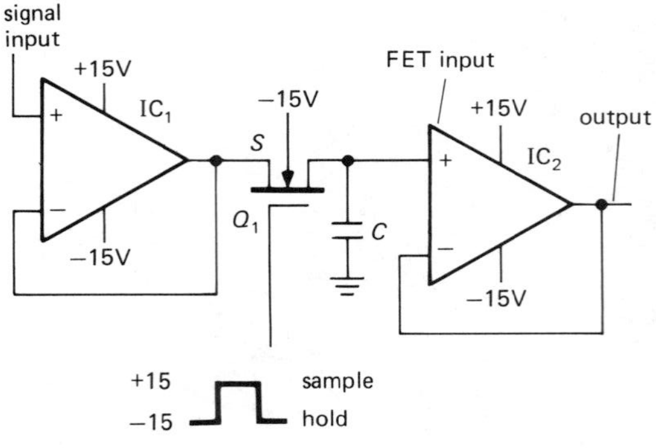

Figure 6.41

scheme. Each of the switches SW0 through SW3 is a CMOS transmission gate of the type discussed in the preceding section. The "select logic" decodes the address and "enables" (jargon for "turns on") the addressed switch only, disabling the remaining switches. Such a multiplexer will usually be used in conjunction with digital circuitry that will generate the appropriate addresses. A typical situation might involve a data-acquisition instrument in which a number of analog input voltages must be sampled in turn, converted to digital quantities, and recorded (or become the input to some on-line computations done by associated computing apparatus).

Since transmission gates are bidirectional, an analog multiplexer such as this is also a "demultiplexer": A signal can be fed into the "output" and will appear on the selected "input." When we discuss digital circuitry in Chapters 8 and 9, you will see that an analog multiplexer such as this can also be used as a "digital multiplexer/demultiplexer," since logic levels are, after all, nothing but voltages that happen to be interpreted as 1s and 0s.

Typical of analog multiplexers are the 506, 507, and 508 series and the IH6108 and IH6116 types, 8- or 16-input MUX circuits that accept TTL or CMOS logic levels for the address inputs and operate with analog voltages up to 35 volts. The 4051–4053 devices in the CMOS digital family are analog multiplexers/demultiplexers with up to 8 inputs, but with 15 volt pp maximum signal levels.

Sample-and-hold circuits

FET switches are the basic ingredients of "sample-and-hold" and "peak-detector" circuits. Figure 6.42 shows the idea. IC_1 is a follower to provide a low-impedance replica of the input. Q_1 passes the signal through

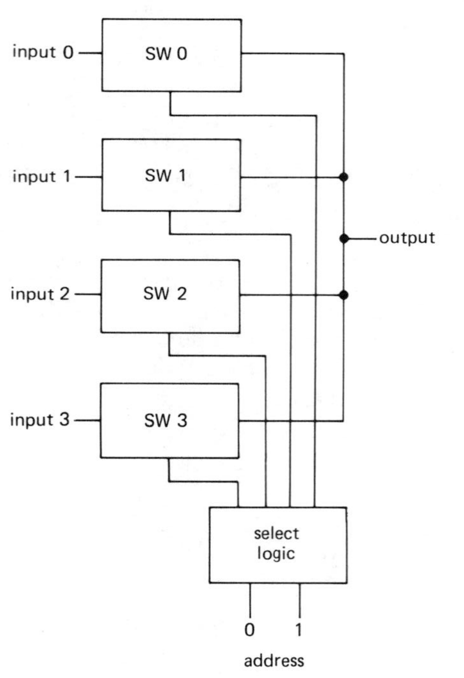

Figure 6.42

during "sample" and disconnects it during "hold." Whatever signal was present when Q_1 was turned OFF is held on capacitor C. IC_2 is a high-input-impedance follower (FET inputs), so that capacitor current during "hold" is minimized. The value of C is a compromise: Leakage currents in Q_1 and the follower cause C's voltage to "droop" during the "hold" interval, according to $dV/dT = I_{leakage}/C$. Thus C should be large to minimize droop. But Q_1's ON resistance forms a low-pass filter in combination with C, so C should be small if high-speed signals are to be followed accurately. IC_1 must be able to supply C's charging current $I = CdV/dT$ and must have sufficient slew rate to follow the input signal. In practice, the slew rate of the whole circuit will be limited by IC_1's output current and Q_1's ON resistance. See Section 6.18 for an improved sample-and-hold circuit.

EXERCISE 6.7

Suppose IC_1 can supply 10mA of output current, and $C = 0.01\mu F$. What is the maximum input slew rate the circuit can accurately follow? If Q_1 has 50 ohms ON resistance, what will be the output error for an input signal slewing at $0.1V/\mu s$? If the combined leakage of Q_1 and IC_2 is 1nA, what is the droop rate during the "hold" state?

FETs are useful in peak-detector circuits, in which the highest value of a waveform is held on a capacitor. Figure 6.43 shows one possibility. IC_1 drives C_1 to the highest point reached by the input waveform since the last reset pulse applied to Q_1's gate. IC_2 buffers the output. IC_1 could take its feedback from

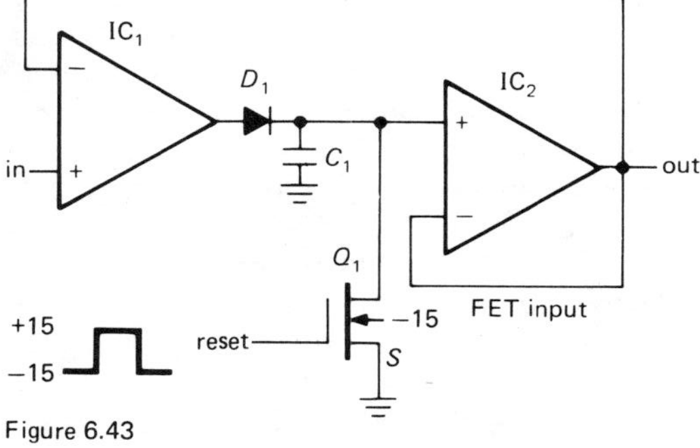

Figure 6.43

C_1, but the connection shown makes it easier to eliminate offsets, at the risk of reduced stability. IC_2 should be a FET-input op-amp, and D_1 should be a low-leakage diode, to minimize droop.

6.15 Limitations of FET switches

Speed

FET switches have ON resistances R_{ON} of 25 to 200 ohms. In combination with substrate and stray capacitances, this resistance forms a low-pass filter that limits operating speeds to frequencies of a few megahertz or less. FETs with lower R_{ON} tend to have larger capacitance (up to 50pF with some MUX switches), so no gain in speed results.

ON resistance

CMOS switches operated from a relatively high supply voltage (15V, say) will have low R_{ON} over the entire signal swing, because one or the other of the transmission FETs will have a forward gate bias at least half the supply voltage. However, when operated with lower supply voltages, the switch's R_{ON} value will rise, the maximum occurring when the signal is about halfway between the supply and ground (or halfway between the supplies, for dual-supply voltages). Figure 6.44 shows why. As V_{DD} is reduced, the

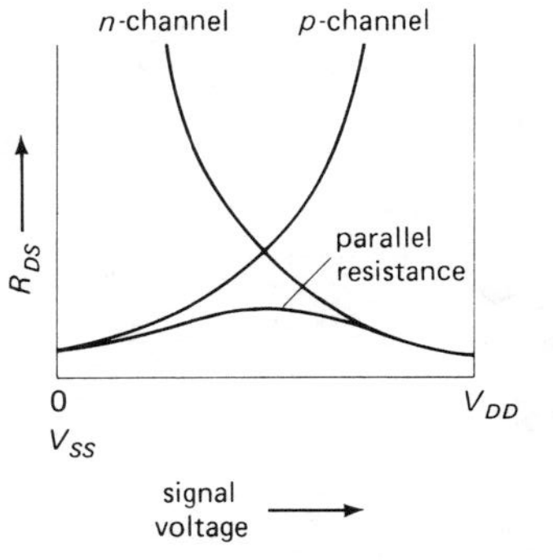

Figure 6.44

FETs begin to have significantly higher ON resistance at $V_{GS} = 0.5V_{DD}$, since for enhancement-mode FETs V_T is at least a few volts, and a gate-source voltage of as much as 5 to 10 volts is required to achieve low R_{ON}. Not only will the parallel resistances of the two FETs rise for signal voltages between the supply voltage and ground, but also the peak at $0.5V_{DD}$ will rise as V_{DD} is reduced, and for sufficiently low V_{DD} the switch will become an open circuit for signals near $0.5V_{DD}$. Figure 6.45 shows the

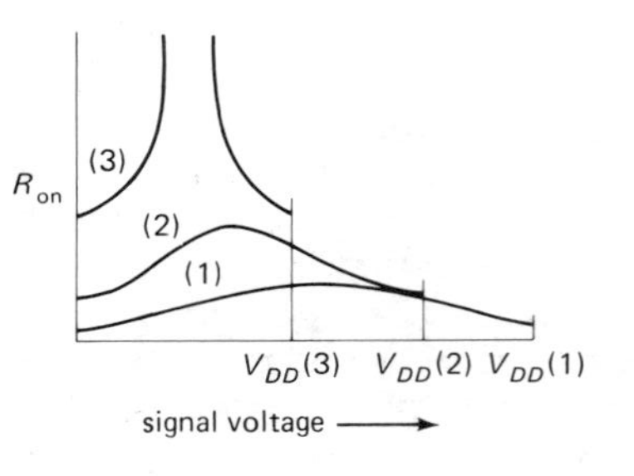

Figure 6.45

sort of behavior you will see for several successively lower values of V_{DD}.

Glitches

During turn-on and turn-off transients, FET transmission gates can do nasty things. The control signal being applied to the gate can couple capacitively to the channel, putting ugly transients on your signal. The situation is most serious if the signal is at high impedance levels. Multiplexers can show similar behavior during transitions of the input address, as well as momentary connection between inputs if turn-off delay exceeds turn-on delay.

Let's look at this problem in a bit more detail. Figure 6.46 shows a typical wave-

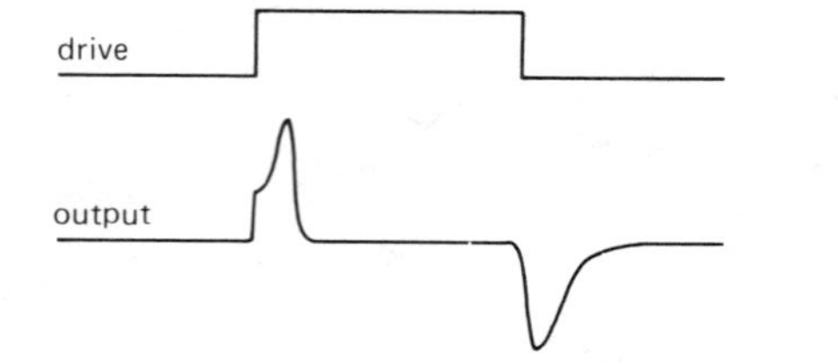

Figure 6.46

form you might see at the output of a MOSFET analog switch circuit similar to Figure 6.33, with an input signal level of zero volts and an output load consisting of 10k in parallel with 20pF, realistic values for an analog switch circuit. The handsome transients are caused by charge transferred to the channel, through the gate-channel capacitance, at the transitions of the gate. The gate makes a sudden step from one supply voltage to the other, in this case between ±15 volt supplies, transferring a slug of charge

$$Q = \pm C_{GC}(V_{G(HIGH)} - V_{G(LOW)})$$

C_{GC} is the gate-channel capacitance, typically around 5pF. Note that the amount of charge transferred to the channel depends only on the total voltage change at the gate, not on its rise time. Slowing down the gate signal gives rise to a smaller-amplitude glitch of longer duration, with the same total area under its graph. Low-pass filtering of the switch's output signal has the same effect. Such measures may help if the peak amplitude of the glitch must be kept small, but in general they are ineffective in eliminating gate feedthrough. In some cases the gate-channel capacitance may be predictable enough for you to cancel the spikes by coupling an inverted version of the gate signal through a small adjustable capacitor.

The gate-channel capacitance is distributed over the length of the channel, which means that some of the charge is coupled back to the switch's input. As a result, the size of the output glitch depends on the signal source impedance and is smallest when the switch is driven by a voltage source. Of course, reducing the load impedance will reduce the size of the glitch, but this also loads the source and introduces error and nonlinearity due to finite R_{ON}. Finally, all other things being equal, a switch with smaller gate-channel capacitance will introduce smaller switching transients, although you pay a price in the form of increased R_{ON}.

Figure 6.47 shows an interesting comparison of gate-induced charge transfers for three kinds of switches. In all cases the gate signal is making a full swing, i.e., a 30 volt swing for PMOS and CMOS, and a swing from −15 volts to the signal level for the *n*-channel JFET switch. The JFET switch shows a strong dependence of glitch size on signal, because the gate swing is propor-

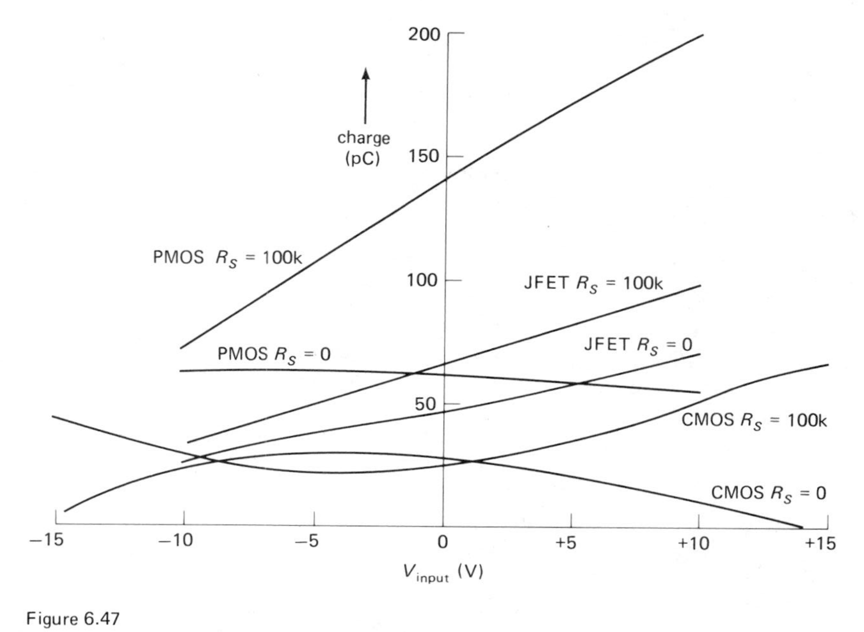

Figure 6.47
Charge transfer for various FET linear switches as a function of signal voltage.

tional to the level of the signal above −15 volts. The CMOS switch has relatively low feedthrough because the charge contributions of the complementary MOSFETs tend to cancel out (one gate is rising while the other is falling). Just to give scale to these figures, it should be pointed out that 30pC corresponds to a 3mV step across a 0.01μF capacitor. That's a rather large filter capacitor, and you can see that this is a real problem, since a 3mV glitch is pretty large when dealing with low-level analog signals.

Handling precautions

MOSFETs have very high gate impedance and moderately high breakdown voltage, and as a result they can be easily damaged by static charge. In a classic situation you have a MOSFET or MOSFET device in your hand. You walk over to your circuit, stick the device into its socket, and turn on the power, only to discover that the FET is dead. You killed it! You should have grabbed onto the circuit with your other hand before inserting the device. This would have discharged your static voltage, which in winter can reach thousands of volts. MOS devices don't take kindly to "carpet shock" (for purposes of static electricity, you are equivalent to 100pF in series with about 1k; in winter your capacitor may charge to 10kV with a bit of shuffling about on a fluffy rug, and even a simple arm motion with shirt or sweater can generate 1kV).

MOS devices should be shipped in conductive foam or bags, and you have to be careful about voltages on soldering irons, etc., during fabrication. It is best to ground soldering irons, table tops, etc., and use conductive wrist straps. Once the device is safely soldered into its circuit, the chances for damage are greatly reduced, especially since many MOS devices have protection diodes in the input gate circuits. Although the internal protection networks of resistors and clamping diodes (or sometimes zeners) compromise performance somewhat, it is often worthwhile to choose those devices because of the greatly reduced risk of damage by static electricity. If a FET comes with a protective metal or rubber ring around the leads, don't remove it until the FET is safely installed in the circuit.

SOME ADDITIONAL FET CIRCUIT IDEAS

□ 6.16 Amplifiers

□ *ac Amplifier with bootstrap*

FETs are often used because of their extremely high input impedance. With dc-coupled signals you get the full advantage of the enormous Z_{in}s ($10^{12}\Omega$ to $10^{14}\Omega$), but with ac-coupled signals the input bias resistor dominates the input impedance. The bias resistor can be quite large (typically many megohms), but it still sets the input impedance, at least at low frequencies. Figure 6.48 shows a simple remedy, using the bootstrap principle.

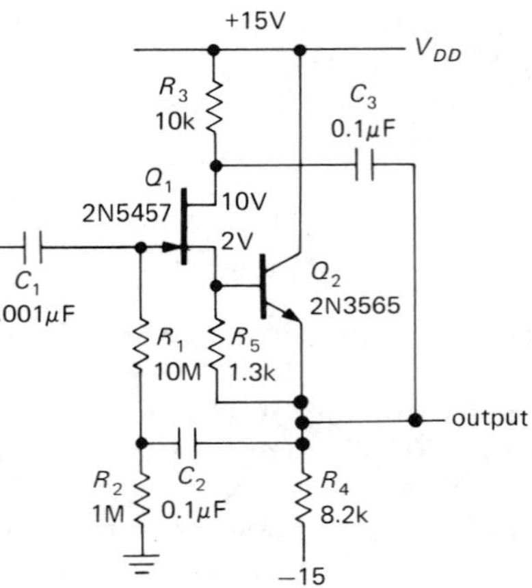

Figure 6.48

As with the transistor bootstrap (see Section 2.16), the bias resistor is split, with the output signal ac-coupled to the junction of R_1 and R_2. Q_2's base resistor provides a current-source load for Q_1's source (as discussed in Section 6.08), resulting in a voltage gain very close to 1.0, and hence effective bootstrapping of R_1 to values of 1000M or so (voltage gain of 0.99). When designing circuits to provide such high input impedances, it is important to have very low input capacitance; otherwise the capacitive reactance at signal frequencies wipes out any advantage gained from bootstrapping the bias network. In this circuit the drain is also bootstrapped, via C_3, to eliminate the effects of gate-drain capacitance. Since gate-source capacitance is automatically bootstrapped by the follower connection, the only remaining capacitance at the input is due to "stray" (wiring) capacitance. If the input signal travels through some cable, a guarded shield, as discussed in Section 6.08, can be used to reduce the effective capacitance of the cable to a negligible value.

□ *FET cascode*

FETs make useful high-frequency amplifiers because of their low feedthrough capacitance. As with bipolar transistors, the Miller effect is the dominant contributor to high-frequency rolloff, and once again it can be eliminated with the cascode configuration. Figure 6.49 shows some possibilities. In Figure 6.49A paired *n*-channel JFETs are connected in the same manner as in the bipolar transistor cascode circuit. Q_2's gate is biased at +5 volts to ensure headroom for Q_1's drain. Q_1 is self-biased, the usual method with depletion-mode FETs.

Dual-gate depletion-mode MOSFETs are available; this is a natural for a cascode. The circuit in Figure 6.49B uses one of these devices, drawn as a MOSFET with two gates. Think of the lower FET's drain as connected to the upper FET's source, as shown in dashed lines (which you never draw, except in textbooks). This circuit is shown with tuned input and output circuits, the usual method in RF amplifiers intended for a limited band of frequencies (for broadband applications you might use untuned RF transformers in the source and drain circuits, rather than resistors). Dual-gate FETs are intended as high-frequency amplifiers (they're good to 250MHz or more) and as mixers. We will have more to say on this subject in Chapter 13.

By using JFETs with tightly controlled I_{DSS}, you can construct a cascode with simplified biasing, as in Figure 6.49C. The 2N5953 has $2.5\text{mA} < I_{DSS} < 5\text{mA}$, whereas the 2N5950 has $10\text{mA} < I_{DSS} < 15\text{mA}$, so the circuit runs at the lower FET's I_{DSS}, with the upper FET's gate back-biased to bring its drain current to the same value.

In Figure 6.49D a cascode JFET, manufactured as a single device, is used in the same configuration as in Figure 6.49C. Once again, no separate bias source is needed for the upper transistor. These devices exhibit

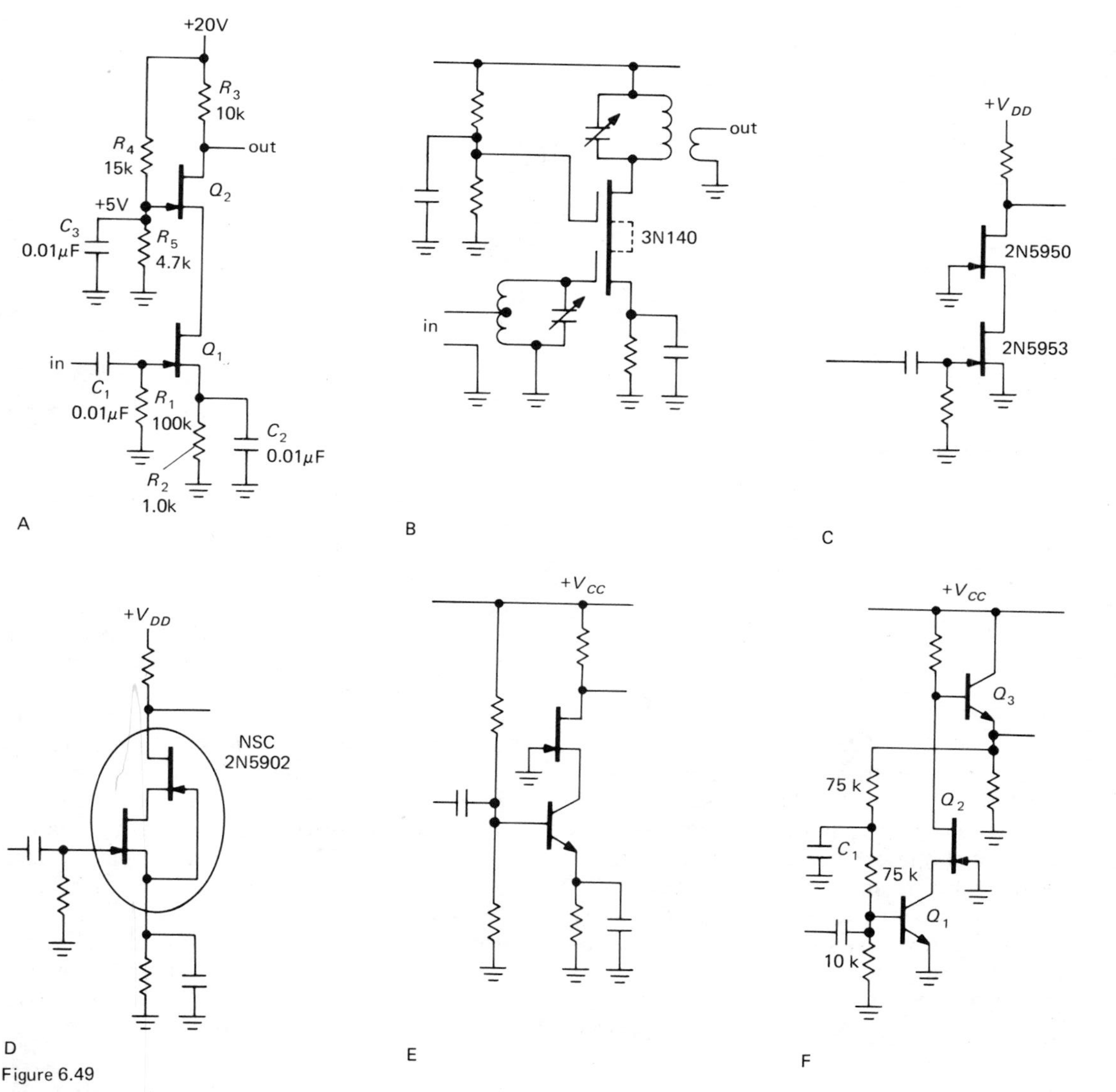

Figure 6.49

FET cascode circuits.

very low dynamic input current because of the reduced drain-gate capacitance. They come as dual cascodes in one package, and they exhibit very low offset voltages because of the matched V_{DS} enforced by the cascode. A similar effect results in good performance of the "no-holds-barred" follower described in Section 6.08.

Figure 6.49E shows a hybrid cascode, attractive because of its simple biasing scheme. Cascodes could also be made with the bipolar transistor on top. In Figure 6.49F the same basic arrangement is used, with feedback at dc to set the quiescent point at $16V_{BE}$, or about 10 volts. C_1 prevents feedback at signal frequencies, for maximum open-loop gain.

□ *CMOS linear amplifiers*

As we mentioned in Section 6.07, a simple technique for getting large single-stage voltage gain from a FET amplifier is to use an active load, instead of a resistor, in the drain circuit. One possibility is to use a current mirror load for a FET differential stage. A nice technique for a single-ended stage is to

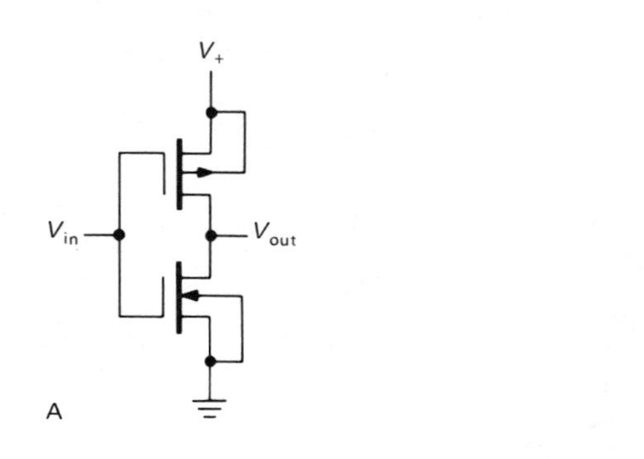

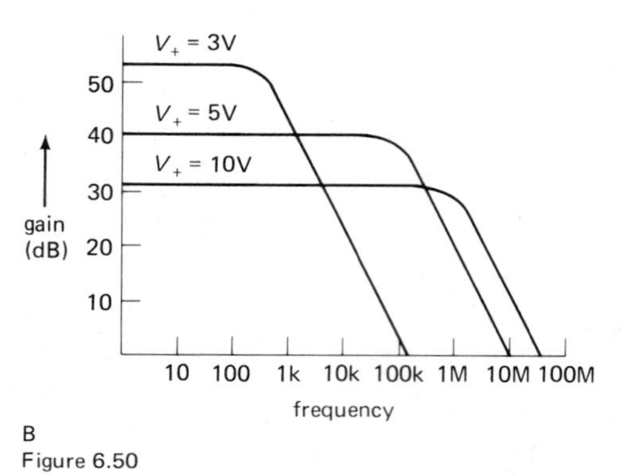

B
Figure 6.50

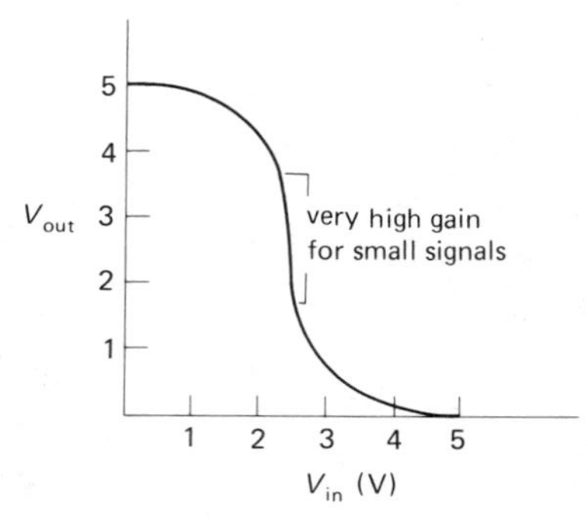

Figure 6.51

use complementary enhancement-mode MOSFETs, i.e., a CMOS inverter (Fig. 6.50). The upper (p-channel) FET serves as active load for the lower (n-channel) FET, and vice versa. This is precisely the classic CMOS inverter circuit you saw earlier as a logic switch, and you will see it again in Chapters 8 and 9 in connection with digital logic. Much of the popularity of this circuit stems from its simplicity and low cost, since CMOS inverters are available 6 to a package for under half a dollar.

This amplifier has very high gain for input levels approximately half the supply voltage. Its transfer characteristic is shown in Figure 6.51. The variation of R_o and g_m with drain current is such that the highest voltage gain occurs for relatively low drain currents, i.e., at low supply voltages (5V is typical). This circuit has the disadvantage of relatively high output impedance, especially when operated with low drain current, poor linearity, and unpredictable gain. However, it is simple and inexpensive, and it is sometimes used to amplify small input signals whose waveforms aren't important. Some examples are proximity switches (which amplify 60Hz pickup), crystal oscillators, and frequency-sensing input devices whose output is a frequency that goes to a frequency counter.

Note that dc feedback is necessary to bring the CMOS amplifier into its active region. Figure 6.52 shows some examples.

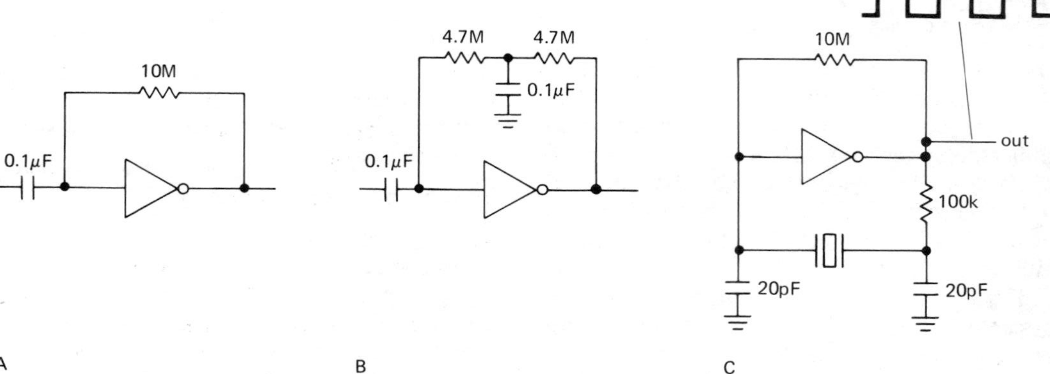

Figure 6.52
CMOS linear amplifier circuits.

The dc feedback resistor in the first circuit lowers the input impedance in the usual way, since it provides negative shunt feedback at signal frequencies also, making the second circuit desirable if a high input impedance at signal frequencies is important. The third circuit is the classic CMOS crystal oscillator discussed in Section 4.16.

□ 6.17 Pinch-off reference

The pinch-off voltage for JFETs is quite stable with temperature, making it feasible to construct a low-power "pinch-off reference" (Fig. 6.53). In the first circuit, the

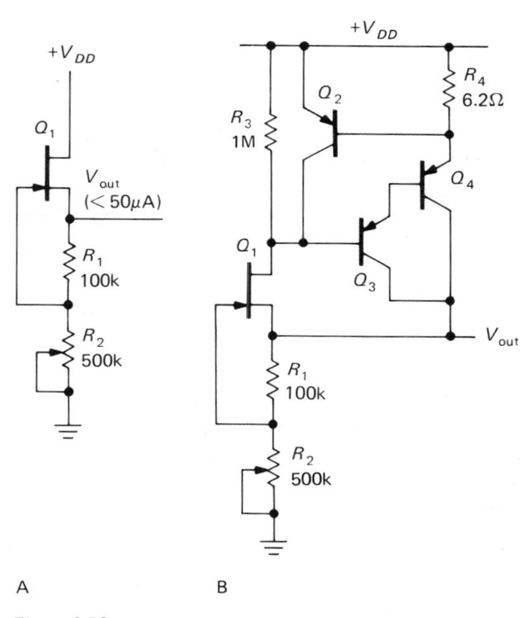

Figure 6.53

FET pinch-off voltage (V_P) reference.

drop across R_1 is essentially Q_1's pinch-off voltage, since the drain current is only a few microamps. R_2 lets you set the output voltage, since the pinch-off voltage appears across the upper leg of the divider formed by the two resistors. This circuit is useful only for very small load currents.

In the second circuit, *pnp* Darlington Q_3Q_4 boosts the output current into the range of milliamps, being biased into conduction by Q_1's drain current of 10μA or so. Q_2 and R_4 provide short-circuit protection, limiting output current to about 100mA. It is important to include such protection in power supplies and references, unless you enjoy replacing a bunch of transistors after you slip while making a measurement with your scope or meter probe. As before, R_2 sets the output voltage. R_2 should provide a wide range of output adjustment in this sort of circuit in order to cope with the disgraceful manufacturing spread in V_P for a given FET type. See Section 5.14 for conventional voltage references.

□ 6.18 Switch circuits

□ *Multiple-stage linear switch*

A nice way of coping with the feedthrough capacitance problem of FET switches is to use several cascaded stages (Fig. 6.54). In the first circuit, capacitive signal feedthrough

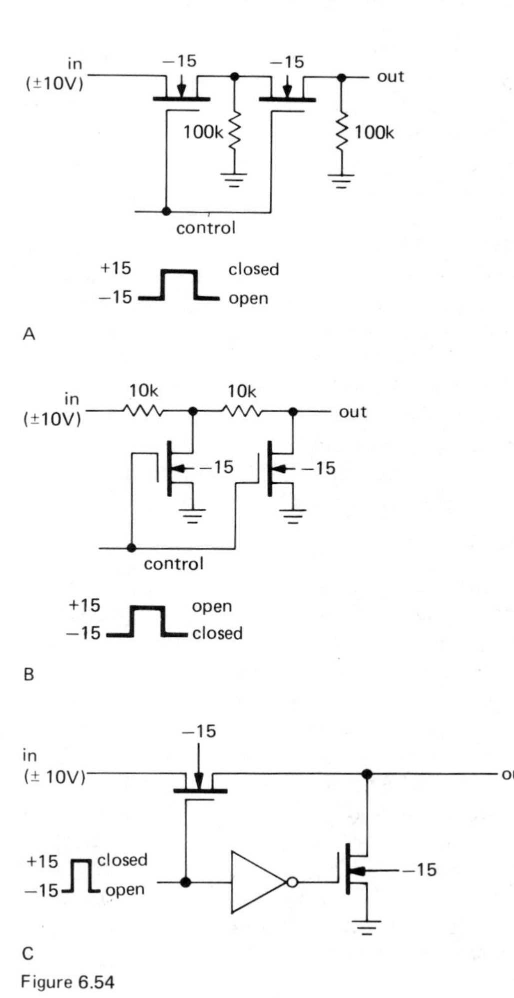

Figure 6.54

of 1% in each section would result in an overall feedthrough of 0.01% (−80dB) in the open state. If R_{ON} is 100 ohms, the signal is essentially unattenuated (99.7% of its input amplitude) when the switch is closed, which means that variations of R_{ON} with signal swing contribute negligible nonlinearity. To get equivalent feedthrough performance with a single stage would require a load resistor of 1.0k, which would result in unsatisfactory attenuation in the closed state (10% attenuation), not to mention nonlinearity due to changing R_{ON} with signal level.

The second circuit does the same sort of thing, using FET switches to short the signal to ground. In this case the use of several stages allows you to keep the series resistors reasonably small.

A third possibility is to use a pair of switches, one in series with the signal and one between the output and ground, as shown in the third circuit in Figure 6.54. By alternately enabling the two switches, you get the best of both worlds, i.e., low feedthrough in the OFF state and good linearity with negligible attenuation in the ON state. CMOS SPDT switches with controlled break-before-make are available commercially in single packages; in fact, you can get a pair of SPDT switches in a single package. Examples are the DG188, IH5042, and IH5142, as well as the DG191, IH5043, and IH5143 (dual SPDT units). Because of the availability of such convenient CMOS switches, it is easy to use this SPDT configuration to achieve excellent performance.

□ *Another look at the integrator*

In the integrator circuits in Section 6.14, drain-source leakage sinks a small current from the summing junction when the FET is OFF. With an ultra-low-input-current op-amp and low-leakage capacitor, this can be the dominant error in the integrator. Figure 6.55 shows a clever circuit solution. Both *n*-channel FETs are switched together, but it may be desirable to switch Q_1 with gate voltages of zero and +15 volts so that all the gate leakage effects are eliminated during the OFF state (zero gate voltage). In the ON state the capacitor is discharged as before, but with twice R_{ON}. In the OFF state,

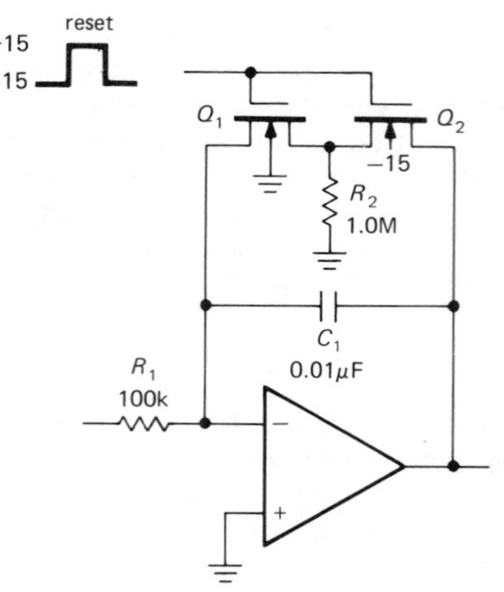

Figure 6.55

Q_2's small leakage passes to ground through R_2 with negligible drop. There is no leakage current at the summing junction because Q_1's source, drain, and substrate are all at the same voltage. Compare this circuit with the "zero-leakage" peak detector in Section 3.15.

EXERCISE 6.8

Assume that Q_1 and Q_2 in Figure 6.55 behave like 10,000M resistors when OFF. Calculate the rate at which the integrator's output will drift, due to FET leakage, for both this configuration and the simple FET reset switch circuit (Fig. 6.40), when the integrator's output is at +10 volts.

□ *Another look at sample-and-hold circuits*

Figure 6.56 shows an improved sample-and-hold circuit. Q_2 and Q_3 are parallel complementary switches, maintaining low ON resistance for all signal levels. Q_1 is turned on during HOLD to prevent saturation in the input op-amp, with its usual problems of slew-rate-limited recovery time. Q_1 could be replaced by a pair of back-to-back diodes. The second op-amp should have a FET input for low droop.

The same trick used to eliminate the effects of drain-source leakage in the integrator (Fig. 6.55) can be used to advantage here. There's a hint in Exercise 6.9.

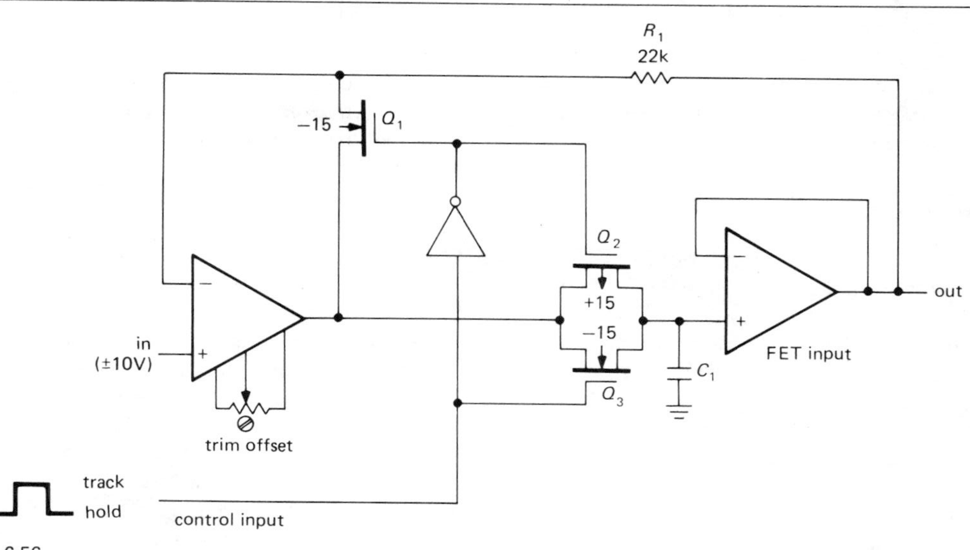

Figure 6.56

EXERCISE 6.9

Modify the sample-and-hold circuit to eliminate droop caused by source-drain leakage in Q_2 and Q_3. Hint: You have to bootstrap the junction of series FETs at the capacitor voltage; however, don't take your bootstrap voltage directly from the capacitor!

6.19 BiFET integrated circuits

Integrated circuits that combine bipolar transistors and FETs are known as BiFET ICs. They offer the best of both worlds, combining traditional linear IC technology (op-amps, comparators, etc.) with the ultralow input current of FETs. We will rather arbitrarily divide BiFETs into "BiJFET" and BiMOS.

p-Channel-JFET-input op-amps

It turns out that *p*-channel JFETs are the easiest to fabricate in the same process with bipolar transistors. They are low-pinch-off FETs, with a V_P of 1 to 2 volts. The 355-357 series of FET op-amps provides popular examples of this kind of chip. The input stage consists of a pair of *p*-channel JFETs with their sources returned to V_+ with a current source (Fig. 6.57). There are a couple of interesting points about the input circuit. First, it turns out that there is a value of drain current for JFETs at which I_D versus V_{GS} has zero temperature coefficient (See Fig. 6.11). This is typically in the range of 25μA to 200μA, and it is the current at which the input FETs are cleverly operated in the 355-357, to keep drifts of offset voltage minimized. The resulting temperature coefficient of offset voltage, 5μV/°C (typ), is quite good even by bipolar transistor standards (it's a lot better than that of the 741, for instance).

The second point to note is that these low values of drain current are really close to pinch-off, which means that the gate is positive by a volt or so relative to the source. As a result, the common-mode input range typically extends to the positive supply rail, and even a bit beyond. This is a useful virtue in a number of applications, e.g., constant-current supplies with sensing at the high side.

The rest of the op-amp is constructed with bipolar transistor circuitry: a differential stage with Miller compensation capacitor,

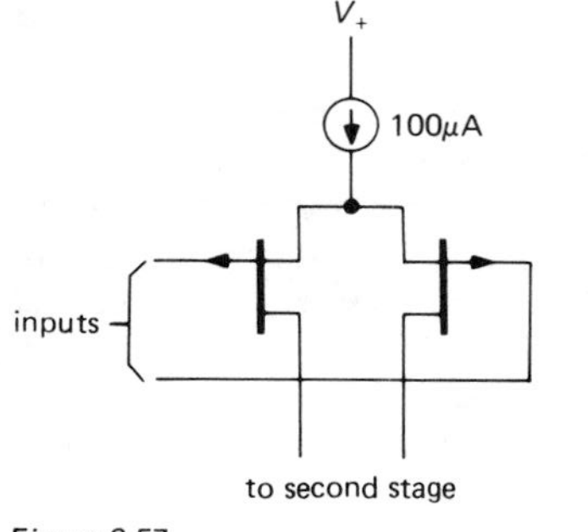

Figure 6.57

followed by the usual unity-gain push-pull follower with protective current limiting.

Another BiJFET chip is the LF311, a FET equivalent of the popular 311 comparator.

MOSFETs in op-amps

p-Channel enhancement-mode MOSFETs are used in a number of BiFET op-amps, among them the 3130/3140/3160 series and the ICH8500 ultra-low-current op-amp. The basic input scheme is again a *p*-channel differential pair, biased by a current source and driving a bipolar transistor current mirror load. Since MOSFETs are easily damaged by static discharge, input-protection diodes are added at the inputs.

The common-mode input situation again has an unusual feature. Since the input transistors are enhancement-mode devices, the inputs can be driven 0.5 volt below the negative supply rail. This is handy in single-supply circuits. MOSFET-input op-amps provide the lowest input-current specification (an astounding 0.01pA over the full temperature range for the ICH8500).

The 3130/3160 use complementary MOSFETs for the *output* stage also. This has the virtue of providing output swing capability all the way to both supply rails, a very useful feature in circuits where you are counting on the output voltage going to the supplies. An example is shown in the triangle generator in Section 3.28. These particular op-amps have a limited range of supply voltage and cannot be operated from ± 15 volts. Their differential input voltage is limited to a scanty ± 8 volts. The inexpensive BiMOS CA080 series will operate from standard ± 15 volt supplies, and it has improved common-mode and differential input ranges. It competes with the popular and inexpensive TL080 series of BiJFET op-amps.

Other linear IC functions can take advantage of BiMOS or BiFET technology. As an example, the 3290 is a dual comparator with MOSFET inputs, and the LF398 is a monolithic sample-and-hold chip that uses *p*-channel JFETs as linear switches.

Linear ICs constructed with MOSFETs can have a few drawbacks, most notably a higher noise level and a tendency for the offset voltage to drift with time if a differential voltage is applied across the input. These effects are treated in Chapter 7 in connection with precision design and low-noise circuits.

6.20 Power MOSFETs

Enhancement-mode MOSFETs capable of high voltage and current levels have recently become available, and at affordable prices. These devices go by names such as VMOS (vertical-groove MOS) and HEXFET, which uses a process developed by International Rectifier (IR). They can handle surprisingly high voltages and currents. The HEXFET line currently includes a device capable of 28 amps maximum drain current, 100 volts maximum drain-source voltage, and 0.055 ohm maximum ON resistance. Other types are available with 400 volts breakdown voltage. In this country there is plenty of activity by companies such as Siliconix, Supertex, and IR, and the larger semiconductor manufacturers (Motorola, RCA, TI) are getting into power MOSFETs. In Japan, where the process was originally developed, power MOSFETs are manufactured by several companies. Small VMOS devices sell for less than a dollar, and they should be considered in many applications traditionally reserved for bipolar power transistors.

Among the advantages of power MOS transistors are high input impedance (watch out for high input capacitance, though, particularly with the high-current devices) and complete absence of second breakdown and thermal runaway. Since MOS drain current drops with increasing temperature, you don't get the junction hot spots that lead to second breakdown in bipolar transistors and hence limit the safe operating area (on a plot of collector current versus collector voltage, see Section 5.07) to a region smaller than that allowed by transistor dissipation limits. Similarly, power amplifiers don't have the nasty runaway tendencies that we've all grown to love in bipolar transistors, and as an added bonus, power MOSFETs can be paralleled without the current-equalizing resistors that are necessary with bipolar transistors.

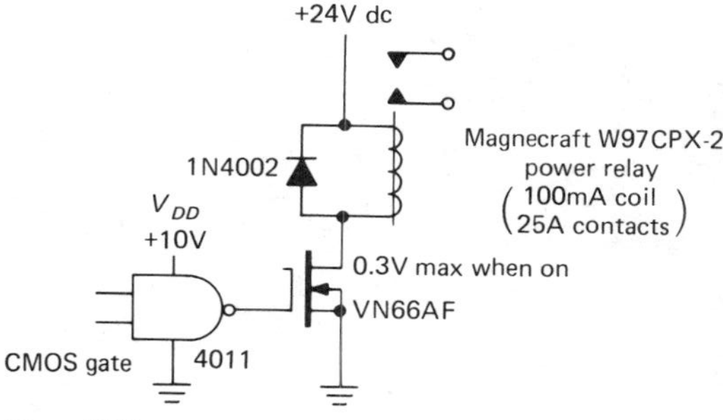

Figure 6.58

As a saturated switch, the power MOSFET is a natural when interfacing from low-power logic such as CMOS, or even plain old TTL. Some devices have specified values for $R_{DS(ON)}$ with only 5 volts of gate drive, and they can therefore be driven directly from TTL (use a pull-up resistor), but most current VMOS devices require the higher drive levels of CMOS (operating from 10–15V supplies) to give really good turn-on. Figure 6.58 shows the simplest example of a CMOS gate driving a high-current load.

Keep in mind that an ON MOSFET behaves like a small resistance, rather than a saturated bipolar transistor, for small values of drain voltage. This can be an advantage, since the "saturation voltage" goes to zero for small drain currents; on the other hand, it means that the drain voltage will vary linearly with drain current, rather than showing the rather stiff saturation characteristic of bipolar power transistors.

Power MOSFETs are also available as *p*-channel devices, although there tends to be a greater variety available among the *n*-channel devices. Low-capacitance devices such as the VMP4 and DV1000 series (in power stripline packages) from Siliconix make it possible to use these handy devices for radiofrequency circuits. The next few years will undoubtedly see a blossoming of this relatively new technology. See Table 6.4 for a sampling of the currently available types.

SELF-EXPLANATORY CIRCUITS

6.21 Circuit ideas

Figure 6.59 presents a sampling of FET circuit ideas.

6.22 Bad circuits

Figure 6.60 presents a collection of bad ideas, some of which involve a bit of subtlety. You'll learn a lot by figuring out why these circuits won't work.

TABLE 6.4. POWER MOSFETs

Type	Gate protect.[a]	BV_{DS} (V)	BV_{GS} rev (V)	BV_{GS} fwd (V)	I_D max (A)	$R_{DS(ON)}$ max (Ω)	@	V_{GS} (V)	$V_{GS(th)}$[b] min (V)	$V_{GS(th)}$[b] max (V)	C_{rss}[c] max (pF)	θ_{JC} (°C/W)	P_{diss} @ $T_c = 75°C$ (W)	Case	Comments
***n*-channel**															
VN10KM	✓	60	0.3	15	0.5	5		10	0.3	2.5	2	125[d]	0.6	TO-92+	Siliconix; small pkg. inexpensive
VN33AJ	—	35	30	30	2	1.8		10	0.8	2	10	5	15		
VN66AJ	—	60	↓	↓	↓	3		↓	↓	↓	↓	↓	↓	TO-3	Siliconix
VN98AJ	—	90				4									
VN64GA	—	60	30	30	12.5	0.4		12	1	4[e]	40	1.6	50	TO-3	Siliconix
VN66AF	✓	60	0.3	15	2	3		10	0.8	—	10	8	9	TO-202	Inexpensive; $R_{ON} = 5\Omega$ @ 5V
VN88AF	✓	80	↓	↓	↓	4		↓	↓	—	↓	↓	↓		Inexpensive
VN1200A	—	120	20	20	20	0.1		10	3	5[e]	100	1.1	70	TO-3	Siliconix; inexpensive; TO-220 available
VN4000A	—	400	20	20	8	1		10	3	5[e]	30	1.2	60	TO-3	
2SK133-	—	120	14	14	7	1.7		12	0	1.5[f]	15	1.25	60	TO-3	Hitachi; 2SJ48-50 compl.
2SK135	—	160	↓	↓	↓	↓		↓	↓	↓	↓	↓	↓	↓	
IRF130-	—	100	20	20	12	0.18		10	1	3	100	1.67	45	TO-3	HEXFET
IRF133	—	60	↓	↓	10	0.25		↓	↓	↓	↓	↓	↓	↓	
IRF150-	—	100	20	20	28	0.055		10	1	3	500	0.83	90	TO-3	Premium; lowest R_{ON}; HEXFET
IRF153	—	60	↓	↓	24	0.08		↓	↓	↓	↓	↓	↓	↓	
IRF530-	—	100	20	20	10	0.18		10	1.5	3.5	100	1.67	45	TO-220	Plastic HEXFET
IRF533	—	60	↓	↓	8	0.25		↓	↓	↓	↓	↓	↓	↓	
IRF330-	—	400	20	20	4	1		10	1	3	20	1.67	45	TO-3	HEXFET
IRF333	—	350	↓	↓	3.5	1.5		↓	↓	↓	↓	↓	↓	↓	
IRF350-	—	400	20	20	11	0.3		10	1	3	200	0.83	90	TO-3	Premium; high voltage; HEXFET
IRF353	—	350	↓	↓	10	0.4		↓	↓	↓	↓	↓	↓	↓	
IRF730-	—	400	20	20	3.5	1		10	1.5	3.5	20	1.67	45	TO-220	Plastic HEXFET
IRF733	—	350	↓	↓	3	1.5		↓	↓	↓	↓	↓	↓	↓	

2N6656	✓	35	0.3	15	2	1.8	10	0.8	2	10	5	15	TO-3	
2N6657	✓	60	↓	↓	↓	3	↓	↓	↓	↓	↓	↓	↓	
2N6658	✓	90				4								
2N6659	✓	35	0.3	15	2	1.8	10	0.8	2	10	15	5	TO-39	
2N6660	✓	60	↓	↓	↓	3	↓	↓	↓	↓	↓	↓	↓	2N6656-58 in TO-39 pkg.
2N6661	✓	90				4								
VN0106N3	—	60	20	20	2	4	10	0.8	2.4	10	125[d]	0.6	TO-92	
VN0206N2	—	60	↓	↓	4	2	↓	↓	↓	20	20	3.75	TO-39	Supertex, Inc.; many voltages and packages avail.; VP series compl.
VN1206N5	—	60			16	0.4			2.4[e]	50	8	9	TO-220	
VN0340N1	—	400			4	3		1	3[e]	25	5	15	TO-3	
IVN5200TND	—	40	30	30	4	0.5	10	0.8	2[g]	60	10	7.5	TO-39	Intersil
IVN5200TNF	—	80	↓	↓	↓	↓	↓	↓	↓	↓	↓	↓	↓	
IVN5200KND	—	40	30	30	5	0.5	10	0.8	2[g]	60	2.5	30	TO-3	Intersil; TO-220 and TO-66 avail.
IVN5200KNF	—	80	↓	↓	↓	↓	↓	↓	↓	↓	↓	↓	↓	
***p*-channel**														
2SJ48-	—	120	14	14	7	1.7	12	0	1.5[f]	15	1.25	60	TO-3	Hitachi; 2SK133-35 compl.
2SJ50	—	160	↓	↓	↓	↓	↓	↓	↓	↓	↓	↓	↓	
VP0106N3	—	60	20	20	1	8	10	1.5	3.5	10	125[d]	0.6	TO-92	Supertex, Inc.; many voltages and packages avail.; VN series compl.
VP0206N2	—	60	↓	↓	2	4	↓	↓	↓	20	20	3.75	TO-39	
VP1206N5	—	60			6	0.6			3.5[e]	50	8	9	TO-220	
VP0800A	—	80	20	20	2	2	10	3	5[e]	—	1.1	70	TO-3	Siliconix; inexpensive
UHF type (*n*-channel)														
VMP4	—	60	30	30	2	—	—	—	—	6.5	5	15	Stripline	Siliconix; 10dB @ 20MHz
DV1006	—	65	40	40	2	—	—	—	—	3.5	4.4	17	Stripline	Siliconix; 25 W 10dB @ 175 MHZ
DV1007	—	↓	↓	↓	4					7	2.2	35	Stripline	Siliconix; 50 W 10dB @ 175MHZ
DV1008	—				8	—	—	—	—	14	1.1	70	Stripline	Siliconix; 100W 10dB @ 175MHz

[a] Gate-protection diodes are losing favor with power MOSFET designers, for several reasons. Power MOSFETs, with their large input capacitance, are quite rugged and are not easily damaged. The input-protection diodes restrict gate swing (in the OFF direction) and have failure problems of their own. Old habits die hard, though, so power MOSFETs with gate-protection diodes are still availabe. [b] $I_D = 1$mA. [c] $V_{DS} = 25$V. [d] θ_{JA}. [e] $I_D = 10$mA. [f] $I_D = 100$mA. [g] $I_D = 5$mA.

Circuit ideas

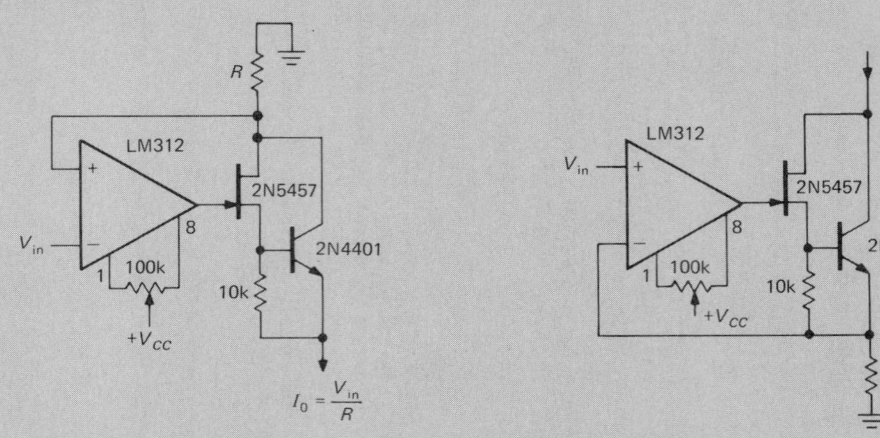

A. precision current source

B. precision current sink

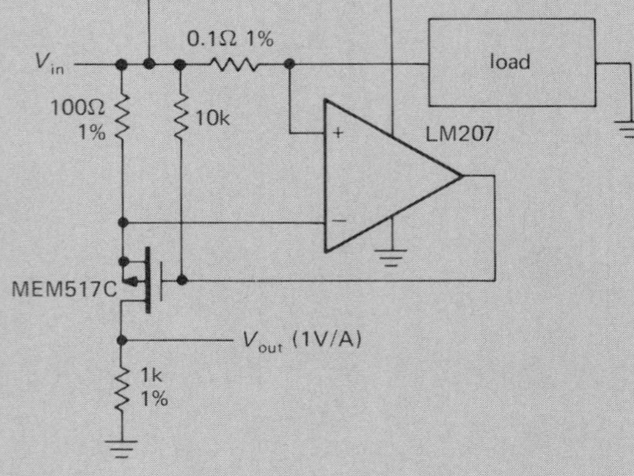

C. current monitor

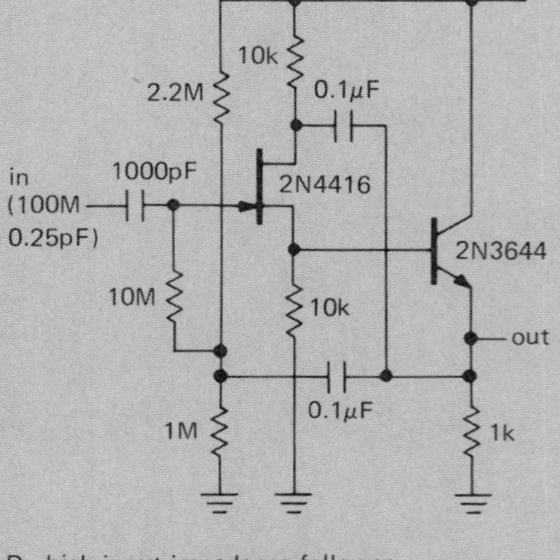

D. high-input-impedance follower

Figure 6.59

Bad circuits

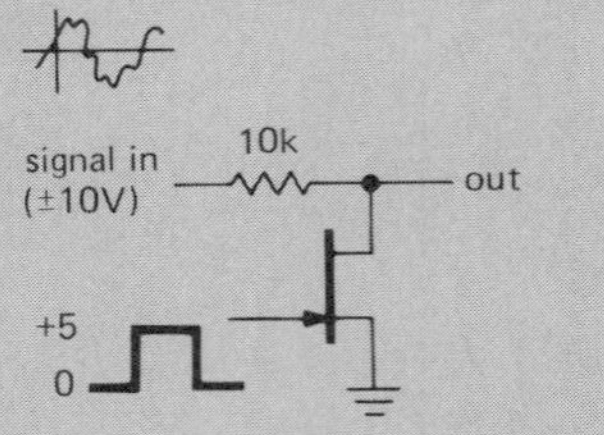

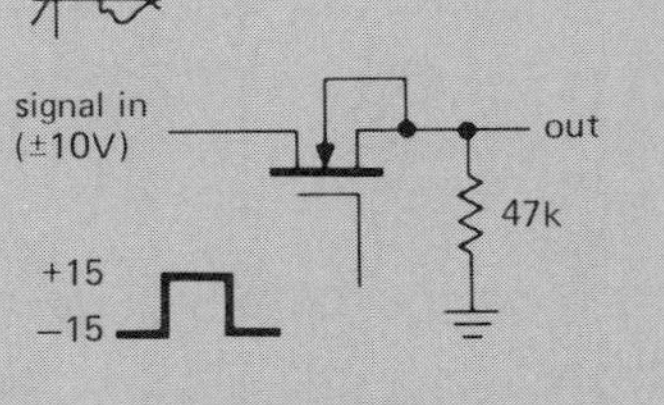

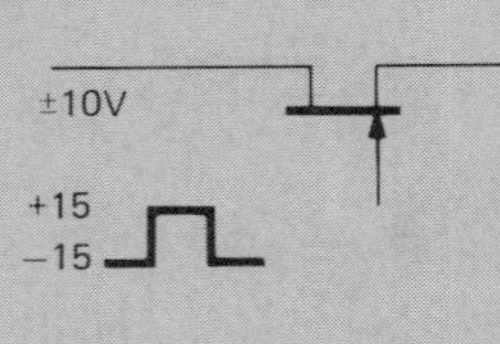

A. analog switches

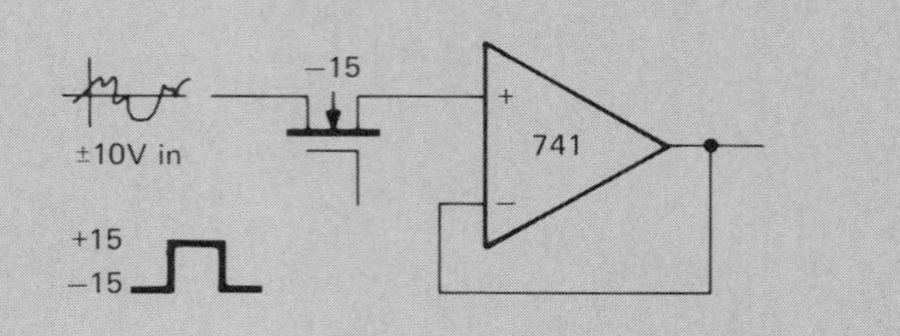

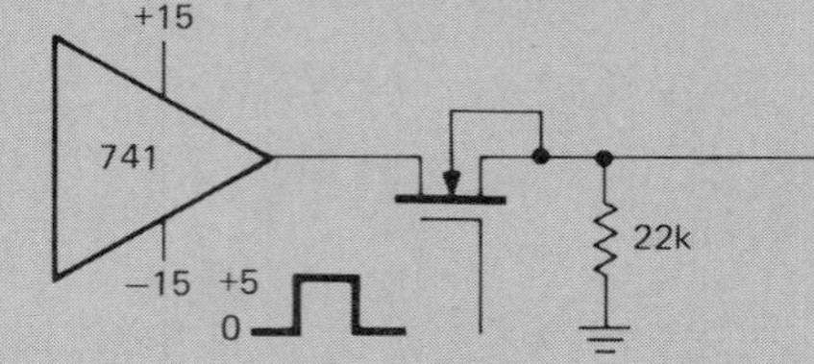

B. op-amps with switches

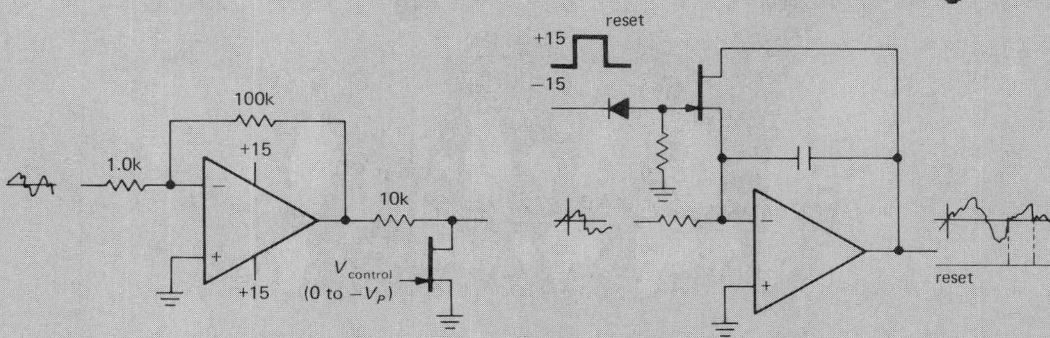

C. FET gain control

D. integrator with reset

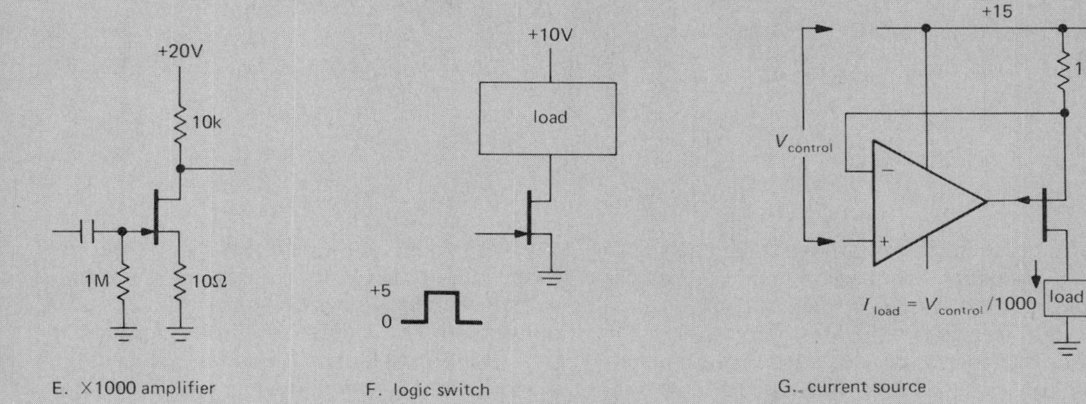

E. ×1000 amplifier

F. logic switch

G. current source

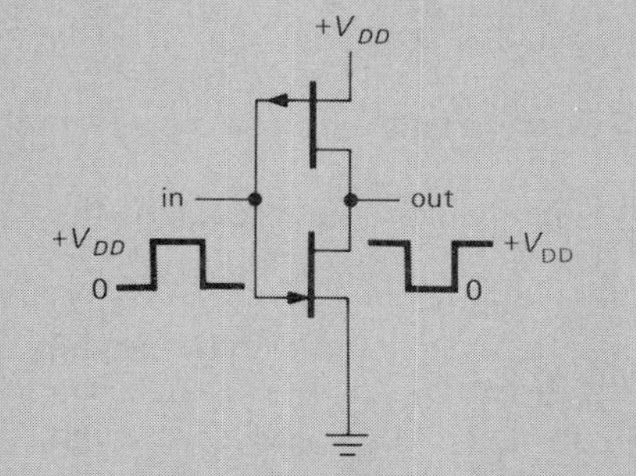

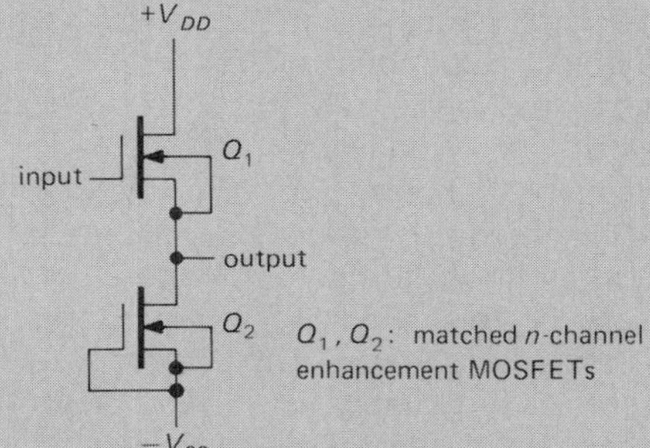

Q_1, Q_2: matched n-channel enhancement MOSFETs

H. complementary JFET inverter

I. zero-offset follower

Figure 6.60

PRECISION CIRCUITS AND LOW-NOISE TECHNIQUES

CHAPTER 7

In the preceding chapters we have dealt with many aspects of analog circuit design, including the circuit properties of passive devices, transistors, FETs, and op-amps, the subject of feedback, and numerous applications of these devices and circuit methods. In all our discussions, however, we have not yet addressed the question of the best that can be done, for example, in minimizing amplifier errors (nonlinearities, drifts, etc.) and in amplifying weak signals with minimum degradation by amplifier "noise." In many applications these are the most important issues, and they form an important part of the art of electronics. In this chapter, therefore, we will look at methods of precision circuit design and the issue of noise in amplifiers. With the exception of the introduction to noise in Section 7.10, this chapter can be skipped over in a first reading. This material is not essential for an understanding of later chapters.

PRECISION OP-AMP DESIGN TECHNIQUES

In the field of measurement and control there is often a need for circuits of high precision. Control circuits should be accurate, stable with time and temperature, and predictable. The usefulness of measuring instruments likewise depends on their accuracy and stability. In almost all electronic subspecialties we always have the desire to do things more accurately – you might call it the joy of perfection. Even if you don't always actually *need* the highest precision, you can still delight in the joy of fully understanding what's going on.

7.01 Precision versus dynamic range

It is easy to get confused between the concepts of *precision* and *dynamic range,* especially since some of the same techniques are used to achieve both. Perhaps the difference can best be clarified by some examples: A 5-digit multimeter has high precision; voltage measurements are accurate to 0.01% or better. Such a device also has wide dynamic range; it can measure millivolts and volts on the same scale. A precision decade amplifier (one with selectable gains of 1, 10, and 100, say) and a precision voltage reference may have plenty of precision, but not necessarily much dynamic range. An example of a device with wide dynamic range but only moderate accuracy might be a 6-decade logarithmic amplifier (log amp) built with carefully trimmed op-amps but with components of only 5% accuracy; even with accurate components a log amp might have limited accuracy because of lack of log conformity (at the extremes of current) of the transistor junction used for the conversion. Another example of a wide-dynamic-range instrument (greater than 10,000:1 range of input currents) with only moderate accuracy (1%) is the coulomb meter described in Section 9.27. It was originally designed to keep track of the total charge put through an electrochemical cell, a quantity that needs to be known only to approximately 5% but that may be the cumulative result of a current that varies over a wide range. It is a general characteristic of wide-dynamic-range design that input offsets must be carefully trimmed in order to maintain good proportionality for signal levels near zero; this is also necessary in precision design, but, in addition, precise components, stable references, and careful attention to every possible source of error must be used to keep the sum total of all errors within the so-called error budget.

7.02 Error budget

A few words on *error budgets.* There is a tendency for the beginner to fall into the trap of thinking that a few strategically placed precision components will result in a device with precision performance. On rare occasions this will be true. But even a circuit peppered with 0.01% resistors and expensive op-amps won't perform to expectations if somewhere in the circuit there is an input offset current multiplied by a source resistance that gives a voltage error of 10mV, say. With almost any circuit there will be errors arising all over the place, and it is essential to tally them up, if for no other reason than to locate problem areas where better devices or a circuit change might be needed. Such an error budget results in rational design, in many cases revealing where an inexpensive component will suffice, and eventually permitting a careful estimate of performance.

7.03 Example circuit: precision amplifier with automatic null offset

In order to motivate the discussion of precision circuits, we have designed an extremely precise decade amplifier with automatic offset. This gadget lets you "freeze" the value of the input signal, amplifying any subsequent changes from that level by gains of exactly 10, 100, or 1000. This might come in particularly handy in an experiment in which you wish to measure a small change in some quantity (e.g., light transmission or radiofrequency absorption) as some condition of the experiment is varied. It is ordinarily difficult to get accurate measurements of small changes in a large dc signal, owing to drifts and instabilities in the amplifier. In such a situation a circuit of extreme precision and stability is required. We will describe the design choices and errors of this particular circuit in the framework of precision design in general, thus rendering painless what could otherwise become a tedious exercise. A note at the outset: Digital techniques offer an attractive alternative to the purely analog circuitry used here. Look forward to exciting revelations in chapters to come! Figure 7.1 shows the circuit.

Circuit description

The basic circuit is a follower (U_1) driving an inverting amplifier of selectable gain (U_2), the latter offsettable by a signal applied to its noninverting input. Q_1 and Q_2 are FETs (see Chapter 6), used in this application as simple analog switches; Q_3–Q_5 generate suitable levels, from a logic-level input, to activate the switches. Q_1 through Q_5 and their associated circuitry could all be replaced by a relay, or even a switch, if desired. For now, just think of it as a simple SPST switch.

When the logic input is HIGH ("hold"), the switch is closed, and U_3 charges the analog "memory" capacitor (C_1) as necessary to maintain zero output. No attempt is made to follow rapidly changing signals, since in the sort of application for which this was designed the signals are essentially dc, and some averaging is a desirable feature. When the switch is opened, the voltage on the capacitor remains stable, resulting in an output signal proportional to the wanderings of the input thereafter.

There are a few additional features that should be described before going on to explain in detail the principles of precision design as applied here: (a) U_4 participates in a first-order leakage-current compensation scheme, whereby the tendency of C_1 to discharge slowly through its own leakage (100,000M, minimum, corresponding to a time constant of 2 weeks!) is compensated by a small charging current through R_{15} proportional to the voltage across C_1. (b) Instead of a single FET switch, two are used in series in a "guarded leakage-cancellation" arrangement. The small leakage current through Q_2, when switched OFF, flows to ground through R_{23}, keeping all terminals of Q_1 within millivolts of ground. Without any appreciable voltage drops, Q_1 hasn't any appreciable leakage! (See Sections 3.15 and 6.18 for discussions of similar circuit tricks.) (c) The offsetting voltage generated at the output of U_3 is attenuated by R_{11}–R_{14}, according to the gain setting. This is done to avoid problems with dynamic range and accuracy in U_3, since drifts or errors in the offset holding circuitry are not amplified by U_2 (more on this later).

7.04 A precision-design error budget

For each category of circuit error and design strategy we will devote a few paragraphs to a general discussion, followed by illustrations from the preceding circuit. Circuit errors can be divided into the categories of (a) errors in the external network components, (b) op-amp (or amplifier) errors associated with the input circuitry, and (c) op-amp errors associated with the output circuitry. Examples of the three are resistor tolerances, input offset voltage, and errors due to finite slew rate, respectively.

Let's start by setting out our error budget. It is based on a desire to keep input errors down to the 10μV level, output drift (from capacitor "droop") below 1mV in 10 minutes, and gain accuracy in the neighborhood of 0.01%. As with any budget, the individual items are arrived at by a process of tradeoffs, based on what can be done with available technology. In a sense the budget represents the end result of the design, rather than the starting point. However, it will aid our discussion to have it now.

Error budget (worst-case values)

1. Buffer amplifier (U_1)
Voltage errors referred to input:

Temperature	2.4μV/4°C
Time	1.0μV/month
Power supply	1.0μV/100mV change
Bias current × R_S	2.0μV/1k of R_S
Load current heating	0.5μV @ full scale (10V)

2. Gain amplifier (U_2)
Voltage errors referred to input:

Temperature	2.4μV/4°C
Time	1.0μV/month
Power supply	1.0μV/100mV change
Bias offset current drift	1.0μV/4°C
Load current heating	1.0μV @ full scale ($R_L \geq$ 10k)

3. Hold amplifier (U_3)
Voltage errors referred to output:

U_3 offset tempco	60μV/4°C
Power supply	10μV/100mV change
Capacitor droop (see current error budget)	100μV/minute

Current errors applied to C_1 (needed for preceding voltage error budget):

Capacitor leakage	
Maximum (uncompensated)	(100pA)
Typical (compensated)	10pA
U_3 input current	0.2pA
U_3 & U_4 offset voltage across R_{15}	1.0pA
FET switch OFF leakage	0.5pA
Printed-circuit-board leakage	5.0pA

The various items in the budget will make sense as we discuss the choices faced in this particular design. We will organize by the categories of circuit errors listed earlier: network components, amplifier input errors, and amplifier output errors.

7.05 Component errors

The degrees of precision of reference voltages, current sources, amplifier gains, etc., all depend on the accuracy and stability of the resistors used in the external networks. Even where precision is not involved directly, component accuracy can have significant effects, e.g., in the common-mode rejection of a differential amplifier made from an op-amp (see Section 3.09), where the ratios of two pairs of resistors must be accurately matched. The accuracy and linearity of integrators and ramp generators depend on the properties of the capacitors used, as do the performances of filters, tuned circuits, etc. As you will see shortly, there are places in the circuit where component accuracy is crucial, and there are other places where the particular component value hardly matters at all.

Components are generally specified with an initial accuracy, as well as the changes in value with time (stability) and temperature. In addition, there are specifications of voltage coefficient (nonlinearity) and bizarre effects such as "memory" and dielectric absorption (for capacitors). Complete specifications will also include the effects of temperature cycling and soldering, shock and vibration, short-term overloads, and moisture, with well-defined conditions of measurement. In general, components of greater initial accuracy will have their other specifications correspondingly better, in order to provide an overall stability comparable with the initial accuracy. However, the overall error due to all other effects combined can exceed the initial accuracy specification. Beware!

As an example, RN55C 1% tolerance metal-film resistors have the following specifications: temperature coefficient (tempco), 50 ppm/°C over the range −55°C to +175°C; soldering, temperature, and load cycling, 0.25%; shock and vibration, 0.1%; moisture, 0.5%. By way of comparison, ordinary 5% carbon-composition resistors (Allen-Bradley type CB) have these specifications: tempco, 3.3% over the range 25°C to 85°C; soldering and load cycling, +4%, −6%; shock and vibration, ±2%; moisture, +6%. From these specs it should be obvious why you can't just select (using an accurate digital ohmmeter) carbon resistors that happen to be within 1% of their marked value for use in a precise circuit, but are obliged to use 1% resistors (or better) designed for long-term stability as well as initial accuracy. For the utmost in precision it is necessary to use wire-wound resistors, available with tolerances of 0.01%.

Nulling amplifier: component errors

In the preceding circuit (Figure 7.1), 0.01% resistors are used in the gain-setting network, R_3–R_9, giving highly predictable gain. As you will see shortly, the value of R_3 is a compromise, with small values reducing offset current error in U_2 but increasing heating and thermal offsets in U_1. Given the value of R_3, the feedback network is forced to take on its complicated form to keep the resistor values below 301k, the maximum value generally available in 1% precision resistors. This trick is discussed in Section 3.18. Note that 1% resistors are used in the offset attenuator network, R_{11}–R_{14}; here accuracy is irrelevant, and metal-film resistors are used only for their good stability.

The largest error term in this circuit, as the error budget shows, is capacitor leakage in the holding capacitor, C_1. Capacitors intended for low-leakage applications give a leakage specification, sometimes as a leakage resistance, sometimes as a time constant (megohm-microfarads). In this circuit C_1 must have a value of at least a few microfarads in order to keep the charging rate from other current error terms small (see budget). In that range of capacitance, polystyrene, polycarbonate, and polysulfone capacitors have the lowest leakage. The unit chosen has a leakage specified as 1,000,000 megohm-microfarads maximum, i.e., a parallel leakage resistance of at least 100,000M. Even so, that's equivalent to a leakage current of 100pA at full output (10V), corresponding to a droop rate of nearly 1mV/min at the output, the largest error term by far. For that reason we have added the leakage-cancellation scheme described earlier. It is fair to assume that the effective leakage can be reduced to 0.1 of the capacitor's worst-case leakage specification (in practice, we can probably do much better). No great stability is required in the cancellation circuit, given the modest demands made of it. As you will see later when we discuss voltage offsets, R_{15} is kept intentionally large so that input voltage offsets in U_3 aren't converted to a significant current error.

While on the subject of errors produced

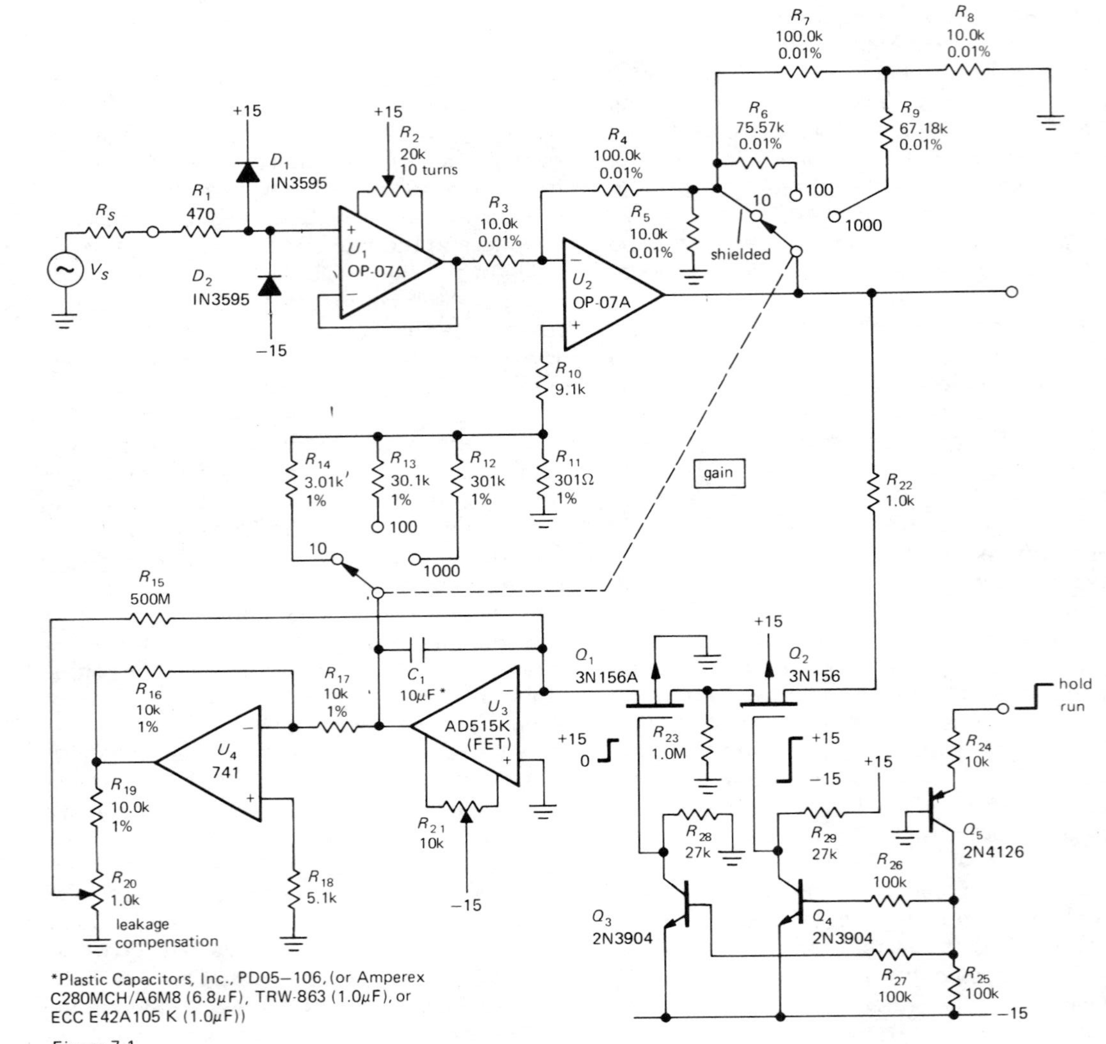

*Plastic Capacitors, Inc., PD05–106, (or Amperex C280MCH/A6M8 (6.8μF), TRW-863 (1.0μF), or ECC E42A105 K (1.0μF))

Figure 7.1

Autonulling dc laboratory amplifier.

by components external to the amplifiers themselves, it should be pointed out that leakage in FET switches is normally in the range of 1nA, a value completely unacceptable in this circuit. The trick of using a pair of series-connected FETs, with Q_2's leakage resulting in only 1mV across Q_1 (with negligible leakage into U_3's summing junction), is elegant and powerful; it is sometimes used in integrator circuits, as discussed in Section 6.18. We have also used it in a novel peak-detector circuit in Section 3.15. As you will see shortly, U_3 is chosen carefully to keep currents through C_1 down in the picoampere range. The philosophy is the same everywhere: Choose circuit configurations and component types as necessary to meet the error budget. At times this involves hard work and circuit trickery, but at other times it falls easily within standard practice.

7.06 Amplifier input errors

The deviations of op-amp input characteristics from the ideal that we discussed in Chapter 3 (finite values of input impedance and input current, voltage offset, common-mode rejection ratio, and power-supply rejection ratio, and their drifts with time and temperature) generally constitute serious obstacles to precision circuit design and force tradeoffs in circuit configuration, component selection, and the choice of a particular op-amp. The point is best made with examples, as we will do shortly. Note that these errors, or their analogs, exist for amplifiers of discrete design as well.

Input impedance

Let's discuss briefly the error terms just listed. The effect of finite input impedance is to form voltage dividers in combination with the source impedance driving the amplifier, reducing the gain from the calculated value. Most often this isn't a problem, because the input impedance is bootstrapped by feedback, raising its value enormously. As an example, the OP-07 precision op-amp (with transistor, not FET, input stage) has a typical "differential-mode input impedance" of 80M. In a circuit with plenty of loop gain, feedback raises the input impedance to the "common-mode input impedance," 200,000M. In any case, some FET-input op-amps have astronomical values of R_{in}, if there's still a problem.

Input bias current

More serious is the input bias current. Here we're talking about currents measured in nanoamps, and this already produces voltage errors of microvolts for source impedances as small as 1k. Again, FET op-amps come to the rescue, but with greatly increased voltage offsets as part of the bargain. Bipolar superbeta op-amps such as the 312 and LM11 can also have surprisingly low input currents. As an example, compare the OP-07 precision bipolar transistor op-amp with the LM11, the AD515K "precision" FET op-amp, and the ICH8500 ultra-low-current MOSFET op-amp:

	Offset voltage @ 25°C V_{OS} max	*Bias current* @ 25°C I_B max	*Tempco of* V_{OS} max
OP-07 (bipolar)	0.025mV	2nA	0.6μV/°C
LM11 (superbeta)	0.3mV	0.05nA	3μV/°C
AD515K (JFET)	1mV	0.00015nA	15μV/°C
ICH8500 (MOSFET)	50mV	0.00001nA	100μV/°C

Well-designed FET amplifiers have extremely low bias current, but with much larger offset voltage, as compared with the precision OP-07. Since the offset voltage can always be trimmed, what matters more is the drift with temperature. In this case the FET amplifiers are 20 to 100 times worse. The op-amps with the lowest input current use MOSFETs for the input stage. They are becoming popular because of the availability of inexpensive units like the 3130, 3140, and 3160, as well as the ultra-low-bias-current devices like the 8500 listed earlier. However, unlike JFETs or bipolar transistors, MOSFETs can have very large drifts of offset voltage with time, an effect that will be discussed shortly. So the improvement in current errors you buy with a FET op-amp can be wiped out by the larger voltage error terms. With any circuit in which bias current can contribute significant error, it is always wise to ensure that both op-amp input terminals see the same dc source resistance, as we discussed in Section 3.12; then the op-amp's *offset current* becomes the relevant specification.

One additional point to keep in mind when using FET-input op-amps is that the input "bias" current is actually gate leakage current, and it rises dramatically with increasing temperature (it roughly doubles for every 10°C increase in chip temperature). Since FET op-amps often run warm (the popular 356 dissipates 150mW quiescent power), the actual input current may be considerably higher than the 25°C figures you see on the data sheet. The input current of a bipolar-transistor-input op-amp, by comparison, is actual base current, and it *drops* with rising temperature. So a FET-input op–amp with impressive input-current specs on paper may not give such an improvement over a good superbeta bipolar unit. As an example, the AD545 with its 1pA input current (at 25°C) will have an input current of about 15pA at 65°C chip temperature, which is as high as the input current of the superbeta LM316A or LM11 at the same temperature. The popular 355 series of FET op-amps has an input current that is comparable to that of the LM316A or LM11 at 25°C and is many times higher at elevated temperatures.

ELECTRICAL CHARACTERISTICS			OP-07A			OP-07			
These specifications apply for $V_s = \pm 15V$, $T_A = 25°C$, unless otherwise noted.									
Parameter	**Symbol**	**Test Conditions**	**Min**	**Typ**	**Max**	**Min**	**Typ**	**Max**	**Units**
Input Offset Voltage	V_{os}	(Note 1)		10	25		30	75	μV
Long Term Input Offset Voltage Stability	V_{os}/Time	(Note 2)		0.2	1.0	-	0.2	1.0	μV/Mo
Input Offset Current	I_{os}		--	0.3	2.0	-	0.4	2.8	nA
Input Bias Current	I_B		--	±.7	±2.0	-	±1.0	±3.0	nA
Input Noise Voltage	$e_{np\text{-}p}$	0.1Hz to 10Hz (Note 3)	--	0.35	0.6	--	0.35	0.6	μV p-p
Input Noise Voltage Density	e_n	f_o = 10Hz (Note 3) f_o = 100Hz (Note 3) f_o = 1000Hz (Note 3)	-- -- --	10.3 10.0 9.6	18.0 13.0 11.0	-- -- --	10.3 10.0 9.6	18.0 13.0 11.0	nV/$\sqrt{Hz}$
Input Noise Current	$i_{np\text{-}p}$	0.1Hz to 10Hz (Note 3)	--	14	30	--	14	30	pA p-p
Input Noise Current Density	i_n	fo = 10Hz (Note 3) fo = 100Hz (Note 3) fo = 1000Hz (Note 3)	-- -- --	0.32 0.14 0.12	0.80 0.23 0.17	-- -- --	0.32 0.14 0.12	0.80 0.23 0.17	pA/$\sqrt{Hz}$
Input Resistance - Differential Mode	R_{in}		30	80	--	20	60	-	MΩ
Input Resistance - Common Mode	R_{inCM}		--	200	--	--	200	--	GΩ
Input Voltage Range	IVR		±13.0	± 14.0	--	±13.0	±14.0	--	V
Common Mode Rejection Ratio	CMRR	$V_{CM} = \pm 13V$	110	126	--	110	126	--	dB
Power Supply Rejection Ratio	PSRR	$V_s = \pm 3V$ to $\pm 18V$	100	110	--	100	110	--	dB
Large Signal Voltage Gain	A_{vo}	$R_L \geq 2k\Omega$, $V_o = \pm 10V$ $R_L \geq 500\Omega$, $V_o = \pm 5V$ $V_s = \pm 3V$	300 150	500 500	-- --	200 150	500 500	-- --	V/mV
Maximum Output Voltage Swing	V_{oM}	$R_L \geq 10k\Omega$ $R_L \geq 2k\Omega$ $R_L \geq 1k\Omega$	± 12.5 ± 12.0 ± 10.5	± 13.0 ± 12.8 ± 12.0	-- -- --	± 12.5 ±12.0 ± 10.5	± 13.0 ±12.8 ± 12.0	-- -- --	V
Slewing Rate	SR	$R_L = \geq 2k\Omega$ (Note 3)	0.1	0.17	--	0.1	0.17	--	V/μsec
Closed Loop Bandwidth	BW	$A_{VCL} = +1.0$ (Note 3)	0.4	0.6	--	0.4	0.6	--	MHz
Open Loop Output Resistance	R_o	$V_o = 0$, $I_o = 0$	--	60	-	--	60	--	Ω
Power Consumption	Pd	 $V_s = \pm 3V$	-- --	75 4	120 6	-- --	75 4	120 6	mW
Offset Adjustment Range		$R_p = 20k\Omega$	--	±4	--	--	±4	--	mV
The following specifications apply for $V_s = \pm 15V$, $-55°C \leq T_A \leq +125°C$, unless otherwise noted.									
Input Offset Voltage	V_{os}	(Note 1)	--	25	60	--	60	200	μV
Average Input Offset Voltage Drift Without External Trim With External Trim	 TCV_{os} TCV_{osn}	 R_p = 20kΩ	 -- --	 0.2 0.2	 0.6 0.6	 -- --	 0.3 0.3	 1.3 1.3	 μV/°C μV/°C
Input Offset Current	I_{os}		--	0.8	4.0	--	1.2	5.6	nA
Average Input Offset Current Drift	TCI_{os}		--	5	25	--	8	50	pA/°C
Input Bias Current	I_B		--	±1.0	±4.0	--	± 2.0	± 6.0	nA
Average Input Bias Current Drift	TCI_B		--	8	25	--	13	50	pA/°C
Input Voltage Range	IVR		± 13.0	± 13.5	--	± 13.0	± 13.5	--	V
Common Mode Rejection Ratio	CMRR	$V_{CM} = \pm 13V$	106	123	--	106	123	--	dB
Power Supply Rejection Ratio	PSRR	$V_s = \pm 3V$ to $\pm 18V$	94	106	--	94	106	--	dB
Large Signal Voltage Gain	A_{vo}	$R_L \geq 2k\Omega$, $V_o = \pm 10V$	200	400	--	150	400	--	V/mV
Maximum Output Voltage Swing	V_{oM}	$R_L \geq 2k\Omega$	± 12.0	± 12.6	--	± 12.0	±12.6	--	V

NOTE 1: Input offset voltage measurements are performed by automated test equipment approximately 0.5 seconds after application of power. Additionally, OP-07A offset voltage is measured five minutes after power supply application at 25°C, -55°C and +125°C.

NOTE 2: Long Term Input Offset Voltage Stability refers to the averaged trend line of Vos vs. Time over extended periods after the first 30 days of operation. Excluding the initial hour of operation, changes in Vos during the first 30 operating days are typically 2.5μV – refer to typical performance curve on Page 5. Parameter is not 100% tested; 90% of units meet this specification.

NOTE 3: Parameter is not 100% tested; 90% of units meet this specification.

Figure 7.2
OP-07 specifications. (Precision Monolithics, Inc., Santa Clara, Calif. 95050.)

Voltage offset

Voltage offsets at the input are obvious sources of error. Since they can be trimmed to zero, what ultimately matters is the drift of offset voltage with time, temperature, and power-supply voltage. A precision bipolar transistor op-amp like the OP-07, LH0044, or μA714 will have superior performance in this regard, but input current may then dominate the error budget.

Another factor to keep in mind is the drift caused by self-heating of the op-amp when it drives a low-impedance load. It is often necessary to keep the load impedance above 10k to prevent large errors from this effect; as usual, that may compromise the next stage's error budget from the effects of bias current! You will see just such a problem in this design example. To get an idea of state-of-the-art performance, look over the OP-07 specifications in Figure 7.2.

Although most good op-amps have offset-adjustment terminals, it is still wise to choose an amplifier with inherently low initial offset voltage V_{OS} max, because offset voltage drift and common-mode rejection are degraded by the unbalance caused by an offset-adjustment trimmer. Figure 7.3 illustrates how a trimmed offset has larger drifts with temperature. We have also shown how the offset adjustment is spread over the trimmer pot rotation, with best resolution near the center, especially for large values of trimmer resistance. (Check out the zero adjust for IC_2 in the coulomb meter circuit in Section 9.27 for another method.)

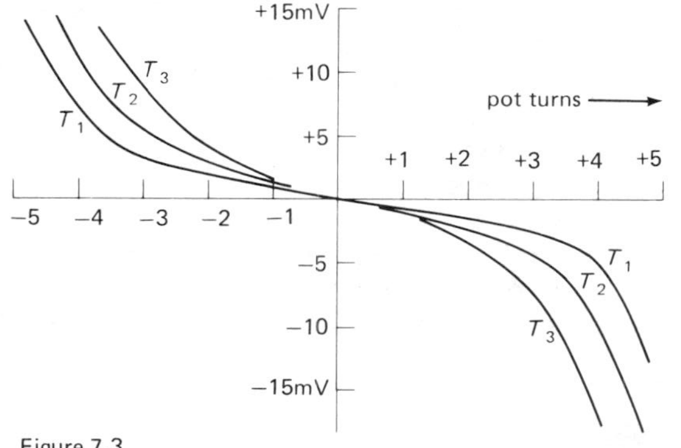

Figure 7.3

Typical op-amp offset versus offset-adjustment potentiometer rotation for several temperatures.

As mentioned earlier, FET-input op-amps suffer from larger initial offsets and much larger drifts of V_{OS} with temperature and time than do bipolar transistor op-amps. MOSFETs, in particular, are subject to a unique debilitating effect that neither JFETs nor transistors have. It turns out that sodium-ion impurities in the gate insulating layer migrate slowly under the influence of the gate's $V_{GS(ON)}$ electric field, resulting in a drift of offset voltage of as much as 0.5mV over the life of the device. The effect is increased for elevated temperatures and for a large applied differential input signal. For example, the data sheet for the RCA 3140 MOSFET-input op-amp shows a typical 5mV change of V_{OS} over 3000 hours of operation at 125°C with 2 volts across the input. Moral: Don't choose MOSFET-input op-amps for circuits where stability of (trimmed) offset voltage is essential. (An exception is the new CAZ amplifier, discussed in Section 7.09 with instrumentation amplifiers.)

Common-mode rejection

Insufficient common-mode rejection ratio (CMRR) degrades circuit precision by effectively introducing a voltage offset as a function of dc level at the input. This effect is usually negligible, since it is equivalent to a small gain change, and in any case it can be overcome by choice of configuration: An inverting amplifier is insensitive to op-amp CMRR, in contrast with a noninverting amplifier. However, in "instrumentation amplifier" applications you are looking at a small differential signal riding on a large dc offset, and a high CMRR is essential. In such cases you have to be careful about circuit configurations and, in addition, choose an op-amp with a high CMRR specification. Once again, a superior op-amp like the OP-07 can solve your problems, with a CMRR (min) of 110dB, compared with the 741's meager specification of 70dB to 80dB. We will discuss high-gain differential and instrumentation amplifiers shortly.

Power-supply rejection

Changes in power-supply voltage cause small op-amp errors. As with most op-amp specifications, the power-supply rejection ratio (PSRR) is referred to a signal at the *input.* For example, the OP-07 has a specified PSRR of 100dB at dc, meaning that a 0.1 volt change in one of the power-supply voltages causes a change at the output equivalent to a change in differential input signal of 1μV.

The PSRR drops drastically with increasing frequency, and a graph documenting this scurrilous behavior is often given on the data sheet. For example, the PSRR of our favorite OP-07 begins dropping at 3Hz and is down to 80dB at 60Hz and 40dB at 10kHz. This actually doesn't present much of a problem, since power-supply noise is also decreasing at higher frequencies if you have used good bypassing. However, 120Hz ripple could present a problem if an unregulated supply is used.

It is worth noting that the PSRR will not, in general, be the same for the positive and negative supplies. Thus the use of dual-tracking regulators (Section 5.19) doesn't necessarily bring any benefits.

Nulling amplifier: input errors

The amplifier circuit in Figure 7.1 begins with a follower, to keep a high input impedance. It is tempting to consider a FET type, but the poor V_{OS} specification more than offsets the advantage of low input current, except with sources of very high impedance. The OP-07's 2nA bias current gives an error of 2μV/1k source impedance; an AD515K, although giving negligible current error, would give voltage offset drifts of 60μV/4°C (4°C is considered standard laboratory ambient temperature range). The input follower is provided with offset trimming, since the initial 25μV spec is too large. As mentioned earlier, feedback bootstraps the input impedance to 200,000M and eliminates any errors from finite source impedance, up to 20M (for gain error less than 0.01%). D_1 and D_2 are included for input overvoltage protection and are low-leakage types (less than 1nA).

U_1 drives an inverting amplifier (U_2), with R_3 being a compromise between heat-produced thermal offsets in U_1 and bias-current offset errors in U_2. The value chosen keeps heating down to 5.6mW (at 7.5V output, the worst case), which works out to a temperature rise of 0.8°C (the op-amp has a thermal resistance of 0.14°C/mW, see Section 5.04), with a consequent voltage offset of 0.5μV. The resultant 10k source impedance seen by U_2 results in an error due to bias-current offset, but since U_2 is inside a feedback loop with U_3 trimming the overall offset to zero, all that matters is the drift in the current error term. The OP-07 has a specification for bias offset drift with temperature (not often specified by manufacturers, incidentally), from which the error result of 1.0μV/4°C in the error budget is calculated. Reducing the value of R_3 would improve this term, at the expense of the heating term in U_1.

As explained in the overall circuit description earlier, the value of R_3 forces the bizarre feedback T network in order to keep the feedback resistor values in the range where precision wire-wound resistors can be manufactured. Using the ordinary inverting amplifier configuration, for example, you would need resistors of 100.0k, 1.0M, and 10.0M for gains of 10, 100, and 1000, respectively.

The input impedance of U_2 comes closer to presenting a problem. At a gain of 1000 its differential input impedance of 80M is bootstrapped by a factor of $A_{vol}/1000$ to 40,000M. Fortunately this exceeds the 9.4k impedance of the gain-setting network by a factor of more than a million, contributing much less than 0.01% error. This is the toughest example we could think of, and even so the op-amp input impedance presents no problem, thus demonstrating that, in general, you can ignore the effects of op-amp input impedances.

Drifts in offset voltage in both U_1 and U_2 over time, temperature, and power-supply variations affect the final error equally and are tabulated in the budget. It is worth pointing out that they are all automatically cancelled at each "zeroing" cycle, and only short-term drifts matter anyway. These errors are all in the microvolt range, thanks to a good op-amp. U_3 has larger drifts, but it must be a FET type to keep capacitor current small, as already explained. Since U_3's output is attenuated according to the gain selected, its error, *referred to the input,* is reduced at high gain. This is an important point, since high gains are used with small input signal levels where high accuracy is needed. U_3's errors are always the same *at the output,* and they are therefore specified as output errors in the error budget.

Note the general philosophy of design that emerges from this example: You work at the problem areas, choosing configurations and components as necessary to reduce errors to acceptable values. Tradeoffs and compromises are involved, with some choices depending on external factors (e.g., the use of a FET-input follower for U_1 would be preferable for source impedances greater than about 50k).

Table 7.1 compares the specifications of op-amps you might choose for precision circuit design.

7.07 Amplifier output errors

As we discussed in Chapter 3, op-amps have some serious limitations associated with the output stage. Limited slew rate, output crossover distortion (see Section 2.14), and finite open-loop output impedance can all cause trouble, and they can cause precision circuits to display astoundingly large errors if not taken into account.

Slew rate: general considerations

As we mentioned in Section 3.11, an op-amp can swing its output voltage only at some maximum rate. This effect originates in the frequency-compensation circuitry of the op-amp, as we will explain in a bit more detail shortly. One consequence of finite slew rate is to limit the output swing at high frequencies, as we showed in Section 3.12: $V_{pp} = S/\pi f$. A second consequence is best explained with the help of a graph of slew rate versus differential input signal (Fig. 7.4). The point to be made here is that a circuit that demands a substantial slew rate operates with a substantial voltage error across the op-amp's input terminals. This can be disastrous for a circuit that pretends to be highly precise.

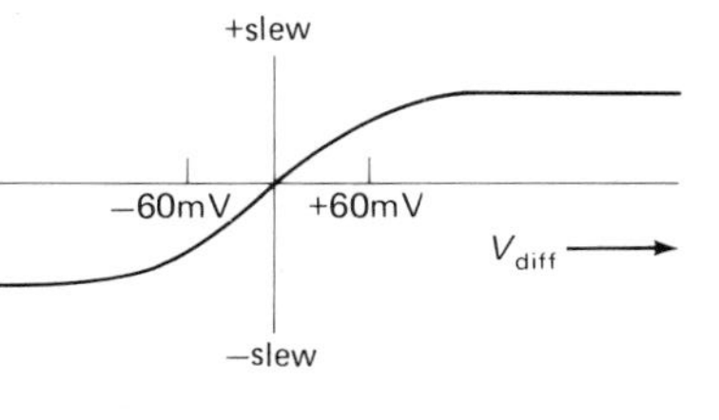

Figure 7.4

Let's look at the innards of an op-amp in order to get some understanding of the origin of slew rate. The vast majority of op-amps can be summarized with the circuit shown in Figure 7.5. A differential input stage, loaded with a current mirror, drives a stage of large voltage gain with a compensation capacitor from output to input. The output stage is a unity-gain push-pull follower. The compensation capacitor C is chosen to bring the open-loop gain of the amplifier to unity before the phase shifts caused by the other amplifier stages have become significant. That is, C is chosen to put f_T, the unity-gain bandwidth, near the frequency of the next amplifier rolloff pole, as described in Section 3.32. The input stage has very high output impedance, and it looks like a current source to the next stage.

The op-amp is slew-rate-limited when the input signal drives one of the differential-stage transistors nearly to cutoff, driving the second stage with the total emitter current I_E of the differential pair. This occurs for a differential input voltage of about 60mV, at which point the ratio of currents in the differential stage is 10:1. At this point Q_5 is slewing its collector as rapidly as possible, with all of I_E going into charging C. Q_5 and C thus form an integrator, with a slew-rate-limited ramp as output. We are now going to derive an expression for the slew rate.

□ *Slew rate: a detailed look*

First, let us write an expression for the open-loop small-signal ac voltage gain, ignoring phase shifts:

$$A_V = g_m X_C = g_m/2\pi fC$$

from which the unity-gain bandwidth product (the frequency at which $A_V = 1$) is

$$f_T = \frac{1}{2\pi}\frac{g_m}{C}$$

Now, the slew rate is determined by a current I_E charging a capacitance C:

$$S = \frac{dV}{dt} = \frac{I_E}{C}$$

TABLE 7.1. PRECISION OP-AMPS

	Input offset				Input current									
	Voltage		Drift		Bias		Offset		Noise	Gain				
Type	typ (μV)	max (μV)	typ (μV/°C)	max (μV/°C)	typ (nA)	max (nA)	typ (nA)	max (nA)	e_n (typ) @ 10Hz (nV/Hz$^{1/2}$)	min (× 10^6)	typ (× 10^6)	Slew rate (V/μs)	f_T (MHz)	Comments
Bipolar														
SSS725A	60	100	0.2	0.6[a]	30	70	0.3	1.0	9	1	3	0.15	0.08	Original precision amp
LM308A-1	300	500	0.6	1.0	1.5	7	0.2	1.0	—	0.08	0.3	0.15	0.3	Superbeta
LM11	100	300	1.0	3.0	25pA	50pA	0.5pA	10pA	—	0.1	0.3	0.3	0.5	Superbeta Darlington
LH0044A	8	25	0.1	0.5	8.5	15	1.5	2.5	11	1	20	0.06	0.4	Hybrid
OP-07C/μA714C	60	150	0.5	1.8	1.8	7	0.8	6	10	0.12	0.4	0.17	0.6	
OP-07E/μA714E	30	75	0.3	1.3	1.2	4	0.5	3.8	10	0.2	0.5	0.17	0.6	
OP-07A/μA714A	10	25	0.2	0.6	0.7	2	0.3	2	10	0.3	0.5	0.17	0.6	Premium
OP-08E	70	150	0.5	2.5	0.08	2	0.05	0.2	22	0.08	0.3	0.12	0.8	Improved 308
OP-12E	70	150	0.5	2.5	0.08	2	0.05	0.2	22	0.08	0.3	0.12	0.8	Compensated OP-08
OP-20B	80	300	1.0	2.0	13	20	0.4	1.5	100	0.5	1	0.025	0.1	Low power
AD510J	—	100	—	3.0	—	25	—	5	18	0.25	—	0.1	0.3	
AD510L	—	25	—	0.5	—	10	—	2.5	18	1	—	0.1	0.3	
AD517J	—	150	—	3.0	—	5	—	1	35	1	—	0.1	0.25	
AD517L	—	25	—	0.5	—	1	—	0.25	35	1	—	0.1	0.25	Premium, low current
3510CM	—	60	—	0.5	—	15	—	10	14	1	—	0.8	0.4	
CAZ														
ICL7600	2	5	0.01	0.1	—	—	0.15	1.5	700	0.05	0.2	0.5	—[b]	Ultralow offset; noisy
Chopper														
HA2905	20	80	0.2	—	0.15	1	0.05	0.5	900	1	50	2.5	3	Noisy
SN72088	70	150	1.0	—	0.6	10	0.2	0.6	—	0.1	10	25	3	Poor CMRR and PSRR
FET														
LFT355	—	500	3	5	30pA	50pA	3pA	10pA	90	0.05	0.2	5	2.5	Precision LF355
LFT356	—	500	3	5	70pA	100pA	3pA	20pA	60	0.05	0.2	12	4.5	Precision LF356
LH0052	100	500	2	5	0.5pA	2.5pA	0.01pA	0.5pA	150	0.1	0.2	3	1	Hybrid
OP-15E	200	500	2	5	15pA	50pA	3pA	10pA	50	0.1	0.24	17	6	Fast precision LF355
OP-16E	200	500	2	5	15pA	50pA	3pA	10pA	50	0.1	0.24	25	8	Fast precision LF356
AD544L	—	500	—	5	—	25pA	—	2pA	35	0.05	—	13	2	Low noise, fast
AD545L	—	500	—	5	—	1pA	—	1pA	55	0.04	—	1	0.7	Low current
3521L	—	250	—	1	—	10pA	—	2pA	—	0.5	—	0.6	1.5	Low drift
3528BM	100	250	2	5	—	0.15pA	—	0.04pA	120	0.5	0.5	0.7	0.7	Low current

[a] With offset nulled. [b] Limited to ~10Hz because of CAZ transients.

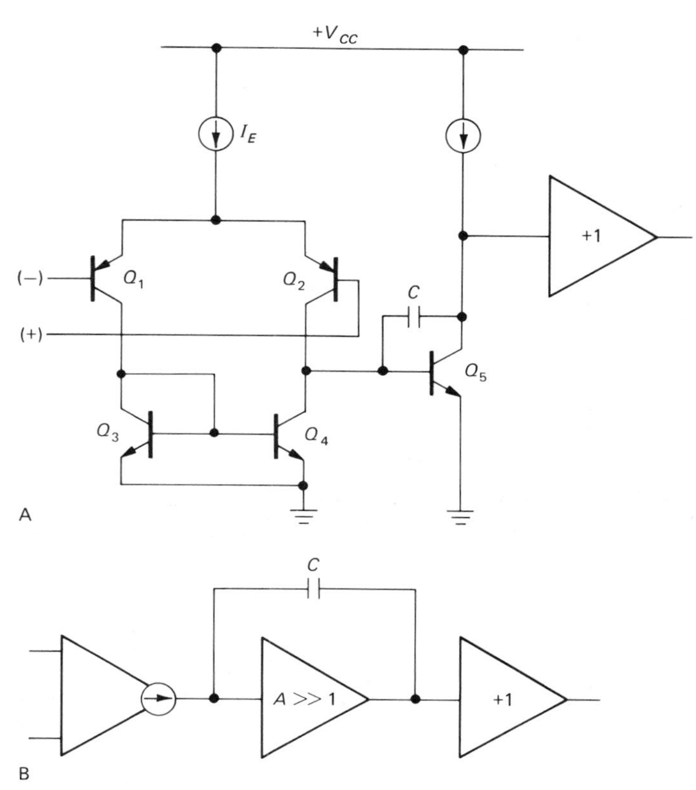

Figure 7.5

For the usual case of a differential amplifier with no emitter resistors, g_m is related to I_E by

$$g_m = \frac{1}{r_e} = \frac{I_E}{2V_T} = \frac{I_E}{50\text{mV}}$$

By substituting this into the slew-rate formula, we find

$$S = 2V_T \frac{g_m}{C}$$

i.e., the slew rate is proportional to g_m/C, just the same as the unity-gain bandwidth! In fact,

$$S = 4\pi V_T f_T = 0.3 f_T,$$

with f_T in MHz and S in V/μs. This is independent of the particular values of C, g_m, I_E, etc., and it gives a good estimate of slew rate. (The 741, with $f_T \approx 1.5$MHz, has a slew rate of 0.5V/μs.) It shows that an op-amp with greater gain-bandwidth product f_T will have a higher slew rate. You can't improve matters in a slow op-amp by merely increasing input-stage current I_E, because the increased gain (from increased g_m) then requires a correspondingly increased value of C for compensation. Adding gain anywhere else in the op-amp doesn't help either.

The preceding result shows that increasing f_T (by raising collector currents, using faster transistors, etc.) will increase the slew rate. A high f_T is, of course, always desirable, a fact not lost on the IC designer, who has already done the best he can with what's on the chip. However, there is a way to get around the restriction that $S = 0.3 f_T$. That result depended on the fact that the transconductance was determined by I_E (through $g_m = I_E/2V_T$). You can use simple tricks to raise I_E (and therefore the slew rate) while keeping f_T (and therefore the compensation) fixed. The easiest is to add some emitter resistance to the input differential amplifier. Let's imagine we do something like that, causing I_E to increase by a factor m while holding g_m constant. Then, by going through the preceding derivation, you would find

$$S = 0.3 m f_T$$

EXERCISE 7.1

Prove that such a trick does what we claim.

□ *Increasing slew rate*

Here, then, are some ways to obtain a high slew rate: (a) Use an op-amp with high f_T. (b) Increase f_T by using a smaller compensation capacitor; of course, this is possible only in applications where the closed-loop gain is greater than unity. (c) Reduce the input-stage transconductance g_m by adding emitter resistors; then reduce C or raise I_E proportionately. (d) Use a different input-stage circuit.

The third technique (reduced g_m) is used in many op-amps. As an example, the HA2607 and HA2507 op-amps are nearly identical, except for the inclusion of emitter resistors in the input stage of the HA2507. The emitter resistors increase the slew rate, at the expense of open-loop gain. The following data demonstrate this tradeoff. FET op-amps, with their lower input-stage g_m, tend to have higher slew rates for the same reason.

	HA2607	HA2507
f_T	12MHz	12MHz
Slew rate	7V/μs	30V/μs
Open-loop gain	150,000	15,000

The fourth technique generally uses the method of "cross-coupled transconductance reduction," which involves having a second set of transistors available at the input stage, biding their time during small signal swings, but ready to help out with some extra current when needed. This has the advantage of improved noise and offset performance, at the expense of some complexity, as compared with the simple emitter resistor scheme. This technique is used in the Harris HA2705, Raytheon 4531, Motorola 1741S, and Signetics 535 and 538 to boost the slew rate for large differential input signals. The resultant graph of slew rate versus input error signal is shown in Figure 7.6.

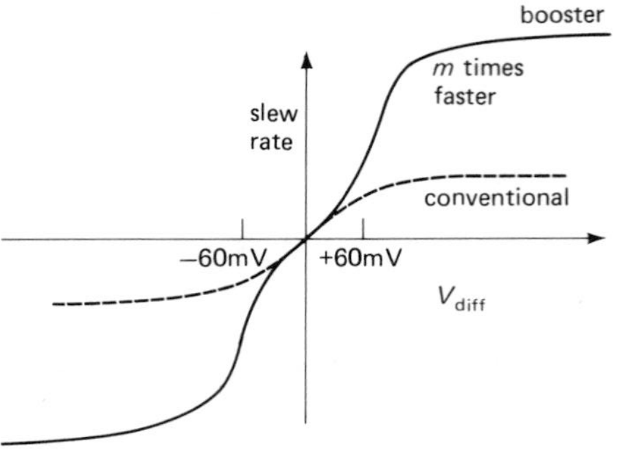

Figure 7.6

Settling time

Slew rate measures how rapidly the output voltage can change. The op-amp slew-rate specification usually assumes a large differential input voltage (60mV or more), which is realistic, since an op-amp whose output isn't where it's supposed to be will have its input driven hard by feedback, assuming a reasonable amount of loop gain. Of perhaps equal importance in high-speed precision applications is the time required for the output to get where it's going following an input change. This *settling-time* specification (the time required to get within the specified accuracy of the final value and stay there) is always given for devices such as digital-to-analog converters, where precision is the name of the game, but it is not normally specified for op-amps.

Settling time depends on the gain and compensation used, and a few waveforms may be given in op-amp sheets. A few points are worth making: (a) Settling time *can* be very rapid (in fact, almost negligible compared with the slew-rate-limited rise time). (b) A fast settling time to 1%, say, doesn't necessarily guarantee a fast settling time to 0.1%, since there may be a long tail. (c) The op-amp will settle more quickly if the frequency-compensation scheme used gives a plot of open-loop phase shift versus frequency that is a nice straight line on a log-log graph. Op-amps with wiggles in the phase-shift graph are more likely to exhibit overshoot and ringing, as in the second waveform shown in Figure 7.7. Note that the slew-rate-limited output tends to begin

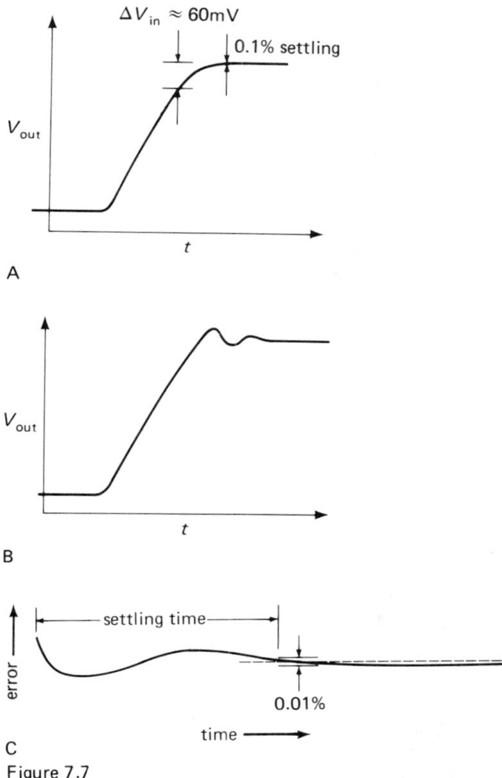

Figure 7.7

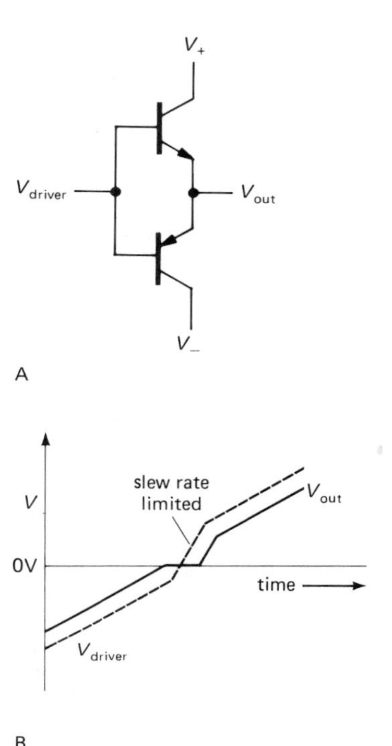

Figure 7.8

bending over when the differential input signal drops below about 60mV.

Table 7.2 lists a selection of high-speed op-amps suitable for applications that demand high f_T, high slew rate, and fast settling time.

Crossover distortion and output impedance

Some op-amps use a simple push-pull output stage, without biasing the bases two diode drops apart, as we discussed in Section 2.14. This leads to class B distortion near zero output, since the driver stage has to slew the bases through $2V_{BE}$ as the output current passes through zero (Fig. 7.8). This crossover distortion can be substantial, particularly at higher frequencies where the loop gain is reduced. It is greatly reduced in op-amp designs that bias the output push-pull pair into slight conduction (class AB). The popular 741 is an example of the latter, whereas its predecessor, the 709, uses the simple class B output-stage biasing. The otherwise admirable 324 can exhibit large crossover distortion for this reason. The right choice of op-amp can have enormous impact on the performance of low-distortion audio amplifiers. Perhaps this problem has contributed to what the audiophiles refer to as "transistor sound."

The open-loop output impedance of an op-amp is highest near zero output voltage, because the output transistors are operating at their lowest current. The output impedance also rises at high frequency as the transistor gain drops off, and it may rise slightly at very low frequencies due to thermal feedback on the chip. It is easy to neglect the effects of finite open-loop output impedence, thinking that feedback will cure everything. But when you consider that some op-amps have open-loop output impedances of a few hundred ohms, it becomes clear that the effects may not be negligible, especially at low to moderate loop gains. Figure 7.9 shows some typical graphs of op-amp output impedance, both with and without feedback.

TABLE 7.2. HIGH-SPEED OP-AMPS

Compensated for gain = 1					Compensated for gain > 1							
Type	FET	f_T (MHz)	Slew rate (V/μs)	Settle to 0.1% (μs)	Type	FET	Min gain	f_T (MHz)	@ Gain	Slew rate (V/μs)	Settle to 0.1% (μs)	Comments
OP-01E[a]		2.5	18	0.7								Precision
OP-15E	✓	6	17	0.9								Precision
OP-16E	✓	8	25	0.7								Precision
LH0024C		70	400	—								
LH0032C	✓	70	500	0.3								
LH0061C		—	70	0.8								TO-3, 0.5A output
LH0062C	✓	15	70	1.0								
LM318		15	70	0.8								
LF356	✓	4.5	12	1.3	LF357	✓	5	20	5	50	1.3	Popular, inexpensive
AD518		12	70	0.8								
AD528J	✓	10	70	0.8								Precision
NE530		3	25	0.9	NE538		5	6	5	60[b]		NE5530/8 dual mini-DIP
MC1741S		1	12	3								High-slew 741
					1435		2	1000	2	300	0.02	Fast settle
					1435 + 2035		2	700	2	270	0.08	120mA output
HA2055A	✓	8	40			✓	3	20	10	120[c]	0.4[c]	Same as 1434
HA2065A	✓	10	5			✓	5	100	10	35[b]	0.8[b]	Same as 1433
HA2505		12	30	0.33								2507 mini-DIP
HA2515[d]		12	60	0.25	HA2525[d]		3	20	10	120	0.2	2517, 2527 mini-DIP
					HA2535[a]		10	70	10	370	0.5	Cable driver
HA2605		12	7	1.5	HA2625[e]		5	100	100	35[b]	0.3	
CA3100		30	25	0.6			10	38	10	70[f]	0.6[g]	
CA3140	✓	4.5	9	1.4								MOSFET, low cost, 3240 dual
3550K	✓	20	100	0.4	3551	✓	10	50	10	250	0.5	
					3554	✓	55	225	10	1200[h]	0.12	TO-3, 100mA output
HA5155	✓	50	60	0.7	HA5165	✓	10	100	10	100	0.5	
					HA5195		5	150	10	200[b]	0.05	Fast settle
					NE5539		7	1200	7	800	0.01[i]	Fast settle
8017C[a]		10	130	1								
9906		100	250	0.15			30	300	100			OEI
9908	✓	100	200	0.3								OEI
9909		100	2500	0.2			100	1000	100			OEI
9912		100	600	0.15			30	800	100			OEI
9914		300	1000	0.2			100	3000	100			OEI
9916		200	300	0.05								OEI
9932	✓	40	600	0.1			3	70	100			OEI

[a] Inverting amp. [b] At $G = 5$. [c] At $G = 3$. [d] Popular, available also as 1322, 3507J, AD509J. [e] Popular, available also as 1321, 3508J, AD507J. [f] At $G = 10$. [g] To 0.5%. [h] Data sheet confusing. [i] At $G = 2$.

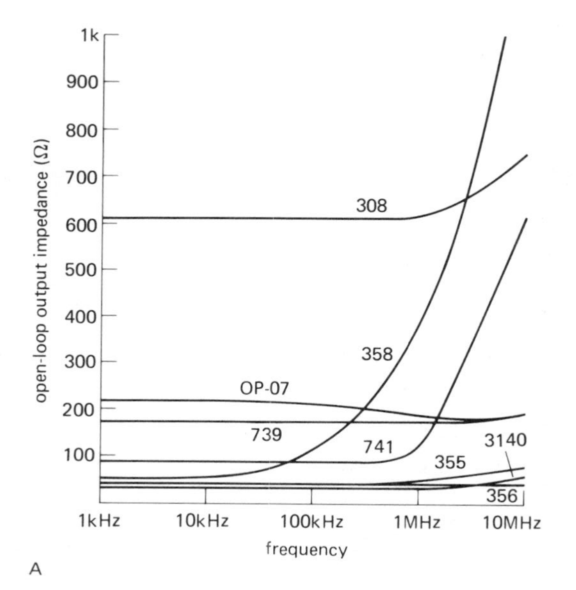

A

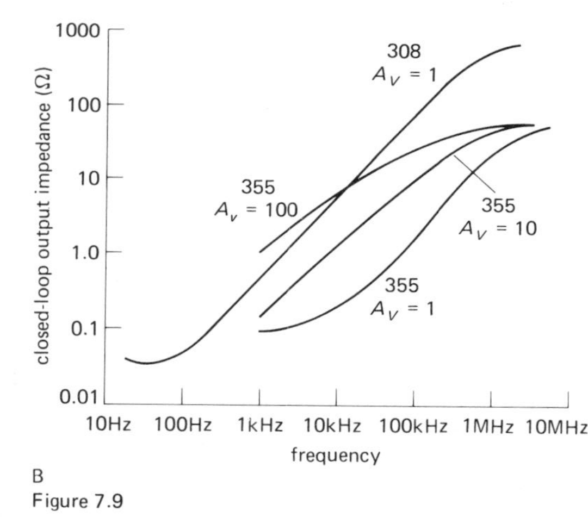

B

Figure 7.9

A: Measured open-loop output impedance for some popular op-amps. B: Closed-loop output impedance for the 308 and 355 op-amps.

Driving capacitive loads

The finite open-loop output impedance of op-amps leads to serious difficulties when you attempt to drive a capacitive load, owing to lagging phase shifts produced by the output impedance in combination with the load capacitance to ground. This can lead to feedback instabilities if the 3dB frequency is low enough, since it adds to the 90° phase shift already present with frequency compensation. As an example, imagine driving a hundred feet of coaxial cable from an op-amp with 200 ohms output impedance. The unterminated coax line acts like a 3000pF capacitor, generating a low-pass *RC* with a 3dB point of 270kHz. This is well below the unity-gain frequency of a typical op-amp, so oscillations are likely at high loop again (a follower, for example).

There are a couple of solutions to this problem. One is to add a series resistor, taking feedback at high frequencies from the op-amp output and feedback at low frequencies and dc from the cable (Fig. 7.10). The parts values shown in the second circuit are specific for that op-amp and circuit configuration, and they give an idea of how large a capacitance can be driven. Of course, this technique degrades the high-frequency performance, since feedback isn't operative at high frequencies on the signal at the cable.

Unity-gain power buffers

If this technique of split feedback paths is unacceptable, the best thing to do is add a unity-gain high-current buffer inside the loop (Fig. 7.11). The devices listed have voltage gain near unity and low output impedance, and they can supply up to 250mA output current. They have no significant phase shifts up to the unity-gain frequency (f_T) of most op-amps, and they can be included in the feedback loop without any additional frequency compensation. Table 7.3 presents a brief listing of buffer amplifiers. These ''power boosters'' can, of course, be used for loads that require high current, regardless of whether or not there are problems with capacitance. Note also that the preceding example would be changed if the cable were terminated in its characteristic impedance. In that case it would look like a pure resistance, somewhere in the range of 50 to 100 ohms, depending on the type of cable. In such a case a buffer would be mandatory, with ±200mA drive capability in order to drive ±10 volt signals into the 50 ohm load impedance. This subject is discussed in greater detail in Section 13.09.

The preceding circuit example does not suffer from any op-amp output-related errors, since it operates essentially at dc.

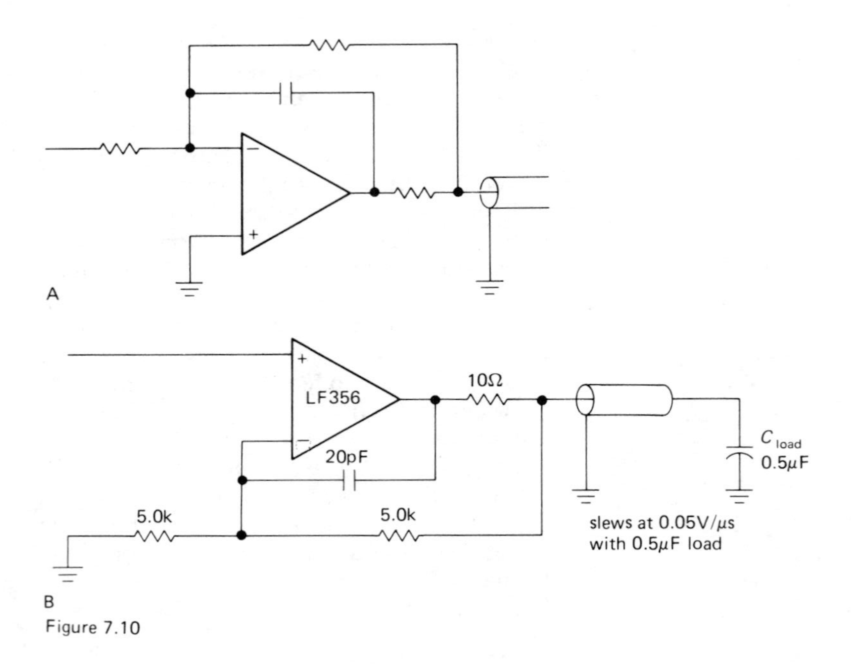

Figure 7.10

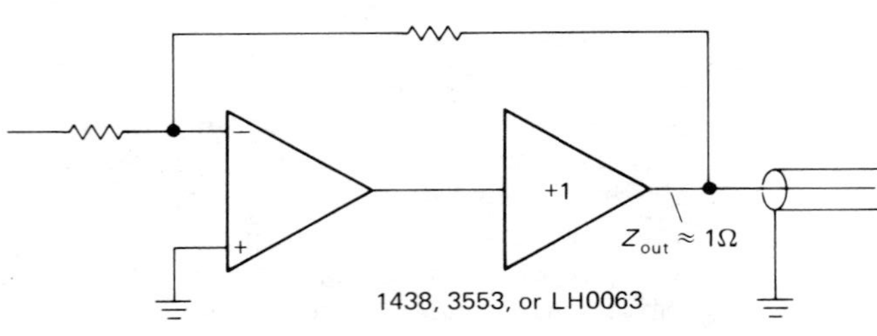

Figure 7.11

TABLE 7.3. FAST BUFFERS

Type	Small signal: Rolloff frequency −3dB (MHz)	Small signal: Rolloff frequency −40° (MHz)	Small signal: Output impedance (Ω)	Supply voltages min (±V)	Supply voltages max (±V)	Large signal: Slew rate (V/μs)	Large signal: Maximum output current (±mA)	Large signal: Output swing V_{out} (±V)	Large signal: Output swing R_{load} (Ω)	Comments
HA2635	8	5	2	5	20	500	600	10	50	
MC1438	30	20	10	5	18	75	300	12	300	Classic, monolithic
SH0002	50	60	6	6	22	200	100	10	50	
9910	60	—	20	6	18	2000	100	10	100	Mini-DIP package
3553	60	60	1	5	20	2000	200	10	50	Insulated metal case
LH0033	100	80	6	5	20	1400	100	10	50	
9911	200	—	6	11	18	1000	500	10	20	
LH0063	200	30	1	5	20	4000	250	10	50	"Damn fast" buffer

DIFFERENTIAL AND INSTRUMENTATION AMPLIFIERS

The term *instrumentation amplifier* is used to denote a high-gain dc-coupled differential amplifier with single-ended output, high input impedance, and high CMRR. They are used to amplify small differential signals coming from transducers in which there may be a large common-mode signal or level.

An example of such a transducer is a strain gauge, a bridge arrangement of resistors that converts strain (elongation) of the material to which it is attached into resistance changes (see Section 14.04); the net result is a small change in differential output voltage when driven by a fixed dc bias voltage (Fig. 7.12). The resistors all have roughly the same resistance, typically 350 ohms, but they are subjected to differing strains. The full-scale sensitivity is typically 3mV per volt, so that the full-scale output is 30mV for 10 volts dc excitation. This small differential output voltage proportional to strain rides on a 5 volt dc level. The differential amplifier must have extremely good CMRR in order to amplify the millivolt differential signals while rejecting variations in the ~5 volt common-mode signal. For example, suppose that a maximum error of 0.1% is desired. Since 0.1% of full scale is 0.03mV, riding on 5000mV, the CMRR would have to exceed 160,000 to 1, i.e., about 140dB.

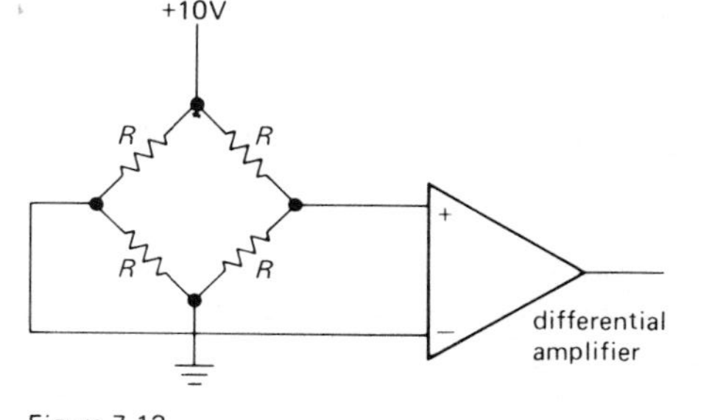

Figure 7.12

The tricks involved in making good instrumentation amplifiers and, more generally, high-gain differential amplifiers are similar to the techniques just discussed. Bias current, offsets, and CMRR errors are all important. Let's begin by discussing the design of differential amplifiers for noncritical applications first, working up to the most demanding instrumentation requirements and their circuit solutions.

7.08 Differencing amplifier

Figure 7.13 shows a typical circuit situation requiring only modest common-mode rejection. This is a current-sensing circuit used as part of a constant-current power supply to generate a constant current in the load. The drop across the precision 4-wire 0.01 ohm power resistor is proportional to load current. Even though one side of R_5 is

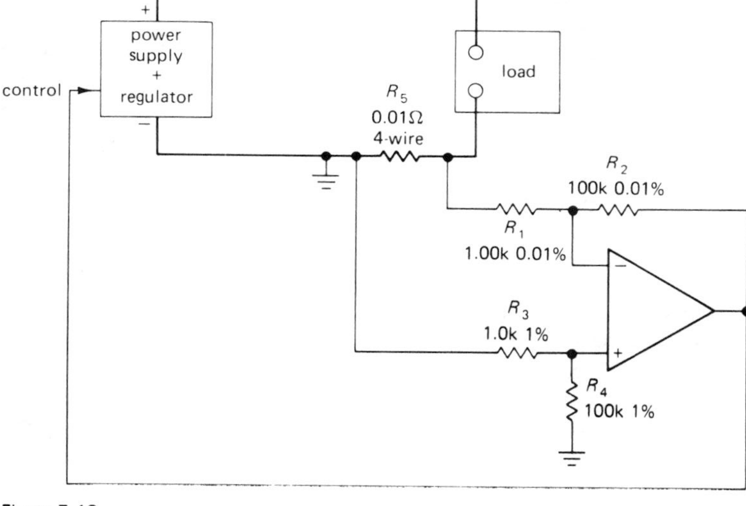

Figure 7.13
Current regulator

connected to ground, it would be unwise to use a single-ended amplifier, since connection resistances of a milliohm would contribute 10% error! A differential amplifier is obviously required, but it need not have particularly good CMRR, since only very small common-mode signals are expected.

The op-amp is connected in the standard differencing amplifier configuration, as discussed in Section 3.09. R_1, R_2, and R_5 are precision wire-wound types for extreme stability of gain, whereas R_3 and R_4, which set the CMRR, can be mere 1% metal-film types. The overall circuit thus has a gain accuracy approaching that of the current-sensing resistor and a CMRR of about 40dB.

Precision differential amplifier

For applications such as strain gauges, thermocouples, and the like, 40dB of common-mode rejection is totally inadequate, and figures more like 100dB to 120dB are often needed. In the preceding example of the strain gauge, for instance, you might have a full-scale differential (unbalance) signal of 2mV per volt. If you want accuracy of 0.05%, you need a common-mode rejection of 110dB, minimum. (Note that this requirement can be relaxed considerably in the special case that the amplifier is zeroed with the common-mode voltage present, as might be done in a laboratory situation.) The obvious first approach to improved CMRR is to beef up the resistor precision in the differencing circuit (Fig. 7.14). The resistor values are chosen to keep the large feedback resistors within the range of available precision wire-wound resistors. With 0.01% resistors, the common-mode rejection is in the range of 80dB (68dB worst case), assuming the op-amp has high CMRR. It takes only one trimmer to null the common-mode sensitivity, as shown. With the values shown, you can trim out an accumulated error up to 0.05%, i.e., a bit more than the worst-case resistor error. The fancy network shown is used because small-value trimmer resistors tend to be somewhat unstable with time and are best avoided.

A point about ac common-mode rejection: With good op-amps and careful trimming, you can achieve 100dB or better

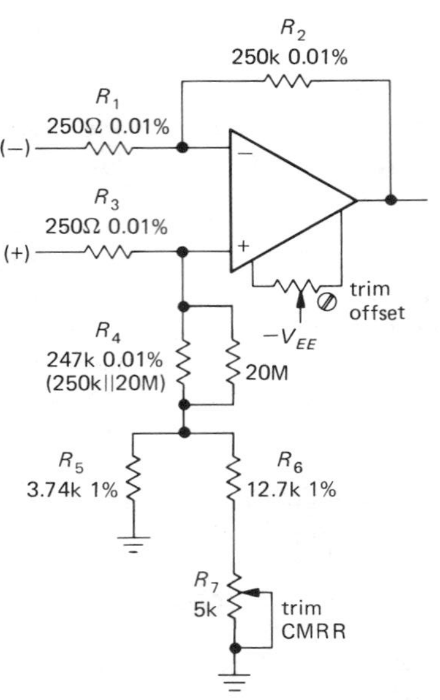

Figure 7.14

CMRR at dc. However, the wire-wound resistors you need for the best stability have some inductance, causing degradation of CMRR with frequency. Noninductive wire-wound resistors (Aryton-Perry type) are available to reduce this effect, which is common to all the circuits we will be talking about. Note also that it is necessary to balance the circuit capacitances to achieve good CMRR at high frequencies. This may require careful mirror-image placement of components.

High-voltage differential amplifier

Figure 7.15 shows a clever method for increasing the common-mode input voltage range of the differencing amplifier circuit beyond the supply voltages, without a corresponding reduction in differential gain. U_2 looks at the common-mode input signal at U_1's input and removes it via R_5 and R_6. Since there is no common-mode signal left at either U_1 or U_2, the CMRR of the op-amps is unimportant. The ultimate CMRR of this circuit is thus set by the matching of resistor ratios $R_1/R_5 = R_3/R_6$, with no great demands made on the accuracy of R_2 and R_4. The circuit shown has a common-mode input range of ±200 volts, a CMRR of 80dB, and a differential gain of 1.0.

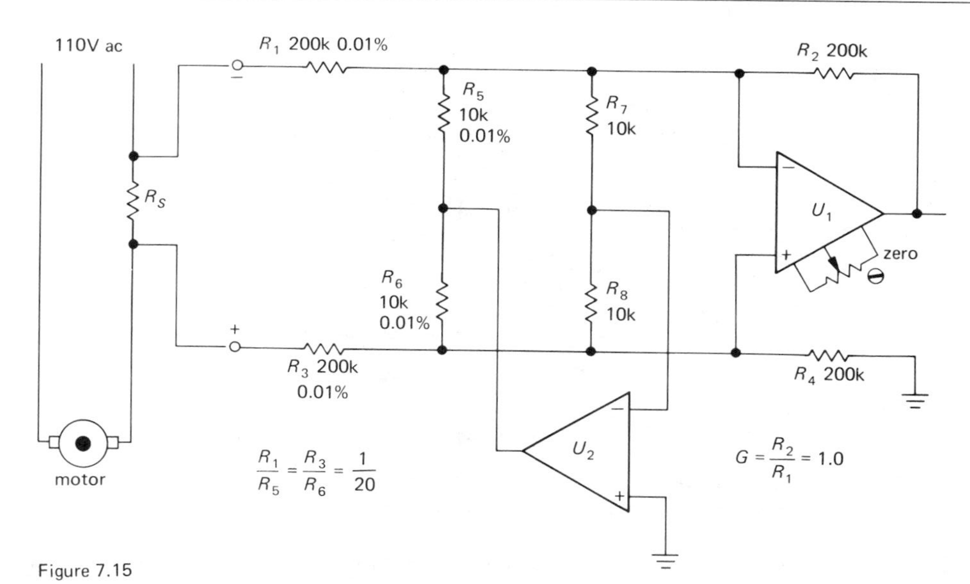

Figure 7.15
High-common-mode-voltage differential amplifier made from low-voltage op-amps.

□ *Raising input impedance*

The differencing circuit with carefully trimmed resistor values would seem to give the performance you want, until you look at the restrictions it puts on allowable source resistances. To get a gain accuracy of 0.1% with the circuit of Figure 7.14, you have to keep the source impedance below 0.25 ohm! Furthermore, the source impedance seen at the two terminals has to be matched to 0.0025 ohm in order to attain a CMRR of 100dB. This last result follows from a look at the equivalent circuit (Fig. 7.16). The

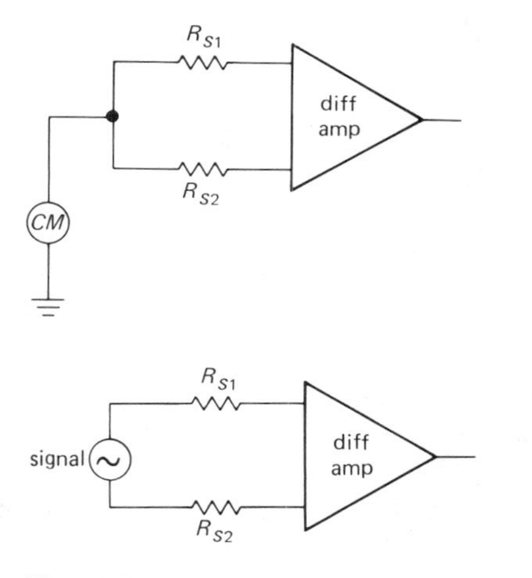

Figure 7.16

triangles represent the whole differential amplifier circuit or, in general, any differential or instrumentation amplifier, and R_{S1} and R_{S2} represent the Thévenin source resistances in each leg. For common-mode signals, the overall amplifier circuit includes the two source impedances in series with the input resistors R_1 and R_3, and so the CMRR now depends on the matching of $R_{S1} + R_1$ with $R_{S2} + R_3$. Obviously the demands this circuit makes on the source impedances as calculated earlier are unreasonable.

Some improvement can be had by increasing the resistor values, using the trick of a T network for the feedback resistors, as in Figure 7.17. This is the differential amplifier version of the T network discussed in Sections 7.06 and 3.18. With the values shown, you get a differential voltage gain of 1000 (60dB). For a gain accuracy of 0.1%, the source impedance must be less than 25 ohms and must be matched to 0.25 ohm for 100dB CMRR. This is still an unacceptable demand on the source in most applications. A strain gauge, for instance, typically has a source impedance of about 350 ohms.

The general solution to this problem involves followers, or noninverting amplifiers, to attain high input impedance. The simplest method would be to add followers to the conventional differential amplifier (Fig.

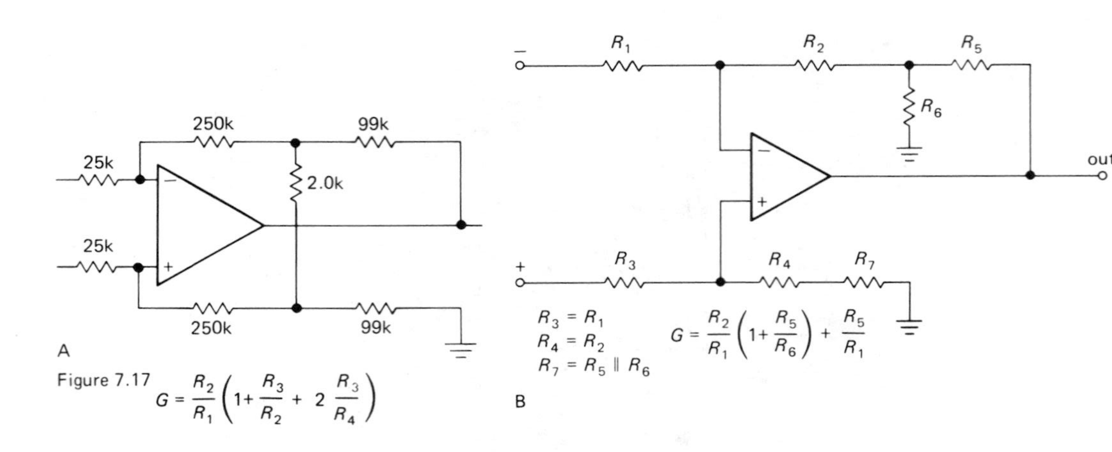

Figure 7.17

7.18). With the enormous input impedances you get, there is no longer any problem with any reasonable source impedance, at least at dc. At higher frequencies it again becomes important to have matched source impedances relative to the common-mode signal, because the input capacitance of the circuit forms a voltage divider in combination with the source resistance. By "high frequencies" we often mean 60Hz, since common-mode ac power-line pickup is a common nuisance; at that frequency the effect of a few picofarads of input capacitance isn't serious.

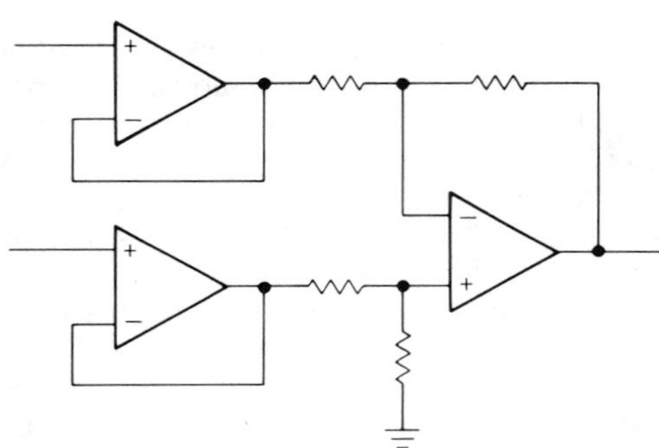
Figure 7.18

7.09 Standard three-op-amp instrumentation amplifier

One disadvantage of the previous follower circuit (Fig. 7.18) is that it requires high CMRR both in the followers and in the final op-amp. Since the input buffers operate at unity gain, all the common-mode rejection must come in the output amplifier, requiring precise resistor matching, as we discussed. The circuit in Figure 7.19 is a significant improvement in this respect. It constitutes the standard instrumentation amplifier configuration. The input stage is a clever configuration of two op-amps that provides high differential gain and unity common-mode gain without any close resistor matching. Its differential output represents a signal with substantial reduction in the comparative common-mode signal, and it is used to drive

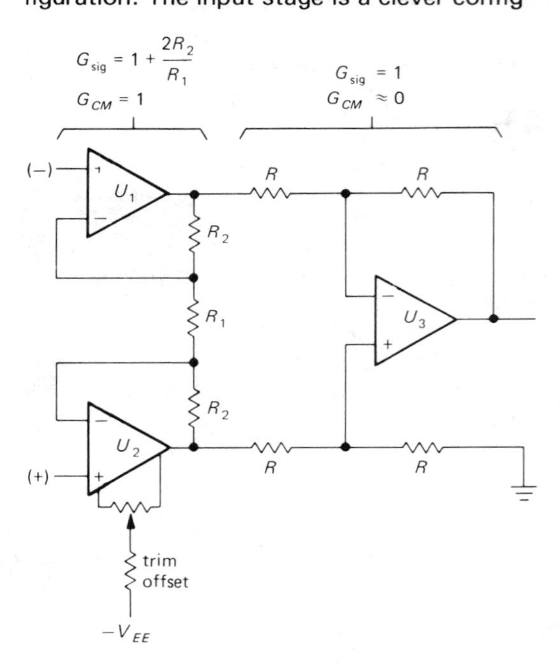

Figure 7.19
Classic instrumentation amplifier.

a conventional differential amplifier circuit. The latter is often arranged for unity gain and is used to generate a single-ended output and polish off any remaining common-mode signal. As a result, the output op-amp, U_3, needn't have exceptional CMRR itself, and resistor matching in U_3's circuit is not terribly critical. Offset trimming for the whole circuit can be done at one of the input op-amps, as shown. The input op-amps must still have high CMRR, and they should be chosen carefully.

Complete hybrid instrumentation amplifier modules containing this standard configuration are available from several manufacturers. All components except R_1 are internal, with gain set by the single external resistor R_1. Typical examples are the micropower LH0036, and AD522, and the high-accuracy 3630. All of these amplifiers offer a gain range of 1 to 1000, CMRR in the neighborhood of 100dB, and input impedances greater than 100M. The LH0036 can run from supply voltages as low as ±1 volt. The 3630 offers gain linearity of 0.002%, initial offset voltage of 25μV, and offset drift of 0.25μV/°C. There are provisions for external trimming of offset voltage and CMRR. Don't confuse these with the 725 "instrumentation operational amplifier," which is nothing more than a good op-amp intended as a building block for instrumentation amplifiers. Figure 7.20 shows the complete instrumentation amplifier circuit that is usually used.

A few comments about these instrumentation amplifier circuits (Fig. 7.20): (a) The buffered common-mode signal at U_4's output can be used as a "guard" voltage to reduce the effects of cable capacitance and leakage. When used this way, the guard output will be tied to the shield of the input cables. If the gain-setting resistor (R_1) is not immediately adjacent to the amplifier (e.g., if it is a panel adjustment, a configuration that should usually be avoided), its connections should be shielded and guarded also. (b) The SENSE and REF terminals allow sensing of output voltage *at the load* so that feedback can operate to eliminate losses in the wiring or external circuit. In addition, the REF terminal also allows you to offset the output

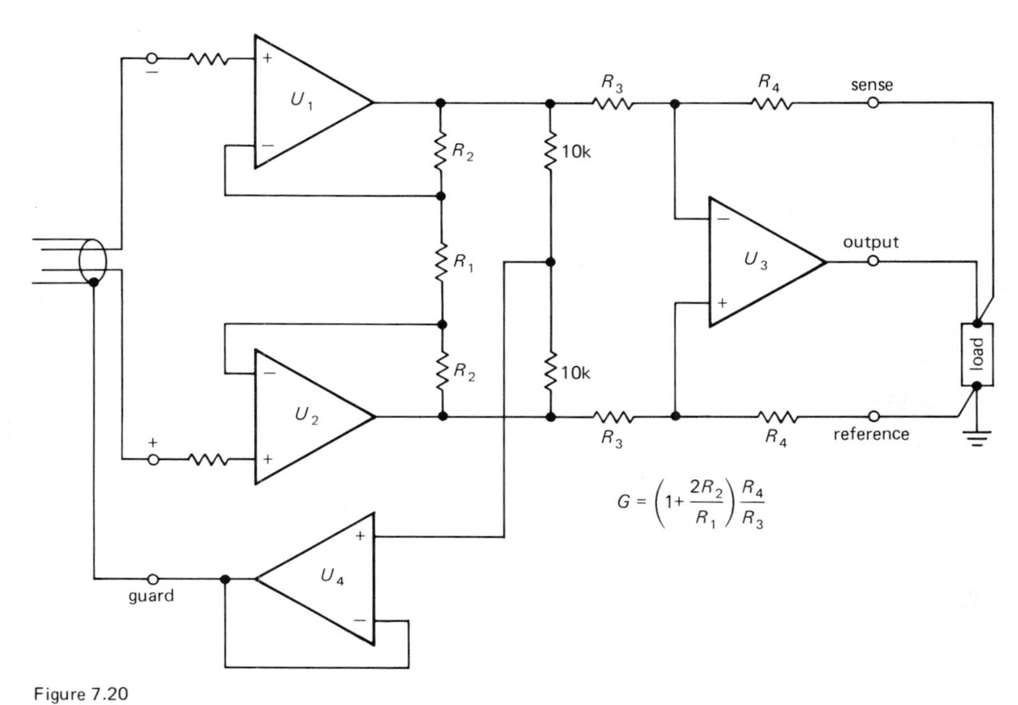

Figure 7.20
Instrumentation amplifier with guard, sense, and reference terminals.

signal by a dc level (or by another signal); however, the impedance from the REF terminal to ground must be kept small, or the CMRR will be degraded. (c) With any of these instrumentation amplifiers there must be a bias path for input current; for example, you can't just connect a thermocouple across the input. Figure 7.21 shows the simple application of an IC instrumentation amplifier with guard, sense, and reference terminals.

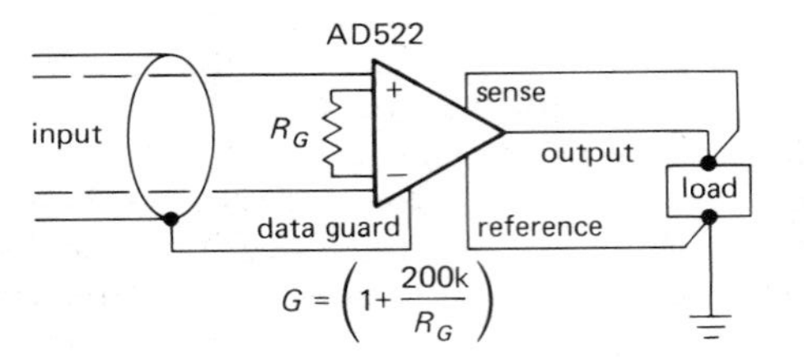

Figure 7.21
Integrated-circuit instrumentation amplifier.

□ *Bootstrapped power supply*

The CMRR of the input op-amps may be the limiting factor in the ultimate common-mode rejection of this circuit. If CMRRs greater than about 120dB are needed, the trick shown in Figure 7.22 can be used. U_4 buffers the common-mode signal level, driving the common terminal of a small floating split supply for U_1 and U_2. This bootstrapping scheme effectively eliminates the input common-mode signal from U_1 and U_2, because they see no swing (due to common-mode signals) at their inputs relative to their power supplies. U_3 and U_4 are powered by the system power supply, as

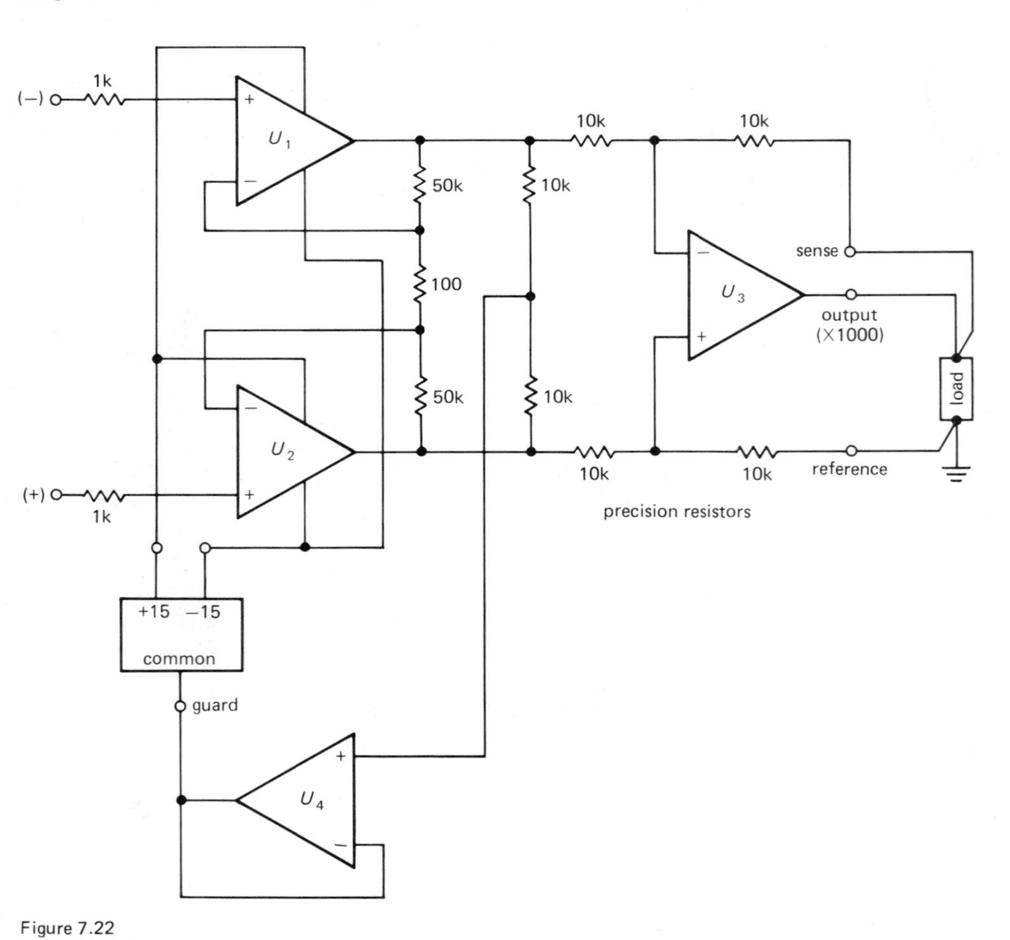

Figure 7.22
Instrumentation amplifier with bootstrapped input power supply for high CMRR.

usual. This scheme can do wonders for the CMRR, at least at dc. At increasing frequencies you have the usual problems of presenting matched impedances to the input capacitances.

Two-op-amp configuration

Figure 7.23 shows another configuration that offers high input impedance with only two op-amps. Since it doesn't accomplish the common-mode rejection in two stages, as in the three-op-amp circuit, it requires precise resistor matching for good CMRR, in a manner similar to that of the standard differencing amplifier circuit.

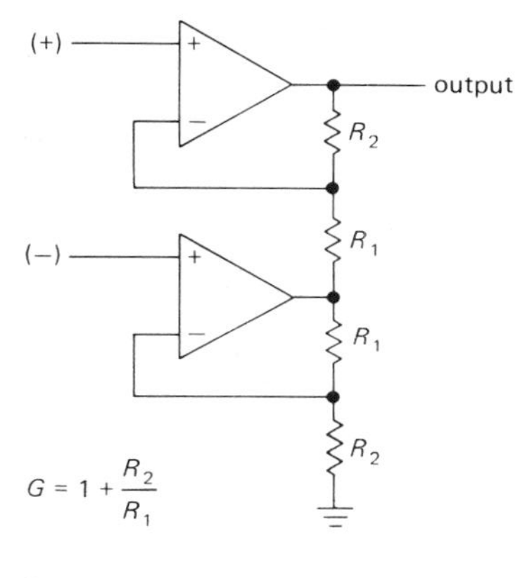

Figure 7.23

Special IC instrumentation amplifiers

There are several interesting instrumentation amplifier configurations available as monolithic (and therefore inexpensive) ICs, some with extremely good performance. They use methods unrelated to the preceding circuits.

□ *Differential transconductance amplifier technique.* This technique, typified by the LF352 and AD521, achieves high CMRR without the need for matched external resistors. In fact, the gain is set by the ratio of a pair of external resistors. Figure 7.24 shows a block diagram of the LF352. The circuit employs two differential transconductance amplifier pairs, with a single external resistor setting the gain in each case. One amplifier is driven by the input signal, and the other is driven by the output signal, relative to the REF terminal. The circuit includes an op-amp that adjusts the output voltage as necessary to make the currents match. The LF352 uses FETs to keep input currents low,

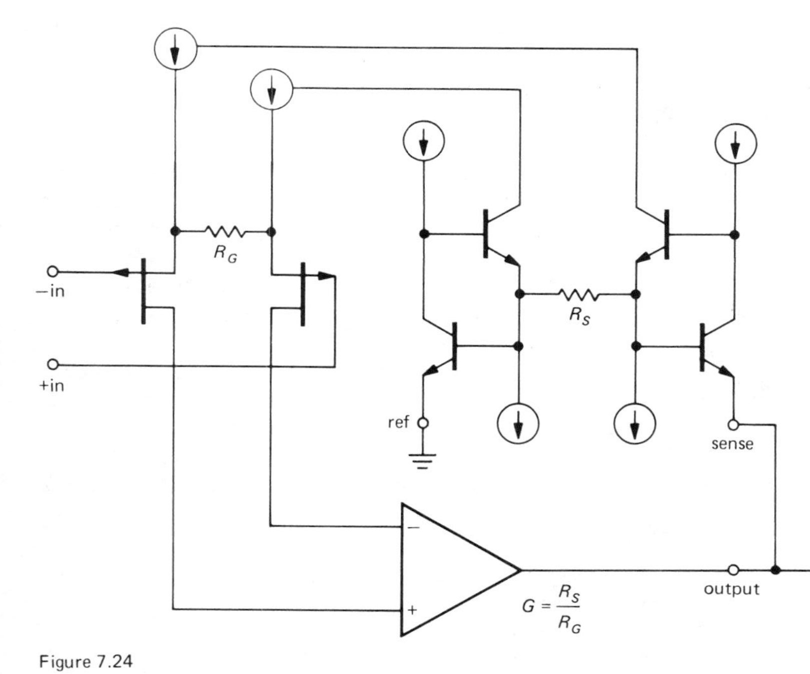

Figure 7.24
Block diagram of the LF352 instrumentation amplifier IC.

whereas the AD521 uses bipolar technology to achieve low offset voltage and drift (Table 7.4).

□ *Commutating auto-zeroing amplifiers.* The recently developed commutating auto-zeroing (CAZ) amplifier technique stands a good chance of revolutionizing precision op-amp and instrumentation amplifier technology. The CAZ technique lets you construct MOSFET op-amps with offsets of a few microvolts and drifts of less than $0.1\mu V/°C$. This is 100 to 1000 times better than ordinary MOSFET op-amps. The basic idea is to package two CMOS op-amps and a bunch of MOSFET switches in a single package; one amplifier is being zeroed (storing its input offset voltage on a capacitor) while the other is acting as an op-amp for the input signal, with its previously stored offset voltage in series with the input. The op-amps reverse roles periodically, via the MOSFET switches, so drifts are effectively cancelled out. Don't confuse this with "chopper-stabilized" amplifiers, which convert the input signal to ac, amplify it, and then convert back to dc. In the CAZ technique the signal is always passing through an op-amp, with normal op-amp frequency response.

Intersil pioneered the CAZ technique with its ICL7600 op-amp, which uses a commutation frequency of 160Hz. It has an initial offset of less than $5\mu V$ and drifts of less than $0.1\mu V/°C$ and $0.2\mu V$/year. The companion ICL7605 instrumentation amplifier uses the basic CAZ op-amp, with a "flying capacitor" switching technique (Fig. 7.25).

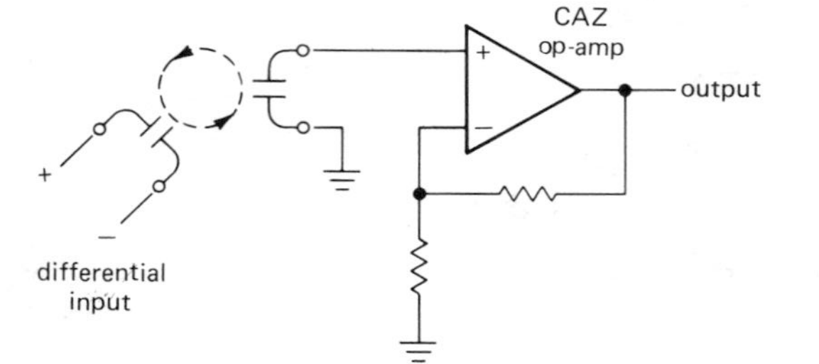

Figure 7.25

MOSFET switches enable you to store the differential input signal across a capacitor, then amplify it with a single-ended noninverting amplifier constructed with the CAZ technique. The 7605 has a common-mode input range extending 0.3 volt beyond both supply rails.

Some minor circuit modifications are generally necessary when substituting CAZ amplifiers into existing op-amp circuits, owing to the presence of switching transients and other artifacts at the output. Furthermore, there are some applications where they are unsuitable altogether, e.g., in a rapidly multiplexed A/D converter system.

AMPLIFIER NOISE

In almost every area of measurement the ultimate limit of detectability of weak signals is set by noise – unwanted signals that obscure the desired signal. Even if the quantity being measured is not weak, the presence of noise degrades the accuracy of the measurement. Some forms of noise are unavoidable (e.g., real fluctuations in the quantity being measured), and they can be overcome only with the techniques of *signal averaging* and *bandwidth narrowing,* which we will discuss in Chapter 14. Other forms of noise (e.g., radiofrequency interference and "ground loops") can be reduced or eliminated by a variety of tricks, including filtering and careful attention to wiring configuration and parts location. Finally, there is noise that arises in the amplification process itself, and it can be reduced through the techniques of low-noise amplifier design. Although the techniques of signal averaging can often be used to rescue a signal buried in noise, it always pays to begin with a system that is free of preventable interference and that possesses the lowest amplifier noise practicable.

We will begin by talking about the origins and characteristics of the different kinds of noise that afflict electronic circuits. Then we will launch into a discussion of transistor and FET noise, including methods for low-noise design with a given signal source, and will present some design examples. After a short discussion of noise in differential and feedback amplifiers, we will conclude with a section on proper grounding and shielding and the elimination of interference and pickup.

TABLE 7.4 INSTRUMENTATION AMPLIFIERS

Type	FET input	Transconductance type	High-Z output ref.	Total supply: Voltage min (V)	Total supply: Voltage max (V)	Total supply: Current max (mA)	Max. input errors, Offset voltage RTI[a] (mV)	Offset voltage RTI[a] (μV/°C)	Offset voltage RTO[a] (mV)	Offset voltage RTO[a] (μV/°C)	Current: Bias (nA)	Current: Offset (nA)	Diffl. Input impedance R (Ω)[b]	Diffl. Input impedance C (pF)	CMRR at dc $G = 1$ (dB, min)	CMRR at dc $G = 1k$ (dB, min)	Slew rate (V/μs)	−3dB bandwidth $G = 1$ (kHz)	−3dB bandwidth $G = 1k$ (kHz)	Bandwidth for 1% error $G = 1$ (kHz)	Bandwidth for 1% error $G = 1k$ (kHz)	Settling time to 0.1% $G = 1$ (μs)	Settling time to 0.1% $G = 1k$ (μs)	Noise Voltage 0.1–10Hz RTI[a] (μV, pp)	Noise Voltage 0.1–10Hz RTO[a] (μV, pp)	Noise Voltage 10Hz–10kHz RTI[a] (μV, rms)	Noise Voltage 10Hz–10kHz RTO[a] (μV, rms)	Noise Current 10Hz–10kHz (pA, rms)
LH0036	—	—	—	2	36	0.6	1	10[f]	5	15[f]	100	40	3E8	—	50	100	0.3	350	0.35	—	—	8	600	—	—	5	—	—
LH0037	—	—	—	—	44	8.4	1	10[f]	5	15[f]	500	100	3E8	—	50	—	0.5	350	0.35	—	—	—	—	—	—	—	—	—
LH0038[c]	—	—	—	10	36	2	0.1	0.25	10	25[f]	100	5	5E6	—	—	114	0.3	—	1.6	—	—	—	80[d]	0.2	—	0.2	—	10
LF352	✓	✓	✓	10	36	2.2	30	10[f]	400	600[f]	0.04	0.02	2E12	2.5	65	105	1	140	7	5	1.5	15	200	1.3	670	8	450	0.01
AD521	—	✓	✓	10	36	5	3	15	400	400	80	20	3E9	1.8	70	100	10	2000	40	75	6	7	35	0.5	150	1.2	30	—
AD522	—	—	—	10	36	10	—	6	0.4	50	25	20	1E9	—	75	100	0.1	300	0.3	—	—	500[d]	20,000[d]	1.5	15	—	15	—
3630B	—	—	—	10	40	14	0.025	0.75	0.2	10	50	50	1E10	3	90	110	0.5	150	2.5	20	0.2	60	500	1.2	—	1.0	—	20
3660K	—	✓	—	14	40	6	1	2	300	500	200	20	2E10/G	9	70	110	1.8	800	22	20	3	17	50	3	900	1	130	50
ICL7605[e]	✓	—	—	2	18	5	0.005	0.1	—	—	1.5	—	—	—	100[f]	100[f]	0.5	0.01	0.01	Slow		Slow		5	—	—	—	—

[a] RTI: referred to the input, RTO: referred to the output. Noise and errors can be separated into components generated at both the input and output. The total input-referred noise (or error) is thus given by RTI + RTO/G. [b] Fortran notation: 3E8 = 3×10^8, for example. [c] Gain range 100–2000. [d] To 0.01%. [e] CAZ type (see Section 7.09). [f] Typical.

7.10 Origins and kinds of noise

Since the term *noise* can be applied to anything that obscures a desired signal, noise can itself be another signal ("interference"); most often, however, we use the term to describe "random" noise of a physical (often thermal) origin. Noise can be characterized by its frequency spectrum, its amplitude distribution, and the physical mechanism responsible for its generation. Let's next look at the chief offenders.

Johnson noise

Any old resistor just sitting on the table generates a noise voltage across its terminals known as Johnson noise. It has a flat frequency spectrum, meaning that there is the same noise power in each hertz of frequency (up to some limit, of course). Noise with a flat spectrum is also called "white noise." The actual open-circuit noise voltage generated by a resistance R at temperature T is given by

$$V_{\text{noise}}\,(\text{rms}) = V_{nR} = (4kTRB)^{1/2}$$

where k is Boltzmann's constant, T is the absolute temperature in degrees Kelvin (°K = °C + 273.16), and B is the bandwidth in hertz. Thus V_{noise} (rms) is what you would measure at the output if you drove a perfect noiseless bandpass filter (of bandwidth B) with the voltage generated by a resistor at temperature T. At room temperature (68°F = 20°C = 293°K),

$$4kT = 1.62 \times 10^{-20}\ \text{V}^2/\text{Hz-}\Omega$$

$$(4kTR)^{1/2} = 1.27 \times 10^{-10}\ R^{1/2}\ \text{V/Hz}^{1/2}$$

$$= 1.27 \times 10^{-4}\ R^{1/2}\ \mu\text{V/Hz}^{1/2}$$

For example, a 10k resistor at room temperature has on open-circuit rms voltage of 1.3μV, measured with a bandwidth of 10kHz (e.g., by placing it across the input of a high-fidelity amplifier and measuring the output with a voltmeter). The source resistance of this noise voltage is just R. Figure 7.26 plots the simple relationship between Johnson noise voltage density (rms voltage per square root bandwidth) and source resistance.

The amplitude of the Johnson noise voltage at any instant is, in general, unpredictable, but it obeys a Gaussian amplitude distribution (Fig. 7.27), where $p(V)\,dV$ is the probability that the instantaneous voltage lies between V and $V + dV$, and v_n is the rms noise voltage, given earlier.

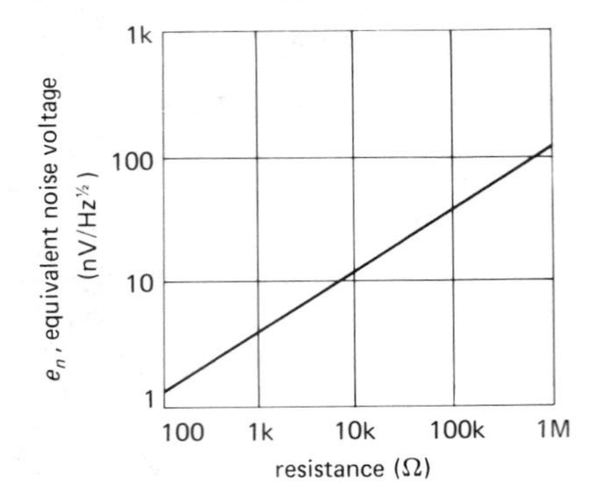

Figure 7.26
Thermal noise voltage versus resistance. (National Semiconductor Corp.)

The significance of Johnson noise is that it sets a lower limit on the noise voltage in any detector, signal source, or amplifier having resistance. The resistive part of a source impedance generates Johnson noise, as do the bias and load resistors of an amplifier. You will see how it all works out shortly.

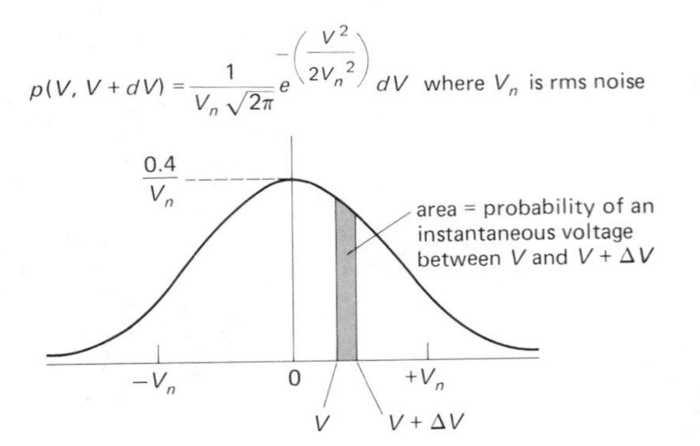

Figure 7.27

It is interesting to note that the physical analog of resistance (any mechanism of energy loss in a physical system, e.g., viscous friction acting on small particles in a liquid) has associated with it fluctuations in the associated physical quantity (in this case, the particles' velocity, manifest as the chaotic Brownian motion). Johnson noise is just a special case of this fluctuation-dissipation phenomenon.

Johnson noise should not be confused with the additional noise voltage created by the effect of resistance fluctuations when an externally applied current flows through a resistor. This "excess noise" has a $1/f$ spectrum (approximately) and is heavily dependent on the actual construction of the resistor. We will talk about it later.

Shot noise

An electric current is the flow of discrete electric charges, not a smooth fluidlike flow. The finiteness of the charge quantum results in statistical fluctuations of the current, given by

$$I_{\text{noise}}\,(\text{rms}) = I_{nR} = (2qI_{dc}B)^{1/2}$$

where q is the electron charge (1.60×10^{-19} coulomb) and B is the measurement bandwidth. For example, a "steady" current of 1 amp actually has an rms fluctuation of 57nA, measured in a 10kHz bandwidth; i.e., it fluctuates by about 0.000006%. The relative fluctuations are larger for smaller currents: A "steady" current of 1μA actually has an rms current noise fluctuation, measured over a 10kHz bandwidth, of 0.006%, i.e., −85dB. At 1pA dc, the rms current fluctuation (same bandwidth) is 56fA, i.e. a 5.6% variation! Shot noise is "rain on a tin roof." This noise, like resistor Johnson noise, is Gaussian and white.

1/f Noise (flicker noise)

Shot noise and Johnson noise are irreducible forms of noise generated according to physical principles. The most expensive and most carefully made resistor has exactly the same Johnson noise as the cheapest carbon resistor (of the same resistance). Real devices have, in addition, various sources of "excess noise." Real resistors suffer from fluctuations in resistance, generating an additional noise voltage (which adds to the ever-present Johnson noise) proportional to the dc current flowing through them. This noise depends on many factors having to do with the construction of the particular resistor, including the resistive material and especially the end-cap connections. Here is a listing of typical excess noise for various resistor types, given as rms microvolts per volt applied across the resistor, measured over one decade of frequency:

Carbon-composition	0.10μV to 3.0μV
Carbon-film	0.05μV to 0.3μV
Metal-film	0.02μV to 0.2μV
Wire-wound	0.01μV to 0.2μV

This noise has approximately a $1/f$ spectrum (equal power per decade of frequency) and is sometimes called "pink noise." Other noise-generating mechanisms often produce $1/f$ noise, examples being base current noise in transistors and cathode current noise in vacuum tubes. Curiously enough, $1/f$ noise is present in nature in unexpected places, e.g., the speed of ocean currents, the flow of sand in an hourglass, the flow of traffic on Japanese expressways, and the yearly flow of the Nile measured over the last 2000 years. If you plot the loudness of a piece of classical music versus time, you get a $1/f$ spectrum! No unifying principle has been found for all the $1/f$ noise that seems to be swirling around us, although particular sources can often be identified in each instance.

Interference

As we mentioned earlier, an interfering signal or stray pickup constitutes a form of noise. Here the spectrum and amplitude characteristics depend on the interfering signal. For example, 60Hz pickup has a sharp spectrum and relatively constant amplitude, whereas car ignition noise, lightning, and other impulsive interferences are broad in spectrum and spiky in amplitude. Other sources of interference are radio and television stations (a particularly serious problem near large cities), nearby electrical equipment, motors and elevators, subways, switching regulators, and television sets. In a slightly different guise you have the same sort of problem generated by anything that puts a signal into the parameter you are measuring. For example, an optical interferometer is susceptible to vibration, and a sensitive radiofrequency measurement (e.g., NMR) can be affected by ambient radiofrequency signals. Many circuits, as well as detectors and even cables, are sensitive to vibration and sound; they are *microphonic,* in the terminology of the trade.

Many of these noise sources can be controlled by careful shielding and filtering, as we will discuss later in the chapter. At other times you are forced to take draconian measures, involving massive stone tables (for vibration isolation), constant-temperature rooms, anechoic chambers, and electrically shielded rooms.

7.11 Signal-to-noise ratio and noise figure

Before getting into the details of amplifier noise and low-noise design, we need to define a few terms that are often used to describe amplifier performance. These involve ratios of noise voltages, measured at the same place in the circuit. It is conventional to refer noise voltages to the input of an amplifier (although the measurements are usually made at the output), i.e., to describe source noise and amplifier noise in terms of microvolts *at the input* that would generate the observed output noise. This makes sense when you want to think of the relative noise added by the amplifier to a given signal, independent of amplifier gain; it's also realistic, because most of the amplifier noise is usually contributed by the input stage. Unless we state otherwise, noise voltages are referred to the input.

Noise power density and bandwidth

In the preceding examples of Johnson noise and shot noise, the noise voltage you measure depends both on the measurement bandwidth B (i.e., how much noise you see depends on how fast you look) and on the variables (R and I) of the noise source itself. So it's convenient to talk about an rms noise-voltage "density" v_n:

$$V_n\,(\text{rms}) = v_n B^{1/2} = (4kTR)^{1/2} B^{1/2}$$

where V_n is the rms noise voltage you would measure in a bandwidth B. White noise sources have a v_n that doesn't depend on frequency, whereas pink noise, for instance, has a v_n that drops off at 3dB/octave. You'll often see v_n^2, too, the mean squared noise density. Since v_n always refers to rms, and v_n^2 always refers to mean square, you can just square v_n to get v_n^2! Sounds simple (and it is), but we want to make sure you don't get confused.

Note that B and the square root of B keep popping up. Thus, for example, for Johnson noise from a resistor R

$$v_{nR}\,(\text{rms}) = (4kTR)^{1/2} \;\; \text{V/Hz}^{1/2}$$
$$v_{nR}^{\;2} = 4kTR \;\; \text{V}^2/\text{Hz}$$
$$V_n\,(\text{rms}) = v_{nR} B^{1/2} = (4kTRB)^{1/2} \;\; \text{V}$$
$$V_n^2 = v_{nR}^{\;2} B = 4kTRB \;\; \text{V}^2$$

On data sheets you may see graphs of v_n or v_n^2, with units like "nanovolts per root Hz" or "volts squared per Hz." The quantities e_n and i_n that will soon appear work just the same way.

When you add two signals that are uncorrelated (two noise signals, or noise plus a real signal), the *squared* amplitudes add:

$$v = (v_s^2 + v_n^2)^{1/2}$$

where v is the rms signal obtained by adding together a signal of rms amplitude v_s and a noise signal of rms amplitude v_n. The rms amplitudes *don't* add.

Signal-to-noise ratio

Signal-to-noise ratio (SNR) is simply defined as

$$\text{SNR} = 10 \log_{10}\left(\frac{V_s^2}{V_n^2}\right) \;\; \text{dB}$$

where the voltages are rms values, and some bandwidth and center frequency are specified; i.e., it is the ratio, in decibels, of the rms voltage of the desired signal to the rms voltage of the noise that is also present. The "signal" itself may be sinusoidal or a modulated information-carrying waveform or even a noiselike signal itself. It is particularly important to specify the bandwidth if the signal has some sort of narrowband spectrum, since the SNR will drop as the bandwidth is increased beyond that of the signal: The amplifier keeps adding noise power, while the signal power remains constant.

Noise figure

Any real signal source or measuring device generates noise because of Johnson noise in its source resistance (the real part of its complex source impedance). There may be additional noise, of course, from other causes. The *noise figure* (NF) of an amplifier is simply the ratio, in decibels, of the output of the real amplifier to the output of a "perfect" (noiseless) amplifier of the same gain, with a resistor of value R_s connected across the amplifier's input terminals in each case. That is, the Johnson noise of R_s is the "input signal."

$$\text{NF} = 10 \log_{10} \left(\frac{4kTR_s + v_n^2}{4kTR_s} \right)$$

$$= 10 \log_{10} \left(1 + \frac{v_n^2}{4kTR_s} \right) \text{dB}$$

where v_n^2 is the mean squared noise voltage per hertz contributed by the amplifier, with a noiseless (cold) resistor of value R_s connected across its input. This latter restriction is important, as you will see shortly, because the noise voltage contributed by an amplifier depends very much on the source impedance (Fig. 7.28).

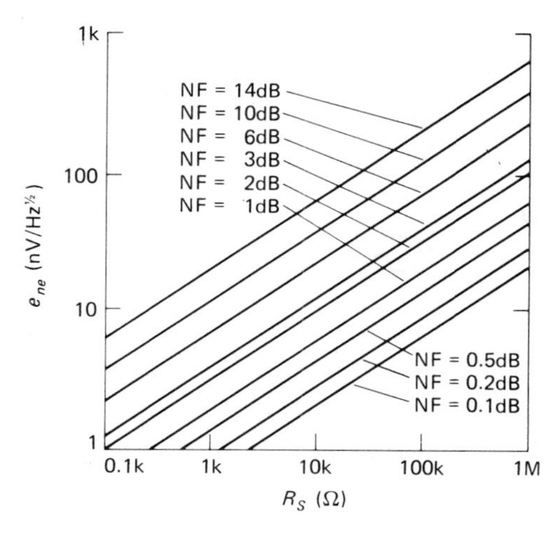

Figure 7.28
Effective noise voltage versus noise figure and source resistance. (National Semiconductor Corp.)

Noise figure is handy as a figure of merit for an amplifier when you have a signal source of a given source impedance and want to compare amplifiers (or transistors, for which NF is often specified). NF varies with frequency and source impedance, and it is often given as a set of contours of constant NF versus frequency and R_s. It may also be given as a set of graphs of NF versus frequency, one curve for each collector current, or a similar set of graphs of NF versus R_s, one for each collector current.

Big fallacy: Don't try to improve things by adding a resistor in series with a signal source to reach a region of minimum NF. All you're doing is making the source noisier to make the amplifier look better! Noise figure can be very deceptive for this reason. To add to the deception, a noise figure specification (e.g., NF = 2dB) for a transistor or FET will always be for the optimum combination of R_s and I_C. It doesn't tell you much about actual performance, except that the manufacturer thinks the noise figure is worth bragging about.

In general, when evaluating the performance of some amplifier, you're probably least likely to get confused if you stick with SNR calculated for that source voltage and impedance. Here's how to convert from NF to SNR:

$$\text{SNR} = 10 \log_{10} \left(\frac{v_s^2}{4kTR_s} \right) - \text{NF (dB) (at}R_s) \quad dB$$

where v_s is the rms signal amplitude, R_s is the source impedance, and NF is the noise figure of the amplifier for source impedance R_s.

After reading the next two sections, we trust you won't ever be confused about noise figure again!

7.12 Transistor amplifier voltage and current noise

The noise generated by an amplifier is easily described by a simple noise model that is accurate enough for most purposes. In Figure 7.29, e_n represents a noise voltage source in series with the input and i_n represents an input noise current. The transistor (or amplifier, in general) is assumed noiseless, and it simply amplifies the input noise voltage it sees. That is, the amplifier contributes a total noise voltage e_a, referred to the input, of

$$e_a \text{ (rms)} = [e_n^2 + (R_s i_n)^2]^{1/2} \text{ V/Hz}^{1/2}$$

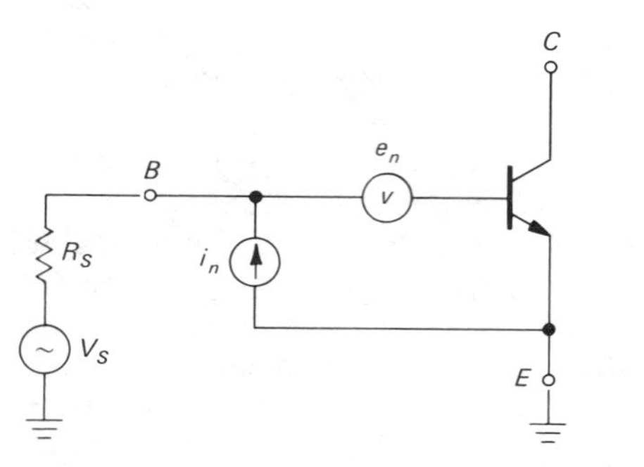

Figure 7.29

The two terms are simply the amplifier input noise voltage and the noise voltage generated by the amplifier's input noise current passing through the source resistance. Since the two noise terms are usually uncorrelated, their squared amplitudes add to produce the effective noise voltage seen by the amplifier. For low source resistances the noise voltage e_n dominates, whereas for high source impedances the noise current i_n generally dominates.

Just to give an idea of what these look like, Figure 7.30 shows a graph of e_n and i_n versus I_C and f, for a 2N4250. We'll go into some detail now, describing these and showing how to design for minimum noise. It is worth noting that voltage noise and current noise for a transistor are in the range of nanovolts and picoamps per root hertz ($Hz^{1/2}$).

Voltage noise, e_n

The equivalent voltage noise looking in series with the base of a transistor arises from Johnson noise in the base spreading resistance, $r_{bb'}$, and collector current shot noise generating a noise voltage across the intrinsic emitter resistance r_e. These two terms look like this:

$$e_n^2 = 4kTr_{bb} + 2qI_C r_e^2$$
$$= 4kTr_{bb} + \frac{2(kT)^2}{qI_C} \text{ V}^2/\text{Hz}$$

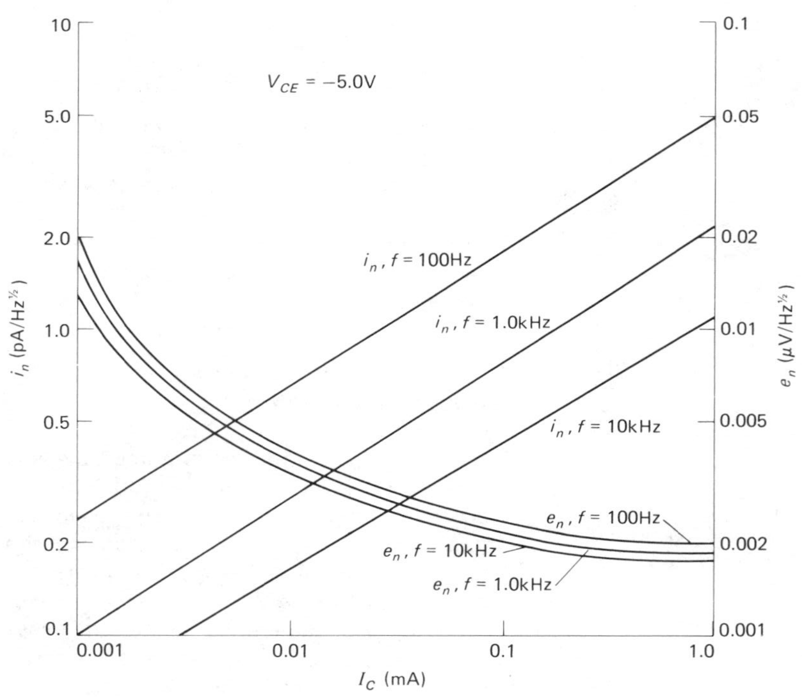

Figure 7.30
Equivalent rms input noise voltage (e_n) and noise current (i_n) versus collector current for a 2N4250 npn transistor. (Fairchild Camera and Instrument Corp.)

Both of these are Gaussian white noise. In addition, there is some flicker noise generated by base current flowing through r_{bb}. This last term is significant only at high base current, i.e., at high collector current. The result is that e_n is constant over a wide range of collector currents, rising at low currents (shot noise through an increasing r_e) and at sufficiently high currents (flicker noise from I_B through r_{bb}). This latter rise is present only at low frequencies, because of its $1/f$ character. As an example, at frequencies above 10kHz the 2N4250 has an e_n of 5nV/Hz$^{1/2}$ at $I_C = 10\mu A$ and 2nV/Hz$^{1/2}$ at $I_C = 100\mu A$. Figure 7.31 shows graphs of e_n versus frequency and current for the low-noise LM394 *npn* differential pair.

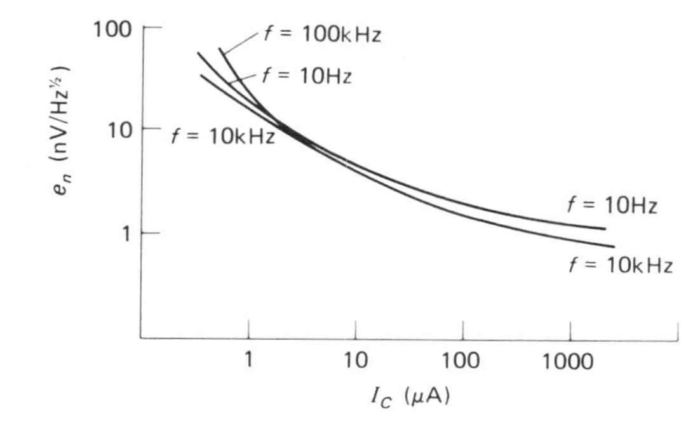

Figure 7.31
Input noise voltage (e_n) versus collector current for the LM394 bipolar transistor.

Current noise, i_n

Noise current is important, because it generates an additional noise voltage across the input signal source impedance. The main source of current noise is shot noise fluctuation in the steady base current, added to the fluctuations caused by flicker noise in r_{bb}. The shot noise contribution is a noise current that increases proportional to the square root of I_B (or I_C) and is flat with frequency, whereas the flicker noise component rises more rapidly with I_C and shows the usual $1/f$ frequency dependence. Taking the example of the 2N4250 again, above 10kHz i_n is about 0.1pA/Hz$^{1/2}$ at $I_C = 10\mu A$ and 0.4pA/Hz$^{1/2}$ at $I_C = 100\mu A$. The noise current increases, and the noise voltage drops, as I_C is increased. In the next section you will see how this dictates operating current in low-noise design. Figure 7.32

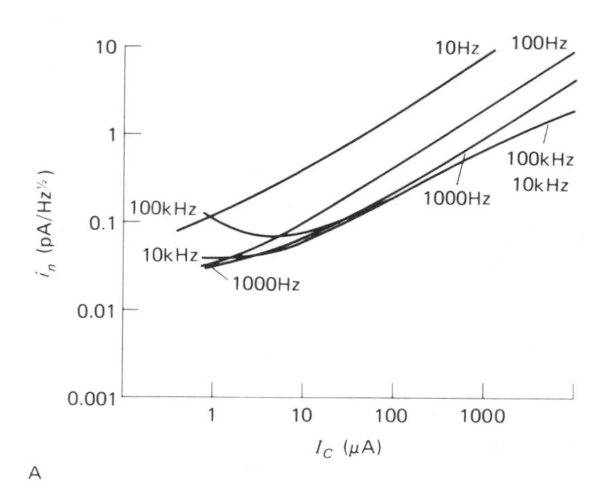

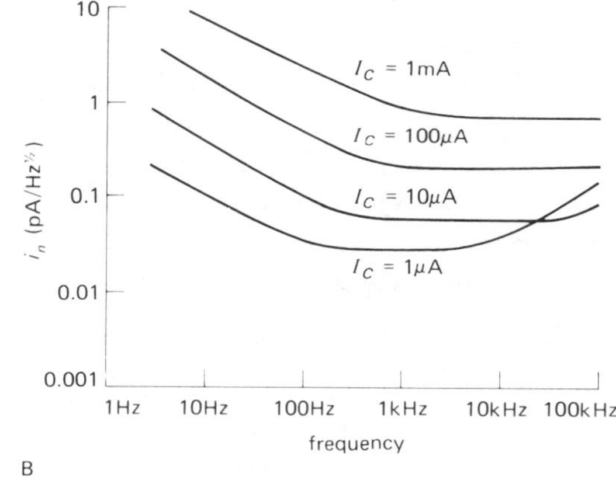

Figure 7.32
Input noise current for the LM394 bipolar transistor. A: Noise current (i_n) versus collector current. B: Noise current (i_n) versus frequency.

shows graphs of i_n versus frequency and current, again for the low-noise LM394.

7.13 Low-noise design with transistors

The fact that e_n drops and i_n rises with increasing I_C provides a simple way to optimize transistor operating current to give lowest noise with a given source. Look at the model again (Fig. 7.33). The noiseless signal source v_s has added to it an irreducible noise voltage from the Johnson noise of its source resistance.

$$e_R^2 \text{ (source)} = 4kTR_s \ \ \text{V}^2/\text{Hz}$$

The amplifier adds noise to its own, namely,

$$e_a^2 \text{ (amplifier)} = e_n^2 + (i_nR_s)^2 \ \ \text{V}^2/\text{Hz}$$

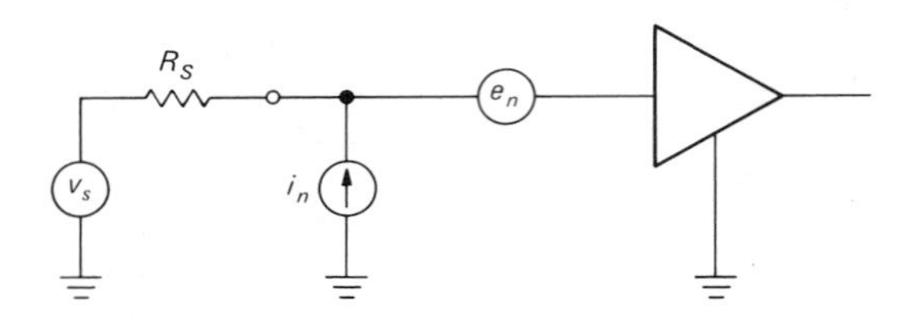

Figure 7.33

Thus the amplifier's noise voltage is added to the input signal, and in addition, its noise current generates a noise voltage across the source impedance. These two are uncorrelated (except at very high frequencies), so you add their squares. The idea is to reduce the amplifier's total noise contribution as much as possible. That's easy, once you know R_s, because you just look at a graph of e_n and i_n versus I_C, in the region of the signal frequency, picking I_C to minimize $e_n^2 + (i_nR_s)^2$. Alternatively, if you are lucky and have a plot of noise figure contours versus I_C and R_s, you can quickly locate the optimum value of I_C.

Noise figure example

As an example, suppose we have a small signal in the region of 1kHz with source resistance of 10k, and we wish to make an amplifier with a 2N4250. From the e_n–i_n graph (Fig. 7.34) we see that the sum of voltage and current terms (with 10k source) is minimized for a collector current of about 10–20μA. Since the current noise is dropping faster than the voltage noise is rising as I_C is reduced, it might be a good idea to use slightly less collector current, especially if operation at a lower frequency is anticipated (i_n rises rapidly with decreasing frequency). We can estimate the noise figure using i_n and e_n at 1kHz:

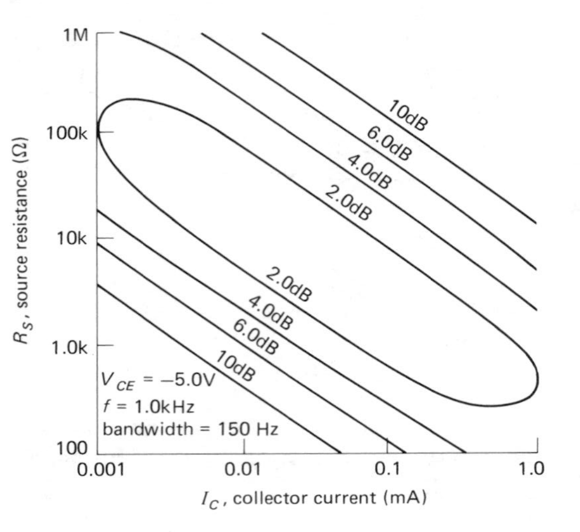

Figure 7.34
Contours of constant narrow-band noise figure for the 2N4250 transistor. (Fairchild Camera and Instrument Corp.)

$$NF = 10 \log_{10}\left(1 + \frac{e_n^2 + (i_nR_s)^2}{4kTR_s}\right) \text{dB}$$

For $I_C = 10\mu A$, $e_n = 3.8nV/Hz^{1/2}$, $i_n = 0.29pA/Hz^{1/2}$, and $4kTR_s = 1.65 \times 10^{-16}$ V^2/Hz for the 10k source resistance. The calculated noise figure is therefore 0.6dB. This is consistent with the graph (Fig. 7.35)

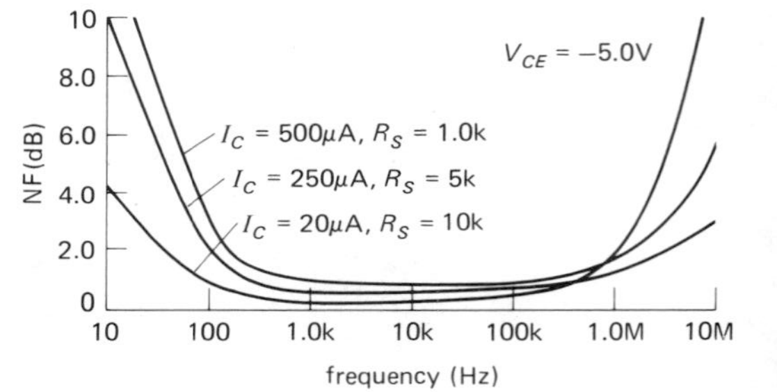

Figure 7.35
Noise figure (NF) versus frequency, for three choices of I_C and R_S, for the 2N4250. (Fairchild Camera and Instrument Corp.)

showing NF versus frequency, in which they have chosen $I_C = 20\mu A$ for $R_s = 10k$. This choice of collector current is also roughly what you would get from the graph in Figure 7.34 of noise figure contours at 1kHz, although the actual noise figure can be estimated only approximately from that plot as being less than 2dB.

EXERCISE 7.2

Find the optimum I_C and corresponding noise figure for $R_s = 100k$ and $f = 1kHz$, using the graph in Figure 7.30 of e_n and i_n. Check your answer from the noise figure contours (Fig. 7.34).

For the other amplifier configurations (follower, grounded base) the noise figure is essentially the same, for given R_s and I_C, since e_n and i_n are unchanged. Of course, a stage with unity voltage gain (a follower) may just pass the problem along to the next stage, since the signal level hasn't been increased to the point that low-noise design can be ignored in subsequent stages.

Charting amplifier noise with e_n and i_n

The noise calculations just presented, although straightforward, make the whole subject of amplifier design appear somewhat formidable. If you misplace a factor of Boltzmann's constant, you suddenly get an amplifier with 10,000dB noise figure! In this section we will present a simplified noise-estimation technique of great utility.

The method consists of first choosing some frequency of interest in order to get values for e_n and i_n versus I_C from the transistor data sheets. Then, for a given collector current, you can plot the total noise contributions from e_n and i_n as a graph of e_a versus source resistance R_s. Figure 7.36 shows what that looks like at 1kHz for a differential input stage using an LM394 matched super-beta transistor running 50μA of collector current. The e_n noise voltage is constant, and the i_nR_s voltage increases proportional to R_s, i.e., with a 45° slope. The amplifier noise curve is drawn as shown, with care being taken to ensure that it passes through a point 3dB (voltage ratio of 1.4) above the crossing point of individual voltage and current noise contributions. Also plotted is the noise voltage of the source resistance, which also happens to be the 3dB NF contour. The other lines of constant noise figure are simply straight lines parallel to this line, as you will see in the examples that follow.

The best noise figure (0.2dB) at this collector current and frequency occurs for a source resistance of 15k, and the noise figure is easily seen to be less than 3dB for all source resistances between 300 ohms and 500k, the points at which the 3dB NF contour intersects the amplifier noise curve.

The next step is to draw a few of these noise curves on the same graph, using different collector currents or frequencies, or maybe a selection of transistor types, in order to evaluate amplifier performance. Before we go on to do that, let's show how we can talk about this same amplifier using a different pair of noise parameters, the noise resistance R_n and the noise figure NF(R_n), both of which pop right out of the graph.

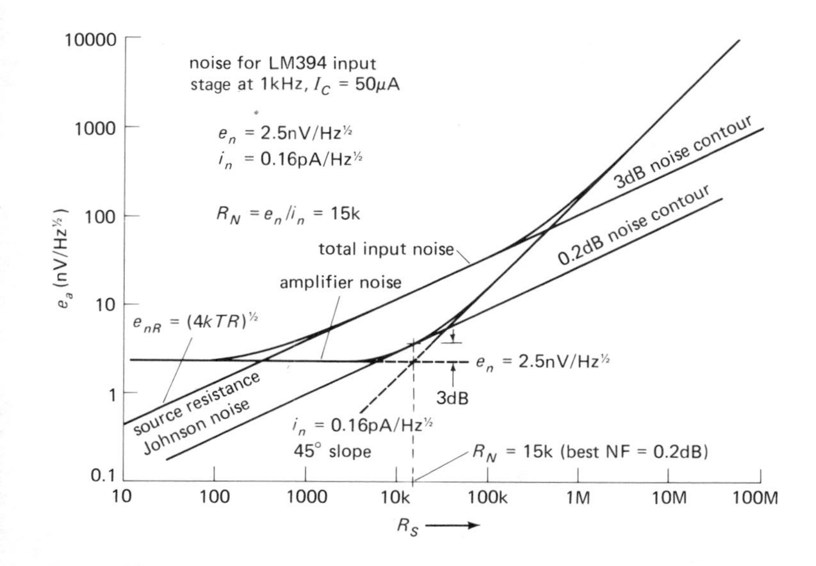

Figure 7.36

Total amplifier input voltage noise (e_a) plotted from the e_n and i_n parameters.

□ *Noise resistance*

The lowest noise figure in this example occurs for a source resistance $R_s = 15\text{k}$, which equals the ratio of e_n to i_n. That defines the noise resistance

$$R_n = \frac{e_n}{i_n}$$

You can find the noise figure for a source of that resistance from our earlier expression for noise figure. It is

$$\text{NF (at } R_n) = 10 \log_{10}\left(1 + 1.23 \times 10^{20} \frac{e_n^2}{R_n}\right) \text{dB} \approx 0.2\text{dB}$$

Noise resistance isn't actually a real resistance in the transistor, or anything like that. It is a tool to help you quickly find the value of source resistance for minimum noise figure, ideally so that you can vary the collector current to shift R_n close to the value of source resistance you're actually using. R_n corresponds to the point where the e_n and i_n lines cross.

The noise figure for a source resistance equal to R_n then follows simply from the preceding equation.

Charting the bipolar/FET shootout

Let's have some fun with this technique. A perennial bone of contention among engineers is whether FETs or bipolar transistors are "better." We will dispose of this issue with characteristic humility by matching two of the best contenders and letting them deliver their best punches. In the interest of fairness, we'll let National Semiconductor intramural teams compete, choosing two game fighters.

In the bipolar corner we have the magnificent LM394 superbeta monolithic matched pair, already warmed up, as described earlier. We'll run it at 1kHz, with collector currents from 1μA to 1mA (Fig. 7.37).

The FET entry is the 2N6483 monolithic *n*-channel JFET matched pair, known far and wide for its stunning low-noise performance, reputed to exceed that of bipolar transistors. According to its data sheet, it was trained only for 100μA and 400μA drain currents (Fig. 7.38).

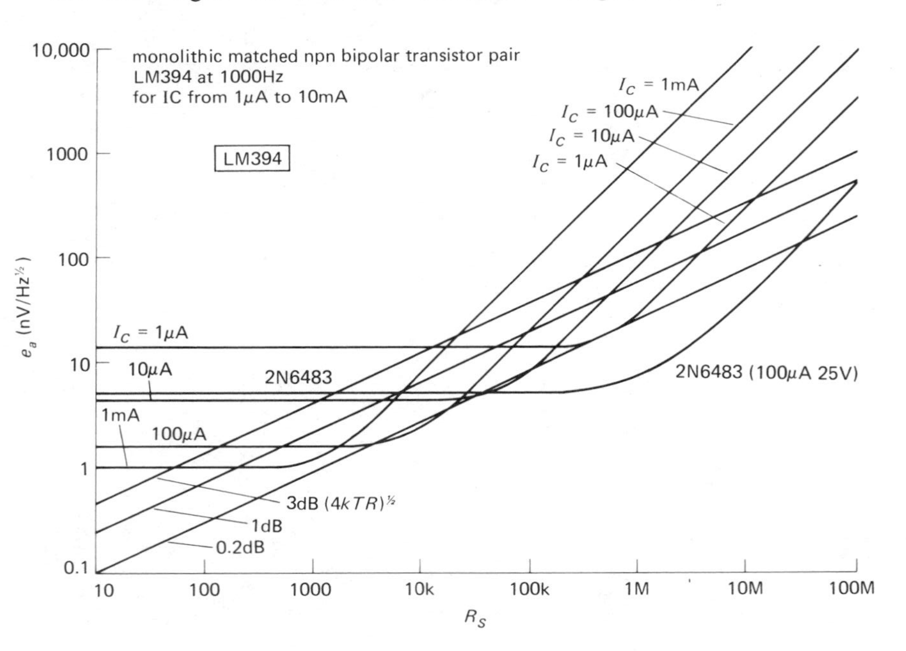

Figure 7.37
Total amplifier input voltage noise (e_a) for the LM394 bipolar transistor under various conditions compared with the 2N6483 JFET.

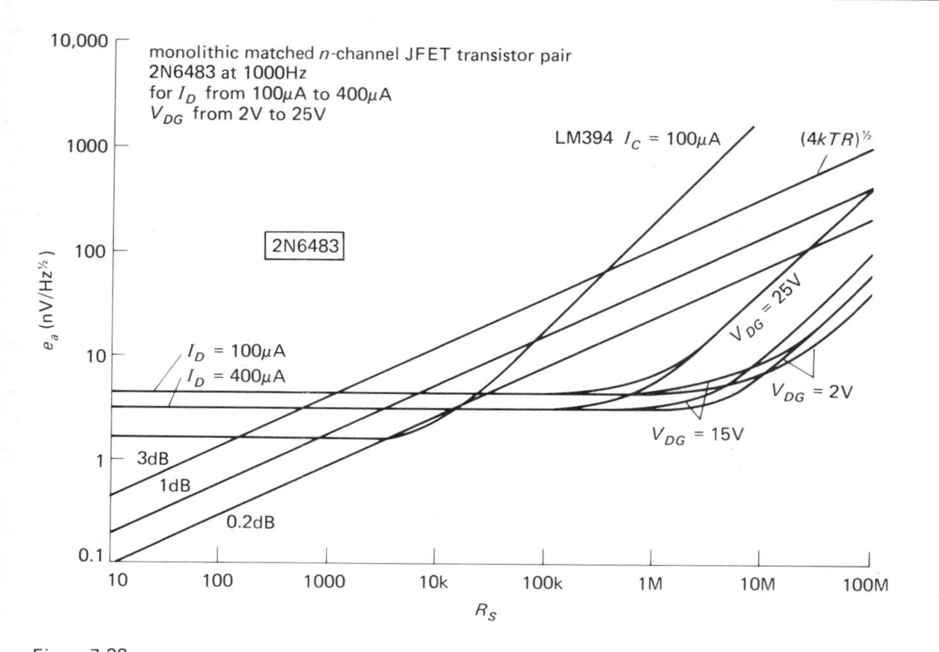

Figure 7.38
Total amplifier input voltage noise (e_a) for the 2N6483 JFET compared with the LM394 bipolar transistor.

And the winner? Well, it's a split decision. The FET won points on lowest minimum noise figure, NF(R_n), reaching a phenomenal 0.05dB noise figure, and dipping well below 0.2dB from 100k to 100M source impedance. For high source impedances, FETs remain unbeaten. The bipolar transistor is best at low source impedances, particularly below 5k, and it can reach a 0.3dB noise figure at R_s = 1k, with suitable choice of collector current. By comparison, the FET cannot do better than 2dB with a 1k source resistance, owing to larger voltage noise e_n.

□ *Low source impedances*

Bipolar transistor amplifiers can provide very good noise performance over the range of source impedances from about 200 ohms to 1M; corresponding optimum collector currents are generally in the range of several milliamps down to a microamp. That is, collector currents used for the input stage of low-noise amplifiers generally tend to be lower than in amplifier stages not optimized for low-noise performance.

For very low source impedances (say 50Ω), transistor voltage noise will always dominate, and noise figures will be poor. The best approach in such cases is to use a transformer to raise the signal level (and impedance), treating the signal on the secondary as before. High-quality signal transformers are available from companies such as James and Princeton Applied Research. As an example, the latter's model 116 FET preamp has voltage and current noise such that the lowest noise figure occurs for signals of source impedance around 1M. A signal around 1kHz with source impedance of 100 ohms would be a poor match for this amplifier, since the amplifier's voltage noise is much larger than the signal source's Johnson noise; the resultant noise figure for that signal connected directly to the amplifier would be 11dB. By using the optional internal step-up transformer, the signal level is raised (along with its source impedance), thus overriding amplifier noise voltage and giving a noise figure of about 1.0dB.

At radiofrequencies (e.g., beginning around 100kHz) it is extremely easy to make good transformers, both for tuned (narrowband) and broadband signals. At these frequencies it is possible to make broadband "transmission-line transformers" of very good performance. We will treat some of these methods in Chapter 13. It is at the very low frequencies (audio and below) that transformers become problematic.

Three comments: (a) The voltage rises proportional to the turns ratio of the transformer, whereas the impedance rises proportional to the square of the ratio. Thus a 2:1 voltage step-up transformer has an output impedance four times the input impedance (this is mandated by conservation of energy). (b) Transformers aren't perfect. They have trouble at low frequencies (magnetic saturation) and at high frequencies (winding inductance and capacitance), as well as losses from the magnetic properties of the core and from winding resistance. The latter is a source of Johnson noise, as well. Nevertheless, when dealing with a signal of very low source impedance, you may have no choice, and transformer coupling can be very beneficial, as the preceding example demonstrates. Exotic techniques such as cooled transformers, superconducting transformers, and SQUIDs (superconducting quantum interference devices) can provide good noise performance at low impedance and voltage levels. With SQUIDs you can measure voltages of 10^{-15} volt! (c) Again, a warning: Don't attempt to improve performance by adding a resistor in series with a low source impedance. If you do that, you're just another victim of the noise figure fallacy.

□ High source impedances

If the source impedance is high, say greater than 100k or so, transistor current noise dominates, and the best device for low-noise amplification is a FET. Although their voltage noise is usually greater than that of bipolar transistors, the gate current (and its noise) can be exceedingly small, making them ideally suited for low-noise high-impedance amplifiers. Incidentally, it is sometimes useful to think of Johnson noise as a current noise $i_n = v_n/R_s$. This lets you compare source noise contributions with amplifier current noise (Fig. 7.39).

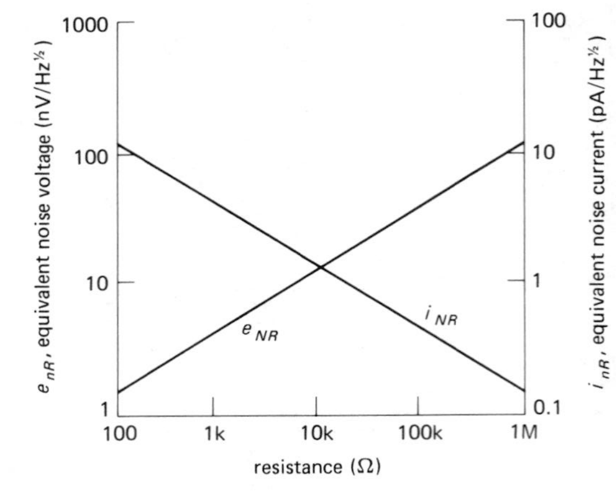

Figure 7.39
Thermal noise voltage density versus resistance at 25°C. The equivalent short-circuit current noise density is also shown. (National Semiconductor Corp.)

7.14 FET noise

We can use the same amplifier noise model for FETs, namely a series noise voltage source and a parallel noise current source. You can analyze the noise performance with exactly the same methods used for bipolar transistors. For example, see the graphs in the section on bipolar/FET shootout.

Voltage noise of JFETs

For JFETs the voltage noise e_n is essentially the Johnson noise of the channel resistance, given approximately by

$$e_n^2 = 4kT\left(\frac{2}{3}\frac{1}{g_m}\right) \ \text{V}^2/\text{Hz}$$

where the inverse transconductance takes the place of resistance in the Johnson noise formula. Since the transconductance rises with increasing drain current, it is best to operate FETs at high current for lowest voltage noise. There is, in addition, some flicker noise associated with the gate leakage current.

When choosing a large drain current in order to minimize voltage noise in FET amplifiers, you have to keep in mind that the performance may suffer in other ways. In particular, the offset voltage drift, CMRR, and voltage gain will be worse at higher drain currents. Obviously you have to compromise.

MOSFETs tend to have much higher voltage noise than JFETs, with $1/f$ noise predominating, since the $1/f$ knee is as high as 10kHz to 100kHz. For this reason you wouldn't normally choose a MOSFET for low-noise amplifiers below 1MHz.

Current noise of JFETs

At low frequencies the current noise i_n is extremely small, arising from the shot noise in the gate leakage current (Fig. 7.40):

$$i_n = (3.2 \times 10^{-19} I_G B)^{1/2} \text{ A (rms)}$$

In addition, there is a flicker noise component in some FETs. The noise current rises

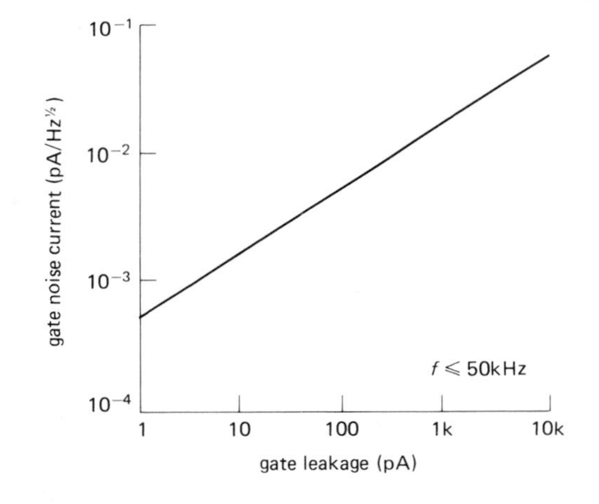

Figure 7.40
Input noise current versus gate leakage current for JFETs. (National Semiconductor Corp.)

with increasing temperature, as the gate leakage current rises. Watch out for the rapidly increasing gate leakage in n-channel JFETs that occurs for operation at high V_{DG} (see Section 6.09).

At moderate to high frequencies there is an additional noise term, namely the real part of the input impedance seen looking into the gate. This comes from the effect of feedback capacitance (Miller effect) when there is a phase shift at the output due to load capacitance; i.e., the part of the output signal that is shifted 90° couples through the feedback capacitance C_{gd} to produce an effective resistance at the input, given by

$$R = \frac{1 + \omega C_L R_L}{\omega^2 g_m C_{gd} C_L R_L{}^2} \text{ ohms}$$

As an example, the 2N5266 p-channel JFET has a noise current of 0.005pA/Hz$^{1/2}$ and a noise voltage e_n of 12nV/Hz$^{1/2}$, both at I_{DSS} and 10kHz. The noise current begins climbing at about 50kHz. These figures are roughly 100 times better in i_n and 5 times worse in e_n than the corresponding figures for the 2N4250 used earlier.

With FETs you can achieve good noise performance for input impedances in the range of 10k to 100M. The PAR model 116 preamp has a noise figure of 1dB or better for source impedances from 5k to 10M in the frequency range from 1kHz to 10kHz. Its performance at moderate frequencies corresponds to a noise voltage of 4nV/Hz$^{1/2}$ and a noise current of 0.013pA/Hz$^{1/2}$.

7.15 Selecting low-noise transistors

As we mentioned earlier, bipolar transistors offer the best noise performance with low source impedances, owing to their lower input voltage noise. Voltage noise, e_n, is reduced by choosing a transistor with low base spreading resistance, $r_{bb'}$, and operating at high collector current (as long as h_{FE} remains high). For higher source impedances the current noise can be minimized instead by operating at lower collector current.

At high values of source impedance, FETs are the best choices. Their voltage noise can be reduced by operating at higher drain currents, where the transconductance is highest. FETs intended for low-noise applications have high k values (see Section 6.04), which usually means high input capacitance. For example, the low-noise 2N6483 has C_{iss} = 20pF, whereas the 2N5902 low-current FET has C_{iss} = 2pF.

Figures 7.41 and 7.42 show comparisons of the noise characteristics of a number of popular and useful transistors.

□ 7.16 Noise in differential and feedback amplifiers

Low-noise amplifiers are often differential, to obtain the usual benefits of low drift and good common-mode rejection. When you

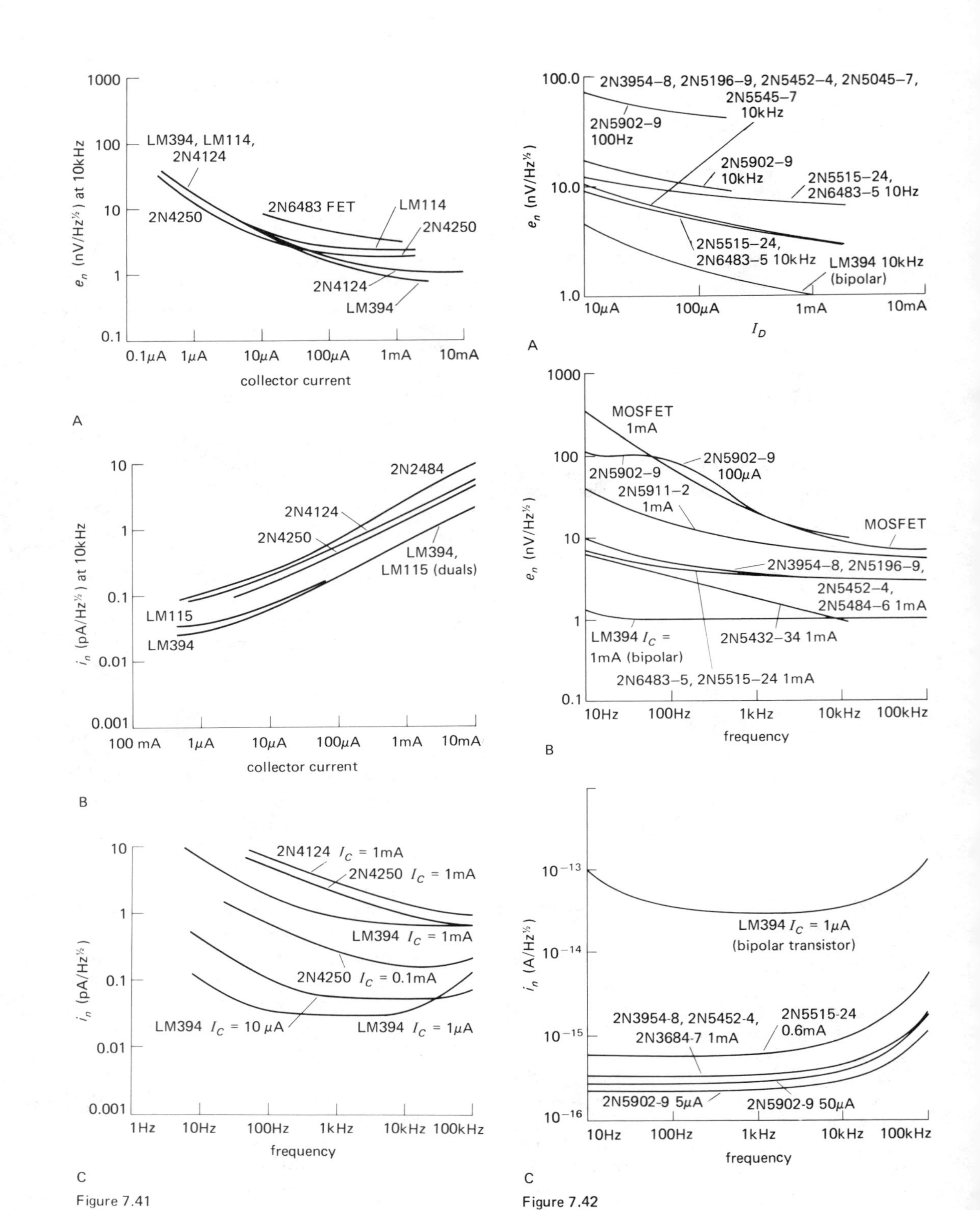

Figure 7.41
Input noise for some popular transistors. A: Input noise voltage (e_n) versus collector current. B: Input noise current (i_n) versus collector current. C: Input noise current (i_n) versus frequency.

Figure 7.42
Input noise for some popular FETs. A: Input noise voltage (e_n) versus drain current (I_D). B: Input noise voltage (e_n) versus frequency. C: Input noise current (i_n) versus frequency.

calculate the noise performance of a differential amplifier, there are three points to keep in mind: (a) Be sure to use the individual collector currents, not the sum, to get e_n and i_n from data sheets. (b) The i_n seen at each input terminal is the same as for a single-ended amplifier configuration. (c) The e_n seen at one input, with the other input grounded, say, is 3dB larger than the single-transistor case, i.e., it is multiplied by 1.414.

In amplifiers with feedback, you want to take the equivalent noise sources e_n and i_n out of the feedback loop, so you can use them as previously described when calculating noise performance with a given signal source. Let's call the noise terms brought out of the feedback loop e_A and i_A, for *amplifier* noise terms. Thus the amplifier's noise contribution to a signal with source resistance R_s is

$$e^2 = e_A^2 + (R_s i_A)^2 \ \ \mathrm{V^2/Hz}$$

Let's take the two feedback configurations separately.

☐ Noninverting

For the noninverting amplifier (Fig. 7.43) the input noise sources become

$$i_A^2 = i_n^2$$

$$e_A^2 = e_n^2 + 4kTR_{\|} + (i_n R_{\|})^2$$

where e_n is the "adjusted" noise voltage for the differential configuration, i.e., 3dB larger than for a single-transistor stage. The additional noise voltage terms arise from Johnson noise and input-stage noise current in the feedback resistors. Note that the effective noise voltage and current are now not completely uncorrelated, so calculations

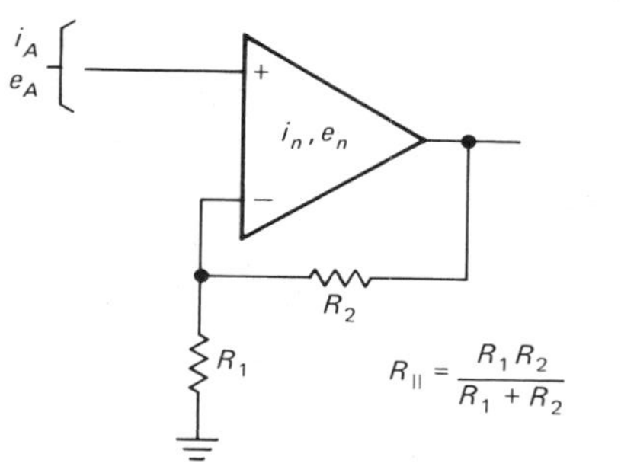

Figure 7.43

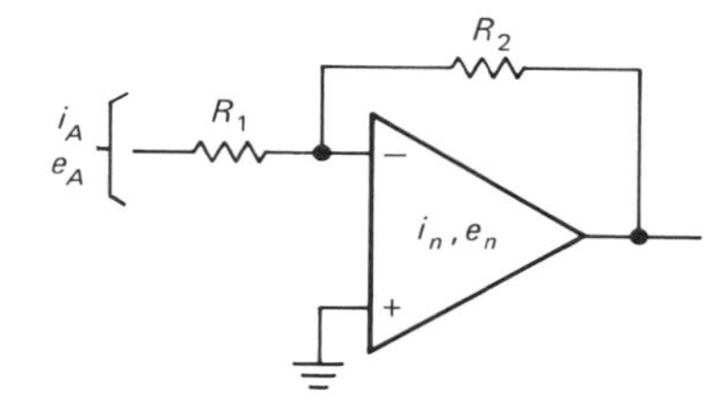

Figure 7.44

in which their squares are added can be in error by a maximum factor of 1.4.

For a follower, R_2 is zero, and the effective noise sources are just those of the differential amplifier alone.

☐ Inverting

For the inverting amplifier (Fig. 7.44) the input noise sources become

$$i_A^2 = i_n^2 + 4kT\frac{1}{R_2}$$

$$e_A^2 = e_n^2 + R_1^2\left(i_n^2 + 4kT\frac{1}{R_2}\right)$$

$$= e_n^2 + R_1^2 i_A^2$$

Op-amp selection curves

You now have all the tools necessary to analyze op-amp input circuits. Their noise is specified in terms of e_n and i_n, just as with transistors and FETs. You don't get to adjust anything, though; you only get to use them. The data sheets really need to be taken with a grain of salt. For example, "popcorn noise" is typified by jumps in offset at random times and durations. It is rarely mentioned in polite company. Figure 7.45 summarizes the noise performance of some popular op-amps.

Wideband noise

Op-amp circuits are generally dc-coupled and extend to some upper frequency limit f_{cutoff}. Therefore it is of interest to know the total noise voltage over this band, not merely the noise power density. Figure 7.46 presents some graphs showing the rms noise voltage in a band extending from dc to the indicated frequency; they were calculated by integrating the noise power curves for the various op-amps.

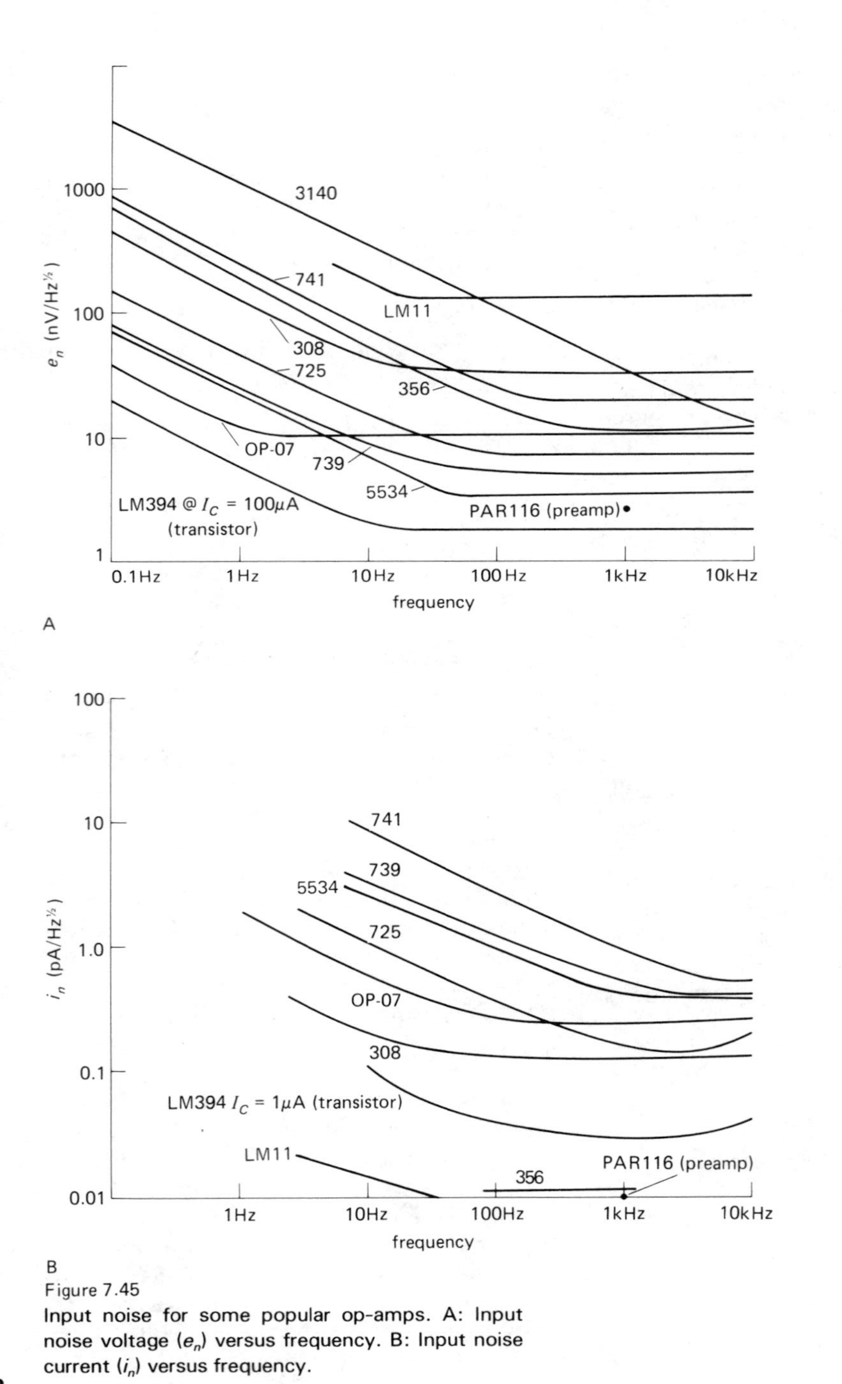

Figure 7.45

Input noise for some popular op-amps. A: Input noise voltage (e_n) versus frequency. B: Input noise current (i_n) versus frequency.

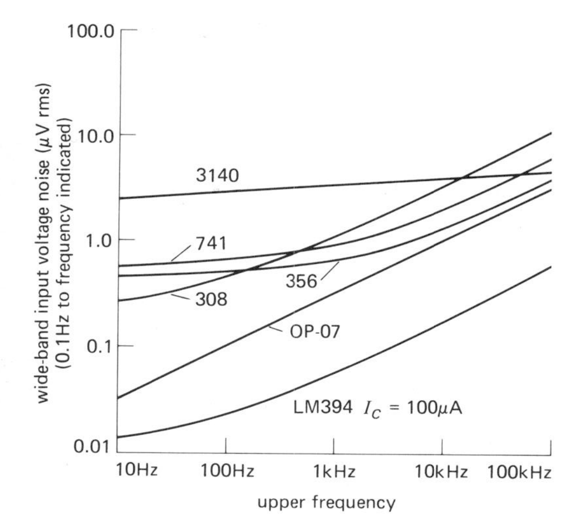

Figure 7.46
Wideband noise voltage for some popular op-amps.

□ NOISE MEASUREMENTS AND NOISE SOURCES

It is a relatively straightforward process to determine the equivalent noise voltage and current of an amplifier, and from these the noise figure and signal-to-noise ratio for any given signal source. That's all you ever need to know about the noise performance of an amplifier. Basically the process consists of putting known noise signals across the input, then measuring the output noise signal amplitudes within a certain bandwidth. In some cases (e.g., a matched input impedance device such as a radiofrequency amplifier) an oscillator of accurately known and controllable amplitude is substituted as the input signal source.

Later we will discuss the techniques you need to do the output voltage measurement and bandwidth limiting. For now, let's assume you can make rms measurements of the output signal, with a measurement bandwidth of your choice.

□ 7.17 Measurement without a noise source

For an amplifier stage made from a FET or transistor and intended for use at low to moderate frequencies, the input impedance is likely to be very high. You want to know e_n and i_n so that you can predict the SNR with a signal source of arbitrary source impedance and signal level, as we discussed earlier. The procedure is simple:

First, determine the amplifier's voltage gain G_V by actual measurement with a signal in the frequency range of interest. The amplitude should be large enough to override amplifier noise, but not so large as to cause amplifier saturation.

Second, short the input and measure the rms noise output voltage, e_s. From this you get the input noise voltage per root hertz from

$$e_n = \frac{e_s}{G_V B^{1/2}} \quad \text{V/Hz}^{1/2}$$

where B is the bandwidth of the measurement (see Section 7.20).

Third, put a resistor R across the input, and measure the new rms noise output voltage, e_r. The resistor value should be large enough to add significant amounts of current noise, but not so large that the input impedance of the amplifier begins to dominate. (If this is impractical, you can leave the input open and use the amplifier's input impedance as R.) The output you measure is just

$$e_r^2 = [e_n^2 + 4kTR_s + (i_n R_s)^2]BG_V^2$$

from which you can determine i_n to be

$$i_n = \frac{1}{R_s}\left[\frac{e_r^2}{BG_V^2} - (e_n^2 + 4kTR_s)\right]^{1/2}$$

With some luck, only the first term in the square root will matter (i.e., if current noise dominates both amplifier voltage noise and source resistor Johnson noise).

Now you can determine the SNR for a signal V_s of source impedance R_s, namely

$$\text{SNR} = 10\log_{10}\left(\frac{V_s^2}{V_n^2}\right)$$

$$= 10\log_{10}\left[\frac{V_s^2}{[e_n^2 + (i_n R_s)^2 + 4kTR_s]B}\right]$$

where the numerator is the signal voltage (presumed to lie within the bandwidth B) and the terms in the denominator are the amplifier noise voltage, amplifier noise current applied to R_s, and Johnson noise in

R_s. Note that increasing the amplifier bandwidth beyond what is necessary to pass the signal v_s only decreases the final SNR. However, if V_s is broadband (e.g., a noise signal itself), the final SNR is independent of amplifier bandwidth. In many cases the noise will be dominated by one of the terms in the preceding equation.

□ 7.18 Measurement with noise source

The preceding technique of measuring the noise performance of an amplifier has the advantage that you don't need an accurate and adjustable noise source, but it requires an accurate voltmeter and filter, and it assumes that you know the gain versus frequency of the amplifier, with the actual source resistance applied. An alternative method of noise measurement involves applying broadband noise signals of known amplitude to the amplifier's input and observing the relative increase of output noise voltage. Although this technique requires an accurately calibrated noise source, it makes no assumptions about the properties of the amplifier, since it measures the noise properties right at the point of interest, at the input.

Again, it is relatively straightforward to make the requisite measurements. You connect the noise generator to the amplifier's input, making sure that its source impedance R_g equals the source impedance of the signal you ultimately plan to use with the amplifier. You first note the amplifier's output rms noise voltage, with the noise source attenuated to zero output signal. Then you increase the noise source rms amplitude V_g until the amplifier's output rises 3dB (a factor of 1.414 in rms output voltage). The amplifier's input noise voltage in the measurement bandwidth, for this source impedance, equals this value of added signal. The amplifier therefore has a noise figure

$$NF = 10 \log_{10}\left(\frac{V_g^2}{4kTR_g} + 1\right) - 3.0 \text{ dB}$$

From this you can figure out the SNR for a signal of any amplitude with this same source impedance, using the formula from Section 7.11

$$SNR = 10 \log_{10}\left(\frac{V_s^2}{4kTR_s}\right) - NF(R_s) \text{ dB}$$

There are nice calibrated noise sources available, most of which provide means for attenuation to precise levels in the microvolt range.

Note that this technique does not tell you e_n and i_n directly, just the appropriate combination for a source of impedance equal to the driving impedance you used in the measurement. Of course, by making several such measurements with different noise source impedances, you could infer the values of e_n and i_n.

□ *Amplifiers with matched input impedance*

This last technique is ideal for noise measurements of amplifiers designed for a matched signal source impedance. The most common examples are in radiofrequency amplifiers or receivers, usually meant to be driven with a signal source impedance of 50 ohms, and which themselves have an input impedance of 50 ohms. We will discuss in Chapter 13 the reasons for this departure from our usual criterion that a signal source should have a small source impedance compared with the load it drives. In this situation e_n and i_n are irrelevant as separate quantities; what matters is the overall noise figure (with matched source) or some specification of SNR with a matched signal source of specified amplitude.

Sometimes the noise performance is explicitly stated in terms of the *narrowband* input signal amplitude required to obtain a certain output SNR. A typical radiofrequency receiver might specify a 10dB SNR with a $0.25\mu V$ rms input signal and 2kHz receiver bandwidth. In this case the procedure consists of measuring the rms receiver output with the input driven by a matched sine-wave source initially attenuated to zero, then increasing the (sine-wave) input signal until the rms output rises 10dB, in both cases with the receiver bandwidth set to 2kHz. It is important to use a meter that reads true rms voltages for a measurement where noise and signal are combined (more about this later). Note that radiofrequency noise measurements often involve output signals that are in the audiofrequency range.

□ 7.19 Noise and signal sources

Broadband noise can be generated from the effects we discussed earlier, namely Johnson noise and shot noise. The shot noise in a vacuum diode is a classic source of broadband noise that is especially useful because the noise voltage can be predicted exactly. More recently, zener diode noise has been used in noise sources. Both of these extend from dc to very high frequencies, making them useful in audiofrequency and radiofrequency measurements.

An interesting noise source can be made using digital techniques, in particular by connecting long shift registers with their input derived from a modulo-2 addition of several of the later bits (see Section 9.35). The resultant output is a pseudorandom sequence of 1's and 0's that after low-pass filtering generates an analog signal of white spectrum up to the low-pass filter's breakpoint, which must be well below the frequency at which the register is shifted. These things can be run at very high frequencies, generating noise up to 100kHz or more. The "noise" has the interesting property that it repeats itself exactly after a time interval that depends on the register length (an n-bit maximal-length register goes through 2^n-1 states before repeating). Without much difficulty that time can be made to be very long (months or years), although most often a period of a second is long enough. For example, a 50-bit register shifted at 10MHz will generate white noise up to 100kHz or so, with a repeat time of 3.6 years. A design for a pseudorandom noise source based on this technique is shown in Section 9.38. Such instruments are available commercially, typified by the Hewlett-Packard model 3722A.

Some noise sources can generate pink noise as well as white noise. Pink noise has equal noise power per *octave,* rather than equal power per hertz. Its power density (power per hertz) drops at 3dB/octave. Since an *RC* filter drops off at 6dB/octave, a more complicated filter is necessary to generate a pink spectrum from a white noise input. The circuit shown in Figure 7.47 uses a 17-bit pseudorandom white noise generator chip to generate pink noise, accurate to ±0.25dB from 10Hz to 40kHz.

Versatile signal sources are available with precisely controlled output amplitude (down to the microvolt range and below) over frequencies from a fraction of a hertz to gigahertz. Some can even be programmed via a digital "bus." An example is the Hewlett-Packard model 8660 synthesized signal generator, with output frequencies from 0.01Hz to 110MHz, calibrated amplitudes from 10nV to 1 volt rms, handsome digital display and bus interface, and nifty accessories that extend the frequency range to 2.6GHz and provide modulation and frequency sweeping. This is a bit more than you usually need to do the job.

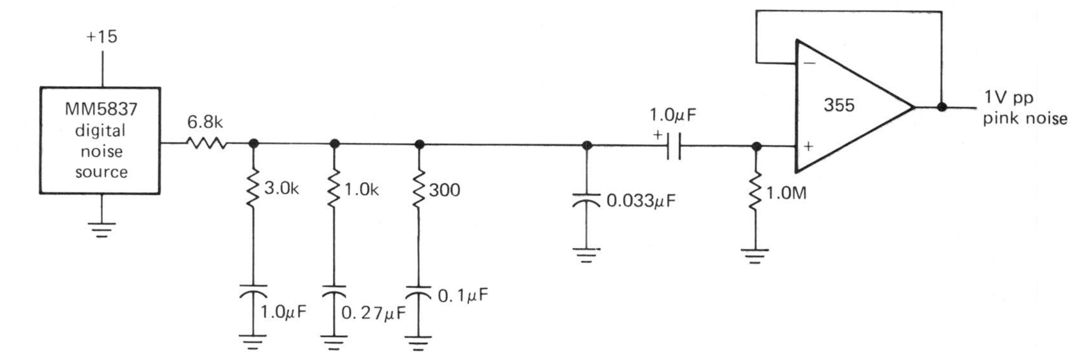

Figure 7.47

Pink noise source (−3dB/octave, ± 0.25dB from 10Hz to 40kHz).

□ 7.20 Bandwidth limiting and rms voltage measurement

□ *Limiting the bandwidth*

All the measurements we have been talking about assume that you are looking at the noise output only in a limited frequency band. In a few cases the amplifier may have provision for this, making your job easier. If not, you have to hang some sort of filter on the amplifier output before measuring the output noise voltage.

The easiest thing to use is a simple *RC* low-pass filter, with 3dB point set at roughly the bandwidth you want. For accurate noise measurements, you need to know the equivalent "noise bandwidth," i.e., the width of a perfect "brick-wall" low-pass filter that lets through the same noise voltage (Fig. 7.48).

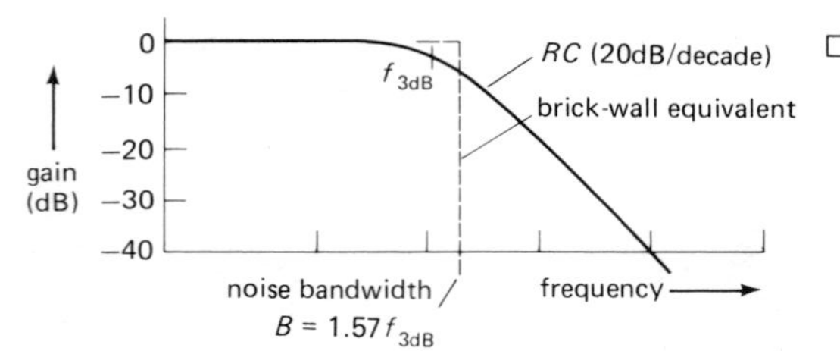

Figure 7.48

This noise bandwidth is what should be used for B in all the preceding formulas. It is not terribly difficult to do the mathematics, and you find

$$B = \frac{\pi}{2} f_{3dB} = 1.57 f_{3dB}$$

For a pair of cascaded *RC*s (buffered so they don't load each other), the magic formula becomes $B = 1.22 f_{3dB}$. For the Butterworth filters discussed in Section 4.05, the noise bandwidth is

$B = 1.57 f_{3dB}$ (1 pole)
$B = 1.11 f_{3dB}$ (2 poles)
$B = 1.05 f_{3dB}$ (3 poles)
$B = 1.025 f_{3dB}$ (4 poles)

If you want to make band-limited measurements up at some center frequency, you can just use a pair of *RC* filters (Fig. 7.49), in which case the noise bandwidth is as indicated.

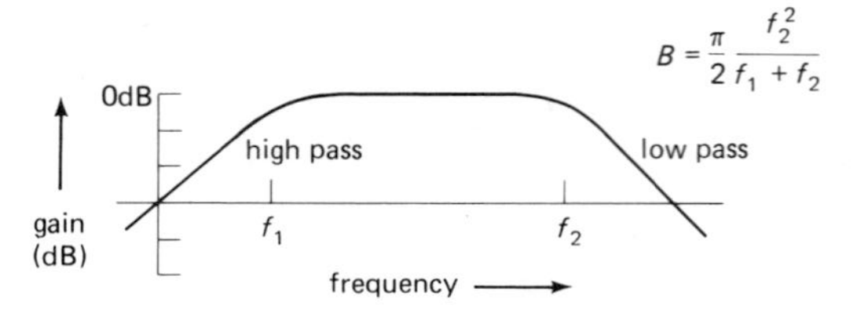

Figure 7.49

If you have had experience with contour integration, you may wish to try the following exercise.

EXERCISE 7.3

Optional exercise: Derive the preceding result, beginning with the response functions of *RC* filters. Assume unit power per hertz input signal, and integrate the output power from zero to infinity. A contour integral then gets you the answer.

□ *Measuring the noise voltage*

The most accurate way to make output noise measurements is to use a true rms voltmeter. These operate either by measuring the heating produced by the signal waveform (suitably amplified) or by using an analog squaring circuit followed by averaging. If you use a true rms meter, make sure it has response at the frequencies you are measuring; some of them only go up to a few kilohertz. True rms meters also specify a "crest factor," the ratio of peak voltage to rms that they can handle without great loss of accuracy. For Gaussian noise, a crest factor of 3 to 5 is adequate.

You can use a simple averaging-type ac voltmeter instead, if a true rms meter is unavailable. In that case, the values read off the scale must be corrected. As it turns out, all averaging meters (VOMs, DMMs, etc.) already have their scales adjusted, so what you read isn't actually the *average,* but rather the rms voltage *assuming a sine-wave signal.* For example, if you measure the power-line voltage in the United States, your meter will read something close to 117 volts. That's fine, but if the signal you're reading is Gaussian noise, you have to apply an additional correction. The rule is as follows: To get the rms voltage of Gaussian noise, multiply the "rms" value you read on an averaging ac voltmeter by 1.13 (or add

1dB). Warning: This works fine if the signal you are measuring is pure noise (e.g., the output of an amplifier with a resistor or noise source as input), but it won't give accurate results if the signal consists of a sine wave added to noise.

A third method, not exactly world-famous for its accuracy, consists of looking at the noise waveform on an oscilloscope: The rms voltage is 1/6 to 1/8 of the peak-to-peak value (depending on your subjective reading of the pp amplitude). It isn't very accurate, but at least there's no problem getting enough measurement bandwidth.

INTERFERENCE: SHIELDING AND GROUNDING

"Noise" in the form of interfering signals, 60Hz pickup, and signal coupling via power supplies and ground paths can turn out to be of far greater practical importance than the intrinsic noise sources we've just discussed. These interfering signals can all be reduced to an insignificant level (unlike thermal noise) with proper layout and construction. In stubborn cases the cure may involve a combination of filtration of input and output lines, careful layout and grounding, and extensive electrostatic and magnetic shielding. In these sections we would like to offer some suggestions that may help illuminate this dark area of the electronic art.

7.21 Interference

Interfering signals can enter an electronic instrument through the power-line inputs or through signal input and output lines. In addition, signals can be capacitively coupled (electrostatic coupling) onto wires in the circuit (the effect is more serious for high-impedance points within the circuit), magnetically coupled to closed loops in the circuit (independent of impedance level), or electromagnetically coupled to wires acting as small antennas for electromagnetic radiation. Any of these can become a mechanism for coupling of signals from one part of a circuit to another. Finally, signal currents from one part of the circuit can couple to other parts through voltage drops on ground lines or power-supply lines.

Eliminating interference

Numerous effective tricks have been evolved to handle most of these commonly occurring interference problems. Keep in mind the fact that these techniques are all aimed at reducing the interfering signal or signals to an acceptable level; they rarely eliminate them altogether. Consequently, it often pays to raise signal levels, just to improve the signal-to-interference ratio. Also, it is important to realize that some environments are much worse than others; an instrument that works just dandy on the bench may perform miserably on location. Some environments worth avoiding are those (a) near a radio or television station (RF interference), (b) near a subway (impulsive interference and power-line garbage), (c) near high-voltage lines (radio interference, frying sounds), (d) near motors and elevators (power-line spikes), (e) in a building with triac lamp and heater controllers (power-line spikes), (f) near equipment with large transformers (magnetic pickup), and (g) near arc welders (unbelievable pickup of all sorts). Herewith a gathering of advice, techniques, and black magic:

Signals coupled through inputs, outputs, and power line

The best bet for power-line noise is to use a combination of RF line filters and transient suppressors on the ac power line. You can achieve 60dB or better attenuation of interference above a few hundred kilohertz this way, as well as effective elimination of damaging spikes.

Inputs and outputs are more difficult, because of impedance levels and the need to couple desired signals that may lie in the frequency range of interference. In devices like audio amplifiers you can use low-pass filters on inputs *and outputs* (much interference from nearby radio stations enters via the speaker wires, acting as antennas). In other situations shielded lines are often necessary. Low-level signals, particularly at high impedance levels, should always be shielded. So should the instrument cabinet.

Capacitive coupling

Signals within an instrument can get around handsomely via electrostatic coupling: Some point within the instrument has a 10 volt signal jumping around; a high-Z input nearby does some sympathetic jumping, too. The best things to do are to reduce the capacitance between the offending points (move them apart), add shielding (a complete metal enclosure, or even close-knit metal screening, eliminates this form of coupling altogether), move the wires close to a ground plane (which "swallows" the electrostatic fringing fields, reducing coupling enormously), and lower the impedance levels at susceptible points, if possible. Op-amp outputs don't pick up interference easily, whereas inputs do. More on this later.

Magnetic coupling

Unfortunately, low-frequency magnetic fields are not significantly reduced by metal enclosures. A turntable, microphone, tape recorder, or other sensitive circuit placed in close proximity to an instrument with a large power transformer will display astounding amounts of 60Hz pickup. The best therapy here is to avoid large enclosed areas within circuit paths and try to keep the circuit from closing around in a loop. Twisted pairs of wires are quite effective in reducing magnetic pickup, because the enclosed area is small, and the signals induced in successive twists cancel.

When dealing with very low level signals, or devices particularly susceptible to magnetic pickup (tape heads, inductors, wirewound resistors), it may be desirable to use magnetic shielding. "Mu-metal shielding" is available in preformed pieces and flexible sheets. If the ambient magnetic field is large, it is best to use shielding of high permeability (high mu) on the inside, surrounded by an outer shield of lower permeability (which can be ordinary iron, or low-mu shielding material), to prevent magnetic saturation in the inner shield. Of course, moving the offending source of magnetic field is often a simpler solution. It may be necessary to exile large power transformers to the hinterlands, so to speak. Toroidal transformers have smaller fringing fields than the standard frame types.

Radiofrequency coupling

RF pickup can be particularly insidious, because innocent-looking parts of the circuit can act as resonant circuits, displaying enormous effective cross section for pickup. Aside from overall shielding, it is best to keep leads short and avoid loops that can resonate. Ferrite beads may help, if the problem involves very high frequencies. A classic situation is the use of a pair of bypass capacitors (one tantalum, one disc ceramic), often recommended to improve bypassing. The pair can form a lovely parasitic tuned circuit somewhere in the HF to VHF region (tens to hundreds of megahertz), with self-oscillations!

7.22 Signal grounds

Ground leads and shields can cause plenty of trouble, and there is a lot of misunderstanding on this subject. The problem, in a nutshell, is that currents you forgot about flowing through a ground line can generate a signal seen by another part of the circuit sharing the same ground. The technique of a ground "mecca" (a common point in the circuit to which all ground connections are tied) is often seen, but it's a crutch; with a little understanding of the problem you can handle most situations intelligently.

Common grounding blunders

Figure 7.50 shows a common situation. Here a low-level amplifier and a high-current driver are in the same instrument. The first circuit is done correctly: Both amplifiers tie to the supply voltages at the regulator (right at the sensing leads), so IR drops along the leads to the power stage don't appear on the low-level amplifier's supply voltages. In addition, the load current returning to ground does not appear at the low-level input; no current flows from the ground side of the low-level amplifier's input to the circuit mecca (which might be the connection to the case near the BNC input connector).

In the second circuit there are two blunders. Supply voltage fluctuations caused by load currents at the high-level stage are impressed on the low-level supply voltages. Unless the input stage has very good supply

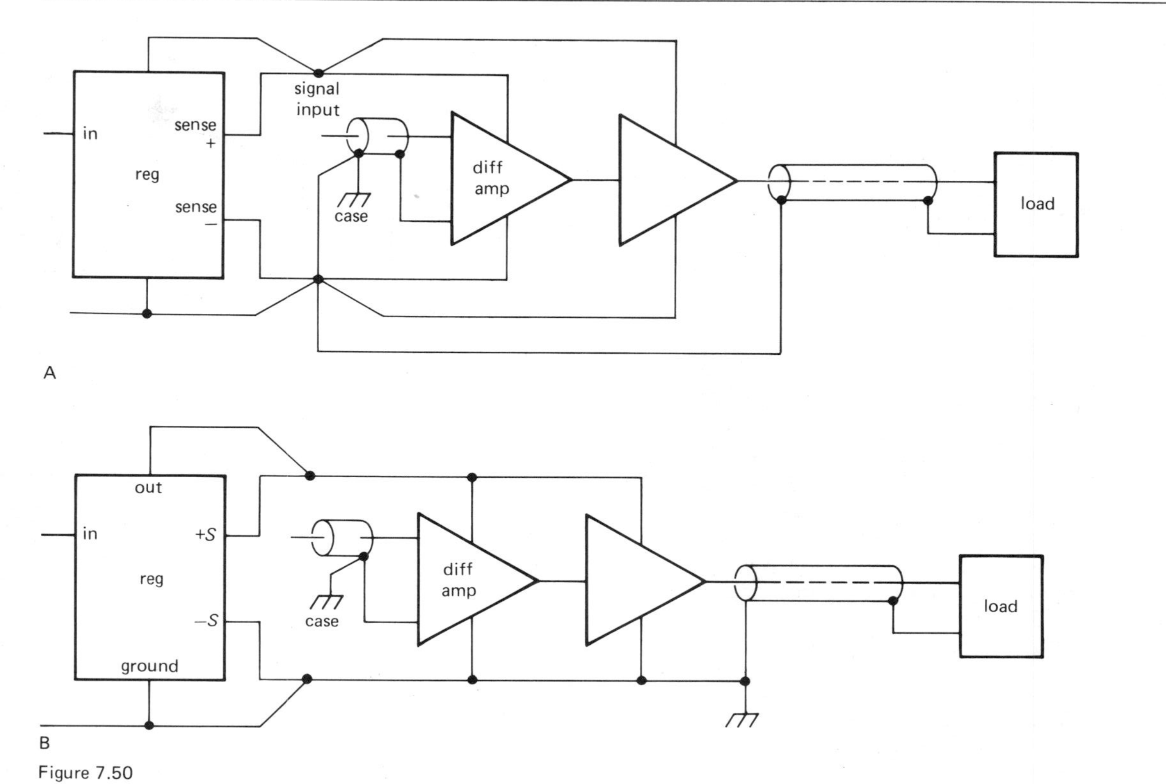

Figure 7.50
Ground paths for low-level signals. A: Right. B: Wrong.

rejection, this can lead to oscillations. Even worse, the load current returning to the supply makes the case "ground" fluctuate with respect to power-supply ground. The input stage ties to this fluctuating ground, a very bad idea. The general idea is to look at where the large signal currents are flowing and make sure their *IR* drops don't wind up at the input. In some cases it may be a good idea to decouple the supply voltages to the low-level stages with a small *RC* network (Fig. 7.51). In stubborn cases of supply

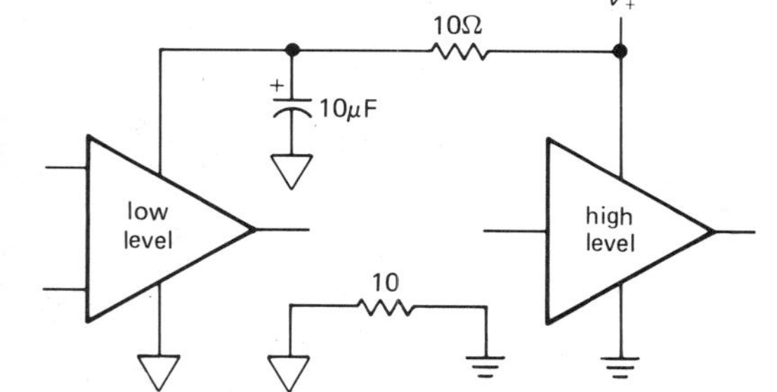

Figure 7.51

coupling it may pay to put a zener or 3-terminal regulator on the low-level-stage supply for additional decoupling.

□ 7.23 Grounding between instruments

The idea of a controlled ground point within one instrument is fine, but what do you do when a signal has to go from one instrument to another, each with its own idea of "ground"? Some suggestions follow.

□ *High-level signals*

If the signals are several volts, or large logic swings, just tie things together and forget about it (Fig. 7.52). The voltage source shown between the two grounds represents the variations in local grounds you'll find on different power-line outlets in the same room or (worse) in different rooms or buildings. It consists of some 60Hz voltage, harmonics of the line frequency, some radio-frequency signals (the power line makes a good antenna), and assorted spikes and

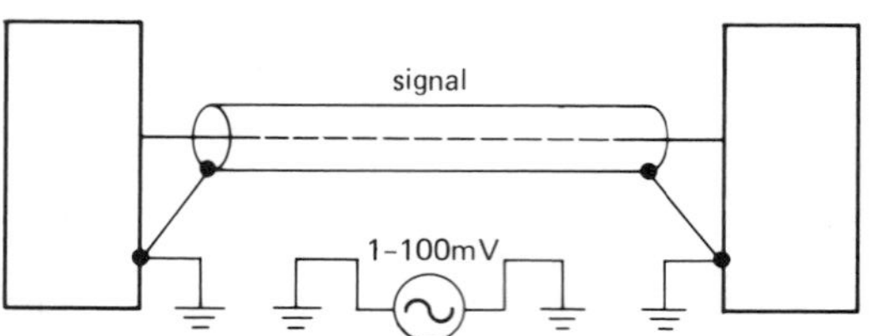

Figure 7.52

other garbage. If your signals are large enough, you can live with this.

☐ *Small signals and long wires*

For small signals this situation is intolerable, and you will have to go to some effort to remedy the situation. Figure 7.53 shows some ideas. In the first circuit, a coaxial shielded cable is tied to the case and circuit ground at the driving end, but it is kept isolated from the case at the receiving end (use a Bendix 4890-1 or Amphenol 31-10 insulated BNC connector). A differential amplifier is used to buffer the input signal, thus ignoring the small amount of "ground signal" appearing on the shield. A small resistor and bypass capacitor to ground is a good idea to limit ground swing and prevent damage to the input stage.

In the second circuit, a shielded twisted pair is used, with the shield connected to the case at both ends. Since no signal travels on the shield, this is harmless. A differential amplifier is used as before on the receiving end. If logic signals are being transmitted, it is a good idea to send a differential signal (the signal and its inverted form), as indicated. Ordinary differential amplifiers can be used as input stages, or if the ground interference is severe, special "isolated amplifiers" are available from manufacturers like Analog Devices and Burr-Brown. The latter permit kilovolts of common-mode signals. So do optoisolator modules, a handy solution for digital signals in some situations.

At radiofrequencies, transformer coupling

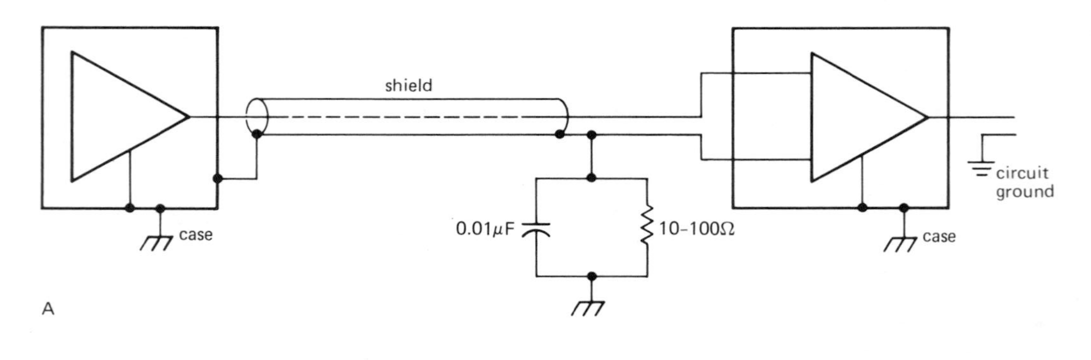

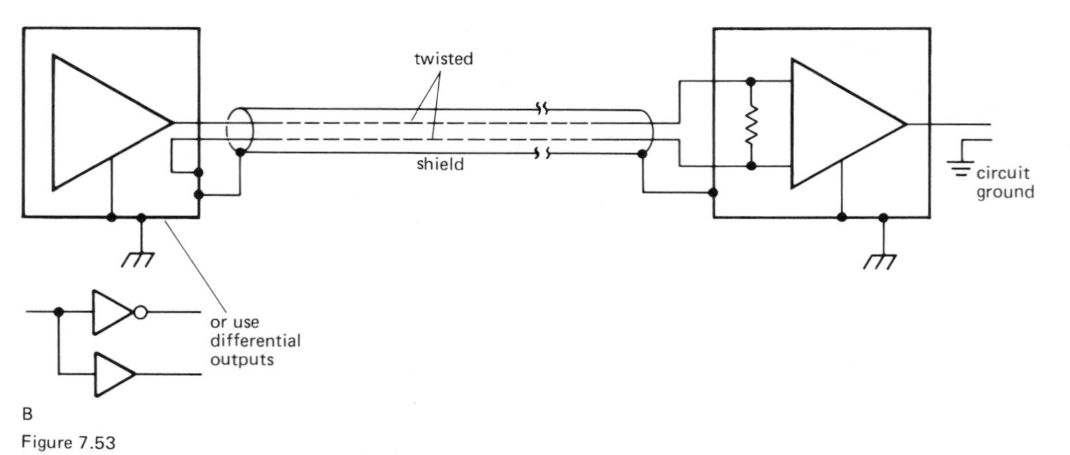

Figure 7.53
Ground connections for low-level signals through shielded cables.

offers a convenient way of removing common-mode signal at the receiving end; this also makes it easy to generate a differential bipolarity signal at the driving end. Transformers are popular in audio applications as well, although they tend to be bulky and lead to some signal degradation.

For very long cable runs (measured in miles) it is useful to prevent large ground currents flowing in the shield at radiofrequencies. Figure 7.54 suggests a method. As before, a differential amplifier looks at the twisted pair, ignoring the voltage on the shield. By tying the shield to the case through a small inductor, the dc voltage is kept small while preventing large radiofrequency currents. This circuit also shows protection circuitry to prevent common-mode excursions beyond ± 10 volts.

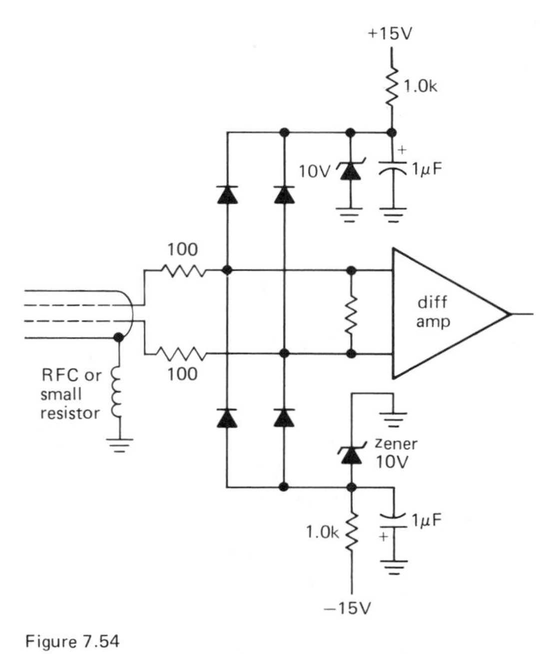

Figure 7.54
Input-protection circuits for use with very long lines.

Figure 7.55 shows a nice scheme to save wires in a multiwire cable in which the common-mode pickup has to be eliminated. Since all the signals suffer the same common-mode pickup, a single wire tied to ground at the sending end serves to cancel the common-mode signals on each of the *n* signal lines. Just buffer its signal (with respect to ground at the receiving end), and use it as the comparison input for each of *n* differential amplifiers looking at the other signal lines.

The preceding schemes work well to eliminate common-mode interference at low to moderate frequencies, but they can be ineffective against radiofrequency interference, owing to poor common-mode rejection in the receiving differential amplifier.

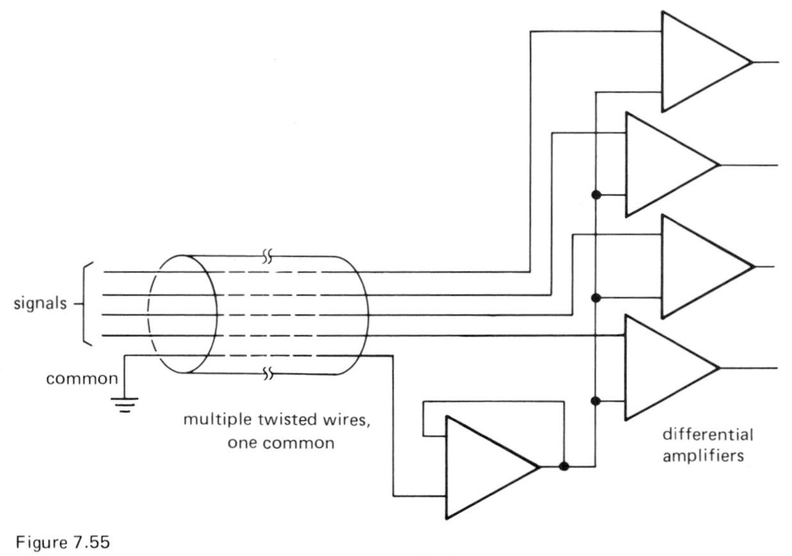

Figure 7.55
Common-mode interference rejection with long multiwire cables.

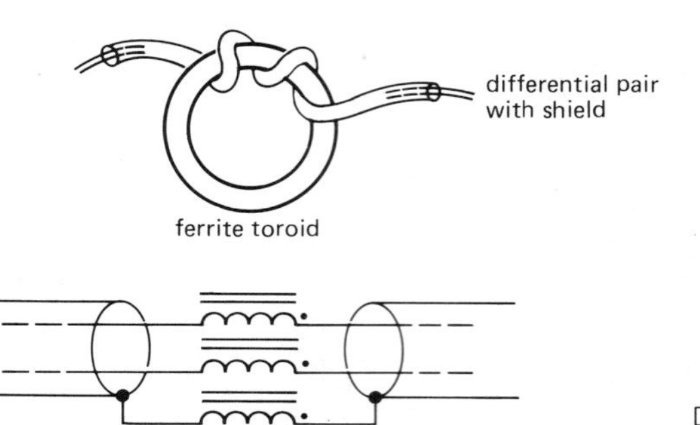

Figure 7.56

One possibility here is to wrap the whole cable around a ferrite toroid (Fig. 7.56). That increases the series inductance of the whole cable, raising the impedance to common-mode signals of high frequency and making it easy to bypass them at the far end with a pair of small bypass capacitors to ground. The equivalent circuit shows why this works without attenuating the differential signal: You have a series inductance inserted into both signal lines and the shield, but since they form a tightly coupled transformer of unit turns ratio, the differential signal is unaffected. This is actually a "1:1 transmission line transformer," as discussed in Section 13.10.

□ *Floating signal sources*

The same sort of disagreement about the voltage of "ground" at separated locations enters in an even more serious way at low-level inputs, just because the signals are so small. As example is a magnetic tape head or other signal transducer that requires a shielded signal line. If you ground the shield at both ends, differences in ground potential will appear as signal at the amplifier input. The best approach is to lift the shield off ground *at the transducer* (Fig. 7.57).

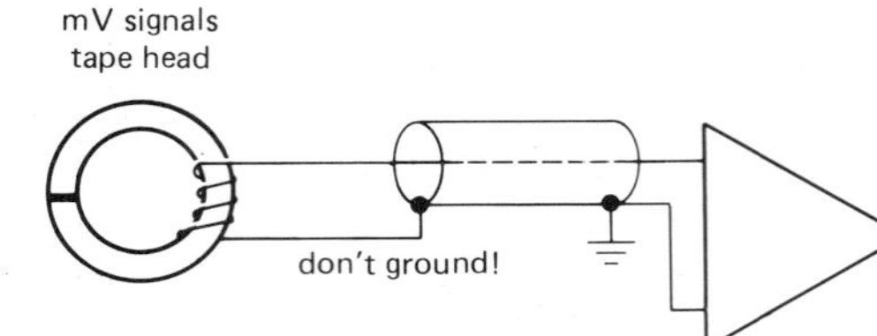

Figure 7.57

□ *Signal guarding*

A closely related issue is signal *guarding,* an elegant technique to reduce the effects of input capacitance and leakage for small signals at high impedance levels. You may be dealing with signals from a microelectrode or a capacitive transducer, with source impedances of hundreds of megohms. Just a few picofarads input capacitance can form a low-pass filter, with rolloffs beginning at a few hertz! In addition, the effects of insulation resistance in the connecting cables can easily degrade the performance of an ultra-low-input current amplifier (bias currents less than a picoamp) by orders of magnitude. The solution to both these problems is a *guard electrode* (Fig. 7.58).

A follower bootstraps the inner shield, effectively eliminating leakage current and capacitive attenuation by keeping zero voltage difference between the signal and its surrounding. An outer grounded shield is a good idea, to keep interference off the guard electrode; the follower has no trouble driving that capacitance and leakage, of course, since it has low output impedance.

You shouldn't use these tricks more than you need to; it would be a good idea to put the follower as close to the signal source as possible, guarding only the short section of cable connecting them. Ordinary shielded cable can then carry the low-impedance output signal out to the remote amplifier.

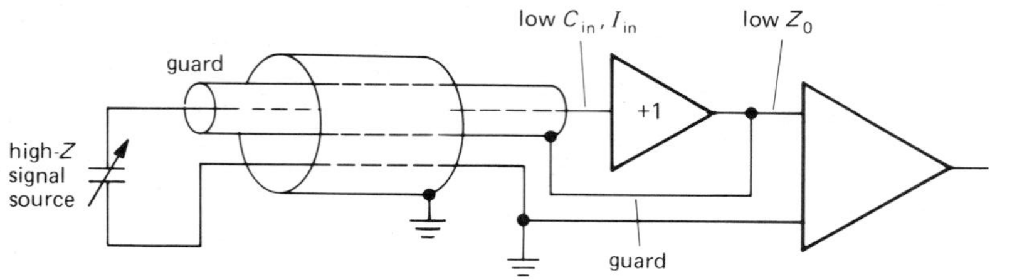

Figure 7.58
Using a guard to raise input impedance.

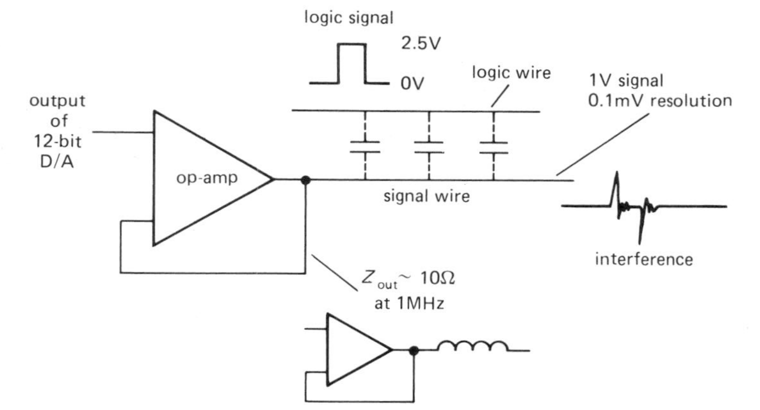

Figure 7.59
Digital cross-coupling interference with linear signals.

We will discuss signal guards in Section 14.08 in connection with high-impedance microelectrodes.

□ *Coupling to outputs*

Ordinarily the output impedance of an op-amp is low enough that you don't have to worry about capacitive signal coupling. In the case of high-frequency or fast-switching interference, however, you have just cause for alarm, particularly if the desired output signal involves some degree of precision. Consider the example in Figure 7.59. A precision signal is buffered by an op-amp and passes through a region containing digital logic signals jumping around with slew rates of 0.5V/ns. The op-amp's closed-loop output impedance rises with frequency, typically reaching values of 10 to 100 ohms or more at 1MHz (see Section 7.07). How large a coupling capacitance is permissible, keeping coupled interference less than the analog signal's resolution of 0.1mV? The surprising answer is 0.02pF.

There are some solutions. The best thing is to keep your small analog waveforms out of the reach of fast-switching signals. A moderate bypass capacitor across the op-amp's output (with perhaps a small series resistor, to maintain op-amp stability) will help, although it degrades the slew rate. You can think of the action of this capacitor as lowering the frequency of the coupled charge bundles to the point where the op-amp's feedback can swallow them. A few hundred picofarads to ground will adequately stiffen the analog signal at high frequencies (think of it as a capacitive voltage divider). Another possibility is to use a low-impedance buffer, such as the 1438 or LH0063. Don't neglect the opportunity to use shielding, twisted pairs, and proximity to ground planes to reduce coupling.

SELF-EXPLANATORY CIRCUITS

7.24 Circuit ideas

Figure 7.60 presents some circuit ideas relevant to the subjects of this chapter.

ADDITIONAL EXERCISES

(*1*) Prove that SNR = 10 $\log_{10}$ $(v_s^2/4kTR_s)$ − NF(dB) (atR_s).

(*2*) A 10μV (rms) sine wave at 100Hz is in series with a 1M resistor at room temperature. What is the SNR of the resultant signal **(*a*)** in a 10Hz band centered at 100Hz and **(*b*)** in a 1MHz band going from dc to 1MHz?

(3) A transistor amplifier using a 2N4250 is operated at 100μA collector current and is driven by a signal source of impedance 2000 ohms. **(*a*)** Find the noise figure at 100Hz, 1kHz, and 10kHz. **(*b*)** Find the SNR (at each of the listed frequencies) for an input signal of 50nV (rms) and an amplifier bandwidth of 10Hz.

(*4*) Measurements are made on a commercial amplifier (with Z_{in} = 1M) in order to

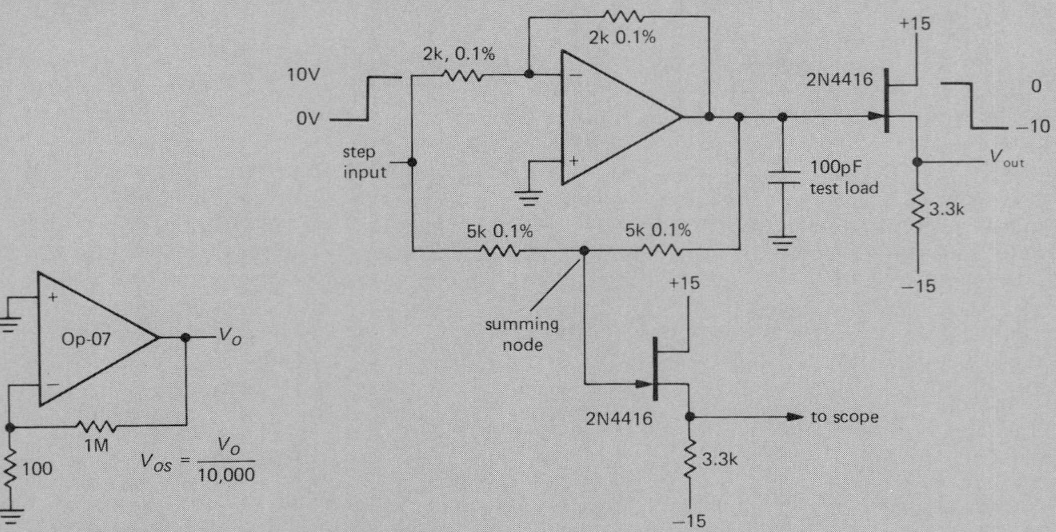

A. offset-voltage test circuit

C. settling-time test circuit

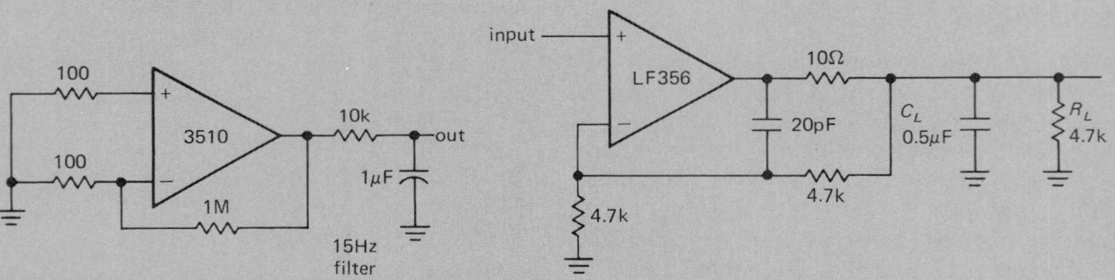

B. low-frequency-noise test circuit

D. driving capacitive loads

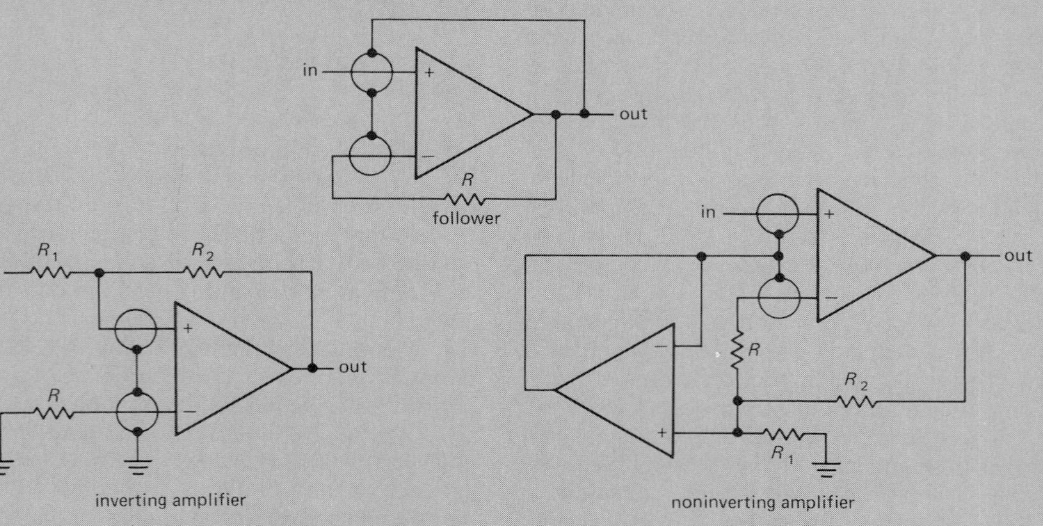

E. input guarding for low-level high-Z inputs
(R compensates source impedance)

Figure 7.60

determine its equivalent input noise e_n and i_n at 1kHz. The amplifier's output is passed through a sharp-skirted filter of bandwidth 100Hz: A 10μV input signal results in a 0.1 volt output. At this level the amplifier's noise contribution is negligible. With the input shorted, the noise output is 0.4mV rms. With the input open, the noise output rises to 50mV rms. (***a***) Find e_n and i_n for this amplifier at 1kHz. (***b***) Find the noise figure of this amplifier at 1kHz for source resistances of 100 ohms, 10k, and 100k.

(***5***) Noise measurements are made on an amplifier using a calibrated noise source whose output impedance is 50 ohms. The generator output must be raised to $2\text{nV/Hz}^{1/2}$ in order to double the output noise power of the amplifier. What is the amplifier's noise figure for a source impedance of 50 ohms?

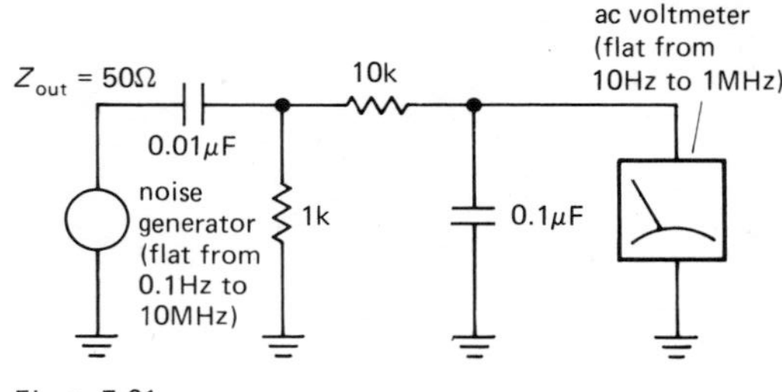

Figure 7.61

(***6***) The output noise voltage of a white noise generator is measured with the circuit shown in Figure 7.61. At a particular setting of the noise generator output level, the ac voltmeter reads 1.5 volts "rms." What is the noise generator's output noise density (i.e., rms volts per root hertz)?

DIGITAL ELECTRONICS

CHAPTER 8

BASIC LOGIC CONCEPTS

8.01 Digital versus analog

Thus far we have been dealing mainly with circuits in which the input and output voltages have varied over a range of values: *RC* circuits, amplifiers, integrators, rectifiers, op-amps, etc. This is natural when dealing with signals that are continuous (e.g., audio signals) or continuously varying voltages from measuring instruments (e.g., temperature-reading or light-detecting devices, or biological or chemical probes).

However, there are instances in which the input signal is naturally discrete in form, e.g., pulses from a particle detector, or "bits" of data from a switch, keyboard, or computer. In such cases the use of digital electronics (circuits that deal with data made of 1's and 0's) is natural and convenient. Furthermore, it is often desirable to convert continuous (analog) data to digital form and vice versa (using D/A and A/D converters) in order to perform calculations on the data with a calculator or computer or to store large quantities of data as numbers. In a typical situation a microprocessor or computer might monitor signals from an experiment or industrial process, control the experimental parameters on the basis of the data obtained, and store for future use the results collected or computed while the experiment is running.

Another interesting example of the power of digital techniques is the transmission of analog signals without degradation by noise: An audio or video signal, for instance, picks up "noise" while being transmitted by cable or radio that cannot be removed. If, instead, the signal is converted to a series of numbers representing its amplitude at successive instants of time, and these numbers are transmitted as digital signals, the analog signal reconstruction at the receiving end (done with D/A converters) will be without error, providing the noise level on the transmission channel isn't high enough to prevent accurate recognition of 1's and 0's. This technique, known as PCM (pulse code modulation), is particularly attractive where a signal must pass through a series of "repeaters," as in the case of a transcontinental telephone call, since digital regeneration at each stage guarantees noiseless transmission. The information and pictures sent back by recent deep space probes were done with PCM.

In fact, digital hardware has become so powerful that tasks that seem well suited to analog techniques are often better solved with digital methods. As an example, an analog temperature meter might incorporate a microprocessor and memory in order to improve accuracy by compensating the instrument's departure from perfect linearity. Because of the wide availability of microprocessors, such applications are becoming commonplace. Rather than attempt to enumerate what can be done with digital electronics, let's just start learning about it. Applications will emerge naturally as we go along.

8.02 Logic states

By digital electronics we mean circuits in which there are only two (usually) states possible at any point, e.g., a transistor that can either be in saturation or be nonconducting. We usually choose to talk about voltages rather than currents, calling a level HIGH or LOW. The two states can represent any of a variety of "bits" (binary digits) of information, such as the following:

one bit of a number
whether a switch is open or closed
whether a signal is present or absent
whether some analog level is above or below some preset limit
whether or not some event has happened yet
whether or not some action should be taken
etc.

HIGH and LOW

The HIGH and LOW states represent the TRUE and FALSE states of Boolean logic, in some predefined way. If at some point HIGH represents TRUE, that signal line is called "positive true," and vice versa. This can be confusing at first. Figure 8.1 shows an example. SWITCH CLOSED is true when the output is LOW; that's a negative-true signal ("LOW-true" might be a better label, since

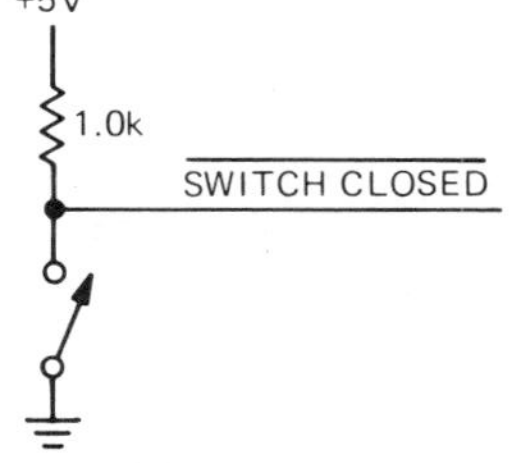

Figure 8.1

no negative voltages are involved), and you might label the lead as shown (a bar over a symbol means NOT; that line is HIGH when the switch is *not* closed). Just remember that the presence or absence of the negation bar over the label tells whether the wire is LOW or HIGH when the stated condition (SWITCH CLOSED) is true.

A digital circuit "knows" what a signal represents by where it comes from, just as an analog circuit might "know" what the output of some op-amp represents. However, added flexibility is possible in digital circuits; sometimes the same signal lines are used to carry different kinds of information, or even to send it in different directions, at different times. In order to do this "multiplexing," additional information must also be sent (address bits, or status bits). You will see many examples of this very useful ability later. For now, imagine that any given circuit is wired up to perform a predetermined function and that it knows what that function is, where its inputs are coming from, and where the outputs are going.

To lend a bit of confusion to a basically simple situation, we introduce 1 and 0. These symbols are used in Boolean logic to mean TRUE and FALSE, respectively, and are sometimes used in electronics in exactly that way. Unfortunately, they are also used in another way, in which 1 = HIGH and 0 = LOW! In this book we will try to avoid any ambiguity by using the word HIGH (or the symbol *H*) and the word LOW (or the symbol *L*) to represent logic states, a method that is in wide use in the electronics industry. We will use 1 and 0 only in situations where there can be no ambiguity.

Voltage range of HIGH and LOW

As we mentioned in Section 1.10, the voltage levels corresponding to HIGH and LOW are allowed to fall in some range. For instance, in TTL logic the LOW state is any voltage between −0.5 volt and +0.4 volt (it's typically a tenth of a volt or so above ground, just the output of a saturated npn transistor with grounded emitter), and the HIGH state is any voltage between +2.4 volts and +5.5 volts (typically around +3.4 V). This is done to allow for manufacturing spread, variations of the circuits with temperature, loading, supply voltage, etc., and the presence of "noise," the miscellaneous garbage that gets added to the signal in its journey through the circuit (from capacitive coupling, external interference, etc.). The circuit receiving the signal decides if it is HIGH or LOW and acts accordingly. As long as noise does not change 1's to 0's, or vice versa, all is well, and any noise is eliminated at each stage, since "clean" 0's and 1's are regenerated. In that sense digital electronics is noiseless and perfect.

The term *noise immunity* is used to describe the maximum noise level that can be added to logic levels (in the worst case) while still maintaining error-free operation. For instance, TTL has 0.4 volt noise immunity, since a TTL *input* is guaranteed to interpret anything less than +0.8 volt as LOW and anything greater than +2.0 volts as HIGH, whereas the worst-case *output* levels are +0.4 volt and +2.4 volts, respectively (see the box on logic levels in Chapter 1). In practice, noise immunity is considerably better than this figure, with typical LOW and HIGH voltages of +0.2 and +3.4 volts and an input decision threshold near +1.3 volts. But always remember that if you are doing good circuit design, you use worst-case values. It is worth keeping in mind that different logic families have different amounts of noise immunity. HNIL (high-noise-immunity logic) and CMOS have greater voltage noise immunity than TTL, whereas the speedy ECL family has less.

8.03 Number codes

Most of the conditions we listed earlier that can be represented by a digital level are self-explanatory. How a digital level can represent part of a number is a more involved, and very interesting, question.

A decimal (base 10) number is simply a string of integers that are understood to multiply successive powers of 10, the individual products then being added together. For instance,

$$137.06 = 1 \times 10^2 + 3 \times 10^1 + 7 \times 10^0 + 0 \times 10^{-1} + 6 \times 10^{-2}$$

Ten symbols (0 through 9) are needed, and the power of 10 each multiplies is determined by its position relative to the decimal point. If we want to represent a number using two symbols only (0 and 1), we use the *binary*, or base 2, number system. Each 1 or 0 then multiplies a successive power of 2. For instance,

$$1101_2 = 1 \times 2^3 + 1 \times 2^2 + 0 \times 2^1 + 1 \times 2^0 = 13_{10}$$

The individual 1's and 0's are called "bits" (binary digits). The subscript (always given in base 10) tells what number system we are using, and often it is essential in order to avoid confusion, since the symbols all look the same.

We convert a number from binary to decimal by the method just described. To convert the other way, we keep dividing the number by 2, and write down the remainders. To convert 13_{10} to binary,

$13/2 = 6$	remainder	1
$6/2 = 3$	remainder	0
$3/2 = 1$	remainder	1
$1/2 = 0$	remainder	1

from which $13_{10} = 1101_2$. Note that the answer comes out in the order LSB (least significant bit) to MSB (most significant bit).

Octal and hexadecimal representations

The binary number representation is the natural choice for two-state systems (however, it is not the only way, and you will see some others soon). Since the numbers tend to get rather long, it is common to write them in the octal (base 8) or hexadecimal (base 16) representation. This is easier than it sounds: To write a binary number in octal, just group it in 3-bit groups, beginning with the LSB, and write the octal equivalent of each group:

$$835_{10} = 1101000011_2 \; (= 1\;101\;000\;011_2) = 1503_8$$

Only the symbols 0–7 are used, of course. To write it as hexadecimal, form 4-bit groups. Each group can have values from 0 to 15; since you want a single symbol for each hex position, the symbols A–F are assigned to the values 10–15:

$$707_{10} = 1011000011_2 \; (= 10\;1100\;0011_2) = 2C3_{16}$$

Hexadecimal representation is best suited to the popular "byte" (8-bit) organization of computers, which are most often organized as 16-bit or 32-bit computer "words"; a word is then 2 or 4 bytes. An alphanumeric character (letter, number, or symbol) is one byte. So in hexadecimal, each byte is 2 hex digits, a 16-bit computer word is 4 hex digits, etc. Unfortunately, 12-bit and 36-bit words were common in earlier computers, which used 6-bit alphanumeric representation. Octal was a natural for representing such 6-bit groups, and it has become entrenched. The result is that the octal system is alive and well for binary number representation, which can be a nuisance.

For example, in computers with 16-bit words, one word is used to store two alphanumeric characters, each 8 bits (one byte) long. Octal representation is a royal pain here, because you can't recognize the individual bytes in the word. Let's say the letters A and B happen to be stored in a particular computer word, in the widely used ASCII representation (more on that in Section 10.17). Here's the situation:

$$A = 101_8 = 01\ 000\ 001$$
$$B = 102_8 = 01\ 000\ 010$$

$$\begin{aligned} \text{word } (AB) &= 01\ 000\ 001 \quad 01\ 000\ 010 \\ &= 0\ 100\ 000\ 101\ 000\ 010 \quad \text{(regroup by 3-bit groups)} \\ &= 040502_8 \quad \text{(write as octal)} \end{aligned}$$

Thus the individual identities of the A and B are lost when the 16-bit word is written in octal, because the higher-order byte gets reshuffled when the entire word is grouped in 3-bit groups. The word would look still different if the two characters were swapped, putting B in the higher-order (left) half and A in the lower-order (right) half.

EXERCISE 8.1

What is the octal representation of the 16-bit word containing BA in ASCII?

Hexadecimal has none of these problems. The character A is 41_{16}, the character B is 42_{16}, the 16-bit word AB is 4142_{16}, and the 16-bit word BA is 4241_{16}. The only problems with hex are getting used to crazy numbers like "3F2A" and doing arithmetic with them. Since octal is also widely used, you really have to get used to both systems.

BCD

Another way to represent a number is to encode each decimal digit into binary. This is called BCD (binary coded decimal), and it requires a 4-bit group for each digit. For instance,

$$137_{10} = 0001\ 0011\ 0111 \quad \text{(BCD)}$$

Note that BCD representation is *not* the same as binary representation, which in this case would be $137_{10} = 10001001_2$. You can think of the bit positions (starting from the right) as representing 1, 2, 4, 8, 10, 20, 40, 80, 100, 200, 400, 800, etc. It is clear that BCD is wasteful of bits, since each 4-bit group could represent numbers 0 through 15 but in BCD never represents numbers greater than 9 (there's a minor exception in 7-track even-parity digital tape recording, but never mind). However, BCD is ideal if you want to display a number in decimal, since all you do is convert each BCD character to the appropriate decimal number and display it. (There are many devices that do exactly this, e.g., a "BCD decoder, driver, display, and latch," which is a little IC with a transparent top. You apply logic levels for your BCD character, and it lights up with the digit.) For this reason BCD is commonly used for input and output of numeric information. Unfortunately, the conversion between pure binary and BCD is complicated, since *each* decimal digit depends on the state of almost every binary bit, and vice versa. Nevertheless, binary arithmetic is so efficient that most computers convert all input data to binary, converting back only when data need to be output. Think how much effort and bother would have been saved if *Homo sapiens* had evolved with 8 (or 16) fingers!

EXERCISE 8.2

Convert to decimal: (a) 1110101.0110_2, (b) $11.01010101\ldots_2$, (c) 25_8, (d) 39_8. Convert to binary: (a) 1023_{10}, (b) 1023_8.

Signed numbers

Sign magnitude representation. Sooner or later it becomes necessary to represent negative numbers in binary, particularly in devices where some computation is done. The simplest method is to devote one bit (the MSB, say) to the sign, with the remaining bits representing the magnitude of the number. This is called "sign magnitude representation," and it corresponds to the way signed numbers are ordinarily written (see Table 8.1). It is used when numbers are displayed, as well as in some A/D conversion schemes. In general, it is not the best method for representing signed numbers, particularly where some computation is done, for several reasons: Computation is awkward; subtraction is different from addition (i.e., addition doesn't "work" for signed numbers). Also, there can be two zeros (+0 and −0), so you have to be careful to use only one of them.

Offset binary representation. A second method for representing signed numbers is "offset binary," in which you subtract half the largest possible number to get the value represented (Table 8.1). This has the advantage that the number sequence from the most negative to the most positive is a simple binary progression, which makes it a natural for binary "counters." The MSB still carries the sign information, and zero appears only once. Offset binary is popular in A/D and D/A conversions, but it is still awkward for computation.

TABLE 8.1. 4-BIT INTEGERS IN THREE SYSTEMS OF REPRESENTATION

Integer	Sign magnitude	Offset binary	2's comp.
+7	0111	1111	0111
+6	0110	1110	0110
+5	0101	1101	0101
+4	0100	1100	0100
+3	0011	1011	0011
+2	0010	1010	0010
+1	0001	1001	0001
0	0000	1000	0000
−1	1001	0111	1111
−2	1010	0110	1110
−3	1011	0101	1101
−4	1100	0100	1100
−5	1101	0011	1011
−6	1110	0010	1010
−7	1111	0001	1001
−8	—	0000	1000
(−0)	1000	—	—

2's complement representation. The method most widely used for integer computation is called "2's complement." In this system, positive numbers are represented as simple unsigned binary. The system is rigged up so that a negative number is then simply represented as the binary number you add to a positive number of the same magnitude to get zero. To form a negative number, first complement each of the bits of the positive number (i.e., write 1 for 0, and vice versa; this is called the "1's complement"), then add 1 (that's the "2's complement"). As you can see from Table 8.1, 2's complement numbers are related to offset binary numbers by having the MSB complemented. As with the other signed number representations, the MSB carries the sign information. There's only one zero, conveniently represented by all bits 0 ("clearing" a counter or register sets its value to zero).

Arithmetic in 2's complement

Arithmetic is simple in 2's complement. To add two numbers, just add bitwise (with carry), like this:

```
5 + (−2):  0101   (+5)
           1110   (−2)
           ----
           0011   (+3)
```

To subtract *B* from *A*, take the 2's complement of *B* and add (i.e., add the negative):

```
2 − 5:  0010 (+2)
        1011 (−5)  (+5 = 0101: 1's
                    comp = 1010, so
                    2's comp = 1011)
        ----
        1101 (−3)
```

Multiplication also "works right" in 2's complement representation. Try the following exercise.

EXERCISE 8.3

Multiply +2 by −3 in 3-bit 2's complement binary arithmetic. Hint: The answer is −6.

EXERCISE 8.4

Show that the 2's complement of −5 is +5.

Because the 2's complement system is natural for computation, it is universally used for integer arithmetic in computers (note, however, that "floating point" numbers are usually represented in a form of "sign magnitude," namely sign-exponent-mantissa).

□ *Excess-3 and 4221 number codes*

There are a few other number codes with interesting and useful properties. Instead of BCD you will occasionally see "XS3" or "4221" used. In each of these a 4-bit group is used for each decimal digit, just as in BCD. To represent a digit (0–9) in excess-3, just add 3 to the number, then write the result as 4-bit binary. In 4221 each digit is again represented as a 4-bit group, but with the bits representing 4, 2, 2, and 1, from the left. To see why these codes are used, look at the representations of the digits 0–9:

Digit	*XS3*	4221
0	0011	0000
1	0100	0001
2	0101	0010
3	0110	0011
4	0111	1000
5	1000	0111
6	1001	1100
7	1010	1101
8	1011	1110
9	1100	1111

These codes have one important feature: The 9's complement of a digit (= 9 − digit) is obtained by taking its 1's complement. This simplifies decimal arithmetic (just as 2's complement representation simplifies binary arithmetic). To subtract a number you just add the number formed by taking the 9's complement and adding one. For instance,

```
27 − 15:     27
            +85      (9's comp of 15 =
                      84; then add 1)
            ___
answer:      12
```

□ *Gray code*

The following code is used for mechanical shaft-angle encoders, among other things. It is called a Gray code, and it has the property that only one bit changes in going from one state to the next. This prevents errors, since there is no way of guaranteeing that all bits will change simultaneously at the boundary between two encoded values. If straight binary were used, it would be possible to generate an output of 15 in going from 7 to 8, for instance. Here is a simple rule for generating Gray code states: Begin with a state of all zeros. To get to the next state, always change the single least significant bit that brings you to a new state.

0000
0001
0011
0010
0110
0111
0101
0100
1100
1101
1111
1110
1010
1011
1001
1000

Gray codes can be generated with any number of bits. They find use also in "parallel encoding," a technique of high-speed A/D conversion that you will see later. We will talk about translation between Gray-code and binary-code representations in the next section.

8.04 Gates and truth tables

Combinational versus sequential logic

In digital electronics the name of the game is generating digital outputs from digital inputs. For instance, an *adder* might take two 16-bit numbers as inputs and generate a 16-bit (plus carry) sum. Or you might build a circuit to multiply two numbers. These are the kinds of operations a computer's processing unit should be able to do. Another task might be to compare two numbers to see which is larger or to compare a set of inputs with the desired input to make sure that "all systems are go." Or you might want to attach a "parity bit" to a number to make the total number of 1's even, say, before transmission over a data link; then the parity could be checked on receipt as a simple check of correct transmission. Another typical task is to take some numbers expressed in binary and display, print, or punch them as decimal characters. All of these are tasks in which the output or outputs are predetermined functions of the input or inputs. As a class, they are known as "combinational" tasks. They can all be performed with devices called *gates,* which perform the operations of Boolean algebra applied to two-state (binary) systems.

There is a second class of problems that cannot be solved by forming a combinational function of the inputs alone, but require knowledge of past inputs as well. Their solution requires the use of "sequential" networks. Typical tasks of this type might be converting a string of bits in serial form (one after another) into a parallel set of bits, or keeping count of the number of 1's in a sequence, or recognizing a certain pattern in a sequence, or giving one output pulse for each four input pulses. All these tasks require digital memory of some sort. The basic device here is the "flip-flop" (the fancy name is "bistable multivibrator").

We will begin with gates and combinational logic, since they're basic to everything. Digital life will become more interesting when we get to sequential devices, but there will be no lack of fun and games with gates alone.

OR gate

The output of an OR gate is HIGH if either input (or both) is HIGH. This can be expressed in a "truth table," as shown in Figure 8.2. The gate illustrated is a 2-input OR gate. In general, gates can have any number of inputs, but the standard packages

A, B, Q
OR

inputs A	inputs B	output Q
0	0	0
0	1	1
1	0	1
1	1	1

Figure 8.2

usually contain four 2-input gates, three 3-input gates, or two 4-input gates. For instance, a 4-input OR gate will have a HIGH output if any one input (or more) is HIGH.

The Boolean symbol for OR is $+$. "A OR B" is written $A + B$.

AND gate

The output of an AND gate is HIGH only if both inputs are HIGH. The logic symbol and truth table are as shown in Figure 8.3. As

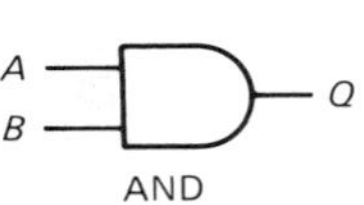

inputs		output
0	0	0
0	1	0
1	0	0
1	1	1

Figure 8.3

with OR gates, AND gates are available with 3 or 4 (sometimes more) inputs. For instance, an 8-input AND gate will have a HIGH output only if *all* inputs are HIGH.

The Boolean symbol for AND is a dot ($\cdot$); this can be omitted; and usually is. "A AND B" is written $A \cdot B$, or simply AB.

Inverter (the NOT function)

Frequently we need the complement of a logic level. That is the function of an inverter, a "gate" with only one input (Fig. 8.4).

The Boolean symbol for NOT is a bar over the symbol, or sometimes a prime symbol. "NOT A" is written $\overline{A}$, or A'.

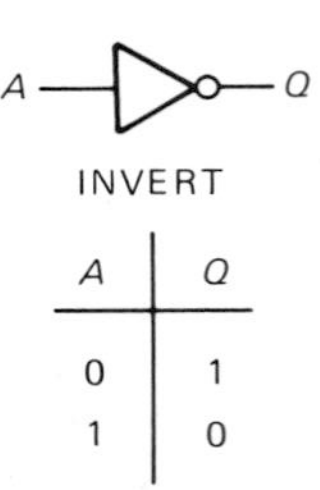

A	Q
0	1
1	0

Figure 8.4

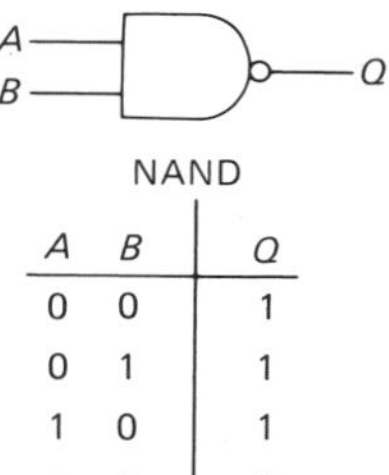
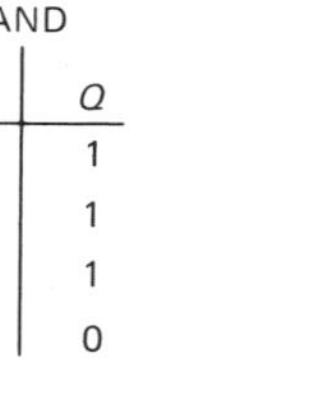

NAND

A	B	Q
0	0	1
0	1	1
1	0	1
1	1	0

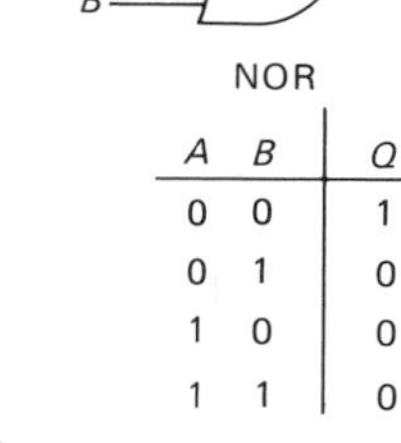

NOR

A	B	Q
0	0	1
0	1	0
1	0	0
1	1	0

Figure 8.5

NAND and NOR

The INVERT function can be combined with gates, forming NAND and NOR (Fig. 8.5). These are actually more popular than AND and OR, as you will see shortly.

Exclusive-OR

Exclusive-OR is an interesting function, although less fundamental than AND and OR (Fig. 8.6). The output of an exclusive-OR gate is HIGH if one or the other (but not both) input is HIGH (it never has more than two inputs). Another way to say it is that the output is HIGH if the inputs are different. The exclusive-OR gate is identical to modulo-2 addition of two bits.

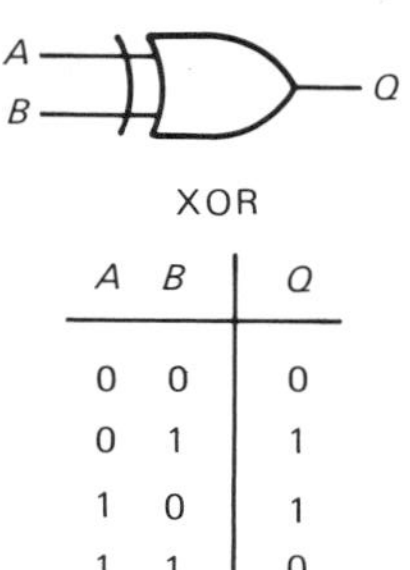

XOR

A	B	Q
0	0	0
0	1	1
1	0	1
1	1	0

Figure 8.6

EXERCISE 8.5

Show how to use the exclusive-OR gate as an "optional inverter," i.e., it inverts an input signal or buffers it without inversion, depending on the level at a control input.

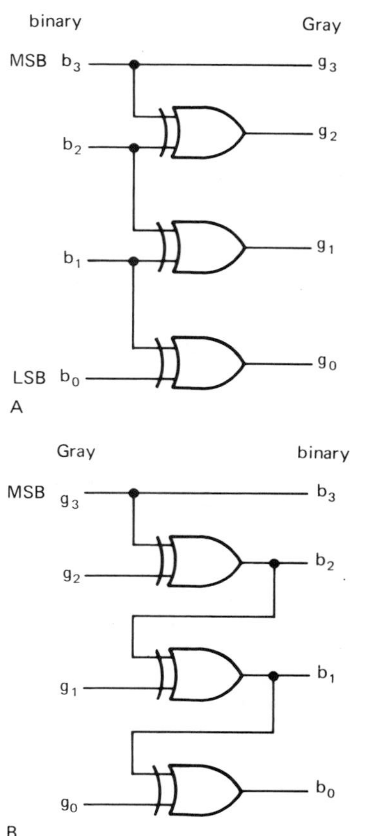

Figure 8.7

Parallel code converters: binary to Gray and Gray to binary.

EXERCISE 8.6

Verify that the circuits in Figure 8.7 convert binary code to Gray code, and vice versa.

8.05 Discrete circuits for gates

Before going on to discuss gate applications, let's see how to make gates from discrete components. Figure 8.8 shows a diode AND

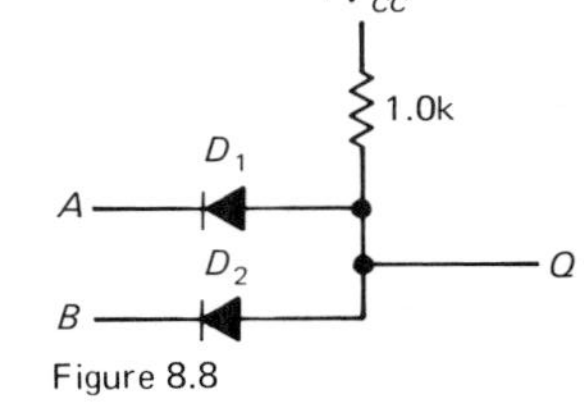

Figure 8.8

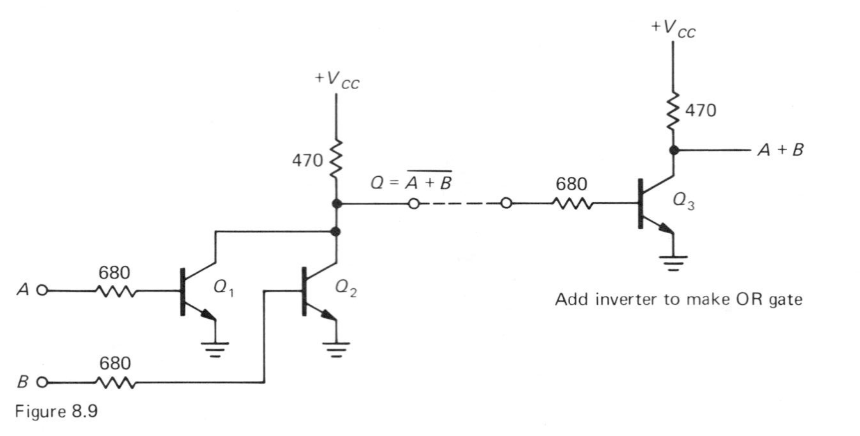

Figure 8.9

gate. If either input is held LOW, the output is LOW. The output can go HIGH only when both inputs go HIGH. This circuit has many disadvantages, in particular: (a) Its LOW output is a diode drop above the signal holding the input LOW. Obviously you couldn't use very many of these in a row! (b) There is no "fanout" (the ability of one output to drive several inputs), since any load at the output is seen by the signal at the input. (c) It is slow, because of resistive pull-up. As a general rule, you cannot do as well with logic constructed from discrete components as with IC logic. Part of the superiority of IC logic lies in the use of special techniques (e.g., gold doping) to achieve excellent performance.

Figure 8.9 shows the simplest form of transistor NOR gate. This circuit was used in the family of logic known as RTL (resistor-transistor logic), which was popular in the 1960s because of its low price, but is now obsolete. A HIGH at either input (or both) turns on at least one transistor, pulling the output LOW. Since this gate is intrinsically inverting, you would have to add an inverter, as shown, to make an OR gate.

8.06 Gate circuit example

Let's work out a circuit to perform the logic we gave as an example in Chapters 1 and 2: the task to sound a buzzer if either car door is open and the driver is seated. The answer is obvious if you restate the problem as "output HIGH if either the left door OR the right door is open, AND driver is seated,"

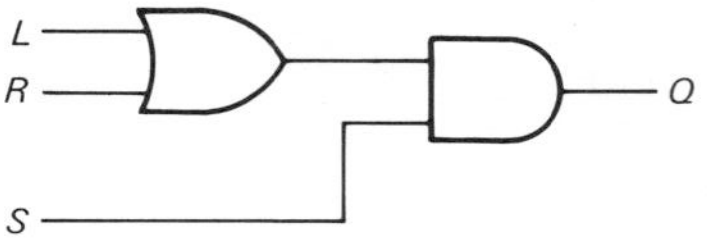

Figure 8.10

i.e., $Q = (L + R)S$. Figure 8.10 shows it with gates. The output of the OR gate is HIGH if one OR the other door (or both) is open. If that is so, AND the driver is seated, Q goes HIGH. With an additional transistor, this could be made to sound a buzzer or close a relay.

In practice, the switches generating the inputs will probably close a circuit to ground, to save extra wiring (there are additional reasons, particularly in the case of the popular TTL logic, and we will get to them shortly). This means that the inputs will go LOW when a door is opened, for example. In other words, we have "negative-true" inputs. Let's rework the example with this in mind, calling the inputs L', R', and S'.

First, we need to know if either door input (L', R') is LOW; i.e., we must distinguish the state "both inputs HIGH" from all others. That's an AND gate. So we make L' and R' the inputs to an AND gate. The output will be LOW if either input is *LOW;* call that EITHER′. Now we need to know when EITHER′ is LOW and S' is LOW; i.e., we must distinguish the state "both inputs LOW" from all others. That's an OR gate. Figure 8.11 shows the circuit. We have used a NOR gate, instead of an OR gate, to get the same output as earlier: Q HIGH when the

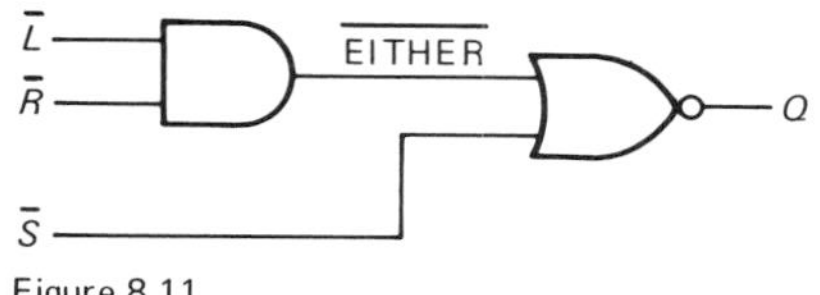

Figure 8.11

desired condition is present. Something strange seems to be going on here, though. We have used AND instead of OR (and vice versa), as compared with the earlier circuit. Section 8.07 should clarify the matter. First, consider the following exercise.

EXERCISE 8.7

What do the circuits shown in Figure 8.12 do?

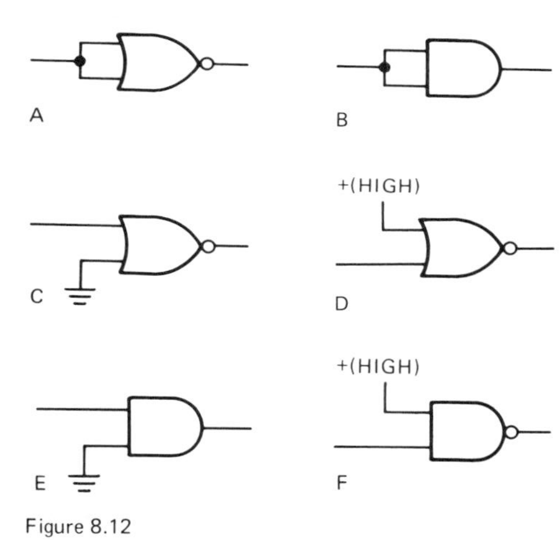

Figure 8.12

Gate interchangeability

When designing digital circuits, keep in mind that it is possible to form one kind of gate from another. For example, if you need an AND gate, and you have half of a 7400 available (quad 2-input NAND), you can substitute as shown in Figure 8.13. The second NAND functions as an inverter, making AND. The following exercises should help you explore this idea.

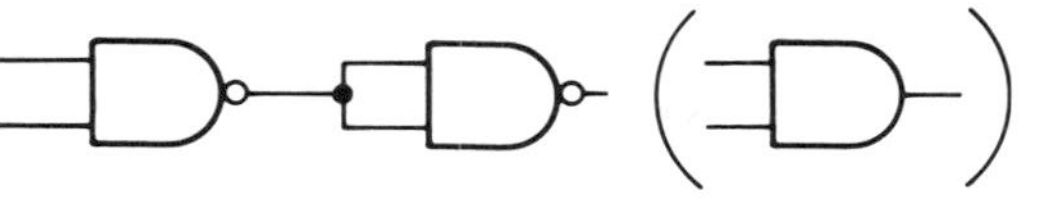

Figure 8.13

EXERCISE 8.8

Using 2-input gates, show how to make (a) INVERT from NOR, (b) OR from NORs, and (c) OR from NANDs.

EXERCISE 8.9

Show how to make (a) a 3-input AND from 2-input ANDs, (b) a 3-input OR from 2-input ORs, (c) a 3-input NOR from 2-input NORs, and (d) a 3-input AND from 2-input NANDs.

In general, multiple use of one kind of inverting gate (e.g., NAND) is enough to make any combinational function. However, this isn't true for a noninverting gate, since there's no way to make INVERT. This probably accounts for the greater popularity of NAND and NOR in logic design.

8.07 Assertion-level logic notation

An AND gate has a HIGH output if both inputs are HIGH. So, if HIGH means "true," you get a true output only if all inputs are true. In other words, with positive-true logic, an AND gate performs the AND function. The same holds for OR.

What happens if LOW means "true," as in the last example? An AND gate gives a LOW if either input is true (LOW): It's an OR function! Similarly, an OR gate gives a LOW only if both inputs are true (LOW): It's an AND function! Very confusing.

There are two ways to handle this problem. The first way is to think through any digital design problem as we did earlier, choosing the kind of gate that gives the needed output. For instance, if you need to know if any of three inputs is LOW, use a 3-input NAND gate. This method is probably used by the majority of designers. When designing this way, you would draw a NAND gate, even though the gate is performing a NOR function on the (negative-true) inputs. You would probably label the inputs as in Figure 8.14. In this example, CLEAR', MR' (master reset), and RESET' might be nega-

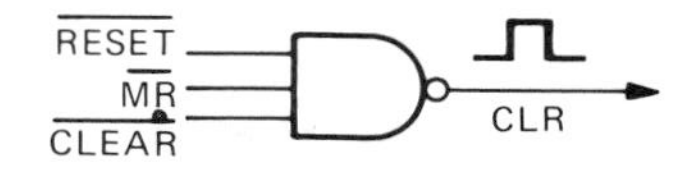

Figure 8.14

tive-true levels coming from various places in a circuit. The output, CLR, is positive-true and will go to the devices that are to be cleared if *any* of the reset signals goes LOW (true).

The second way to handle the problem of negative-true signals is to use "assertion-level logic." If a gate performs an OR function on negative-true inputs, draw it that way, as in Figure 8.15. The 3-input OR gate

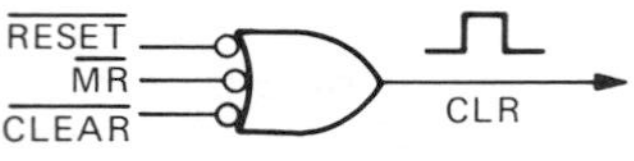

Figure 8.15

with negated inputs is functionally identical to the preceding 3-input NAND. That equivalence turns out to be an important logical identity, as stated in DeMorgan's theorem, and we will spell out a number of such useful identities shortly. For now, it is enough to know that you can change AND to OR (and vice versa) if you negate the output and all inputs. Assertion-level logic looks forbidding at first, because of the proliferation of funny-looking gates. It is better, though, because the logical functions of the gates in the circuit stand out clearly. You'll find it "friendly" after you've used it for a while, and you won't want to use anything else. Let's rework the car door example again with assertion-level logic (Fig. 8.16). The

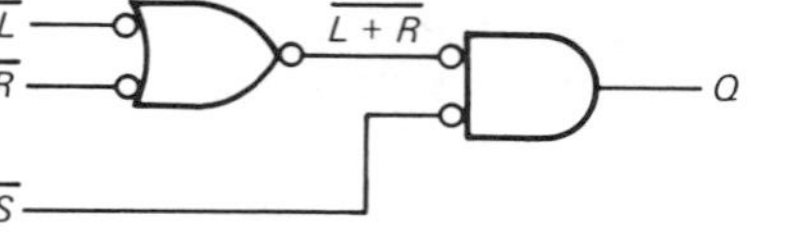

Figure 8.16

gate on the left determines if L or R is true, i.e., LOW, giving a negative-true output. The second gate gives a HIGH output if both ($L + R$) and S are true, i.e., LOW. From DeMorgan's theorem (after a while you won't even need that, you'll recognize these gates as equivalent) the first gate is AND, the second is NOR, just as in the circuit drawn earlier. Two important points:

1. Negative-true doesn't mean that the logic levels are negative *polarity.* It means that the lower of the two states (LOW) stands for TRUE.

2. The symbol used to draw the gate itself assumes positive-true logic. A NAND gate used as an OR for negative-true signals can be drawn as a NAND or, using assertion-level logic, as an OR with negation symbols (little circles) at the inputs. In the latter case you think of the circles as indicating inversion of the input signals, followed by an OR gate operating on positive-true logic as originally defined.

Postscript: Logical AND and OR shouldn't be confused with the *legal* equivalents. The weighty legal tome known as *Words and Phrases* has over 40 pages of situations in which AND can be construed as OR. For example: "OR will be construed AND, and AND will be construed OR, as the necessities of the case may require. . . ." This isn't the same as DeMorgan's theorem!

TTL AND CMOS

TTL (transistor-transistor logic) and CMOS (complementary MOS) are the two most popular logic families in current use, with at least 10 manufacturers of integrated circuits offering an enormous variety of functions in both families. These families should satisfy your needs for all digital design, with the exceptions of large-scale integration (LSI), which is dominated by MOS logic, and ultra-high-speed logic, where emitter-coupled logic (ECL) reigns supreme. Throughout the rest of the book we will rely heavily on these families.

8.08 Catalog of common gates

Table 8.2 shows the common gates you can get in the TTL and CMOS families of digital logic. Each gate is drawn in its normal (positive-true) incarnation, and also the way it looks for negative-true logic. The last entry in the table is an AND-OR-INVERT gate, sometimes abbreviated AOI.

A word of explanation: TTL is available in a half dozen subfamilies, all offering the same functions with the same logic levels. The differences have to do with speed and power dissipation (see Section 9.01). The best type for most applications is currently the low-power Schottky variety, specified by

TABLE 8.2 COMMON GATES IN THE TTL AND CMOS FAMILIES

Name	Expression	Symbol	Negative true symbol	Type	No. per chip	Part number							
						CMOS	TTL	→	L	LS	H	S	C
AND	AB	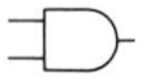	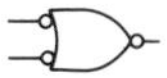	2-input	4	4081	7408			✓		✓	✓
				3-input	3	4073	7411			✓		✓	
				4-input	2	4082	7421			✓	✓		
NAND	$\overline{AB}$	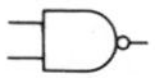	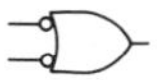	2-input	4	4011	7400		✓	✓	✓	✓	✓
				3-input	3	4023	7410		✓	✓	✓	✓	✓
				4-input	2	4012	7420		✓	✓	✓	✓	✓
				8-input	1	4068	7430		✓	✓	✓	✓	✓
				13-input	1	—	74133				✓		
OR	$A + B$	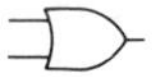	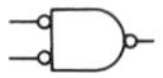	2-input	4	4071	7432		✓	✓		✓	✓
				3-input	3	4075	—						
				4-input	2	4072	—						
NOR	$\overline{A + B}$		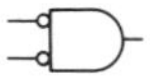	2-input	4	4001	7402		✓	✓		✓	✓
				3-input	3	4025	7427			✓			
				4-input	2	4002	7425						
				5-input	2	—	74260					✓	
				8-input	1	4078	—						
INVERT	$\bar{A}$	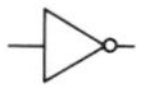	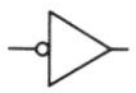		6	4069/4049	7404		✓	✓	✓	✓	✓
BUFFER	A	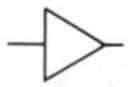	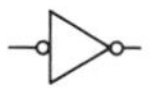		6	4503/4050	74365						
XOR	$A \oplus B$	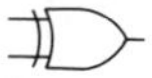	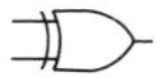	2-input	4	4070	7486/386 (−135)		✓	✓		✓	✓
XNOR	$\overline{A \oplus B}$		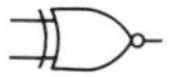	2-input	4	4077	74266 (−135)			✓			
AOI		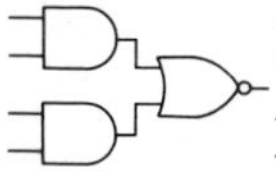	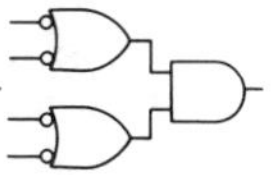	2-2-input	2	4085	7450/51		✓	✓	✓	✓	
				2-2-2-2-input	1	4086	7453/54		✓	✓	✓	✓	

adding the letters LS after the digits 74, e.g., 74LS00. For simplicity we will omit such letters routinely in this book, with the understanding that LS should usually be used. "Standard" TTL (no letters) has now become essentially obsolete.

8.09 IC gate circuits

Although a NAND gate, for instance, performs identical logic operations in both TTL and CMOS versions, the logic levels and other characteristics (speed, power, input current, etc.) are quite different. In general, you can't mix the two logic family types. To understand the differences, look at the schematics of a NAND gate in Figure 8.17.

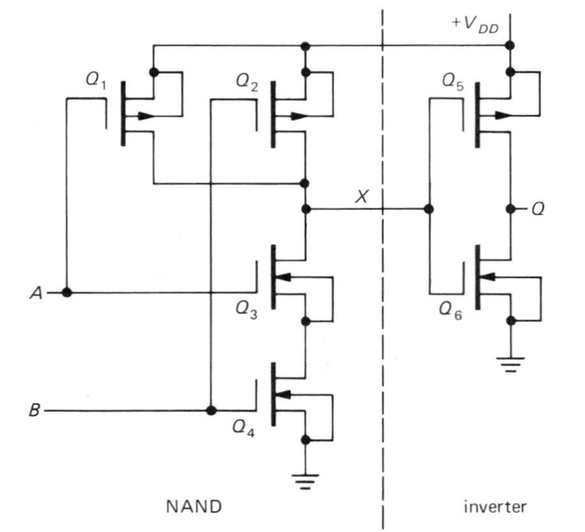

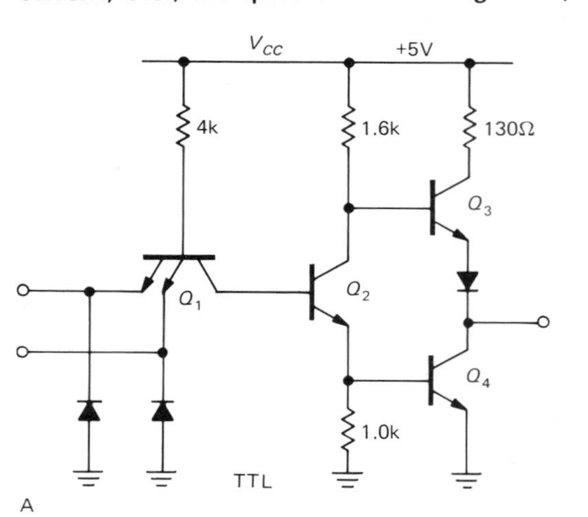

Figure 8.17
A: TTL NAND gate. B: CMOS AND gate.

The CMOS gate is constructed from enhancement-mode MOSFETs of both polarities, connected as switches rather than followers. An ON FET looks like a low resistance to whichever supply rail it is connected. Both inputs must be HIGH to turn on the series pair Q_3Q_4 and to turn off both of the pull-up transistors Q_1Q_2. That produces a LOW at the output, i.e., it is a NAND gate. Q_5 and Q_6 constitute the standard CMOS inverter, to generate an AND gate. From this example it should be evident how to generalize to AND, NAND, OR, and NOR with any number of inputs.

EXERCISE 8.10

Draw the circuit of a 3-input CMOS OR gate.

A word of explanation about the TTL input circuit: Q_1 has two emitters. You can think of it as two transistors with their bases and collectors in parallel. With either input LOW, Q_1 is saturated, Q_2 is off, and the output is HIGH. When both inputs are HIGH, Q_1 functions as an inverted transistor, with $h_{FE} \ll 1$. You can think of its base-collector junction as a diode in conduction, turning on Q_2 and pulling the output LOW.

Note that both CMOS and TTL gates have an output circuit with "active pull-up" to the positive supply rail, unlike our discrete gate circuit examples.

8.10 TTL and CMOS characteristics

Let's compare family characteristics:

Supply voltage. TTL requires +5 volts, ±5%, whereas CMOS operates with a supply voltage from +5 to +15 volts; +5 and +12 are popular values.

Input. A TTL input held in the LOW state sources current into whatever drives it (1.0mA typ for "standard" TTL), so to pull it LOW you must sink current. Since the TTL output circuit is good at sinking current, this presents no problems when TTL logic is wired together, but you must keep it in mind when driving TTL with other circuitry.

By contrast, CMOS has no input current. The TTL logic threshold is two diode drops above ground (about 1.4V); CMOS has its threshold at about one-half the supply voltage, but it can vary quite a bit (from one-third to two-thirds the supply voltage). CMOS is susceptible to damage from static electricity during handling. In both families, unused inputs should be tied HIGH or LOW, as necessary (more on this later).

Output. The TTL output stage is a saturated transistor to ground in the LOW state, and a follower in the HIGH state (about two diode drops below V_{CC}); with CMOS the output is a turned-on MOSFET, either to ground or to V_{DD}.

Speed and power. TTL is fast (25–50 MHz), but it consumes lots of power; new low-power subfamilies offer considerable improvement in this respect. CMOS is slower, but it has negligible *quiescent* power dissipation. However, its power dissipation rises linearly with increasing frequency (switching capacitive loads requires current), and CMOS operated near its upper frequency limit dissipates nearly as much power as TTL.

In general, TTL is used where speed is important, whereas CMOS dominates low-power applications. Other families are used for special requirements (ECL for highest speed; HNIL for highest noise immunity, NMOS and PMOS for high density).

Within any one logic family, outputs are designed to drive other inputs easily, so you don't often have to worry about thresholds, input current, etc. For instance, with TTL or CMOS any output can drive at least 10 other inputs (the official term for this is *fanout:* TTL has a fanout of 10), so you don't have to do anything special to ensure compatibility. In the next chapter we will go into the issue of interfacing between logic families and between logic circuits and the outside world.

8.11 Three-state and open-collector devices

The TTL and CMOS gates we have just discussed have push-pull output circuits: The output is held either HIGH or LOW by an ON transistor or MOSFET. Nearly all digital logic uses this sort of circuit (called active pull-up; in TTL it's also called a totem-pole output) because it provides low output impedance in both states, giving faster switching time and better noise immunity, as compared with an alternative such as a single transistor with passive collector pull-up resistor. In the case of CMOS, it also results in lower power dissipation.

However, there are a few situations for which active pull-up output is unsuitable. As an example, imagine a computer system in which several functional units have to exchange data. The central processor (CPU), memory, and various peripherals all need to be able to send and receive 16-bit words. It would be awkward (to put it mildly) to have separate 16-wire cables connecting each device to all others. The solution is the so-called *data bus,* a single set of 16 wires accessible to all devices. It's like a telephone party line: Only one device at a time may "talk" (assert data), but all may "listen" (receive data). With a bus system there must be agreement as to who may talk, and words like "bus arbitrator," "bus master," and "control bus" pop up.

You can't use gates (or any other devices) with active pull-up outputs to drive a bus, since you couldn't disconnect your output from the shared data lines (you're holding it either HIGH or LOW at all times). What's needed is a gate whose output can be "open." Such devices are available, and they come in two varieties, "three-state devices" and "open-collector devices." Let's start with open-collector devices, with the understanding that the discussion applies also to three-state devices.

In an open-collector gate the output circuit simply omits the active pull-up transistor of the output stage (Fig. 8.18). The name "open collector" is a good one. When using such gates, you must supply an external pull-up resistor somewhere. Its value isn't critical; a small-value resistor gives increased speed and improved noise immunity at the expense of increased power dissipation and loading of the driver. Values of a few hundred to a few thousand ohms are typical with TTL. As we will explain shortly, everything we are about to say about open-

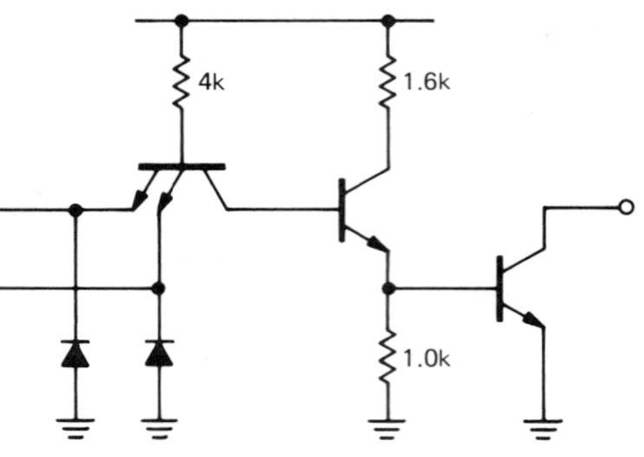

Figure 8.18

collector gates applies also to gates with three-state outputs.

A look ahead: data buses

Open-collector gates can be used to drive a data bus. Each line of the bus has a single pull-up resistor, and every device that needs to put data on the bus ties onto it with open-collector NAND (or three-state) gates. At any instant, all devices except one have their bus drivers disabled (in the HIGH state, i.e., open) by having one input of their NAND inputs held LOW (or by having their "third state" enabled). In a typical situation the selected device "knows" to assert data onto the bus by decoding its particular address from a set of address and control lines (Fig. 8.19). In this case the device is wired as device 6. It decodes the address on address lines A_0–A_2 and asserts data (negative-true) onto data bus D'_0 through D'_3 when it sees its address on the address lines *and* it sees a READ pulse. Such a bus protocol is adequate for many simple systems. Something like this is used in most minicomputers and microprocessors, as you will see in Chapters 10 and 11. Many ordinary gates, e.g., the 7400 quad NAND, are available in open-collector versions (7403). Open-collector gates with high current-sinking capability, e.g., the 7438 with a fanout of 30 TTL loads, are available for bus driving applications. An asterisk next to the output pin, or a vertical line across the width of the gate symbol, signifies open collector; sometimes you see OC written within the gate symbol instead.

Three-state logic

One problem with open-collector logic is that speed and noise immunity are degraded, when compared with logic constructed with ordinary active pull-up devices, because of the resistive pull-up circuit. A large bus can have significant capacitance and potential for noise pickup. The so-called three-state or TRI-STATE logic (a trademark of National

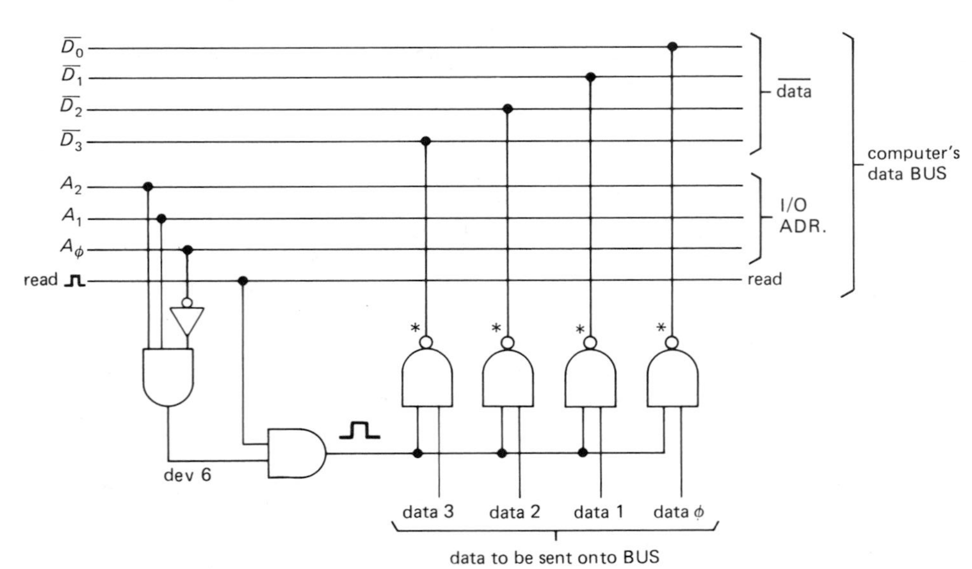

Figure 8.19
Open-collector NAND gates driving a computer data bus.

Semiconductor Corporation) provides an elegant solution. The name is misleading; it is not digital logic with three voltage levels. It's just ordinary logic, with a third output state: open circuit. With three-state logic you have the best of both worlds – the speed and noise immunity of active pull-up outputs and the busing capability of open-collector logic. A separate *enable* input is used to put the output into the third state, regardless of the logic levels present at the gate's input. Three-state outputs are available on many digital chips, including counters, latches, registers, etc., as well as on gates and inverters. Three-state devices are generally replacing open-collector gates for busing applications because of their superior performance. As a bonus, you don't need a whole bunch of pull-up resistors.

Driving external loads

Another application for open-collector logic is driving external loads that are returned to a higher-voltage positive supply. You might want to drive a low-current lamp that requires 12 volts, or perhaps just generate a 15 volt logic swing by running a resistor from gate to +15 volts, as in Figure 8.20.

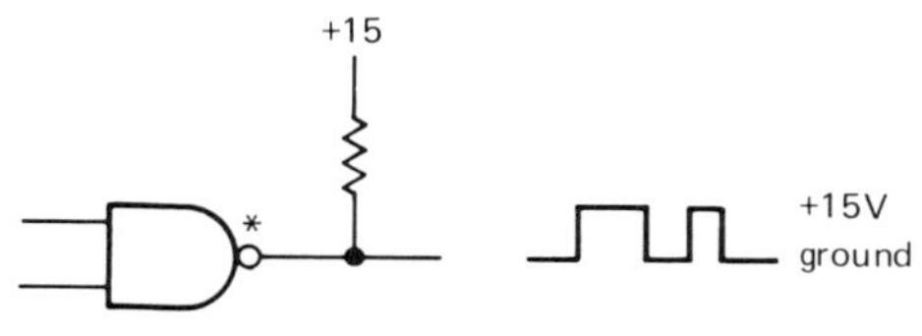

Figure 8.20

In the TTL family, for instance, the 7416 is an open-collector hex inverter with 15 volt breakdown, whereas the 7406 has 30 volt breakdown rating. The 75450 series of "dual peripheral drivers" can sink up to 300mA from loads returned to +30 volts. In CMOS the 40107 is a dual-NAND "open-drain" buffer with up to 120mA sink capability; in CMOS that's really something! More on these subjects in Chapters 9, 10, and 11.

Wired-OR

There is one other important application of open-collector logic: wired-OR. This is just gate expansion done the simplest way, namely by tying the outputs of a few gates together. You can't do that with active pull-up outputs (there would be a contest of wills, if all the gates didn't agree on what the output should be), but with open collectors it's easy (Fig. 8.21). In this case the output goes LOW if any input is HIGH. Three 2-

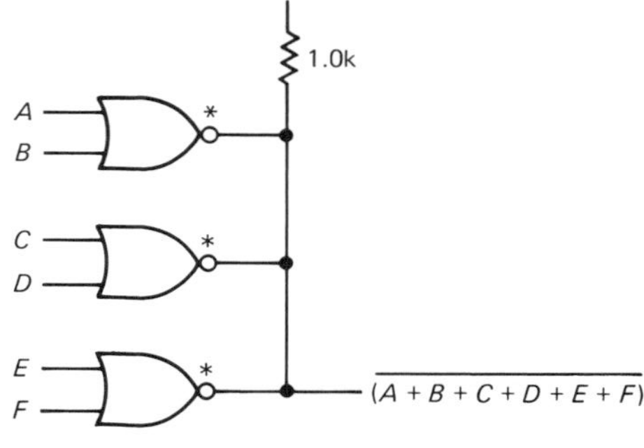

Figure 8.21

input NORs have been expanded to a 6-input NOR. NORs, NANDs, etc., may be combined. The output will be LOW if any gate asserts a LOW output. This connection is sometimes called "wired-AND," since the output is HIGH only if all gates have HIGH (open) outputs. Both names are describing the same thing: It's wired-AND for positive-true logic and wired-OR for negative-true logic.

Wired-OR has a rather significant liability, namely the difficulty of debugging circuits in which some malfunctioning wired-OR output is holding a point LOW. The problems you face when trying to track down such "stuck nodes" takes a lot of the fun out of wired-OR. Wired-OR was given a long trial run with the now-defunct DTL logic, but it didn't survive the changeover to TTL.

COMBINATIONAL LOGIC

As we discussed earlier in Section 8.04, digital logic can be divided into *combinational* and *sequential*. Combinational circuits are those in which the output state depends only on the present input states in some predetermined fashion, whereas in sequential circuits the output state depends both on the input states and on the previous history. Combinational circuits can be constructed with gates alone, whereas sequential circuits require some form of memory (flip-flops). In these sections we will explore the possibilities of combinational logic before entering the turbulent world of sequential circuits.

8.12 Logic identities

No discussion of combinational logic is complete without the identities shown in Table 8.3. Most of these are obvious. The last two compose DeMorgan's theorem, the most important for circuit design.

TABLE 8.3 LOGIC IDENTITIES

$ABC = (AB)C = A(BC)$
$AB = BA$
$AA = A$
$A1 = A$
$A0 = 0$
$A(B + C) = AB + AC$
$A + AB = A$
$A + BC = (A + B)(A + C)$

$A + B + C = (A + B) + C = A + (B + C)$
$A + B = B + A$
$A + A = A$
$A + 1 = 1$
$A + 0 = A$

$\overline{1} = 0$
$\overline{0} = 1$
$A + \overline{A} = 1$
$A\overline{A} = 0$
$\overline{\overline{A}} = A$
$A + \overline{A}B = A + B$
$\overline{A + B} = \overline{A}\,\overline{B}$
$\overline{AB} = \overline{A} + \overline{B}$

Example: exclusive-OR gate

We will illustrate the use of the identities with an example: making the exclusive-OR function from ordinary gates. Figure 8.22 shows the XOR truth table. From studying

A	B	$A \oplus B$
0	0	0
0	1	1
1	0	1
1	1	0

Figure 8.22

this, and by realizing that the output is 1 only when $(A,B) = (0,1)$ or $(1,0)$, we can write

$$A \oplus B = \overline{A}B + A\overline{B}$$

from which we have the realization shown in Figure 8.23. However, this realization is not

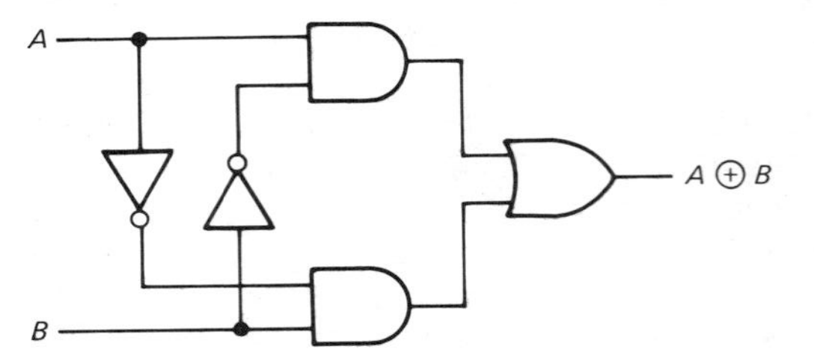

Figure 8.23

unique. Applying the identities, we find

$$A \oplus B = A\overline{A} + A\overline{B} + B\overline{A} + B\overline{B} \qquad (A\overline{A} = B\overline{B} = 0)$$
$$= A(\overline{A} + \overline{B}) + B(\overline{A} + \overline{B})$$
$$= A(\overline{AB}) + B(\overline{AB})$$
$$= (A + B)(\overline{AB})$$

(In the first step we used the trick of adding two quantities that equal zero; in the third step we used DeMorgan's theorem.) This has the realization shown in Figure 8.24.

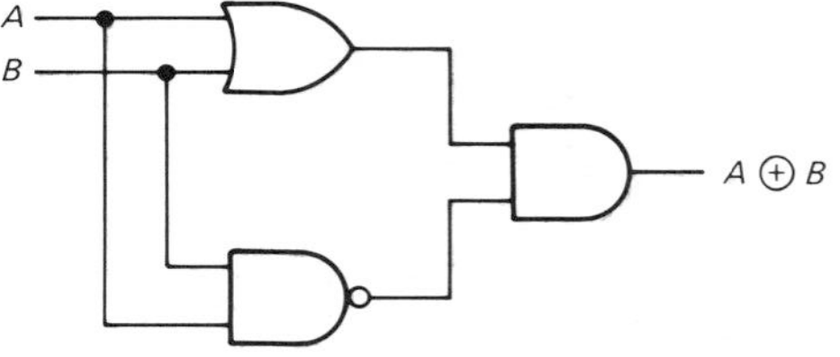

Figure 8.24

There are still other ways to construct XOR. Consider the following exercise.

EXERCISE 8.11

Show that

$$A \oplus B = \overline{AB + \overline{A}\,\overline{B}}$$
$$A \oplus B = (A + B)(\overline{A} + \overline{B})$$

by logic manipulation. You should be able to convince yourself that these are true by inspection of the truth table, combined with suitable handwaving.

EXERCISE 8.12

What are the following: (a) $0 \cdot 1$, (b) $0 + 1$, (c) $1 \cdot 1$, (d) $1 + 1$, (e) $A(A + B)$, (f) $A(A' + B)$, (g) A XOR A, (h) A XOR A'?

8.13 Minimization and Karnaugh maps

Since a realization of a logic function (even one as simple as exclusive-OR) isn't unique, it is often desirable to find the simplest, or perhaps most conveniently constructed, circuit for a given function. Many good minds have worked on this problem, and there are several methods available, including algebraic techniques that can be coded to run on a computer. For problems with four or fewer inputs, a Karnaugh map provides one of the nicest methods; it also enables you to find a logic expression (if you don't know it) once you can write down the truth table.

We will illustrate the method with an example. Suppose you want to generate a logic circuit to count votes. Imagine that you have three positive-true inputs (each either 1 or 0) and an output (0 or 1). The output is to be 1 if at least two of the inputs are 1;

Step 1. Make a truth table:

A	*B*	*C*	*Q*
0	0	0	0
0	0	1	0
0	1	0	0
0	1	1	1
1	0	0	0
1	0	1	1
1	1	0	1
1	1	1	1

All possible permutations must be represented, with corresponding output(s). Write an X (= "don't care") if either output state is OK.

Step 2. Make a Karnaugh map. This is somewhat akin to a truth table, but the variables are represented along two axes. Furthermore, they are arranged in such a way that only one input bit changes in going from one square to an adjacent square (Fig. 8.25).

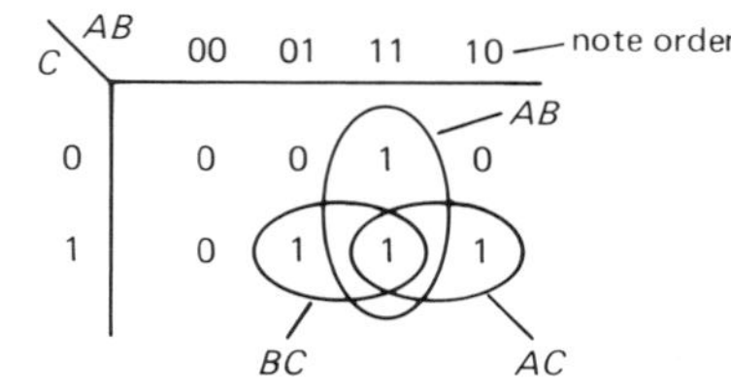

Figure 8.25

Step 3. Identify on the map groups of 1's (alternatively, you could use groups of 0's): The three blobs enclose the logic expressions AB, AC, and BC. Finally, read off the required function, in this case

$$Q = AB + AC + BC$$

with the realization shown in Figure 8.26. The result seems obvious, in retrospect. We

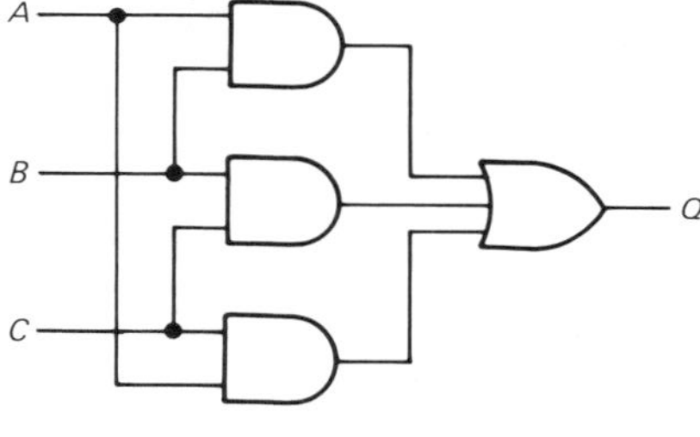

Figure 8.26

could have read off the patterns of 0's to get instead

$$Q' = A'B' + A'C' + B'C'$$

which might be useful if the complements A', B', and C' already exist somewhere in the circuit.

Some comments on Karnaugh maps

1. Look for groups of 2, 4, 8, etc., squares; they have the simple logic expressions.
2. The larger the block you describe, the simpler the logic.
3. The edges of the Karnaugh map connect up. For instance, the map in Figure 8.27 is described by $Q = B'C$.
4. A block of 1's with only one or two 0's may be best described by the grouping illustrated in Figure 8.28, which corresponds to the logic expression

$$Q = A(BCD)'$$

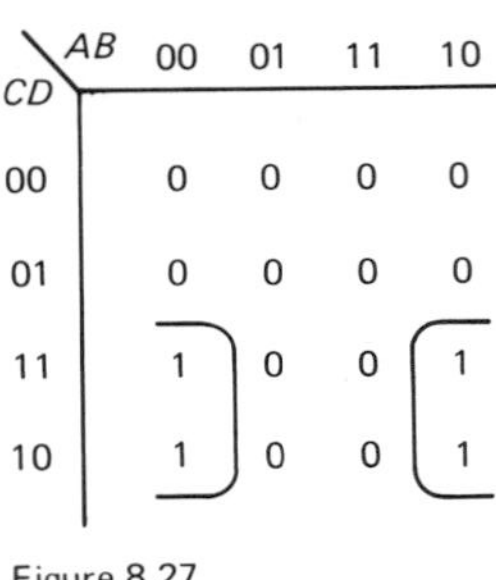

Figure 8.27

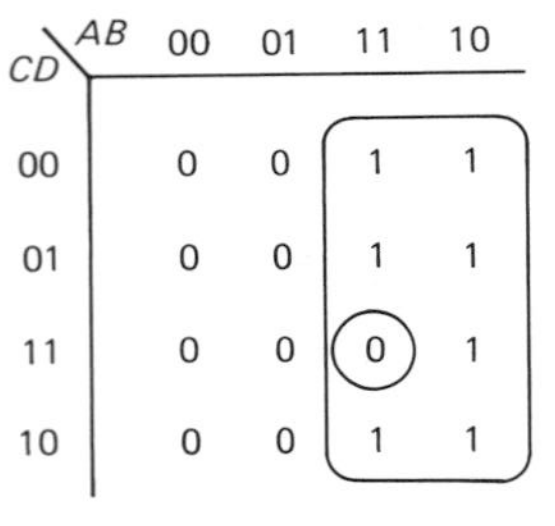

Figure 8.28

5. *X*s (don't care) are "wild cards." Use them as 1's or 0's to generate the simplest logic.

6. A Karnaugh map may not lead directly to the best solution: A more complicated logic expression may sometimes have a simpler realization in gates, e.g., if some of its terms already exist as logic in your circuit, and you can exploit intermediate outputs (from other terms) as inputs. Furthermore, exclusive-OR realizations are not obvious from Karnaugh maps. Finally, package constraints (e.g., the fact that four 2-input gates come in a single IC) also figure into the choice of logic used in the final circuit realization.

EXERCISE 8.13

Draw a Karnaugh map for logic to determine if a 3-bit integer (0 to 7) is prime (assume 0, 1, and 2 are not primes). Show a realization with 2-input gates.

EXERCISE 8.14

Find logic to perform multiplication of two 2-bit unsigned numbers (i.e., each 0 to 3), producing a 4-bit result. Hint: Use a separate Karnaugh map for each output bit.

8.14 Combinational functions available as ICs

Using Karnaugh maps, you can construct logic to perform rather complicated functions such as binary addition or magnitude comparison, parity checking, multiplexing (selecting one of several inputs, as determined by a binary address), etc. In the real world the most frequently used complex functions are available as single MSI functions (medium-scale integration, upward of 100 gates on one chip). Although many of the MSI functions involve flip-flops, which we will get to shortly, lots of them are combinational functions involving only gates. Let's see what animals live in the MSI combinational zoo.

Quad 2-input select

The quad 2-input select is a very useful chip. It is basically a 4-pole 2-position switch for logic signals. Figure 8.29 shows the basic

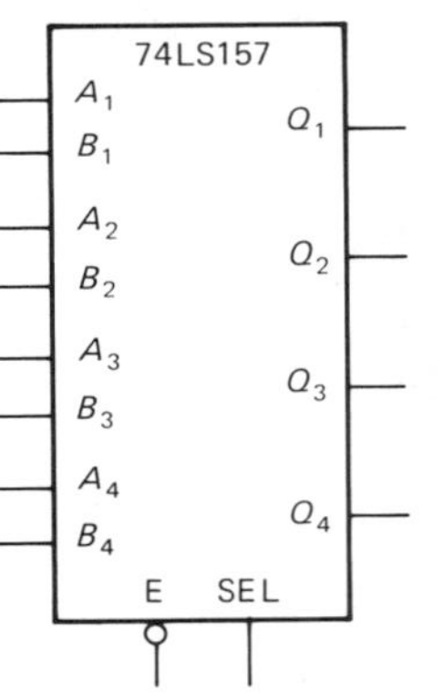

Figure 8.29

idea. When SELECT is LOW, the *A* inputs are passed through to their respective *Q* outputs. For SELECT HIGH, the *B* inputs appear at the outputs. ENABLE' HIGH disables the device by forcing all outputs LOW. This is an important concept you will see more of later. Here's the truth table, which illustrates the *X* (don't care) entry:

Inputs				*Outputs*
E'	SEL	A_n	B_n	Q_n
H	*X*	*X*	*X*	*L*
L	*L*	*L*	*X*	*L*
L	*L*	*H*	*X*	*H*
L	*H*	*X*	*L*	*L*
L	*H*	*X*	*H*	*H*

Figure 8.29 and the preceding table correspond to the 74157 TTL select chip. It is also available with inverted outputs (74158). In CMOS the 4019 is similar, but it requires the equivalent of SEL and SEL′ (it is actually a quad AND/OR gate). The CMOS 74C157 is functionally identical to the TTL 74157, an equivalence that characterizes the entire 74Cxx family.

EXERCISE 8.15

Show how to make a 2-input select from an AND-OR-INVERT gate.

Although the function of a select gate can be performed by a mechanical switch in some cases, the gate is a far better solution, for several reasons: (a) it is cheaper; (b) all channels are switched simultaneously and rapidly; (c) it can be switched, nearly instantaneously, by a logic level generated elsewhere in the circuit; (d) even if the select function is to be controlled by a front-panel switch, it is better not to run logic signals around through cables and switches, to avoid capacitive signal degradation and noise pickup. With a select gate actuated by a dc level, you keep logic signals on the circuit board and get the bonus of simpler off-board wiring (a single line with pull-up switched to ground by an SPST switch). Controlling circuit functions with externally generated dc levels in this manner is known as "cold switching," and it is a much better approach than controlling the signals themselves with switches, potentiometers, etc. Besides its other advantages, cold switching lets you bypass control lines with capacitors to eliminate interference, whereas signal lines cannot generally be bypassed. You will see some examples of cold switching later.

Transmission gates

As we discussed in Section 6.12, with CMOS it is possible to make "transmission gates," simply a pair of complementary MOSFET switches in parallel, so that an input (analog) signal between ground and V_{DD} is either connected through to the output through a low resistance (a few hundred ohms) or open-circuited (essentially infinite resistance). As you may remember, such a device is bidirectional and doesn't

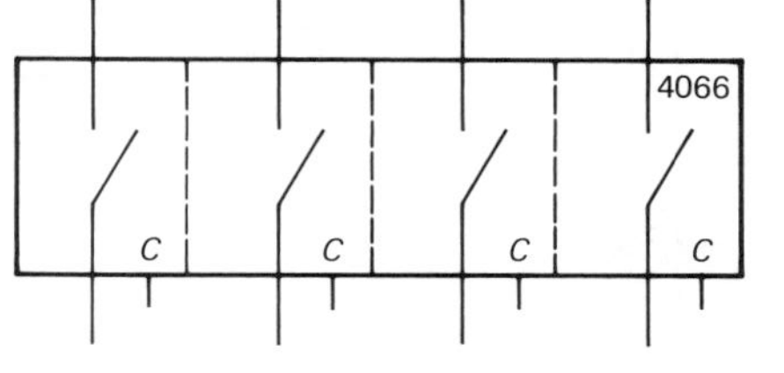

Figure 8.30

know (or care) which end is input and which end is output. Transmission gates work perfectly well with digital CMOS levels and are used extensively in CMOS design. Figure 8.30 shows the layout of the popular 4066 CMOS "quad bilateral switch." Each switch has a separate "control" input; input HIGH closes the switch, and LOW opens it.

With transmission gates you can make 2-input (or more) select functions, usable with CMOS digital levels or analog signals. To select among a number of inputs, you can use a bunch of transmission gates (generating the control signals with a "decoder," as will be explained later). This is such a useful logic function that it has been institutionalized as the "multiplexer," which we will discuss next.

EXERCISE 8.16

Show how to make a 2-input select with transmission gates. You will need an inverter.

Multiplexers

The 2-input select gate is also known as a 2-input multiplexer. Multiplexers are also available with 4, 8, and 16 inputs (the 4-input variety comes as a dual unit, 2 in one package). A binary address is used to select which of the input signals appears at the output. For instance, an 8-input multiplexer has a 3-bit address input to address the selected data input (Fig. 8.31). The digital multiplexer illustrated is a 74151. It has a STROBE (another name for ENABLE) input (negative-true), and it provides true and complemented outputs. When the chip is disabled (STROBE held HIGH), Q is LOW and Q' is HIGH, independent of the states of the address and data inputs.

In CMOS, two varieties of multiplexer are available. One type is for digital levels only, with an input threshold and "clean" regen-

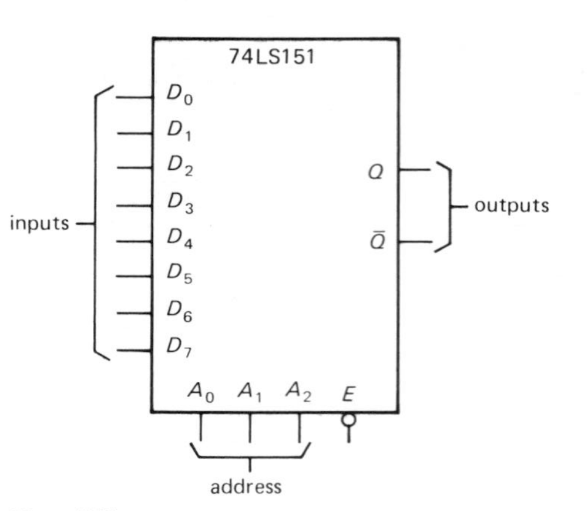

Figure 8.31

eration of output levels according to the input state; that's also the way all TTL functions work. As an example, the 4539 is a CMOS dual 4-input multiplexer, just like the 74153 TTL multiplexer. The other kind of CMOS multiplexer is analog and bidirectional; it's really just an array of transmission gates. The 4051-4053 CMOS multiplexers work this way. Since transmission gates are bidirectional, these multiplexers can be used as "demultiplexers," or decoders. We will discuss them next.

EXERCISE 8.17

Show how to make a 4-input multiplexer using gates.

You might wonder what to do if you want to select among more inputs than are provided in a multiplexer. This question comes under the general category of chip "expansion" (using several chips that have small individual capabilities to generate a larger capability), and it applies to decoders, memories, shift registers, arithmetic logic, and many other functions as well. In this case the job is easy (Fig. 8.32). Here we have expanded two 74LS151 8-input multiplexers into a 16-input multiplexer. There's an additional address bit, of course, and you use it to enable one chip or the other. The disabled chip holds its Q LOW, so an OR gate at the output completes the expansion. With three-state outputs the job is even simpler, since you can connect the outputs directly together.

Demultiplexers and decoders

A demultiplexer takes an input and routes it to one of several possible outputs, according to an input binary address. The other outputs are either held in the inactive state or open-circuited, depending on the type of demultiplexer.

A decoder is similar, except that the address is the only input, and it is "decoded" to assert one of n possible outputs. Figure 8.33 shows an example. This is the 7442 "BCD-to-decimal decoder." The output corresponding to (addressed by) the BCD character at the input is LOW; all others are HIGH. It can be used to drive a decimal display, for instance, or to determine the state of a counter.

In CMOS logic the multiplexers that use transmission gates are also demultiplexers,

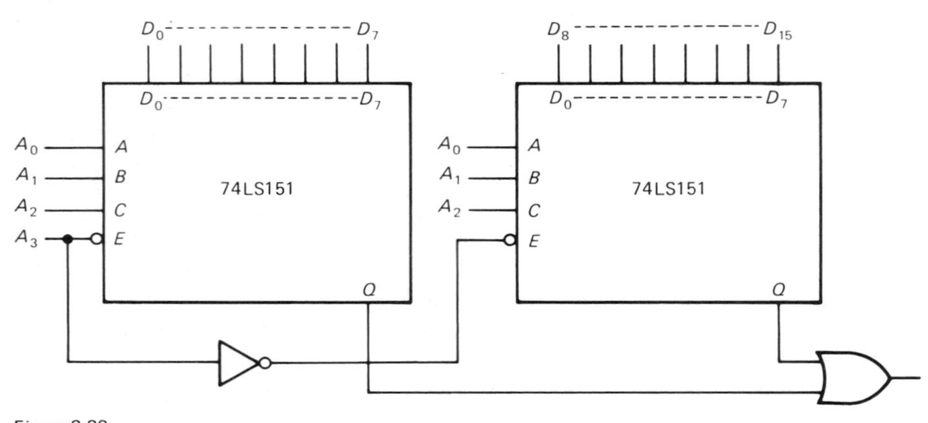

Figure 8.32
Multiplexer expansion.

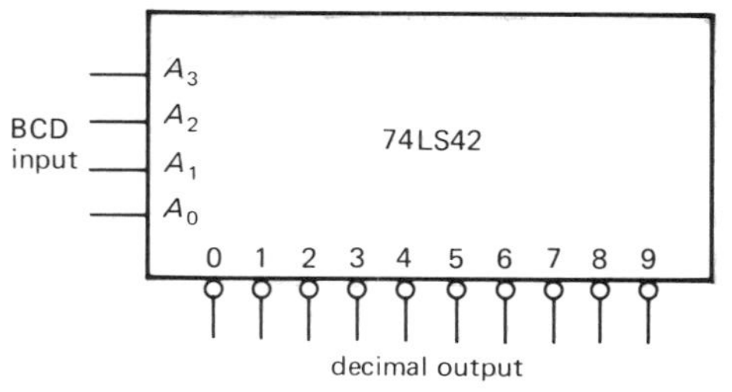

Figure 8.33

since transmission gates are bidirectional. When they are used that way, it is important to realize that the outputs that aren't selected are open-circuited. A pull-up resistor, or equivalent, must be used to assert a well-defined logic level on those outputs (the same requirement as with TTL open-collector gates).

There is another kind of decoder generally available in all logic families. An example is the 7447 "BCD-to-7-segment decoder/driver." It takes a BCD input and generates outputs on 7 lines corresponding to the segments of a "7-segment display" that have to be lit to display the decimal character. This type of decoder is really an example of a "code converter," but in common usage it is called a decoder.

EXERCISE 8.18

Design a BCD-to-decimal deccder using gates.

Priority encoder

The priority encoder generates a binary code giving the address of the highest-numbered input that is asserted. It is particularly useful in "parallel-conversion" A/D converters (see next chapter) and in microprocessor system design. Examples are the CMOS 4532 and the TTL 74148 8-input (3 output bits) priority encoders.

EXERCISE 8.19

Design a "simple" encoder: a circuit that outputs the (2-bit) address telling which of 4 inputs is HIGH (all other inputs must be LOW).

Adders and other arithmetic chips

Figure 8.34 shows a "4-bit full adder." It adds the 4-bit number A_i to the 4-bit

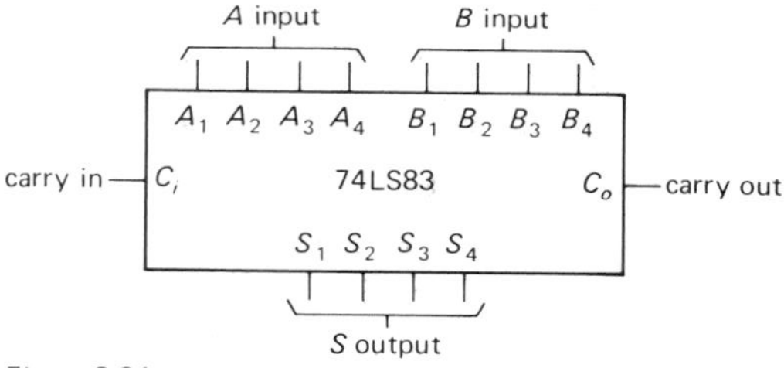

Figure 8.34

number B_i, generating a 4-bit sum S_i plus carry bit C_o. Adders can be "expanded" to add larger numbers: The "carry-in" input C_i is provided, to accept the carry out of the next lower adder. Figure 8.35 shows how you would add two 8-bit numbers.

A device known as an arithmetic logic unit (ALU) is often used as an adder. It actually has the capability of performing a number of different functions. For instance, the 74181 4-bit ALU (expandable to larger word lengths) can do addition, subtraction, bit shifts, magnitude comparison, and a few other functions. Adders and ALUs do their arithmetic in times measured in tens of nanoseconds (TTL types) to hundreds of nanoseconds (CMOS).

Multiplier chips are now available in configurations such as 8 bits times 8 bits, or 16 bits times 16 bits. At the present time they are expensive, but they give excellent performance. Typical speeds are around 100ns for 16×16 multiply.

Magnitude comparators

Figure 8.36 shows a 4-bit magnitude comparator. It determines the relative sizes of the 4-bit input numbers A and B and tells you via outputs whether $A < B$, $A = B$, or $A > B$. Inputs are provided for expansion to numbers larger than 4 bits.

EXERCISE 8.20

Construct a magnitude comparator, using XOR gates, that tells whether or not $A = B$, where A and B are 4-bit numbers.

Parity generator/checker

This chip is used to generate a parity bit to be attached to a "word" when transmitting (or recording) data and to check the received parity when such data is recovered. Parity

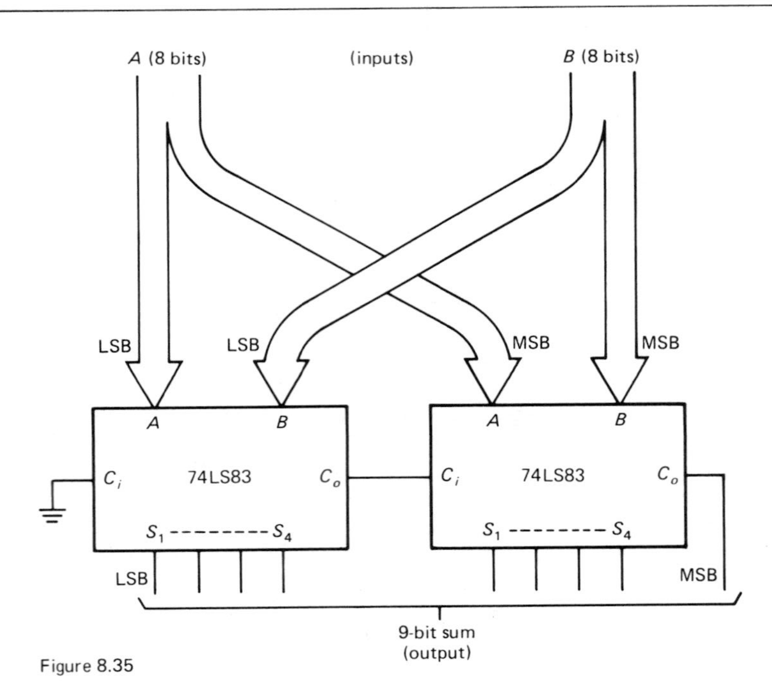

Figure 8.35

Figure 8.36

can be even or odd (e.g., with odd parity the number of 1 bits in each character is odd). The 74180 parity generator, for instance, accepts an 8-bit input word and an even/odd control input, giving an even and an odd parity bit output. The basic construction is an array of exclusive-OR gates.

EXERCISE 8.21

Figure out how to make a parity generator using XOR gates.

□ *Some other strange functions*

There are many other interesting MSI combinational chips worth knowing about. For example, in CMOS you can get a "majority logic" IC that tells you whether or not a majority of n inputs are asserted. Also available is a BCD "9's complementer," whose function is obvious. Signetics has a TTL chip called an 8243 that is a "bit position scaler." It shifts an 8-bit input over by n (selectable) bits; it can be expanded to any width. With time the selection of exotic functions available can only become greater as new requirements require solutions.

8.15 Implementing arbitrary truth tables

The number of gates you need to wire up some complicated truth table can become awfully large. You may begin to ask yourself if there isn't some other way. Fortunately, there are several. In this section we will look at the use of multiplexers and demultiplexers to implement arbitrary truth tables. Then we will talk briefly about ROMs and PLAs, the ultimate combinational functions.

Multiplexers as generalized truth tables

It should be obvious that an n-input multiplexer can be used to generate any n-entry truth table, without any external components, by simply connecting its inputs to HIGH or LOW as required. For example, Figure 8.37 shows a circuit that tells if a 3-bit binary input is prime.

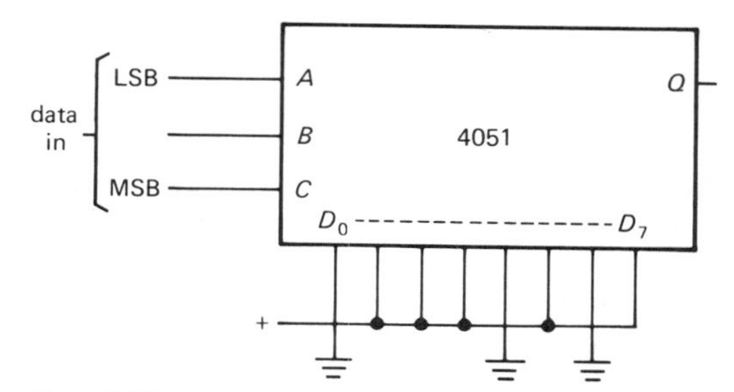

Figure 8.37

What is not so obvious is that an N-input multiplexer can be used to generate any $2N$-entry truth table, with at most one external inverter. The trick is this: Organize the truth table for each permutation of $n - 1$ bits. The output, for each permutation of $n - 1$ bits, can only depend on the state of the nth bit in one of these four ways:

$$Q(D_1, D_2, \ldots, D_{n-1}; D_n) = 0$$
$$Q(D_1, D_2, \ldots, D_{n-1}; D_n) = 1$$
$$Q(D_1, D_2, \ldots, D_{n-1}; D_n) = D_n$$
$$Q(D_1, D_2, \ldots, D_{n-1}; D_n) = D_n'$$

That function of D_n is then applied to the input addressed by $D_1, D_2, \ldots, D_{n-1}$. To illustrate, let's build a circuit that tells whether or not a given month of the year has 31 days, where the month is specified by a 4-bit input. First, we make a truth table (Table 8.4). Then we just read off the appropriate multiplexer inputs by looking at how Q depends on the fourth input bit, D. The circuit realization is shown in Figure 8.38, where we have used a 4051 CMOS 8-input multiplexer. In general, the individual inputs may have to connect to LOW, HIGH, D, or D'.

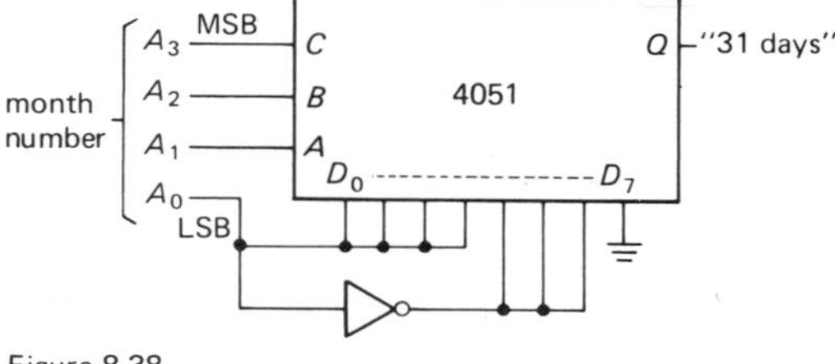

Figure 8.38

EXERCISE 8.22

Verify that the preceding circuit does indeed tell you if a month has 31 days.

Amusing postscript: It turns out that this truth table can be implemented with a single XOR gate, if you take advantage of X's (don't care) for the months that don't exist! Try your hand at this challenge. It will give you a chance to exercise Karnaugh map skills.

TABLE 8.4. DAYS/MONTH LOGIC CALCULATION

Month	No. of month	MUX address A_3	A_2	A_1	Selected input	Address LSB (A_0)	31 days? Q	Q as function of D
—	0	0	0	0	D_0	0	0	D
Jan	1	0	0	0	D_0	1	1	
Feb	2	0	0	1	D_1	0	0	D
Mar	3	0	0	1	D_1	1	1	
Apr	4	0	1	0	D_2	0	0	D
May	5	0	1	0	D_2	1	1	
Jun	6	0	1	1	D_3	0	0	D
Jul	7	0	1	1	D_3	1	1	
Aug	8	1	0	0	D_4	0	1	D'
Sep	9	1	0	0	D_4	1	0	
Oct	10	1	0	1	D_5	0	1	D'
Nov	11	1	0	1	D_5	1	0	
Dec	12	1	1	0	D_6	0	1	D'
—	13	1	1	0	D_6	1	0	
—	14	1	1	1	D_7	0	0	0
—	15	1	1	1	D_7	1	0	

Decoders as generalized truth tables

Decoders also provide a nice shortcut for combinational logic, particularly in situations where you need several simultaneous outputs. As an example, let's generate a circuit to convert BCD to excess-3. Here's the truth table:

Decimal	BCD	XS3	Decimal	BCD	XS3
0	0000	0011	5	0101	1000
1	0001	0100	6	0110	1001
2	0010	0101	7	0111	1010
3	0011	0110	8	1000	1011
4	0100	0111	9	1001	1100

We use the 4-bit (BCD) input as an address to the decoder, then use the (negative-true) decoded outputs as inputs to several OR gates, one for each output bit, as shown in Figure 8.39. Note that with this scheme the output bits don't have to be mutually exclusive. You might use something like this as a cycle controller for a washing machine, in which you turn on several functions (pump out water, fill, spin, etc.) at each input state. You will see shortly how to generate equally timed consecutive binary codes. The individual outputs from the decoder are known as *minterms,* and they correspond to positions on a Karnaugh map.

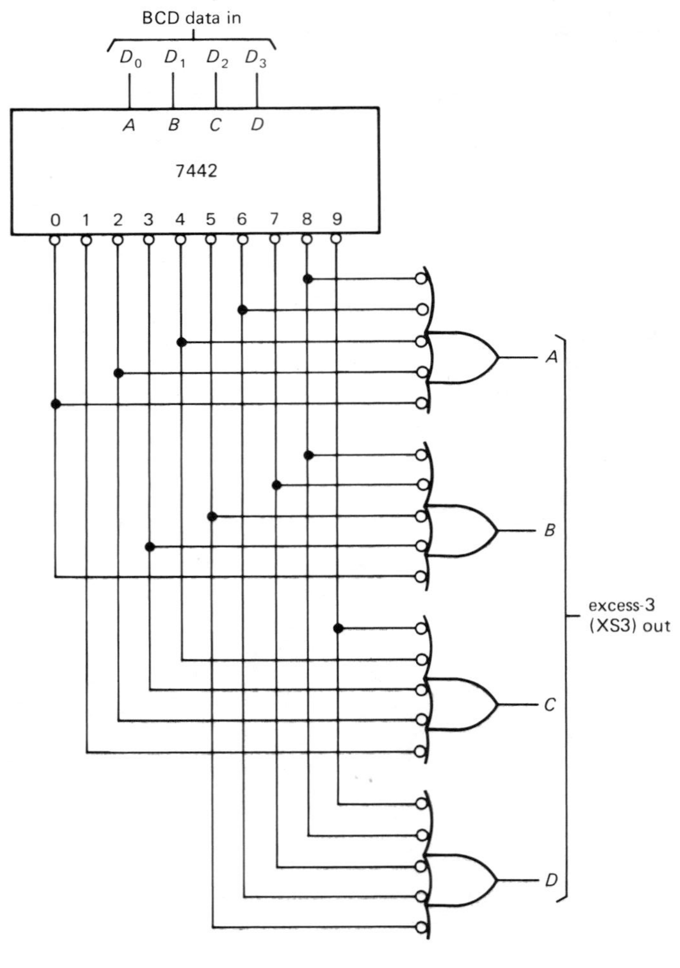

Figure 8.39
Minterm code conversion: BCD to excess-3.

□ *ROM and PLA: the ultimate combinational functions*

The "read-only memory," or ROM, is really an example of a device with memory and should properly be discussed later, along with flip-flops, registers, etc. The special feature of the ROM is that it is permanently programmed with a particular bit pattern, and it retains that information even when the power is removed. Thus, a 32 × 8 ROM, for instance, gives 8 output bits for each of 32 input states, specified by a 5-bit input address. Any combinational truth table can therefore be programmed into a ROM. For example, the 2-bit by 2-bit multiplier in Exercise 8.14 could be implemented with a single 74188 programmable ROM, using 4 of the 5 input "address" lines and 4 of the 8 output "data" lines, if suitably programmed. ROMs are available in various configurations, such as 1024 × 4 (4 output bits for each 10-bit address), up to memories as large as 8K × 8. More on them in Chapter 11.

ROMS come in several varieties, according to their method of programming. "Mask-programmed ROMs" have their bit pattern built in at the time of manufacture. "Programmable ROMs" (PROMs) are programmed by the user; once programmed, their pattern is permanent. "Erasable programmable ROMs" (EPROMs) are PROMs that can be erased if necessary (they have a transparent quartz window, and they are erased by exposing them to intense ultraviolet light for a few minutes). "Electrically alterable ROMs" (EAROMs) can be programmed and erased electrically, while connected in the circuit.

ROMs find extensive use in computer and microprocessor applications, where they are used to store finished programs. However, you should keep the smaller ROMs in mind as replacements for complicated arrays of gates.

The "programmable logic array," or PLA, is another device worth knowing about if you need an elaborate array of combinational logic. It is a chip with many gates that can be interconnected to form the desired logic functions. PLAs come in mask-programmable and field-programmable varieties. The former are programmed at the factory to meet the user's specifications; the latter are user-programmed by fusing internal links that initially interconnect the internal gates. As an example, the 82S100 is a field-programmable PLA that can generate 8 outputs, each of which can be a programmed logic combination of 16 inputs. The 82S200 is the mask-programmable version with the same capabilities.

SEQUENTIAL LOGIC

8.16 Devices with memory: flip-flops

All our work with digital logic so far has been with combinational circuits (e.g., arrays of gates), for which the output is determined completely by the existing state of the inputs. There is no "memory," no history, in these circuits. Digital life gets really interesting when we add devices with memory. This makes it possible to construct counters, arithmetic accumulators, and circuits that generally do one interesting thing after another. The basic unit is the flip-flop, a colorful name to describe a device that, in its simplest form, looks as shown in Figure 8.40.

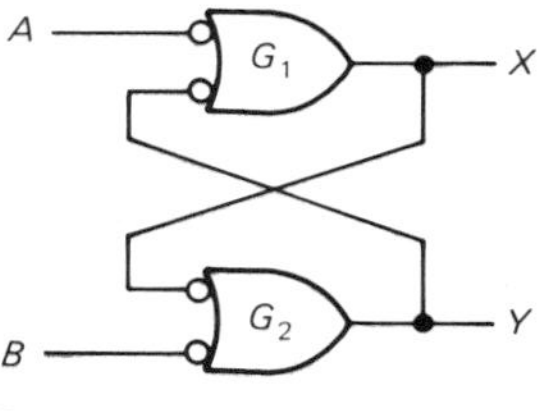

Figure 8.40

Assume that both A and B are HIGH. What are X and Y? If X is HIGH, then both inputs of G_2 are HIGH, making Y LOW. This is consistent with X being HIGH, so we're finished. Right?

X = HIGH
Y = LOW

Wrong! The circuit is symmetrical, so an equally good state is

X = LOW
Y = HIGH

The states X, Y both LOW and X, Y both HIGH are not possible (remember, $A = B =$ HIGH). So the flip-flop has two stable states (it's sometimes called a "bistable"). Which state it is in depends on past history. It has memory! To write into the memory, just bring one of the inputs momentarily LOW. For instance, bringing A LOW momentarily guarantees that the flip-flop goes into the state

$X =$ HIGH
$Y =$ LOW

no matter what state it was in previously.

Switch debouncing

This kind of flip-flop (with a SET and RESET input) is quite useful in many applications. Figure 8.41 shows a typical example. This circuit is supposed to enable the gate and pass input pulses when the switch is

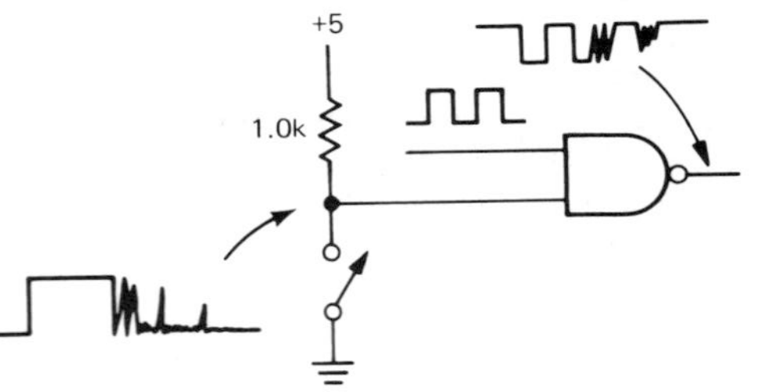

Figure 8.41

opened. The switch is tied to ground (not +5), because you must *sink* current from a TTL input in the LOW state, whereas in the HIGH state you only need to overcome leakage current (40μA maximum, for standard TTL). Besides, ground is generally available as a convenient return for switches and other controls. The problem with this circuit is that switch contacts "bounce." When the switch is closed, the two contacts actually separate and reconnect, typically 10 to 100 times over a period of about 1ms. You would get waveforms as sketched; if there were a counter or shift register using the output, it would faithfully respond to all those extra "pulses" caused by the bounce.

Figure 8.42 shows the cure. The flip-flop changes state when the contacts first close. Further bouncing against that contact makes no difference (SPDT switches never bounce

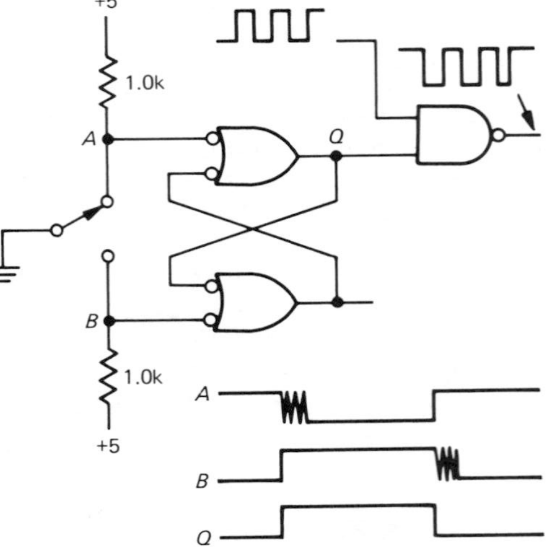

Figure 8.42

all the way back to the opposite position), and the output is a "debounced" signal, as sketched. This debouncer circuit is widely used, and a "quad switch debouncer" (with internal pull-up resistors and three-state outputs) is available in a single package (National DM8544). Incidentally, the preceding circuit has a minor flaw: The first pulse after the gate is enabled may be shortened, depending on when the switch is closed relative to the input pulse train; the same holds for the final pulse of a sequence (of course, a switch that is not debounced has the same problem). A "synchronizer" circuit (see Section 8.19) can be used to prevent this from happening, for applications where it makes a difference.

Multiple-input flip-flop

Figure 8.43 shows another simple flip-flop. Here NOR gates have been used; a HIGH input forces the corresponding output LOW. Multiple inputs allow various signals to set or

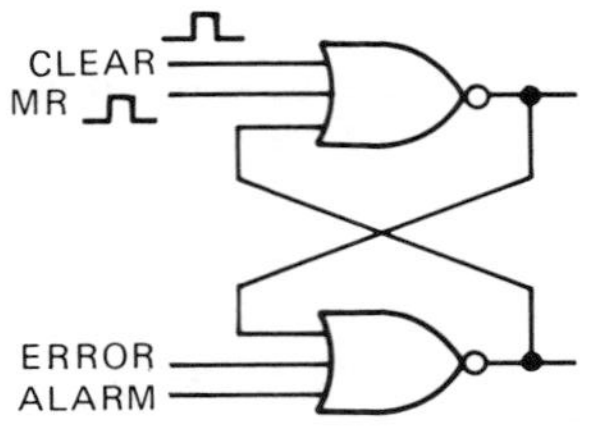

Figure 8.43

clear the flip-flop. In this circuit fragment, no pull-ups are used, since logic signals generated elsewhere (by standard active pull-up outputs) are used as inputs.

8.17 Clocked flip-flops

Flip-flops made with two gates, as in Figure 8.43, are known generically as *SR* (set-reset), or jam-loaded, flip-flops. You can force them into one state or the other whenever you want by just generating the right input signal. They're handy for switch debouncing and many other applications. But the most widely used form of flip-flop looks a little different. Instead of a pair of jam inputs, it has one or two "data" inputs and a single "clock" input. The outputs can change state or stay the same, depending on the levels at the data inputs when the clock pulse arrives.

The simplest clocked flip-flop looks as shown in Figure 8.44. It's just our original

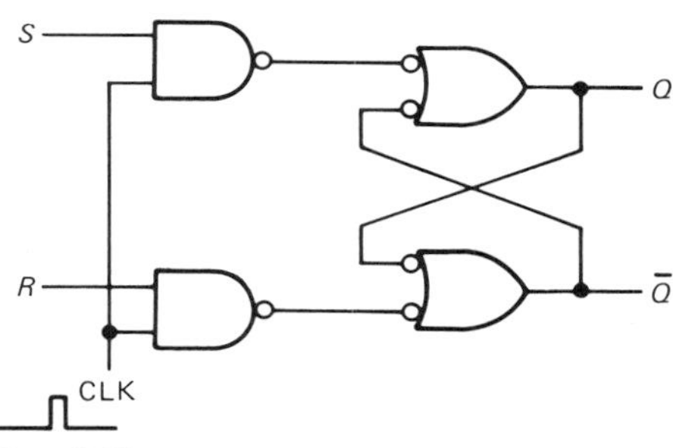

Figure 8.44

flip-flop, with a pair of gates (controlled by the clock) to enable the set and reset inputs. It is easy to verify that the truth table is

S R	Q_{n+1}
0 0	Q_n
0 1	0
1 0	1
1 1	indeterminate

where Q_{n+1} is the Q output after the clock pulse and Q_n is the output before the clock pulse. The basic difference between this and the previous flip-flops is that R and S should now be thought of as data inputs. What is present on R and S when a clock pulse comes along determines what happens to Q.

This flip-flop has one awkward property, however. The output can change in response to the inputs during the time the clock is HIGH. In that sense it is still like the jam-loaded *SR* flip-flop (it's also known as a "transparent latch," since the output "sees through" to the input when the clock is HIGH). The full utility of clocked flip-flops comes with the introduction of slightly different configurations, the master-slave flip-flop and the edge-triggered flip-flop.

Master-slave and edge-triggered flip-flops

These are by far the most popular flip-flops. The data present on the input lines just before a clock transition, or "edge," determine the output state after the clock has changed. These flip-flops are available as inexpensive packaged ICs and are always used in that form. But it is worth looking at their innards in order to understand what is going on. Figure 8.45 shows the schematics. Both are known as type D flip-flops. Data present at the D input will be transferred to the Q output after the clock pulse. The master-slave configuration is probably easier to understand. Here's how it works:

While the clock is HIGH, gates 1 and 2 are enabled, forcing the master flip-flop (gates 3 and 4) to the same state as the D input: $M = D$, $M' = D'$. Gates 5 and 6 are disabled, so the slave flip-flop (gates 7 and 8) retains its previous state. When the clock goes LOW, the inputs to the master are disconnected from the D input, while the inputs of the slave are simultaneously coupled to the outputs of the master. The master thus transfers its state to the slave. No further changes can occur at the output, because the master is now stuck. At the next rising edge of the clock, the slave will be decoupled from the master and will retain its state, while the master will once again follow the input.

The edge-triggered circuit behaves the same externally, but the inner workings are different. It is not difficult to figure it out. The particular circuit shown happens to be the popular 7474 TTL positive-edge-trig-

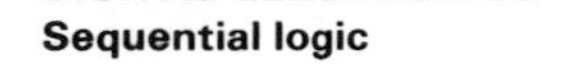

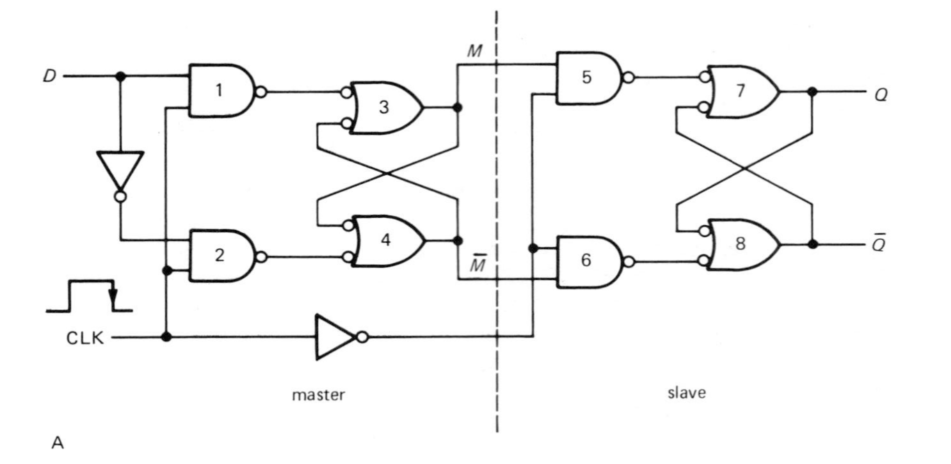

A

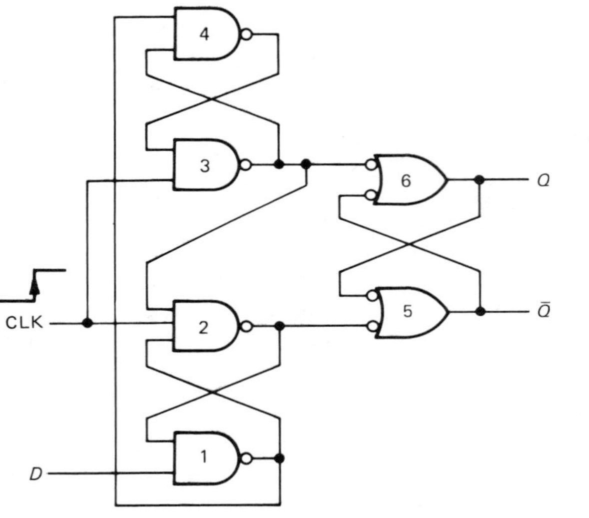

B

Figure 8.45
Edge-triggered type-*D* flip-flops.

gered type *D* flip-flop. The preceding master-slave circuit transfers data to the output on the negative edge. Flip-flops are available with either positive or negative edge triggering. In addition, most flip-flops also have SET and CLEAR jam-type inputs. They may be set and cleared on HIGH or on LOW, depending on the type of flip-flop. Figure 8.46 shows a few popular flip-flops. The wedge means "edge-triggered," and the little circles mean "negation," or

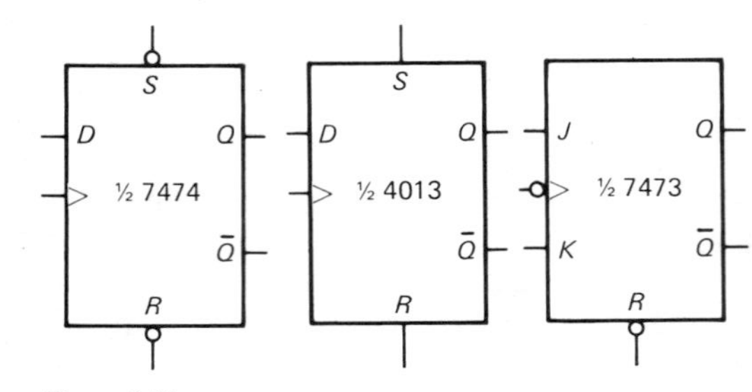

Figure 8.46

complement. Thus, the 7474 is a TTL dual type D positive-edge-triggered flip-flop with active LOW jam-type SET and CLEAR inputs. The 4013 is a CMOS dual type D positive-edge-triggered flip-flop with active HIGH jam-type SET and CLEAR inputs. The 7473 is a TTL dual JK master-slave flip-flop with data transfer on the negative edge and with an active LOW jam-type CLEAR input.

The JK flip-flop. The JK flip-flop works on principles similar to those of the type D flip-flop, but it has two data inputs. Here's the truth table:

J	K	Q_{n+1}
0	0	Q_n
0	1	0
1	0	1
1	1	Q_n'

Thus, if J and K are complements, Q will go to the value of the J input at the next clock edge. If J and K are both LOW, the output won't change. If J and K are both HIGH, the output will "toggle" (reverse its state after each clock pulse).

Warning: Some older JK flip-flops are "ones-catching," a term you won't find in the data sheet, but an effect that can have dire consequences for the unsuspecting. This means that if either J or K (or both) changes state momentarily while the slave is enabled by the clock, then returns to its previous state before the clock makes its transition, the flip-flop will "remember" that momentary state and behave as if that state had persisted. Thus the flip-flop may change state at the next clock transition even if the J and K inputs existing at that transition should cause the flip-flop to remain in its present state. This can lead to peculiar behavior, to put it mildly. The problem arises because such flip-flops were designed with short clock pulses in mind, whereas in common usage you clock flip-flops with just about anything. Be careful when using master-slave flip-flops, or avoid them altogether and use true edge-triggered flip-flops instead.

Divide by 2

It is easy to make a divide-by-2 circuit by just exploiting the toggling capability of flip-flops. Figure 8.47 shows two ways. The JK flip-flop toggles when both inputs are HIGH, producing the output shown. The second circuit also toggles, since with the D input tied to its own Q' output, the type D flip-flop always sees the complement of its existing output at its D input as the time of the clock pulse. The output signal in either case is at half the frequency of the input.

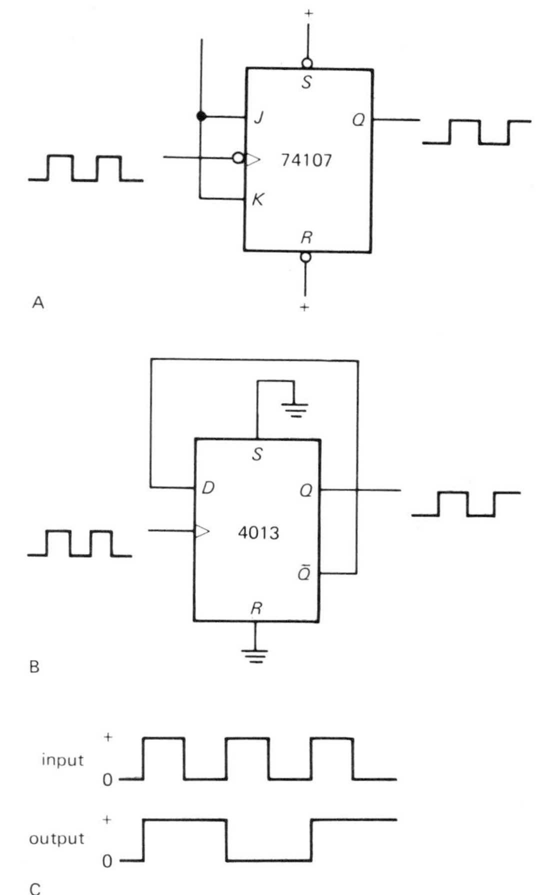

Figure 8.47

Data and clock timing

This last circuit raises an interesting question: Will the circuit fail to toggle, since the D input changes almost immediately after the clock pulse? In other words, will the circuit get confused, with such crazy things happening at its input? You could, instead, ask this question: Exactly *when* does the D flip-flop (or any other flip-flop) look at its

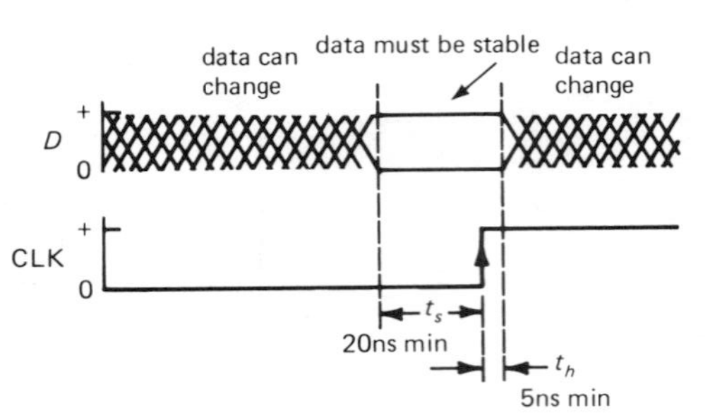

Figure 8.48
Data setup and hold times.

input, relative to the clock pulse? The answer is that there is a specified "setup time" t_s and "hold time" t_h for any clocked device. Input data must be present and stable from at least t_s before the clock transition until at least t_h after it, for proper operation. For the 7474, for instance, t_s = 20ns and t_h = 5ns (Fig. 8.48). So, for the preceding toggling connection, the setup-time requirement is met if the output has been stable for at least 20ns before the next clock rising edge. It may look as if the hold-time requirement is violated, but that's OK, also. The minimum "propagation time" from clock to output is 10ns, so a *D* flip-flop connected to toggle as described is guaranteed to have its *D* input stable for at least 10ns after the clock transition. Most devices nowadays have a zero hold-time requirement.

An interesting thing can happen if the level at the *D* input changes during the setup-time interval, namely a so-called *metastable* state in which the flip-flop can't make up its mind which state to go into. We will have more to say about this shortly.

Divide by more

By cascading several toggling flip-flops (connect each *Q* output to the next clock input), it is easy to make a divide-by-2^n, or binary, counter. Figure 8.49 shows a four-stage "ripple counter" and its waveforms. Note that flip-flops that clock on the falling edge

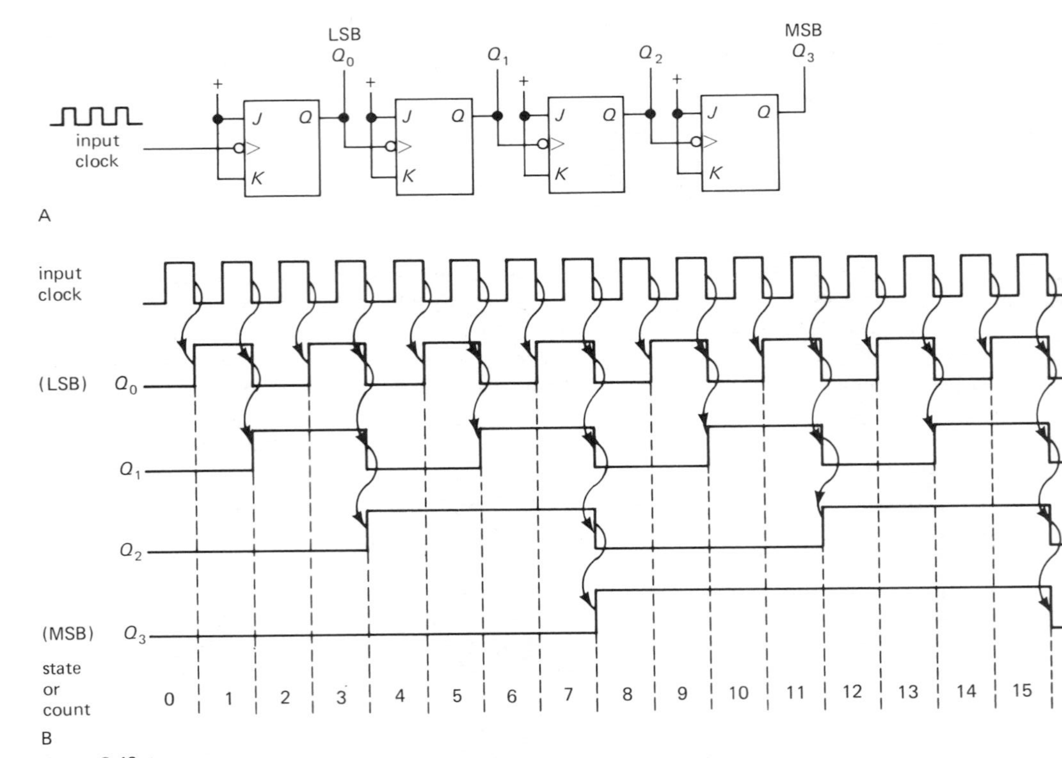

Figure 8.49
Four-bit counter. A: Schematic. B: Timing diagram.

(indicated by the negation circle) must be used if each Q output drives the next clock input. This circuit is a divide-by-16 counter: The output waveform from the last flip-flop is a square wave whose frequency is 1/16 of the circuit's input clock frequency. Such a circuit is called a *counter* because the data present at the four Q outputs, considered as a single 4-bit binary number, go through a binary sequence from 0 to 15, incrementing after each input pulse. The waveforms in Figure 8.49 demonstrate this fact. In the figure the abbreviation MSB is used to mean "most significant bit," and LSB means "least significant bit"; the curved arrows are used to indicate what causes what, to aid in understanding.

As you will see in Section 8.25, the counter is such a useful function that many versions are available integrated onto single chips, including 4-bit, BCD, and multidigit counting formats. By cascading several such counters and displaying the count on a numeric display device (e.g., an LED digital display) you can easily construct an event counter. If the input pulse train to such a counter is gated for exactly 1 second, you've got a frequency counter, which displays frequency (cycles per second) by actually counting the number of cycles in a second. Section 14.10 shows diagrams of this simple and highly useful scheme. In fact, single-chip frequency counters are available, complete with oscillator, counter, control, and display circuitry; see Figure 8.65 for an example.

In practice, the simple scheme of cascading counters by connecting each Q output to the next clock input has some interesting problems related to the cascaded delays as the signal "ripples" down through the chain of flip-flops, and a "synchronous" scheme (in which all clock inputs see the same clocking signal) is usually better. Let's look into this question of synchronous clocked systems.

8.18 Combining memory and gates: sequential logic

Having explored the properties of flip-flops, let's see what can be done when they are combined with the combinational (gate) logic we discussed earlier. Circuits made with gates and flip-flops constitute the most general form of digital logic.

Synchronously clocked systems

As we hinted in the last section, sequential logic circuits in which there is a common source of clock pulses driving all flip-flops have some very desirable properties. In such a *synchronous system* all action takes place just after each clocking pulse, based on the levels present just before each clock pulse. Figure 8.50 shows the general scheme.

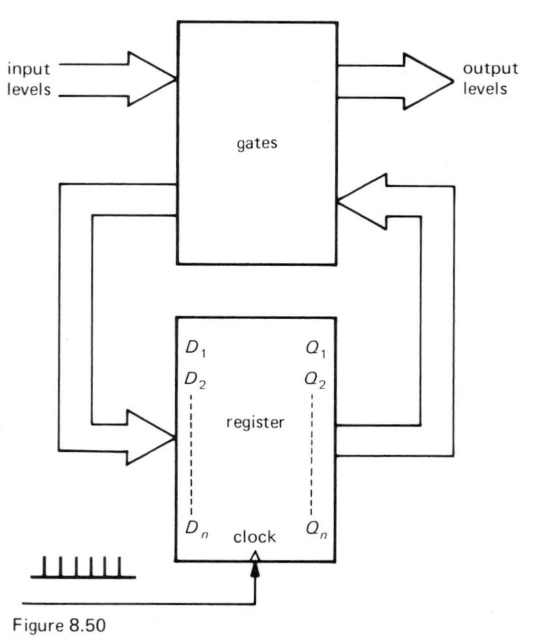

Figure 8.50
The classic sequential circuit: a set of memory registers combined with combinational logic.

The flip-flops have all been combined into a single *register,* which is nothing more than a set of type D flip-flops with their clock inputs all tied together and their individual D inputs and Q outputs brought out; i.e., each clock pulse causes the levels present at the D inputs to be transferred to the respective Q outputs. The box full of gates looks at both the Q outputs and whatever input levels are applied to the circuit and generates a new set of D inputs and logic outputs. This simple-looking scheme is extremely powerful. Let's look at an example.

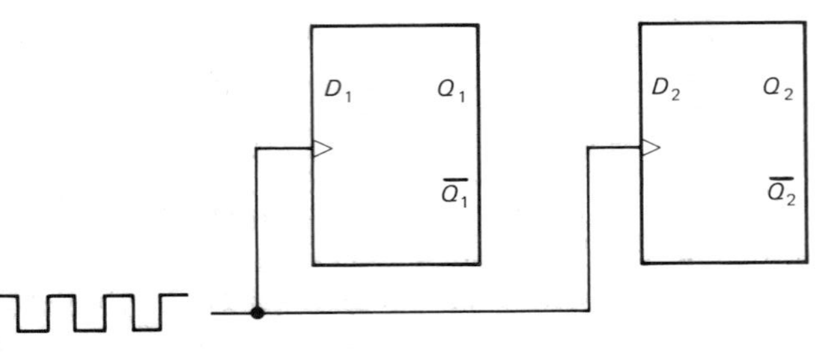

Figure 8.51

Example: divide by 3

Let's design a synchronous divide-by-3 circuit with two type *D* flip-flops, both clocked from the input signal. In this case, D_1 and D_2 are the register inputs, Q_1 and Q_2 are the outputs, and the common clock line is the master clocking input (Fig. 8.51).

1. Choose the three states. Let's use

Q_1	Q_2	
0	0	
0	1	
1	0	
0	0	(i.e., first state)

2. Find the combinational logic network outputs necessary to generate this sequence of states, i.e., figure out what the *D* inputs have to be to get those outputs:

Q_1	Q_2	D_1	D_2
0	0	0	1
0	1	1	0
1	0	0	0

3. Concoct suitable gating (combinational logic), using available outputs, to produce those *D* inputs. In general, you can use a Karnaugh map. In this simple case you can see by inspection that

$$D_1 = Q_2$$

$$D_2 = (Q_1 + Q_2)'$$

from which the circuit of Figure 8.52 follows.

It is easy to verify that the circuit works as planned. Since it is a synchronous counter, all outputs change simultaneously (when you feed one output to the next clock, you've got a *ripple* counter instead). In general, synchronous (or "clocked") systems are desirable, since susceptibility to noise is improved: Things have settled down by the time of the clock pulse, so circuits that only look at their inputs at clock edges aren't troubled by capacitively coupled interference from other flip-flops, etc. A further advantage of clocked systems is that transient states (caused by delays, so that all outputs don't change simultaneously) don't produce false outputs, since the system is insensitive to what happens just *after* a clock pulse. You will see some examples later.

Excluded states

What happens to the divide-by-3 circuit if the flip-flops somehow get into the state

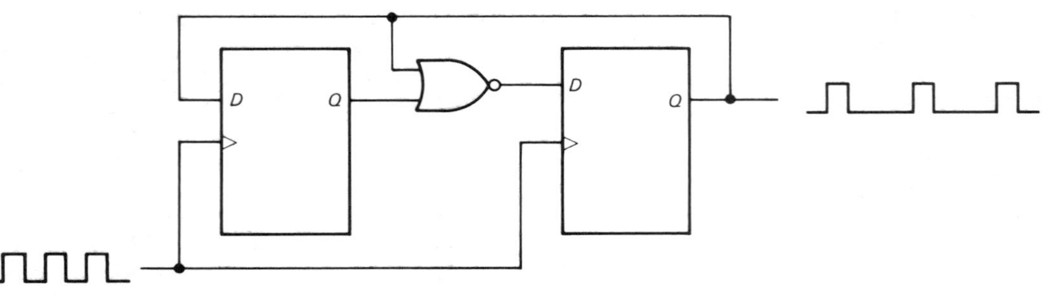

Figure 8.52

$(Q_1,Q_2) = (1,1)$? This can easily happen when the circuit is first turned on, since the initial state of a flip-flop is anyone's guess. From the diagram, it is clear that the first clock pulse will cause it to go to the state (1,0), from which it will function as before. It is important to check the excluded states of a circuit like this, since it is possible to be unlucky and have it get stuck in one of those states. (Alternatively, the initial design procedure can include a specification of all possible states.) A useful diagnostic tool is the *state diagram,* which for this example looks like Figure 8.53. Usually you write the

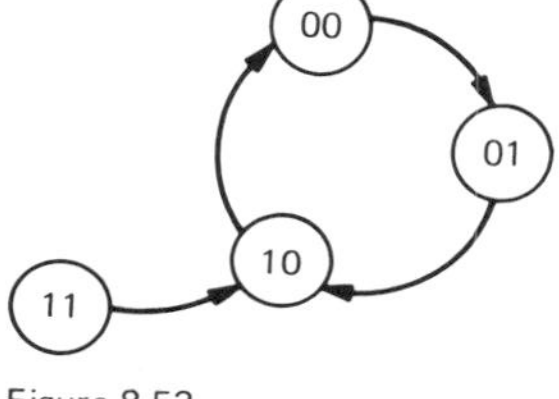

Figure 8.53

conditions for each transition next to the arrows, if other variables of the system are involved. Arrows may go in both directions between states, or from one state to several others.

EXERCISE 8.23

Design a synchronous divide-by-3 circuit using two *JK* flip-flops. It can be done (in 16 different ways!) without any gates or inverters. One hint: When you construct the table of required J_1,K_1 and J_2,K_2 inputs, keep in mind that there are two possibilities for J,K at each point. For instance, if a flip-flop output is to go from 0 to 1, $J,K = 1,X$ (X = don't care). Finally, check to see if the circuit will get stuck in the excluded state (of the 16 distinct solutions to this problem, 4 will get stuck and 12 won't).

State diagrams as design tools

The state diagram can be very useful when designing sequential logic, particularly if the states are connected together by several paths. In this design approach, you begin by selecting a set of unique states of the system, giving each a name (i.e., a binary address). You will need a minimum of n flip-flops, or bits, where n is the smallest integer for which 2^n is equal to or greater than the number of distinct states in the system. Then you set down all the rules for moving between states, i.e., all possible conditions for entering and leaving each state. From there it is a straightforward (but perhaps tedious) job to generate the necessary combinational logic, since you have all possible sets of Qs and the set of Ds that each leads to. Thus you have converted a sequential design problem into a combinational design problem, always soluble through techniques such as the Karnaugh map. Figure 8.54 shows a real-world example. Note that there may be states that don't lead to others, e.g., "receive diploma."

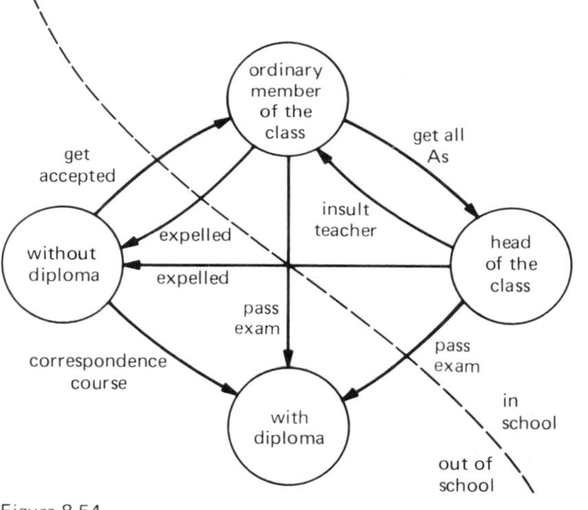

Figure 8.54
State diagram.

8.19 Synchronizer

An interesting application of flip-flops in sequential circuits is their use in a *synchronizer.* Suppose you have some external control signal coming into a synchronous system that has clocks, flip-flops, etc., and you want to use the state of that input signal to control some action. For example, a signal from an instrument or experiment might signify that data is ready to be sent to a computer. Since the experiment and the computer march to the beats of different drummers, so to speak (in fancy language you would say they are *asynchronous*

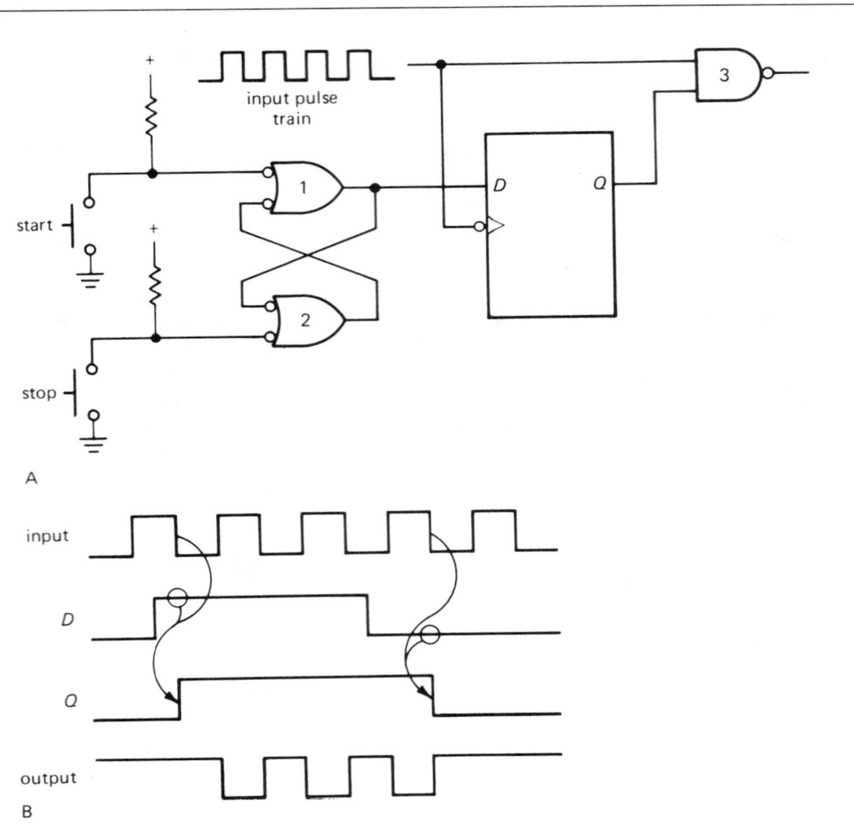

Figure 8.55
Pulse-train synchronizer.

processes), you need a method to restore order between the two systems.

Example: pulse synchronizer

As an example, let's reconsider the circuit in which a debouncer flip-flop gated a pulse train (Section 8.16). That circuit enables the gate whenever the switch is closed, regardless of the phase of the pulse train being gated, so that the first or last pulse may be shortened. The problem is that the switch closure is asynchronous with the pulse train. In some applications it is important to have only *complete* clock cycles, and that requires a synchronizer circuit like that in Figure 8.55. Pushing START brings the output of gate 1 HIGH, but Q stays LOW until the next falling edge of the input pulse train. In that way only complete pulses are passed by NAND gate 3. Figure 8.55 shows some waveforms. The curved arrows are drawn to show exactly what causes what. You can see, for instance, that the transitions of Q occur slightly *after* the falling edges of the input.

Logic races and glitches

This example brings up a subtle but extremely important point: What would happen if a positive-edge-triggered flip-flop were used instead? If you analyze it carefully, you'll find that START still works OK, but if STOP is pushed while the input is LOW, a bad thing happens (Fig. 8.56). A short spike, or "glitch," gets through because the final NAND gate isn't disabled until the flip-flop output has a chance to go LOW, a delay of about 20ns for TTL or 100ns (or more) for CMOS. This is a classic example of a "logic race." With some care these situations can be avoided, as the example shows. Glitches are terrible things to have running through your circuits. Among other things, they're hard to see on an oscilloscope, and

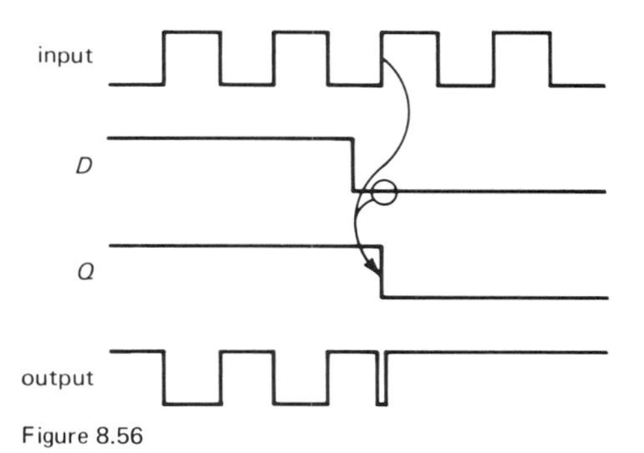

Figure 8.56

you may not know they are present. They can clock subsequent flip-flops erratically, and they may be widened or narrowed to extinction by passage through gates and inverters.

EXERCISE 8.24

Demonstrate that the preceding pulse synchronizer circuit (Fig. 8.55) does not generate glitches.

A few comments about synchronizers: The input to the *D* flip-flop can come from other logic circuitry, rather than from a debounced switch. There are applications in computer interfacing, etc., where an asynchronous signal must communicate with a clocked device; in such cases clocked flip-flops or synchronizers are ideal. In this circuit, as in all logic, unused inputs must be handled properly. For instance, SET and CLEAR must be connected so that they are not asserted (for a 7474, tie them HIGH; for a 4013, they are grounded). Unused inputs that have no influence on the outputs can be left unconnected (e.g., inputs to unused gates), except in CMOS, where they should be grounded to prevent output-stage current (more on that in Chapter 9). A dual synchronizer is available as the 74120, although it has not been widely used.

MONOSTABLE MULTIVIBRATORS

The monostable multivibrator, or "one-shot" (emphasis on the word "one"), is a variation of the flip-flop (which is sometimes called a bistable multivibrator) in which the output of one of the gates is capacitively coupled to the input of the other gate. The result is that the circuit sits in one state. If it is forced to the other state by a momentary input pulse, it will return to the original state after a delay time determined by the capacitor value and the circuit parameters (input current, etc.). It is very useful (some would say *too* useful!) for generating pulses of selectable width and polarity. Making one-shots with gates and *RC*s is tricky, and it depends on the details of the gate's input circuit, since, for instance, you wind up with voltage swings beyond the supply voltages. Rather than encourage bad habits by illustrating such circuits, we will just treat the one-shot as an available functional unit. In actual circuits it is best to use a packaged one-shot; you construct your own only if absolutely necessary, e.g., if you have a gate available and no room for an additional IC package (even then, maybe you shouldn't).

8.20 One-shot characteristics

Inputs

One-shots are triggered by a rising or falling edge at the appropriate inputs. The only requirement on the triggering signal is that it have some minimum width, typically 25ns to 100ns. It can be shorter or longer than the output pulse. In general, several inputs are provided so that several signals can trigger the one-shot, some on positive edges and some on negative edges (remember, a negative edge means a HIGH-to-LOW transition, not a negative polarity). The extra inputs can also be used to inhibit triggering. Figure 8.57 shows four examples.

Each horizontal row of the table represents a valid input triggering transition. For example, the 74121 will trigger when one of the *A* inputs makes a HIGH-to-LOW transition, if the *B* input and the other *A* input are both HIGH. The 9602 is a dual monostable with OR gating at the input; if only one input is used, the other must be disabled, as shown. The 74121 has three inputs, with a combination of OR and AND gating (and triggering), as shown. Its *B* input is a Schmitt trigger, more forgiving with slowly rising or noisy input signals. This monostable also includes a not-too-good internal

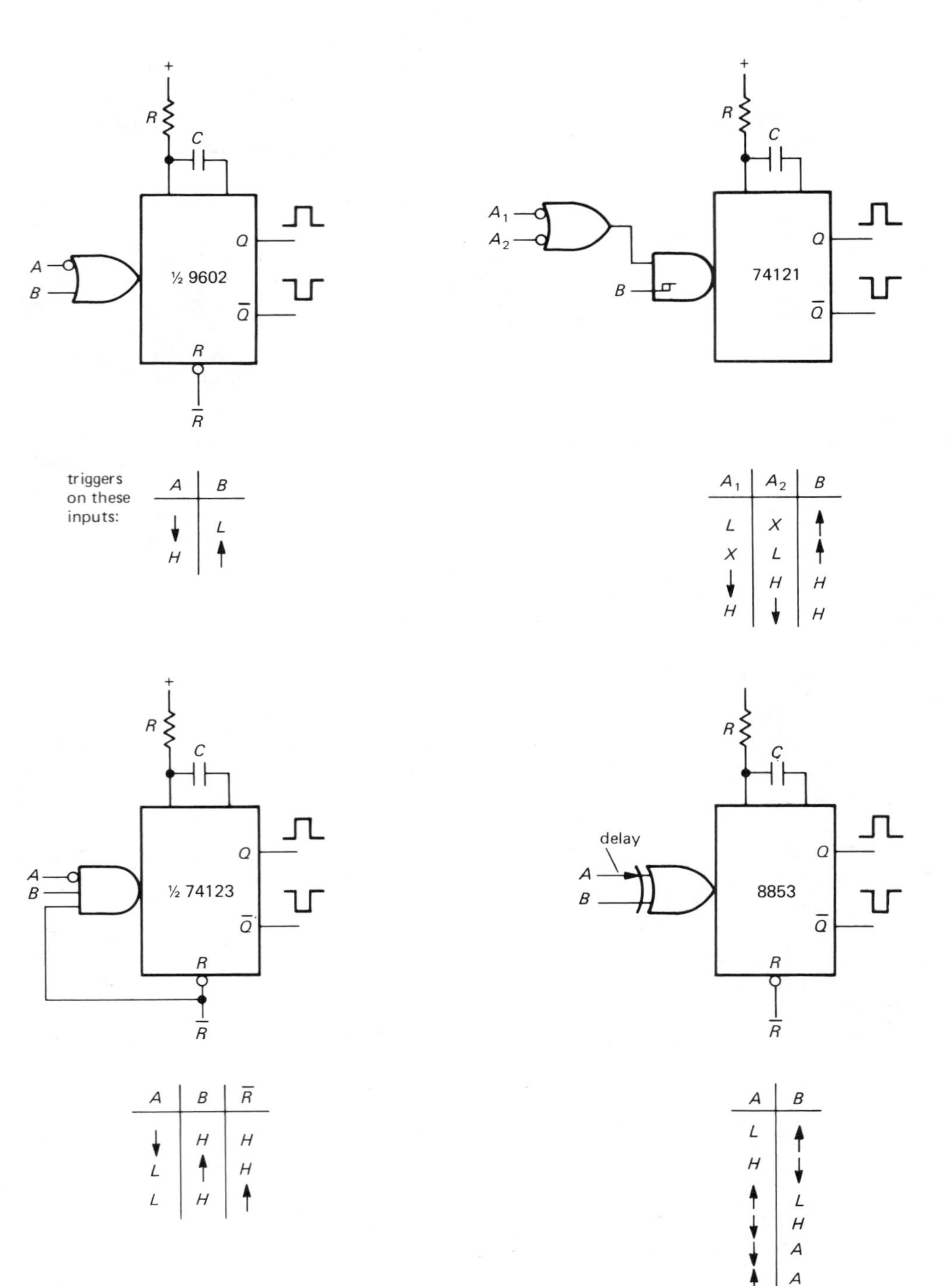

A	B
↓	L
H	↑

A_1	A_2	B
L	X	↑
X	L	↑
↓	H	H
H	↓	H

A	B	$\overline{R}$
↓	H	H
L	↑	H
L	H	↑

A	B
L	↑
H	↓
↑	L
↓	H
↓	A
↑	A

Figure 8.57
Four popular one-shots with their truth tables.

timing resistor you can use instead of *R*, if you're feeling lazy. The popular 74123 is a dual monostable with AND input gating; unused inputs must be enabled. Note particularly that it triggers when RESET is disabled if both trigger inputs are already asserted. This is not a universal property of monostables, and it may or may not be desirable in a given application (it's usually not). The 8853 has an unusual input circuit consisting of a delay and an exclusive-OR. A logic level on one input determines the edge polarity for triggering at the other input. With the two inputs connected together, it triggers on *both* edge polarities.

When drawing monostables in a circuit diagram, the input gating is usually omitted, saving space and creating a bit of confusion.

Retriggerability

Most monostables, e.g., the 9602, 74123, and 8853 mentioned earlier, and the 4098 (CMOS), will begin a new timing cycle if the input triggers again during the duration of the output pulse. They are known as retriggerable monostables. The output pulse will be longer than usual if they are retriggered during the pulse, finally terminating one pulse width after the last trigger. The 74121 is nonretriggerable; it ignores input transitions during the output pulse. Most retriggerable one-shots can be connected as nonretriggerable one-shots. Figure 8.58 shows an example that's easy to understand.

Resettability

Most monostables have a jam RESET input that overrides all other functions. A momentary input to the RESET terminal terminates the output pulse. The RESET input can be used to prevent a pulse during power-up of the logic system; however, see the preceding comment about the 74123.

Pulse width

Pulse widths from 40ns up to milliseconds (or even seconds) are attainable with standard monostables, set by an external capacitor and (usually) resistor combination. A device like the 555 (Section 4.13) can be used to generate longer pulses, but its input properties are sometimes inconvenient. Very long delays are best generated digitally (see Section 8.23).

8.21 Monostable circuit example

Figure 8.59 shows a square-wave generator with independently settable rate and duty cycle (ratio of HIGH to LOW) and an input that permits an external signal to "hold" the output following a negative edge. Current mirror Q_1-Q_3 generates a ramp at C_1. When it reaches the threshold of the upper comparator at two-thirds V_+, the one-shot is triggered and generates a 2μs positive pulse, putting *n*-channel VFET Q_4 into conduction and discharging the capacitor. C_1 therefore has a sawtooth waveform going from ground to +8 volts, with rate set by potentiometer R_2. The lower comparator generates an output square wave from the sawtooth, with duty cycle adjustable linearly between 1% and 99% via R_5. Both comparators have a few millivolts of hysteresis (R_8 and R_9) to prevent noise-induced multiple transitions. The LM393 is a low-power dual comparator with input common-mode range right down to ground and open-collector outputs.

A feature of this circuit is its ability to synchronize (start/stop) to an externally applied control level. The HOLD input lets

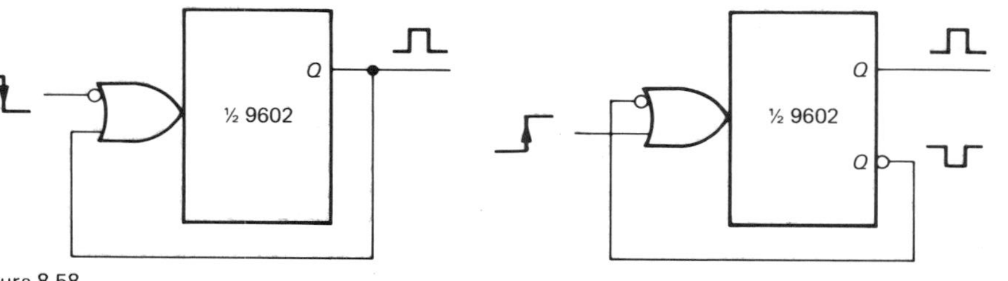

Figure 8.58
Nonretriggerable one-shot circuits.

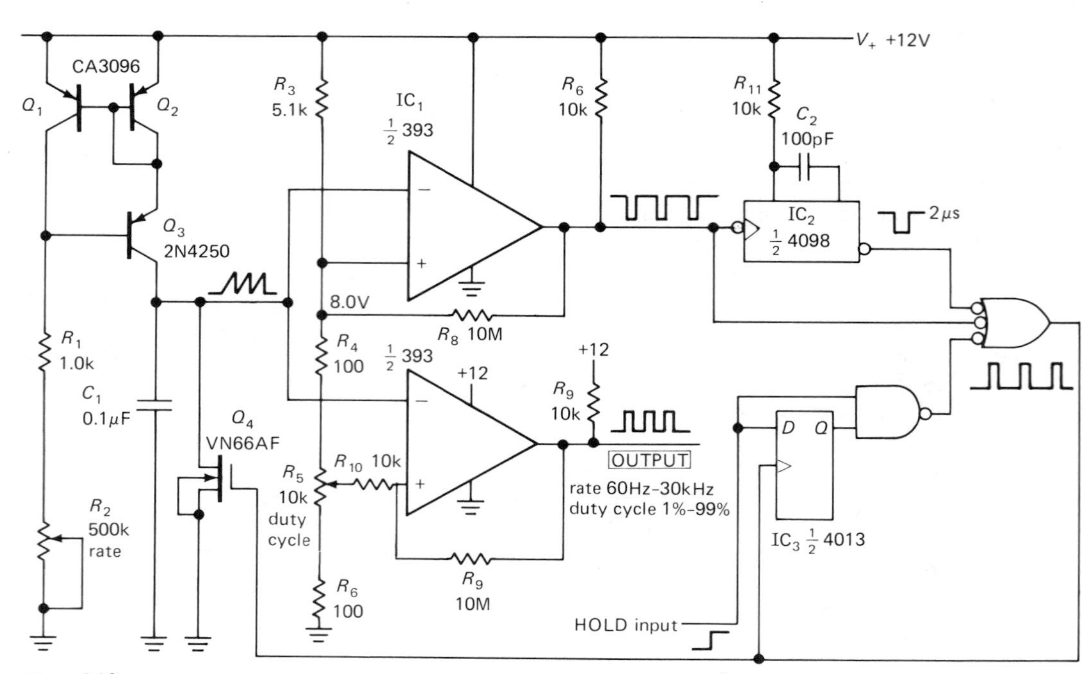

Figure 8.59
Autosynchronizing triggerable pulse generator.

the driven circuit stop the oscillator at the next negative transition at the output. When HOLD is again brought LOW, the oscillator immediately resumes full cycles as if a falling edge had occurred at the time HOLD was released. The additional input to the 3-input NAND from the comparator output ensures that the circuit won't get stuck with C_1 charged up. In this circuit the one-shot pulse width has been chosen long enough to ensure that C_1 is fully discharged during the pulse.

8.22 Cautionary notes about monostables

Monostables have some problems you don't see in other digital circuits. In addition, there are some general principles involved in their use. First, a rundown on monostable pathology.

Some problems with monostables

Timing. One-shots involve a combination of linear and digital techniques. Since the linear circuits have the usual problems of V_{BE} and h_{FE} variation with temperature, etc., one-shots tend to exhibit temperature and supply voltage sensitivity of output pulse width. A typical unit like the 9602 will show pulse width variations of a few percent over a 0–50°C temperature range and over a ±5% supply voltage range. In addition, unit-to-unit variations give you a ±10% prediction accuracy for any given circuit. The 74121, 74221, and 74C221 are superior in this regard, with variations of a few tenths of a percent over the same temperature and voltage ranges and typical unit-to-unit spread of about 1%. When looking at temperature and voltage sensitivity, it is important to remember that the chip may exhibit self-heating effects and that supply voltage variations *during the pulse* (e.g., small glitches on the V_+ line) may affect the pulse width seriously.

Long pulses. When generating long pulses, the capacitor value may be a few microfarads or more; in that case electrolytic capacitors are necessary. You have to worry about leakage current (which is insignificant with the smaller capacitor types), especially since most monostable types apply voltage of both polarities across the capacitor during the pulse. It may be necessary to add a diode or transistor to prevent this problem, or to use a digital delay method instead (involving a clock and many cascaded flip-flop stages, as in Section 8.23). The use of an external diode or transistor will degrade temperature and voltage sensitivity and pulse width predictability; it may also degrade retriggerable operation.

Duty cycle. With some one-shots the pulse width is shortened at high duty cycle. A typical example is the TTL 9600-9602 series, which has constant pulse width up to 60% duty cycle, decreasing about 5% at 100% duty cycle. The otherwise admirable 74121 is considerably worse in this respect, with erratic behavior at high duty cycles.

Triggering. One-shots can produce substandard or jittery output pulses when triggered by too short an input pulse. There is a minimum trigger pulse width specified, e.g., 50ns for the 74121, 140ns for the 4098 with +5 volt supply, and 40ns for the 4098 with +15 volt supply (CMOS is faster and has more output drive capability when operated at higher supply voltages).

Noise immunity. Because of the linear circuits in a monostable, the noise immunity is generally poorer than in other digital circuits. One-shots are particularly susceptible to capacitive coupling near the external R and C used to set the pulse width. In addition, some one-shots are prone to false triggering from glitches on the V_+ line or ground.

Specsmanship. Be aware that monostable performance (predictability of pulse width, temperature and voltage coefficients, etc.) may degrade considerably at the extremes of its pulse width range. Specifications are usually given in the range of pulse widths where performance is good, which can be misleading.

Output isolation. Finally, as with any digital device containing flip-flops, outputs should be buffered (by a gate, inverter, or perhaps an interface component like a line driver) before going through cables or to devices external to the instrument. If a device like a one-shot drives a cable directly, the load capacitance and cable reflections may cause erratic operation to occur.

General considerations for using monostables

Be careful, when using one-shots to generate a train of pulses, that an extra pulse doesn't get generated at the "ends." That is, make sure that the signals that enable the one-shot inputs don't themselves trigger a pulse. This is easy to do by looking carefully at the one-shot truth table, if you take the time.

Don't overuse one-shots. It is tempting to put them everywhere, with pulses running all over the place. Circuits with lots of one-shots are the mark of the neophyte designer. Besides the sort of problems just mentioned, you have the added complication that a circuit full of monostables doesn't allow much adjustment of the clock rate, since all the time delays are "tuned" to make things happen in the right order. In many cases there is a way to accomplish the same job without a one-shot, and that is to be preferred. Figure 8.60 shows an example.

The idea is to generate a pulse and then a second delayed pulse following the falling edge of an input signal. These might be used to set up and initiate operations that require that some previous operation be completed, as signaled by the input falling edge. Since the rest of the circuit is probably controlled by a "clock" square wave, let's assume that the signal at the D input falls synchronous with a clock rising edge. In the first circuit the input triggers the first one-shot, which then triggers the second one-shot at the end of its pulse.

The second circuit does the same thing with type D flip-flops, generating output pulses with width equal to one clock cycle. This is a synchronous circuit, as opposed to the asynchronous circuit using cascaded flip-flops. The use of synchronous methods is generally preferable from several standpoints, including noise immunity. If you wanted to generate shorter pulses, you could use the same kind of circuit, with the system clock divided down (via several toggling flip-flops) from a master clock of higher frequency. The master clock would

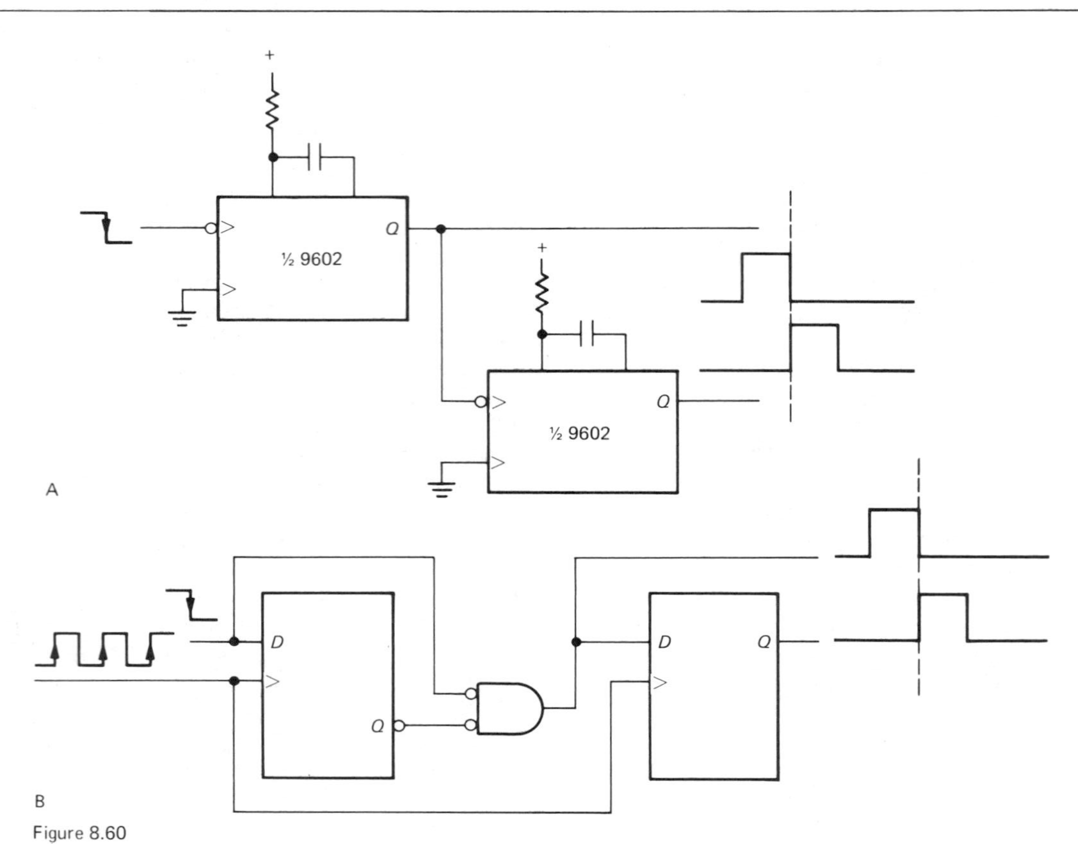

Figure 8.60

A digital delay can replace one-shot delays.

then be used to clock the D flip-flops in this circuit. The use of several subdivided system clocks is common in synchronous circuits.

8.23 Timing with counters

As we have just emphasized, there are many good reasons for avoiding the use of monostables in logic design. Figure 8.61 shows another case where flip-flops and counters (cascaded toggling flip-flops) can be used in place of a monostable to generate a long output pulse. The 4060 is a 14-stage CMOS binary counter (14 cascaded flip-flops). A rising edge at the input brings Q HIGH, enabling the counter. After 2^{n-1} clock pulses,

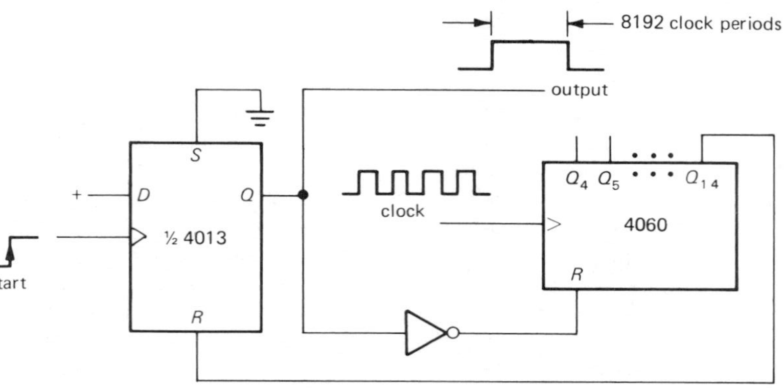

Figure 8.61

Digital generation of long pulses.

Q_n goes HIGH, clearing the flip-flop and the counter. This circuit generates an accurate long pulse whose length may be varied by factors of 2. The 4060 also includes internal oscillator circuitry that can substitute for the external clock reference.

SEQUENTIAL FUNCTIONS AVAILABLE AS ICs

As with the combinational functions we described earlier, it is possible to integrate various combinations of flip-flops and gates onto a single chip. In the following sections we will present a survey of the most useful types, listed according to function. We will refer to TTL chips by their 7400 numbers for simplicity, although in practice you will probably be using the 74LS00 subfamily or one of its relatives.

8.24 Latches and registers

Latches and registers are used to "hold" a set of bits, even if the inputs change. A set of *D* flip-flops constitutes a register, but it has more inputs and outputs than necessary. Since you don't need separate clocks, or SET and CLEAR inputs, those lines can be tied together, requiring fewer pins and therefore allowing 4 or 6 flip-flops to fit in the standard 16-pin package. For instance, the 74174 (TTL) and 4174 (CMOS) are hex (6 per package) *D* registers, whereas the 74175 and 4175 are quad *D* registers with true and complemented outputs. Both have

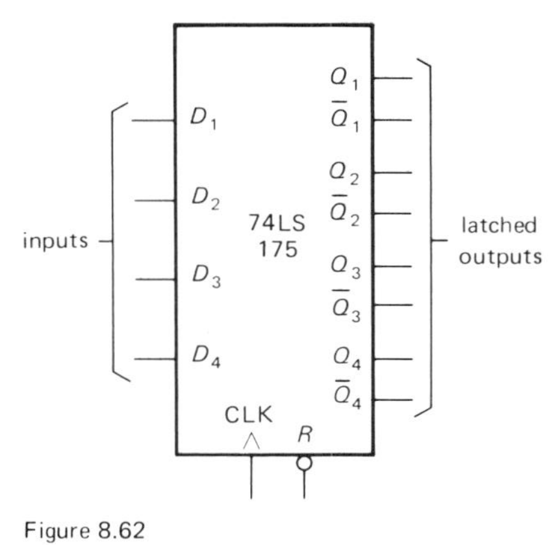

Figure 8.62
74LS175 4-bit D register.

a common clock line and a common clear, and they come in 16-pin packages (Fig. 8.62).

The term "latch" is usually reserved for a special kind of register: one in which the outputs follow the inputs when enabled and hold the last value when disabled. Since the label "latch" has become ambiguous with use, the terms "transparent latch" and "type *D* register" are often used to distinguish these closely related devices. Examples of the first are the 7475 and 4042 quad transparent latches; the 7475 has two ENABLE inputs, one for each pair of input bits. Both transparent latches and type *D* registers are available for as many as 8 input bits. Examples are the 74373 and 74374, respectively.

Latches and registers are available with three-state outputs, an example being the 4-bit 8551 type *D* register. These are particularly convenient for multiplexing 4-bit groups onto a single 4-bit "bus," as in display multiplexing (a subject we'll consider shortly). The 74373 and 74374, mentioned earlier, have three-state outputs.

A variation on the latch is the "addressable latch," a multibit latch that lets you address any of the bits for updating, while keeping the others held. The 74259 and 4099, for instance, are 8-bit addressable latches in 16-pin packages. They have a single data input, a 3-bit address input, and 8 output bits. They can also be used as demultiplexers.

A memory device known as a "random-access memory" (RAM) is sometimes what's needed in place of a latch or register. It works like a ROM, except that data can be both written and read. They come in many varieties, from the 16-bit 7481 and 16-word (4 bits each) 7489 up to the giant 64K-bit RAMs. RAMs will be discussed in more detail in connection with microprocessors in Chapter 11.

Finally, the latch function is frequently implemented in a chip with other functions. For instance, the 74399 is a quad 2-input multiplexer with output *D*-type register, the 8554 is a 4-bit counter with transparent latch and three-state outputs, and the HP 5082-7300 is a decimal display chip with BCD decoder, transparent latch, and driver built in. It pays to look for combined functions when you need them, to save space, cost, and complexity.

8.25 Counters

As we mentioned earlier, it is possible to make a "counter" by connecting flip-flops together. There is available an amazing variety of such devices, integrated onto single chips. We will try to point out some of the features to look for and give an idea of what's available.

Size

There are many BCD chips (divide by 10) and binary chips (or hexadecimal, divide by 16), and even a few divide-by-12 chips, available in the popular 4-bit counter category. In TTL you can get 8-bit counters (74390, BCD; 74393, binary, see Fig. 8.63), and in CMOS there are binary coun-

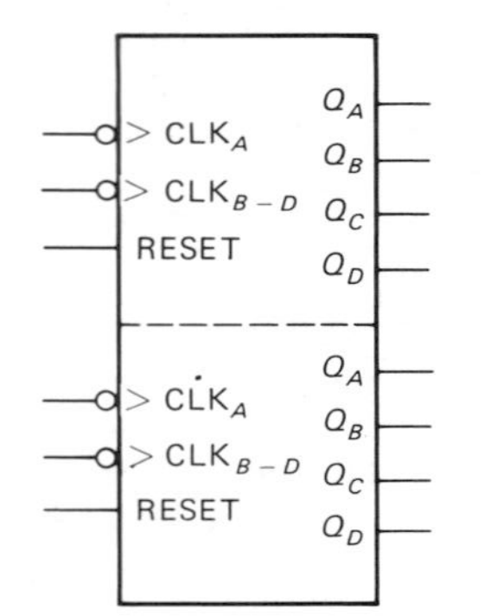

Figure 8.63
74LS390 dual BCD ripple counter.

ters with 7, 12, and 14 stages with most of the outputs accessible (4024, 4020, 4040, 4060), as well as larger chips like the 4536, a 24-stage counter with a selectable number of stages (the output stage is set by a 5-bit input) and an internal oscillator. "Modulo-*n*" counters are available in both families; they allow you to divide by an integer *n*, specified as an input word. Counters (including synchronous types) can always be cascaded to get more stages.

Clocking

Counters are available with positive or negative edge clocking. A much more important distinction is whether the counter is a "ripple" counter or a "synchronous" counter. The latter clocks all flip-flops simultaneously, whereas in a ripple counter each stage is clocked by the output of the previous stage. Ripple counters generate transient states, since the earlier stages toggle slightly before the later ones. For instance, a ripple counter going from a count of 7 (0111) to 8 (1000) goes through the states 6, 4, and 0 along the way. This doesn't cause trouble in well-designed circuits, but it would in a circuit that used gates to look for a particular state (this is a good place to use something like a *D* flip-flop, so that the state is examined only at the clock edge). Ripple counters are slower than synchronous counters, because of the accumulated propagation delays. Ripple counters must have falling-edge clocking inputs for easy expandability (by connecting the Q output of one counter directly to the clock input of the subsequent counter).

As examples, the 7490 and 7493 are 4-bit BCD and binary ripple counters, respectively, and the 74160 and 74161 are synchronous counters.

Up/down

Some counters can count in either direction. Examples are the 74190 (BCD) and 74191 (binary) synchronous up/down counters, which use a separate UP/DOWN input control line, and the 74192 (BCD) and 74193 (binary), which have separate UP and DOWN clock inputs (Fig. 8.64). The 4029 is a nifty CMOS counter that can count either up or down, in either BCD or binary. Cape Canaveral, move over!

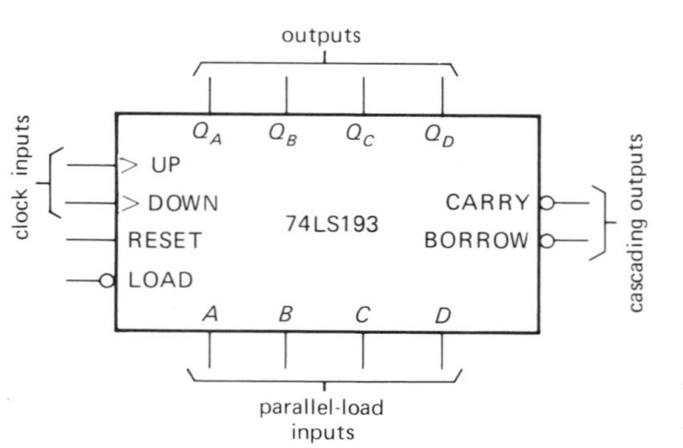

Figure 8.64
74LS193 4-bit synchronous up/down counter.

Presetting

Some counters have data inputs so that they can be preset to a given count. This is handy if you want to make a modulo-n counter, for instance, as well as for a number of other applications. The preset, or "load" function, can be synchronous or asynchronous. The 74160-74163 have synchronous load, which means that data on the input lines is transferred to the counter coincident with the next clock edge, if the LOAD line is also enabled. The 74190–74193 are asynchronous, or jam load; data on the input lines is transferred to the counter when LOAD is enabled, regardless of the state of the clock. The term "parallel load" is sometimes used, since all bits are loaded at the same item.

The CLEAR (or RESET) function is a form of presetting. The majority of counters have a jam-type CLEAR function, although some have a synchronous CLEAR (e.g., the 74162 and 74163).

Other counter features

Some counters feature integral latches on the output lines; these are always of the transparent type, so the counter can be used as if no latch were present. (It should be kept in mind that any counter with parallel-load inputs can function as a latch, but you can't count at the same time as data are held, as you can with a counter/latch chip.) The combination of counter plus latch is sometimes very convenient, e.g., if you want to display or output the previous count while beginning a new counting cycle. In a frequency counter this will allow a stable display, with updating after each counting cycle, rather than a display that repeatedly gets reset to zero and then counts up (see Section 14.10).

There are also available counters with latch and three-state outputs. These are great for applications where the digits (or 4-bit groups) are multiplexed onto a bus for display or transfer to some other device. Examples are the 8552 (BCD) and 8554 (binary) 4-bit counters. For bus applications there are even counters with combined input/output terminals. The 8555 (BCD) and 8556 (binary) counters have three-state outputs that, when disabled, are used as parallel-load inputs. You will see some applications later.

If you want a counter to use with a display, there are several that combine counter, latch, 7-segment decoder, and driver on one chip. The 74142–74144 series, for example, includes a counter in a 16-pin package with 7-segment output only, as well as some counters in 24-pin packages that have binary outputs as well. There are even counters with display in one package (e.g., the TIL 306/307): You just look at the IC, which lights up with a digit telling the count!

In the domain of LSI (large-scale integration) there are chips with several stages of counters, integral latch, decoder, and multiplexed display drivers. The 74C925–74C928 series of 4-digit CMOS counters combines these functions in a single 16-pin package. With them you can make a 4-digit counter with display with one IC plus one 4-digit LED display. Figure 8.65 shows a very nice LSI counter circuit that doesn't require a lot of support components.

8.26 Shift registers

If you connect a series of flip-flops so that each Q output drives the next D input, and all clock inputs are driven simultaneously, you get what's called a "shift register." At each clock pulse the pattern of 0's and 1's in the register shifts to the right, with the data at the first D input entering from the left. As with flip-flops, the data present at the serial input just prior to the clock pulse is entered, and there is the usual propagation delay to the outputs. Thus they may be cascaded without fear of a logic race. Shift registers are very useful for conversion of parallel data (n bits present simultaneously, on n separate lines) to serial data (one bit after another, on a single data line), and vice versa. They're also handy as memories, particularly if the data are always read and written in order. As with counters and latches, shift registers come in a pleasant variety of prefab styles. The important things to look for are the following.

Size

The 4-bit (e.g., 7495) and 8-bit (e.g., the 74164 and 4021) registers are standard.

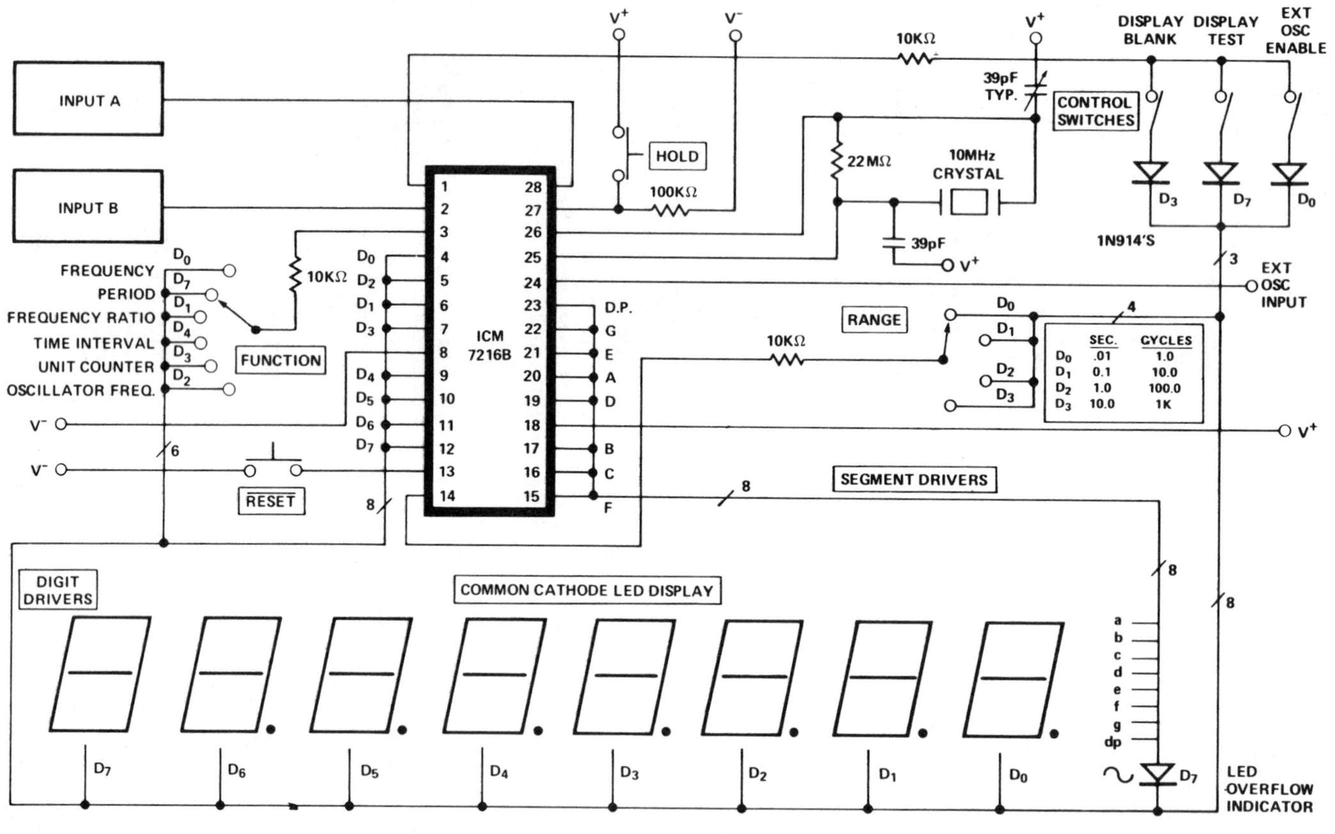

Figure 8.65
Intersil 7216 8-digit 10MHz universal counter on a chip. (Courtesy of Intersil, Inc.)

Registers as large as 1024 bits (e.g., the 2533, in mini-DIP package) or larger are available. In CMOS you can get variable-length shift registers, e.g., the 4557, whose length is selectable from 1 to 64, as determined by a 6-bit control input. Figure 8.66

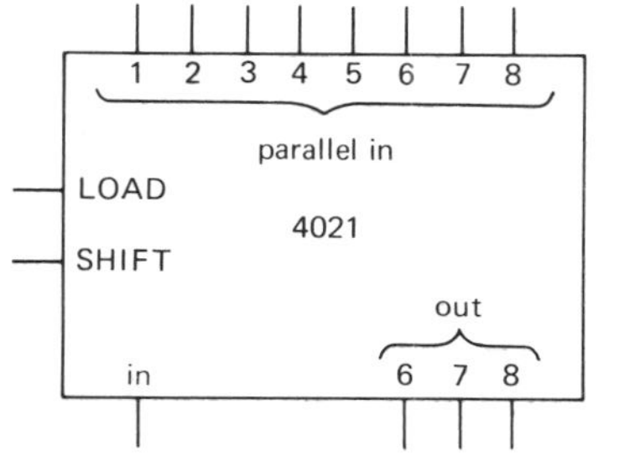

Figure 8.66

4021 CMOS 8-bit parallel-input shift register.

shows the popular CMOS 4021 8-bit parallel-input shift register.

Organization

Most shift registers consist of a single register, but duals, quads, and even hex registers are made. The 4517 is a dual 64-bit register with separate clock inputs; the 2518/2519 registers are hex 32-bit registers with common clock input.

Static/dynamic

A static register holds its data indefinitely, like a flip-flop. A dynamic register must be clocked at some minimum rate at all times; otherwise it forgets. Small shift registers are always of the static type (all the registers mentioned so far are static, as are all TTL and CMOS shift registers), but the larger MOS registers may be dynamic. The latter are easier to make and consume less power, but they may be inconvenient to use.

Inputs and outputs

Shift registers with more than 8 bits or so usually have only serial inputs and outputs, i.e., only the input to the first flip-flop and the output from the last are accessible. In some cases a few selected intermediate outputs are provided (e.g., the 18-bit 4006 gives outputs every 4 bits, and the 128-bit 4562 has outputs every 16 bits). Small shift registers can provide parallel inputs or outputs, and usually do. Examples are the 7495, a 4-bit parallel-input/output shift register, and the 74166, an 8-bit parallel-in serial-out shift register. The 74199 is an 8-bit parallel-in parallel-out shift register; since that already commits 16 pins, it has to be put in a larger package (24 pins). One way to put such a device in a standard 16-pin package is to share input and output on the same lines. The 8546 does that by using its three-state output lines as parallel-input lines also.

As with counters, inputs and clear lines can be either synchronous or jam load. The 74166 8-bit shift register has synchronous load and jam clear, the 7496 5-bit shift register has jam load and clear (but see the following explanation), and the 7495 4-bit register has synchronous load with a separate load clock input, but no clear input at all. The 7496 has one peculiar property that you better know about: Its parallel-load inputs can only *set*, not *clear*, the corresponding flip-flop; one of the authors fell into that little trap, once!

Shift left/shift right

A few shift registers are bidirectional, i.e., they can shift to the right or left, as determined by an input control line (or lines). Examples are the 74194 and 4194 (4-bit) and 74198 (8-bit) bidirectional shift registers; all have parallel load (the 74198 is a 24-pin package), separate serial inputs, and a pair of "mode" inputs to determine function (left shift, right shift, parallel load, or disable). The 7495 is sometimes called a bidirectional shift register, but it's not the genuine article; it's a 4-bit parallel-load right-shift register, and by connecting the output of each stage to the load input of the previous stage, you get a left shift at the next transition of the load clock input. A true bidirectional shift register requires no external connections and shifts left with the same effortless ease with which it shifts right (Fig. 8.67).

□ *RAMs as shift registers*

A random-access memory can always be used as a shift register (but not vice versa) by using an external counter to generate successive addresses. Figure 8.68 shows

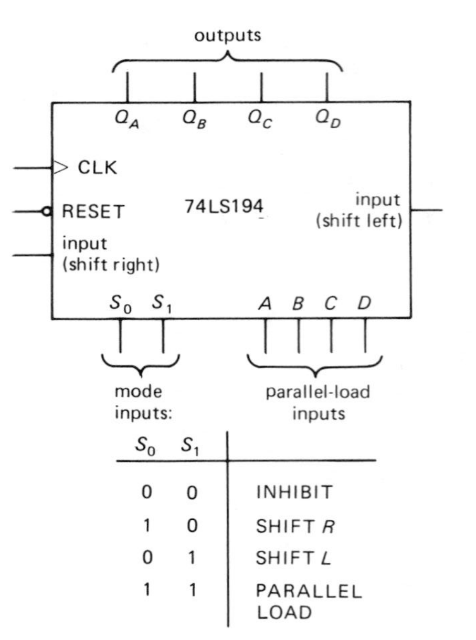

Figure 8.67
74LS194 4-bit bidirectional shift register.

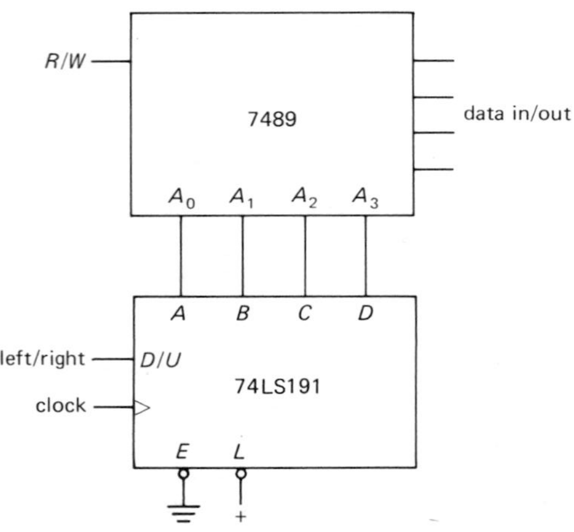

Figure 8.68

the idea. The 74191 synchronous up/down counter generates successive addresses for the 7489 4-bit × 16-word RAM. The combination behaves like a quad 16-bit shift register, with left/right direction of shift selected by the 74191's DOWN/UP control line. All other inputs of the counter are shown enabled for counting.

8.27 Miscellaneous sequential functions

With the increasing availability of large-scale integration (LSI) chips (equivalent to 1000 gates or more on a single chip), you can get weird and wonderful gadgets all on one chip. This brief section will present just a sampling.

First-in first-out memory

A first-in first-out (FIFO) memory is somewhat akin to a shift register in that data entered at the input appear at the output in the same order. The important difference is that with a shift register the data get "pushed along" as additional data are entered and clocked, but with a FIFO the data "fall through" to the output queue with only a small delay. Thus if the output queue is empty, data just entered are available almost immediately. Input and output are controlled by separate clocks, and the FIFO keeps track of what data have been entered and what data have been removed. A helpful analogy might be a bowling alley, in which black and white bowling balls (bits) are returned to the bowling station: The bits are input by the pin-setting machine, and the time it takes for a ball to roll the length of the alley is the "ripple-through delay time" of the FIFO (typically 1–25 μs), whereupon the bits are available at the output to be removed at the whim (asynchronously) of the user.

FIFOs are extremely useful for buffering asynchronous data. The classic application is buffering a keyboard (or other input device, such as magnetic tape) to a computer or slow data-computation instrument. By this method no data are lost if the computer isn't ready for each word as it is generated, provided the FIFO isn't allowed to fill up completely. Typical FIFOs are the 3341 (MOS, 64 words of 4 bits each), the 74S225 (TTL, 16 words of 5 bits each), and the 40105 (CMOS, 16 words of 4 bits each). The newer 9423 is particularly suited to buffering data from a high-density magnetic disk, owing to its combination of high speed (10MHz) and internal serial/parallel conversion at input and output.

The application of FIFOs is illustrated in Section 8.31 with a buffered keyboard.

Universal asynchronous receiver/transmitter

The universal asynchronous receiver/transmitter (UART) is a nifty gadget that converts parallel data (usually 8-bit words) to a serial data stream for transmission over a single-wire cable and simultaneously converts a received serial bit stream to parallel words. The serial data include self-synchronizing START and STOP pulses, so no extra timing lines are needed. They are used extensively when computers or microprocessors communicate with serial devices such as terminals or printers, but they can be used whenever it is convenient to reduce the number of connecting lines between a source of digital data and its recipient, at the expense of speed.

Typical UART type numbers are the 6011 and the 1602, constructed with PMOS technology. For new designs you might prefer the IM6402/6403, versatile CMOS UARTs that even include crystal oscillator circuitry. Note that a UART is a "stand-alone" chip that is easy to use in any digital system, large or small, whereas the newer software-programmable serial microprocessor peripheral chips (called USARTs) require a resident microprocessor for control (see Section 11.9).

Rate multiplier

Rate multipliers are used to generate output pulses at a frequency that is related to the clock frequency by a rational fraction. For instance, a 3-decade BCD rate multiplier allows you to generate output frequencies of *nnn*/1000 of the input frequency, where *nnn* is a 3-digit number specified as three BCD input characters. This isn't the same as a modulo-*n* counter, since, for instance, you cannot generate an output frequency equal to 3/10 of the input frequency with a modulo-*n* divider. One important note: The output pulses generated by a rate multiplier are not, in general, equally spaced. They coincide with input clock pulses, and therefore they come in funny patterns whose *average* rate is as previously described. The 7497 and 4089 are binary rate multipliers, and the 74167 and 4527 are BCD rate multipliers. (You may recognize that the rate multiplier is not an LSI chip. We beg your forgiveness. We just couldn't find any other pigeonhole for it in this chapter.)

Digital voltmeters

You can get complete digital voltmeters (DVMs) on a single chip. They include analog/digital conversion circuitry and the necessary timing, counting, and display circuitry. Examples are the 7107 and the 74C936-74C938 series of "3½-digit" voltmeters.

Consumer chips

There has recently been extensive development of ICs for use in large-market consumer products. You can get single chips to make watches, clocks (including "snooze alarm"), calculators, smoke detectors, and push-button telephone dialers. In addition, many radio and TV circuits have been integrated onto chips, and the development of effective automobile circuits (for engine functions, collision-avoidance systems, etc.) seems to be the next big frontier.

Games and toys

Video games have hit the big time, and semiconductor manufacturers are competing to produce the world's best Ping-Pong or space war. These chips or chip sets typically accept some analog ("joystick") and digital (push-button) inputs and generate a video output. You sit in front of the Tube and watch things blow up.

There are interesting chips being produced for use in electronic organs. A nice example is the MM5871 "rhythm pattern generator." It has an on-chip tempo oscillator, and it produces waltz, swing, country/western, march, Latin, and rock rhythms, with others available on special order.

Another example of a far-out consumer chip is the 76477, a sound-effects circuit from TI. It contains a number of clocks and noise generators, filters, mixers, modulators, and logic circuitry. With it you can mimic the sounds of a locomotive, a sports car race

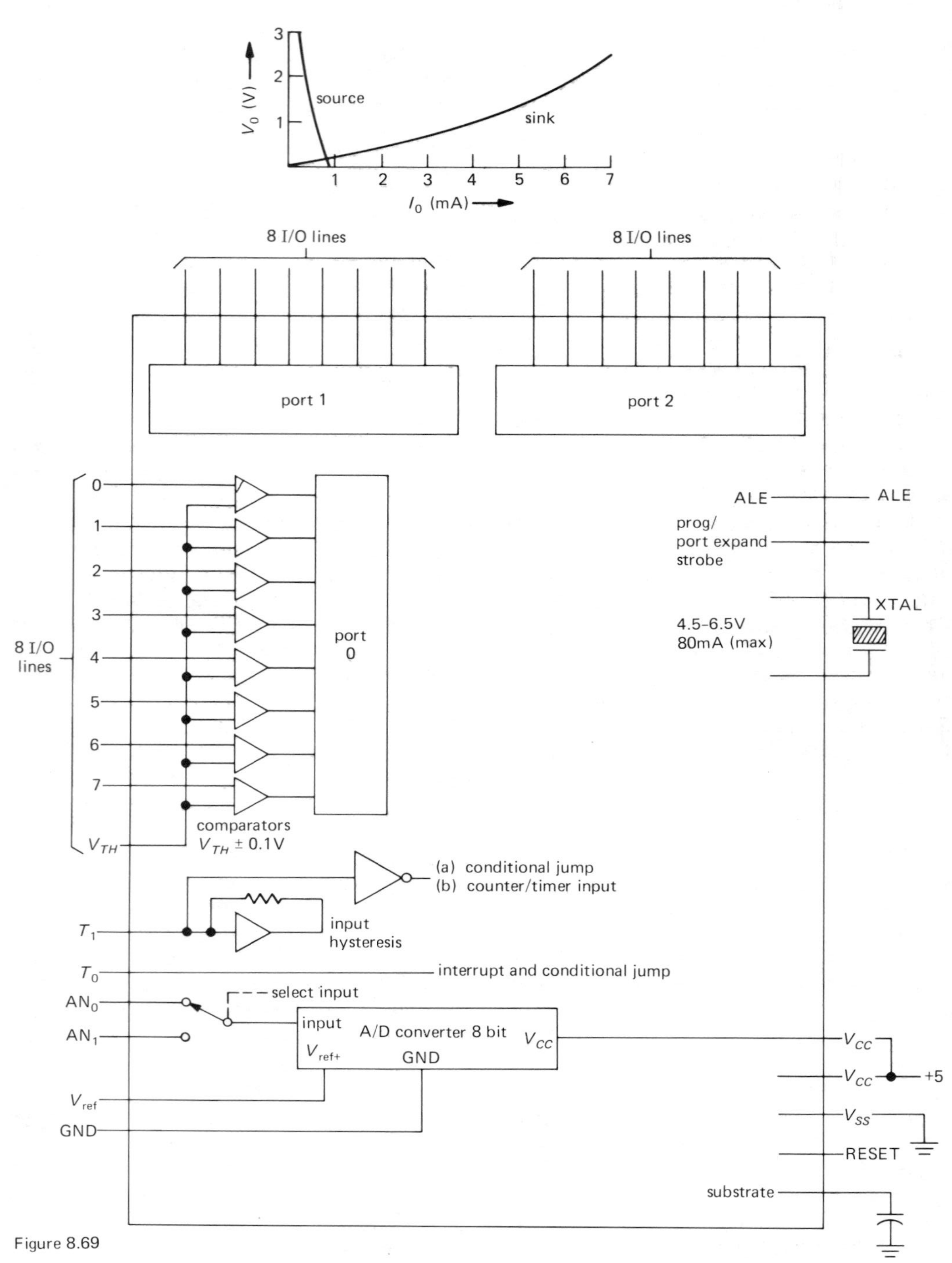

Figure 8.69

Intel 8022, the first of the microprocessors incorporating on-chip A/D converters and analog comparators.

(with crash), a bird chirping, and a gun with silencer (for use with bird, above), to name a few!

Those who frequent toy stores have probably noticed the talking boxes that you can buy to help speed your infant on toward college and success. They contain some pretty sophisticated examples of speech-synthesis technology and large memories.

The outstanding characteristic of these consumer chips is their incredibly low cost. The sound-effects generator, for instance, sells for less than three dollars in single quantities.

Microprocessors

The most stunning example of the wonders of LSI is the microprocessor, a computer on a chip. At one extreme there are powerful "number crunchers" like the 8086 (a 16-bit processor with features like instruction look-ahead, hardware multiply and divide, and stack and string operations) and chips like the LSI-11 and micro-Nova that emulate existing minicomputers. At the other extreme are single-chip processors with various input, output, and memory functions included on the same chip, for "stand-alone" use. An example of the latter is the 8022, which has 8 comparator inputs, an 8-bit A/D converter, and 2048 8-bit words of memory on the same chip as the 8-bit processor (Fig. 8.69). The latter types are intended as dedicated controllers in instruments, rather than as versatile computation devices. We expectantly await the continued development of exciting LSI functions.

SOME TYPICAL DIGITAL CIRCUITS

In the following sections we will present some digital circuits that illustrate a number of standard techniques. Some of these functions can be performed with LSI circuits, but the implementations shown are reasonably efficient and illustrate the kind of circuit design being done with what's now available.

8.28 Modulo-*n* counter

The circuit in Figure 8.70 produces one output pulse for every *n* input clock pulses. The 74LS190s are BCD synchronous up/down counters, here enabled to count down by tying the DOWN input HIGH. The idea is to count down until the "ripple clock" output goes LOW, which happens when the count has reached zero and the input clock goes LOW. The counter is then preset back to *n* via the LOAD′ input, to begin a new cycle of *n* clock pulses. The value of *n* is set by a pair of BCD thumbwheel switches; since pull-up resistors are used, the switch has to have inverted BCD bits, a commonly available form of digital switch. The two inverters are used as delay elements to make sure the LOAD′ pulse is long enough (it's specified as 35ns minimum), since parallel loading takes the counter out of the zero state, thereby killing the "ripple clock" output that initiated loading in the first place. A "runt pulse" might otherwise result (see Section 8.34). The combination of inverter plus counter loading delays results in a 50 ns pulse.

The output is taken from the MAX/MIN output (HIGH when the counter is at zero), giving output pulses of width equal to the HIGH part of the input clock. This scheme can easily be expanded for larger *n* by using more counter chips cascaded synchronously as shown. If you build a circuit like this with fully synchronous cascaded high-speed counters like 74S160 or 74S162, it will work with input clock rates up to 50MHz or more.

8.29 Multiplexed LED digital display

This example illustrates the technique of display multiplexing: displaying an *n*-digit number by displaying successive digits rapidly on successive 7-segment LED displays (of course, the characters need not be numbers, and the displays can have a different organization than the popular 7-segment arrangement). Display multiplexing is done for reasons of economy and simplicity: Displaying each digit continuously requires separate decoders, drivers, and current-limiting resistors for each digit, as well as separate connections from each register to its corresponding decoder (4 lines) and from each driver to its corresponding display (7 wires); it's a mess!

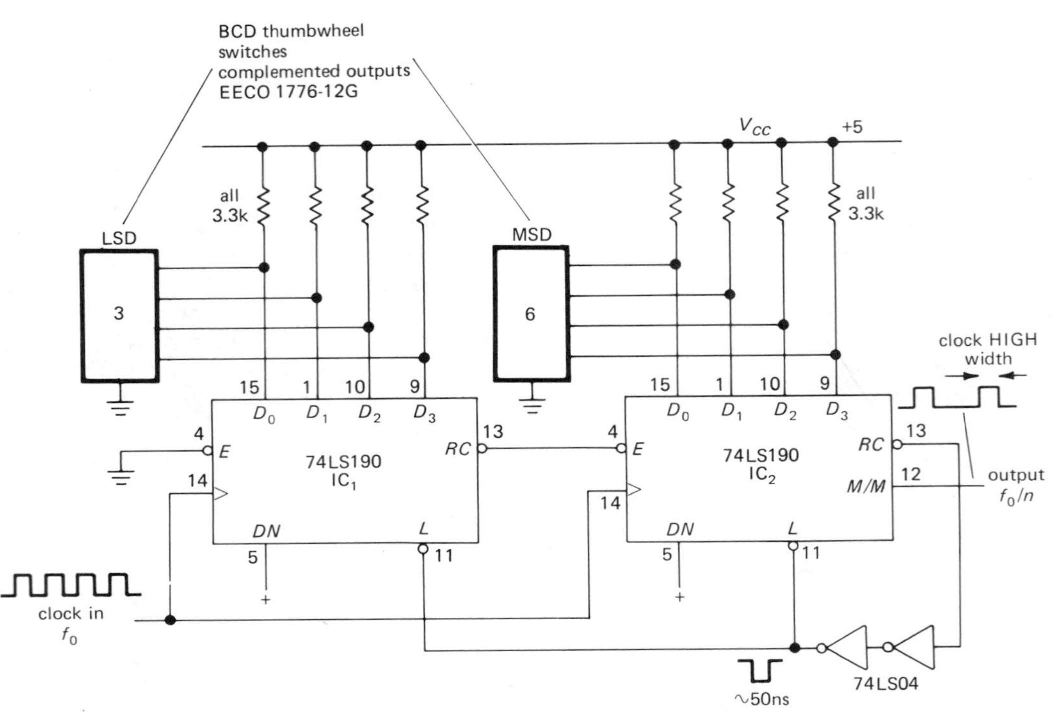

Figure 8.70

Two-digit modulo-n divider.

With multiplexing, there's only one decoder/driver and one set of current-limiting resistors. Furthermore, since LED displays come in n-character "sticks" with the corresponding segments of all characters tied together, the number of interconnections is enormously reduced. An 8-digit display requires 15 connections when multiplexed (7-segment inputs, common to all digits, plus one cathode or anode return for each digit), rather than the 57 required for continuous display. An interesting bonus of multiplexing is that the subjective brightness perceived by the eye is greater than if all digits were illuminated continuously with the same average brightness.

Figure 8.71 shows the schematic diagram. The digits to be displayed are resident in registers IC_1–IC_4; they could be counters, if the device happened to be a frequency counter, or perhaps a set of latches receiving data from a computer, or possibly the output of an A/D converter, etc. In any case, the technique is to assert each digit successively onto an internal 4-bit "bus" (in this case with 4503 CMOS three-state buffers) and decode and display it while on the bus (4511 BCD-to-7-segment decoder/driver).

In this circuit a pair of inverters is used to form a classic CMOS oscillator operating at about 1kHz, driving a 4022 octal counter/decoder. As each successive output of the counter goes HIGH, it enables one digit onto the bus and simultaneously pulls the corresponding digit's cathode LOW via the high-current open-drain 40107 buffers. The 4022 is rigged up to cycle through the states 0–3 by resetting when the count reaches 4. Display multiplexing will work with greater numbers of digits, and it is universally used in instruments with multidigit LED displays. Try waving a calculator around in front of your eyes – you get numeric alphabet soup!

Many LSI display-oriented chips, such as counters, timers, and watches, include on-chip display multiplexing (and even driver)

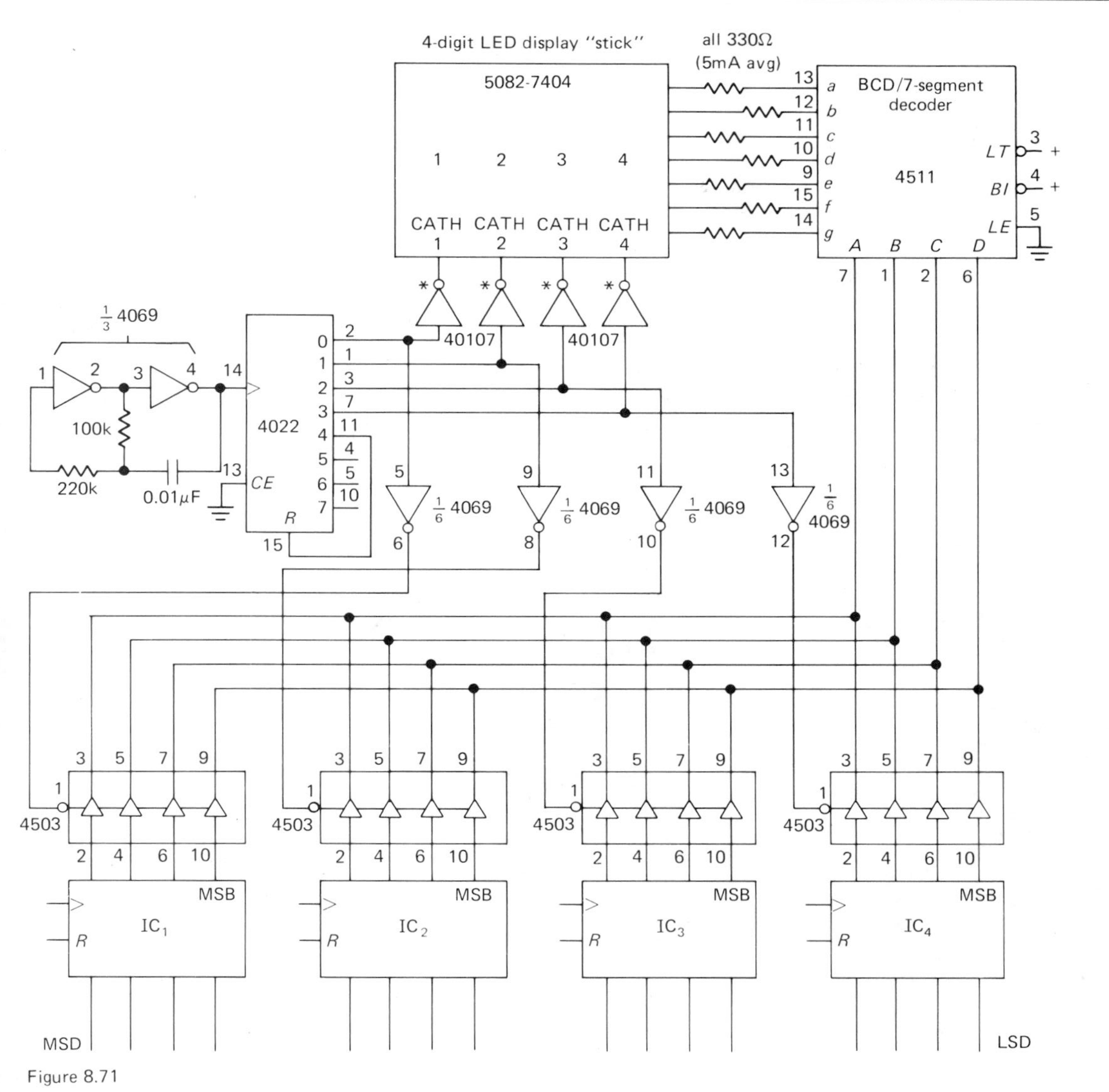

Figure 8.71

Four-digit multiplexed display.

circuitry. In addition, you can get LSI display controller chips (e.g., the 74C911 and 74C912) for handling the kind of job done earlier with MSI circuits.

□ 8.30 Siderial telescope drive

The circuit in Figure 8.72 was designed to drive Harvard's 61-inch optical telescope. We needed a 60Hz power source for the equatorial drive motor (1 revolution/day), accurately settable to any frequency near 60Hz (55Hz to 65Hz, say). You wouldn't want exactly 60Hz, for several reasons: (a) stars move at the siderial rate, not the solar rate, so you would want 60.1643Hz, approximately; (b) starlight gets bent, traveling obliquely through the atmosphere; this "refraction" depends on zenith angle, so the apparent motion is at a slightly different rate; (c) sometimes you want to look at the moon, planets, or comets, which have different rates. The solution here was to use a 5-digit rate multiplier to generate output

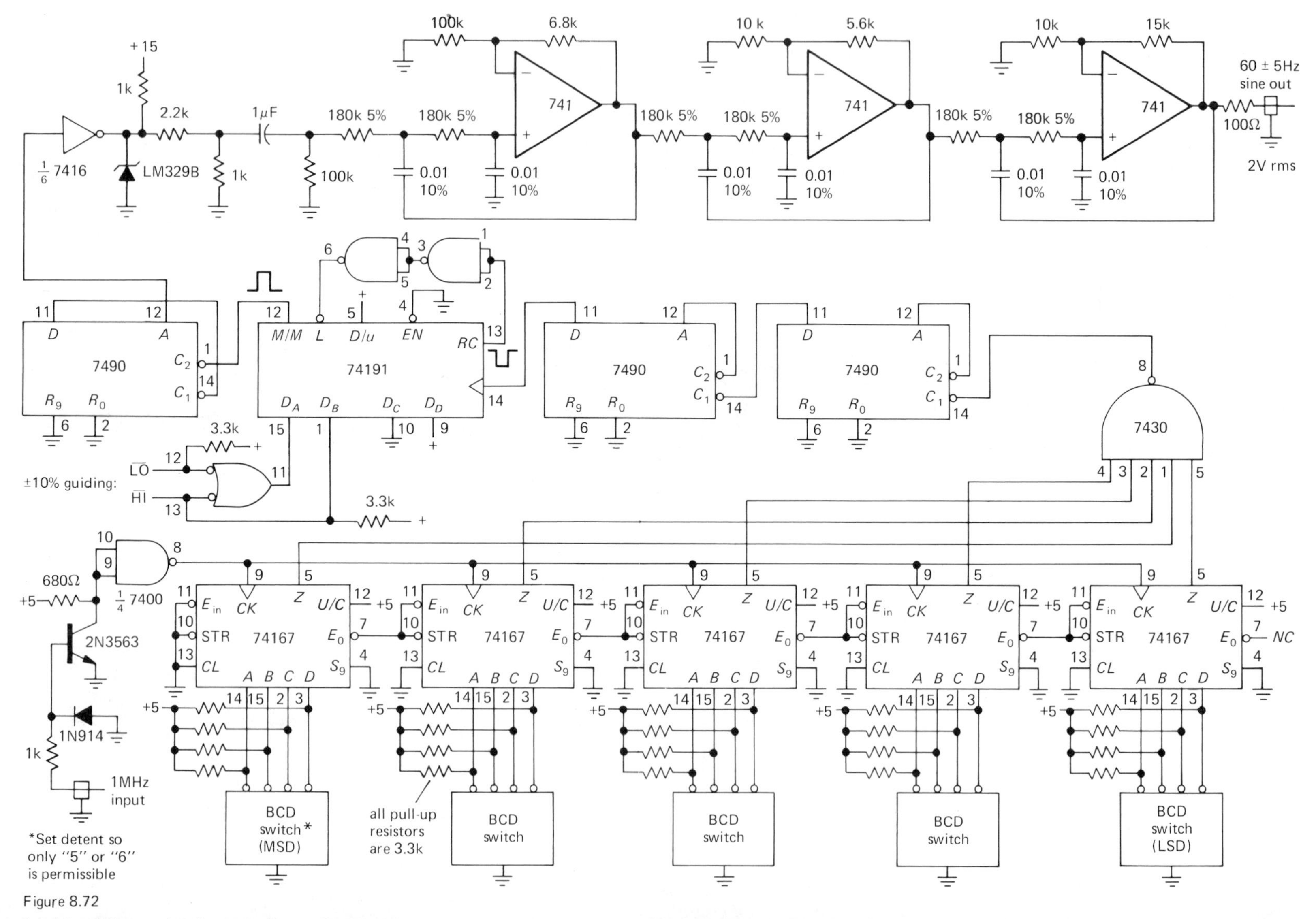

Figure 8.72

Precision 60Hz ac signal source. Output frequency = xx.xxx; e.g., to generate sidereal rate, set switches to 60165

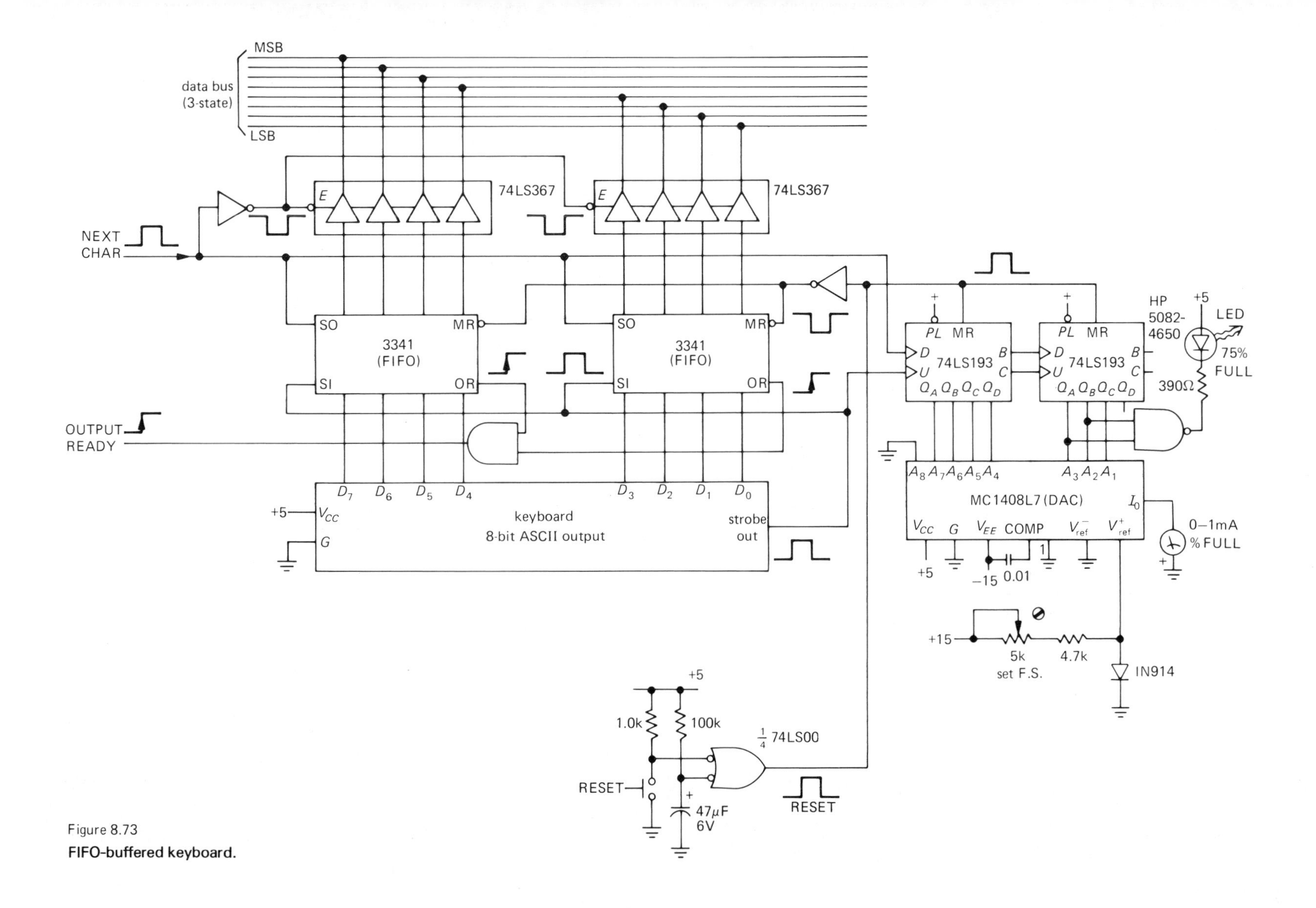

Figure 8.73

FIFO-buffered keyboard.

pulses at a rate $f_{in}n/10^5$, where n is a 5 digit number set by front-panel BCD thumbwheel switches.

The output is then near 600kHz, since f_{in} is an accurate 1MHz generated by a stable crystal oscillator. The output of the rate multiplier is divided by 10^4 by four decade counters, with the last counter arranged as a divide-by-5 followed by a divide-by-2 for symmetrical square waves at 60Hz. The output is clamped by a zener for stable square-wave amplitude and filtered by a 6-pole Butterworth low-pass filter (f_0 = 90Hz) to generate a good sine wave (you can think of the filter as stripping away the higher Fourier components, or "overtones," of the square wave). Then 115 volts ac is generated by the "overcompensated" amplifier illustrated in Section 3.33. The output of the Butterworth looks "perfect" on a scope, as it should, since in this case a 6-pole Butterworth reduces the largest overtone to 1.5% of its unfiltered amplitude; this means that the distortion is more than 35dB down. Note that this technique of sine-wave generation is convenient only if the input frequency is confined to a narrow range.

The ±10% guiding inputs alter the synthesized output frequency 10% by changing the third divider to divide-by-9 or divide-by-11. That stage is a modulo-n divider constructed along the lines of Figure 8.70.

8.31 FIFO-buffered keyboard

The subcircuit in Figure 8.73 interfaces an alphanumeric keyboard to a data-using instrument (a computer, say) by interposing a 64-character FIFO memory (Section 8.27). Most keyboards encode keystrokes into an 8-bit binary code known as ASCII (see Chapters 10 and 11), with a "strobe" pulse output as each 8-bit character is ready. Good keyboards debounce the contacts, and they usually include "2-key rollover," meaning that if a second key is struck while the first key is still held down, the keyboard will generate the two output codes, in correct sequence.

In this circuit the successive ASCII codes are loaded into the 3341 64 × 4 FIFOs, expanded to hold 64 × 8, by SI (shift in) pulses generated by the keyboard. If anything is in the FIFO it puts OR (output ready) HIGH, signaling the external instrument that there are some data available (in Chapter 10 you will see how such a message is sent to a computer, and what it does about it). The external instrument obtains a character by issuing a NEXT CHAR pulse, which unloads a character from the FIFO via the SO (shift out) input, simultaneously enabling the three-state 74LS367 bus drivers.

In order to keep track of what's in the buffer, the circuit includes an up/down counter and digital/analog converter (of which lots more in the next chapter), driving a "% FULL" meter; each SI increments the counter, and each SO decrements it. The FIFO and counters are cleared by a reset command generated either by a push-button or at power-up; the 47μF capacitor holds one of the NAND inputs LOW for about 0.15 second after initial turn-on. It is a good idea to use some sort of power-up reset circuit in any device with sequential logic; otherwise you get nonsense at first when you turn it on, since all flip-flops come up in random states. This circuit also includes an LED indicator to signal when the buffer is 75% full; it lights when the count reaches 48.

Synchronizer for up/down counters

There is a minor flaw in the preceding circuit. Although the FIFO has completely asynchronous input and output (i.e., data may be input and output at random times, even overlapping if data are available), the 74LS193s will not clock properly if the UP clock and the DOWN clock have overlapping pulses. With 1μs strobe pulses this happens, on the average, less than once in 100,000 keystrokes. However, if you want to be compulsive about it, you can improve your odds with a synchronizer circuit like that shown in Figure 8.74. SI and SO may overlap, but UP and DOWN clock pulses are separated by clocking them on opposite phases of the internal ~1MHz clocking oscillator. The operation of this subcircuit should be self-explanatory.

The 3341 FIFO, by the way, is a MOS LSI device with TTL-compatible inputs and

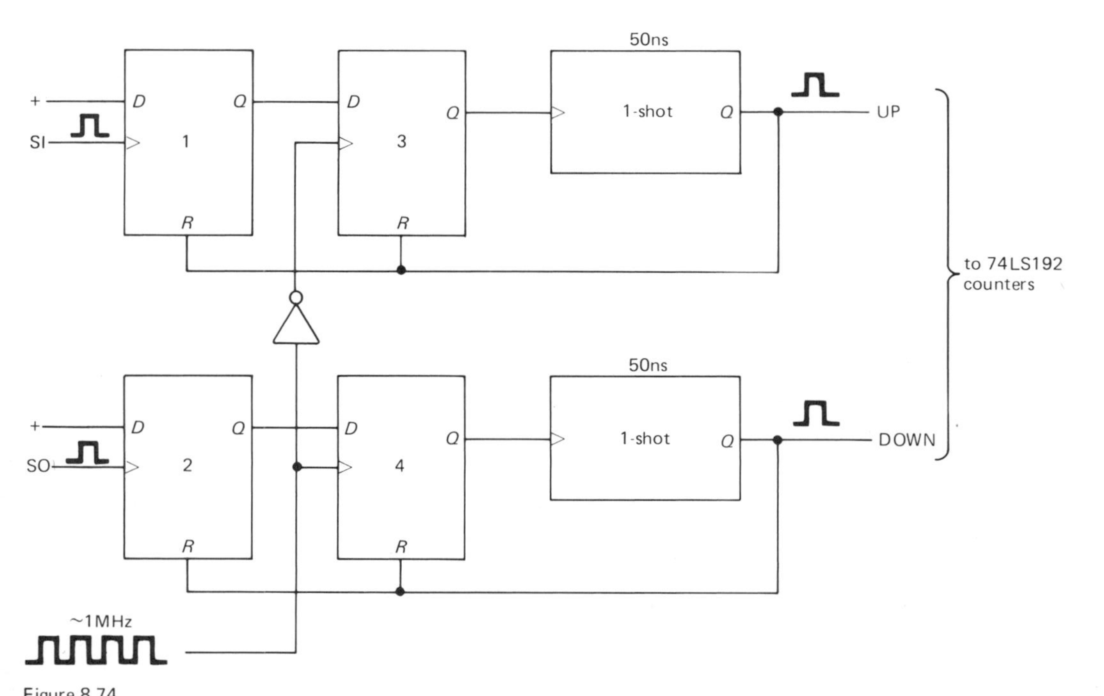

Figure 8.74

Synchronizer for use with up/down counters.

outputs. That's common practice with many MOS LSI devices, since they frequently find application in TTL circuits.

8.32 *n*-Pulse generator

The *n*-pulse generator is a useful little test instrument. It generates a burst of *n* output pulses following an input trigger signal (or you can push a button), with a few selectable pulse repetition rates. Figure 8.75 shows the circuit. The 40102s are CMOS 2-decade down counters, clocked continuously by the free-running 555 oscillator, but disabled by having both APE (asynchronous preset enable) asserted and CI (carry in) disasserted. When a trigger pulse comes along, flip-flop 1 enables the counter, and flip-flop 2 synchronizes counting following the next rising edge of the clock. Pulses are passed by NAND gate 3 until the counter reaches zero, at which time both flip-flops are reset; this parallel-loads the counter to *n* from the BCD switches, disables counting, and readies the circuit for another trigger. Note that the use of pull-down resistors in this circuit means that true (rather than complemented) BCD switches must be used. Note also that the manual trigger input must be debounced, since it clocks a flip-flop. That is not necessary for the free-run/*n*-pulse switch, which simply enables a continuous stream of output pulses.

Output-stage options

The output stage consists of 40107 CMOS high-current open-drain NAND, capable of driving TTL or CMOS. The driven circuit supplies the pull-up voltage in order to generate a proper HIGH logic level. For convenience it might be nice to have TTL levels available without the need to tie into the V_+ supply of the driven circuit. In that case the output circuit shown in Figure 8.76 might be preferable.

Here the CMOS outputs are generated by a 40109 "level shifter," a CMOS chip with active pull-ups to a second supply terminal connected (in this circuit) to the driven circuit's V_{DD} line, which can be higher or lower than the pulse generator's V_{DD}. In that way you get clean CMOS output levels at the

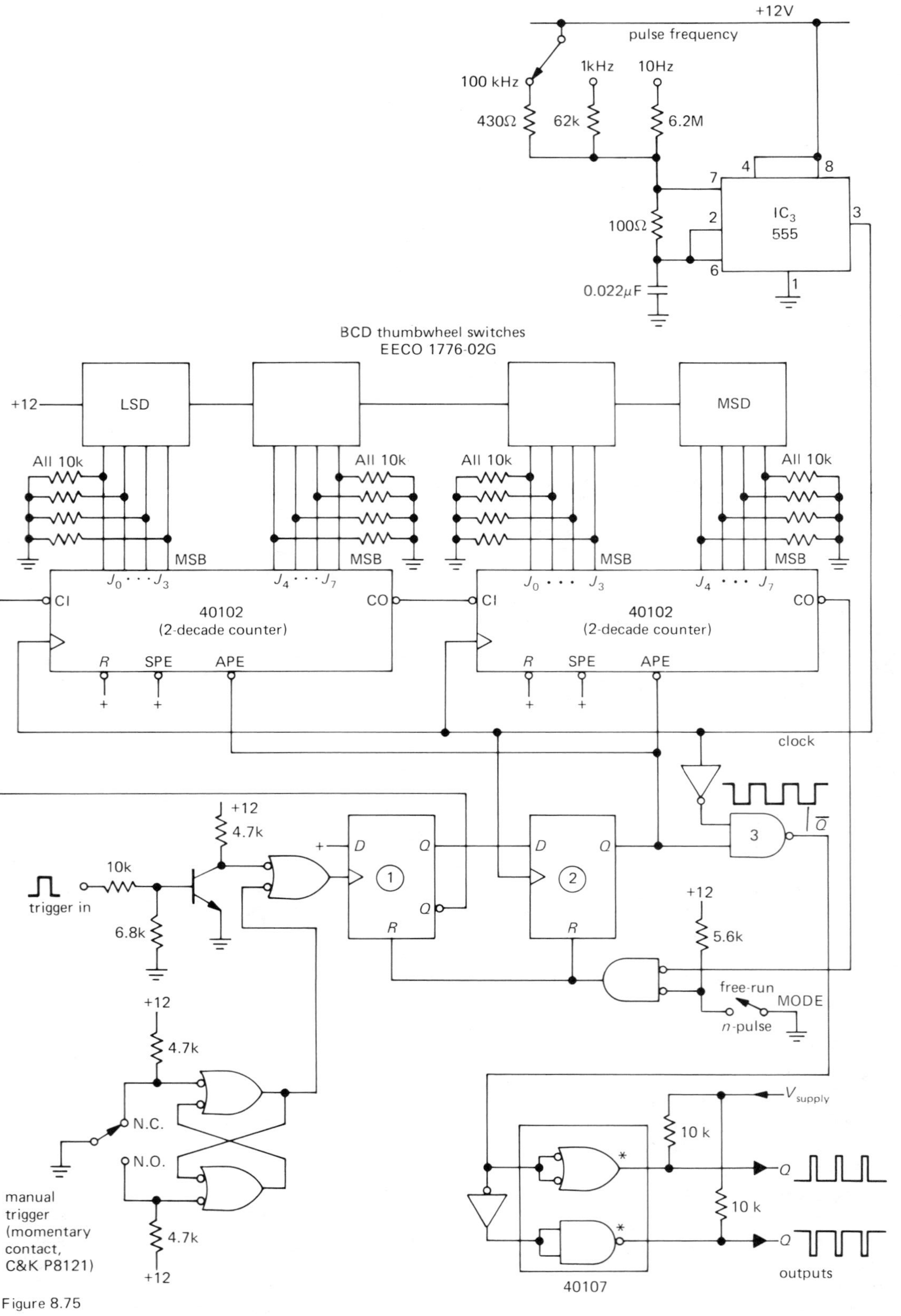

Figure 8.75

Laboratory *n*-pulse generator.

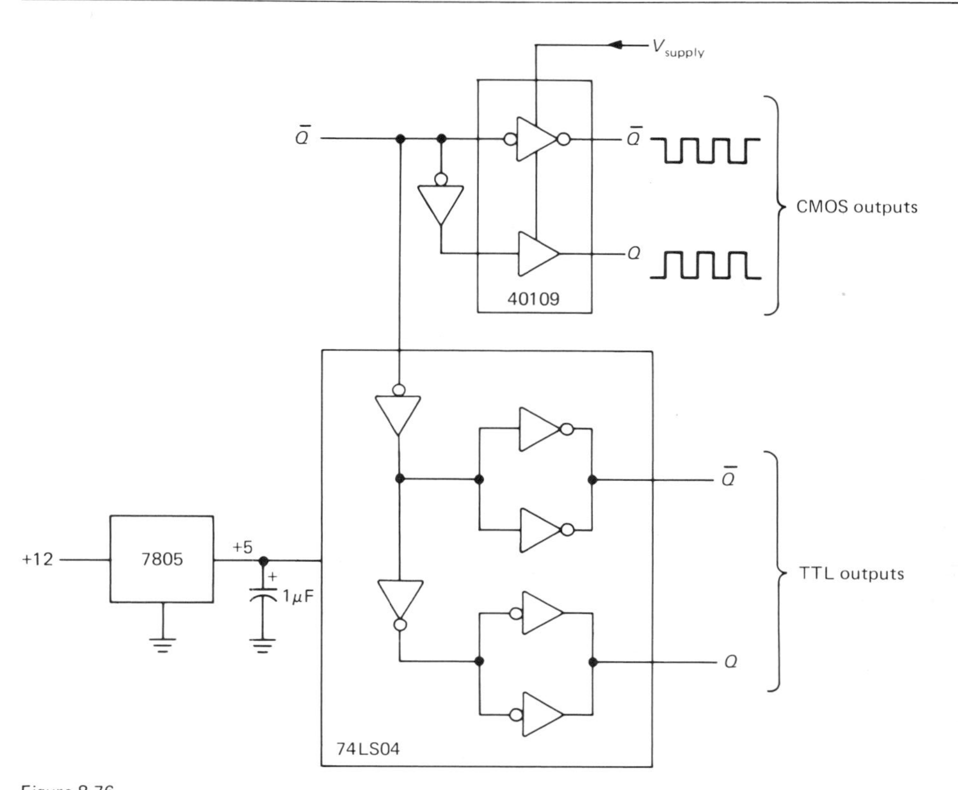

Figure 8.76

right voltage for the circuit under test. Unfortunately, the 40109 doesn't have enough drive compatability for TTL loads, so a 74LS04 hex inverter is used to generate TTL levels. That means another supply voltage, unless you run the CMOS from +5 volts also. Two interesting points about the TTL circuit: (a) It's OK to connect inverters in parallel, to get increased drive capability, if they're on the same chip. (b) Most low-power Schottky TTL gates use a diode input circuit that allows input levels up to +15 volts, so you can drive them directly from CMOS, assuming you have sufficient fanout. However, don't try this with other TTL subfamilies!

This last output circuit illustrates the kind of compromise you often face in circuit design. It has improved properties (active pull-up on the CMOS outputs, and TTL levels without connection to any external supplies), but it also requires additional supply voltages, making operation from a single 9 volt battery impractical. In such situations you have to decide what's most important, although you can sometimes have the best of both worlds by going to discrete circuitry of greater complexity. In situations like these, a good knowledge of transistor and FET circuit design may be invaluable. Chapter 9 treats this area in detail.

LOGIC PATHOLOGY

There are interesting, and sometimes amusing, pitfalls awaiting the unsuspecting digital logician. Some of these, such as logic races and lockup conditions, can occur regardless of the logic family in use. Others (e.g., "SCR latch-up" in CMOS chips) are "genetic abnormalities" of one logic family or another. In the following sections we have collected our bad experiences in the hope that such anecdotes can help others avoid such problems.

8.33 dc Problems

Lockup

It is easy to fall into the trap of designing a circuit with a lockup state. Suppose you have some gadget with a number of flip-flops, all going through their proper states. Everything seems to work fine. Then one day it just stops dead. The only way you can get it to work is to turn the power off and back on again. The problem is that there is a lockup state (an excluded state of the system that you can't escape from), and you got into it because of some power-line transient that sent the system into the forbidden state. It is very important to look for such states when you design the circuit and rig up logic so that the circuit recovers automatically. At a minimum, things should be arranged so that a RESET signal (generated manually, at start-up, etc.) brings the system to a good state. This may not require any additional components (e.g., Exercise 8.23).

Start-up clear

A related issue is the state of the system at start-up. It is always a good idea to provide some sort of RESET signal at start-up. Otherwise the system may do weird things when first turned on. Figure 8.77 shows a

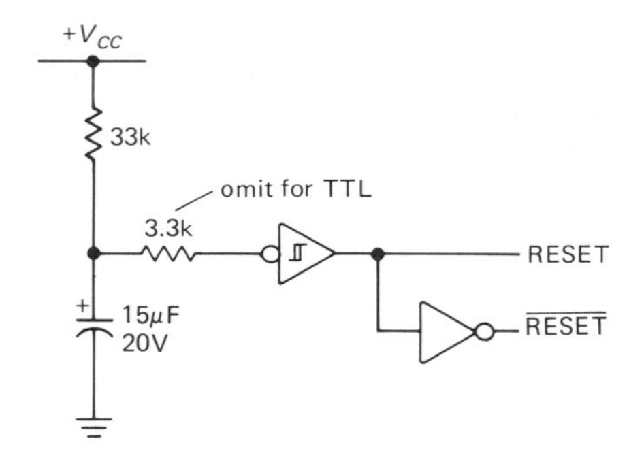

Figure 8.77
Power-on clear circuit.

suitable circuit. The series resistor is necessary with CMOS to prevent damage when power is removed from the circuit, since otherwise the electrolytic capacitor will try to power the system via the CMOS input-gate protection diode. A Schmitt trigger (4093, 74LS14) may be a good idea, to make the RESET signal switch off cleanly. The hysteresis symbol shown in the figure indicates an inverter with Schmitt trigger input, e.g., the TTL 74LS14 (hex inverter) or CMOS 4093 (quad 2-input NAND) or 40106 (hex inverter).

8.34 Switching problems

Logic races

Lots of subtle traps lurk here. The classic race was illustrated with the pulse synchronizer in Section 8.19. Basically, in any situation where gates are enabled by signals coming from flip-flops (or any clocked device), you must be sure that a gate doesn't get enabled and then disabled a flip-flop delay time later. Likewise, make sure that signals appearing at flip-flop inputs aren't delayed with respect to the clock (another plus for synchronous systems!). In general, delay the clock rather than the data. It is surprisingly easy to overlook a race condition.

□ *Metastable states*

As we remarked earlier, a flip-flop (or any clocked device) can get confused if the data input changes during the setup-time interval preceding a clock pulse. In the worst case the flip-flop's output can hover at the logic thresholds literally for microseconds (by comparison, normal propagation delays are around 20ns for LS TTL). This is not usually of concern to the logic designer, but it can create problems in fast systems where asynchronous signals must be synchronized. It has been blamed for mysterious computer crashes, although we are skeptical. The cure involves a set of concatenated synchronizers, or a "metastable state detector" that resets the flip-flop.

Clock skew

Clock skew affects CMOS more than TTL. The problem arises when you have a clocking signal of slow rise time driving several interconnected devices (Fig. 8.78). In this case two shift registers are clocked by a slowly rising edge, caused by capacitive loading of a relatively high impedance CMOS output (around 500Ω, when operating from +5V). The problem is that the first register may have its threshold at a lower voltage

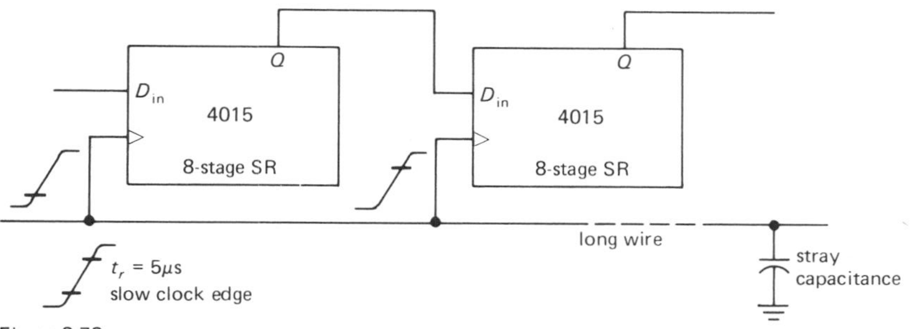

Figure 8.78

Slow rise times cause clocking skew.

than that of the second register, and this causes it to shift earlier than the second register. The last bit of the first register is then lost. CMOS devices can display quite a spread of input threshold voltages, which compounds the problem (the threshold is specified only to be between one-third and two-thirds of V_{DD}, and they mean it!). The best cure is to use a nearby chip without much capacitive loading to drive clock inputs in this sort of situation.

Runt pulses

In Section 8.28 (modulo-*n* counter) we remarked that some delay should be added if a counter's output clears itself, in order to prevent a pulse of substandard width. The same comment goes for LOAD pulses when using counters or shift registers. Runt pulses will make your life miserable, since you may have marginal operation or intermittent failures. Use the worst-case propagation delay specifications when designing.

8.35 Congenital weaknesses of TTL and CMOS

We will divide this section into nuisance problems and really bizarre behavior.

Nuisance problems

TTL. You have to remember the TTL inputs *source* current when held in the LOW state. That makes it difficult to use *RC* delays, etc., because of the low impedances necessary, and in general you have to give some thought when interfacing linear levels to TTL inputs.

The TTL threshold is too close to ground, making the whole logic family somewhat noise-prone (more on this in Chapter 9). Because TTL is fast, it recognizes short spikes on the ground line. These spikes are easily generated by the fast output transition speeds, making the problem worse.

TTL makes demands on the power supply: +5 volts ±5%, with relatively high power dissipation. Power-supply current spikes generated by the active pull-up output circuitry generally require liberal use of power-supply bypassing, ideally one capacitor per chip.

Finally, "standard" TTL inputs have a low base-emitter breakdown voltage, making it inadvisable to connect inputs to the +5 volt supply. (If the power-supply pass transistor fails, the voltage will soar to the crowbar limit, typically around 6.5–7V. This is below the TTL 7V supply-voltage limit, but above the 5.5V input breakdown.) You're obliged to use resistor pull-ups, or some other subterfuge.

CMOS. CMOS inputs are prone to damage from static electricity. The mortality rate really climbs in winter! The inputs show a large spread in logic threshold voltage, which in combination with the high output impedance (200–500Ω) generates problems of clock skew (see Section 8.34). You can get double output transitions on slowly rising inputs. CMOS requires *all* unused inputs, even those of unused gate sections, to be connected to HIGH or LOW. Finally, CMOS is just plain slow, when operated from a +5 volt supply.

Bizarre behavior

TTL. TTL doesn't do many really weird things. However, some TTL monostables will trigger on a glitch on the supply (or ground) line, and they generally behave somewhat fidgety. And a circuit that works well with standard TTL may malfunction when replaced by LS TTL. The latter seems more sensitive to short ground-line spikes. Most weird TTL operation can be traced to noise problems.

CMOS. CMOS can drive you crazy! For example, the chip can go into "SCR latch-up" if the input is driven beyond the supply momentarily. A 50mA input current through the input-protection diodes is what it takes. Then the chip gets hot, and you have to turn the power supply off before it will behave itself again. If you let this happen for more than a few seconds, you'll have to replace the chip.

CMOS has some strange and subtle failure modes. One of the output FETs can open up, giving pattern-sensitive failures that are difficult to detect. An input may begin to source or sink current. Or the whole chip may start drawing substantial supply current. Putting a 10 ohm resistor in series with each chip's V_{DD} lead makes it easy to locate faulty CMOS chips that are drawing quiescent supply current.

Circuit ideas

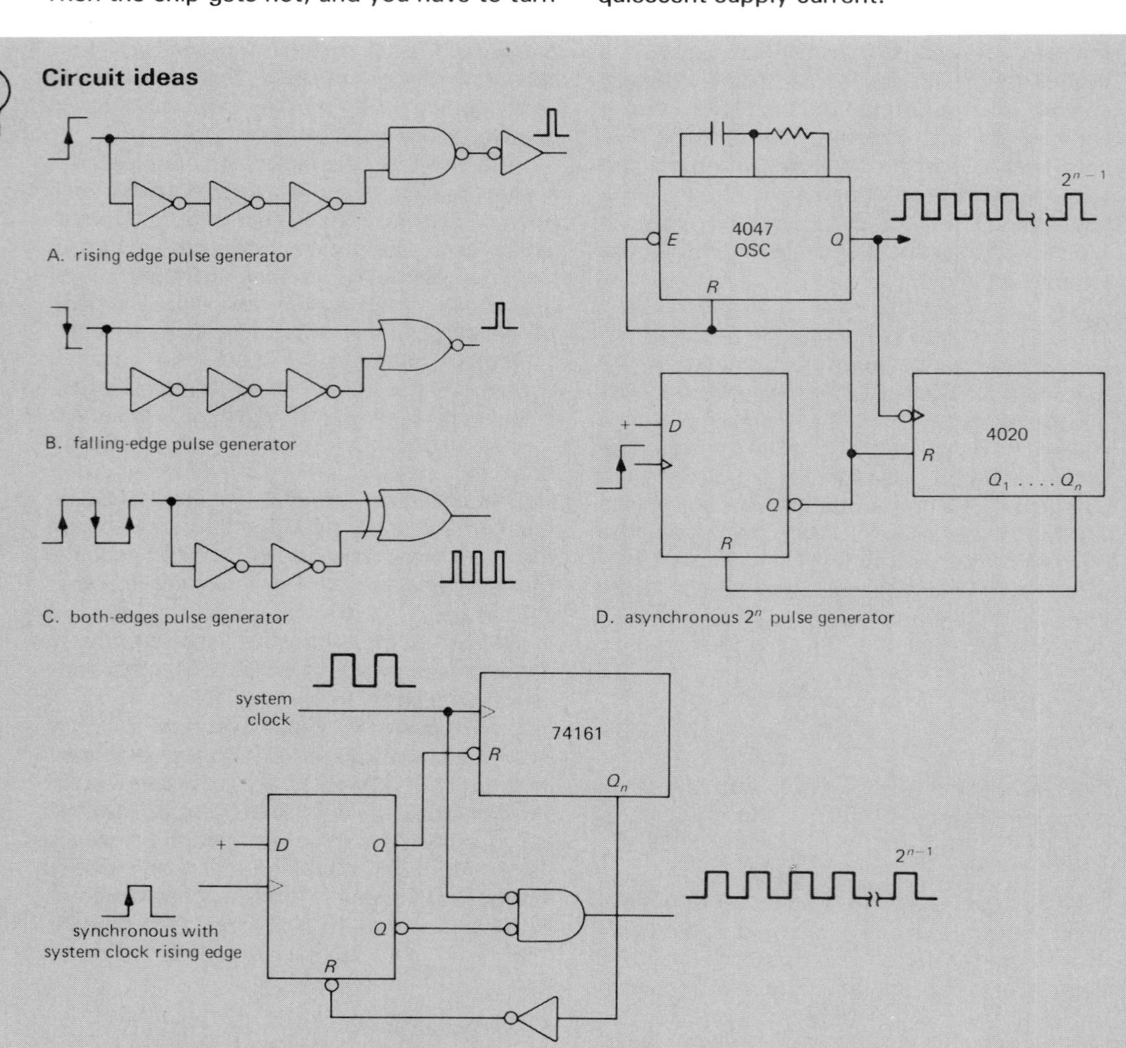

A. rising edge pulse generator

B. falling-edge pulse generator

C. both-edges pulse generator

D. asynchronous 2^n pulse generator

E. synchronous 2^n pulse generator

Figure 8.79

Besides the input threshold skew between chips, a single chip can exhibit different thresholds for several functions driven from a single input. For example, the RESET input of a 4013 brings *Q'* HIGH before it brings *Q* LOW. This means that you should not terminate a reset pulse based on the output at *Q'*, since the runt pulse that will be generated may actually fail to clear the flip-flop. For example, the digital delay circuit in Figure 8.61 won't work if you try to eliminate the inverter by clearing the counter directly from the *Q'* output.

Open inputs on CMOS chips are bad news! You might have a circuit that intermittently misbehaves. You put your scope probe onto a point in the circuit, and it shows zero volts, as it should. Then the circuit works fine for a few minutes! What happened is that the scope discharged the open input, and it took a long time to charge back up to the threshold.

Here's the craziest of them all: You forgot to wire up the V_{DD} pin on a CMOS chip, but the circuit works just fine! That's because it is being powered by one of its logic inputs (via the input-protection diode from the input to the V_{DD} line in the chip). You might get away with this for a long time, but suddenly the circuit reaches a state where all the logic inputs to the chip are simultaneously LOW; the chip loses its power and forgets its state. Of course, this is a bad situation anyway, since the output stage isn't adequately powered and can't source much current. The trouble is that this situation may produce symptoms only occasionally, and it can have you running around in circles until you figure out what's going on.

SELF-EXPLANATORY CIRCUITS

8.36 Circuit ideas

Figure 8.79 shows some digital circuit ideas.

8.37 Bad circuits

Figure 8.80 shows some classic digital circuit blunders.

ADDITIONAL EXERCISES

(1) Show how to make a *JK* flip-flop using a type *D* flip-flop and a 4-input multiplexer. Hint: Use the address inputs for *J* and *K*.

(2) Design a circuit that reads out, on 7-segment digits, how many milliseconds you've held a button down. The device should be smart enough to reset itself each time. Use a 1.0MHz oscillator.

(3) Design a reaction timer. "A" pushes his button; an LED goes on, and a counter begins counting. When "B" pushes her button, the light goes out and an LED display reads the time, in milliseconds. Be sure to design the circuit so that it will function properly even if A's button is still held down when B's button is pushed.

(4) Design a period counter: a device to measure the number of microseconds in one period of an input waveform. Use a Schmitt comparator to generate TTL; use a 1.0MHz clock frequency. Make it work so that pushing a button initiates the next measurement.

(5) Add latches to the period counter, if you haven't already.

(6) Now make it measure the time for 10 periods. Also, have it light an LED while it's counting.

(7) Design a true electronic stopwatch. Button *A* starts and stops the count. Button *B* resets the count. The output should be of the form *xx.x* (seconds and tenths); assume that you have a 1.0MHz square wave.

(8) Some stopwatches use a single button (start, stop, reset, start, etc., each time it is pushed). Design an electronic equivalent.

(9) Design a nice frequency counter to measure the number of cycles per second of an input waveform. Include lots of digits, latched count while counting the next interval, and choice of 1 second, 0.1 second, or 0.01 second counting interval. You might add a good input circuit with several sensitivities, a Schmitt trigger with adjustable hysteresis and trigger point (use a fast comparator), and a logic signal input for TTL signals. How about a BCD output? Multiplex the digits on output, as well as parallel output? Spend some time on this one.

Bad circuits

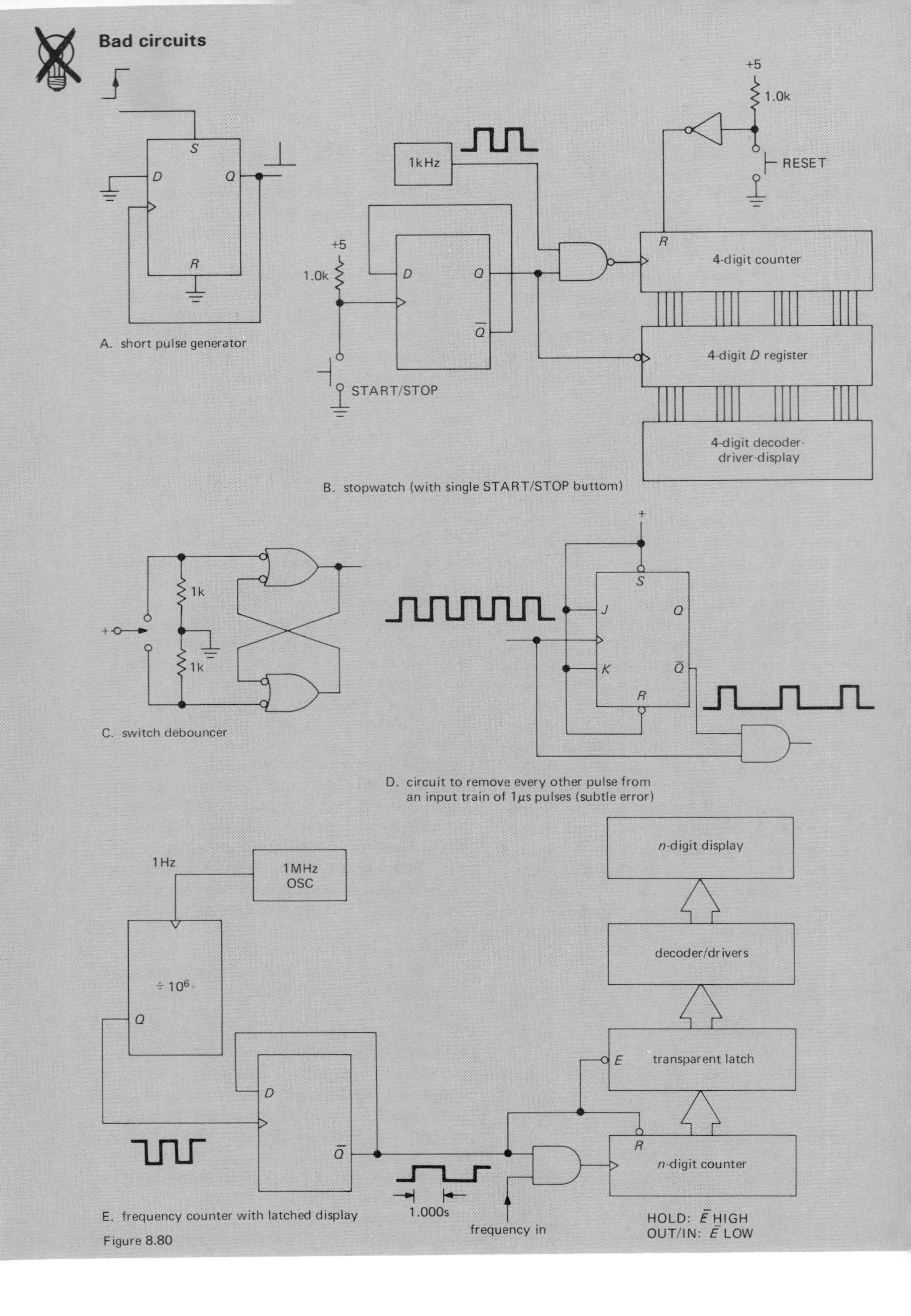

A. short pulse generator

B. stopwatch (with single START/STOP buttom)

C. switch debouncer

D. circuit to remove every other pulse from an input train of 1μs pulses (subtle error)

E. frequency counter with latched display

Figure 8.80

(10) Design a circuit, using TTL logic, to time a speeding bullet. The projectile breaks a thin wire stretched across its path; then, some measured distance farther along its path, it breaks a second wire. Beware of problems like "contact bounce." Assume that you have a 10MHz TTL square wave, and design your circuit to read out, in microseconds (4 digits), the time interval between breaking the two wires. A push button should reset the circuit for the next shot.

(11) Make a 1-of-16 decoder from two 74LS42s (1-of-10). The input is a 4-bit binary number. Output will be negative-true (as with the 74LS42). Hint: MSB input to the 74LS42 can be used as an "ENABLE."

(12) Imagine that you have four 256-bit ROMs, TTL style, each of which has an 8-bit parallel address input, a three-state positive-true output, and an output enable (negative-true); i.e., the ROM asserts the selected data bit at its output if the enable is LOW. Show how to "expand" these into a 1024-bit ROM, using whatever else you need. (A 74LS42 might be handy, or you can do it with gates. Try both.)

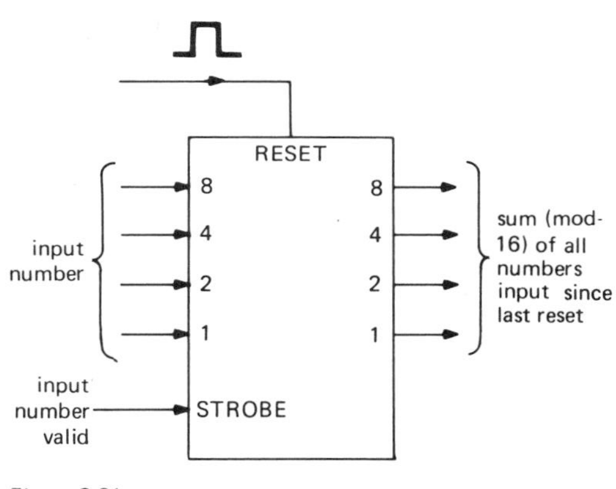

Figure 8.81

(13) Invent a circuit to keep a running sum of successive 4-bit numbers that are input to it. Only keep your result to 4 bits (i.e., perform a sum modulo-16). (Such a sum is useful as a "check sum" to be written on inherently error-prone data-recording media, e.g., punched paper tape.) Assume that a positive 1μs TTL pulse occurs once during the time that each input number is valid. Provide a reset input. Thus, your circuit is as shown in Figure 8.81.

Now add another feature to the circuit, namely an output bit that is 1 if the *total number* of 1 bits in all the input numbers (since the last reset input) is odd, 0 if even. Hint: An XOR "parity tree" will tell you if the sum of 1's in each number is odd; figure it out from there.

(14) In Exercise 8.14 you designed a 2 × 2 multiplier by using a Karnaugh map for each output bit. In this problem you are to accomplish the same task by the process of "shift and add." Begin by writing out the product the way you would in elementary school:

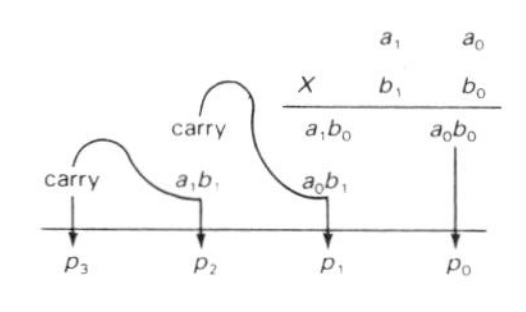

This process has a simple repeating pattern, requiring 2-input gates (what kind?) to generate the intermediate terms a_0b_0, etc., and 1-bit "half adders" (adders that have a carry out but no carry in) to sum the intermediate terms.

(15) Now design a 4 × 4 multiplier along the same lines, this time using three 4-bit full adders (74LS83) and 16 2-input gates.

Although sheer "number crunching" is an important application of digital electronics, the real power of digital techniques is seen when digital methods are applied to analog (or "linear") signals and processes. In this chapter we will begin by reviewing the input and output properties of the TTL and CMOS families in order to understand how to interface the two logic families to each other and to digital input devices (switches, keyboards, comparator outputs, etc.) and output devices (indicator lamps, relays, etc.). We will also look at *p*-channel and *n*-channel MOS logic, since this is widely used in LSI functions. We will continue with the important subject of bringing digital signals on and off circuit boards, in and out of instruments, and through cables. Then we will discuss the important subject of conversion between analog and digital signals. Finally, with an understanding of these techniques, we will look at a number of applications in which combined analog and digital techniques provide powerful solutions to interesting problems.

TTL AND CMOS LOGIC INTERFACING

9.01 TTL and CMOS logic families

An understanding of the input and output properties of a logic family is essential for any interconnection to the outside world. As usual, we will consider TTL and CMOS in detail, since they are the logic families of choice for nearly all applications.

Varieties of TTL

TTL itself comes in a number of varieties. For most purposes the "low-power Schottky" series is currently the best choice. Table 9.1 presents a quick summary of the TTL family tree.

Most popular functions are available in most of the TTL families. For instance, a 7474 is a dual type *D* flip-flop, with identical pinouts in all families. You would specify a 7474, 74L74, 74S74, 74LS74, etc. Input and output logic levels are the same, resulting in general compatibility among families, restricted in some cases by output drive limitations (e.g., a 74L*xx* series device can drive only two 74*xx* loads, compared with the usual 10-load "fanout" within any one family). Important note: There are some other TTL families, such as the 2500 series from AMD, the 8000 series from National Semiconductor, a *different* 8000 series from Signetics, and the Fairchild 9000 series. These TTL families also offer LS, S, and L versions. To simplify the discussion, we will refer only to the relevant 7400 series; for instance, when we mention the 74LS*xx* series, the LS functions from these other TTL families are included as well.

The 74*xx* series has been the standard for a decade, but it is now being replaced by the low-power Schottky series (74LS*xx*), which offers slightly greater speed at about one-fourth the power. The recently introduced "advanced low-power" families (74ALS*xx*) promise even better speed/power ratios. "Schottky" refers to the use of metal-to-semiconductor junction diodes (Schottky diodes) as clamps to prevent transistor saturation, with consequent reduction of switching time (see Section 13.22)

The "Schottky" series (74S*xx*) has been the traditional speed leader, but the newly introduced "fast" (74F*xx*) and "advanced Schottky" families offer even better speed and speed/power ratios. When broad selections of these new series of TTL become available at competitive prices, they may well replace the 74S*xx* and 74LS*xx* series. The "low-power" series (74L*xx*) offers lowest power, but because of its very low speed it has also been replaced by 74LS*xx*. The 74H*xx* series was the original "high-speed" family, offering moderate increase in speed over the standard series, with correspondingly higher power dissipation; it has been rendered nearly obsolete by the 74S*xx* series, and it is now used occasionally where high output drive current is needed.

As a guide for new TTL designs, we would recommend (a) 74LS*xx* for nearly everything, with 74*xx* used where additional drive current is needed (e.g., driving lamps), and (b) 74S*xx* (or 74F*xx* or 74AS*xx*) where very high speeds are needed. The recent proliferation of new TTL subfamilies appears to represent a transient situation that will probably resolve itself in a few years, with perhaps three or four families ultimately remaining.

Varieties of CMOS

CMOS is the family of choice for lowest power, where speed is not important. It also imposes fewest requirements on the power supply (+3V to +18V at very low current), and it is therefore a good general-purpose logic family, especially for small simple circuits where a power supply suitable for TTL would be more complicated than the logic circuit itself.

There are a few choices available in CMOS. The original 4000 series was pioneered by RCA, then supplanted by the 4000A series. The newer B series is now

TABLE 9.1. TTL FAMILIES COMPARED

	Gate					
Series	Delay	Power	Flip-flop max f	I_{in} (low) (max)	I_{out} (sink) (min @ 0.4V)	Comments
74*xx*	9ns	10mW	15MHz	1.6mA	16mA	The original TTL; good speed, but high power
74LS*xx*	9.5ns	2mW	25MHz	0.4mA	4mA	Best for new designs; as fast as standard TTL, much less power
74S*xx*	3ns	19mW	75MHz	2.0mA	16mA	High speed; may be supplanted by 74F*xx*, when full family is available
74F*xx*	2.7ns	4mW	115MHz	0.6mA	16mA	New high-speed low-power series; full family not yet available
74L*xx*	33ns	1mW	2.5MHz	0.2mA	3.6mA	Lowest power, slow; obsolete
74H*xx*	6ns	22mW	35MHz	2.0mA	20mA	Obsolete high-speed family; replaced by 74S*xx*

the standard 4000 series CMOS part. It claims to operate with supply voltages from 3 to 18 volts. In reality, operation even from 5 volts is frequently unsatisfactory, because of the low speed and high output impedance. The Fairchild 4000B series "isoplanar CMOS" offers somewhat better performance at low supply voltages. But if you want speed and low output impedance, it is best to operate CMOS from a supply voltage of 10 or 12 volts.

The 74Cxx series of CMOS is the main alternative to the 4000 series. It is *functionally* identical to the other TTL series (with identical pinouts), but it is actually CMOS, with the same power-supply range, logic levels, and input/output characteristics as CMOS. It is completely compatible with the standard 4000 series of CMOS, but it has only limited compatibility with TTL. We will discuss the important issue of CMOS/TTL interfacing shortly.

There are also some specialty low-threshold metal-gate CMOS chips (e.g., wristwatch circuits) with unusual (and low) supply voltage specifications. As an example, there is a nice oscillator/divider chip that requires a 1.0 to 2.1 volt supply. It is incompatible with both TTL and standard CMOS.

9.02 Input and output characteristics of TTL and CMOS

Digital logic families are designed so that the output from a chip can properly drive many inputs within the same logic family. A typical fanout capability is 10 loads, meaning that an output from a gate or flip-flop, for example, can be connected to 10 inputs and still perform within specifications. In other words, in normal digital design practice you can get by without knowing anything about the electrical properties of the chips you're using, as long as your circuit consists only of digital logic driving more digital logic of the same type. In practice, this means that you don't often have to worry about what's actually going on at logic inputs and outputs.

However, as soon as you attempt to drive digital circuitry with externally generated signals, whether digital or analog, or whenever you use digital logic outputs to drive other devices, you must face the realities of what it takes to drive a logic input and what a logic output can drive. Furthermore, when mixing logic families it is essential to understand the circuit properties of logic inputs and outputs. Interfacing between logic families is not an academic question. In order to take advantage of the advanced MOS LSI chips that are increasingly becoming available, you must know how to mix logic types. In the next few sections we will consider the circuit properties of logic inputs and outputs in detail, with examples of interfacing between logic families and between logic devices and the outside world.

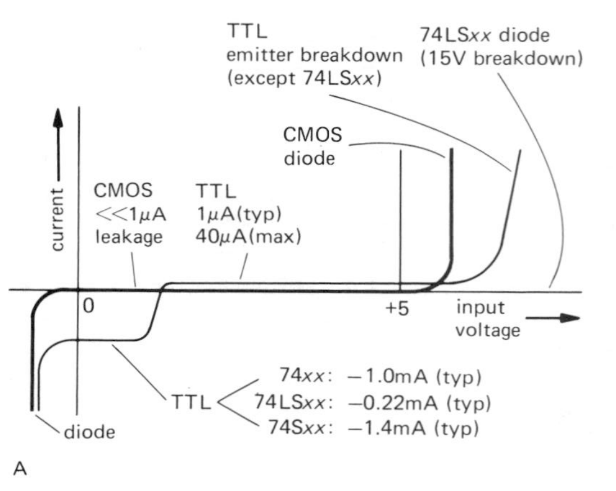

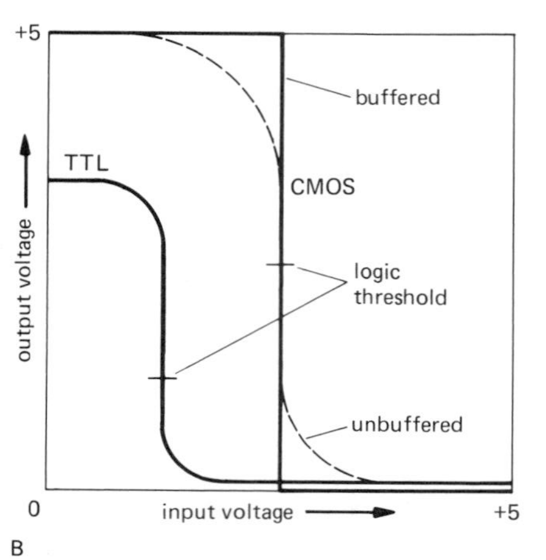

Figure 9.1
Logic gate characteristics. A: Input current. B: Transfer function.

Input characteristics

The graphs in Figure 9.1 show the important properties of TTL and CMOS inputs: input current and output voltage (for an inverter) as a function of input voltage. We have extended the graphs to input voltages beyond the range encountered in purely digital circuits, since in interface situations the input signals might easily exceed the power-supply voltages. As the graphs imply, both TTL and CMOS are normally operated with the negative supply pin connected to ground.

A TTL input sources a sizable current when held LOW, and it draws only a small current when HIGH (typically a few microamps, never more than 40μA); the HIGH input current is actually the effective collector current of the "inverted input transistor," and not leakage current, as is commonly believed (Fig. 9.2). To drive a 74*xx* TTL input, you must be able to sink a milliamp or so while holding the input below 0.4 volt. Failure to understand this may lead to widespread circuit malfunction in interfacing situations! For input voltages below ground, a TTL input looks like a clamping

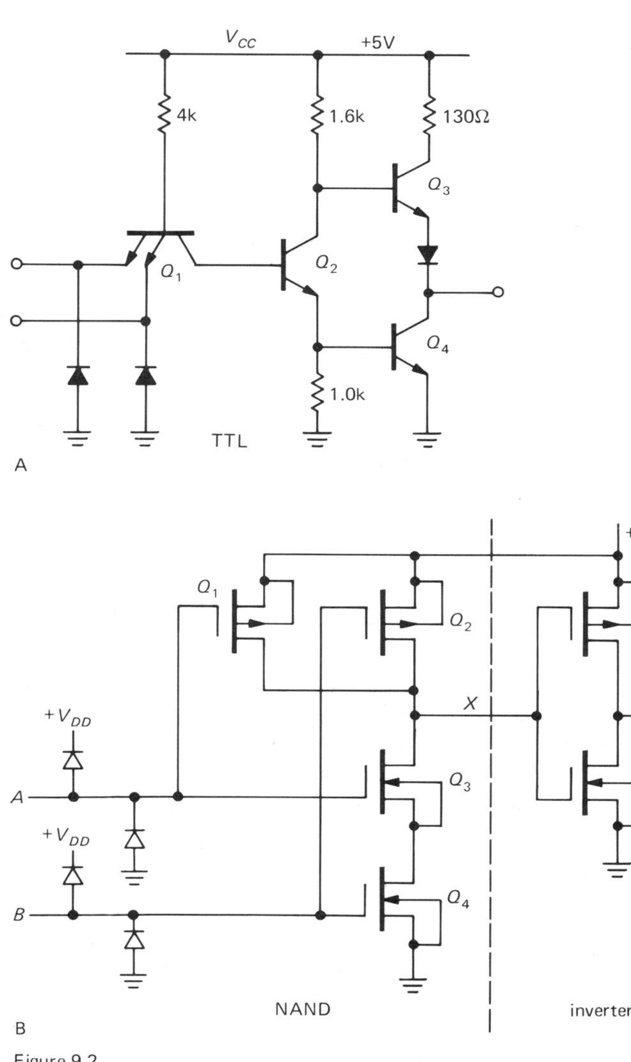

Figure 9.2
A: TTL NAND gate. B: CMOS AND gate.

diode to ground, and for inputs above +5 volts it looks like a transistor with low breakdown voltage somewhere above (but not much above) +5.5 volts. The input logic transition threshold is typically around +1.3 volts, but it can range from +0.8 volt to +2.0 volts in the worst case. TTL gates with Schmitt-trigger inputs are available (7413, 7414, 74132), with ±0.4 volt hysteresis; they are indicated by a hysteresis symbol within the gate outline (e.g., Fig. 9.30). V_{supply} (usually called V_{CC}) is +5.0 volts, ±5%.

A CMOS input draws no current (except for leakage current, typically $10^{-5}\mu A$) for input voltages between ground and the supply voltage. For voltages beyond the supply range, the input looks like a pair of clamping diodes to the positive supply and ground (Fig. 9.2). The current through these diodes must never exceed 10mA, even momentarily! These are the famous input-protection diodes, without which CMOS would be extremely susceptible to damage from static electricity during handling (as it is, CMOS is still relatively fragile). The input logic threshold voltage is typically at half the supply voltage, but it can range from one-third to two-thirds of V_+ (V_+ is called V_{DD}). CMOS gates with Schmitt-trigger inputs are available (4093, 40106, 4584) with 1 to 2 volts hysteresis; they are indicated by a hysteresis symbol within the gate outline (e.g., Fig. 9.8). V_{supply} can range from +3 to +18 volts, with +5 and +12 volts being popular.

Output characteristics

The TTL output circuit is an *npn* transistor to ground and an *npn* follower to V_+, with a current-limiting resistor in its collector and (sometimes) a diode in series with its emitter (Fig. 9.2). One transistor is saturated and the other is off. As a result, a TTL device can sink a large current (16mA for 74*xx*) to ground with a small (saturation) voltage drop and can source a few milliamps with its output HIGH (about +3.5V). The output circuit is designed to drive TTL inputs, with a fanout of 10 (i.e., 1 output can drive 10 inputs).

The CMOS output circuit is a push-pull pair of complementary MOSFETs, one ON and the other OFF (Fig. 9.2). The output looks like a few hundred ohms to ground or to V_+ for small output currents, becoming something like a current source for output currents at which the output voltage has moved more than a volt or so away from the supply rails. Figure 9.3 shows a summary of the output characteristics in graphic form.

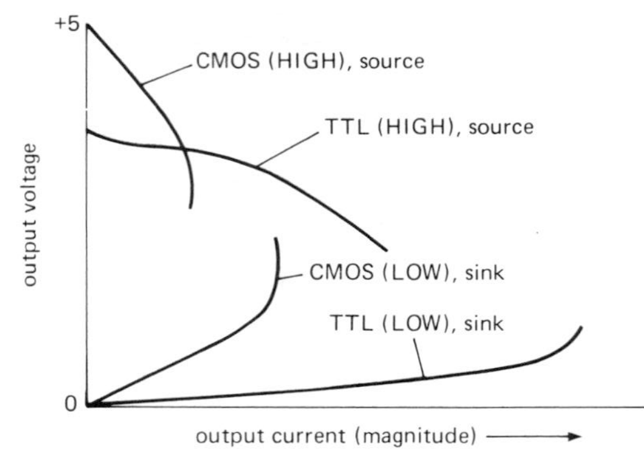

Figure 9.3
Logic gate output characteristics.

We have plotted the output voltage, for both HIGH and LOW output states, against output current. To simplify the graph, output current is drawn positive. Note that CMOS pulls its outputs all the way to V_+ or ground, generating a full supply swing if not heavily loaded. In normal use, CMOS outputs drive CMOS inputs. Since there is no input current (except for currents to charge the small input capacitance), the outputs swing fully to V_+ or ground. TTL levels, by comparison, are typically at 50mV (LOW) or +3.5 volts (HIGH) when driving other TTL devices as loads. With a pull-up resistor (almost any size will do), HIGH TTL outputs go to +5.

9.03 Interfacing between TTL and CMOS

It is important to know how to make CMOS and TTL talk to each other, in order to feel at home in either family. There are some very nice functions available in CMOS that you can't get in TTL. If you have a TTL system that isn't running at high speeds, you can mix in some CMOS functions without any trouble. Likewise, it is sometimes a good idea to put some TTL buffers at the inputs or outputs of a CMOS system, for easy connection to external TTL-compatible signals and for driving cables.

Driving CMOS from TTL

If the CMOS is run from +5 volts, the levels are almost compatible. The only problem is that the TTL HIGH (3.4V typical) is marginal for CMOS, which would like to see at least +4.3 volts. Just put a pull-up resistor to V_+ (3.3k, equivalent to one TTL load, is a possible value) on the TTL outputs, and all is well. This will work on both active pull-up and open-collector outputs.

For CMOS running from a higher supply voltage, you can still use a pull-up resistor, but only with open-collector "high voltage" TTL chips. Examples are the 7406 (hex inverter), 7407 (hex buffer), and 7426 (quad 2-input NAND). A second possibility is to use a CMOS "level-shifter" chip like the 40109; it accepts inputs relative to a V_{CC} supply (CMOS levels) and has CMOS outputs relative to a second V_{DD} supply. To drive CMOS operating at $V_{DD} > 5$ volts from TTL, the V_{CC} pin is connected to the TTL 5 volt supply, the V_{DD} pin is connected to the CMOS supply, and you use pull-ups on the standard TTL outputs, as previously. A third possibility is to use an *npn* transistor. Figure 9.4 presents schematics, with supply voltages shown explicitly.

In this transistor circuit a pair of resistors is used in the base circuit in order to put the input "threshold" at about two diode drops above ground (the same as a real TTL input) for good noise immunity. The "speedup" capacitor improves switching speed (see the discussion in Section 13.22). You sometimes see R_2 omitted; without it the transistor turns on at about 0.7 volt input, providing inadequate noise immunity, especially since 0.5 volt spikes on the ground line are not uncommon in TTL systems (see Section 9.14). Note that the *npn* transistor acts as an inverter. The pull-up resistor value in the preceding open-collector circuit can be much larger if speed is not important; smaller values can be used for improved noise immunity.

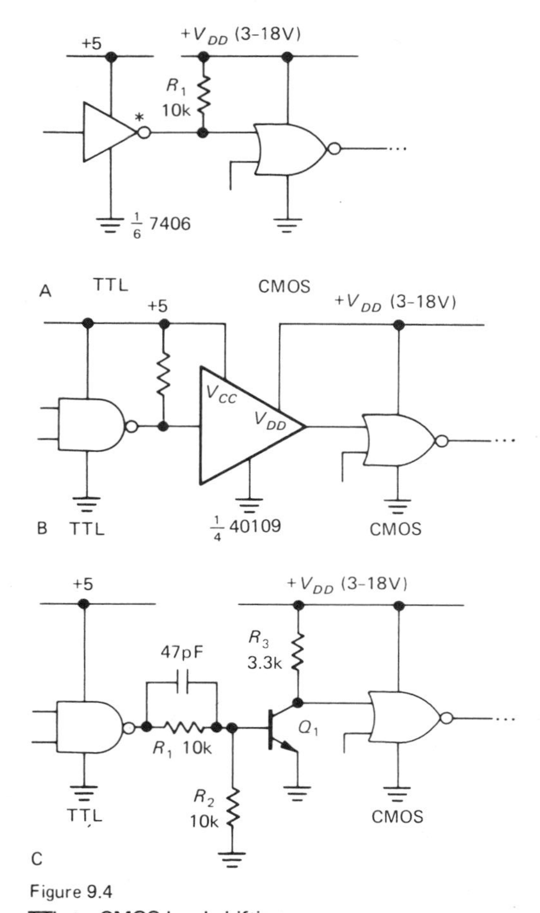

Figure 9.4
TTL-to-CMOS level shifting.

Driving TTL from CMOS

If the CMOS is running from +5 volts, it can drive a 74LS*xx* load or two 74L*xx* loads directly. CMOS buffers like the 4049 (hex inverter) or 4050 (hex buffer) can drive two 74*xx* loads or eight 74LS*xx* loads directly, and an "open-drain" buffer like the 40107 (with pull-up resistor to +5V) can drive 10 74*xx* loads or 40 74LS*xx* loads.

For CMOS running from a higher supply voltage, there are several good methods. First, you can still use a 4049/4050. The inputs of these two chips are allowed to exceed the supply voltage, so you connect their V_{DD} pins to +5 volts. This gives an output swing from zero to +5 volts capable of driving two 74*xx* loads or eight 74LS*xx*

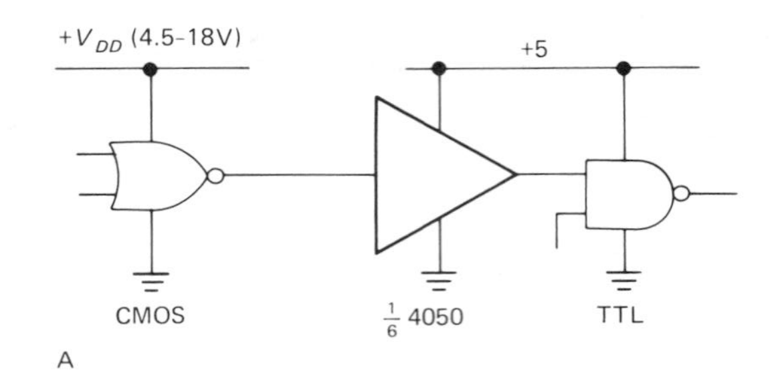

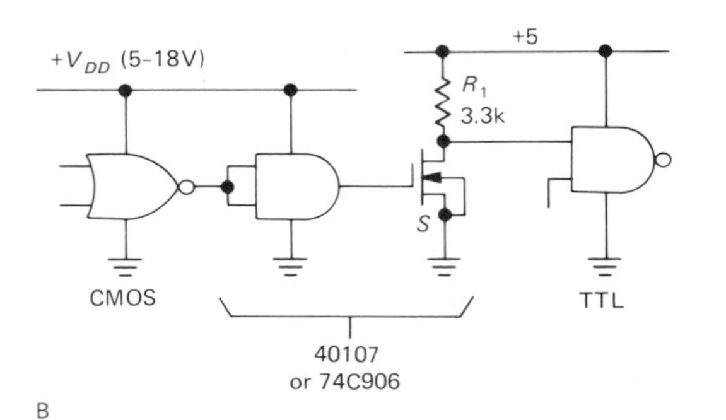

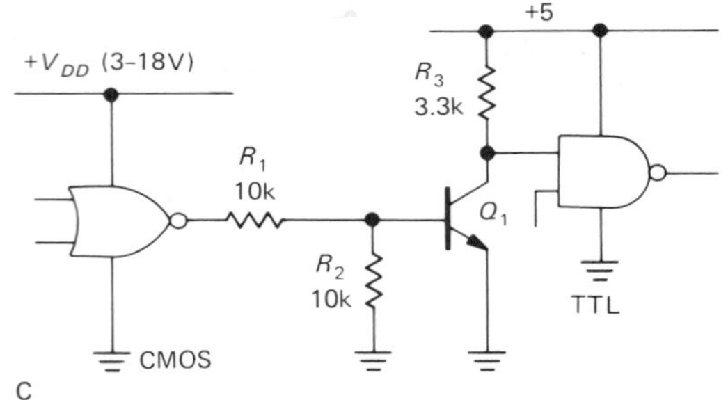

Figure 9.5
CMOS-to-TTL level shifting.

loads, with the input threshold around +2.5 volts. An alternative method is to use a 40107 or 74C906 operating from the CMOS supply, with pull-up to +5 volts. A third method is to use an *npn* transistor as previously. Figure 9.5 shows the schematics. As before, the transistor stage inverts.

9.04 Driving TTL and CMOS inputs

Switches as input devices

It is easy to drive digital inputs from switches, keyboards, comparators, etc., if you keep in mind the input characteristics of the logic you're driving. The simplest way is with a pull-up resistor (Fig. 9.6). When driving TTL, a pull-up resistor with switch closure to ground is by far the best method, because of the input properties of TTL. The switch easily sinks the LOW-state input current, and the pull-up brings the HIGH state to +5 volts, giving good noise immunity; in addition, it is convenient to have the switch return to ground.

The alternative possibility, a pull-down resistor to ground with switch closure to +5 volts, is undesirable because it requires a small pull-down resistor (such as 220Ω) to guarantee a TTL LOW of a few tenths of a volt, which means you have rather large currents with the switch closed. In the pull-up circuit, the noise immunity with the switch open (the worst case, from the standpoint of noise pickup) is at least 3 volts, whereas in the undesirable pull-down circuit it could be as little as 0.4 volt (stan-

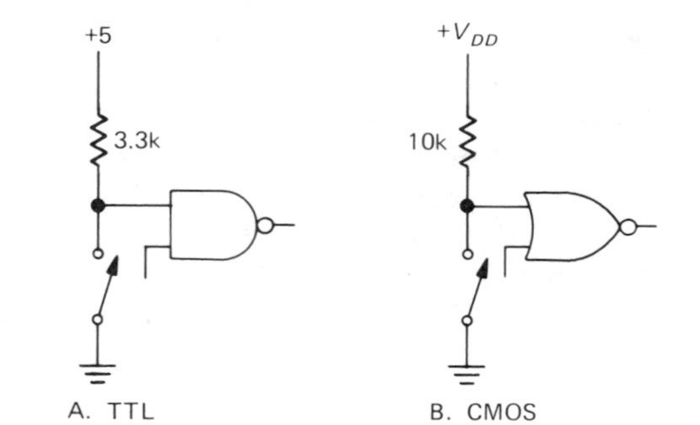

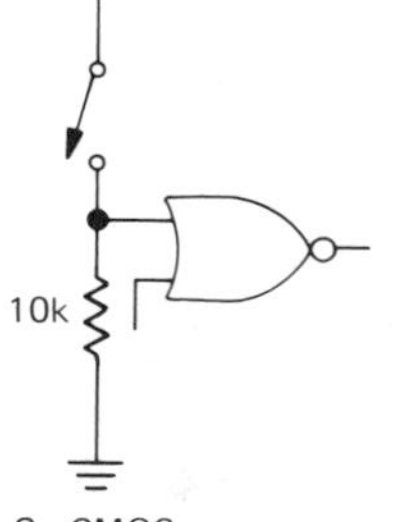

Figure 9.6
Mechanical switch to logic-level circuits (not debounced).

dard TTL, −1.6mA input current, LOW threshold at +0.8V). In addition, it is not a good idea to connect TTL inputs directly to the +5 volt supply, as you will see shortly.

Either pull-up or pull-down with CMOS is fine, since the inputs draw no current and the threshold is typically halfway from V_{DD} to ground. It is usually more convenient to ground one side of the switch, but if the circuit is simplified by having a HIGH input when the switch is closed, the method with pull-down resistor will be perfectly OK. Figure 9.6 shows these three methods.

Switch bounce

As we remarked in Chapter 8, mechanical switch contacts usually bounce for about a millisecond after closure. With large switches the bounce can last for as much as 50ms. This can wreak havoc with circuits that are sensitive to changes of state, or "edges" (a flip-flop or counter would toggle many times if clocked directly from a switch input, for instance). In such cases it is essential to debounce the switch electronically. Here are a few methods:

1. Use a pair of gates to make a jam-type *RS* flip-flop. Use pull-ups at the inputs to the debouncer, of course (Fig. 9.7). You could use a flip-flop with SET and CLEAR inputs instead (e.g., the 7474); in that case ground the clocking input.

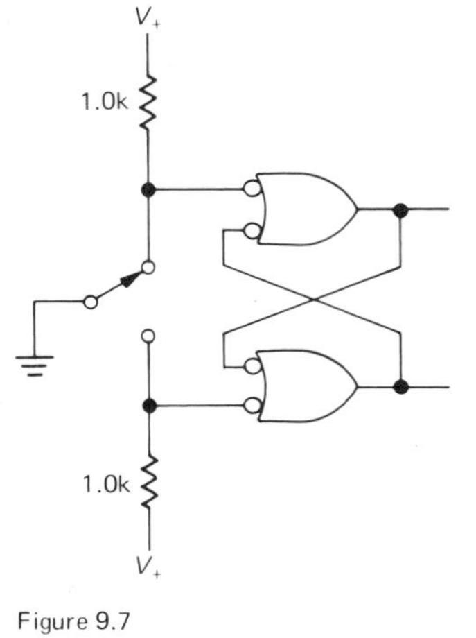

Figure 9.7
Switch debouncing circuit.

2. Use an integrated version of the preceding arrangement. The 74279, 4043, and 4044 are quad *RS* latches, and the 8544 is a "quad debouncer" with internal pull-ups, input "STROBE" (ENABLE), and three-state outputs.

3. Use an *RC* slowdown network to drive a CMOS Schmitt trigger (Fig. 9.8). The low-

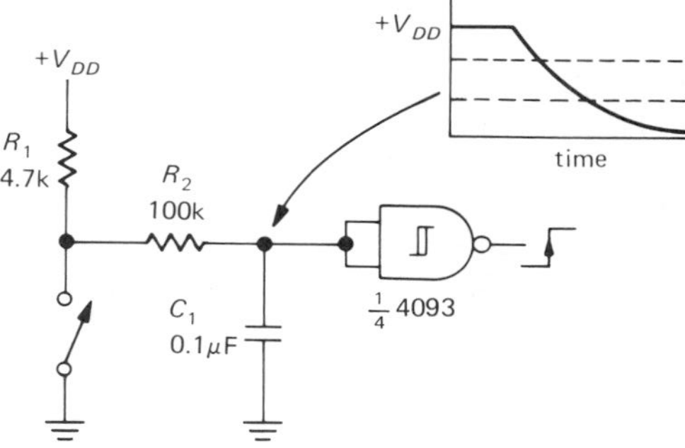

Figure 9.8

pass filter R_2C_1 smooths the bouncy waveform so that the Schmitt-trigger gate makes only one transition. A 10ms to 25ms *RC* time constant is generally long enough. This method isn't well suited to TTL because of the low driving impedance that TTL inputs require.

4. Use a chip like the 4490 "hex contact bounce eliminator," a nifty device that uses a digital delay (a 5-bit shift register for each switch) as a sort of digital low-pass filter. It includes internal pull-up resistors and clock circuitry. The user supplies a timing capacitor that sets the oscillator frequency, thus determining the delay time. There isn't an equivalent TTL device at present.

5. Use the circuit shown in Figure 9.9. The CMOS buffer can be a noninverting gate, as shown, or one section of a buffer chip like the 4050 or 4502. It's OK to short a CMOS

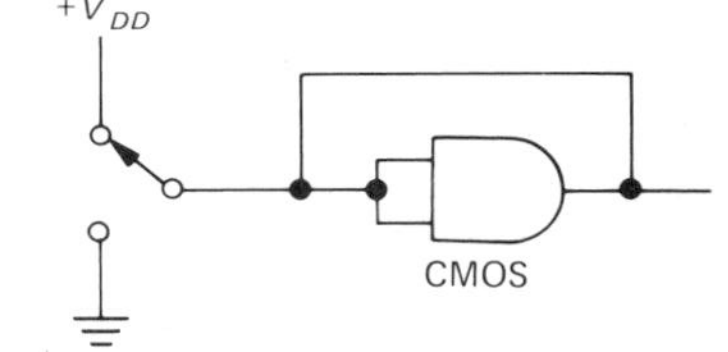

Figure 9.9

output to V_{DD} or ground momentarily, unlike the situation with TTL, which is protected only against shorts to ground, not to V_{CC}.
6. Use a device with built-in debouncer. Keyboard encoders, for instance, are designed with mechanical switches in mind as input devices, and they usually include debouncing circuitry.
7. Use a Hall-effect switch. These are magnetically operated solid-state switches, available as panel switches or keyboard switches. In either case the magnet and switch come as a complete assembly. They require +5 volts and produce a debounced logic output suitable for driving TTL or CMOS (operated from +5V). Since they have no mechanical contact assembly to wear out, Hall-effect switches last just about forever.

A few general comments on switches as input devices: Note that SPST switches (sometimes called "form A") can be used with methods 3 and 4 (and usually 6), whereas SPDT switches (form B) are necessary with the other methods. Keep in mind, also, that often it isn't necessary to debounce switch inputs, since they aren't always used to drive edge-sensitive circuitry. Another point: Well-designed switches are usually "self-wiping" to maintain a clean contact surface (take one apart to see what that means), but it's a good idea to choose circuit values so that a current of at least a few milliamps flows through the switch contacts to clean them.

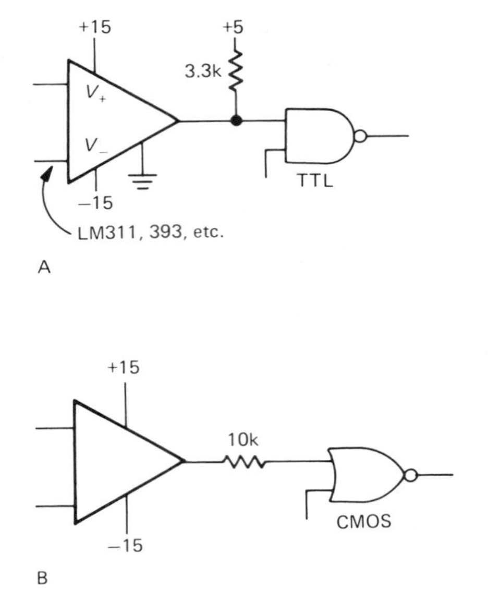

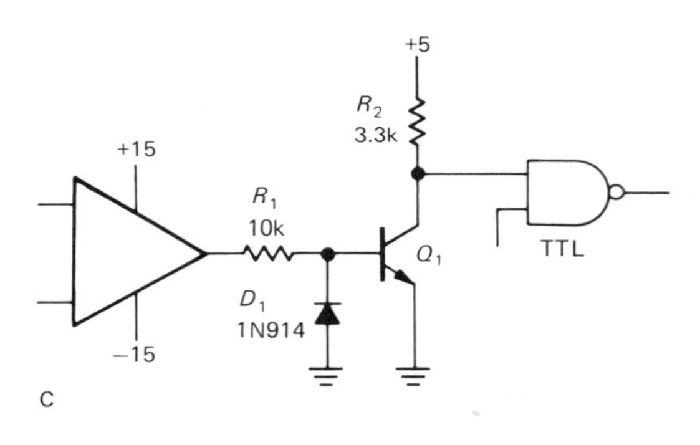

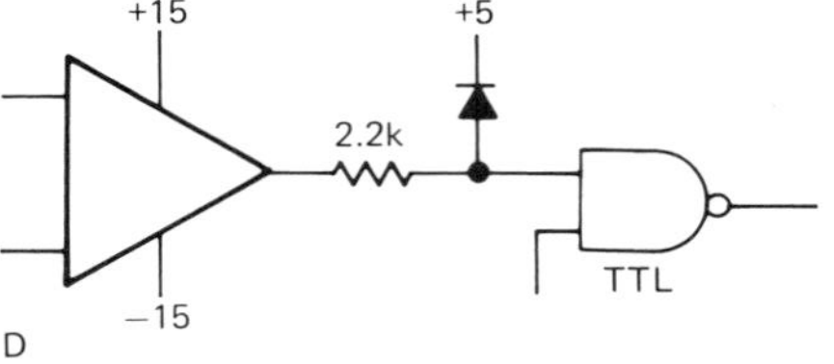

Figure 9.10

9.05 Driving digital logic from comparators and op-amps

Comparators and op-amps, together with analog-to-digital converters, are the common input devices by which analog signals can drive digital circuits. Figure 9.10 shows some examples. In the first circuit, a comparator drives TTL directly. Since most comparators have an *npn* output transistor with open collector and grounded emitter, all that's needed is a pull-up to +5 volts. The same scheme works for CMOS as well, with pull-up to $+V_{DD}$. The comparator may not have to run from split supply, since many are designed for single-supply operation ($V-$ grounded), and some will even operate with a single 5 volt supply (e.g., the 311, 339, 393, or 329).

The second circuit shows an op-amp driving CMOS with only a series current-limiting resistor. The input-protection diodes of the

CMOS device form effective clamps to V_{DD} and ground, provided that input current is kept below 10mA. In the third circuit an op-amp switches an *npn* transistor into saturation to drive a TTL load; the diode prevents base-emitter reverse breakdown (~6V). The last circuit is not highly recommended, but it does work. The input clamping diode of the TTL device limits negative swing to a diode drop below ground, and the external diode limits positive swing. The series resistor prevents damage if base-emitter reverse breakdown occurs in the TTL input transistor. The series resistor is chosen small enough to sink the TTL LOW-state input current when the op-amp output is a few volts negative.

Clock inputs: hysteresis

A general comment about driving digital logic from op-amps: Don't try to drive clock inputs from these op-amp interfaces; the transition times are too long, and you may get glitches as the input signal passes through the logic threshold voltage. If you intend to drive clocking inputs (of flip-flops, shift registers, counters, monostables, etc.), it is best to use a comparator with hysteresis or to buffer the input with a gate (or other logic device) with Schmitt-trigger input. The same comment goes for signals derived from transistor analog circuitry. Figure 9.11 shows the idea. R_2 is chosen for 50mV hysteresis. A small capacitor (C_2) across the

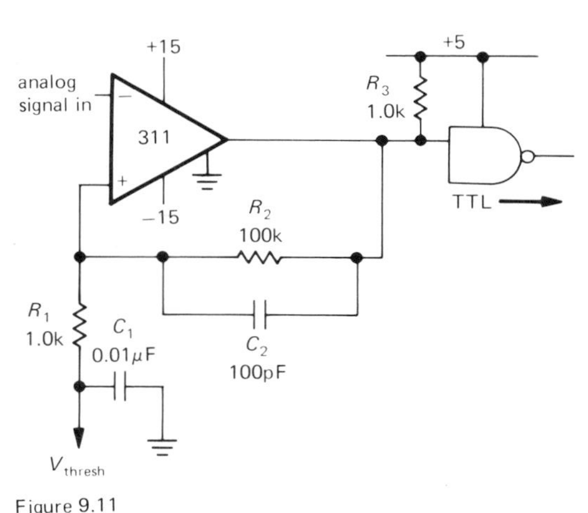

Figure 9.11
Threshold detector with hysteresis.

feedback resistor is a good idea to ensure fast transitions and to prevent multiple pulses as the input passes through the threshold (the 311, in particular, has a tendency to do this). Bypass capacitor C_1 is also important in preventing glitches on the reference supply during transitions. In many cases the reference voltage will be ground, in which case C_1 is omitted.

9.06 Some comments about logic inputs

TTL chips tend to be designed with active LOW control inputs; for instance, flip-flops generally have SET and RESET inputs that are enabled when LOW. External digital signals used as inputs will therefore nearly always have a pull-up resistor and will pull LOW (sinking current) when active, a convenient arrangement because switches, etc., can use a common ground return. It also leads to greater noise immunity, since a line held near +5 volts has 3 volts of noise immunity, as compared with the 0.8 volt of noise immunity of a line held near ground. This intrinsic weakness of TTL logic (poor noise immunity in the LOW state) is particularly apparent when you realize that a negative 0.5 volt spike on the chip's ground line can be interpreted as a HIGH input by the chip. Such spikes are not uncommon, since they can be generated by short current spikes through the ground-line inductance. See Section 9.14 for further discussion of this troublesome problem.

With CMOS the situation is reversed, with inputs generally being active HIGH. Since the noise immunity is the same either way, you can use pull-up or pull-down resistors as input terminations when driving from devices that have an open state. Pull-down resistors tend to be used more often, although you will see pull-ups when the driving device is something like a switch, with a ground return.

An open TTL input is "barely HIGH." It sits at the logic threshold (~1.3V), but since no current is being sunk, it does not turn on the input transistor. You may occasionally see "designs" in which inputs that should be tied HIGH on TTL devices are left open. Don't *ever* do this! It is foolish and dangerous: There is zero noise immunity on an open input, so capacitive coupling from nearby signals can generate short LOW spikes at the input. This generates glitches at the ouputs of combinational devices (gates), which is bad enough, but in the case of a flip-flop or a register it can be devastating, since an open RESET' input can clear the device at unpredictable times. The offending glitches may be impossible to see on a scope, since they may be "single-event" pulses of about 20ns duration. Although you may be able to get away with this unsavory practice most of the time, especially if there is low capacitance between the open pin and neighboring pins, it is still a very poor idea; if you try to troubleshoot the circuit by clipping a logic analyzer or test clip over the IC, you've got a new circuit, since the additional capacitance of the test device is almost certain to cause transient LOWs at the open pin(s). Besides, why build unreliable circuits when you know how to make them reliable with a few simple connections? (End of tirade.)

Unused inputs

Unused inputs that affect the logic state of a chip (e.g., a RESET input of a flip-flop) must be tied HIGH or LOW, as appropriate. Inputs that have no effect (e.g., inputs of unused gate sections in the same package) may be left unconnected in TTL, but not in CMOS devices. Here's what you should do with unused logic inputs:

CMOS inputs may be tied directly to ground or V_{DD}, as required. Since CMOS SET and RESET inputs are usually enabled when HIGH, they would be grounded, whereas unused inputs of a multiple-input NAND gate would be tied HIGH, for instance. As mentioned earlier, an unused *section* of a CMOS chip should not be left unconnected, as you would do with any other logic family. An unused gate that is left with inputs open, for instance, will bring its output to half the supply voltage, with both MOS output transistors forward-biased, thus drawing considerable class A current. This results in excessive supply current and can even lead to failure in devices with hefty output stages (4049, 4050, etc.) when operated at 10 volts or above. It is best to ground all inputs of unused functions on every chip.

TTL inputs can be connected to ground, and the inputs of nearly all the Schottky families (74S*xx*, 74LS*xx*, etc.) can be connected directly to the 5 volt supply. However, inputs of the older 74*xx*, 74L*xx*, 74H*xx*, etc., as well as a few members of the Schottky families (e.g., the 74LS74; check the data sheets for devices with emitter inputs), should not be connected directly to the +5 volt supply. The reason is that TTL inputs are restricted to an absolute maximum of +5.5 volts, whereas the supply voltage can rise to +7.0 volts without damage. If inputs are connected directly to the supply, a power-supply fault or transient can produce widespread damage in the circuit via input breakdown, since power-supply "crowbars" typically do not go into action until the supply voltage exceeds 6 to 6.5 volts (see Section 5.06). For inputs that need to be held HIGH, there are two alternatives: (a) Connect them through a resistor to +5 volts: 1k is a good value, and many inputs may be held HIGH by the same resistor (remember that TTL inputs draw essentially no current in the HIGH state). The resistor prevents damage in the event of supply transients. (b) Ground the input of an inverter (or inverting gate), and use the output to hold unused inputs HIGH; many inputs can be tied to the same inverter output.

With TTL you can ignore unused sections of a chip, as well as irrelevant inputs of circuits you are using. For instance, you can leave the parallel-load data lines of a counter unconnected if you never enable the LOAD line.

9.07 Comparators

Comparators were introduced briefly in Section 3.22 to illustrate the use of positive feedback (Schmitt trigger) and to show that special-purpose comparator ICs deliver considerably better performance than general-purpose op-amps used as comparators. These improvements (short delay times, high output slew rate, and relative immunity to large overdrive) come at the expense of the properties that make op-amps useful (in particular, careful control of phase shift versus frequency). Comparators are not frequency-compensated (Section 3.31) and cannot be used as linear amplifiers.

Comparators provide an important interface between analog (linear) input signals and the digital world. For example, the 8022 microprocessor illustrated in Section 8.27 includes eight of them on the same chip as the processor itself, in addition to an 8-bit analog-to-digital converter. In this section we would like to look at comparators in some detail, with emphasis on their output properties, their flexibility regarding power-supply voltages, and the care and feeding of input stages.

Supply voltages and outputs

Most comparators have open-collector outputs, suitable for driving logic inputs (with a pull-up resistor, of course) or higher current/voltage loads. The 311, for instance, can drive loads connected to supplies up to 40 volts with currents up to 50mA, and the 306 can handle even more current. These comparators have a ground pin in addition to the negative and positive supply pins, so the load is pulled to ground regardless of supply voltages. Comparators intended for very high speed (521, 527, 529, 360, 361, Am686, 760, HA4925, and HA4950) often come with active pull-up output circuits. These are meant to drive 5 volt digital logic, and they usually have 4 power-supply pins: V_+, V_-, V_{CC} (+5), and ground.

One thing to watch out for is the fact that many comparators require positive and negative supplies, even if the inputs never go negative. Examples are the 306, 710, and 711, in addition to the comparators with active pull-up listed previously. It is inconvenient to have to generate a negative supply just to run a comparator in an instrument that otherwise uses only positive supplies, so it is important to know about comparators that will run on a single positive supply (e.g., the 311, 319, 339, 393, 775, CA3290, HA4905, CMP-01, and CMP-02). In fact, these will all operate with a single 5 volt supply, a great advantage in a digital system. When operated from a single +5 volt supply, the 339, 393, 775, CA3290, and HA4905 have an input common-mode range all the way down to ground. These last comparators are intended specifically for single-supply operation and, with the exception of the 4905, have only two supply pins (V_+ and ground); if operated from split supplies, the output will be pulled to V_-. The first three of these also have the unusual property of being able to operate from a single supply of as little as +2 volts (up to +36V maximum).

While on the subject of power supplies, it should be mentioned that there are special comparators designed to run on low supply currents, generally less than 0.5mA: examples are the Intersil 8001, the Analog Devices AD351, the Siliconix L161, and the Motorola MC14574. The last two are quad comparators with programmable operating current; the L161 can operate on as little as 1.0μA total supply current! However, at that skimpy supply current the output rise time is 1ms. The 393 dual comparator also runs on low supply current, although not as low as that of the comparators just mentioned.

Inputs

There are some general cautions concerning input circuits of comparators. Hysteresis should be used whenever possible, because erratic switching is otherwise likely to result. To see why, imagine a comparator without hysteresis in which the differential input voltage has just passed through zero volts, slewing relatively slowly, since it is an analog waveform. A mere 2mV input differential causes the output to change state, with switching times of 50ns or less. Suddenly you have 3000mV fast digital logic transitions in your system, with current pulses impressed on the power supplies, etc. It would be a miracle if some of these fast waveforms didn't couple into the input signal, at least to the extent of a few millivolts, overcoming the 2mV input differential and thus causing multiple transitions and oscillations. This is why generous amounts of hysteresis (including a small capacitor across the feedback resistor), combined with careful layout and bypassing, are generally required to get sensitive comparator circuits to function well. It is a good idea to avoid high-speed comparators, which only aggravate these problems, if speed is not needed. Then, too, some comparators are more troublesome in this regard than others; we have had plenty of headaches using the otherwise admirable 311.

Another caution on inputs: Some comparators have a very limited differential input voltage range, as little as +5 volts for some types (e.g., the 710, 711, and 306). In such cases it may be necessary to use diode clamps to protect the inputs, since excessive differential input voltage will degrade h_{FE} and cause permanent input offset errors and may even destroy the base-emitter junctions of the input stage. More recent comparators are generally better in this respect, with typical differential input voltage ranges of ±30 volts (e.g., the 311, 393, 775, etc.).

An important feature of comparator inputs is the bias current at the input terminals and the way it changes with differential input voltage. Nearly every comparator uses bipolar transistors for its input stage, with input bias currents ranging from tens of nanoamps to tens of microamps. Since the input stage is just a high-gain differential amplifier, the bias current changes as the input signal takes the comparator through its threshold. In addition, internal protection circuitry may cause a larger change in bias current a few volts away from threshold. The curve shown in Figure 9.12 (for the CMP-02) is typical. The small current "step" at zero volts (differential) is actually a smooth transition, taking place over 100mV or so, and represents the voltage change necessary to switch the input differential amplifier stage fully from one state to the other.

For comparator applications where extremely low input current is necessary, there are FET-input comparators available. Examples are the CA3290, a MOSFET-input dual comparator, and the LF311, a FET version of the popular 311. The latter offers maximum input currents of 50pA, compared with 250nA for the 311, with essentially no sacrifice in offset voltage or speed. In situations where the properties of a particular comparator are needed, but with lower input current, one solution is to add a matched-pair FET follower at the input, as illustrated in Section 6.09.

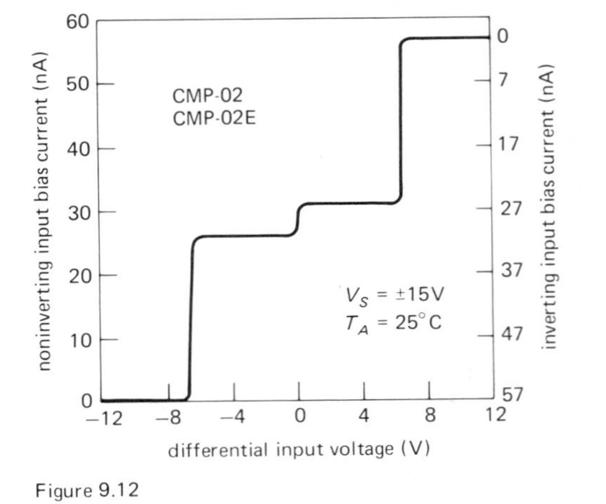

Figure 9.12
Input bias current versus differential input voltage for the CMP-02 comparator. (Precision Monolithics, Inc., Santa Clara, Calif. 95050)

One last comment on input properties: Thermal gradients set up on the chip from dissipation in the output stage can degrade input offset voltage specifications. In particular, it is possible to have "motorboating" (a slow oscillation of the output state) take place for input signals near zero volts (differential), since the state-dependent heat generated at the output can cause the input to switch.

Overall speed

It is convenient to think of a comparator as an ideal switching circuit for which any reversal in the differential input voltage, however small, results in a sudden change at the output. In reality, a comparator behaves like an amplifier, for small input signals, and the switching performance depends on the gain properties at high frequencies. As a result, a smaller input "overdrive" (i.e., more than enough signal to cause saturation at dc) causes a greater propagation delay and (often) a slower rise or fall time at the output. Comparator specifications usually include a graph of "response times for various input overdrives." Figure 9.13 shows some for the 311. Note particularly the reduced performance in the configuration where the output transistor is used as a follower, i.e., with less gain. Increased input

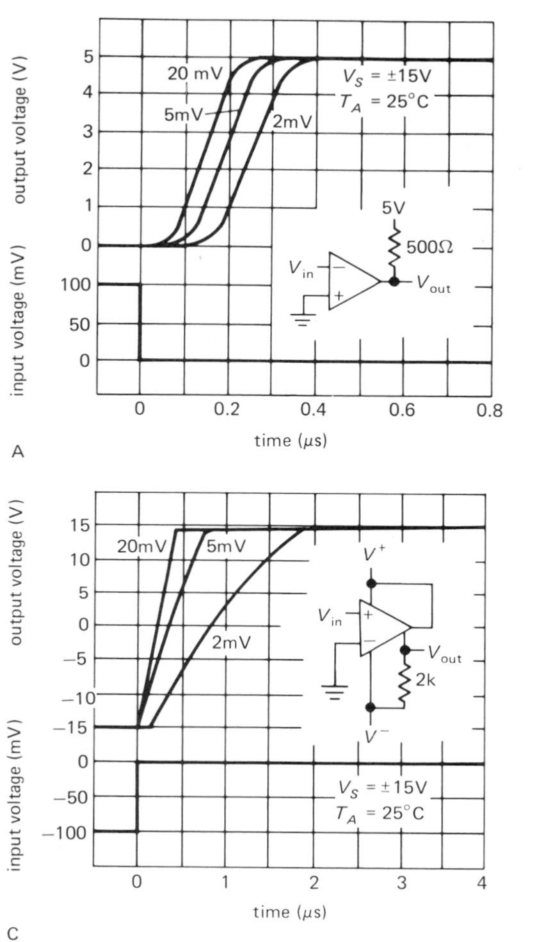

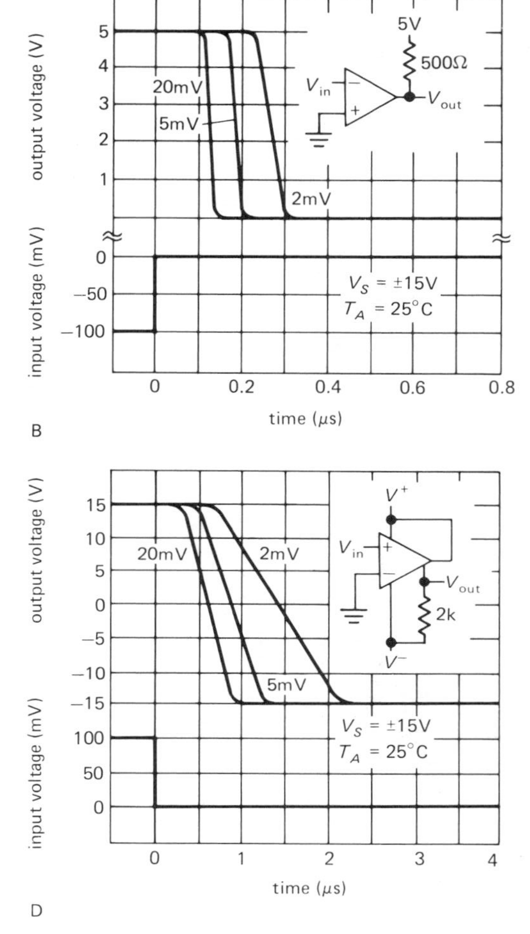

Figure 9.13
LM311 comparator response times for various input overdrives. (National Semiconductor Corp.)

drive speeds things up because the amplifier's reduced gain at high frequencies is overcome by a larger signal. In addition, larger internal amplifier currents cause internal capacitances to charge faster.

Table 9.2 compares the properties of most of the comparators currently available.

9.08 Driving external digital loads from TTL and CMOS

It is easy to drive on/off devices like lamps (LEDs), relays, displays, and even ac loads from the outputs of TTL or CMOS logic. Figure 9.14 shows some methods. Circuit A shows the standard method of driving LED indicator lamps from 74*xx* TTL; since TTL is designed to sink large currents, the LED is returned to +5 volts. The LED behaves like a diode with a forward drop of 1.5 to 2.5 volts at typical operating currents of 5mA to 20mA (with some of the newer-efficiency GaAsP LEDs you get plenty of light with just a few milliamps). Circuit B is a shortcut that we're not sure we should be telling you about; the TTL gate sources current, limited by the internal current-limiting circuitry. The gate output cannot be used to drive other TTL inputs, since the LED clamps the output below a safe HIGH value when ON.

Circuit C shows how to drive a 5 volt

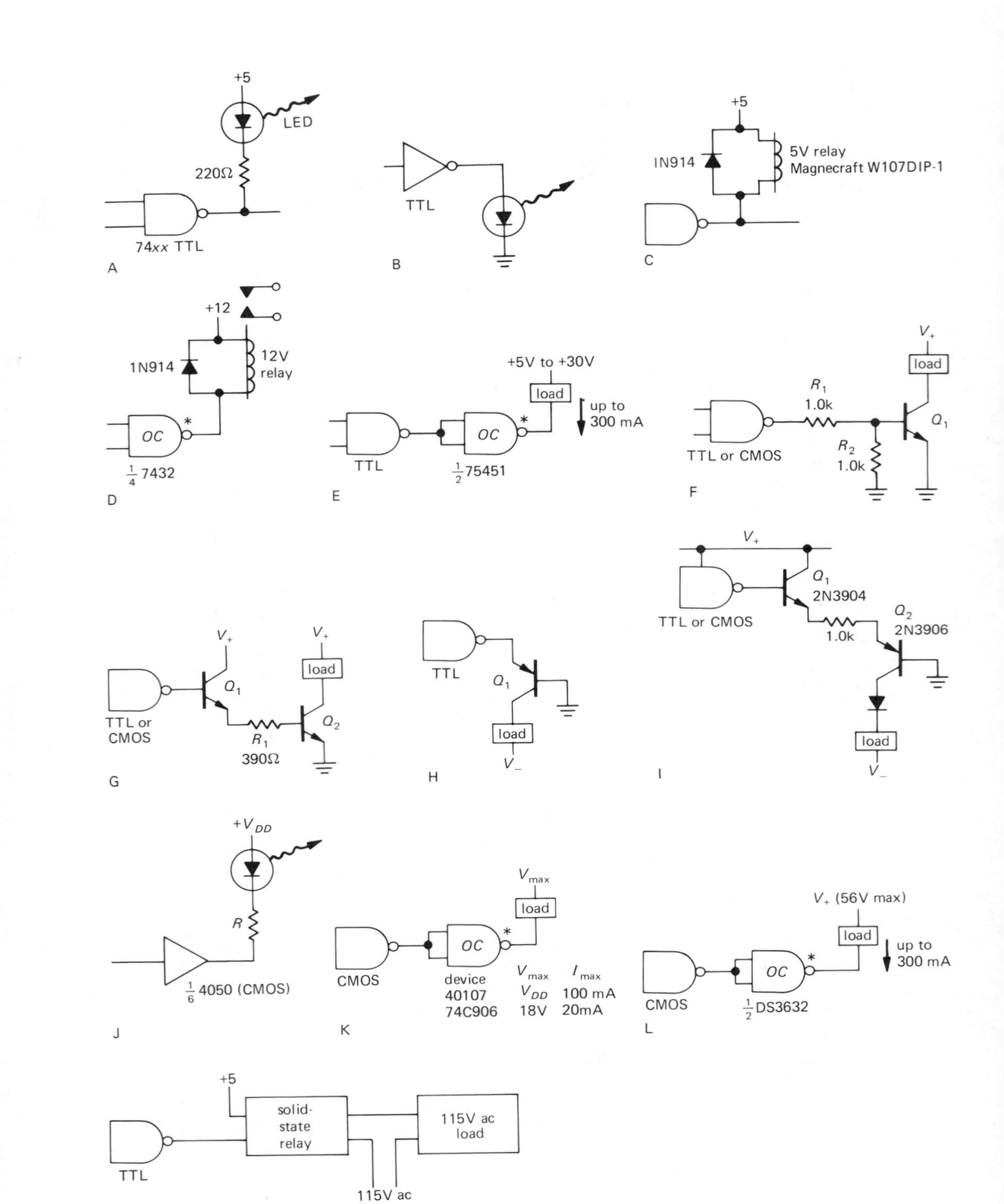

+5
LED
220Ω
74xx TTL
A
TTL
B
+5
IN914
5V relay
Magnecraft W107DIP-1
C
+12
1N914
12V
relay
OC
*
1/4 7432
D
+5V to +30V
load
up to
300 mA
TTL
OC
*
1/2 75451
E
V+
load
R1
1.0k
Q1
TTL or CMOS
R2
1.0k
F
V+
Q1
TTL or CMOS
load
R1
390Ω
Q2
G
TTL
Q1
load
V−
H
V+
Q1
2N3904
TTL or CMOS
Q2
2N3906
1.0k
load
V−
I
+VDD
R
1/6 4050 (CMOS)
J
Vmax
load
CMOS
OC
*
device
40107
74C906
Vmax
VDD
18V
Imax
100 mA
20mA
K
V+ (56V max)
load
up to
300 mA
CMOS
OC
*
1/2 DS3632
L
+5
solid-
state
relay
115V ac
load
TTL
115V ac
M

Figure 9.14

low-current relay directly from TTL, sinking current as in circuit A; the diode is essential to clamp the inductive spike. The relay shown comes in a standard DIP IC package and has 500 ohms coil resistance (10mA, or six standard TTL loads). For driving higher-voltage loads, the schemes shown in circuits D and E are effective. In circuit D, a 7432 open-collector gate with 15 volt rating drives a 12 volt relay, and in circuit E a 75451 "dual peripheral driver" chip rated at up to 300mA and 30 volts drives an unspecified load. Open-collector devices like these are available with 80 volt ratings (e.g., the DS3611–3614) and with even higher current capability.

The schemes mentioned previously may not be practical with 74LS*xx* TTL, which can sink only 8mA in the LOW state; when using 74LS*xx* TTL, it may be a good idea to set aside a chip like a 7404 (or even a 74H04) for LED driving. When driving high-current loads directly from logic chips, take particular care to use a hefty ground wire to the chip, since the current from the load returns through the chip into the logic supply ground. In some cases a separate ground return path to the supply may be desirable.

Circuit F shows how to use an *npn* transistor to switch a high-current load from TTL or CMOS. R_1 could be smaller, or you could use a second transistor, as in circuit G, if higher output currents are needed. The circuits H and I show how to drive loads returned to a negative supply. A HIGH output state turns on the *pnp* transistor, pulling the collector into saturation at one diode drop above ground. In circuit H the TTL current limit sets the emitter current, and therefore the maximum collector (load) current, whereas in the improved circuit I an *npn* follower is used as a buffer. In the latter a diode in series with the output keeps the load from swinging above ground. In both cases the maximum load current is equal to the drive current to the *pnp* transistor's emitter. Note that with circuit H you cannot use the gate's output to drive other logic, since it is clamped one diode drop above ground. Similar IC circuits are available (e.g., the DS3687 "negative relay driver") with CMOS/TTL-compatible inputs and 300mA −56 volt output ratings (the companion DS3686 drives positive loads with the same ratings).

Standard CMOS logic devices have considerably lower output sink capability, less than 1mA with 5 volt supplies, so it is generally necessary to use some sort of power driver to light LEDs, etc. Circuit J shows the trusty 4050 hex buffer driving an LED. It can typically sink 5mA to 50mA with 5 to 15 volt supplies, respectively (output drive capability increases with increasing supply voltage). The circuits K and L show the application of even heftier drivers: The 40107 has a large "open-drain" *n*-channel MOS output transistor good for sinking 16mA to 50mA (5–15V supply voltages, respectively), and the DS3632 uses an *npn* Darlington driver, good for 300mA. Of course, discrete external transistors can always be used, as in circuits G and I, but they will be limited to approximately 0.25mA base drive current.

For driving ac loads, the easiest method is to use a "solid-state relay," as in circuit M. It is an optically coupled triac with TTL-compatible input and 10 to 40 amps load-current capability when switching a 115 volt ac load. Alternatively, you can use an ordinary relay, energized from logic. However, be sure to check the specifications, because most small logic-driven relays cannot drive heavy ac loads, and you may have to use the logic relay to drive a second larger relay. Many solid-state relays use "zero-crossing" switching, a desirable feature that prevents spikes and noise from being impressed onto the power line. Much of the "garbage" on the ac power line comes from triac controllers that don't switch at zero crossings, e.g., phase-controlled dimmers used on lamps, thermostatic baths, motors, etc.

When driving 7-segment numeric displays, it is usually simplest to take advantage of one of the decoder/driver combinations available. They come in a bewildering variety, including current-sinking (common anode) and current-sourcing (common cathode) types. Typical types are the tried-and-true TTL 7447 BCD-to-7-segment decoder/driver (current sinking) and the CMOS 4511 BCD-to-7-segment latch/decoder/driver (current sourcing). We will have some examples of their use shortly.

TABLE 9.2. COMPARATORS

Type	Qty. per pkg.	Response time[a] typ (ns)	V_{os} max (mV)	I_B max (μA)	CM to V_-?[b]	Abs. max common-mode range[c] min (V)	Abs. max common-mode range[c] max (V)	Abs. max diff'l input V_{diff}[d] (V)	Gain typ
Advanced Micro Devices (AMD)									
Am685	1	6	2.0	10	—	−4	4	6	1600
Am686	1	9	2.0	10	—	−4	4	6	
Am687	2	7	2.0	10	—	−4	4	6	
Fairchild (FSC)									
μA734	1	200	5.0	0.1	—	V_-	V_+	10	60k
μA760	1	16	6.0	60	—	V_-	V_+	5	5k
μA775	4	1300	5.0	0.25	√	−0.3	V_+	V_+	200k
Harris									
HA4905	4	150	7.5	0.15	√	V_-	V_+	15	400k
HA4925	4	33	6.0	8	—	V_-	V_+	6	25k
HA4950	1	40	2.3	3	—	V_-	V_+	6	
Intersil									
8001	1	250	5.0	0.05	—	V_-	V_+	15	60k
Motorola									
MC14574	4	1000	50	0.001	√	−0.5	$V_+ + 0.5$	V_+	100k
MC14574-1	4	1000	10	0.001	√	−0.5	$V_+ + 0.5$	V_+	100k
National (NSC)									
LM306	1	28	6.5	5	—	−7	7	5	40k
LM311/2311	1/2	200	3.0	0.1	—	$V_+ - 30$	$V_- + 30$	30	200k
LF311	1	200	4.0	0.00005	—	$V_+ - 30$	$V_- + 30$	30	200k
LM319	2	80	4.0	0.5	—	V_-	V_+	5	40k
LM339/393	4/2	1300	5.0	0.25	√	−0.3	36	36	200k
LM360	1	14	5.0	20	—	V_-	V_+	5	3k
LM361	1	14	5.0	30	—	−6	6	5	3k
Plessey									
SP1650B	2	3.5[i]	20	10	—	−3	2.5	5	
SP1651B	2	3.0[i]	20	40	—	−2.5	3	5	
SP9685	1	2.3	5	20	—	−5	3	5	300
SP9687	2	2.8	5	20	—	−5	5	5	300
SP9750	1	5.0	5	25	—	−4	4	6	60
Precision Monolithics (PMI)									
CMP01C	1	110	2.8	0.9	—	V_-	V_+	11	500k
CMP02	1	190	0.8	0.003	—	V_-	V_+	11	500k
CMP04	4	1300	1.0	0.1	√	−0.3	30	36	200k
RCA									
CA3290A	2	1000	10.0	0.00004	√	$V_- - 5$	$V_+ + 5$	36	150k
Signetics									
NE521	2	11	7.5	20	—	−5	5	6	
NE522	2	14	7.5	20	—	−5	5	6	
NE527	1	33	6.0	2	—	−6	6	5	
NE529	1	20	6.0	20	—	−6	6	5	
Texas Instruments (TI)									
TL510C/514C	1/2	30	3.5	20	—	−7	7	5	33k
TL810C/820C	1/2	30	3.5	20	—	−7	7	5	33k

[a] 100mV step with 5mV overdrive. [b] Operating input common-mode range includes negative supply. [c] Maximum range without input breakdown; will not operate properly over entire range. [d] Maximum allowable voltage betwen input terminals. [e] Ability to accept signals of both polarities and drive unipolar logic. [f] G: Output pulled to GND. R: Low output is

Power supplies								Output parameters									
Positive		Negative		Total supply													
min (V)	max (V)	min (V)	max (V)	min (V)	max (V)	GND?	Single +5V OK?	TTL-compatible	+/− to logic[e]	Open collector	Active pull-up	Complementary	Strobe	Latch	Output low[f]	Max ext. pull-up[g]	Comments
6[h]		−5.2[h]		9.7	14	√	—	—	√	—	—	√	√	√	E		ECL
5[h]		−6[h]		9.7	14	√	—	√	√	—	√	—	√	—	G		Fastest TTL comp.
5[h]		−5.2[h]		9.7	14	√	—	—	√	—	—	√	√	√	E		ECL
5	18	−5	−18	10	36	—	—	√	√	√	—	—	—	—	G		
4.5	8	−4.5	−8	9	16	√	—	√	√	—	√	√	—	—	G		
2	36	—	—	2	36	√	√	√	—	√	—	—	—	—	G	30	Low power; sim. to 339
5	30	0	−30	5	33	—	√	√	√	—	√	—	—	—	R		Flexible output stage
15[h,i]		−15[h]				—	—	√	√	—	√	—	—	√	R		
15[h,i]		−15[h]				√	—	√	√	—	√	—	√	—	R		
5	15	−15[h]			36	√	—	√	√	√	—	—	—	—	G		
3	15	—	—	3	18	√	√	√	—	—	√	—	—	—	G	V_+	100% CMOS, programmable
3	15	—	—	3	18	√	√	√	—	—	√	—	—	—	G	V_+	specs @ I_{set} = 50μA
12[h]		−3	−12		30	√	—	√	√	—	√	—	√	—	G	24	High output current
5	30	0	−30	4.5	36	—	√	√	√	√	—	—	—	—	R	40	May oscillate; popular
5	30	0	−30	4.5	36	—	√	√	√	√	—	—	—	—	R	40	JFET 311
5	30	0	−30	4.5	36	—	√	√	√	√	—	—	—	—	R	36	
2	36	—	—	2	36	√	√	√	—	√	—	—	—	—	G	30	Our favorite; low power
4.5	6.5	−4.5	−6.5	9	13	√	—	√	√	—	√	√	—	—	G		Similar to 760
5	15[j]	−6	−15	11	30	√	—	√	√	—	√	√	√	—	G	7	Similar to 529
5[h]		−5.2[h]				√	—	—	√	—	—	—	√	√	E		ECL
5[h]		−5.2[h]				√	—	—	√	—	—	—	√	√	E		ECL
5[h]		−5.2[h]			12	√	—	—	√	—	—	√	√	√	E		ECL; fast Am685
5[h]		−5.2[h]			12	√	—	—	√	—	—	√	√	√	E		ECL; fast Am687
5[h]		−5.2[h]			11	√	—	—	√	—	—	—	√	—	E		ECL & current outputs
5	30	0	−30	5	36	—	√	√	√	—	√	—	—	—	R	32	
5	30	0	−30	5	36	—	√	√	√	—	√	—	—	—	R	32	Precision; recommended
3	36	0	−30	3	36	√	√	√	—	√	—	—	—	—	G	30	Precision 339
4	36	—	—	4	36	√	√	√	—	√	—	—	—	—	G	36	MOSFET input
5[h]		−5[h]		9.5	10.5	√	—	√	√	—	√	—	√	—	G		
5[h]		−5[h]		9.5	10.5	√	—	√	√	—	√	—	√	—	G		
5	10	−6	−10	10	20	√	—	√	√	—	√	√	√	—	G	15	529 with Darlington
5	10	−6	−10	10	20	√	—	√	√	—	√	√	√	—	G	15	
10	14	−5	−7	15	21	√	—	√	√	—	√	—	√	—	G		
10	14	−5	−7	15	21	√	—	√	√	—	√	—	—	—	G		510/514 without strobe

saturated *npn* transistor to terminal that may be operated at voltages other than GND. E: Output (open *npn* emitter) is designed to drive ECL logic. [g]Maximum voltage to which output can be externally pulled up. [h]Nominal. [i]And additional +5 volt logic supply. [j]100mV overdrive.

N-CHANNEL AND *P*-CHANNEL MOS LSI INTERFACING

Large-scale integrated circuits (LSI) are often constructed from enhancement MOSFETs of one polarity only. This has the advantage over CMOS of fewer steps in the manufacturing process, and higher densities are possible than with either CMOS or bipolar technologies. Because of the widespread use of such chips, it is important to understand how to interface between MOS and TTL (or CMOS) and how to connect between MOS inputs/outputs and external discrete circuitry. Most MOS LSI chips are designed for TTL compatibility, since they are used extensively within TTL systems; however, there are some subtle points to be considered.

9.09 NMOS inputs

The input circuit of an *n*-channel MOS IC intended for use with TTL driving circuitry is shown in Figure 9.15. Q_1 is an inverter, and

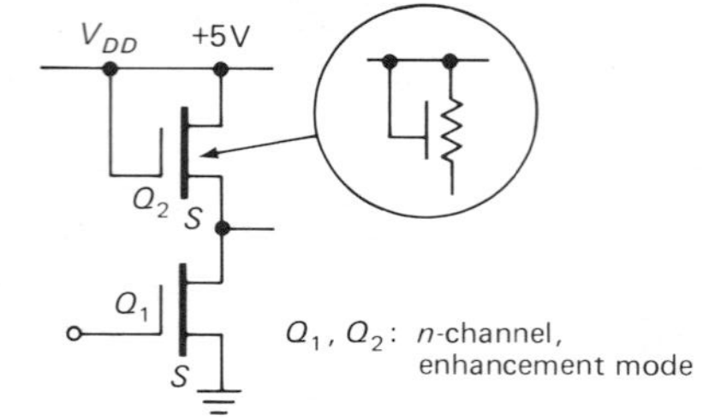

Figure 9.15
NMOS logic input circuit.

Q_2 is a small-geometry "source follower" providing pull-up current (resistors take up too much space, so MOSFETs are universally used as drain loads); the alternative symbol shown for Q_2 is widely used. The threshold voltage of the input transistor can be as low as 0.5 to 1.5 volts in the newer "silicon-gate" low-threshold designs, but in older devices the threshold voltage can be in the range of 2 to 3 volts. Thus, in order to generate a safe universal HIGH input, it is best to use a pull-up resistor on TTL outputs driving NMOS inputs. The TTL LOW output is perfectly fine as is. As usual, the pull-up resistor should be chosen small enough to give adequate speed in a LOW-to-HIGH transition into whatever load capacitance the TTL output sees; values in the 1k to 10k range are usually good.

When using TTL outputs with a few feet of connecting cable to drive NMOS with high input threshold voltage, you may get into speed problems, since the load capacitance can reach a few hundred picofarads or more. The TTL output waveform will look something like that in Figure 9.16. Although the

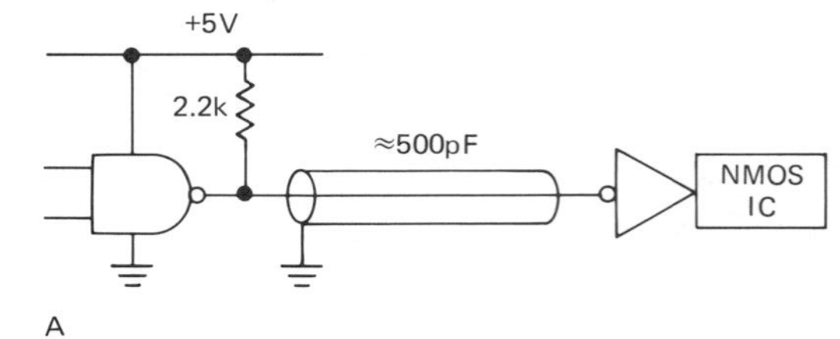

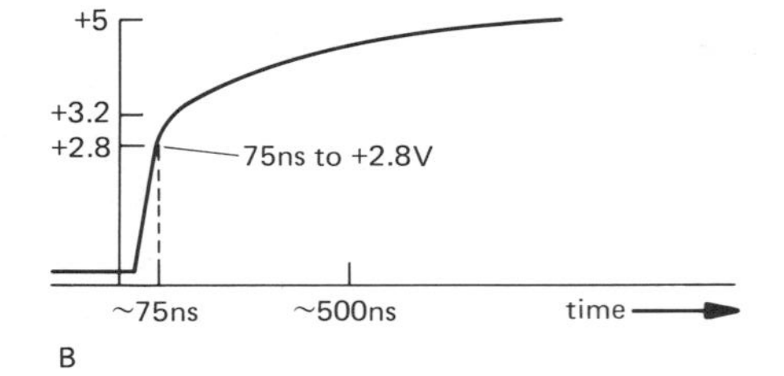

Figure 9.16
Capacitive load slows pull-up resistor rise time.

output reaches +2.8 volts in 50ns to 75ns, it will take perhaps another 500ns to reach +3.5 volts (necessary to provide reasonable noise immunity for a 3V MOS HIGH logic threshold), even with a 2.2k pull-up resistor. One solution to this problem is to use specially-designed MOS driver chips, e.g., the 74LS363 and 74LS364 three-state transparent-latch bus drivers. These have a minimum HIGH output voltage of 3.65 volts, as compared with the usual 2.4 volts for most other TTL devices, and the active pull-up output circuit guarantees rapid transition to that more positive HIGH level. With newer MOS devices with low input threshold, such special output circuits are often unnecessary, and ordinary TTL devices will usually suffice.

By comparison, CMOS logic operating on +5 volts can be directly connected to TTL-compatible MOS LSI inputs; pull-up resistors aren't needed, since the CMOS output goes all the way to +5 volts.

Some NMOS chips include internal-input pull-up FETs, in some cases controllable with an external input. They can be driven directly from TTL outputs without difficulty, unless there is excessive cable capacitance, as described previously.

9.10 NMOS outputs

The output stage of the popular low-voltage NMOS logic (single +5V supply) is shown in Figure 9.17. Q_1 is a switch, and Q_2

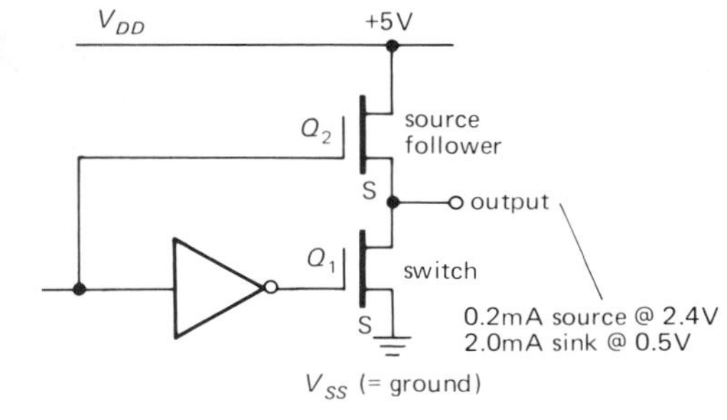

Figure 9.17
NMOS logic output circuit.

is a source follower. The circuit is designed so that with the output in the LOW state, Q_1 can sink a few milliamps while keeping the output below 0.5 volt. This is possible because Q_1's gate can be driven to V_{DD} (usually +5V for TTL compatibility), resulting in a low $R_{DS(ON)}$. The situation in the HIGH output state is dismal by comparison: At a minimum TTL HIGH output of +2.4 volts, Q_2 finds only 2.6 volts from gate to source, resulting in relatively high R_{ON}, with the situation rapidly deteriorating for higher output voltages.

The graph in Figure 9.18 illustrates the situation. As a result, the NMOS output drive capability might be only 0.2mA (sourcing) at +2.4 volts output. This is fine for driving TTL inputs, but it is a disaster in a circuit like that shown in Figure 9.19.

To operate the LED with multiplexed-display currents (25–50mA while ON), the output would have to put out about 1mA at +4.1 volts. That's impossible, since V_{GS} would be only 0.9 volt, possibly even below

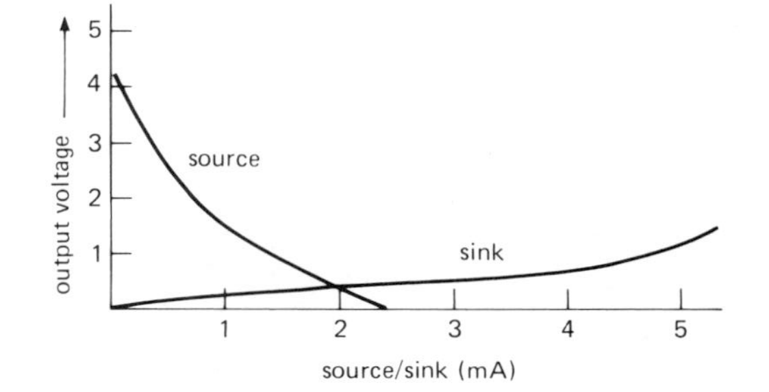

Figure 9.18
Typical NMOS output source and sink current capability.

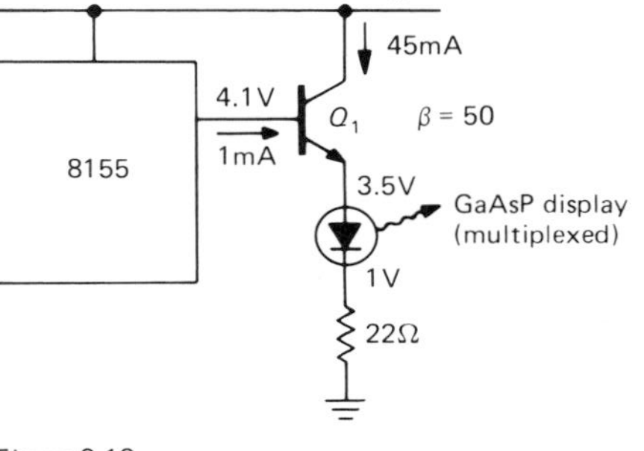

Figure 9.19

the threshold voltage for the FET. Remember, too, that all circuits designed to operate on TTL +5 volt supplies should work with supply voltages as low as +4.5 to allow for ±10% tolerance of supply voltage. To drive LEDs (or other high-current devices), the circuits in Figure 9.20 are good.

In the first circuit, the NMOS output LOW sinks 2mA, driving the *pnp* transistor into hard conduction. In the second circuit, a Darlington *npn* transistor is switched ON by the small output current of the NMOS device while in the HIGH state. This circuit clamps the HIGH output two diode drops above ground, which might seem a bit unfriendly, but it turns out that NMOS outputs are designed so that they can be shorted to ground in this manner, with output currents controlled at small enough values so that they can drive the base of a grounded-emitter Darlington without damage (but don't try this with a high-voltage NMOS device, described in the next paragraph). A typical low-voltage NMOS output would source 2mA into the Darlington's base at

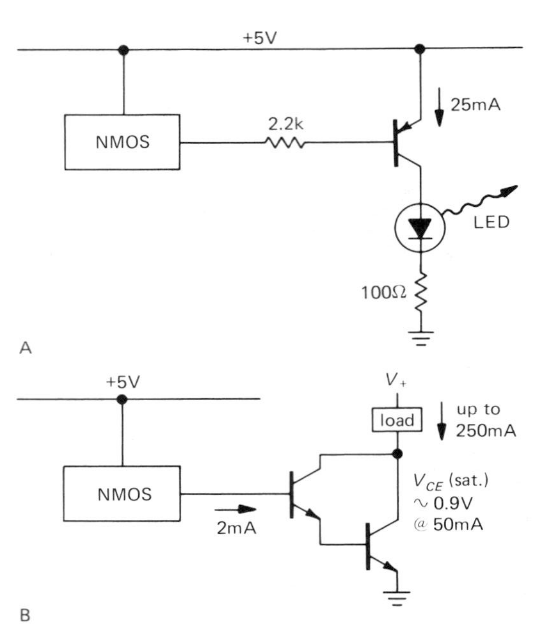

Figure 9.20
NMOS logic outputs driving loads.

+1.5 volts, giving an output sink capability of 250mA at 1 volt with an IC like the 75492 hex Darlington array.

□ *High-voltage NMOS*

A pleasant variety of NMOS LSI from the standpoint of output interfacing is provided by the so-called high-voltage low-threshold types. These chips operate with two positive supplies, typically +5 and +12 volts, and have output circuits like that in Figure 9.21.

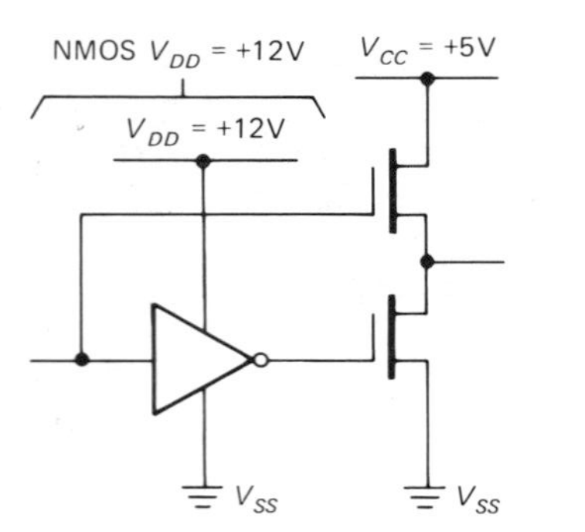

Figure 9.21

Most of the internal circuitry operates from a V_{DD} supply of +12V, with only the upper output transistor returned to +5 volts. As a result, the output stage can swing to +5 volts or to ground with excellent drive capability; the upper NMOS transistor (source follower) has 7 volts of gate bias when the output is at +5 volts. With these devices all your problems are over (well, almost all; you've got to figure out how to get +12V).

From the preceding discussion it should be clear why NMOS devices can have TTL-compatible outputs while running from a single +5 volt supply (low-voltage NMOS); this is not the case with PMOS, which we will discuss next. In practice, NMOS is sometimes run with an additional supply voltage for greater speed (high-voltage NMOS), since the limiting factor in the speed of a MOS circuit is ultimately the ratio of circuit capacitances to drive currents. With additional supply voltages it is possible to achieve lower R_{ON} by driving the gates to higher voltages.

□ 9.11 PMOS inputs

Figure 9.22 shows the input circuit of a PMOS chip designed for TTL input compatibility. Q_1 is an inverter (common-source

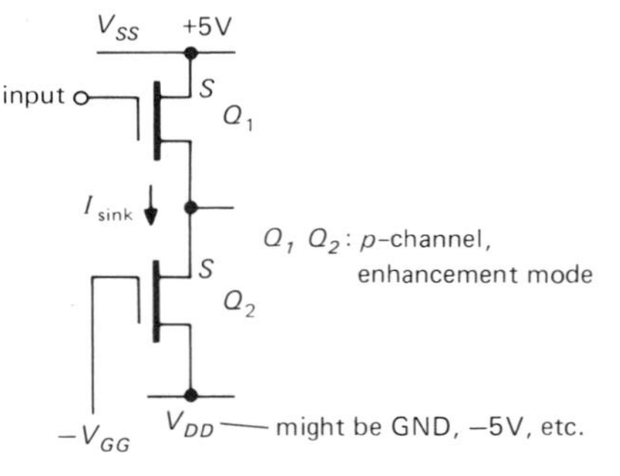

Figure 9.22

configuration), with Q_2 acting as the drain load. As you will see shortly, the negative V_{GG} supply (in addition to the +5V supply) is unavoidable with TTL-compatible PMOS chips, unlike the situation with NMOS, where it is possible to make compatible circuits that operate with a single +5 volt supply. Although V_{DD} may be connected to ground with some PMOS chips, many others use a negative V_{DD} supply (−5V, −9V, and −12V are common), with no connection to ground whatsoever! In some cases V_{GG} and V_{DD} may be the same negative supply.

To drive PMOS inputs it is necessary to pull the input far enough toward ground (LOW) to turn Q_1 ON and high enough (HIGH) to turn it OFF. The *p*-channel MOSFETs have intrinsically higher threshold voltages and ON resistances than *n*-channel devices of the same size (the charge carriers are holes rather than electrons); typical threshold voltages are in the range of 1.5 to 3 volts. For driving this sort of PMOS input stage, TTL outputs are fine in the LOW state, but they should be pulled up with a resistor to +5 volts to ensure an adequate HIGH voltage.

□ *Active pull-up*

There is another solution to the general problem of pulling up TTL HIGH outputs to levels sufficient to drive PMOS inputs reliably, namely the use of switched active pull-ups within the PMOS device itself. Figure 9.23 shows the general idea. Q_1

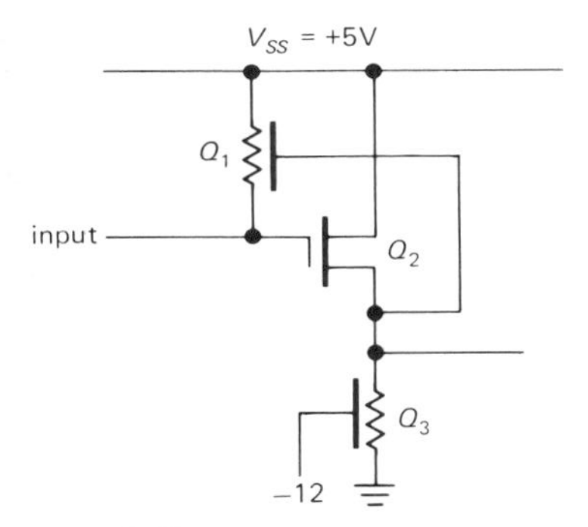

Figure 9.23

sources current when inverter Q_2Q_3 detects a HIGH input state; thus the input is pulled up once it reaches the MOS logic threshold, but not during the LOW state. In this way the pull-up is effective in maintaining a good MOS HIGH input state, without actually having to source any quiescent current in either state. The time-honored 3341 FIFO Memory uses this sort of input circuit to achieve TTL-compatible inputs.

□ 9.12 PMOS outputs

A standard PMOS output stage designed for driving TTL inputs is shown in Figure 9.24. Q_3 and Q_4 form a push-pull output stage,

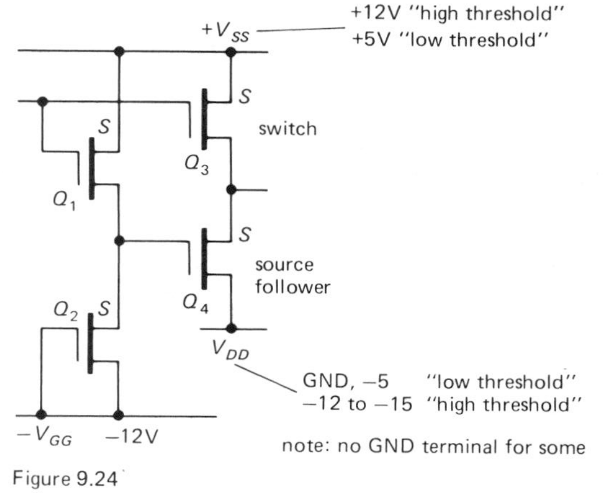

Figure 9.24

PMOS logic output circuits.

with Q_3 acting as a switch to V_{SS} (usually +5V) and Q_4 acting as a follower. Remember, both are *p*-channel devices. Doesn't CMOS, with its complementary switches to ground and $+V_{DD}$, seem elegant by comparison? Q_3 has no problem pulling the output HIGH, since its gate can be brought to ground (or even negative), turning it fully on.

Pulling a TTL input LOW is another story, since Q_4 is a follower; with the relatively high threshold voltages of *p*-channel (enhancement) MOS, the output could not be brought below a couple of volts with any current-sinking capability if the gate were brought only to ground. That would be totally inadequate for driving TTL, so a negative gate supply (V_{GG}, usually −12V) is used so that negative gate drive can hold Q_4 in good conduction when its source (the output) is near ground.

In the preceding circuit, Q_1, with drain load Q_2 returned to −12 volts (V_{GG}), acts as an inverter with swing down to −10 volts or so, providing good bias for follower Q_4. For PMOS devices with V_{DD} at ground, this brings the output close to ground with good current-sinking capability (like CMOS), whereas for PMOS devices with negative V_{DD} and no ground terminal, the output sinks substantial current and can pull its output well below ground. In the latter case the input clamping diodes of a driven TTL input are expected to limit negative swings to one diode drop below ground. Q_4 is designed to sink current while driving a TTL input, and it can be used to light LED indicators, etc., via a *pnp* transistor, as illustrated with NMOS; since the outputs are generally designed to sink only one TTL load, they can't drive high-current loads directly. Unlike the situation with NMOS, the PMOS output can *source* significant current (although generally less than it can sink). It can drive an *npn* switch with grounded emitter, for a load returned to +5 volts (or higher).

An important caution regarding PMOS operating with negative V_{DD} supply: In the LOW state a PMOS output may be able to sink substantial current while pulling a load below ground. TTL inputs can take this kind of treatment, because they have clamping diodes to ground on all inputs. However, CMOS devices may be damaged, and you may be in for some unpleasant surprises when driving other MOS LSI devices with such nasty outputs, which, after all, were really designed for TTL compatibility.

One last comment about PMOS: The trend is toward greater use of TTL-compatible NMOS in newer IC designs. Nevertheless, there are many very nice PMOS chips, and a few of them are totally incompatible with TTL. As an example, National has a series of music-synthesizer chips that includes a chromatic equal-tempered synthesizer, a digital noise source, and the rhythm-pattern generator mentioned in Section 8.27. These chips operate with supplies of +14 volts and −13 volts. In addition, many calculator chips are designed for operation from a single 9 volt battery. In most cases the outputs will interface to CMOS directly. If not, an interface using bipolar transistors or MOSFETs can usually be rigged up, once you understand the output properties of the chips involved; unfortunately that information is not always provided on the data sheets.

□ 9.13 Summarizing MOS family characteristics

Figure 9.25 summarizes the MOS LSI supply voltage situation, and Figure 9.26 presents a graph that shows output drive capabilities. The supply voltages and resulting input HIGH and LOW thresholds are shown separately for each of six MOS families.

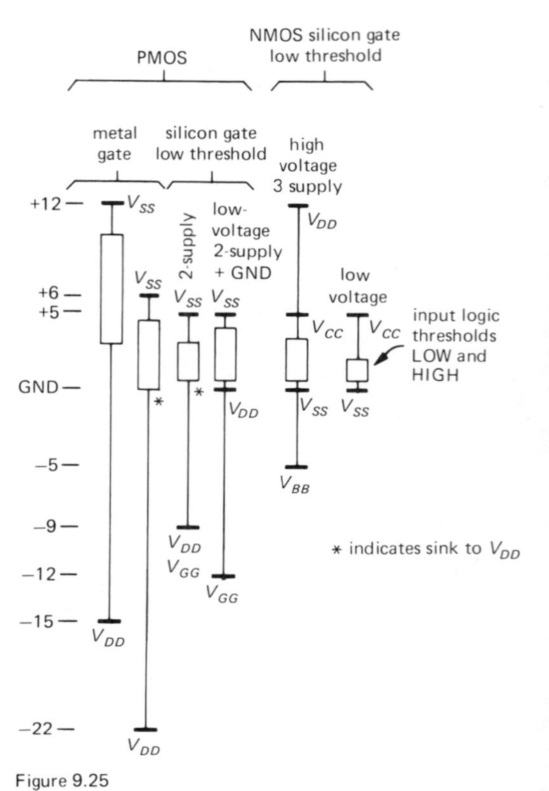

Figure 9.25
Chart of PMOS and NMOS supply voltages and logic thresholds showing high and low threshold processes.

There are a few points to note here: Often the V_{DD} and V_{GG} supply voltages are flexible; for example, you can substitute −12 volts for −15 volts. The supply voltages can be shifted, if convenient, to put the logic thresholds or output logic levels where you want them. Notice that PMOS outputs can pull their loads below ground while sinking current, since the outputs are followers with a negative supply voltage. Low-voltage NMOS (+5V supply) can't pull its load all the way to +5 volts, but it can sink plenty of current to ground. It is ideal for driving TTL, for which it was designed.

DIGITAL SIGNALS AND LONG WIRES

Special problems arise when you try to send digital signals through cables or between instruments. Effects such as capacitive loading of the fast signals, common-mode interference pickup, and "transmission-line"

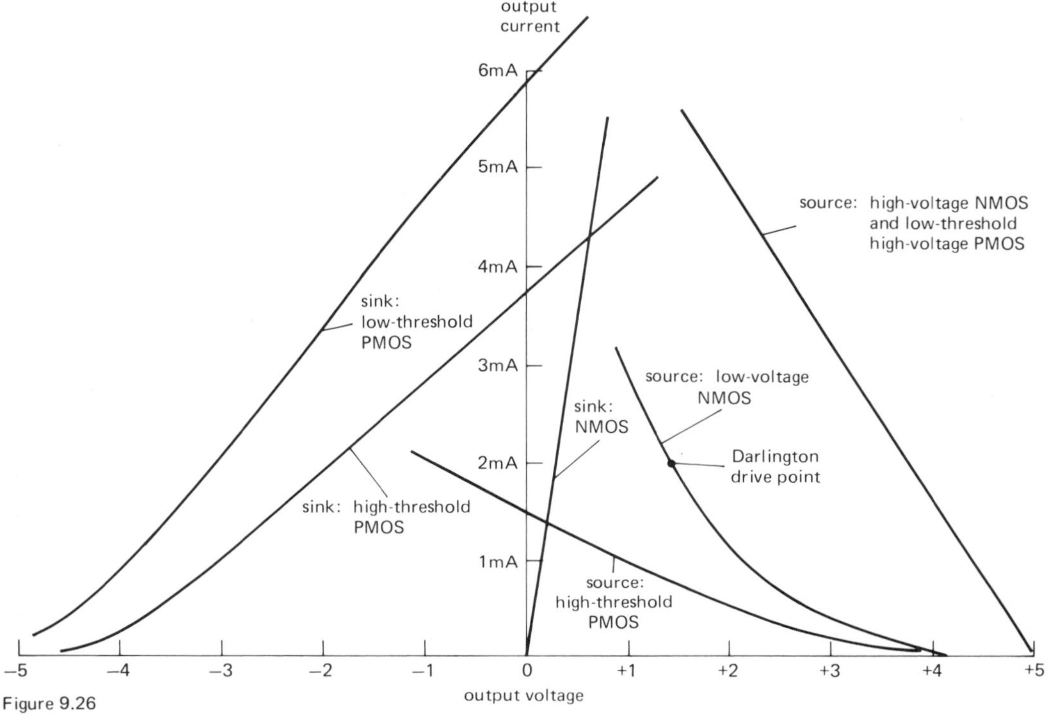

Figure 9.26

Output current characteristics for TTL-compatible NMOS and PMOS logic.

effects (reflections from impedance mismatching, see Section 13.09) become important, and special techniques and interface ICs are often necessary to ensure reliable transmission of digital signals. Some of these problems arise even on a single circuit board, so a knowledge of digital transmission techniques is generally handy. We will begin by considering on-card problems. Then we will go on to consider the problems that arise when signals are sent between cards, on data buses, and finally between instruments via twisted-pair or coaxial cables.

9.14 On-board interconnections

Output-stage current transient

The push-pull output circuit for TTL and CMOS ICs consists of a pair of transistors going from V_+ to ground. When the output changes state, there is a brief interval during which both transistors are ON; during that time a pulse of current flows from V_+ to ground, putting a short negative-going spike on the V_+ line and a short positive spike on the ground line. The situation is shown in Figure 9.27. Suppose that IC_1 makes a transition, with a momentary large current from +5 to ground along the paths as indicated (with 74S*xx* circuits the current might reach 100mA). This current, in combination with the inductance of the ground and V_+ leads, causes short voltage spikes relative to the reference point, as shown. These spikes may be only 5ns to 20ns long, but they can cause plenty of trouble: Suppose that IC_2, an innocent bystander located near the offending chip, has a steady LOW output that drives IC_3, located some distance away. The positive spike at IC_2's ground line appears also at its output, and if it is large enough, it gets interpreted by IC_3 as a short HIGH spike. Thus at IC_3, some distance away from the troublemaker IC_1, a full-size bona fide TTL output pulse appears, ready to mess up an otherwise well-behaved circuit. It doesn't

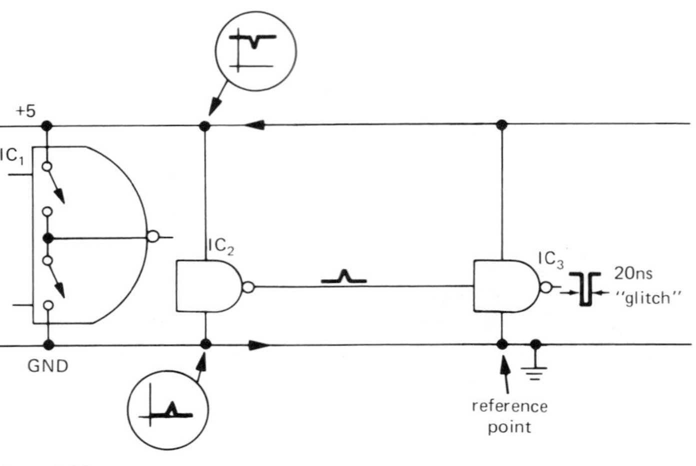

Figure 9.27
Ground current noise.

take very much to toggle or reset a flip-flop, and this sort of ground current spike can do the job nicely.

The best therapy for this situation consists of (a) using hefty ground lines throughout the circuit, even to the extent of using a large "ground plane" (one whole side of a double-sided printed-circuit board devoted to ground), and (b) using bypass capacitors liberally throughout the circuit. Large ground lines mean smaller current-induced spikes (lower inductance and resistance), and bypass capacitors from V_+ to ground sprinkled throughout the circuit mean that current spikes travel only over short paths, with the reduced inductance resulting in much smaller spikes (the capacitor acts as a local voltage source, since its voltage does not change appreciably during the brief current spikes). It is best to use a $0.05\mu F$ to $0.1\mu F$ capacitor near each TTL IC, although one capacitor for every two or three ICs may suffice. In addition, a few larger tantalum capacitors ($6.8\mu F$ 35V is a good value) scattered throughout the circuit for energy storage is a good idea. Incidentally, bypass capacitors from power-supply lines to ground are recommended in any circuit, digital or linear. They help make the supply lines low-impedance voltage sources at high frequencies, and they prevent signal coupling between circuits via the power supply. Unbypassed power-supply lines can cause peculiar circuit behavior, oscillations, and headaches.

Spikes caused by driving capacitive loads

Even with the supplies bypassed, your problems aren't over. Figure 9.28 shows why. A

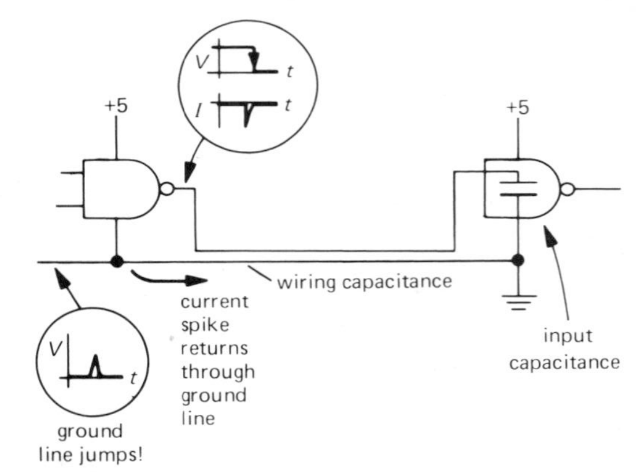

Figure 9.28
Capacitive-load ground current noise.

digital output sees the stray wiring capacitance and the input capacitance of the chip it drives (5–10pF, typically) as part of its overall load. To make a fast transition between states it must sink or source a large current into such a load, according to $I = C(dV/dT)$. For instance, consider a 74S*xx* chip (3V output swing in 3ns) driving a total load capacitance of 25pF (equivalent to three or four TTL loads connected with short leads). The current during the logic transition is 25mA, nearly the maximum output capability of the driving chip! This current returns through ground (HIGH-to-LOW transition) or the +5 volt line (LOW-to-HIGH transition), producing those cute little spikes at the receiving end, as before. [Just to get an idea of the magnitude of this effect, consider the fact that wiring inductance is roughly 5nH/cm. An inch of wire with this logic transition current occurring in 2ns would have a spike of $V = L(dI/dt) = 0.15V$!] A similar ground spike is generated near the driven chip, where the drive current spikes return to ground through stray capacitance and the input capacitance of the driven device.

In a synchronous system, with a number of devices making output transitions simultaneously, the noise spike situation can become serious. In a large printed-circuit card, with long interconnections and long ground runs, these current spikes can lead to real trouble. The best approach is to use massive ground runs (for low inductance) and short runs wherever possible. Because of its slower transition time, CMOS is less troublesome in this regard (but see Section 8.34 on logic "diseases" for a discussion of the related problem of "clocking skew"). 74S*xx* is worse in this respect than the all-purpose 74LS*xx*, which itself seems more susceptible to this problem than "standard" 74*xx* TTL, and the ultrafast ECL logic can be downright infuriating.

□ 9.15 Intercard connections

With logic signals going between circuit cards, the opportunities for trouble multiply rapidly. There is greater wiring capacitance, as well as longer ground paths through cables, connectors, card extenders, etc., so the ground spikes induced by drive currents during logic transitions are generally larger and more troublesome. It is best to avoid sending clocking signals with large fanout between boards, if possible, and ground connections to the individual cards should be robust. If clocking signals are sent between boards, it is important to use a gate as input buffer on each board. In extreme cases it may be necessary to use line driver and receiver chips, as we will discuss later. In any case, it is best to try to keep critical circuits together on one card, where you can control the inductance of the ground paths and keep wiring capacitance at a minimum. The problems you will encounter in sending fast signals around through several cards should not be underestimated; they can turn out to be the major headache of an entire project!

□ 9.16 Data buses

Where many subcircuits are connected together by a data bus (more on this in Chapters 10 and 11), the sort of problems already mentioned become more severe. In addition, a new factor comes into the picture: transmission-line effects due to the length and inductance of the signal lines themselves. With the fastest ECL chips ("MECL III," with rise times of less than 1ns), these effects are so severe that signal runs of more than 1 inch must be treated as transmission lines and properly terminated!

The best approach with data buses of any substantial length (a foot or more) is probably the use of a "motherboard" with ground plane. As we will explain in Chapter 12, a motherboard is simply a printed-circuit board containing a row of printed-circuit edge connectors to accept the individual circuit cards that make up the logic circuit. Motherboards are economical solutions to the problem of interconnecting cards, and if they are properly laid out, they are electrically superior as well. Wires that run close to a ground plane have lower inductance and less tendency to couple capacitively to adjacent signal lines, so a good way to arrange a motherboard is with all the signal lines on one side, with the other side being a solid ground plane (double-sided printed-circuit boards are standard; multiple-layer boards are even used on occasion for complicated circuits).

One last note on this problem: In desperation you may be tempted to use the common trick of putting a capacitor directly across the input of a gate driven from a long line, when transmission-line effects like "ringing" or ground spikes are driving you up the wall. In spite of the fact that we have done it ourselves, we don't recommend this inelegant fix, since it only compounds the problem of large ground currents during logic transitions (Section 9.14).

□ *Bus termination*

With buses of substantial length, it is common to "terminate" the signal lines at the far end with a resistive pull-up or pull-down. As we will discuss in Chapter 13, long pairs of wires or coaxial cables have a "characteristic impedance" Z_o. For a cable terminated with that impedance (which is always a resistance), a signal traveling down the cable will be entirely absorbed, with no reflection whatsoever. Any other value of termination, including an open circuit, will produce reflected waves, with amplitude and phase depending on the impedance mismatch. The sorts of conductor width and spacing used on printed-circuit boards give a characteristic impedance in the neighborhood of 100 ohms, which is also close to the characteristic impedance of twisted pairs made from ordinary insulated wire of 24 gauge or so.

A popular method of terminating a TTL bus is to use a voltage divider from +5 volts to ground. In that way the logic HIGH is kept around +3 volts, which means that less swing (and therefore less current into the load capacitance) is necessary during logic transitions. The combination of a 180 ohm resistor to +5 and a 390 ohm resistor to ground is typical.

□ *Bus drivers*

When driving bus lines of substantial length or fanout, it is necessary to use special TTL functions with high current-sinking capability. Here are the most popular types:

Type	*Output can sink*	*Description*
7438	48mA	Quad O/C 2-NAND
8880	48mA	Quad O/C 2-NAND
8095–8098 or 74365-8	32mA	Hex 3-state buffer/inverter
8T95–8T98	48mA	Hex 3-state buffer/inverter
75450-4	300mA	Dual O/C 2-AND, NAND, OR, NOR
3440 series	48mA	O/C drivers and transceivers, some with internal pull-ups
8836-8	16mA	3-state "unified" drivers and transceivers, high-Z inputs
74LS240-5	24mA	3-state drivers and transceivers

The sinking capability is specified at 0.4 or 0.5 volt maximum in the LOW state. The transceivers connect both the output of a driver and the input of a buffer to each bus line, all in one IC package.

9.17 Driving cables

Twisted pairs or coaxial cables are used to connect signals to a remote instrument, since single conductor lines are prone to pickup of interference. We will deal with coaxial cables (affectionately called "coax") in Chapter 13 in connection with radiofrequency techniques; here we would like to mention some of the techniques used to send digital signals over such lines, since these methods constitute an important part of digital interfacing. In most cases there are special-purpose line driver/receiver chips available to make your job easier.

RS-232C

For relatively slow (thousands of bits per second) transmission, the RS-232C signal standard is probably the best. It specifies bipolarity levels (positive and negative supply voltages are needed for the driver chip, but not generally for the receiver), and the receivers usually allow control of hysteresis and response time. Figure 9.29 shows a

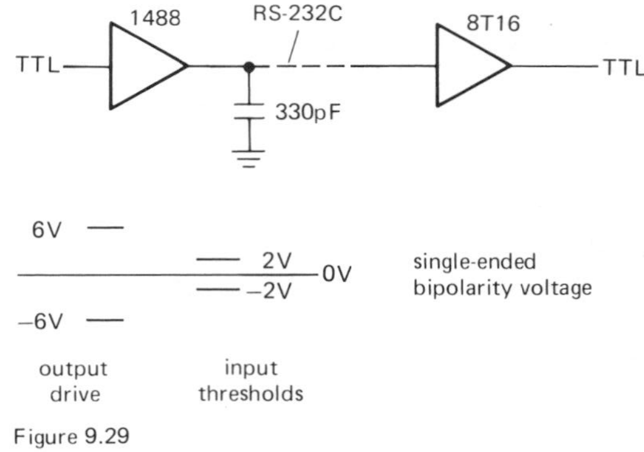

Figure 9.29
RS-232C high-noise-immunity cable driver and receiver.

typical setup. Note the 330pF load capacitor to keep rise and fall times slower than 1μs. RS-232C is used extensively for communication between computers and terminals, at standardized data rates ranging from 110 to 19,200 bits/s. The full standard even specifies the pin connections to be used with a 25-pin type D subminiature connector and is usually used for transferring ASCII-encoded data, as we will discuss in Section 10.17. A new standard (IEEE 422) allows considerably faster data transfer rates, but it is gaining acceptance slowly.

Direct TTL drive

Just as with data buses, it is possible to drive lines of moderate length directly with TTL devices; in general, high-current-capability gates should be used. Figure 9.30

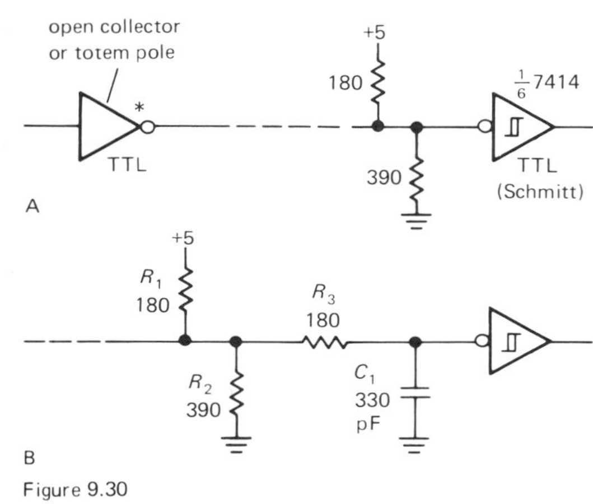

Figure 9.30
Use of termination on long TTL lines.

shows the usual method. In the first circuit, a buffer (which could be an open-collector type) drives a terminated line, with a TTL Schmitt trigger as a receiver for improved noise immunity. If noise is a serious problem, the *RC* slowdown network shown in the second circuit can be used, with the *RC* time constant (and the transmitted bit rate!) adjusted to give good noise immunity. With this circuit the Schmitt trigger is essential.

Important note: *Never* attempt to drive long lines from the output of clocked devices (flip-flops, monostables, shift registers); the capacitive loading and transmission-line effects can cause erratic or erroneous behavior, unless the chip happens to have "buffered" output signals.

□ *Differential TTL*

Much better noise immunity can be obtained by using differential signals (i.e., *Q* and *Q'*) on a twisted pair, with a differential receiver (Fig. 9.31). In this example, paired TTL inverters drive a terminated twisted pair with true and complemented signals, and a 75115 differential line receiver regenerates clean TTL levels from the mess. This scheme gives very good common-mode rejection and generates clean logic levels from line signals that can look just awful. The waveform shown gives an idea of what you might see on the individual signal lines in a relatively clean system; the individual signals tend to acquire bumps and wiggles, although they generally remain monotonic through the transition (they don't reverse direction).

The 75115 is an example of a line receiver with adjustable response time; the 75152, another differential receiver, allows control of hysteresis instead. For peace of mind, it is a nice idea to use a receiver with hysteresis (and adjustable time constant),

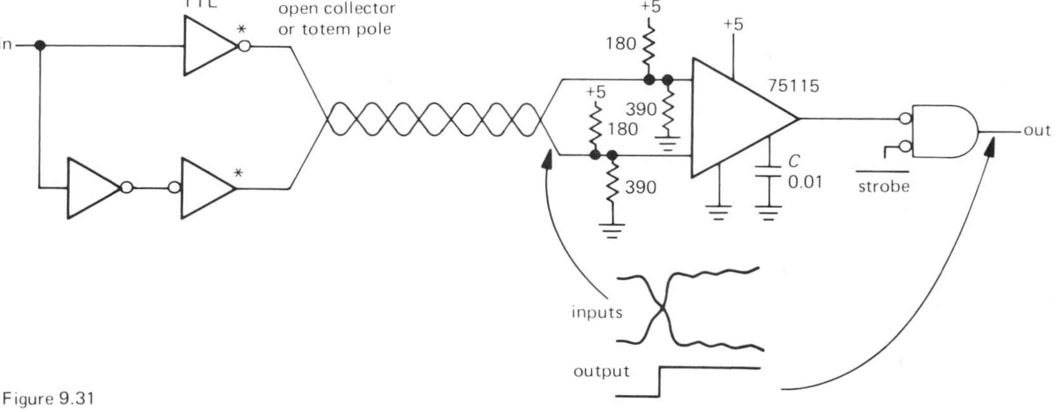

Figure 9.31
High-speed differential TTL cable driver and receiver.

given the bizarre waveforms these receivers are called on to recognize.

□ *Current-sinking drivers*

Chips like the 75110 have switched current-sinking outputs, which can be used as single-ended outputs or in a differential mode as shown previously. Figure 9.32

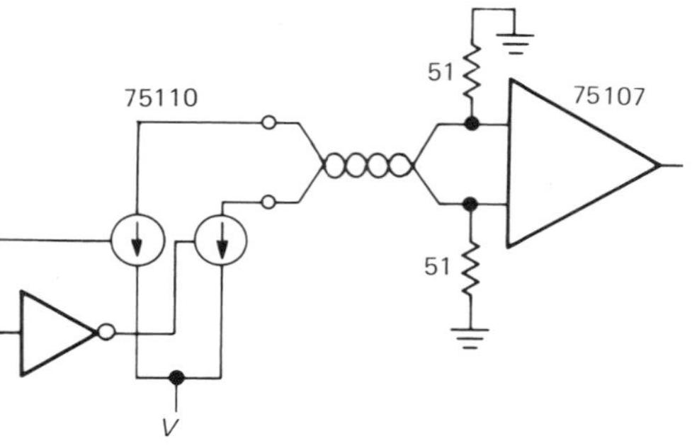

Figure 9.32
Differential current drive with terminated receivers.

shows a differential connection. The 75107 is a companion differential receiver chip. It would normally be used with a termination as shown, or perhaps with matched terminations at both ends of the transmission line. This simple data transmission scheme is claimed to allow data rates in excess of 1 megabit per second (Mb/s) over line lengths of 2000 feet and rates up to 10Mb/s over lengths of several hundred feet or less.

□ *Coaxial-cable drivers*

Coaxial cable gives very good protection against interference because of its completely shielded geometry. There are several driver/receiver pairs available that are suited to coax; Figure 9.33 shows an example. The cable is terminated in its characteristic impedance, in this case 51 ohms. The 8T23 can drive a 50 ohm load directly, and the

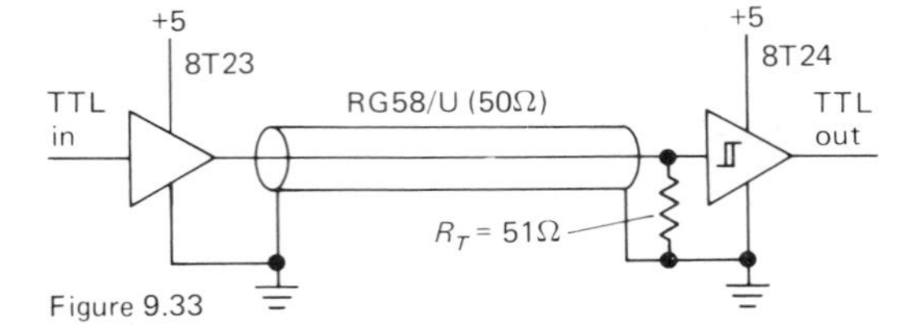

Figure 9.33
Fifty ohm cable driver and receiver.

8T24 has a fixed amount of hysteresis for noise immunity and fast output transition time. Other driver/receiver pairs are the 8T13/8T14, 75123/75124, and a number of other members of the 75*xxx* interface family. Be sure to use the specified receivers when driving 50 ohm coaxial lines, since the logic-level swings may be less than ordinary TTL levels. The circuit illustrated in Figure 9.33 will allow bit rates in excess of 100kb/s over cables a mile long and up to 20Mb/s over shorter lines.

TTL outputs can drive a 50 ohm cable if buffered by a single *npn* emitter follower (Fig. 9.34). The 2N4401 is a husky little

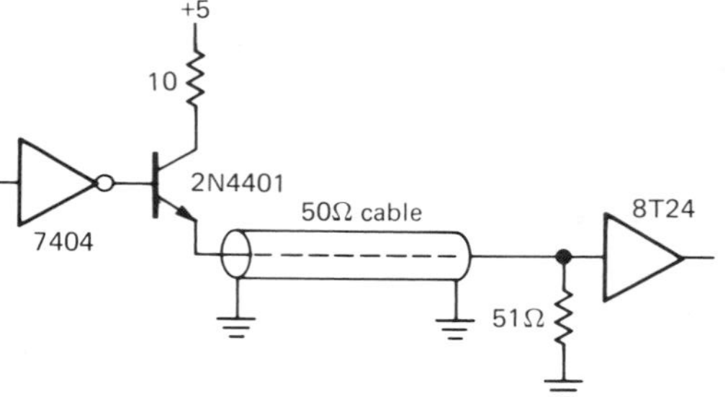

Figure 9.34

transistor with plenty of beta at high currents ($h_{FE} > 100$ at $I_C = 150$mA). The 10 ohm resistor is for short-circuit protection. In comparison with the carefully designed and expensive 50 ohm cable driver chips that are available, this circuit is embarrassingly simple. Note that the "open-emitter" output must be loaded with a low-resistance return to ground in order to work, which is also true of some IC cable driver chips, (e.g., the 8T13 mentioned earlier).

ANALOG/DIGITAL CONVERSION

9.18 Introduction to A/D conversion

In addition to applying the purely "digital" interfacing (switches, lights, etc.) discussed in the last few sections, it is often necessary to convert an analog signal to an accurate digital number proportional to its amplitude, and vice versa. This is essential in any application in which a computer or processor is logging or controlling an experiment or process, or whenever digital techniques are used to do a "normally" analog job. Applications in which analog information is converted to an intermediate digital form for error- and noise-free transmission [e.g., "digital audio" or pulse-code modulation (PCM)] make heavy use of analog/digital conversion. It is necessary in a wide variety of measurement instruments (including ordinary bench instruments such as digital multimeters and more exotic instruments such as transient averagers, "glitch catchers," and digital memory oscilloscopes), as well as signal-generation and processing instruments such as digital waveform synthesizers and data encryption devices.

Finally, conversion techniques are essential ingredients for the generation of analog displays by a digital instrument, e.g., a meter indication or xy display (or plot) created by a computer. Even in a relatively unsophisticated electronic apparatus there

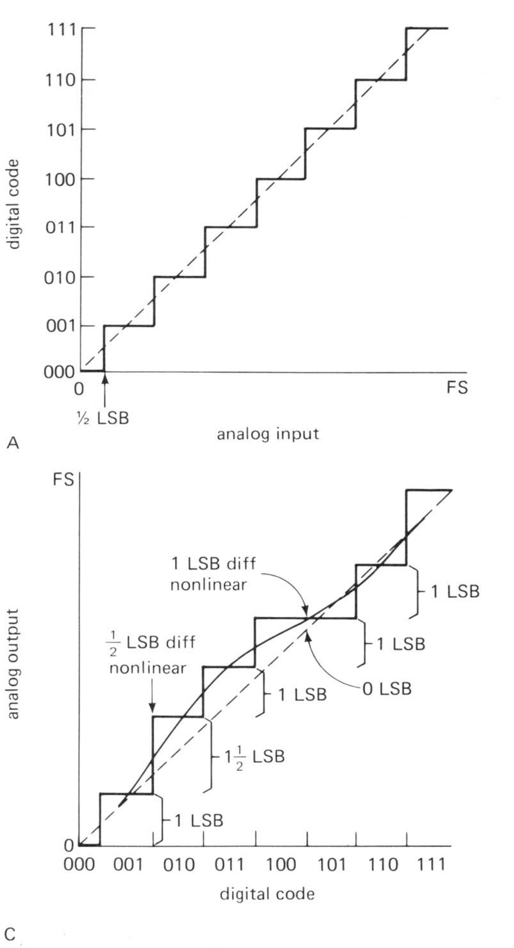

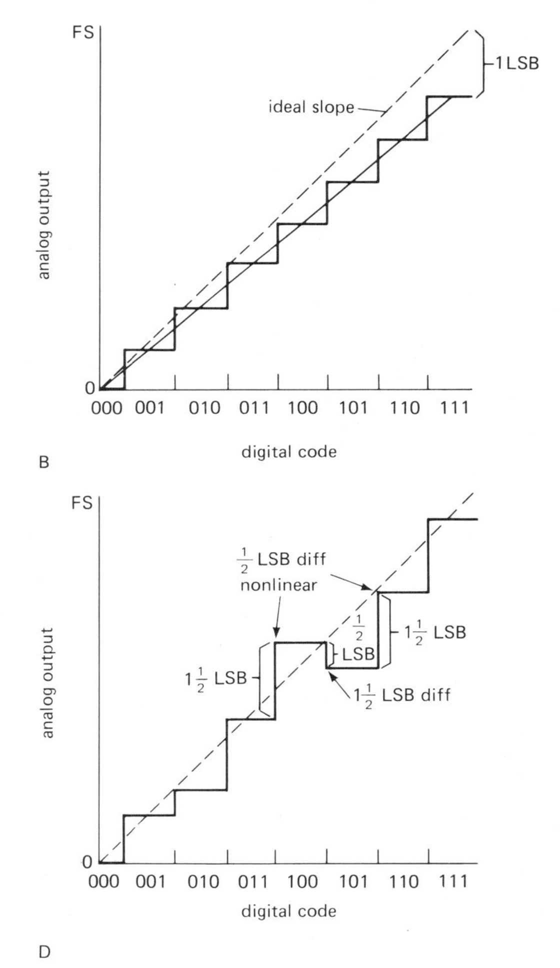

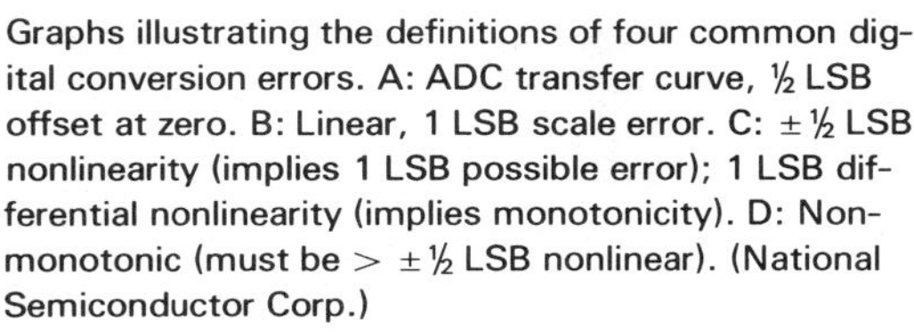

Figure 9.35

Graphs illustrating the definitions of four common digital conversion errors. A: ADC transfer curve, ½ LSB offset at zero. B: Linear, 1 LSB scale error. C: ±½ LSB nonlinearity (implies 1 LSB possible error); 1 LSB differential nonlinearity (implies monotonicity). D: Nonmonotonic (must be > ±½ LSB nonlinear). (National Semiconductor Corp.)

are plenty of nice applications that call for A/D and D/A conversion, and it is worth developing familiarity with the various techniques and available modules used in analog/digital conversion, especially since the day of the $5 A/D and D/A converter is nearly here.

Our treatment of the various conversion techniques will not be aimed at developing skill in converter design itself. Rather, we will try to point out the advantages and disadvantages of each method, since in most cases the most sensible thing is to buy commercially available chips or modules, rather than build the converter from scratch. An understanding of conversion techniques and idiosyncrasies will guide you in choosing among the hundreds of available units.

Codes

At this point you should review Section 8.03 on the various number codes used to represent signed numbers. Offset binary and 2's complement are commonly used in A/D conversion schemes, with sign-magnitude and Gray codes also popping up from time to time. Here is a reminder:

	Offset binary	2's Complement
+ Full scale	11111111	01111111
+ Full scale − 1	11111110	01111110
↓	↓	↓
0 + 1 LSB	10000001	00000001
0	10000000	00000000
0 − 1 LSB	01111111	11111111
↓	↓	↓
− Full scale + 1	00000001	10000001
− Full scale	00000000	10000000

Converter errors

The subject of A/D and D/A errors is a complicated one, about which whole volumes could be written. According to Bernie Gordon at Analogic, if you think a high-accuracy converter system lives up to its claimed specifications, you probably haven't looked closely enough. We won't go into the application scenarios necessary to support Bernie's claim, but we will show the four most common types of converter errors. Rather than bore you with a lot of complicated talk, we'll just present self-explanatory graphs of the four most common errors: offset error, scale error, nonlinearity, and nonmonotonicity (Fig. 9.35).

9.19 Digital-to-analog converters (DACs)

The goal is to convert a quantity specified as a binary (or multidigit BCD) number to a voltage or current proportional to the value of the digital input. There are several popular methods:

Scaled resistors into summing junction

As you saw in Section 3.09, by connecting a set of resistors to an op-amp summing junction you get an output proportional to a weighted sum of the input voltages (Fig. 9.36). This circuit generates an output from

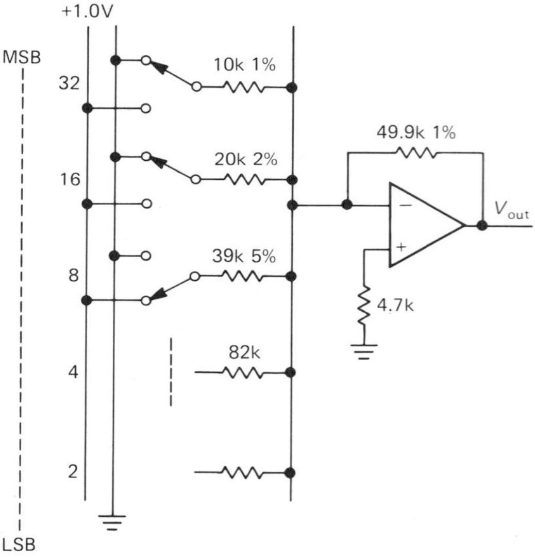

Figure 9.36

zero to −10 volts, with full output corresponding to an input count of 64. Actually, the maximum input count is always $2^n - 1$, i.e., all bits set to 1. In this case the maximum input count is 63, with output voltage of $-10 \times 63/64$. By changing the feedback resistor, you can generate an output of zero to −6.3 volts (i.e., output in volts numerically equal to −1/10 of input count), or you can add an inverting amplifier, or dc offset at the summing junction, to get positive outputs. By changing the input resistor values you can properly convert a multidigit BCD input code, or any other weighted code. The input voltages must be clamped to an accurate reference, and the smaller input resistors must be of correspondingly higher precision. Of course, the switch resistance must be smaller than $1/2^n$ of the smallest resistor, an important consideration since in actual circuits the switching is done with transistor or FET switches.

EXERCISE 9.1

Design a 2-digit BCD D/A converter. Assume that the inputs are zero or +1 volt; the output should go from zero to 9.9 volts.

R-2R ladder

The *R*-2*R* ladder is an interesting variation of the preceding circuit, generating a voltage output directly, using a set of resistors of two values only (Fig. 9.37). This circuit generates an output of zero to +10 volts, with "full output" corresponding to an input count of 16 (again, the maximum input count is 15, with output voltage of 10 × 15/16). With a few modifications, an *R*-2*R* scheme can be used for BCD conversion.

Scaled current sources

In this method the input binary code turns on a selection of current sources wired to generate appropriately scaled currents (Fig. 9.38). The output currents are summed and are either brought out directly as a current output or converted internally to a voltage via an op-amp summing junction circuit ("transresistance" configuration, see Section 3.09). The scaled currents are generated internally by transistors in conjunction

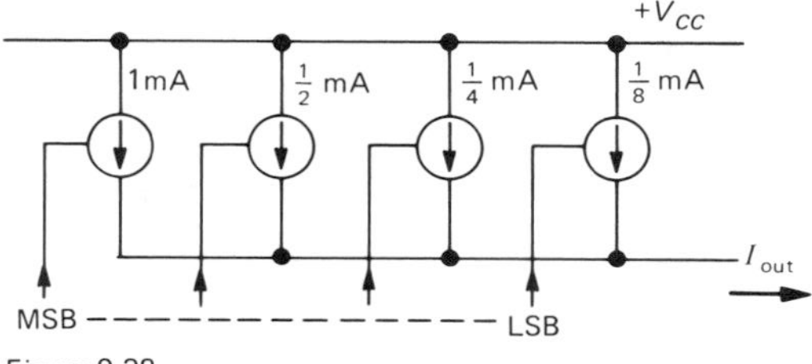

Figure 9.38

with a set of suitably scaled resistors or a ladder-like resistor array, depending on the particular converter. For devices with current output, the output voltage compliance can range from as little as 0.5 volt for some types (e.g., the 1406) to as much as 25 volts or more (e.g., the DAC-08). Outputs can be either current-sourcing or current-sinking.

In most converters of this type, the current sources are actually ON all the time, and their output current is switched to the output terminal or to ground, under control of the digital input code. This gives improved speed and accuracy, and the switching itself is easily accomplished with transistors or diodes (Fig. 9.39). In the first circuit, the current sources are scaled with a ladder network, with their outputs switched to V_+ or the current output terminal, as dictated by the digital input. The emitter areas are scaled as indicated, for constant emitter current density. The op-amp, in combination with Q_{ref}, generates a negative voltage reference with suitable V_{BE} dependence for biasing the current sources: The stable positive reference voltage $+V_{ref}$, which may be internally generated or supplied from an external reference, is used to generate a collector current

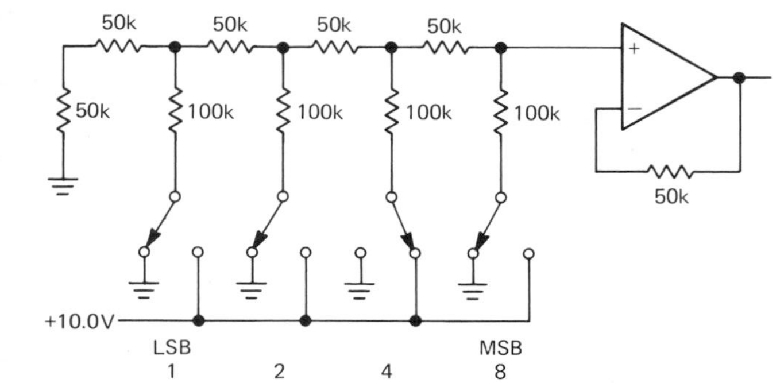

Figure 9.37

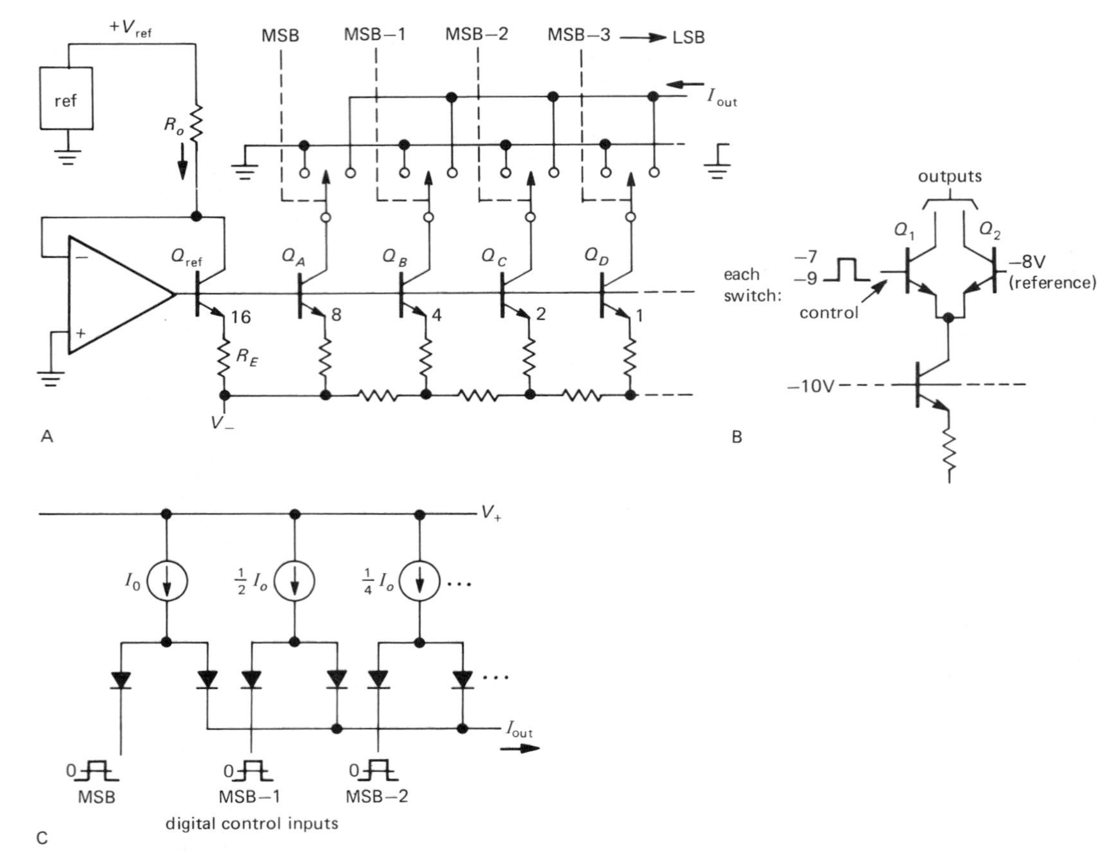

Figure 9.39
Classic current-switched DAC.

in Q_{ref} of $I_C = V_{ref}/R_o$, and hence a stable emitter voltage relative to V_-; the base sits one V_{BE} drop above, generating the desired bias voltage for the current sources $Q_A - Q_D$, which therefore generate the necessary binary scaled currents.

The switch circuitry itself is extremely simple. Each switch is a pair of *npn* transistors connected as shown, with one base tied to a reference voltage a few volts above the base voltage of the current-source transistors. You can think the switch as a pair of cascode transistors, with the emitter current passing to the collector of the transistor with the higher base voltage. The control voltage to the switch (Q_1's base) has to swing only a fraction of a volt above or below Q_2's base voltage for complete switching of the current. This sort of switch is fast and has large output compliance. The second circuit shows an even simpler diode switching arrangement in which the output current from the individual scaled current sources either goes to the output terminal (through the right-hand diode of each pair) or gets diverted by the left-hand diode, according to the control voltage applied to the cathodes of the left-hand diodes. This scheme tends to have a relatively small output compliance range, since the control voltage must swing beyond the possible range of output voltages, but it is extremely fast. Even inexpensive (less than $10) current-output DACs have settling times of 100ns or less.

Generating a voltage output

There are a few ways to generate an output voltage from a current DAC. Figure 9.40 shows some ideas. If the load capacitance is low, and large voltage swings aren't needed,

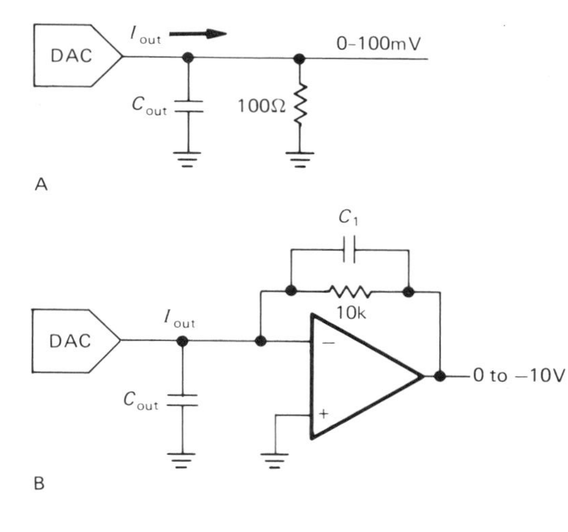

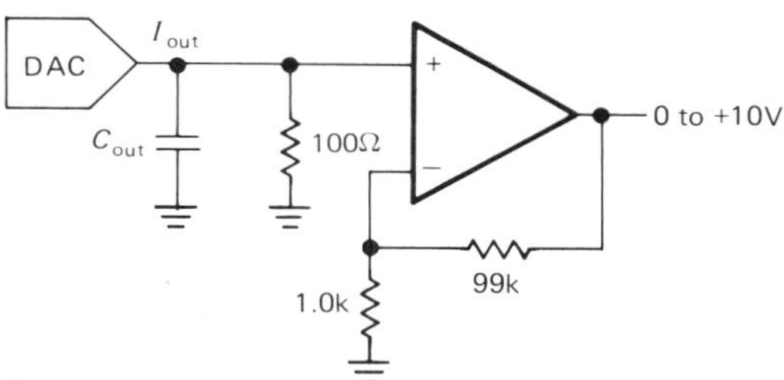

Figure 9.40

Generating voltages from current-output DACs.

a simple resistor to ground will do nicely. With the usual 1mA full-scale output current, a 100 ohm load resistor will give 100mV full-scale output with 100 ohms output impedance. If the capacitance of the DAC's output combined with the load capacitance doesn't exceed 100pF, you will get 100ns settling time in the preceding example, assuming the DAC is that fast. When worrying about the effect of *RC* time constants on DAC output response, don't forget that it takes quite a few *RC* time constants for the output to settle to within 1/2 LSB of the final voltage. It takes 7.6 *RC* time constants, for instance, to settle to within 1 part in 2048, which is what you would want for a 10-bit converter output.

To generate large swings, or to buffer into small load resistances or large load capacitances, an op-amp can be used in the transresistance configuration (current-to-voltage amplifier), as shown. The capacitor across the feedback resistor may be necessary for stability, because the DAC's output capacitance in combination with the feedback resistance introduces a lagging phase shift; unfortunately, that compromises the speed of the amplifier. It is an irony of this circuit that a relatively expensive high-speed (fast-settling) op-amp may be necessary to maintain the high speed of even an inexpensive DAC. In practice the last circuit may give better performance, since no compensation capacitor is needed. Watch out for offset voltage error, since the op-amp's input offset voltage is amplified 100 times.

Commercial D/A converter modules are available with precision ranging from 6 bits up to 18 bits, with settling times of 25ns to 100μs (for converters of the highest precision). Prices range from a few dollars to several hundred dollars. A typical popular unit is the DAC-80, a 12-bit converter with internal reference, settling time of 3μs for voltage output (0.3μs for current output), and a price of about $20.

□ 9.20 Time-domain (averaging) DACs

□ *Frequency-to-voltage converters*

In conversion applications the "digital" input may be a train of pulses or other waveform of some frequency; in that case, direct conversion to a voltage is sometimes more convenient than the alternative of counting for a predetermined time, then converting the binary count via the preceding methods. In direct F/V conversion, a standard pulse is generated for each input cycle; it may be a voltage pulse or a pulse of current (i.e., a fixed amount of charge).

An *RC* low-pass filter or integrator then averages the pulse train, giving an output voltage proportional to the average input frequency. Of course, some output ripple results, and the low-pass filter necessary to keep this ripple less than the D/A precision (e.g., ½ LSB) will, in general, cause a slow output response. To ensure less than ½ LSB output ripple, the time constant T of a simple *RC* low-pass filter must be at least $T = 0.69(n + 1)T_o$, where T_o is the output period of the n-bit F/V converter corresponding to maximum input frequency. The output of this *RC* network will settle to ½

LSB, following a full scale change at the input, in $0.69(n + 1)$ filter time constants. In other words, the output settling time to ½ LSB will be approximately $t = 0.5(n + 1)^2T_o$. A 10-bit F/V converter with 100kHz maximum input frequency, smoothed with an *RC* filter, will have an output voltage settling time of 0.6 ms. With more complicated low-pass filters (sharp cutoff) you can get improved performance. Before you get carried away with fancy filter design, however, you should remember that F/V techniques are most often used when a voltage output is not what's needed. For some perspective, we will next comment about intrinsically slow loads in connection with pulse width modulation.

□ *Pulse width modulation*

In this technique the digital input code is used to generate a train of pulses of fixed frequency, with width proportional to the input count; this can be easily done with a counter, magnitude comparator, and high-frequency clock (see Exercise 9.2). Once again, a simple low-pass filter can be used to generate an output voltage proportional to the average time spent in the HIGH state, i.e., proportional to the digital input code. More often, this sort of D/A conversion is used when the load is itself a slowly responding system; the pulse width modulator then generates precise parcels of energy, averaged by the system connected as a load. For example, the load might be capacitive (as in a switching regulator, see Chapter 5), thermal (a thermostated bath with heater), mechanical (a tape speed servo), or electromagnetic (a large electromagnet controller).

EXERCISE 9.2

Design a circuit to generate a 10kHz train of pulses of width proportional to an 8-bit binary input code. Use counters and magnitude comparators (suitably expanded).

□ *Averaged rate multiplier*

The rate-multiplier circuit described in Section 8.27 can be used to make a simple D/A converter. A parallel binary or BCD input code is converted to a train of output pulses of average frequency proportional to the digital input; simple averaging, as in the preceding F/V converter, can be used to generate a dc output proportional to the digital input code, although in this case the resulting output time constant will be intolerably long, since the rate-multiplier output will have to be averaged for a time equal to the longest output period it can generate. Rate multipliers are most advantageously used as D/A converters when the output is intrinsically averaged by the slowly responding characteristics of the load itself, as described earlier.

Perhaps the nicest application is in digital temperature control, where complete cycles of ac power are switched across the heater for each rate-multiplier output pulse. In this application the rate multiplier is arranged so that its lowest output frequency is an integral submultiple of 120Hz, and a solid-state relay (or triac) is used to switch the ac power (at zero crossings of its waveform) from logic signals.

Note that the last three conversion techniques involve some time averaging, whereas the resistor-ladder and current-source methods are "instantaneous," a distinction that also exists in the various methods of analog-to-digital conversion. Whether a converter averages the input signal or converts an instantaneous sample of it can make an important difference, as you will see shortly in some examples.

9.21 Multiplying DACs

Most of the preceding techniques can be used to construct a "multiplying DAC," in which the output equals the product of an input voltage (or current) and an input digital code. For instance, in a scaled current-source DAC you can scale all the internal current sources by an input programming current. Multiplying DACs can be made from DACs that have no internal reference by using the reference input for the analog input signal. However, not all DACs are optimized for use in this way, so it is best to check the data sheets of the converters you're considering for details. A DAC with good multiplying properties (wide analog input range, high speed, etc.) will usually be called a "multiplying DAC" right at the top of the data sheet. The AD7521, DAC348, DAC921, and 562 are all examples of 12-bit multiplying DACs, priced around $10 to $20.

Multiplying DACs (and the A/D equivalent) open the possibility of *ratiometric* measurements and conversions. If a sensor of some sort (e.g., a variable-resistance transducer like a thermistor) is powered by a reference voltage that also supplies the reference for the A/D or D/A converter, then variations in the reference voltage will not affect the measurement. This concept is very powerful, since it permits measurement and control with accuracy greater than the stability of voltage references or power supplies; conversely, it relaxes the requirements on supply stability and accuracy. The ratiometric principle is used in its simplest form in the classic *bridge* circuit, in which two ratios are adjusted to equality by nulling the differential signal taken between the two voltage divider outputs (see Section 14.02). Devices like the 555 (see Section 4.13) achieve good stability of output frequency with large variations of supply voltage by using essentially a ratiometric scheme: The capacitor voltage, generated by an *RC* network from the supply, is compared with a fixed fraction of the supply voltage ($\frac{1}{3}V_{CC}$ and $\frac{2}{3}V_{CC}$), giving an output frequency that depends only on the *RC* time constant. We will have more to say on this important subject in connection with A/D converters later in this chapter, as well as in Chapter 14, when we discuss scientific measurement techniques.

9.22 Analog-to-digital converters

There are half a dozen basic techniques of A/D conversion, each with its peculiar advantages and limitations. Since you usually use a commercial A/D module or chip rather than build your own, we will describe the various conversion techniques somewhat briefly, mainly to serve as a guide for intelligent selection in a given application. The next section of the chapter will illustrate some typical A/D applications. In Chapter 11 we will discuss some A/D converters that use exactly the same conversion methods, but that have outputs designed for simplified interfacing to microprocessors.

Parallel encoder

In this method the input signal voltage is fed simultaneously to one input of each of *n* comparators, the other inputs of which are connected to *n* equally spaced reference voltages. A priority encoder generates a digital output corresponding to the highest comparator activated by the input voltage (Fig. 9.41).

Parallel encoding (sometimes called "flash" encoding) is the fastest method of A/D conversion. The delay time from input to output equals the sum of comparator plus encoder delays; with NE521 comparators and a 74148 encoder, the delay will typi-

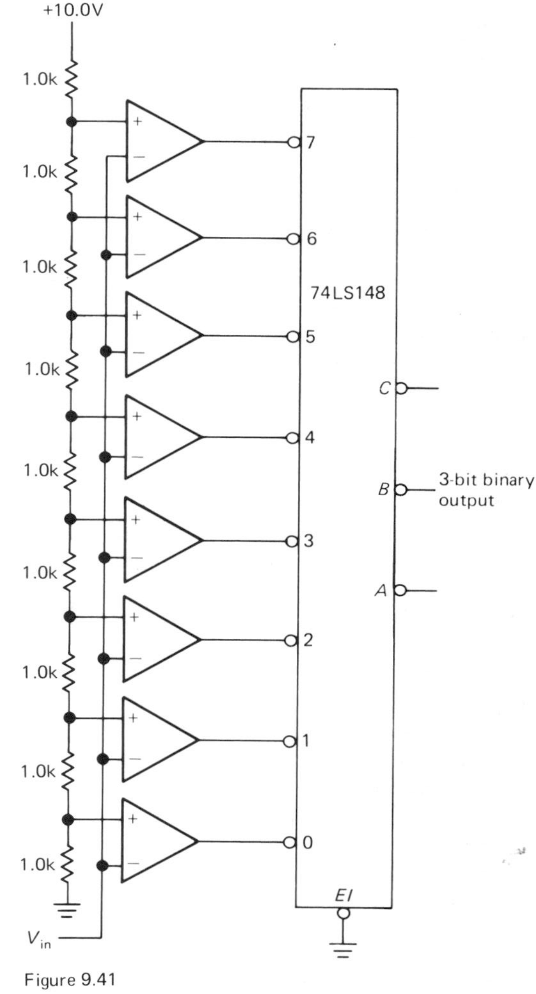

Figure 9.41

Parallel-encoded A/D converter (ADC).

cally be less than 20ns. Commercial parallel encoders are available with 16 to 256 levels (4-bit to 8-bit outputs). Beyond that they become prohibitively expensive and bulky. The 256-level TDC1007J manufactured by TRW, for instance, converts in 33ns and is currently priced in the hundreds of dollars. The design of this high-performance converter requires extremely fast comparator response (e.g., 15ns) with only 1/256 full-scale voltage overdrive, and that leaves just 15ns for digital encoding. In reality, the priority encoder should generate Gray code output in order to avoid incorrect codes for input levels near the comparator thresholds.

Successive approximation

In this popular technique you try various output codes by feeding them into a D/A converter and comparing the result with the analog input via a comparator. The way it's usually done is to set all bits initially to 0. Then, beginning with the most significant bit, each bit in turn is set provisionally to 1. If the D/A output does not exceed the input signal voltage, the bit is left as a 1; otherwise it is set back to 0. For an n-bit A/D, n such steps are required. What you're doing could be described as a binary search, beginning at the middle. A successive-approximation A/D module has a BEGIN CONVERSION input and a CONVERSION DONE output. The digital output is always provided in parallel format (all bits at once, on n separate output lines) and usually in serial format as well (n successive output bits, MSB first, on a single output line).

Successive-approximation A/D converters are relatively accurate and fast, requiring only n settling times of the DAC for n-bit precision. Typical conversion times range from 1μs to 50μs, with accuracies of 8 to 12 bits commonly available; prices range from about $10 to upwards of $400. This type of converter operates on a brief sample of the input voltage, and if the input is changing during the conversion, the error is no greater than the change during that time; however, spikes on the input are disastrous. Although generally quite accurate, these converters can have strange nonlinearities and "missing codes."

A variation known as a "tracking A/D converter" uses an up/down counter to generate successive trial codes; it is slow in responding to jumps in the input signal, but it follows smooth changes somewhat more rapidly than a successive-approximation converter. For large changes its slew rate is proportional to its internal clocking rate.

□ Voltage-to-frequency conversion

In this method an analog input voltage is converted to an output pulse train whose frequency is proportional to the input level. This can be done simply by charging a capacitor with a current proportional to the input level and discharging it when the ramp reaches a preset threshold. For greater accuracy, a feedback method is generally used. In one technique you compare the output of an F/V circuit with the analog input level and generate pulses at a rate sufficient to bring the comparator inputs to the same level. In the more popular methods, a "charge-balancing" technique is used, as will be described in greater detail later (in particular, the "capacitor-stored charge-dispensing" method).

Typical V/F output frequencies are in the range 10kHz to 1MHz for maximum input voltage. Commercial V/F converters are available with the equivalent of 12-bit resolution (0.01% accuracy). They are inexpensive, and they are handy when the output is to be transmitted digitally over cables or when an output frequency (rather than digital code) is desired. If speed isn't important, you can get a digital count proportional to the average input level by counting the output frequency for a fixed time interval. This technique is popular in moderate-accuracy (3-digit) digital panel meters.

□ Single-slope integration

In this technique an internal ramp generator (current source + capacitor) is started to begin conversion, and at the same time a counter is enabled to count pulses from a stable clock. When the ramp voltage equals the input level, a comparator stops the counter; the count is proportional to the input level, i.e., it's the digital output. Figure 9.42 shows the idea.

At the end of the conversion the circuit discharges the capacitor and resets the

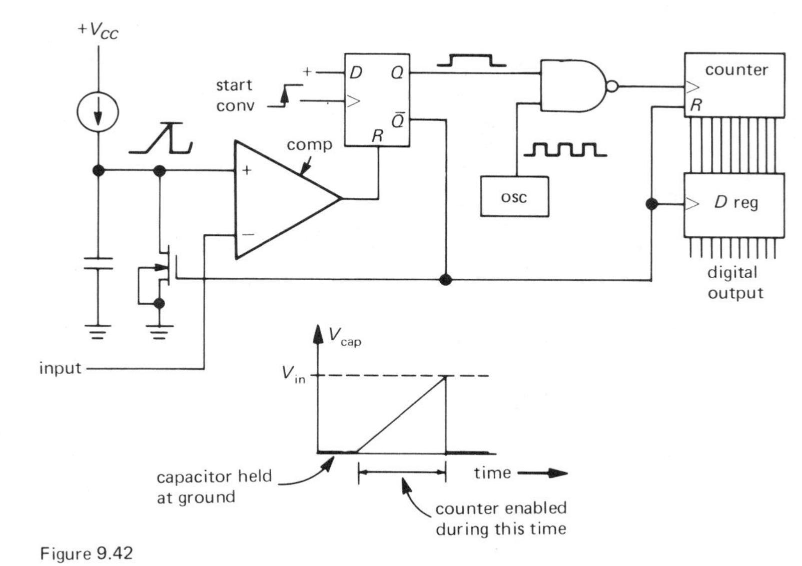

Figure 9.42

Single-slope ADC.

counter, and the converter is ready for another cycle. Single-slope integration is simple, but it is not used where high accuracy is required because it puts severe requirements on the stability and accuracy of the capacitor and comparator. The method of ''dual-slope integration'' eliminates that problem (and several others as well) and is now generally used where precision is required.

Single-slope integration is still alive and well, particularly in applications that don't require absolute accuracy, but rather need conversion with good resolution and uniform spacing of adjacent levels. A good example is pulse-height analysis (see Section 14.16), where the amplitude of a pulse is held (peak detector) and converted to an address. Channel width equality is essential for this application, for which a successive-approximation converter would be totally unsuitable. The technique of single-slope integration is also used in time-to-amplitude conversion (TAC).

9.23 Charge-balancing techniques

There are several techniques that have in common the use of a capacitor to keep track of the ratio of an input signal level to a reference. These methods all average (integrate) the input signal for a fixed time interval for a single measurement. There are two important advantages:

1. Because these methods use the same capacitor for the signal and reference, they are relatively forgiving of capacitor stability and accuracy. These methods also make fewer demands on the comparator. The result is better accuracy for equivalent-quality components, or equivalent accuracy at reduced cost.

2. The output is proportional to the *average* input voltage over the (fixed) integration time. By choosing that time inverval to be a multiple of the power-line period, the converter becomes insensitive to 60Hz ''hum'' (and its harmonics) on the input signal. As a result, the sensitivity to interfering signals as a function of frequency is as shown in Figure 9.43 (0.1 s integration).

This nulling of 60Hz interference requires accurate control of the integration time, since an error of a fraction of a percent in the clock timing will result in incomplete cancellation of hum. One possibility is to use a crystal oscillator. You will see in Section 9.31 an elegant method of synchronizing the workings of a charge-balancing con-

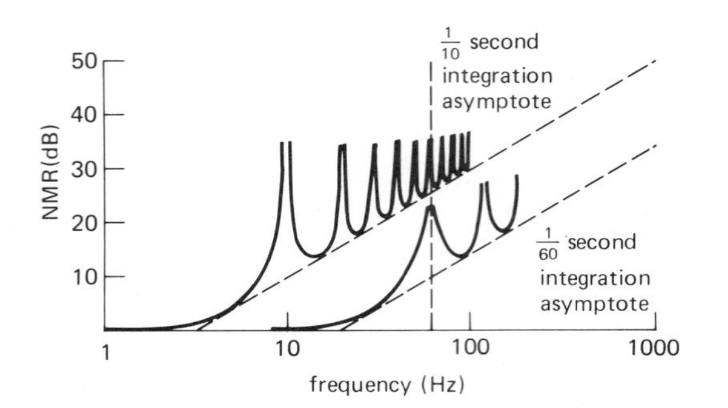

Figure 9.43

Normal-mode rejection with integrating A/D converters.

verter to a multiple of the line frequency in order to make this rejection perfect.

Charge-balancing techniques have the disadvantage of slow speed, as compared with successive approximation.

Dual-slope integration

This elegant and very popular technique eliminates most of the capacitor and comparator problems inherent in single-slope integration. Figure 9.44 shows the

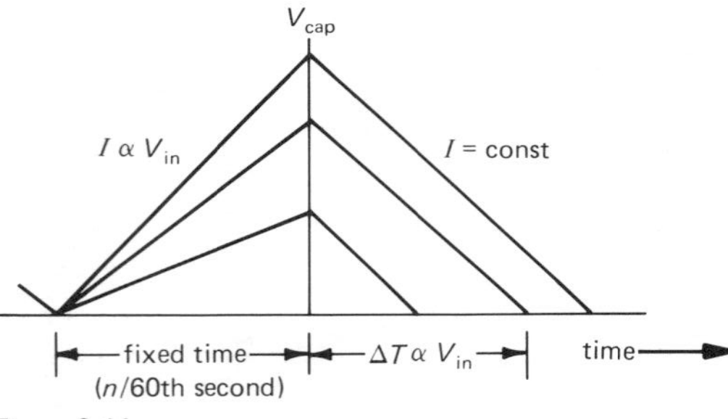

Figure 9.44

Dual slope conversion cycle.

idea. First, a current accurately proportional to the input level charges a capacitor for a fixed time interval; then the capacitor is discharged by a constant current until the voltage reaches zero again. The time to discharge the capacitor is proportional to the input level and is used to enable a counter driven from a clock running at a fixed frequency. The final count is proportional to the input level, i.e., it's the digital output.

Dual-slope integration achieves very good accuracy without putting extreme requirements on component stability. In particular, the capacitor value doesn't have to be particularly stable, since the charge cycle and the discharge cycle both go at a rate inversely proportional to C. Likewise, drifts or scale errors in the comparator are cancelled out by beginning and ending each conversion cycle at the same voltage and, in some cases, at the same slope. In the most accurate converters, the conversion cycle is preceded by an "auto-zeroing" cycle in which the input is held at zero. Since the same integrator and comparator are used during this phase, subtracting the resulting "zero-error" output from the subsequent measurement results in effective cancellation of errors associated with measurements near zero; however, it does not correct for errors in overall scale.

Note that even the clock frequency does not have to have high stability in dual-slope conversion, because the fixed integration time during the first phase of the measurement is generated by subdivision from the same clock used to increment the counter. If the clock slows down by 10%, the initial ramp will go 10% higher than normal, requiring 10% longer ramp-down time. Since that's measured in clock ticks that are 10% slower than normal, the final count will be the same! Only the discharge current has to be of high stability in a dual-slope converter with internal auto-zeroing. Precision voltage and current references are relatively easy to produce, and the (adjustable) reference current sets the scale factor in this type of converter.

Dual-slope integration is used extensively in precision digital multimeters, as well as in conversion modules of 10-bit to 18-bit resolution. It offers good accuracy and high stability at lowest cost, combined with excellent rejection of power-line (and other) interference, for applications where speed is not important. For a fixed amount of money, you will get greatest precision with a module that uses this technique. The digital output codes are strictly monotonic with increasing input.

Delta-sigma convertors

There are several methods of A/D conversion that involve cancellation of the (average) signal input current with a switched internal source of current or charge. Figure

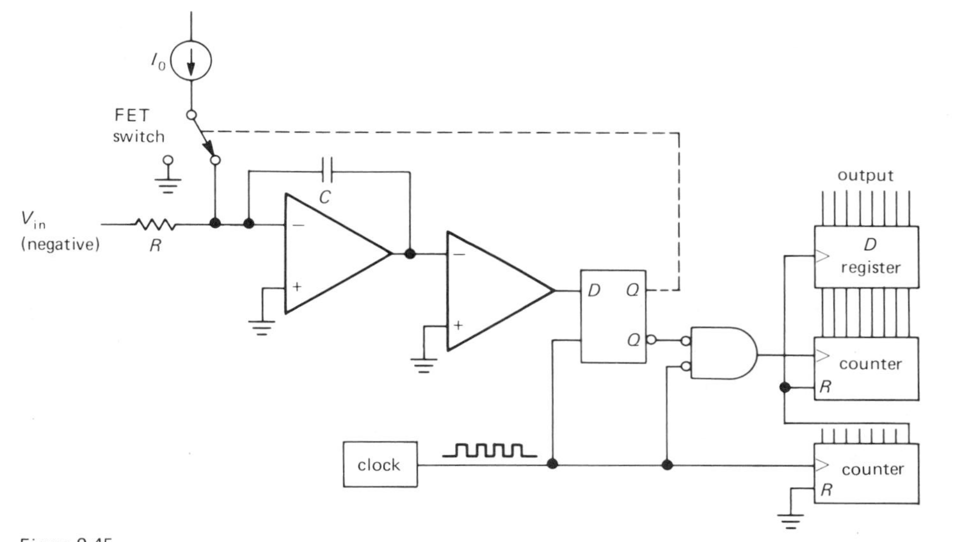

Figure 9.45

Delta-sigma charge-balancing ADC.

9.45 shows a functional diagram of a "delta-sigma" converter.

The input voltage drives an integrator, whose output is compared with any fixed voltage, such as ground. Depending on the comparator output, pulses of current of fixed length (i.e., fixed increments of charge) are switched into the summing junction or to ground at each clock transition, with the effect of maintaining zero average current into the summing junction. This is the balancing concept. A counter keeps track of the number of charge pulses switched into the summing junction for a given number of clock pulses, say 4096. That count is proportional to the average input level during the 4096 clock pulses, i.e., it's the output.

Delta-sigma converters can also be constructed with the current pulses generated with a resistor from a stable reference voltage, since the summing junction is a virtual ground. In that case you have to make sure that the switch ON resistance is small compared with the series resistor, so that variations of R_{ON} don't cause drifts.

□ *Switched-capacitor A/D*

A closely related charge-balancing method uses the "capacitor-stored charge-dispensing A/D," or "switched-capacitor" A/D. In this technique, fixed amounts of charge are created by repeatedly charging a capacitor from a stable reference voltage, then discharging it into the summing junction. The comparator looks at the integrator output, as previously, and controls the rate at which the capacitor is switched. That rate is counted for a fixed time interval to generate the digital output. This method has advantages for circuits that are meant to operate from a single supply voltage, since the effective polarity of charge transferred from the capacitor to the summing junction can be reversed by suitably connected FET switches (i.e., by switching both sides of the capacitor).

An example of this technique is the LM331 voltage/frequency converter, which has the unique advantage of operation from a single +5 volt supply. We showed its application as a VCO in Section 4.13.

Comments on integrating A/D converters

As with the dual-slope integration A/D, these charge-balancing methods all average the input over fixed time intervals, so they can be made insensitive to 60Hz pickup and its harmonics. In general, charge-balancing methods are inexpensive (they don't require a particularly good comparator, for instance) and accurate, and they produce a strictly monotonic output. However, they are slow, as compared with successive approximation. The ADC100 from Burr-Brown gives 16-bit resolution with 200ms conversion time, with a price of about $200. By comparison, the ADC-16Q from Analog Devices is a 16-bit successive-approximation converter with 400μs conversion time and $1720 price tag. Compared with dual-slope converters, delta-sigma and switched-capacitor methods can be characterized by low-accuracy comparators following the integrators, but with precise charge-switching circuits, whereas dual-slope methods use highly repeatable endpoint comparators, with somewhat simpler switch requirements, at least from the viewpoint of speed and charge injection.

One interesting point to keep in mind with any of the integrating techniques (single- and dual-slope integration and charge balancing) is that the input to the integrator can be either a current, or a voltage in series with a resistor. In fact, some converters have two input terminals, one tied directly to the summing junction for use with devices that source a current directly. When used with a current input, the offset voltage of the integrator becomes unimportant, whereas with a voltage input (with internal series resistor) the integrator op-amp produces an error equal to its input offset voltage. Current input is useful, therefore, to get wide dynamic range, particularly if the A/D is used with a device that has a current output anyway; examples are photomultipliers and photodiodes. Beware of this specmanship "gotcha," though: A/D accuracy may be specified assuming input current, for converters that accept either current or voltage input; don't expect as good performance for small inputs when using such a converter with voltage input.

Note that these charge-balancing methods all include a highly accurate V/F converter and that they can be used as such if an output *frequency* is desired, as indicated in Figure 9.46.

SOME A/D CONVERSION EXAMPLES

9.24 16-Channel A/D data-acquisition system

Figure 9.47 shows a circuit to digitize any of 16 analog inputs on command, with a 12-bit digital output. This could form the "front end" of a microprocessor-controlled data-taking experiment.

The HI-506 is a 16-channel CMOS analog multiplexer with CMOS-compatible digital inputs. This particular multiplexer has some very nice properties. In particular, its switches are of the "break-before-make" variety, which means that the various input channels don't find themselves shorted together during address changes in the MUX. In addition, the inputs may swing beyond the supply rails, and when they do there is no "SCR latch-up" or cross-talk between inputs. Watch out for consider-

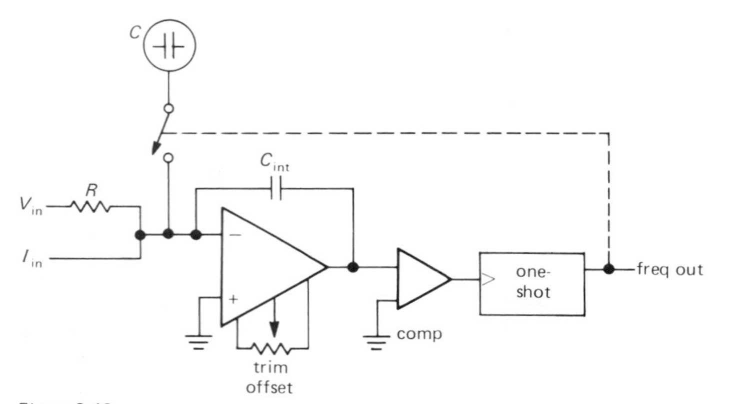

Figure 9.46

Charge-balancing V/F converter.

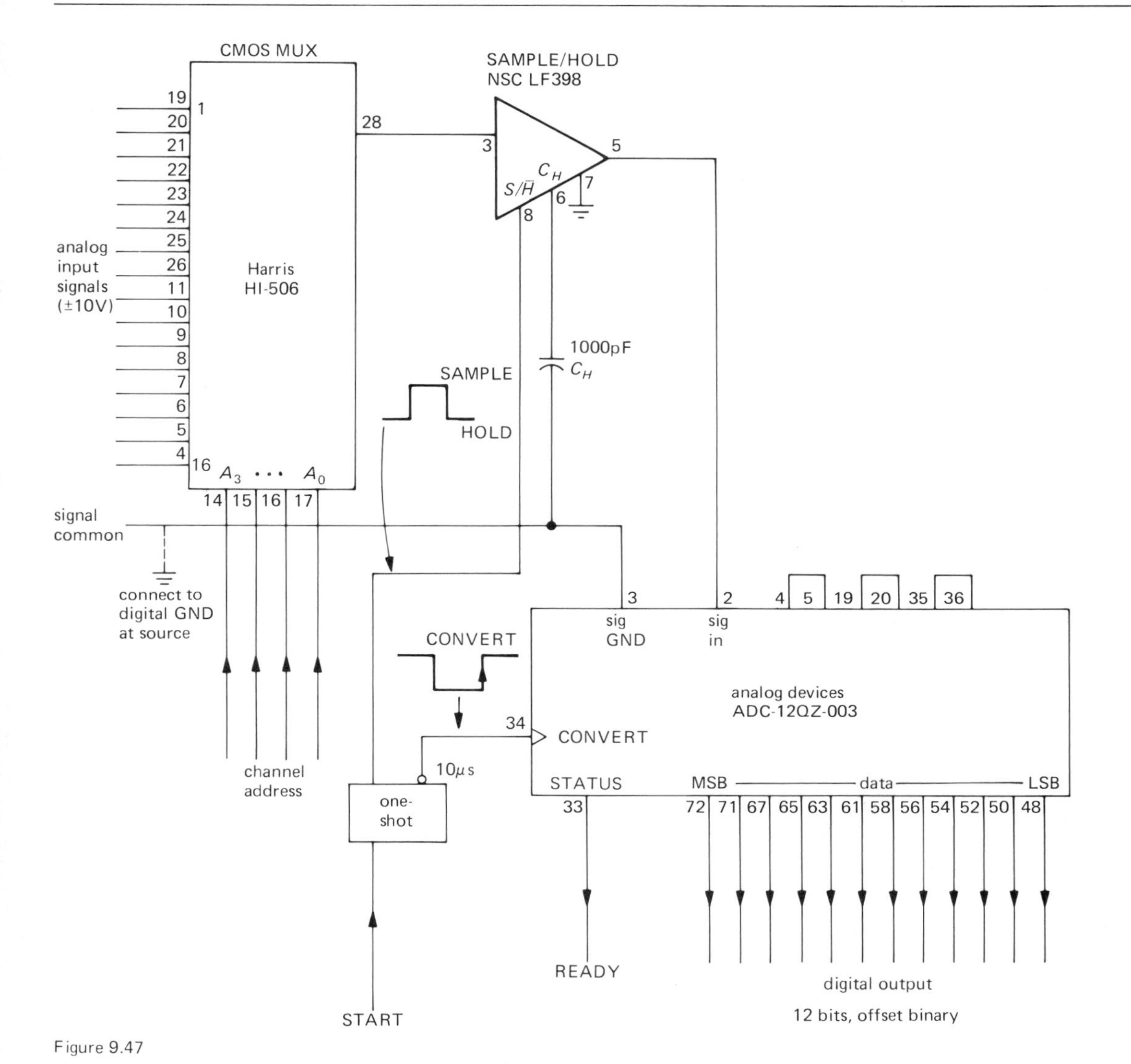

Figure 9.47

A 12-bit 16-channel successive-approximation A/D converter system (50μs/conversion).

ations of this type when shopping for linear switches. They sometimes involve a compromise. For example, "break-before-make" results in a slower switching-time specification, since the "make" must be delayed to allow the switch to open.

The multiplexer's single analog output drives an LF398 sample/hold IC. The LF398 settles in 4μs with a 1000pF hold capacitor and droops less than 10μV during the subsequent 40μs A/D conversion. The device controlling this circuit normally asserts an address and gives a START pulse. The one-shot then generates a 10μs SAMPLE pulse (long enough for the S/H output to stabilize), at the end of which the A/D begins its conversion. The conversion is complete 40μs later, at which time READY goes HIGH. Total elapsed time is 50μs/input, or 20,000 conversions per second. Since the A/D module has been strapped for 20 volt offset mode, the input range is −10 to +10 volts, with −10 volts input giving an output of 0 counts, zero volts input giving

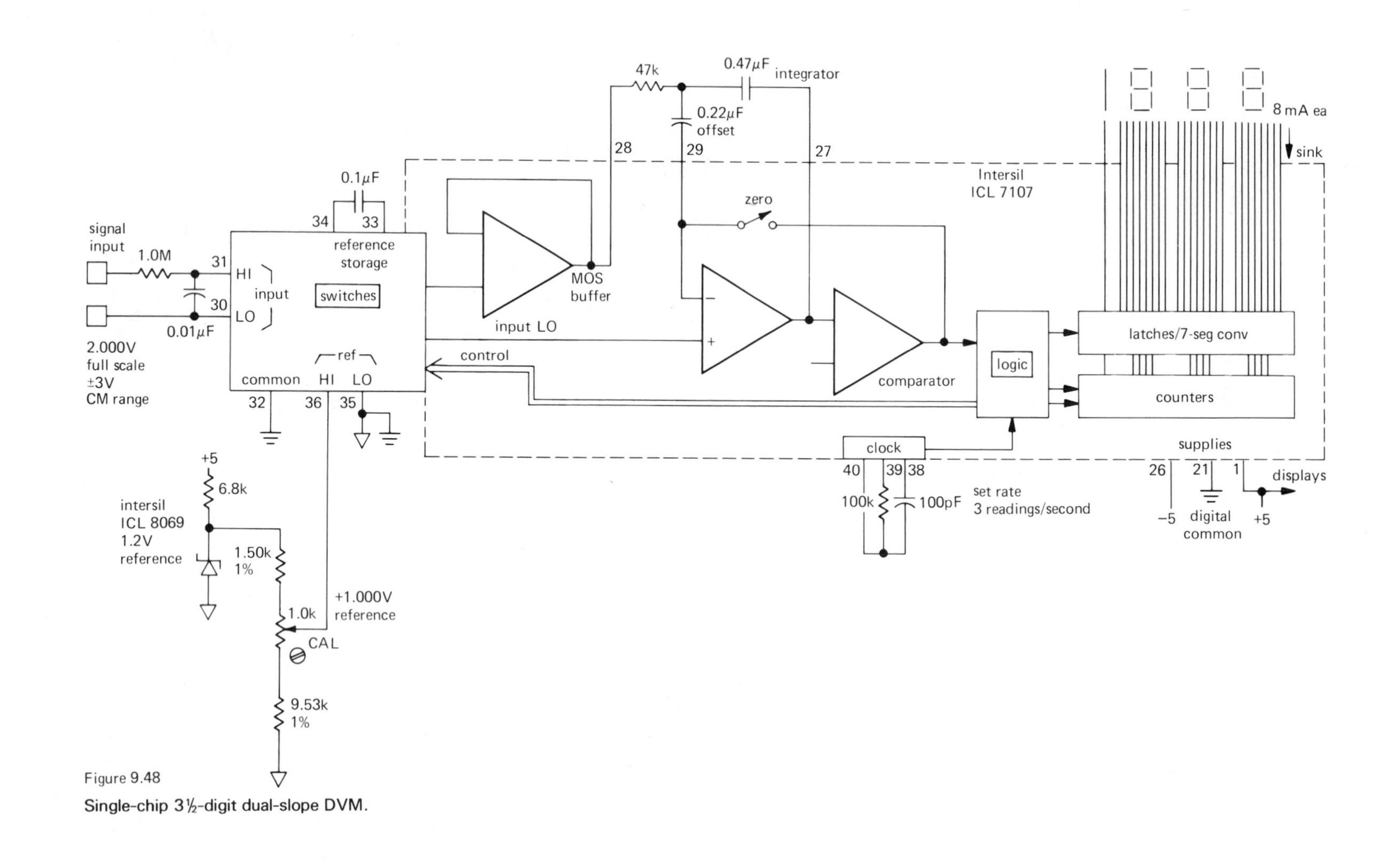

Figure 9.48

Single-chip 3½-digit dual-slope DVM.

an output of 2048 counts, and +9.995 volts input giving a full-scale count of 4095 ($2^{12} - 1$).

A successive-approximation A/D is the natural choice here, since speed is important when jumping from one input to another. At today's prices this circuit will cost about $150, dominated by the A/D converter.

9.25 3½-Digit voltmeter

Figure 9.48 shows a circuit that exploits the advantages of dual-slope integration. Almost the entire circuit is included in the digital-voltmeter CMOS LSI chip, the only external components being the integrator and clock *RC*s, an accurate voltage reference, and the display itself. The ICL 7107 includes an automatic zeroing cycle in its operation, and it even generates all the 7-segment multiplexed outputs to drive a 4-digit LED display directly. By using an external attenuator at the input (or a reference of a different voltage), you can generate other full-scale voltage ranges. Dual-slope conversion is well suited to DVM operation because it provides good accuracy (including auto-zeroing) and 60Hz rejection in an averaging instrument at low cost; the converter chip used here costs less than $20.

□ 9.26 Delta-sigma continuous-integrating converter

Figure 9.49 shows a complete delta-sigma converter, built with CMOS logic and dual op-amps and comparators. It features the ability to perform continuous-integrating A/D conversion. The input signal is initially offset with a calibrated current into IC_{1a}'s summing junction, so that an input voltage range of −5 to +5 volts is converted to a current of zero to −100.0μA at the summing junction of the integrator, IC_{1b}. A precision voltage reference is necessary, since any drifts in the offsetting current will appear as drifts at the output. Offsetting the input in this way results in an offset binary output code, with zero volts at the input corresponding to an output of half the largest count (100000000000).

The output of the integrator drives a comparator whose output controls a *D* flip-flop, clocked at a fixed 16,384Hz rate. The flip-flop output will be HIGH or LOW for full clock cycles, depending on whether the integrator output is above or below ground at the beginning of each clock cycle; its output controls an analog SPDT switch, adding an accurate current into the integrator summing junction when HIGH. In this way the average current into the integrator due to the input signal is cancelled by the switched reference current. IC_7 counts the number of 16,384Hz clock cycles during which current is switched into the integrator during each 0.25 second measurement interval. Its count is then latched and cleared, and the DATA READY output flag is set by monostable IC_5. The contents of the three-state latch are enabled (typically, onto a data bus) by a READ DATA′ input, which also clears the DATA READY flip-flop IC_6.

□ *Design highlights*

□ *Clock frequency.* By using a 4Hz clock frequency, 60Hz pickup (and its harmonics) is effectively nulled, since integral numbers of full cycles of hum are averaged in each measurement. In Section 9.31 we will show how to lock the system clock to a multiple of the power-line frequency in order to make this cancellation complete.

□ *Comparator.* No great precision or stability is required of the comparator, since all it does is keep the integrator output constant. It is entirely irrelevant whether that output is at zero volts or at some other voltage. In fact, the integrator output could be fed directly into the *D* flip-flop, using the logic threshold as a "comparator!"

□ *Capacitor.* By the same token, the integrating capacitor does not have to be particularly stable, since the particular value of the integral is not read, only cancelled at the input. Only capacitance drifts on the time scale of the individual measurements (0.25s) matter.

□ *Switch.* Finally, note that an SPDT switch is used for the reference current, switching between a precise +5 volt reference and ground. This is done to reduce the effect of switch capacitance, since the residual charge stored in the ~9pF capacitance of the 4053 will otherwise be integrated by IC_{1b}

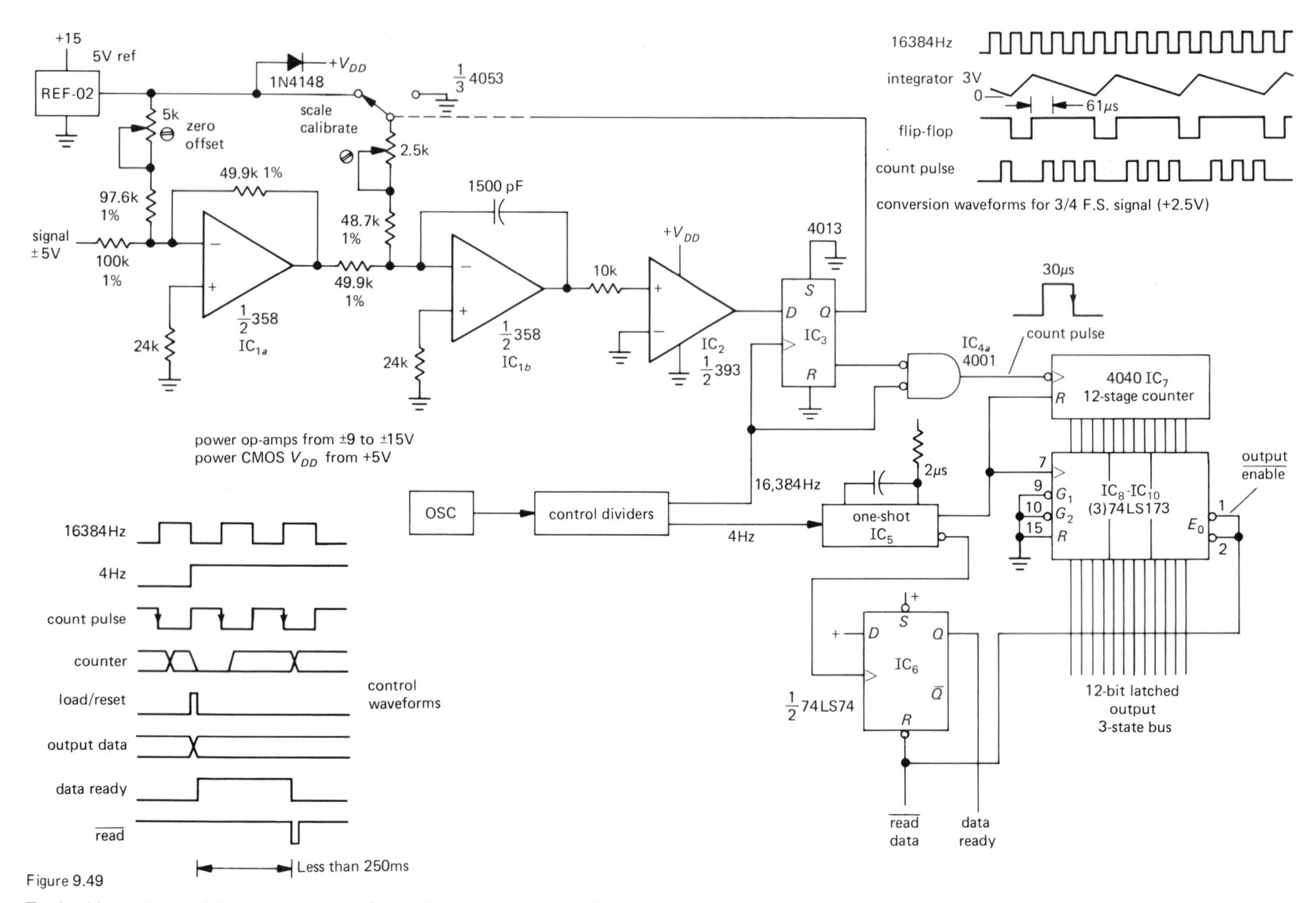

Figure 9.49

Twelve-bit continuous-integrating converter (delta-sigma charge balancing).

each time the switch is opened if an SPST switch is used instead (we'll have more to say on this important point in the next example).

□ *Calibration*

To calibrate this converter, the "zero-offset" trimmer is adjusted to give zero count with an accurate −5.000 volt input. The "scale-calibrate" trimmer is then adjusted to give full count (4095) with an accurate input of +4.9976 volts. In this way, two adjustments suffice, without having to go back and "iterate." If accuracy near zero volts input is particularly important, it may be desirable to ground the input and make slight calibration adjustments (*after* the normal calibration has been carried out) to bring the zero volt output exactly to 2048. In this way you can obtain improved accuracy as a percentage of *reading* rather than as a percentage of *full scale,* since you may wish to minimize one or the other error, depending on the particular application.

□ *Measurement continuity and input signal filtering*

Note that this circuit is a *continuous*-integrating converter: The counter is cleared and counting is resumed without loss of any 16,384Hz clocking pulses, so no count intervals are lost during readout. The converter is even "continuous" from one readout cycle to another, since the integrator holds leftover charge from the previous conversion, and thus successive converter outputs can be added together, if desired, to give increased measurement resolution and integration time. This is a valuable property not obtainable with A/D conversion schemes that hold and convert an instantaneous sample of the input waveform. With the latter, it is necessary to use a low-pass filter to match the signal's bandwidth to the converter in order not to lose data or generate "aliases" in the frequency spectrum. The optimum choice of the low-pass-filter response depends on the repetitive sampling interval used (you usually use a sharp cutoff filter, with the passband extending to half the sampling frequency, or Nyquist frequency), which means that you can't change your mind later and "clump" successive samples (to obtain improved accuracy or signal/noise ratio) without some penalty, owing to improper input filtering. With continuous-integrating measurements, on the other hand, the sample interval can be modified later in software (to a multiple of the interval actually used during the measurement), while retaining the optimum properties of correctly sampled data.

Spend a bit of time looking at the waveforms in Figure 9.49; they speak volumes.

□ 9.27 Coulomb meter

The circuit in Figure 9.50 is a charge-balancing current integrator, or "coulomb meter." This instrument can be used to measure the integrated current (total charge) over some time period; it might find application in electrophoresis or electrochemistry. The action begins in the lower left-hand corner, where the current to be integrated flows through a precision 4-wire power resistor, generating a proportional voltage. IC_2 is a relatively low cost (under $5) precision op-amp with low initial voltage offset (0.15mV max) and low drift of offset with time and temperature (less than 2μV per degree and per month). It generates an output current, programmed by the current being measured, to drive the charge-balancing integrator, IC_3. Five decades of input sensitivity are selectable via the rotary switch at the input, with 200μA collector current in Q_1 corresponding to full-scale input in any range. Q_1 is a high-beta transistor (h_{fe} = 1000 minimum at 100μA) for low base current error.

The charge-balancing circuitry is a standard delta-sigma scheme, with *p*-channel enhancement-mode MOSFET Q_2 doling out parcels of charge as directed by the state of flip-flop IC_{5a} after each clock cycle. IC_{5b} acts as a monostable, incrementing binary scaler IC_7 for each clock cycle during which Q_2 conducts. This circuit doesn't count for a fixed number of clock cycles, but simply integrates until it is stopped. The 4-digit counters IC_9 and IC_{10} keep track of the total charge, driving an 8-digit LED display.

If the current being measured ever exceeds the full-scale current of the range selected, Q_2 will be unable to balance Q_1's

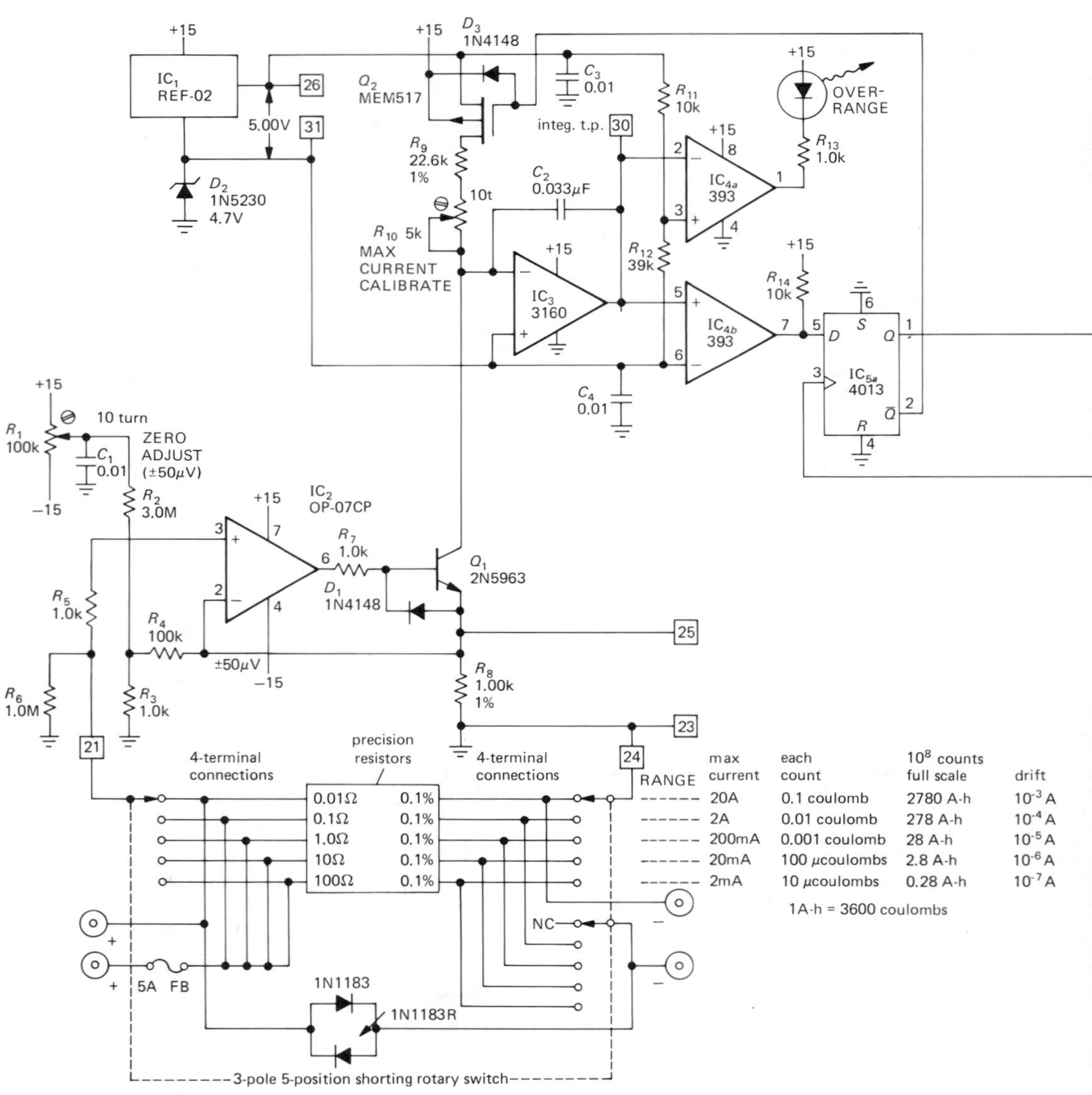

RANGE max current	each count	10^8 counts full scale	drift
20A	0.1 coulomb	2780 A-h	10^{-3} A
2A	0.01 coulomb	278 A-h	10^{-4} A
200mA	0.001 coulomb	28 A-h	10^{-5} A
20mA	100 μcoulombs	2.8 A-h	10^{-6} A
2mA	10 μcoulombs	0.28 A-h	10^{-7} A

1A-h = 3600 coulombs

Figure 9.50

Coulomb meter: accumulated charge counter.

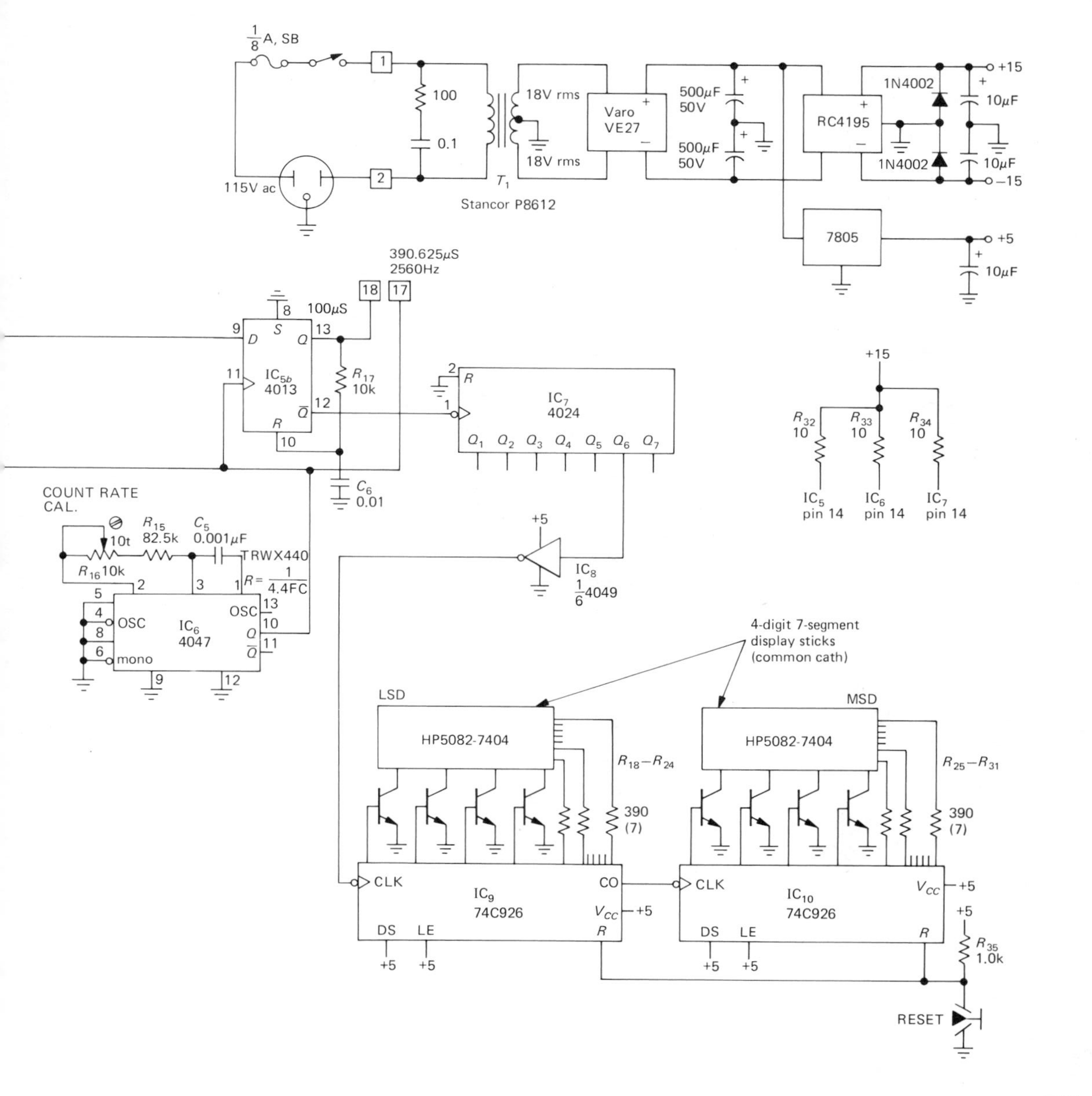

1/8 A, SB
115V ac
100
0.1
T1
Stancor P8612
18V rms
18V rms
Varo VE27
500μF 50V
500μF 50V
RC4195
1N4002
1N4002
10μF
10μF
+15
−15
7805
+5
10μF
390.625μS 2560Hz
100μS
IC5b 4013
R17 10k
C6 0.01
IC7 4024
+15
R32 10
R33 10
R34 10
IC5 pin 14
IC6 pin 14
IC7 pin 14
COUNT RATE CAL.
10t
R15 82.5k
C5 0.001μF
TRWX440
R16 10k
R = 1/4.4FC
IC6 4047
OSC
mono
IC8 1/6 4049
4-digit 7-segment display sticks (common cath)
LSD
MSD
HP5082-7404
HP5082-7404
R18−R24
R25−R31
390 (7)
390 (7)
CLK
IC9 74C926
CO
DS
LE
IC10 74C926
R35 1.0k
RESET

current even if it is ON continuously, and the measured charge registered in the counters will be in error. IC_{4a} checks for this overrange condition, lighting the LED if the integrator output rises past a fixed reference voltage (chosen comfortably larger than the output excursions of the integrator under normal conditions).

□ *Design calculations*

There are several interesting decisions that have to be made in designing a circuit like this. For instance, most of the CMOS logic is operated from +15 volts in order to simplify switching of Q_2. Since the 4-digit counters require +5 volts, a 4049 is used to interface the high-level CMOS logic signals to the counters. IC_4 is operated with single supply so that its output goes between ground and +15, for simple connection to IC_{5a}. The reference voltage for the integrator and comparator is put at about +4.7 volts by zener D_2, in order to allow headroom for Q_1; a simple zener is fine, since no accuracy is needed here. Note that a precision reference rides on the +4.7 volt level used to scale the current switched into the integrator. The REF-02's operating current is conveniently used to bias the zener.

The choice of switch (Q_2) can critically affect the overall precision of the instrument. If it has too much capacitance, the additional charge residing on its drain will cause error. The scheme used in the previous circuit example (switching to ground during current-OFF cycles) is not used here because offset voltage errors in IC_3 will cause a fixed error at very low currents. You get increased dynamic range at the expense of some accuracy (owing to residual charge on Q_2's drain that gets integrated each cycle) by using an SPST switch configuration as shown. Note that the integrator op-amp chosen is a low-bias-current MOSFET type for negligible current error (10pA typ). Since FET op-amps tend to have larger voltage offsets than bipolar transistor types, this choice of op-amp will aggravate the dynamic-range problem just discussed if SPDT switching is used.

□ *Dynamic range*

It is important to understand that this instrument is designed to have a large dynamic range, accurately integrating a current that may vary over several orders of magnitude during the measurement. That is why great care has been exercised in the design of the "front end," using a precision op-amp with an external voltage offset scheme capable of precise adjustment (the internal circuitry provided for external trimming usually has a total range of a few millivolts, making it difficult to adjust the offset precisely to zero). With IC_2 trimmed to $10\mu V$ offset or better, the instrument's dynamic range exceeds 10,000:1.

□ PHASE-LOCKED LOOPS

□ 9.28 Introduction to phase-locked loops

The phase-locked loop (PLL) is a very interesting and useful building block, available from several manufacturers as a single integrated circuit. A PLL contains a phase detector, amplifier, and voltage-controlled oscillator (VCO) and represents a blend of digital and analog techniques all in one package. A few of its applications, which we will discuss shortly, are tone decoding, demodulation of AM and FM signals, frequency multiplication, frequency synthesis, pulse synchronization of signals from noisy sources (e.g., magnetic tape), and regeneration of "clean" signals.

There has traditionally been some reluctance to use PLLs, partly because of the complexity of discrete PLL circuits and partly because of a feeling that they cannot be counted on to work reliably. With inexpensive and easy-to-use PLLs now widely available, the first barrier to their acceptance is

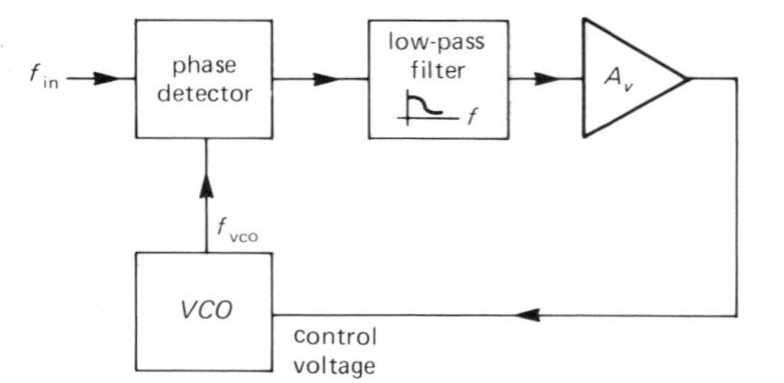

Figure 9.51
Phase-locked loop.

rapidly vanishing. And with proper design and conservative application, the PLL is as reliable a circuit element as an op-amp or flip-flop.

Figure 9.51 shows the classic PLL configuration. The phase detector is a device that compares two input frequencies, generating an output that is a measure of their phase difference (if, for example, they differ in frequency, it gives a periodic output at the difference frequency). If f_{IN} doesn't equal f_{VCO}, the phase-error signal, after being filtered and amplified, causes the VCO frequency to deviate in the direction of f_{IN}. If conditions are right (lots more on that soon), the VCO will quickly "lock" to f_{IN}, maintaining a fixed phase relationship with the input signal.

At that point the filtered output of the phase detector is a dc signal, and the control input to the VCO is a measure of the input frequency, with obvious applications to tone decoding (used in digital transmission over telephone lines) and FM detection. The VCO output is a locally generated frequency equal to f_{IN}, thus providing a clean replica of f_{IN}, which may itself be noisy. Since the VCO output can be a triangle wave, sine wave, or whatever, this provides a nice method of generating a sine wave, say, locked to a train of input pulses.

In one of the most common applications of PLLs, a modulo-*n* counter is hooked between the VCO output and the phase detector, thus generating a multiple of the input reference frequency f_{IN}. This is an ideal method for generating clocking pulses at a multiple of the power-line frequency for integrating A/D converters (dual-slope, charge-balancing), in order to have infinite rejection of interference at the power-line frequency and its harmonics. It also provides the basic technique of frequency synthesizers.

□ 9.29 PLL components

□ *Phase detector*

Let's begin with a look at the phase detector. There are actually two basic types, sometimes referred to as type I and type II. The type I phase detector is designed to be driven by analog signals or digital square-wave signals, whereas the type II phase detector is driven by digital transitions (edges). They are typified by the type I 565 (linear) and the type II 4044 (TTL); the CMOS 4046 contains both.

The simplest phase detector is the type I (digital), which is simply an exclusive-OR gate (Fig. 9.52). With low-pass filtering, the graph of the output voltage versus phase difference is as shown, for input square waves of 50% duty cycle. The type I (linear) phase detector has similar output-voltage-versus-phase characteristics, although its internal circuitry is actually a "four-quadrant multiplier," also known as a "balanced mixer." Highly linear phase detectors of this type are essential for *lock-in detection,* a lovely technique we will discuss in Section 14.15.

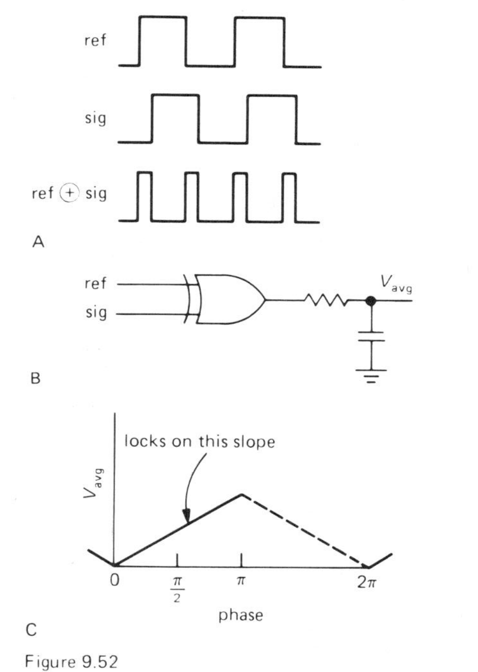

Figure 9.52
Exclusive-OR-gate phase detector (type I).

The type II phase detector is sensitive only to the relative timing of *edges* between the signal and VCO input, as shown in Figure 9.53. The phase comparator circuitry generates either *lead* or *lag* output pulses, depending on whether the transitions of the VCO output occur before or after the transitions of the reference signal, respectively. The width of these pulses is equal to the time between the respective edges, as

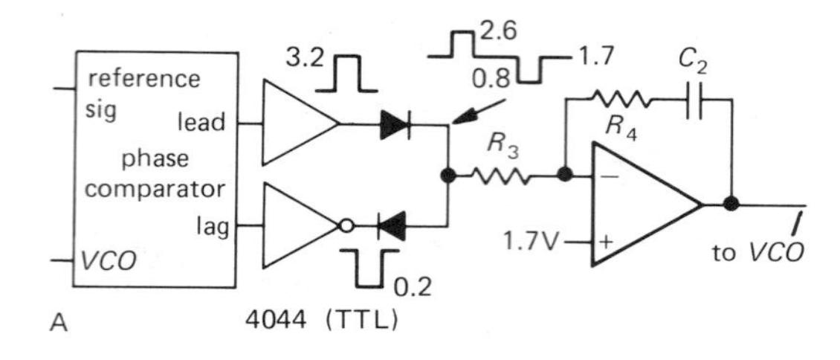

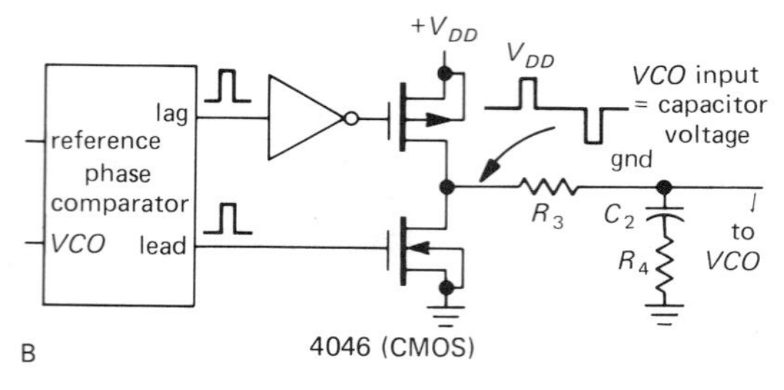

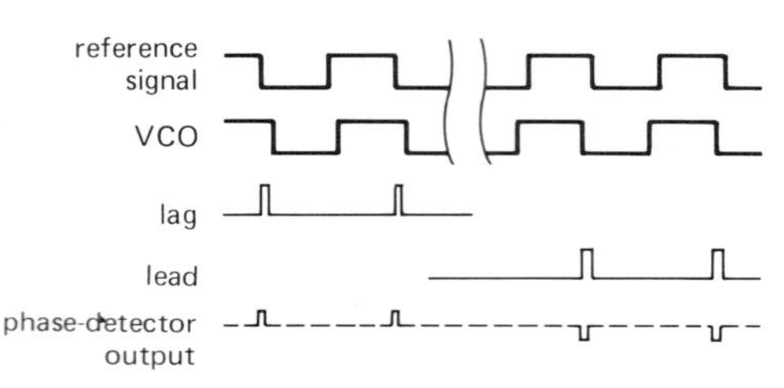

Figure 9.53
Edge-sensitive lead-lag phase detector (type II).

shown. The output circuitry then either sinks or sources current (respectively) during those pulses and is otherwise open-circuited, generating an average output voltage versus phase difference like that in Figure 9.54. This is completely independent of the

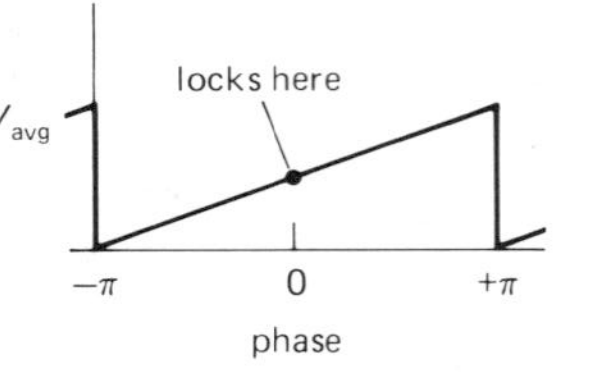

Figure 9.54

duty cycle of the input signals, unlike the situation with the type I phase comparator discussed earlier. Another nice feature of this phase detector is the fact that the output pulses disappear entirely when the two signals are in lock. This means that there is no "ripple" present at the output to generate periodic phase modulation in the loop, as there is with the type I phase detector.

Here is a comparison of the properties of the two basic types of phase detector:

	Type I Exclusive-OR	*Type II edge-triggered ("charge pump")*
Input duty cycle	50% optimum	irrelevant
Lock on harmonic?	yes	no
Rejection of noise	good	poor
Residual ripple at $2f_{IN}$	high	low
Lock range (*L*)	full VCO range	full VCO range
Capture range	fL $(f < 1)$	L
Output frequency when out of lock	f_{center}	f_{min}

There is one additional point of difference between the two kinds of phase detectors. The type I detector is always generating an output wave, which must then be filtered by the loop filter (much more on this later). Thus in a PLL with type I phase detector the loop filter acts as a low-pass filter, smoothing this full-swing logic-output signal. There will always be residual ripple, and consequent periodic phase variations, in such a loop. In circuits where phase-locked loops are used for frequency multiplication or synthesis, this adds "phase-modulation sidebands" to the output signal (see Section 13.17).

By contrast, the type II phase detector generates output pulses only when there is a phase error between the reference and VCO signal. Since the phase detector output otherwise looks like an open circuit, the loop filter capacitor then acts as a voltage-storage device, holding the voltage that gives the right VCO frequency. If the reference signal moves away in frequency, the phase detector generates a train of short pulses, charging (or discharging) the capacitor to the new voltage needed to put the VCO back into lock.

□ *VCOs*

An essential component of a phase-locked loop is an oscillator whose frequency can be controlled by the phase detector output. Some PLL ICs include a VCO (e.g., the linear

565 and the CMOS 4046). Then there are separate VCO chips, such as the 4024 (a companion chip to the 4044 TTL phase detector illustrated earlier) and various members of the 74*xx* TTL series (e.g., the 74S124 and the 74LS324–327). Another interesting class of VCOs is composed of the sine-wave-output types (8038, 2206, etc.), since they let you generate a clean sine wave locked to some horrendous input waveform. Table 9.3 presents a brief summary.

TABLE 9.3. SELECTED VCOs

Type	Family	f_{max}	Output
566	Linear	1MHz	Square, triangle
2206	Linear	0.5MHz	Square, triangle, sine
2207	Linear	0.5MHz	Square, triangle
4024	TTL	25MHz	TTL
4046	CMOS	1MHz	CMOS
8038	Linear	0.1MHz	Square, triangle, sine
74LS124	TTL	20MHz	TTL
74S124	TTL	60MHz	TTL
74LS324	TTL	20MHz	TTL

One thing to keep in mind is the fact that the VCO doesn't have to be restricted to logic speeds. You could, for instance, use a radiofrequency oscillator tuned with a varactor (variable capacitor) diode (Fig. 9.55).

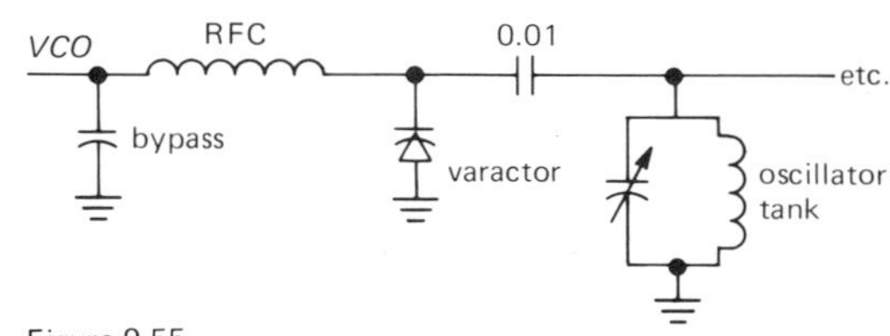

Figure 9.55

Carrying this idea one step further, you could even use something like a reflex klystron, a microwave (gigahertz) oscillator that is electrically tuned by varying the voltage on the *repeller.* Of course, a phase-locked loop built with such oscillators would require a radiofrequency phase detector.

A VCO for use in a phase-locked loop doesn't have to be particularly linear in its frequency-versus-control-voltage characteristic, but if it is highly nonlinear, the loop gain will vary according to the signal frequency, requiring better loop stability.

□ 9.30 PLL design

□ *Closing the loop*

The phase detector gives us an error signal related to the phase difference between the signal and reference inputs. The VCO allows us to control its frequency with a voltage input. It would seem straightforward to treat this like any other feedback amplifier, closing the loop with some gain, just as we did with op-amp circuits.

However, there is one essential difference. Previously, the quantity adjusted by feedback was the same quantity measured to generate the error signal, or at least a proportional quantity. For example, in a voltage amplifier we measured output voltage and adjusted input voltage accordingly. In a PLL there's an integration; we measure *phase,* but adjust *frequency,* and phase is the integral of frequency. This introduces a 90° phase shift in the loop.

This integrator included within the feedback loop has important consequences, since an additional 90° of lagging phase shift at a frequency where the loop gain is unity can produce oscillations. A simple solution is to avoid any further lagging components within the loop, at least at frequencies where the loop gain is close to unity. After all, op-amps have a 90° lagging phase shift over most of their frequency range, and they work quite nicely. This is one approach, and it produces what is known as a "first-order loop." It looks just like the PLL block diagram shown earlier, with the low-pass filter omitted.

Although they are useful in many circumstances, first-order loops don't have the desirable property of acting as a "flywheel," allowing the VCO to smooth out noise or fluctuations in the input signal. Furthermore, a first-order loop will not maintain a fixed phase relationship between the reference and VCO signals, since the phase detector output drives the VCO directly. A "second-order loop" has additional low-pass filtering within the feedback loop (as drawn earlier), carefully designed to prevent instabilities. This provides flywheel action and also reduces the "capture range" and increases the capture time. Furthermore, with type II phase detectors, a second-order loop guarantees phase lock with zero phase difference between reference and VCO, as will be explained later. Second-order loops are used almost universally, since the applications of phase-locked loops usually demand an output frequency with low phase noise and some "memory," or flywheel action. Second-order loops permit high loop gain at low frequencies, resulting in high stability (in analogy with the virtues of high loop gain in feedback amplifiers). Let's get right down to business, illustrating the use of phase-locked loops with a design example.

□ 9.31 Design example: frequency multiplier

Generating a fixed multiple of an input frequency is one of the most common applications of PLLs. This is done in frequency synthesizers, where an integer multiple n of a stable low-frequency reference signal (1Hz, say) is generated as an output; n is settable digitally, giving you a flexible signal source that can even be controlled by a computer. In more mundane applications, you might use a PLL to generate a clock frequency locked to some other reference frequency already available in the instrument. For example, suppose we generate a 61,440Hz clock signal for a dual-slope A/D converter. That particular choice of frequency permits 7.5 measurement cycles per second, allowing 4096 clock periods for the ramp-up (remember that dual-slope conversion uses a constant time interval) and 4096 counts full scale for the constant-current ramp-down. The unique virtue of a PLL scheme is that the 61.440kHz clock can be locked to the 60Hz power line (61,440 = 60 × 1024), giving infinite rejection of 60Hz pickup present on any signals input to the converter, as we discussed in Section 9.23.

We begin with the standard PLL scheme, with a divide-by-n counter added between the VCO output and the phase detector (Fig. 9.56). In this diagram we have indicated the units of gain for each function in the loop. That will be important in our stability calculations. Note particularly that the phase detector converts phase to voltage and that the VCO converts voltage to the time derivative of phase (i.e., frequency). This has the important consequence that the VCO is actually an integrator, thinking of phase as the variable in the lower part of the diagram; a fixed input voltage error produces a linearly rising phase error at the VCO output. The low-pass filter and the divide-by-n counter both have unitless gain.

□ *Stability and phase shifts*

The trick to a stable second-order phase-locked loop is shown in the Bode plots of loop gain in Figure 9.57. The VCO acts as an integrator, with $1/f$ response and 90° lagging phase shift (i.e., its response is proportional to $1/j\omega$, a current source driving a capacitor). In order to have a respectable phase margin (the difference between 180° and the phase shift around the loop at the frequency of unity loop gain), the low-pass filter has an additional resistor in series with the capacitor to stop the rolloff at some frequency (fancy name: a "zero"). The combination of these two responses produces the loop gain shown. As long as the loop gain rolls off at 6dB/octave in the

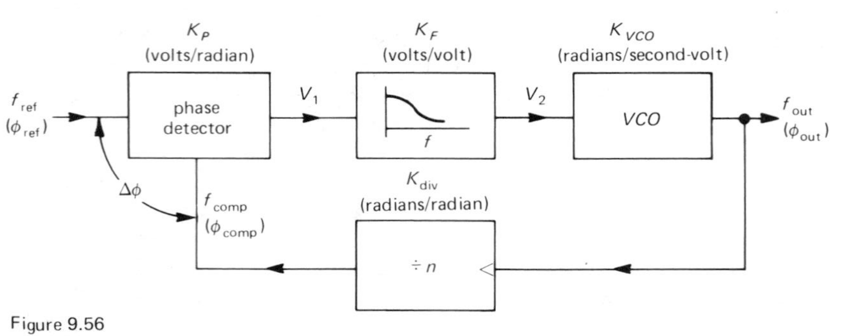

Figure 9.56
Frequency multiplier block diagram.

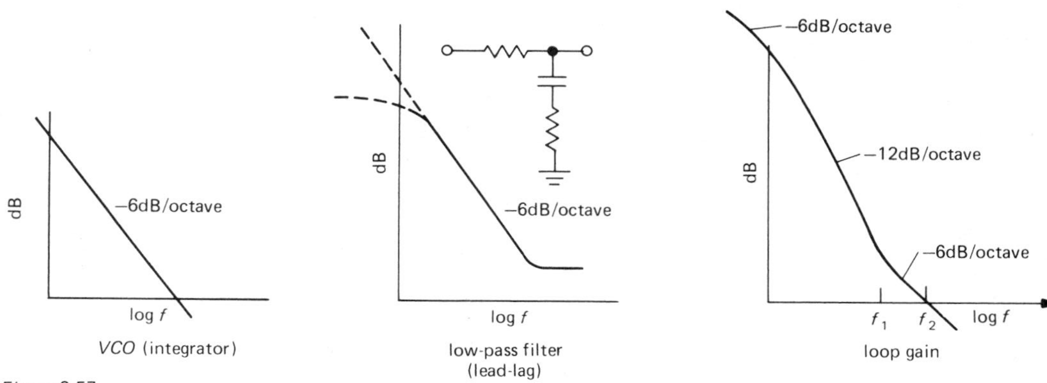

Figure 9.57

neighborhood of unity loop gain, the loop will be stable. The "lead-lag" low-pass filter does the trick, if you choose its properties correctly (this is exactly the same as lead-lag compensation in op-amps). Next you will see how it is done.

□ *Loop gain calculations*

Figure 9.58 shows the schematic of the 61.440Hz PLL synthesizer. Both the phase detector and VCO are parts of a 4046 CMOS PLL. We have used the edge-triggered type of phase detector in this circuit (the 4046 actually contains both kinds). Its output comes from a pair of CMOS transistors generating saturated pulses to V_{DD} or ground. It is really a three-state output as explained earlier, since it is in the high-impedance state except during actual phase-error pulses.

The VCO allows you to set the minimum and maximum frequencies corresponding to control voltages of zero and V_{DD}, respectively, by choosing R_1, R_2, and C_1 according to some design graphs. We have made the choices shown. Note: The 4046 has chronic severe supply sensitivity disease; check the graphs on the data sheet. The rest of the loop is standard PLL procedure.

Having rigged up the VCO range, the remaining task is the low-pass-filter design. This part is crucial. We begin by writing down the loop gain, as in Table 9.4, considering each component (refer to Fig. 9.56). Take special pains here to keep your units consistent; don't switch from f to ω or (worse) from hertz to kilohertz. The only gain

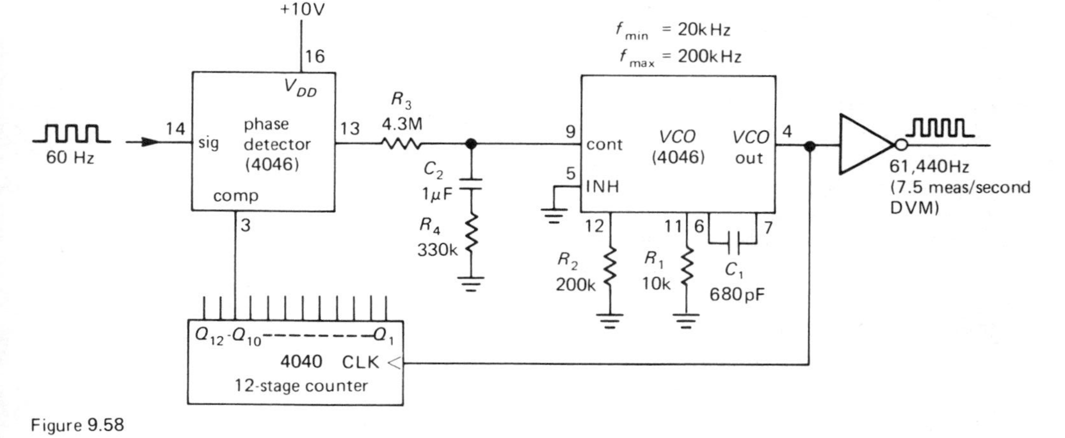

Figure 9.58
Using a PLL multiplier to generate a clock locked to the 60Hz ac line.

TABLE 9.4. PLL GAIN CALCULATION

Component	Function	Gain	Gain calculation ($V_{DD} = +10\text{V}$)
Phase detector	$V_i = K_P \Delta\phi$	K_P	0 to $V_{DD} \leftrightarrow 0°$ to $360°$ $\rightarrow K_P = 10\text{V}/360° = 1.59$ volts/radian
Low-pass filter	$V_2 = K_F V_1$	K_F	$K_F = \dfrac{1 + j\omega R_4 C_2}{1 + j\omega(R_3 C_2 + R_4 C_2)}$ volts/volt
VCO	$\dfrac{d\phi_{out}}{dt} = K_{VCO} V_2$	K_{VCO}	20 kHz ($V_2 = 0$) to 200 kHz ($V_2 = 10\text{V}$) $\rightarrow K_{VCO} = 18\text{kHz/volt} = 1.13 \times 10^5$ radians/second-volt
Divide-by-n	$\phi_{comp} = \dfrac{1}{n}\phi_{out}$	K_{div}	$K_{div} = \dfrac{1}{n} = \dfrac{1}{1024}$

term still to be decided is K_F. We do this by writing down the overall loop gain, remembering that the VCO is an integrator:

$$\phi_{out} = \int V_2 K_{VCO}\, dt$$

The loop gain is therefore given by

$$\text{Loop gain} = K_P K_F \frac{K_{VCO}}{j\omega} K_{div}$$

$$= 1.59 \times \frac{1 + j\omega R_4 C_2}{1 + j\omega(R_3 C_2 + R_4 C_2)} \times \frac{1.13 \times 10^5}{j\omega} \times \frac{1}{1024}$$

Now comes the choice of frequency at which the loop gain should pass through unity. The idea is to pick a unity-gain frequency high enough so that the loop can follow input frequency variations you want to follow, but low enough to provide flywheel action to smooth over noise and jumps in the input frequency. For example, a PLL designed to demodulate an FM input signal, or decode a rapid sequence of input tones, needs to have rapid response (for the FM input signal, the loop should have as much bandwidth as the input signal, i.e., response up to the maximum modulating frequency, while to decode input tones, its response time must be short compared with the time duration of the tones). On the other hand, a loop such as this one, designed to generate a fixed multiple of a stable and slowly varying input frequency, should have a low unity-gain frequency. That will reduce phase noise at the output and make the PLL insensitive to noise and glitches on the input. It will hardly even notice a short dropout of input signal, because the voltage held on the filter capacitor will instruct the VCO to continue producing the same output frequency.

In this case, we choose the unity-gain frequency f_2 to be 2Hz, or 12.6 radians/second. This is well below the reference frequency, and you wouldn't expect genuine power-line frequency variations on a scale shorter than this (remember that the 60Hz power is generated by enormous generators with lots of mechanical inertia). As a rule of thumb, the breakpoint of the low-pass filter (its "zero") should be lower by a factor of at least 3 to 5, for comfortable phase margin. Remember that the phase shift of a simple *RC* goes from 0° to 90° over a frequency range of roughly 0.1 to 10 times the −3dB frequency (its "pole"), with a 45° phase shift at the −3dB frequency. In this case we put the frequency of the zero, f_1, at 0.5Hz, or 3.1 radians/second (Fig. 9.59). The breakpoint f_1 determines the time constant

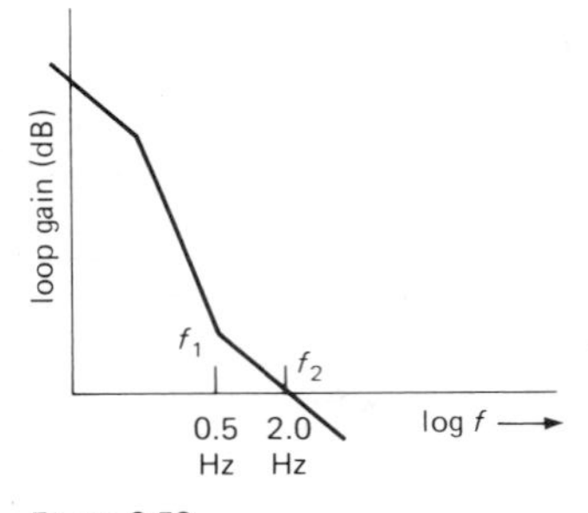

Figure 9.59

R_4C_2: $R_4C_2 = 1/2\pi f_1$. Tentatively, take $C_2 = 1\mu F$ and $R_4 = 330k$. Now all we do is choose R_3 so that the magnitude of the loop gain equals 1 at f_2. In this case that works out to $R_3 = 4.3M\Omega$.

EXERCISE 9.3

Show that these choices of filter components actually give a loop gain of magnitude 1.0 at $f_2 = 2.0Hz$.

Sometimes the filter values are inconvenient, so you have to readjust them, or move the unity-gain frequency somewhat. With a CMOS phase-locked loop these values are acceptable (the VCO input terminal has a typical input impedance of $10^{12}\Omega$). With bipolar transistor PLLs (the 4044, for example), you might want to use an external op-amp to buffer the impedances.

We used an edge-triggered (type II) phase detector in this circuit example because of its simplified loop filter; in practice that might not be the best choice for a PLL locked to the 60Hz power line because of the relatively high noise level present on the 60Hz signal. With careful design of the analog input circuit (e.g., a low-pass filter followed by a Schmitt trigger) it would probably perform well; otherwise an exclusive-OR (type I) phase detector should be used.

□ *"Cut and try"*

For some people, the art of electronics consists in fiddling with filter component values until the loop "works." If you are one of those, we will oblige you by looking the other way. We have presented these loop calculations in detail because we suspect that much of the PLL's bad reputation is the result of too many people "looking the other way." Nevertheless, we can't resist supplying a hot tip for cut-and-try addicts: R_3C_2 sets the smoothing (response) time of the loop, and R_4/R_3 determines the damping, i.e., absence of overshoot for step changes in frequency. You might begin with $R_4 = 0.2R_3$.

□ *Video clock generation*

Another nice application of a high-frequency oscillator locked to the 60Hz power line is in video signal generation, as in alphanumeric computer terminals. The standard video display rate is 30 pictures per second. Since a small amount of 60Hz pickup is almost inevitable, the picture will "weave" slowly sideways unless the vertical video sync rate is locked exactly to the power-line frequency. PLLs provide a nice way. You just lock a high-frequency VCO (around 15MHz) to a predetermined multiple of 60Hz, so that subdivisions of that high-frequency clock generate (successively) the dots for each displayed character, the number of characters on each line, and the number of horizontal lines in each picture.

□ 9.32 PLL capture and lock

Once locked, it is clear that a PLL will stay locked as long as the input frequency doesn't wander outside the range of the feedback signal. An interesting question to ask is how PLLs get locked in the first place. After all, an initial frequency error results in a periodic output from the phase detector at the difference frequency. After filtering by the low-pass filter, it is reduced to small-amplitude wiggles, rather than a nice clean dc error signal.

□ *Capture transient*

The answer is a little complicated. First-order loops will always lock, because there is no low-pass attenuation of the error signal. Second-order loops may or may not lock, depending on the type of phase detector and the bandpass of the low-pass filter. In addition, the exclusive-OR (type I) phase detector has a limited *capture* range that depends on the filter time constant (this fact can be used to advantage, if you want a PLL that will lock to signals only within a certain frequency range).

The capture transient goes like this: As the (phase) error signal brings the VCO frequency closer to the reference frequency, the error-signal waveform varies more slowly, and vice versa. So the error signal is asymmetric, varying more slowly over that part of the cycle during which f_{VCO} is closer to f_{ref}. The net result is a nonzero average, i.e., a dc component that brings the PLL into lock. If you look carefully at the VCO control

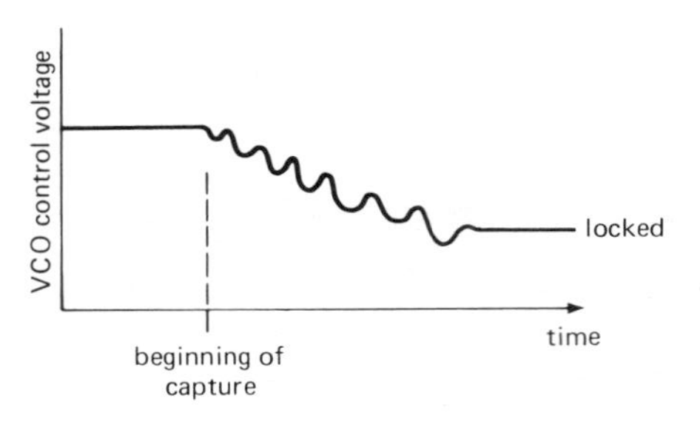

Figure 9.60

voltage during this *capture transient,* you'll see something like what is shown in Figure 9.60. That final overshoot has an interesting cause. Even when the VCO *frequency* reaches its correct value (as indicated by correct VCO control voltage), the loop isn't necessarily in lock, since the *phase* may be wrong. So it may overshoot. Each capture transient is an individual – it looks a bit different each time!

□ *Capture and lock range*

For the exclusive-OR (type I) phase detector, the capture range is limited by the low-pass-filter time constant. This makes sense, since if you begin sufficiently far away in frequency, the error signal will be attenuated so much by the filter that the loop will never lock. It should be evident that a longer filter time constant results in narrower capture range, as does reduced loop gain. It turns out that the edge-triggered phase detector does not have this limitation. Both types have a lock range extending to the limits of the VCO, given the available control input voltage.

□ 9.33 Some PLL applications

We have spoken already of the common use of phase-locked loops in frequency synthesizers and frequency multiplication. The latter application, as in the preceding example, is so straightforward that there should be no hesitation about using these mysterious PLLs. In simple frequency-multiplication applications (e.g., the generation of higher clock frequencies in a digital system) there isn't even any problem of noise on the reference signal, and a first-order loop may suffice.

We would like to point out some other interesting applications, just to give an idea of the diversity of PLL uses.

□ *FM detection*

In frequency modulation, information is encoded onto a "carrier" signal by varying its frequency proportional to the information waveform. We will talk about FM and other modulation techniques in some detail in Chapter 13. There are two methods of recovering the modulating information using phase detectors or PLLs. The word *detection* is used to mean a technique of demodulation.

In the simplest method, a PLL is locked to the incoming signal. The voltage setting the VCO frequency is proportional to the input frequency and is therefore the desired modulating signal (Fig. 9.61). In such a system

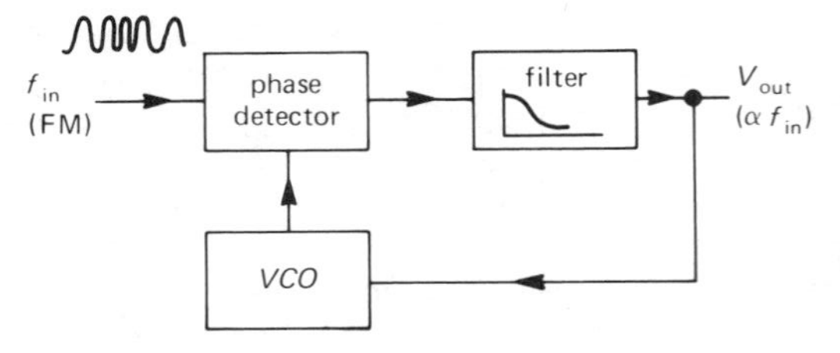

Figure 9.61

you would choose the filter bandwidth wide enough to pass the modulating signal, i.e., the response time of the PLL must be short compared with the time scale of variations in the signal being recovered. As you will see in Chapter 13, the signal applied to the PLL does not have to be the actual transmitted waveform; it can be an "intermediate frequency" generated in the receiving system by the process of *mixing.* A high degree of linearity in the VCO is desirable in this method of FM detection, for low audio distortion.

The second method of FM detection involves a phase detector, although not in a phase-locked loop. Figure 9.62 shows the idea. Both the input signal and a phase-shifted version of the signal are applied to a phase detector, generating some output voltage. The phase-shifting network is diabolically arranged to have a phase shift varying linearly with frequency in the region of the input frequency (this is usually done with resonant *LC* networks), thus generating

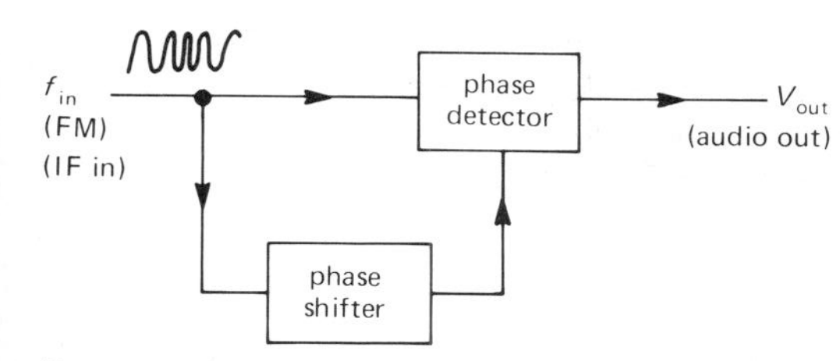

Figure 9.62

an output voltage with linear dependence on input frequency. That is the demodulated output. This method is called doubly balanced quadrature FM detection, and it is used in many IF amplifier/detector integrated circuits (e.g., the CA3089).

□ *AM detection*

Wanted: a technique to give an output signal proportional to the instantaneous *amplitude* of a high-frequency signal. The usual method involves rectification (Fig. 9.63). Figure 9.64

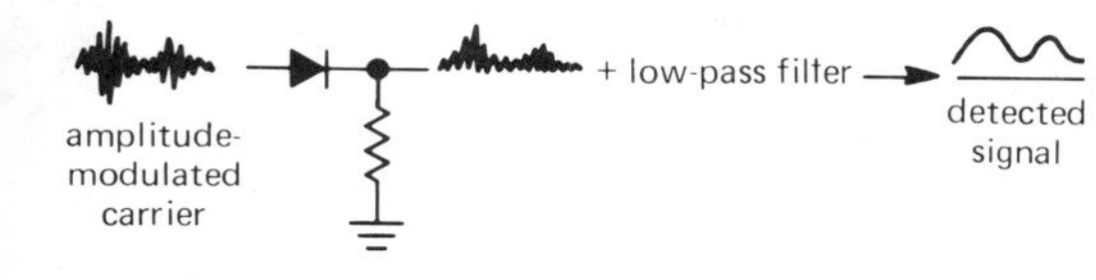

Figure 9.63

shows a fancy method ("homodyne detection") using PLLs. The PLL generates a square wave at the same frequency as the modulated carrier. Multiplying the input signal by this square wave generates a full-wave-rectified signal that only needs some low-pass filtering to remove the remnants of the carrier frequency, leaving the modulation *envelope.* If you use the exclusive-OR type of phase detector in the PLL, the output is shifted 90° relative to the reference signal, so a 90° phase shift would have to be inserted in the signal path to the multiplier.

□ *Pulse synchronization and clean-signal regeneration*

In digital signal transmission, a string of bits containing the information is sent over a communications channel. The information may be intrinsically digital, or it may be digitized analog signals, as in "pulse-code modulation" (PCM, see Section 13.19). A closely related situation is the decoding of digital information from magnetic tape or disk. In both cases there may be noise or variations in pulse rate (e.g., from tape stretch), and it is desirable to have a clean clock signal at the same rate as the bits you are trying to read. PLLs work very nicely here. The low-pass filter would be chosen to eliminate the jitter and noise in the input synchronizing signal while following slow variations in tape speed, for example.

Another example of signal synchronization might be the circuit in Section 8.30, in which an accurate digitally generated "60Hz" signal (actually, anything from 50Hz to 70Hz) is used to generate a nice output sine wave. In that circuit we used a 6-pole Butterworth low-pass filter to convert the square wave to a sine. An attractive alternative would be to use a sine-wave VCO chip (e.g., the 8038) phase-locked to the precision 60Hz square wave. That would guarantee a constant sine-wave amplitude, permit a wide range of frequency variation, and allow you to eliminate jitter in the rate-multiplier output.

□ PSEUDO-RANDOM BIT SEQUENCES AND NOISE GENERATION

□ 9.34 Digital noise generation

An interesting blend of digital and analog techniques is embodied in the subject of pseudo-random-bit sequences (PRBS). It turns out to be remarkably easy to generate sequences of bits (or words) that have good randomness properties, i.e., a sequence that has the same sort of probability and correlation properties as an ideal coin-flipping machine. Since these sequences are gener-

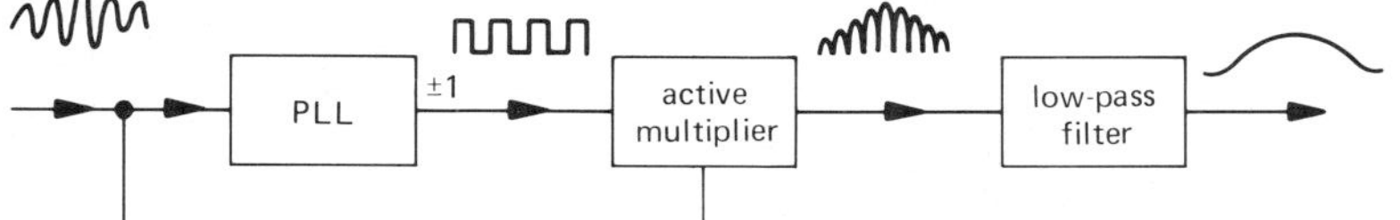

Figure 9.64

ated by standard deterministic logic elements (shift registers, to be exact), the bit sequences generated are in fact predictable and repeatable, although any portion of such a sequence looks for all the world just like a random string of 0's and 1's. With just a few chips you can generate sequences that literally go on for centuries without repeating, making this a very accessible and attractive technique for the generation of digital bit sequences or analog noise waveforms. In fact, there is even an inexpensive "digital noise source" chip available in a mini-DIP package (National MM5837), and the 76477 sound-effects chip mentioned in the last chapter includes one of them in its innards.

Analog noise

Simple low-pass filtering of the output bit pattern of a PRBS generates band-limited white Gaussian noise, i.e., a noise voltage with a flat power spectrum up to some cutoff frequency (see Chapter 7 for more on noise). Alternatively, a weighted sum of the shift register contents (via a set of resistors) performs *digital filtering,* with the same result. Flat noise spectra out to several megahertz can easily be made this way. As you will see later, such digitally synthesized analog noise sources have many advantages over purely analog techniques such as noise diodes or resistors.

Other applications

Besides their obvious applications as analog or digital noise sources, psuedo-random-bit sequences are useful in a number of applications that have nothing to do with noise. They can be used for encipherment of messages or data, since an identical PRBS generator at the receiving end provides the key. They are used extensively in error-detecting and error-correcting codes, since they allow the transcription of blocks of data in such a way that valid messages are separated by the greatest "Hamming distance" (measured by the number of bit errors). Their good autocorrelation properties make them ideal for radar ranging codes, in which the returned echo is compared (cross-correlated, to be exact) with the transmitted bit string. They can even be used as compact modulo-n dividers.

9.35 Feedback shift register sequences

The most popular (and the simplest) PRBS generator is the feedback shift register (Fig. 9.65). A shift register of length m bits is

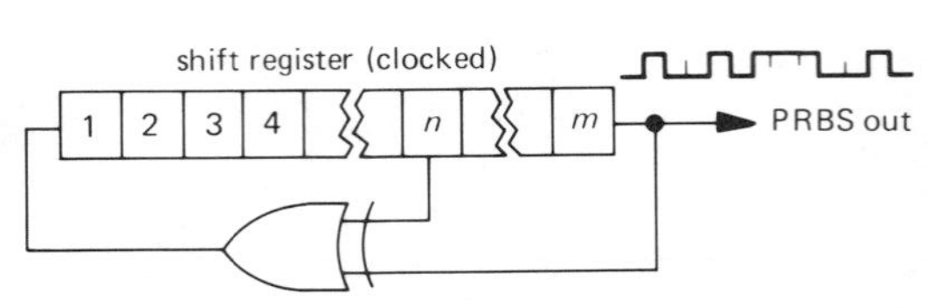

Figure 9.65
Pseudo-random bit sequence generator.

clocked at some fixed rate, f_0. An exclusive-OR gate generates the serial input signal from the exclusive-OR combination of the nth bit and the last (mth) bit of the shift register. Such a circuit goes through a set of states (defined by the set of bits in the register after each clock pulse), eventually repeating itself after K clock pulses; i.e., it is cyclic with period K.

The maximum number of conceivable states of an m-bit register is $K = 2^m$, i.e., the number of binary combinations of m bits. However, the state of all 0's would get "stuck" in this circuit, since the exclusive-OR would regenerate a 0 at the input. Thus the maximum-length sequence you can possibly generate with this scheme is $2^m - 1$. It turns out that you can make such "maximal-length shift register sequences" if m and n are chosen correctly, and the resultant bit sequence is pseudorandom. (The criterion for maximal length is that the polynomial $1 + x^n + x^m$ be irreducible and prime over the Galois field.) As an example, consider the 4-bit feedback shift register in Figure 9.66. Beginning with the state 1111

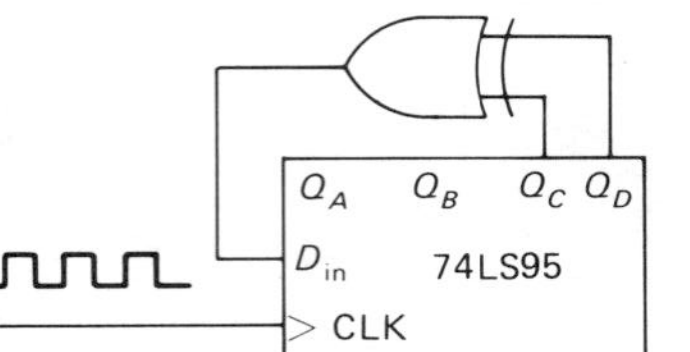

Figure 9.66

(we could start anywhere except 0000), we can write down the states it goes through:

1111
0111
0011
0001
1000
0100
0010
1001
1100
0110
1011
0101
1010
1101
1110

We have written down the states as 4-bit numbers $Q_AQ_BQ_CQ_D$. There are 15 distinct states $(2^4 - 1)$, after which it begins again; therefore it is a maximal-length register.

EXERCISE 9.4

Demonstrate that a 4-bit register with feedback taps at the second and fourth bits is not maximal length. How many distinct sequences are there? How many states within each sequence?

□ *Feedback taps*

Maximal-length shift registers can be made with exclusive-OR feedback from more than two taps (in these cases you use several exclusive-OR gates in the standard parity-tree configuration, i.e., modulo-2 addition of several bits). In fact, for some values of m, a maximal-length register can only be made with more than two taps. Here is a listing of all values of m up to 33 for which maximal-length registers can be made with just two taps, i.e., feedback from the nth bit and the mth (last) bit, as previously. A value is given for n and for the cycle length K, in clock cycles. In some cases there is more than one possibility for n, and in every case the value $m - n$ can be used instead of n; thus the earlier 4-bit example could have used taps at $n = 1$ and $m = 4$.

m	*n*	*Length*
3	2	7
4	3	15
5	3	31
6	5	63
7	6	127
9	5	511
10	7	1023
11	9	2047
15	14	32767
17	14	131071
18	11	262143
20	17	1048575
21	19	2097151
22	21	4194303
23	18	8388607
25	22	33554431
28	25	268435455
29	27	536870911
31	28	2147483647
33	20	8589934591

Since shift register lengths of multiples of 8 are common, you may want to use one of those lengths. In that case, more than two taps are necessary. Here are the magic numbers:

m	*Taps*	*Length*
8	4, 5, 6	255
16	4, 13, 15	65535
24	17, 22, 23	16777215

The MM5837 IC noise-generator chip uses a 17-bit register with a tap at stage 14. Its internal clock runs at about 80kHz, generating white noise output up to about 35kHz (3dB down) with a cycle time of about 1.6 seconds. With a 33-bit register clocked at 1MHz, the cycle time would be over 2 hours. A 100-bit register clocked at 10MHz would have a cycle time a million times longer than the age of the universe!

□ *Properties of maximal-length shift register sequences*

We generate a string of pseudorandom bits from one of these registers by clocking it and looking at successive output bits. The output can be taken from any position of the register; it is conventional to use the last (mth) bit as the output. Maximal-length shift register sequences have the following properties:

1. In one full cycle (K clock cycles), the number of 1's is one greater than the number of 0's. The extra 1 comes about because of the excluded state of all 0's. This says that heads and tails are equally likely (the extra 1 is totally insignificant for any reasonable-length register; a 17-bit register will produce 65,536 1's and 65,535 0's in one of its cycles).

2. In one full cycle (K clock cycles), half the runs of consecutive 1's have length 1, one-fourth the runs have length 2, one-eighth have length 3, etc. There are the same numbers of runs of 0's as of 1's, again with the exception of a missing 0. This says that the probability of heads and tails does not depend on the outcome of past flips, and therefore the chance of terminating a run of successive 1's or 0's on the next flip is 1/2 (contrary to the man-in-the-street's understanding of the "law of averages").
3. If one full cycle (K clock cycles) of 1's and 0's is compared with the same sequence shifted cyclically by any number of bits n (where n is not 0 or a multiple of K), the number of disagreements will be one greater than the number of agreements. In fancy language, the autocorrelation function is a Kronecker delta at zero delay, and $-1/K$ everywhere else. This absence of "side lobes" in the autocorrelation function is what makes PRBSs so useful for radar ranging.

EXERCISE 9.5

Show that the 4-bit shift register sequence listed earlier (taps at $n = 3$, $m = 4$) satisfies these properties, considering the Q_A bit as the "output": 100010011010111.

□ 9.36 Analog noise generation from maximal-length sequences

□ *Advantages of digitally generated noise*

As we remarked earlier, the digital output of a maximal-length feedback shift register can be converted to band-limited white noise with a low-pass filter whose cutoff frequency is well below the clock frequency of the register. Before getting into the details, we will point out some of the advantages of digitally generated analog noise. Among other things, it allows you to generate noise of known spectrum and amplitude, with adjustable bandwidth (via clock frequency adjustment), using reliable and easily maintained digital circuitry. There is none of the variability of diode noise generators, nor are there the interference and pickup problems that plague the sensitive low-level analog circuitry used with diode or resistor noise generators. Finally, it generates repeatable "noise" and, when filtered with a weighted digital filter (more about this later), repeatable noise waveforms independent of clocking rate (output noise bandwidth).

□ 9.37 Power spectrum of shift register sequences

The output spectrum generated by maximal-length shift registers consists of noise extending from the repeat frequency of the entire sequence, f_{clock}/K, up to the clock frequency and beyond. It is flat within ±0.1dB up to 12% of the clock frequency (f_{clock}), dropping rather rapidly beyond its −3dB point of 44% f_{clock}. Thus a low-pass filter with a high-frequency cutoff of 5%–10% of the clock frequency will convert the unfiltered shift register output to a band-limited analog noise voltage. Even a simple *RC* filter will suffice, although it may be desirable to use active filters with sharp cutoff characteristics (see Chapter 4) if a precise frequency band of noise is needed.

To make these statements more precise, let's look at the shift register output and its power spectrum. It is usually desirable to eliminate the dc offset characteristic of digital logic levels, generating a bipolarity output with 1 corresponding to $+a$ volts and 0 corresponding to $-a$ volts (Fig. 9.67). This

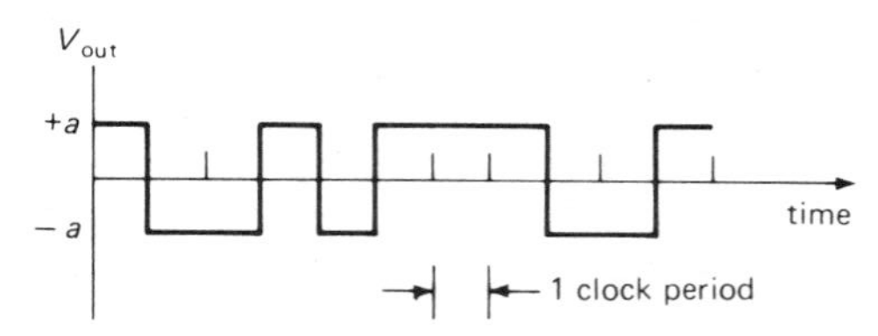

Figure 9.67

can be easily done with the sort of transistor push-pull stage shown in Figure 9.68. Alternatively, you can use MOS transistors, a circuit with clamping diodes to stable reference voltages, or a fast op-amp with adjustable dc offset current into a summing junction.

As we remarked earlier, the string of output bits has a single peak in its autocorrelation. If the output states represent +1 and −1, the digital autocorrelation (the sum of the product of corresponding bits, when the

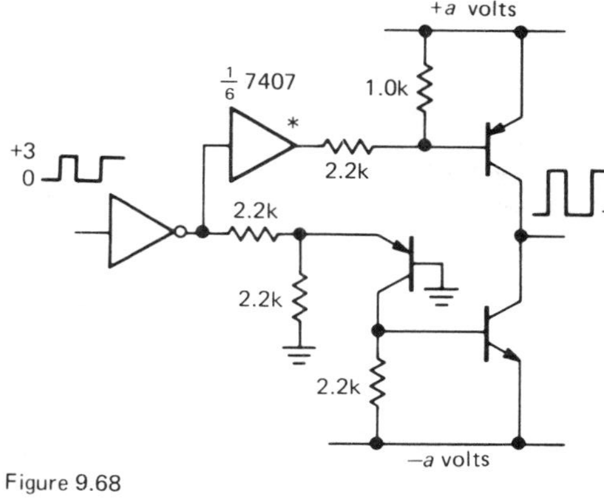

Figure 9.68

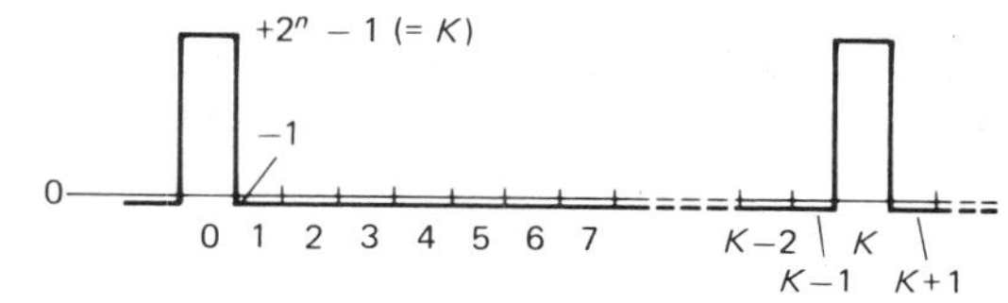

Figure 9.69
Full-cycle discrete autocorrelation for a maximal-length shift-register sequence.

bit string is compared with a shifted version of itself) is as shown in Figure 9.69.

Don't confuse this with a *continuous* autocorrelation function, which we will consider later. This graph is defined only for shifts corresponding to a whole number of clock cycles. For all shifts that aren't zero or a multiple of the overall period K, the autocorrelation function has a constant -1 value (because there is an extra 1 in the sequence), negligible when compared with the zero-offset value of K. Likewise, if we consider the unfiltered shift register output as an *analog* signal (whose waveform happens to take on values of $+a$ and $-a$ volts only), the normalized autocorrelation becomes a continuous function, as shown in Figure 9.70. In other words, the waveform

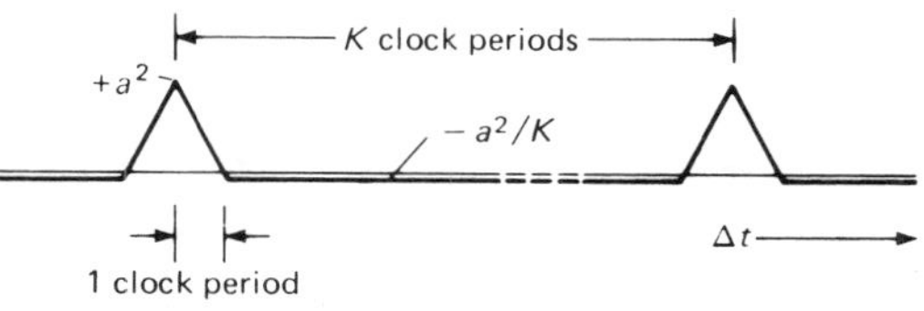

Figure 9.70

is totally uncorrelated with itself when shifted more than one clock period forward or backward.

The power spectrum of the unfiltered digital output can be obtained from the autocorrelation by standard mathematical techniques. The result is a set of equally spaced series of spikes (delta functions), beginning at the frequency at which the whole sequence repeats, f_{clock}/K, and going up in frequency by equal intervals f_{clock}/K. The fact that the spectrum consists of a set of discrete spectral lines reflects the fact that the shift register sequence eventually (and periodically) repeats itself. Don't be alarmed by this funny spectrum; it will look continuous for any measurement or application that takes less time than the cycle time of the register. The envelope of the spectrum of the unfiltered output is shown in Figure 9.71. The envelope is proportional to

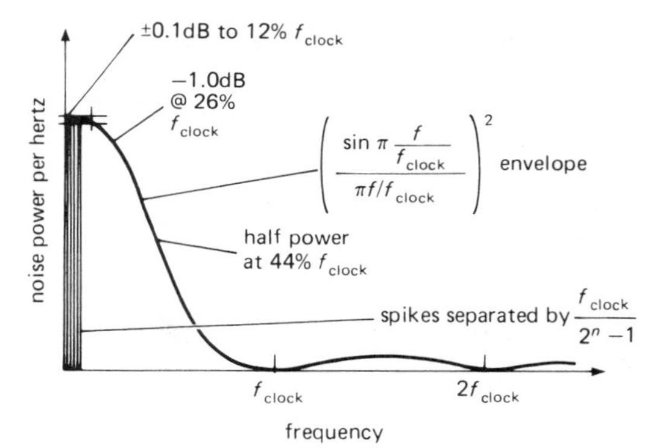

Figure 9.71
Power spectrum of unfiltered digital shift register output signal.

the square of $(\sin x)/x$. Note the peculiar property that there is *no* noise power at the clock frequency or its harmonics.

□ *Noise voltage*

Of course, for analog noise generation you use only a portion of the low-frequency end of the spectrum. It turns out to be easy to calculate the noise power per hertz in terms of the half-amplitude (a) and the clock frequency (f_{clock}). Expressed as an rms noise *voltage*, the answer is

$$V_{rms} = a\left(\frac{2}{f_{clock}}\right)^{1/2} \quad \text{V/Hz}^{1/2} \quad (f \leq 0.2 f_{clock})$$

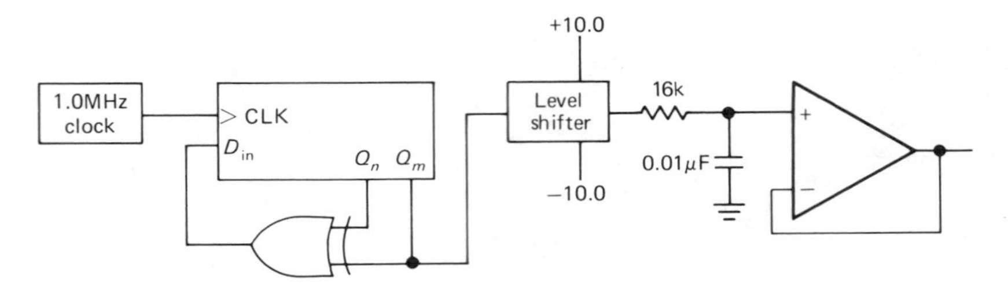

Figure 9.72
Simple pseudorandom noise source.

This is for the bottom end of the spectrum, the part you usually use (you can use the envelope function to find the power density elsewhere).

For example, suppose we run a maximal-length shift register at 1.0MHz and arrange it so that the output voltage swings between +10.0 and −10.0 volts. The output is passed through a simple *RC* low-pass filter with 3dB point at 1kHz (Fig. 9.72). We can calculate the rms noise voltage at the output exactly. We know from the preceding equation that the output from the level shifter has an rms noise voltage of 14.14mV per root hertz. From Section 7.20 we know that the noise bandwidth of the low-pass filter is $(\pi/2)(1.0\text{kHz})$, or 1.57kHz. So the output noise voltage is

$$V_{rms} = 0.01414(1570)^{1/2} = 560\text{mV}$$

with the spectrum of a single-section *RC* low-pass filter.

□ 9.38 Low-pass filtering

□ *Analog filtering*

The spectrum of useful noise from a pseudo-random-sequence generator extends from a low-frequency limit of the reciprocal repeat period (f_{clock}/K) up to a high-frequency limit of perhaps 20% of the clock frequency (at that frequency the noise power per hertz is down by 0.6dB). Simple low-pass filtering with *RC* sections, as illustrated in the earlier example, is adequate provided that its 3dB point is set far below the clock frequency (e.g., less than 1% of f_{clock}). In order to use the spectrum closer to the clock frequency, it is advisable to use a filter with sharper cutoff, e.g., a Butterworth or Chebyshev. In that case the flatness of the resultant spectrum depends on the filter characteristics, which should be measured, since component variations can produce ripples in the passband gain. Likewise, the filter's actual voltage gain should be measured if the precise value of noise voltage per root hertz is important.

□ *Digital filtering*

A disadvantage of analog filtering is the need to readjust the filter cutoff if the clock frequency is changed by large factors. In situations where that is desirable, an elegant solution is provided by digital filtering, in this case performed by taking an analog weighted sum of successive output bits (nonrecursive digital filtering). In this way the effective filter cutoff frequency changes to match changes in the clock frequency. In addition, digital filtering lets you go to extremely low cutoff frequencies (fractions of a hertz) where analog filtering becomes awkward.

In order to perform a weighted sum of successive output bits simultaneously, you can simply look at the various parallel outputs of successive shift register bits, using resistors of various values into an op-amp summing junction. For a low-pass filter the weights should be proportional to $(\sin x)/x$; note that some levels will have to be inverted, since the weights are of both signs. Since no capacitors are used in this scheme, the output waveform consists of a set of discrete output voltages.

The approximation to Gaussian noise is improved by using a weighting function over many bits of the sequence. In addition, the analog output them becomes essentially a continuous waveform. For this reason it is desirable to use as many shift register stages as possible, adding additional shift register stages outside the exclusive-OR feedback if necessary. As before, pull-ups or MOS switches should be used to set stable digital voltage levels (CMOS logic is ideal for this application, since the outputs saturate cleanly at V_{DD} and ground).

The circuit in Figure 9.73 generates pseudorandom analog noise, with bandwidth selectable over an enormous range, using this technique. A 2.0MHz crystal oscillator drives a 14536 24-stage programmable divider, generating clock frequencies going from 1.0MHz down to 0.12Hz by factors of 2. A 32-bit shift register is connected with feedback from stages 31 and 18, generating a maximal-length sequence with 2 billion states (at the maximum clock frequency the register completes one cycle in a half hour). In this case we have used a $(\sin x)/x$ weighted sum over 32 successive values of the sequence. U_1 and U_2 amplify the inverted and noninverted terms, respectively, driving differential amplifier U_3. The gains are chosen to generate a 1.0 volt rms output with no dc offset into a 50 ohm load impedance (2.0V rms open circuit). Note that this noise amplitude is independent of the clock rate, i.e., the total bandwidth. This digital filter has a cutoff at about $0.05f_{clock}$, giving a white noise output spectrum extending from dc to 50kHz (maximum clock frequency) down to dc to 0.006Hz (minimum clock frequency), in 24 steps of bandwidth. The circuit also provides an unfiltered output waveform, going between +1.0 volt and −1.0 volt.

There are a few interesting points about this circuit. Note that an exclusive-NOR gate is used for feedback so that the register can be simply initialized by bringing it to the state of all zeros. This trick of inverting the serial input signal makes the excluded state the state of all 1's (rather than all 0's as with the usual exclusive-OR feedback), but it leaves all other properties unaffected.

A weighted sum of a finite number of bits cannot ever produce truly Gaussian noise, since the peak amplitude is limited. In this case it can be calculated that the peak output amplitude (into 50Ω) is ±4.34 volts, giving a "crest factor" of 4.34. That calculation is important, by the way, because you must keep the gain of U_1 through U_3 low enough to prevent clipping. Look carefully at the methods used to generate an output of zero dc offset from the CMOS levels of +6.0 volts average value (LOW = 0V, HIGH = 12.0V).

This method of digital low-pass filtering of maximal-length shift register sequences is used in the Hewlett-Packard 3722A noise generator.

□ 9.39 Wrap-up

A few comments about shift register sequences as analog noise sources: You might be tempted to conclude from the three properties of maximal-length shift registers listed earlier that the output is "too random," in the sense of having exactly the right number of runs of a given length, etc. A genuine random coin-flipping machine would not generate exactly one more head than tail, nor would the autocorrelation be absolutely flat for a finite sequence. To put it another way, if you used the 1's and 0's that emerge from the shift register to control a "random walk," moving forward one step for a 1 and back one step for a 0, you would wind up exactly one step away from your beginning point after the register had gone through its entire cycle, a result that is anything but "random"!

However, the shift register properties mentioned earlier are only true of the entire sequence of $2^n - 1$ bits, *taken as a whole.* If you use only a section of the entire bit sequence, the randomness properties closely approximate a random coin-flipper. To make an analogy, it is as if you were drawing red balls and blue balls at random from an urn initially containing K balls in all, half red and half blue. If you do this *without replacing them,* you would expect to find approximately random statistics at first. As the urn becomes depleted, the statistics are modified by the requirement that the total numbers of red and blue balls must come out the same.

You can get an idea how this goes by thinking again about the random walk. If we assume that the only nonrandom property of

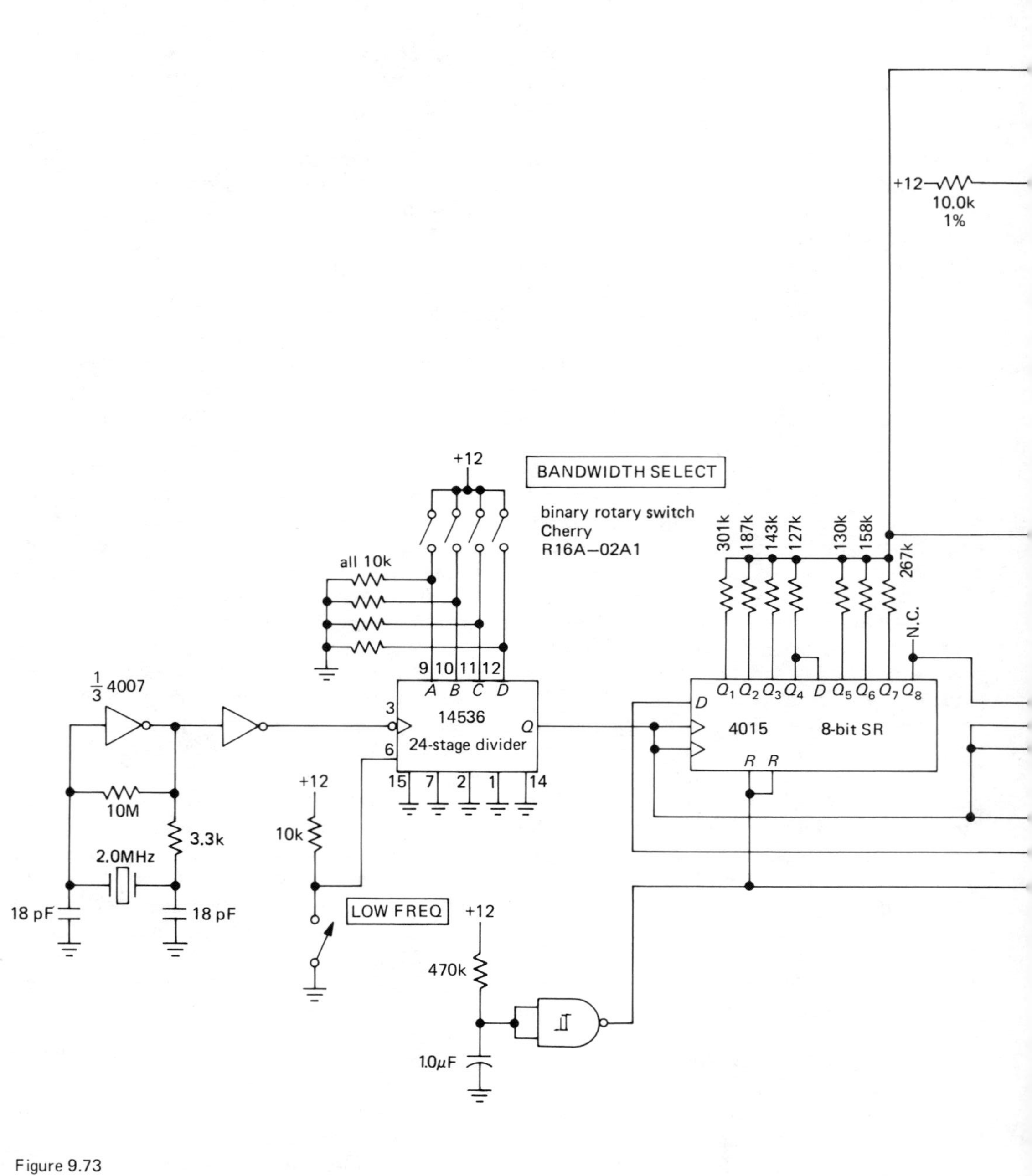

Figure 9.73

Wide-frequency-range laboratory noise source.

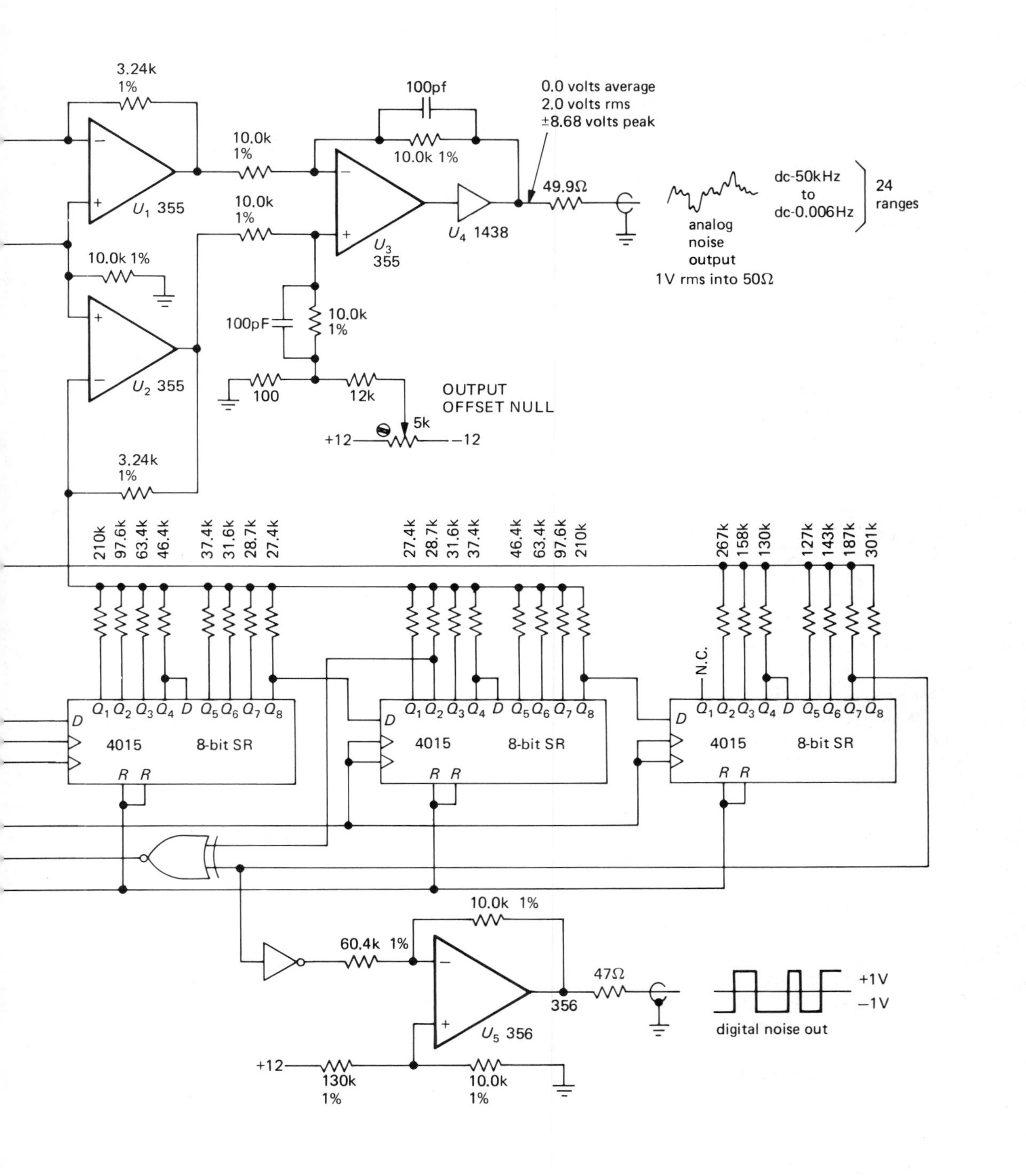
3.24k 1%
U1 355
10.0k 1%
U2 355
100pf
10.0k 1%
U3 355
U4 1438
0.0 volts average
2.0 volts rms
±8.68 volts peak
49.9Ω
analog noise output
1V rms into 50Ω
dc-50kHz to dc-0.006Hz
24 ranges
100pF
100
12k
OUTPUT OFFSET NULL
5k
+12
−12
210k 97.6k 63.4k 46.4k 37.4k 31.6k 28.7k 27.4k
27.4k 28.7k 31.6k 37.4k 46.4k 63.4k 97.6k 210k
267k 158k 130k 127k 143k 187k 301k
N.C.
4015
8-bit SR
60.4k 1%
U5 356
356
47Ω
130k 1%
digital noise out
+1V
−1V

the shift register sequence is the exact equality of 1's and 0's (ignoring the single excess 1), it can be shown that the random walk as described should reach an average distance from the starting point of

$$X = [r(K - r)/(K - 1)]^{1/2}$$

after r draws from a total population of $K/2$ 1's and $K/2$ 0's. Since in a completely random walk X equals the square root of r, the factor $(K - r)/(K - 1)$ expresses the effect of finite urn contents. As long as $r \ll K$, the randomness of the walk is only slightly reduced from the completely random case (infinite urn contents), and the pseudo-random-sequence generator is indistinguishable from the real thing. We tested this with a few thousand PRBS-mediated random walks, each a few thousand steps in length, and found that the randomness was essentially perfect, as measured by this simple criterion.

Of course, the fact that PRBS generators pass this simple test does not guarantee that they would satisfy some of the more sophisticated tests of randomness, e.g., as measured by higher-order correlations. Such correlations also affect the properties of analog noise generated from such a sequence by filtering. Although the noise amplitude distribution is Gaussian, there may be higher-order amplitude correlations uncharacteristic of true random noise. Current thinking on this subject is that the use of many (preferably about $m/2$) feedback taps (using an exclusive-OR parity-tree operation to generate the serial input) generates "better" noise in this respect.

Noise generator builders should be aware of the 4557 CMOS variable-length shift register (1 to 64 stages); of course, you have to use it in combination with a parallel-output register (such as the 4015 or 74C164) in order to get at the n tap.

In Section 7.19 there is a discussion of noise, with an example of a "pink noise" generator using the MM5837 maximal-length shift register IC.

□ 9.40 Digital filters

The last example brought up the interesting topic of digital filtering, in that case the generation of a low-pass-filtered analog output signal by taking a weighted sum of 32 samples of pseudorandom bits, each corresponding to a voltage level of 0 volts or +12 volts. The "filter" accepted as its input a waveform that happened to have only two voltage levels. In general, the same thing can be done with an analog waveform as input, forming a weighted sum over its values (x_i) at equally spaced times

$$y_i = \sum_{k=-\infty}^{\infty} h_k x_{i-k}$$

The x_is are the discrete input signal samples, the h_ks are the weights, and the y_is are the output of the filter. In real life a digital filter will sum only over a finite set of input values, as for example in the noise generator circuit where we used 32 terms. Figure 9.74 shows schematically what is happening.

Note that such a filter can have the interesting property of being symmetrical in time, i.e., averaging past and future "history" to arrive at its present output. Of course, real analog filters can only look backward in time, corresponding to a digital filter with nonzero weighting factors h_k only for $k \geq 0$.

□ *Symmetrical filter frequency response*

For a symmetrical filter ($h_k = h_{-k}$) it can be shown that the frequency response is given by

$$H(f) = h_0 + 2 \sum_{k=1}^{\infty} h_k \cos 2\pi k f t_s$$

where t_s is the time between samples. Thus the individual h_ks are recognizable (to those who know about such things!) as the Fourier series components of the desired frequency response curve, which explains why the weights in the noise generator circuit shown earlier were chosen proportional to $(\sin x)/x$, the Fourier components of a "brick-wall" low-pass filter. For such a symmetrical filter the phase shift at any frequency is either zero or 180°.

□ *Recursive filters*

An interesting class of digital filters can be made if you allow the filter to use as its inputs the value of its own output, in addi-

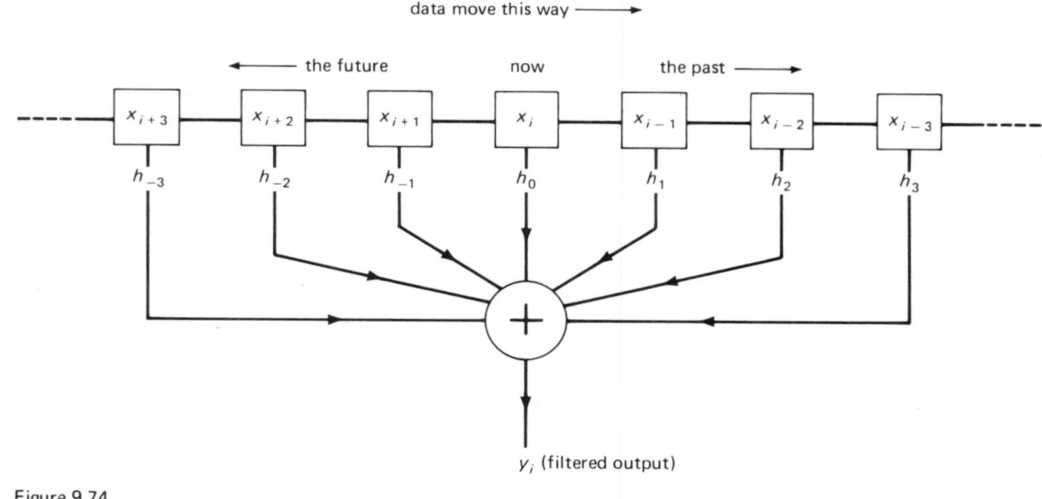

Figure 9.74
Nonrecursive digital filter.

tion to the value of the signal being filtered. You can think of the filter as having "feedback." The fancy name for such a filter is a *recursive* digital filter, as opposed to the nonrecursive filters just discussed. For example, you could form the outputs y_i according to

$$y_i = Ay_{i-1} + (1 - A)x_i$$

This happens to give you a low-pass response, equivalent to that of a simple *RC* low-pass filter, according to

$$A = e^{-t_s/RC}$$

where t_s is the time between successive samples x_i of the input waveform. Of course, the situation is not *identical* to an analog low-pass filter operating on an analog waveform because of the discrete nature of the sampled waveform.

□ Low-pass-filter example

As a numerical example, suppose you want to filter a set of numbers representing a signal, with a low-pass 3dB point at $f_{3dB} = 1/20t_s$. Thus the time constant equals the time for 20 successive samples. Then $A = 0.95123$, and so the output is given by

$$y_i = 0.95123y_{i-1} + 0.04877x_i$$

The approximation to a real low-pass filter becomes better as the time constant becomes long compared with the time between samples, t_s.

You would probably use a filter like this to process data that are already in the form of discrete samples, e.g., an array of data in a computer. In that case the recursive filter becomes a trivial arithmetic pass once through the data. Here is what the low-pass filter would look like in a FORTRAN program:

```
      A = EXP (-TS/TC)
      B = 1. - A
      DO 1Ø I = 2, N
1Ø    X(I) = A*X(I - 1) + B*X(I)
```

where X is the array of data, TS is the time between samples of the data (i.e., TS = $1/f_s$), and TC is the desired filter time constant. Ideally, TC ≫ TS. This little program does the filtering *in place,* i.e., it replaces the original data by the filtered version. You could, of course, put the filtered data into a separate array.

□ Low-pass commutating filter

This same filter can be built with hardware, using the circuit shown in Figure 9.75. The FET switches S_1 and S_2 are toggled at some clock rate f_s, repeatedly charging C_1 to the input voltage, then transferring its charge to C_2. If C_2 has voltage V_2, and C_1 charges up to

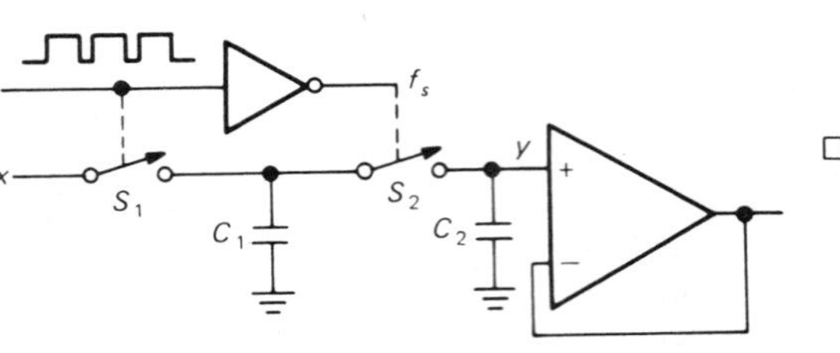

Figure 9.75
Recursive switched-capacitor filter.

the input level V_1, then when C_1 is connected to C_2, the new voltage will be

$$V = \frac{C_1V_1 + C_2V_2}{C_1 + C_2}$$

i.e., it is identical to the preceding low-pass recursive filter, with

$$y_i = \frac{C_2}{C_1 + C_2} y_{i-1} + \frac{C_1}{C_1 + C_2} x_i$$

Equating these coefficients to the value of *A*, given earlier, gives us

$$f_{3dB} = \frac{1}{2\pi} f_s \log_e \left(\frac{C_1 + C_2}{C_2}\right)$$

EXERCISE 9.6

Prove that this result is correct.

This filter is perfectly practicable, and it offers the nice feature of electronic tuning via the clock rate f_s. In practice, you would use CMOS switches, and C_1 would probably be much larger than C_2. Consequently, the switch-driving waveform should be unsymmetrical, spending most of its time closing S_1.

The preceding circuit is a simple example of a *commutating filter,* which includes filters made from arrays of switched capacitors. They have periodic frequency-response properties that make them particularly suitable for "comb" and notch filters.

It is possible to synthesize discrete approximations to all the classic filters (Butterworth, Chebyshev, etc.) in high-pass, low-pass, bandpass, and band-reject forms, either symmetrical in time or with the genuine "lagging" time response. Such filters are extremely useful when processing digitally quantized data, which is clearly the way of the future.

□ *Digital sine-wave generation*

An interesting technique related to nonrecursive digital filtering is the synthesis of sine waves by taking weighted sums from the outputs of a Johnson counter (or "twisted-ring" or "walking-ring" counter). The circuit in Figure 9.76 shows the way. The 4015 is

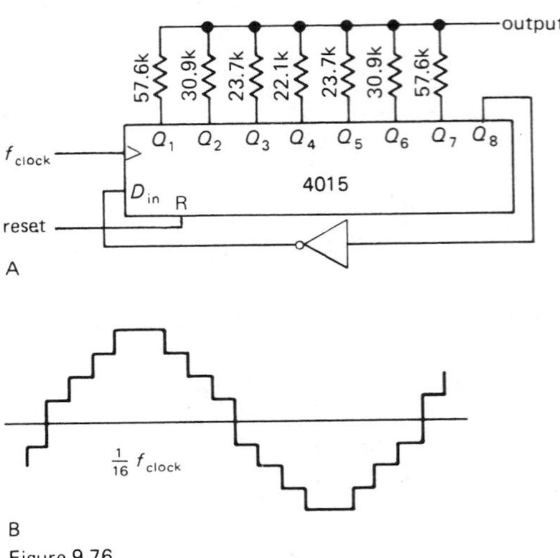

Figure 9.76
Digital sine-wave generation.

an 8-stage parallel-output shift register. By driving the input with the complement of the last stage, you get a Johnson counter, which goes through 16 states ($2n$ states, in general, for an n-stage shift register). Beginning with the state of all 0's, 1's begin marching in from the left until the register contains all 1's, at which point 0's march in again, and so on. The weighting shown generates an 8-level approximation to a sine wave, as shown, with a frequency 1/16th that of the clock, and with the first nonzero distortion term (assuming perfect resistor values) being the 15th harmonic, which is down by 24dB.

SELF-EXPLANATORY CIRCUITS

9.41 Circuit ideas

Figure 9.77 shows a few examples of interfacing between logic and linear signals.

Circuit ideas

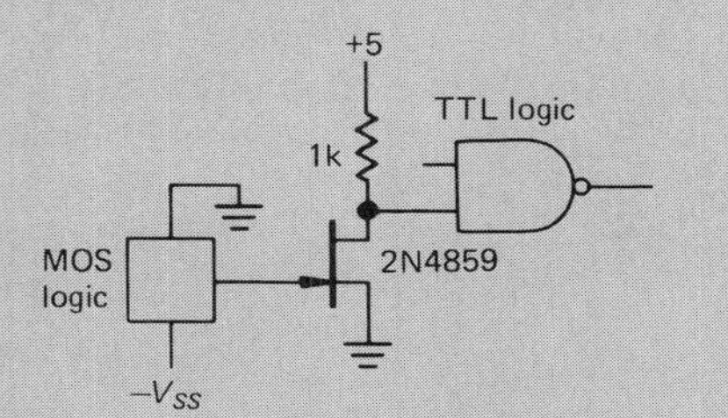

A. negative logic to TTL shifter

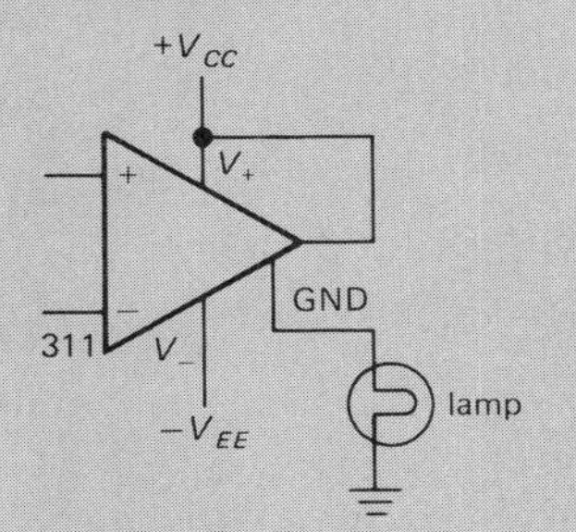

B. driving ground-returned load

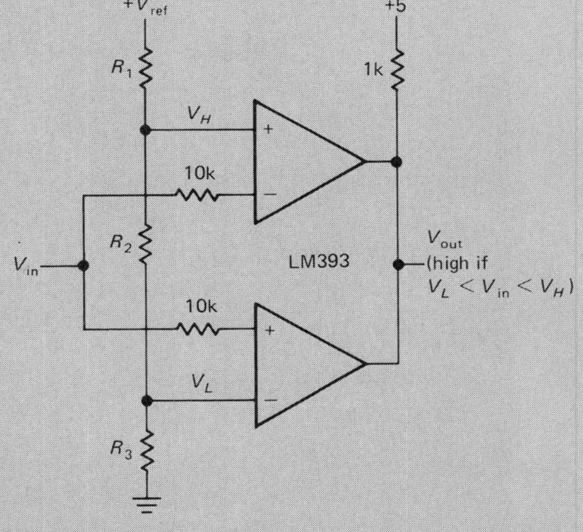

C. window discriminator

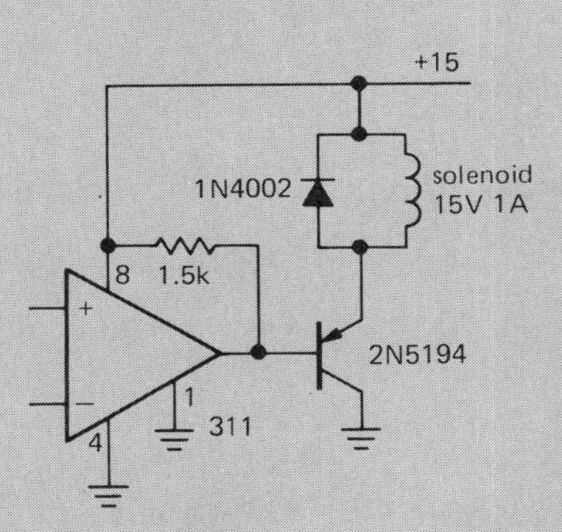

D. solenoid driver

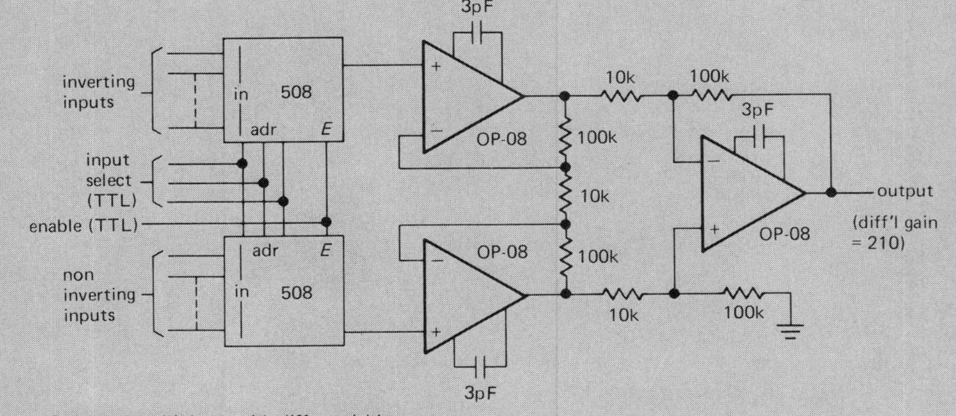

E. 8-channel multiplexer with differential input

Figure 9.77

Bad circuits

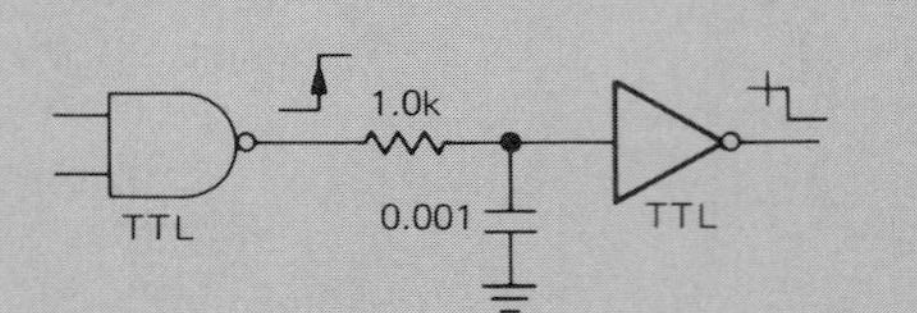

A. delayed edge generator

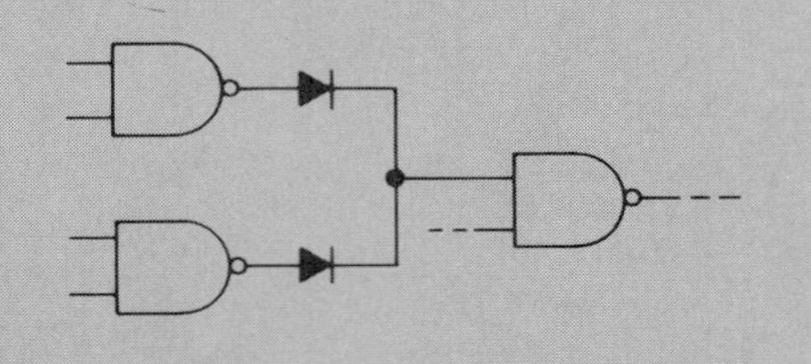

E. wired-OR from active pull-up gates

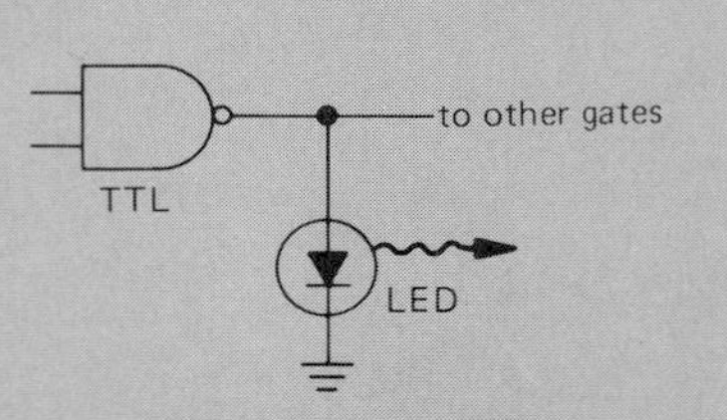

B. logic-state indicator

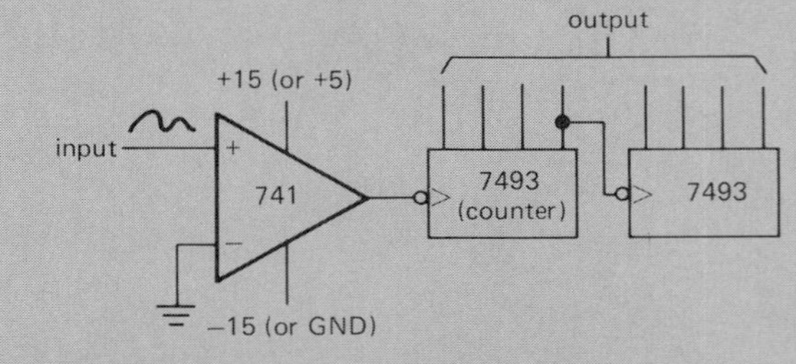

F. zero-crossing counter

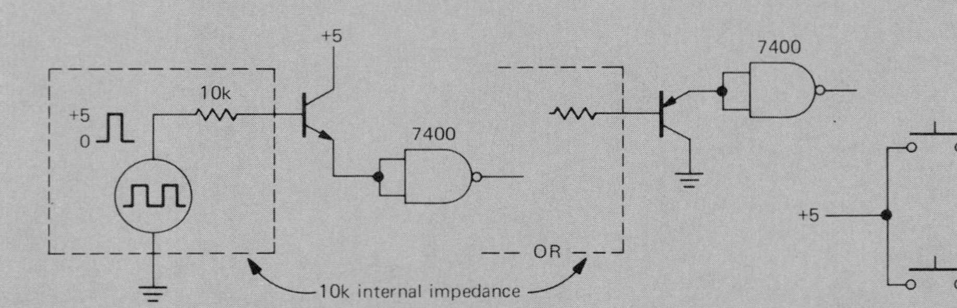

C. high-Z-to-TTL interface (2 bad circuits)

G. SR flip-flop

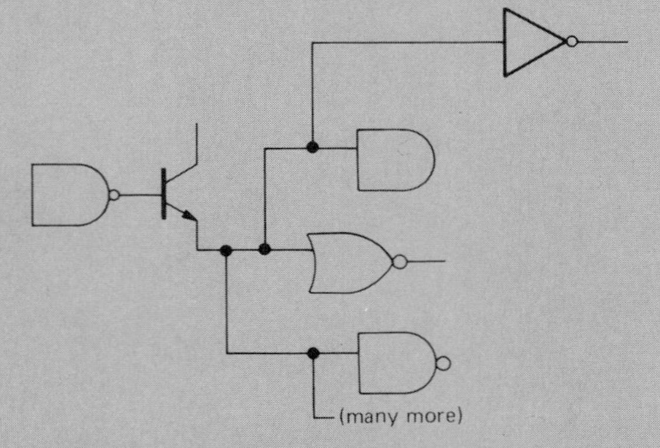

D. increasing TTL fanout with a follower

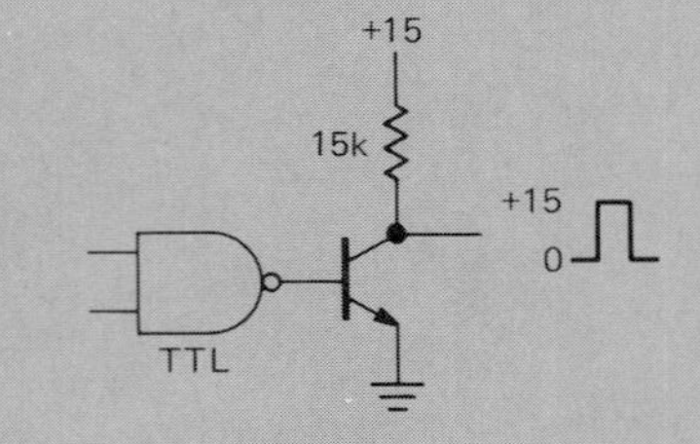

H. TTL-to-high-level interface

Figure 9.78

9.42 Bad circuits

The circuits in Figure 9.78 illustrate some basic interfacing blunders; in each case try to figure out what is wrong and how to fix it.

ADDITIONAL EXERCISES

1. Design a circuit to indicate if the logic power (+5V) has failed momentarily. It should have a push button to RESET it and a LED lamp that indicates CONTINUOUS POWER; make it operate from the +5 volt logic supply.

2. Why can't two n-bit DACs be used to make a $2n$-bit DAC by just summing their outputs proportionally ($OUT_1 + OUT_2/2^n$)?

3. Verify that the peak output of the pseudo-random-noise generator (Fig. 9.73) is ±8.68 volts.

4. An experiment is being controlled by a programmable calculator interfaced to various stimulus and measurement devices. The calculator increments a variable under its control (e.g., the wavelength of light coming from a monochromator) and processes the corresponding measurement (e.g., the amount of transmitted light, corrected for the known sensitivity curve of the detector). The result is a set of x,y pairs. Your job is to design a circuit so that they can be plotted on an analog x,y plotter.

The calculator outputs each x,y pair as two 3-digit BCD characters. To reduce the number of connections necessary, the numbers are presented one digit at a time ("bit parallel, character serial"), along with an address (2 bits). A CHARACTER VALID pulse signifies that the data and address are valid and can be latched, for example. An x'/y level tells whether the character being presented belongs to the x or y number. Figure 9.79 presents a summary.

The data are sent in the order x_n(LSD) ... x_n(MSD), y_n(LSD) ... y_n(MSD), so you know you've got one complete x,y pair after receiving the MSD of a y value ($A_1 = 0$, $A_2 = 1$, $x'/y = 1$). At *that* point you should update the digits seen by your D/A converters (don't update them one at a time).

In your circuit you needn't use particular

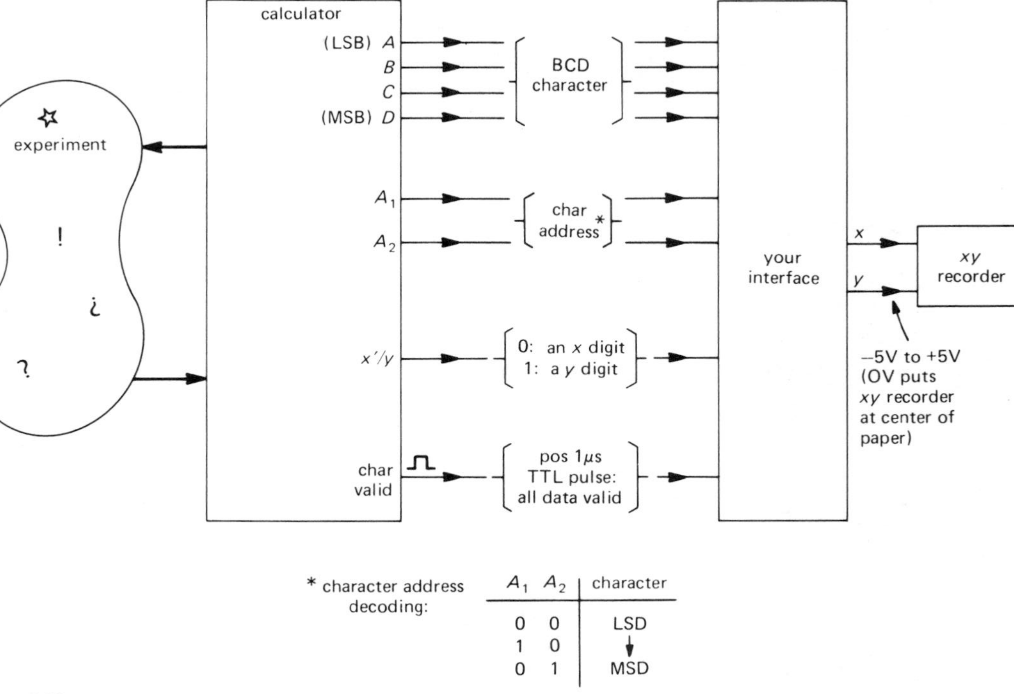

Figure 9.79

device numbers; just label them generically, e.g., a type *D* flip-flop, or a 1-of-10 decoder. Be sure to indicate where inputs or outputs are inverted (by showing small circles). Assume that you have some D/A converters that accept 3-digit BCD inputs, TTL-compatible, and whose outputs are a current, zero to 1mA corresponding to inputs of 000 to 999. Since the x,y plotter has 10 volt full-scale sensitivity, you'll have to convert the current to a voltage. As an additional obstacle, so that you can exercise your ingenuity, assume that the D/A converters have an output compliance of only 1 volt.

MINICOMPUTERS

CHAPTER 10

MINICOMPUTERS, MICROCOMPUTERS, AND MICROPROCESSORS

The availability of relatively inexpensive ($5k) small computers has made it feasible to control experiments and processes, collect data, and perform computation directly under the control of a computer. Small computers are commonly used in laboratory and industrial settings, and a knowledge of their capabilities, program languages, and interfacing requirements is becoming an essential part of electronics know-how.

We will use the term *minicomputer* to refer to a small computer whose central processing unit (CPU) is constructed from SSI and MSI circuits, usually occupying one or more large printed-circuit boards. A *microcomputer* is one whose CPU is constructed from just a few (often only one) LSI microcircuits; the CPU chip (or chip set) constitutes a *microprocessor.* The capabilities of minicomputers and microprocessors overlap considerably, although, generally speaking, a microprocessor with a small number of memory and I/O support chips is likely to be used for dedicated control of a process or instrument, whereas a minicomputer will be used where powerful (and fast) computational and input/output (I/O) capabilities are needed. A microprocessor plus a few assorted chips and some ROM (read-only memory) can replace a complicated logic circuit of gates, flip-flops, and analog/digital conversion functions and should be considered whenever embarking on a large design project. And microprocessors, when teamed up with large amounts of memory and peripherals, form microcomputing systems that have become so powerful that they are now serious contenders for large ''number crunching'' (computational) applications, an area previously dominated by minicomputers and larger computers. In fact, the words minicomputer and microcomputer are coming to be used almost interchangeably, in some cases with the distinction referring more to the physical size of the computer or the number of peripherals than to the scale of integration used in the construction of the CPU.

In this chapter we will describe minicomputer architecture, programming, and interfacing, with some examples of useful and simple interfacing to peripherals. Most of the ideas introduced in this chapter will carry over to the next chapter, where we will get into a detailed discussion of the selection and construction of microprocessor-based circuits and systems. Generally speaking, with minicomputers (and, to some extent, microcomputers) the design of the computer itself, including integration of memory and I/O control, as well as system programming and utility program development, is taken care of by the manufacturer. The user need worry only about special-purpose interfaces and the job of user programming. By contrast, in a dedicated microprocessor system, the choices of memory types, system interconnection, and programming generally have to be made by the designer. Minicomputer manufacturers are generally committed to providing extensive system and utility software as part of a complete computing system (including peripherals), whereas the microprocessor manufacturers (semiconductor manufacturers) generally see the design and marketing of microprocessor and support chips as their central tasks. In this chapter, then, we will describe computer architecture and programming and will concentrate on the details of intercommunication and interfacing.

10.01 Computer architecture

Although different computers do things in different ways, Figure 10.1 summarizes the organization typical of most computers. Let's take it from left to right:

CPU

The central processing unit, or CPU, is the heart of the machine. Computers do their computation in the CPU on chunks of data organized as computer "words." Word size can range from 4 bits to 32 bits or more, with a 16-bit word size being the most popular in current minicomputers. A "byte" is 8 bits (half a byte, or 4 bits, is sometimes called a "nibble"). The CPU has an *arithmetic unit,* which can perform operations such as add, complement, compare, shift, move, etc., on quantities contained in *registers* (and sometimes in *memory*). The *program counter* keeps track of the current location in the executing program. It normally increments after each instruction, but it can take on a new value after a "jump" or "branch" instruction. The *bus control circuitry* handles communication with memory and I/O. Most computers also have a *stack pointer register* (more on that later), and a few *flags* (carry, zero, and sign) that get tested for conditional branching.

Memory

All computers have some fast random-access memory, called RAM, "core," or "semiconductor memory." In a large minicomputer this may include 128K or more words of 16 bits each, although 32K words is a more usual number. (When used to describe memory sizes, K doesn't mean 1000, but rather 1024, or 2^{10}; thus 64K words of memory is actually 65,536 words. The lower-case symbol k is sometimes used to mean 1000.) This memory can be read and written at speeds of 1μs or less. Magnetic-core memory is "nonvolatile" (it

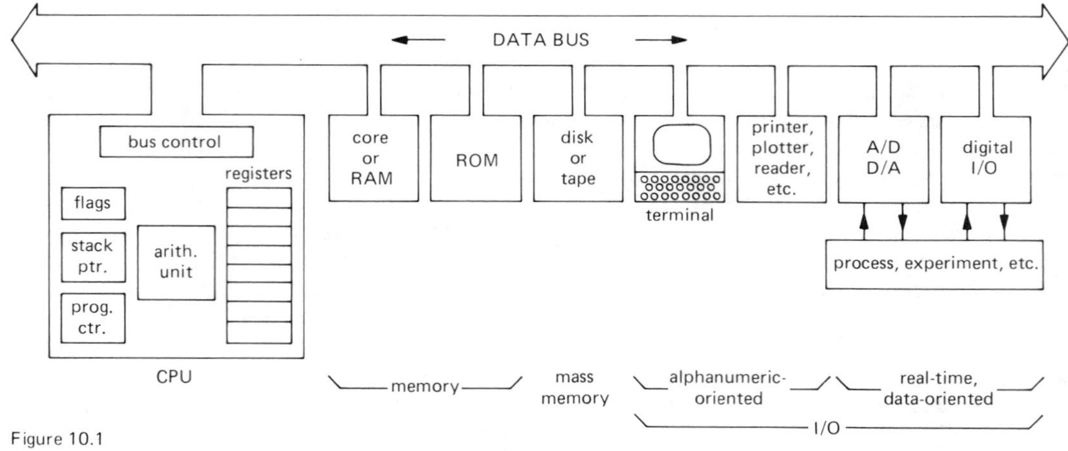

Figure 10.1
Block diagram of a laboratory computer.

retains its information when power is shut off), whereas semiconductor memory is volatile (its information evaporates when power is removed). Some computers, particularly dedicated microprocessor systems, will include some ROM (read-only memory), which is always nonvolatile, preprogrammed with frequently used routines. Nearly all computers now include at least one ROM program to "bootstrap" the computer, i.e., get it started from the state of total amnesia when power is first turned on.

To get or store information in memory, the CPU "addresses" the desired word. Most computers address memory by words, beginning at word 0 and going sequentially through to the last word in memory. In some machines the words are addressed by *bytes* (8-bit groups), which has advantages for character manipulations (letters, numbers, etc.), since one character usually occupies one byte. In a computer with lots of memory, two (or even three) bytes may be required to specify an arbitrary memory address anywhere in the machine. Since most memory references in an actual program are usually "nearby," all computers provide for simplified addressing modes: "Relative" addressing specifies an address according to its distance from the present instruction, "indexed" addressing uses the contents of a register to point to a location in memory, "paged" addressing uses a shortened address to refer to a memory location within a small area (a page), and "immediate" instructions refer to the next word in memory.

Both programs and data are kept in memory during program execution. The CPU fetches instructions from memory, figures out what they mean, and does the appropriate things, often involving data stored somewhere else in memory. General-purpose computers store programs and data in the same memory, and in fact the computer doesn't even know one from the other. Amusing things start to happen if a program goes awry and you "execute data"!

Mass memory

Computers intended for program development or computation, as opposed to dedicated control processors, usually have one or more mass-storage media. Disks ("hard" or "floppy") and magnetic tapes are the usual ones, with storage capabilities going from a few hundred thousand bytes (floppy disks) to several million bytes (hard disks, tapes) and up to as much as 200 million bytes (large hard disks). Mass-storage media are generally slow, magnetic tape being the slowest and cheapest, with access times of several seconds or more, and hard disks being the fastest and most expensive, with access time of milliseconds. Once the data has been located, data transfer is rapid, typically 10,000 to 100,000 bytes per second. You generally keep programs, data files, plot files, etc., on some sort of mass-storage device and bring these into RAM only when doing computation. Many users can simultaneously fit their programs on one disk; a large disk can hold the contents of the *Encyclopaedia Britannica.*

Alphanumeric I/O

It is nice to have a powerful computer, capable of millions of smart computations per second, but it doesn't do you any good if it keeps all its results to itself. Peripherals such as an alphanumeric terminal (keyboard plus screen), line printer, graphic plotter or terminal, etc., let man and machine communicate, and these are essential in any "friendly" computer system. These peripherals are mostly oriented toward programming and numbers; you use them when writing programs, debugging, listing, inputting data, playing space war, and producing output data. These peripherals, together with suitable interfaces, are available from many sources, including the computer ("mainframe") manufacturer.

Real-time I/O

For experiment or process control and data logging, or for exotic applications such as speech or music synthesis, you need A/D and D/A devices that can communicate with the computer in "real time," i.e., while things are happening. The possibilities are almost endless here, although a general-purpose set of multiplexed A/D converters, a few fast D/As, and some digital "ports" for exchange of digital data will permit many interesting applications. Such general-purpose peripherals are manufactured commercially. If you want something fancier, such as improved performance (higher speed, more channels) or special-purpose functions (tone generation, frequency synthesis, time-interval generation, etc.), you've got to build it yourself. This is where a knowledge of bus interfacing and programming techniques is essential, although it's helpful in any case.

Data bus

For communication between the CPU and various peripherals, all computers use a "bus," a set of shared lines for exchange of digital words. (In principle, many buses also allow communication *between* peripherals, although this capability isn't often used.) This same bus is often used for communication between the CPU and memory, as well. The use of a shared bus vastly simplifies interconnections, and if you use a little care in bus design and implementation, it doesn't cause any problems.

The bus contains a set of DATA lines (generally as many lines as there are bits in a computer word – 4, 8, or 16 for microcomputers, 16, 24, or 32 for most minicomputers), some ADDRESS lines for determining who should "talk" or "listen" on the line, and a bunch of CONTROL lines that specify what action is going on (data going to or from the CPU, interrupt handling, DMA transfers, etc.). All the DATA lines, as well as a number of others, are "bidirectional" – they're driven either by open-collector gates, with resistor pull-ups somewhere (usually at the end of the bus, where they also serve as terminators to minimize reflections, see Section 13.09), or by three-state devices, in which case no pull-ups are necessary (although line termination may still be necessary if the bus is physically long).

Don't confuse this use of open-collector/three-state with wired-OR; in normal operation only one device is asserting data onto the bus at any time, and open-collector or three-state devices are used only so that the other devices can disable their bus drivers. Each computer has a well-defined protocol for determining who asserts data at any time. If it didn't, total chaos would result, with everyone shouting at once (so to speak). (Computer people can't resist personalizing their machines, peripherals, etc. Engineers are even worse, with flip-flops and even gates coming to life. Naturally, we follow the trend.)

There is one interesting distinction in computer buses. They can be either *synchronous* or *asynchronous,* with examples of each in currently popular minicomputers. You will see what that means when we get into the details of communication via the bus.

A COMPUTER INSTRUCTION SET

10.02 Assembly language and machine language

In order to make the rest of the chapter sensible, we're going to introduce a mythical computer with a 16-bit word. Let's call it the "MC-16." We'll endow it with a simple instruction set and bus and then show some examples of interfacing and programming. These examples will help convey the idea of programming at the "machine language" level, something quite different from programming in a high-level language like FORTRAN. Since real machines work in closely similar ways, careful reading of an actual computer manual should allow you to carry on from where this chapter leaves off.

First, a word on "assembly language." As we mentioned earlier, the computer's CPU is designed to interpret certain words as instructions and carry out the appointed tasks. This "machine language" consists of a set of instructions, each of which may occupy one or more computer words. Incrementing the contents of a register would be a single-word instruction, for example, whereas adding the contents of a memory location to the contents of a register would usually require at least two words (the first would specify the operation and the second would specify the memory location). It is a sad fact of life that different computers have different machine languages, and there is no standard whatsoever.

Programming in machine language is

extremely tedious, since you wind up dealing with columns of binary numbers, each bit of which has to be bit-perfect, so to speak. For this reason a program called an *assembler* is provided by the computer manufacturer. It allows you to write programs using easily remembered mnemonics for the instructions and symbolic names of your own choosing for memory locations and variables. This *assembly-language* program, really nothing more than a number of cryptic-looking lines of letters and numbers, is massaged by a program called an *assembler* to produce as its output a finished program in machine-language "object code" that the computer can execute. Each line of assembly code gets turned into a few (one to four) machine-language words. The computer cannot execute assembly-language instructions directly. To make these ideas concrete, let's look at the MC-16 assembly language and do a few examples.

□ 10.03 MC-16 instruction set

The MC-16 has a 16-bit word, a single 16-bit accumulator (or "register") plus a carry bit, and the following assembly-language instruction set:

Instruction		*What you call it*	*What it does*
STA	m	store accum	AC → (m); AC unchanged
LDA	m	load accum	(m) → AC; (m) unchanged
ADD	m	add	AC + (m) → AC; (m) unchanged
SUB	m	subtract	AC − (m) → AC; (m) unchanged
AND	m	and	AC AND (m) → AC (bitwise); (m) unchanged
OR	m	or	AC OR (m) → AC (bitwise); (m) unchanged
ISZ	m	increment, and skip if zero	(m) + 1 → (m); skip next instr if zero result
DSZ	m	decrement, and skip if zero	(m) − 1 → (m); skip next instr if zero result
COM		complement	1's complement of AC → AC
NEG		negate	negative (2's comp) of AC → AC
CLR		clear	0 → AC
SHL		shift left	AC "circular" left shift one bit
SHR		shift right	AC "circular" right shift one bit
JMP	label	jump	jump to instr "label"
JZ	label	jump on zero	jump to instr "label" if AC = 0
JNZ	label	jump nonzero	jump to instr "label" if AC not zero
JPL	label	jump plus	jump to instr "label" if AC ≥ 0
JMI	label	jump minus	jump to instr "label" if AC < 0
JCA	label	jump on carry	jump to instr "label" if carry bit = 1
JNC	label	jump nocarry	jump to instr "label" if carry bit = 0
IN	dev	input	read from device "dev" to AC
OUT	dev	output	write to device "dev" from AC
JSR	sub	jump subroutine	jump to subroutine "sub"
IQ		interrupt query	interrupting device → AC
IE		interrupt on	enable interrupts
ID		interrupt off	disable interrupts

A few explanations: The symbol (m) means "contents of the memory location whose address is m," as opposed to the *address* of that memory location. For example, the instruction

```
LDA 1000
```

means "get the number stored in the memory location numbered 1000, and put it into the accumulator." A "circular" shift moves each bit over, with the last one wrapping around; the carry bit is included. It is as if the accumulator and carry bit were all part of a 17-bit shift register. Thus, a left circular shift moves each bit of the accumulator over one notch to the left, with the carry bit going to the right-hand bit (the LSB) and the MSB going into the carry bit position. JSR jumps to the subroutine labeled "sub" and puts the return location into the first location of sub, i.e., into location "sub" (in most machines a "stack" is used instead to store return addresses during a subroutine call, and another stack may be used for variables; this has the advantage that a subroutine is *re-entrant* – it may call itself, or call a subroutine that calls it). Finally, any memory reference (m, label, sub) can be preceded by the symbol "@," signifying an "indirect" reference. For instance, the command

```
LDA @const
```

means "go to location 'const,' read its contents, then go to *that* location and load the accumulator with *its* contents"; i.e., ((const)) ⟶ AC.

10.04 A programming example

To get an idea of what programming looks like in assembly language, as compared with the more familiar "higher-level" languages like BASIC and FORTRAN, let's do an example. Suppose you want to increment a number, N, if it equals another number M. This will typically be a tiny step in a larger program, and in FORTRAN it will be a simple instruction:

IF (N .EQ. M) N = N + 1

In MC-16 language it will look like Program 10.1. The assembler program will convert this set of mnemonics to machine language, generally using one or two machine-language words per line, and the resultant machine-language code will get loaded into successive locations in memory before being executed. Note that it is necessary to define locations somewhere in the program (by preceding the instruction with a symbolic name and colon) to hold the two variables and the constant 1. Symbolic labels can also be used to tag instructions, for instance the label NEXT in this example; this is usually done only if there is a jump to that location (JNZ NEXT). Giving some locations understandable (to you!) names and adding comments (separated by a semicolon) makes the job of programming simpler; it also means that you have a chance of understanding what you've written a few weeks later! Programming in assembly language can still be a nuisance, but it is often necessary to write short routines in it, callable from a higher-level language, to handle I/O. Assembly-language programs run faster than programs compiled from a higher language, so it is also used where speed is important (e.g., the innermost loop of a long numerical calculation). In any case, you can't really understand computer interfacing without understanding the nature of assembly-language I/O. The correspondence between mnemonic assembly language and executable machine language is explored further in Section 11.4, in that case illustrated by 8085 microprocessor programming.

BUS SIGNALS AND INTERFACING

A typical 16-bit minicomputer bus has about 50 signal lines, devoted to the transfer of data, addresses, and control signals. We will approach the subject by building up a typical bus, beginning with the signal lines necessary for the simplest kind of data interchange (programmed I/O) and adding additional signal lines as they become necessary. We will give some useful interface examples

```
        LDA N          ;get n
        SUB M          ;subtract m
        JNZ NEXT       ;test N=M?
        LDA N          ;yes. get n again
        ADD C1         ;add one
        STA N          ;store result
NEXT:     •            ;program continues
          •
          •            ;last executible statement

N:      Ø              ;the variable n stored here
M:      Ø              ;the variable m stored here
C1:     1              ;the constant 1
```

Program 10.1

as we go along, to keep things comprehensible and interesting.

10.05 Fundamental bus signals: data, address, strobe

A minimum I/O bus must have DATA lines (for the data to be transferred), ADDRESS lines (to identify the I/O device), and some STROBE lines (which tell when data is being transferred). To speed things up, there are usually as many DATA lines as bits in the computer word, so a whole word can be transferred at once. The number of ADDRESS lines depends on whether or not the bus is also used to address memory. In a machine like the PDP-11, in which a single bus is used for everything, there are 18 ADDRESS lines, whereas with a machine like the NOVA, with a separate memory bus, the I/O bus has only 6 ADDRESS lines, permitting up to 64 external I/O device codes. Data transfer itself is synchronized by pulses on additional "strobing" bus lines. There are two ways this can be done: by having separate IN and OUT lines, with a pulse on one or the other synchronizing data transfer; or by having one SYNC line and one IN/OUT line, with a pulse on SYNC synchronizing data transfer in a direction specified by the level on the IN/OUT line.

10.06 Programmed I/O: data out

The simplest method of data exchange on a computer bus is known as "programmed I/O," meaning that data is transferred via an IN or OUT statement in the program (the directions for IN and OUT are among the few things on which all computer manufacturers agree: IN always means *toward* the CPU, and OUT always means *from* the CPU). The whole process of data OUT is extremely simple and logical: The ADDRESS of the recipient and the DATA to be sent are put onto the respective bus lines by the CPU. After several hundred nanoseconds (to allow for address decoding by the intended recipient's hardware), an OUT pulse is sent by the CPU, with DATA and ADDRESS guaranteed to be valid during the pulse, and for 250ns on either side (in some computers the DATA and ADDRESS are not guaranteed valid on *both* sides of the strobe pulse). To play the game, the peripheral (in this case, an XY scope display) looks at the ADDRESS and DATA lines. When it sees its own address, it latches the information on the DATA lines, using the OUT pulse as a clocking signal. That's all there is to it.

Let's look at the example shown in Figure 10.2. In this example we have designed an XY scope display, giving the X register address 112 and the Y register address 113. The 8131 is a 6-bit address comparator, giving a LOW output when the six high-order bits A_2–A_7 match the fixed comparison inputs, in this case when the ADDRESS bus contains address 112–115 (you could use gates, but an address comparator is more compact). In a typical interface you generally assign several consecutive addresses to the same peripheral, using each address for different register, as we've done here. The 3-input NORs then decode the remaining ADDRESS bits to give a HIGH output on individual addresses 112 and 113 (another method will be described later). These outputs enable AND gates that clock the 74LS174 hex D registers, thus latching the 12 low-order bits of the DATA lines when (a) the correct address is present and (b) an OUT pulse is sent. A pair of monostables generates a 5μs "unblanking" pulse after the Y coordinate has been latched, to intensify the selected spot on the scope (all scopes have a "*Z* input" for that purpose). To draw a graph or set of characters on the screen, all you do is output successive XY coordinates repetitively, fast enough so the eye doesn't see the flicker. Minicomputers are fast enough to display a few thousand XY pairs repetitively without annoying flicker. Even this simple interface enhances the power of a small computer enormously.

You can save a few parts by using a strobed decoder in the address decoding circuitry, as indicated. Notice the trick of using the high-order input bits as enable lines, for a decoder (like the 7442) that has no separate enable inputs. Decoders like the 74138 (3-line to 8-line) and 74139 (dual 2-line to 4-line) include one or more enable inputs, and they are handy in this sort of application.

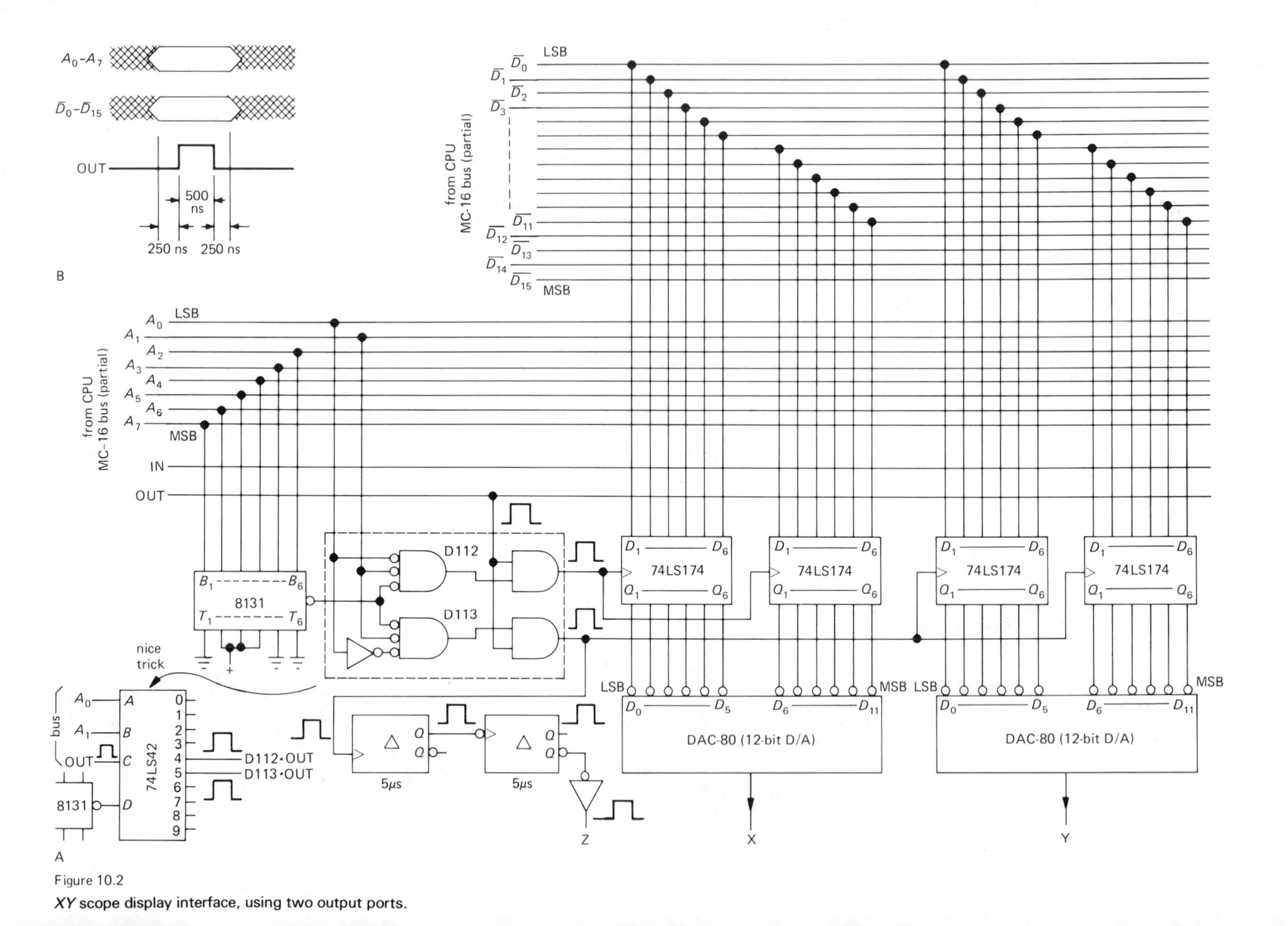

Figure 10.2

XY scope display interface, using two output ports.

```
                            ;routine to drive XY display
INIT:     LDA XPOINT
          STA X             ;initialize X pointer
          LDA YPOINT
          STA Y             ;initialize Y pointer
          LDA NPOINT
          STA N             ;initialize counter

LOOP:     LDA @X            ;get X value
          OUT 112           ;output it
          LDA @Y            ;get Y value
          OUT 113           ;output it
          ISZ X             ;advance X pointer. never "skips"
          ISZ Y             ;advance Y pointer. never "skips"
          DSZ N             ;decrement counter.  done yet?
          JMP LOOP          ;nope.  display more stuff
          JMP INIT          ;yup.  start over

XPOINT:  (address of first X value)
YPOINT:  (address of first Y value)
NPOINT:  (number of points to plot)
X:        Ø                 ;X pointer stored here
Y:        Ø                 ;Y pointer stored here
N:        Ø                 ;counter stored here
```

Program 10.2

□ *Programming the scope display*

The programming to run this interface is straightforward. Program 10.2 shows what you do.

The addresses of the first X and Y, and the number of points to be plotted, have to be available to the program. This will probably be a subroutine, with those parameters as passed arguments in the subroutine call. The program puts the initial values into locations X, Y, and N, then enters a loop in which successive XY pairs are sent to devices 112 and 113. The X and Y pointers are advanced each time around, and the counter is decremented and checked for 0, which means that the last pair has been displayed; the pointers are then reinitialized, and the process begins again.

A couple of important points: Once started, this program displays the XY array forever. In real life the program would probably check the keyboard, or perhaps a computer "sense switch," to see if the operator wants the plot terminated. Alternatively, the display could be terminated by an "interrupt," which we will discuss shortly. With this sort of "refreshed" display, there usually isn't time to do much computing while displaying. A display device refreshed from its own memory takes that burden off the computer, and this is generally a better method. Nevertheless, a simple refreshed display is adequate for many tasks. This book is being written on a computer with such a display.

10.07 Programmed I/O: data in

The other direction of programmed I/O is equally simple. The interface looks at the ADDRESS lines as before. If it sees its own address, it puts a word onto the DATA lines coincident with the IN pulse. Figure 10.3 shows an example. This interface lets the computer read a 6-bit word latched in the 74LS174 D-type register. Since the clock input and data inputs of the register are accessible to an external device, the register could hold just about any sort of digital information (the output of a digital instrument, A/D converter, etc.). For variety, the device address (213) has been decoded with an 8-input NAND gate, which enables the 74366 inverting three-state buffer when an IN pulse comes along. Many minicomputers use open-collector bus drivers, with resistive pull-ups on the bus lines; in that case you could drive the bus with open-collector high-current NANDs, as shown. This is the reason most minicomputers use negative-

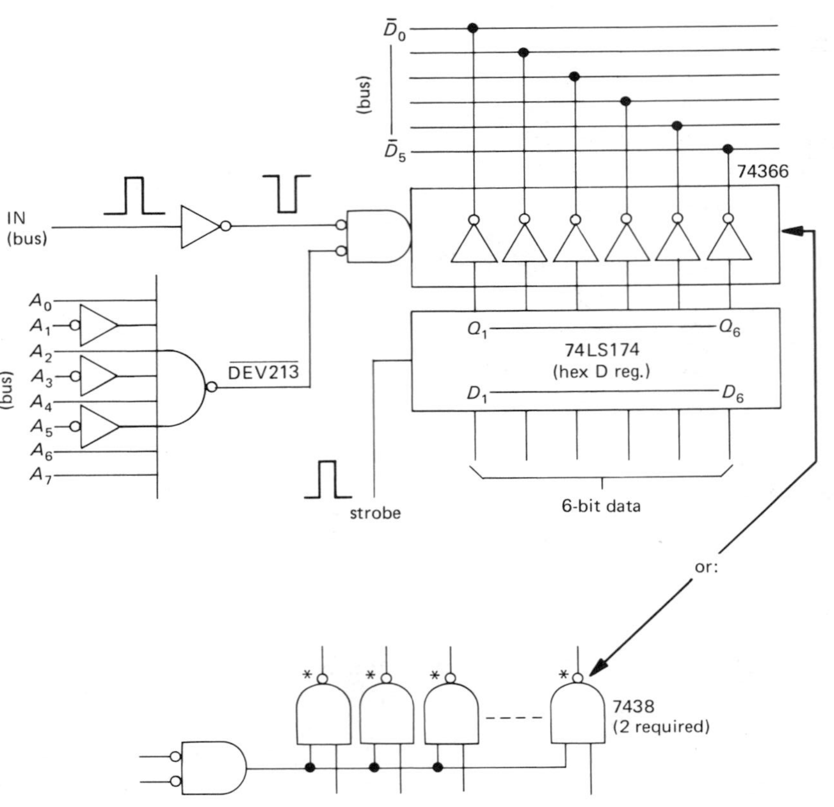

Figure 10.3
Data input port.

true DATA lines; DATA lines not driven are held HIGH by the pull-ups and hence are interpreted as 0. This is convenient in situations where you don't want to use all 16 DATA lines (as in the preceding examples or, as you will see shortly, in the use of "status registers"), since you don't have to include drivers for the unused lines just to assert zeros on them. In this example, as in all further interface circuits, we will omit the tangle of bus lines and simply call them out by name.

10.08 Programmed I/O: status registers

In the last example, the computer can read a word from the interface any time it wants to. That's nice, but how does it know when there's something worth reading? In some situations you may want the computer to read data at equally spaced intervals, as determined by its "real-time clock." Perhaps the computer instructs an A/D to begin conversion at regular intervals (via an OUT command), then reads the result a few microseconds later (via an IN command). That might suffice in a data-logging application. However, it is often the case that the external device has a mind of its own, and it would be nice if it could communicate what's happening to the computer without having to wait around.

A classic example is an alphanumeric input terminal, with someone banging away at a keyboard. You don't want characters to get lost; the computer has to get every character, without too much delay. With a fast storage device like disk or tape the situation is even more serious; data must be moved at rates up to 100,000 bytes per second without delay. There are actually three ways to handle this general problem: status registers, interrupts, and direct mem-

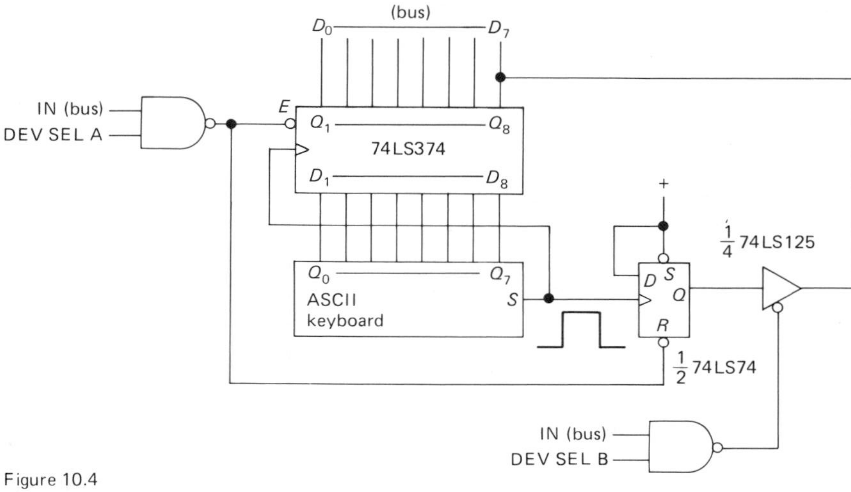

Figure 10.4
Keyboard input port.

ory access. Let's begin with the simplest method (status registers), illustrated by the keyboard interface in Figure 10.4.

In this example, an ASCII keyboard drives a 74LS374 8-bit D-type register, clocking in a character via the keyboard's STROBE output pulse when a key is struck. This octal latch happens to have three-state outputs, so it's easy to rig up a programmed data IN circuit, as shown. The input labeled DEV SEL A comes from an address decoding circuit of the sort shown explicitly in the previous examples, and it goes HIGH when the particular address chosen for this interface appears on the ADDRESS lines of the bus.

What's new in this example is the flip-flop, which gets set when a character is struck and cleared when a character is read by the computer. It's a 1-bit "status" register, HIGH if there's a new character available, LOW otherwise. The computer can query the status bit by doing a data IN from the other address of this device, decoded (with gates, decoders, or whatever) as DEV SEL B. You only need one bit to convey the status information, so the interface only drives the most significant bit, in this case with a three-state buffer. The line coming into the side of the buffer symbol is the three-state OUTPUT ENABLE; the 74LS125's output is enabled when that line is brought LOW, as indicated by the negation bubble.

□ *Program example: keyboard terminal*

The computer now has a way to find out when data is ready. Program 10.3 shows how.

This is a routine to get characters from a keyboard terminal (device code = KBDCHAR; the actual numerical device codes will be defined in some statements near the beginning of the program, omitted here for simplicity), echoing each character on the printer (device code = OUTCHAR). When it has gotten a whole line, it transfers control to a line handling routine, which might do just about anything, based on what the line says. When it's ready for another line, it types an asterisk. The program loops on the input flag (device code = KBDFLAG), checking the sign of the word brought IN from the status register; it goes negative when the MSB is set (remember, the MSB is the sign bit in 2's complement arithmetic), i.e., when the interface has a new character ready. The program then gets a character (this clears the status flag flip-flop), stores it consecutively in the line buffer, increments the pointer, echoes the character to the printer, and checks to see if the line has been terminated by a carriage return character. If so, it transfers control to the line handler; otherwise it goes back and loops on the keyboard status flag again.

A subroutine has been used to type a

```
                                ;keyboard handler -- uses flags
INIT:    LDA BEGIN              ;get beginning address of character buffer
         STA POINT              ;initialize pointer

LOOP:    IN  KBDFLAG            ;new character available?
         JPL LOOP               ;nope
         IN  KBDCHAR            ;yup. get character
         STA @POINT             ;store it in line buffer
         ISZ POINT              ;advance pointer
         JSR TYPE               ;echo last character to printer
         SUB CR                 ;was it a carriage return?
         JNZ LOOP               ;if not, get next character
LINE:     •                     ;if so, do something with the line
          •                     ;keep at it
          •                     ;don't quit now
          •                     ;done at last!
         LDA STAR
         JSR TYPE               ;type a "prompt" -- asterisk
         JMP INIT               ;get another line

POINT:   Ø                      ;pointer to character buffer goes here
CR:      15                     ;ASCII for carriage return
STAR:    52                     ;ASCII for "*"
BEGIN:   (first address of line buffer goes here)

                                ;routine to type character in AC
                                ;preserves contents of AC
TYPE:    Ø                      ;return address gets put here
         STA TEMP               ;save character to type
CHECK:   IN  OUTFLAG            ;check printer busy?
         JMI CHECK              ;if so, check again
         LDA TEMP               ;if not, get char to type
         OUT OUTCHAR            ;type it
         JMP @TYPE              ;return via address in first loc of subr.

TEMP:    Ø                      ;temporary storage
```

Program 10.3

character, since even that simple operation requires some flag checking (device code = OUTFLAG; MSB is set if the printer is still active). Note the method of return to the main program: an indirect jump through location TYPE (remember that the MC-16 puts the return address in the first location of a subroutine, beginning execution at the second). The awkwardness of a single-register machine is well illustrated in this example, where the AC has to be saved just to allow checking of the printer status via an IN command.

In an actual keyboard interface, there will usually be provision for clearing the status flag(s) independently of reading a character. In most interfaces there will be several flags to signal various conditions. For instance, in a magnetic-tape interface you will normally have status bits for beginning of tape, end of reel, parity error, tape in motion, etc. The usual procedure is to put all the status bits into one word, so that a data IN command from the status register gets all bits at once. Typically you will put an error bit as the MSB of the status word, so that a simple test of the sign tells if there are any errors; if there are, you test specific bits of the word (by ANDing with a ''mask'' word) to find out what's wrong.

In some computers the use of status flags is so widespread that several lines are added to the bus just for status information. As an example, the NOVA computers use BUSY and DONE lines: An interface is BUSY if it is involved in some action initiated by the computer; it is DONE when that action is complete. The interface pulls down the corresponding line, according to its BUSY/DONE state, when it sees its address on the ADDRESS lines. The instruction set is arranged so that a single instruction checks

BUSY or DONE of a selected device, with conditional jump based on the outcome.

10.09 Interrupts

The use of status flags just illustrated is one of three ways for a peripheral device to "tell" the computer when some action needs to be taken. Although it will suffice for some peripherals, it has the serious drawback that the peripheral cannot "announce" that some action needs to be taken – it has to wait to be "asked" by the CPU, via a data IN command from its status register. Devices that need quick action (such as disks, tape, card readers, or real-time I/O) would have to have their status flags queried often, and with a few such devices in a computer system the CPU would soon find itself spending most of its time checking status flags, as in the earlier example. What is needed is a mechanism for a peripheral to interrupt the normal action of the CPU when something needs to be done. The CPU can then check the status register to find out what the trouble is, take care of what needs to be done, and go back to its normal business.

To add the interrupt capability to the MC-16 computer, it is necessary to add at least one line to the bus, a shared INT′ line that is pulled LOW by any interrupting device. For reasons you will see shortly, two other lines are also used: INTQ and INTP. Let's look at a typical interrupt circuit (Fig. 10.5).

The DATA READY flip-flop in the interface is set (by the interface, not by the computer) when the device wants to interrupt; it might indicate the presence of new data, as in the previous keyboard example, or it might signal completion of a task, as in the case of a printer or plotter. Whatever the cause, the peripheral pulls the INT′ line LOW, a signal to the CPU that something needs to be done. If interrupts are enabled (the usual state of affairs), hardware in the CPU causes a jump to a special memory location. In some machines the jump is "vectored" to a location in the program that depends on which device interrupted, whereas in simpler machines the jump is always to a fixed location. The MC-16 is simple. It jumps to location 1, putting the present value of the program counter (i.e., the address of the next instruction that would have been exe-

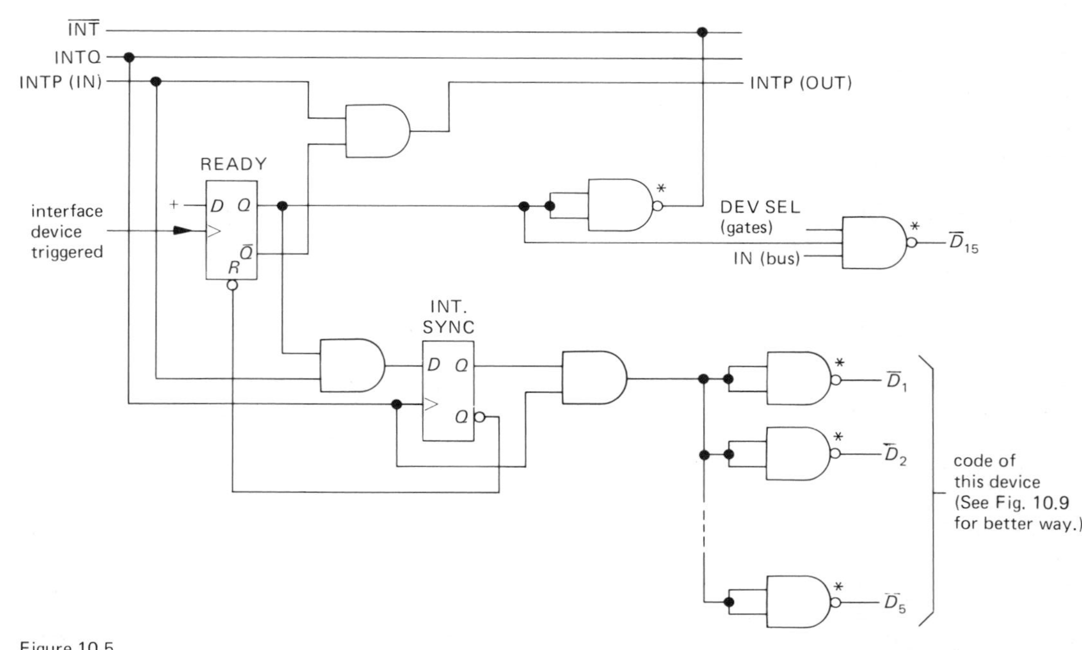

Figure 10.5
Priority interrupt circuit.

cuted) in location 0, so the task that was interrupted can be resumed after the interrupt is serviced. It also disables interrupts, to prevent your interrupt routine from being interrupted before you're ready.

At this point the program has to figure out which device interrupted, then jump to an interrupt handling routine (specific for each device that can interrupt) to take the appropriate action. For instance, if the program determines that the keyboard interrupted, it will jump to a keyboard handler, which will read in a character to a line buffer; that clears the keyboard's DATA READY flip-flop, releasing the INT′ line. For a more complicated device, the interrupt might signal one of a number of possible conditions. In that case the handler will begin by looking at the status register, via a data IN command, to find out what action to take. To write the (device-specific) handlers, you have to know what the device does and what information it puts in the various bits of its status register, as well as what information is needed by the program from the device (via data IN commands) and what commands need to be sent to the device (via data OUT commands to a "command register" in the device).

□ 10.10 Interrupt handling method I: device polling

There are two ways to figure out which device(s) have interrupted (more than one can interrupt at one time). The simplest method is interrupt *polling,* the process of asking each device in turn if it interrupted. This you do by simply checking the status register of each device in the system. When you find one with a status bit requiring action, you simply jump to the corresponding handler as shown in Program 10.4.

The instruction pair IN STATx, JMI DEVx constitutes the actual polling of each device. These instructions check the status of each device's READY flip-flop via the sign bit (DATA bit 15), the status flag asserted by the interface (look at the interrupt circuitry in Fig. 10.5). Note that it is essential that interrupts be left disabled during handling in this example; otherwise you will lose your return address if interrupted while handling an interrupt. By putting return addresses on a *stack,* you can allow an interrupt to be interrupted without losing track of things. Note also that the state of the machine (registers, carry bit) has to be saved, since the interrupt can occur at any place in a running program, and the contents of the registers must not be lost.

Device polling is perfectly OK, provided you don't have a large number of possible interrupting devices, all requiring rapid response to an interrupt; you generally poll the latency-sensitive (impatient) devices first, i.e., the ones that can't wait, eventually getting to the devices that don't require rapid response. This effectively sets up a "software priority" among the interrupting devices. You will see shortly that there is, in addition, a hardware priority built into the bus protocol itself.

□ 10.11 Interrupt handling method II: vectored interrupt

A more efficient (and therefore faster) way of handling interrupts, especially when many devices that can interrupt are connected to a system, is the use of *vectored interrupts,* in which the interrupting device "tells" the CPU its name. As mentioned earlier, this is automatic in some machines, in which the hardware generates a jump to a location specified by the interrupting device; the PDP-11 works this way. In machines with simpler architecture (e.g., the NOVA computers) the program must ask which device interrupted, by executing an IQ (interrupt query) type of instruction. This puts a pulse on the INTQ line of the bus, and the interrupting device responds by asserting its own device code onto the DATA lines. That device code is then available to the program in one of the registers. Let's suppose it gets put in the AC (where else?) by the hardware in the MC-16. Program 10.5 shows the software.

The programming begins with temporary storage of the AC and carry. An interrupt query command then gets the address of the interrupting device and puts it in the AC. Now we have to jump to an interrupt service routine specific for that device. This is done

```
                             ;interrupt handling code -- device polling
Ø:        Ø                  ;value of PC at interrupt (Ø is a dummy)
1:        JMP INTS           ;go to interrupt service routine
            •                ;other parts of the program
            •
            •
INTS:     STA ACSAVE         ;save contents of AC
          SHL
          STA CASAVE         ;save carry bit

          IN  STATa          ;check device A's status reg
          JMI DEVa           ;if set, jump to handler
          IN  STATb          ;check device B's status reg
          JMI DEVb           ;if set, jump to handler
            •
            •
            •
          IN  STATx          ;check last device's status reg
          JMI DEVx           ;if set, jump to handler
          JMP ERROR          ;none set! unknown device interrupting

DEVa:       •                ;beginning of specific code for device a
            •
          JMP RET            ;prepare to return to interrupted task
DEVb:       •                ;beginning of specific code for device b
            •
          JMP RET            ;prepare to return to interrupted task
            •
            •
            •
          JMP RET
DEVx:       •                ;beginning of specific code for last device
            •
RET:      LDA CASAVE         ;get archived carry bit
          SHR                ;rotate it into carry potition
          LDA ACSAVE         ;restore old AC contents
          IE                 ;enable interrupts again
          JMP @Ø             ;return via PC, stored in location Ø


ACSAVE:   Ø
CASAVE:   Ø                  ;locations for temporary storage

          (end of interrupt handling routines)
```

Program 10.4

with an indirect jump through a table of starting addresses of the various routines. The device code is used as an "offset," added to the first address of the table; the sum of device code plus beginning table address is therefore the address of the table entry for that device, and an indirect jump puts you at the beginning of the specific routine. The routine will usually begin by reading the status register, then doing whatever needs to be done, and finally returning to location RET, at which point the AC and carry bit are restored and the interrupted program is resumed.

□ *Interrupt priority*

What happens if several devices interrupt at the same time? In particular, what prevents all interrupting devices from asserting their device codes simultaneously when an interrupt query is issued by the CPU? Look again at the circuit in Figure 10.5, which shows the essential part of the interface interrupt circuitry. The device code is asserted onto the bus coincident with INTQ if the device has requested an interrupt AND if a line called INTP (interrupt priority) is HIGH. The signal called INTP is special, in that it is not

```
                                ;interrupt handling code -- vectored
Ø:        Ø                     ;value of PC at interrupt
1:        STA ACSAVE            ;stash current AC
          SHL
          STA CASAVE            ;save carry bit
          IQ                    ;code of interrupting device?
          ADD TABADR            ;add to first address of handler table
          STA TEMP              ;points into table
          LDA @TEMP             ;points to handler
          STA TEMP              ;prepare for indirect jump to handler
          JMP @TEMP             ;jump to handler via table entry

RET:      LDA CASAVE
          SHR                   ;restore carry bit
          LDA ACSAVE            ;restore AC
          IE                    ;enable interrupts again
          JMP @Ø                ;return to interrupted program

TABADR:   TABLE                 ;address of beginning of table
TEMP:     Ø                     ;temporary storage
ACSAVE:   Ø
CASAVE:   Ø

TABLE:    (starting address of handler for device Ø)
          (starting address of handler for device 1)
          (starting address of handler for device 2)
          (etc.)
```

Program 10.5

shared by devices on the bus, but rather is passed along *through* each device's interface circuit, beginning at the CPU and threading along through each interface (that's called a "daisy chain" in the colorful language of electronics). The rule for INTP hardware logic is as follows: If you have not requested an interrupt, pass INTP through to the next device unchanged; if you *have* interrupted, hold your INTP output LOW. In that way only the device electrically closest to the CPU that has interrupted will answer the interrupt query.

This sets up a "serial priority" chain, with devices closest to the CPU (i.e., highest priority) getting serviced frist; it also prevents chaos. In some computers there are several such priority chains, providing a hierarchy of parallel priorities, each of which has its own serial priority. The PDP-11, for example, has five parallel levels of "bus requests." Important point: When removing an interface card from a computer, it is important to jumper the INTP-type lines across the unused slots you create.

As we mentioned earlier, computers with more sophisticated hardware often provide automatic vectored interrupt, in which the CPU automatically performs a jump to a "vector area" of memory when an interrupt occurs, the exact location in that area depending on the device code of the interrupting device. The current location is stored in a stack, to allow interrupts within interrupts. With such computers it is not necessary to execute the equivalent of the MC-16's "interrupt query" command. This is sometimes called a "fully vectored interrupt."

10.12 Direct memory access

There are situations in which data must be moved very rapidly to or from a device; the classic examples are fast mass-storage devices like disk or tape and on-line data-acquisition applications such as multichannel pulse-height analysis. Interrupt processing of each data transfer in these examples would be awkward, and probably too slow. For example, data comes off a standard 800 bytes/inch 75 inches/second tape drive at the rate of one byte every 16μs (remember, a byte is 8 bits). With all the bookkeeping involved in handling an interrupt, data would probably be lost, even if the tape drive were the only interrupting device in the system; with a few such devices, the situation becomes hopeless. Devices like disks and tapes can't stop in midstream, so a method must be provided for reliably fast response and high overall word transfer rates. Even with peripherals with low average data transfer rates, there are sometimes requirements for short *latency* time, the time from initial request to actual movement of data.

The solution to these problems is direct memory access, or DMA, a method for direct communication from peripheral to memory. In many minicomputers the communication is actually handled by the CPU hardware, but that doesn't really matter. The important point is that no programming is involved; the data is moved to memory via the bus, without program intervention. The only effect on the executing program is some slowing down of execution time, because the DMA activity "steals" bus cycles that would otherwise be used to access memory for program execution. DMA involves more hardware complexity in the interface itself, and it should not be used unless necessary. However, it is useful to know what can be done, so we will describe briefly what you need to make a DMA interface. The details tend to be machine-dependent and complicated, so go to the manual of the particular computer for more details.

In DMA transfers the peripheral requests access to the bus via special lines included in the bus. The CPU gives permission. The peripheral then asserts memory addresses onto the bus and either sends or receives data, one location at a time, depending on the direction of transfer as specified by the peripheral on another bus control line. Data transfer is synchronized by control signals generated by both CPU and peripheral, and words like "handshaking" are used to describe the interlocked sort of communication that results. The interface is responsible for generating addresses (usually a block of successive addresses, generated with a binary counter) and keeping track of the number of words moved. The usual way to do this is to have a word counter and an address counter in the interface. These are loaded from the CPU, via programmed I/O, to set up the DMA transfer desired. On command from the CPU (via a command bit, written with programmed I/O), the interface makes its DMA request and begins to move the data. It notifies the program that it is finished by setting a status bit and requesting an interrupt, whereupon the CPU can decide what to do next.

Getting data or programs from disk is a common example of DMA transfer: The executing program asks for some "file" by name; the "operating system" (more about this soon) translates this into a set of programmed data OUT commands to the disk interface's control (or "command") register, word count register, and address register (specifying where to go on the disk, how many words to read, and where to put it in memory). Then the disk interface finds the right place on the disk, makes a DMA request, and begins moving blocks of data to the specified place in memory. When it's done, it sets bits in its status register to signify completion and then makes an interrupt. The CPU, which has meanwhile been executing other instructions (or possibly just waiting for the data from disk), responds to the interrupt, finds out from the status register of the disk interface that the data is now in memory, and then goes on to the next task. Thus, programmed I/O to the interface (the simplest level of I/O) was used to set up the DMA transfer, DMA itself was used for rapid transfer of data, and an interrupt was used to let the computer know the task was done. This sort of I/O hierarchy is extremely common, especially with mass-storage devices. Maximum DMA transfer rates are usually in the vicinity of a million words per second.

10.13 Synchronous versus asynchronous communication

The data IN/OUT protocol we described earlier constitutes a *synchronous* exchange of data; data is asserted onto or retrieved from the bus synchronously with strobing signals generated in the CPU. Such a scheme has the virtue of simplicity, but it does open the possibility of problems with long buses, since the long propagation delays you get mean that data is not asserted soon enough during a data IN operation for reliable transmission. In fact, with a synchronous bus the device sending the data never even knows if it was received!

This sounds like a serious disadvantage, but in reality it doesn't seem to matter; computer systems with synchronous buses seem to work just fine.

The alternative is an asynchronous bus, in which a data IN transfer will go something like this: The CPU will assert the device ADDRESS and a *level* (not a pulse) on a control line (call it "IN") that signifies data IN to the addressed device. The addressed device then asserts the DATA and a level signifying that the DATA is valid (call it "DATA READY"). When the CPU sees DATA READY, it latches the DATA and then releases its IN level. When the interface sees the IN line go LOW, it releases the DATA READY line and DATA lines. In other words, the CPU says "Give me data." The peripheral then says "OK, here it is." The CPU then says "OK, got it." And the peripheral finally says "Great! I'll go back to sleep again."

Asynchronous bus protocol allows any length (and speed) of bus and gives the communicating devices assurance that data is being moved. If a remote device is switched off, the CPU will know about it! Actually, that information is available via status registers with any kind of bus, and the chief advantage of asynchronous protocol is the flexibility of using any length of bus, bought at the expense of greater hardware complexity. (In computers that address memory with the same bus as used for I/O, an asychronous bus has the additional advantage of letting you mix memories that operate at different speeds.) Asynchronous buses are used in some minicomputers, e.g., the PDP-11, whereas synchronous buses are used in microcomputers and in the majority of minicomputers. In some cases a synchronous bus includes a READY line that can be pulled LOW to halt further bus activity in order to permit a slow operation to finish.

It should be kept in mind that a peripheral interfaced to a computer must communicate with its particular interface, which itself communicates with the CPU via the computer's bus. This peripheral-interface communication involves transfer of data, flags, strobe signals, etc., and usually consists of an asynchronous transfer of data to permit the peripheral to operate at its own speed. We will have more to say on this in the next chapter.

10.14 Connecting peripherals to the computer

Interface circuits are usually built on printed-circuit cards or Wire-Wrap cards (see Chapter 12) designed to plug right into the computer's mainframe. Computers generally contain a number of unused slots for just this purpose (or they can be "expanded" to accommodate extra cards), with power-supply voltages and bus signals already connected to some of the pins and with other uncommitted pins available for bringing out signals from the interface card to the peripheral being controlled by that particular card. Each computer has a standard card size (or sizes), ranging from 5 × 10 inches or so up to 15 inches square or larger. Each card has a set of plated edge connections so that it can be plugged directly into a multipin edge connector containing from 50 to 200 connections.

Commercially available interfaces designed to drive common peripherals such as tape drives, disks, and terminals are usually built on printed-circuit cards that are compatible with the particular computer's mainframe. Cables from the interface card to the peripheral either plug into sockets mounted right on the interface card or plug onto the unused pins of the backplane that connect to the interface card. In either case it is common to use flat *ribbon* cable, with some care being taken to prevent cross-coupling of strobing signals with data. One good method is to ground every other wire in the ribbon; another technique uses ribbon cable bonded to a flexible metal groundplane to reduce inductance and coupling, at the same time maintaining a nearly constant cable impedance. In both cases you can get nice multipin connectors that attach to the cable with one simple crimping operation. An alternative to ribbon cable is a cable made from a set of twisted pairs, each pair consisting of one signal line and one ground wire. Twisted-pair cable is available in many configurations, including a nifty ribbonlike flat cable in which there is a flat untwisted region every 18 inches or so for easy connection to crimp-on connectors of the type used for ordinary ribbon cable. Because of the strobed data-transfer protocol used on computer buses and in connections to peripherals, it generally isn't necessary to use twisted pairs for *all* signal lines, just for the synchronizing pulses and other strobing or enabling lines. Suitable terminations and driver/receiver combinations should be used for long lines, as discussed in Sections 9.15–9.17.

Custom interfaces are best handled in the same way, either laying out large circuit-boards or using one of the general-purpose interfacing cards available commercially from companies such as Douglas or MDB or from the computer manufacturer. These blank cards have places for ICs and other components, and they come in solder and Wire-Wrap styles (more in Chapter 12). Some of them include built-in circuitry to handle bus communication, including interrupts and even DMA. With these boards you have the choice of bringing small cables directly off the card or making connections to the peripheral via backplane pins; with some computers backplane connection is inconvenient, so direct cabling to the card must be used, preferably with flat cable connectors.

It is possible to design an interface that connects to the computer via an external I/O bus cable. Nearly all computers provide for bus extension outside the mainframe. In such cases it is particularly important to follow the manufacturer's recommendations on cable type, length, and termination, since you may cause trouble elsewhere in the computer if I/O bus signals become degraded.

Another possibility is to build an interface that resides partly in the computer and partly outside. In such cases the interface circuitry that goes in the computer will probably be a simple digital input/output "port," as suggested in Figure 10.6.

The interface in the computer mainframe latches a word during data OUT, setting a DATA AVAILABLE flag and soon thereafter receiving a DATA ACCEPTED flag from the external part of the interface. Data IN is handled similarly, with a flag to indicate that the buffer can accept data and an input STROBE line to enable the latch when data is valid. The cable connecting the two parts of

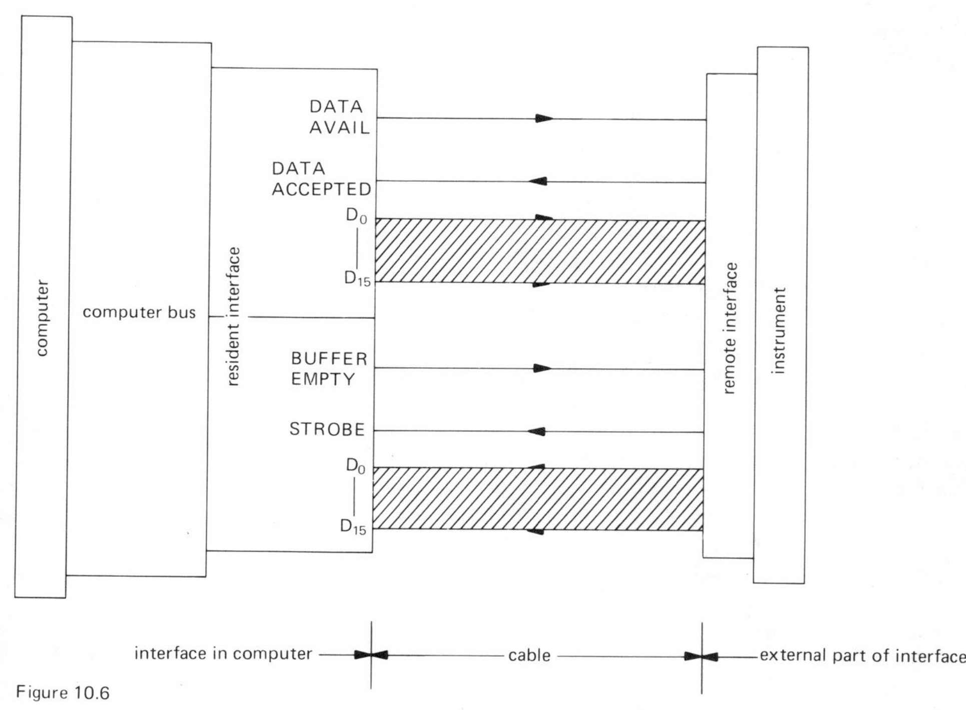

Figure 10.6

the interface is considerably less than a full I/O bus cable, particularly if less than a full 16-bit word is needed. This sort of scheme may be particularly useful for interfaces that handle low-level analog signals, since the noise-susceptible linear circuitry can be kept away from the general roar of digital interference present in the computer; this also allows you to pay careful attention to maintaining clean analog signal ground lines.

One further note on interconnections: When connecting to a peripheral device via backplane pins, it may be awkward to plug connectors directly onto the backplane itself. In that case it may be a good solution to mount a connector somewhere nearby on the computer housing, then wirewrap a set of wires from the backplane pins you're using to the connector. The peripheral will then be connected by a twisted-pair cable plugged into the connector. The so-called type D subminiature connectors (Fig. 1.98) are often used for this.

SOFTWARE SYSTEM CONCEPTS

In this section we will discuss some general aspects of small-computer programming, since a knowledge of computer interfacing is of limited value without an understanding of the hierarchy of programs that actually make the computer come to life. In particular, we would like to discuss the important areas of programming, operating systems, files, and use of memory. It is easy to get carried away admiring the beauty of computer hardware and underestimate the importance of good software. Software is what makes the computer fly, and a good operating system and package of "utilities" can make all the difference. Computer manufacturers spend prodigious sums of money developing software to go with their hardware, and this attention to software support has contributed to keeping the minicomputer companies competitive in the face of widespread availability of inexpensive microcomputers.

Following our discussion of software and systems, we will end the chapter with a return to input/output concepts, with emphasis on bit-serial I/O, specifically RS-232C ASCII.

10.15 Programming

Assembly language

As we mentioned earlier in the chapter, the computer's CPU is designed to recognize certain instructions and carry out the appointed tasks. It is extremely rare to program directly in this machine language (you might do it when debugging via front-panel switches and lights, or when checking for a hardware fault). Instead, you write programs in a mnemonic assembly language (like the MC-16 programming examples), which a program called an assembler converts into machine language. Assembly language is very close to machine language; each instruction is converted directly into one line or a few lines of machine code. Assembly-language programming produces the most efficient code and allows you to do special operations, such as I/O commands, that are not accessible from higher-level languages. But it is tedious programming, as the MC-16 examples demonstrate, and for most computing jobs (especially those involving plenty of numerical computation) it pays to use a compiled or interpreted high-level language, such as FORTRAN or BASIC, with calls to assembly-language routines only where necessary.

Compilers

FORTRAN, ALGOL, PL1, COBOL, and PASCAL are popular examples of high-level compiled programming languages. You write a program with algebraic types of commands. For instance:

X = (−B + SQRT (B*B − 4*A*C))/2A

This is called the *source* code, which a program called a *compiler* converts into assembly code. From there it's business as usual, with the assembler converting that intermediate assembly language to machine language. In some cases the compiler includes a built-in assembler, for direct conversion to machine-language code. In another variation, typified by PASCAL, the compiler generates an intermediate code (called p-code, in the case of PASCAL), which is then run by an interpreter-like run-time package (more about this later).

Loaders and libraries

The assembler produces machine code (well, almost; it's actually called "relocatable machine code") from the assembly code produced by the compiler and from separate subroutines written in assembly code. In addition, there are usually routines needed by the particular commands in the high-level program. For example, a FORTRAN program might need function subroutines like SQRT or a host of I/O subroutines needed for a READ or WRITE statement. A program called a "linker" (or "relocatable loader") handles the bureaucratic nightmare of getting the appropriate subroutines (in relocatable form) from a "library," then rigging up all the linking jumps and references so the whole mess fits together in memory. It is the linker's job to put final numerical values into the memory references and variable addresses of the assembled code, and it can do this only when it knows how long each program is and which program calls which. That's why the code produced by the assembler must be in relocatable form, as must the assembled subroutines that sit in the various libraries (there are usually several, e.g., a FORTRAN library, a system library, and perhaps a home-grown library of useful subroutines such as I/O handlers, random-number generators, etc.). In some microcomputer systems you don't have relocatable loaders, and you have to assign memory areas for each subroutine, or assemble the entire program simultaneously. Needless to say, this is unspeakably primitive.

Editors

How do you actually write and enter the programs to be compiled, assembled, and eventually run? Life has improved considerably since the days when all programs were punched on cards, then gobbled up by a huge card reader connected to the computer. A program called an "editor" allows you to type in commands at a terminal, creating a "file" of text. Good editors let you see (on a screen) what's in the file as you type, and they provide nifty commands that let you search for words, change text, move stuff around, etc. An editor doesn't know, or care, what you are writing; it could be a program, a sonnet, or a book (the one you are reading is an example). It just creates the text file according to your keyboard instructions. Afterward you may choose to have it printed out, compiled, assembled (or published). You can recall the text file and edit it later, if you want, to make program changes, etc. A really first-class editor is worth its weight in gold.

Interpreters

Languages like BASIC and APL work in a different manner than the compiled languages mentioned earlier. Instead of compiling an assembly-language program from the source program, they "look at" the statements and execute appropriate computer instructions. With such an *interpreter* program, the source code is kept intact and is easily modified. In fact, languages like BASIC include a primitive editor for program entry and modification.

In general, interpreted languages run much slower than compiled languages. However, since there's no compilation, assembly, or linking, there's no delay after entering a program before it can run, and the resulting simplicity of commands makes an interpreter easier to use. Interpreters are favorites in systems without rapid mass storage (usually a disk), since there's no need to get editors, compilers, assemblers, loaders, and the various intermediate outputs from mass storage into the computer's memory, and vice versa. BASIC, for example, is the darling of the microcomputer hobbyist.

10.16 Operating systems, files, and use of memory

Operating systems

As you might guess from the preceding discussion, you frequently want to run different programs at different times, trading data back and forth between them. For instance, in writing and running a program you begin by running the editor program, creating a text file from the keyboard. After temporarily storing the text file, you bring in the compiler program and compile the stored text file to form an assembly-language file. You store that, bring in the assembler, and produce a relocatable machine-language file from the stored assembly-language file. Finally, the linker combines the relocatable machine code with other assembled subroutines and library routines to produce the executable machine-language program, which (at last!) you run. For all these operations you need some sort of super program to juggle things around, getting programs from disk, putting them into memory, and transferring control to the relevant programs. In addition, it would be nice if each program didn't have to contain all the commands necessary to do disk reads and writes (including interrupt handling, loading of status and command registers, etc.).

These are some of the tasks of the *operating system,* a vast program that oversees the loading and running of user programs (the ones you write) and utility programs (editor, compiler, assembler, linker, debugger, etc.), as well as the handling of I/O and interrupts, and file creation and manipulation. The system includes a *monitor* for operator communication (you tell it to run the editor, compile a program, or run a program) and many "system calls" that permit a running program to read or write a line of text from some device, find out the time of day, swap control to another program, bring in a program "overlay," etc. Good systems handle all the busywork of I/O handling, including "spooling" (the buffering of input or output data so that the program can run at the same time that data is being read or written to some device). When running under a system, a user program doesn't have to worry about interrupts; an interrupt is taken care of by the system, and it affects the running program only if it wants to take part in the handling of a particular device's interrupts. The whole business of successful "time sharing" (using one computer to handle many users at once) is system programming at its finest.

Files

Data stored by a system on mass storage (disks, tape) is organized into *files.* Since text, user programs, utility programs (e.g., editor, assembler, compiler, debugger, linker), libraries, etc., are stored in the same ways, they also constitute files. Although the mass-storage medium may be divided into physical blocks or sectors of well-defined size (256 words/block is common), the files themselves may have any length. The operating system takes care of block/sector addressing, etc.; it gets the data you want, if you know the file name. There are all sorts of interesting details having to do with file organization that we don't have space to describe here. What is important is to understand that all those programs (editor, compiler, etc., as well as user source text, compiled programs, and even data) reside on some mass-storage device as named files, and the system can get them for you. In the normal course of its duties, the system does enormous amounts of file handling.

Use of memory

Files are stored in some mass-storage device, but a program must reside in memory while being executed. Simple stand-alone programs (those run without benefit of a system) can be loaded almost anywhere in memory, but in a computer of any complexity there are special areas reserved for special functions. For instance, the bottom part of memory (lowest-numbered locations) is often used as a vector area for interrupts or as a "page-zero" area for direct addressing with a single-byte address. Some computers have special "auto-incrementing" locations. And in some machines an area of memory addresses is actually used to address peripherals ("memory-mapped I/O"). Systems make additional demands. The system itself often uses the top of memory (the highest-numbered locations) for most of the system, including its buffers and stacks as well as the system program itself; it may also use an area at the bottom of memory. When operating under a system, the allocation of memory for user programs will be handled by the system. It is essential to realize that memory allocation isn't entirely arbitrary, especially if you get involved in DMA interfaces, in which data will be moved to memory under external hardware control.

DATA COMMUNICATIONS CONCEPTS

A small computer system will usually be configured with some mass-storage devices, such as tapes or disks, and some "hard-copy" or interactive devices, such as alphanumeric terminals, printers, plotters, etc. All these devices are commercially available in great variety, and they can be interfaced to nearly any computer. Because each computer has a unique bus structure and interfacing protocol, it is necessary to construct (or buy) interfaces for the particular computer involved. The interface card is usually designed to fit into the computer's mainframe, with cables going from the interface to the peripheral itself.

Incompatibility

As luck would have it, this general lack of compatibility among computers extends to the peripherals themselves. You can't hook a tape drive to a disk interface, or a terminal to a plotter interface, etc. To make matters worse, the peripherals offered by different manufacturers may use different signals and data-transfer conventions, and in general they won't be "plug-compatible" (you will see some pleasant exceptions to this generally bleak picture shortly). To maximize performance, different peripherals transfer their data to and from the interface differently, contributing to the incompatibility. For instance, a magnetic tape moves words in parallel binary format (all bits of one word at a time) for highest speed, and the corresponding interface uses DMA transfer of the data to or from memory. By contrast, a keyboard terminal nearly always uses a standardized alphanumeric bit-serial format, moving each character as a coded string of bits on a single line. The corresponding interface invariably communicates with the CPU by interrupt and programmed I/O. A new high-speed parallel-bus standard (IEEE 488-1975, also known as GPIB or HPIB) could lead to standardized interfacing conventions for high-speed peripherals, but for the time being the situation with high-speed peripherals with parallel data transfer is one of general incompatibility (with the possible exception of magnetic-tape drives, which all use nearly the same conventions), with interfacing being handled on a case-by-case basis.

10.17 Alphanumeric codes and serial communication

As mentioned earlier, alphanumeric communication between devices of moderate speed and a computer is most frequently done using the 8-bit ASCII code (American Standard Code for Information Interchange), with bit-serial transmission over a single line. There are several other alphanumeric codes in use, e.g., the 8-bit "BCD code" EBCDIC (pronounced "ebsedick," accent on the first syllable) used primarily on the larger computers, the Hollerith code used for punched cards, and the 5-bit Baudot code that used to be standard on teleprinters, but that is now becoming a dusty museum piece. In addition, there are several alphanumeric codes used exclusively for communication with certain I/O devices, e.g., the IBM Selectric code (which exists in two varieties, "correspondence" and "BCD"). In these cases you do software code conversion on input and output data.

ASCII code

ASCII is by far the most common alphanumeric code; it is universally used in minicomputers and microcomputers. Figure 10.7 presents a listing of the 7-bit codes (the 8th bit is either used for parity or set to zero). The tabulation is arranged by vertical columns (or "sticks") according to the most significant 3 bits, with 16 rows for the least significant 4 bits. For example, the letter M (upper case) is 1001101 (binary), 115 (octal), or 4D (hex). From this convenient arrangement you can see some nice regularities: The code for a lower-case letter is just the corresponding code for an upper-case letter with bit 6 set to 1; the ASCII value for an integer is just the integer plus 48 (decimal); most punctuation is in stick 2; the nonprinting "control characters" comprise the first 32 ASCII characters. The latter are used to control printing, or they can be interpreted as commands by programs that otherwise expect to receive alphanumeric characters, e.g., text editors. Among the frequently used control characters are the

<table>
<tr><td colspan="5">Bits b7 b6 b5</td><td>0 0 0</td><td>0 0 1</td><td>0 1 0</td><td>0 1 1</td><td>1 0 0</td><td>1 0 1</td><td>1 1 0</td><td>1 1 1</td></tr>
<tr><td>b4</td><td>b3</td><td>b2</td><td>b1</td><td>COLUMN / ROW</td><td>0</td><td>1</td><td>2</td><td>3</td><td>4</td><td>5</td><td>6</td><td>7</td></tr>
<tr><td>0</td><td>0</td><td>0</td><td>0</td><td>0</td><td>NUL</td><td>DLE</td><td>SP</td><td>0</td><td>@</td><td>P</td><td>`</td><td>p</td></tr>
<tr><td>0</td><td>0</td><td>0</td><td>1</td><td>1</td><td>SOH</td><td>DC1</td><td>!</td><td>1</td><td>A</td><td>Q</td><td>a</td><td>q</td></tr>
<tr><td>0</td><td>0</td><td>1</td><td>0</td><td>2</td><td>STX</td><td>DC2</td><td>"</td><td>2</td><td>B</td><td>R</td><td>b</td><td>r</td></tr>
<tr><td>0</td><td>0</td><td>1</td><td>1</td><td>3</td><td>ETX</td><td>DC3</td><td>#</td><td>3</td><td>C</td><td>S</td><td>c</td><td>s</td></tr>
<tr><td>0</td><td>1</td><td>0</td><td>0</td><td>4</td><td>EOT</td><td>DC4</td><td>$</td><td>4</td><td>D</td><td>T</td><td>d</td><td>t</td></tr>
<tr><td>0</td><td>1</td><td>0</td><td>1</td><td>5</td><td>ENQ</td><td>NAK</td><td>%</td><td>5</td><td>E</td><td>U</td><td>e</td><td>u</td></tr>
<tr><td>0</td><td>1</td><td>1</td><td>0</td><td>6</td><td>ACK</td><td>SYN</td><td>&</td><td>6</td><td>F</td><td>V</td><td>f</td><td>v</td></tr>
<tr><td>0</td><td>1</td><td>1</td><td>1</td><td>7</td><td>BEL</td><td>ETB</td><td>'</td><td>7</td><td>G</td><td>W</td><td>g</td><td>w</td></tr>
<tr><td>1</td><td>0</td><td>0</td><td>0</td><td>8</td><td>BS</td><td>CAN</td><td>(</td><td>8</td><td>H</td><td>X</td><td>h</td><td>x</td></tr>
<tr><td>1</td><td>0</td><td>0</td><td>1</td><td>9</td><td>HT</td><td>EM</td><td>)</td><td>9</td><td>I</td><td>Y</td><td>i</td><td>y</td></tr>
<tr><td>1</td><td>0</td><td>1</td><td>0</td><td>10</td><td>LF</td><td>SUB</td><td>*</td><td>:</td><td>J</td><td>Z</td><td>j</td><td>z</td></tr>
<tr><td>1</td><td>0</td><td>1</td><td>1</td><td>11</td><td>VT</td><td>ESC</td><td>+</td><td>;</td><td>K</td><td>[</td><td>k</td><td>{</td></tr>
<tr><td>1</td><td>1</td><td>0</td><td>0</td><td>12</td><td>FF</td><td>FS</td><td>,</td><td><</td><td>L</td><td>\</td><td>l</td><td>|</td></tr>
<tr><td>1</td><td>1</td><td>0</td><td>1</td><td>13</td><td>CR</td><td>GS</td><td>-</td><td>=</td><td>M</td><td>]</td><td>m</td><td>}</td></tr>
<tr><td>1</td><td>1</td><td>1</td><td>0</td><td>14</td><td>SO</td><td>RS</td><td>.</td><td>></td><td>N</td><td>^</td><td>n</td><td>~</td></tr>
<tr><td>1</td><td>1</td><td>1</td><td>1</td><td>15</td><td>SI</td><td>US</td><td>/</td><td>?</td><td>O</td><td>_</td><td>o</td><td>DEL</td></tr>
</table>

0/0	NUL	Null	1/0	DLE	Data Link Escape
0/1	SOH	Start of Heading	1/1	DC1	Device Control 1
0/2	STX	Start of Text	1/2	DC2	Device Control 2
0/3	ETX	End of Text	1/3	DC3	Device Control 3
0/4	EOT	End of Transmission	1/4	DC4	Device Control 4
0/5	ENQ	Enquiry	1/5	NAK	Negative Acknowledge
0/6	ACK	Acknowledge	1/6	SYN	Synchronous Idle
0/7	BEL	Bell	1/7	ETB	End of Transmission Block
0/8	BS	Backspace	1/8	CAN	Cancel
0/9	HT	Horizontal Tabulation	1/9	EM	End of Medium
0/10	LF	Line Feed	1/10	SUB	Substitute
0/11	VT	Vertical Tabulation	1/11	ESC	Escape
0/12	FF	Form Feed	1/12	FS	File Separator
0/13	CR	Carriage Return	1/13	GS	Group Separator
0/14	SO	Shift Out	1/14	RS	Record Separator
0/15	SI	Shift In	1/15	US	Unit Separator
			7/15	DEL	Delete

Figure 10.7.
ASCII code. (Reproduced with permission from American National Standard X3.4-1977, copyright 1977 by the American National Standards Institute. Copies may be purchased from the American National Standards Institute, 1430 Broadway, New York, New York, 10018.)

following: NUL (null), a character of all zeros used to delimit character strings, among other things; CR (carriage return) and LF (line feed), used to begin a new line of printing; FF (form feed), used to begin a new page; ESC (escape), often used as a command delimiter; ETX (end of text, affectionately called "control C"), which many operating systems interpret as a command to abort a running program. ASCII keyboards usually have a key marked CTRL. When an alphabetic key (columns 4 and 5) is struck with CTRL held down, the keyboard generates the corresponding control character from columns 0 and 1. In addition, it is common to have separate keys that generate CR, LF, ESC, and (sometimes) NUL.

Unfortunately, ASCII doesn't provide for subscripts, exponents, or any Greek letters. It is unfortunate that there aren't characters for pi, mu, omega, and the degree symbol, since these crop up frequently in technical writing. It is possible, of course, to use a control character to indicate a change of font or alphabet. A special word-processing program can then interpret subsequent ASCII characters differently, printing them on a "hard-copy" device (printer or plotter) with any alphabet you wish.

Bit-serial transmission

ASCII (or any other alphanumeric code) can be transmitted as a parallel 8-bit group (8 separate wires) or as a serial string of 8 bits, one after the other. An ASCII keyboard itself will usually generate an 8-bit parallel TTL-compatible output. For reasons of convenience, serial format is usually used to send ASCII to and from terminals. Devices called MODEMs (modulator demodulator) convert a serial-bit string to audio tones (and vice versa) for transmission via telephone lines (with an acoustic coupler, an ordinary telephone handpiece can be used for transmission). For that application, serial format makes the most sense. Serial transmission has a standard bit-transmission protocol and fixed bit rates: The data is sent *asynchronously,* with a *start* bit and a *stop* bit (sometimes two) attached at the ends of each 8-bit ASCII character, forming a 10-bit group. The sender and receiver use a fixed bit rate, which may be 110 (2 stop bits), 150, 300, 600, 1200, 2400, 4800, 9600, or 19,200 baud (= clock periods per second). Figure 10.8 shows the idea.

When no information is being sent, the transmitter sits in the "marking" state (the language comes from the teletypewriter days, with "mark" and "space"). Every character begins with a START bit, followed by the eight ASCII bits and a final STOP bit; the latter must be at least one clock period long (two, for transmission at the 110 baud rate), but may extend any amount longer. At the receiving end a UART (or equivalent device, see Sections 8.27 and 11.9) operating at the same baud rate synchronizes to each 10-bit group, generating 8-bit parallel data groups from the input serial string. By resynchronizing on the START and STOP bits of each character, the receiver doesn't require a highly accurate clock; it only has to be accurate and stable enough for the transmitter and receiver to stay synchronized to a fraction of a bit period over the time of one character, i.e., an accuracy of a few percent. The STOP bit terminates the character and is the resting state if no new characters are sent immediately. Programmable baud-rate generators (i.e., programmable dividers) are available that generate any of the standard

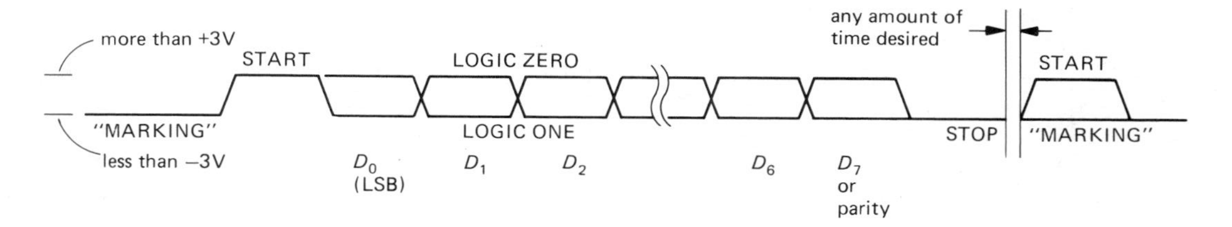

Figure 10.8
RS-232C serial data-byte timing waveform.

baud rates from a single clock input frequency, with the output baud rate selected by a binary input code.

Half and full duplex

Alphanumeric terminals (a terminal has both a keyboard and display or printer) can be connected to a computer in either "full duplex" or "half duplex" mode. In full duplex mode the keyboard and printer act as completely independent devices, each connected to the computer by a separate serial ASCII line. This is the usual setup, with the computer "echoing" each character as it is received. In half duplex mode the keyboard sends to the receiver and simultaneously prints at the terminal. Characters sent by the computer are printed when received at the terminal. With half duplex the computer does not echo received characters.

Current loop and RS-232C

The actual serial ASCII signals can be sent in one of two ways. The method used with teleprinters, and still available on most terminals as an option, consists of switching a 20mA (or sometimes 60mA) current at the selected baud rate. The more popular method consists of sending the data as bipolar EIA RS-232C standard levels, as discussed in Section 9.17. Even the connector type (25-pin type D subminiature) and pin connections are standardized, making RS-232C devices all plug-compatible. In the jungle of incompatible devices that is the computer industry, that's really something!

10.18 Numeric data interfacing

Example: Hardware data packing

When you actually begin designing an interface to bring in numbers from some external instrument, you discover that life is full of variety; different instruments output their data in different ways. Instruments with just a few digits (or bits) of resolution (e.g., a 3½-digit DVM) usually give you all the digits at once (character-parallel, bit-parallel), whereas something like an 8-digit frequency counter is likely to give you one digit after another (character-serial, bit-parallel), often at the display's internal refresh rate. To illustrate the sort of circuits involved, Figure 10.9 shows an example of an interface to get data from an 8-digit counter with character-serial output.

This is a complete interface to the MC-16 bus, including a status flag, interrupt capability, and selectable device code. The action begins at the lower left, where the counter is busy putting out successive digits, their addresses (0–7), and a STROBE pulse when the data is valid. The counter goes from the least significant digit (LSD) to the most significant digit (MSD), so a complete output cycle ends with receipt of the MSD (digit 7). The eight 74LS173s (4-bit D registers with three-state outputs) latch the successive digits, being driven in parallel and separately clocked via the decoded digit address. Note the use of a strobed 1-of-8 decoder (74LS138) to generate the digit clocking signals from the address and strobe.

The counter output is thus latched in the eight registers, with the outputs connected as two 4-digit groups (16 bits each). The computer can bring in all eight digits with two data IN commands, one from the device code set in the set of six switches (a "DIP switch," eight SPST switches in the same package as a standard IC), and one from the next higher device code. Note the address decoding: The 8131 6-bit comparator generates a LOW output when the high-order six address bits match: its output, combined with a data IN pulse, enables the 74LS138 1-of-8 strobed decoder, which decodes the low-order two address bits to generate the separate data IN enabling pulses corresponding to successive device codes. This is a common method of handling address decoding, since it is common to assign a few contiguous device codes to the various registers of a single interface.

The status flag (READY flip-flop) is set when the last digit of each group is received from the counter; it can be read with a data IN from device $DC + 2$, where DC is the device code set with the DIP switch. READY also initiates an interrupt request, with the usual interrupt hardware. Note that the same levels generated by the device select switches are used to assert the device code onto the DATA lines during an interrupt

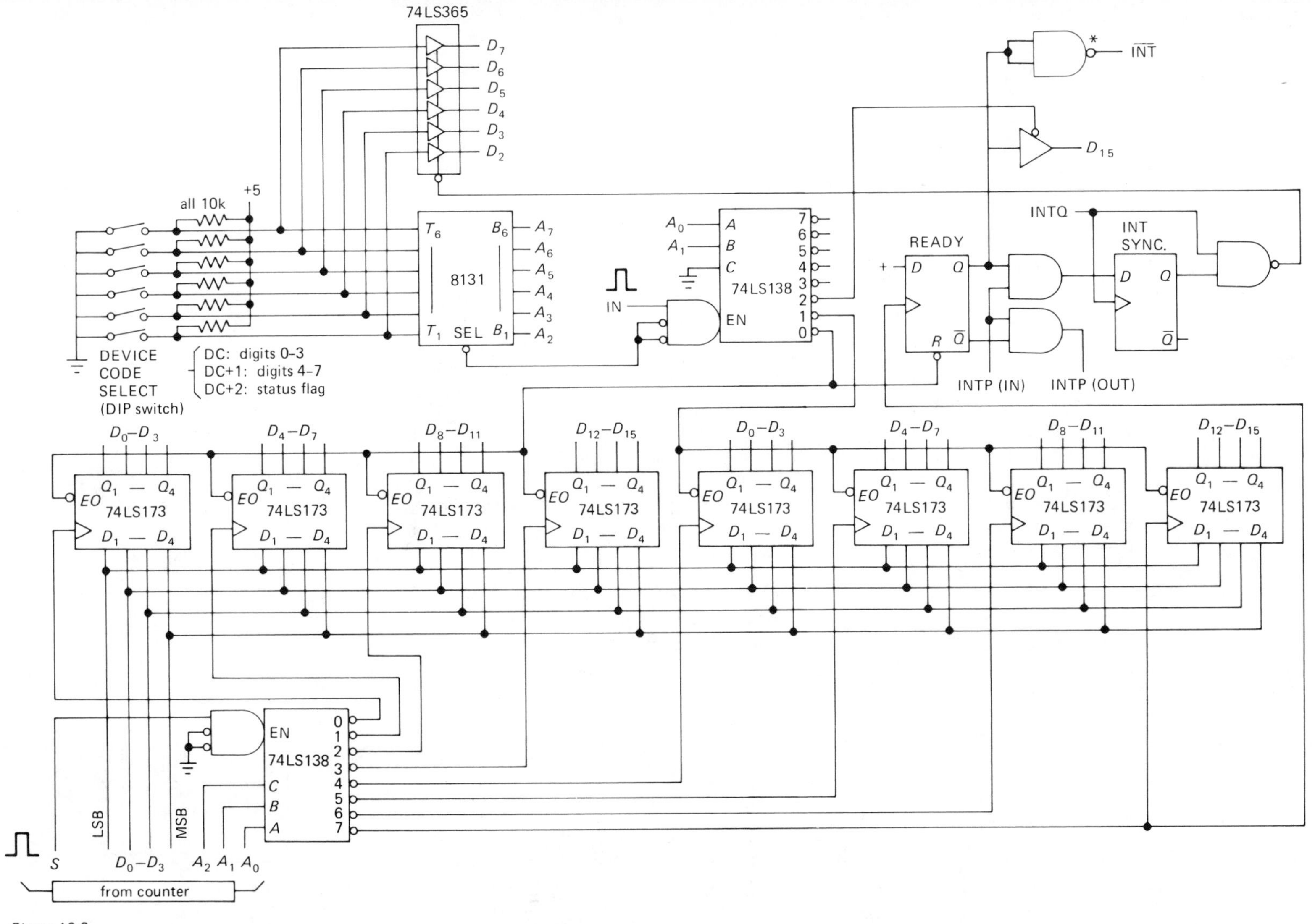

Figure 10.9

DVM interface: serial digits to parallel words.

query, via the 74LS365 three-state buffer. This arrangement is considerably more flexible than the "hard-wired" interrupt address assertion and decoding circuitry shown in the interrupt example earlier (Fig. 10.5).

This interface is an example of "packing" data, the process by which several numbers are stuffed into one computer word. If the "numbers" happen to consist of single bits, you can pack 16 of them into a 16-bit word. This isn't as crazy as it sounds; for the highest I/O throughput rate of a periodically sampled "hard-clipped" waveform, you would do exactly that. Of course, if speed is not important, the simplest thing is to bring in the data with the least hardware and do the packing and conversion in software. In the preceding example, for instance, you might latch and transfer to the computer one digit at a time if you can be sure that the latency time of the computer is short enough so that no digits will be lost.

Number-format conversion

In the preceding example the two words brought in are not in the computer's internal binary number format; they're really BCD, packed four digits per word. To do meaningful computation, it is necessary to convert them into an integer or a floating-point number. Let's take a look at the usual number formats used in computers (Fig. 10.10), a subject we touched on briefly at the beginning of Chapter 8. In this figure we've grouped the bits in 16-bit words, the most popular word size in minicomputers; in machines with other word sizes, say 8 bits, you just use twice as many computer words to store each number. Integers are represented as 2's complement 16-bit groups. Thus the most significant bit (MSB) tells the sign. A 16-bit integer can represent numbers from $-32,768$ to $+32,767$. For larger numbers you sometimes see "double-precision" (32-bit) integers (they would be standard on a computer with 32-bit word), with an extra 16 bits added on. With 32 bits you can represent integers in the range $\pm 2.15 \times 10^9$.

Floating-point numbers have several common representations, the most popular using 32 bits: 1 sign bit, 7 bits of exponent, and 24 bits of fraction. In this representation the exponent tells the power of 16 that the fraction should be multiplied by. The exponent is "biased" (offset binary) so that the exponent field 1000000_2 corresponds to an exponent of 0; exponents thus go from -64 to $+63$. Numbers of either sign from about 5.4×10^{-79} to 7.2×10^{75} can be represented with this format. Since the exponent field of the number specifies a power of 16 (rather than 2), the fraction may have up to three leading zeroes; i.e., a normalized fraction has a nonzero most significant hex digit.

EXERCISE 10.1

In order to understand the meaning of this last statement, write out the floating-point representation of the number 1.0. Now

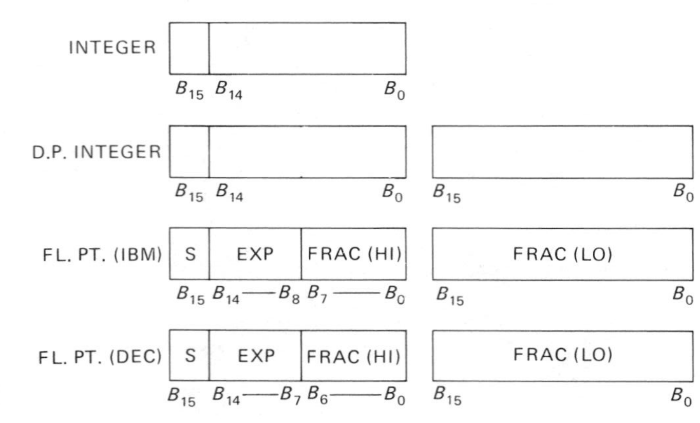

Figure 10.10

Computer number representation.

write the next smaller number that can be represented in this format.

EXERCISE 10.2

Show that the range of numbers that can be represented in this floating-point format is as stated. Hint: The fraction portion of the number must be "normalized."

There is another common floating-point format (typified by the PDP-11 and other DEC machines) used with 32-bit words: 1 sign bit, 8 bits of exponent, and 23 bits of fraction. In this format the exponent is again offset binary, this time specifying powers of 2; numbers of either sign from about 2.9×10^{-39} to 1.7×10^{38} can be represented. An interesting feature of this representation is the "hidden bit" of precision: A normalized fraction always has a nonzero MSB; therefore it would be redundant to display it in the word. Consequently, the MSB of the fraction field is actually the second most significant bit of the represented number's fraction. This floating-point number representation sacrifices some dynamic range to give improved accuracy, compared with the more popular power-of-16 representation. In either representation double precision is achieved by adding 32 additional bits of fraction, with no change in the exponent or sign field.

EXERCISE 10.3

Show that the range of numbers that can be represented in this floating-point format is as stated. See hint in Exercise 10.2.

Depending on the type of input data, the number of digits, its range of variation, etc., it may be best to convert the incoming data to floating point (greatest dynamic range) or to integer (best resolution) or to do some other sort of numerical massaging (e.g., taking differences from the average value). This might be done in the particular device's software "driver," the section of program that handles the actual input of data. In this sense the software cannot be optimized without an understanding of the hardware and what its data means. Just another reason why it is important to know your way around the wonderful world of electronic hardware!

ADDITIONAL EXERCISES

(1) Design an interface that loads a 16-bit parallel-load counter (e.g., four 74LS193s) on data OUT and sends its count back on data IN. The counter is clocked by an external TTL input pulse train. Provide a DIP switch for device code selection.

This simple peripheral is enough to make a "multichannel scaler." The computer is programmed to write the contents of successive memory locations to the counter at the beginning of each time interval and read them back at the end, going on to the next memory location. In that way a histogram of pulse arrivals versus time can be built up; repetitive signals can be averaged by repeating the whole process synchronously with the signal's period (see Section 14.13).

(2) An algorithm known as the Cooley-Tukey fast Fourier transform (FFT) enormously reduces the time to compute a discrete Fourier transform on a set of data, and it is frequently used to search for periodicities in scientific data. The FFT performs a Fourier transform *in place,* i.e., it replaces the data array by the array of frequency components. However, there's one peculiarity: The frequencies are not in order; they're scrambled, and it is necessary to swap the contents of each memory location X with the contents of the memory location that is X read *backward* in binary (i.e., MSB $\leftrightarrow$ LSB, etc.).

To speed up this process (computers have no simple way to read binary numbers backward), design a peripheral for an 8-bit computer that will generate the reversed version of an 8-bit number that the CPU sends to it. The peripheral is to have device code 13 (decimal) and use programmed I/O only. (This is an easy problem.)

(3) Design the hardware for a "software" 16-bit successive-approximation A/D converter. The computer outputs a 16-bit trial number to your device via data OUT; it then reads from your device (via data IN) a bit telling whether or not this trial number ex-

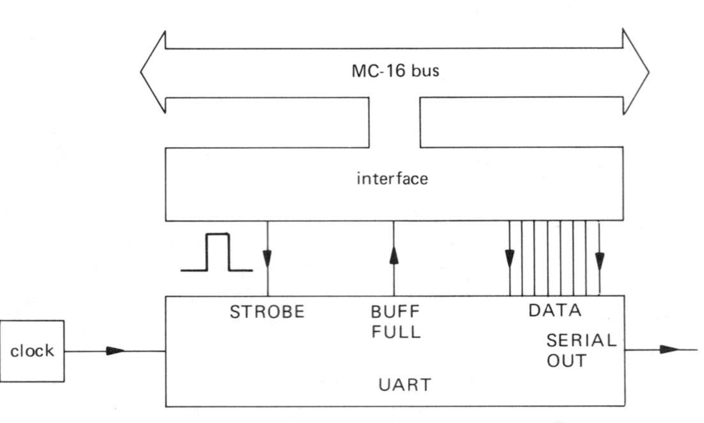

Figure 10.11

ceeds the analog input level. The computer software (see next exercise) generates the appropriate set of trial numbers, based on the results of preceding trials.

Some details: Use device code 99 (decimal). Data OUT should start a monostable with pulse width as long as the time the D/A takes to convert; at the end of that pulse a status flag is set telling the computer it's OK to read the output of the comparator. Put the status bit in the MSB of the status register, and put the comparator output in bit 7 (MSB of lower-order byte), with a 1 signifying that the D/A output exceeds the analog input.

(4) Now for the fun. Write an MC-16 program to run your A/D converter.

(5) Design a programmable timer. It interrupts (and sets a status bit) every *n* milliseconds, where *n* is set by an 8-bit binary number sent to the peripheral via a data OUT command; i.e., time intervals from 1ms to 255ms can be programmed.

(6) Now add a programmable scaler circuit to the timer so that you can specify what factor of 10 to multiply the period by; i.e., an additional two bits specifies an "exponent," giving you the capability of multiplying the programmed period by 1, 10, 100, or 1000.

(7) Design a serial ASCII output peripheral. It should accept an 8-bit character from the CPU via programmed I/O, then send it out using a UART. The peripheral should have two status bits; one (BUSY) tells if it is in the process of sending a character, and the second (DONE) tells if it has completed sending a character. DONE should also initiate an interrupt. Assume that the UART has an 8-bit parallel data input, a STROBE input, and a BUFFER FULL output flag (set when it is in the process of sending something out), as well as a clock input. Figure 10.11 shows the idea.

(8) Write a subroutine ("WRL" – write a line) to drive your peripheral from the preceding exercise. It should be passed (in the AC) the address of the first character to be sent; the subroutine should then output characters from successive 8-bit memory locations and return to the calling program after finding and typing a carriage return (after which it should append a "line feed" output character).

(9) It is possible to create a stack in a computer that does not have special stack manipulation instructions. In particular, if we choose a particular memory location (call it POINT) to serve as a stack pointer, and if we initially load it with the first address in memory where the stack is to reside (actually, that address minus one), then the MC-16 instruction pair

```
ISZ   POINT
STA   @POINT
```

"pushes" the contents of the accumulator onto the stack, and the instruction pair

```
LDA   @POINT
DSZ   POINT
```

''pops'' the last number from the stack into the accumulator. ***(a)*** Explain how these instructions work and how such a stack could be used to store return addresses and variables during subroutine calls, in order to make them re-entrant. ***(b)*** Explain why the two instructions of each pair must be in the order shown, and not interchanged, if the stack is to function properly during interrupts. ***(c)*** What would happen if you were to pop a stack more times than you had pushed it?

MICROPROCESSORS

CHAPTER 11

A microprocessor is an integrated-circuit computer, a computer on a chip, and it typifies the most advanced kind of integrated circuit available. As we discussed in Chapter 10, these microprocessors are genuine computer central processing units (CPUs), including an arithmetic unit, several registers, hardware stack implementation, on-chip memories (both RAM and ROM), and analog input/output (I/O). Not all microprocessors include memory and I/O, of course; some have been optimized for computational speed and elegance, whereas others have been designed for simple applications where a minimum of "support chips" is desirable. The most inexpensive microprocessor chips cost as little as $2 in quantity, and considerable computational power is available for under $10. We will continue to use the terms *microprocessor* and *microcomputer* as defined in Chapter 10. A microprocessor is the CPU chip itself; a microcomputer is a computational system built around a microprocessor, usually including mass storage (disks), terminals, printers, etc.

Most of the interfacing and programming concepts we introduced in Chapter 10 in connection with minicomputers are directly applicable to microprocessors, and this chapter assumes a familiarity with the contents of Chapter 10. We will have little to say about microcomputers, since they are similar to minicomputers in most essential respects. However, microprocessors can be designed right into a piece of equipment, and this has revolutionized the design of electronic instruments. Microprocessor-based instruments generally deliver better performance at lower cost and with simpler construction than do equivalent instruments implemented with discrete logic chips. As a consequence, no competent engineer can afford to ignore these versatile devices. If any further incentive is needed, we might point out that microprocessors are fun. A new hobby, complete with a half dozen magazines, has sprung up around the microcomputer.

Because microprocessors are designed into an instrument as "dedicated" devices, the designer must play a greater role in the design and programming than he would with a minicomputer system. In particular, design with microprocessors includes tasks such as choosing the type of memory (RAM, EPROM, floppy disk) and deciding where in "memory space" to put it, deciding the form that input/output takes (including the choice of I/O hardware, whether constructed from the conventional MSI functions we dealt with in Chapters 8 and 9 or constructed from custom LSI "peripheral support" chips), and writing and debugging the dedicated software (programming) in the context of the instrument it controls. In general, designers of microprocessor-based instruments must have a thorough knowledge of both hardware and machine-language software techniques in order to be successful.

In this chapter we will begin by looking in detail at a particular microprocessor (the Intel 8085). This processor is convenient for dedicated-instrument design, having simple power-supply and clock requirements (a single +5V supply, and on-chip oscillator circuitry requiring only a single external crystal), as well as an adequate instruction set. We will look first at its internal operation and instruction set. Then we will discuss programming and will show a simple application: a complete multichannel counter using only standard TTL functions plus a microprocessor with its memory. No microprocessor system is complete without software, and we will show the programming necessary for this example.

Having gone through one complete concrete example, we will go on to discuss the essential components for microprocessor applications: MSI and LSI support chips and the various forms of memory. We will illustrate the use of these chips with a circuit for a general-purpose microprocessor system, needing only programming to make it do a specific task. The chapter will continue with a discussion of timing, data buses, and other popular processors. Finally, we will step back and look at the overall process of electronic design with microprocessors–development systems, evaluation boards, and emulators.

A DETAILED LOOK AT THE 8085

11.1 Architecture

The abundance of different microprocessor types can present a real problem to the circuit designer. Incompatibility between different microprocessor chips, both in their hardware implementation (signal lines, interfacing protocol, etc.) and in their instruction sets, is the rule in the microprocessor world, just as it is in the world of the minicomputer and large computer. Rather than attempt to choose precisely the best microprocessor for each job, it is perhaps best to settle on a sufficiently good microprocessor, then build a good development system and gain expertise with it. This is especially true in view of the fact that software development costs and effort often exceed those of the hardware design in microprocessor-based designs.

In this chapter we will concentrate on the Intel 8085, a popular microprocessor that is a refined version of the famous original 8080. It features the same instruction set, but with hardware refinements that simplify circuit design. For instance, the 8080 requires three power-supply voltages and two externally supplied clocking signals at 12V levels with a carefully timed delay between them. As an example, the awkwardness of designing around the 8080 caused designers to fall into circuit traps such as the so-called clock problem that beset the MITS Altair 8800 (the original hobbyist microcomputer), wherein the precise requirements for the two clock inputs were not carefully met. Several companies had advertisements in the hobbyist computer magazines for kits to solve the "clock problem." In those days we were happy to get anything working. Today's designers demand, and get, far greater simplicity of design in their microprocessor chips. Although the 8085 has been outclassed by more recent microprocessors, it is entirely adequate for most tasks, and it has remained popular because of its low price and the widespread familiarity with the original 8080 instruction set.

8085 block diagram

Figure 11.1 shows an 8085 CPU block diagram. Note the organization around an internal 8-bit data bus, with connections to an accumulator, arithmetic logic unit, instruction register, and a register array consisting of both 8-bit and 16-bit registers. The 8-bit registers can be combined into 16-bit register pairs, and there are instructions to operate on them individually and as register pairs. Even though the 8085 is basically an 8-bit (byte-oriented) computer,

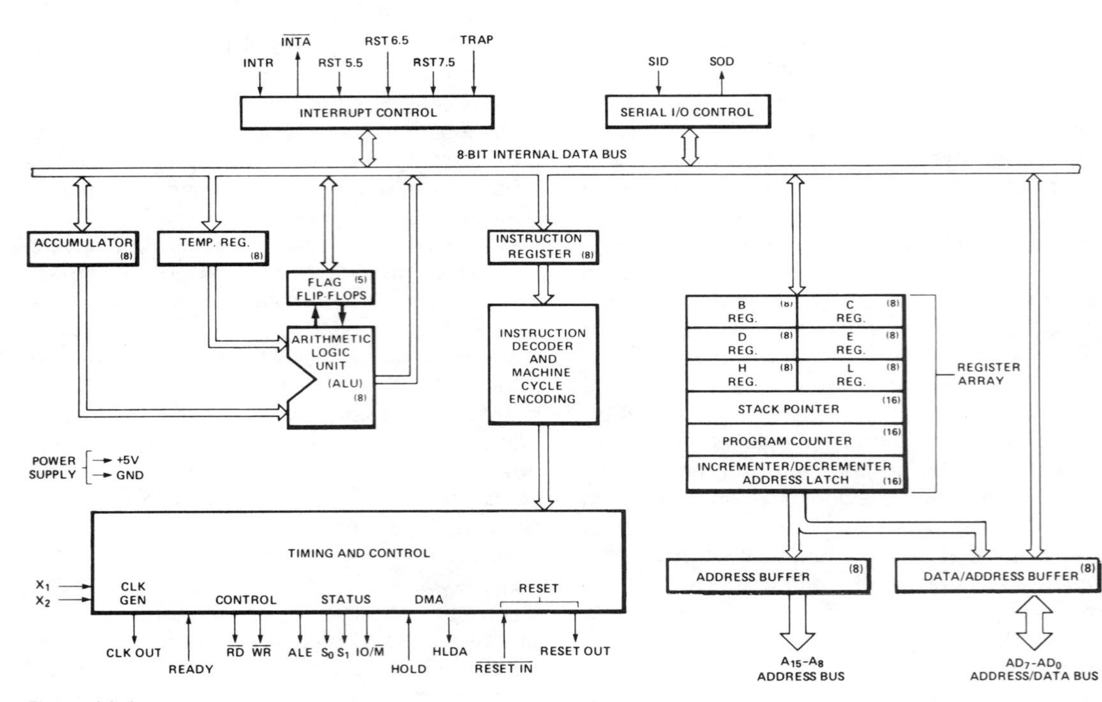

Figure 11.1.

Block diagram of the 8085 8-bit microprocessor. (Reprinted by permission of Intel Corporation, copyright 1978.)

16-bit registers are needed to address memory. With 8 bits you can address only 256 locations, but with 16 bits you can address 65,536 locations. The microprocessor contains a timing and control section to orchestrate the activity on the internal data bus and on the external control lines in response to the output of the instruction decoder. Note that the microprocessor's internal data bus is its own responsibility; you communicate with the microprocessor via the external signals shown (data bus, address bus, etc.) and never talk directly to the internal bus.

To help you get oriented, we would like to point out that the accumulator is the place where you do most of your arithmetic (add, subtract, complement, etc.) and logic (AND, OR, compare, etc.) operations. The register pair HL serves a very special function in being a "pointer" to memory; the memory location whose address is in HL can substitute for a register in most arithmetic operations. The register pairs, in general, are used for 16-bit arithmetic and to handle addresses (16 bits needed), and double-register arithmetic is used to calculate addresses in order to do indexing (calculating successive addresses for elements of an array, table, string of text, etc.). The flag register is vital for all forms of testing and branching.

The address bus of the 8085 deserves a bit of explanation. In order to fit the 8085 into a 40-pin package and still have enough pins available for desirable input lines such as the interrupts, the designers decided to use the 8-bit data bus pins to do double duty as half of the address bus also. Since an address is 16 bits wide, the top 8 bits of the address are brought out on their own pins, with the bottom 8 bits sharing the same pins as the data, but multiplexed in time. In other words, to send out an address the microprocessor uses both the address bus (high-order byte) and the "address/data" bus (low-order byte), whereas to send or receive data it uses the address/data bus only. As you will see shortly, sharing the same bus for both address and data is an advantage because it reduces external wiring.

11.2 Internal operation

Let's begin with the program counter, known as the PC. This 16-bit register always contains the memory address of the next instruction in the program to be executed. Except for jumps, successive instructions are consecutive in memory, so the PC is advanced after each instruction. When the processor is ready for the next instruction, it sends the contents of the PC to the external memory via the address bus and reads the resulting byte of data from that memory location into its instruction register via the data bus. It then acts on the instruction and may access the external memory one or more times in the process of completing the instruction.

An important point to understand about microcomputers, or indeed almost any computer with a small word size, is that a single computer instruction may be more than one word long. For example, any instruction that designates a location in memory (such as "store accumulator at such-and-such a place in memory") must have 16 bits just to specify the address in memory, not to mention some bits to specify what the instruction is supposed to be doing. In the 8085, an instruction is from one to three words (bytes) long; the first byte specifies the operation (e.g., "store accumulator") and is known as the operation code ("opcode"), whereas additional bytes specify whatever else is needed (address, data) to complete the instruction. As you will see, all instructions are written as a single assembly-language mnemonic, which often confuses the beginner into thinking that one instruction is the same as one word in memory. A consequence of this variable instruction length (1 to 3 bytes) is that the PC will have advanced by 1, 2, or 3 following each instruction execution, depending on what instruction was just executed. (Of course, it is altered to some other value following a jump instruction.)

□ *Anatomy of a single instruction cycle*

In order to give a feeling for the kind of operations the microprocessor is carrying out while performing its duties, let's look at the detailed steps it goes through while executing a particular instruction. Although an understanding at this level of detail is not necessary in order to use microprocessors, it will help to understand how they perform their particular magic, and it may aid in debugging. Figure 11.2 shows the timing diagram for a typical instruction cycle, in this case the instruction MOV M,B. This instruction simply moves the contents of one of the CPU registers (B) to a specific memory location, whose address happens to be specified by the pair of bytes in registers H and L (this is called indirect addressing; H and L contain the high and low halves of the address, respectively). Details for the timing and the address and data lines are shown, along with three control lines: the ADDRESS LATCH ENABLE (ALE) line, the READ line (RD'), and the WRITE line (WR'). The ALE is needed because data is shared with the lower half of the address on the address/data bus.

This particular instruction cycle is broken up into two machine cycles, M_1 and M_2, each of which involves a transfer to or from memory. Each of these is further divided into machine states, T_1 through T_4, etc., that correspond to cycles of the system clock. During M_1 the CPU obtains the instruction opcode from memory (in this case the instruction is only one byte long), and during M_2 it actually performs the data movement from the B register to memory. In general, an instruction may require up to five machine cycles, M_1 through M_5, and each cycle may itself contain up to six machine states, T_1 through T_6. Moving data to or from memory always takes up one full machine cycle, with the ALE pulse present at the beginning of each T_1. During T_1 the low-order bits of the address are available on the address/data bus AD_0–AD_7, and the trailing edge of ALE should be used to latch them into an external register. During the remainder of each machine cycle the address/data lines are available for data operations, with timing provided by the READ and WRITE lines.

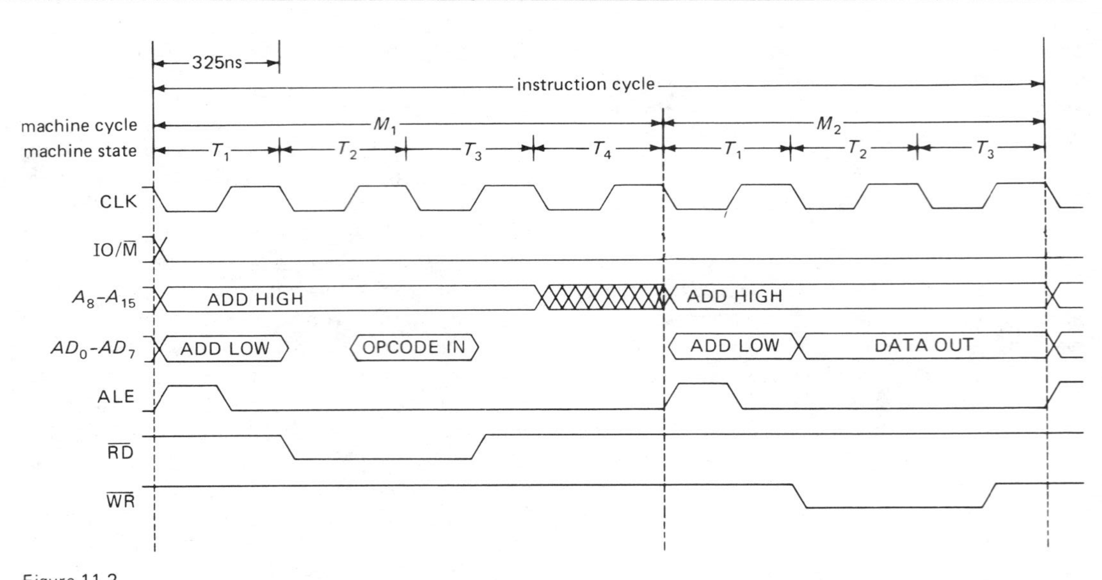

Figure 11.2
Timing diagram for the instruction MOV M,B.

Every instruction cycle must begin with an M_1 cycle that gets the opcode of the instruction and therefore includes a memory READ operation during machine states T_2 and T_3. The processor reads this opcode from the address/data lines into the instruction register during the middle of the T_3 machine state and initiates execution of the instruction during the T_4 machine state. Many instructions can be completed without further memory references, and therefore they terminate during T_4 or T_5. However, in the timing diagram we have illustrated, the processor realizes (during T_4) that a transfer from register B to memory has been requested. It therefore begins the second machine cycle, M_2, by outputting the address of the destination in memory, which in the case of this instruction comes from its register pair HL.

During the T_2 and T_3 machine states of cycle M_2, the processor uses the 8-bit internal data bus to connect the B register to the data bus, simultaneously asserting the WRITE pulse. The data is kept stable until long after the WRITE pulse has been removed, and data should be jam-loaded by the external memory during the WRITE pulse (or latched on the trailing edge). The memory has at least two machine states to realize that it has been addressed and to accept the data being written to it. This is called the access time requirement, t_A, of the memory. This completes the execution of this instruction. The CPU will then initiate the next M_1 cycle to get the opcode of the next instruction, which sits in the next memory location unless the last instruction was a jump.

The user has no control over the sequence of the processor's machine states, with one exception. Pulling the READY line LOW causes extra T states to be inserted after T_2; this "wait state" is useful for synchronizing data transfer from external devices (e.g., floppy disks) or for accessing slow memory.

Instruction speed

The processor clock frequency for the 8085 may range from 0.5MHz to 5MHz (for the 8085A-2). The CPU contains the oscillator circuitry, so an external crystal is simply connected between pins 1 and 2. A clock output is provided from the CPU, since the crystal oscillates at twice the clock frequency and is divided internally by a flip-flop to ensure a 50%-duty-cycle system clock signal. Intel recommends a standard operating frequency for the 8085 of 3.072MHz (6.144MHz crystal). The time for each machine state is then about 325ns, and the memory access requirement is about 525ns, a relaxed specification for MOS memory (see Sections 11.10 and 11.11 for further details). The MOV M,B instruction previously illustrated requires seven machine states, or 2.2μs, to execute with this clock frequency. This particular clock frequency was selected because it is an integral multiple of the clock frequency needed for a serial I/O chip (see Section 11.9). Since serial communication to a terminal is often needed, this choice of the system clock frequency makes it particularly easy to add the necessary hardware.

11.3 Instruction set

A data sheet for a microprocessor looks different than most data sheets, partly because of an extensive section on the "instruction set." To use a microprocessor it is essential to understand exactly what the instructions do: how to address memory ("addressing modes"), how to access registers, where the results of an arithmetic operation wind up, how to make a condi-

TABLE 11.1. 8085 INSTRUCTION SET

Register-to-register operations

INR r	increment r: (r) ← (r) + 1
DCR r	decrement r: (r) ← (r) − 1
MOV r1,r2	copy byte into r1 from r2: (r1) ← (r2)
MVI r,d[d]	copy byte into r from d

Accumulator operations

load-store[a]

LDA nn	load byte into A from address nn: (A) ← (nn)
STA nn	store byte into address nn from A: (nn) ← (A)
MOV A,M	load A from HL indirect address: (A) ← ((HL))
MOV M,A	store byte to HL indirect from A: ((HL)) ← (A)
LDAX rp[b]	load A with byte from (rp)address: (A) ← ((rp))
STAX rp[b]	store byte into (rp)address from A: ((rp)) ← (A)
IN port	read port into A
OUT port	write A to port

register	immediate	
MOV r,A		copy data into r from A
MOV A,r	MVI A,d	load A with data from r
ADD r	ADI d	add data from r to A
ADC r	ACI d	add with carry
SUB r	SUI d	subtract r: (A) ← (A) − (r)
SBB r	SBI d	subtract with carry
ORA r	ORI d	(A) ← (A) bitwise OR with (r)
ANA r	ANI d	(A) ← (A) bitwise AND with (r)
XRA r	XRI d	(A) ← (A) bitwise XOR with (r)
CMP r	CPI d	sets flags: Z if (A) = (r), C if (A) < (r)

TABLE 11.1. *(cont.)*

rotate A

RAL	RAR	rotate A left or right, looping through carry
RLC	RRC	rotate A into itself; copy extreme bit to carry

miscellaneous

CMA	1's complement A
DAA	decimal-adjust A[c]
INR A	increment A
DCR A	decrement A
XRA A	zero A
STC	set carry bit
CMC	complement carry bit

Register-pair "extended" operations[e]

LXI rp,dd[d]	load register pair with immediate data
DAD rp	HL "double add," (HL) ← (HL) + (rp)
INX rp	increment register pair, (rp) ← (rp) + 1
DCX rp	decrement register pair, (rp) ← (rp) − 1
XCHG	exchange (HL) with (DE)
LHLD nn	load HL direct: (HL) ← (nn) and (nn + 1)
SHLD nn	store HL direct: (nn) and (nn + 1) ← (HL)

Stack operations

PUSH rp[f]	store register pair on stack[g]
POP rp[f]	retrieve register pair from stack
LXI SP, dd	set the stack pointer (SP) ← dd
DAD SP	(HL) ← (HL) + (SP)
INX SP	increment stack: (SP) ← (SP) + 1
DCX SP	decrement stack: (SP) ← (SP) − 1
SPHL	(SP) ← (HL)
XTHL	exchange top of stack with HL

JUMP	**Call**	**Return**	
JMP	CALL nn	RET	unconditional
JNZ	CNZ	RNZ	not zero (not equal)
JZ	CZ	RZ	zero flag (equal)
JNC	CNC	RNC	not carry
JC	CC	RC	carry
JPO	CPO	RPO	odd parity
JPE	CPE	RPE	even parity
JP	CP	RP	plus or zero
JM	CM	RM	minus

Misc. control

NOP	no-operation
HLT	halt (wait for interrupt)

Note: Abbreviations are as follows: A, accumulator; d, 8-bit data or label;[d] dd, 16-bit data or label;[d] (*x*), contents of *x*; ((*x*)), contents of memory at (*x*); M, memory byte addressed by (HL); nn, 16-bit address or label; p, 8-bit port address or label;[d] r, registers A, B, C, D, E, H, L, and M (at HL); rp, register pairs B, D, H, for BC, DE, HL.
The following notes apply as indicated: [a]Move and load instructions don't affect any flags. [b]Register-pairs B (BC) and D (DE) only; HL and SP disallowed. [c]Use after an ADD or ADC instruction on packed BCD arguments. [d]Numbers are assumed decimal, except that 0nnnnH is hex, ddddddddB is binary. [e]16-bit operations. None affects any flags, except that DAD affects the C flag. [f]Also PUSH PSW (processor status word: A + flags) and POP PSW. [g]((SP)) and ((SP) − 1) ← (rp), (SP) ← (SP) − 2 (the stack pointer decrements).

tional branch, etc. In the 8085, each instruction consists of one to three bytes, with the first byte giving the opcode. The opcode determines the nature of the instruction; the CPU determines from the opcode whether or not additional bytes are necessary, and if so it will obtain them in subsequent cycles M_2, M_3. Since the opcode byte consists of 8 bits, there are just 256 possible opcodes, of which the 8085 uses 244. Table 11.1 lists the complete 8085 instruction set.

The instructions in the table are given in their mnemonic assembly-language form, rather than as hexadecimal machine language, because the processor is normally programmed in the more readable assembly language, with addresses and data labeled with understandable names rather than as actual numbers. After a program has been written in assembly language, a standard program known as the assembler converts these names and labels into numbers. The assembly language is called the "source code," and the actual machine-language program generated by the assembler is called the "object code." During this process a single assembly-language instruction will be converted into 1, 2, or 3 bytes of object code, depending on the particular instruction. We will have more to say later about the relationship between assembly language and machine language.

To figure out how the 8085 works, there's no substitute for plunging right in with the instruction set.

Register-oriented operations

The 8085 has a set of registers, indicated in the mnemonic instruction list by lower-case "r" and specified in actual assembly-language code by one of the eight possibilities A, B, C, D, E, H, L, and M. Letter "A" means the accumulator, and M always means the memory byte addressed by the contents of HL. Thus, for example, MOV A,M means "move to the accumulator the contents of the memory location whose address is now in the register pair HL." You can think of this use of the HL register as making a particular memory location into a register substitute.

Any of these registers may be incremented or decremented, and data may be moved (copied) from one register to another. In addition, "immediate" data (data that is part of the instruction) may be copied into a register. Note that a command like "MOV r1,r2" means "copy the contents of r2 to r1, while leaving r2 unchanged." In other words,

MOV (destination),(source)

Typical register instructions might be MOV A,B ("copy contents of register B to the accumulator") and MVI A,4 ("load the accumulator with the constant 4").

In these four register-oriented commands one of the symbols r1,r2 may be replaced in the assembly-language instruction by upper-case M. This causes the instruction to use the location in memory addressed by the contents of HL ("indirect addressing"). An example is MOV B,M ("load register B with the contents of the memory location whose address is in HL"). As you will see soon, memory can also be addressed directly, without the assistance of HL.

Accumulator-oriented commands

The next commands in the table refer to the accumulator register only. (For completeness, we have duplicated the register-oriented commands, with the accumulator acting as one of the registers.)

The load and store commands (LDA, STA) copy data between memory and the accumulator, with the memory location being specified directly in the instruction ("direct addressing"). The MOV A,M and MOV M,A instructions can, of course, do the same thing, but with indirect addressing through HL. Similarly, LDAX and STAX let you move data between memory and the accumulator, with indirect addressing through the other register pairs BC or DE. Finally, IN and OUT let you move data between the accumulator and an input/output port. Note that names like IN, OUT, load, and store are always with reference to the CPU: OUT means "move data *from* the CPU *to* the outside world." This CPU-centric convention extends to hardware labels like READ and WRITE: READ means that the CPU reads data *from* memory or I/O.

The instruction list continues with a group of arithmetic commands (ADD, SUB, etc.)

that refer to the accumulator. In all cases the operation is performed on data in the accumulator plus the register (or immediate), and the result is stored in the accumulator. Thus, SUB r subtracts the register contents from the accumulator, leaving the register contents unchanged. When performing these operations, the CPU sets "condition flags" (actually individual bits in a "flag register") that can be tested later to determine if the result was positive, negative, or zero or caused overflow. The CMP instruction merely sets flags and does not affect the accumulator contents. The condition flags are essential for conditional jumping, which we will discuss soon. Two instructions are available to affect the carry flag: STC sets carry, and CMC complements (toggles) carry. To clear the carry flag you could use an STC,CMC sequence (or, quicker, a trick like ANA A).

EXERCISE 11.1

Write a set of instructions to double the number in memory location nn; i.e., (nn) ← (nn)*2

The ANA, ORA, XRA instructions perform bitwise logic operations (AND, OR, exclusive-OR) on the accumulator. For instance, use logical AND to look at a single bit of a data byte:

```
ANI Ø4
JNZ address
```

This code looks at accumulator bit D2 (D0 is the LSB), jumping to the address location if D2 is a 1, otherwise continuing on to the next instruction.

Use logical OR to set an individual bit and exclusive-OR to toggle it. For example, the instruction XRI 80H toggles the most significant bit of the accumulator (H after a constant means hexadecimal, B means binary; otherwise decimal is assumed. Thus 80H = 80_{16} = 10000000_2). Exclusive-OR can be used to zero the accumulator ("XRA A"), as can the instruction "SUB A."

The rotate instructions RAL and RAR do a circular shift of the accumulator through the carry bit by one bit location; i.e., the accumulator plus carry behave like a 9-bit circular register. The alternate rotate instructions RLC and RRC do not rotate through the carry, but instead copy the wraparound bit of the original accummulator into the carry flag. The complement instruction CMA performs a 1's complement on the accumulator, i.e., it toggles each bit. There is no 2's complement instruction (negation), so to negate the accumulator you use the sequence

```
CMA
INR A
```

since the 2's complement of a number is the 1's complement plus 1.

Register-pair operations

A pair of registers (BC, DE, HL) holds a 16-bit word, essential for indexing and addressing, and therefore a few operations are provided for doing arithmetic on them. The LXI command loads a register pair with an immediate 2-byte number. The DAD "double-add" operation adds a register pair to HL; for example, DAD D would add the contents of double register DE to HL. This is the only double-register operation that sets any flags (DAD can set carry). INX and DCX increment and decrement register pairs, and they are very important for index operations (accessing consecutive locations in memory). Program 11.1 moves a table of 50 bytes

```
        LXI H,1ØØØH     ;first address of input list
        LXI D,2ØØØH     ;first address of output list
        MVI B,5Ø        ;initialize loop counter
LOOP:   MOV A,M         ;fetch a number
        STAX D          ;store it
        INX H           ;increment pointer to input list
        INX D           ;increment pointer to output list
        DCR B           ;decrement loop count
        JNZ LOOP        ;loop again if not finished
          •
          •
          •
```

Program 11.1

```
        LXI H,2000H
        LXI D,3000H
        LXI B,-1000H      ;initialize loop count
LOOP:   MOV A,M
        STAX D
        INX H
        INX D             ;move byte and increment pointers
        INR C
        JNZ LOOP          ;loop if low-order counter not zero
        INR B
        JNZ LOOP          ;exit if both counter bytes zero
         •
         •
         •
```

Program 11.2

starting at memory location 1000 hex to 50 memory locations starting at 2000 hex. Here we have used the single register B as a loop counter, checking the zero flag. If more than 256 bytes are to be moved, it is necessary to use a double register as a loop counter, as illustrated in the following example (Program 11.2), where 4096 bytes are moved from 2000 hex to 3000 hex. In this example we have used the register pair BC as a loop counter, initializing it with the negative of the loop count, and incrementing each half and checking for zero bytes explicitly (the double-register decrement DCX doesn't set flags). After the first move BC will contain −0FFF, etc.

Because the register pair HL is often used as a pointer for data transfers between the accumulator and memory, the LHLD (load HL direct) and SHLD (store HL direct) instructions are particularly useful. These load and store the contents of double register HL from (to) the memory location specified in the instruction (direct addressing). You might use these instructions to bring pointers into HL, as in this piece of code:

```
LHLD POINT
MOV A,M
INX H
SHLD POINT
```

Here we assume that the memory location POINT (actually two words) contains a pointer to an array; the code retrieves the pointer, gets a word from the array at the place pointed to, increments the pointer, and saves it in memory for later use. Another useful technique for getting data from arrays or tables is to keep the base address (first address of the array) in DE, then put an "offset" (position up from the bottom of the array) in HL, like this:

```
LXI H,n
DAD D
MOV A,M
```

The code adds the base address to the offset, n, to give the final address in HL for indirect addressing. The base address that was in DE is not destroyed, so other offsets can be brought into HL for additional addressing.

Stack operations

Unlike the MC-16 mythical computer we introduced in Chapter 10, most modern microprocessors include hardware "stack" operations. A stack is a set of sequential locations in memory that works just like the familiar XYZT stack used in Hewlett-Packard calculators: You can "push" numbers onto the stack and "pop" them in reverse order, i.e., last in, first out; it's a bus driver's coin dispenser (or a lunchroom tray gizmo). In the 8085 the stack is always used for double words, either the contents of a double register or the accumulator plus status byte (a byte containing all the condition flags). The stack expands downward in RAM, with the stack pointer (SP) automatically decremented by 2 each time a push is done and incremented by 2 for each pop.

The major use of the stack is for subroutine calls: The processor pushes the return address each time a subroutine is called, so it can return without getting lost (see the discussion of the CALL statement that follows). The last-in/first-out organization of the stack means that subroutines can call other subroutines without trouble. In fact, subroutines can even call themselves, if the passed arguments are also put on the stack.

The stack can be used for temporary storage of data, arguments, and the like. This allows routines to be re-entrant (i.e., they can call themselves, or they can call other routines that may, in turn, call the original routine), and it also economizes on memory space, as compared with allocating specific memory locations for temporary storage. As with any use of the stack, the only caution is that you must keep track of what you have put there, and in what order. If you pop more times than you push, you're in trouble.

The stack is useful when going off to a subroutine, since it can be used to save the "state of the machine" (the contents of accumulator, condition flags, and registers). Here's an example:

```
SUBR:    PUSH PSW
         PUSH B
         PUSH D
         PUSH H
           •
           •
           •
         POP H
         POP D
         POP B
         POP PSW
         RET
```

The command PUSH PSW pushes the accumulator and status byte. This routine saves the state of the machine on the stack, restoring everything at the end.

The stack pointer must be initialized at power-on reset, usually with the immediate instruction LXI SP. Occasionally, when branching to a subroutine, it may be desirable to save the stack pointer and relocate it to its original position at the end of the routine. Here's code to do that:

```
LXI H,Ø
DAD SP            ;get SP into HL
SHLD OLDSTK       ;stash it
LXI SP,NEWSTK     ;define new stack
  •
  •
  •
LHLD OLDSTK
SPHL              ;restore original stack
RET
```

EXERCISE 11.2

Create a software stack; i.e., write instruction sequences for PUSH and POP. Hint: Use a double register as a stack pointer. For PUSH, you must decrement (by how much?), then store indirect.

Jump instructions and condition flags

The normal instruction sequence can be altered with jumps and calls. A jump instruction causes a branch to the (immediate) address specified: The unconditional jump (JMP) always branches, but the conditional jumps (JC, JNC, JZ, JNZ, etc.) only jump if the appropriate condition flag is set. Thus, to loop 100 times, you could use this code:

```
         MVI B,1ØØ
LOOP:      •
           •
           •
         DCR B
         JNZ LOOP
           •
           •
           •
```

You have to be somewhat careful, because not all operations set flags. As a rough guide, (a) arithmetic operations affect all flags, (b) logic operations clear carry and affect all other flags, (c) increment and decrement don't affect carry, but affect all other flags, and (d) double operations don't affect flags (with the exception of DAD). For example, you can't use the carry to check for underflow on decrement.

The CALL statement causes unconditional branching to the immediate address, just like a JMP statement, but in addition it pushes the contents of the program counter (the next instruction location) onto the stack. A return statement (RET) at the end of a subroutine then causes a return to the original sequence by popping the stack to the PC, as illustrated in Program 11.3. Subroutine ADDAB just adds B to the accumulator; it doesn't deserve to be a subroutine. However, it does illustrate the essential features of a subroutine call. Here arguments are passed through the registers A and B. You could also use agreed-upon memory locations, or pass a pointer through registers or the stack.

Call and return instructions also come in eight conditional varieties, just like jumps.

```
        MOV B,M
        INX H
        MOV A,M
        CALL ADDAB      ;jump to subroutine; push PC
          •             ;program goes here after doing ADDAB
          •
          •
ADDAB:  ADD B           ;adds B to accumulator
        RET             ;pop PC
```

Program 11.3

□ *Restarts and interrupts*

At power-on, execution always begins at location 0. In addition, the instruction "RST n" transfers control to location 8∗n, pushing the return address onto the stack. Thus, "RST 3" causes a call-like jump to location 18 hex (24 decimal), whereas "RST 0" mimicks a power-on restart. The 8085 also has two varieties of hardware interrupts: One form has a protocol similar to the minicomputer interrupt protocol described in Chapter 10 (namely, an interrupt line is asserted, followed by an interrupt acknowledge sequence). A more convenient interrupt procedure is provided by the three pins "RST 5.5," "RST 6.5," and "RST 7.5." When one of these lines is asserted (the first two are enabled HIGH, the last is edge-triggered), the processor does a call-like jump (return address pushed on stack) to location 2C, 34, or 3C, respectively. As with the MC-16, there is an interrupt mask and an interrupt disable, accessible with the SIM, EI, and DI commands.

EXERCISE 11.3

Figure out why the vectored interrupts "RST 5.5," etc., are numbered that way.

Finally, the NOP instruction is a space (and time) waster, whereas HLT stops execution and waits for an interrupt. Instead of HLT, you could just use a jump loop to wait for an interrupt.

□ *Memory-mapped I/O*

An interesting alternative to the special I/O instructions IN and OUT is the use of "memory-mapped I/O," in which input/output devices hook onto the address bus and mimic memory locations, usually high in address space. You can always do this with processors like the 8085, but with some other processors, it is the *only* way to do input/output, because they have no separate I/O commands. Examples are the 6800, 6809, and 68000 and the PDP-11 minicomputers.

□ *No relative addressing in the 8085*

The 8085 instruction set is typical of fairly simple microprocessors. However, one omission in the 8085 language that is common in other microprocessors is the use of *relative addressing.* Here is an example:

JZ offset

This would usually be a two-byte instruction, with "offset" being a single-byte signed number telling how many locations to jump, *relative to the next instruction in the sequence.* For example, in this program

```
          •
          •
          •
        IN port
        RLC
        JNC -5
NEXT:     •
          •
          •
```

the jump instruction keeps testing the byte brought in from the I/O port, until it receives a nonzero MSB, at which point it continues on to the next instruction. The use of relative addressing is economical in program memory, since a single-byte offset usually suffices (jumps up to ± 128 locations are possible); jumps with two-byte offsets are usually allowed in addition, to reach any address. Relative addressing is also elegant, and it can yield relocatable code without reassembly. In practice, you would probably write this example with a jump to a label (at the IN instruction); the assembler, in its wisdom, would then assemble the jump using relative addressing, assuming, of course, that the microprocessor has a relative addressing mode.

11.4 Machine-language representation

As we mentioned earlier, the assembly language that we have been using is not the "object code" actually executed by the microprocessor, but rather a mnemonic representation convenient for writing programs. The set of assembly-language instructions that constitutes a program must be converted to a set of bytes that the processor actually executes. This can be done by hand, looking up the opcode bytes for each instruction and filling in the correct addresses and data for instructions that are longer than one byte. Alternatively, a standard *assembler* program can be used to convert a stored file containing the alphanumeric *source code* (assembly-language program listing) to a set of bytes constituting the *object code.*

Short programs are easily assembled by hand, and you would probably do just that when working with an "evaluation board" (see Section 11.14). For more complicated programs the job is tedious and prone to errors, and a more elaborate "development system" (see Section 11.14), complete with disk storage, operating system, text editor, and assembler, is highly desirable. In either case it is essential for debugging purposes to understand the correspondence between assembly-language source code and machine-language object code.

□ *Opcodes*

Each instruction in the 8085 language gets converted to a machine-language instruction one to three bytes in length. Here are a few examples:

Mnemonic	*Opcode (hex)*	*Operand*		*Length (bytes)*
JMP nn	C3	n(high)	n(low)	3
JZ nn	CA	n(high)	n(low)	3
OUT p	D3	p		2
IN p	DB	p		2
CPI d	FE	d		2
MOV C,A	4F			1
MOV C,M	4E			1
ADD E	83			1
RET	C9			1
RZ	C8			1
STC	37			1
ANI d	E6	d		2

As in Table 11.1, nn stands for a two-byte address, p is a port number (one byte), and d stands for one-byte immediate data. The instructions for which the operands are tacitly understood are a single byte in length, whereas instructions that need information such as immediate data or an address are several bytes long; in such cases the single assembly-language mnemonic will translate to several consecutive bytes of program storage, and their execution will require several memory accesses just for the processor to obtain the complete instruction.

□ *Execution times and instruction lengths*

Table 11.2 lists all the 8085 instructions, giving the length of the assembled instruction (in bytes) and the execution time (in *T* states). A *T*-state equals one clock cycle; thus, for example, the instruction "IN port" takes 3.25μs (325ns clock period), whereas "JNC adr" takes 2.28μs if the carry is set and 3.25μs otherwise. The lengths of the instructions are generally understandable in terms of the information needed for execution. For example, ADD C takes only a byte, whereas ADI 2CH requires an additional byte for the immediate constant. You will see shortly that the specification of which register is involved in a command like ADD is coded into the opcode byte itself as three of the opcode bits and thus doesn't require an additional byte in the instruction. Note that any instruction that includes an address in memory (CALL, JMP, LDA, STA) requires a total of three bytes, unless, of course, a double register holds the address, as in ADD M (a one-byte instruction).

Assembly example

In order to make clear the operation of the assembler and the nature of machine-language instructions, we will attempt to assemble by hand the machine-language code for a section of a simple program. For this purpose, let's consider the following assembly-language (Program 11.4), used to output characters to an alphanumeric screen and retrieve characters typed at a keyboard. The combination of keyboard plus alphanumeric output device is usually called a *terminal;* the master terminal in a computer system is often called the *system console.*

TABLE 11.2. AFFECTED FLAGS, MEMORY SPACE, AND EXECUTION TIMES FOR 8085 INSTRUCTIONS

Mnemonic	Flags	Bytes	*T* states	Exceptions	*T* states
Register					
INR	Z, S,P,AC	1	4	INR M	10
DCR	Z, S,P,AC	1	4	DCR M	10
MOV		1	4	MOV M	7
MVI		2	7	MVI M	10
Accumulator					
LDA, STA		3	13		
LDAX, STAX		2	10		
IN, OUT		2	10		
ADD, ADI	Z,C,S,P,AC	1,2	4,7	ADD M	7
ADC, ACI	Z,C,S,P,AC	1,2	4,7	ADC M	7
SUB, SUI	Z,C,S,P,AC	1,2	4,7	SUB M	7
SBB, SBI	Z,C,S,P,AC	1,2	4,7	SBB M	7
ANA, ANI	Z,C,S,P,AC	1,2	4,7	ANA M	7[a]
ORA, ORI	Z,C,S,P,AC	1,2	4,7	ORA M	7[b]
XRA, XRI	Z,C,S,P,AC	1,2	4,7	XRA M	7[b]
CMP, CPI	Z,C,S,P,AC	1,2	4,7	CMP M	7[c]
RAL, RAR	C	1	4		
RLC, RRC	C	1	4		
CMA		1	4		
DAA	Z,C,S,P,AC	1	4		
CMC, STC	C	1	4		
Register pair					
LXI		3	10		
DAD	C	1	10		
INX, DCX		1	6		
XCHG		1	4		
LHLD, SHLD		3	16		
Stack					
PUSH		1	12		
POP[d]		1	10		
SPHL		1	6		
XTHL		1	16		
Jumps, calls, returns					
JMP		3	10		
J (condition)		3	7/10[e]		
CALL		3	18		
C (condition)		3	9/18		
RET		1	10		
R (condition)		1	6/12		
PCHL		1	6		
Interrupts, etc.					
RST		1	12		
DI, EI		1	4		
RIM, SIM		1	4		
NOP		1	4		
HLT		1	5		

Note: C = carry, Z = zero, S = sign, P = parity, and AC = decimal-adjust carry. [a]The C flag is cleared and the AC flag is set. [b]The C flag is cleared and the AC flag is cleared. [c]The Z flag is set for equal; the C flag is set if A < operand. [d]The POP PSW instruction restores all the flags from the stack. [e]Ten states if condition met and jump occurs; 7 states otherwise.

```
                        ;console in/out routine

        CONFLG = ØØH    ;console status input port
        CONCHR = Ø1H    ;console data port -- in/out
        KEYST  = Ø1H    ;keyboard status mask
        PRIST  = 8ØH    ;printer status mask

        ORG 1ØØH        ;starting address
                        ;vectors --
START:  JMP INIT        ;initialization
CHECK:  JMP CONST       ;returns carry if ready
READ:   JMP CONIN       ;input character returned in accum
WRITE:  JMP CONOUT      ;character passed in reg C

INIT:     •             ;initialization routine
          •
          •

CONST:  IN CONFLG       ;get console status
        ANI KEYST       ;check input status bit, DØ
        RZ              ;not ready: return with no carry
        STC
        RET             ;ready: return with carry set

CONIN:  CALL CONST      ;check status
        JNC CONIN       ;wait for ready
        IN CONCHR       ;get byte from keyboard
        ANI 7FH         ;strip parity bit, D7
        RET             ;return with byte in accum

CONOUT: IN CONFLG       ;get console status
        ANI PRIST       ;check printer status bit, D7
        JZ CONOUT       ;wait for ready
        MOV A,C         ;put output byte in accum
        OUT CONCHR      ;send to printer
        RET

        END
```

Program 11.4

Remember that an assembler is a computer program that operates on the program source code, such as the preceding lines of program statements; these have been generated by the user, typically with the assistance of the computer's editor program, and stored temporarily on disk or other mass-storage device. The assembler converts the source code into machine-language code, which can be executed directly by the processor. This is done by subsequent loading into RAM or by burning an EPROM for permanent accessibility by the microprocessor. The assembler understands the mnemonics of the standard instruction set, but it does not understand the labels or names used in the program until they are defined. The labels used in jump or call commands are defined by the assembler as address locations, and the assembler searches for these locations in the program.

In this example the first four lines define symbols to be used in the assembly-language program; this makes the code more readable and makes it easy to change hardware bit and port assignments without having to make a lot of changes in the program. For example, the printer status mask is 80H, meaning that we have assumed that the MSB of the console status byte is set if the printer is ready for data. The ORG statement tells the assembler the memory location where we want the program to begin; the assembler will therefore assign the value 100 hex to the symbolic location START, the first executable statement in the program. Since the instruction JMP INIT requires three bytes, the next label, CHECK, will be assigned the address 103 hex. These four locations are jump vectors, just an extra jump into the program from outside, which we will explain later.

The program then continues with an initialization routine, beginning at symbolic address INIT. Note that you don't have to give locations symbolic addresses unless you reference those addresses, e.g., with a JMP or CALL statement. We have omitted the details of the uninteresting initialization routine and then written out the three sequences of instructions to check the input status, bring in a byte from the keyboard, and send a byte out to the printer or screen. Again, locations are given symbolic addresses only where necessary.

At this point you should go through the preceding program to make sure its operation is entirely clear. These routines would be semipermanent residents of your computer and would be called from other programs if input or output via the terminal were needed. For example, an executing program might get a byte from the keyboard with the code in Program 11.5. This use of a common handler routine for devices like the terminal is convenient, since each individual program that uses it doesn't have to bother with the details of status flags and input/output protocol.

□ *Hand assembly*

Let's try assembling a section of the terminal handler code. We will assume for purposes of illustration that the initialization routine requires 23 bytes of code. When you add to that the 12 bytes needed for the jump vectors, you've used up 35 bytes from START, putting the first statement of CONST at 123 hex (check our arithmetic). At that location we put a DB byte, the opcode for IN from the partial opcode listing given earlier. The instruction requires a second byte giving the port, in this case 00. The opcode byte E6 for the next instruction, ANI, goes into the next memory location, followed by its immediate byte 01, which is the value of KEYST, and so on. Using Table 11.2 and the preceding list of opcodes, you can see that the next few bytes are

Address (hex)	*Data (hex)*
0123	DB
0124	00
0125	E6
0126	01
0127	C8
0128	37
0129	C9
012A	CD
etc.	

Locations 123 through 129 constitute the whole CONST routine, with the opcodes for the instructions winding up at locations 123, 125, 127, 128, and 129. You should go through this example in detail, verifying the preceding code.

One possible point of confusion is the convention of listing the contents of successively higher locations in memory as a list going down the page, while showing diagrams of memory usage going *up,* as in Figure 11.3. In such a memory map a program like this might reside near the bottom, with execution proceeding upward.

Assembler

You could perform hand assembly of all your programs. However, this is very tedious, and any change in the program can alter many addresses when reassembled. Life is improved enormously by using an assembler program, which can assemble pages of assembly language in seconds. Since you may need a detailed listing of the object code for debugging and single-stepping, you can instruct the assembler to produce a nice listing showing both the original source code and the assembled object bytes. Program 11.6 shows the actual assembler output listing. Note that the assembler has condensed the machine-language code by

```
        CALL CHECK      ;check if keyboard has a character
        JNC NEXT        ;nothing there -- continue
        CALL CONIN      ;keyboard has a character -- get it
           •
           •
NEXT:      •
           •
```

Program 11.5

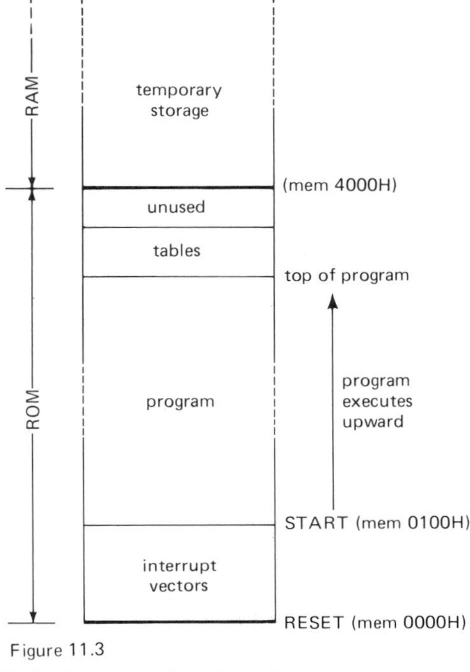

Figure 11.3

Typical memory allocation scheme.

putting one complete instruction on a line, even if it requires more than one byte. Note also that destination addresses are printed "backward," an idiosyncrasy of the microprocessor, which expects operands to have the least significant byte first.

□ *Jump vectors*

The four jump vectors at the beginning of this program deserve some explanation. These merely jump to the appropriate routine at a later location in the program, which might seem like a useless exercise. (For example, whenever you need to send a byte to the printer, wouldn't it be better just to go directly to CONOUT, rather than going through a middleman called WRITE?) However, there is a good reason for this "vectoring" procedure. As a program evolves through various versions and additional code is added to the program, the actual location of a routine may move up or down in memory. Since this I/O routine may be accessed by other programs that are not assembled at the same time, the use of the four jump vectors at the beginning of the program ensures that their locations always remain fixed despite the possible changing memory locations for the actual subroutines in later versions of the I/O program. Thus external programs that use the console I/O routines will always know where to find the fixed entry points, and any changes made in the body of the I/O routines only require reassembly of the I/O program itself. The jump vectors are a good use of 12 bytes of memory, and they illustrate the kind of feature you want to have in good software.

Alternatively, you could accomplish the same goal without jump vectors by calling the routine through a single entry point, passing arguments to indicate which function is to be performed.

□ *8085 opcode assignments*

Before leaving the subject of the assembler and its ability to convert source code to machine-language bytes, we can get some additional insight into machine language by looking at the way opcodes have been assigned. Here are the opcodes for add and subtract:

Mnemonic	*Opcode*
ADD r	10000*xxx*
ADC r	10001*xxx*
SUB r	10010*xxx*
SBB r	10011*xxx*

Note that these instructions all have their highest-order bits, D7 to D5, equal to 100. This indicates to the processor that the instruction is either add or subtract. D4 distinguishes between add and subtract, and D3 tells whether or not the carry bit should be included. The last three bits specify which of the eight possible registers should be used; for example, 000 indicates the B register, and 110 indicates the M "register." For the most part, the 256 possible opcode bytes are assigned to instructions in a systematic manner similar to the preceding four instructions.

Programming example

We will conclude this section with one brief example of 8085 programming (Program 11.7): a routine to search through a list of 100 bytes, beginning at location 1000 hex, looking for one that matches the byte in the accumulator. A nice technique that avoids the need for a loop counter is to load the

```
                                         ;console in/out routine

0000 =                   CONFLG = 00H    ;console status input port
0001 =                   CONCHR = 01H    ;console data port -- in/out
0001 =                   KEYST  = 01H    ;keyboard status mask
0080 =                   PRIST  = 80H    ;printer status mask

0100                     ORG 100H        ;starting address
                                         ;vectors --
0100 C3 0C 01   START:   JMP INIT        ;initialization
0103 C3 23 01   CHECK:   JMP CONST       ;returns carry if ready
0106 C3 2A 01   READ:    JMP CONIN       ;input char returned in accum
0109 C3 35 01   WRITE:   JMP CONOUT      ;character passed in reg C

010C            INIT:     •              ;initialization routine
 •                        •
 •                        •

0123 DB 00      CONST:   IN CONFLG       ;get console status
0125 E6 01               ANI KEYST       ;check input status bit, D0
0127 C8                  RZ              ;not ready: ret with no carry
0128 37                  STC
0129 C9                  RET             ;ready: return with carry set

012A CD 23 01   CONIN:   CALL CONST      ;check status
012D D2 2A 01            JNC CONIN       ;wait for ready
0130 DB 01               IN CONCHR       ;get byte from keyboard
0132 E6 7F               ANI 7FH         ;strip parity bit, D7
0134 C9                  RET             ;return with byte in accum

0135 DB 00      CONOUT:  IN CONFLG       ;get console status
0137 E6 80               ANI PRIST       ;check printer status bit, D7
0139 CA 35 01            JZ CONOUT       ;wait for ready
013C 79                  MOV A,C         ;put output byte in accum
013D D3 01               OUT CONCHR      ;send to printer
013F C9                  RET

0140                     END
```

Program 11.6

byte being sought in the location immediately following the list. Then the loop will always exit normally, with the address in HL indicating whether or not there was a match within the input list.

EXERCISE 11.4

Modify the program to search an array with 500 elements.

EXERCISE 11.5

Modify the program to search the original array for a byte that matches in bits D6, D5, and D4, disregarding the other bits.

EXERCISE 11.6

Modify the program to find all occurrences of the comparison byte, storing their addresses as an array of two-byte addresses beginning in memory at location 2000H.

EXERCISE 11.7

Write a program to go through a string of ASCII bytes (i.e., alphanumeric characters) in memory, changing all lower-case characters to upper-case characters. Lower-case ''a'' is 61 hex, with the alphabet sequential through lower-case ''z'' (7A hex). The upper-case alphabet goes sequentially from 41 hex through 5A hex. Let the starting address and size of the array be specified by registers HL and B, respectively.

EXERCISE 11.8

Write a program to search for a ''string'' (a particular sequence of bytes, e.g., the ASCII characters in a particular word). Assume

```
                        ;routine to find matching byte
                        ;enter with comparison byte in accum
        LXI H,1000H     ;starting address of table into HL
        MVI B,100       ;array size into loop counter

LOOP:   CMP M
        JZ MATCH        ;a match! exit via match, with addr in HL
        INX H           ;next byte
        DCR B           ;decrement counter
        JNZ LOOP        ;done yet?
          •             ;exit here if no match
          •
MATCH:    •             ;exit here if match. addr in HL
          •
          •
```

Program 11.7

that HL points to the beginning of the table being searched, that BC contains the length of the table, and that DE points to the beginning of the comparison string. You don't know the length of the comparison string, but its end is signaled by a null terminator; i.e., the string ends with a zero byte. This procedure is a valuable capability in text editors, letting you look for all occurrences of some word or word fragment.

A COMPLETE DESIGN EXAMPLE: 6-CHANNEL EVENT COUNTER

In the following sections we will design a complete instrument, based around the 8085 microprocessor. For simplicity we will use only simple TTL MSI chips, in addition to the essential CPU and memory, saving the discussion of LSI peripheral chips for later in the chapter. Our example, a 6-channel event counter, will include the complete hardware design and all necessary software.

11.5 Circuit design

The basic idea of a microprocessor-based event counter is that we don't actually need counter chips (like a half dozen 7490s) to count incoming pulses; instead, we can keep the count in a memory location and use the microprocessor to monitor the input, incrementing the count in memory when an input pulse occurs. Such a scheme will not be as fast as a hardware counter. However, the microprocessor-based design has the potential of easy expansion to hundreds of channels, and it provides the ability to perform arithmetic operations on the accumulated counts and even take action based on these counts.

Memory

To begin the design, we need some permanent memory for the program, for which we choose a small EPROM (an erasable programmable read-only memory; see Section 11.10), the 2758. We also need some "scratch-pad" memory in which to store the counts and a few other pieces of temporary information. For this we choose the smallest NMOS RAM (random-access memory) we can find, the 68B10. Both of these memories are 8 bits wide and are intended for use with 8-bit microprocessors.

Now we need to configure the circuit, using ordinary TTL chips to handle functions such as low-order address latching, memory chip enabling, obtaining input counts, examining the panel switches, and driving the numeric display. Figure 11.4 shows the complete block diagram.

Figure 11.5 shows the schematic of the CPU, memory, and control portions of the finished design. We use a 74LS373 octal transparent (jam-type) latch to obtain $A_{0\text{-}7}$ from $AD_{0\text{-}7}$. This low-order half of the address is latched off the data bus by the ALE control line from the CPU. The EPROM needs 10 address bits, $A_{0\text{-}9}$, to address its 1K bytes, and the RAM needs 7 address bits, $A_{0\text{-}6}$, for its 128 bytes. In any microprocessor system you have to provide a way to enable the individual memory chips, since any one chip occupies only a small part of

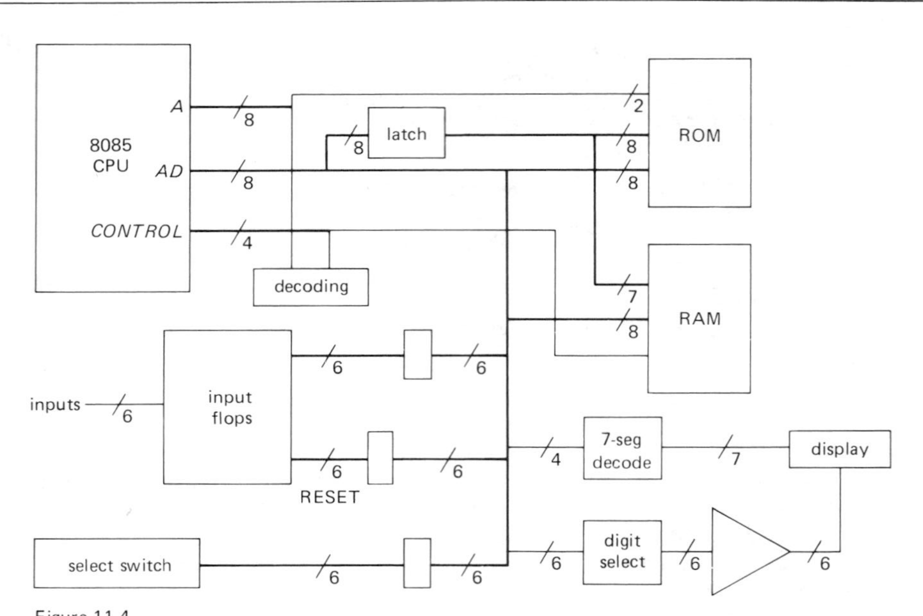

Figure 11.4

Block diagram of a 6-channel event counter.

the 64K locations that the CPU can address with its 16 address bits. Because the power-on reset starts program execution at memory location 0, you always put a ROM containing program at the bottom of memory space. In this case we have enabled the EPROM when address bit A_{10} is LOW and the IO/M′ line is also LOW, putting it at the bottom 1K of memory space (and repeating every 2K further up, which won't matter here). The RAM is enabled when address bit A_{10} is HIGH and the IO/M′ line is LOW, putting its 128 bytes into memory space beginning at location 1K (and repeating every 128 bytes higher, which again won't matter here, see Fig. 11.6).

The EPROM is enabled only during read cycles, since otherwise there would be a conflict for the data bus. The RAM is enabled during read and write, with gating to convert the RD′,WR′ pair of control signals from the CPU to the R/W′ and ENABLE pair required by the 6810 RAM. This completes the circuitry needed for the memories.

Input and output

We need circuitry to obtain input counts, and we need to look at panel controls to see which counter channel should be displayed and when to clear the counters. Then we need circuitry to drive a 6-digit multiplexed 7-segment LED display. The 8085 can address 256 I/O device codes (ports), signaled by IO/M′ being HIGH, and addressed by the 8 bits $AD_{0\text{-}7}$ (duplicated in $A_{8\text{-}15}$ without the ALE strobing requirement). Programmed I/O transfers are strobed by RD′ and WR′, in the same manner as memory transfers.

Input circuitry. Let's begin with the circuitry to read the displayed-channel selector switch, shown in Figure 11.7. As you saw in the last chapter, a simple programmed I/O for data IN requires a three-state buffer for each bit, enabled by the simultaneous combination of RD, IO/M′, and the appropriate device code on the address lines. The 8085 commands "IN port" and "OUT port" do programmed I/O between the accumulator and the selected port.

The 74LS155 dual 1-of-4 decoder (Fig. 11.5) generates an enabling signal for the 74LS367 three-state buffer when RD′ is LOW, IO/M′ is HIGH, and port 0 is selected (A_8 and A_9 LOW). Thus the command "IN 0" loads the accumulator with a byte, six bits of which tell the position of the rotary selector switch. That's all the processor needs; software will take care of the rest.

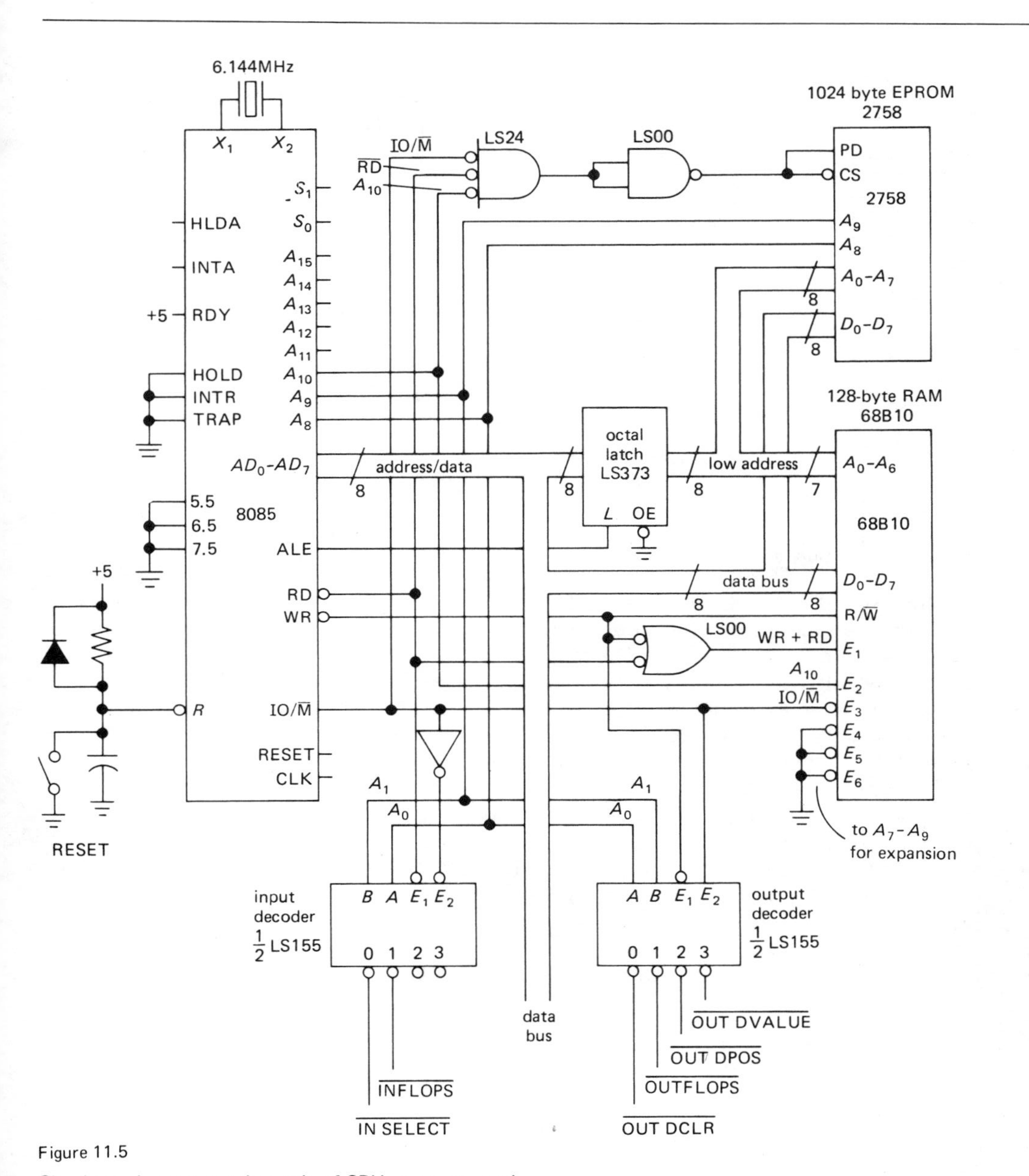

Figure 11.5

Six-channel counter: schematic of CPU, memory, and control.

The input events are detected with the circuit shown in Figure 11.8. Since most input pulses will occur while the processor program is occupied with other tasks and is not looking at the inputs, a flip-flop is used on each input line to allow the program to determine that a pulse has occurred. The state of these six flip-flops is read via programmed data IN from port 1 ("IN 1"), buffered by another 74LS367. The flip-flops can be selectively cleared via a data word output to port 1, which is latched by the 74LS174.

Display circuitry. We need to display the contents of the selected counting channel, with the displayed digits being sent to the

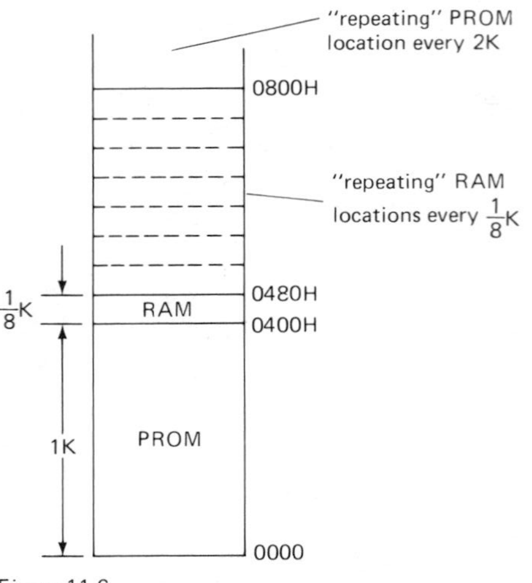

Figure 11.6

Six-channel counter: memory allocation.

6-digit multiplexed display via programmed data OUT from the processor. The program will handle this by outputting the count in memory one digit at a time to the display, making sure not to ignore the register of input flip-flops any longer than necessary in order to obtain the highest maximum counting rate. Of course, the processor will also need to check the channel selector switch occasionally, in order to know where to look in memory for the displayed data. Since the display is multiplexed (one digit illuminated at a time), the whole cycle must repeat rapidly to avoid annoying flicker in the display. The whole display should be refreshed at least 50 times per second; because there are six digits, this leaves only 3.3ms per digit.

The display circuitry is shown in Figure 11.9. The digit value is latched in the 9368, addressed as output port 3. The 9368 is an unusual chip that combines a BCD latch, decoder, and 7-segment current-limited anode driver. The digit position to be displayed is output by setting one bit at a time in a byte sent to the 74LS174 connected as port 2. (A decoder could have been used here, but it would have required more chips.) Thus, the sequence

```
OUT 3
MVI A,2
OUT 2
```

will display the digit originally in the accumulator as the second digit from the right in the display. Note that whatever digit you send out stays on until you send something else; if the program crashes, the last digit you send out really shines brightly.

There is one subtlety in the display driving circuitry. The decoder/driver is specified for 20mA segment drive (source) into a load at +2V maximum. Since the display has a forward drop of 1.7V at 20mA, the digit position drivers (sink) must be able to pull the selected cathode line down to 0.3V, maximum, at a current of 140mA (all 7 segments lit, 20mA/segment). This means you can't use a Darlington driver chip (e.g., the popular 75492), but are forced to use

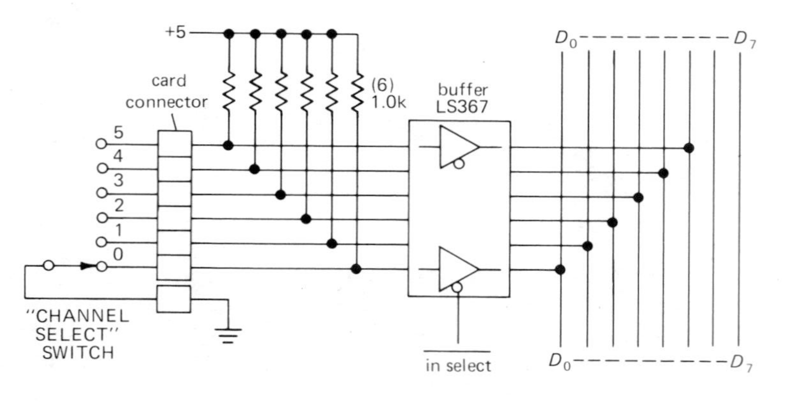

Figure 11.7

Six-channel counter: switch-reading circuit.

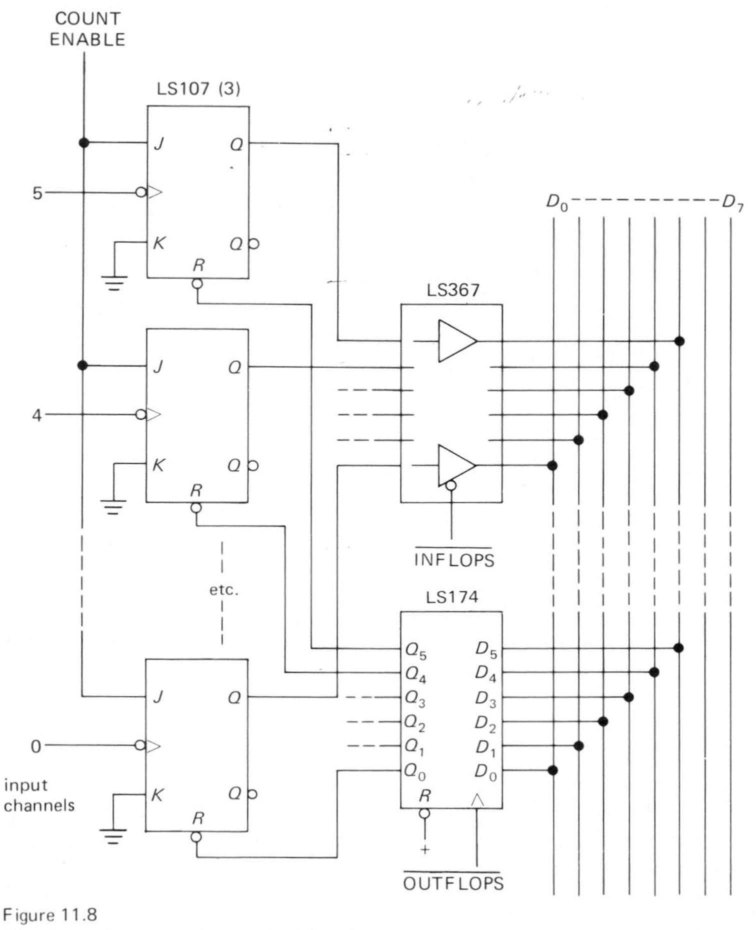

Figure 11.8
Six-channel counter: input circuit.

the 75494, which saturates at 0.25V, max, at 0.25A.

11.6 Programming the 6-channel event counter

We will divide up the programming task in a "top-down" fashion, which means that we begin by identifying the major tasks of the job, the order in which they will be handled, and the protocol for interchange of information between major modules of the program. As part of this approach, we will make a flow chart to organize the task. Next, we will write the individual program modules (these parts will often correspond to individual subroutines) and test them individually. Finally, we will hook them together and test the completed program.

Organizing the task

Our basic jobs are the following: (a) Periodically check the input flip-flops to see if any counts have come in; if so, increment the count in the corresponding locations in memory. (b) Periodically check the displayed-channel selector switch. (c) Refresh the display. These tasks can be done in just about any order, but there are two constraints. First, the display digits must be refreshed at a regular rate, to keep the displayed brightness steady. Furthermore, the refresh process must spend equal time on each digit. Second, the maximum rate of

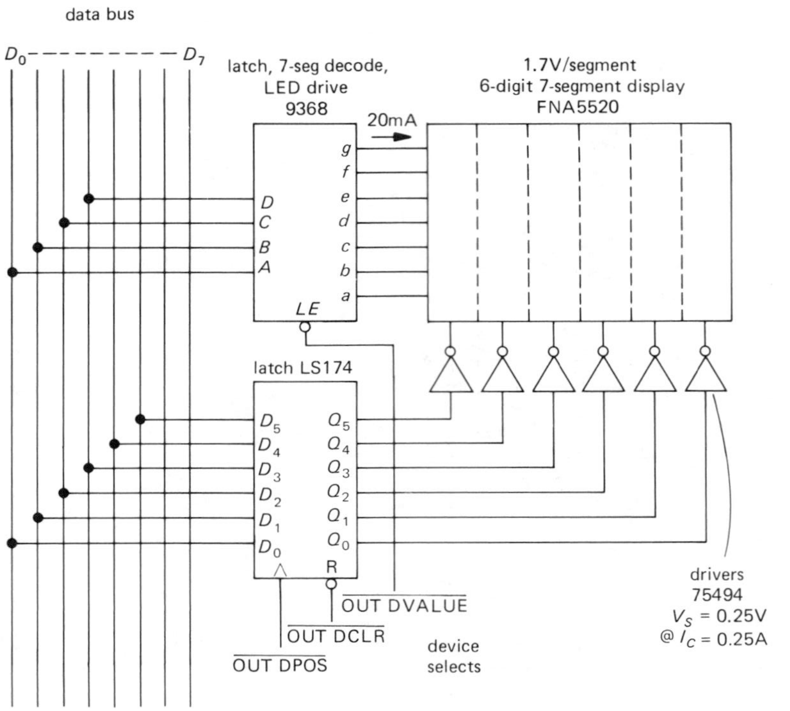

Figure 11.9
Six-channel counter: display circuit.

input pulses is set by the longest delay in checking the input flip-flops. In programming, a common constraint is limited memory space, but in this example these timing considerations are the only constraints.

Figure 11.10 suggests an efficient way to handle the task. We update the counters as often as possible, doing only the smallest piece of display refreshing possible in between. Figure 11.11 expands on this procedure with a simple flow chart.

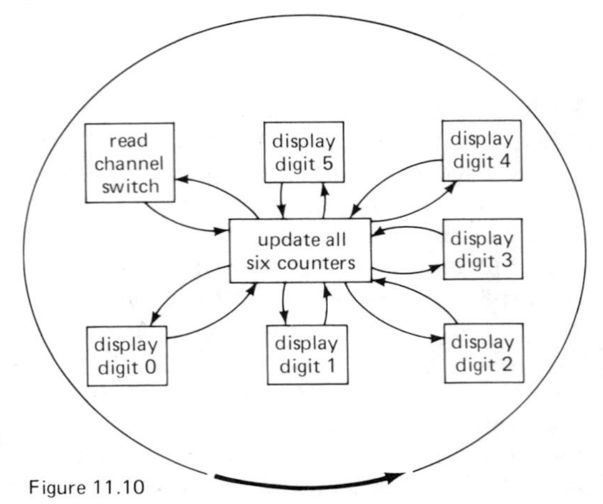

Figure 11.10
Program sequence for 6-channel counter.

With our top-down approach, we can immediately write the program (Program 11.8). The maximum input counting rate is determined by the longest time it takes to get around the loop. In operation the instrument does nothing else except count and display. The counters are cleared by restarting the program.

We are ready to define protocols between modules, so we can code the main program and subroutines. First we allocate the use of RAM memory: We need room for the counts in the six counting channels, and we need places to keep pointers to the channel selected for display and the digit being displayed. Figure 11.12 shows our choice, allowing four bytes (six digits plus overflow) of RAM for each of the six counter channels. The digit pointer sits in the next available byte, followed by the two-byte channel

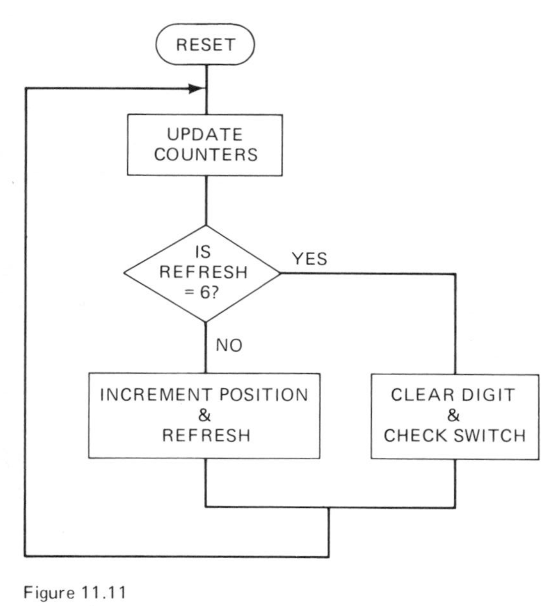

Figure 11.11
Six-channel counter: overall flow chart.

pointer. These locations are given symbolic names, CNTRØ, POINTER, CHANNEL, for use by the assembler. The individual routines pass information back and forth through these locations; this is the protocol for communication between program modules.

Main program

Program 11.9 is the main program. Symbolic locations are equated with their actual addresses in RAM in the first four lines. The next five lines equate the I/O symbolic names with their port addresses. Having written these, we are now free to use the symbolic tags in our assembly code, to make it more understandable.

The ORG statement tells the assembler the starting address in memory for this program. We use location 0000 because the power-on reset initiates program execution at location 0000; that is built into the 8085 hardware and cannot be changed. The program first initializes the stack pointer, then clears RAM (clearing the counters), and finally proceeds to the count and display loop. Work through the loop that clears RAM, to be sure you understand how it works.

Unlike most computer programs, this program has no exit. This is typical of programs that control dedicated instruments – life is unending work, work, work!

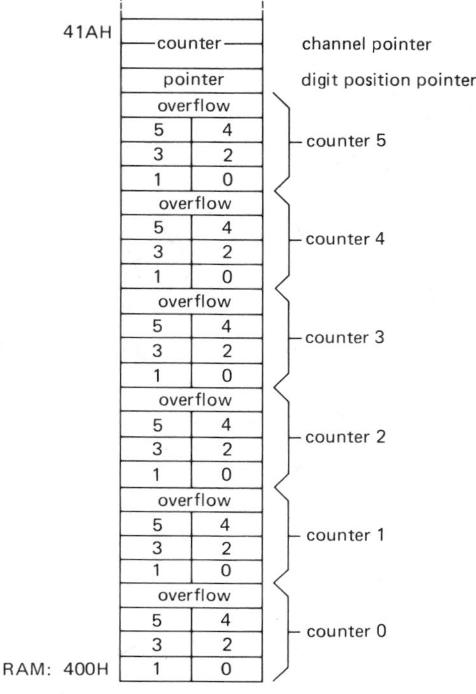

Figure 11.12
Six-channel counter: data storage in RAM.

EXERCISE 11.9

Write a loop to clear 1K locations in RAM, beginning at location 2000 hex. Be careful with double-register operations that don't set flags.

```
        (reset routines)
           •
           •
           •
LOOP:   CALL COUNT      ;flip-flop check and memory increment
        CALL DISPLA     ;display digit or check switch
        JMP LOOP
```

Program 11.8

```
                            ;program for 6-channel event counter
                            ;with 7-segment readout

                            ;symbolic assignments to RAM:
RAM = Ø4ØØH                 ;initial RAM location
CNTRØ = RAM                 ;space for counters
POINTER = CNTRØ + 4*6       ;pointer to displayed digit (1 byte)
CHANNEL = POINTER + 1       ;pointer to addr of displayed chan (2 bytes)
NEXT = CHANNEL + 2          ;next available location in RAM

                            ;symbolic assignments for I/O ports
SELECT = ØØ                 ;input port for reading chan select switch
DCLR = ØØ                   ;output port for clearing display driver
FLOPS = Ø1                  ;in/out port for reading and clearing flops
DPOS = Ø2                   ;output port for digit position driver
DVALUE = Ø3                 ;output port for BCD digit to display

        ORG = ØØØØ          ;reset location, bottom of ROM
RESET:  LXI SP,RAM+128      ;set stack pointer
        LXI H,RAM           ;first RAM location to clear
        MVI C,8ØH           ;number of RAM locations to clear (128)
        XRA A               ;clear accumulator

LOOP1:  MOV M,A             ;clear one RAM location
        INX H               ;increment memory pointer
        DCR C               ;decrement counter
        JNZ LOOP1           ;keep clearing RAM until done

LOOP:   CALL COUNT          ;update all 6 counters
        CALL DISPLA         ;display a new digit
        JMP LOOP            ;don't ever stop
```

Program 11. 9

Counting module

Figure 11.13 is a flow chart for the subroutine that checks the input flip-flops and increments the RAM locations accordingly. The routine reads a byte from input port 1, for which each bit corresponds to a count in the corresponding channel. It then clears the flip-flops that were set. Only flip-flops that have received a count since the last pass should be reset, because you can't be sure there won't be another count coming into the other flip-flops during the reset pulse (this assumes that input pulses are not coming at greater than the maximum count rate the program can tolerate). Once the FLOPS byte has been brought in, the program continues by adding to the count in each channel the corresponding bit of the FLOPS byte. This means it adds 0 or 1. This is better than checking for 1's and incrementing only those channels, because this way the time required to execute the subroutine is independent of how many counts were received. This ensures a uniform time interval for each displayed character.

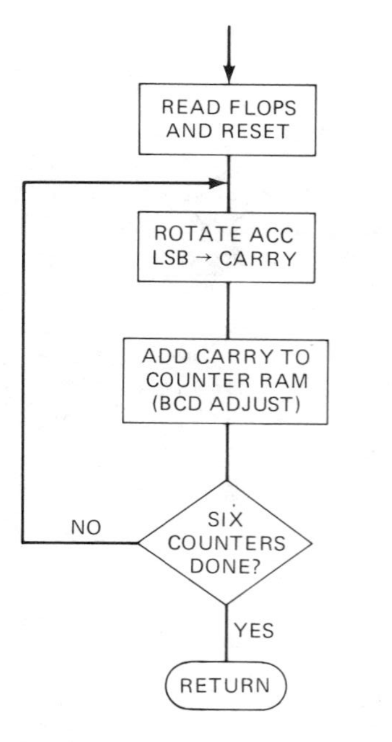

Figure 11.13

Six-channel counter: input-checking flow chart.

The actual program code is given in Program 11.10. This subroutine is called from the loop in the main program. It begins by reading in the byte from FLOPS, then clearing the set flip-flops by outputting the complement. This jams the reset inputs of those flip-flops. Because we want a reset *pulse,* rather than the level you get from a latch, we've got to send out a byte of all 1's to end the pulse. You might call this a "software pulse." Because input/output requires the accumulator, the FLOPS byte was temporarily stored in register B.

The routine then initializes the C register with the loop count (6) and HL with the first counter memory location. Then it enters the memory adding loop. Each pass of the loop updates all the memory locations for one channel. The memory update is a multiple-precision decimal add, with 2 digits stored in each of three bytes (6 digits per channel). The loop consists of three repetitions of the sequence

```
MVI A,ØØ
ADC M
DAA
MOV M,A
INX H
```

in each case entered with the carry set on overflow from the previous operation (initially, the carry comes from the input flip-flop). The sequence adds the contents of memory to the carry bit, then does a decimal adjustment (DAA) to maintain the format of 2 BCD digits per byte before sending the sum back to memory. The last step prepares

```
                        ;routine to read all 6 channels
                        ;and increment counters

COUNT:  IN FLOPS        ;read input counts
        MOV B,A         ;save flops
        CMA             ;invert positive true
        OUT FLOPS       ;reset flops that have counts
        XRA A
        CMA             ;accumulator all 1's
        OUT FLOPS       ;terminate the reset pulses
        MVI C,6         ;counter: 6 channels to update
        LXI H,CNTRØ     ;first channel, first byte

LOOP2:  MOV A,B         ;get flops
        RRC             ;get count bit into carry
        MOV B,A         ;save flops for next channel
        MVI A,ØØ        ;clear A, keep carry
        ADC M           ;add digits Ø and 1, with carry
        DAA             ;correct to packed BCD addition
        MOV M,A         ;put answer back
        INX H           ;look for next byte
        MVI A,ØØ
        ADC M           ;add digits 2 and 3, with carry
        DAA
        MOV M,A         ;put them back
        INX H
        MVI A,ØØ
        ADC M           ;add digits 4 and 5
        DAA
        MOV M,A         ;put them back
        INX H           ;look for overflow
        MVI A,ØØ        ;clear A, keep carry (overflow)
        RAL             ;A = 1 if overflow
        ORA M           ;set overflow byte
        MOV M,A         ;store it
        INX H           ;first digit of next channel
        DCR C           ;do for all 6 counters
        JNZ LOOP2
        RET             ;finished -- back to main loop
```

Program 11.10

for the next repetition by incrementing the pointer to the next byte of the 3-byte number.

Display module

Figure 11.14 is the flow chart for the display and switch-reading subroutine. It is called from the main program's loop immediately after the count routine, and it is expected to refresh the next digit of the display, unless all six digits have been refreshed, in which case it reads the channel selector switch instead. It begins by retrieving the pointer to the digit that was displayed last. If that was the sixth digit, it jumps to the switch-reading routine. Otherwise, it fetches the CHANNEL pointer, figures out which half byte should be displayed, and displays it.

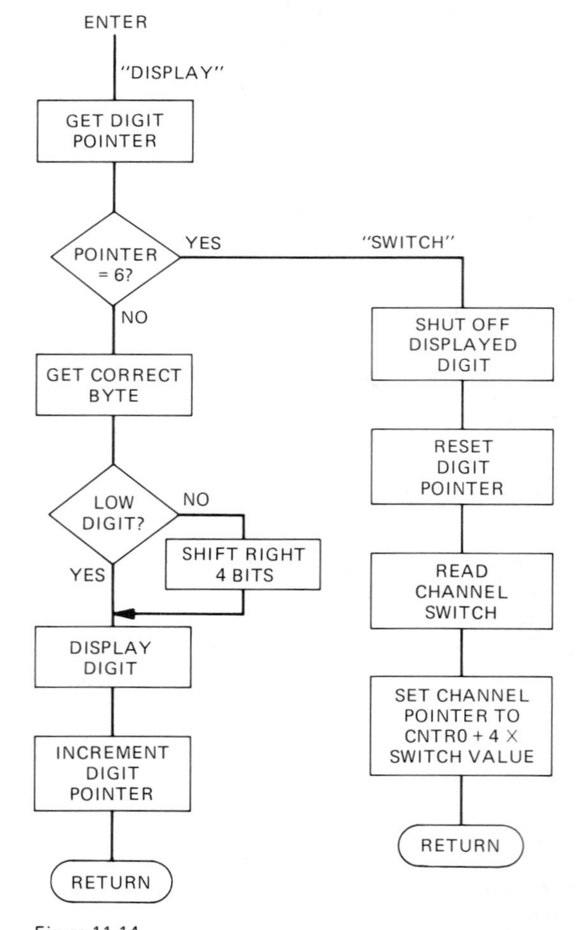

Figure 11.14
Six-channel counter: display and switch-reading flow chart.

Program 11.11 shows the actual program code. There are some interesting subtleties. CHANNEL already contains a two-byte pointer to the first RAM location of the selected channel. You will see the code that does this in the switch-reading routine. POINTER is a number from 0 to 5, giving the digit position, starting from the right, that is to be refreshed. Note the code sequence to determine both the correct byte and the correct half of that byte for the displayed digit (look again at Fig. 11.12, to remind yourself how the digits are stored). Another interesting point is the two-step "time-wasting" code beginning at address WAIT. This approximately equalizes the time spent when the digit happens to be located in the low-order half of the data byte, and therefore doesn't need four shifts.

An interesting problem crops up in this routine. The digit position is stored in POINTER as a value from 0 to 5, but the form of hardware I/O demands a single bit set and output to port 2 (DPOS) (see Fig. 11.9). We could at this point go back and add a 74LS42 BCD decoder, along with another hex inverter. This illustrates the classic situation wherein you discover a hardware incompatibility that can be fixed either by adding additional hardware or by changing the software. In this case we'll take the software route, saving two chips (and the redrawing of the schematic!). We do it by setting up a 6-entry table, beginning at location TABLE, containing the binary numbers 00000001, 00000010, etc. The program adds the digit position (an "offset") to the table's first address to find the location of the byte to be sent to port DPOS.

Switch-reading module

Figure 11.14 also shows the flow chart of the switch-reading module, called from the display module if the sixth digit has already been refreshed. It begins by shutting off the displayed digit, in order to ensure that the sixth digit is displayed for the same amount of time as the others. It then resets the digit pointer (POINTER), in preparation for the

```
                                ;display new digit on each pass
                                ;CHANNEL points to RAM where selected
                                ;channel is stored, and POINTER tells
                                ;which digit to display

TABLE:   DB 00000001B           ;digit lookup table, for fast conversion
         DB 00000010B           ;to format needed by display hardware
         DB 00000100B
         DB 00001000B
         DB 00010000B
         DB 00100000B

DISPLA:  LDA POINTER            ;get digit pointer (0-5)
         CPI 06
         JNC SWITCH             ;look at switch on 7th pass
         MOV B,A                ;save the digit pointer
         ANA A                  ;clear carry
         RAR                    ;A = pointer/2, carry = pointer odd
         PUSH PSW               ;save carry on stack
         LHLD CHANNEL           ;base address of selected channel
         MOV E,A                ;want to add A to base address
         MOV D,0
         DAD D                  ;HL now points to CHANNEL + 1/2 POINTER
         POP PSW                ;get carry back (tells which half has digit)
         MOV A,M                ;get byte from channel memory
         JNC WAIT               ;right digit of byte; go waste some time
         RAR
         RAR
         RAR
         RAR                    ;get left digit in right side

OUT:     MOV C,A                ;save digit
         XRA A
         OUT DCLR               ;temporarily turn off display
         MOV A,C                ;get digit back
         OUT DVALUE             ;update BCD/7-segment driver
         MOV A,B                ;get digit pointer back again
         LXI D,TABLE            ;lookup table starting location
         ADD E                  ;add offset (digit number)
         MOV E,A                ;table index in DE
         LDAX D                 ;get mask to enable that driver
         OUT DPOS               ;update digit position driver
         MOV A,B
         INR A                  ;increment digit pointer for next pass
         STA POINTER
         RET                    ;finished -- go back to main loop

WAIT:    NOP                    ;waste 4 states
         JMP OUT                ;waste 10 more, then go back
```

Program 11.11

next DISPLAY loop. A byte is then brought in from the selector switch and converted to a number from 0 to 5, telling which counter's contents are to be displayed. This it does by successively checking each bit, incrementing a counter until it finds the single 1 bit. That counter then contains the selected channel number. Finally, it finishes by figuring the first address of the selected channel's data in RAM, by multiplying the channel number by 4 and adding the address of the first channel, CNTRØ.

Program 11.12 shows the actual program code. An interesting feature of this routine is the "safety bit" OR'd into the byte retrieved from the selector switch port. In the event of a switch malfunction (no closure to ground), the program would otherwise loop forever, searching for an asserted bit. The safety bit ensures that the instrument

```
                        ;program to look at "display channel"
                        ;selector switch, and compute pointer
                        ;to RAM address where that channel's
                        ;contents are stored

SWITCH: OUT DCLR        ;turn off display
        XRA A
        STA POINTER     ;reset digit pointer to Ø
        LXI H,CNTRØ     ;counter RAM location base value
        IN SELECT       ;read the select switch
        CMA             ;complement: switch closed = logic LOW
        ORI 2ØH         ;safety bit (default to 6th channel)
        MVI B,-1        ;initialize counter

LOOP3:  INR B
        RAR             ;is it this switch position?
        JNC LOOP3       ;if not, try next channel

DONE:   MOV A,B         ;counter in B holds switch position (Ø-5)
        ADD A
        ADD A           ;mult by 4, since 4 bytes of RAM per chan
        ADD L           ;add base location to get RAM pointer
        MOV L,A         ;move to HL
        SHLD CHANNEL    ;store pointer for use by DISPLA routine
        RET             ;our job is done -- back to main loop
```

Program 11.12

won't go dead, but will instead display the contents of channel 6, so you'll be able to troubleshoot it more easily.

EXERCISE 11.10

Make software changes so that the counter will read all zeros if the switch is set to an open contact.

EXERCISE 11.11

Rewrite the routine SWITCH so that a binary-coded selector switch can be used instead of the 1-of-6 rotary switch shown in Figure 11.7. Note that this simplifies the program and speeds execution.

11.7 Program timing and performance

Recall that the maximum input counting rate without loss of data for this 6-channel counter is set by the largest latency time before the program checks the input flip-flops. Also, in evaluating the performance of the counter we must check to see that the display is refreshed often enough to avoid annoying flicker. In doing these timing calculations we use the execution times listed in Table 11.2, taking all the worst-case branching possibilities within the program.

The flip-flops are checked once during each loop of the main program. That loop occupies 38 clock cycles, 18 for each CALL and 10 for the JMP. The COUNT routine requires 889 clock cycles. Finally, the DISPLAY routine takes 218 clock cycles if it refreshes a digit and 282 cycles (worst case) if it instead jumps to SWITCH to check the position of the channel selector switch. These figures are obtained simply by adding the execution timings for the various instructions, being sure to include all the passes around each loop. For example, in the COUNT routine the loop LOOP2 is executed six times, with 138 clock cycles each pass. In SWITCH, the loop LOOP3 is executed a variable number of times, depending on the switch position. It is obvious that loops that are executed many times can dominate the total execution time.

In this example the worst-case execution time for the main loop is 1209 clock cycles, or 393.5μs. Thus the display is completely refreshed 350 times per second, and the maximum guaranteed count rate is 2.54kHz. This has got to be one of the slowest counting circuits in the book.

Because the execution of LOOP2 costs us 798 out of the total 1209 clock cycles, any improvements we make in other areas of the program won't have a dramatic effect on the performance. It is worth trying to optimize

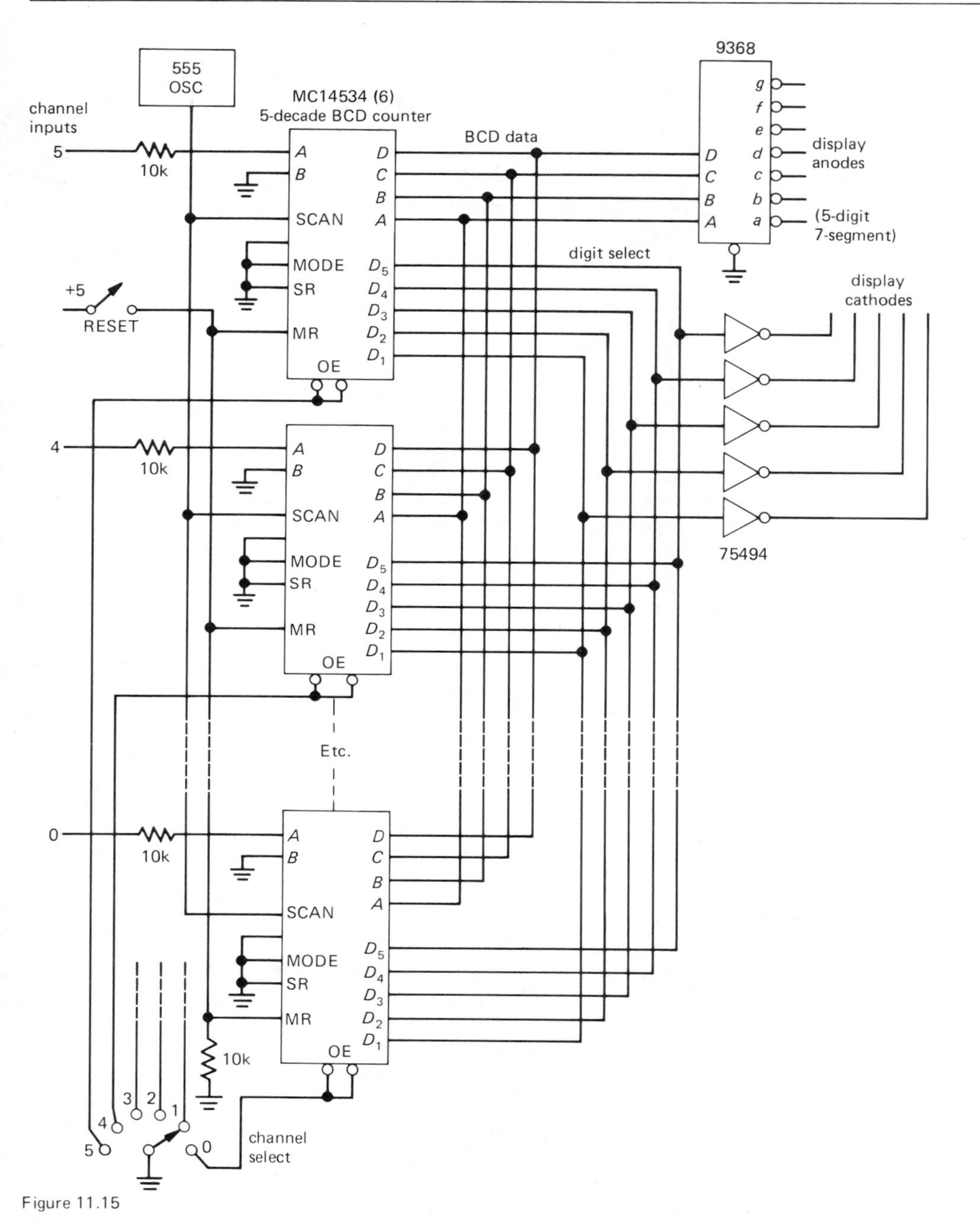

Figure 11.15

Six-channel counter constructed with discrete LSI logic, instead of a microprocessor.

that loop, something we've already tried hard to do. One alternative you might consider is the use of binary number format, rather than "packed BCD," for storing the counts in RAM. This trims the execution time by 72 cycles, but it expands the DISPLAY routine enormously. Going in the other direction (one BCD digit per byte of RAM) speeds up the DISPLAY routine, but further worsens the LOOP2 bottleneck.

Another figure of merit for an instrument is the chip count and total chip cost. Our design uses 2 gates, 11 MSI chips, and a microprocessor plus two memory chips. The total IC parts cost is about $50 (not to mention the software development costs).

Comparison with discrete logic

In this case it is tempting to compare our finished design with a 6-channel counter built without a microprocessor. For this purpose we have designed a humble 6-channel counter using MC14534 CMOS LSI counters. The circuit is shown in Figure 11.15. We have cheated a bit, changing the specification to match the 5-digit capability of these chips. These pleasant counter chips have three-state outputs, a virtue in this circuit because the outputs can be tied together, with the displayed counter enabled by the selector switch. Its display outputs are easily buffered to drive the usual LED digit stick. This circuit requires an oscillator and two MSI chips, in addition to the six counters. The maximum count rate is 500kHz, and the IC parts cost is about the same as that of the microprocessor circuit.

Among other things, this comparison illustrates that the use of microprocessors doesn't necessarily enhance a circuit's performance. You may even wind up with an inferior design, because you got bogged down in the details of just making it work at all. In defense of microprocessors, however, we should point out that even in a simple case like this the microprocessor adds a great deal of potential. For example, you could activate an alarm at a preset count, do statistics on the distribution of arrival times, subtract average values, or add serial RS-232C output capability, all of which would be extremely awkward to add to the discrete circuit. In addition, an instrument designed with microprocessors can be reprogrammed to do some other task without hardware changes. Furthermore, in the realm of highly complex instrumentation, the case for microprocessors is clear-cut.

The hardware/software tradeoff can take still another form, since we could have used discrete counter chips in combination with a microprocessor. This simplifies the programming and improves the performance, at the expense of hardware complexity and speed. Such a combination generally yields the advantages of both: maximum performance, combined with the flexibility and computational power of microprocessors. Microprocessor manufacturers generally provide compatible support chips to perform often-needed tasks, e.g., the 8253 triple 1MHz counter from Intel.

MICROPROCESSOR SUPPORT CHIPS

In the example we just finished, a chip count revealed that the majority of the ICs we used were medium-scale integration (MSI) chips, such as latches, three-state buffers, display decoder-driver, and power buffers. The actual microprocessor and its memories (RAM and EPROM) accounted for only 3 of the 16 chips. Latches and three-state buffers, in particular, are used extensively in all microprocessor systems, and so it is important to take a more detailed look at some of the choices you have.

□ 11.8 Medium-scale integration

□ *Transparent and edge-triggered latches*

We mentioned latches briefly in Section 8.24. A latch whose outputs follow the respective inputs while enabled is called a *transparent latch* or jam-loaded latch, in contrast with an *edge-triggered latch,* which is just an array of *D* flip-flops with a single common clock input. The difference has important consequences when latching data from a bus, because of the relative timing of DATA and WRITE strobe signals. The situation is summarized in Figure 11.16. In a typical microprocessor bus protocol, the data is not necessarily valid at the leading edge of the WRITE strobe pulse, but it is guaranteed to be valid (and to have been valid for some minimum setup time) at the trailing edge of the WRITE strobe. As a result, the output of a transparent latch, which is enabled during the entire WRITE pulse, may pass through a transient state, as shown in Figure 11.16A. The edge-triggered latch, by comparison, changes state at the end of the WRITE pulse, and it is guaranteed not to have any glitches at its output (Fig. 11.16B).

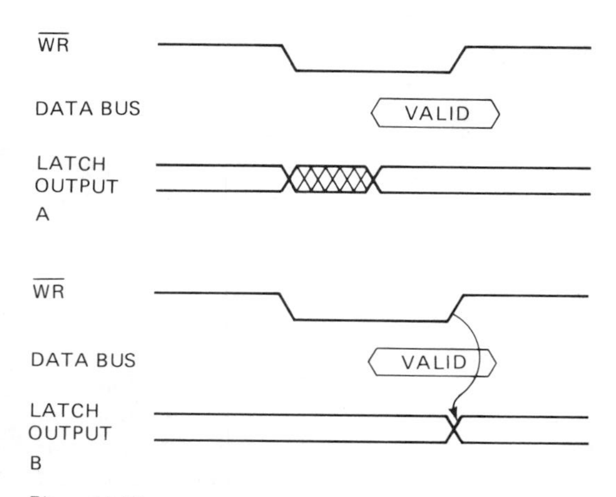

Figure 11.16
Write-cycle timing. A: Transparent latch. B: Edge-triggered latch.

There is a nice variety available in both transparent and edge-triggered TTL latches, with features such as RESET input, convenient pinout (inputs and outputs on opposite sides), inverted outputs, complemented outputs (both true and inverted outputs), three-state outputs (useful for driving buses), and separate input enable. The latter lets you simplify external gating by driving the clock input with the WRITE signal and the INPUT ENABLE with the device select logic. Note that there aren't enough pins to provide all these features on a single chip, since you want to use the pins for data. Because of the organization of most processors by bytes, it is desirable to be able to buffer 8-bit chunks of data in a single chip. You can't do this with a 16-pin package (count 'em), so the ever-versatile semiconductor industry has responded with a nice slim (0.3-inch width, same as the 16-pin package) 20-pin DIP. We will have more to say about packages later (look for our all-new "PEA" index).

Tables 11.3 and 11.4 summarize available TTL latches. At this writing the 74LS377 is a pretty good all-purpose output latch, and the 74LS374 is a good input latch (for driving buses).

□ *Buffers*

Another chip used by the bucketful in microprocessor system design is the three-state buffer. You use them for asserting data and address information onto the bus. Most often you're simply sending data to the CPU. As with latches, there are 20-pin versions with 8-bit width. Options include hysteresis, inverted output, opposite-side pinout, and pinout with separate enable inputs for bidirectional use. Figure 11.17 shows a bidirectional three-state buffer used to send data in either direction. These are used to buffer the feeble microprocessor input/output data bus to the relatively high current and capacitive system bus, a process that is handled by high-current drivers on the CPU board in a

TABLE 11.3. EDGE-TRIGGERED LATCHES

	Bits	Pins	Input enable	Opposite side pinout	Output inverted	Three-state output	Complementary output	Reset
74LS175	4	16	—	—	—	—	√	√
379	4	16	√	—	—	—	√	—
173	4	16	√	√	—	√	—	√
376	4	16	—	—	—	—	—	√
276	4	20	—	—	—	—	—	√
174	6	16	—	—	—	—	—	√
378	6	16	√	—	—	—	—	—
273	8	20	—	—	—	—	—	√
374	8	20	—	—	—	√	—	—
364	8	20	—	—	—	√	—	—
574	8	20	—	√	—	√	—	—
534	8	20	—	—	√	√	—	—
564	8	20	—	√	√	√	—	—
377	8	20	√	—	—	—	—	—
25LS2520	8	22	√	—	—	√	—	√

TABLE 11.4. TRANSPARENT LATCHES

	Bits	Input enable	Opposite-side pinout	Output inverted	Three-state output	Complementary outputs	Reset
74LS375	4	—	—	—	—	√	—
373	8	—	—	—	√	—	—
363	8	—	—	—	√	—	—
573	8	—	√	—	√	—	—
533	8	—	—	√	√	—	—
563	8	—	√	√	√	—	—

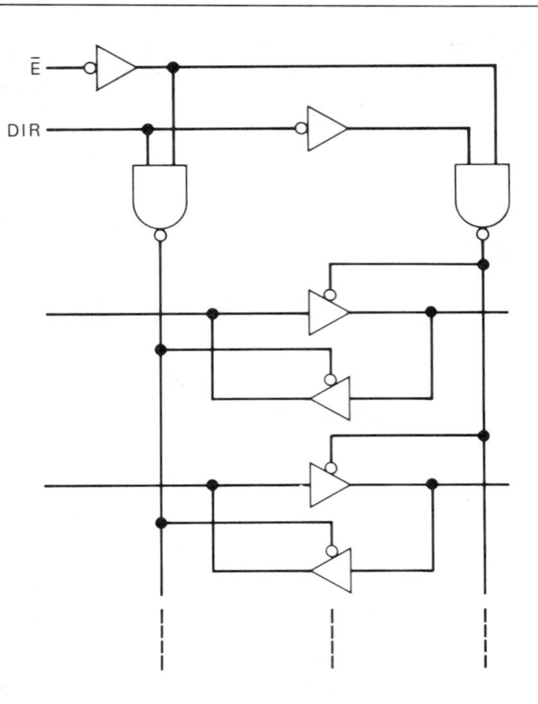

Figure 11.17
Bidirectional three-state buffer.

minicomputer, as we discussed in Chapter 10.

Note that we're really talking about two different kinds of uses here. You always need a three-state driver to get data into a microprocessor, even in a simple application like the counter we illustrated; in addition, a buffer with substantial output current capability is needed to drive the heavily loaded backplane bus typical of a larger system. For the latter you often avoid the low-power Schottky types, with their 24mA maximum sinking capability, in favor of "standard" TTL buffers with up to 48mA sinking capability.

Table 11.5 lists an impressive array of three-state buffers. The 74365-368 (identical to 8095–8) are the classics, but they suffer from having only 6 bits (16-pin package). The Signetics 8T95-8 are pin-compatible, with greater sinking capability (48mA). The 74245 is a popular 8-bit bidirectional buffer.

□ *Packages*

The 20-pin DIP arose from the needs of 8-bit microprocessors, as we mentioned earlier. Unlike the earlier 24-pin DIP with 0.6-inch width, the newer packages economize on printed-circuit board "real estate" by squeezing into the standard 0.3-inch width. Since a typical circuit board is dominated by peripheral chips, as we demonstrated earlier, it is important to raise the issue of what sort of packages you are going to use to populate your board. An important alternative to these MSI 20-pin chips is the use of NMOS LSI peripheral chips to handle tasks like latching, buffering, counting and timing, serial I/O, disk and video control, DMA transfers, and interrupts. These "system components" are usually introduced along with the CPU chips by the same manufacturer and are specifically designed to mate with the host CPU.

Before getting carried away with the CPU manufacturer's persuasive literature proclaiming the virtues of these peripheral chips, you want to consider their tradeoffs, particularly the high price tag, large package size, and excessive flexibility. Often humble MSI components can do the job at least as well. To illustrate the space tradeoff, we have created Table 11.6, listing the package

TABLE 11.5. THREE-STATE BUFFERS

	Bits	Pins	Bidirectional	Pinout for bidirectional	Opposite-side pinout	Hysteresis	Inverting	Enables	Alternates
74LS125	4	14	—	—	—	—	—	4	8093
126	4	14	—	—	—	—	—	4	8094
242	4	14	✓	✓	✓	✓	—	2	
243	4	14	✓	✓	✓	✓	—	2	
365	6	16	—	—	—	—	—	2	8T95
366	6	16	—	—	—	—	✓	2	8T96
367	6	16	—	—	—	—	—	2	8T97
368	6	16	—	—	—	—	✓	2	8T98
8T37	6	16	—	—	—	✓	✓	2	8T38
241	8	20	—	✓	—	✓	—	2	
541	8	20	—	✓	✓	✓	—	2	
244	8	20	—	✓	—	✓	—	2	
240	8	20	—	✓	—	✓	✓	2	
540	8	20	—	✓	✓	✓	✓	2	
245	8	20	✓	✓	✓	✓	—	1	8T125
645	8	20	✓	✓	✓	✓	—	1	
640	8	20	✓	✓	✓	✓	✓	1	
641	8	20	✓	✓	✓	✓	—	1	
642	8	20	✓	✓	✓	✓	✓	1	

TABLE 11.6. PEA[a] FOR VARIOUS DIP IC SIZES

Pins	Width	Chip area	PC board area	PEA[a]
14	0.3	0.18	0.54	0.9
16	0.3	0.21	0.60	1.0
18	0.3	0.24	0.66	1.1
20	0.3	0.27	0.72	1.2
22	0.3	0.3	0.78	1.3
22	0.4	0.40	0.91	1.5
24	0.4	0.44	0.98	1.6
24	0.6	0.66	1.26	2.1
28	0.6	0.78	1.44	2.4
40	0.6	1.14	1.98	3.3

Note: Effective PC board areas calculated for 0.15 inch clearance all around the chip, i.e., 0.3 inch between chips.
[a] 16-pin package equivalent area.

equivalent area (PEA) in units of a 16-pin DIP. The popular 40-pin LSI package uses the space of 2.8 20-pin packages. Having sounded this caution, let's take a look at some of these LSI wonders.

11.9 Peripheral LSI chips

General characteristics

As we mentioned earlier, LSI chips intended for microprocessor support are usually constructed with NMOS technology, and they are usually supplied in large packages with 28 or 40 pins. They tend to be designed with lots of flexibility, often with programmable parameters of operation. Although they are usually designed for specific microprocessors, their generality allows you to use one manufacturer's support chips with another's CPU. They are expensive at the time of their introduction, often costing more than the CPU itself, but they show the usual exponential decay in price that is characteristic of IC technology (and precious little else in this world!). Figure 11.18 illustrates this apparently universal law of "Silicon Valley" (a section of the San Andreas Fault between San Francisco and San Jose).

Although we cast some disparaging remarks in their direction, many LSI support chips are virtually indispensable: Disk and video controllers are obvious examples.

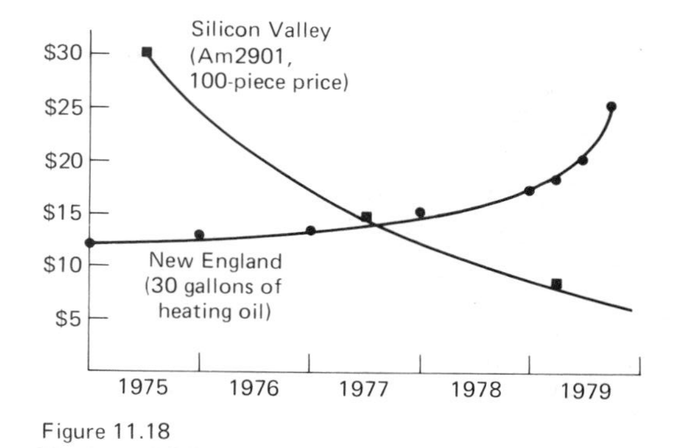

Figure 11.18
The law of Silicon Valley: learning curve.

Another widely used type of support chip is the universal synchronous/asynchronous receiver transmitter, USART for short.

□ *How to use a USART*

A USART is a microprocessor-controlled version of the UART, which we discussed in Section 8.27. Although conventional UARTs can be used with a microprocessor, the USARTs offered with the various microprocessor families are more convenient, with control of operating mode programmable through the bus. Most CPU families have their own USART, e.g., the Motorola 6850 (6800 CPU), the Zilog SIO (Z80 CPU), the MOS Technology 6550 (6502 CPU), and the Intel 8251 (8085 CPU). With the possible exception of the 6850 (more about this later), these chips are essentially compatible. We will illustrate with the popular 8251A.

USARTs are most often used to send data to and from terminals or hard-copy devices (printers, plotters), where the major requirement is universal compatibility and simplicity of interconnection. The usual method is to use serial ASCII transmitted via RS-232C levels, as described in Section 10.17. For this sort of communication the USART is operated in the asynchronous mode, with each 8-bit character sandwiched between a START and STOP bit, and transmitted as a 10-bit serial string at a standard baud rate (see Section 10.17).

The 8251 comes in a 28-pin package, and it communicates with the CPU via the 8 DATA lines of the CPU's bus, five additional control lines, and a single CHIP SELECT input. Figure 11.19 shows the simple

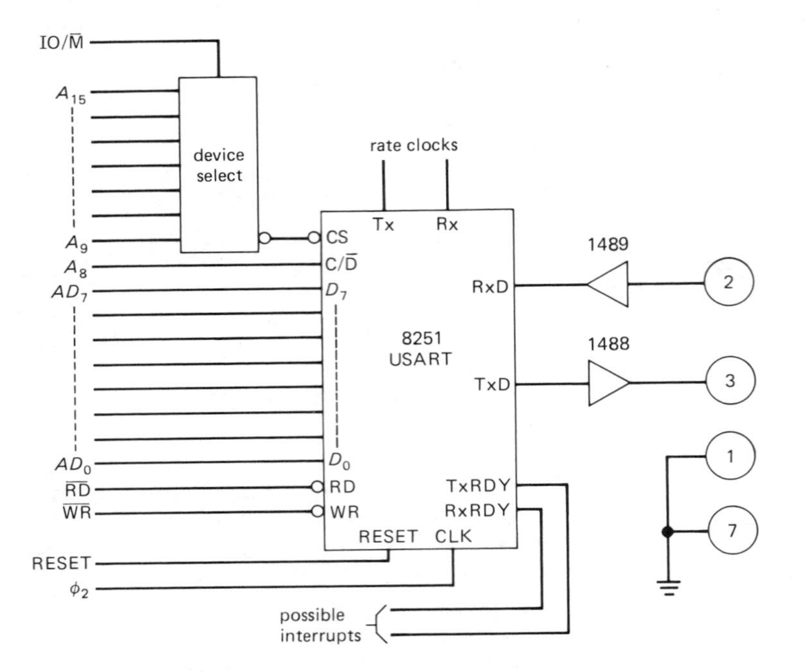

Figure 11.19
Interfacing a USART to a microprocessor bus.

connection to a typical bus, along with the data-rate clock, serial data input/output, and external control lines of the RS-232C protocol.

□ *CPU lines.* The RESET input initializes the chip to an idle mode, awaiting software mode selection. It is usually tied to the power-on clear line of the bus. The CLK line is tied to any clock signal that is at least 30 times the send/receive bit rate; the frequency and phase are not critical. The chip select (CS′) is used to activate the chip for processor communication via the bus. RD′ and WR′ strobe parallel data via the bus if the C/D′ line is LOW, and they strobe status and control information if the C/D′ line is HIGH. The C/D′ is usually connected to address line A_0 (LSB), which means that data and status/control are treated as two I/O ports.

□ *Serial I/O lines.* The serial data input and output lines and the various serial control lines (CTS, RTS, DTR, DSR) all conform to the standard RS-232C protocol. On these four lines we'll make no comment. RxD and TxD are the received and transmitted data lines, respectively.

□ *Other lines.* TxRDY, RxRDY, and TxEMPTY let you monitor the status of the USART's registers for troubleshooting or for interrupts. TxC and RxC are the transmitter and receiver reference rate inputs, supplied from some external oscillator, system clock (e.g., the 6.144MHz clock of the 8085 CPU), or baud-rate generator chip. These are normally set to 16 times the desired serial baud rate, e.g., 19.2kHz for 1200 baud. (As we discussed in Chapter 10, 1200 baud is 1200 bits/second, corresponding to 120 10-bit words/second, i.e., 120 ASCII characters/second.)

□ *Software.* As we said at the outset, the operating modes of the USART are controlled by software commands. In other words, a byte sent to the USART in C mode (C/D′ HIGH) is interpreted by the USART as a control command, setting the mode of operation. You can choose, for example, synchronous versus asynchronous operation, the number of STOP bits, and even versus odd parity (or none at all). Here are the control command bytes for three common serial-data modes:

Data	*Parity*	*Command byte*
8 bits	none	01001110
7 bits	odd	01011010
7 bits	even	01111010

```
CDATA = 80H                ;address of the USART data register
CSTAT = 81H                ;address of the USART control/status register
  •
  •
  •
ORG     0                  ;CPU reset starting address
        MVI A, 01001110B   ;command word for 8 bits, no parity
        OUT CSTAT          ;send to USART control register
          •
          •
          •
```

Program 11.13

Note that there are numerous other possible operating modes under your control, and you are forced to set every bit of the command word correctly. These commands must be sent by the processor to initialize the chip, and they require a few lines of program code. In other words, a software command operation must precede any hardware serial-data transmission. This sort of complexity is the price you pay for the extreme flexibility of these microprocessor support chips. In this case you would have to execute the lines of 8085 code shown in Program 11.13.

After setting up the USART operating mode via the command register, actual bytes of data are sent and received by CPU I/O commands in the D mode (C/D′ LOW). The status register must also be interrogated (C mode again) to determine when serial data received by the USART is ready to be picked up by the CPU and when it can accept data to be transmitted. In addition, other bits of the status register tell if a parity error was detected, if incoming data was lost, etc. You usually ignore these latter dire indications of doom and plunge boldly ahead. Program 11.14 is an example.

Note that these are the simplest form of handlers, using programmed I/O to check for status information (see Sections 10.06–10.08). These hold up CPU operation by looping on the status flags. Input, in particular, would benefit from an interrupt-driven routine, or at least a loop that checks the status flags less frequently.

Parallel I/O (PIO) chips

Another really important class of LSI peripheral chips is composed of the multiple parallel-port I/O chips. These let you latch outgoing data and buffer incoming data in parallel format (typically 8 bits wide); they sometimes include programmable timers and may even have parallel/serial conversion capability. Like the USART, they have programmable operating modes: You can select size and direction of data transfer, perform individual bit manipulations, and set timing

```
                   ;transmit routine
                   ;enter here, with outgoing data byte
                   ;in register C
LOOP1:  IN CSTAT
        ANI 04     ;trans buff mask
        JZ LOOP1   ;wait until trans buff empty
        MOV A,C
        OUT CDATA  ;send data byte
        RET


                   ;receive routine
                   ;incoming data byte returned in A
LOOP2:  IN CSTAT
        ANI 02     ;receive rdy mask
        JZ LOOP2   ;wait for data byte
        IN CDATA   ;bring it into A
        ANI 7FH    ;strip MSB
        RET
```

Program 11.14

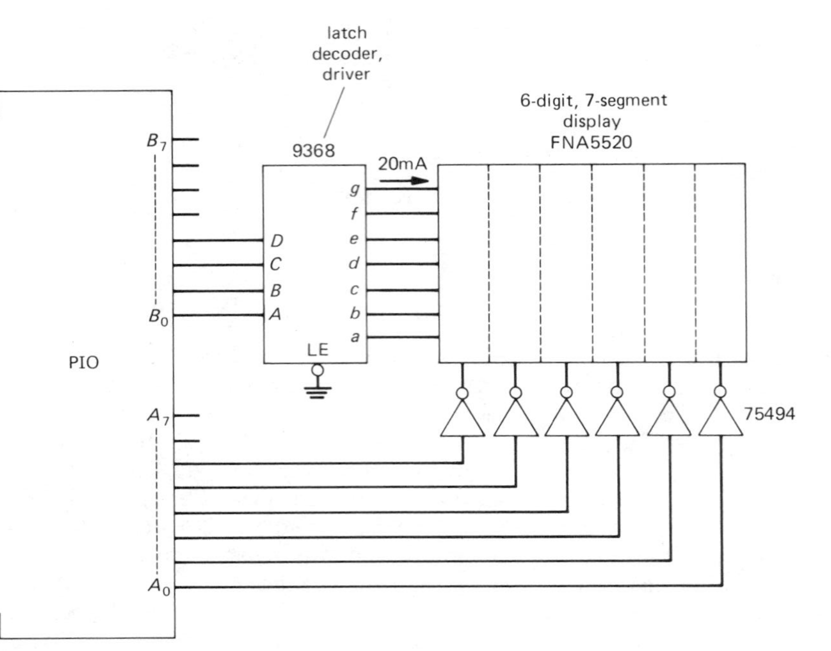

Figure 11.20

Driving a multiplexed display with a PIO.

intervals. The PIO's data-transfer protocol can also be programmed; for example, you can enable automatic interrupts and select a handshake mode (see Section 10.13 and the next section for a brief discussion of handshakes). Table 11.7 summarizes typical PIOs.

Nearly all PIO chips are constructed with NMOS technology, which means that they cannot source much current and can typically fan out into four or five low-power Schottky (LS) TTL loads. As a result, they are usually used in conjunction with power driver chips to drive loads that require significant amounts of current. Don't try to turn on a relay directly with a PIO output.

Figure 11.20 suggests the sort of circuit you might use, in this case to refresh a 6-digit display. It is worth noting that these NMOS outputs can be used to drive a grounded-emitter Darlington switch directly, as we discussed in Section 9.10. The output (sourcing) current with the HIGH output clamped at 1.5 volts is often specified on

TABLE 11.7. LSI PIO CHIPS

Part	RAM bytes	I/O (bits)	DDR[a]	Bit S,R[b]	Input latching	Handshakes	Interrupts	Timer counter (bits)	Address lines
6522[d]	—	16	√	—	√	2[c]	1	16	4
6532[d]	128	16	√	—	—	—	1	8	7
6820,21[d]	—	16	√	—	—	2	2	—	2
8154	128	16	—	√	—	1	1	—	7
8155,56	256	22	—	—	√	2	2	14	0[e]
8254	—	16	—	√	—	1	1	—	7
8255	—	24	—	—	√	2	2	—	2

[a] Individual-bit data direction register. [b] Individual-bit set/reset with one instruction. [c] Serial-to-parallel I/O conversion available. [d] The 65*xx* and 68*xx* PIO chips require Φ_2 clock. [e] Uses $AD_{0\text{-}7}$ lines for both data and address.

the data sheet under the heading "Darlington drive current." For the 8255, for example, it is 1.5mA (min) at 1.5 volts.

□ *Handshaking.* The business of handshaking deserves a bit more explanation. Suppose you are sending bytes of data to a processor via a PIO port. You want to know when the PIO is ready to accept the next byte, i.e., when the previous byte has been picked up by the processor. The natural way to handle this is to have an output READY signal from the PIO. It is asserted when the PIO is able to take a byte, and it is disasserted from the time you have strobed data in until that data is picked up by the processor, at which time READY is again asserted. In other words, you can strobe your data any time READY is asserted. The actual data should be valid at the trailing edge of the strobe pulse, when the PIO accepts the data.

Figure 11.21 illustrates this simple concept, using the terminology of the 8255 PIO: IBF is "input buffer full," the complement of READY; STB′ is the input strobe signal; INTR is the PIO's signal to the processor that it has some data that needs fetching. The strobe pulse is specified as 500ns minimum, and the data must remain valid for 180ns after the strobe's trailing edge. The diagram also shows what happens after the processor reads the data from the PIO.

Although it is usual to use a strobe pulse of fixed width, you could simply hold the strobe level until you see IBF (input buffer full) asserted, as suggested by an arrow on the diagram. This kind of interlocked data exchange ensures that no data is lost.

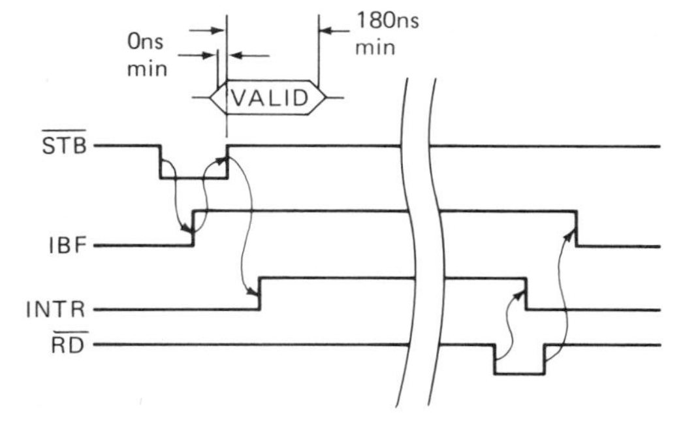

Figure 11.21
PIO handshaking.

Microprocessor-bus A/D converters

Because there is often a need for A/D converters in microprocessor-based instruments, several manufacturers have introduced ADCs designed for simplified connection to a data bus. In particular, these converters have byte-wide three-state outputs that can be connected directly to the bus. For example, the ICL7109 from Intersil is an integrating 12-bit A/D converter whose output can be asserted onto the bus as two successive bytes, without the need for the usual interface paraphernalia such as external latches, three-state buffers, etc. START and STATUS lines on the converter permit the microprocessor to do other things while the A/D is carrying out its relatively slow conversion (7.5 conversions per second).

The ICL7109 typifies bus-oriented stand-alone converters; it is a complete A/D converter, whether used with or without a microprocessor. Chips of another class, also called microprocessor A/D converters, *require* a microprocessor for their operation; they cannot operate as stand-alone converters. Such chips contain some linear-conversion circuitry, but they must be driven by the microprocessor software in order to function. In other words, the combination of microprocessor software plus converter chip is capable of doing A/D conversion, and software is an inherent part of the ADC. For example, one such converter chip contains a DAC and analog comparator; to bring the converter to life the software would have to generate trial codes and check the comparator's output, thus forming a complete successive-approximation converter.

Note that D/A converters, as opposed to A/D converters, are inherently easy to use with microprocessors. In particular, those with latches at the data inputs can be connected directly to a data bus.

11.10 Memory

In both minicomputer and microcomputer systems, as discussed in Section 10.16, allocation of memory is handled by an operating system, and it can vary from moment to moment; program space (instructions) at one time may become an array of data at another time. The memory itself is completely general, with some minor exceptions. In dedicated microprocessor applications, on the other hand, the assignment of memory is part of the design, with blocks of nonvolatile (permanent) ROM being used for program storage, and read/write volatile RAM being used for temporary storage of data, stacks, and program workspace. Nonvolatile program storage is universally used in dedicated instruments so that it isn't necessary to load a program each time the instrument is turned on.

In this section we will discuss the various kinds of memory: static versus dynamic RAM, RAM versus ROM, mask ROM versus PROM and EPROM (not to mention EAROM).

Static and dynamic RAM

A static RAM stores bits in an array of flip-flops, whereas a dynamic RAM stores bits as charged capacitors. A bit once written in a static RAM stays there until rewritten, unless the power is turned off. In a dynamic RAM the data will disappear in less than a second, typically, unless "refreshed." In other words, a dynamic RAM is always busy forgetting data, and it is only rescued by a refresh clock, which periodically accesses the rows of data. Typically, you have to access each of 64 row addresses every 2ms.

You might wonder what would possess anyone to choose a dynamic RAM. By not using flip-flops, the dynamic RAM saves space, giving you more data on a chip, at lower cost. For example, the classic 2114 static RAM stores 4K (1K × 4) bits and costs about $5, whereas the 4116 dynamic RAM stores 16K (16K × 1) for about the same price. Since the package sizes are similar, you can get four times as much memory on a large memory board, and at the same price, by using dynamic RAM.

Now you might wonder why anyone would choose static RAM (fickle, aren't you?). For small systems, a pair of static 2114s might suffice, whereas you are forced to use dynamic RAMs in multiples of eight because of their nearly universal single-bit width. The other advantage of static RAM is its simplicity of use, with no refresh clocks or timing contention to worry about (the refresh cycle competes with normal memory access cycles and must be properly synchronized).

Another kind of RAM combines the best features of static and dynamic types. These are actually dynamic RAMs with built-in refresh circuitry that is transparent to the user. They are 4 or 8 bits wide and are easy to use. They cost considerably more than comparable dynamic memories and may be more difficult to obtain.

RAMs are available in sizes from 1K bits to at least 64K bits, with access times from 750ns down to about 50ns. The largest memories are dynamic, whereas the convenient byte-wide and other small memories (e.g., the 8185 1K-by-8-bit static) are invariably static.

Ready-only memory (ROM)

ROMs are used as nonvolatile program storage, for bootstrap programs, lookup tables (including character generators), sequence controllers, etc. They come in several varieties: mask-programmed ROMs, fusible-link ROMs, UV-erasable EPROMs, and electrically alterable EAROMs.

Mask-programmed ROMs. These are born with your bit pattern built in. The semiconductor house converts your bit specification into a custom metallization mask used to process the chip. Sizes range from less than 1K bytes to 8K bytes and above. These are for large production jobs, and you wouldn't dream of having a mask-programmed ROM designed for prototyping. Typical costs are $1000 to $3000 setup charge, with the manufacturer strongly discouraging you from buying fewer than 1000 ROMs at a time. In those quantities the chip may cost less than $5.

□ *Fusible-link ROMs.* These typically begin life with all bits set, and you give them electrical shock treatments until you've blown out the offending bits. For example, the Intel 3624 uses the "time-proven polycrystalline silicon fuse" technology. It contains 512 bytes in a 24-pin package, and it is programmed by dumping a large current into the appropriate output pins while running the ROM at an elevated supply voltage: You check to see if the links have blown; if not, you just keep zapping it.

Fusible-link ROMs are good where you want small and inexpensive ROMs.

UV-erasable ROMs (EPROM). These ROMs can be programmed (and erased, if necessary) by the user. They employ an array of MOSFETs with floating gates that can be charged by "avalanche injection," a fancy name for breakdown of the gate insulating layer by an applied 30 volt pulse. These memories store data by retaining indefinitely a tiny charge on these insulated gates. You can think of them as capacitors with time constants of centuries. You read out the state of an individual capacitor by allowing it to be the gate of an associated MOSFET channel. Since the gate is not electrically accessible, it can only be erased by exposing it to intense ultraviolet radiation for 10 to 30 minutes, which causes the stored charge to leak off by photoconduction. A UV-erasable ROM has a quartz window over the chip so that you can erase it.

The classic 2716 is a popular EPROM (erasable programmable ROM), with 2K bytes of storage, operation from a single +5V supply, and a price tag of about $25. It is programmed by applying +25V dc to a programming pin, then writing your data at the rate of 50ms per byte, checking to see if it has been properly written. It takes 100 seconds to complete the job. The older 2708 has half the capacity, requires three supply voltages (+12, +5, and −5), and costs about $5. The EPROM is the basic ROM form for small systems and small-level production. They're great for prototyping.

□ *Electrically alterable ROMs (EAROM).* These ROMs can be programmed and erased electrically, avoiding the lengthy erasing procedure of the EPROM. The EAROM is an example of some of the new ROM technology making its appearance. This is a hot field, with great advances promised in density and convenience.

11.11 Designing a system with LSI

We have been discussing MSI and LSI peripheral chips, RAMs, and EPROMs. At this point it should be instructive to see how to combine these memory types and support chips into a small system. We will show how RAM and ROM can be mixed, how peripheral chips are integrated into the system, and how to add memory chips not specifically tailored to the microprocessor.

A general-purpose microprocessor module

Look at Figure 11.22. This drawing could be the design of an instrument soon to be marketed. However, since we haven't divulged the contents of the ROM, you can't tell what it does. This circuit is more general than the example in Figure 11.4, including software-programmable uncommitted multiple I/O ports, expandable ROM space, serial I/O, and room for additional RAM.

We have begun with our favorite 8085. There are two 8155 parallel I/O (PIO) chips, used as parallel ports for external I/O and also as a timer to program the baud rate of the 8251 USART. They also contain 256 bytes of RAM each, making possible a small system without any additional RAM. A 2716 EPROM sits at the bottom of memory and holds initialization routines as well as the dedicated program for this instrument. The entire system is small and compact: three 40-pin packages, a 28-pin DIP, a 24-pin DIP, two small DIPs, and one each resistor, capacitor, and crystal. If you saw it all on a printed-circuit board, it would look unimpressive. The whole system could have been made with LSI, except that we needed a latch to hold the low-order part of the EPROM address (remember, they're multiplexed on the data bus) and a decoder to partition memory according to the high-order address bits.

The EPROM is put at the bottom of memory because power-on reset starts the processor at location 0, as previously mentioned. Also, interrupts force calls at locations low down in memory (2C, 34, and 3C, all hexadecimal), so you want program code down there to handle them. The 74LS138 1-of-8 decoder puts the EPROM in the bottom 2K of memory, with additional decoded select outputs for the other three ROMs you're absolutely sure you'll never need, but inevitably will. Address bit A_{13} must be LOW to enable the ROMs; when A_{13}

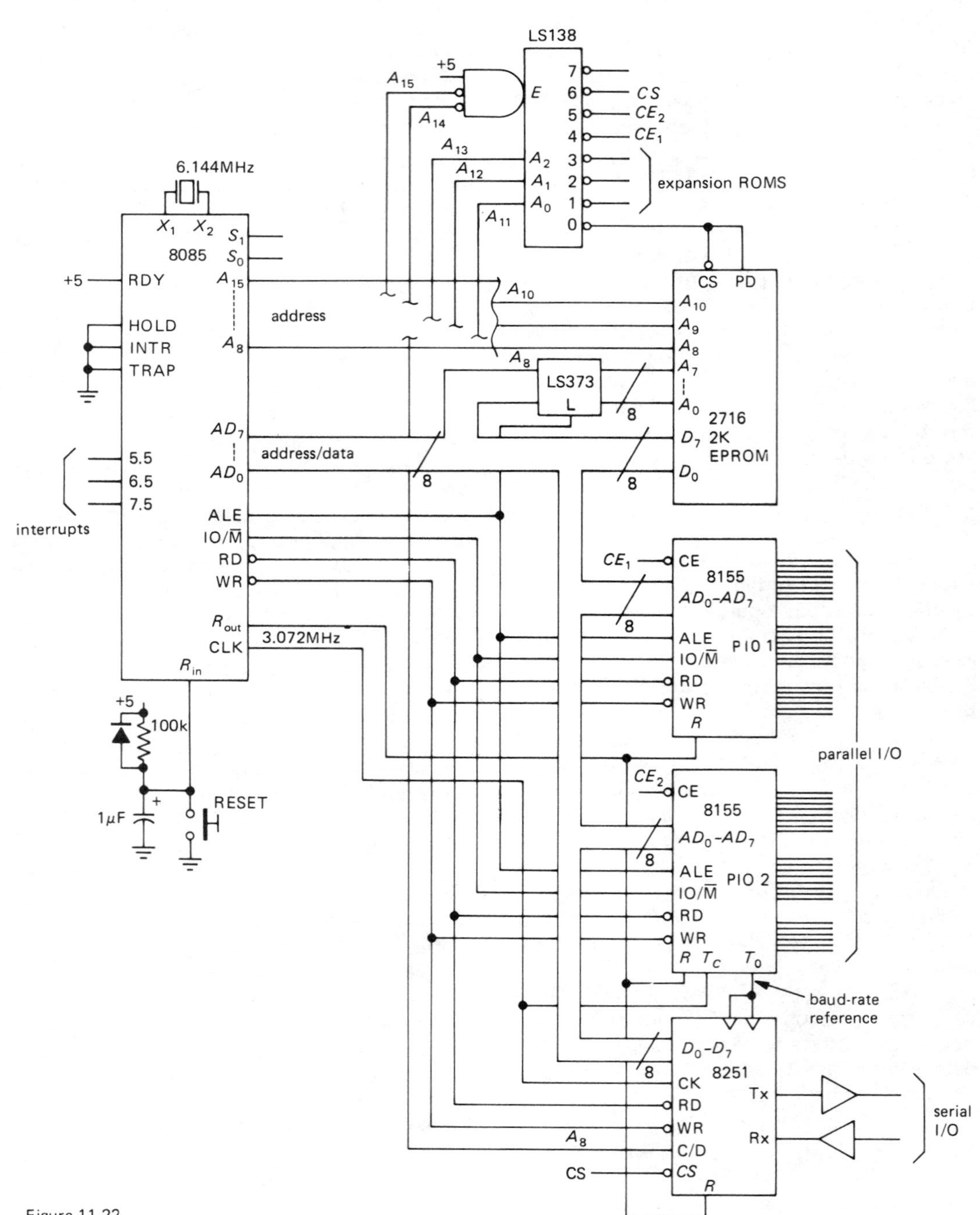

Figure 11.22

General-purpose microprocessor circuit.

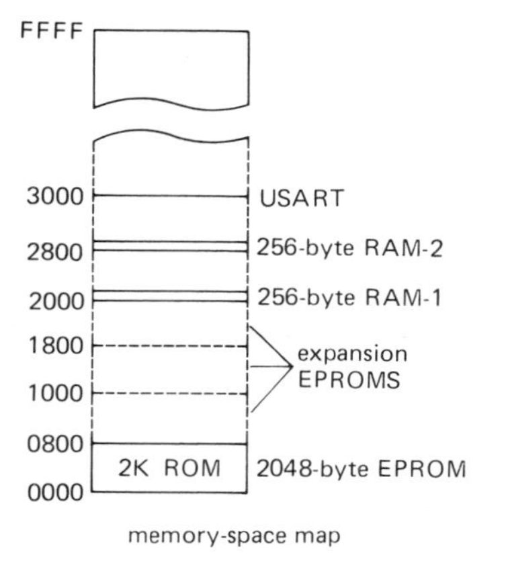

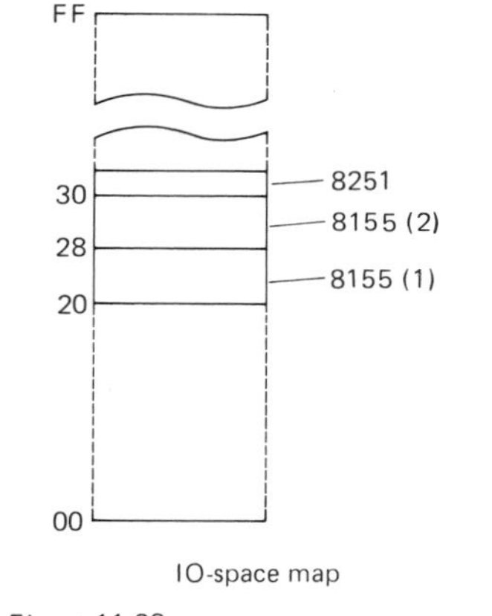

Figure 11.23
Memory allocation for the general-purpose microprocessor circuit.

is HIGH the ROMs are disabled, and the 74LS138 decoder outputs then select the peripherals (USART and PIOs). See Figure 11.23 for the "memory map" of this system.

Some additional points: Note that it isn't necessary to fill up memory space contiguously, beginning at the bottom. It often simplifies the circuitry to assign blocks of memory according to available high-order address lines, as we've done. The resulting gaps don't hurt anything. Note also that by saving a few gates and not fully decoding the address bits from the 8085, we've allowed the peripheral chips to be accessed at multiple places in high memory, just as with the 6-channel counter example (see Fig. 11.6). An additional subtlety of the addressing in this example involves the peripheral chips. Since the select gates (74LS138) don't look at the IO/M′ line from the processor, they enable the peripheral chips during both I/O and memory instructions. This doesn't cause any trouble, however, since the 8155 looks at the IO/M′ line (selecting its I/O or memory function internally), and the USART simply appears in both memory and I/O space at the same locations.

EXERCISE 11.12

Add gates to do full address decoding, so that RAM and EPROM don't make multiple appearances in address space.

□ *Adding RAM: timing considerations*

In many applications the 512 bytes of RAM that are available in the PIOs in the preceding example simply don't provide enough storage for the data you're handling. An example might be a device to buffer bursts of incoming data, say from a tape or disk unit. In such an application you need enough storage to hold one entire block of data in RAM, which you then process and send out before accepting the next burst of incoming data. It is tempting to use a popular low-cost RAM like the ubiquitous 2114, because typically the RAMs designed to mate with a given microprocessor cost considerably more. For example, 1K bytes made with a pair of 2114s cost about $8, whereas the 8185 1K-byte RAM intended for use with the 8085 costs about $25.

In general, when adding memory not specifically designed for the CPU, you will have to add some external circuitry for the strobe and enabling signals. You also have to check to see that the timing requirements are met (setup and hold times, for example). Figure 11.24 shows how to add 2K bytes to our previous circuit. The 74LS138 is enabled only when address bit A_{14} is HIGH, putting the RAM into address space beginning at location 4000 hex. The other enabling inputs require a memory access (IO/M′ LOW) and a READ or WRITE strobe, in order for the 74LS138 outputs to enable the chip select of the appropriate pair of 2114s. With this circuit, up to 8K of RAM can be added.

The timing situation is shown in Figure

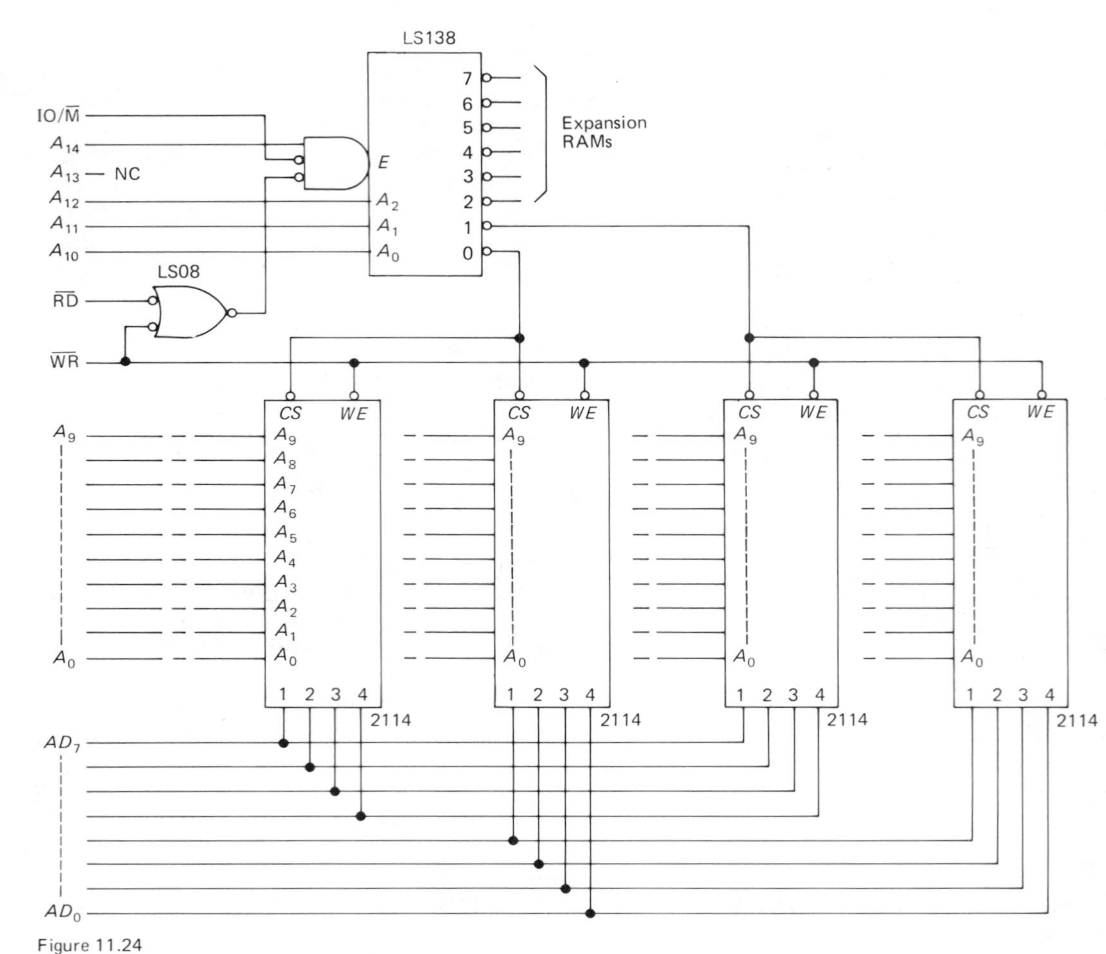

Figure 11.24
Adding 2K bytes of 2114 RAM to the general-purpose microprocessor circuit.

11.25, with the CPU and memory requirements shown on the same scale. The entire address is latched and presented to the RAM by the end of ALE (earlier if a jam-type register is used to latch the low-order byte of the address), and the CPU expects data back from the RAM to be asserted onto the bus within 460ns after the trailing edge of ALE in order to allow 100ns of setup time before the end of RD′. The timing diagram for the 2114 is also shown in Figure 11.25; the essential feature is that data is valid 450ns (t_A, the access time) after the address is asserted. In this case the setup time requirement is barely missed if an edge-triggered latch is used for the low-order address byte, since the memory takes 450ns to access its data, whereas the CPU imposes a deadline of 460ns. Since the latch has a propagation delay of 34ns (max), the setup requirement is actually violated by 24ns in the worst case. Fortunately, we've used a jam-loaded (transparent) register, which passes through a valid address approximately 100ns before the end of ALE, giving us 75ns of safety margin. A similar calculation for the WRITE cycle would also show adequate timing margins. Note that the 2114 is guaranteed to disassert its data before the CPU again asserts address information on the data bus at the beginning of the next machine cycle, thus avoiding overlapping contention for the data bus. In this case there is a 50ns margin.

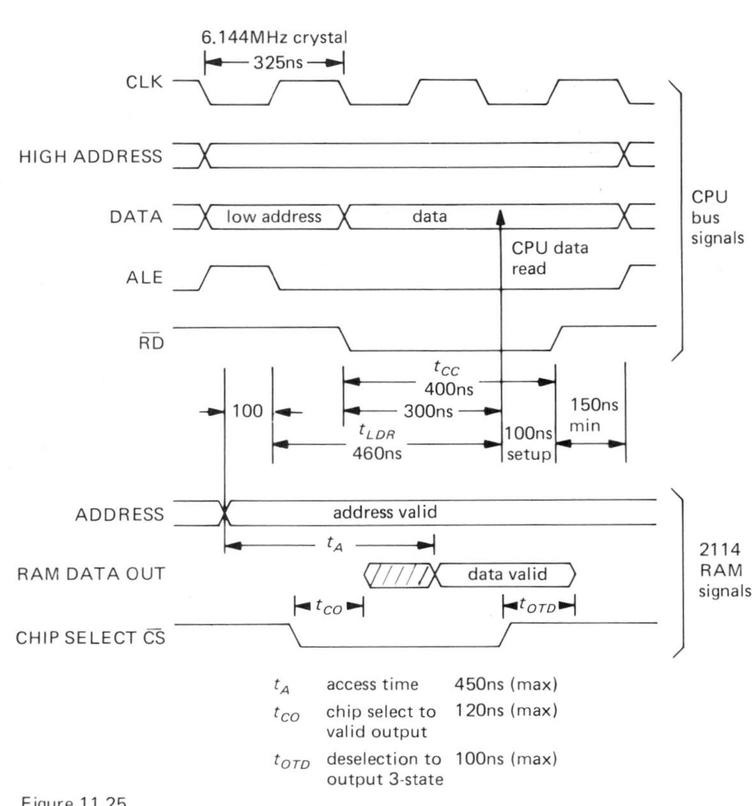

Figure 11.25
Memory timing diagram for an 8085 READ machine cycle, with 2114 RAM.

Two comments: Our timing calculations use worst-case values throughout. In reality, the circuit with edge-triggered address latching would probably work fine. Furthermore, we've used the slowest members of the 2114 and 8085 families; extra timing margin can always be gained by going to the faster versions or by reducing the system clock frequency.

□ *Capacitance*

Data sheets always give timing specifications with a certain assumed load capacitance. In the preceding example the worst-case values given all assume 150pF of load capacitance on all lines. In a system with a dozen chips or more, it is worth checking to make sure that the total load capacitance is less than 150pF; otherwise the timing specifications must be adjusted according to a formula given in the CPU data book (microprocessors never have data *sheets;* often you get a pretty thick book).

Here is the capacitance situation for the preceding circuit:

Quantity	*Device*	*Address lines load (pF)*	*Data lines load (pF)*	*Conrol lines load (pF)*
16	2114	80	40	40
2	8155	—	40	20
1	74LS373	5	5	—
1	8251	10	20	10
1	74LS155	10	—	—
Total		105	105	70

In this case the capacitive loading comes close to the specified value of 150pF, when the wiring capacitance is added in. It is clear that this small system cannot be expanded much further without using buffers to increase the drive capability of the CPU signals. However, with the extra chips that requires, the circuit may expand to the point that it may not fit conveniently on a single printed-circuit board. At that point we should begin to think about putting the system onto several boards, plugged into sockets on a motherboard (or backplane), with the signals bused along to every socket. We will discuss some popular microprocessor buses in the next section.

Buffering the CPU signals has one interesting subtlety: Since the data bus (AD_0–AD_7) is bidirectional, the buffer must be able to transmit signals in both directions, although not simultaneously. This is easily handled with a chip such as the 74LS245, an octal bus transceiver (see Table 11.5) that consists simply of a pair of back-to-back three-state buffers on each signal line. Figure 11.26 shows the circuit of an 8085

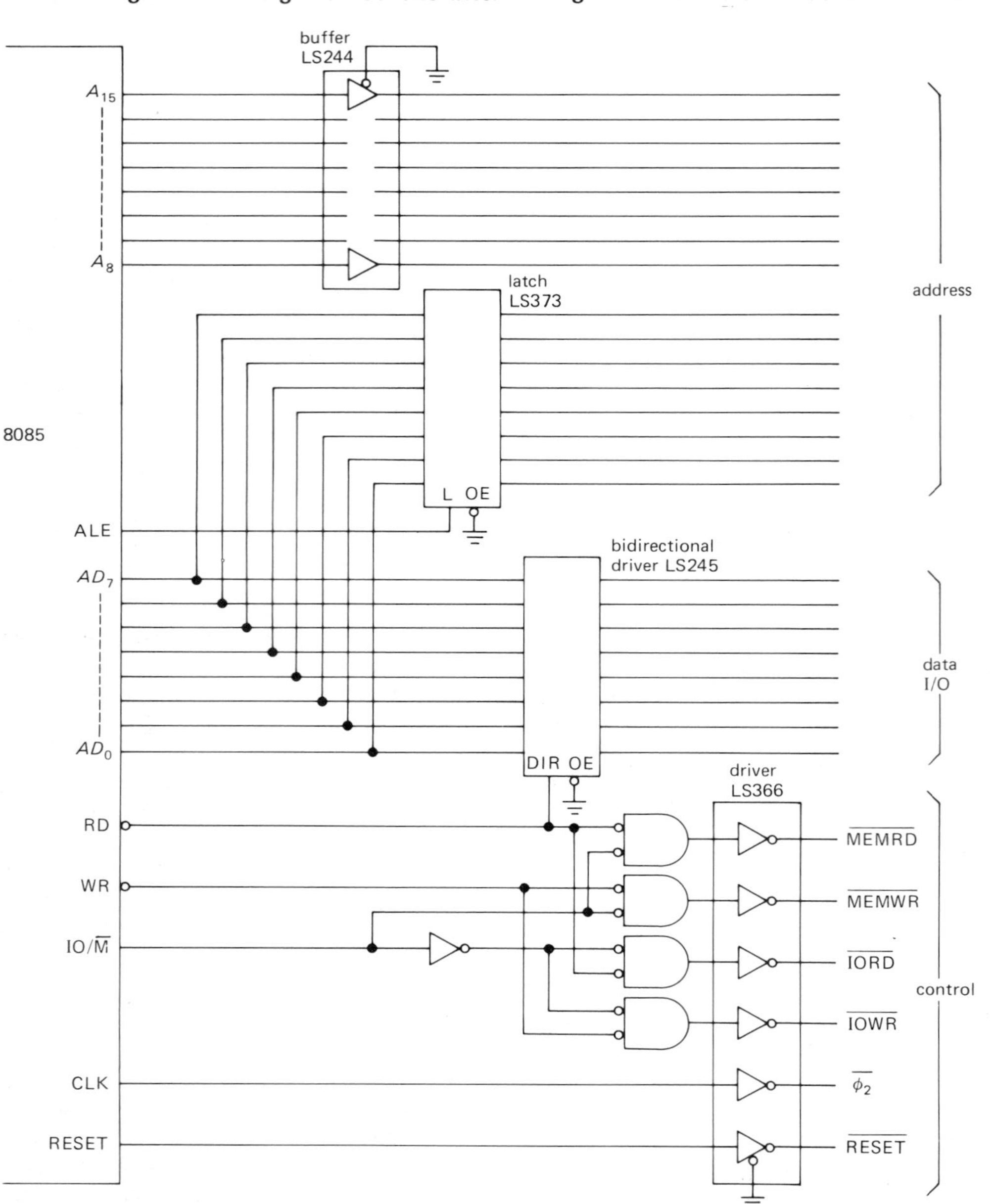

Figure 11.26
Buffering the 8085 bus.

CPU with bus buffers. The AD_0–AD_7 lines are normally enabled to drive the bus (74LS245 in transmit mode), and the buffer is switched to the receive mode only during the CPU RD pulse. The remaining lines are unidirectional and are buffered with ordinary hex and octal buffers. Note that the RD′, WR′, and IO/M′ have been converted to a set of four control lines, so that decoding on individual boards isn't necessary.

Of course, once you have a large system with plenty of capacitance loading the motherboard bus lines, you're forced to buffer the signals at the other end also. A memory card would use a bidirectional buffer transceiver on the data lines, with the transceiver direction controlled by the MEMRD′ line, and it would probably buffer the incoming address lines also, because of the large capacitive load it presents to the bus. It is usual to have some specification for the maximum loading a card presents to any bus line, e.g., two 74LS*xx* loads.

FURTHER TOPICS IN MICROPROCESSOR SYSTEM DESIGN

11.12 The S100 bus

History

The designers of the first microcomputer kit, the MITS Altair 8800, were faced with the problems we have just discussed. In their historic article in *Popular Electronics* in 1975, the photographs and schematic showed a stack of printed-circuit boards screwed together with spacers and interconnected by a set of wires soldered through aligned holes in the individual boards. Luckily, MITS didn't stick with this unserviceable arrangement, and purchasers of the Altair 8800 received a kit with a motherboard populated with 100-pin card-edge connectors. In the hectic weeks between their announcement and first shipments, the now-famous S100 bus was born.

MITS's 100-line bus quickly became the de facto standard for the exploding hobbyist computer market, with new manufacturers jumping on the bus almost weekly. MITS called their bus the Altair bus, but their competitors renamed it the S100 bus (for standard, 100 pins), and that name stuck. Unfortunately, it wasn't really a standardized bus. MITS did their memory write line decoding on the front panel, but not all microcomputers had a front panel. There was confusion about the meaning of certain bus signals, leading to bus contention. Then other companies created their own definitions for various unused lines (pin 67, the PHANTOM′ line, for instance). Manufacturers were supplying CPU boards with missing synchronization lines, etc. You could never be sure that two "S100-compatible" boards would in fact work together in the same system.

In spite of the problems with the S100 bus, many manufacturers and individuals made relatively powerful and inexpensive computers using it. All this investment made it worthwhile to fix up the S100's problems. A committee of interested engineers worked out a standard (with the blessing of IEEE) that included signal names and functions and timing specifications. This new standard has legitimized the S100 bus. Manufacturers can now design CPUs and peripherals to this standard, knowing what specifications their boards must meet to work with any other standardized cards.

S100 bus signals and timing

Table 11.8 shows the pin assignments for the IEEE standard S100 bus. Five ground lines and four power-supply lines distribute unregulated dc to the cards, each of which has on-board regulators to generate the regulated voltages needed, usually +5V and ±12V. There are 16 address lines (A_0–A_{15}), with provision for expansion to 24. Two sets of 8-bit data lines are used, one for writing (DO_{0-7}) and one for reading (DI_{0-7}); there is provision for using them together as a 16-bit bidirectional data bus. A set of status lines, labeled with lower-case s (e.g., sMEMR), indicates the nature of processor activity. Strobing pulses from the CPU are labeled with lower-case p (e.g., pDBIN, "data bus in"). An exception is MWRITE, which is generated somewhere in the system by NORing pWR′ and sOUT (Fig. 11.27).

There are two CPU clock signals available on the bus: a T-state clock, called Φ_2, typically 2MHz or 4MHz, and an M-cycle sync

TABLE 11.8. S100 BUS SIGNALS

Pin	Function	Pin	Function	Pin	Function
79	A_0	95	DI_0	36	DO_0
80	A_1	94	DI_1	35	DO_1
81	A_2	41	DI_2	88	DO_2
31	A_3	42	DI_3	89	DO_3
30	A_4	91	DI_4	38	DO_4
29	A_5	92	DI_5	39	DO_5
82	A_6	93	DI_6	40	DO_6
83	A_7	43	DI_7	90	DO_7
84	A_8	4[c]	VI_0'*	76	pSYNC
34	A_9	5	VI_1'*	77	pWR'
37	A_{10}	6	VI_2'*	78	pDBIN
87	A_{11}	7	VI_3'*	75	RESET'*
33	A_{12}	8	VI_4'*	74	HOLD*
85	A_{13}	9	VI_5'*	26	pHLDA
86	A_{14}	10	VI_6'*	25	sSTVAL'
32	A_{15}	11	VI_7'*	96	sINTA
16[a]	A_{16}	55	DMA_0'*	97	sWO'
17	A_{17}	56	DMA_1'*	44	sM1
15	A_{18}	57	DMA_2'*	45	sOUT
59	A_{19}	14	DMA_3'*	46	sINP
61	A_{20}	73	INT'*	47	sMEMR
62	A_{21}	12	NMI'*	48	sHLTA
63	A_{22}	13	PWR FAIL'	18	STAT DSB'*
64	A_{23}	3	XRDY	19	C/C DSB'*
49	2MHz clock	72	RDY*	22	ADD DSB'*
24	Φ_2	60	SIXTN'*	23	DO DSB'*
54	SLAVE CLR'*	100	GND	67[d]	PHANTOM'*
58[b]	sXTRQ'	50	GND	99	POC'
68	MWRITE	20	GND	98	ERROR'*
1	+8V	53	GND	2	+16V
51	+8V	70	GND	52	−16V

Notes: Asterisk indicates open collector lines with pull-up. [a]Addresses A_{16} to A_{23} are optional. [b]Data OUT lines DO_0 to DO_1 are used for (a) data output and (b) 16-bit I/O transfers if established by SXTRO request and SIXTN response. [c]Eight vectored-interrupt lines. [d]Bootstrap enable line, open collector.

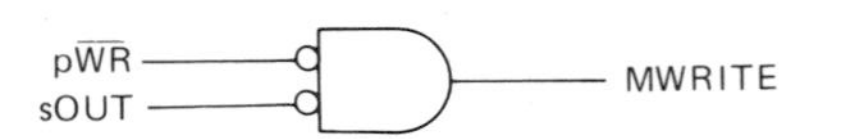

Figure 11.27
Generating MWRITE for the S100 bus.

pulse, called pSYNC. There is, in addition, a 2MHz clock line available for timing; it doesn't have to be related to the CPU clock. Figure 11.28 shows the standardized S100 timing specification for read and write operations, with relative timing referenced to the Φ_2 clock signal.

Example

Figure 11.29 shows how to use the S100 bus signals to read and write data. The peripheral asserts data onto the DI_{0-7} lines during a CPU read operation by enabling its three-state buffers, and it accepts data from the DO_{0-7} lines during a CPU write operation. The peripheral decodes its address from A_{0-15} to generate SELECT. PHANTOM is asserted by the CPU to inhibit normal memory transfer; it is used, for example, to fetch the start-up program from the power-on bootstrap ROM. Programmed I/O on the S100 bus is very simple.

Other buses

The S100 bus is the hobbyist's bus, although IEEE standardization has given it wider applicability. There are, in addition, a few other bus standards, used commonly for industrial designs. They are generally intended for dedicated processor designs rather than for general-purpose microcomputer systems.

Intel Multibus. Intel has promulgated a bus standard, known as the Multibus, that is currently being given an IEEE standard specification. It is a general-purpose bus and can be used with any microprocessor. The connectors are 86 pins (43 pins, dual-readout card-edge connectors) with 0.156-inch spacing, and the cards measure 6.75 × 12 inches. A number of manufacturers already offer compatible CPU cards, memory, and input/output. The bus has 20 address lines (1 Mbyte) and supports multiprogramming. Cards for this bus generally cost more than the equivalent function on the S100 bus.

STD BUS. This bus has been promoted by Pro-Log and Mostek, and it also is the subject of an IEEE standard. Its special feature is its small size: Cards are 4.5 × 6 inches, plugging into a 72-pin socket (36 pins, dual-readout) with 0.1-inch spacing. Many manufacturers already offer compatible cards.

11.13 Other microprocessors

Like any Darwinian process, the evolution of microprocessors has proceeded in several divergent directions. In the contest for survival, some of the less fit have become endangered species. As an example of different evolutionary paths, there is the division

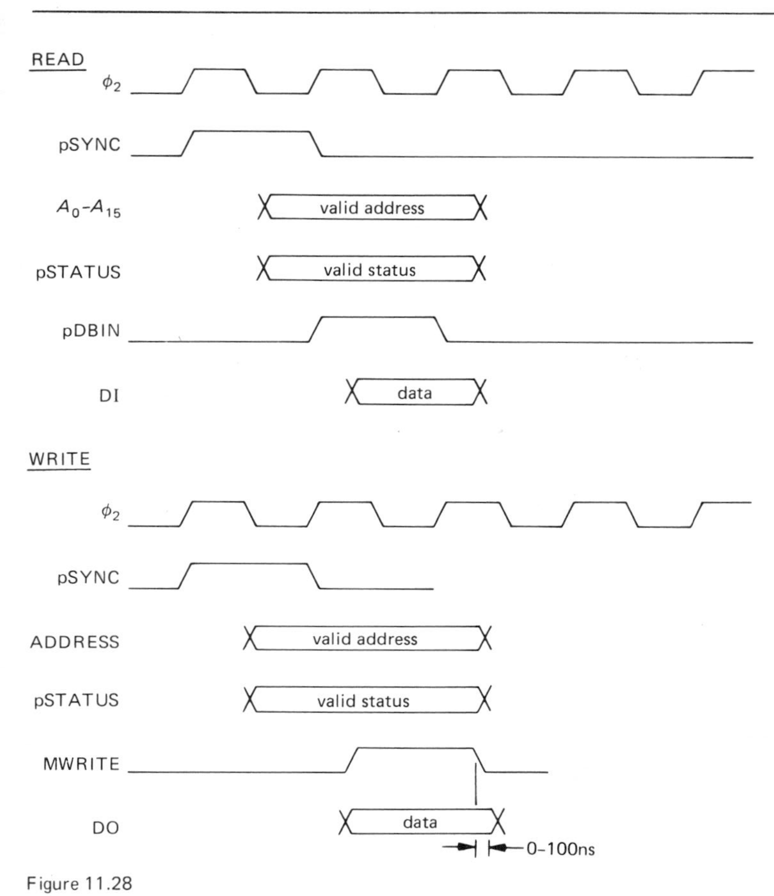

Figure 11.28
Ideal S100 bus timing: READ and WRITE.

between processors with separate I/O instructions and those that require "memory-mapped I/O," in which peripheral device registers simply look like individual locations in memory. Then there is the division created by machines that use memory instead of registers for most arithmetic operations. Another choice involves the use of available pinouts: Some CPUs make multiple uses of pins, to allow more flexibility within the 40-pin package constraint. Then there is the question of word size (4, 8, or 16 bits), stacks, and elegance of the instruction set. Each microprocessor has its own assembly language, another stumbling block for the beginner. Microprocessors are fabricated with technologies other than the usual NMOS (e.g., integrated-injection logic, ECL, PMOS, and CMOS) in order to optimize considerations of speed, density, or power.

There are even greater differences among processors that have to do with their intended applications. There are the "single-chip" processors with RAM and ROM, and even analog/digital converters on the chip (e.g., the 8022 we mentioned in Section 8.27). At the other extreme, powerful 16-bit CPUs like the Z8000 and MC68000 overlap the computational finesse of minicomputers, but they require extensive hardware and software support to take full advantage of their advanced features. Microprocessors with an astounding 32-bit word size are even available as single chips.

We have used the 8085 for all our examples in this chapter, but we don't want to leave the impression that other microprocessors are less useful. Table 11.9 lists a small selection of the most popular microprocessors now available. This table is not intended to be a comprehensive listing of all microprocessors.

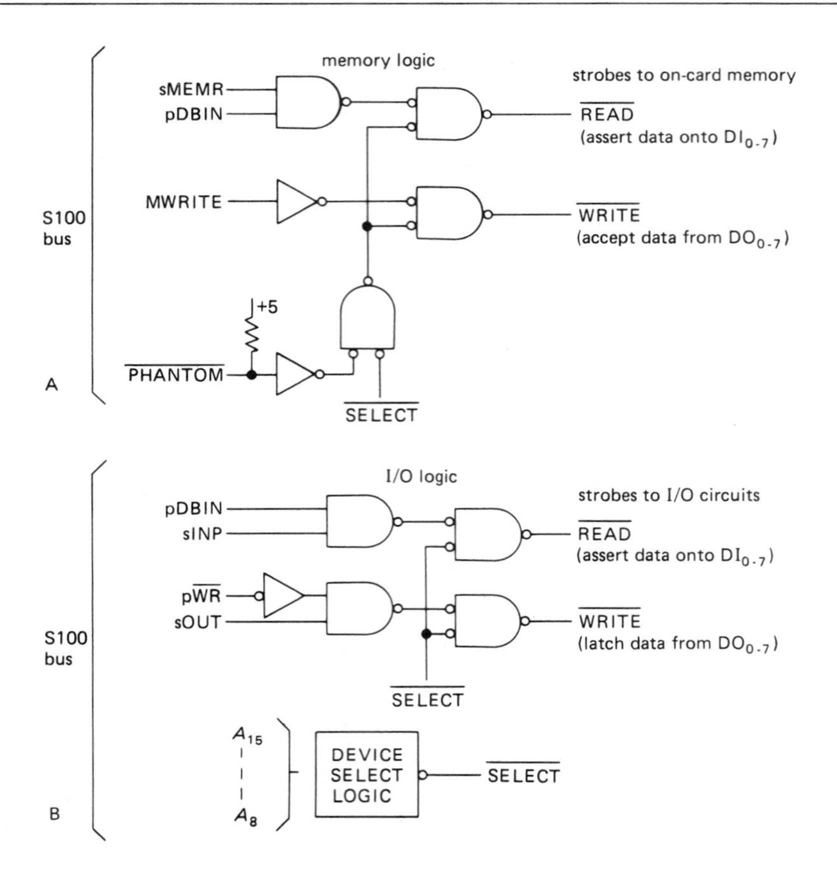

Figure 11.29
Using the S100 bus. A: Memory. B: Input/output.

11.14 Evaluation boards, development systems, and emulators

How are you going to get a program written, assembled, debugged, and loaded into a ROM for use in some instrument you've designed? This is a major problem, especially for someone beginning to work with microprocessors. There are a few techniques available, involving methods that range from simple stand-alone "evaluation boards" up to elaborate development systems and high-level language emulators, complete with terminals and disk storage. In this section we will try to describe what is available and how it can be useful in designing instruments with microprocessors.

Evaluation boards

One of the easiest ways to get started is to buy an evaluation board for the processor you're using. These are generally available from the manufacturer, and sometimes from other suppliers as well. For example, for the 8085 Intel makes the SDK-85 kit, and for the 6502 MOS Technology makes the KIM-1. Evaluation boards for the 6502 are also offered by Synertek and by Rockwell (the AIM-65). These boards have a small keypad, hexadecimal display, RAM, ROM, some parallel I/O ports, and a breadboard area for adding custom circuitry of your own choosing. Often they also include serial I/O for attachment to a terminal or to an inexpensive cassette tape unit for program storage. The ROM you get comes programmed with routines to allow you to enter instructions into RAM from the keypad, display the contents of RAM memory locations and registers, and execute programs you have entered. Figure 11.30 shows the SDK-85 evaluation board for the 8085.

An evaluation board is a good way to learn the basics of a microprocessor. It is also the least expensive way to get started. With most evaluation boards you are forced to hand-assemble the programs you write, since only a rudimentary substitute for an assembler may be provided. With such an evaluation board you can actually develop a complete program in RAM, test it in conjunction with the rest of the instrument you're designing, and have an EPROM "burned" for use in the final dedicated application. EPROM burning presents a problem, since these boards don't provide any way to do it. EPROM burners can be rented, although you'll want to buy or build one if you plan to continue in this line of work. Of course, you will have to have I/O circuitry on your evaluation board similar to that in your dedicated circuit.

In practice, finished software is not often generated this way. The reason is that the process of hand assembling and entering programs is slow and awkward; the whole process is streamlined enormously by the use of a development system.

Development systems

A much better way to develop programs for use in dedicated applications is with a full-

TABLE 11.9. TYPICAL MICROPROCESSOR CPUs

		Data word size (bits)	Accumulators	Gen.-purpose registers	Index registers	Register-register add (μs)	Cycle clock freq. (MHz)	Addressable Memory space (bytes)	DMA?	Ready?	Hardware multiply
Traditional general-purpose 8-bit bus											
*8080	Intel, original design, 1973	8	1	6	(1)	2.0	2	64K	√	√	—
*6800	Motorola, original design	8	2	0	1	1.0	2	64K	√	—	—
2650B	Signetics	8	0	7	(7)	1.5	2	32K	√		—
8060	National, SC/MP	8	1	3	4	7.0	4	64K	√		—
*6502	MOS Technology	8	1	0	2	1.0	2	64K	—	√	—
F8	Fairchild, unique architecture	8	1	65	2	2.0	2	64K	√	—	—
1802	CMOS	8	1	16	(16)	1.3	6.4	64K	√	√	—
Modern general-purpose 8-bit bus											
8070	Improved version of SC/MP	8	3	3	4	7	4	64K	√		√
*8085	8080 instr., improved hdwre.	8	1	7	1	0.8	5	64K	√	√	—
6802	6800 with improvements	8	2	0	1	1.0	2	64K	√	√	—
*Z80	8080 set + added instr.	8	1	14	2	1.0	4	64K	√	√	—
*6809	greatly improved 6800	8	2	4	4	1.0	2	64K	√	√	√
9980	8-bit version of 9900	16	16	(16)	15	8.8	2	64K	√	—	√
*8088	8-bit version of 8086	16	8	14	2	0.3	5	1024K	√	√	√
16008	8-bit version of 16000	16	(8)	8	3	—	—	64K	√		√
NSC800	CMOS Z80 with 8085 bus	8	1	14	2	1.0	4	64K	√	√	—
Traditional 16-bit bus											
9900	TI minicomputer CPU	16	16	(16)	15	4.7	3	64K	√		√
mN601	Data General micro-NOVA	16	(4)	4	(2)	2.4	8.3	32K	√	—	√
9440	Fairchild, Nova inst. set	16	(4)	4	(2)	1.5	10	32K	√		√
LSI-11	DEC chip set	16	(6)	6	(6)	7.0	2.9	64K	√	√	√
Advanced 16-bit CPUs											
CP9008	Executes PASCAL p-code	16		Stacks		—	3	64K	√	—	√
*8086	Intel (to 48-bit words)	16	8	14	2	0.3	5	1024K	√		√
*68000	Motorola (to 80-bit words)	16	8	8	16	0.5	8	16384K	√		√
*Z8001	Zilog (to 64-bit words)	16	(16)	16	(16)	1.0	4	8192K	√		√
16000	National	16	(8)	8	3	—	—	16384K	√		√

Note: An asterisk indicates that the microprocessor is historically important or specially recommended. Parentheses indicate multiple use of a register, rather than additional registers.

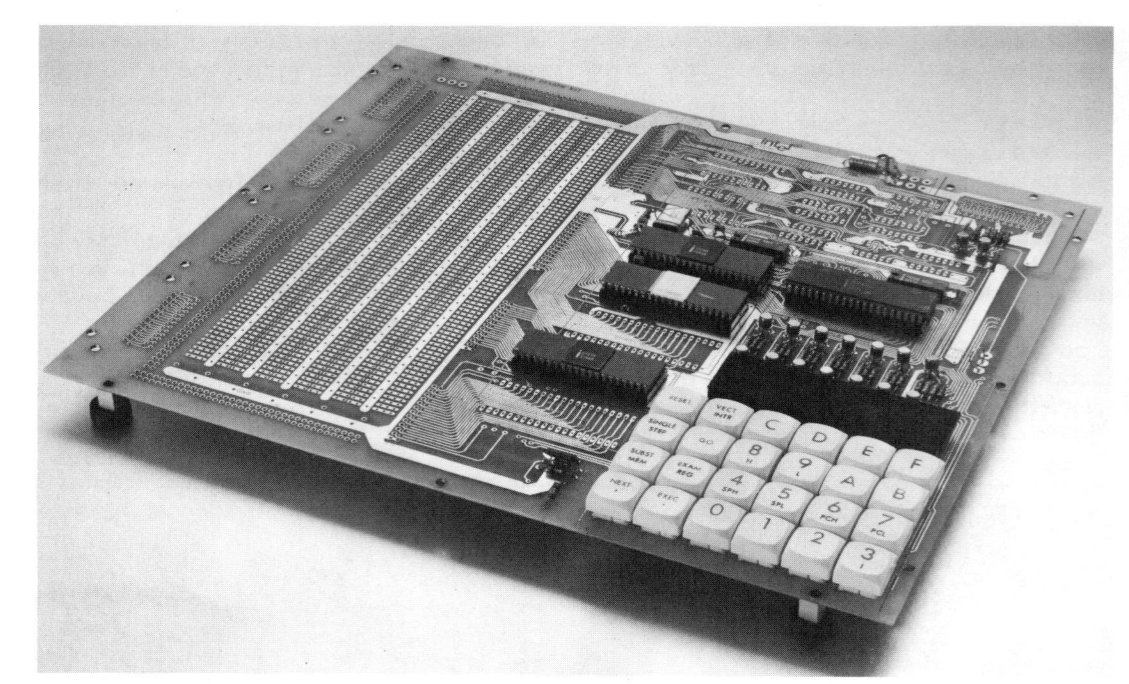

Figure 11.30
The SDK-85 evaluation board for the 8085 (courtesy Intel Corporation).

blown development system. This is a real microcomputer system, complete with video screen, disks, and lots of memory. Software is available for text and program editing, assembly, and debugging. These systems also typically let you do computation, with high-level languages like BASIC and FORTRAN. Development systems are available from the microprocessor manufacturer, as well as from other companies like Tektronix and Future Data.

With a development system you can write your program in assembly language, have it assembled, then burn a ROM to try in the actual instrument you're building. In practice, this can be awkward, because it takes a long time to erase and reprogram a ROM. Furthermore, it gives you no easy way to figure out what is going wrong.

A better way to test your assembled program is to use a "ROM emulator" technique. With this method you get a card to go in the development system, with a cable and 24-pin DIP plug at the end. Through this plug the development system, with the aid of special software, emulates a ROM with the finished program in it. In this way it is easy to test a finished program in circuit, without going through the procedure of erasing and burning ROMs. You can easily make changes and reassemble the program, as you attempt to flush out the bugs. It is even possible to add breakpoints and halts to help you diagnose problems. When things are working to your satisfaction, you can burn a ROM and do final testing. A virtue of the ROM emulator technique is that it is universal and is not processor-dependent. When you switch to a different processor, you don't have to buy new hardware. On the other hand, it is not as powerful as the method of "in-circuit emulation," which will be discussed next.

A powerful technique for testing microprocessor programs in the final circuit environment is the method of in-circuit emulation. This involves another special board that goes into the development system, this time emulating the *microprocessor* in the dedicated circuit via a cable with a 40-pin (or whatever) DIP plug. The development system then emulates the microprocessor in the dedicated circuit itself, with the ability to see what's happening in the various registers, etc. This provides the most powerful in-circuit debugging method. Since you need a special card for each kind of processor, this method can be expensive.

Development systems made by microprocessor manufacturers usually support development of that family of microprocessors only. However, there is no reason why a development system cannot contain a "cross-assembler," generating finished programs for any given microprocessor. The object-code output can then be burned in ROM or used in a ROM emulator mode. However, it still takes special hardware to do in-circuit emulation, and there are often long delays before the manufacturers of development systems bring out new in-circuit emulator hardware. The development systems available from companies other than the semiconductor manufacturers usually support the development of programs for several popular microprocessors.

The kind of testing we have just described assumes that the CPU and memory hardware, at a minimum, is working properly. When there are problems at that basic a level, additional methods are usually needed to find the trouble. Development systems usually include options for integral logic analyzers. These enhancements typically let you trigger on some condition, seeing the states of 24 lines (data and address) on the bus for 1000 consecutive states. Memory lets you look backward in time, seeing the states that preceded the trigger event.

Finally, it should be pointed out that a good way to track down problems during software development is to analyze carefully the symptoms of the instrument, deducing the probable cause of the trouble. With ROM emulation it is easy to make changes, and trial breakpoints can be added if it isn't clear where the program is going astray.

ELECTRONIC CONSTRUCTION TECHNIQUES

CHAPTER 12

Between completing a circuit design and testing a finished product, you've got a number of decisions to make: Is the instrument going in a benchtop case, a "relay rack" enclosure, or perhaps some sort of modular "bin" chassis? Should the circuit itself be constructed with point-to-point wiring, on a "breadboard" card, with Wire-Wrap connections, or on a printed-circuit board? Are connections to the circuit board made with solder lugs, flat ribbon-cable connectors, or edge connectors? Should individual circuit cards be housed in a card "cage," plug into a motherboard, or what? Does it pay to design a printed-circuit motherboard or use hand-wired backplane connections? Which adjustments should be on the circuit board and which on the front (or rear) panel? Decisions like these are important for the appearance, reliability, and serviceability of the finished product, not to mention cost and ease of construction and testing. In this chapter we will try to give some information and guidance on this important subject, one that tends to be overlooked in electronics courses where circuit work is usually done on plug-in breadboarding gadgets. We will begin with the circuit construction itself, treating interconnections, controls, and enclosures last.

Because this chapter does not deal with circuit design, it could be skipped in an abbreviated reading.

PROTOTYPING METHODS

12.01 Breadboards

The unusual name breadboard seems to have arisen from the early practice of building radios on handsome slabs of varnished wood, with tubes, coils, condensers, etc., and the interconnecting wires all fastened to the topside of the board. Later, radios of greater refinement and elegance (to be used by stiff-skirted ladies in the parlor) were built with holes near each component so that the wiring would be hidden from view underneath the board. The practice of testing circuits by constructing trial versions on some sort of board or jig is still called breadboarding.

Wooden breadboards are no longer used (except in the kitchen). Instead, you can get handy plastic blocks with rows of holes spaced to accommodate ICs or other components and (usually) some extra rows for distributing the power-supply voltages. These breadboards are typified by the Super Strip manufactured by AP, and more elaborate breadboarding boxes are made by E&L

Instruments and Hewlett-Packard, among others. These are intended for *testing* circuits, not for constructing permanent (even semipermanent) versions.

12.02 PC prototyping boards

To construct one-of-a-kind circuits of some permanence, probably the best approach is to use a different kind of breadboard, one of the many printed-circuit (PC) prototyping cards available with predrilled pads for ICs and other components, but with no interconnections laid out on the board itself. Instead, each component pad is connected to two or three uncommitted pads nearby, and you wire the circuit by soldering lengths of insulated wire from pad to pad. There are usually some additional lines running around the board for power-supply distribution and ground. These boards, made by companies such as Douglas Electronics, Artronics, Vector, Triad, and many others, usually include a card-edge connector – gold-plated "fingers" of copper board material aligned

Figure 12.1
A "solder breadboard" prototyping card is useful for wiring up small circuits, especially those involving both discrete parts and ICs. This particular specimen will accommodate 12 dual in-line (DIP) packages, and it includes common lines for ground and power-supply voltages. An "edge-connector" foil pattern is standard, so the card can be plugged into a card cage or supporting connector, as shown. This particular circuit uses a variety of components, including single-turn and multiturn trimmers, inductors, a crystal, a DIP switch, a miniature relay, and a logic-state indicator, in addition to both transistor and IC circuitry.

at one end of the card for easy plug-in connection to a mating PC card-edge socket.

There are several standard connector configurations, a very common one being 22 connections on each side of the card, with 0.156 inch spacing (spacings of 0.125 and 0.1 inch are also common). A connector to mate with such a card is called a 44-pin "dual-readout" PC card-edge connector. Prototyping cards are available in various sizes, accommodating from 12 to 36 or more IC packages, with some larger computer-compatible cards available that will accept 100 or more ICs and will plug directly into small-computer mainframes. Some of these boards are single-sided, and others are double-sided with plated-through holes, a subject we will discuss shortly in connection with custom PC design. Figure 12.1 shows a small PC prototyping card (Douglas Electronics 11-DE-3) plugged into a 44-pin dual-readout socket that includes card-supporting guides (Elco 6022).

Another form of breadboard that has enjoyed considerable popularity is the so-called perfboard, a thin sheet of laminated insulating material manufactured with regularly spaced holes ($\frac{3}{16}$ inch is a common spacing) designed to accept little metal pins. To wire up a circuit, you shove in dozens of the little pins wherever you want, then stick in the components. You then solder wires from pin to pin to complete the circuit. Perfboard works OK, but it is awkward to use with the tight spacings of IC packages (0.1 inch between pins). Figure 12.2 shows an example.

12.03 Wire-Wrap panels

A variation on the PC breadboard is the Wire-Wrap panel, a circuit card festooned with IC sockets (or pads), with a pin 0.3 to

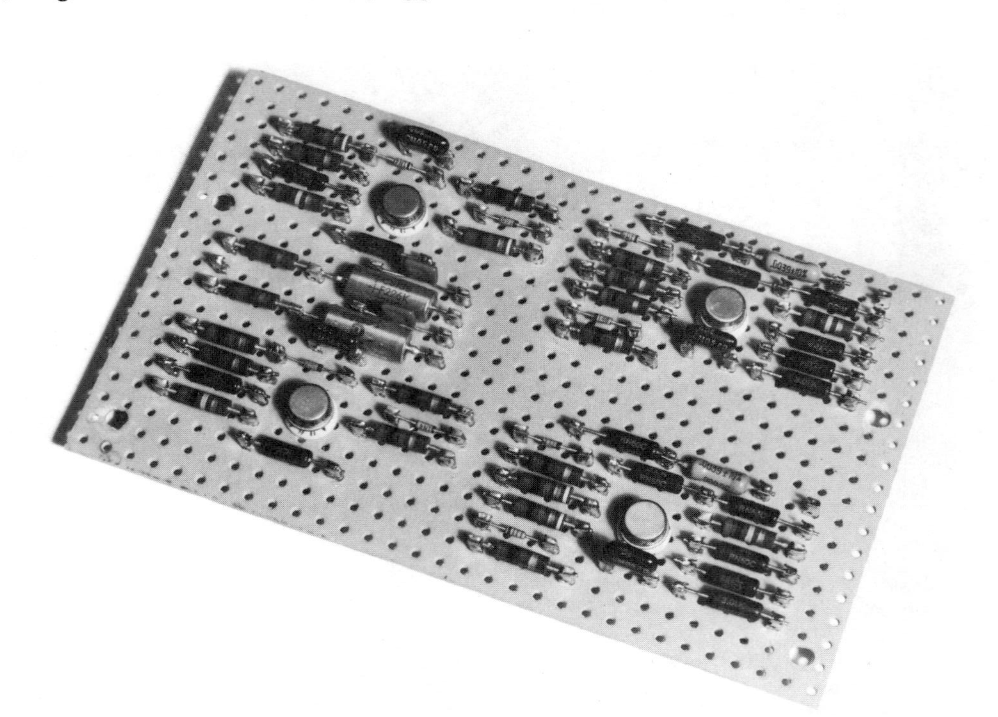

Figure 12.2
"Perfboard" can be handy for prototype circuits constructed with discrete components, although it is not particularly good for ICs. The terminals are press-fit into the holes (or flared with a special tool) and wired underneath.

0.6 inch long protruding from each socket connection. These pins are square in cross section, typically 0.025 inch on a side, and made of a hard metal with sharp corners, plated with gold or tin. Instead of soldering to the pin, you wrap an inch of bare wire tightly around it, using an electric wire-wrapping tool (there's an inexpensive variation known euphemistically as a "hand-operated Wire-Wrap tool"). Wire-wrapping is very fast. You just stick the stripped end of the wire into the Wire-Wrap tool, put the tool over the Wire-Wrap pin, and zip, you're done. The standard wire used for this purpose is silver-plated copper wire of 26 or 30 gauge, with Kynar insulation. There are special tools available for stripping the insulation from the thin wire without nicking it. The wire is stretched tightly around the sharp corners during the wrapping process, forming a few dozen gas-tight cold welds. As a result, wire-wrapped connections are as reliable as well-soldered connections, and it is extremely easy to do them rapidly. For logic circuits, where you have few discrete components, wire-wrapping is probably the best technique for constructing one or two custom circuits of reasonable complexity. Because Wire-Wrap panels are laid out primarily for IC packages, the technique is less convenient for linear circuits with many resistors, capacitors, etc., and the soldered prototype breadboard technique described earlier will usually be better.

It is possible to use discrete components with a Wire-Wrap panel. You just mount

Figure 12.3
Wire-Wrap boards provide a neat and fast construction method particularly good for circuits made with digital ICs. This board uses a printed-circuit pattern to bring out the Wire-Wrap pins on the component side, an alternative to the usual underside pin configuration. Its peculiar shape is dictated by the interior size of the oceanographic pressure cell it fits into.

them on little "headers" that plug into IC sockets, then do the wire-wrapping as usual from the socket pins. Some Wire-Wrap panels have extra solderable pads (rather than IC sockets) available for discrete components. A very nice kind of Wire-Wrap board is available with the pins on the *component* side of the board (the usual procedure is to have the pins stick out the other side). Although this type of board is less dense (fewer ICs per square inch), it is easier to use with discrete components, since you can see them while wrapping, and it allows closer spacing between adjacent circuit boards, since both components and Wire-Wrap pins take space on the same side. This kind of board without sockets is actually quite convenient for construction of linear or digital circuits. Figure 12.3 shows an example. In Figure 12.4 we have compared a prototype circuit built on a Wire-Wrap panel with the final printed-circuit version used in production. Printed circuits are much easier to produce in quantity; they are superior electrically and less cluttered than Wire-Wrap panels. We will talk about PC cards next.

PRINTED CIRCUITS

12.04 PC board fabrication

The best method of constructing any electronic circuit in quantity is to use a printed circuit, a stable insulating sheet of material with thin plated copper lines bonded to the sheet forming the circuit paths. Although early printed circuits were associated with poor reliability (Remember the advertisements stressing the superior quality that only handcrafted television sets without printed circuits could provide?), the process of manufacturing board material and producing finished boards has been perfected to the point that printed-circuit boards now have very few problems. In fact, PC boards offer the most reliable fabrication technique. They are routinely used in computers, spacecraft, and military electronics where high reliability is essential.

Layout and taping

To make a printed circuit, you begin by laying out a pattern on Mylar. There are lots of rules and tricks in this business, but the

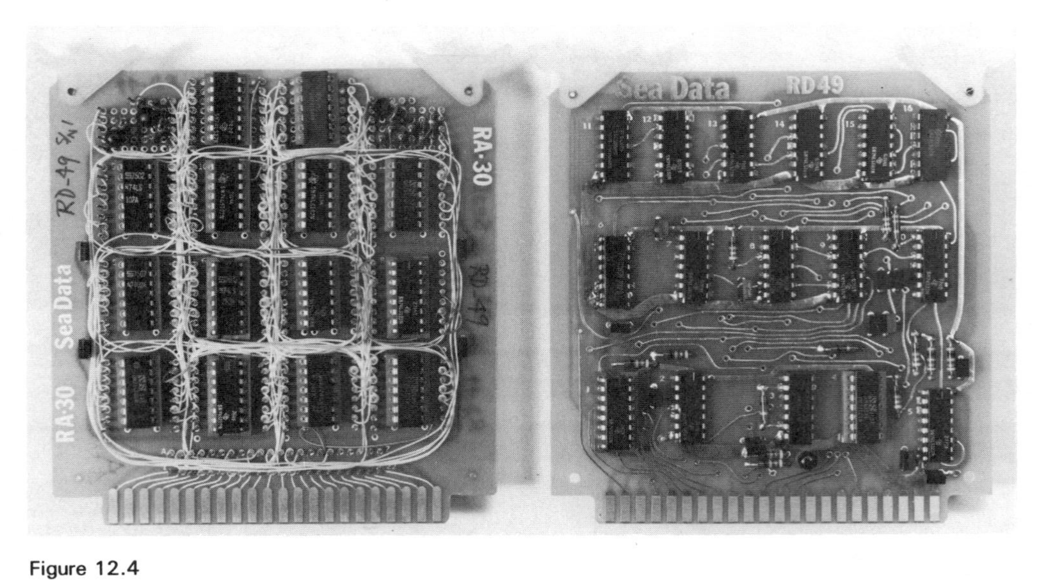

Figure 12.4
A Wire-Wrap prototype board and its printed-circuit successor. PC boards are less cluttered and far easier to fabricate in quantity. They eliminate wiring errors, too.

basic idea is to figure out how to make all the interconnections the circuit demands by running lines around a board. The use of a double-sided board is nearly universal (although often a simple circuit can be laid out using one side only), and this allows you to minimize the number of wire "jumpers" in the finished board. Furthermore, most boards use plated-through holes (the holes in the board are actually conductive clear through from one side of the board to the other). Besides ensuring a superior solder joint to the component leads, this type of hole lets you eliminate jumpers entirely, since a "run" can switch sides when it bumps into a dead end.

The Mylar pattern is made by sticking special adhesive symbols onto clear polyester film, using a precision gridded sheet of Mylar or glass underneath for accurate alignment. There are patterns for IC packages, transistors, edge connectors, etc., and you use black tape (available in many widths) for the interconnections. The usual procedure is to make the Mylar double size, then photographically reduce it to an actual-size negative. Accurate alignment of the patterns is important for double-sided boards to ensure good registration of the two sides when the board is made.

Manufacturing

The board material (usually 1/16 inch of so-called FR-4 board, a fire-resistant epoxy-bonded fiberglass) comes clad on both sides with copper ("2 ounce" thickness is standard; the copper is 0.0027 inch thick). The first step is to drill the holes, using a template or automated drilling machine keyed to the full-size Mylar pattern. The holes are then "plated through" by a tricky multistep copper plating process, creating continuous conducting paths from one side of the board to the other. Then a tough "resist" material is silk-screened onto both sides of the board, everywhere except where the foil for the circuit is to remain. The screen is generated photographically from the Mylar, of course (this step can be done instead by coating the board with a light-sensitive photo-resist, then exposing the board to light with the full-size negative sandwiched on top). Then the board is immersed into a solder-plating bath, plating solder (a tin/lead alloy) everywhere the foil pattern is to remain, including the insides of the holes.

Next the resist is removed chemically, exposing the copper that is to be removed, and the board is treated with a copper-etching compound. That leaves the desired pattern of solder-plated copper, complete with plated-through holes. At this point it is important to carry out a step known as "reflow soldering," which consists of heating the board to make the thin solder plating flow. This prevents the formation of tiny slivers of metal (from the undercutting action of the etching bath) that could otherwise cause conductive bridges. Reflow soldering also improves the solderability of the finished board; a reflow-soldered board is a delight to "stuff" with components. The final step in board manufacture is to electroplate the edge-connector fingers with gold. In industrial board manufacture, the board might then be stuffed by automatic machines, with all joints soldered in a few seconds in a "wave soldering" machine. The alternative is to stuff and solder by hand.

There is a simpler process of board manufacture that is sometimes used, especially in small production situations or for single-sided boards, where plated-through holes aren't needed. In this method you begin by coating the board with photo-resist, then expose it through a full-size negative of the desired pattern, i.e., a negative that is transparent wherever you want foil to remain. The resist is "developed," and then the unexposed resist is dissolved away just as in conventional photography. This board then has a layer of tough resist covering the copper that is to remain, so you simply expose it to the etching compound. After the superfluous copper has been etched away, the remaining resist is washed off with solvent, leaving the desired pattern in copper. At this point it is best to treat the board with an "electroless" tin-plating bath in order to cover the copper with a metal less susceptible to corrosion. As before, the edge-connector fingers will then be gold-plated. The final step in this process consists of drilling the holes by hand, using the actual conductive pattern as a guide (each pad has a small opening in the center to aid in drilling the finished board).

□ 12.05 PC board design

There are several important decisions you have to make during PC board design, during component "stuffing," and finally when the board is used in an instrument. In this section we will try to touch on the most important of these.

□ *PC board layout*

There are several stages along the way from a schematic diagram to a final printed circuit. Beginning with the diagram, you generally work out trial pencil sketches of component layouts and interconnections, eventually working these together into a final pencil layout drawing. From this you make the "Mylar," consisting of accurately aligned "pads" (terminal areas for component connections) and taped interconnections. Precut patterns are used for IC and transistor pads and for ribbon and edge connectors, since these have standard spacings and dimensions. The pencil sketch and Mylar are usually made double size to allow greater accuracy (and to keep your eyes from popping out!). When the Mylar (two Mylars for double-sided boards) is completed, it is photographically reduced to an actual-size negative, from which a trial board is made as described previously. You generally "stuff" the prototype board with components, turn on the power, and then hunt down the errors; this lets you correct the Mylar artwork to produce final boards. The following subsections provide some further details and hints.

□ *Initial sketch*

We recommend doing the initial layout with pencil on grid paper (5 lines/inch), with two colors to indicate foil patterns on the top and bottom (assuming it is a double-sided board). We usually use black pencil for runs on the bottom and green or red for the top (component) side. Since you're likely to do plenty of erasing, it is best to use vellum graph paper. The 0.2 inch gridding corresponds to 0.1 inch final size, the universal measure for IC pin spacings, transistor pinouts, edge connectors, etc. Your drawing should be the view from the component side; i.e., the sketch of the component-side (top-side) foil pattern looks like the final pattern, and the sketch of the bottom foil pattern is what you would see looking down through the finished board with x-ray eyes. While working on the layout, indicate component outlines with a pencil of a third color. All this work should be freehand. Don't waste time with an outline template; just use the grid lines as a guide to draw IC and component pinouts.

It is generally best to work up some trial layouts on a piece of scratch paper, particularly for sections of the circuit that may require special layout to minimize long lines or capacitive coupling. It may take some experimentation to arrive at good component arrangements. A trial layout might consist of a block of the circuit with two or three op-amps, or perhaps the input or output section of the circuit. These blocks should then be worked together onto the large gridded vellum, with adjustments being made as you go. Don't hesitate to do lots of erasing!

□ *Layout dimensions and hints*

Try to have all ICs pointing in the same direction, preferably in straight rows. Likewise, resistors should be in even rows, not staggered. We use 0.040 black tape for signal runs, with wider tape for power supplies (0.05 or 0.062 inch) and very wide ground runs (0.1 to 0.2 inch, or even wider; it's common to broaden the ground runs with lots of tape). Be sure to include plenty of bypass capacitors, one 0.1μF for every two to four ICs. As you scratch your head, trying to juggle the tangled maze of interconnections, don't forget that components act as "jumpers" – they can hop over runs on the board.

Dimensions and spacings: On the actual-size PC board, we recommend holes spaced 0.4 inch for resistors (¼W size), with spacing of 0.1 or 0.15 inch between resistors (with 0.15 inch spacing you can get a tape run between adjacent pads). We favor the CK05 and CK06 types of ceramic capacitors, with their controlled 0.2 inch lead spacing; they can also be spaced 0.1 inch from other capacitors or resistors. Leave some room around ICs for logic clips: a minimum of 0.3 inch to the next IC pads and a minimum of 0.15 inch to the nearest resistor or capacitor pads. Leave 0.040 inch spacing between tape runs, and don't run anything closer than 0.25 inch from the edge of the board, to allow room for card lifters, guides, standoffs, etc. Avoid running lines between the 0.1 inch spaced pads of an IC, unless absolutely necessary. You can fit four tape runs (five, if you squeeze) lengthwise between the pads of a standard DIP IC pattern (they're spaced 0.3 inch).

□ *Connections to the board*

For the majority of boards it is probably best to bring out all connections through "edge-connector" contacts, which mate directly with sockets available in a variety of contact configurations. The most commonly used spacings are 0.156 inch, 0.125 inch, and 0.100 inch between fingers. Generally you'll put an edge-connector pattern at one end of the card, bringing power-supply voltages and signals through that connector. The card is mechanically supported, and it plugs in at that end (more on that shortly).

Often you see an edge-connector pattern at the other end of the card also, used instead of a flat ribbon connector to bring some other signals off the board or to other boards. Another method for bringing out signals is to use flat ribbon cable terminated in DIP plugs; such cables plug right into IC sockets on the board. You can buy these cables prefabricated in various lengths, or you can make them yourself with a kit consisting of flat cable, unassembled DIP plugs, and a crimping tool. Ribbon cables can also connect to the board via in-line or "mass-termination" connectors, which use one or two rows of pins on 0.1 inch centers.

For simple boards the best method of connection may be to use swage-solder terminals or PC-type barrier strips with screw terminals. Avoid the use of large pads alone for connection of external wires to PC boards.

Figure 12.5 illustrates a variety of PC board connection techniques.

□ *Odds and ends*

With plated-through boards, use several holes to join ground foils on opposite sides of the board. Try to avoid using multiple passes through the board to reach your destination, since plated-through connections where no component is mounted can give trouble. The layout of a double-sided board generally winds up with most tape runs going horizontally on one side, vertically on the other.

General philosophy: Use smooth curves rather than right-angle turns. Bring lines into pads as if heading for the center of the pad, rather than coming in at an oblique angle. Don't mount heavy components on boards (a couple of ounces ought to be the limit); assume that the instrument will be dropped 6 feet onto a hard surface sometime during its life! Put polarity markings on the component side for diodes and electrolytic capacitors, and label IC numbers and pin 1 location (if there's room). It is always nice to label test points, trimmer functions (e.g., "ZERO ADJ"), inputs and outputs, indicator light functions, etc., if you have room.

□ *Taping the Mylar*

General advice: Use an illuminated "light table" with a piece of precision gridded Mylar taped to it. Don't confuse this with the inexpensive gridded plastic films that are neither accurate nor dimensionally stable; a piece of precision gridded film will set you back at least $20. Put your clear Mylar over, and stick down the IC pads accurately on it. Use the pencil sketch for guidance while taping. Wash your hands often to prevent deposition of oily film on the Mylar, and use alcohol to wipe any areas that might become oily. Use an Xacto knife with curved blade for tape and outline cutting, and learn not to cut through the Mylar. Press the tape down firmly after positioning; otherwise it will eventually curl up. Allow generous overlap where tape meets pad, etc. When laying out tape, don't hold it under tension; it will

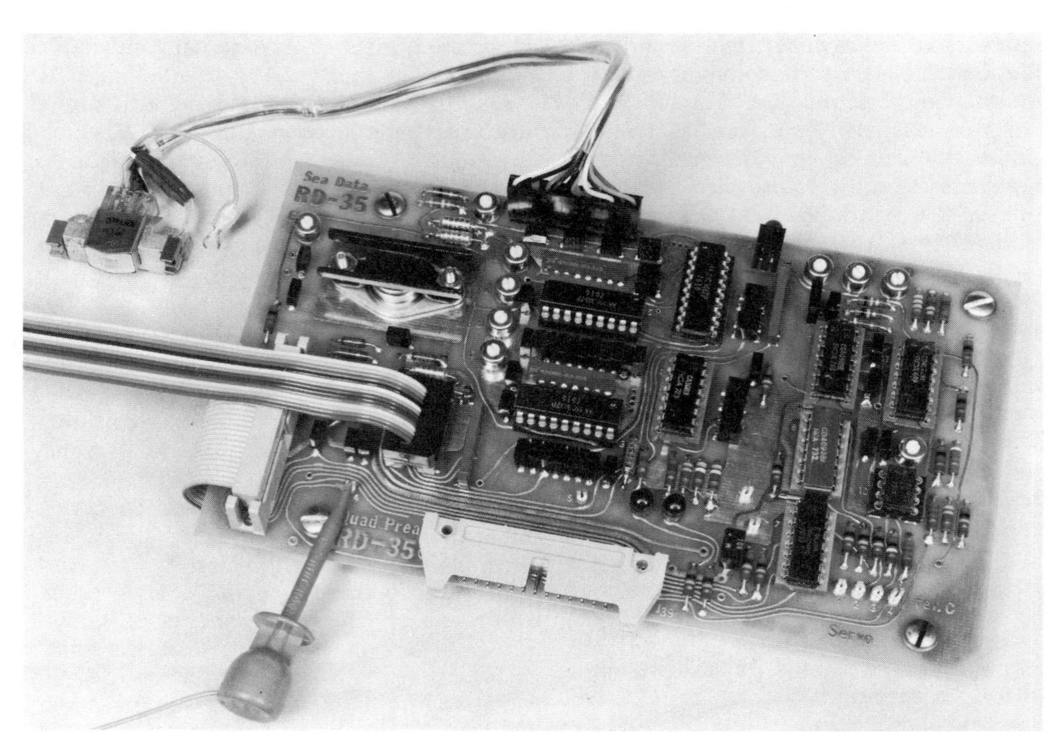

Figure 12.5
Several connection techniques are illustrated in this digital recorder printed-circuit card. The tape head connects via an in-line connector (which mates with a row of Wire-Wrap-type pins), and other signals are brought out with "mass-termination" ribbon connectors and a dual in-line ribbon connector. A test lead is shown clipped onto a "test point" terminal. This board also illustrates PC board heat sinking (upper left), a logic-state indicator (upper right), miniature single-turn trimmers, and single in-line (SIP) resistor networks.

shrink and pull away from pads. Use precut bends and circles for the larger tape widths (0.062 inch or wider) when navigating tight turns. After the Mylar is completely taped, check it against the schematic by going over each connection on the diagram with a red pencil. When all seems OK, seal up flaws on the Mylar with a black felt-tip pen.

Precut PC graphics patterns are available from several manufacturers. Table 12.1 shows some recommended types. The Bishop Graphics catalog (20450 Plummer Street, Chatsworth, CA 91311) includes extensive information on PC board layout and execution.

12.06 Stuffing PC boards

Your worries aren't over when you've got a finished board. You've got some decisions to make (e.g., whether or not to use IC sockets) and some important things to do (e.g., defluxing and lead trimming). Herewith, some thoughts on these subjects:

Sockets

There is great temptation to use IC sockets everywhere, for ease of troubleshooting. However, if you're not careful, the sockets may well cause more trouble than they prevent. In general, sockets are a good idea at the prototyping phase, where IC substitution may be necessary to convince you that the trouble you're having is a *design* error, not a bad component. They should also be used for expensive ICs (e.g., a D/A converter, microprocessor, or the like), ICs that you're likely to want to change from time to time (e.g., a program ROM), and ICs that have a good chance of being damaged sooner or later (e.g., chips that buffer input or output signals from outside the instrument).

The problem is that a poorly designed socket may prove unreliable over extended time periods. A nonsoldered joint must have a gas-tight seal, such as that created by a mechanical metal-to-metal wiping action, with the seal then being left undisturbed. PC edge connectors, for example, used to be somewhat unreliable; with time, manufacturers learned some good tricks: bifurcated contacts (two independently sprung contacts for each finger), gold plating on the socket and on the edge fingers, and good mechanical design to ensure firm contact pressure during wiping and afterward. Joints that aren't gas-tight can be expected to fail after some time, perhaps a year or so. This sometimes happens inadvertently, e.g., by inserting a component in a PC board and then forgetting to solder it. Such connections have the maddening property of working fine at first, then becoming intermittent months or years later, owing to the formation of corrosion. A different problem can arise when heavy ICs (24 pins or more) are held in sockets. They can work their way out of the sockets after repeated vibration or shock.

We have found that the pin-and-jack type of IC socket (popularized by the Augat 5*xx*-AG series), although expensive compared with many other socket types, gives good reliability.

Soldering and defluxing

The usual procedure is to insert some components, turn the board over and bend the leads aside to hold the components in place, then solder them using a thermostated soldering iron and fine solder. ICs can be inserted easily with an insertion tool (highly recommended), and it is best to use a lead bender on resistor leads, etc., in order

TABLE 12.1. SELECTED PC GRAPHICS PATTERNS

	Bishop	Centron	Chartpak
Small pads (0.150″ OD)	D203	P7180	TPCC 628
Standard pads (0.187″ OD)	D104	P7260	TPCC 71
Large pads (0.250″ OD)	D108	P7400	TPCC 76
Giant pads (0.300″ OD)	D111	P7460	TPCC 572
14-pin DIP	6402	2070	TPCL 6402
16-pin DIP	6404	2071	TPCL 6404
16-pin DIP with in-betweens	6764		TPCL 6764
TO-5 transistor	6077	2015	TPCL 6077
TO-18 transistor	6274	2001	TPCL 6274
0.100″ connector pads	5004	1501	TPGP 6185
0.100″ edge-connector strip	6714	2079P	TPCP 6714
0.156″ edge-connector strip	6722	2081P	TPCP 6722
0.031″ black tape	201-031-11	CT100-.031	CP3101
0.040″ black tape	201-040-11	CT100-.040	CP0401
0.050″ black tape	201-050-11	CT100-.050	CP0501
0.062″ black tape	201-062-11	CT100-.062	CP6201
0.100″ black tape	201-100-11	CT100-.100	CP1001
0.200″ black tape	201-200-11	CT100-.200	CP2001
0.062″ universal corners	CU601	UC 812	TPUC 2482
0.100″ universal corners	CU607	UC 820	TPUC 2485
0.200″ universal corners	CU609	UC 830	TPUC 2489

to prevent slivers of solder being shaved off the leads during insertion. After soldering, the leads should be trimmed with a snipper.

Now comes a very important step: Solder flux should be removed from the board. If it isn't, the board will look just terrible in a few years, when you're not around to defend it! Rules for defluxing:

1. Do it.
2. Do it soon. The stuff gets much harder to remove with time.
3. Use a solvent such as Freon, alcohol, or some other organic solvent recommended for this purpose. Use a small brush to help dislodge stubborn globs of flux.

12.07 Some further thoughts on PC boards

The solderability of PC boards tends to decrease with time, owing to oxide formation, so it is best to stuff the components soon after the board is made. For the same reason, you should keep unstuffed boards in plastic bags, away from corrosive fumes. Good circuit boards should be made from 1/16 inch FR-4-type board material (sometimes referred to as "epoxy fiberglass") clad with 2 ounce copper. Remember that a circuit on a PC board is basically sitting on a piece of glued-together stuff; the board can absorb moisture and develop electrical leakage. Another pathology of PC board material is "hook," the variation of dielectric constant with frequency; the consequent variation of stray capacitance can make it impossible to build an amplifier with flat frequency response, for instance. Oscilloscope manufacturers are very aware of this bizarre effect.

PC runs with large currents passing through them have to be widened to prevent excessive heating and voltage drops. As a rough guide, here is a table of approximate conductor widths that give temperature rises of 10°C or 30°C for the currents listed, for 2 ounce copper PC boards. For other foil thicknesses, just scale the widths accordingly.

	0.5A	1.0A	2.0A	5.0A	10.0A	20.0A
10°C rise	0.004"	0.008"	0.020"	0.070"	0.170"	0.425"
30°C rise	0.002"	0.004"	0.010"	0.030"	0.080"	0.200"

Tools

As a starting point, we lined up the most heavily used tools on our bench and came up with the following part numbers:

Long-nose pliers	Erem 11d, Utica 321-4½ or 775-5½
Snippers	Erem 90E or 71AE
Soldering iron	Weller WCTP-N, Ungar "Ungarmatic"
Solder	Ersin Multicore 22 ga, Sn60 alloy
IC inserter	Dipsert 880
Lead bender	Production Devices PD801
Solvent dispenser	Menda 613
Solder sucker	Edsyn Soldapullt DS017

Lots of useful gimmicks for PC assembly are listed in the Contact East catalog (7 Cypress Drive, Burlington, MA 01803) and the Marshall Claude Michael catalog (9674 Telstar Avenue, El Monte, CA 91731).

INSTRUMENT CONSTRUCTION

12.08 Housing circuit boards in an instrument

Circuit cards, whether printed circuits, Wire-Wrap panels, or breadboarding cards, have to be mounted in some sort of enclosure and connected to power supplies, panel controls and connectors, and other circuitry. In this section we will discuss some of the popular methods of putting instruments together so that circuits are neatly mounted and accessible for testing and repair. We will begin with methods for holding the circuit cards themselves, then discuss the business of cabinets, front and rear controls, power-supply mounting, etc.

Circuit card mounting

In simple instruments you may have only a single circuit card, whether printed circuit, Wire-Wrap, breadboard card, or whatever. In that case a simple solution is to drill holes near the corners and mount the card with screws (and standoff bushings) to a flat surface, component side up. Connections can then be made with a card-edge connector socket (if the card has plated fingers), with flat cable terminated with a connector to mate with a plug on the board, or with individual soldered connections to swaged terminals on the board. With edge or ribbon connectors, the card will support the connector adequately, so no extra connector supports have to be used. Whatever the method of connection, it is wise to arrange the wiring such that the board can be tipped upward for access to the underside so that you can make modifications or repairs.

In a system with several circuit boards, the best way to arrange things is with some sort of card "cage," a rigid assembly with guides for individual cards to slide into and aligned holes along the rear so that you can mount edge connectors to mate with the cards. There's lots of flexibility in card width, spacing, and number of cards that can fit into a card cage. A very common size accommodates cards 4.500 inches wide with 44-pin dual-readout (22 pins each side) edge connections on 0.156 inch centers. Cards can be spaced as little as 0.5 inch apart, if necessary, although 0.6 inch is a more comfortable spacing; if space is no problem, 0.75 inch spacing will allow plenty of room, even with Wire-Wrap pins and bulky components. It is best to look at some catalogs to see what's available. You can get variations with plastic card guides or just dimples in the metal sides to align the cards, and there are various types of card ejectors (attached to the PC cards) to help remove a card. Card cages are available with simple flange mounting to a flat surface parallel to the cards, as well as in various configurations that fit nicely into rack enclosures, etc. You can even get modular enclosures that include an integral card cage, with some additional room for power supplies, panel controls, etc.

Backplane connections

Card-edge connectors are available with lugs for solder connections, with Wire-Wrap pins, and with small pins for insertion into PC boards. In many cases it is best to wire up the intercard connections with point-to-point wiring between card-edge connector pins, using the edge connectors with solder lugs. A neat job requires some cabling of wire bundles, with the wires running in straight lines parallel to the card cage dimensions. In other cases it may be preferable to use Wire-Wrap connections on this backplane, especially if there are many connections between backplane pins with relatively few connections to other points in the instrument, and if there is no need for shielded-cable connections to the backplane.

The third possibility is to use a motherboard backplane, a PC board designed just to hold the card-edge sockets. Motherboards are popular in bused systems (they are nearly universal in computers) and should be considered in any case if the instrument is intended for production in significant quantities. With double-sided motherboards, you can have the advantage of a groundplane (lower inductance and coupling of signal lines), or you can use both sides for signals if the intercard wiring is complicated. With bused systems the backplane is usually simple, with lines connecting corresponding bus pins on all cards. In computer backplanes you sometimes see a motherboard used with Wire-Wrap pins sticking through. This is very handy if you want the motherboard to do all the bus and power-supply connections, leaving the unbused pins to be connected in a custom configuration by wire-wrapped connections. Figure 12.6 shows a simple PC motherboard.

12.09 Cabinets

Depending on the intended use, an electronic instrument might be housed in a benchtop cabinet (complete with rubber feet and hinged front "bail"), in a cabinet or panel designed for mounting in a standard "relay rack" 19 inches wide (either screwed directly to the rack flanges that run vertically up from the floor or mounted on ball-bearing rack slides for simplified access), in a modular instrument case designed to plug into slots in a larger rack-mounted "bin," "cage," or "crate" (the latter usually provide dc power connections through standardized connectors at the rear), or perhaps in some other format such as a free-standing pedestal-mounted case.

There are many cabinet configurations available in both benchtop and rack-mounting formats. Among the most popular are cabinets 17 inches wide, available in various heights and depths, that accept optional

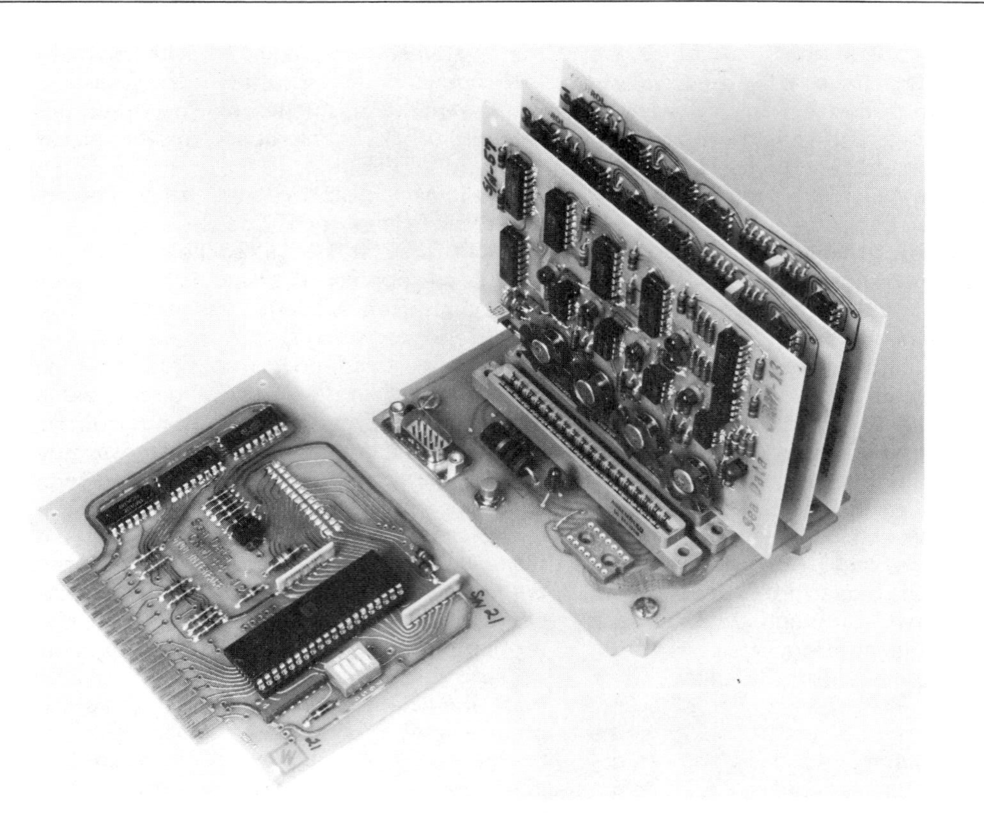

Figure 12.6
A "motherboard" provides a convenient method of interboard connection, reducing hand wiring and the possibilities for error considerably, while simultaneously providing superior electrical performance. In large systems the motherboard and its connectors would probably be mounted rigidly at the rear of a card cage.

rack-mounting flanges or slides (a rack 19 inches wide has about 17½ inches of clearance between the flanges). That way you can convert an instrument from rack-mounted to benchtop format, or vice versa, by just changing a bit of cabinet hardware. A point to check: Some of these convertible cabinets require removal of the outer case for rack mounting, whereas others let you keep the cabinet intact.

In the category of modular instrumentation, the NIM bin is popular in nuclear and atomic measurements, the CAMAC crate has established itself in computer interfacing, and several manufacturers have defined modules and bins (e.g., the TM500 series from Tektronix and the EFP series of blank modules from Vector). Blank chassis are available in each of these formats, complete with rear connectors to mate with the dc power receptacle of the mother bin.

12.10 Construction hints

Rather than attempt to list the enormous variety of manufactured cabinets by name or style, we will simply offer some general comments on construction of instruments. These suggestions, together with the figures in this chapter, should help you choose electronics enclosures wisely and fill them up with circuitry in a sensible way.

In general, you use the front panel for indicators, meters, displays, etc., as well as controls and frequently used connectors. It is common to put seldom-used adjustments and connectors that don't require frequent access on the rear panel, along with large connectors, line cord, fuses, etc.

Perhaps the most important thing to remember when laying out an instrument is the need for good accessibility to circuit cards and controls. It should be possible to replace any component in the instrument without great pain. This means neat cabling of wiring, so that modular units can be raised up without having to unsolder anything, and careful planning, so that circuit cards can be tested while operating in the instrument. For instance, a card cage might be mounted with the cards vertical; to get at them, you remove the top panel from the cabinet, then plug in extender cards to make the circuit cards accessible. If the cards are mounted horizontally, you might make the front panel removable, or hinged, to provide access. At all costs, fight the temptation to lay down the circuit in "layers," with circuitry nicely covering other circuitry. Figure 12.7 shows an example of neat and accessible front-panel cabling built in a cabinet with a removable front panel.

12.11 Cooling

Instruments that consume more than a few watts will usually require some sort of forced air cooling. As a rule of thumb, a small instrument running more than 10 watts, or a larger (rack-width) instrument consuming more than about 25 watts, will probably benefit from a blower. It is important to keep in mind that a box full of electronics may run at a nice temperature when sitting on the bench with the top cover removed, but when installed in a rack with other heat-producing equipment (where the ambient temperature may reach 50°C), complete

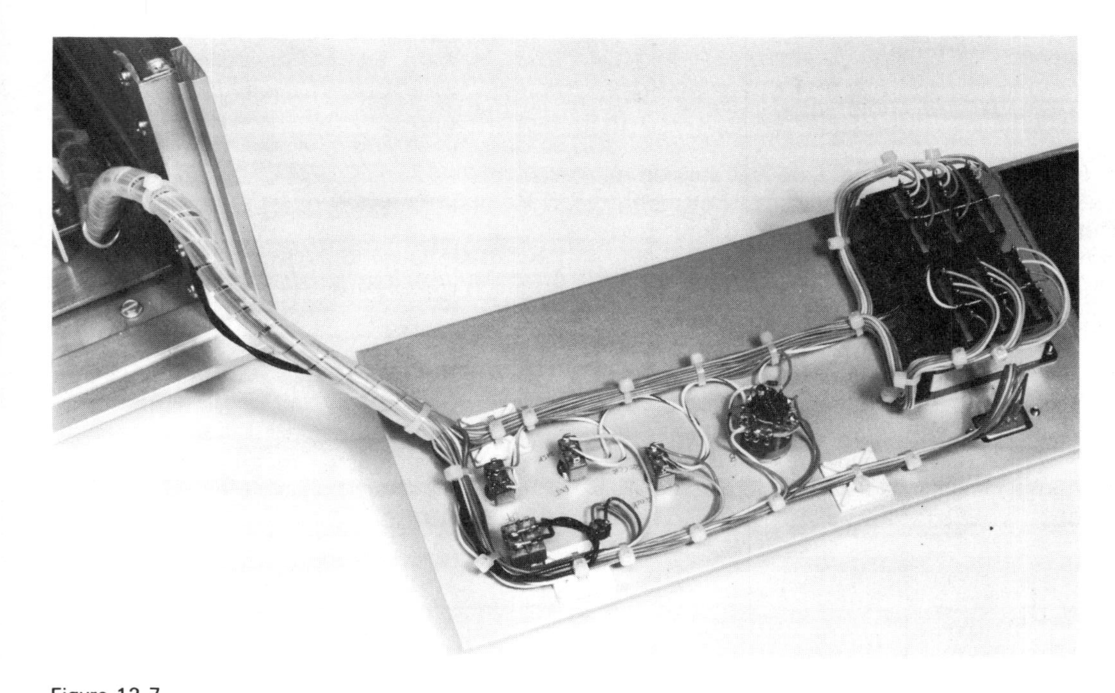

Figure 12.7
One way to ensure good accessibility to panel controls is to bring all wiring away at one end, so the panel can be hinged or otherwise detached from the instrument. In this example the panel slides into a slotted instrument case. Note the use of cable "ties" and self-adhesive supports to keep the wiring tidy.

with its outer cover, it is likely to run very hot, leading to early failure of components and generally unsatisfactory operation.

Instruments running at moderately low power, say at the figures mentioned earlier, can often be cooled adequately with simple convective cooling. In such cases you might perforate the top and bottom covers, paying attention to the location of major heat-producing components (power resistors and transistors). It may be best to mount the high-power components on the rear panel, using heat sinks with their fins aligned vertically (see Section 5.04). Circuit boards will also be better ventilated if mounted vertically, although heat dissipation in circuit cards is often negligible. If simple convective cooling doesn't keep things cool enough, you have to resort to a blower. The simple "Muffin-type" venturi instrument blower, with flow rate of about 100 cubic feet per minute (CFM) assuming relatively unimpeded air flow, will adequately cool instruments running 100 watts or more. Here's the relevant formula:

$$\text{air temperature rise}(^\circ\text{C}) = \frac{1.6 \times P\ (\text{watts})}{\text{airflow (CFM)}}$$

If less airflow will suffice, a quieter version of the venturi fan is available from most manufacturers. Table 12.2 shows some part numbers. The airflow of these fans is greatly reduced when operating against high back-pressure. Figure 12.8 shows a graph.

When laying out an instrument designed for forced air cooling, try to arrange things so that the air enters the box at one end, flows around the components, and exits at the far end. In an instrument with an interior horizontal chassis partition, for example, you might punch some inlet perforations at the bottom rear, perforate the internal chassis near the front of the instrument, and mount the exhaust fan at the top rear, thus forcing the airflow to pass through all parts of the instrument. Keep in mind that a circuit board will block airflow, and plan accordingly. If there is significant impedance to the flow of air (high back-pressure), a centrifugal blower will work better than the propeller type. The blades of the latter go into "stall" when the back-pressure exceeds about 0.3 inch of water, rendering the fan totally ineffective. Finally, in any cooling situation it is a good idea to design conservatively; failure rates for electronic instruments rise dramatically when equipment is operated hot. Figure 12.9 illustrates good instrument design in regard to cooling and accessibility.

12.12 Some electrical hints

Unreliable components

The most unreliable components in any electronic system will be the following (worst first):

1. Connectors and cables
2. Switches
3. Potentiometers and trimmers

Keep them in mind as your brainchild proliferates in complexity.

RF line filters

As we mentioned earlier, it is a good idea to use RF filters on the ac power-line inputs. These are manufactured by a number of companies, including Corcom, Cornell-Dubilier, and Sprague. They are available as

TABLE 12.2. VENTURI FANS

	Standard, 4½" square 105–120 CFM	Quiet, 4½" square 70 CFM	Very quiet, 4½" square 50 CFM	Mini, 3⅛" square 35 CFM
Rotron	MU2A1 MU2B1	WR2H1	WR2A1	SP2A2
IMC	WS2107FL BS2107FL	WS2107FL-2 BS2107FL-2	WS2107FL-9	PWS2107FL PBS2107FL
Pamotor	4500C 4600	4800		8500C 8506

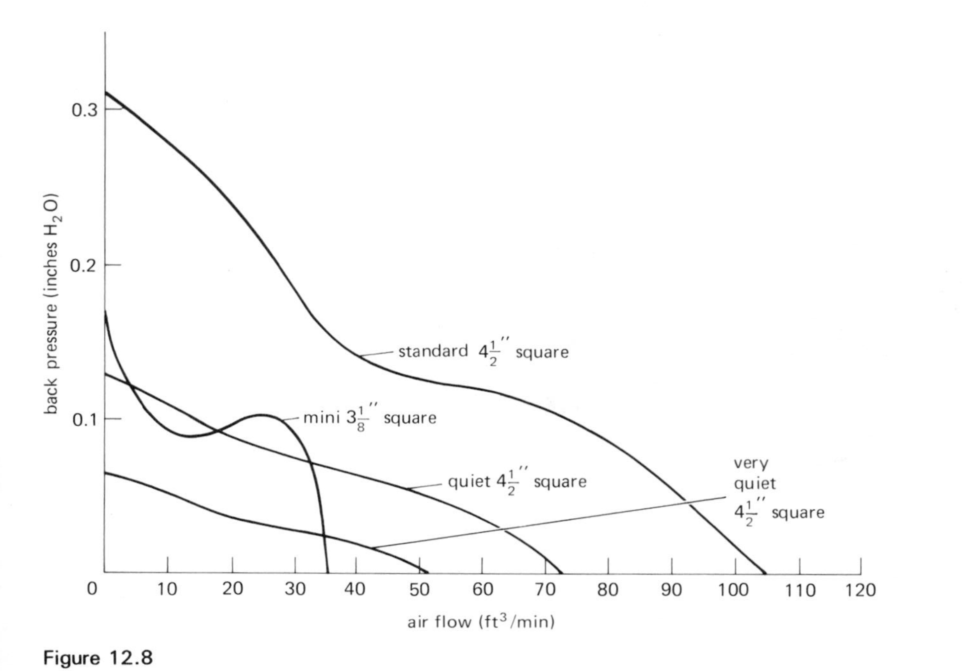

Figure 12.8
Air flow versus back pressure for the Venturi blowers listed in Table 12.2.

simple modules with solder-lug terminals or in configurations that include an integral chassis-mounting ac line plug to mate with the standard IEC line cord. These filters provide excellent rejection of RF signals on the power line (as well as preventing their generation by the instrument itself), and they are also partially effective in reducing line transients. As an example, the Corcom 3R1 filter (rated at 3A, 115V) has 50dB rejection of RF at 200kHz and more than 70dB rejection for signals above 0.5MHz (see Section 5.10 and Table 5.3).

Transient suppressors

Power-line transient suppressors are also a good idea in any instrument to prevent malfunction (or even damage) from the occasional 1kV to 5kV spikes that occur on everyone's ac power line. You just put them across the power-line terminals; they act like bidirectional zeners with enormous peak current capability. They come in packages that look like disc ceramic capacitors or power diodes; the inexpensive and small GE V130LA10A, for instance, costs about a dollar, begins conducting at 185 volts, and can handle peak currents of 4000 amps (see Section 5.10 and Table 5.2 for more details).

Fusing

A line fuse is *mandatory* in every line-operated electronic instrument, without exception. As we indicated in Section 5.10, the wall socket is fused at a current designed to prevent fire hazard in the wall wiring, typically 15 or 20 amps. That won't prevent a major disaster in a malfunctioning instrument, such as one with a power-supply capacitor failure, where the instrument may begin to draw 10 amps or so (more than 1kW heating in the power transformer). Important fact (learned the hard way by the authors): The lead from the power line goes to the innermost terminal of the fuse holder, so when you insert a new fuse you can't get your fingers onto the ''hot'' terminal. Use a slow-blow fuse with rating 50%–100% greater than worst-case current drain of the instrument.

Cold-switching philosophy

Whenever possible, it is a good idea to avoid running logic signals or analog waveforms to panel controls; this is to prevent cross-

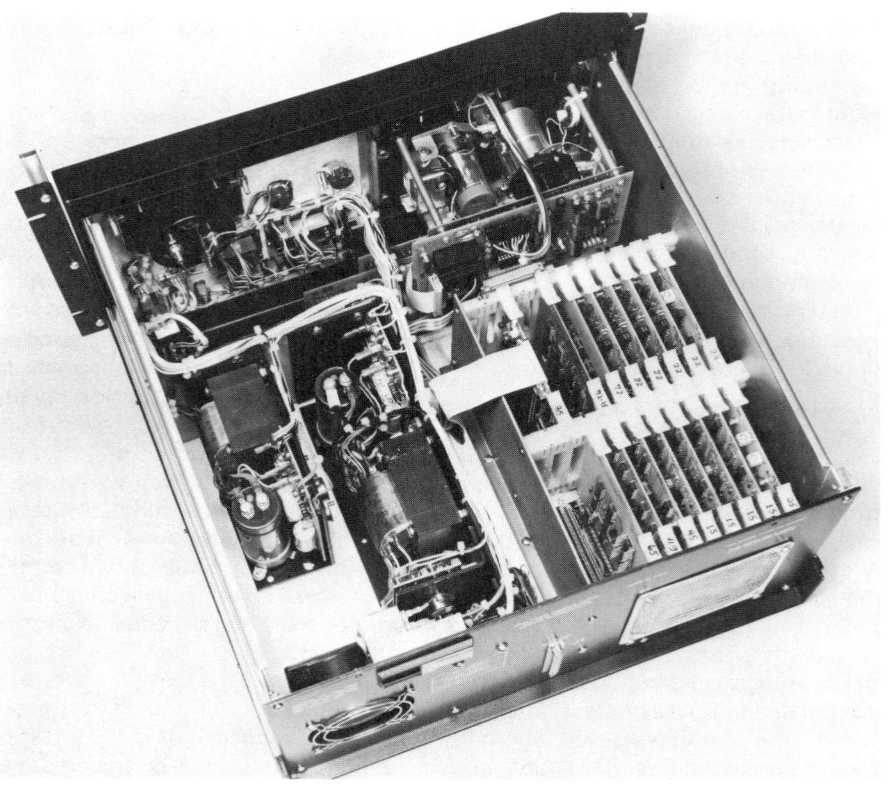

coupling and signal degradation that otherwise might occur. Instead, run dc control levels to the panel switches and pots, and use on-board circuitry to handle the actual signal switching, etc. This is especially important in noisy environments, or when dealing with high-speed or low-level signals, since the dc control signals can be thoroughly bypassed, whereas fast signals cannot. For example, use select gates (multiplexers) rather than routing the logic signals through a switch, and use a voltage-controlled oscillator rather than an *RC* oscillator for panel control of frequency. The few extra components you need to do things this way will buy you increased reliability and simplified assembly (no shielded cable, for example).

12.13 Where to get components

Getting the parts you need to build some piece of electronic equipment can present some real difficulties. Most of the large electronics distributors have abandoned over-the-counter sales, making it nearly impossible for the small purchaser to go down to the store and buy a few parts. Fortunately, the large (well-stocked) distributors will still take an order by telephone, with cash pickup at the "will call" counter. When playing this game, it is essential to know exactly what you want, by part number and manufacturer (for ICs, you may have to know the full part number, prefixes, suffixes, and all).

Many distributors are extremely hesitant to sell small quantities, so you're often forced to buy at least five or ten of each item. Add to this the fact that a given distributor handles only a fraction of the brands you may need, and you're faced with a major chore. The consumer-oriented electronics stores (Radio Shack, Lafayette, etc.) will deal in small quantities and do have counter sales, but they tend to stock an extremely limited range of parts. The parts distribution system seems to be aimed at the industrial user, with his quantity ordering. Electronics manufacturers are treated well by the distributors, who pay them frequent visits and engage in competitive pricing.

Some special cautions are in order for buying ICs. Many kinds of ICs are manufactured without 100% testing; instead, a sample of each batch is tested, with the whole batch being rejected if the sample shows excessive failure rate. As a result, you can, on occasion, get a perfectly worthless chip straight from a reputable manufacturer's production line. As a rough guide, you can expect something in the neighborhood of 1% of new chips to be defective. That's not too serious, and in any case you can have your chips tested if a lower reject rate is necessary. Furthermore, all manufacturers test their LSI chips, and some manufacturers (AMD, for example) perform 100% testing of all their ICs.

A more serious problem arises when those rejected batches find their way into the hands of a "relabeler," a kind term for a junk peddler. Labeling machines are inexpensive; consequently, "counterfeit ICs" are all too common. Our experience has been that the large distributors (Arrow, Hamilton/Avnet, Harvey, Newark, and Schweber, to name a few) are reliable, at least for the brands for which they are authorized distributors. Surprisingly, most mail-order houses seem to ship good merchandise, often at very good prices, but there is an element of risk involved. Be suspicious of any IC without a date code. Because of the extra time and annoyance involved in finding bad ICs in a circuit, we recommend using regular distributors for all IC buying, in spite of the generally higher prices.

Figure 12.9 (facing page)
Views of a complete instrument (a digital cassette tape reader) illustrating several techniques of support and interconnection. Most of the electronics are housed in a card cage (with hand-wired backplane and mass-termination connections), and the electronics associated with the tape drive are on two boards near the drive motor (with mass-termination, in-line, and DIP plug connections). Adjustments and test points are accessible near the edges of each circuit card. Note the flow of cooling air: Air is sucked in behind the card cage; it flows between the cards, then around the central partition and back over the power supplies before being blown out by the exhaust fan at the right rear.

HIGH-FREQUENCY AND HIGH-SPEED TECHNIQUES

CHAPTER 13

HIGH-FREQUENCY AMPLIFIERS

In this chapter we will discuss the important subject of high-frequency and radiofrequency techniques, as well as the digital equivalent, high-speed switching. High-frequency techniques find wide application in communications and broadcasting and in the domain of radiofrequency laboratory measurements (resonance, plasmas, particle accelerators, etc.), whereas high-speed switching techniques are essential for the fast digital instrumentation used in computers and other digital applications. High-frequency and high-speed techniques are extensions of our ordinary linear and digital techniques into the domain where the effects of interelectrode capacitance, wiring inductance, stored charge, and short wavelength begin to dominate circuit behavior. As a result, circuit techniques depart radically from those used at lower frequencies, with such bizarre incarnations as stripline and waveguide and devices like Gunn diodes, klystrons, and traveling-wave tubes. To give an idea of what is possible, there are now commercially available digital ICs (counters, etc.) that operate at pulse rates of 1GHz and higher and linear circuit elements (amplifiers, etc.) that operate at frequencies in excess of 100GHz.

We will begin with a discussion of high-frequency transistor amplifiers, complete with simple transistor and FET models. After a few examples, we will move to the important subject of radiofrequency techniques, followed by a discussion of communications concepts and methods, including modulation and detection. Finally, we will look at high-speed switching techniques in some detail. Because of the specialized nature of these subjects, this chapter could be passed over in a first reading.

13.01 Transistor amplifiers at high frequencies: first look

Amplifiers of the type we discussed earlier (e.g., common-emitter amplifiers with resistive collector load) show a rolloff of gain with increasing signal frequency, mostly owing to the effects of load capacitance and junction capacitance. Figure 13.1 shows the situation in its simplest form (we'll complicate things soon enough!). C_L represents the effective capacitance from collector to ground and forms a low-pass filter of time constant R_LC_L in combination with the amplifier's collector load resistance R_L. Remember that at signal frequencies V_+ is the same as ground; hence the equivalent circuit shown. C_L includes collector-to-emitter and collec-

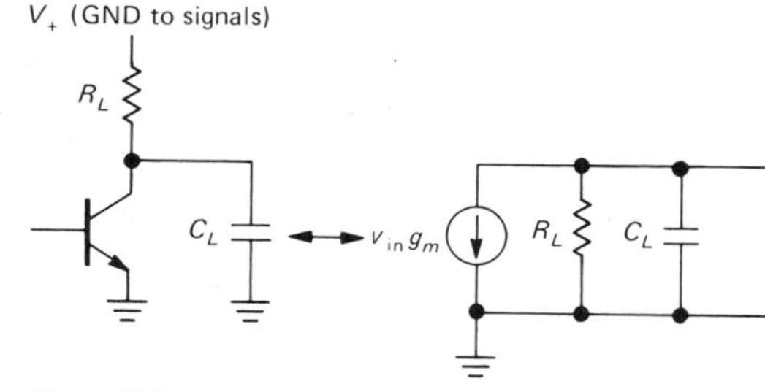

Figure 13.1

tor-to-base capacitances, as well as load capacitance. At frequencies approaching $f \approx 1/R_LC_L$ the amplifier's gain begins dropping rapidly.

Reducing load capacitance effects

The simplest therapy consists of measures to reduce the product R_LC_L. For example:

1. Choose a transistor (or FET) with low interelectrode (junction and lead) capacitance; these are usually designated as RF or switching transistors.
2. Isolate the load with an emitter follower, thus reducing the capacitive load seen at the collector.
3. Reduce R_L. If you keep I_C constant, the gain drops, owing to reduced $g_m R_L$. Remember that for a transistor, $g_m = 1/r_e$, or I_C(mA)/25 for an amplifier with bypassed emitter. To keep the gain constant with decreasing R_L, you have to raise the collector current by keeping V_+ constant. Thus

$$f_{max} \approx 1/R_LC_L \propto I_C/C_L$$

which accounts for the rather high currents often used in high-frequency circuits.

□ 13.02 High-frequency amplifiers: the ac model

Load capacitance is not the only effect reducing amplifier gain at high frequencies. As we mentioned earlier (see the discussion of Miller effect in Chapter 2), the feedback capacitance (C_{cb}) from output to input can dominate the high-frequency rolloff, especially if the input signal source impedance is not low. In order to determine where an amplifier will roll off, and what to do about it, it is necessary to introduce a relatively simple ac model of transistors and FETs. We will do that now, with a worked example of a high-frequency amplifier to illustrate how to use it.

□ *ac Model*

The common-emitter (or source) models diagrammed in Figure 13.2 are just about the simplest possible; yet they are reasonably useful in estimating the performance of high-speed circuits. Both models are straightforward. In the bipolar transistor model, C_{ie} (also called C_{ib} or C_{be}; note the alternative naming of input and output capacitances) is the input junction capacitance, r_b is the impedance looking into the base, C_{cb} is the feedback (Miller) capacitance, and C_{ce} is the capacitance from collector to emitter. The current source models the transistor's gain at signal frequencies. The FET model is similar, but with different names for the capacitances and with the simplification of infinite input resistance.

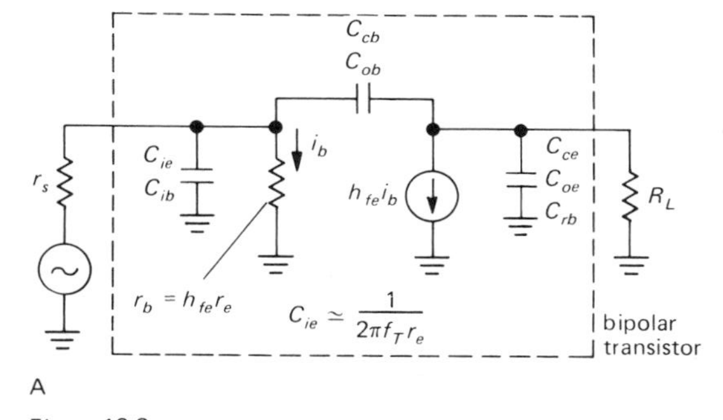

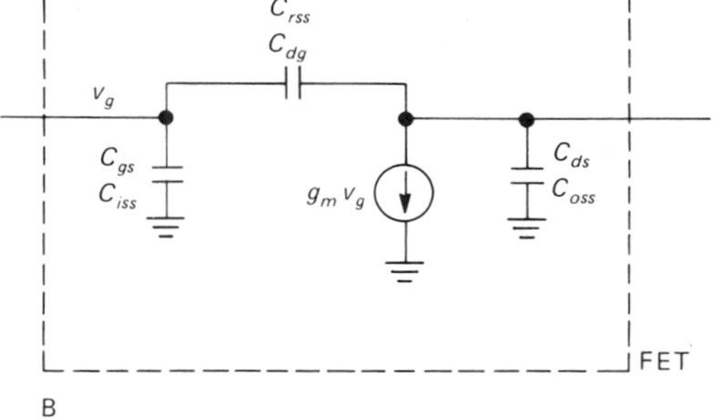

Figure 13.2
Bipolar transistor and FET high-frequency equivalent circuits.

☐ *Effects of collector voltage and current on transistor capacitances*

The feedback and output capacitances (C_{cb}, C_{rss}, C_{ce}) consist of a combination of the small capacitance of the transistor leads and the larger capacitances of the semiconductor junctions. The latter behave like reverse-biased diodes, with a capacitance that decreases gradually with increasing back-bias, as shown in Figure 13.3 (this effect is exploited in the voltage-variable capacitors known as "varactors"). The capacitance varies with voltage approximately as $C = k(V - V_d)^n$, where n is in the range of $-\frac{1}{2}$ to $-\frac{1}{3}$ for transistors and V_d is a "built-in" voltage of about 0.6 volt.

The input capacitance C_{ie} is different, since you're dealing with a forward-biased junction. In this case the effective capacitance rises dramatically with increasing base current, since V is near V_d, and it would make little sense to specify a value for C_{ie} on a transistor data sheet. However, it turns out that the effective C_{ie} increases with increasing I_E (and therefore decreasing r_e) in such a way that the RC product (r_bC_{ie}) remains roughly constant. As a result, the transistor's gain at a particular frequency depends primarily on the ratio between current lost into C_{ie} and current that actually "drives the base" and is not strongly dependent on collector current. Therefore, instead of attempting to specify C_{ie}, the transistor manufacturer usually specifies f_T, the frequency at which the current gain (h_{fe}) has dropped to unity. It is easy to show that f_T is given by

$$f_T = \frac{1}{2\pi C_{ie} r_e}$$

or, equivalently,

$$C_{ie} = \frac{1}{2\pi f_T r_e}$$

for particular values of C_{ie} and r_e at some collector current. Transistors intended for radiofrequency applications have f_Ts in the

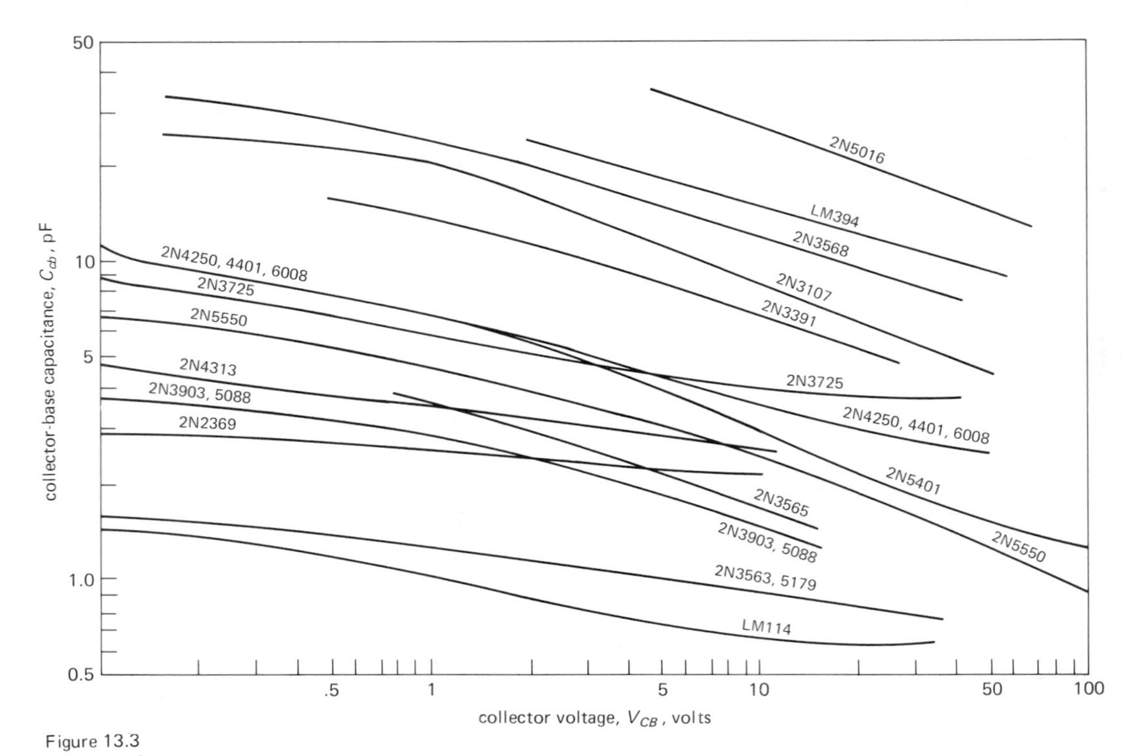

Figure 13.3
Collector-to-base capacitance versus voltage for some popular bipolar transistors.

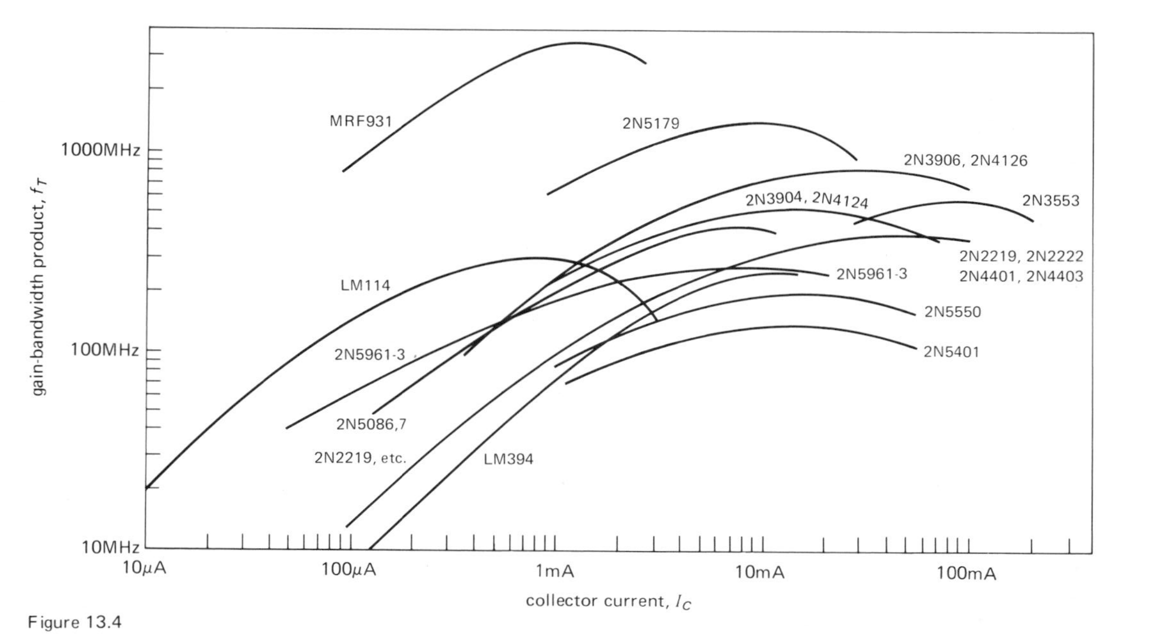

Figure 13.4
Gain-bandwidth product, f_T, versus collector current for some popular bipolar transistors.

range of 500MHz to 2000MHz, whereas "general-purpose" transistors have f_Ts in the range of 50MHz to 250MHz. Figure 13.4 shows the variation of f_T with collector current for typical transistors.

□ 13.03 A high-frequency calculation example

Let's apply our simple model to the design of a high-frequency broadband amplifier. We will show the driving stage also, so the driving (source) impedance is known. As it will turn out, the amplifier will exhibit poor performance and severe loading of the driving stage. The sort of performance problems you will see are characteristic of real-life circuit design, and we will talk about ways to improve performance by changes in circuit configuration and operating points. Figure 13.5 shows the circuit fragment. This subcircuit is assumed to lie within an overall amplifier circuit with feedback at dc to stabilize the quiescent point at $\frac{1}{2}V_{CC}$; it would not be biased stably as shown. Since we are interested in high-frequency performance, we won't worry further about how the biasing is accomplished. Note that the

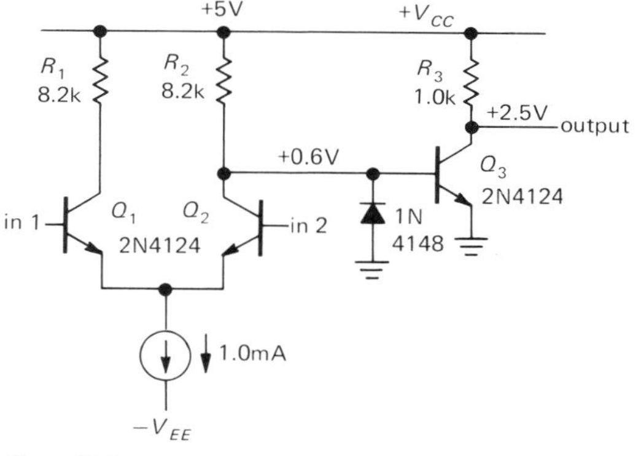

Figure 13.5

differential stage has very little common-mode input range, extending to perhaps +0.25 volts or so and limited in the negative direction by the compliance of the emitter current source.

□ *Analysis of high-frequency rolloff*

The differential stage has calculable gain and output impedance, allowing us to analyze the output-stage rolloff in detail. Our analysis of the gain of the Q_3 amplifier stage consists of the following:

1. Find the low-frequency gain with zero source impedance. Then find the 3dB point, owing to output capacitance and feedback capacitance in combination with load resistance:

$$f_{-3\text{dB}} = \frac{1}{2\pi R_L(C_L + C_{cb})}$$

2. Find the input impedance, the combination of base input impedance (r_b and C_{ie}) and effective feedback capacitance ($G_V C_{cb}$).
3. Compute the 3dB point due to input loading of the source; compare with the "output 3dB point" calculated in step 1, to see where the high-frequency bottleneck is.
4. Improve performance, if necessary, by alleviating the problem that dominates the high-frequency rolloff.

Note that the feedback capacitance C_{cb} appears in both the output and input circuit calculations, in the latter multiplied by the voltage gain (Miller effect).

Let's try our simple method of analysis on this circuit, modelled in Figure 13.6. The

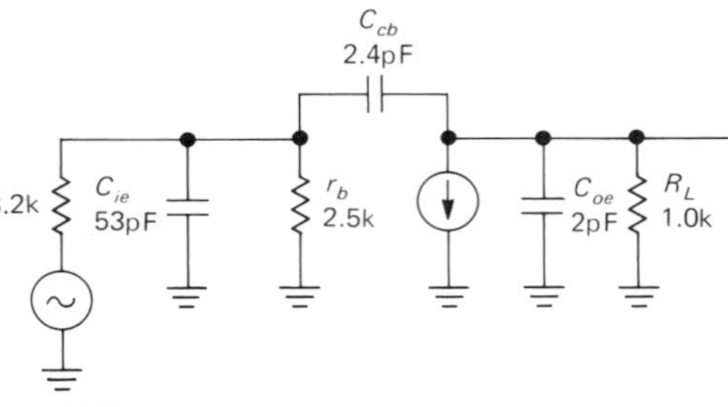

Figure 13.6

2N4124 is parametrized by $C_{cb} = 2.4$pF at 2.5 volts, $h_{fe} \approx 250$, and $f_T = 300$MHz.

1. Assuming Q_3 is driven by a voltage source, its low-frequency voltage gain is 100, since $r_e = 10$ ohms at 2.5mA collector current. The 3dB point set by output capacitance is at roughly 40MHz (2.4pF in parallel with 2pF, driven by 1.0k). Note that in this simple calculation we have ignored the load capacitance and stray wiring capacitance.
2. The input resistance is roughly 2.5k ($h_{fe}r_e$), paralleled by the Miller capacitance (240pF) and by C_{ie}; the latter works out to about 53pF, using the formula given earlier.
3. The 3dB point due to input capacitances comes out roughly at 280kHz ($R = 8.2$k paralleled with 2.5k; $C = 240\text{pF} + 53\text{pF}$), dominated by the Miller-effect capacitance $C_{cb}G_V$ in combination with the relatively high impedance at the base. Note that the low-frequency gain is actually less than 100, considering the input signal to be the *unloaded* output of the differential stage, owing to the loading of the previous stage by the low input resistance; when you include this effect, the gain at low frequencies is actually $100 \times 2.5/(2.5 + 8.2)$, or about 23.

The combination of excessive previous-stage loading and low 3dB rolloff frequency makes this a poor circuit, included here to illustrate real-life problems of high-frequency amplifier design. In practice, you would improve performance by using greatly reduced collector impedances or by going to a different amplifier configuration. In the next section we will discuss some popular high-frequency amplifier configurations intended to reduce or eliminate the effects of input capacitance (f_T) and feedback capacitance ($C_{cb}G_V$, Miller effect).

13.04 High-frequency amplifier configurations

As the preceding example illustrates, the Miller effect can dominate the high-frequency performance of an amplifier driven by moderately high source impedance. In that example, an f_T of 300MHz and an output time constant corresponding to a 3dB point of 40MHz were swamped by an input time constant giving a 3dB point of 280kHz.

Three cures for Miller effect

Besides the brute-force approach of reducing collector resistances enormously, there are several interesting configurations that aim to reduce driving (source) impedance or reduce feedback capacitance or both. Figure 13.7 shows these configurations, drawn in their simplest forms, without regard to bias or power supplies (i.e., the signal-frequency circuit alone is drawn).

In the first circuit, an emitter follower reduces the driving impedance seen at the

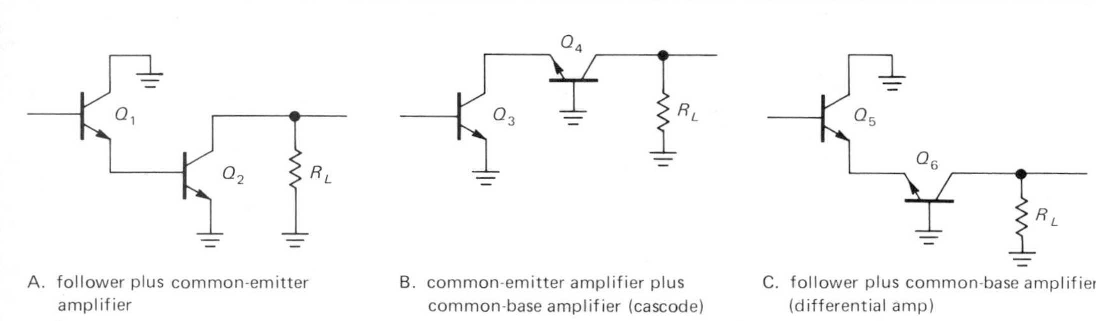

A. follower plus common-emitter amplifier
B. common-emitter amplifier plus common-base amplifier (cascode)
C. follower plus common-base amplifier (differential amp)

Figure 13.7
Simplified high-frequency amplifier configurations.

input of a common-emitter amplifier. This greatly reduces the degradation of high-frequency performance caused by f_T and $C_{cb}G_V$. The second circuit is the popular cascode, in which a common-emitter stage drives a common-base stage, eliminating $C_{cb}G_V$ Miller effect (Q_4's emitter is pinned by the fixed base voltage; it just passes Q_3's collector current through to R_L). In the third circuit a follower drives a common-base stage, eliminating Miller effect and reducing the driving impedance at the same time; this circuit is the familiar differential amplifier, with unbalanced collector resistors and one input grounded.

More techniques

In addition to these circuit configurations, there are two other approaches to the input and feedback capacitance problem, namely (a) the use of a simple grounded-base amplifier alone, if the driving impedance is low enough, and (b) the use of tuned circuits at the input and output of a common-emitter (or other) amplifier, to "tune away" the effects of interelectrode capacitance. Note that such a tuned amplifier does not have broadband response, but amplifies only a narrow range of frequencies (which may be an advantage, depending on the application). In addition, *neutralization* may be necessary. We will discuss narrowband tuned amplifiers in a later section of the chapter. An in-between approach involves the use of "peaking" inductances of a few microhenrys in series with collector load resistances to cancel some of the effects of capacitance and hold up the gain at frequencies somewhat above the normal high-frequency rolloff (Fig. 13.8).

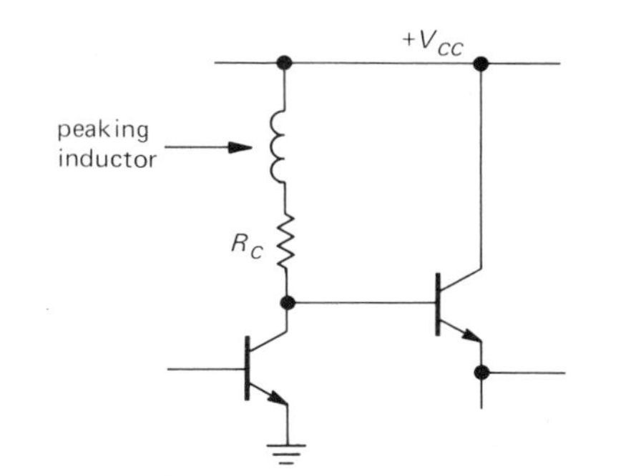

Figure 13.8

In order to be able to estimate the high-frequency performance of circuits involving followers and grounded-base stages, we will need simple ac transistor models for these configurations (Fig. 13.9). Note that in the emitter follower model the impedances depend on source and load impedances (reactance as well as resistance). We will apply these models in the next example.

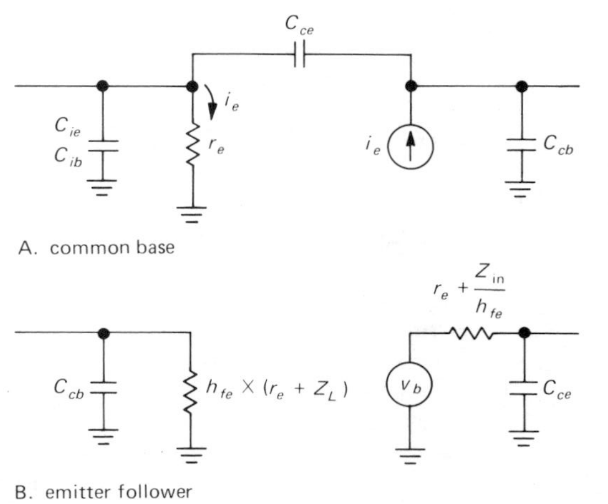

Figure 13.9

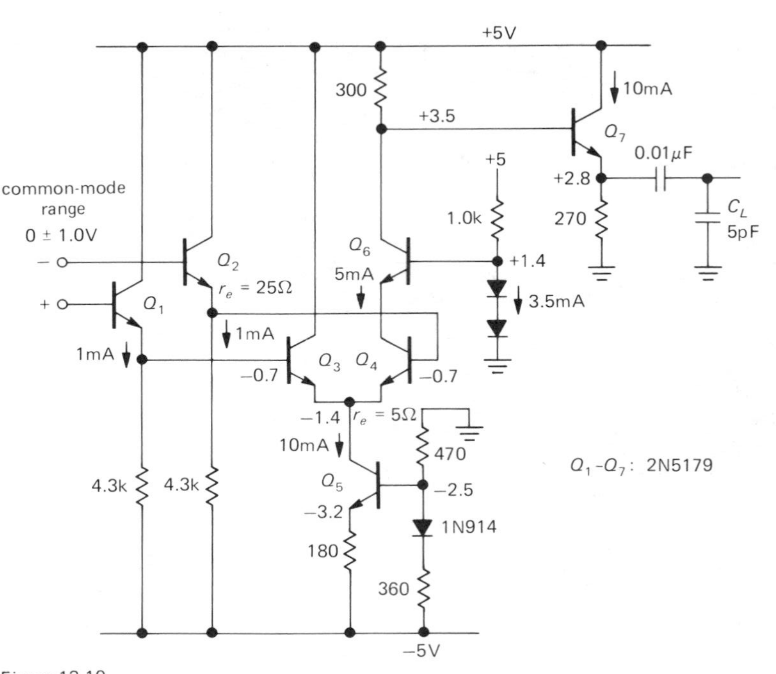

Figure 13.10
Wide-band differential amplifier.

□ 13.05 A wideband design example

As an example of an improved broadband amplifier design, consider the circuit in Figure 13.10, configured to eliminate almost entirely the rolloff caused by Miller effect. It uses emitter follower inputs (high input impedance) to a differential amplifier; the output is isolated by an emitter follower from the cascode-connected output section of the differential amplifier. The design is based on the use of a good high-frequency transistor such as the 2N5179, with an f_T of 1000MHz (specified as $h_{fe} = 10$ at 100MHz) and a C_{cb} of 0.5pF at 2 volts. The approximate equivalent circuit, in terms of junction and stray capacitances and their shunt resistances, is shown in Figure 13.11.

To determine the high-frequency rolloff point of this amplifier, you have to go through each stage, analyzing the various *RC*s by substituting the appropriate equivalent circuits. There is usually one stage that sets the lowest limit, and with some intuition and guesswork you can often put your finger right on it. In this case the limiting performance is set by the finite driving impedance (300Ω) to Q_7's base, in combination with the capacitance of Q_7 and the load capacitance C_L as seen buffered at Q_7's base (remember that h_{fe} drops approximately as $1/f$, so at very high frequencies the isolating effect of an emitter follower is seriously degraded).

The simple method we used to figure the 3dB point goes something like this: Apply the emitter follower equivalent circuit to Q_7 to get the impedance looking into the base, knowing the load capacitance, junction capacitances, and wiring capacitance (we used $C_{cb} = 0.5$pF, $C_{ce} = 0.2$pF, and $C_{stray} = 0.3$pF). Since the impedance looking into the base depends on h_{fe}, you have to do the calculation as a function of frequency (assuming $h_{fe} \approx 1/f$ at high frequency); we chose instead to do the calculation at a few high frequencies, guessing that the 3dB point would be in the neighborhood of a few hundred megahertz. Figure 13.12 summarizes the process. At frequencies of 100MHz, 200MHz, and 400MHz we took the load impedance, multiplied it by the transistor beta (of Q_7, assuming $h_{fe} \approx 1/f$),

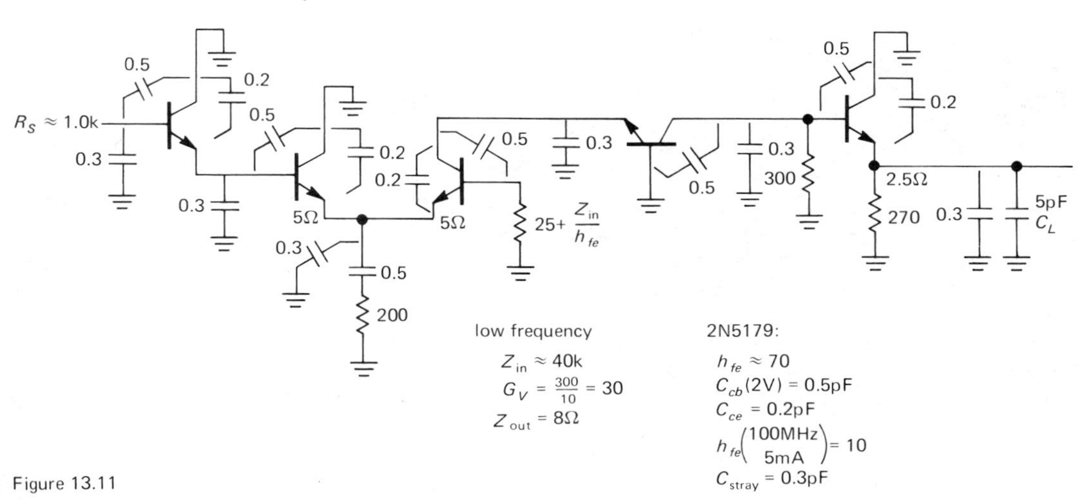

Figure 13.11
An ac equivalent circuit for Figure 13.10.

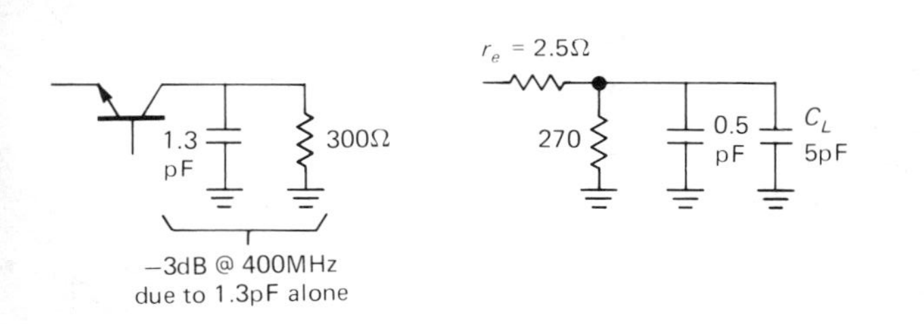

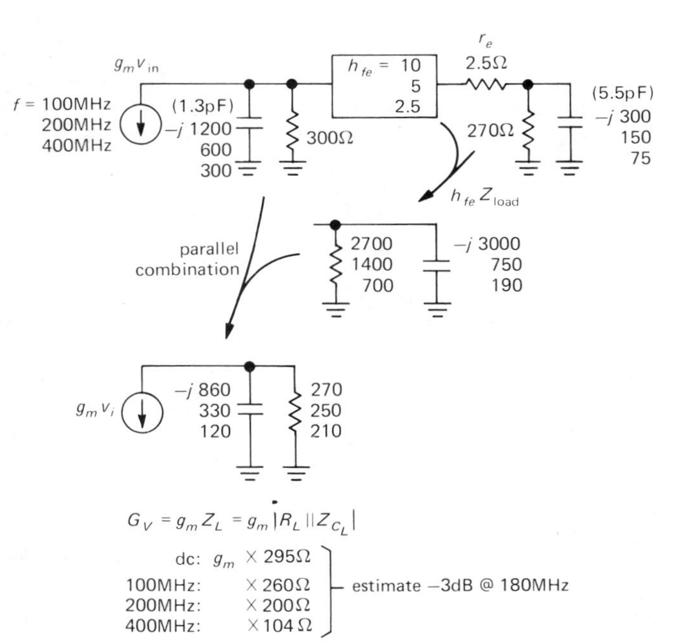

Figure 13.12
Calculating amplifier response rolloff.

combined it with the other impedances already present at the base, and then calculated the resultant magnitude of impedance in order to get the relative output swing as a function of frequency. As can be seen, the output was down 3dB somewhere around 180MHz.

Now, using this as an estimate of the 3dB point, we went through the rest of the circuit, checking to see if other *RC*s gave significant attenuation at this particular frequency. As an example, Q_4's collector circuit would be down 3dB at about 1000MHz, *using the value of transistor beta at 180MHz* ($h_{fe} \approx 5$); in other words, the cascode portion of the circuit does not degrade overall performance.

In a similar manner, it is a relatively straightforward process to verify that no other portion of the circuit sets as low a 3dB point. When dealing with the input stage, you have to assume some value of driving (source) impedance. If you assume Z_s = 1000 ohms (rather high for a video circuit like this), you find, finally, that the combination of source resistance in combination with input capacitance (1.0k, 0.8pF) contributes a 3dB point at about 200MHz. Thus, the overall circuit has good performance up to about 200MHz, for source impedances somewhat less than 1k, but this will be degraded for source impedances comparable to or exceeding 1k. This is a considerable improvement over the earlier circuit we analyzed.

□ 13.06 Some refinements to the ac model

□ *Base spreading resistance*

It is worth noting that the models we have been using are somewhat simplistic, and they neglect some important effects, e.g., the finite resistance r'_b of the base contact. Transistors intended for high-frequency use often specify the parameter $r'_b C_{cb}$, the "collector-base time constant." For the 2N5179 it is 3.5ps (typ), implying a base contact "spreading" resistance of about 7 ohms. When analyzing performance at extremely high frequencies, it is necessary to include such effects in the calculations; in this example it has no effect on the conclusions we reached earlier.

□ *Pole splitting*

Another simplification in the preceding treatment is the assumption that each *RC* rolloff acts independent of the others. It is intuitively easy to see that there must be some interaction, by the following argument: The Miller effect itself is a form of high-frequency negative feedback. Since it samples output *voltage,* it must therefore act to lower the output impedance of the transistor stage, particularly at higher frequencies, where its "loop gain" is high (of course, it also lowers the voltage gain, which is the whole problem). The resultant reduced impedance at the collector drives the $R_L C_L$ rolloff up to a higher frequency, since the collector impedance parallels R_L. Thus, lowering the frequency of the Miller-effect rolloff (by raising G_V or C_{cb}) raises the rolloff due to collector and load capacitance. This is known as "pole splitting."

□ 13.07 The shunt-series pair

A popular circuit for broadband low-gain amplifiers is the shunt-series pair (Fig. 13.13). The idea is to make an amplifier of relatively low gain (perhaps 10dB) with flat response over a broad range of frequencies. That sounds like an application for negative feedback. However, negative feedback can

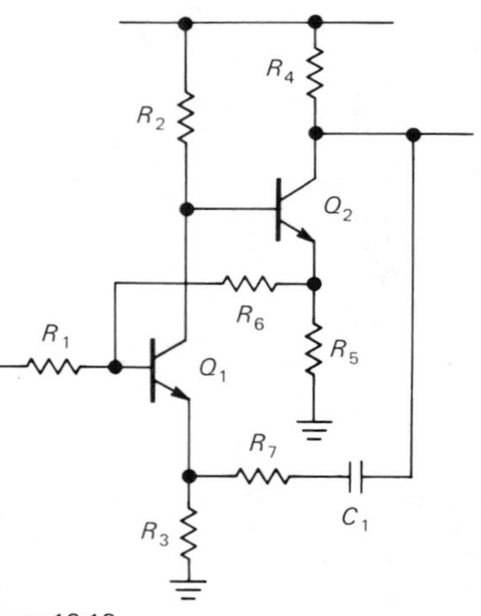

Figure 13.13

TABLE 13.1 RF TRANSISTORS

Type	Case	P_{diss} $T_C = 25°C$ (W)	V_{CEO} (V)	V_{CBO}[a] (V)	I_C max (A)	h_{FE} typ	@ I_C (mA)	C_{cb} 10V (pF)	f_T (MHz)	@ I_C (mA)	Power gain (dB)	@ f (MHz)	Output power (W)	@ f (MHz)	Comments
2N3375	TO-60	12	40	65	1.5	10[b]	250	12	550	120	6	175	10	100	Low cost, easy to mount
2N3553	TO-39	7	40	65	1.0	10[b]	250	7	600	100	15	175	7	100	Low cost, popular
2N3866	TO-39	5	30	55	0.4	50	50	3	900	40	15	400	1[b]	400	
2N4427	TO-39	2	20	40	0.4	50	100	4	800	50	15	175	1[b]	175	
2N5016	TO-60	30	30	65	4.5	10[b]	500	25	600	500	8	400	30	100	
2N5109	TO-39	2.5	20	40	0.4	60	50	3	1500	70	15	200	—	—	Low noise, popular
2N5179	TO-72	12	20	20	0.05	70	3	0.7	1500	10	20	200	0.02	500	Low noise, $r'_b C_c = 7ps$
2N5994	strip	35	30	65	5	—	—	70	—	—	10	100	35	175	VHF power
2N6267	strip	20	50	50	1.5	—	—	13	—	—	10	2000	10	2000	Microwave power
2N6604	strip	0.5	15	25	0.05	—	—	—	5500	30	16	1000	—	—	Microwave small signal
MRF931	strip	0.05	5	10	0.005	70	0.25	0.4[c]	3000	1	12	1000	—	—	Battery-powered, telemetry

(a) Since the base is reverse-biased when the collector tuned circuit goes high, V_{CBO} is often the relevant breakdown voltage. (b) Minimum. (c) At $V_{CB} = 1V$.

be troublesome at radiofrequencies, owing to uncontrolled phase shifts in a high-loop-gain feedback path. The shunt-series pair overcomes this difficulty by having several feedback paths, each with relatively low loop gain.

In the preceding circuit, both Q_1 and Q_2 operate as low-gain voltage amplifiers, since their emitter resistors are not bypassed. R_6 provides feedback around Q_1 alone, since Q_2 is used as a follower for that loop. Once Q_1's overall voltage gain is set (R_6/R_1), R_4 is chosen to set Q_2's open-loop gain (R_4/R_5). Finally, the feedback to Q_1's emitter is added to reduce the gain to its design value.

The shunt-series pair is a convenient amplifier building block because it is extremely stable and easy to design. Amplifiers with bandwidth to 300MHz or so are easily constructed with this technique. You can get gains of 10dB to 20dB per amplifier, cascading several stages, if necessary, to obtain greater gain.

In Section 13.11 we will discuss the techniques used to construct tuned (narrow-band) amplifiers, as contrasted with the broadband design we have been talking about so far. Since signals of interest are often confined to a narrow band of frequencies in instruments that operate at radiofrequencies, tuned amplifiers are extremely useful.

13.08 Modular amplifiers

From the foregoing discussion of RF amplifiers it might seem that any project at high frequencies would become a formidable design effort, with messy calculations and numerous trial designs. Luckily, there are complete packaged amplifier *modules* available from more than a dozen suppliers, in configurations to meet almost any need. In fact, nearly every RF component can be obtained as a module, including oscillators, mixers, modulators, voltage-controlled attenuators, power combiners and dividers, circulators, hybrids, directional couplers, etc. We will describe some of these other circuit elements in Section 13.12.

In its most basic form, the prepackaged RF amplifier comes as a thin-film hybrid circuit with gain over a wide band, packaged in a 4-pin transistor package. Two of the pins are input and output terminals, with convenient 50 ohm impedance levels, and the remaining pins are for ground and the dc supply. There are dozens of different amplifiers available, some optimized for low noise and others for high power or large dynamic range. Individual amplifiers may be designed for operation over a very wide frequency range, or for a particular band of frequencies used in communications. As an example, the UTO-514 from Avantek has 15dB of gain over the frequency range of 30MHz to 200MHz, with a noise figure of 2dB (maximum) and a gain flatness of ±0.75dB. It is packaged in a 4-pin TO-8 transistor package.

These hybrid amplifiers can be used singly or in cascade, usually as part of a stripline (Section 13.20). To make life even easier, the amplifier manufacturers have thoughtfully provided complete amplifier building blocks as prepackaged modules. These beasts typically occupy a small metal box, perhaps 2 × 2 × 1 inch, with SMA-type RF coax connectors for input and output. To give an idea of what you can get, we have thumbed through the impressive Avantek catalog and come up with the following: The UTC2-102 is a nice low-noise amplifier with 29dB gain and a 1.5dB noise figure over the frequency range 30MHz to 200MHz. For wider bandwidth the AMG-502 spans 5MHz to 500MHz with a 2.8dB noise figure and 27dB gain. For even wider bandwidth you might choose the AWL-500, which spans 0.001MHz to 500MHz with a 5dB noise figure and 25dB of gain. All these amplifiers are flat to ±1dB. Wideband amplifiers are available all the way to 18GHz, using the recent GaAs FET technology.

Amplifiers for use over a narrow band of frequencies can be optimized for low-noise performance; extremely good amplifiers are available for the communications bands. For example, for your own backyard satellite-downlink receiver, Avantek's AW-4286 has 60dB of gain (±0.5dB) in the 3.7–4.2GHz band, with a phenomenal 1.5dB noise figure. In the 7.25–7.75GHz band, their AM-7724 delivers a gain of 35dB

(±0.25dB) with an astounding 1.8dB noise figure.

There is plenty of commercial competition in these amplifier modules, as well as other RF modular components. For complete amplifier modules, some of the larger suppliers are Aertech/TRW, Avantek, Aydin Vector, Hewlett-Packard, Narda, Scientific Communications, and Watkins-Johnson. In practice, when designing an RF system you might well choose to thumb through catalogs of available (and custom) modules in order to assemble a system. Screw them all down to a plate, connect them together with coaxial cable, and off you go!

RADIOFREQUENCY CIRCUIT ELEMENTS

13.09 Transmission lines

Before proceeding to the subject of communications circuits, it is necessary to deal briefly with the interesting subject of transmission lines. You have met these earlier in connection with digital signal communications in Chapter 9, where we introduced the ideas of characteristic impedance and line terminations. Transmission lines play a central role in radiofrequency circuits, where they are used to pipe signals around from one place to another within a circuit, and often to an antenna system. Transmission lines provide one of the most important exceptions to the general principles (see Chapter 1) that a signal source ideally should have a source impedance small compared with the impedance of the load being driven and that the load should present an input impedance large compared with the source impedance driving it. The equivalent rule for transmission lines is that the load (and possibly the source) should present an impedance equal to the characteristic impedance of the line. The line is then "matched."

Transmission lines for signals of moderate frequency (up to 1000MHz, say) come in two major types: parallel conductors and coaxial line. The former is typified by the inexpensive molded 300 ohm "twin lead" used to bring the signal from a television antenna to the receiver, and the latter is widely used in short lengths with BNC fittings to carry signals between instruments (Fig. 13.14).

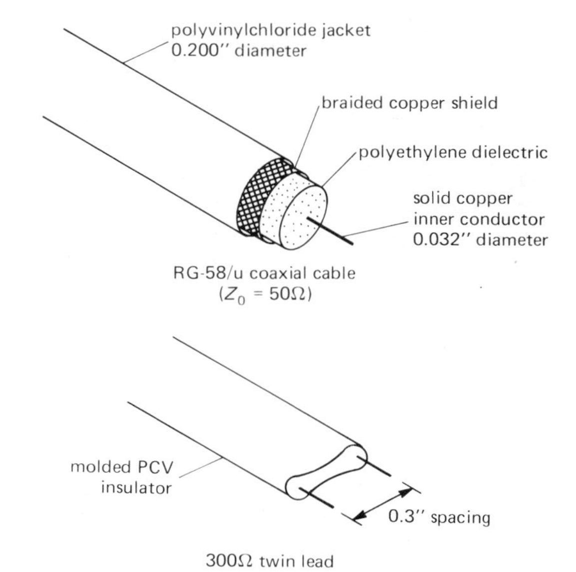

Figure 13.14

In the domain of ultra-high-frequency circuitry there are "stripline" techniques that involve parallel-conductor transmission lines as part of the actual circuit, and at the higher "microwave" frequencies (upwards of 2GHz, say) conventional lumped circuit elements and transmission lines are replaced by cavity and waveguide techniques, respectively. Except at these extremes of frequency, the familiar coaxial cable is probably the best choice for most radiofrequency applications. Compared with parallel-conductor line, a properly matched coax line has the advantage of being totally shielded, i.e., there is no radiation or pickup of external signals.

Characteristic impedance and matching

A transmission line, whatever its form, has a "characteristic impedance" Z_0, meaning that a wave moving along the line has a ratio of voltage to current equal to Z_0. For a lossless line, Z_0 is resistive and equal to the square root of L/C, where L is the inductance per unit length and C is the capacitance per unit length. Typical coaxial lines have impedances in the range of 50 to 100 ohms, whereas parallel-conductor lines have impedances in the range of 300 to 1000 ohms.

When used with high-frequency (or short-rise-time) signals, it is important to ''match'' the load to the characteristic impedance of the line. The important facts are the following: (a) A transmission line terminated with a load equal to its characteristic impedance (resistance) will transfer an applied pulse to the termination without reflection. In that case all the power in the signal is transferred to the load. (b) The impedance looking into such a terminated line, at any frequency, is equal to its characteristic impedance (Fig. 13.15).

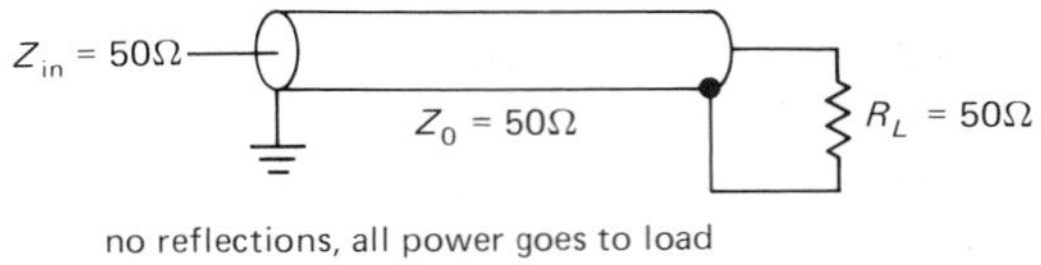

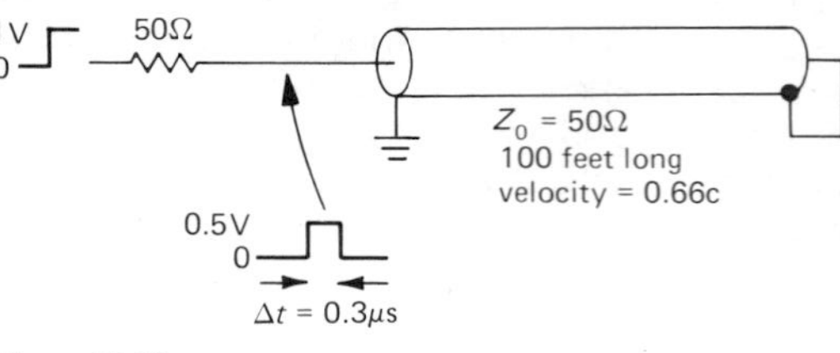

Figure 13.15

This is surprising at first, since at low frequencies you tend to think of a length of coax as a small capacitive load, generally a pretty high (capacitive) impedance. Also, at low frequencies (wavelength ≫ length of cable) there is no need to match the line's impedance, provided you can handle the capacitance (typically 30pF per foot). If the cable is terminated with a resistor, on the other hand, it magically becomes a pure resistance at all frequencies.

Mismatched transmission lines

A mismatched transmission line has some interesting, and occasionally useful, properties. A line terminated in a short circuit produces a reflected wave of opposite polarity, with the delay time of the reflected wave determined by the electrical length of the line (the speed of wave propagation in coax lines is about two-thirds the speed of light, because of the solid dielectric spacing material). You can see the reason for this, since the short circuit enforces a point of zero voltage at the end; the cable produces this obligatory boundary condition by creating a wave of opposite phase at the short. In similar manner, an open-circuited cable (boundary condition of zero current at the end) produces a noninverted reflection of amplitude equal to the applied signal.

This property of a shorted cable is sometimes exploited to generate a short pulse from a step waveform. The step input is applied to the cable input through a resistance equal to Z_0, with the other end of the cable shorted. The waveform at the input is a pulse of width equal to the round-trip travel time, since the reflected step cancels the input (Fig. 13.16).

Figure 13.16
Pulse generation with shorted transmission lines (inverted reflection).

Cables terminated with a resistance R unequal to Z_0 also produce reflections, although of lesser amplitude. The reflected wave is inverted if $R < Z_0$ and uninverted if $R > Z_0$. The ratio of reflected wave amplitude to incident wave amplitude is given by

$$A_r/A_i = (R - Z_0)/(R + Z_0)$$

Transmission lines in the frequency domain

Looked at in the frequency domain, a transmission line matched at the far end looks like a load of impedance Z_0, i.e., a pure resistance if line losses are neglected. That makes sense, since it just swallows any wave you apply, all the power going into the matching resistor. This is true independent of cable length or wavelength. It is when you deal with mismatched lines that things begin to get interesting in the frequency domain. Since, for a given line length, the reflected wave arrives back at the input with a phase (relative to the applied signal) that depends on applied frequency, the impedance seen looking into the input depends on the mismatch and on the electrical length of the transmission line, in wavelengths.

As an example, a line that is an odd number of quarter wavelengths long terminated in an impedance Z_{load} at the far end presents an input impedance $Z_{in} = Z_0^2/Z_{load}$. If the load is resistive, the input will look

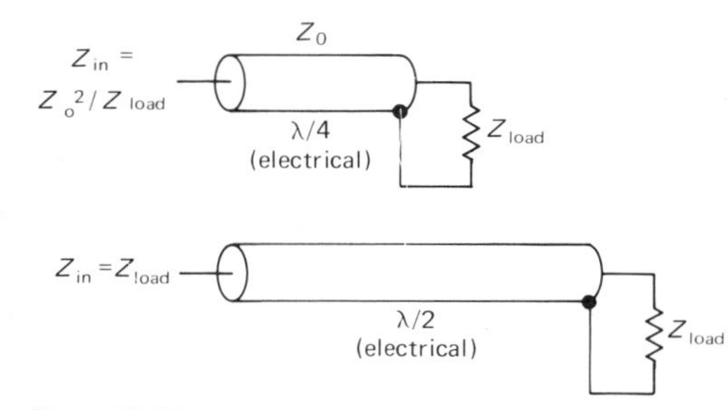

Figure 13.17

resistive. On the other hand, a line that is an integral number of half wavelengths long presents an input impedance equal to its terminating impedance (Fig. 13.17).

The presence of reflected signals on a transmission line is not necessarily bad. For operation at a single frequency, a mismatched line can be driven (through a line tuner) in such a way as to match its resultant input impedance, often with only negligibly greater line losses (due to higher voltages and currents for the same forward power) than with a matched load. But a mismatched line has different properties at different frequencies (the famous "Smith chart" can be used to determine transmission-line impedances and "standing-wave ratio," or SWR, a measure of the amplitude of reflected waves), making it undesirable for broadband or multifrequency use. In general, strive to terminate a transmission line in its characteristic impedance, at least at the receiving end.

□ 13.10 Stubs, baluns, and transformers

There are some interesting applications of transmission lines that exploit the properties of mismatched sections or generally use sections of line in an unconventional way. The simplest is the quarter-wave matching section, which exploits the relationship $Z_{in} = Z_0^2/Z_{load}$. This can be rearranged to read $Z_0 = (Z_{in}Z_{load})^{1/2}$. In other words, a quarter-wave section can be used to match any two impedances by choosing the characteristic impedance of the matching section appropriately.

In a similar manner, a short length of transmission line (a "stub") can be used to "tune" a mismatched load by simply putting the stub across or in series with the mismatched line, choosing the stub length and termination (open or shorted) and its position along the mismatched line correctly. In this sort of application the stub is really functioning as a circuit element, not a transmission line. At very short wavelengths the use of sections of transmission line as circuit elements is common (Fig. 13.18).

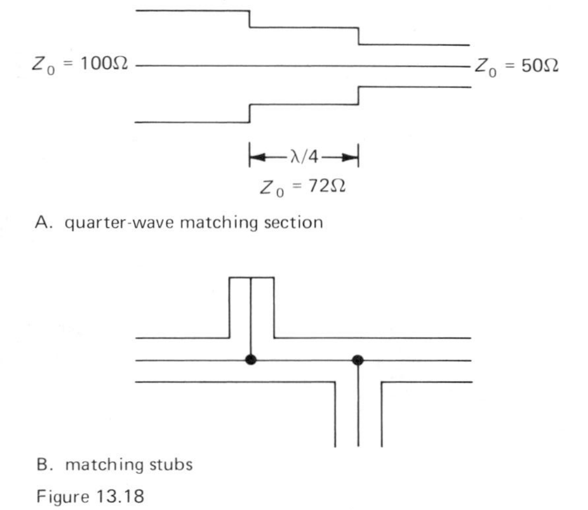

A. quarter-wave matching section

B. matching stubs

Figure 13.18

Sections of transmission line (or a transformer made with several interconnected windings) can be used to construct a "balun," a device for matching an unbalanced line (coax) to a balanced load (e.g., an antenna). There are simple configurations for making fixed-impedance transformations at the same time (1:1 and 4:1 are common). Perhaps the nicest circuit element made from transmission line is the broadband transmission-line transformer. These gadgets consist simply of a few turns of miniature coax or twisted pair wound on a ferrite core, suitably interconnected. They avoid the high-frequency limitations of conventional transformers (caused by the resonant combination of "parasitic" winding capacitance and inductance) because the coils are arranged so that the winding capacitance and inductance form a transmission line, free of resonances. They can provide various impedance transformations with astounding broadband performance (e.g., less than 1dB loss from 0.1MHz to 500MHz), a property not shared by transformers constructed from simple coupled inductors. Transmis-

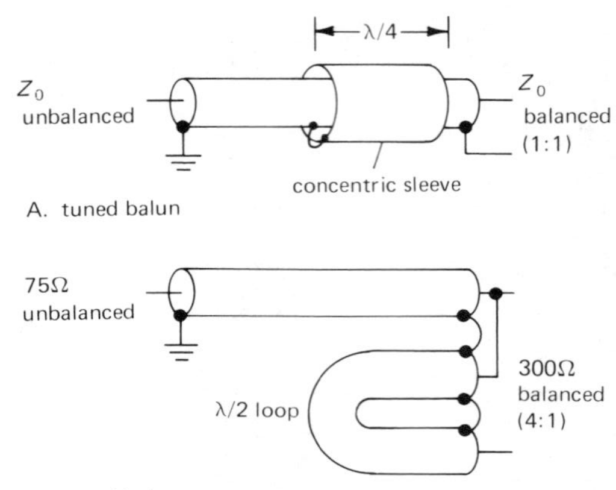

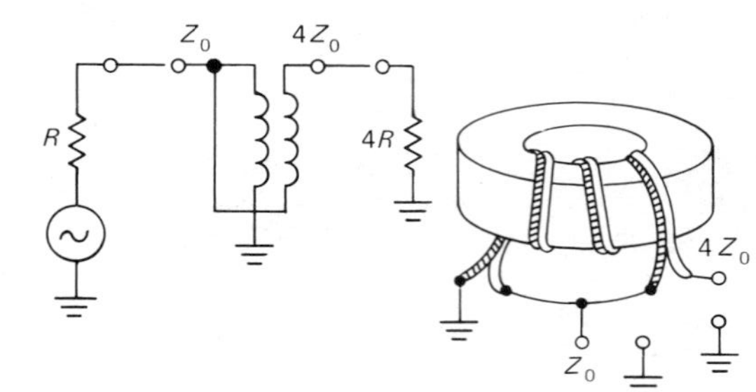

C. 4:1 unbalanced transmission-line transformer

Figure 13.19
Transmission-line transformers.

sion-line transformers are available from the Vari-L Co. and Mini-Circuits Laboratory, among others, as packaged modules. Figure 13.19 shows a few examples of baluns and a transmission-line transformer.

13.11 Tuned amplifiers

In radiofrequency circuits intended for communications, or for other applications where the operating frequency is confined to a narrow range, it is common to use tuned LC circuits as collector or drain loads. This has several advantages: (a) higher single-stage gain, since the load presents a high impedance at the signal frequency ($G_V = g_m Z_{load}$) while allowing arbitrary quiescent current; (b) elimination of the undesirable loading effects of capacitance, since the LC circuit "tunes out" any capacitance by making it part of the tuned circuit capacitance; (c) simplified interstage coupling, since an LC circuit can be tapped or transformer-coupled (or even configured as a resonant matching network, as in the popular pi network) to achieve any desired impedance transformation; (d) elimination of out-of-band signals and noise owing to the frequency selectivity of the tuned circuits.

Examples of tuned radiofrequency circuits

You will see tuned RF amplifiers in their natural element when we discuss communications circuits shortly. At this point we would simply like to illustrate the use of tuned circuits in oscillators and amplifiers with a few examples. Figure 13.20 illustrates the classic tuned amplifier. A dual-gate depletion-mode FET is used to eliminate the problems of Miller effect, since the input is untuned. By operating the lower gate at dc ground, the stage runs at I_{DSS}. The parallel tuned LC sets the center frequency of amplification, with output buffering via Q_2. Since the drain sits at +10 volts, the output follower requires a higher collector voltage. This sort of circuit has quite high voltage gain at resonance, limited by the LC circuit Q and loading by the follower.

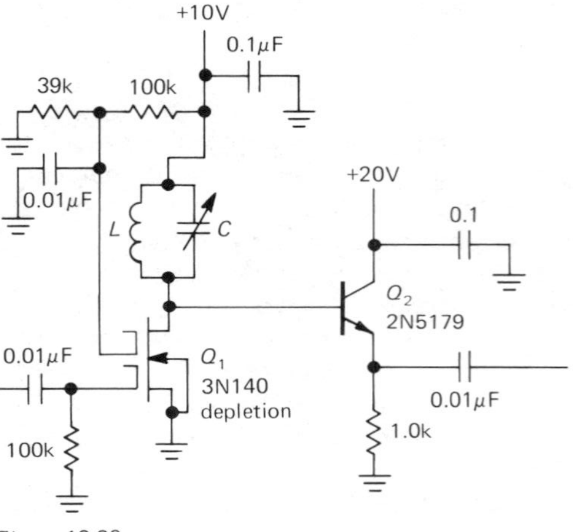

Figure 13.20
Dual-gate MOSFET (cascode) tuned amplifier.

In the circuit shown in Figure 13.21 a carefully constructed and tunable LC circuit is used to set the frequency of an oscillator. This is known as a VFO (variable-frequency oscillator); it is used as the tunable element in some transmitters and receivers, as well

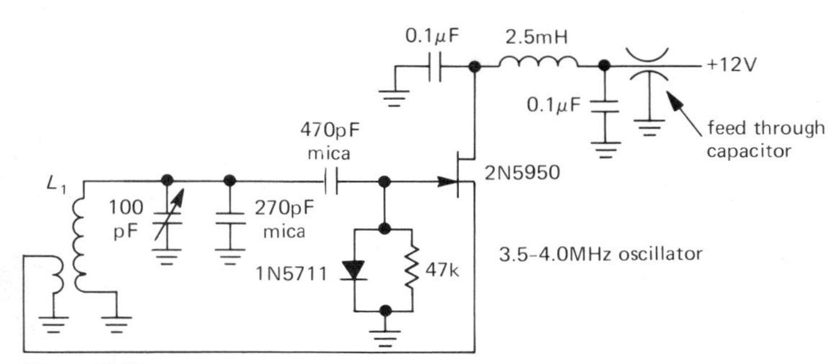

Figure 13.21
JFET *LC* oscillator.

as variable-frequency RF signal sources. In this circuit a JFET provides the necessary power gain, with positive feedback from the source coupled into a "link" on L_1. The link has fewer turns than the inductor, providing voltage gain and therefore oscillation. By adding a varactor diode, which acts as a voltage-variable capacitor, you can make such an oscillator voltage-tunable. Note the use of a feedthrough capacitor and decoupling RF chokes on the power-supply leads; this is nearly universal practice in radiofrequency circuits.

The circuit in Figure 13.22 is a 200MHz common-emitter transistor amplifier stage. This circuit illustrates neutralization, the technique of canceling out capacitively coupled signal from output to input by adding a current of the opposite phase. C_N is the neutralizing capacitor, driven from the bottom of the collector "tank" circuit where the phase is opposite to that at the collector. This circuit matches the output impedance to the line by tapping down on the "tank" (the collector *LC* circuit), a simple but inflexible method.

This last circuit (Fig. 13.23) is a 25kW RF amplifier, using a zero-bias grounded-grid triode. Vacuum tubes are still used in high-power radiofrequency amplifiers, because no solid-state device can match their performance. The grounded-grid configuration requires no neutralization. The output circuit is the popular pi network, driven by blocking capacitor C_8. C_9, L_4, and C_{10} form the actual network, with their values determined by the desired resonant frequency, impedance transformation, and loaded Q (Q, or quality factor, is a measure of the sharpness of resonance, see Section 1.22). The RF choke at the output prevents dc voltage appearing there, and the plate RF choke is used to apply plate voltage while allowing signal swing at the operating frequency.

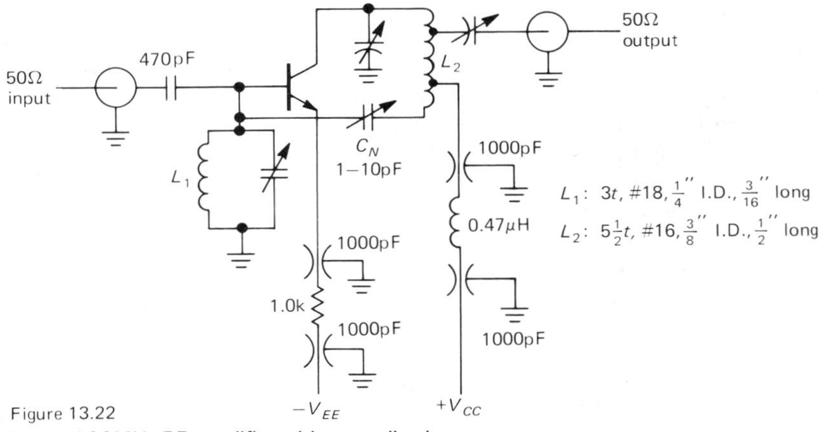

Figure 13.22
Tuned 200MHz RF amplifier with neutralization.

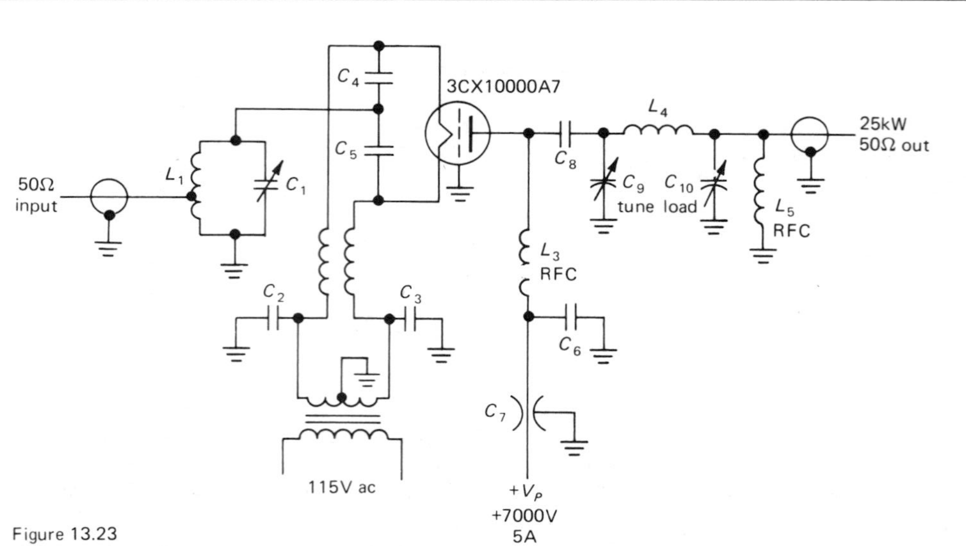

Figure 13.23
High-power grounded-grid triode RF amplifier (25kW output).

13.12 Radiofrequency circuit elements

In radiofrequency circuits you meet several kinds of specialized circuit modules that have no equivalents in low-frequency circuitry. Before going on to consider RF communications circuits, we will mention some of these, since they enjoy widespread use in the generation and detection of radiofrequency signals.

Oscillators

If great stability is not important, a simple *LC* oscillator of the type just illustrated will generate a radiofrequency signal with adjustability over an octave or more by varying either *C* or *L* (the latter is sometimes called a "permeability-tuned oscillator," or PTO). With careful design and attention to detail during construction, such a VFO (variable-frequency oscillator) can have drifts of less than a few parts per million measured over hours and will be quite satisfactory for use in receivers and noncritical transmitting applications. *LC* oscillators can be operated at frequencies ranging from audiofrequencies to hundreds of megahertz.

Just as with the amplifier modules we mentioned in Section 13.08, prepackaged oscillator modules with excellent performance are readily available. Tunable oscillator modules use varactors (voltage-variable capacitors) to adjust the operating frequency of an *LC* oscillator in response to an externally applied voltage. A fancier version of tunable oscillator for gigahertz frequencies uses a yttrium iron garnet (YIG) sphere as a magnetically tunable resonant cavity; YIG-tuned oscillators provide high spectral purity and tuning linearity.

For high stability, the best kind of oscillators use quartz crystals to set the operating frequency. With off-the-shelf garden-variety crystals, you can expect overall stabilities of a few parts per million, with tempco of order 1 ppm/degree or better. A temperature-compensated crystal oscillator (TCXO), which uses capacitors of controlled tempco to offset the crystal's frequency variation, can deliver frequency stability of 1 ppm over a temperature range of 0°C to +50°C or better. For the utmost in performance, oscillators with the crystal maintained in a constant-temperature "oven" are available, with stabilities of a few parts per billion over time and temperature. Even the so-called atomic oscillators (rubidium, cesium) actually use a high-stability quartz oscillator as the basic oscillating element, with frequency adjusted as necessary to agree with a particular atomic transition frequency.

Crystal oscillators are commercially available in frequencies ranging from about 10kHz to about 100MHz, in all of the variations just mentioned. There are even little DIP and transistor can (TO-5) oscillators, with logic outputs. Only a slight electrical adjustment of frequency is possible, so the frequency must be specified when the oscillator or crystal is ordered.

To get both adjustability and high stability, a frequency synthesizer is the best choice. It uses tricks to generate any desired frequency from a single source of stable frequency, typically a 10MHz crystal oscillator. A synthesizer driven from a rubidium standard (stability of a few parts in 10^{12}) makes a nice signal source.

Mixers/modulators

A circuit that forms the product of two analog waveforms is used in a variety of radiofrequency applications and is called, variously, a modulator, mixer, synchronous detector, or phase detector. The simplest form of modulation, as you will see shortly, is amplitude modulation (AM), in which the high-frequency *carrier* signal is varied in amplitude according to a slowly varying *modulating* signal. A multiplier obviously performs the right function. Such a circuit can also be used as a variable gain control, thinking of one of the inputs as a dc voltage. There are convenient ICs to do this job, e.g., the MC1495 and MC1496.

A mixer is a circuit that accepts two signal inputs and forms an output signal at the sum and difference frequencies. From the trigonometric relationship

$$\cos \omega_1 t \cos \omega_2 t = \tfrac{1}{2} \cos (\omega_1 + \omega_2)t + \tfrac{1}{2} \cos (\omega_1 - \omega_2)t$$

it should be clear that a "four-quadrant multiplier," i.e., one that performs the product of two input signals of any polarity, is in fact a mixer. If you input two signals of frequency f_1 and f_2, you will get out signals at $f_1 + f_2$ and $f_1 - f_2$. A signal at frequency f_0 mixed with a band of signals near zero frequency (band-limited to a maximum frequency of f_{max}) will produce a symmetrical band of frequencies around f_0, extending from $f_0 - f_{max}$ to $f_0 + f_{max}$ (the spectrum of amplitude modulation, see Section 13.14).

It is not necessary to form an accurate analog product in order to mix two signals. In fact, any nonlinear combination of the two signals will produce sum and difference frequencies. Take, for instance, a "square law" nonlinearity applied to the sum of two signals:

$$(\cos \omega_1 t + \cos \omega_2 t)^2 = 1 + \tfrac{1}{2} \cos 2\omega_1 t + \tfrac{1}{2} \cos 2\omega_2 t + \cos (\omega_1 + \omega_2)t + \cos (\omega_1 - \omega_2)t$$

This is the sort of nonlinearity you would get (roughly) by applying two small signals to a forward-biased diode. Note that you get harmonics of the individual signals, as well as the sum and difference frequencies. The term "balanced mixer" is used to describe a circuit in which only the sum and difference signals, not the input signals and their harmonics, are passed through to the output. The four-quadrant multiplier is a balanced mixer, whereas the nonlinear diode is not.

Among the methods used to make mixers are the following: (a) simple nonlinear transistor or diode circuits, often using Schottky diodes; (b) dual-gate FETs, with one signal applied to each gate; (c) multiplier chips like the MC1495, MC1496, μA796, or 5596; (d) balanced mixers constructed from transformers and arrays of diodes, generally available as packaged "double-balanced mixers." The latter are typified by the inexpensive SBL-1 double-balanced mixer from Mini-Circuits, spanning the frequency range from dc to 500MHz with 30dB to 50dB of signal isolation and just a few decibels of conversion loss. Mixers are widely used in heterodyne receivers, as well as in the generation of radiofrequency signals at arbitrary frequencies; they let you shift a signal up or down in frequency without changing its spectrum. You will see how it all works shortly.

Frequency multipliers

A nonlinear circuit is often used to generate a signal at a multiple of the input signal's frequency. This is particularly handy if a signal of high stability is required at a very high frequency, above the range of good oscillators. One of the most common methods is to bias an amplifier stage for highly nonlinear operation, then use an *LC* output circuit tuned to some multiple of the input signal; this can be done with bipolar transistors, FETs, or even tunnel diodes. A multiplier like the 1496 can be used as an efficient doubler at low radiofrequencies by connecting the input signal to both inputs, thus forming the square of the input waveform. The square of a sine wave contains frequencies at the second harmonic only. Prepackaged frequency doublers that use balanced mixers are commercially available; they are very broadband (50kHz to 150MHz is typical). Exotic devices such as SNAP diodes and varactors are also used as multipliers. A frequency-multiplier circuit should include a tuned output circuit or should be followed by tuned amplifiers, since, in general, many harmonics of the input signal are generated in the nonlinear process.

Attenuators, hybrids, circulators

There are some fascinating passive devices that are used to control the amplitude and direction of radiofrequency signals passing between circuit modules. All of these are broadband transmission-line (or waveguide) components, meant to be inserted in a line of fixed impedance, usually 50 ohms. They are all widely available as modules.

The simplest is the attenuator, a device to reduce the amplitude of a signal. They come with a big knob and accurately calibrated steps of attenuation, or as voltage-controlled attenuators. The latter are simply balanced mixers with the control current serving as one of the multiplying inputs.

A hybrid (also known as a "rat race," magic-T, 3dB coupler, or iso-T) is a clever transmission-line configuration with four ports. A signal fed into any port emerges from the two closest ports, with specific phase shifts (usually 0° or 180°). A hybrid that has one port terminated in its characteristic impedance is a 3-port "power splitter/combiner." Splitter/combiners can be cascaded to make multiport splitter/combiners. A close cousin of the hybrids is the directional coupler, a 3-port device that couples a small fraction of a passing wave out to a third port. Ideally there is no output at the third port for a wave passing through in the opposite direction.

The most magical devices in this general category are the circulators and isolators. They employ exotic ferrite materials and magnetic fields to achieve the impossible: a device that will transmit waves in only one direction. The isolator has two ports and allows transmission in one direction only. Circulators have three or more ports, and they transmit an incoming signal at any port to the next port in succession.

Filters

As you will see, frequency selectivity is often needed in the design of radiofrequency circuits. The simple tuned *LC* amplifier provides a good measure of selectivity, with the peakiness of the response adjustable via the Q factor of the *LC* circuit. The latter depends on losses in the inductor and capacitor, as well as loading by the associated circuitry. Qs as high as several hundred can be easily obtained. At very high frequencies, lumped *LC* circuits are replaced by stripline techniques, and at microwave frequencies you use cavity resonators, but the basic idea remains the same. Tuned circuits can also be used to *reject* a particular frequency, if desired.

For applications where it is necessary to have a filter that passes a very narrow band of frequencies relatively unattenuated, with a sharp dropoff outside the limits of the band, a superior bandpass filter can be made from a set of piezoelectric (ceramic or quartz-crystal) or mechanical resonators. There are commercially available 8-pole and 16-pole Butterworth crystal-lattice filters with center frequencies in the range of 1MHz to 50MHz and bandwidths ranging from as little as a few hundred hertz to several kilohertz. These filters are extremely important in setting receiver selectivity and in the generation of certain kinds of modulated signals. Recently, surface acoustic wave (SAW) filters have become popular and inexpensive; these, too, can have level passband characteristics with extremely steep skirts. This desirable characteristic is usually expressed as a "shape factor," e.g., the ratio of −3dB bandwidth to −40dB bandwidth, with values as small as 1.1. In a typical application SAW filters are used in television receivers and cable systems to limit the received passband.

Of course, in situations where such narrow passbands are not needed, filters can be designed with multiple resonant *LC* sections. Appendix H shows some low-pass and high-pass *LC* filter examples.

Detectors

The bottom line in the extraction of information from a modulated radiofrequency signal involves *detection,* the process of stripping the modulating signal from the "carrier." There are several methods, depending on the form of modulation (AM, FM, SSB, etc.), and we will discuss this important topic next, along with communications concepts.

RADIOFREQUENCY COMMUNICATIONS: AM

Since radiofrequency techniques find their greatest application in communications, it is important to understand how signals can be modulated and demodulated, i.e., how radiofrequencies are used to carry information from one place to another. Besides, how would you feel if, after taking a course in electronics, someone asked you how a radio works and you didn't know?

13.13 Some communications concepts

In communications theory we speak of a communications "channel," a means of conveying information from A to B. For example, the channel might consist of a cable or an optical-fiber link. In radiofrequency communications the channel is the electromagnetic frequency spectrum, which, roughly speaking, extends from very low frequencies (VLF) of a few kilohertz, through the "short waves" of a few megahertz to a few tens of megahertz, the very high frequencies (VHF) and ultrahigh frequencies (UHF) extending up to several hundred megahertz, and the microwave region beginning at about 1GHz.

A signal, consisting of speech, say, is sent on a radiofrequency channel by having it modulate a radiofrequency "carrier." It is important to understand why this is done at all, rather than transmitting the speech directly. There are basically two reasons. First, if the information were transmitted at its natural band of frequencies with radio waves (in this case, in the VLF portion of the spectrum), any two signals would overlap and jam each other; i.e., by encoding the information onto carriers in separate portions of the spectrum, it is possible to "frequency-multiplex" the signal, and thereby maintain many channels simultaneously. Second, some wavelengths are more conveniently generated and propagated than others. For instance, in the region from 5MHz to 30MHz, signals can travel around the world by multiple reflections from the ionosphere, and at microwave frequencies antennas of modest size can form narrow beams. Consequently, the HF (short-wave) region is used for over-the-horizon communication, whereas microwaves are used for line-of-sight repeaters and radar.

There are several ways to modulate a carrier. Roughly speaking, all methods have in common the property that the modulated signal occupies a bandwidth at least comparable to the bandwidth of the modulating signal, i.e., the bandwidth of the information being sent. Thus, a high-fidelity audio transmission will occupy 20–40kHz of spectrum, regardless of the carrier frequency. A perfect unmodulated carrier has zero bandwidth and conveys no information. A transmission of low information content, e.g. telegraphy, occupies a relatively narrow slice of spectrum (perhaps 50–100Hz), whereas something like a television picture requires several megahertz. For completeness it should be pointed out that more information can be sent on a channel of given bandwidth if there is sufficiently high signal/noise ratio (SNR). Such "frequency compression" takes advantage of the fact that "channel capacity" equals bandwidth times $\log_2$ SNR.

13.14 Amplitude modulation

Let's begin with the simplest form of modulation (AM), taking a look at its frequency spectrum and methods of detection. Imagine a simple carrier, $\cos \omega_c t$, varied in amplitude by a modulating signal of much lower frequency, $\cos \omega_m t$, in the following way:

$$\text{signal} = (1 + m \cos \omega_m t) \cos \omega_c t$$

with m, the "modulation index," less than or equal to 1. Expanding the product, you get

$$\text{signal} = \cos \omega_c t + \tfrac{1}{2} m \cos (\omega_c + \omega_m) t + \tfrac{1}{2} m \cos (\omega_c - \omega_m) t$$

i.e., the modulated carrier has power at frequency ω_c and at frequencies on either side ω_m away. Figure 13.24 shows the

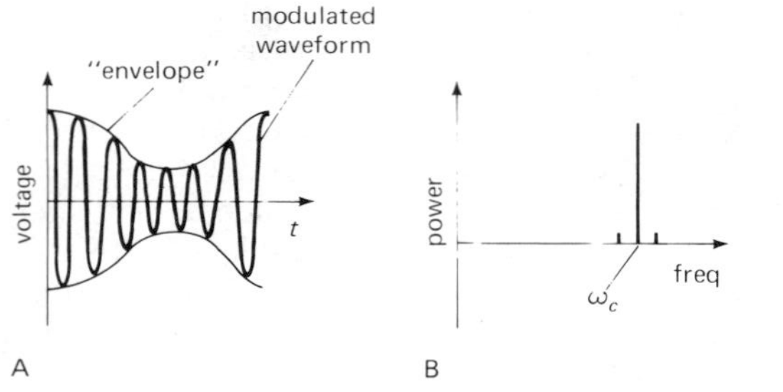

Figure 13.24
Amplitude modulation.

signal and its spectrum. In this case the modulation (m) is 50%, and the two "sidebands" each contain 1/16 of the power contained in the carrier.

If the modulating signal is some complex waveform [$f(t)$], like speech, the amplitude-modulated waveform is given by

$$\text{signal} = [A + f(t)] \cos \omega_c t$$

with the constant A large enough so that $A + f(t)$ is never negative. The resulting spectrum simply appears as symmetrical sidebands around the carrier (Fig. 13.25).

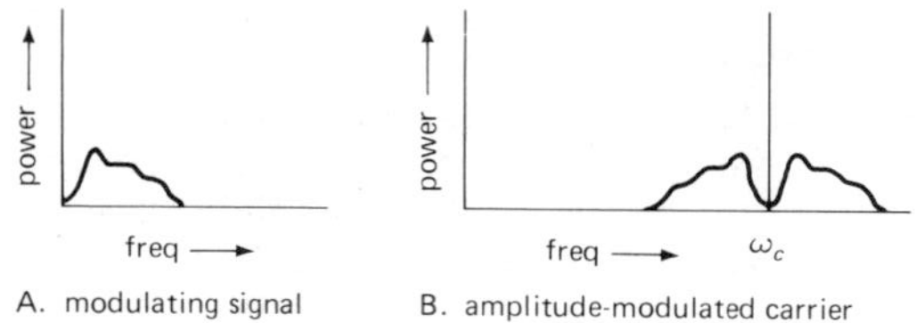

Figure 13.25
AM spectrum for a band of modulating frequencies (speech).

AM generation and detection

It is easy to generate amplitude-modulated RF. Any technique that lets you control the signal amplitude with a voltage in a linear manner will do. Common methods involve varying the RF amplifier supply voltage (if the modulation is done at the output stage) or using a multiplier chip such as the 1496. When the modulation is done at a low-level stage, all following stages of amplification must be linear. Note that in AM the modulating waveform must be biased up so that it never assumes negative values. Look at the graphs in Figure 13.26.

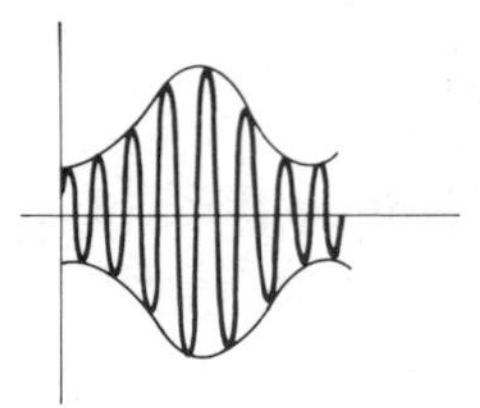

A. 50% modulation

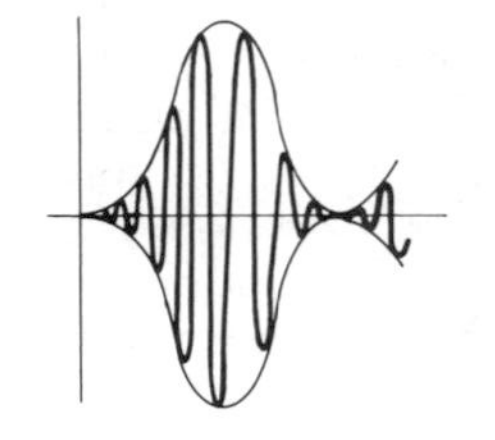

B. 100% modulation

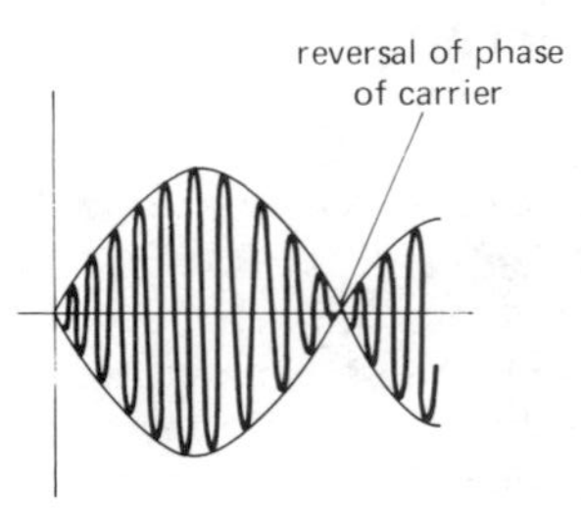

C. overmodulation

Figure 13.26

The simplest receiver of AM consists of several stages of tuned RF amplification, followed by a diode detector (Fig. 13.27). The amplifier stages provide selectivity against signals nearby in frequency, and they amplify the input signals (which may be at the microvolt level) for the detector. The latter simply rectifies the RF waveform, then recovers the smooth "envelope" with low-pass filtering. The low-pass filter should

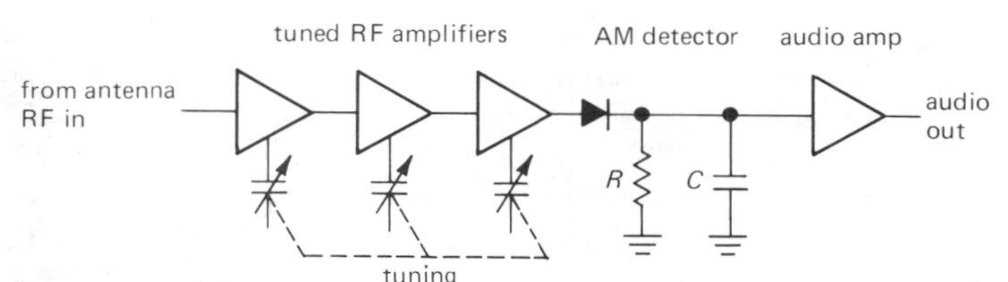

Figure 13.27

reject RF while passing the audiofrequencies unattenuated. This simple scheme leaves much to be desired, as you will see. It is really just a glorified crystal set.

13.15 Superheterodyne receiver

A receiver consisting of a set of tuned RF amplifiers is undesirable for several reasons. First of all, the individual amplifiers must be tuned to the same frequency, requiring either great coordination by someone with a lot of hands or extremely good tracking of a set of simultaneously tuned *LC* circuits. Second, since the overall frequency selectivity is determined by the combined responses of the individual amplifiers, the shape of the passband will depend on the accuracy with which the individual amplifiers are tuned; the individual amplifiers cannot have as sharp a response as would be desirable, since tuning would then be practically impossible. And since the signal being received can be at any frequency within the tuning range of the amplifiers, it isn't possible to take advantage of crystal-lattice filters to generate a flat passband with steep falloff on either side (steep "skirts"), a very desirable passband characteristic.

A nice solution to these problems is the superheterodyne ("superhet") receiver shown in Figure 13.28. The incoming signal is amplified with a single stage of tuned RF amplification, then mixed with an adjustable local oscillator (LO) to produce a signal at a fixed intermediate frequency (IF), in this case 455kHz. From then on the receiver consists of a set of fixed-tuned IF amplifiers, including selective elements such as crystal or mechanical filters, finally terminating in a detector and audio amplifier. Changing the LO frequency tunes the receiver, since a different input frequency then gets mixed to the IF passband frequency. The input RF amplifier must be gang-tuned with the LO, but the alignment is not critical. Its purposes are (a) to improve the sensitivity with a stage of low-noise amplification prior to mixing and (b) to reject signals at the "image" frequency, in this case input signals at a frequency of 455kHz *above* the LO (remember that a mixer generates sum

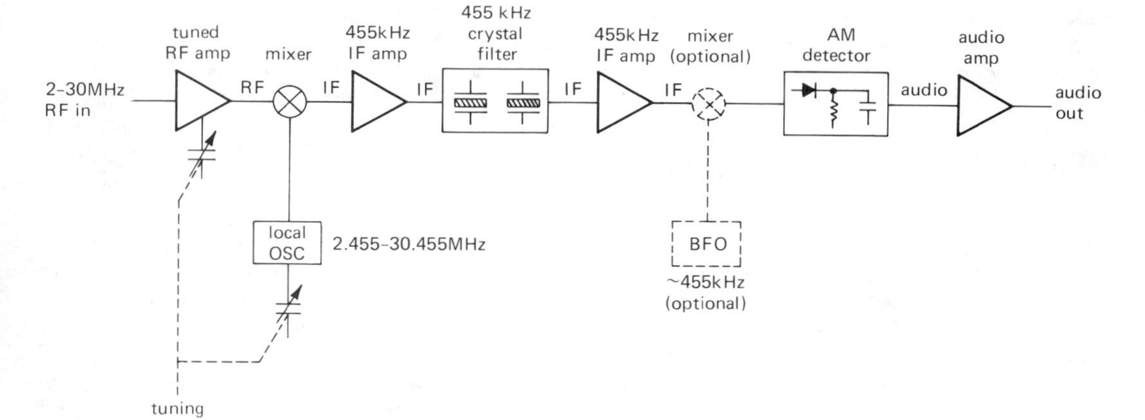

Figure 13.28
Superheterodyne receiver.

and difference frequencies). In other words, the superheterodyne receiver uses a mixer and local oscillator to shift a signal at the (variable) input frequency over to a fixed intermediate frequency where most of the gain and selectivity are concentrated.

Superhet potpourri

There are some additional features often added to a superheterodyne receiver. In this example a beat frequency oscillator (BFO) is shown; it is used in the detection of some signals with modulation other than AM (telegraphy, suppressed carrier telephony, frequency-shift keying, etc.). It can even be used for AM detection in what is known as a "homodyne" or "synchronous" detector. Receivers often have more than one mixing stage (they're called "multiple-conversion" receivers). By using a high first IF, image rejection is improved (the image is twice the IF frequency away from the actual received signal). A lower second IF makes it easier to use sharp-cutoff crystal filters, and a third IF is sometimes generated to allow the use of audio-type notch filters, low-frequency ceramic or mechanical filters, and "product detectors." Recently, the use of direct up-conversion (an IF higher than the input signal frequency) in a front-end balanced mixer, with crystal filters at the ~40MHz IF, followed by detection with no further mixing, has become popular. Such a single-conversion scheme offers better performance in the presence of strong interfering signals, and it has become practical with the availability of good VHF crystal-lattice filters and low-distortion wide-range balanced mixers with good noise performance.

ADVANCED MODULATION METHODS

□ 13.16 Single sideband

From a glance at the spectrum of an AM signal it is obvious that things can be improved. Most of the power (67%, to be exact, at 100% modulation) is in the carrier, conveying no information. AM is at most 33% efficient, and that only when the modulation index is 100%. Since voice waveforms generally have a large ratio of peak amplitude to average amplitude, the modulation index of an AM signal carrying speech is generally considerably less than 100% (although speech-waveform "compression" can be used to get more power into the sidebands). Furthermore, the symmetrical sidebands, by conveying the identical information, cause the signal to occupy twice the bandwidth actually necessary.

With a bit of trickery it is possible to eliminate the carrier [a balanced mixer does the job; note that $\cos A \cos B = \frac{1}{2} \cos (A + B) + \frac{1}{2} \cos (A - B)$], creating what is known as "double-sideband-suppressed carrier," or DSBSC. (This is just what you will get if the audio signal multiplies the carrier directly, without first being biased so that the audio waveform is always positive, as in normal AM.) Then, either by using sharp crystal filters or by using a method known as "phasing," one of the remaining sidebands can be eliminated. The "single-sideband" (SSB) signal that remains forms a highly efficient mode of voice communication and is widely used by radio amateurs and commercial users for long-range high-frequency telephony channels. When you're not talking, there's nothing being transmitted. To receive SSB, you need a BFO and product detector, as shown in the last block diagram, to reinsert the missing carrier.

□ *Modulation spectra*

Figure 13.29 shows representative spectra of voice-modulated AM, DSBSC, and SSB. When transmitting SSB, either sideband can be used. Note that SSB consists simply of the audio spectrum translated upward in frequency by f_c. When SSB is being received, the BFO and mixer combine to translate the spectrum down to audiofrequencies again. If the BFO is slightly mistuned, all audiofrequencies will be offset by the amount of mistuning. This dictates good stability for the LO and BFO in a receiver used for single sideband.

Note that a mixer (modulator) can always be thought of as a frequency translator, especially when combined with suitable filters to eliminate the undesired outputs: When used as a modulator, a low-frequency band of frequencies is shifted up by the

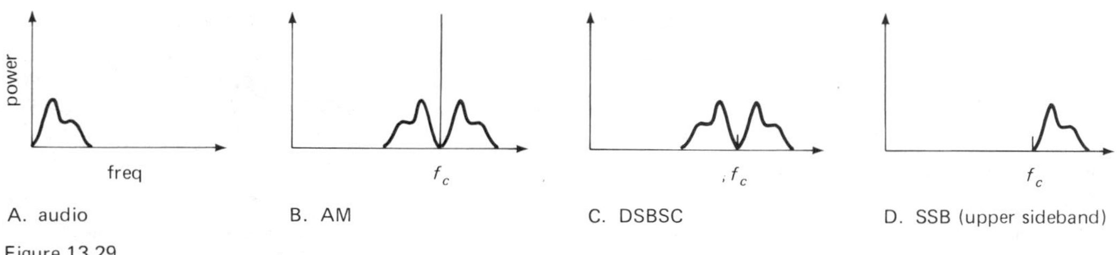

Figure 13.29
Suppressed-carrier spectra.

carrier frequency, to form a band centered around f_c. When used as a mixer, a band of frequencies around f_c is shifted down to audiofrequencies ("baseband"), or to a band centered around the IF frequency, by the action of a high-frequency LO.

13.17 Frequency modulation

Instead of modulating the amplitude of a carrier, as in AM, DSBSC, and SSB, it is possible to send information by modulating the frequency or phase of the carrier:

signal $= \cos\,[\omega_c + kf(t)]\,t$ frequency modulation (FM)

signal $= \cos\,[\omega_c t + kf(t)]$ phase modulation (PM)

FM and PM are closely related and are sometimes referred to as "angle modulation." FM is familiar as the mode used in the 88–108MHz VHF broadcast band, and AM is used in the 0.54–1.6MHz broadcast band. Anyone who has tuned an FM receiver has probably noticed the "quieting" of background noise characteristic of FM reception. It is this property (the steep rise of recovered SNR with increasing SNR of the channel) that makes wideband FM preferable to AM for high-quality transmission.

Some facts about FM: When the frequency *deviation* $kf(t)/2\pi$ is large compared with the modulating frequency [highest frequency present in $f(t)$], you have "wideband FM" as used in FM broadcasting. The modulation index, m_f, equals the ratio of frequency deviation to modulating frequency. Wideband FM is advantageous because under the right conditions the received SNR increases 6dB per doubling of FM deviation. The price you pay is increased channel bandwidth, since a wideband FM signal occupies approximately $2f_{dev}$ of bandwidth, where f_{dev} is the peak deviation of the carrier. FM broadcasting in the 88–108MHz band uses a peak deviation f_{dev} of 75kHz, i.e., each station uses about 150kHz of the band. This explains why wideband FM is not used in the AM band (0.54–1.6MHz): There would be room for only six stations in any broadcasting area.

FM spectrum

A carrier that is frequency-modulated by a sine wave has a spectrum similar to that shown in Figure 13.30. There are numerous

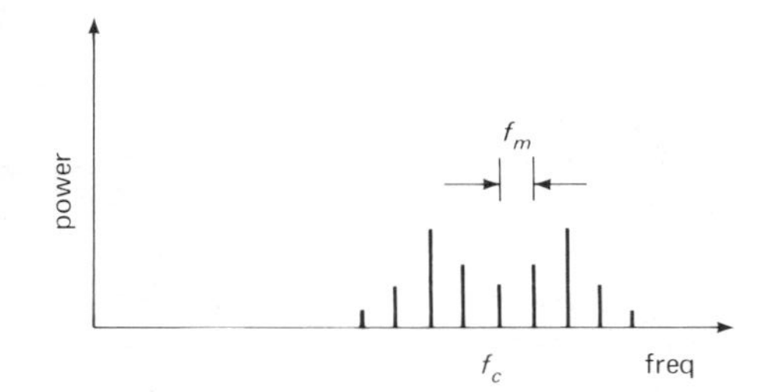

Figure 13.30
Wideband FM spectrum.

sidebands spaced at multiples of the modulating frequency from the carrier, with amplitudes given by Bessel functions. The number of significant sidebands is roughly equal to the modulation index. For narrowband FM (modulation index < 1), there is only one component on either side of the carrier. Superficially this looks the same as AM, but when the phase of the sidebands is taken into account, you have a waveform of constant amplitude and varying frequency (FM), rather than a waveform of varying amplitude and constant frequency (AM). With wideband FM the carrier amplitude may be very small, with correspondingly high efficiency, i.e., most of the transmitted power goes into the information-carrying sidebands.

□ *Generation and detection*

FM is easily produced by varying an element of a tuned circuit oscillator; a varactor (a diode used as a voltage-variable capacitor) is ideal. Another technique involves integrating the modulating signal, then using the result to do phase modulation. In either case it is often best to modulate at low deviation, then use frequency multiplication to increase the modulation index. This works because the *rate* of frequency deviation is not changed by frequency multiplication, whereas the deviation is multiplied along with the carrier.

To detect FM, an ordinary superheterodyne receiver is used, with two differences. First, the final stage of IF amplification includes a "limiter," a stage run at constant (saturated) amplitude. Second, the subsequent detector (called a discriminator) has to convert frequency deviation to amplitude. There are several popular methods of detection:

1. A "slope detector," which is nothing more than a parallel *LC* circuit tuned off to one side of the IF frequency; as a result, it has a rising curve of response versus frequency across the IF bandwidth, thereby converting FM to AM. A standard envelope detector converts the AM to audio. There are improved versions of the slope detector involving a balanced pair of *LC* circuits tuned symmetrically to either side of the IF center frequency.
2. The Foster-Seely detector, or its variant, the "ratio detector," using a single tuned circuit in a fiendishly clever diode arrangement to give a linear curve of amplitude output versus frequency over the IF bandpass. These discriminators are superior to the simple slope detector (Fig. 13.31).
3. A "phase-locked loop" (PLL). This is a device that varies the frequency of a voltage-controlled oscillator to match an input frequency, as we discussed in Section 9.33. If the input is the IF signal, the control voltage generated by the PLL is linear in frequency, i.e., it is the audio output.
4. An averaging circuit, in which the IF signal is converted to a train of identical pulses at the same frequency. Averaging this pulse train generates an output proportional to IF frequency, i.e., the audio output plus some dc.
5. A "balanced quadrature detector," which is a combination of a phase detector (see Sections 9.29 and 9.33) and a phase-shifting network. The IF signal is passed through a network that produces a phase shift varying linearly with frequency across the IF passband (an *LC* circuit would do nicely). The resultant signal and the original signal are compared in a phase detector, giving an output that varies with relative phase. That output is the desired audio signal (Fig. 13.31).

It is often pointed out that FM provides essentially noise-free reception if the channel has sufficient SNR, as compared with AM,

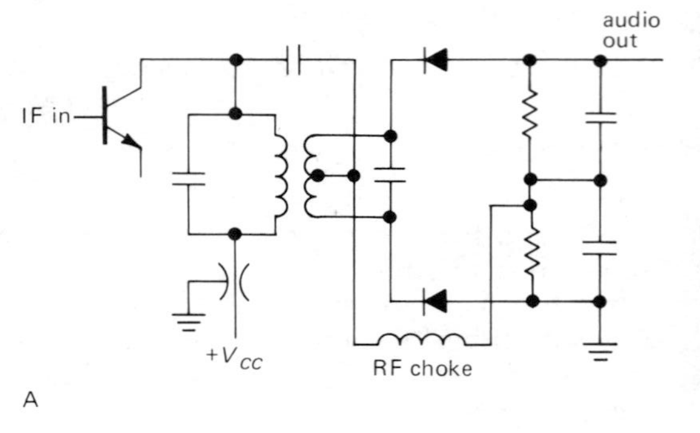

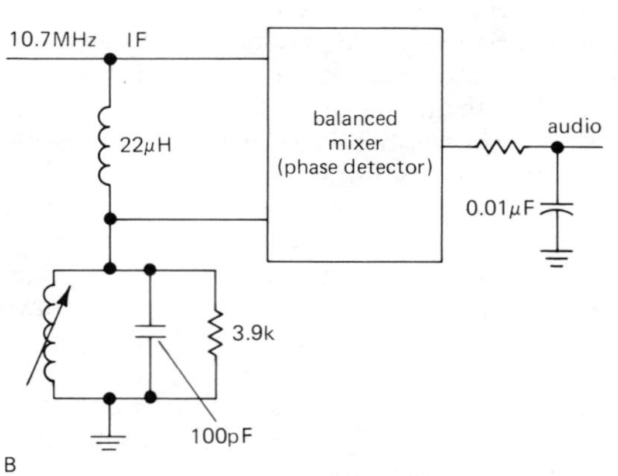

Figure 13.31

FM discriminators. A: Foster-Seeley. B: Balanced quadrature detector.

where the rejection of interference improves only gradually with increasing signal power. This makes sense when you remember that FM signals pass through a stage of amplitude limiting before detection. As a result, the system is relatively insensitive to interfering signals and noise, which appear as amplitude variations added to the transmitted signal.

□ 13.18 Frequency-shift keying

Transmission of digital signals (radioteletype, RTTY) is usually done by shifting a continuous-running carrier in frequency between two closely spaced frequencies according to the 1's and 0's being transmitted; 850Hz of shift is a typical value. The use of frequency-shift keying (FSK), rather than on/off modulation, is extremely effective in the presence of large signal fading from changing propagation conditions. To demodulate FSK, you simply use a differential amplifier looking at the outputs from a pair of filters set at the two detected audiofrequencies. You can think of FSK as digital FM. Narrow-shift FSK has been used to circumvent selective fading between the two signal frequencies. However, the shift cannot be reduced below the information bandwidth of the keyed signal itself, roughly the "baud" rate (number of bit cells per second), or about 100Hz for ordinary radioteletype.

□ 13.19 Pulse-modulation schemes

There are several methods whereby analog signals can be transmitted as pulses. The basic fact that makes digital transmission of analog signals possible is expressed in the Shannon sampling theorem, which states that a band-limited waveform is fully described by sampling its amplitude at a rate equal to twice the highest frequency present. Thus a method that conveys the amplitude of a waveform, by digital methods or whatever, at instants of time separated by $1/2f_{max}$ can be used instead of a continuous modulation scheme. Several methods are shown in Figure 13.32.

In pulse-amplitude modulation (PAM), a train of pulses of amplitude equal to the signal is transmitted at regular intervals. This scheme is useful for *time multiplexing* of several signals on one information channel, since the time between samples can be used to transmit the samples of another signal (with an increase of bandwidth, of course). In pulse-width modulation (PWM), the width of fixed-amplitude pulses is proportional to the instantaneous signal amplitude. PWM is easy to decode, using simple averaging. In pulse-position modulation (PPM), pulses of fixed width and amplitude are delayed or advanced relative to a set of fixed times, according to the amplitude of the signal.

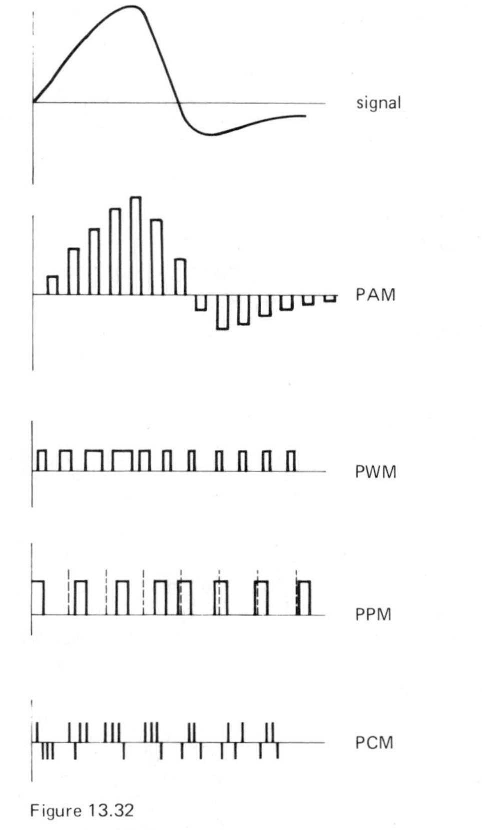

Figure 13.32
Pulse modulation schemes.

□ *Pulse-code modulation*

Finally, in pulse-code modulation (PCM) the instantaneous amplitude of the signal is converted to a binary number and transmitted as a serial string of bits. In the illustration, a 4-bit offset binary code corresponding to 16 levels of quantization has been used. PCM excels when error-free transmission is required over noisy channels. As long as 1's and 0's can be identified unambiguously, the correct digital code, and hence a replica of the original signal, can be recovered. PCM is particularly useful in repeater applications, e.g., transcontinental telephone channels, where the signal must pass through many stations and be amplified along the way. With any of the linear modulation schemes (AM, FM, SSB), noise accumulated in transit cannot be removed, but with PCM the digital code can be correctly regenerated at each station. Thus the signal starts anew at each station.

There are variations of PCM (known as coded PCM) in which techniques other than simple serial binary sequences are used to encode the quantized samples; for instance, a burst of one of 16 tones could be used in the preceding example. PCM is routinely used for telemetry of images from space vehicles, owing to its error-free properties. In any PCM application the bit rate must be chosen low enough to ensure a low probability of error in bit recognition. In general, this limits transmission on a given channel to speeds much below what could be used with direct analog modulation techniques.

□ RADIOFREQUENCY CIRCUIT TRICKS

In this chapter we are attempting to highlight some of the motivation and techniques of circuitry operated at radiofrequencies. In such limited space it is not possible to consider circuit design and construction in as much detail as we have generally attempted in the other chapters, nor would that even be desirable in a book intended as a broad introduction to electronics. In keeping with this philosophy, we would like to give some idea of the techniques that are ordinarily used in RF circuits. These are generally aimed at reducing stray inductance and capacitance and coping with circuitry whose dimensions are often comparable with a wavelength. There will be no attempt to weave these together into a coherent methodology; just think of them as a bag of tricks.

□ 13.20 Special construction techniques

RF "chokes" (small inductors, in the range of microhenrys to millihenrys) are used extensively as signal-blocking elements. Power-supply voltages will usually be brought into a shielded enclosure with shielded "feedthrough capacitors" (bypass to ground combined with a mechanical feedthrough terminal), with an RF choke in series. A variation is to use ferrite beads on leads of transistors, FETs, etc. These are used because of the tendency of RF circuits toward "parasitic" oscillations, encouraged by unintentional tuned circuits at UHF formed by the wiring itself. Stringing a few beads on a base or collector lead here and there raises the inductance enough to prevent the oscillations (if you're lucky, that is!).

Inductors play a major role in RF design, and you see plenty of open coils and "slug-tuned" inductors and transformers (such as the little metal IF transformer cans you see everywhere in receiver circuits). Small-value air-variable capacitors are equally popular.

As suggested earlier, RF circuits are constructed in shielded enclosures, often with internal grounded partitions between sections of the circuit to prevent coupling. It is common to build circuits on double-sided PC board, with one side used as a ground-plane. Alternatively, a circuit may be constructed immediately adjacent to a shield or other grounded surface. Grounds can't be wishy-washy at RF; you've got to solder a shield along its whole length, and you have to use a lot of screws to mount a partition or cover.

When building circuits at higher radiofrequencies, it is absolutely essential to keep component leads as short as possible. That means snipping off leads right at the resistor or capacitor and soldering them with no visible lead showing (the components get plenty hot, but they seem to survive). At VHF and UHF you often use ceramic capacitor "chips," soldered to PC strips, etc., without leads at all. The use of wide straps or metal ribbon, rather than ordinary wire, reduces inductance and is a favorite at UHF. At these frequencies you get into stripline and microstrip techniques, where every lead is itself a transmission line, complete with impedance matching. In fact, strips of sheet metal can be used as parts of tuned circuits; here's a specification for an inductor in a 440MHz circuit (ARRL handbook, 1978, p. 447): "L_1–L_3, incl. – 2 ⅝ × ¼-inch strip of brass, soldered to the enclosure on one end and to the capacitor at the other. Input and output taps are ½-inch up from the ground end." Of course, at microwave frequencies all such techniques give way to waveguide and cavity circuits, complete with exotica such as circulators and "magic T's" (Fig. 13.33).

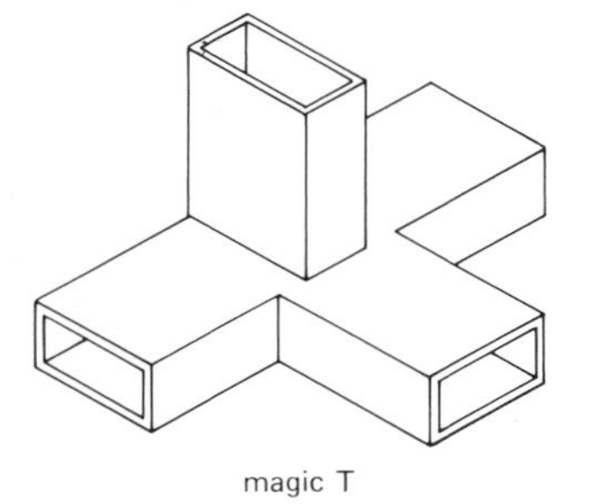

Figure 13.33

A facet of RF design that surprises beginners is the use of test instruments combined with "cut-and-try" techniques. You see widespread use of sweep generators (RF signal sources that sweep repetitively through a range of frequency), grid dip meters (for measuring resonances), SWR bridges, and spectrum analyzers, with plenty of circuit experimentation. At these frequencies you just can't predict everything; it takes some trial and error (and lots of experience) to make things work well.

□ 13.21 Exotic RF amplifiers and devices

Familiar devices such as bipolar transistors and FETs are used at radiofrequencies, although often in somewhat different incarnations. Transistors intended for use at VHF and above come in strange-looking packages, with flat strips radiating out from the center for connection to stripline or PC board (Fig. 13.34). There are also devices and circuits with no low-frequency analog, such as the following.

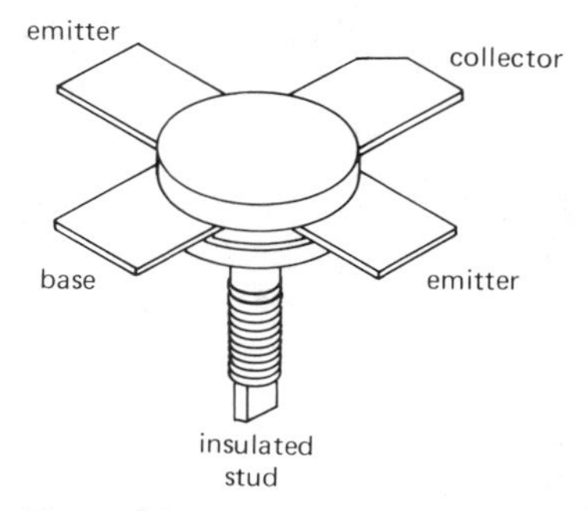

Figure 13.34

Parametric amplifiers. These devices amplify by varying a *parameter* of a tuned circuit. An analogy is a pendulum formed by hanging a weight on a length of string. Imagine that the motion represents output signal. You can build up the swing by gently shoving the weight at the resonant frequency; this is analogous to an ordinary amplifier, with a transistor or other active device providing the "shove." But there's another completely different way to get the thing swinging, namely by pulling up and down on the string (varying its length, a parameter of the system) at *twice* the natural resonant frequency. Try it (Fig. 13.35). The pendulum is closely analogous to the Adler parametric amplifier. In a paramp you can vary the capacitance of a tuned circuit with a varactor (voltage-variable capacitor) by driving it with a "pump" signal. Paramps are used for low-noise amplification.

Masers. Maser is an acronym for microwave amplification by stimulated emission of radiation. These things are basically atomic or molecular amplifiers, tricky to make and use but delivering the lowest noise of any amplifier.

GaAs FETs. The latest word in simple microwave amplifiers. Performance is comparable to that of paramps, without the fuss

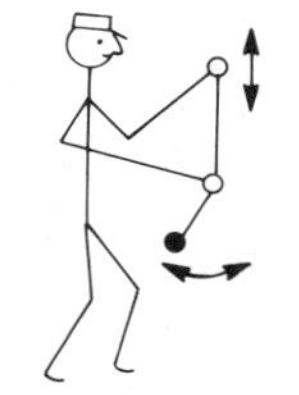
Figure 13.35

Pendulum analogy to the parametric amplifier.

and bother. Present-day commercial GaAs FETs will deliver 11dB of gain at 4GHz with a 1.3dB noise figure.

Klystrons and traveling-wave tubes. Vacuum-tube amplifiers used at microwave frequencies, klystrons and TWTs take advantage of transit-time effects within the tube. A variation known as a reflex klystron works as an oscillator by bouncing its electron beam back into its guts. There are klystrons available that can continuously deliver 0.5MW of RF output at 2000MHz.

Magnetrons. The heart of radar and microwave ovens. A high-power oscillator tube, full of little resonant cavities, and operated in a large magnetic field to make the electrons spiral around inside.

Gunn diodes, IMPATT diodes, PIN diodes. These exotic devices are used extensively at UHF and microwave frequencies. Gunn diodes are used as low-power oscillators in the 5–100GHz range, delivering output powers of 100mW or so. IMPATT diodes are analogous to klystrons, with capabilities of a few watts at a few gigahertz. PIN diodes behave as voltage-variable resistances and are used to switch microwave signals on and off by becoming a short circuit across a waveguide.

Varactors, SNAP diodes. Varactors are reverse-biased diodes used as variable capacitances for tuning purposes, or in paramps. Because of their nonlinearity they are also used for harmonic generation, i.e., as frequency multipliers. SNAP diodes are also popular for harmonic generation, since they exhibit sub-picosecond rise times.

Schottky diodes, back-diodes. You have seen Schottky diodes earlier as high-speed diodes with low forward drop. They're often used as mixers, as are back-diodes, variations of tunnel diodes.

HIGH-SPEED SWITCHING

The same effects that limit linear amplifier performance at high frequencies (the combination of junction capacitance, feedback capacitance, with its Miller effect, and stray capacitance in combination with finite source and load resistance) impose speed limitations on high-speed digital circuits. Many of these problems don't affect the designer directly, since they've been handled well in the design of digital ICs themselves. The average circuit designer would have a difficult time even coming close to the performance of TTL circuits, for example, using discrete transistor design.

Nevertheless, there are plenty of occasions when you've got to know how to design fast switching circuits. For example, in driving some external high-voltage or high-current load (or a load of opposite polarity) from a logic output, it is quite easy to lose a factor of 100 in switching speed through careless design. Furthermore, there are situations in which no packaged digital logic is used at all, and you're on your own all the way.

In this section we will begin with a simple transistor model useful for switching-circuit calculations. We will apply it to a few example circuits to show how it goes (and how important the choice of transistor can be). We will conclude by illustrating transistor switching design with a complete high-speed circuit (a photomultiplier preamp/discriminator).

13.22 Transistor model and equations

Figure 13.36 shows a saturated transistor switch, connected as an inverter, driven from a source of pulses with extremely fast rise and fall times. R_S represents the source impedance, r_b' is the relatively small intrinsic transistor base "spreading" resistance (of the order of 5Ω), C_{cb} is the all-important feedback capacitance, and R_C is the load resistance, paralleled by load capacitance C_L.

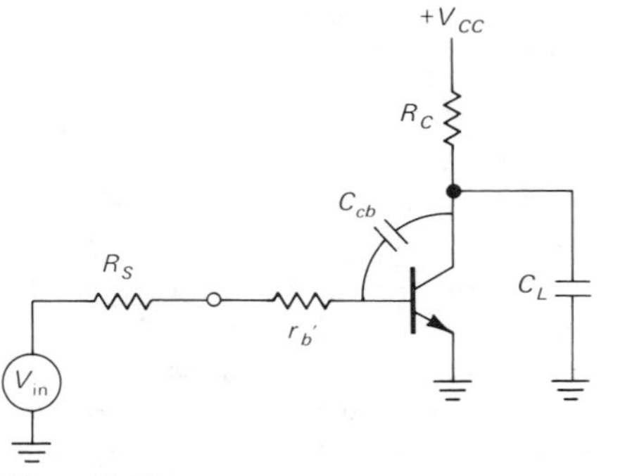

Figure 13.36

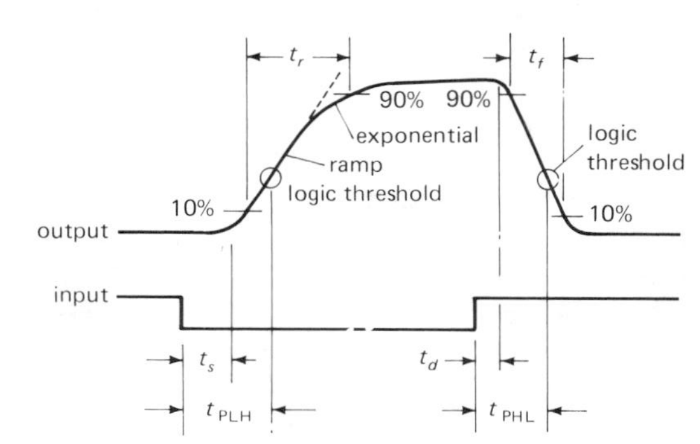

Figure 13.37

Waveform parameters of a transistor switch.

The effects of finite load resistance can be included by letting R_C represent the Thévenin equivalent of the combined resistances, with V_{CC} suitably modified. The collector-to-emitter capacitance has been absorbed into C_L, and C_{be} has been ignored, since C_{cb} always dominates owing to Miller effect.

Figure 13.37 shows a typical output waveform from this circuit when driven by a clean negative-going input pulse. The rise time, t_r, is usually defined as the time to go from 10% to 90% of the final value, with the corresponding definition of fall time, t_f. Note especially the relatively long storage time, t_s, that is required for the transistor to come out of saturation, compared with the correspondingly shorter delay time, t_d, to bring the transistor into conduction. These definitions are all conventionally taken between the 10% and 90% points. More useful for digital logic purposes are the propagation times, t_{PLH} and t_{PHL}, defined as the time from the input transition until the output passes through the logic threshold (rising or falling, respectively). Other symbols are in common use, e.g., t_{pd1} or t_{pr} is often used for what we've called t_{PLH}.

Let's use the circuit model to estimate rise and fall times for a given circuit. In the process you will even come to understand why the rising portion of the output waveform sometimes ends with an exponential.

□ *Estimation of rise time*

After the input signal has dropped to its LOW state and t_s has elapsed (more on that later), the collector begins to rise. Two effects limit the rate of rise: (a) R_C in combination with C_{cb} and C_L sets a time constant, generating an exponential rise toward V_{CC}, but (b) if that rate of collector rise is great enough, the resulting current through C_{cb} generates forward base bias across the source impedance ($R_S + r_b'$), and it can turn on the base with the resultant effect of slowing the collector rise through negative feedback. What you have in the latter case is an integrator, and the collector waveform is a ramp. In general (depending on circuit values and transistor parameters), the collector waveform may begin as a ramp and change over to an exponential, as shown previously.

A simple way to estimate circuit behavior is the following:

1. Compute the "integrator-limited" rate of rise of collector voltage, according to

$$\frac{dV_C}{dt} = \frac{V_{BE} - V_{in}(\text{LOW})}{C_{cb}(R_S + r_b')}$$

2. Find the collector voltage V_X at which the output waveform changes from a ramp to an exponential, according to

$$V_X = V_{CC} - \left(\frac{V_{BE} - V_{in}(\text{LOW})}{R_S + r_b'} + C_L\frac{dV_C}{dt}\right)R_C$$

This allows you to determine the collector waveform and rise time, as we will illustrate with examples presently. If V_X comes out negative, that means the entire collector rise is exponential: The capacitive load dominates, and the base is never turned on via current through the feedback capacitor. The term r_b' is usually negligible.

EXERCISE 13.1

Derive the two preceding formulas. Hint: For the second formula, equate the feedback current needed to bring the base into conduction with the available collector pull-up current less the current needed to drive the (capacitive) load.

□ *Estimation of fall time*

Following the short delay time after the input has gone HIGH, the collector begins to drop, on its way to saturation. With a little bookkeeping of currents, it is easy enough to see that the collector current is given by

$$I_C = \left(\frac{V_{in}(\text{HIGH}) - V_{BE}}{R_S + r'_b} + C_{cb}\frac{dV_C}{dt}\right) h_{fe}$$
$$= \frac{V_{CC} - V_C}{R_C} - (C_L + C_{cb})\frac{dV_C}{dt}$$

where the first line is the net base current multiplied by h_{fe} and the second line is the available collector current through R_C, less the current needed to drive the capacitance seen at the collector. Remember that dV_C/dt is negative. Rearranging, we get

$$-\frac{dV_C}{dt} = \frac{1}{C_L + (h_{fe} + 1)C_{cb}} \times \left(\frac{V_{in}(\text{HIGH}) - V_{BE}}{R_S + r'_b} h_{fe} - \frac{V_{CC} - V_C}{R_C}\right)$$

where the first term in parentheses is recognizable as $h_{fe}i_{\text{drive}}$ and the second is $i_{\text{pull-up}}$. You are now licensed to try some circuits; you will be able to see what sort of rise times and fall times can be expected and which capacitance dominates. First, however, a word on delay and storage times.

□ *Delay and storage times*

In general, delay times are very short. The main effect is the time constant involved in moving the base capacitance up to V_{BE}, a time constant of order

$$T \sim (R_S + r_b)(C_{cb} + C_{be})$$

When working at extremely high speeds, transistor transit-time effects may also become important.

Storage times are another matter. A transistor in saturation has charge stored in the base region, and even after the base drive signal has gone close to ground (or even a bit negative), it requires a relatively long time for the extra injected minority carriers from the emitter to be swept from the base region by the collector current. Transistors differ widely in storage time; it can be shortened by using less base overdrive during saturation and by reverse-biasing the base to reverse the base current when switching the transistor OFF. This equation for storage time, t_s, makes these points:

$$t_s = K \ln \frac{I_B(\text{ON}) - I_B(\text{OFF})}{\dfrac{I_C}{h_{FE}} - I_B(\text{OFF})}$$

where I_B(OFF) is negative for "discharge" reverse base currents. The constant K includes a "minority-carrier lifetime" term, which is greatly reduced by gold doping. However, such doping reduces h_{FE} and increases the leakage current. This explains the good speed performance of TTL, along with its low breakdown voltage (7V).

Storage times can be as long as several hundred nanoseconds, and they are typically an order of magnitude longer than delay times. The popular general-purpose 2N3904, for instance, has a specified maximum delay time of 35ns and a storage time of 200ns under standardized test conditions, which include driving the base negative by two diode drops.

Since storage times can turn out to be a severe limitation on the performance of high-speed switching circuits, there are several measures that can be taken to circumvent the problem. One solution is to avoid saturation altogether. A Schottky clamping diode (a "Baker clamp") from base to collector will accomplish this by robbing current from the base when the collector is nearing saturation. It prevents transistor saturation, since its forward voltage drop is less than that of the collector-base junction. The Schottky families of TTL logic use this trick. A small "speedup" capacitor (25–100pF) across the base driving resistor is often a good idea in addition, since it can reduce storage time by providing a pulse of current to remove base charge at turn-off, and in addition it increases base drive current during turn-on transitions. Figure 13.38 illustrates these methods.

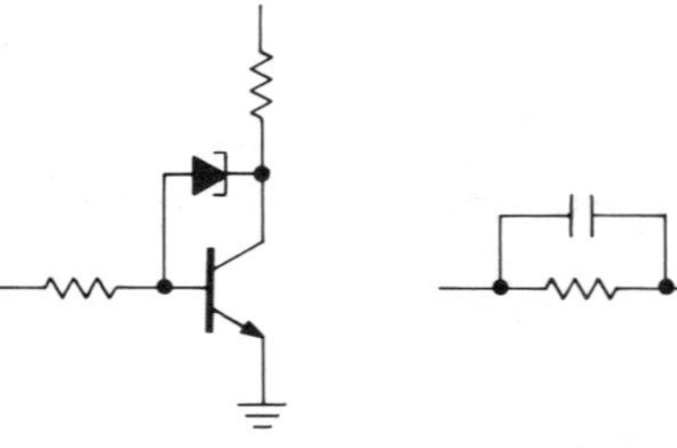

Figure 13.38

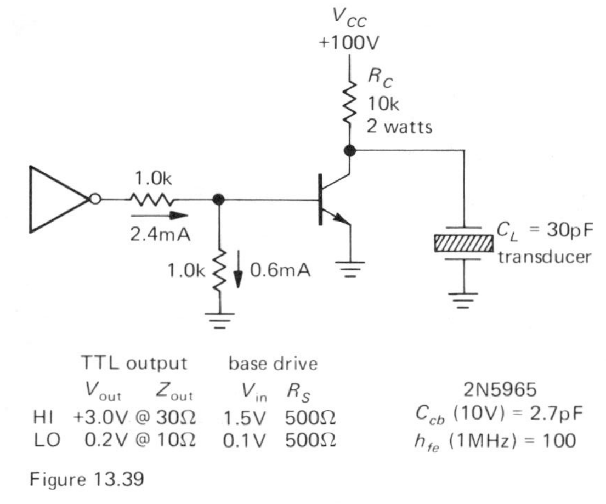

Figure 13.39

SOME SWITCHING-SPEED EXAMPLES

In this section we will analyze the performance of a few simple circuits, based on the methods just discussed.

13.23 High-voltage driver

Let's begin with the circuit in Figure 13.39. It is a simple inverting stage intended for driving a piezoelectric crystal with 100 volt pulses, generated originally with TTL logic. The TTL output, and therefore the base driving signal, is roughly as indicated. In these calculations we will ignore r_b', which is small compared with the source impedance.

Rise time

We begin by calculating the integrator-limited collector rise:

$$\frac{dV_C}{dt} = \frac{V_{BE} - V_{in}(\mathrm{LOW})}{C_{cb}R_S} \approx 450\mathrm{V}/\mu\mathrm{s}$$

from which the estimated rise time will be

$$t_r = \frac{0.8V_{CC}}{dV_C/dt} \approx 180\mathrm{ns}$$

Now we find the collector voltage at which the rise changes from an integrator-limited ramp to an exponential:

$$V_X = V_{CC} - R_C\left(\frac{V_{BE} - V_{in}(\mathrm{LOW})}{R_S} + C_L\frac{dV_C}{dt}\right) \approx -50\mathrm{V}$$

This means that the collector rising waveform is exponential the whole way, with the feedback current ($C_{cb}dV_C/dt$) insufficient to pull the base up into conduction, given the source impedance. The collector time constant is $R_C(C_L + C_{cb})$, or 0.33μs, with a rise time (10% to 90%) of 2.2 time constants, or 0.73μs. It is clear that the combination of collector resistor and load capacitance dominates the rise.

Fall time

To analyze the fall time, we use the formula derived earlier to find

$$-\frac{dV_C}{dt} = \frac{1}{C_L + (h_{fe}+1)C_{cb}} \times \left\{h_{fe}\left(\frac{V_{IN}(\mathrm{HIGH}) - V_{BE}}{R_S}\right) - \frac{V_{CC} - V_C}{R_C}\right\} \approx 530\mathrm{V}/\mu\mathrm{s}$$

$$t_f = \frac{0.8V_{CC}}{dV_C/dt} \approx 0.15\mu\mathrm{s}$$

The last term depends on V_C, but is negligible compared with the first term in parentheses. If it weren't, you would have to evaluate it at several values of collector voltage to get a good picture of the falling waveform. At this point it should be noted that the calculated fall time corresponds to a frequency of about 3MHz, and therefore the value of $h_{fe} = 100$ we used is realistic ($f_T = 300$). If a calculated rise time or fall time corresponds to a frequency much higher than originally assumed, it is generally necessary to go back and recompute the transition time, using a new h_{fe} based on a better estimate of the transition time. This iteration process will usually give a satisfactory answer on the second pass.

Switching waveform

For this circuit, then, the collector waveform is as shown in Figure 13.40. The rise is dominated by the time constant of the load

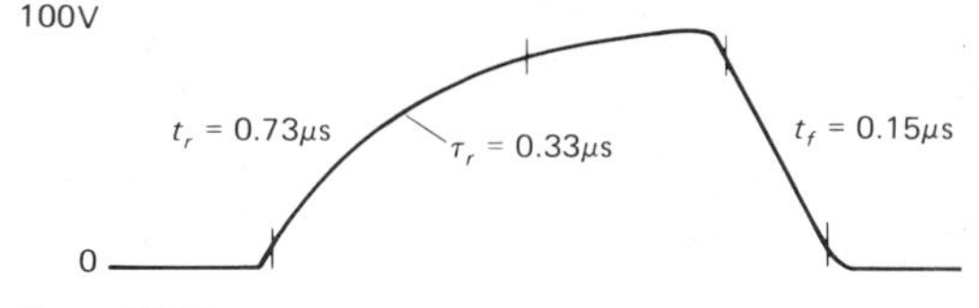

Figure 13.40

capacitance and collector resistor, whereas the fall is dominated by the feedback capacitance in combination with the source impedance. To put it another way, the collector voltage falls at a rate such that the current through the feedback capacitance is almost sufficient to cancel the base drive current and bring the base out of conduction. Note that we have assumed throughout that the TTL output waveform is much faster than the output of our circuit. With typical rise and fall times of about 5ns, that is a good approximation.

□ 13.24 Open-collector bus driver

Suppose we want to drive an open-collector TTL bus from the output of an NMOS circuit. We can do it by interposing an *npn* inverting stage, as in Figure 13.41. The base resistors

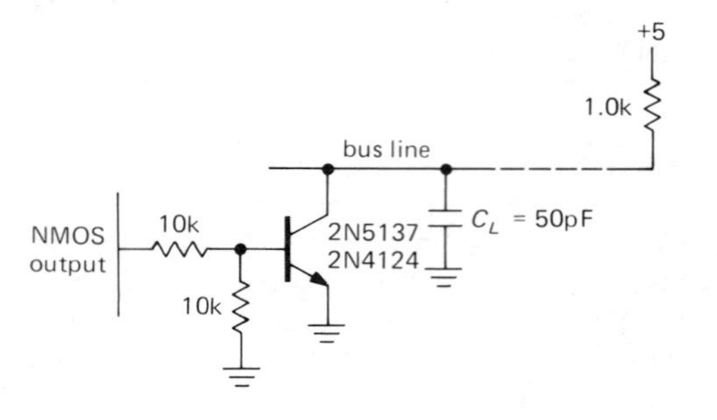

	NMOS output		base drive	
	V_{out}	Z_{out}	V_{in}	R_S
HI	+3.5 @ 1kΩ		1.7V	5.5k
LO	0.0V @ 200Ω		0.0V	5.1k

	2N5137	2N4124
C_{cb} (10V)	16pF	1.8pF
h_{fe} (1MHz)	100	100

Figure 13.41

are necessarily large because of the low output sourcing capability of NMOS operating from +5 volts (see Section 9.10). We have chosen two popular transistors in order to illustrate the effect that parameters like C_{cb} can have.

For rise time, we calculate as before, and we find the following integrator-limited rise times:

	2N5137	2N4124
dV_C/dt	8.5V/μs	76V/μs
t_r	470ns	53ns

The crossover to exponential is calculated as:

	2N5137	2N4124
V_X	4.4V	1.1V
time const	66ns	52ns

The falling waveform comes out:

	2N5137	2N4124
dV_C/dt	−11V/μs	−78V/μs
t_f	360ns	51ns

□ *Choice of transistor*

The situation is as shown in Figure 13.42. The inferior performance of the 2N5137 is

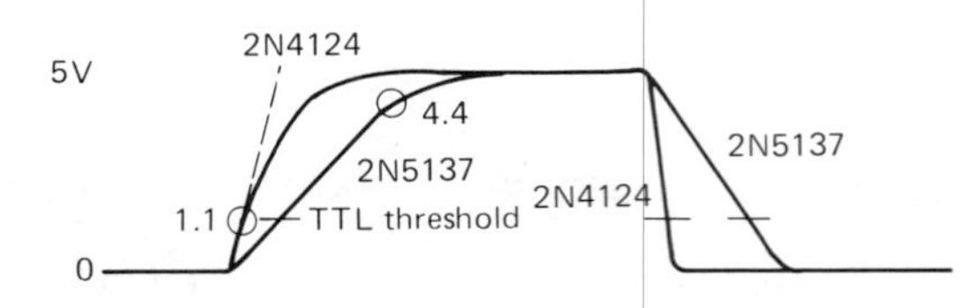

Figure 13.42

due entirely to the effects of feedback capacitance, aggravated in this example by the relatively high value of source impedance. The transition times of the 2N4124 are probably a bit optimistic, since they correspond to a frequency of about 10MHz, at which h_{fe} is somewhat lower than assumed.

It is interesting to look at the time required to reach the TTL threshold voltage of about 1.3 volts, the relevant parameter in a system in which TTL gates are driven by the bus signals. Ignoring storage and delay times, the times to reach the TTL threshold voltage are the following:

	2N5137		2N4124	
	calc	*meas*	*calc*	*meas*
rise (t_{PLH})	150ns	130ns	17ns	30ns
fall (t_{PHL})	340ns	360ns	47ns	52ns

The rise and fall times we measured are in reasonable agreement with the predictions of our somewhat simplistic model, except perhaps for the rise time of the 2N4124 circuit. There are a few possible explanations why the rise time we predicted is too small in that case: The calculation used the value for h_{fe} at 10MHz, whereas a 17ns rise time implies somewhat higher frequencies, and hence a lower value of h_{fe}. Also, by actual measurement the particular transistor has

C_{cb} = 2.2pF at 10 volts and C_{cb} = 3pF at 2 volts. Curiously, the 2N5137 we used had a much lower C_{cb} (about 5pF) than specified on the data sheet, and so we added a small external capacitor to bring it up to specifications. This probably represents a change in the manufacturing process since the original data sheet was published.

EXERCISE 13.2

Verify the results just calculated for dV_C/dt (rise and fall) and V_X.

□ *Pull-up to +3 volts*

Note that the times to reach TTL threshold from a HIGH state are much longer than the times from a LOW state, even though the output slew rates (in the case of the 2N4124 circuit) are almost the same. That's because the TTL threshold voltage is not symmetrically positioned between +5 and ground, forcing the collector to slew through a larger voltage on the way down. For this reason TTL buses are often terminated to a source of +3 volts (a pair of diodes tied to +5 is one trick sometimes used), or each line of the bus can be terminated with a voltage divider, as in Figure 13.43.

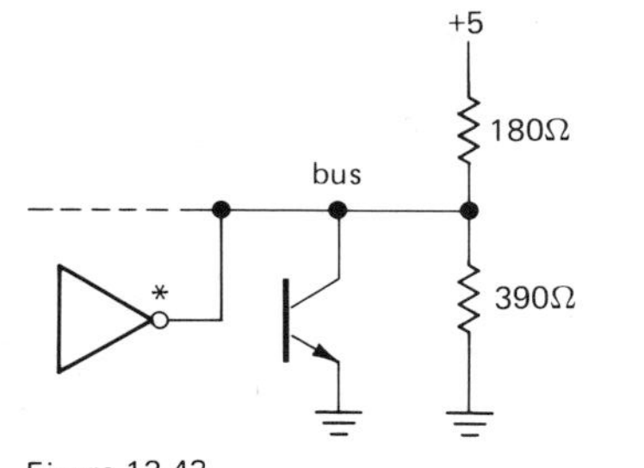

Figure 13.43

EXERCISE 13.3

Calculate the rise and fall times and the propagation delays for a 2N4124 driving the preceding bus with C_L = 100pF. Show your work.

□ 13.25 Example: photomultiplier preamp

As we will discuss in Chapter 14, a device called a photomultiplier tube (PMT) is an extremely useful light detector, combining high sensitivity with high speed. Photomultipliers are also useful in applications where the quantity being directly measured isn't light, e.g., high-energy-particle detectors in which a scintillator crystal generates light flashes in response to particle bombardment. To take advantage of a photomultiplier's properties, it is necessary to use a charge-sensitive high-speed discriminator, a circuit that generates an output pulse when an input pulse of charge exceeds some threshold corresponding to the detection of a photon of light.

Figure 13.44 shows a circuit of a high-speed photomultiplier preamp and discriminator that illustrates the high-frequency and switching techniques discussed in this chapter. The output of the photomultiplier tube consists of negative pulses of charge (electrons are negative), each pulse having a width of perhaps 10–20ns. The larger pulses correspond to detected photons (quanta of light), but there are also lots of smaller pulses that arise from noise within the photomultiplier tube itself and that should be rejected by the discriminator.

□ *Circuit description*

The circuit begins with an inverting input amplifier (Q_A–Q_C) with current and charge feedback via C_1 and R_1. The input follower presents a low driving impedance to Q_B (which provides the voltage gain) to reduce the effects of feedback capacitance (C_{cb}). The follower at the output of the gain block, Q_C, provides a low output impedance while allowing Q_B to have a reasonable amount of gain. The signal at this point is a small positive pulse corresponding to the negative-charge input from the PMT; dc feedback stabilizes Q_C's output at about $2V_{BE}$. Q_1 is biased as a class A emitter follower, giving a low-impedance "monitor" output of the amplified photomultiplier pulses before discrimination.

Differential amplifier Q_2 and Q_3 form the discriminator; the threshold is set by R_{22}, referenced to a voltage (set by Q_E, operating as an "adjustable diode") that tracks the input amplifier's $2V_{BE}$ quiescent point. This diode-drop tracking occurs because the transistors Q_A–Q_E are in a monolithic transistor array (CA3046) and are all at the same temperature. Q_4 forms an inverted cascode

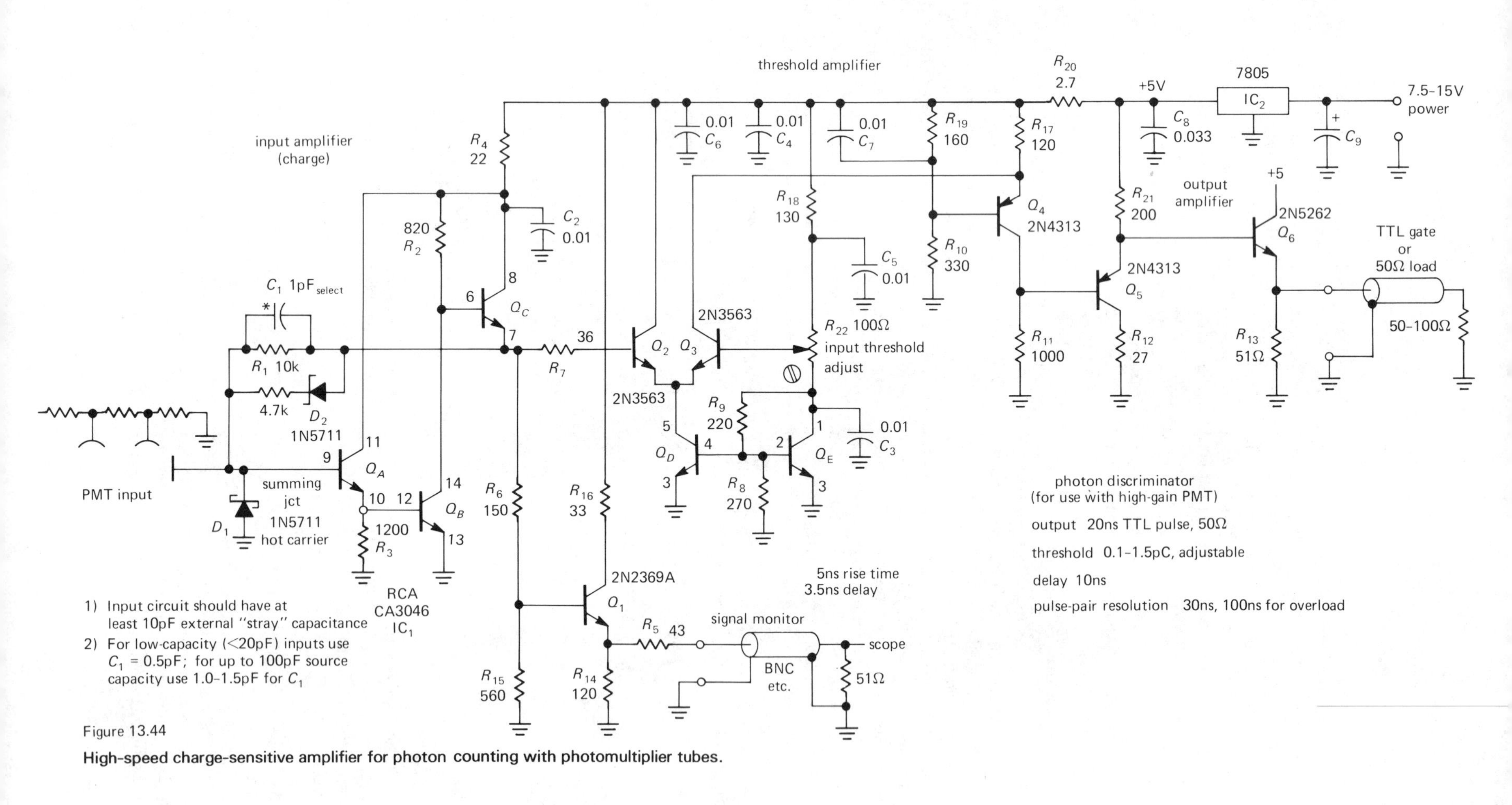

Figure 13.44

High-speed charge-sensitive amplifier for photon counting with photomultiplier tubes.

with Q_3, both for high speed and level shifting. Two stages of follower, arranged with opposite-polarity transistors to cancel V_{BE} offsets, complete the circuit.

There are several interesting points in this circuit. Transistor quiescent currents are set rather high (the differential pair Q_2Q_3 has 11mA emitter current, Q_5 idles at 20mA, and the output transistor has to source 120mA to drive a 50Ω load) in order to get good high-speed performance. Note that the cascode base (Q_4) is bypassed to V_+, not ground, since its input signal is referenced to V_+ via R_{17}. The comparator's emitter current source is a current mirror, convenient since Q_E is already used for the threshold reference. D_1 and D_2 are used to improve overload performance. Although it complicates the circuit, clamping diode D_1 can be returned to Q_E's collector (bypassed to ground) to put a tighter limit on negative (overload) swings at the input.

□ *Performance*

Figure 13.45 shows a graph of output pulse shape and timing versus input pulse size (measured as quantity of charge). The output pulses are stretched by large input overloads, but the overall performance is quite good, measured by usual photomultiplier preamp standards.

13.26 Circuit ideas

Figure 13.46 shows a few wideband circuit ideas.

ADDITIONAL EXERCISES

(1) In this problem you are to work out in detail the high-frequency behavior of the circuit in Figure 13.10, summarized briefly in Section 13.05. ***(a)*** Begin by repeating the calculation of driver/output-stage rolloff diagrammed in Figure 13.12. Be careful as you combine complex impedances. Write to one of the authors if you find an error! ***(b)*** Now check to see that the high-frequency rolloffs of the previous stages are significantly higher in frequency than the ~180MHz 3dB frequency of the output stage and its driver.

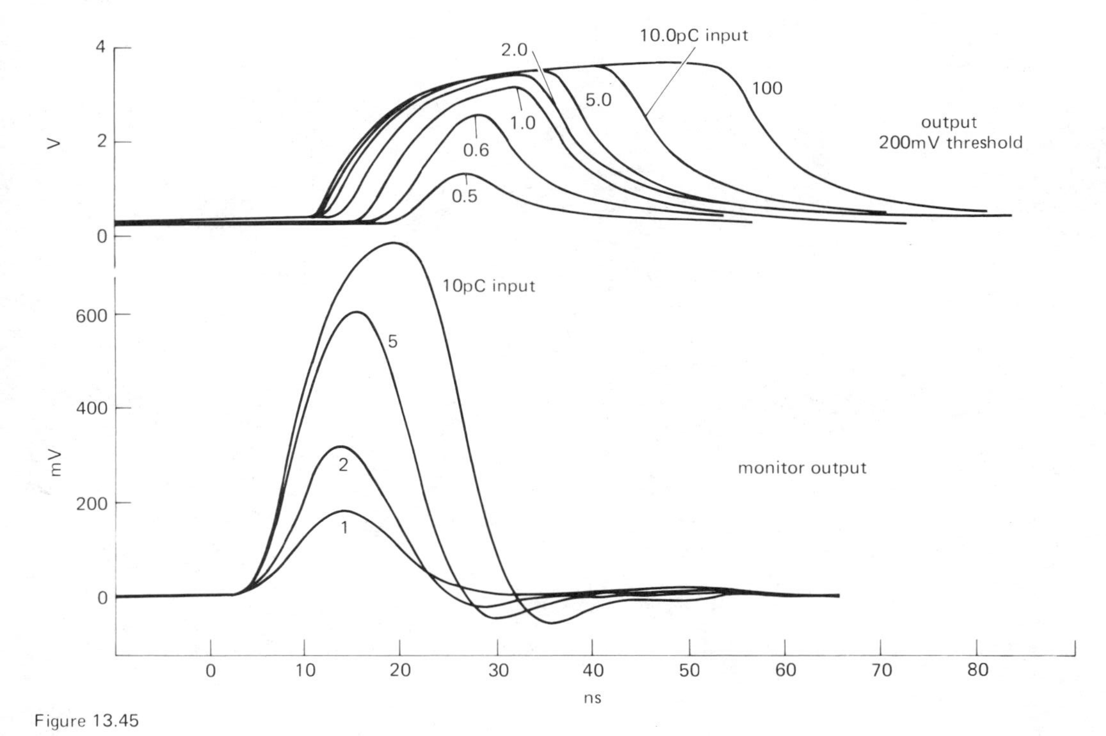

Figure 13.45

Pulse response of the amplifier in Figure 13.44.

In particular, check the following rolloffs: Q_1's output (emitter) impedance driving a capacitive load (see Fig. 13.11); Q_2's output driving a slightly different capacitive load (because Q_4's collector is not grounded); Q_3 and Q_4's emitters driving a capacitive load; Q_4's collector driving a capacitive load.

(2) What is the impedance looking into a length of coaxial cable that is ***(a)*** open-circuited at the far end, and a quarter wave long, electrically, at the frequency of interest, ***(b)*** short-circuited at the far end, and a quarter wave long, electrically, at the frequency of interest, ***(c)*** same as (a), but a half wave long, ***(d)*** same as (b), but a half wave long? The result in (d) is the basis of the so-called "choke joint" used in waveguides.

(3) Work out in detail the rise time and fall time of the high-voltage switching circuit in Figure 13.39, as summarized in Section 13.23. Use $V_{BE} = 0.7V$.

(4) "The Rise and Fall of a Bus Driver": Calculate the rise and fall times for the TTL bus driver circuit of Figure 13.41, as summarized in Section 13.24. Use $V_{BE} = 0.7V$.

(5) Design a video amplifier with a gain of +5 and a rolloff of 20MHz or more. The input impedance should be 75 ohms, and the output should be able to drive a 75 ohm load with 1 volt pp output capability. A nice way to achieve the noninverting gain is to use a common-base input stage with an emitter follower output, as suggested in Figure 13.47. If you like the circuit, finish the design by choosing operating currents, resistor values, and biasing components. You can, of course, use something like a differential amplifier/cascode/follower combination, if you prefer. Note that the gain must be noninverting, or the image will be reversed.

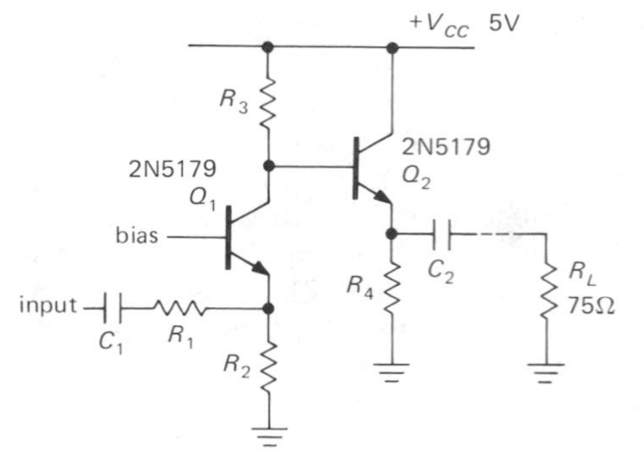

Figure 13.47

Circuit ideas

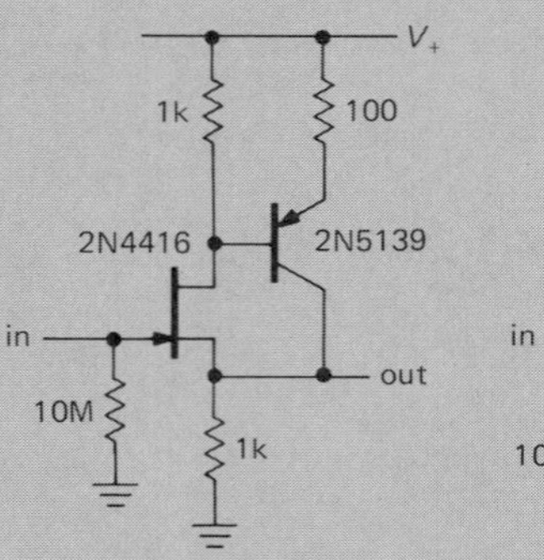

A. high-Z low-C wide-band follower

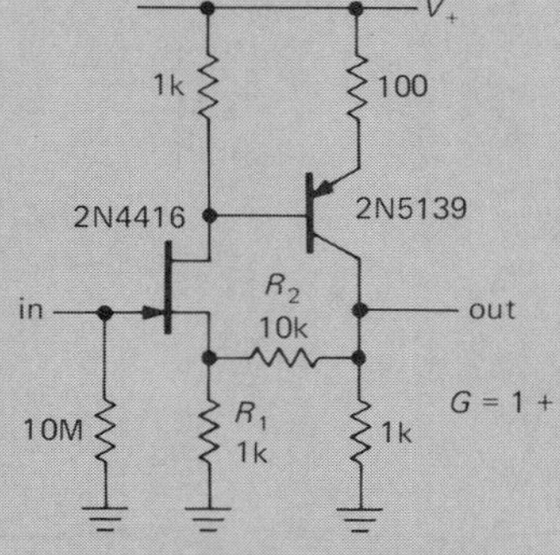

B. high-Z low-C amplifier

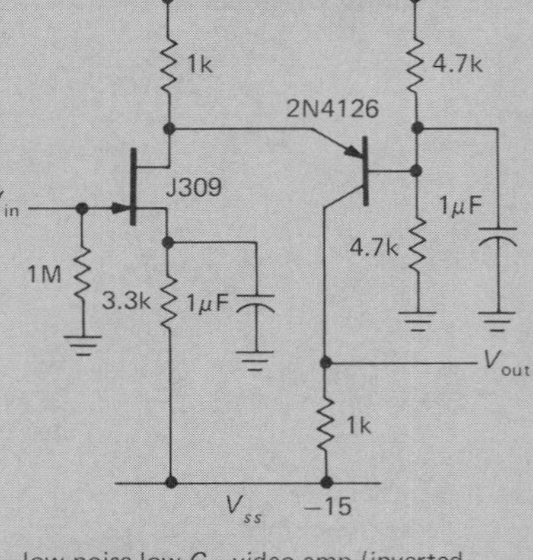

C. low-noise low-C_{in} video amp (inverted cascode)

Figure 13.46

MEASUREMENTS AND SIGNAL PROCESSING

CHAPTER 14

OVERVIEW

Perhaps the most exciting (and most useful) area of electronics involves the gathering and manipulation of data from an industrial process or a scientific experiment. Generally speaking, *transducers* (devices that convert some physical quantity, such as temperature or light level, to a voltage or some other electrical quantity) are used to generate signals that can be manipulated by electronic circuits, quantified by analog-to-digital converters, and logged and analyzed by computers. If the signal you're looking for is masked by noise or interference, powerful "bandwidth-narrowing" techniques such as lock-in detection, signal averaging, multichannel scaling, and correlation and spectral analysis can magically retrieve the sought-after signal. Finally, the results of such physical measurements can be used to control the experiment or process itself, with "on-line" control usually provided by a small computer or microprocessor dedicated to the task. The recent development of powerful and inexpensive microprocessors and support chips has brought about an explosion in the use of electronics to control and log processes that would not have seemed likely candidates only a decade ago.

In this chapter we will begin with a sampling of quantities that can be measured and the transducers that are normally used for the job. In this area there is plenty of room for ingenuity, and the catalog of transducers we will describe should therefore be considered representative, not exhaustive. We will go into some detail describing the particular problems some of these measurement transducers present and the circuit solutions you might use with them. We will try to cover the most common difficulties, dealing with ultrahigh source impedances (hundreds of megohms in the case of microelectrodes or ion-specific probes), low-level low-impedance transducers (e.g., thermocouples, strain gauges, magnetic pickups) high-impedance ac sensors (capacitance transducers), and others.

The chapter will continue with a look at precision standards (standards of frequency and time, as well as voltage and resistance) and some of the techniques of precision measurements. We will then describe in some detail the whole business of bandwidth narrowing, "pulling the signal from the noise." These techniques are extremely powerful, and they are mysterious to the uninitiated. Finally, we will conclude the chapter with a brief look at spectrum analysis and Fourier techniques. Readers interested primarily in electronic circuit design may wish to skip this chapter.

MEASUREMENT TRANSDUCERS

In some situations the quantity you want to measure is itself an electrical quantity. Examples might be nerve impulses (voltage), seawater conductivity (resistance), charged-particle fluxes (current), etc. In these cases measurement techniques tend to be relatively straightforward, with most of the difficulties centering around the kind of collection electrode to use and how to handle the signals once they have been collected. You might encounter very high impedances (e.g., with microelectrodes) or very small signals (e.g., a current generated from radioactive decay).

More often a "transducer" of some sort is necessary to convert some physical quantity to a quantifiable electrical quantity. Examples are measurements of temperature, light level, magnetic field, strain, acceleration, sound intensity, etc. In the following sections we will take a look at some of the more common input transducers to give an idea of what can be measured and how accurately. We will go into somewhat greater detail when describing the more common measurements, such as heat and light, but in a book of this scope we can cover only a fraction of the measurement possibilities.

14.01 Temperature

Temperature transducers illustrate a nice variety of performance tradeoffs. Temperature range, accuracy, repeatability, conformity to a universal curve, size, and price are all involved.

Thermocouples

A junction between two dissimilar metals generates a small voltage (with low source impedance!), typically in the millivolt range, with coefficients of about $50\mu V/^{\circ}C$. This junction is called a *thermocouple,* and it is useful for measuring temperatures over a broad range. By using various pairs of alloys it is possible to span temperatures from −270°C to +2500°C with reasonable accuracy (0.5–2°C). The thermoelectric properties of different alloys are well known, so thermocouple probes in different formats (rods, washers, armored probes, etc.) made from the same alloys can be interchanged without affecting calibration.

The classic thermocouple circuit is shown in Figure 14.1. The particular choice of

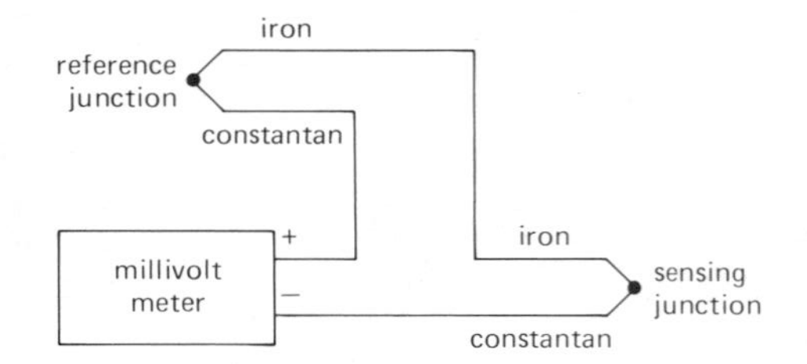

Figure 14.1
Classic thermocouple circuit.

metals in this figure constitutes what is known as a type J thermocouple (look at Table 14.1 for a listing of the standard choices and their properties). Each couple is made by welding the two dissimilar metals together to make a small junction. (People have been known to get away with twisting the wires together, but not for very long!) The reference junction is absolutely necessary, since otherwise you wind up with additional dissimilar thermocouples where the dissimilar metals join the meter terminals. Those extra thermoelectric voltages produced at uncontrolled places in the circuit would result in erratic and inaccurate results. Even with a pair of thermocouple junctions, you still have thermocouples formed where the leads join the metal terminals. However, this seldom causes problems, since those junctions are at the same temperature.

The thermocouple circuit gives you a voltage that depends on the temperatures of both junctions. Roughly speaking, it is proportional to the *difference* between the two junction temperatures. What you actually want is the temperature at the sensing junction. There are two ways to handle the problem of the reference: (a) Classically, you put the reference junction at a fixed temperature, usually 0°C. They used to use ice baths, and you still can, but you can also buy nice little stabilized cold boxes to do the same job. If you are measuring very high temperatures, you may not even care about small errors caused by having the reference junction at "room" temperature. (b) A more modern technique is to build a compensation circuit that corrects for the difference caused by having the reference junction at a temperature other than 0°C.

Figure 14.2 shows how this is done. The basic idea is to use a temperature-sensing chip and circuitry that adds in a voltage that makes up for the difference between the actual reference junction temperature and the standard 0°C. The AD590 (see the subsequent section on IC temperature sensors) produces an output current in microamps equal to the temperature in degrees Kelvin. R_1 is chosen according to the thermoelectric coefficient, in this case converting 1μA/°C to 51.5μV/°C (see Table 14.1), and the AD580 3-terminal reference (in combination with R_2 and R_3) is used to subtract the AD590 offset of 273μA at 0°C (273.16°K). Thus there is no correction made when the reference junction is at 0°C, and 51.5μV/°C (the thermoelectric coefficient of a type J junction at room temperature) is added to the net output voltage of the pair of junctions when the reference junction is at some other temperature.

The metering circuit deserves a few words of comment. The circuit problems you have with thermocouples stem from their low output voltage (50μV/°C or thereabouts), combined with large common-mode ac and radiofrequency interference. The amplifier must have good common-mode rejection at 60Hz and stable differential gain. In addition, the input impedance must be moderately high (of order 10k or more) in order to prevent error from loading, since the thermocouple leads do have some resistance (5 feet of 30 gauge type K junction wire has a resistance of 30Ω, for example).

The circuit shown in Figure 14.3 is a good solution. It is just the standard differencing amplifier, with the T connection in the feedback path to get high voltage gain (200 in this case) while keeping the input impedance large enough so that loading of the source impedance doesn't contribute error. The op-amp is a precision low-offset type, with drift of less than 1μV/°C to keep its contribution to the measurement error much less than the 50μV that corresponds to a 1°C error.

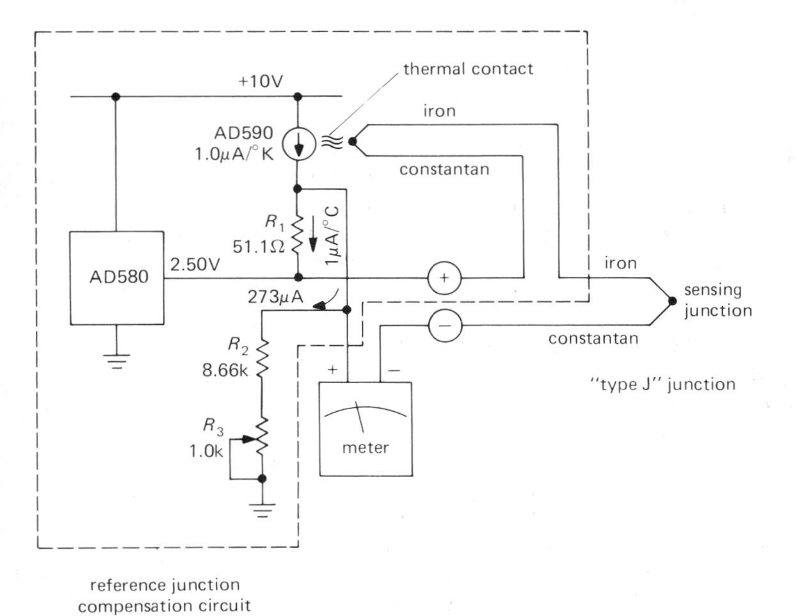

Figure 14.2
Thermocouple reference junction compensation.

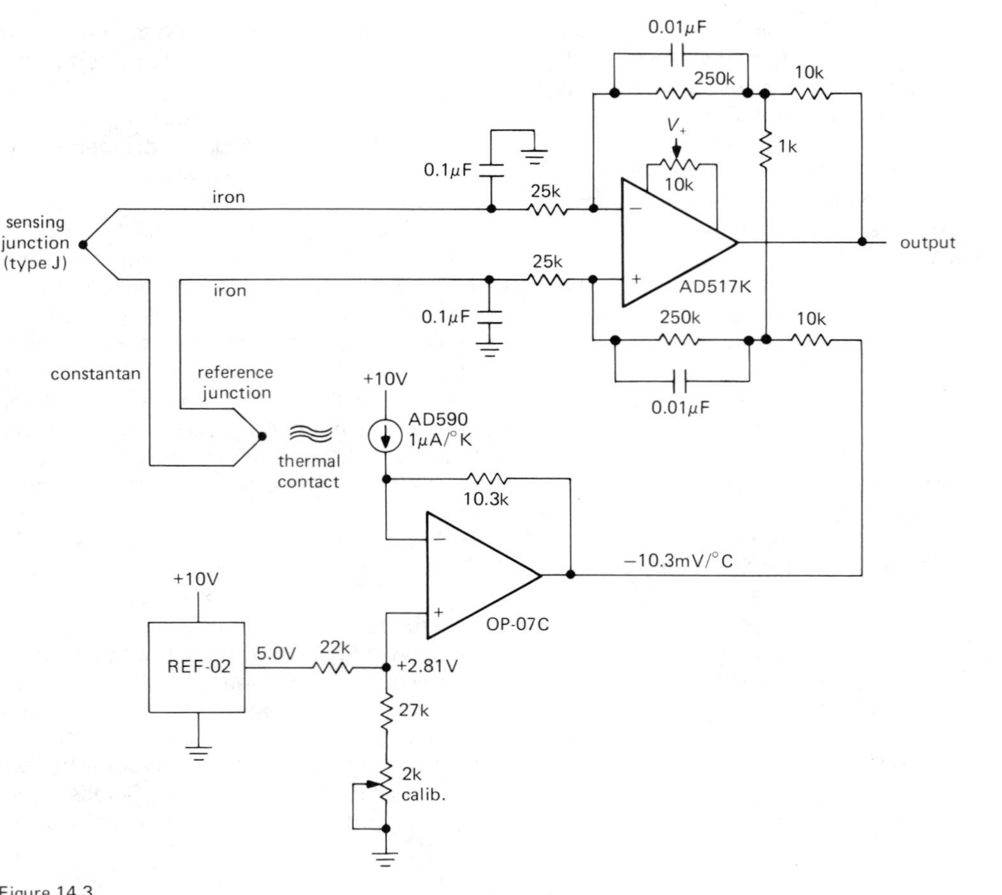

Figure 14.3
Balanced thermocouple amplifier with compensation at output.

The input bypass capacitors are a good idea to reduce the common-mode interference at 60Hz and at radiofrequencies (thermocouples and their long connecting cables tend to behave like radio antennas). Since thermocouples respond slowly anyway, you can limit the bandwidth with capacitors across the feedback resistors, as shown. In cases of extreme radiofrequency problems, it may be necessary to shield the input leads and add RF chokes before the input bypass capacitors.

Note that the reference junction compensation circuit in Figure 14.3 acts at the *output,* rather than the usual method of compensating the voltage from the thermocouple at the input, as in Figure 14.2. This is done to keep the input truly differential, in order to preserve the advantages of the good common-mode rejection of the differencing amplifier. Since the amplifier has a voltage gain of approximately 200, the compensation circuit has to add 200 × 51.5μV/°C, or 10.3mV/°C at the output.

An instrumentation amplifier, as in Figure 7.19, could be used instead of the differencing amplifier we've shown; in that case be sure to provide a dc bias path at the input.

Complete "smart" temperature-measuring instruments configured for various thermocouple pairs are available commercially. These instruments include computational circuitry to convert the theromelectric voltage to temperature. For instance, the digital thermometers manufactured by Analog Devices and Omega Engineering achieve an accuracy of about 1°C over a temperature

TABLE 14.1. THERMOCOUPLES

Type	Alloy	Max temp[a] (°C)	Tempco @ 20°C (μV/°C)	Output voltage[b] 100°C (mV)	400°C (mV)	1000°C (mV)	30 Gauge lead resistance[c] (Ω)
J	Iron / Constantan[d]	760	51.45	5.268	21.846	—	3.6
K	Chromel[e] / Alumel[f]	1370	40.28	4.095	16.395	41.269	6.0
T	Copper / Constantan[d]	400	40.28	4.277	20.869	—	3.0
E	Chromel[e] / Constantan[d]	1000	60.48	6.317	28.943	76.358	7.2
S	Platinum / 90%Pt-10%Rh	1750	5.88	0.645	3.260	9.585	1.9
R	Platinum / 87%Pt-13%Rh	1750	5.80	0.647	3.407	10.503	1.9
B	94%Pt-6%Rh / 70%Pt-30%Rh	1800	0.00	0.033	0.786	4.833	1.9

[a]Thermocouple life is shortened by prolonged operation near maximum temperature. [b]Reference junction at 0°C. [c]Per double foot (for 24 gauge, multiply values by 0.25). [d]55%Cu-45%Ni. [e]90%Ni-10%Cr. [f]96%Ni-2%Mn-2%Al.

range from −200°C to +1000°C and an accuracy of 3°C at temperatures up to +2300°C.

When compared with other methods of temperature measurement, thermocouples have the advantages of small size and wide temperature range, and they are particularly good for measuring high temperature.

Thermistors

Thermistors are semiconductor devices that exhibit a negative coefficient of resistance with temperature, typically in the neighborhood of −4%/°C. They are available in all sorts of packages, ranging from tiny glass beads to armored probes. Thermistors intended for accurate temperature measurement (they can also be used as temperature-compensation elements in circuits, for instance) typically have a resistance of a few thousand ohms at room temperature, and they are available with tight conformity (0.1–0.2°C) to standard curves. Their large coefficient of resistance change makes them easy to use, and they are inexpensive and stable. Thermistors are a good choice for temperature measurement and control in the range of −50°C to +300°C. It is relatively easy to design a simple and effective circuit for "proportional temperature control" using a thermistor sensing element; see, for example, RCA application note ICAN-6158 or the Plessey SL445A data sheet.

Because of their large resistance change with temperature, thermistors make no great demands on the circuitry that follows. Some simple ways to generate an output voltage are shown in Figure 14.4. The circuit in part A expands the low-temperature end of the range because of the thermistor's exponential resistance change, whereas the circuit in part B produces a somewhat more linear variation of output voltage with temperature. The circuit in part C is the classic Wheatstone bridge, balanced when $R_T/R_2 = R_1/R_3$; since it is ratiometric, the null doesn't shift with variations in supply voltage. The bridge circuit, with a high-gain amplifier, is particularly good for detecting small changes about some reference temperature; for small deviations the (differential) output voltage is linear in the unbalance. With all thermistor circuits you have to be careful about self-heating effects. A typical small thermistor probe might have a dissipation constant of 1mW/°C, meaning that I^2R heating should be kept well below 1mW if you would like your reading accurate to better than 1 degree.

Complete "smart" temperature-measuring instruments using curve-conforming thermistors are available commercially.

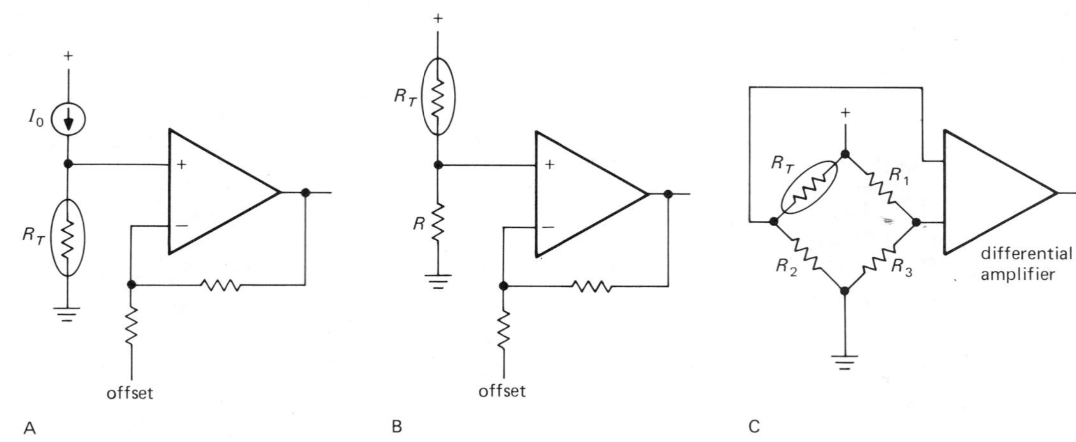

Figure 14.4
Thermistor circuits.

These devices include internal computational circuitry to convert resistance readings directly to temperature. As an example, the Hewlett-Packard 3467A "logging multimeter" includes a temperature mode with 0.3°C accuracy over the range −80°C to +80°C, with reduced accuracy to +150°C.

When compared with other methods of temperature measurement, thermistors provide simplicity and accuracy, but they suffer from self-heating effects, fragility, and a narrow temperature range.

Platinum resistance thermometers

These devices consist simply of a coil of very pure platinum wire, which has a positive temperature coefficient of about 0.4%°C. Platinum thermometers are extremely stable with time and conform very closely (0.02–0.2°C) to a standard curve. They are usable over a temperature range of −200°C to +1000°C. They aren't terribly cheap.

IC temperature sensors

As we remarked in Section 5.14, a bandgap voltage reference can be used to generate a temperature-sensing output voltage proportional to absolute temperature, as well as its usual stable zero-tempco reference output. The REF-02, for example, provides a "temp" output with a linear coefficient of +2.1mV/°C. If you buffer this output with an amplifier of adjustable gain and offset for calibration, you can achieve accuracies of about 0.5°C over the range −55°C to +125°C. The LM335 is a convenient 2-terminal temperature sensor that behaves like a zener diode with a voltage of +10mV/°K; e.g., at 25°C (298.2°K) it acts like a 2.982 volt zener. It comes with an initial accuracy as good as ±1°C, and it can be externally trimmed.

Another approach to IC temperature sensors is the AD590, a two-terminal device that acts as a constant-current element, passing a current in microamps equal to the absolute temperature; e.g., at 25°C (298.2°K) it behaves like a constant-current regulator of 298.2μA (±0.5μA). With this simple device you get 1°C accuracy (best grade) over the range −55°C to +150°C. The simplicity of external circuitry required makes this a very attractive device.

Quartz thermometer

The change of resonant frequency of a quartz crystal with temperature can be exploited to make an accurate and repeatable thermometer. Although the usual objective in quartz-crystal oscillator design is lowest possible temperature coefficient, in this case you choose a crystal cut with a large coefficient and take advantage of the high precision possible in frequency measurements. A good example of a commercial instrument is the Hewlett-Packard 2804A, a microprocessor-controlled thermometer with an absolute accuracy of 40 millidegrees over the range −50°C to +150°C (reduced accuracy over a wider range) and temperature *resolution* of 100 microdegrees. To get this kind of performance, the instrument contains calibration data for the individual sensor that it uses in the temperature calculation.

Pyrometers and thermographs

An interesting method of "noncontacting" temperature measurement is exemplified by the classic pyrometer, a gadget that lets you sight through a telescope at an incandescent object, comparing its glowing color with that of a filament inside the pyrometer. You adjust the filament's current until its brightness matches the object's, both being viewed through a red filter, then read off the temperature. This is a handy method of measuring the temperatures of very hot objects, objects in inaccessible places like ovens or vacuum chambers, or objects in oxidizing or reducing atmospheres where thermocouples cannot be used. Typical optical pyrometers cover a range of 750°C to 3000°C, with an accuracy of about 4°C at the low end and 20°C at the high end.

The development of good infrared detectors has extended this sort of measurement technique down to ordinary temperatures. By measuring the intensity of infrared radiation, perhaps at several infrared wavelengths, you can determine with good accuracy the temperature of a remote object. Such "thermography" has recently become popular in quite diverse fields: in medicine, for the detection of tumors, and in the energy business, where a thermograph of your house can tell you where your energy dollars are evaporating.

Low-temperature measurements

Cryogenic (very cold) systems pose special problems when it comes to accurate temperature measurement. What matters there is how close to absolute zero (0°K = −273.16°C) you are. Two popular methods involve measuring the resistance of ordinary carbon-composition resistors, which soars at low temperatures, and measuring the degree of paramagnetism of some salt. These are really specialty measurement techniques that will not be dealt with here.

Measurement allows control

If you have a way of adjusting some quantity, then the availability of a good measurement technique lets you control that quantity accurately. Thermistors, in particular, provide a nice method for controlling the temperature of a bath or oven.

14.02 Light level

The measurement, timing, and imaging of low light levels are parts of a well-developed field, thanks to the existence of amplification methods that do not rely on conventional circuit techniques. Photomultipliers, channel-plate intensifiers, CCDs (charge-coupled devices), and ISITs (intensifier silicon intensifier target) are included in the catalog of high-performance optical detection devices. We will begin with the simplest detectors (photodiodes and phototransistors) and then go on to discuss the exotic and the wonderful.

Photodiodes and phototransistors

A diode junction acts as a photodetector: Light creates electron-hole pairs, and therefore a current through the external circuit. Diodes intended as photodetectors (photodiodes and PIN diodes) are packaged in a transparent case and are designed for high speed, high efficiency, low noise, and low leakage current. In the simplest mode of operation, a photodiode can be connected directly across a resistive load, or current/voltage converter, as shown in Figure 14.5. You get faster response (and

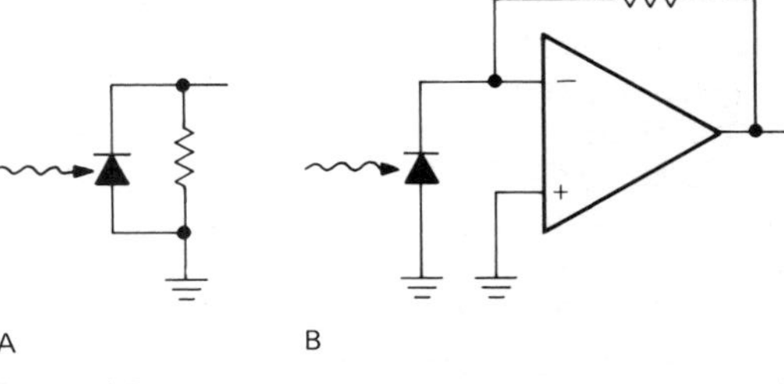

Figure 14.5

the same photocurrent) by reverse-biasing the junction, as in Figure 14.6. High-speed PIN diodes have response times of a nanosecond or less (1GHz bandwidth) when loaded into a low impedance. It should be

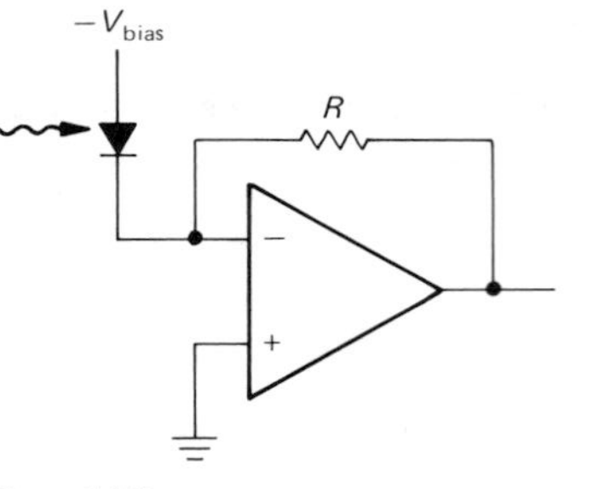

Figure 14.6

noted that the leakage current of good PIN diodes is so low (less than a nanoamp) that the Johnson noise in the load resistance dominates for load resistances less than 100MΩ or so, implying a speed/noise tradeoff. An additional problem to be aware of is the error caused by the amplifier's input offset voltage, or the applied bias voltage, in combination with the photodiode's "dark resistance," when working at low light levels.

Photodiodes are pretty good light detectors when there is plenty of light around, but the output signal can be inconveniently small at low light levels. Typical sensitivities are of the order of 1 microamp per microwatt of incident light. A flux of 1000 photons per second, quite visible with the unaided eye, would cause a photocurrent of 4×10^{-16} amps when focused onto a PIN diode, totally undetectable when compared with the leakage current and noise. No silicon photodetectors are sensitive at the photon level (see the subsequent section on photomultipliers for that), but a device known as a *phototransistor* has considerably more output current than a photodiode at comparable light levels, bought at the expense of speed. It works like an ordinary transistor, with the base current provided by the photocurrent produced in the base-collector junction. Inexpensive phototransistors like the FPT120 have output currents of a few milliamps at an illumination of 1mW/cm^2, with rise and fall times of 18μs, and photo-Darlingtons like the FPT400 have even higher photocurrents, but with rise times of 100μs. Note, however, that the additional current gain of a phototransistor or photo-Darlington doesn't improve its ability to detect extremely low light levels (its "detectivity"), since the ultimate limit is set by the detector diode's "dark current."

Opto-isolators

Photodiodes and phototransistors, in combination with an LED, form the basis of the optically coupled isolator, a handy device for isolating a signal source from its load. They come in various incarnations, from simple LED/phototransistor pairs like the popular Monsanto MCA-2 to high-speed logic couplers with TTL-compatible inputs and outputs. Typically these devices provide 3kV of isolation, 10^{12} ohms insulation resistance, and less than a picofarad coupling between input and output.

An interesting variant is the optical interrupter, an LED phototransistor pair with a 1/8 inch slot. It can sense the presence of an opaque strip, for example, or the rotation

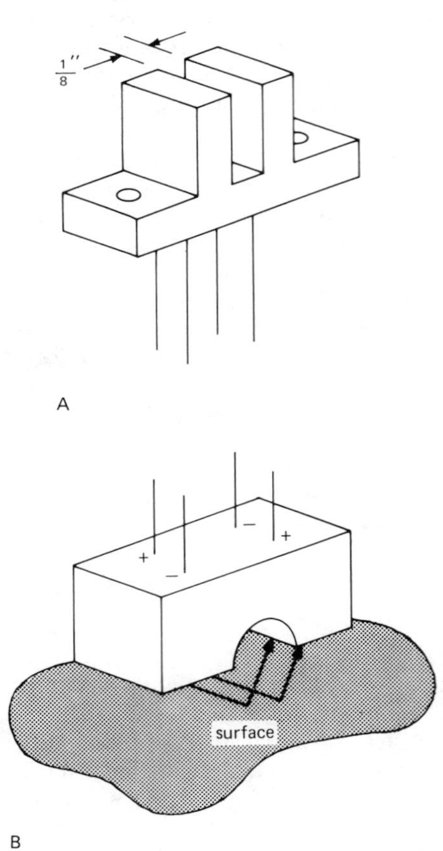

Figure 14.7
A: Optical interrupter. B: Reflective object sensor.

of a slotted disk. An alternative form has the LED and photodetector looking in the same direction, and it senses the presence of a reflective object nearby (most of the time, anyway!). Take a look at Figure 14.7.

Photomultipliers

For low-light-level detection and measurement (and, incidentally, for nanosecond resolution), you can't beat the photomultiplier. This clever device allows a photon (the smallest unit of light) to eject an electron from a photosensitive alkalai metal "photocathode." The photomultiplier then amplifies this feeble photocurrent by accelerating the electron onto successive surfaces (dynodes), from which additional electrons are easily ejected. Figure 14.8 suggests the

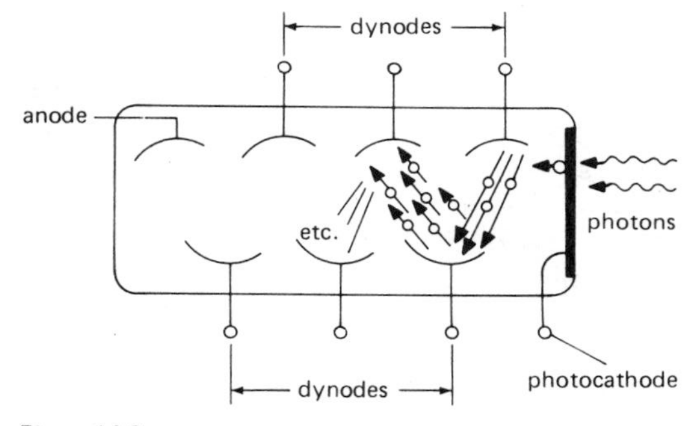

Figure 14.8
Photomultiplier multiplication process.

process. This use of "electron multiplication" yields extremely low noise amplification of the initial photocurrent signal. Typically, you use a voltage divider to put about 100 volts between successive dynodes, for a gain of about 10 per stage, or 1 million overall. The final current is collected by the anode, usually run near ground potential (look at Fig. 14.9), and is large enough so that subsequent amplifier noise is negligible.

The most efficient photocathode materials have quantum efficiencies exceeding 25%, and with the large gain provided by the dynodes, individual photoelectron events are easily seen. At low light levels that's how you would use it, following the PMT (photomultiplier tube) with charge-integrating pulse amplifiers, discriminators (as detailed in Fig. 13.44), and counters. At higher light levels the individual photoelec-

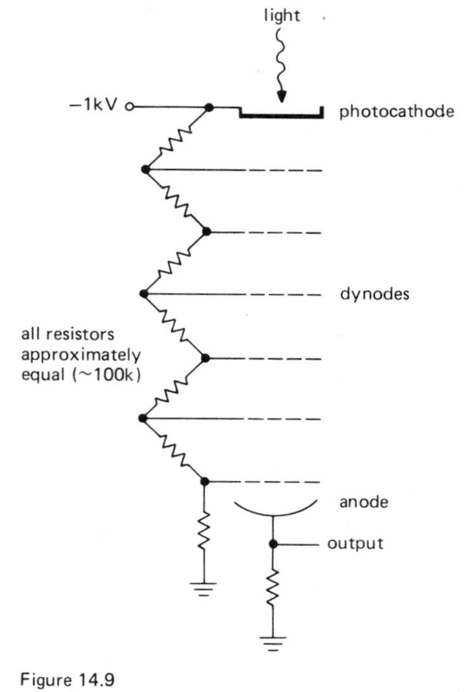

Figure 14.9
Photomultiplier biasing.

tron count rate becomes too high, so you measure the anode *current* as a macroscopic quantity, instead. PMTs have typical sensitivities of an ampere per microwatt, although you could never operate them at such a high current; maximum PMT anode currents are limited to a milliamp or less. It should be noted that the practical limit to photon counting, something like 1 million counts per second, corresponds to roughly 2 micro-microwatts incident power!

Convenient electronic packages are available for both pulse-counting and output-current-measurement modes. An example is the "quantum photometer" from PAR, a handy gadget with built-in high-voltage supply and both pulse and current electronics. It has 11 ranges of pulse counting (10 pps to 10^6 pps, full scale) and 11 ranges of anode-current readout (10nA to 1mA, full scale).

Even in total darkness you get a small anode current from a photomultiplier. This is caused by electrons thermally excited from the photocathode and dynodes, and it can be reduced by cooling the PMT down to

$-25\,°C$ or so. Typical dark currents for a sensitive "bialkali" cathode PMT are in the neighborhood of 30 counts per second per square centimeter of cathode area, at room temperature. A cooled PMT with a small cathode can have dark currents of less than a count per second. It should be pointed out that a powered PMT should never be exposed to ordinary light levels; a PMT that has seen the light of day, even without power applied, may require 24 hours or more to "cool down" to normal dark-current levels.

When compared with photodiodes, PMTs have the advantages of high quantum efficiency while operating at high speed (2ns rise time, typically). They are bulky, though, and they require a stable source of high voltage, since the tube's gain rises exponentially with applied voltage.

It should be emphasized that PMTs are to be used with extremely low light levels. You run them with typical anode currents of a microamp or less, and they can easily see light that you cannot. Photomultipliers are used not only for the detection of light directly, as in astronomy (photometry) and biology (bioluminescence, fluorescence), but also in conjunction with scintillators as particle detectors and x-ray/gamma-ray detectors, as we will discuss in Section 14.07. Photomultipliers find wide use in spectrophotometry, where they are combined with prisms, gratings, or interferometers to make precise measurements of optical spectra.

CCDs, intensifiers, SITs, ISITs, and image dissectors

It is possible to do *imaging* at the light quantum level, thanks to some clever recent technology; i.e., you can form an image with the same sort of sensitivity to low light levels that you get with the (nonimaging) photomultiplier. These recent inventions are amazing to see. You can sit in what appears to be a completely dark room, then peer into a television monitor in which are imaged, albeit with plenty of "snow," all the objects in the room.

The key to all this is the image intensifier, an incredible device that produces as its output a brightened replica of an input image. You begin with either an ordinary silicon target vidicon (TV camera) or a CCD array. These are light-sensitive two-dimensional targets that accumulate an image and can be read out electronically by scanning with an electron beam or by shifting the image along as an analog shift register, respectively. At this point all you have is a television camera whose sensitivity is far below the individual photon level; it is the two-dimensional analog of a photodiode. To bring about the miraculous, you simply put an imaging intensifier tube in front. Figure 14.10 suggests the process schematically.

Intensifiers come in two varieties. The

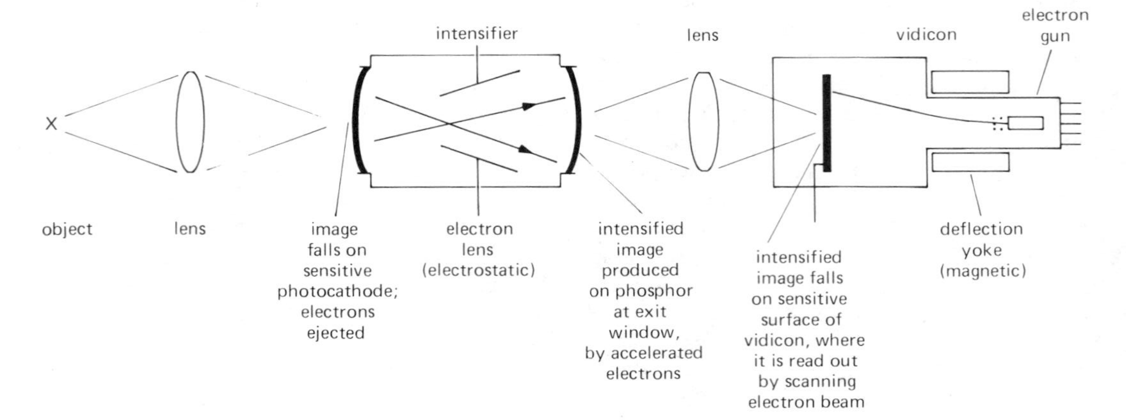

Figure 14.10
Vidicon with single-stage intensifier.

first-generation type consists of a sensitive photocathode surface of the type used in photomultipliers, with electron-focusing optics and a phosphor screen arranged behind so that photoelectrons from the cathode are accelerated by high applied voltages and hit the phosphor with enough energy to give off a bright flash of light. With this kind of intensifier you can get single-stage light amplification of about 50, with resolution of about 50 lines/mm. Popular types cascade two, three, or four such stages of light amplification to achieve overall light amplifications of a million or more. The input and output may simply be glass surfaces with their internal photosensitive and phosphor coatings, or they may be paved with a dense fiber-optic bundle. Fiber optics are nice because they let you match a perfectly flat entrance or exit surface to a curved tube surface, and they simplify the external optical system, since you can cascade these devices by just sticking things together, without any lenses.

Second-generation intensifiers using "microchannel plates" allow you to achieve much higher values of single-stage light amplification, and they are better at really low light levels because of fewer "ion events," the result of positive ions being ejected from the phosphor and returning to the cathode, where they make a big splash. In these channel-plate intensifiers the space from cathode to phosphor contains a bundle of microscopic hollow tubes whose insides are coated with a dynode-type multiplication surface. Photoelectrons from the cathode bounce their way down these channels, ejecting secondary electrons to give light amplification of about 10,000 (Fig. 14.11). You can get resolutions of about 20 lines/mm, and with special configurations ("J-channel," "C-channel," "chevron") the ion-event problem can be eliminated almost entirely. The result is an imaging intensifier with the same sort of quantum efficiency as photomultipliers (20%–30%). The use of nearly noiseless electron multiplication results in light amplification to a level that the vidicon or CCD can see.

Such an intensifier combined with a silicon-target vidicon in a single tube is called a "SIT" (silicon intensifier target). An ISIT is a SIT with an additional intensifier placed externally in front (Fig. 14.12); this is the sort of gadget that lets you see in the dark. These things are very popular with astronomers and with night-warfare people.

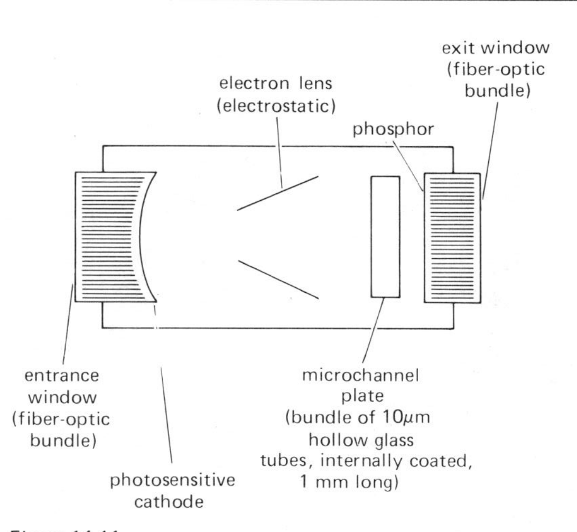

Figure 14.11
Electrostatically focused channel plate intensifier.

An interesting variation of the imaging intensifier is the so-called image dissector, a clever device that actually preceded the devices just described. It consists of a sensitive photocathode area, followed by the usual photomultiplier dynode chain. In between is a small aperture and some deflection electrodes, so that any spot on the photocathode can become the active area for electron multiplication by the dynode system. You can think of an image dissector as a photomultiplier with an electronically movable photocathode area. It has the quantum efficiency and gain of a conventional PMT, but it differs from the intensified vidicons, CCDs, and SITs (which are all image-integrating devices) in that it does not accumulate the image over the entire field in between readouts.

14.03 Strain and displacement

The field of measurements of physical variables such as position and force has its own bag of tricks, and any accomplished measurer should be aware of things like

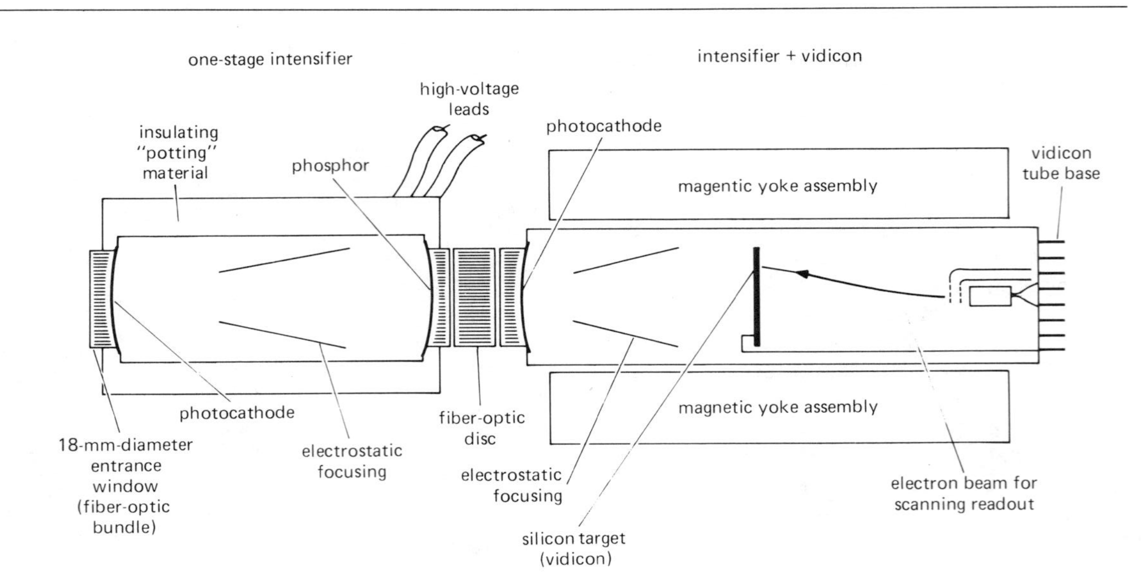

Figure 14.12
Intensified silicon intensified target (ISIT).

strain gauges, LVDTs, and the like. The key to all of these measurements is the measurement of displacement.

There are several nice ways to measure position, displacement (changes in position), and strain (relative elongation).

LVDT. A popular method is the LVDT (linear variable differential transformer), which is almost self-explanatory. You construct a transformer with a movable core, excite one set of coils with ac, and measure the induced voltage in the second set. The secondary is center-tapped (or brought out as two separate windings) and arranged symmetrically with respect to the primary, as shown in Figure 14.13. LVDTs come in an enormous variety of sizes, with full-scale displacements ranging from 0.005 inch to 25 inches, excitation frequencies from 50Hz to 25kHz, and accuracies of 1% down to 0.1% or better.

Strain gauge. A strain gauge measures elongation or flexure by subjecting an array of four metal thin-film resistors to deformation. They come as complete assemblies, in sizes from 1/64 of an inch to several inches, and they generally have impedances in the neighborhood of 350 ohms/leg. Electrically they look like a Wheatstone bridge; you apply dc across two of the terminals and look at the voltage difference between the other two, as discussed in Section 7.08. The output voltages are very small, typically 3mV per volt of excitation for full-scale deformation, with accuracies ranging from 1% down to 0.1% of full scale (see Fig. 14.13D).

It is not easy to measure small relative elongations, and strain-gauge specifications are notoriously unreliable. Small differences in the temperature coefficients of the bridge elements are responsible for the temperature sensitivity, which limits the performance of the strain gauge. This is a problem, even in controlled-temperature environments, because of self-heating. For example, 10 volts of dc excitation on a 350 ohm bridge will produce 300mW dissipation in the sensor, with a temperature rise of 10°C (or more), causing errors corresponding to a real signal of 0.1% to 0.5% of full scale.

Recently, semiconductor strain gauges have become popular. They have outputs that are 10 times higher than those of the metal-film variety and impedances of a few thousand ohms. It is often necessary to use a current source as excitation, rather than a voltage source, to minimize temperature sensitivities.

Capacitance transducers. Very sensitive

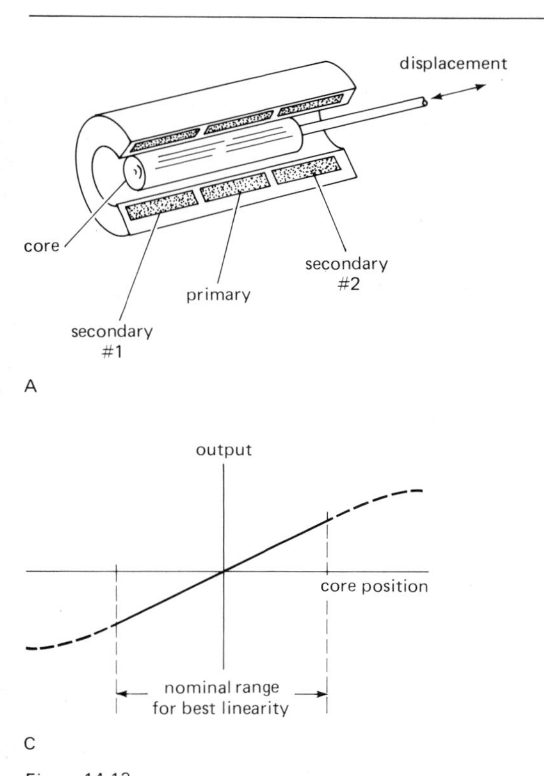

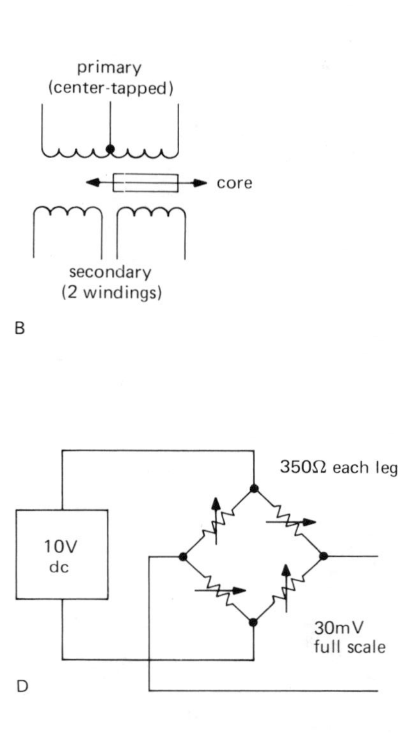

Figure 14.13
Displacement transducers. A: Linear variable differential transformer (LVDT) cutaway view. B: LVDT schematic. C: LVDT output versus displacement. D: Strain gauge schematic.

measurements of displacement can be made with a transducer consisting simply of two closely spaced plates, or a plate suspended between a pair of outer plates. By making the capacitor part of a resonant circuit, or by using a high-frequency ac bridge, you can sense or control very small changes in position. Capacitor microphones use this principle to convert acoustic pressure or velocity to an audio signal.

The amplifiers used with capacitor microphones illustrate some interesting circuit ideas, and they are of practical importance, since many of the best recording microphones are capacitive position transducers, made by supporting a thin metallized plastic foil in close proximity to a fixed plate. You charge the capacitor through a large resistor with a bias of 50 to 100 volts, and you look at the changes of voltage as the diaphragm moves in the sound field.

Capacitor microphones have enormously high source impedances (a typical capsule has about 20pF of capacitance, or a reactance of about 400M at 20Hz), which means that you don't have a chance of running the signal through any length of cable whatsoever without putting a preamp right at the capsule. Figure 14.14 shows two ways to buffer the voltage from the capsule, which might have an amplitude of 1mV to 100mV (rms) for typical ranges of program material. In the first circuit, a low-noise FET op-amp provides 20dB of gain and the low impedance necessary to drive a single-ended shielded line. Since the amplifier has to be located close to the microphone capsule (within a few inches), it is necessary to supply the operating voltages (bias for the capsule, as well as op-amp power) through the microphone cable, in this case on additional wires. Note the trick of floating the microphone capsule in order to simplify biasing of the op-amp. R_1 and C_1

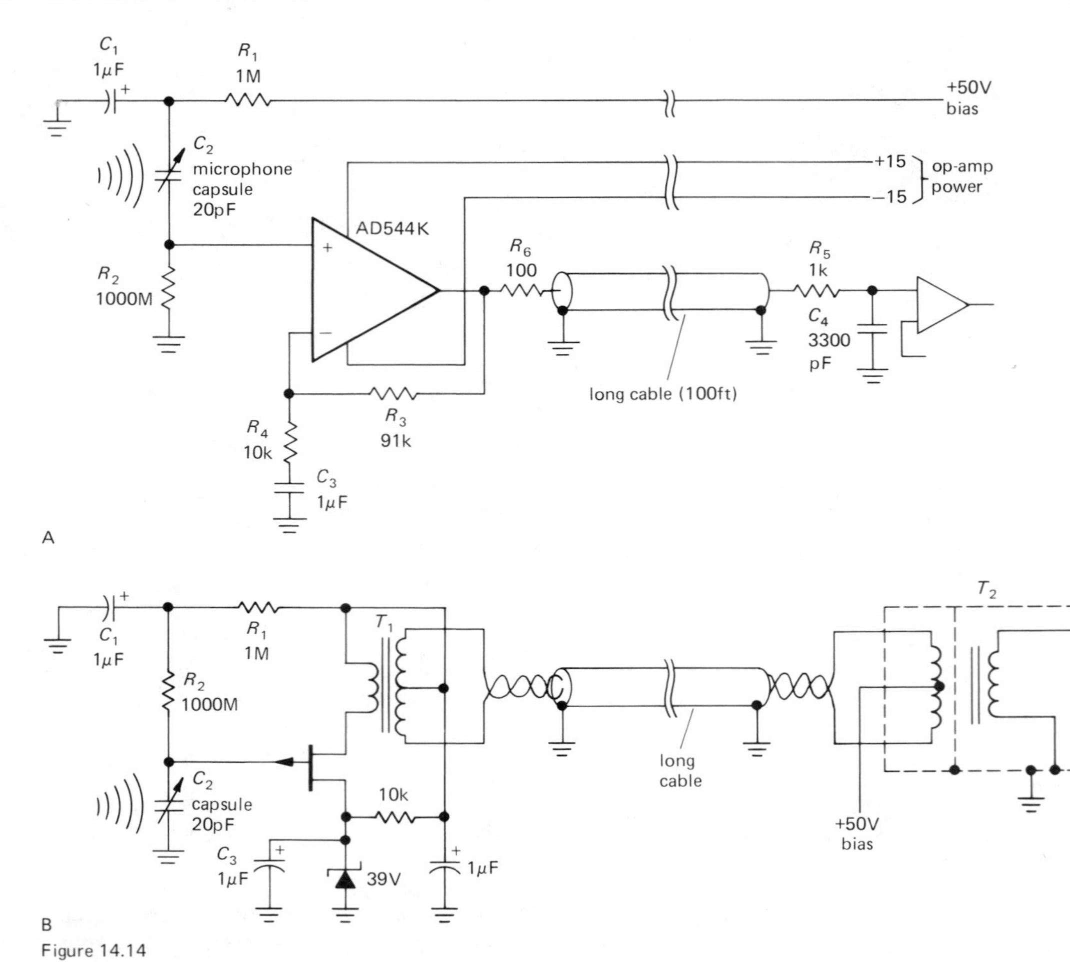

Figure 14.14
Capacitor microphone amplifiers.

filter the bias supply, and R_2 must be chosen to have a high impedance compared with the capsule at all audio frequencies. R_5 and C_4 form an RF filter, since the line is unbalanced and therefore prone to radiofrequency interference.

There are a few bad features about this circuit. It requires a 4-wire cable, rather than the industry-standard shielded pair. Also, the floating capsule can create mechanical problems. These drawbacks are remedied in the second circuit, in which the capsule bias is sent on the same lines as the audio itself, which happens to be a balanced 200 ohm pair. One side of the capsule is grounded, and a *p*-channel JFET is used as a source follower to drive a small audio transformer. A single-ended output is recovered at the far end, where bias is applied via the transformer's center tap. Some would complain that the proliferation of transformers is a poor idea, but in practice they perform admirably.

□ *Angle.* It is possible to convert angles to electronic signals with pretty good precision. There are angular versions of the LVDT, for instance, and popular devices known as resolvers. In both cases you use an ac excitation, and you can easily measure angular position down to an arc-minute. With great care it is possible to measure angles at the arc-second level. There are other techniques, e.g., using light beams looking at a glass disc with gray coded radial stripes.

□ *Interferometry.* Highly accurate position measurements can be made by bouncing laser beams off mirrors attached to the object and counting interference fringes. The ultimate accuracy of such methods is set by the wavelength of light, so you have to work hard to do much better than a half micron (1 micron, or micrometer, is 1/1000 of a millimeter, or 1/25000 of an inch). An example of a commercial laser measurement instrument is the 5526A from Hewlett-Packard, which claims resolutions approaching a microinch and 0.1 arc-second. Laser interferometer systems are now used routinely in surveying, in flatness measurements, and in various tasks around research laboratories.

The most precise distance measurements have been made interferometrically by Deslattes at the National Bureau of Standards. Deslattes is a real wizard when it comes to precise physical measurement, routinely measuring spacings to milliangstroms (one ten-millionth of a micron) and angles to milliseconds of arc.

□ *Quartz oscillators.* A quartz crystal responds to deformation with a change in its resonant frequency, thus providing a very accurate method for measuring small displacements or changes in pressure. Quartz-oscillator pressure transducers provide the highest resolution presently available (more about this later).

14.04 Acceleration, pressure, force, velocity

The techniques just mentioned allow you to measure acceleration, pressure, and force. Accelerometers consist of strain gauges attached to a test mass, or capacitance-sensing transducers that sense the change of position of the test mass. There are various tricks to damp the system to prevent oscillation, in accelerometers that simply measure the displacement of the test mass to provide an output signal; alternatively, some systems use feedback to prevent the test mass from being displaced relative to the body of the accelerometer, the amount of applied feedback force then being the accelerometer's output signal.

LVDT, strain gauges, capacitance transducers, and quartz oscillators are used for pressure measurements, along with special devices such as a Bourdon gauge, a spiral hollow quartz tube that unwinds when inflated. LVDT transducers, for example, are available with full-scale ranges going from 1 psi to 100,000 psi or more. Quartz-crystal oscillator types provide the highest resolution and accuracy. The types available from Paroscientific, for example, will deliver accuracies of 0.01% and stabilities of 0.001%. Hewlett-Packard has a quartz pressure gauge with 11,000 psi full-scale sensitivity and claimed resolution of 0.01 psi.

LVDT transducers are often used to measure force or weight, although any of the displacement techniques can be used. Full-scale sensitivites go from 10 grams to 250 tons for one popular series, with accuracies of 0.1%. For highly precise laboratory measurements of small forces, you will find quartz-fiber torsion balances, electrostatic balances, and the like. An interesting example of the latter is the clever gravimeter developed by Goodkind and Warburton. It uses a superconducting sphere levitated approximately to zero weight by a persistent magnetic field, then balanced the rest of the way by electrostatic sensing and levitating plates. It can measure changes in gravitational field of one part in a billion, and it easily sees barometric pressure variations because of the effect of the changing overhead air mass on local gravity!

Magnetic velocity transducers

The position transducers we have been talking about can also be used to keep track of velocity, which is just the time derivative of position. However, it is possible to make a direct velocity measurement by exploiting the fact that the voltage induced in a loop of wire moving through a magnetic field is proportional to the rate of change of magnetic flux linked by the loop. There are velocity-measurement gadgets available that consist of long coils of wire with magnetic rods moving through the central bore.

Much more prevalent are the magnetic velocity transducers used in the audio industry: microphones (and their inverse, loudspeakers), phono cartridges, and analog tape recorders. These devices typically generate signals at very low levels (a few millivolts is typical), and they present unique and interesting circuit challenges. For high-quality sound you have to keep noise and interference down 60dB or more, i.e., at the microvolt level. Since these signals get piped around over large distances in recording studios and radio stations, the problem can become serious.

Figure 14.15 shows how low-level signals from microphones and phono cartridges are usually handled. A dynamic microphone is a loudspeaker in reverse: A coil moves in a magnetic field, propelled by the sound pressure. Typically these things have output impedances of 200 ohms, with signals of 50μV to 5mV (rms) for quiet speech and concert-hall sound levels, respectively. For any significant length of connecting cable, you always use a balanced and shielded twisted pair, terminated in the industry-standard Cannon 3-pin audio connector.

At the far end you transform to a terminated impedance level of about 50k with a high-quality audio matching transformer, as indicated. Signal levels are then in the range of 1mV to 100mV (rms) and should be amplified by a low-noise preamp, as shown. Although you will see preamps with 40dB front-end gain, for good overload performance it is best to stick with a gain of 20dB. This is especially true for popular-music recording, where singers often wind up hollering into the microphone at close range.

The use of balanced 200 ohm microphone cable pretty much eliminates interference because of its good common-mode

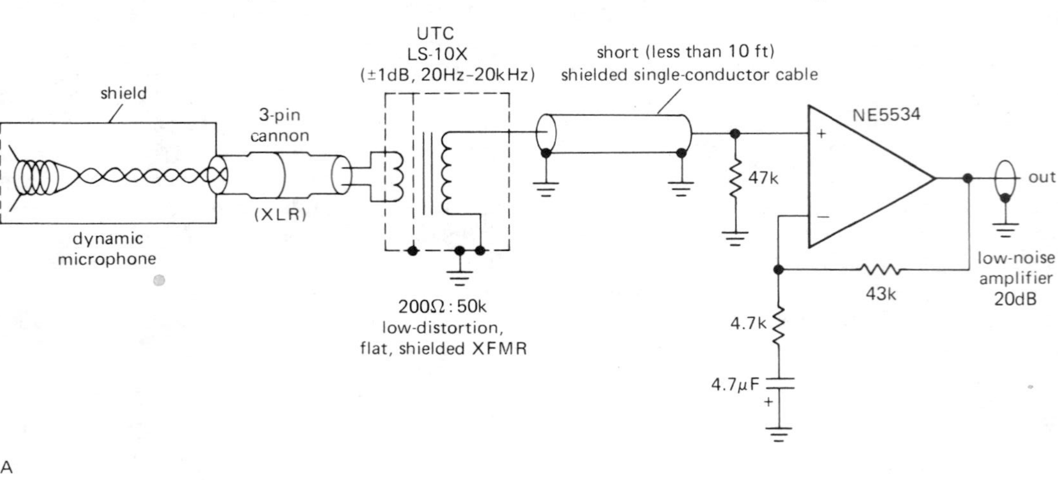

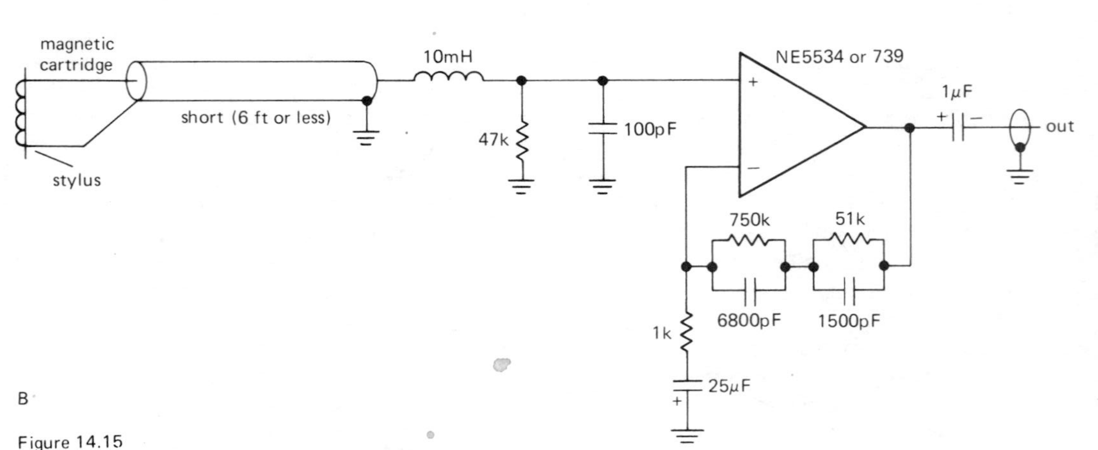

Figure 14.15
Dynamic microphone and phono cartridge amplifiers.

rejection. Good audio transformers for this kind of application have electrostatic shielding between the windings, which further reduces sensitivity to RF pickup. If radiofrequency interference is not suppressed enough with this scheme, as may be the case near transmitting stations, you can add a low-pass filter at the preamp input. A 1k resistor or small RF choke in series at the input, followed by a 100pF capacitor to ground, will usually tame the beast.

Phono cartridges don't require balanced lines, because the cable run to the amplifier is usually very short. The standard method is simply to use single-conductor shielded cable, terminated with the 47k to ground that the cartridge requires for proper frequency response (Figure 14.15B). We have also shown an input filter to reduce RF interference, because that is such a common problem in urban areas. RF signals at the inputs of audio equipment pose a particularly insidious problem, because the audio amplifier's nonlinearities at radiofrequencies produce rectification, with consequent interference (audio detection) and distortion. When designing RF filters, be sure to keep the load capacitance small (a maximum of 300pF, including cable capacitance), since otherwise the cartridge's frequency response is changed. The series impedance should not have a resistance greater than a few hundred ohms, to keep the noise low. Quite large values of inductance can be safely used, since the cartridge's inductance is typically 0.5 henry. The amplifier circuit shown has the standard RIAA response used for recording in the United States.

14.05 Magnetic field

Accurate magnetic-field measurements are important in the physical sciences, in connection with instrumentation that uses a magnetic field (magnetic resonance, magnetrons, magnetically focused electron devices, etc.), and in geology and prospecting. For measurements at the 1% level, a Hall-effect probe is adequate. The Hall effect is the production of a transverse voltage in a current-carrying conductor (usually a semiconductor) in a magnetic field, and commercial Hall-effect magnetometers cover a range of about 1 gauss to 10kG full-scale. To give an idea of scale, the earth's field is about 0.5 gauss, whereas the field of a strong permanent magnet is a few thousand gauss. Hall magnetometers are inexpensive, simple, small, and reliable. The Hall effect is also used to make noncontacting keyboard and panel switches, as we remarked in Section 9.04.

A method with considerably greater roots into the past is the flip coil, a multiturn coil of wire that is either rotated in the magnetic field at some fixed speed or simply pulled out; you measure the induced ac voltage or the integrated current, respectively. A flip coil is simplicity itself, and it has the elegance of pure electromagnetic theory, but it tends to be a bit bulky and old-fashioned-looking.

For measurements of miniscule magnetic fields you can't beat the exotic SQUID (superconducting quantum interference device), a clever arrangement of superconducting junctions that can easily measure a single quantum of magnetic flux (0.2 microgauss-cm^2). A SQUID can be used to measure the magnetic fields set up in your body when you drink a glass of cold water, for whatever that's worth. These are fancy devices that require a considerable investment in cryogenic hardware, liquid helium, etc., and shouldn't be considered ordinary circuit items.

For precision magnetic-field measurements in the kilogauss range you can't do better than an NMR (nuclear magnetic resonance) magnetometer, a device that exploits the precession of nuclear (usually hydrogen) spins in an external magnetic field. This is the physicist's magnetometer, and it effortlessly yields values of magnetic field accurate to a part in a million or better. Since the output is a *frequency,* all the precision of frequency/time measurements can be used (more about this later).

Devices known as flux-gate magnetometers and transductors provide yet another way to measure magnetic fields. They work by exciting a piece of ferrite with an ac excitation field, with the response, as modified by the ambient field, being observed.

14.06 Vacuum gauges

The measurement of vacuum presents no great obstacles, which is fortunate, since it is a quantity of great importance in processes such as transistor and IC manufacture, thin-film evaporation, and the preparation of freeze-dried coffee. The basic device here is the ionization gauge, which looks like an inside-out triode vacuum tube (Fig. 14.16). A hot filament emits electrons

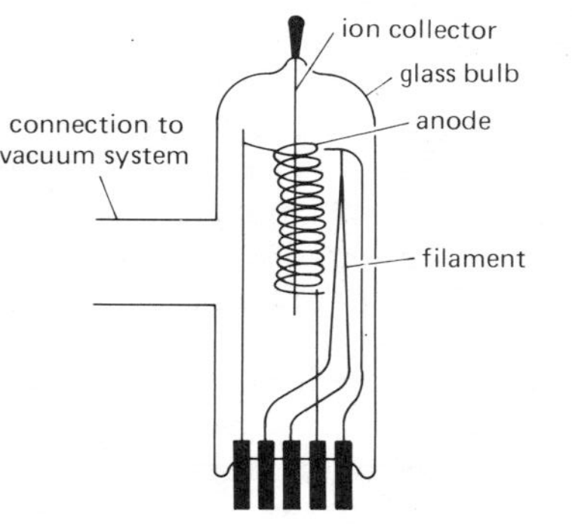

Figure 14.16
Ionization gauge.

that are collected at a positive wire anode. Along the way the electron beam scatters from residual gas molecules, creating positive ions that are collected at a central wire electrode held near ground. The ion current is accurately proportional to the density of gas molecules, i.e., the pressure. Ion gauges are usable at pressures (vacuums!) from about 10^{-4} to 10^{-10}mm Hg (1mm Hg is also known as 1 torr; atmospheric pressure is 760mm Hg). It takes great care to maintain a vacuum of 10^{-10}mm Hg; even a fingerprint on the inside of the chamber will frustrate your efforts.

At more mundane levels of vacuum (1mm Hg down to 1 micron Hg, which you get with mechanical "roughing" pumps) the popular choice of measurement device is the thermocouple gauge. It is simply a thermocouple bonded to a small heater; you run some current through the heater, then measure the temperature with the thermocouple. Residual gas cools the contraption, lowering the thermocouple's output voltage. Thermocouple gauges are usually used so that you know when it's safe to turn on the high vacuum (diffusion or ion) pumps.

14.07 Particle detectors

The detection, identification, spectroscopy, and imaging of charged particles and energetic photons (x rays, gamma rays) constitute an essential part of the fields of nuclear and particle physics, as well as numerous fields that make use of radioactivity (medical radiography tracers, forensic science, industrial inspection, etc.). We will treat x-ray and gamma-ray detectors first, then charged-particle detectors.

X-ray and gamma-ray detectors

The classic uranium prospector was a slightly grizzled and shriveled character who went poking around the desert with clicking Geiger counter in hand. The detector situation has now improved considerably. These detectors all have in common the property that they use the energy of an incoming photon to ionize an atom of something, giving off an electron via the photoelectric effect. What they do with the electron depends on the particular detector.

Ionization chamber, proportional counter, geiger counter. These detectors consist simply of a cylindrical (usually) chamber, typically a few inches in size, with a thin wire running down the center. They are filled with some gas or mixture of gases. There's a thin "window" on one side, made of some material that the desired x rays can penetrate (plastic, beryllium, etc.). The central wire is held at a positive potential and connected to some electronics. Figure 14.17 shows a typical configuration.

When an x ray enters, it ionizes an atom by ejecting a photoelectron, which then loses energy by ionizing gas atoms until it is brought to rest. It turns out that the electron loses about 20 volts of energy per electron-ion pair it creates, so the total free charge left after the photoelectron is brought to rest is proportional to the x ray's initial energy. In an ionization chamber that charge is collected and amplified by a charge-sensitive (integrating) amplifier, just as with a photo-

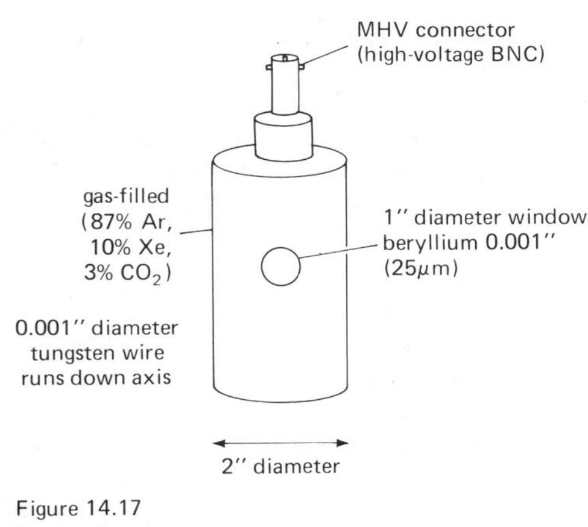

Figure 14.17
Proportional counter.

multiplier. Thus, the output pulse is proportional to the x ray's energy. The proportional counter works the same way, but with the central wire held at a higher voltage, so that electrons drawn toward it cause additional ionization, resulting in a larger signal. This charge-multiplication effect makes proportional counters useful at low x-ray energies (down to a kilovolt, or less) where an ionization counter would be useless. In a Geiger counter the central wire is at a high enough voltage that *any* amount of initial ionization causes a single large (fixed-size) output pulse. This gives a nice large output pulse, but in the process you lose all information about the x ray's energy.

As you will see in Section 14.16, a clever device known as a pulse-height analyzer lets you convert an input stream of pulses of assorted heights into a histogram. If the pulse heights are a measure of particle energy, you wind up with an energy spectrum! Thus with a proportional counter (but not with a Geiger counter) you are doing x-ray energy spectroscopy.

These gas-filled counters are usable in the energy range from about 1keV to 100keV. Proportional counters have an energy resolution of about 15% at 5.9keV (a popular x-ray calibration energy provided by decay of iron 55). They're inexpensive and can be made in very large or very small sizes, but they require a well-regulated power supply (the multiplication rises exponentially with voltage) and are not terribly fast (25,000 counts/s is a rough practical maximum counting rate).

Scintillators. Scintillators work by converting the energy of the photoelectron, Compton electron, or electron-positron pair to a pulse of light, which is then detected by an attached photomultiplier. A popular scintillator is crystalline sodium iodide (NaI) doped with thallium. As with proportional counters, the output pulse is proportional to the incoming x-ray (or gamma-ray) energy, which means that you can do spectroscopy, with the help of a pulse-height analyzer (see Section 14.16). Typically, an NaI crystal will give an energy resolution of about 7% at 1.3MeV (a popular gamma-ray calibration energy provided by decay of cobalt 60) and is usable in the energy range of 10keV to many GeV. The light pulse is about 1μs long, making these detectors reasonably fast. NaI crystals come in various sizes up to a few inches; they absorb water, though, so they have to be sealed. Since you must keep light out anyway, they're usually supplied in a metal package with thin aluminum or beryllium entrance window and integral photomultiplier tube.

Plastic (organic) scintillator materials are also popular, being very inexpensive. They have poorer resolution than sodium iodide and are used primarily at energies above 1MeV. Their light pulses are very short, roughly 10ns. Liquid scintillation "cocktails" are routinely used in biological studies. In such applications the material being examined for radioactivity is mixed into the scintillator cocktail, and the whole works is put into a dark chamber with a photomultiplier. You'll see handsome instruments in biology labs that automate the whole process, passing one vial after another through the counting chamber and recording the results.

Solid-state detectors. As with the rest of electronics, the great revolution in x-ray and gamma-ray detection has come about through advances in silicon and germanium technology. "Solid-state" detectors work just like the classic ionization chamber, but with the active volume filled with a nonconducting (intrinsic) semiconductor. An applied potential of about 1000 volts sweeps the ionization out, generating a pulse of charge.

In silicon an electron loses only about 2eV per electron-ion pair created, so many more ions are created for the same incident x-ray energy, as compared with a gas-filled proportional detector, giving better energy resolution through improved statistics. Other subtle effects also contribute to improved performance.

Solid-state detectors come in several varieties, Si(Li), Ge(Li), and intrinsic germanium, or IG (the first two are pronounced "silly" and "jelly"), according to the semiconductor material and dopants used to make it insulating. They are all operated at liquid-nitrogen temperature (−196°C), and the lithium-drifted types must be kept cold at all times (if allowed to warm up, they decay, permanently, with about the same time constant as fresh fish). Typical Si(Li) detectors come in diameters from 4mm to 16mm and are usable for x-ray energies from about 1keV to 50keV. Ge(Li) and IG detectors are used at higher energies, 10keV to 10MeV. Good Si(Li) detectors have energy resolutions of 150eV at 5.9keV (2.5%, six to eight times better than proportional counters), and the germanium detectors have energy resolutions of about 1.8keV at 1.3MeV (0.14%).

In order to illustrate what that extra resolution buys you, we bombarded a random hunk of stainless steel with 2MeV protons and measured the x-ray spectrum produced. This is called PIXE (proton-induced x-ray emission), and it is a powerful technique for determining spatially resolved trace-element distribution. Figure 14.18 shows the energy spectrum (made with a pulse-height analyzer), with two x-ray lines visible for each element, at least with the Si(Li) detector. You can see iron, chromium, and nickel. A few additional elements are visible if you expand the lower part of the graph. With the proportional counter all you get is mush.

Figure 14.19 shows the same kind of comparison for gamma detectors. This time it's an NaI scintillator versus a Ge(Li). (We ran out of steam, so we cribbed this one from the friendly folks at Canberra Industries. Many thanks, Mr. Tench.) As before, solid-state detectors win, hands down, for resolution.

Solid-state detectors have the best energy resolution of all the x-ray and

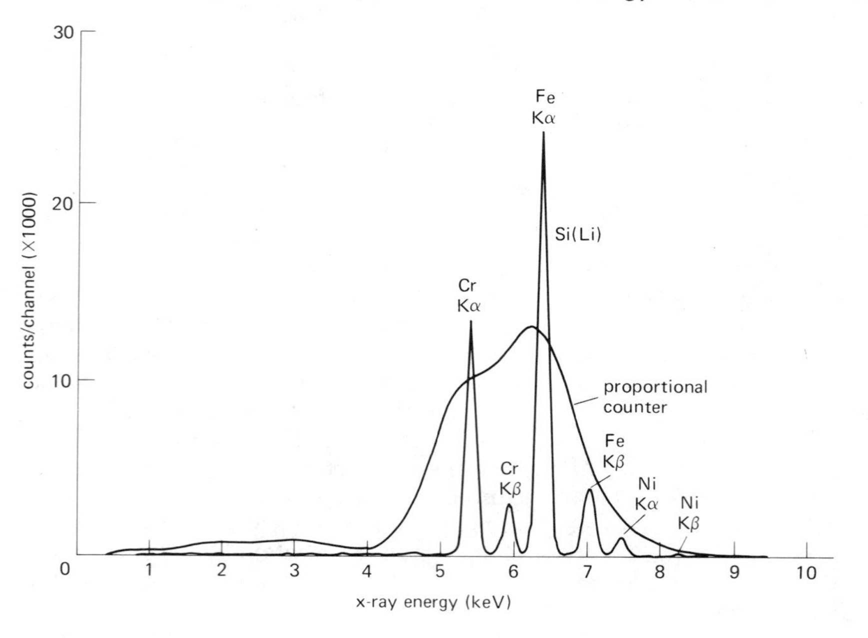

Figure 14.18
X-ray spectrum from a piece of stainless steel, as seen by an argon proportional counter and a Si(Li) detector.

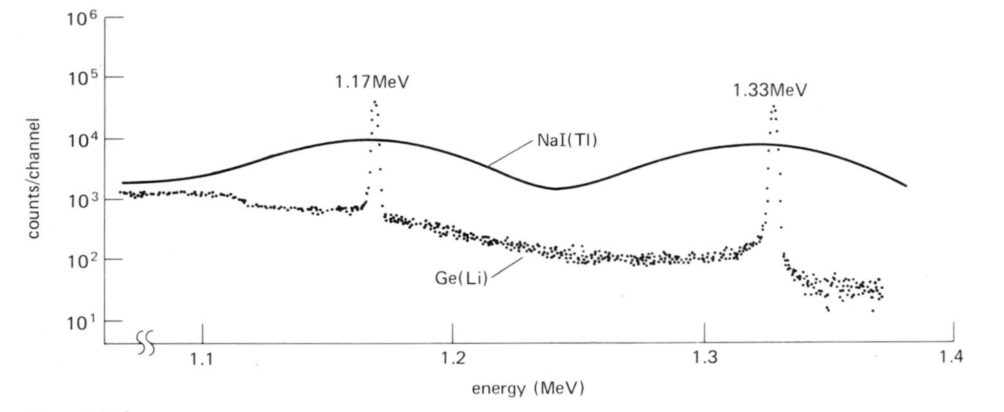

Figure 14.19
Cobalt 60 gamma-ray spectrum, as seen by a sodium iodide scintillator and a Ge(Li) detector. (From Canberra Ge(Li) Detector Systems Brochure, Canberra Industries, Inc., Meriden, CT., U.S.A.)

gamma-ray detectors, but they have the disadvantages of a small active area in a large clumsy package (see Fig. 14.20 for an example), relatively slow speed (50μs or longer recovery time), high price, and high nuisance value (unless you enjoy being a full-time baby-sitter to a liquid-nitrogen guzzler).

Charged-particle detectors

The detectors we've just described are intended for energetic *photons* (x rays and gamma rays), not particles. Particle detectors have somewhat different incarnations; in addition, charged particles are deflected by electric and magnetic fields according to their charge, mass, and energy, making it much easier to measure particle energies.

Surface-barrier detectors. These germanium and silicon detectors are the analogs of the Ge(Li) and Si(Li) detectors. They don't have to be cooled, which simplifies the packaging enormously. (It also lets you take an occasional vacation!) Surface-barrier detectors are available in diameters from 3mm to 50mm. They are usable at particle energies of 1MeV to hundreds of MeV, and they have energy resolutions of 0.3% to 1% for 5.5MeV alpha particles (a popular alpha calibration energy provided by decay of americium 241).

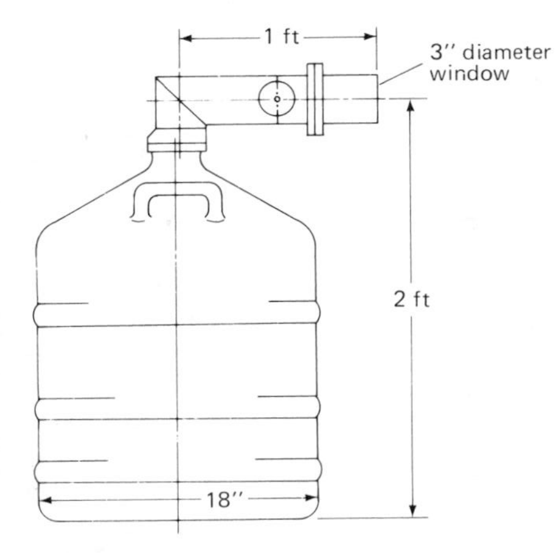

Figure 14.20
Ge(Li) cryostat. (From Canberra Ge(Li) Detector Systems Brochure, Canberra Industries, Inc., Meriden, CT., U.S.A.)

☐ *Cerenkov detectors.* At very high energies (1GeV and above) even a heavy charged particle can outrun light in material media, giving rise to Cerenkov radiation, a "visible sonic boom." They are used extensively in high-energy physics experiments.

☐ *Ionization chambers.* The classic gas-filled ionization chamber previously described in connection with x-ray detection can also be used as a detector of energetic charged particles. In its simplest form it consists of a single collecting wire running the length of an argon-filled chamber. Depending on the particle energies involved, the chamber may range from inches to feet in size; variations include the use of multiple collecting wires or plates, and other filling gases.

- ☐ *Shower chambers.* A shower chamber is the electron equivalent of an ionization chamber. An energetic electron enters a box full of liquid argon, where it generates a "shower" of charged particles that are subsequently collected at charged plates. High-energy physicists like to call these things "calorimeters."
- ☐ *Scintillation chambers.* A charged particle can be detected with very good energy resolution by using photomultipliers to detect the ultraviolet-rich scintillations caused by the particle's ionized path in a chamber filled with argon or xenon in gas or liquid form. Scintillation chambers are delightfully fast, in contrast with the more leisurely response of ionization and shower chambers.
- ☐ *Drift chambers.* These are the latest rage in high-energy physics, and they are made possible by advances in high-speed on-line computing. They're simple in conception: a box filled with gas at atmospheric pressure (an argon-ethane mixture is typical) and crisscrossed by hundreds of wires with an applied voltage. The box is full of electric fields, and when a charged particle goes in and ionizes the gas, the ions are swept out by the array of wires. You keep track of the signal amplitudes and timing on all the wires (that's where the computer comes in), and from that information you deduce the particle's path. With an applied magnetic field, that also tells you the momentum.

The drift chamber has become the universal imaging charged-particle detector for high-energy physics. It can deliver spatial resolution of 0.2mm or better over a volume large enough for you to climb into.

14.08 Biological and chemical voltage probes

In the biological and chemical sciences there are many examples of measurement wizardry: electrochemical methods such as electrochemistry with ion-specific electrodes, electrophoresis, voltametry, and polarography, as well as techniques like chromatography, IR and visible spectroscopy, NMR, mass spectroscopy, x-ray spectroscopy, nuclear quadrupole spectroscopy, ESCA, etc. It is hopeless in a volume this size to attempt any kind of comprehensive catalog of these sophisticated techniques. Furthermore, these techniques can be characterized as less fundamental than the direct physical measurements cataloged earlier in this chapter.

In order to give an idea of the special problems that arise in chemical and biological measurements, we will describe only the simplest sort of measurement: the determination of the potentials generated by a microelectrode (used to explore nerve and muscle signals in biological systems), by an ion-specific electrode (used to measure the concentration of some specific ionic species in solution), and by a voltametric electrochemical probe. As usual, there are some interesting electronic challenges you face, if you want to get anything meaningful out of your measurement.

Microelectrodes

In order to look at the voltages on nerves or in the interiors of cells, it is standard practice to make electrodes that are just a few hundred angstroms in tip diameter (1 angstrom = 10^{-8}cm, approximately the size of a hydrogen atom). That turns out to be easily done by drawing a glass capillary, then filling it with a conductive solution. You wind up with a nice probe, but with interesting circuit problems arising from the electrode's source impedance of 100MΩ or more. Interference pickup, loading by the circuit, and high-frequency rolloffs of a few hertz due to cable and stray capacitances plague the unwary.

In order to see nerve or muscle signals, you want to have decent high-frequency performance, at least out to a few kilohertz, or so (this isn't exactly high frequency in the sense of Chapter 13!). The amplifier must have very high input impedance, and preferably low input noise. In addition, it must be insensitive to common-mode interference.

The circuit in Figure 14.21 represents a good solution. The use of a reference electrode connected near the point of actual measurement keeps interference from appearing as normal (differential) mode signal. The inputs are buffered as close as possible

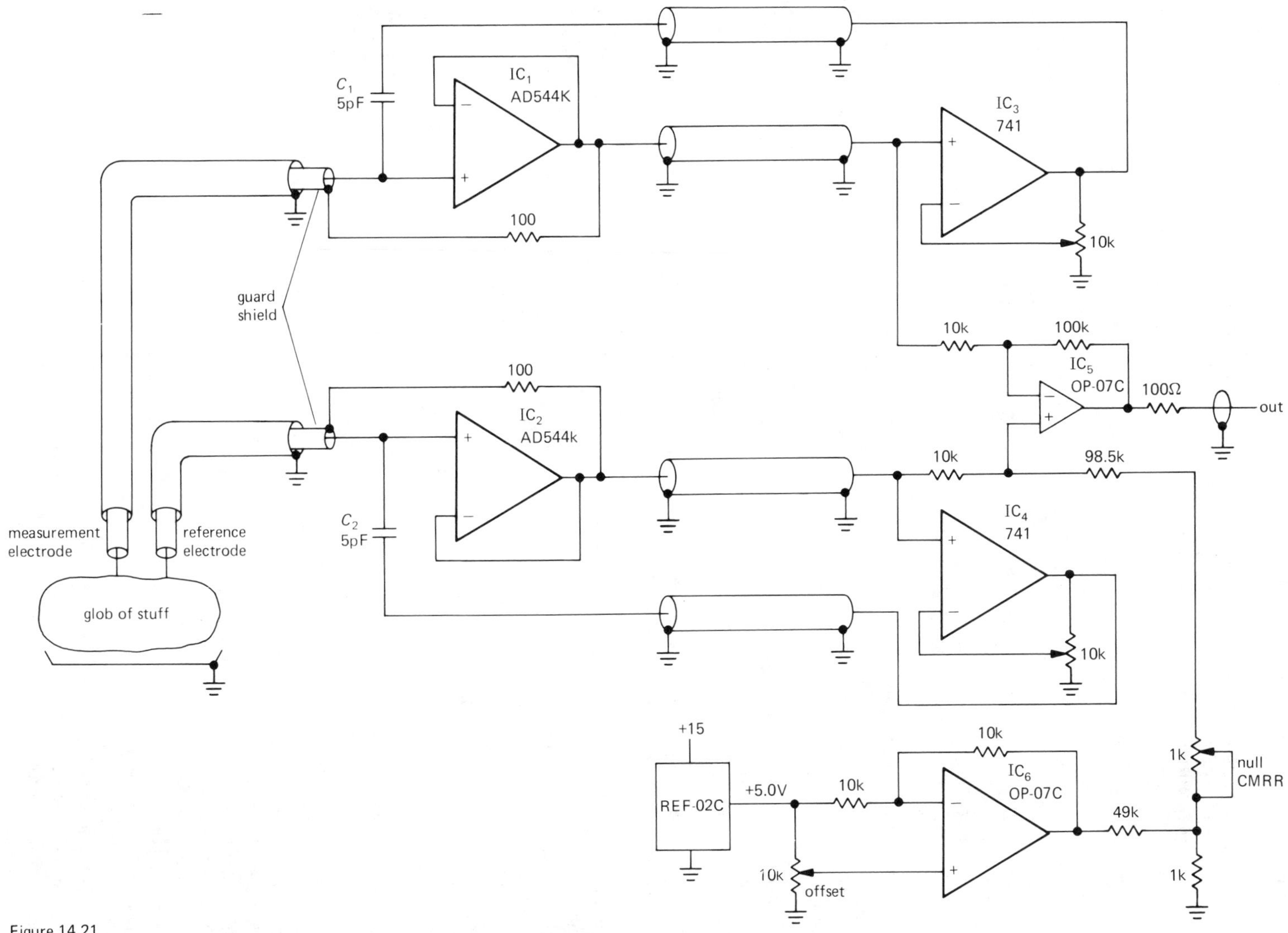

Figure 14.21
Compensating microelectrode amplifier with guarding and reference channel.

to the actual microelectrode by low-noise FET-input op-amps IC_1 and IC_2, which also bootstrap the guard electrode in order to reduce the effective cable capacitance. Note that the guard is itself shielded. You've got to use FET amplifiers in order to get high input impedance and low input current noise; the particular types shown were chosen for their low input noise voltage (2μV pp. max. 0.1–10Hz), often a problem with FET or MOSFET amplifiers. The pair of buffered signals is applied to the standard differencing amplifier configuration with a low-noise-voltage low-drift op-amp, with 100mV of stable adjustable output offset added via IC_6.

At this point you have an amplifier with a differential gain of 10, suitable noise performance, good common-mode rejection, and low input current (25pA). However, even with input guarding the residual input capacitance at the input buffers and at the microelectrode tip will result in poor speed performance. For example, a 100MΩ source impedance driving 20pF has a high-frequency 3dB point of only 80Hz. The solution is active compensation via positive feedback, provided by IC_3 and IC_4 through C_1 and C_2. In practice, you adjust the voltage gain of amplifiers IC_3 and IC_4 for decent high-frequency performance (or transient response), with response to several kilohertz possible.

Ion-specific electrodes

The classic example of an ion-specific electrode is the pH meter, which measures the voltage developed between a reference electrode and a thin-walled glass electrode through which hydrogen ions can diffuse. Once again you're dealing with very high source impedances, although the problems here are less severe than with microelectrodes, because you don't often care about frequency response.

There are more than 20 kinds of ion-specific electrode systems available, e.g., to measure activities of K^+, Na^+, NH_4^+, CN^-, Hg^{++}, SCN^-, Br^-, Cl^-, F^-, I^-, Ca^{++}, or Cu^{++}. In general, you have two electrodes: a reference electrode, typically silver-coated with silver chloride and immersed in a concentrated solution of potassium chloride that communicates with the solution you want to measure via a porous plug or gel, and an ion-specific electrode, typically consisting of an electrode immersed in a concentrated solution of the ion you're interested in and separated from the solution under test by a membrane that is selectively permeable to the ion of interest. The membrane is commonly an ion-selective glass or an organic liquid containing mobile ion-transporting organic molecules. Your task is to measure a voltage that is in the range of 0 to 2 volts, with an accuracy of a millivolt, while drawing less than 100pA. The situation is complicated by a temperature coefficient of as much as a few percent change in voltage per degree centigrade, which you can attempt to cancel automatically with thermistor-driven compensation circuitry. Conversion from measured ionic activity to concentration requires attention to the total ionic strength of the sample and to the crossover sensitivity of the ion-specific electrode to other ions present. In any case, chemists say you get best results with this sort of black art if you calibrate on some standard solutions just before and after making your measurements. With care, you can see concentrations of 0.1ppm and achieve measurement accuracies of about 1% in solutions of moderate concentration.

Electrochemical measurements

In the area of electrochemistry, it is possible to make very sensitive analytical measurements of the concentration of specific ions by measuring electrode currents (reaction rates) versus applied voltage in a solution. By scanning the applied voltage, you pass through the potentials at which specific reactions occur, giving rise to steps or peaks. Terms such as cyclic voltametry, polarography, and anodic stripping voltametry (ASV) are used to describe various ways of doing such analytical measurements. Among the most sensitive of these techniques is ASV, which uses a hanging-drop mercury (hdm) electrode, a renewable electrode onto which you electroplate at a relatively high potential for a while, then reverse the current and strip off each element sequentially. This technique can detect elements like lead and cadmium at the parts-per-billion level, and it should be considered on a par with other trace-element techniques such as neutron activation, flame spectroscopy, and x-ray and ion microprobes.

The technique of measuring a small current while subjecting a system to a fixed voltage is called a "voltage clamp," and it finds application also in nerve and cell physiology. Nerve membranes have voltage-dependent channels through which specific ions can diffuse, and nerve physiologists like to measure the voltages at which such channels open. Again, voltage clamps are used, this time with microelectrodes.

In preparative electrochemistry, the same techniques are used, but with currents measured in amperes rather than microamperes. Once again the idea is to drive a specific reaction and produce palpable quantities of reaction product by applying the right voltage.

Figure 14.22 shows a simple potentiostat (or voltage clamp) circuit. The electrolytic cell consists of an electrode to inject current (the counter electrode), a common return electrode (the working electrode), and a small probe to measure the voltage in the solution near the working electrode (the reference electrode). IC_1 maintains a voltage equal to V_{ref} between the reference and working electrodes by varying the current into the counter electrode appropriately (in measurements of membrane potential the upper two electrodes would be inside the cell and the working electrode outside). IC_2 holds the working electrode at virtual ground, converting the current to an output voltage. The range of voltages encountered is typically ±1 volt; currents in the range of 1nA to 1mA are typical of analytical measurements, whereas currents of 1mA to 10 amps are used in preparative electrochemistry.

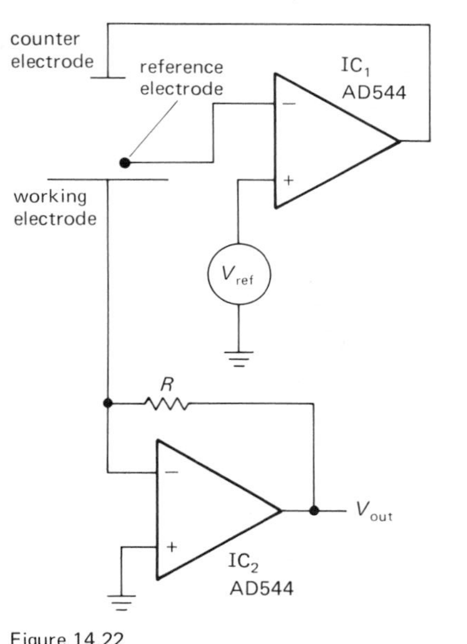

Figure 14.22
Potentiostatic electrochemistry circuit (voltage clamp).

In order to do scanning measurements, V_{ref} would be replaced with a ramp generator. For low-current membrane measurements you would have to shield the input leads carefully, and perhaps use guarding and positive feedback with a capacitor in the manner of Figure 14.21 to maintain some frequency response.

PRECISION STANDARDS AND PRECISION MEASUREMENTS

In Chapter 7 we talked about the circuit methods that are necessary in high-accuracy instruments to maintain small voltage offsets and drifts, e.g., when amplifying very small voltages. We dealt there only with analog electronics, the amplification of continuously varying voltages and currents. For a number of reasons, it turns out that measurements of digital quantities such as frequency, period, and time intervals can be made to far greater precision than any analog measurement. In the following sections, we will explore the accuracy of electronic standards (time, voltage, resistance), and you will see how to make measurements of high precision using such standards as references. We will devote the bulk of the discussion to time/frequency measurements, because of their greater inherent precision and because we have already treated precision analog circuitry in some detail in Chapter 7.

□ 14.09 Frequency standards

Let's take a look at the high-stability frequency standards you can get, and then discuss how you set and maintain their frequency.

☐ *Quartz-crystal oscillators*

Back in Sections 4.11–4.16 we described briefly the stabilities you can expect from frequency standards, going from the simplest *RC* relaxation oscillator to the atomic standards based on rubidium and cesium. For any serious timing you wouldn't consider anything less stable than a quartz-crystal oscillator. Fortunately, the cheapest crystal oscillators cost only a few dollars and can deliver stabilities of a few parts per million. For about $50 you can buy a good TCXO (temperature-compensated crystal oscillator), stable to one part per million from 0°C to 50°C. For better performance you need ovenized crystals, with price tags from a couple of hundred dollars to more than $1000. Once you begin talking about stabilities of a few parts per billion, you have to worry also about "aging," the tendency of crystal oscillators to drift in frequency at a more or less constant rate once they are initially "broken in." The 10544 series from Hewlett-Packard typifies good crystal oscillator modules (no power supplies), with stabilities of better than 1 part in 100 million over the full temperature range and aging rates of less than 0.5 part per billion per day.

Uncompensated crystal oscillators, and even TCXOs, are logical choices as part of a small instrument. The fancier ovenized oscillators are usually rack-mounted standards with an identity all their own.

☐ *Atomic standards*

There are three standards in use today: rubidium, cesium, and hydrogen. Rubidium has a microwave absorption at 6,834,682,608Hz, cesium has an absorption at 9,192,631,770Hz, and hydrogen has an absorption at 1,420,405,751.768Hz. A frequency standard based on one of these is considerably more complicated (and expensive) than a good crystal oscillator.

☐ *Rubidium.* In the rubidium standard you have a glass bulb with rubidium vapor, heated and contained in a microwave cavity with glass end windows. A rubidium lamp shines through the cavity, with a photocell detecting the transmitted light. Meanwhile, a modulated microwave signal referenced to a stable crystal oscillator is introduced into the cavity. By using lock-in detection (see Section 14.15) of the transmitted light, you can bring the microwave signal exactly to the rubidium resonance frequency, since the optical absorption of the rubidium gas is altered when its microwave resonance is excited. The crystal's frequency is then related in a known way to the rubidium resonance, so it is straightforward to generate a standard frequency like 10MHz. (There are actually several additional complications that we have glossed over.)

Rubidium standards have better stability than ovenized crystal oscillators, although they do exhibit a form of aging. Commercial units are available with stabilities of a few parts in 10^{11} over the full temperature range and long-term stabilities of 1 part in 10^{11} per month. Rubidium standards make sense in a laboratory situation, and you find them at observatories and other places where extremely accurate observations are made. It should be pointed out that a rubidium standard, just like a crystal oscillator, must be calibrated, because changing conditions within the resonance cell affect the frequency at the part-per-billion level.

☐ *Cesium.* A cesium standard is practically a small atomic-beam laboratory, in which cesium atoms are launched from an oven into a vacuum chamber, where they pass through spin-state selector magnets and oscillatory electric fields before being detected with a hot-wire ionization detector. As with the rubidium standard, a microwave signal referenced to a stable crystal oscillator is locked to the resonance with feedback from a phase-sensitive detector, and the output frequency is synthesized from the crystal.

Cesium standards aren't small, and they aren't cheap. But they are *primary* standards; you don't have to calibrate them. In fact, by international agreement, cesium *defines* the second: "the duration of exactly 9192631770 periods of the radiation corresponding to the transition between the two hyperfine levels of the ground state of the cesium-133 atom." Cesium clocks are used to keep official time in this country and to calibrate time transmissions (more on this shortly). The cesium clocks used to keep time are elaborate devices, but even commercially built cesium standards keep exceptional time: long-term stability and reproducibility of 3 parts in 10^{12} for the model 5061 from Hewlett-Packard (priced at $24,000).

□ *Hydrogen.* Neutral hydrogen atoms have a hyperfine resonance at about 1420MHz, and in contrast to the situation with the other atomic standards, it is possible to make an actual oscillator with them. As with cesium, you make an atomic beam and run it through magnetic state selectors, then into a Teflon-coated quartz bulb in a microwave cavity. The atoms bounce around inside this "storage bulb" for about 1 second and give off enough radiofrequency energy to sustain an oscillation in the cavity. That makes it easy to lock a crystal oscillator, using PLLs and mixers. You call this object a hydrogen *maser* (microwave amplification by stimulated emission of radiation).

Hydrogen masers are extraordinarily stable over short times (up to a few hours), with stabilities of 1 part in 10^{15}. They have not replaced cesium-beam apparatus for primary timekeeping, however, because the problem of determining the frequency-pulling effect of the cavity has not been solved, and because of long-term drifts caused by the changing properties of the storage bulb wall surface.

□ *Methane laser.* A fourth atomic standard is used at infrared wavelengths, namely the methane-stabilized helium-neon laser. It has a frequency stability comparable to that of the other atomic standards, but at its frequency of 8.85×10^{13}Hz (3.39μm wavelength) it is not a usable radiofrequency standard.

□ *Calibrating a clock*

Unless you happen to own a cesium-beam standard, you've got to have access to a stable calibration signal to keep your oscillator on frequency. In addition, you may wish to keep accurate absolute time as well as frequency, i.e., you have to set your clock, after you have it running at the right rate. There are several services to help you keep time. On the East Coast of the United States, and in several other areas, you can receive Loran-C, a navigational signal at 100kHz, from which you can determine frequency and time. Loran-C is generated by cesium clocks and is compared with the cesium-beam master clock at the Naval Observatory, which publishes corrections each month. Another time service is WWVB, from the National Bureau of Standards in Colorado. This is a 60kHz signal that you can receive most anywhere in the United States. For both these low-frequency transmissions you can synchronize to 1μs or better if you are within range of the "ground wave" signal (a few hundred miles), but ionospheric effects (day/night shifts, solar winds, etc.) make synchronization via the "sky wave" less accurate (10–50μs). A more recent network known as Omega transmits at very low frequencies (around 10kHz) and can be received anywhere.

If you can receive one of these time services, you can compare your oscillator frequency with the real thing. There are nice commercial gadgets that will take care of all the fuss and bother and even generate pretty graphs of the results. It is a bit more difficult to set your clock's time. The most reliable way is to carry it (or some portable clock) to one of the standards, set it, then carry it back. As soon as you get home you make observations of Loran-C, or whatever, to determine the time delay from the transmitter to you. Save that number! (We still remember the magic number 53,211μs for the delay from Loran-C in Nantucket to Harvard's 60-inch telescope dome.) As long as no one builds a new mountain between you and the transmitter, you're all set to tell time.

Some other methods of time and frequency synchronization being used or talked about include microwave repeaters, commercial television signals, satellites, and pulsar time-of-arrival observations.

14.10 Frequency, period, and time-interval measurements

With an accurate reference oscillator and just a small amount of digital electronics, it is disarmingly easy to make frequency and period measurements of high precision.

Frequency

Figure 14.23 shows the basic circuit of a frequency counter. A Schmitt trigger converts the analog input signal to logic levels, at which point it is gated by an accurate 1 second pulse derived from a crystal oscillator. The frequency in hertz is the number of pulses counted by the multidigit BCD counter. It is best to latch the count and reset the counter between counting intervals.

In practice, you would arrange the clock circuit so that shorter or longer intervals can be selected, with a choice of 0.1 second, 1 second, and 10 seconds as a minimum. Also, you can eliminate the 1 second interval between measurements. Additional features might include the following: an adjustable preamp, with selectable trigger point and hysteresis, and perhaps a front-panel output from the discriminator so you can see the trigger point on an oscilloscope; BCD output for readout into a computer or logger; provision for an external oscillator, when a precision standard is available; a manual start/stop input for simple counting (totalization).

□ *Microwave counting.* You can go to frequencies of 1GHz with the digital ICs available today. In particular, Plessey Semiconductor manufactures a series of astoundingly fast counters, going to 1.3GHz. For higher frequencies you can use heterodyne techniques to mix the microwave input signal down to a directly countable frequency, or you can use a so-called transfer oscillator technique, in which you phase-lock the nth harmonic of a VCO to the input signal, then measure the VCO frequency and multiply the result by n.

□ *The ±1 count ambiguity.* One disadvantage of this simple frequency counting scheme is that low frequencies cannot be measured to high precision, because of the ±1 count error. For example, if you were to measure a signal near 10Hz with a gate time of 1 second, your answer would be accurate only to 10%, since the result would be either 9, 10, or 11. You could measure for a longer interval, but it would take a whole day's counting to get the relative accuracy (1 part per million) that you would get in 1 second when measuring a 1MHz signal, for example. There are several solutions to this problem: period (or reciprocal) counting, interpolation methods, and phase-locked-loop frequency-multiplication techniques. We will deal with the first two in the next sections, since they aren't really direct frequency measurements.

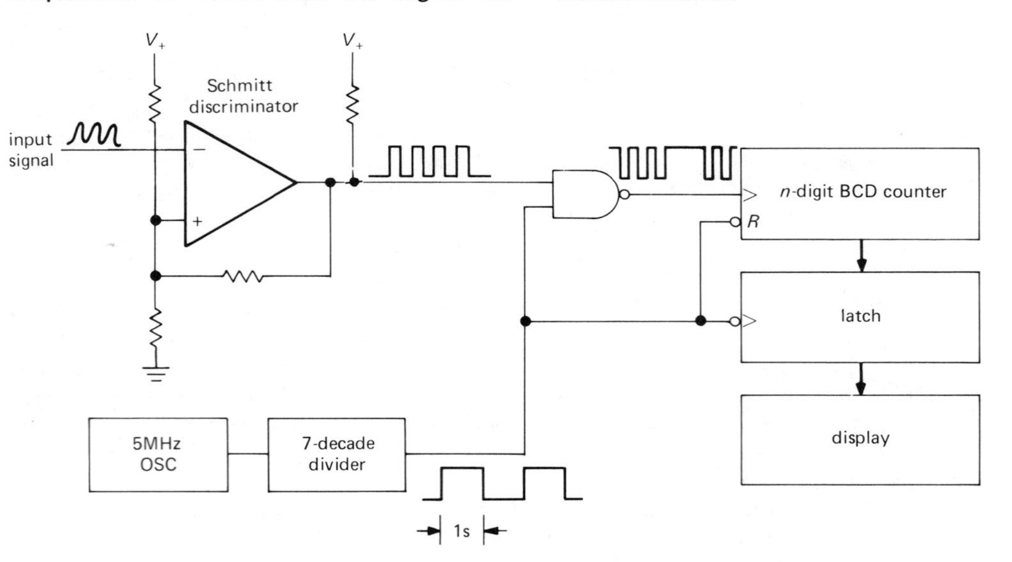

Figure 14.23
Frequency counter.

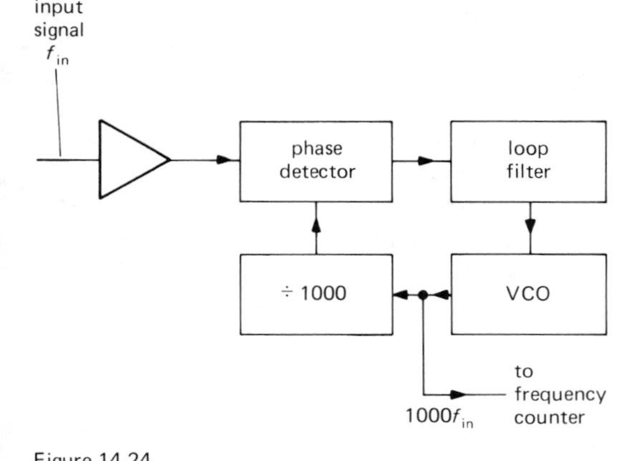

Figure 14.24
PLL resolution multiplication for low-frequency counting.

Figure 14.24 shows the PLL "resolution multiplication" technique. A standard phase-locked loop is used to synthesize a frequency of 1000 times the input signal, say, which is then counted as described earlier. The accuracy of this technique is limited by the phase jitter in the PLL phase detector and the loop compensation parameters. For example, if a 100Hz signal is multiplied by 1000 and counted for 1 second, and the jitter in the phase detector is 1% of a cycle (3.6°), or 100μs, then the *accuracy* of the measurement will be 1 part in 10,000, even though the *resolution* is 1 part in 100,000.

We will now mention two other ways to improve frequency-measurement accuracy: period measurement and interpolated time-interval measurement.

Period (reciprocal counting)

A good way to handle the problem of resolution when measuring low frequencies is to turn things around and use the input signal (or some subdivision of it) to gate the clock. Figure 14.25 shows the standard configuration for such a period counter. The number of periods measured is normally switch-selectable to some power of 10 (1, 10, 100, etc.). You will usually pick a number of periods such that the measurement takes a convenient length of time, typically a second, giving an answer to about seven significant figures. Of course, that answer is in units of time, not frequency, so you have to take the reciprocal to recover the frequency. Luckily, you soon won't even have to know how to divide, since modern counters use dedicated microprocessors to do the period-to-frequency conversion.

Note that the accuracy of period measurement is critically dependent on stable triggering and requires good signal/noise ratios. Figure 14.26 indicates the problem here.

The main advantage of reciprocal counting is that you get a constant resolution $\Delta f/f$ for a given length of measurement, independent of the input frequency. The graph in Figure 14.27 compares the resolution of frequency and reciprocal frequency (period) measurements of duration 1 second, using a 10MHz clock. The period graph should actually be somewhat jagged, since you normally have to live with the closest power of 10 for the number of periods averaged.

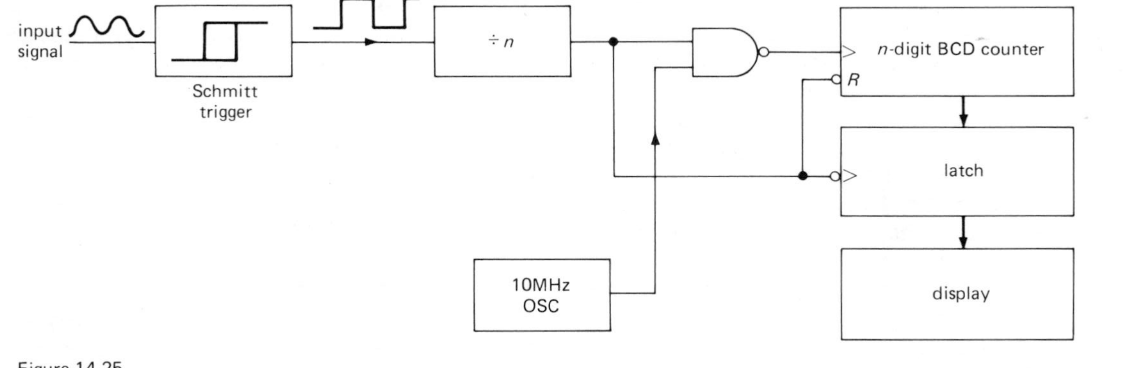

Figure 14.25
Period counter.

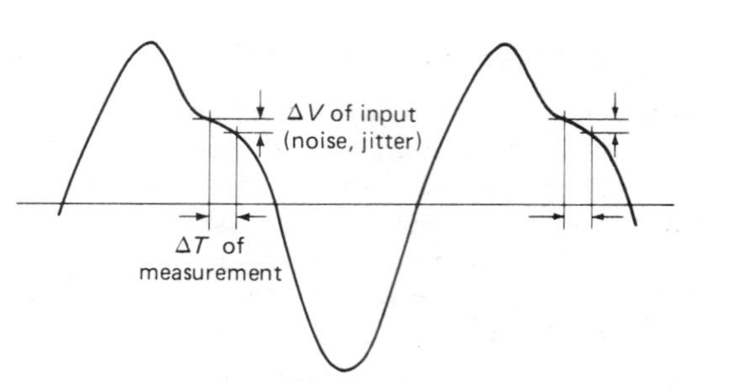

Figure 14.26

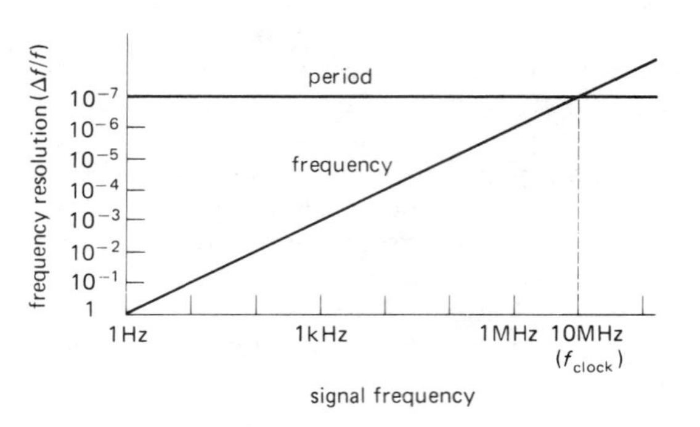

Figure 14.27
Fractional resolution of frequency and period counting.

Even this restriction is evaporating, with the advent of "smart" microprocessor counters (e.g., the Hewlett-Packard 5315) that have *continuous* adjustment of gate time; they know how many periods were averaged, and they divide the answer accordingly. They also oblige you by switching from period to frequency mode for input frequencies greater than the clock frequency, in order to get optimum resolution at any input frequency.

A second advantage of reciprocal frequency measurement is the ability to control externally the time at which the gating occurs. This is advantageous if you wish to measure the frequency of a tone burst, for example, a situation in which a simple frequency counter would give incorrect results, since its internally controlled gating interval might not coincide with the burst. With period counting you can gate the measurement externally and can even make a set of measurements at various points along the burst, given the generally superior resolution of period measurement.

You might wonder if it is possible to do better than the "uncertainty-principle" resolution limit of $\Delta f/f \approx 1/f_{clock}T$ (period measurement) or $1/f_{input}T$ (frequency counting), for the relative error $\Delta f/f$ of a frequency measurement made by counting for time T. The answer is yes. In fact, several clever schemes have been invented. We will discuss them in the next subsection (time-interval measurement), but just to show that it can be done, we've drawn in Figure 14.28

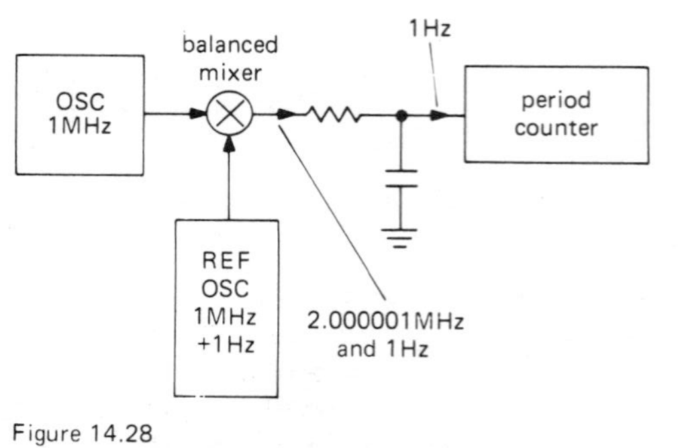

Figure 14.28
High-resolution frequency comparison.

a method of measuring the frequency of a 1MHz oscillator to a resolution of 1 part in 10^{12} in 1 second of measurement time. The unknown oscillator is mixed with a stable reference offset slightly from 1.0MHz, say 1.000001MHz (this could be synthesized with a PLL). The mixer output contains the sum and difference frequencies. After low-pass filtering, you've got a 1Hz signal that is the difference between the two oscillators, and that can be easily measured with a period counter to one part per million in 1 second. In other words, you've measured 1MHz to 1μHz in 1 second.

This technique assumes that you have extremely good signal/noise ratios, and in practice you would have to worry about low-frequency noise, settling time of the filter, etc., so you might not do better than 1 part in 10^{10} in 1 second. Still, this is considerably better than simple frequency (or period) counting. In addition, the *accuracy* will be less than the resolution unless the reference oscillator is also accurate to 1 part in 10^{12} (possible, but not easy, with today's technology). You can think of this scheme as a relative way of comparing the frequencies of two oscillators, if you prefer.

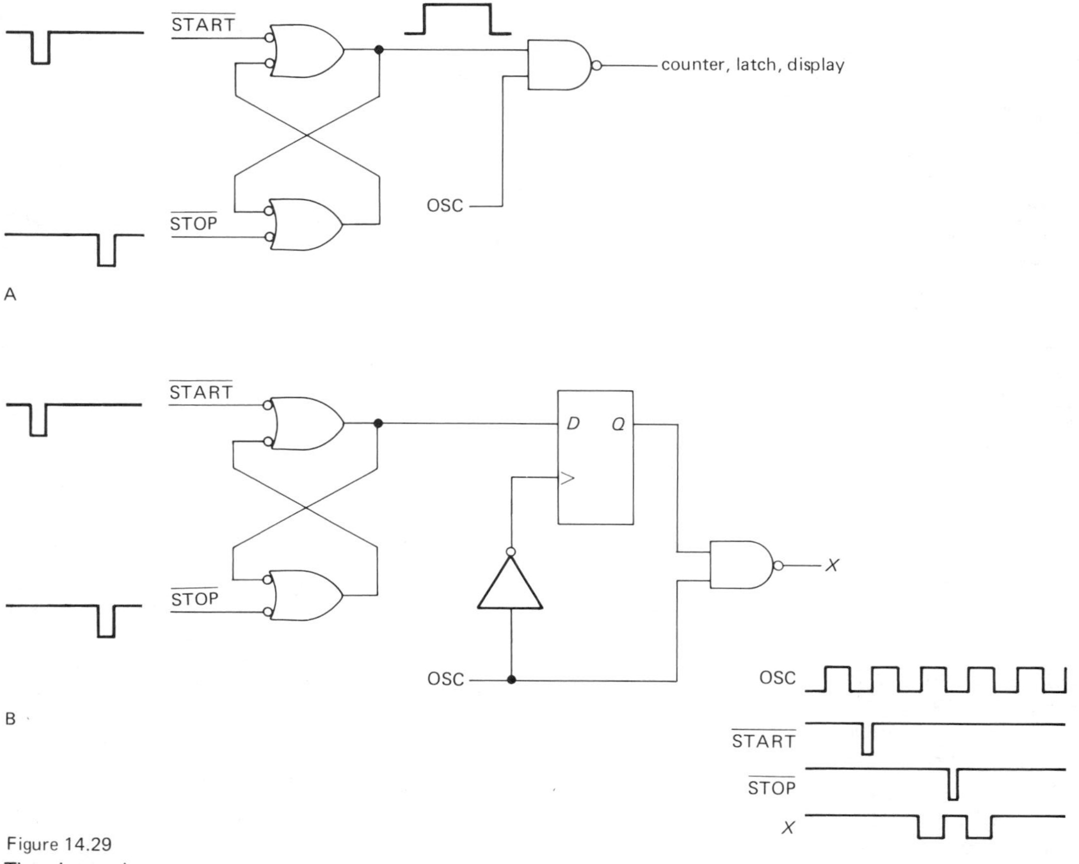

Figure 14.29
Time-interval measurement.

Time-interval measurement

With a trivial change in the circuitry of the period counter, you can measure the time interval between two events. Figure 14.29 shows how. In practice it may be better to add a synchronizer, as shown in the second circuit, to prevent the generation of runt pulses. The best resolution is obviously obtained by running the oscillator at the highest possible frequency, and commercial counters use local oscillator references as high as 500MHz, phase-locked from a stable crystal at 5MHz or 10MHz. With a 500MHz reference, you have a resolution of 2ns.

As we hinted earlier, there are ways to beat the reciprocal frequency resolution limit when making time-interval measurements, essentially by exploiting the extra information you have about the position of zero crossings of the input signal relative to the reference. The oscillator comparison scheme we showed earlier really exploited that same information, but in a more subtle way. For these schemes you must have a clean signal with very low noise level. There are two interpolation methods in use in commercial instruments: linear interpolation and vernier interpolation.

□ *Linear interpolation.* Suppose you wish to measure the time interval between the start and stop pulses in Figure 14.30. You begin by measuring the number of clock pulses, *n*, during the interval τ, as shown (with a synchronizer you would start and stop with the first clock pulse after the respective input signal, as shown). To improve the resolution of the measurement, all you need to know are the time intervals T_0 and T_1, the time

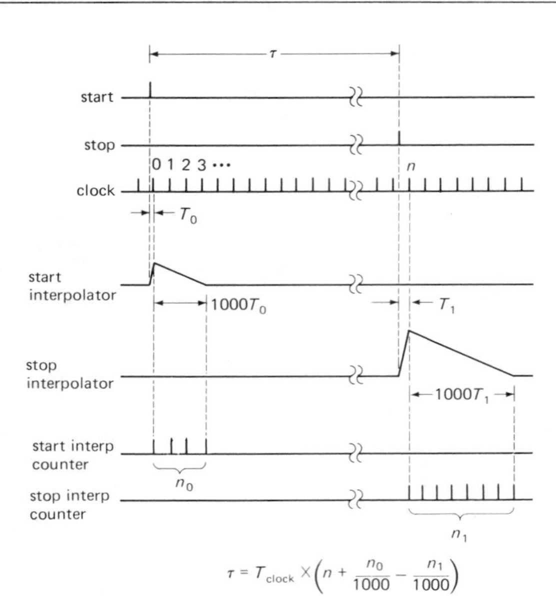

Figure 14.30
Linear interpolation (time interval measurement).

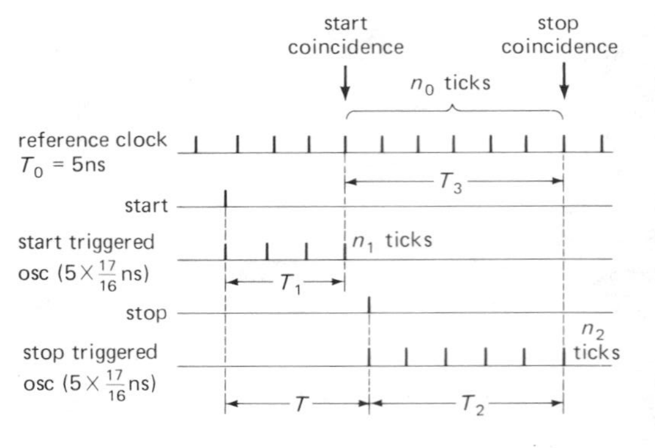

Figure 14.31
Vernier interpolation (time interval measurement).

elapsed from the occurrence of each input pulse to the next clock pulse. Assuming you're already running the system clock at the highest rate you can conveniently count, you have to expand those unknown intervals in order to measure them. A variation of the dual-slope principle works here: Integrate charge onto a capacitor during those intervals, then ramp down at a small fraction (say 1/1000th) of the charging rate, thus expanding the unknown time intervals by a factor of 1000. During those expanded intervals, you count the system clock, generating counts of n_0 and n_1. The unknown time interval is therefore given by

$$t = T_{clock} \times (n + n_0/1000 - n_1/1000)$$

with the obvious improvement in resolution. The ultimate accuracy of this method is limited by the accuracy of the interpolators and the system clock.

☐ *Vernier interpolation.* Vernier interpolation is a digital technique that lets you find out where in the clock cycle the input pulse occurred. Figure 14.31 shows the method schematically. There are three clocks involved: The master reference clock runs continuously, with a period T_0 of, say, 5ns; the input START pulse triggers a second oscillator with a period greater than the reference by a factor $1 + 1/n$ (we've set $n = 16$ for this example); the input STOP pulse triggers a third oscillator of the same period as the other triggered oscillator. Fast circuitry then looks for coincidences between the triggered oscillators and the master clock, while counting the number (n_1, n_2) of ticks of each before coincidence. The arithmetic is shown in the figure; the net result is to determine to within $1/n$th of a master clock pulse the duration between START and STOP.

The Hewlett-Packard 5370A uses this technique, with $T_0 = 5$ns and $n = 256$. The result is time-interval resolution of 20ps. The technique can clearly be used for period measurements, since period is simply the time interval for one cycle of an input wave. When used this way, the counter just mentioned determines frequency to 11-digit resolution in 1 second!

☐ *Time-interval averaging.* There is a third way to improve the resolution of a time-interval measurement, namely by repeating the measurement many times and taking the average. The ± 1 count ambiguity gets averaged that way, and the result converges to the true interval, provided only that the repetition rate of START pulses is not commensurate with the master clock. Some counters include a "jittered clock" to make sure this doesn't happen.

Spectrum analysis

A powerful technique that must be mentioned in connection with the measurement of frequency is *spectrum analysis,* looking at signals in the frequency domain. Spectrum analyzers can measure frequency (and in fact they are very useful when you need to know the frequency of a weak signal in the presence of other stronger signals), but in addition they can do a lot more. We will talk about them in Section 14.18.

□ 14.11 Voltage and resistance standards and measurements

As we hinted earlier, analog standards and measurements do not have anything like the precision we have just been talking about. Here you're lucky to get accuracy of a part per million. The analog standards are voltage and resistance; from them you can determine current, if need be.

The traditional standard of voltage is the Weston cell, an electrochemical device with reproducible output voltage, intended for use as a reference only (no more than $10\mu A$, and preferably no current at all, should be drawn from it). Its terminal voltage is 1.018636 volts at 20°C. Unfortunately, Weston cells are fussy gadgets. They have to be maintained at a precise temperature because of their large temperature coefficient ($40\mu V/°C$, far worse than good IC voltage references) and even larger sensitivity to temperature gradients (the individual "limbs" of the cell have tempcos of about $350\mu V/°C$). Standard cells are carefully maintained by the National Bureau of Standards (NBS) for comparison with secondary standards. Nowadays there are very stable solid-state references with controllable output voltage. They can be used to transfer a measurement from a finicky standard cell to an actual measurement situation. Typical specifications are 10 ppm stability for a month after calibration and 30 ppm stability in a year.

To make an accurate voltage measurement, you use precision voltage dividers (known as Kelvin-Varley dividers), available with linearities in the 0.1ppm range. The divider is used to generate a precise fraction of the unknown voltage, for comparison with the voltage standard. Accurate null detectors and instruments for compensation of wiring resistance are available for this task. Routine calibrations at accuracies of just a few parts per million are possible.

Recently a measurement based on a superconducting Josephson junction has replaced the standard cell as the definition of voltage. With care it is possible to measure voltages reproducibly to 1 part in 10^7. The method has the pleasant simplicity of requiring only a measurement of a frequency and knowledge of the physical constants h (Planck's constant) and e (the electron charge). Unfortunately, the technology is too complicated to be considered for standard laboratory use, involving cryogenic apparatus, liquid helium, and the like.

As with voltage, standards of resistance are carefully maintained by the NBS. By comparison in a Wheatstone bridge circuit, you can calibrate a secondary standard and maintain accuracies of a few parts per million.

We should point out some of the limitations that prevent analog measurements from having the same high accuracy as time measurements. Analog measurements rely on physical properties such as electrochemical potentials, breakdown voltages, and resistances, and these all vary with temperature and time. Interfering effects such as Johnson and $1/f$ noise, leakage currents, and thermoelectric potentials (thermocouple effect) complicate any measurement. To measure a voltage with precision comparable to state-of-the-art time or frequency measurements would require a measurement accuracy of a picovolt at a voltage of 1 volt. Think of this not as an indictment of analog methods, but merely as a celebration of the incredible precision attainable in the time/frequency domain. And, in practice, choose time/frequency transducers and measurements, rather than voltage/resistance measurements, whenever possible.

BANDWIDTH-NARROWING TECHNIQUES

14.12 The problem of signal-to-noise ratio

Up to this point we have been talking about the various experimental quantities that can be detected, how you might measure them, and what sort of tradeoffs you face. As luck would have it, the signals you often want to measure are buried in noise or interference, frequently to the extent that you can't even see them on an oscilloscope. Even when external noise isn't a problem, the statistics of the signal itself may make detection difficult, as, for example, when counting nuclear disintegrations from a weak source, with only a few counts detected per minute. Finally, even when the signal is detectable, you may wish to improve the detected signal strength in order to make a more accurate measurement. In all these cases some tricks are needed to improve the signal/noise ratio; as you will see, they all amount to a narrowing of the detection bandwidth in order to preserve the desired signal while reducing the total amount of (broadband) noise accepted.

The first thing you might be tempted to try when thinking of reducing the bandwidth of a measurement is to hang a simple low-pass filter on the output, in order to average out the noise. There are cases where that therapy will work, but most of the time it will do very little good, for a couple of reasons. First, the signal itself may have some high frequencies in it, or it may be centered at some high frequency. Second, even if the signal is in fact slowly varying or static, you invariably have to contend with the reality that the density of noise signal usually has a $1/f$ character, so as you squeeze the bandwidth down toward dc you gain very little. Electronic and physical systems are twitchy, so to speak.

In practice, there are a few basic techniques of bandwidth narrowing that are in widespread use. They go under names like signal averaging, transient averaging, boxcar integration, multichannel scaling, pulse-height analysis, lock-in detection, and phase-sensitive detection. All these methods assume that you have a repetitive signal; that's no real problem, since there is almost always a way to force the signal to be periodic, assuming it isn't already. Let's see what is going on.

14.13 Signal averaging and multichannel averaging

By forming a cumulative sum of a repetitive signal versus time, you can improve the signal/noise ratio enormously. This usually goes under the heading of "signal averaging," and it is often applied to analog signals. We will consider first what may seem to be an artificial situation, namely a signal consisting of pulses whose rate is proportional to the amplitude of some sought-after waveform versus time. We begin with this example because it makes our calculations easier. In reality, it isn't even an artificial situation, since it is the rule when using pulse-counting electronics such as particle detectors or photomultipliers at low light levels.

Multichannel scalers

We begin with multichannel scaling because it typifies all these techniques and, in addition, is easy to understand and quantify. The multichannel scaler (MCS) is a piece of hardware that contains a set of memory registers (typically 1024 or more), each of which can store a number up to 1 million (20 bits binary or 24 bits BCD) or so. The MCS accepts pulses (or continuous voltages, as will be described later) as its input; in addition, it accepts either a channel-advance signal (a pulse) or a parallel multibit channel address. Each time there is an input pulse, the MCS increments the count in the memory channel currently being addressed. Additional inputs let you reset the address to 0, clear the memory, etc.

To use an MCS you need a signal that repeats itself at some interval. Let's suppose for the time being that the phenomenon you're observing is itself periodic, with period T; although this is not the case most of the time (you usually have to coax the experiment into periodicity), there are good examples in the real world of strictly periodic phenomena, e.g., the light output of a pulsar. Let's suppose that the input consists of pulses, with rate proportional to the signal plus a large background rate of noise pulses, i.e., pulses randomly distributed in time (again, realistic for pulsars, where the actual signal is swamped by light from the night sky). By sending timing pulses to the channel advance and reset inputs, we arrange to sweep the MCS repetitively through its 1024 channels once every T seconds, accumulating additional input (signal plus background) counts into the memory channels each sweep. As time goes on, the signal will keep adding counts to the same subgroup of channels, with the background noise adding counts in all channels, because the sweep through the entire set of channels is timed to coincide with the signal's periodicity. Thus the signal keeps adding on top of itself, the accumulated sum getting larger after each repetition.

Signal-to-noise computation

Let's see what happens. To be specific, let the background pulse rate have an average value that contributes n_b pulses per channel each sweep, with the signal contributing an additional n_s pulses into the channel where its peak lies (Fig. 14.32). Let's give ourselves a poor signal/background ratio, i.e., $n_s \ll n_b$, meaning that most of the counts added during each sweep through the memory are contributed by background, rather than signal. Now, when the memory contents are graphed, the signal should be recognizable as a bump above the background. You might think the criterion is that the number of signal counts in a channel with signal should be comparable with the number of counts contributed to that channel by the background noise. That would be wrong, since the *average* value contributed by noise is quite irrelevant; all that matters is the level of *fluctuations* of that average value about the mean.

Thus, a poor input signal/noise ratio is actually characterized by $n_s \ll n_b^{1/2}$, meaning that in one sweep the signal will not be recognizable above the "noise" consisting of an undulating graph of accumulated random background pulses. For purposes of computation, let's let $n_s = 10$ and $n_b = 1000$. Therefore in one sweep an initially

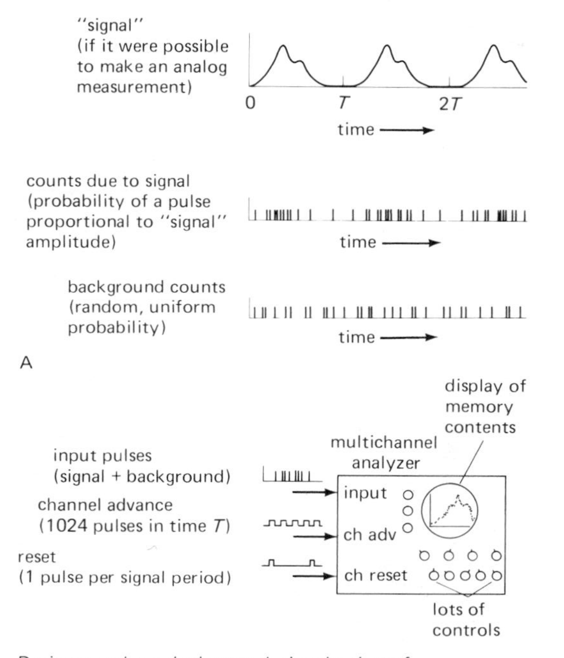

Figure 14.32
Multichannel signal averaging (pulse input).

cleared MCS will acquire an average of 1000 counts in each channel, with an additional 10 counts in the channels where the signal peaked. Since the fluctuations in the channel totals equal about 31 (square root of 1000), the actual signal bump is left pretty much buried in the noise after only one sweep. But after 1000 sweeps, say, the average count in any channel is about 1,000,000, with fluctuations of 1000. The channels where the signal peaks have an additional 10,000 counts (1000 sweeps × 10 counts/sweep), for a signal/noise ratio of 10. In other words, the signal has emerged from the background.

Example: Mössbauer resonance

Figure 14.33 shows the results of just such an analysis, in this case a Mössbauer resonance signal consisting of six dips in the

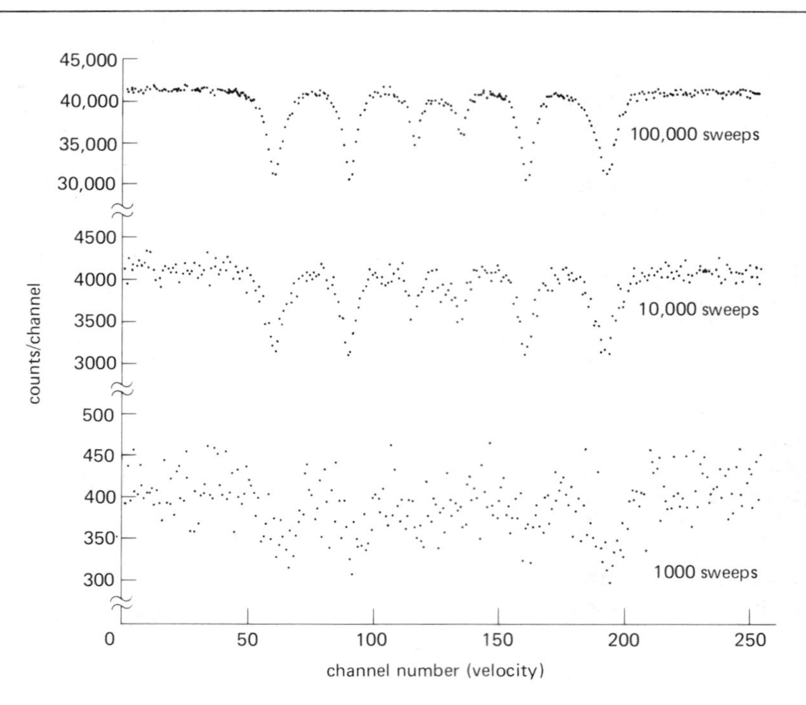

Figure 14.33
Mössbauer absorption spectrum, showing effect of signal averaging.

transmission of an enriched iron-57 foil to gamma radiation from a cobalt-57 radioactive source. In this case $n_b = 0.4$ and $n_s = 0.1$, approximately, for a situation of poor signal/noise ratio. The Mössbauer signal is totally swamped by noise even after 10 or 100 sweeps; it becomes visible only after 1000 sweeps or so. The results are shown after 1000, 10,000, and 100,000 sweeps, with each graph scaled to keep the signal size the same. Note the rise of the "baseline" caused by the steady background, as well as the nice enhancement of SNR with time.

It is easy to see by what factor the ratio of signal amplitude to background fluctuation ("noise") increases as time goes on. The signal amplitude increases proportional to t; the average background count ("baseline") also increases proportional to t, but the *fluctuations* in the background count ("noise") rise only proportional to the square root of t. Therefore, the ratio between signal and fluctuations in background increases as t divided by the square root of t. In other words, the signal-to-noise ratio improves in proportion to the square root of time.

Multichannel analysis of analog signals (signal averaging)

You can play the same game with analog signals by simply using a voltage-to-frequency converter at the input. Commercial MCSs often provide the electronics for you, giving you a choice of analog or pulse input modes. In this form you often hear these gadgets called signal averagers or transient averagers. One company (TMC) called theirs a "CAT" (computer of averaged transients), and the name has stuck, in some circles at least.

It is possible to make a completely analog MCS by using a set of integrators to store the accumulated signal. A simpler device, known as a boxcar integrator, is an analog signal averager with a single "sliding channel." With the enormous reductions in digital memory prices that have taken place in the last decade, such analog signal averagers are becoming impractical, except perhaps for specialized applications.

Multichannel analysis as bandwidth narrowing

We suggested at the beginning of this discussion that there was an equivalence between these magical SNR-reduction methods and a reduction in effective measurement bandwidth. It is not hard to see how that goes in this case. Imagine another (interfering) signal added into the input, but with periodicity T' slightly different from the desired signal of period T. After just a few sweeps, its signal will also begin to accumulate, causing trouble. But wait – as time goes on, its "bump" will gradually drift along through the channels, successively contributing counts through all the channels. It will have drifted all the way around through all the channels once after a time

$$t = 1/\Delta f$$

where Δf is the frequency difference $1/T - 1/T'$ between the desired signal and the interfering signal.

EXERCISE 14.1

Derive this result.

In other words, by accumulating data for a time t (as given in the preceding equation), the interfering signal has been spread equally through all the channels. Another way to say the same thing is that the measurement's bandwidth is reduced roughly to

$$\Delta f = 1/t$$

after accumulating data for time t. By running for a long time, you reduce the bandwidth and exclude nearby interfering signals! In fact, you also exclude most of the noise, since it is spread evenly in frequency. Viewed in this light, the effect of multichannel analysis is to narrow the accepted bandwidth, thereby accepting the signal power but squeezing down the amount of noise power.

Let's see how the calculation goes. After time t, the bandwidth is narrowed to $\Delta f = 1/t$. If the noise power density is p_n watts per hertz, and the signal power P_s stays within the measurement bandwidth, then the SNR after time t is

$$\text{SNR} = 10 \log (P_s t/p_n)$$

The signal amplitude improves proportional to the square root of t (3dB for each doubling of t), just as we found in the analysis we did earlier by considering the number of counts per channel and its fluctuations.

14.14 Making a signal periodic

We mentioned initially that all signal-averaging schemes require a signal that repeats many times in order to realize significant reduction in signal/noise ratio. Since most measurements don't involve intrinsically periodic quantities, it is usually necessary to force the signal to repeat. There are many ways to do this, depending on the particular measurement. It is probably easiest to give a few examples, rather than attempt to set down rules.

A measurable quantity that depends on some external parameter can easily be made periodic – just vary the external parameter. In NMR (nuclear magnetic resonance) the resonance frequency varies linearly with the applied field, so it is standard to modulate the current in a small additional magnet winding. In Mössbauer studies you vary the source velocity. In quadrupole resonance you can sweep the oscillator.

In other cases an effect may have its own well-defined transient, but allow external triggering. A classic example is the pulse of depolarization in a nerve fiber. In order to generate a clean graph of the waveform of such a pulse, you can simply trigger the nerve with an externally applied voltage pulse, starting the MCS sweep at the same time (or even "anticipating" the trigger by starting the sweep, then triggering the nerve with a delayed pulse); in this case you would pick a repetition period long enough so that the nerve has fully recovered before the next pulse. This last case illustrates graphically the importance of a repeatable phenomenon as fodder for signal averaging; if the frog whose leg is twitching chances to expire, your experiment is over, whatever the signal/noise ratio!

It should be pointed out that cases where the phenomenon you're measuring has its

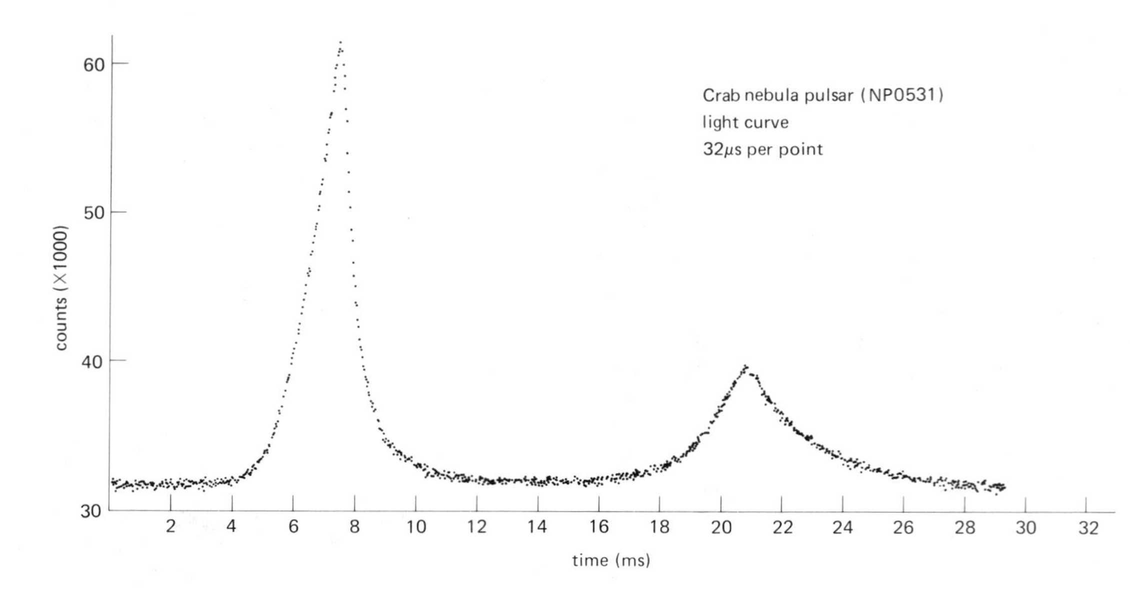

Figure 14.34
Crab nebula pulsar brightness versus time (light curve).

own well-defined periodicity may in fact be the most difficult to work with, since you have to know the periodicity precisely. The graph of the "light curve" (brightness versus time) in Figure 14.34 is an example. We made this curve by using an MCS on the output of a photomultiplier stationed at the focus of a 60-inch telescope, run exactly in synchronism with the pulsar's rotation. Even with that size telescope it required an average of approximately 5 million sweeps to generate such a clean curve, since the average number of detected photons for each entire pulsar pulse was about 1. With such a short period, that puts enormous accuracy requirements on the MCS channel-advance circuitry, in this case requiring clocks of part-per-billion stability and frequent adjustment of the clock rate to compensate for the earth's motion.

It is worth saying again that the essence of signal averaging is a reduction in bandwidth, gained by running an experiment for a long period of time. The bottom line here is the total length of the experiment; the particular rate of scanning, or modulation, is usually not important, as long as it takes you far enough from the $1/f$ noise present near dc. You can think of the modulation as simply shifting the signal you wish to measure from dc up to the modulating frequency. The effect of the long data accumulation is then to center an effective bandwidth $\Delta f = 1/T$ at f_{mod}, rather than at dc.

14.15 Lock-in detection

This is a method of considerable subtlety. In order to understand the method, it is necessary to take a short detour into the phase detector, a subject we first took up in Section 9.29.

Phase detectors

In Section 9.29 we described phase detectors that produce an output voltage proportional to the phase difference between two digital (logic-level) signals. For purposes of lock-in detection, you need to know about linear phase detectors, since you are nearly always dealing with analog voltage levels.

The basic circuit is shown in Figure 14.35. An analog signal passes through a linear amplifier whose gain is reversed by a square-wave "reference" signal controlling a FET switch. The output signal passes through a low-pass filter, RC. That's all there is to it. Let's see what you can do with it.

Phase-detector output. To analyze the

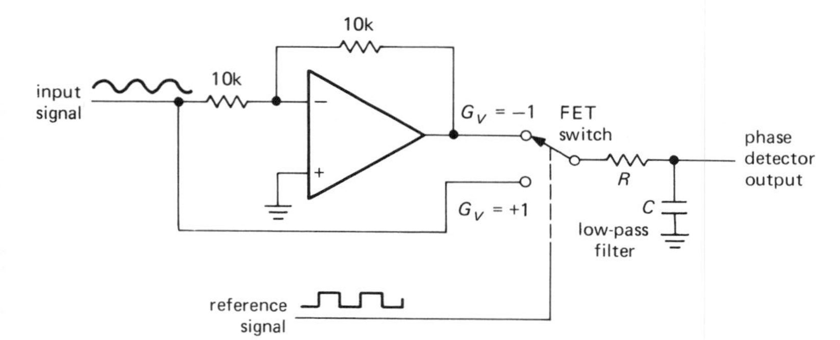

Figure 14.35
Phase detector for linear input signals.

phase detector operation, let's assume we apply a signal

$$E_s \cos(\omega t + \phi)$$

to such a phase detector, whose reference signal is a square wave with transitions at the zeros of sin ωt, i.e., at $t = 0, \pi/\omega, 2\pi/\omega$, etc. Let us further assume that we average the output, V_{out}, by passing it through a low-pass filter whose time constant is longer than one period:

$$\tau = RC >> T = 2\pi/\omega$$

Then the low-pass-filtered output is

$$\langle E_s \cos(\omega t + \phi)\rangle \Big|_0^{\pi/\omega} - \langle E_s \cos(\omega t + \phi)\rangle \Big|_{\pi/\omega}^{2\pi/\omega}$$

where the brackets represent averages, and the minus sign comes from the gain reversal over alternate half cycles of V_{ref}. As an exercise, you can show that

$$\langle V_{out}\rangle = -(2E_s/\pi)\sin\phi$$

EXERCISE 14.2

Perform the indicated averages by explicit integration to obtain the preceding result for unity gain.

Our result shows that the averaged output, *for an input signal of the same frequency as the reference signal*, is proportional to the amplitude of V_s and sinusoidal in the relative phase.

We need one more result before going on: What is the output voltage for an input signal whose frequency is close to (but not equal to) the reference signal? This is easy, since in the preceding equations the quantity ϕ now varies slowly, at the difference frequency:

$$\cos(\omega + \Delta\omega)t = \cos(\omega t + \phi) \quad \text{with} \quad \phi = t\Delta\omega$$

giving an output signal that is a slow sinusoid:

$$V_{out} = (2E_s/\pi)\sin(\Delta\omega)t$$

which will pass through the low-pass filter relatively unscathed if $\Delta\omega < 1/\tau = 1/RC$ and will be heavily attenuated if $\Delta\omega > 1/\tau$.

The lock-in method

Now the so-called lock-in (or phase-sensitive) amplifier should make sense. First you make a weak signal periodic, as we've discussed, typically at a frequency in the neighborhood of 100Hz. The weak signal, contaminated by noise, is amplified and phase-detected relative to the modulating signal. Look at Figure 14.36. You need an experiment with two "knobs" on it, one for fast modulation in order to do phase detection and one for a slow sweep through the interesting features of the signal (in NMR, for example, the fast modulation might be a small 100Hz modulation of the magnetic field, and the slow modulation might be a frequency sweep of 10 minutes' duration through the resonance). The phase shifter is adjusted to give maximum output signal, and the low-pass filter is set for a time constant long enough to give good signal/noise ratio. The low-pass-filter rolloff

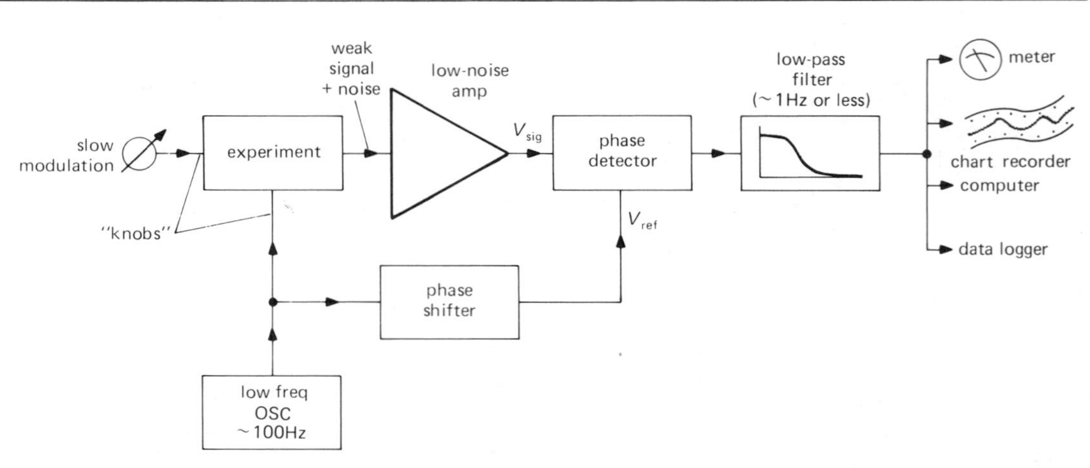

Figure 14.36
Lock-in detection.

sets the bandwidth, so a 1Hz rolloff, for example, gives you sensitivity to spurious signals and noise only within 1 Hz of the desired signal. The bandwidth also determines how fast you can adjust the "slow modulation," since now you must not sweep through any features of the signal faster than the filter can respond. People use time constants of fractions of a second up to tens of seconds and often do the slow modulation with a geared-down clock motor turning an actual knob on something!

Note that lock-in detection amounts to bandwidth narrowing again, with the bandwidth set by the postdetection low-pass filter. As with signal averaging, the effect of the modulation is to center the signal at the fast modulation frequency, rather than at dc, in order to get away from $1/f$ noise (flicker noise, drifts, and the like).

Two methods of "fast modulation"

There are two ways to do the fast modulation: The modulation waveform can be either a very small sine wave or a very large square wave compared with the features of the sought-after signal (line shape versus magnetic field, for example, in NMR), as sketched in Figure 14.37. In the first case the output signal from the phase-sensitive detector is proportional to the *slope* of the line shape (i.e., its derivative), whereas in the second case it is proportional to the line shape itself (providing there aren't any other

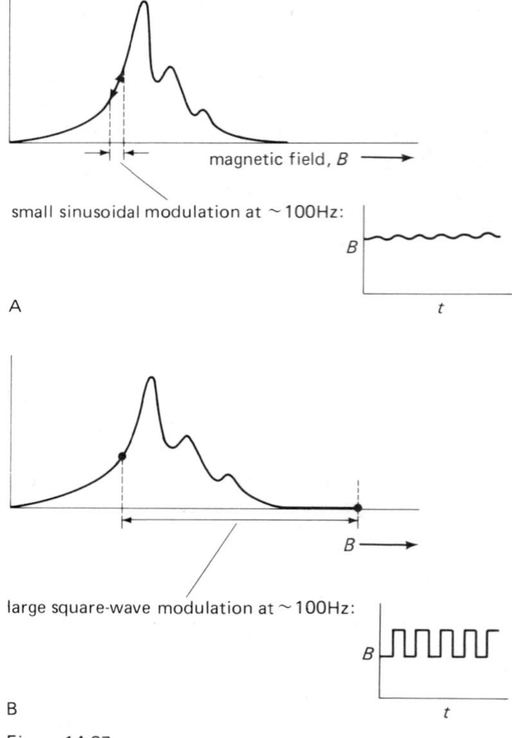

Figure 14.37
Lock-in modulation methods. A: Small sinusoid. B: Large square wave.

lines out at the other endpoint of the modulation waveform). This is the reason all those simple NMR resonance lines come out looking like dispersion curves (Fig. 14.38).

For large-shift square-wave modulation

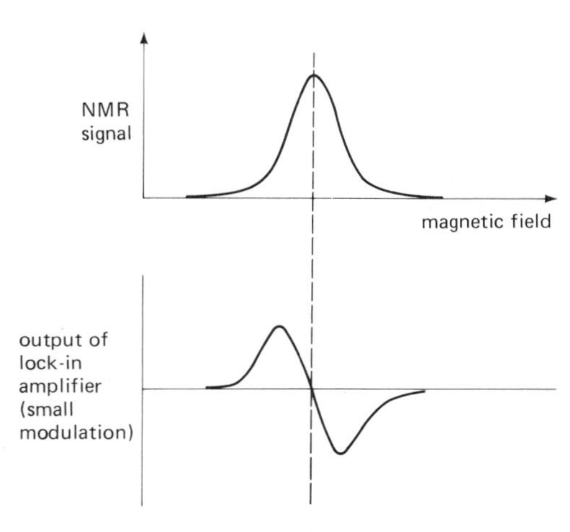

Figure 14.38
Line shape differentiation resulting from lock-in detection.

there's a clever method for suppressing modulation feedthrough, in cases where that is a problem. Figure 14.39 shows the modulation waveform. The offsets above and below the central value kill the signal, causing an on/off modulation of the signal at *twice* the fundamental of the modulating waveform. This is a method for use in

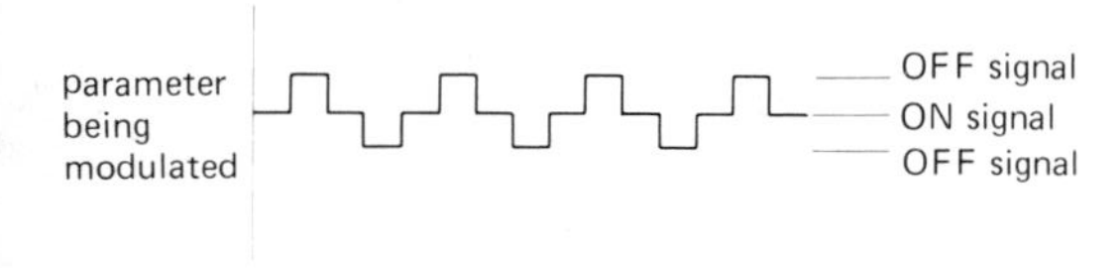

Figure 14.39
Modulation scheme for suppressing modulation feedthrough.

special cases only; don't get carried away by the beauty of it all!

Large-amplitude square-wave modulation is a favorite with those dealing in infrared astronomy, where the telescope secondary mirrors are rocked to switch the image back and forth on an infrared source. It is also popular in radioastronomy, where it's called a Dicke switch.

Commercial lock-in amplifiers have a variable-frequency modulating source and tracking filter, a switchable time-constant post-detection filter, a good low-noise wide-dynamic-range amplifier (you wouldn't be using lock-in detection if you weren't having noise problems), and a nice linear phase detector. They also let you use an external source of modulation. There's a knob that adjusts the phase shift, so you can maximize the detected signal. The whole item comes packed in a handsome cabinet, with a meter to read output signal. Typically these things cost a few thousand dollars.

In order to illustrate the power of lock-in detection, we usually set up a small demonstration for our students. We use a lock-in to modulate a small LED of the kind used for panel indicators, with a modulation rate of a kilohertz or so. The current is very low, and you can hardly see the LED glowing in normal room light. Six feet away a phototransistor looks in the general direction of the LED, with its output fed to the lock-in. With the room lights out, there's a tiny signal from the phototransistor at the modulating frequency (mixed with plenty of noise), and the lock-in easily detects it, using a time constant of a few seconds. Then we turn the room lights on (fluorescent), at which point the signal from the phototransistor becomes just a huge messy 120Hz waveform, jumping in amplitude by 50dB or more. The situation looks hopeless on the oscilloscope, but the lock-in just sits there, unperturbed, calmly detecting the same LED signal at the same level. You can check that it's really working by sticking your hand in between the LED and detector. It's darned impressive.

14.16 Pulse-height analysis

A pulse-height analyzer (PHA) is a simple extension of the multichannel scaler principle, and it is a very important instrument in nuclear and radiation physics. The idea is simplicity itself: Pulses with a range of amplitudes are input to a peak-detector/ADC circuit that converts the relative pulse height to a channel address. A multichannel scaler then increments the selected address. The result is a graph that is a histogram of pulse heights. That's all there is to it.

The enormous utility of pulse-height

analyzers stems from the fact that many detectors of charged particles, x rays, and gamma rays have output pulse sizes proportional to the energy of the radiation detected (e.g., proportional counters, solid-state detectors, surface-barrier detectors, and scintillators, as we discussed in Section 14.07). Thus a pulse-height analyzer converts the detector's output to an energy spectrum.

Pulse-height analyzers used to be designed as dedicated hardware devices, with buckets of ICs and discrete components. Nowadays the standard method is to use an off-the-shelf minicomputer, preceded by a fast pulse-input ADC. That way you can build in all sorts of useful computational routines, e.g., background subtraction, energy calibration and line identification, disk and tape storage, and on-line control of the experiment. We have an apparatus that scans a proton microbeam over a specimen in a two-dimensional raster pattern, detects the emitted x rays, sorts them by chemical element, and stores a picture of the distribution of each element in the sample, all the while letting you view the x-ray spectrum and images as the picture accumulates. The whole operation is handled by a pulse-height analyzer that doesn't realize that it's really a computer.

There is an interesting subtlety involving the ADC front end of a pulse-height analyzer. It turns out that you can't use something like a successive-approximation A/D converter, in spite of its superior speed, because you wouldn't get exact equality of channel widths, with the disastrous effect of producing a lumpy baseline from a smooth continuum of input radiation. All PHAs use a so-called Wilkinson converter, a variation on single-slope conversion whereby an input pulse charges a capacitor, which is then discharged by a constant current while a fast counter (200MHz is typical) counts up the address. This has the disadvantage of giving an analyzer "dead time" that depends on the height of the last pulse, but it gives absolute equality of channel widths.

Most pulse-height analyzers provide inputs so that you can use them as multichannel scalers. Why shouldn't they? All the electronics are already there.

14.17 Time-to amplitude converters

In nuclear physics it is often important to know the distribution of decay times of some short-lived particle. This turns out to be easy to measure, by simply hooking a time-to-amplitude converter (TAC) in front of a pulse-height analyzer. The TAC starts a ramp when it receives a pulse at one input and stops it when it receives a pulse at a second input, discharging the ramp and generating an output pulse proportional to the time interval between pulses. It is possible to build these things with resolution down in the picoseconds. Figure 14.40

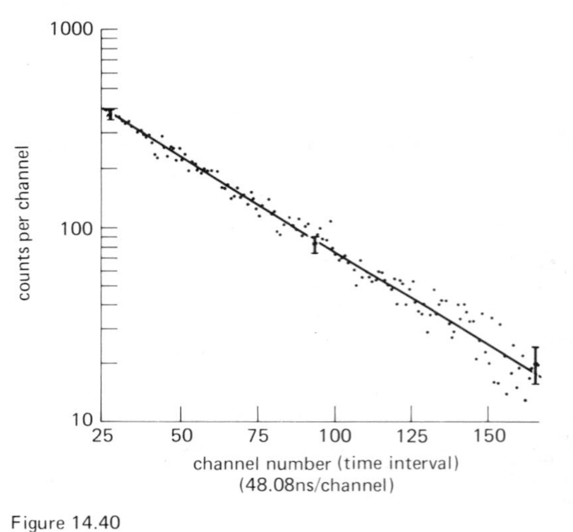

Figure 14.40
Muon lifetime measurement from time-interval spectrum (TAC + PHA).

shows a measurement of the muon lifetime made by a student by timing the delay between the capture of a cosmic-ray muon in a scintillator and its subsequent decay. Each event creates a flash of light, and a TAC is used to convert the intervals into pulses. A cosmic-ray muon decayed in this student's apparatus once a minute on the average, so he accumulated data for 18 days to determine a lifetime of 2.198 ± 0.02μs (accepted value is 2.197134 ± 0.00008μs). Note the use of log-lin axes to plot data that should be an exponential, and the systematic shift of $n^{1/2}$ (counting) error bars. The line plotted is the decay according to the accepted value, $n(t) = n_0 \exp(-t/\tau)$.

SPECTRUM ANALYSIS AND FOURIER TRANSFORMS

14.18 Spectrum analyzers

An instrument of considerable utility, particularly in radiofrequency work, is the spectrum analyzer. These devices generate an *xy* oscilloscope display, with *y* representing signal strength (usually logarithmic, i.e., in decibels), but with *x* representing frequency. In other words, a spectrum analyzer lets you look in the *frequency domain*, plotting the amount of input signal versus its frequency. You can think of it as a Fourier decomposition of the input waveform (if you know about such things), or as the response you would get as you tuned the dial of a broad-range high-performance (wide dynamic range, stable, sensitive) receiver through its frequency range. This ability can be very handy when analyzing modulated signals, looking for intermodulation products or distortion, analyzing noise and drift, trying to make accurate frequency measurements on weak signals in the presence of stronger signals, and making a host of other measurements.

Spectrum analyzers come in two basic varieties: swept-tuned and real-time. Swept analyzers are the most common variety, and they work as shown in Figure 14.41. What you have is basically a superheterodyne receiver (see Section 13.15), with a local oscillator (LO) that can be swept by an internally generated ramp waveform. As the LO is swept through its range of frequencies, different input frequencies are successively mixed to pass through the IF amplifier and filter. For example, suppose you have a spectrum analyzer with an IF of 200MHz and an LO that can sweep from 200MHz to 300MHz. When the LO is at 210MHz, input signals at 10MHz ($\pm$ the IF filter bandwidth) pass through to the detector and produce vertical deflection on the scope. Signals at 410MHz (an "image" frequency) would also pass through, which is the reason for the low-pass filter at the input. At any given time, input frequencies 200MHz lower than the LO are detected.

Real spectrum analyzers allow lots of flexibility as to sweep range, center frequency, filter bandwidth, display scales, etc. Typical input frequency ranges go from hertz to gigahertz, with selectable bandwidths ranging from hertz to megahertz. In addition, sophisticated spectrum analyzers have convenience features such as absolute amplitude calibration, storage of spectra to prevent flicker during sweeping, additional storage for comparison and normalization, and display of digital information on the screen. Fancy spectrum analyzers let you analyze phase versus frequency, generate frequency markers, program the operation via the IEEE 488 bus, include tracking oscillators (for stimulation) and tracking preselectors (for increased dynamic range), make precise frequency measurements of features in the spectrum, generate tracking noise voltages for system stimulus, and even do signal averaging (particularly useful for noisy signals).

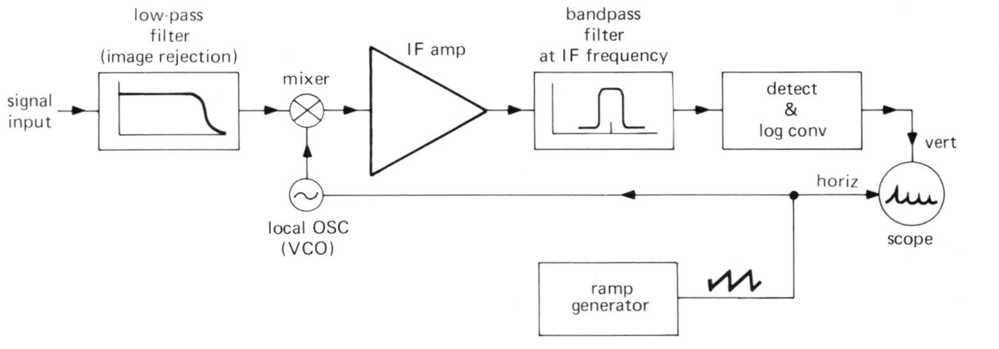

Figure 14.41

Swept-LO spectrum analyzer.

Note that this sort of swept spectrum analyzer looks at only one frequency at a time and generates a complete spectrum by sweeping in time. This can be a real disadvantage, since you can't look at transient events. In addition, when scanning with narrow bandwidth, the sweep rate must be kept slow. Finally, only a small portion of the input signal is being used at any one time.

These disadvantages of swept spectrum analysis are remedied in real-time spectrum analyzers. Again, there are several approaches. The clumsy method employs a set of narrow filters to look at a range of frequencies simultaneously. More recently, sophisticated analyzers based on digital Fourier analysis (in particular, the famous Cooley-Tukey fast Fourier transform, FFT for short) are becoming popular. These instruments convert the analog input signal (after mixing, etc.) to numbers, using a fast analog/digital converter. Then a special-purpose computer turns the crank on the FFT, generating a digital frequency spectrum. Since the method looks at all frequencies simultaneously, it has excellent sensitivity and speed, and it can be used for analysis of transients. It is particularly good for low-frequency signals, where swept analyzers are too slow. In addition, it can perform correlations between signals. Since the data comes out in digital form, it is natural to apply the full power of signal averaging, a feature available in some commercial instruments.

A clever real-time spectrum analyzer can also be constructed using the so-called chirp/Z transform. In this method a dispersive filter (delay time proportional to frequency) replaces the IF bandpass filter in the swept-LO analyzer (Fig. 14.41). By matching the LO sweep rate to the filter's dispersion, you get an output that superficially resembles the swept analyzer output, namely a linear scan of frequency versus time during each sweep. However, in constrast to the swept-LO analyzer, this scheme gathers signals from the entire band of frequencies continuously. Another interesting technique for real-time spectral analysis is the Bragg cell, in which the IF signal is used to generate acoustic waves in a transparent crystal. These deformations diffract a laser beam, generating a real-time display of the frequency spectrum as light intensity versus position. An array of photodetectors completes the analyzer output. When choosing a spectrum analyzer type, be sure to consider tradeoffs among bandwidth, resolution, linearity, and dynamic range.

Figure 14.42 shows the sort of radiofrequency spectra that endear spectrum analyzers to people who earn their living above 1MHz. The first four spectra show oscillators: A is just a pure sine-wave oscillator, B is distorted (as indicated by its harmonics), C has noise sidebands, and D has some frequency instability (drifting or residual FM). You can measure amplifier intermodulation products, as in E, where second-, third-, and

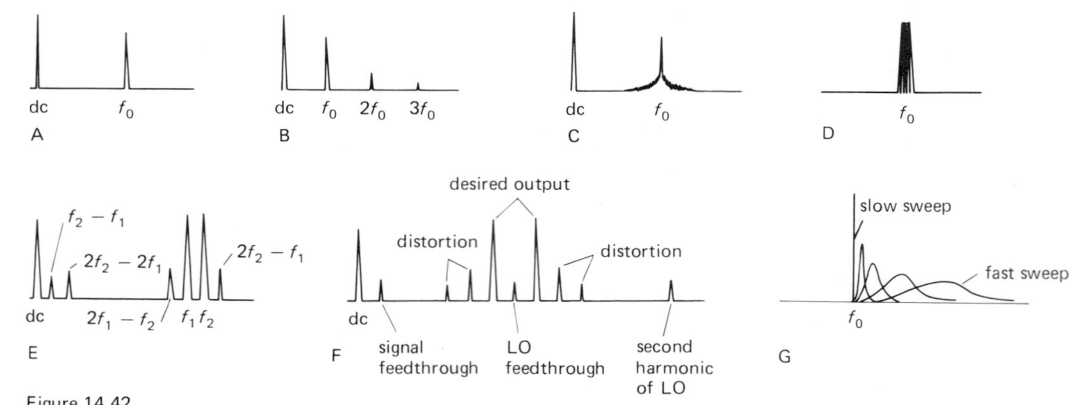

Figure 14.42

Spectrum analyzer displays.

fourth-order intermodulation frequencies are visible in the output of an amplifier driven by a "two tone" test signal consisting of pure sine waves at frequencies f_1 and f_2. Finally, in F you can see the uncouth behavior of a double-balanced mixer; there is feedthrough of both the LO and input signal, as well as distortion terms ($f_{LO} \pm 2f_{sig}$, $f_{LO} \pm 3f_{sig}$). This last spectrum may actually indicate quite respectable mixer performance, depending on the vertical scale shown. Spectrum analyzers are designed with enormous dynamic range (internally generated distortion products are typically down by 70dB or more; with a "tracking preselector" they're down by 100dB) so that you can see the failings of even a very good circuit.

The last graph G in Figure 14.42 shows what happens when you sweep the LO too fast in a swept analyzer. If the LO sweep causes a signal to pass through the filter bandwidth Δf in a time shorter than $\Delta t \approx 1/\Delta f$, it will be broadened, roughly to $\Delta f' \approx 1/\Delta t$.

14.19 Off-line spectrum analysis

The fast Fourier transform applied to digitized data from an experiment provides a very powerful method of signal analysis, particularly the recognition of weak signals of well-defined periodicity buried in interfering signals or noise, or the recognition of vibrations or oscillatory modes. For instance, we have used the FFT to search for pulsars, perform audio analysis, enhance the resolution of astronomical images (speckle imaging), and look for signals from intelligent life in space (SETI). In the last experiment, narrowband signals 35dB below receiver noise could be detected in one minute's time, corresponding to a radio flux of less than 1 micro-microwatt total over the entire earth's disk!

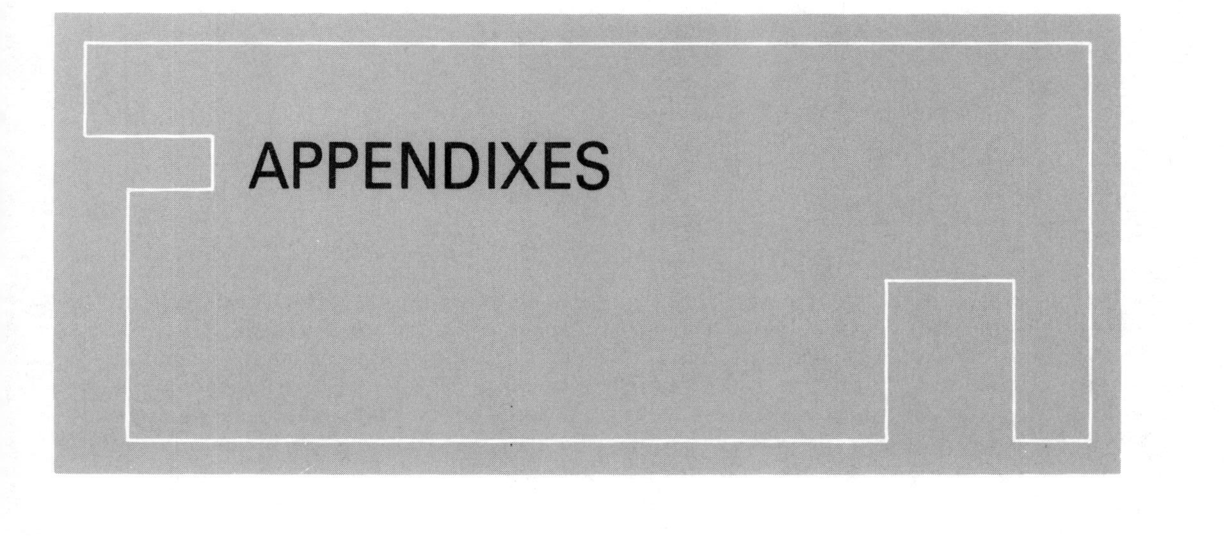

APPENDIXES

THE OSCILLOSCOPE
APPENDIX A

The oscilloscope (scope for short) is the most useful and versatile electronic test instrument. As usually used, it lets you "see" voltages in a circuit as a function of time, triggering on a particular point of the waveform so that a stationary display results. We've drawn a block diagram (Fig. A1) and typical front panel (Fig. A2) to help explain how it works. The scope we will describe is usually called a dc-coupled dual-trace triggered scope. There are special-purpose scopes used for TV servicing and the like, and there are scopes of an older vintage that don't have the features needed for circuit testing.

VERTICAL

Beginning with the signal inputs, most scopes have two channels; that's very useful, since you often need to see the relationship between signals. Each channel has a calibrated gain switch, which sets the scale of VOLTS/DIVISION *on the screen.* There's also a VARIABLE gain knob (concentric with the gain switch) in case you want to set a given signal to a certain number of divisions. Warning: Be sure the variable gain knob is in the "calibrated" position when making voltage measurements! It's easy to forget. The better scopes have indicator lights to warn you if the variable gain knob is out of the calibrated position.

The scope is dc-coupled, an essential feature: What you see on the screen is the signal voltage, dc value and all. Sometimes you may want to see a small signal riding on a large dc voltage, though; in that case you can switch the input to ac coupling, which capacitively couples the input with a time

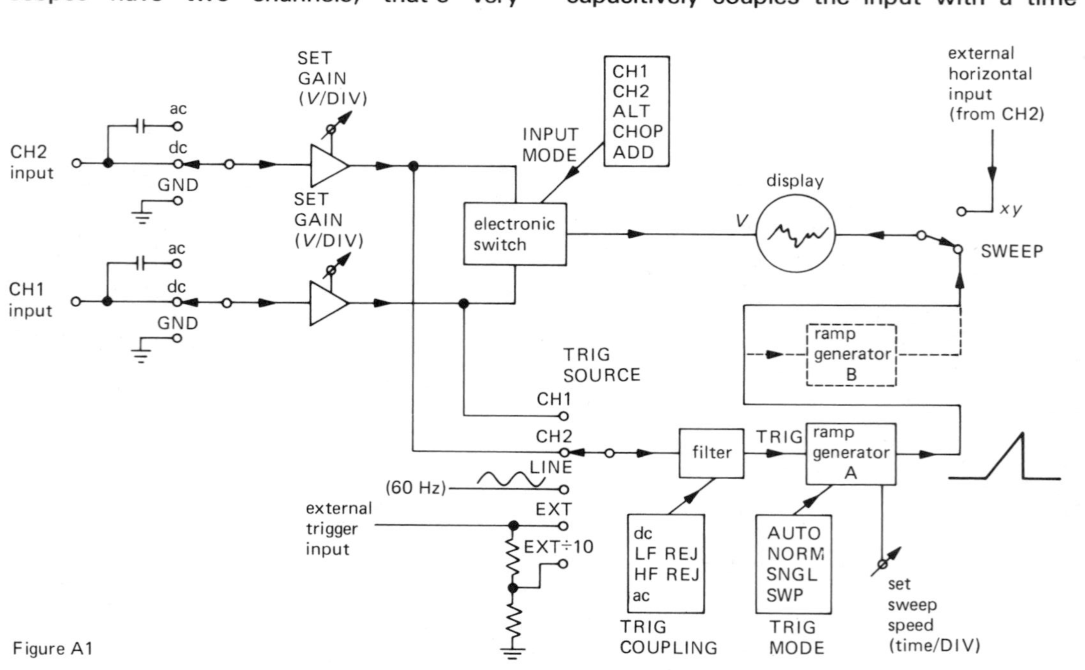

Figure A1

constant of about 0.1 second. Most scopes also have a grounded input position, which lets you see where zero volts is on the screen. (In GND position the signal isn't shorted to ground, just disconnected from the scope, whose input is grounded.) Scope inputs are usually high impedance (a megohm in parallel with about 20pF), as any good voltage-measuring instrument should be. The input resistance of 1.0 megohm is an accurate and universal value, so that high-impedance attenuating probes can be used (as will be described later); unfortunately, the parallel capacitance is not standardized, which is a bit of a nuisance when changing probes.

The vertical amplifiers include a vertical POSITION control, an INVERT control on at least one of the channels, and an INPUT MODE switch. The latter lets you look at

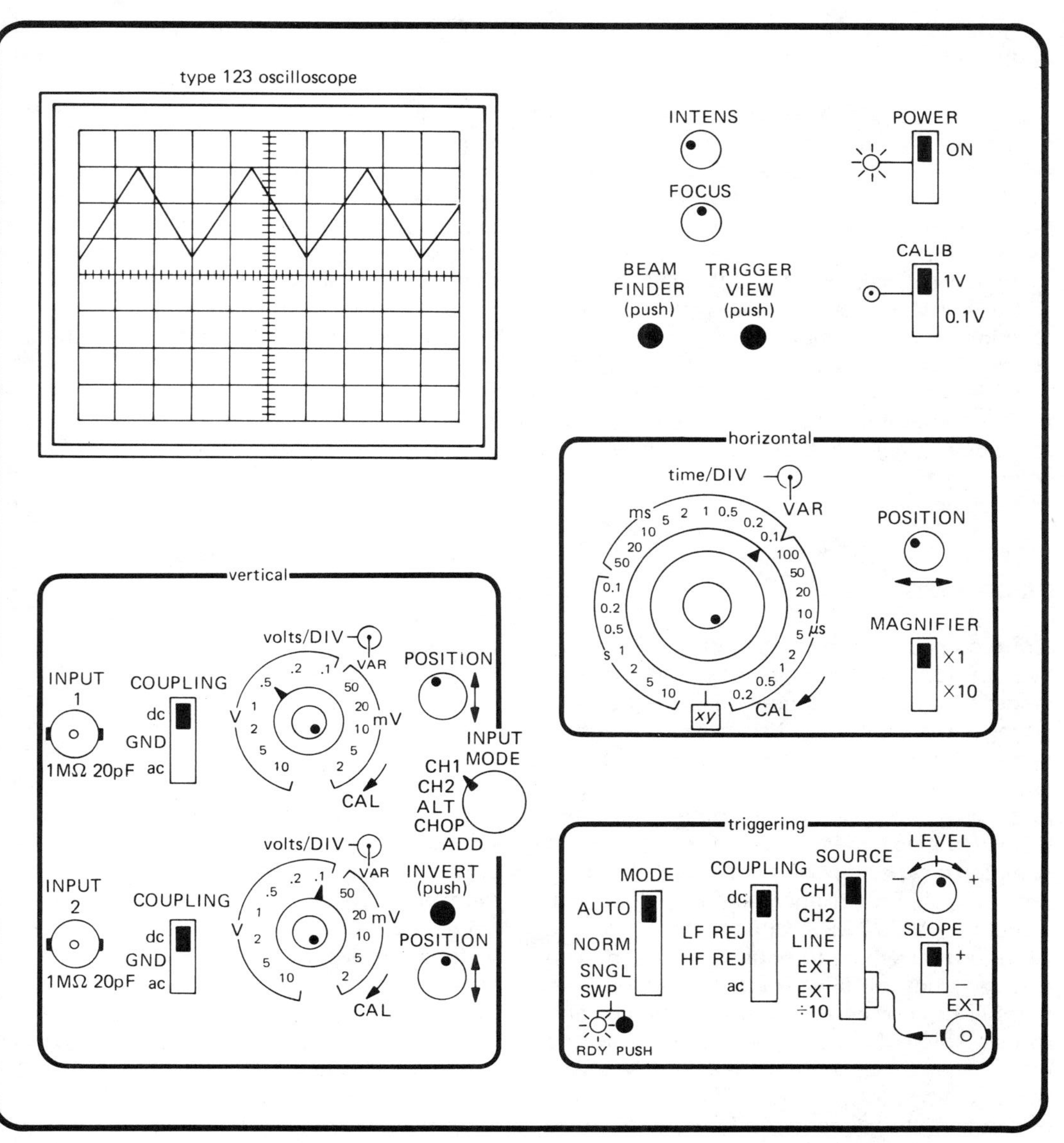

Figure A2

either channel, their sum (their difference, when one channel is inverted), or both. There are two ways to see both: ALTERNATE, in which alternate inputs are displayed on successive sweeps of the trace; CHOPPED, in which the trace jumps back and forth rapidly (0.1–1MHz) between the two signals. ALTERNATE mode is generally better, except for slow signals. It is often useful to view signals both ways, to make sure you're not being deceived.

HORIZONTAL

The vertical signal is applied to the vertical deflection electronics, moving the dot up and down on the screen. The horizontal sweep signal is generated by an internal ramp generator, giving deflection proportional to time. As with the vertical amplifiers, there's a calibrated TIME/DIVISION switch and a VARIABLE concentric knob; the same warning stated earlier applies here. Most scopes have a 10× MAGNIFIER and also allow you to use one of the input channels for horizontal deflection (this lets you generate those beloved but generally useless ''Lissajous figures'' featured in elementary books and science fiction movies).

TRIGGERING

Now comes the trickiest part: triggering. We've got vertical signals and horizontal sweep; that's what's needed for a graph of voltage versus time. But if the horizontal sweep doesn't catch the input signal at the same point in its waveform each time (assuming the signal is repetitive), the display will be a mess – a picture of the input waveform superimposed over itself at different times. The trigger circuitry lets you select a LEVEL and SLOPE (+ or −) on the waveform at which to begin the sweep. You can see from the front panel that you have a number of choices about trigger sources and mode. NORMAL mode produces a sweep only when the source selected crosses through the trigger point you have set, moving in the direction (SLOPE) you have selected. In practice, you adjust the level control for a stable display. In AUTO the sweep will ''free run'' if no signal is present; this is good if the signal sometimes drops to small values, since the display won't disappear and make you think the signal has gone away. It's the best mode to use if you are looking at a bunch of different signals and don't want to bother setting the trigger each time. SINGLE SWEEP is used for nonrepetitive signals. LINE causes the sweep to trigger on the ac power line, handy if you're looking at hum or ripple in a circuit. The EXTERNAL trigger inputs are used if you have a clean signal available at the same rate as some ''dirty'' signal you're trying to see; it's often used in situations where you are driving some circuit with a test signal, or in digital circuits where some ''clock'' signal synchronizes circuit operations. The various coupling modes are useful when viewing composite signals; for instance, you may want to look at an audio signal of a few kilohertz that has some spikes on it. The HF REJ position (high-frequency reject) puts a low-pass filter in front of the trigger circuitry, preventing false triggering on the spikes. If the spikes happen to be of interest, you can trigger on them instead in LF REJ position.

Many scopes now have BEAM FINDER and TRIGGER VIEW controls. The beam finder is handy if you're lost and can't find the trace; it's a favorite of beginners. Trigger view displays the trigger signal; it's especially handy when triggering from external sources.

HINTS FOR BEGINNERS

Sometimes it's hard to get *anything* to show on the scope. Begin by turning the scope on; set triggering for AUTO, DC COUPLING, CH1. Set sweep speed at 1ms/div, cal., and the magnifier off (×1). Ground the vertical inputs, turn up the intensity, and wiggle the vertical position control until a horizontal line appears (if you have trouble at this point, try the beam finder). Warning: Some scopes (the popular Tektronix 400 series, for example) don't sweep on AUTO unless the trigger level is adjusted correctly. Now you can apply a signal, unground the input, and fiddle with the trigger. Become familiar with the way things look when the vertical gain is far too high, when the sweep speed is too fast or slow, and when the trigger is adjusted incorrectly.

PROBES

The oscilloscope input capacitance seen by a circuit under test can be quite high, especially when the necessary shielded connecting cable is included. The resulting input impedance (1 megohm in parallel with 100 picofarads or so) is often too low for sensitive circuits and loads it by the usual voltage-divider action. Worse yet, the capacitance may cause some circuits to misbehave, sometimes even to the point of going into oscillation! In such cases the scope obviously is not acting like the "low-profile" measurement instrument we expect; it's more like a bull in a china shop.

The usual solution is the use of high-impedance "probes." The popular 10× probe works as shown in Figure A3. At dc it's just a 10× voltage divider. By adjusting C_1 to 1/9th the parallel capacitance of C_2 and C_3, the circuit becomes a 10× divider at all frequencies, with input impedance of 10 megohms in parallel with a few picofarads. In practice, you adjust the probe by looking at a square wave of about 1kHz, available on all scopes as CALIB, or PROBE ADJ, setting the capacitor on the probe for a clean square wave without overshoot. Sometimes the adjustment is cleverly hidden; on some probes you twist the body of the probe and lock it by tightening a second threaded part. One drawback: A 10× probe makes it difficult to look at signals of only a few millivolts; for these situations use a "1× probe," which is simply a length of low-capacitance shielded cable with the usual probe hardware (wire "grabber," ground clip, handsome knurled handle, etc.). The 10× probe should be the standard probe, left connected to the scope, with the 1× probe used when necessary. Some probes feature a convenient choice of 1× or 10× attenuation, switchable at the probe tip.

GROUNDS

As with most test instruments, the oscilloscope input is referred to the instrument ground (the outer connection of the input BNC connectors), which is usually tied electrically to the case. That, in turn, connects to the ground lead of the ac power line, via the 3-wire power cord. This means that you cannot measure voltages between two arbitrary points in a circuit, but are forced to measure signals relative to this universal ground.

An important caution is in order here: If you try to connect the ground clip of an oscilloscope probe to a point in the circuit that is at some voltage relative to ground, you will end up shorting it to ground. This can have disastrous consequences to the circuit under test; in addition, it can be downright dangerous with circuits that are "hot to ground" (transformerless consumer electronics like television sets, for example). If it is imperative to look at the signal between two points, you can either "float" the scope by lifting the ground lead (not recommended, unless you know what you're doing) or make a differential measurement by inverting one input channel and switching to ADD (some plug-in modules permit direct differential measurements).

Another caution about grounds when

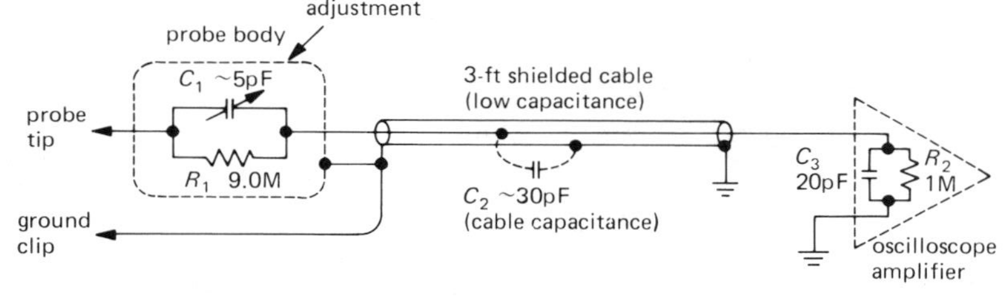

Figure A3

you're measuring weak signals or high frequencies: Be sure the oscilloscope ground is the same as the circuit ground where you're measuring. The best way to do this is by connecting the short ground wire on the probe body directly to the circuit ground, then checking by measuring the voltage of "ground" with the probe, observing no signal. One problem with this scheme is that those short ground clips are usually missing, lost! Keep your probe accessories in a drawer somewhere.

OTHER SCOPE FEATURES

Many scopes have a DELAYED SWEEP; that lets you see a segment of a waveform occurring some time after the trigger point. You can dial the delay accurately with a multiturn adjustment and a second sweep-speed switch. A delay mode known as A INTENSIFIED BY B lets you display the whole waveform at the first sweep speed, with the delayed segment brightened; this is handy during setup. Scopes with delayed sweep sometimes have "mixed sweep," in which the trace begins at one sweep speed, then switches to a second (usually faster) speed after the selected delay. Another option is to begin the delayed sweep either immediately after the selected delay, or at the next trigger point after the delay; there are two sets of trigger controls, so the two trigger points can be set individually. (Don't confuse delayed sweep with "signal delay." All good scopes have a delay in the signal channel, so you can display the event that caused the trigger; it lets you look a little bit backward in time!) Many scopes now have a TRIGGER HOLDOFF control; it inhibits triggering for an adjustable interval after each sweep, and it is very useful when viewing complicated waveforms without the simple periodicity of a sine wave, say. The usual case is a digital waveform with a complicated sequence of 1's and 0's, which won't generate a stable display otherwise (except by adjustment of the sweep-speed vernier, which means you don't get a calibrated sweep). There are also scopes with "storage" that let you see a nonrepetitive event, and scopes that accept plug-in modules. These let you do just about anything, including display of eight simultaneous traces, spectrum analysis, accurate (digital) voltage and time measurements on waveforms, etc. Digital-storage analog oscilloscopes of a new generation are becoming available; they let you catch a one-shot waveform, and even let you look backward in time (before the trigger event) up to 3/4 of the whole screen.

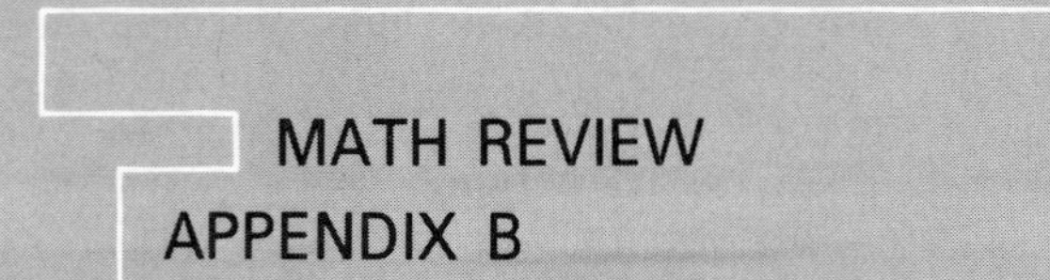

MATH REVIEW
APPENDIX B

Some knowledge of algebra and trigonometry is essential to understand this book. In addition, a limited ability to deal with complex numbers and derivatives (a part of calculus) is helpful, although not entirely essential. This appendix is meant as the briefest of summaries of complex numbers and differentiation. It is not meant as a textbook substitute. For a highly readable self-help book on calculus, we recommend *Quick Calculus,* by D. Kleppner and N. Ramsey (John Wiley & Sons, 1972).

COMPLEX NUMBERS

A complex number is an object of the form

$$\mathbf{N} = a + bi$$

where a and b are real numbers and i (called j in the rest of the book, to avoid confusion with small-signal currents) is the square root of -1; a is called the real part, and b is called the imaginary part. Boldface letters or squiggly underlines are sometimes used to denote complex numbers. At other times you're just supposed to *know!*

Complex numbers can be added, subtracted, multiplied, etc., just as real numbers:

$$(a + bi) + (c + di) = (a + c) + (b + d)i$$

$$(a + bi) - (c + di) = (a - c) + (b - d)i$$

$$(a + bi)(c + di) = (ac - bd) + (bc + ad)i$$

$$\frac{a + bi}{c + di} = \frac{(a + bi)(c - di)}{(c + di)(c - di)} = \frac{ac + bd}{c^2 + d^2} + \frac{bc - ad}{c^2 + d^2} i$$

All these operations are natural, in the sense that you just treat i as something that multiplies the imaginary part, and go ahead with ordinary arithmetic. Note that $i^2 = -1$ (used in the multiplication example) and that division is simplified by multiplying top and bottom by the *complex conjugate,* the number you get by changing the sign of the imaginary part. The complex conjugate is sometimes indicated with an asterisk. If

$$\mathbf{N} = a + bi$$

then

$$\mathbf{N}^* = a - bi$$

The magnitude (or *modulus*) of a complex number is

$$|\mathbf{N}| = |a + bi| = [(a + bi)(a - bi)]^{1/2} = (a^2 + b^2)^{1/2}$$

i.e.,

$$|\mathbf{N}| = (\mathbf{NN}^*)^{1/2}$$

simply obtained by multiplying by the complex conjugate and taking the square root. The magnitude of the product (or quotient) of two complex numbers is simply the product (or quotient) of their magnitudes.

The real (or imaginary) part of a complex number is sometimes written

$$\text{real part of } \mathbf{N} = \mathcal{R}e(\mathbf{N})$$

$$\text{imaginary part of } \mathbf{N} = \mathcal{I}m(\mathbf{N})$$

You get them by writing out the number in the form $a + bi$, then taking either a or b. This may involve some multiplication or division, since the complex number may be a real mess.

Complex numbers are sometimes represented on the complex plane. It looks just like an ordinary x,y graph, except that a complex number is plotted by taking its real part as x and its imaginary part as y; i.e., the axes represent REAL (x) and IMAGINARY (y), as shown in Figure B1. In keeping with

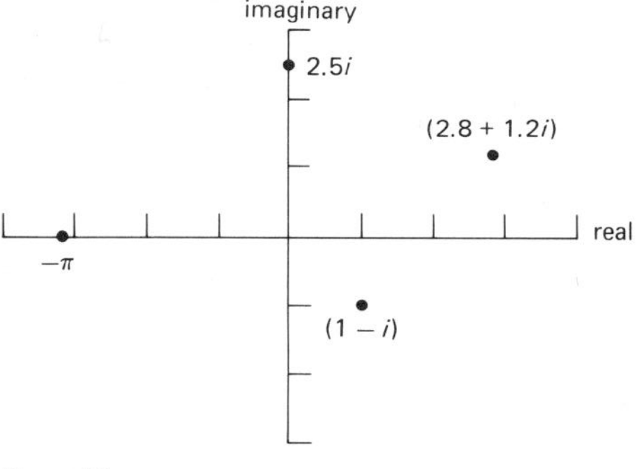

Figure B1

this analogy, you sometimes see complex numbers written just like x,y coordinates:

$$a + bi \leftrightarrow (a, b)$$

Just as with ordinary x, y pairs, complex numbers can be represented in polar coordinates; that's known as "magnitude, angle" representation. For example, the number $a + bi$ can also be written (Fig. B2)

$$a + bi = (R, \theta)$$

where $R = (a^2 + b^2)^{1/2}$ and $\theta = \tan^{-1}(b/a)$.

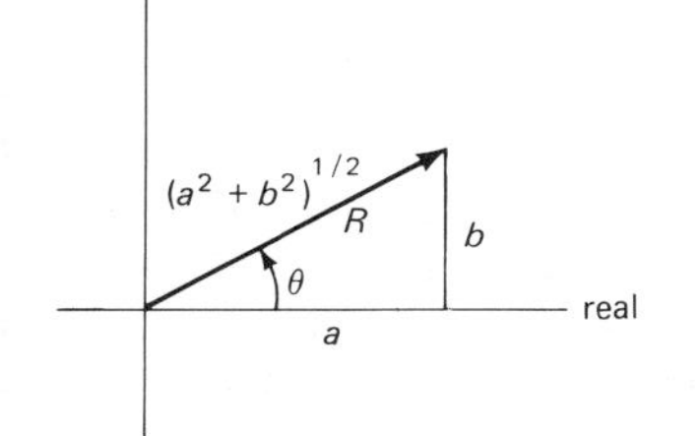

Figure B2

This is usually written in a different way, using the fact that

$$e^{ix} = \cos x + i \sin x$$

(You can easily derive the preceding result, known as Euler's formula, by expanding the exponential in a Taylor series.) Thus we have the following equivalents:

$$\mathbf{N} = a + bi = Re^{i\theta}$$
$$R = |\mathbf{N}| = (\mathbf{NN^*})^{1/2} = (a^2 + b^2)^{1/2}$$
$$\theta = \tan^{-1}(b/a)$$

i.e., the modulus R and angle θ are simply the polar coordinates of the point that represents the number in the complex plane. Polar form is handy when complex numbers have to be multiplied (or divided); you just multiply (divide) their magnitudes and add (subtract) their angles:

$$(ae^{ib})(ce^{id}) = ace^{i(b+d)}$$

Finally, to convert from polar to rectangular form, just use Euler's formula:

$$ae^{ib} = a \cos b + ia \sin b$$

i.e.,

$$\mathcal{Re}(ae^{ib}) = a \cos b$$

$$\mathcal{Im}(ae^{ib}) = a \sin b$$

If you have a complex number multiplying a complex exponential, just do the necessary multiplications. If

$$\mathbf{N} = a + bi$$

$$\mathbf{N}e^{ix} = (a + bi)(\cos x + i \sin x) = (a \cos x - b \sin x) + i(b \cos x + a \sin x)$$

DIFFERENTIATION (CALCULUS)

We start with the concept of a *function* $f(x)$, i.e., a formula that gives a value $y = f(x)$ for each x. The function $f(x)$ should be *single-valued,* i.e., it should give a single value of y for each x. You can think of $y = f(x)$ as a

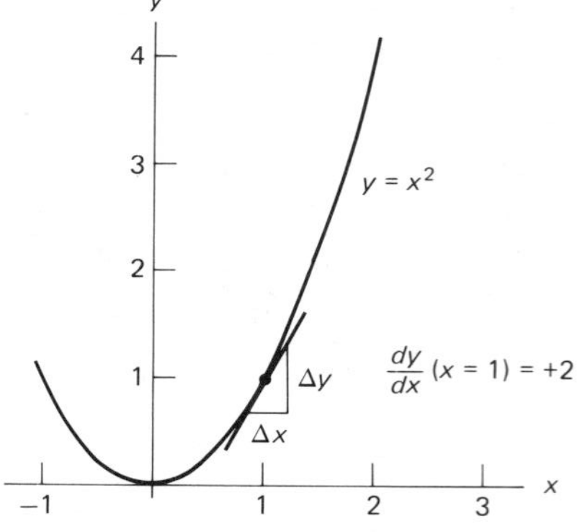

Figure B3

graph, as in Figure B3. The derivative of y with respect to x, written dy/dx ("dee y dee x"), is the *slope* of the graph of y versus x. If you draw a tangent to the curve at some point, its slope is *dy/dx at that point;* i.e., the derivative is itself a function, since it has a value at each point. In Figure B3, the slope at the point (1, 1) happens to be 2, whereas the slope at the origin is zero (you will see shortly how to compute the derivative).

In mathematical terms, the derivative is the limiting value of the ratio of the change in y (Δy) to the change in x (Δx) corresponding to a small change in x (Δx), as Δx goes to zero. To quote a song once sung in the hallowed halls of Harvard (by Tom Lehrer and Lewis Branscomb),

You take a function of *x*, and you call it *y*
Take any *x*-nought that you care to try
Make a little change and call it delta *x*
The corresponding change in *y* is what you find nex'
And then you take the quotient, and now, carefully
Send delta *x* to zero, and I think you'll see
That what the limit gives us (if our work all checks)
Is what you call *dy/dx*
It's just *dy/dx*.
(to the tune of "There'll Be Some Changes Made," W. Benton Overstreet).

Differentiation is a straightforward art, and the derivatives of many common functions are tabulated in standard tables. Here are some rules (u and v are arbitrary functions of x):

Some derivatives

$$\frac{d}{dx}x^n = nx^{n-1}$$

$$\frac{d}{dx}\sin x = \cos x$$

$$\frac{d}{dx}e^x = e^x$$

$$\frac{d}{dx}au(x) = a\frac{d}{dx}u(x) \qquad (a = \text{constant})$$

$$\frac{d}{dx}(u + v) = \frac{du}{dx} + \frac{dv}{dx}$$

$$\frac{d}{dx}\left(\frac{u}{v}\right) = \frac{v\dfrac{du}{dx} - u\dfrac{dv}{dx}}{v^2}$$

$$\frac{d}{dx}\{u[v(x)]\} = \frac{du}{dv}\frac{dv}{dx}$$

The last one is very useful and is called the chain rule.

Once you have differentiated a function, you often want to evaluate the value of the derivative at some point. Other times you may want to find a minimum or maximum of the function; that's the same thing as having a zero derivative, so you can just set the derivative to zero and solve for x. Here are some examples:

$$\frac{d}{dx}x^2 = 2x \text{ (Fig. B3: } \begin{matrix}\text{slope} = 2 \text{ at } x = 1,\\ \text{slope} = 0 \text{ at } x = 0)\end{matrix}$$

$$\frac{d}{dx}xe^x = xe^x + e^x \qquad \text{(product rule)}$$

$$\frac{d}{dx}\sin(ax) = a\cos(ax) \qquad \text{(chain rule)}$$

$$\frac{d}{dx}a^x = \frac{d}{dx}(e^{x\log a}) = a^x \log a \qquad \text{(chain rule)}$$

$$\frac{d}{dx}\left(\frac{1}{x^{1/2}}\right) = -\frac{1}{2}x^{-3/2}$$

THE 5% RESISTOR COLOR CODE
APPENDIX C

Low-power axial-lead carbon-composition and film resistors with 2% to 20% tolerances have a standard set of values and a standard color-band marking scheme. Although it may seem diabolical to the beginner, the practice of color banding makes it easy to recognize resistor values in a circuit or parts bin, without having to search for a printed legend. The standard resistor values are chosen so that adjacent values have relative ratios of about 10% for the 2% and 5% tolerance types and 20% for the 10% and 20% tolerance types. Thus there are many values that could be described by the color code but that are not available.

Two digits and a multiplier digit determine the resistor value, and resistors are color-banded in that order starting from one end of the resistor (Fig. C1). A fourth tolerance band is usually present, and occasionally you'll see a fifth band for other parameters (such as a yellow or orange band for MIL spec reliability rating).

Here is the set of standard values for the first two digits (lightface type indicates 2% and 5% only):

10	16	**27**	43	**68**
11	**18**	30	**47**	75
12	20	**33**	51	**82**
13	**22**	36	**56**	91
15	24	**39**	62	**100**

Carbon-composition resistors range in

digit	color	multiplier	number of zeros
	silver	0.01	−2
	gold	0.1	−1
0	black	1	0
1	brown	10	1
2	red	100	2
3	orange	1k	3
4	yellow	10k	4
5	green	100k	5
6	blue	1 M	6
7	violet	10 M	7
8	gray		
9	white		

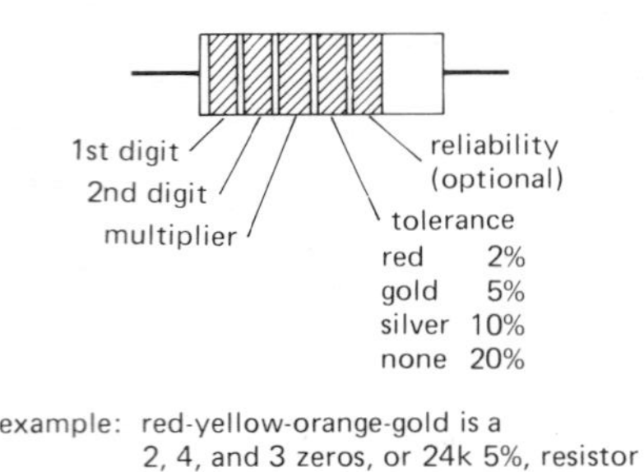

example: red-yellow-orange-gold is a 2, 4, and 3 zeros, or 24k 5%, resistor

Figure C1

price from 3 cents each (in quantities of 1000) to 15 cents (quantities of 25). Distributors may be unwilling to sell less than 25 to 50 pieces of one value; thus an assortment box (made by Stackpole or Ohmite) may be a wise purchase.

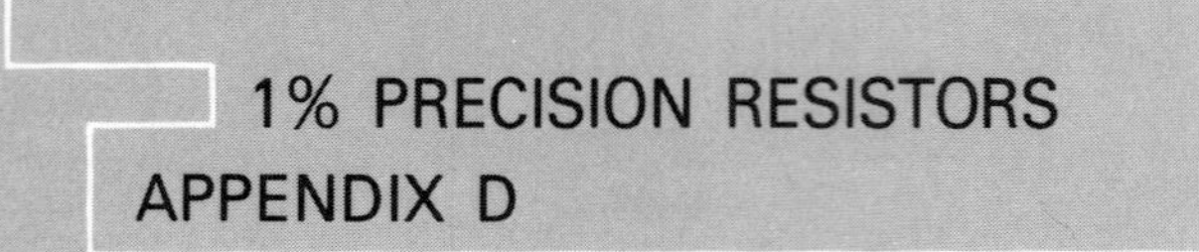

APPENDIX D
1% PRECISION RESISTORS

Metal-film precision resistors with ½% and 1% tolerance ratings have seen sufficient use in the industry to have attractively low prices. In particular, the RN55D and RN60D resistors are often available for as little as 8 cents each in quantities of 100, and a distributor may be willing to sell an assortment of mixed values at a quantity discount. The RN55D resistors are the same size as ordinary ¼ watt "composition" resistors (although their military rating will be ⅒ or ⅛ watt at 70°C ambient temperature), whereas the RN60D are the size of ½ watt composition resistors. The RN55D resistors have a temperature coefficient of 100ppm/°C, and RN55C resistors (same size) have a 50ppm/°C rating.

Metal-film precision resistors use a four-digit code printed on the resistor body, rather than the ordinary color-banding scheme. The first three digits denote a value, and the last digit is the "number-of-zeros" multiplier. For example, 1693 denotes a 169k resistor, and 1000 denotes a 100 ohm resistor. (Note that the color bands work the same way, but with only three digits altogether. Many capacitor types use this same printed number scheme.) If the resistor's value is too small to be described this way, an R will be used to indicate the decimal point; for example, 49R9 is a 49.9 ohm resistor, and 10R0 is 10.0 ohms.

The standard values range from 10.0 ohms to 301kΩ by approximately 2% ratios, although some companies may offer similar (non-MIL-spec) resistors with values from 4.99 ohms to 2.00 MΩ. Standard values in each decade are given in the table that follows.

One percent resistors are often used in applications that require excellent stability and accuracy; a small adjustable "trimmer" resistor may be connected in series to set a precise resistance value. But it's important to realize that, from a worst-case standpoint, 1% resistors are only guaranteed to be within 1% of their rated value under a specified set of conditions. Resistance variation due to temperature change, high humidity, and operation at full rated power can easily exceed 1%. Resistance drift with time can approach 0.5%, particularly if the resistors are used at rated power. Circuits that require extremely accurate or stable performance (good to 0.1% or better, say) should use precision wire-wound resistors or some of the special metal-film resistors designed for such stability. This advice goes for composition resistors, as well. Resist, if you will, the temptation to regard the manufacturer's specifications as being overly conservative.

100	140	196	274	383	536	750
102	143	200	280	392	549	768
105	147	205	287	402	562	787
107	150	210	294	412	576	806
110	154	215	301	422	590	825
113	158	221	309	432	604	845
115	162	226	316	442	619	866
118	165	232	324	453	634	887
121	169	237	332	464	649	909
124	174	243	340	475	665	931
127	178	249	348	487	681	953
130	182	255	357	499	698	976
133	187	261	365	511	715	
137	191	267	374	523	732	

HOW TO DRAW SCHEMATIC DIAGRAMS
APPENDIX E

A well-drawn schematic makes it easy to understand how a circuit works and aids in troubleshooting; a poor schematic only creates confusion. By keeping a few rules and suggestions in mind, you can draw a good schematic in no more time than it takes to draw a poor one. In this appendix we dispense advice of three varieties: general principles, rules, and hints. We have also drawn some real knee-slappers to illustrate habits to avoid.

GENERAL PRINCIPLES

1. Schematics should be unambiguous. Therefore, pin numbers, parts values, polarities, etc., should be clearly labeled to avoid confusion.

2. A good schematic makes circuit functions clear. Therefore, keep functional areas distinct; don't be afraid to leave blank areas on the page, and don't try to fill the page. There are conventional ways to draw functional subunits; for instance, don't draw a differential amplifier as in Figure E1, because the function won't be easily recognized. Likewise, flip-flops are usually drawn with clock and inputs on the left, set and clear on top and bottom, and outputs on the right.

RULES

1. Wires connecting are indicated by a heavy black dot; wires crossing, but not connecting, have no dot (don't use a little half-circular ''jog''; they went out in the 1950s).

2. Four wires must not connect at a point; i.e., wires must not cross *and* connect.

3. Always use the same symbol for the same device; e.g., don't draw flip-flops in two different ways (exception: assertion-level logic symbols show each gate in two possible ways).

4. Wires and components are aligned horizontally or vertically, unless there's a good reason to do otherwise.

5. Label pin numbers on the outside of a symbol, signal names on the inside.

6. All parts should have values or types indicated; it's best to give all parts a label, too, e.g., R_7 or IC_3.

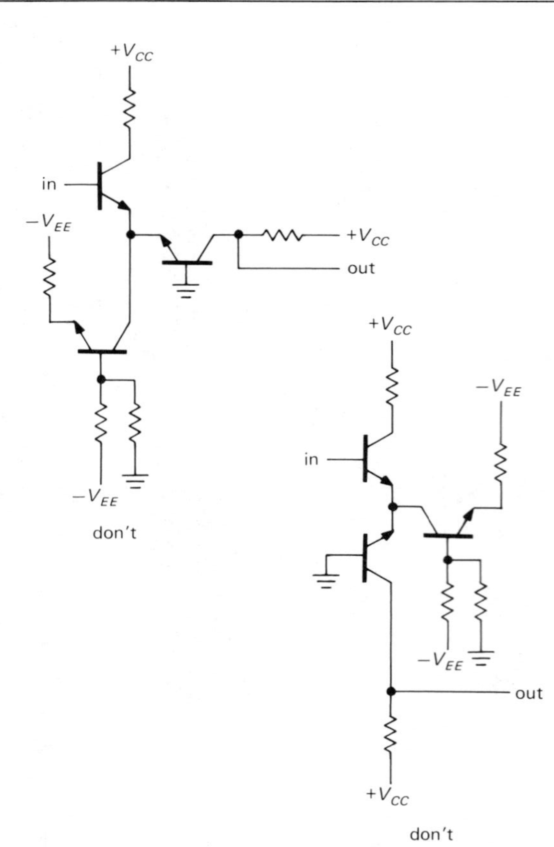

Figure E1

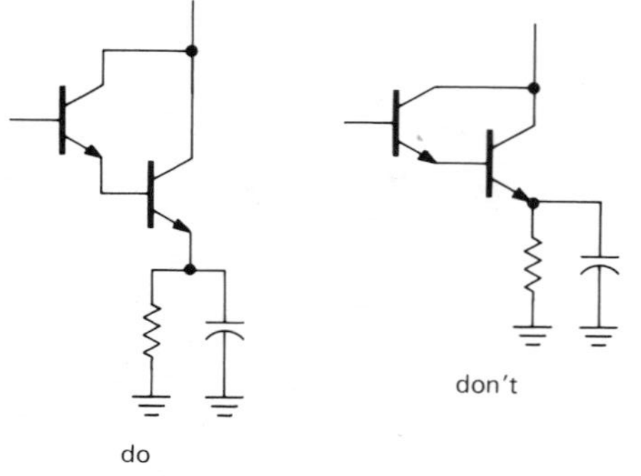

Figure E2

HINTS

1. Identify parts immediately adjacent to the symbol, forming a distinct group giving symbol, label, and type or value.
2. In general, signals go from left to right; don't be dogmatic about this, though, if clarity is sacrificed.
3. Put positive supply voltages at the top of the page, negative at the bottom. Thus, *npn* transistors will usually have their emitter at the bottom, whereas *pnps* will have emitter topmost.
4. Don't attempt to bring all wires around to the supply rails, or to a common ground wire. Instead, use the ground symbol(s) and labels like $+V_{CC}$ to indicate those voltages where needed.
5. It is helpful to label signals and functional blocks and show waveforms; in logic diagrams it is especially important to label signal lines, e.g., RESET' or CLK.
6. It is helpful to bring leads away from components a short distance before making connections or jogs. For example, draw transistors as in Figure E2.
7. Leave some space around circuit symbols; e.g., don't draw components or wires too close to an op-amp symbol. This keeps the drawing uncluttered and leaves room for labels, pin numbers, etc.
8. Label all boxes that aren't obvious: comparator versus op-amp, shift register versus counter, etc. Don't be afraid to invent a new symbol.
9. Use small rectangles, ovals, or circles to indicate card-edge connections, connector pins, etc. Be consistent.
10. The signal path through switches should be clear. Don't force the reader to follow wires all over the page to find out how a signal is switched.
11. Power-supply connections are normally assumed for op-amps and logic devices. However, show any unusual connections (e.g., an op-amp run from a single supply, where V_{-} = ground) and the disposition of unused inputs.
12. It is very helpful to include a small table of IC numbers, types, and power-supply connections (pin numbers for V_{CC} and ground, for instance).
13. Include a title area near the bottom of the page, with name of circuit, name of instrument, by whom drawn, by whom designed or checked, date, and assembly number. Also include a revision area, with columns for revision number, date, and subject.
14. We recommend drawing schematics freehand on coarse graph paper (nonreproducing blue, 4 to 8 lines per inch) or on plain

paper on top of graph paper. This is fast, and it gives very pleasing results. Use dark pencil or ink; avoid ball-point pen.

As an illustration, we've drawn a humble example (Fig. E3) showing "awful" and "good" schematics of the same circuit; the former violates nearly every rule and is almost impossible to understand. See how many bad habits you can find illustrated. We've seen all of them in professionally drawn schematics! (Drawing the "bad" schematic was an occasion of great hilarity; we laughed ourselves silly.)

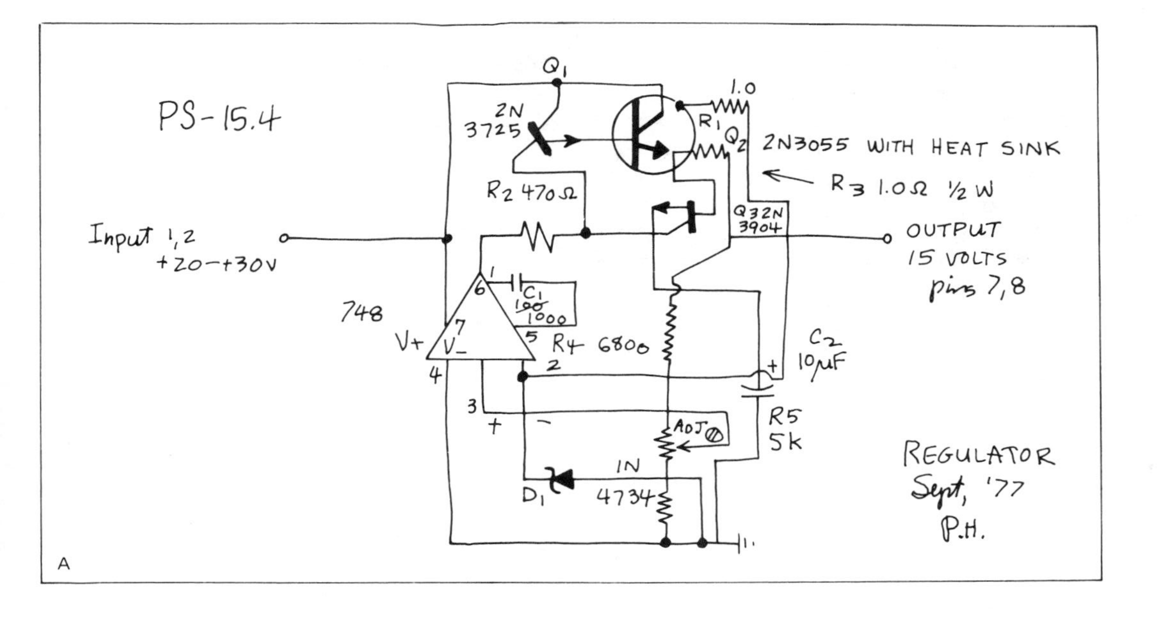

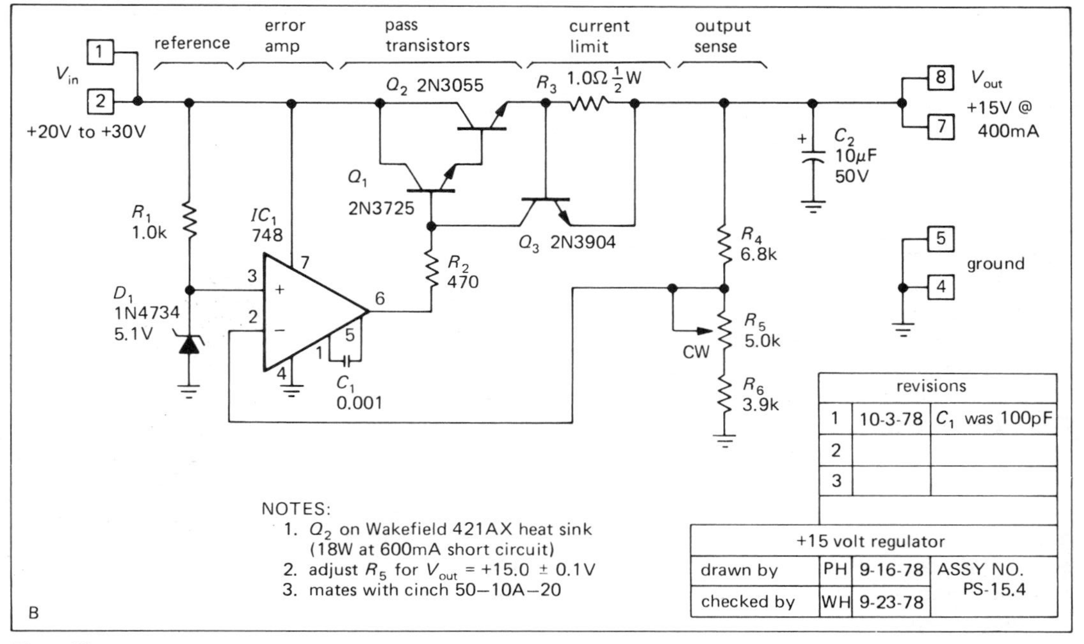

Figure E3

APPENDIX F
LOAD LINES

The graphic method of ''load lines'' usually makes an early appearance in electronic textbooks. We have avoided it because, well, it just isn't useful in transistor design, the way it was in vacuum-tube circuit design. However, it is of use in dealing with some nonlinear devices (tunnel diodes, for example), and in any case it is a useful conceptual tool.

Let's start with an example. Suppose you want to know the voltage across the diode in Figure F1. Assume that you know the voltage versus current (*VI*) curve of the particular diode (of course, it would have a manufacturing ''spread,'' as well as depending on ambient temperature); it might look something like the curve drawn. How would you figure out the quiescent point?

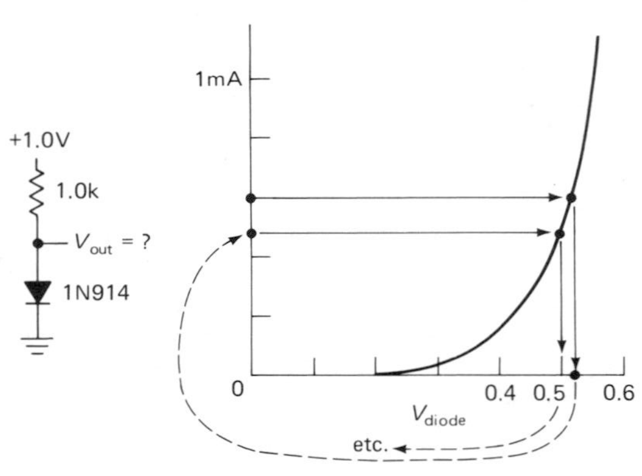

Figure F1

One method might be to guess a rough value of current, say 0.6mA, then use the curve to get the drop across the resistor, from which you get a new guess for the current (in this case, 0.48mA). This iterative method is suggested in Figure F1. After a few iterations, this method will get you an answer, but it leaves a lot to be desired.

The method of load lines gets you the answer to this sort of problem immediately. Imagine *any* device connected in place of the diode; the 1.0k resistor is still the load. Now plot, on a *VI* graph, the curve of resistor current versus device voltage. This turns out to be easy: at zero volts the current is just V_+/R (full drop across the resistor); at V_+ volts the current is zero; points in between fall on a straight line between the two. Now, on the same graph, plot the *VI* curve of the device. The operating point lies on both curves, i.e., at the intersection, as shown in Figure F2.

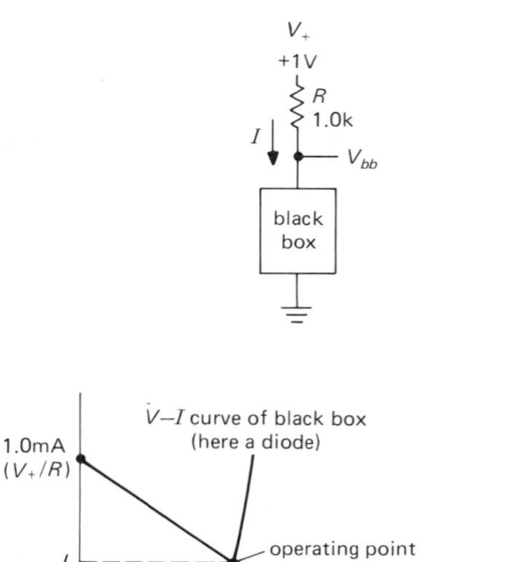

Figure F2

Load lines can be used with a 3-terminal device (tube or transistor, for example) by plotting a family of curves for the device. Figure F3 shows what such a thing would look like for a depletion-mode FET, with the curve family parametrized by gate-source

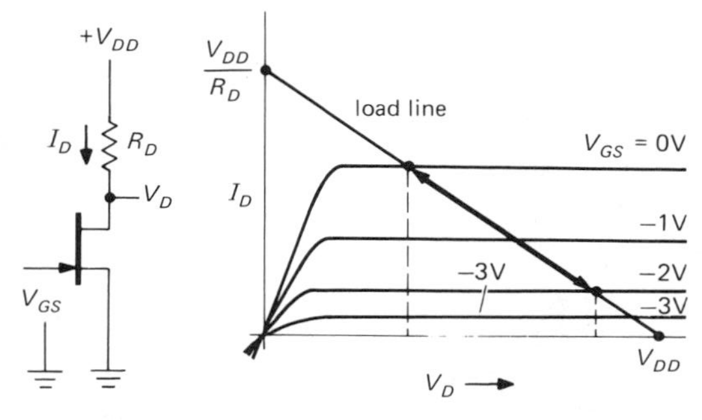

Figure F3

voltage. You can read off the output for a given input by sliding along the load line between appropriate curves corresponding to the input you've got, then projecting onto the voltage axis. In this example we've done this, showing the drain voltage (output) for a gate swing (input) between ground and −2 volts.

As nice as this method seems, it has very limited use for transistor or FET design, for a couple of reasons. For one thing, the curves published for semiconductor devices are "typical," with manufacturing spread that can be as large as a factor of 5. Imagine what would happen to those nice load-line solutions if all the curves shrank by a factor of 4! Another reason is that for an inherently logarithmic device like a diode junction, a linear load-line graph can be used to give accurate results only over a narrow region. Finally, the nongraphic methods we've used in this book are adequate to handle solid-state design. In particular, these methods emphasize the parameters you can count on (r_e, I_C vs. V_{BE} and T, etc.), rather than the ones that are highly variable (h_{FE}, V_p, etc.). If anything, the use of load lines on published curves for transistors only gives you a false sense of security, since the device spread isn't also shown.

Load lines turn out to be very useful in understanding the circuit behavior of highly nonlinear devices. The example of tunnel diodes illustrates a couple of interesting points. Let's analyze the circuit in Figure F4.

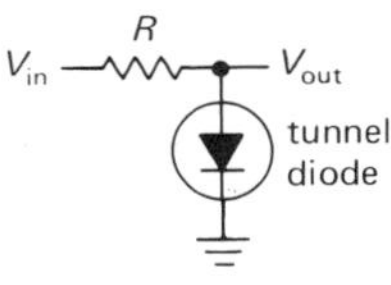

Figure F4

Note that in this case, V_{IN} takes the place of the supply voltage in the previous examples. So a signal swing will generate a family of parallel load lines intersecting with a single device VI curve (Fig. F5A). The values shown are for a 100 ohm load resistor. As can be seen, the output varies most rapidly as the input swing takes the load line across the negative-resistance portion of the tunnel-diode curve. By reading off values of V_{out} (projection on the x axis) for various values of V_{in} (individual load lines), you get the "transfer" characteristics shown. This particular circuit has some voltage gain for input voltages near 0.2 volt.

An interesting thing happens if the load

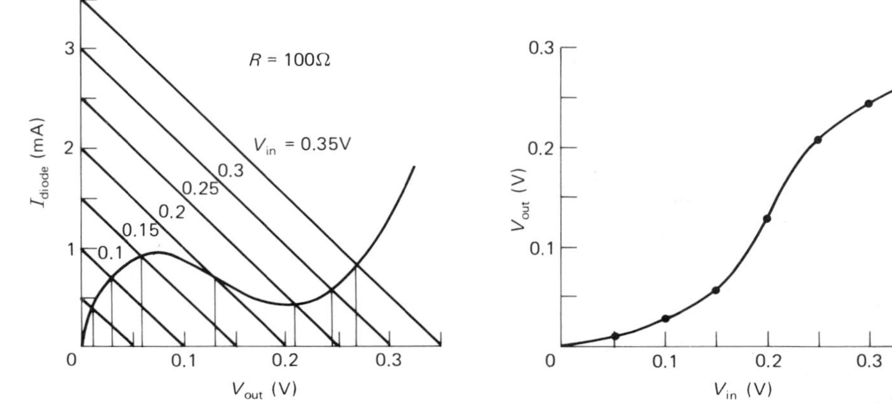

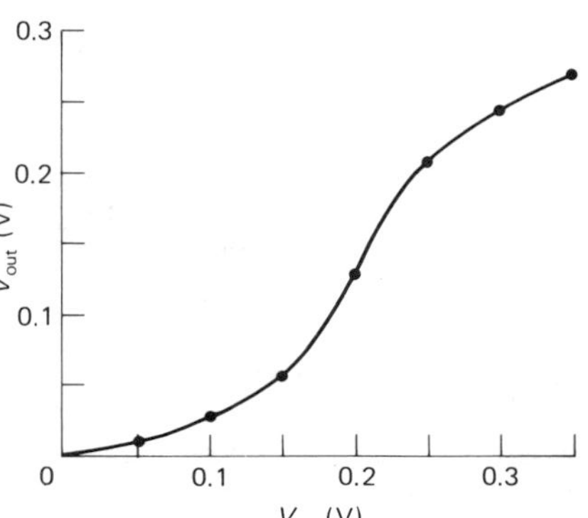

Figure F5

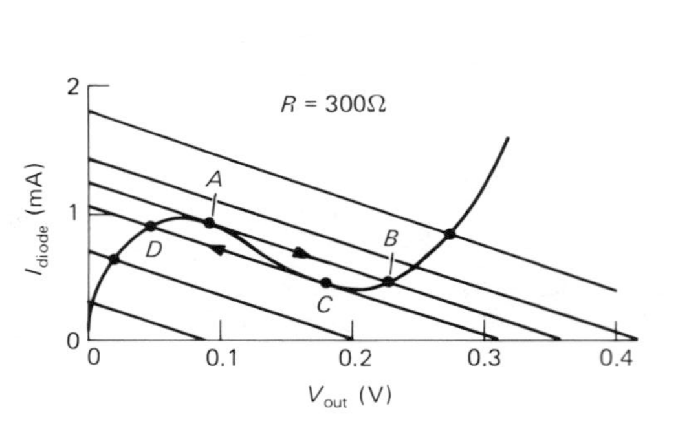

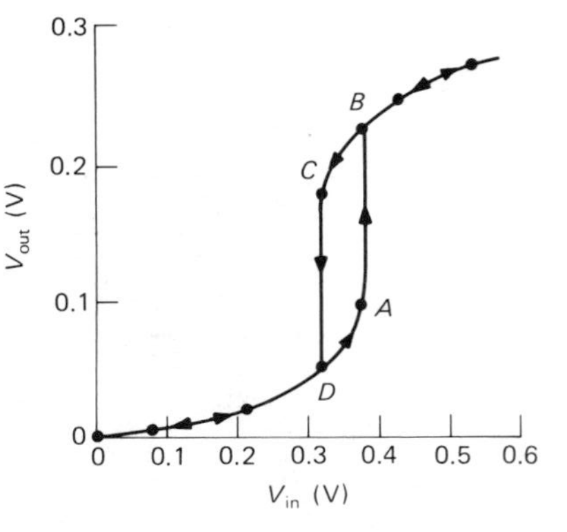

Figure F6

lines become flatter than the middle section of the diode curve. That happens when the load resistance exceeds the magnitude of the diode's negative resistance. It is then possible to have *two* intersection points, as in Figure F6. A rising input signal carries the load lines up until the intersection point has nowhere to go and has to jump across to a higher V_{out} value. On returning, the load lines similarly carry the intersection point down until it must again jump back. The overall transfer characteristic has hysteresis, as shown. Tunnel diodes are used in this manner as fast switching devices (triggers).

TRANSISTOR SATURATION
APPENDIX G

The subtitle of this appendix might be "transistor man defeated by the collector-base diode." With a simple model we can see the reason for the finite "saturation voltage" exhibited by bipolar transistors. The basic idea is that the collector-base junction is a big diode, with a high I_S (Ebers-Moll equation), so that it has a lower ON voltage for a given current than the base-emitter diode. Therefore at small values of collector-to-emitter voltage (typically 0.25 volt or less), some of the base current will be "robbed" by conduction of the collector-base diode (Fig. G1). This lowers the effective h_{FE} and makes it necessary to supply relatively large base currents to bring the collector close to

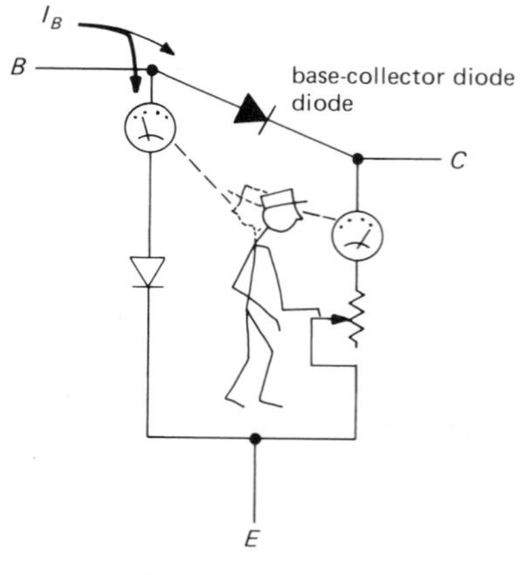

Figure G1

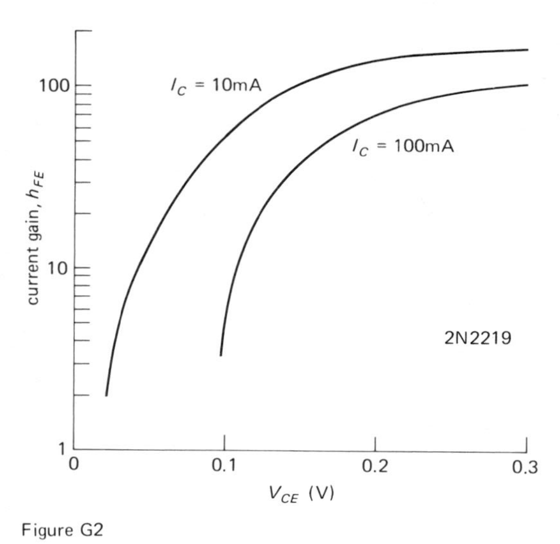

Figure G2

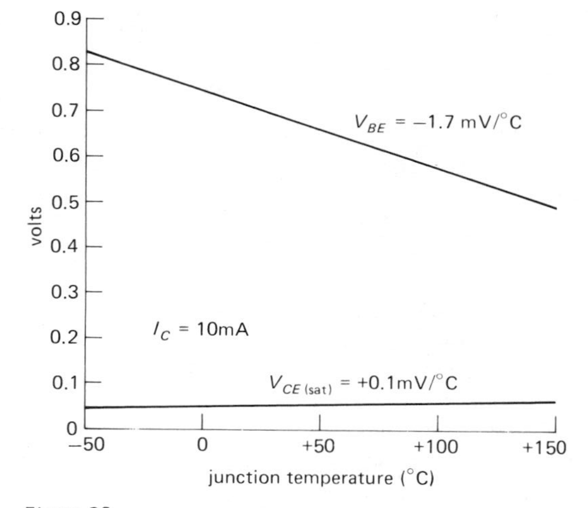

Figure G3

the emitter, as shown in the measured data of Figure G2.

V_{CE}(sat), the collector saturation voltage at a particular value of base current and collector current, is also relatively independent of temperature because of cancellation of the temperature coefficients of the two diodes (Fig. G3). This is of interest because a saturated transistor is frequently used to switch large currents and may get hot (e.g., 10A at a saturation voltage of 0.5V is 5W, enough to bring the junction of a small power transistor to 100°C or more).

In saturated switching applications you usually provide generous amounts of base current (typically 1/10 or 1/20 of the collector current) to achieve values of V_{CE}(sat) of 0.05V to 0.2V. If the load inadvertently demands much greater collector currents, the transistor will go out of saturation, with

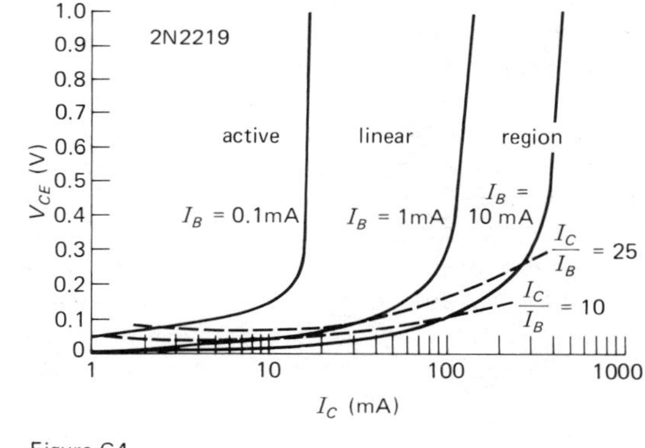

Figure G4

greatly increased power dissipation. The measured data in Figure G4 show that it is hard to define exactly when a transistor is "saturated"; you might use some arbitrary criterion such as $I_C = 10 I_B$.

APPENDIX H
LC BUTTERWORTH FILTERS

Active filters, as discussed in Chapter 4, are very convenient at low frequencies, but they are impractical at radiofrequencies because of the slew-rate and bandwidth requirements they impose on the operational amplifiers. At frequencies of 100kHz and above (and often at lower frequencies), the best approach is to design a passive filter with inductors and capacitors. (Of course, at UHF and microwave frequencies these "lumped-component" filters are replaced by stripline and cavity filters.)

As with active filters, there are many methods and filter characteristics possible with *LC* filters. For example, you can design the classic Butterworth, Chebyshev, and Bessel filters, each in low-pass, bandpass, high-pass, and band-reject varieties. It turns out that the Butterworth filter is particularly easy to design, and we can present in just a page or two all the essential design information for low-pass and high-pass Butterworth *LC* filters, and even a few examples. For further information we recommend the excellent handbook by Zverev cited in the Bibliography.

Table H1 gives the values of normalized inductances and capacitances for low-pass filters of various orders, from which actual circuit values are obtained by the frequency and impedance scaling rules:

Low-pass scaling rules

$$L_n(\text{actual}) = \frac{R_L L_n(\text{table})}{\omega}$$

$$C_n(\text{actual}) = \frac{C_n(\text{table})}{\omega R_L}$$

where R_L is the load impedance and ω is the angular frequency ($\omega = 2\pi f$).

Table H1 gives normalized values for 2-pole through 8-pole low-pass filters, for the two most common cases, namely (a) equal source and load impedances and (b) either source or load impedance much larger than the other. To use the table, first decide how many poles you need, based on the Butter-

TABLE H1. BUTTERWORTH LOW-PASS FILTERS[a] ($R_L = 1\Omega$)

π / T	R_S / $1/R_S$	C_1 / L_1	L_2 / C_2	C_3 / L_3	L_4 / C_4	C_5 / L_5	L_6 / C_6	C_7 / L_7	L_8 / C_8
$n = 2$	1	1.4142	1.4142						
	∞	1.4142	0.7071						
$n = 3$	1	1.0	2.0	1.0					
	∞	1.5	1.3333	0.5					
$n = 4$	1	0.7654	1.8478	1.8478	0.7654				
	∞	1.5307	1.5772	1.0824	0.3827				
$n = 5$	1	0.6180	1.6180	2.0	1.6180	0.6180			
	∞	1.5451	1.6944	1.3820	0.8944	0.3090			
$n = 6$	1	0.5176	1.4142	1.9319	1.9319	1.4142	0.5176		
	∞	1.5529	1.7593	1.5529	1.2016	0.7579	0.2588		
$n = 7$	1	0.4450	1.2470	1.8019	2.0	1.8019	1.2470	0.4450	
	∞	1.5576	1.7988	1.6588	1.3972	1.0550	0.6560	0.2225	
$n = 8$	1	0.3902	1.1111	1.6629	1.9616	1.9616	1.6629	1.1111	0.3902
	∞	1.5607	1.8246	1.7287	1.5283	1.2588	0.9371	0.5776	0.1951

[a] Values of L_n, C_n for 1Ω load resistance and cutoff frequency (−3dB) of 1 rad/s. See text for scaling rules.

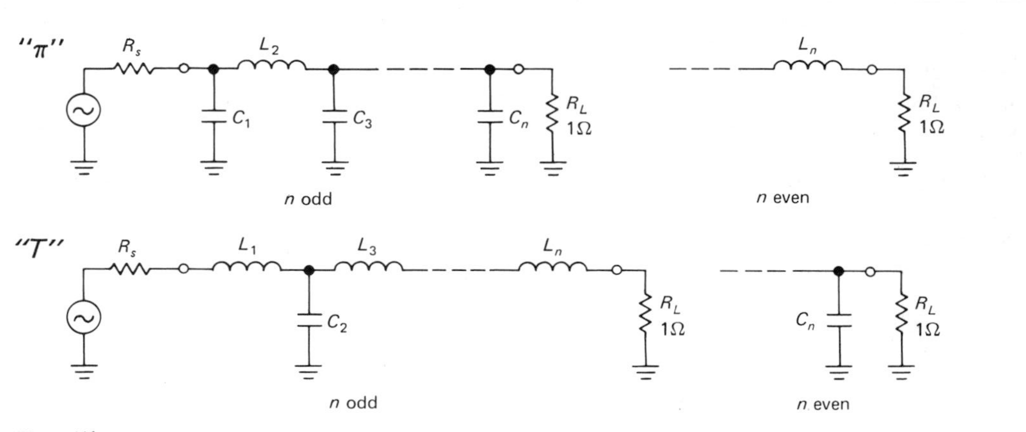

Figure H1

worth response (graphs are plotted in Sections 4.05 and 4.07). Then use the preceding equations to determine the filter configuration (T or π; see Figure H1) and component values. For equal source and load impedances, either configuration is OK; the π configuration may be preferable because it requires fewer inductors. For load impedance much higher (lower) than the source impedance, use the T (π) configuration.

To design a high-pass filter, follow the procedure outlined to determine which filter configuration to use and how many poles are necessary. Then do the universal low-pass to high-pass transformation shown in Figure H2, which consists simply of replacing

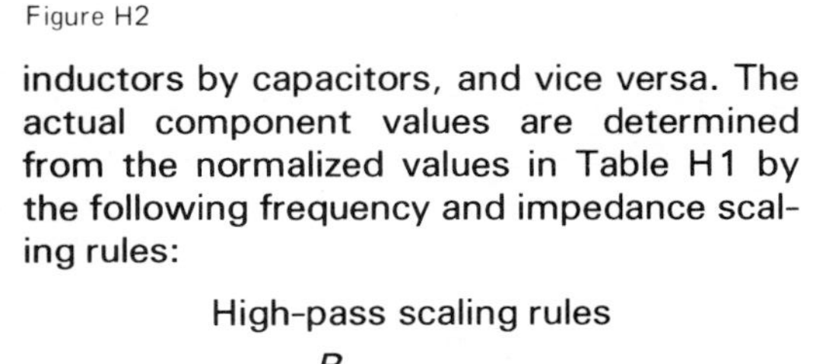

L (table) → C (actual) $\quad C\text{ (actual)} = \dfrac{1}{R_L\,\omega L\text{ (table)}}$

C (table) → L (actual) $\quad L\text{ (actual)} = \dfrac{R_L}{\omega C\text{ (table)}}$

normalized low-pass — actual high-pass

Figure H2

inductors by capacitors, and vice versa. The actual component values are determined from the normalized values in Table H1 by the following frequency and impedance scaling rules:

High-pass scaling rules

$$L_n\text{(actual)} = \frac{R_L}{\omega C_n\text{(table)}}$$

$$C_n\text{(actual)} = \frac{1}{R_L \omega L_n\text{(table)}}$$

We will show how to use the table to design both low-pass and high-pass filters with a few examples.

EXAMPLE I

Design a 5-pole low-pass filter for source and load impedances of 75 ohms, with a cutoff frequency (−3dB) of 1MHz.

We use the π configuration to minimize the number of required inductors. The scaling rules give us

$$C_1 = C_5 = \frac{0.618}{2\pi \times 10^6 \times 75} = 1310\text{pF}$$

$$L_2 = L_4 = \frac{75 \times 1.618}{2\pi \times 10^6} = 19.3\mu\text{H}$$

$$C_3 = \frac{2}{2\pi \times 10^6 \times 75} = 4240\text{pF}$$

The complete filter is shown in Figure H3.

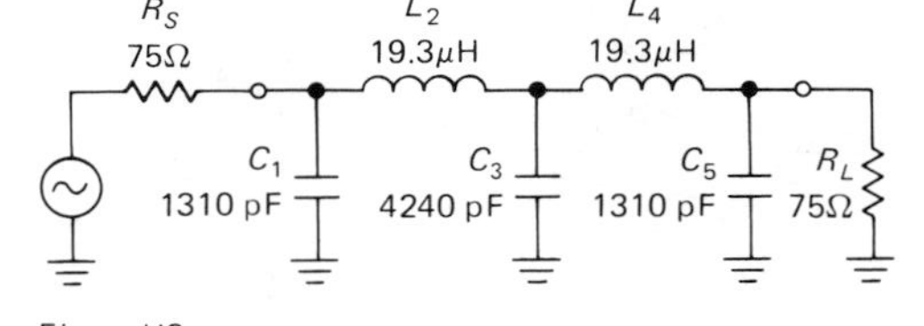

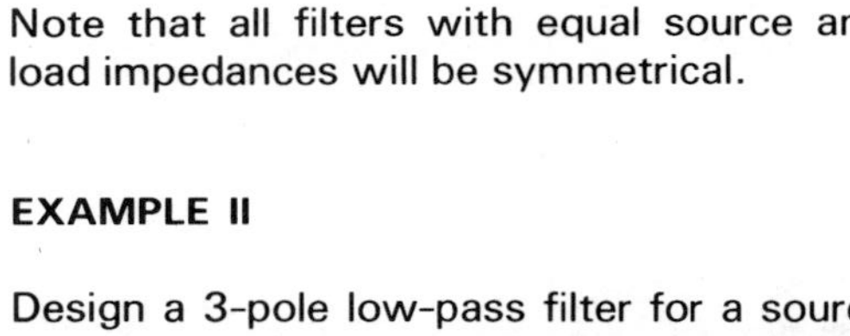

Figure H3

Note that all filters with equal source and load impedances will be symmetrical.

EXAMPLE II

Design a 3-pole low-pass filter for a source impedance of 50 ohms and a load impedance of 10k, with a cutoff frequency of 100kHz.

We use the T configuration, because $R_S \ll R_L$. For $R_L = 10\text{k}$, the scaling rules give

$$L_1 = \frac{10^4 \times 1.5}{2\pi \times 10^5} = 23.9\text{mH}$$

$$C_2 = \frac{1.3333}{2\pi \times 10^5 \times 10^4} = 212\text{pF}$$

$$L_3 = \frac{10^4 \times 0.5}{2\pi \times 10^5} = 7.96\text{mH}$$

The complete filter is shown in Figure H4.

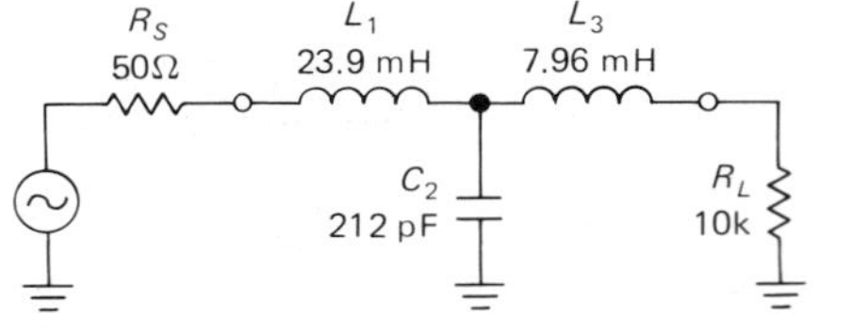

Figure H4

EXAMPLE III

Design a 4-pole low-pass filter for a zero-impedance source (voltage source) and a 75 ohm load, with cutoff frequency of 10MHz.

We use the T configuration, as in the previous example, because $R_S \ll R_L$. The scaling rules give

$$L_1 = \frac{75 \times 1.5307}{2\pi \times 10^7} = 1.83\mu\text{H}$$

$$C_2 = \frac{1.5772}{2\pi \times 10^7 \times 75} = 335\text{pF}$$

$$L_3 = \frac{75 \times 1.0824}{2\pi \times 10^7} = 1.29\mu\text{H}$$

$$C_4 = \frac{0.3827}{2\pi \times 10^7 \times 75} = 81.2\text{pF}$$

The complete filter is shown in Figure H5.

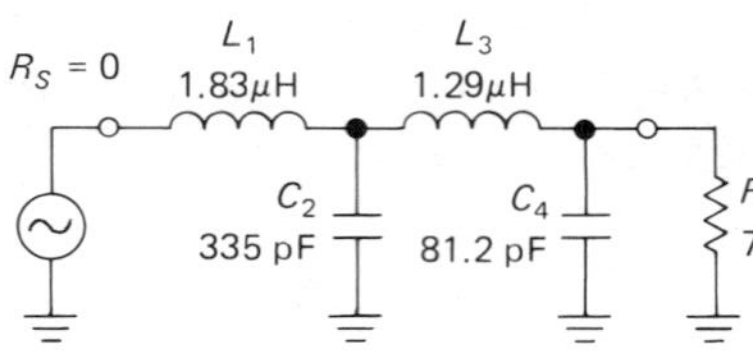

Figure H5

EXAMPLE IV

Design a 2-pole low-pass filter for current-source drive and 1k load impedance, with cutoff frequency of 10kHz.

We use the π configuration, because $R_S \gg R_L$. The scaling rules give

$$C_1 = \frac{1.4142}{2\pi \times 10^4 \times 10^3} = 0.0225\mu\text{F}$$

$$L_2 = \frac{10^3 \times 0.7071}{2\pi \times 10^4} = 11.3\text{mH}$$

The complete filter is shown in Figure H6.

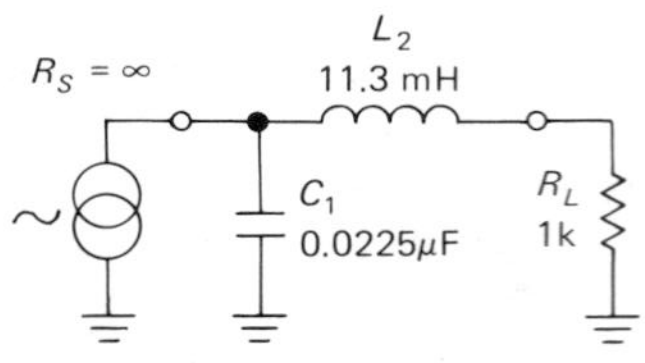

Figure H6

EXAMPLE V

Design a 3-pole high-pass filter for 52 ohm source and load impedances, with cutoff frequency of 6MHz.

We begin with the T configuration, then transform inductors to capacitors, and vice versa, giving

$$C_1 = C_3 = \frac{1}{52 \times 2\pi \times 6 \times 10^6 \times 1.0} = 510\text{pF}$$

$$L_2 = \frac{52}{2\pi \times 6 \times 10^6 \times 2.0} = 0.690\mu\text{H}$$

The complete filter is shown in Figure H7.

We would like to emphasize that the field of passive filter design is rich and varied and that this simple table of Butterworth filters doesn't even begin to scratch the surface.

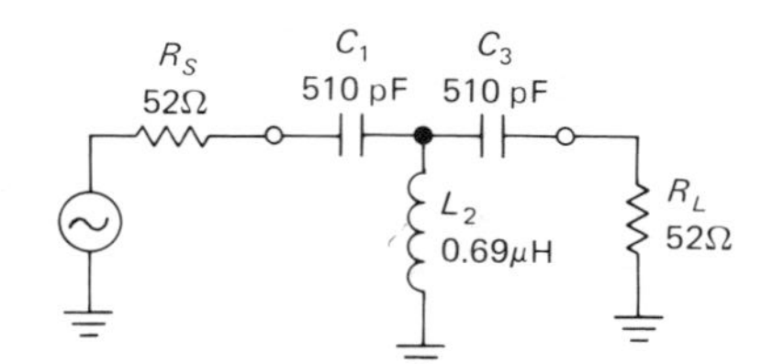

Figure H7

ELECTRONICS MAGAZINES AND JOURNALS

APPENDIX I

In this appendix we've gathered together under various headings a collection of magazines that are worth knowing about. We've marked with an asterisk the magazines that in our opinion are the major ones in their fields. All of the trade magazines are well supplied with advertisements proclaiming the specifications and virtues of new ICs, instruments, etc., and these ads are a good source of information about new products; they don't have the nuisance value of ordinary newspaper and magazine ads. There is always a "reader information card" (or "bingo" card) at the back of the magazine on which you can circle the numbers corresponding to advertisements you're interested in. More information then arrives in the mail in a few weeks. The system works well.

TRADE JOURNALS, GENERAL

*****Electronics***
****EDN***
****Electronic Design***
Electronic Products. At least one of these four magazines should be considered required reading to keep up with new components and design ideas. The advertisements are as important as the articles. *Electronics* is available only as a paid subscription; the others are free to qualifying engineers and are available as expensive subscriptions otherwise (one of the authors tried for more than a year to obtain a free subscription to *Electronic Design,* without success).
Electronic Business. The *Business Week* of electronics.
****Electronic News.*** Best newspaper for the electronics industry.
Spectrum (IEEE). General-interest electronics magazine, put out by IEEE. Good review articles covering broad range of subjects.

TRADE JOURNALS, COMMUNICATIONS

CATJ. Emphasis on cable and satellite TV systems.
****Microwave Journal.*** Emphasis on components for VHF and above. Includes good technical articles.
Microwave Systems News. Similar to *Microwave Journal.*
Microwaves. Similar to *Microwave Journal.*
Telecommunications. Emphasis on communications systems.

TRADE JOURNALS, COMPUTING

Computer (IEEE). The *Scientific American* of computing.
****Computer Design.*** Definitive magazine on digital hardware and software techniques for large computer systems.
Datamation. Oriented toward large computer systems.
Digital Design. Oriented toward peripherals.
Mini-micro Systems. The professional microprocessor and minicomputer magazine.

TRADE JOURNALS, INSTRUMENTATION

Actuator Systems. Emphasis on hydraulic, pneumatic, electrical, and mechanical actuators.
****Industrial Research and Development.*** General-interest technology monthly.
****Instrument and Apparatus News.*** Gen-

eral-purpose instrument and sensor newspaper.
Measurement and Control News. Includes biomedical and chemical instrumentation.

TRADE JOURNALS, MANUFACTURING

Circuits Manufacturing. For the serious manufacturer of electronic systems and components.
Electronics Test. Emphasis on production test equipment.
Insulation Circuits. Emphasis on printed-circuit boards and interconnections.

TRADE JOURNALS, OPTICS

Electro-optical Systems Design (EOSD). Optical technology.
Laser Focus. Optical technology.
Optical Spectra. Optical technology.

HOBBY MAGAZINES, AUDIO

Audio. Some technical articles.
The Audio Amateur. Devoted to the experimenter/builder.

HOBBY MAGAZINES, COMMUNICATIONS

CQ. Amateur radio, general interest.
****Ham Radio.*** The most technical of the amateur publications.
****QST.*** Broad interest, published by ARRL.
73. Amateur radio, general interest.

HOBBY MAGAZINES, COMPUTERS

****Byte.*** The first large-circulation personal computing magazine. Good tutorials on wide range of topics to which computers can be applied.
Creative Computing. Emphasis on computers in education; a "fun" magazine – games, etc.
Dr. Dobbs Journal. For programmers. Emphasis on software systems and design.
****Interface Age.*** Good coverage of new peripherals and products.
****Kilobaud Microcomputing*** (formerly *Kilobaud*). Oriented toward smallest microcomputer systems (e.g., the TRS-80). Includes hardware and assembly-language articles.
Personal Computing. Oriented toward computer basics. Features BASIC.

HOBBY MAGAZINES, GENERAL

Popular Electronics. Wide-circulation magazine for the hobbyist.
Radio Electronics. Wide-circulation magazine for the hobbyist.
Wireless World. British all-around electronics magazine for hobbyists and professionals.

PROFESSIONAL JOURNALS

****Journal of the Audio Engineering Society.*** Definitive journal on audio and acoustics.
****Journal of Solid State Circuits*** (IEEE). Circuit design and new ICs.
****Nuclear Instruments and Methods.*** Scientific instrumentation, with emphasis on particle, x-ray, gamma-ray, and nuclear science.
*****Proceedings of the IEEE.*** Good review articles and other technical articles for engineers.
****Review of Scientific Instruments.*** Scientific instrumentation.
Studio Sound. British professional audio journal.

In addition, there are the many *IEEE Transactions,* such as *Acoustics, Speech, and Signal Processing; Audio; Biomedical Engineering; Broadcasting; Circuit Theory; Circuits and Systems; Communications; Computers; Electron Devices; Instrumentation and Measurement;* and *Microwave Theory and Techniques.*

APPENDIX J
IC GENERIC TYPES

Typical problem: You need to replace an integrated circuit, or at least find some data on it. It says

DM8095N

7410 NS

and lives in a 16-pin DIP. What is it? The 7410 has a familiar sound, so you order a few. A week later they arrive, in 14-pin DIPs! Banging your head on the nearest wall (since you should have known this all along), you realize you're back where you were a week ago, but with a handful of spare 3-input NANDs for consolation.

What's needed is a master list of IC prefixes, suffixes, and number series for all ICs. This appendix is our attempt to bring some order out of chaos. We make no pretense of accuracy or completeness, especially since the list is proliferating every day.

That mystery chip is a National Semiconductor 8095 hex three-state TTL buffer, by the way, manufactured in the 10th week of 1974.

COMPANY PREFIXES

The various semiconductor manufacturers use distinctive (usually) prefixes in front of the IC number, even if it is an IC type made by many different companies. The DM in the preceding example indicates a digital monolithic IC made by National Semiconductor (also indicated by the NS logo). Here is a list of most of the prefixes now in use:

Prefix(es)	*Manufacturer*
AD	Analog Devices
Am	Advanced Micro Devices (AMD)
AY, GIC, GP	General Instrument (GI)
C, I	Intel
CA, CD, CDP	RCA
CA, TDC, MPY	TRW
CMP, DAC, MAT, OP, PM, REF, SSS	Precision Monolithics
DM, LF, LFT, LH, LM, NH	National Semiconductor (NSC)
F, μA, μL, U*nx*	Fairchild (FSC)
FSS, ZLD	Ferranti
GEL	GE
HA	Harris
HEP, MC, MCC, MCM, MFC, MM, MWM	Motorola
ICH, ICL, ICM, IM	Intersil
ITT, MIC	ITT
L, LD	Siliconix
MB	Fujitsu
MCS	MOS Technology
MIL	Microsystems International
MK	Mostek
MM	Teledyne-Amelco, Monolithic Memories, Motorola
MN, SL, SP	Plessey
N, NE, S, SE, SP	Signetics
PD	Philco-Ford (old)
R, RAY, RC, RM	Raytheon
SFC	ESMF
SG	Silicon General
SN, TMS	Texas Instruments (TI)
SW	Stewart Warner
TAA, TBA	AEG, Amperex, SGS, Siemens, Telefunken
TVR	Transitron
UC	Union Carbide, Solitron
ULN, ULS	Sprague
μPB, μPD	NEC
WC, WM	Westinghouse
5082-*nnnn*	Hewlett-Packard (HP)

LINEAR GENERIC NUMBER SERIES

These are most of the common linear series; *n* means a digit, *x* means a letter. Parentheses mean an optional digit or letter.

AD5*nn*	Analog Devices
CA3*nnn*	RCA
CMP*nn*, MAT*nn*, OP*nn*, PM*nnn*, REF*nn*	Precision Monolithics

HA2*nnn*, HA4*nnn*, HA5*nnn*	Harris
LM1*nnn*, LM2*nn*, LM3*nn*, LF1*nnn*, LF2*nnn*, LF3*nnn*	National Semiconductor
MC1*nnn*	Motorola
NE5*nn*, NE5*nnn*, SE5*nn*, SE5*nnn*	Signetics
RC4*nnn*	Raytheon
TLO*nn*	Texas Instruments
μA7*nn*	Fairchild
1*nnn*	Teledyne-Philbrick
3*nnn*	Burr-Brown
8*nn*	Teledyne-Amelco
9*nnn*	Optical Electronics (OEI)

DIGITAL GENERIC NUMBER SERIES

Chaos reigns supreme in the digital world, especially with the rapid introduction of new devices and microprocessor (μP) series.

10*nnn*	ECL
14*nnn*(*n*)	Motorola CMOS
18*nn*	RCA μP
2*nnn*	memory
25*nn*(*n*), 25LS*nn*(*n*)	AMD TTL
26*nn*	Signetics μP
29*nn*	AMD μP
3*nn*	Signetics Utilogic
3*nn*	Teledyne HNIL logic
34*nnn*	Motorola CMOS (old)
38*nn*	Fairchild μP
4*nnn*	Motorola TTL (old)
4*nnn*(*n*)	CMOS
54---	MIL version of 74---
65*nn*	MOS Technology μP
68*nn*(*n*)	Motorola μP
7*nn*	Motorola RTL
7*nnn*, 7L*nnn*, 7LS*nnn*	National TTL
74*nn*(*n*)	TTL
74ALS*nn*(*n*)	"Advanced low-power Schottky" TTL
74AS*nn*(*n*)	"Advanced Schottky" TTL
74F*nn*(*n*)	Fast (Fairchild advanced Schottky technology) TTL
74H*nn*(*n*)	High-speed TTL
74L*nn*(*n*)	Low-power TTL
74LS*nn*(*n*)	Low-power Schottky TTL
74S*nn*(*n*)	Schottky TTL
75*nnn*	Interface
8*nn*	Motorola RTL
8*nn*	DTL
8*nnn*, 8L*nnn*, 8LS*nnn*	National TTL
8*nnn*	Intel μP
8*nnn*	Signetics TTL
8T*nn*	Signetics Interface
9*nn*	DTL
9*nnn*	Signetics TTL (MIL)
9*nnn*, 9*n*L*nn*, 9*n*LS*nn*	Fairchild TTL
95*nn*	AMD μP

SUFFIXES

Suffix letters indicate package type and temperature range. There are three standard temperature ranges: "Military" (−55°C to +125°C), "Industrial" (−25°C to +85°C), and "Commercial" (0°C to +70°C). Commercial is adequate for anything intended for use in normal building environments. As luck would have it, each manufacturer has its own set of suffixes, subject to frequent modification. Since it wouldn't help you identify an IC, we won't attempt to list them here. (However, it's essential to look up the correct suffix before you order, or ask the distributor for assistance.)

DATE CODES

Most ICs and transistors, and many other electronic components, are stamped with a simple four-digit code giving date of manufacture: the first two digits are the year, the last two are the week of the year. In the example given earlier, 7410 means the second week of March 1974. They're sometimes useful, such as for estimating the age of components that have a finite useful life (electrolytic capacitors, for instance); unfortunately the components with the shortest life (batteries) are purposely coded so you can't figure out the date. If you get a batch of ICs with an abnormally high failure rate (most manufacturers test only a sample of each batch; typically 1% of the ICs you buy will not meet specifications), avoid replacements with the same date code. Date codes can also help you estimate the date of manufacture of commercial electronic equipment. Since ICs don't become stale, there's no reason to avoid an IC with an old date code.

Warning: As the preceding example illustrates, 1974 was an unfortunate vintage because of confusion of date codes with the popular 7400 series of TTL ICs. For a couple of years after that, even some electronic parts distributors were shipping the wrong stuff!

DATA SHEETS
APPENDIX K

In this appendix we have reproduced six data sheets just as they were printed by the manufacturer. We chose representative or popular devices, looking especially for data sheets that were comprehensive and clear.

On the following pages you will find data sheets for these devices:

1N914 The universal signal diode (from the 3rd edition of the *GE Semiconductor Handbook*). (Courtesy of General Electric Semiconductor Products Department, Auburn, N.Y.)

2N4400-4401 A popular signal transistor (from the *Motorola Semiconductor Library, Vol. 1,* 1974). (Courtesy of Motorola Semiconductor Products Inc.)

LM194-394 An excellent superbeta matched transistor pair (from the *National Semiconductor Linear Data Book,* 1978). (Courtesy of National Semiconductor Corp.)

LF355-357 A popular series of JFET operational amplifiers (from the *National Semiconductor Linear Data Book,* 1978). (Courtesy of National Semiconductor Corp.)

LM317 A popular 3-terminal adjustable positive voltage regulator (from the *National Semiconductor Linear Data Book,* 1978). (Courtesy of National Semiconductor Corp.)

96LS02 A good data sheet for a good dual TTL one-shot. (Courtesy of Fairchild Camera and Instrument Corp.)

Silicon
Diodes

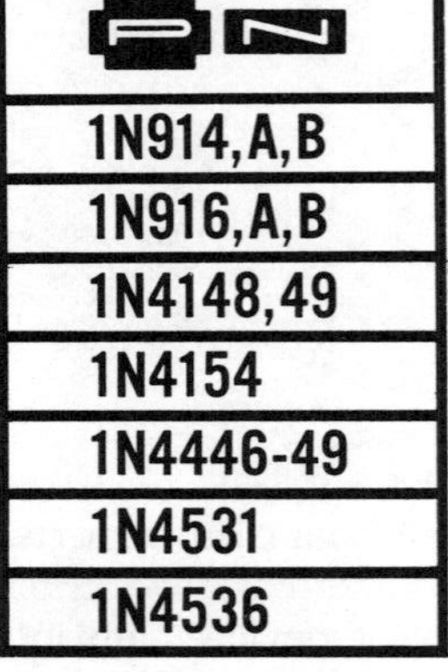

This family of General Electric silicon signal diodes are very high speed switching diodes for computer circuits and general purpose applications. These diodes incorporate an oxide passivated planar structure. This structure makes possible a diode having high conductance, fast recovery time, low leakage, and low capacitance combined with improved uniformity and reliability. These diodes are contained in two different packages; double heat sink miniature package, and milli-heat sink package.
They are electrically the same as their equivalent types in each of the two different packages (see page two for groupings of electrically equivalent types in each of the two packages).

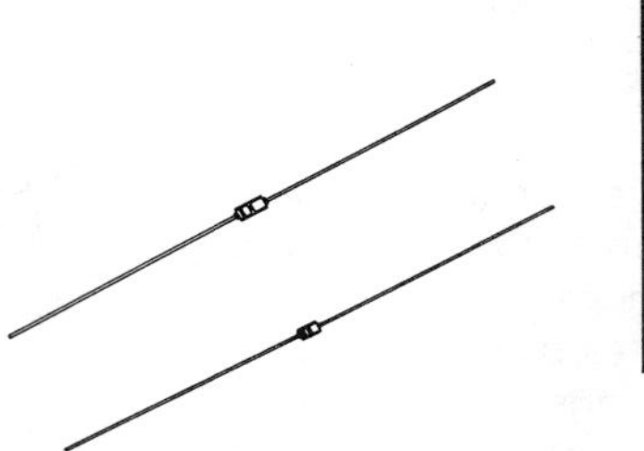

PLANAR EPITAXIAL PASSIVATED
with Controlled Conductance

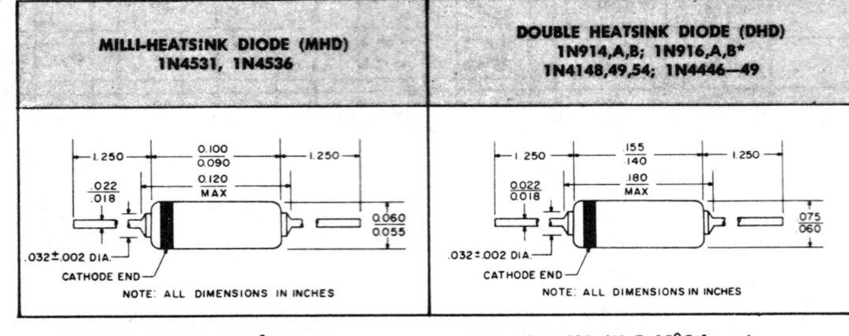

Dissipation: 500mW @ 25°C free air
Derate: 2.85mW/°C for temp. above 25°C amb. based on max. $T_J = 200°C$

Dissipation: 500mW @ 25°C free air
Derate: 2.85mW/°C for temp. above 25°C amb. based on max. $T_J = 200°C$

FEATURES	1N914 1N914A 1N914B	1N4148 1N4446 1N4448 1N4531	1N916 1N916A 1N916B	1N4149 1N4447 1N4449	1N4536 1N4154
Reverse Recovery Time of 2 nanoseconds maximum					●
Reverse Recovery Time of 4 nanoseconds maximum	●	●	●	●	
Capacitance of 2 pF maximum			●	●	
Capacitance of 4 pF maximum	●	●			●
Power Dissipation to 500 mW		●		●	●
Power Dissipation to 250 mW					
Meets all MIL-S-19500C requirements	●	●	●	●	●

Figure 1

HEATSINK SPACING FROM END OF DIODE BODY	STEADY STATE THERMAL RESISTANCE °C/mW (NOTE 1)			POWER DISSIPATION AT 25°C mW (NOTE 2)		
	MHD		DHD	MHD		DHD
.062″	.230		.250	760		700
.250″	.319		.319	550		550
.500″	.438		.438	400		400

NOTE 1 See Figure 7 for thermal resistance for short pulses.
NOTE 2 This power rating is based on a maximum junction temperature of 200°C.

absolute maximum ratings: (25°C)

(unless otherwise specified)

	1N914,A,B 1N916,A,B 1N4148–49 1N4446–47 1N4448–49 1N4531	1N4154 1N4536	
	MHD & DHD	MHD & DHD	
Voltage			
Reverse	75	25	Volts
Current			
Average Rectified	150	150	mA
Recurrent Peak Forward	450	450	mA
Forward Steady-State DC	200	200	mA
Peak Forward Surge (1μsec. pulse)	2000	2000	mA
Power			
Dissipation	500	500	mW
Temperature			
Operating	←—65 to +200—→		°C
Storage	←—65 to +200—→		°C

electrical characteristics: (25°C) (unless otherwise specified)

Type	Minimum Breakdown Voltage @ 100μA	Forward Voltage		Maximum Reverse Current, I_R			C_o (1)	t_{rr} (2)	V_f (3)
		I_F	V_F	20V		75V			
				25°C	150°C	25°C			
	Volts	mA	V	nA	μA	μA	pF	ns	V
1N914 1N4148 1N4531	100	10	1.0	25	50	5	4	4	
1N914A 1N4446	100	20	1.0	25	50	5	4	4	
1N914B 1N4448	100	5 100	0.62–0.72 1.0	25(4)	50	5	4	4	2.5
1N916 1N4149	100	10	1.0	25	50	5	2	4	
1N916A 1N4447	100	20	1.0	25	50	5	2	4	
1N916B 1N4449	100	5 30	0.63–0.73 1.0	25	50	5	2	4	2.5
1N4154 1N4536	35 @ 5μA	30	1.0	100 @ 25V	100 @ 25V		4	2	

*Except as noted.

NOTES (1) Maximum Capacitance is measured on Boonton model 75A capacitance bridge at a signal level of 50 mV at $V_R = 0$

(2) Maximum Reverse Recovery Time, I_f = 10mA, V_R = −6V, R_L = 100Ω, Recovery to 1.0mA (Figure 6)

(3) Maximum Forward Recovery Voltage, −50mA peak square wave, 0.1 μsec. pulse width, 5 to 100 kHz repetition rate, generator rise time (t_r) ≦ 30nsec.

(4) Also 3μA at 20 V at 100°C

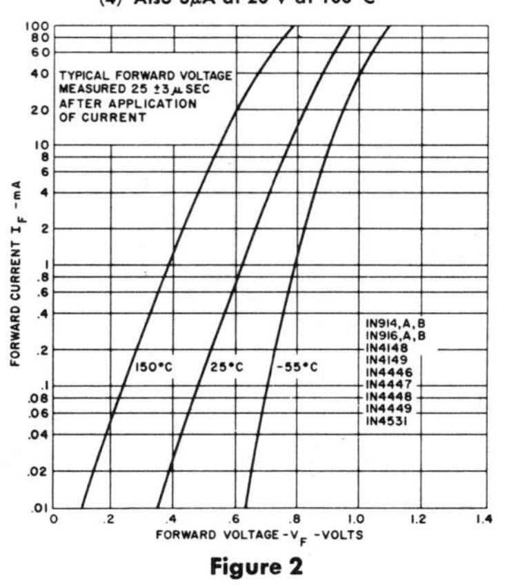

Figure 2

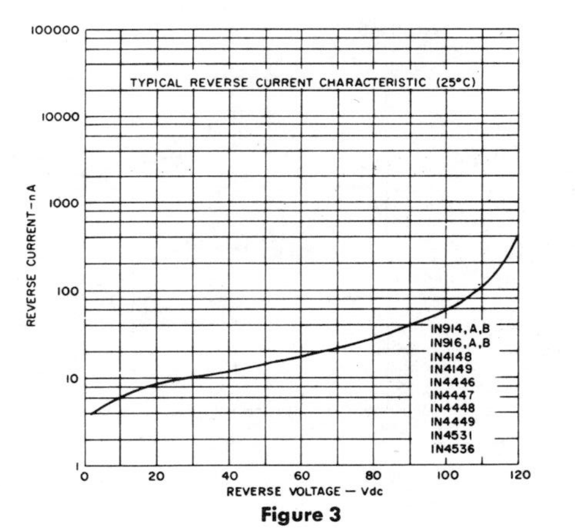

Figure 3

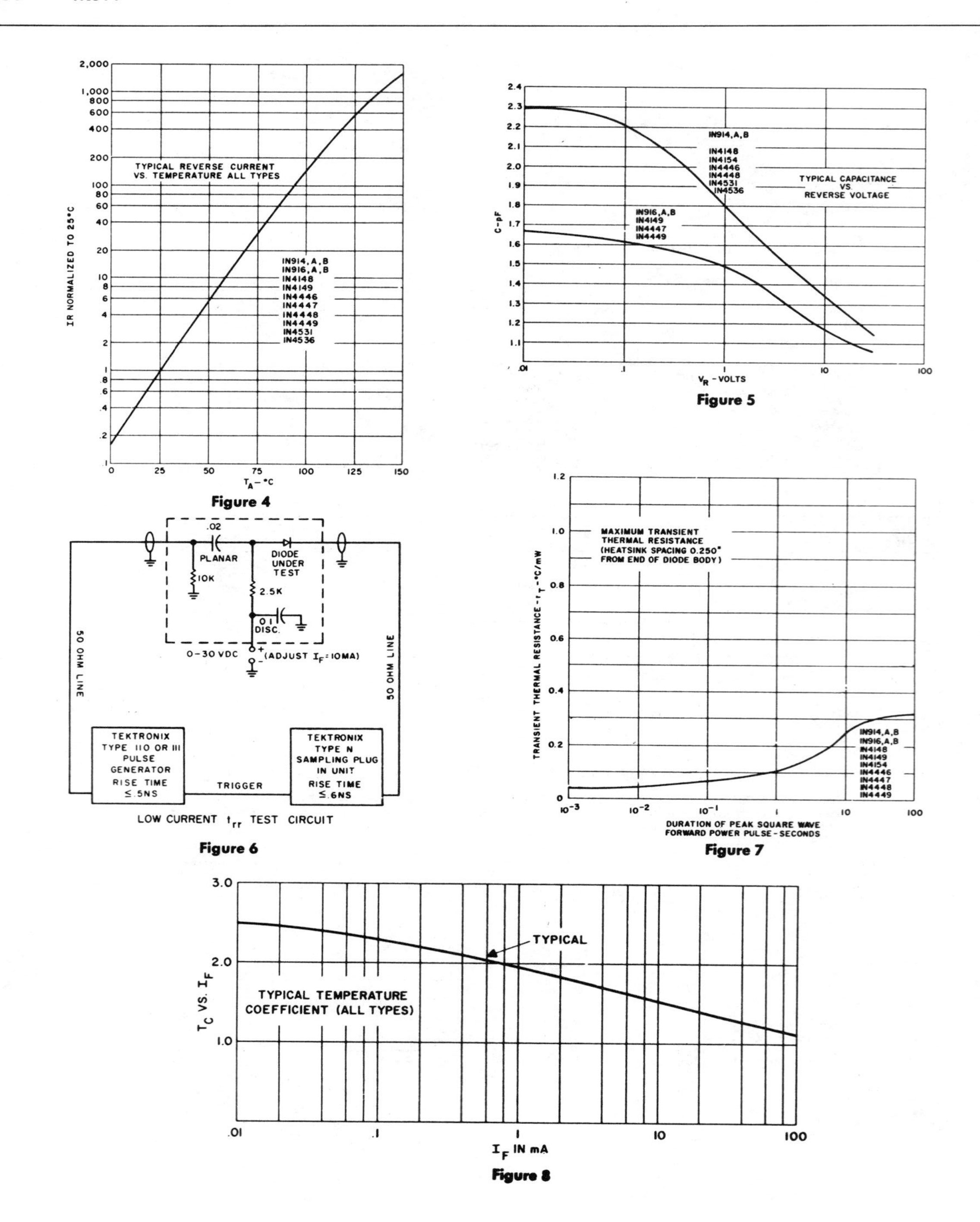

Figure 4

Figure 5

Figure 6

Figure 7

Figure 8

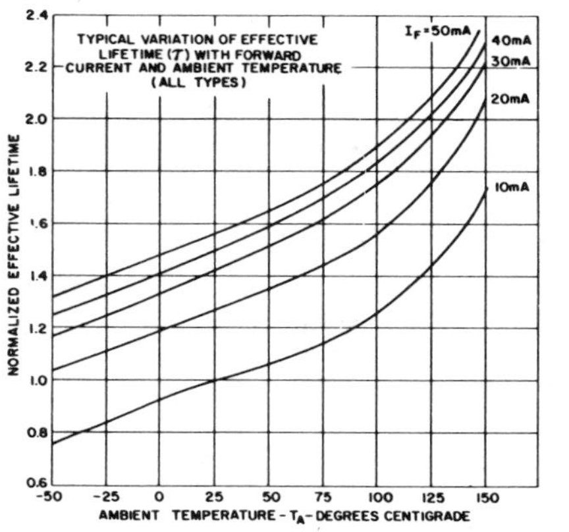

Figure 9

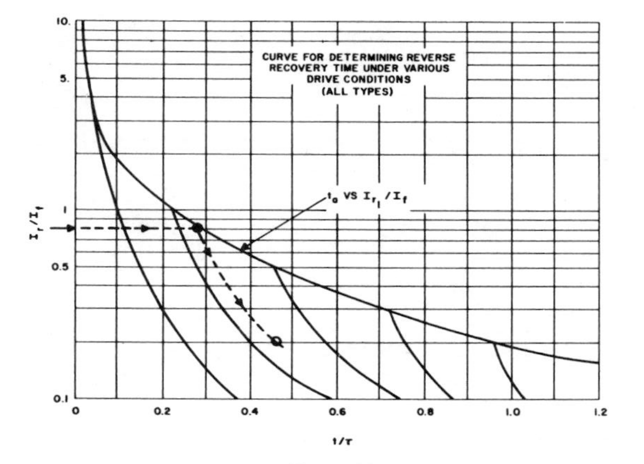

Figure 10

ESTIMATION OF REVERSE RECOVERY TIME UNDER VARIOUS DRIVE CONDITIONS

The reverse recovery time of a silicon signal diode has been shown* to be determined by a quantity called the effective lifetime, τ, and the ratio of forward and reverse current. The exact equations expressing times t_a and t_b (as defined in the sketch at right) are somewhat inconvenient for numerical evaluation, but in many cases an estimation of response time is sufficient. Figure 10 is a graphical solution to the response time equations and its use can best be illustrated by the following example:

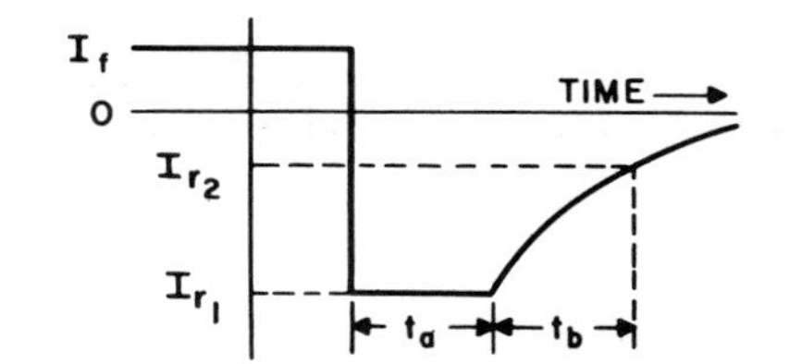

FIND: Recovery time to 5 mA reverse current when the forward current is 25 mA and the maximum reverse current is 20 mA.

SOLUTION: Enter the left side of Figure 10 at $I_{r1}/I_f = 20/25 = 0.8$ and follow horizontally until the t_a vs. I_{r1}/I_f line is reached (see dotted line). From the t/τ scale of the horizontal axis, it is seen that t_a is 0.28τ. The t_b portion of the recovery curve is estimated by moving downward parallel to the general contour lines until the $I_{r2}/I_f = 5/25 = 0.2$ line is reached. The total switching time is thus 0.46τ. The delay time, t_b, is $0.46\tau - 0.28\tau$ or 0.18τ.

The value of τ on the spec sheet should be corrected for current level. Figure 9 shows the typical variation of effective lifetime with forward current. Since the current level of the example is 25 mA, the maximum effective lifetime is approximately (6.8) (1.35) or 9.3 nsec., therefore:

$t_a \approx (9.3)\ (.28) \approx 2.6$ nsec. maximum
$t_b \approx (9.3)\ (.18) \approx 1.7$ nsec. maximum

Total reverse recovery time $\approx$ 4.3 nsec. maximum

Additional information on this method of diode recovery time calculation is contained in a paper entitled "Predicting Reverse Recovery Time of High Speed Semiconductor Junction Diodes" by C. H. Chen, (Publication #90.36) available on request.

*Ko, W. H., "The Reverse Transient Behavior of Semiconductor Junction Diodes," IRE Trans. ED-8, March 1961, pp. 123-131.

2N4400
2N4401

NPN SILICON SWITCHING & AMPLIFIER TRANSISTORS

AUGUST 1966 — DS 5198

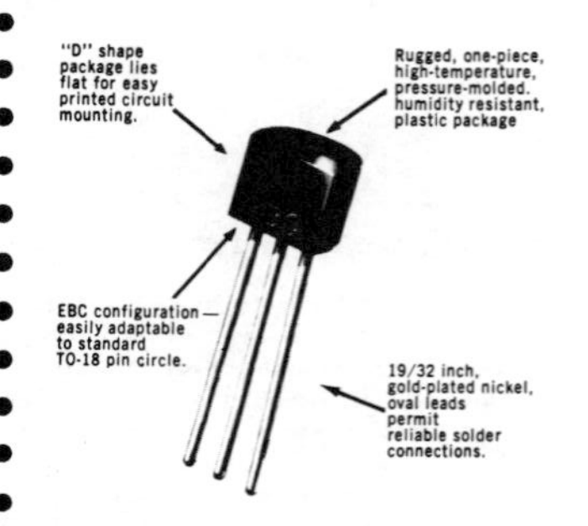

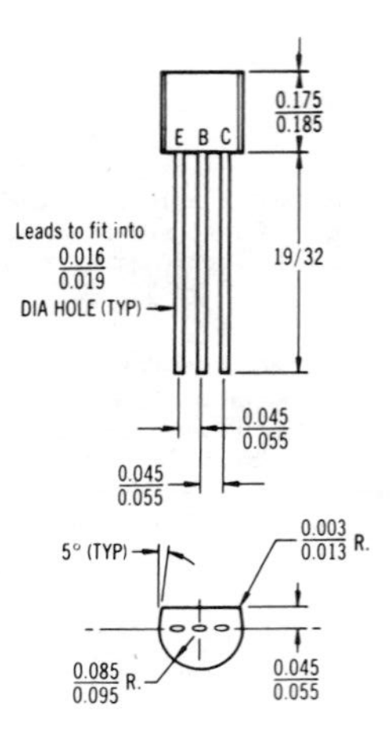

TO-92 OUTLINE

NPN SILICON ANNULAR* TRANSISTORS

. . . designed for general purpose switching and amplifier applications and for complementary circuitry with PNP types 2N4402 and 2N4403.

- High Voltage Ratings — BV_{CEO} = 40 V minimum
- Current Gain Specified from 0.1 mA to 500 mA
- Low Saturation Voltage
  $V_{CE(sat)}$ = 0.4 V maximum @ I_C = 150 mA
- Complete Switching and Amplifier Specifications
- One-Piece, Injection-Molded Unibloc† Package

MAXIMUM RATINGS

Characteristic	Symbol	Rating	Unit
Collector-Emitter Voltage	V_{CEO}	40	Vdc
Collector-Base Voltage	V_{CB}	60	Vdc
Emitter-Base Voltage	V_{EB}	6	Vdc
Collector Current - Continuous	I_C	600	mAdc
Total Device Dissipation T_A = 25°C	P_D	310	mW
Derate above 25°C		2.81	mW/°C
Operating & Storage Junction Temperature Range	T_J, T_{stg}	-55 to +135	°C

THERMAL CHARACTERISTICS

Characteristic	Symbol	Max	Unit
Thermal Resistance, Junction to Case	θ_{JC}	0.137	°C/mW
Thermal Resistance, Junction to Ambient	θ_{JA}	0.357	°C/mW

*Annular Semiconductors patented by Motorola Inc.
†Trademark of Motorola Inc.

ELECTRICAL CHARACTERISTICS (T_A = 25°C unless otherwise noted)

Characteristic	Fig. No.	Symbol	Min	Max	Unit
OFF CHARACTERISTICS					
Collector-Emitter Breakdown Voltage* (I_C = 1 mAdc, I_B = 0)		BV_{CEO}*	40	—	Vdc
Collector-Base Breakdown Voltage (I_C = 0.1 mAdc, I_E = 0)		BV_{CBO}	60	—	Vdc
Emitter-Base Breakdown Voltage (I_E = 0.1 mAdc, I_C = 0)		BV_{EBO}	6	—	Vdc
Collector Cutoff Current (V_{CE} = 35 Vdc, $V_{EB(off)}$ = 0.4 Vdc)		I_{CEX}	—	0.1	μAdc
Base Cutoff Current (V_{CE} = 35 Vdc, $V_{EB(off)}$ = 0.4 Vdc)		I_{BL}	—	0.1	μAdc
ON CHARACTERISTICS					
DC Current Gain (I_C = 0.1 mAdc, V_{CE} = 1 Vdc) 2N4401	15	h_{FE}	20	—	—
(I_C = 1 mAdc, V_{CE} = 1 Vdc) 2N4400 2N4401			20 40	— —	
(I_C = 10 mAdc, V_{CE} = 1 Vdc) 2N4400 2N4401			40 80	— —	
(I_C = 150 mAdc, V_{CE} = 1 Vdc)* 2N4400 2N4401			50 100	150 300	
(I_C = 500 mAdc, V_{CE} = 2 Vdc)* 2N4400 2N4401			20 40	— —	
Collector-Emitter Saturation Voltage* (I_C = 150 mAdc, I_B = 15 mAdc) (I_C = 500 mAdc, I_B = 50 mAdc)	16, 17, 18	$V_{CE(sat)}$	— —	0.4 0.75	Vdc
Base-Emitter Saturation Voltage* (I_C = 150 mAdc, I_B = 15 mAdc) (I_C = 500 mAdc, I_B = 50 mAdc)	17, 18	$V_{BE(sat)}$	0.75 —	0.95 1.2	Vdc
SMALL-SIGNAL CHARACTERISTICS					
Current-Gain — Bandwidth Product (I_C = 20 mAdc, V_{CE} = 10 Vdc, f = 100 MHz) 2N4400 2N4401		f_T	200 250	— —	MHz
Collector-Base Capacitance (V_{CB} = 5 Vdc, I_E = 0, f = 100 kHz, emitter guarded)	3	C_{cb}	—	6.5	pF
Emitter-Base Capacitance (V_{BE} = 0.5 Vdc, I_C = 0, f = 100 kHz, collector guarded)	3	C_{eb}	—	30	pF
Input Impedance (I_C = 1 mAdc, V_{CE} = 10 Vdc, f = 1 kHz) 2N4400 2N4401	12	h_{ie}	500 1.0k	7.5k 15k	ohms
Voltage Feedback Ratio (I_C = 1 mAdc, V_{CE} = 10 Vdc, f = 1 kHz)	13	h_{re}	0.1	8	X 10^{-4}
Small-Signal Current Gain (I_C = 1 mAdc, V_{CE} = 10 Vdc, f = 1 kHz) 2N4400 2N4401	11	h_{fe}	20 40	250 500	—
Output Admittance (I_C = 1 mAdc, V_{CE} = 10 Vdc, f = 1 kHz)	14	h_{oe}	1	30	μmhos
SWITCHING CHARACTERISTICS					
Delay Time — V_{CC} = 30 Vdc, $V_{EB(off)}$ = 2 Vdc,	1, 5	t_d	—	15	ns
Rise Time — I_C = 150 mAdc, I_{B1} = 15 mAdc	1, 5, 6	t_r	—	20	ns
Storage Time — V_{CC} = 30 Vdc, I_C = 150 mAdc,	2, 7	t_s	—	225	ns
Fall Time — I_{B1} = I_{B2} = 15 mAdc	2, 8	t_f	—	30	ns

*Pulse Test: Pulse Width ≤ 300 μs, Duty Cycle ≤ 2%

SWITCHING TIME EQUIVALENT TEST CIRCUITS

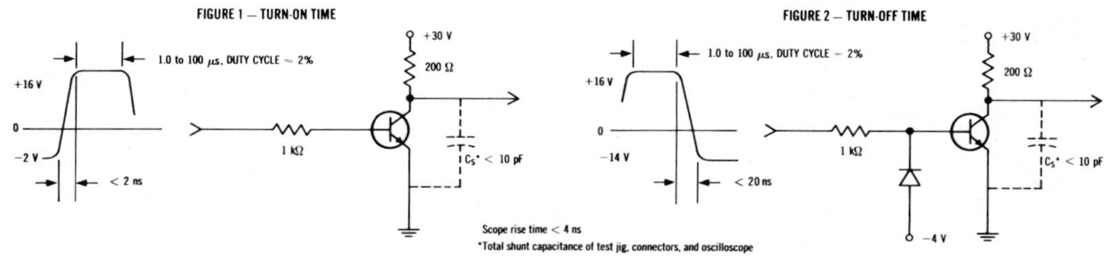

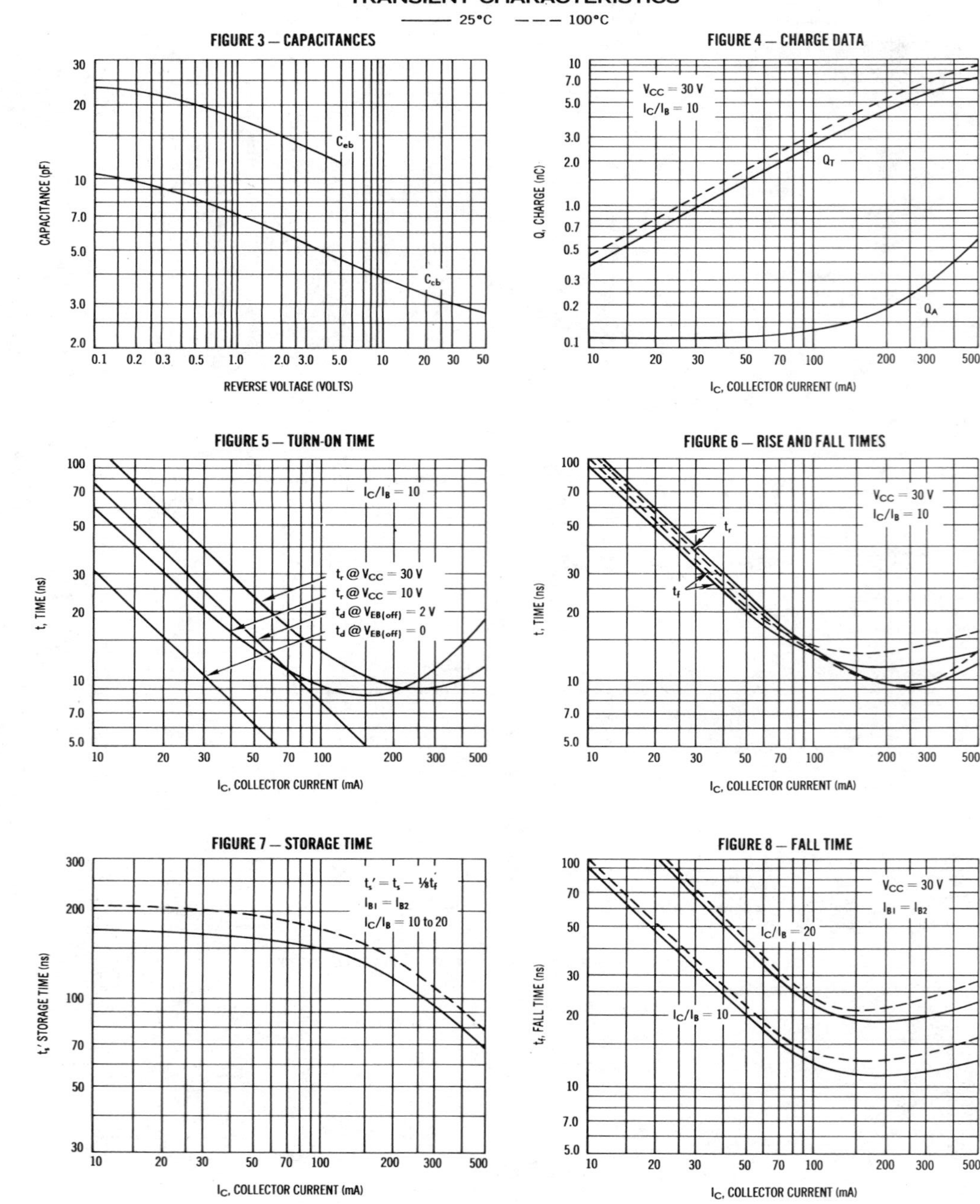

TRANSIENT CHARACTERISTICS

——— 25°C – – – 100°C

FIGURE 3 — CAPACITANCES

CAPACITANCE (pF)
30
20
10
7.0
5.0
3.0
2.0
C_{eb}
C_{cb}
0.1 0.2 0.3 0.5 1.0 2.0 3.0 5.0 10 20 30 50
REVERSE VOLTAGE (VOLTS)

FIGURE 4 — CHARGE DATA

Q, CHARGE (nC)
10
7.0
5.0
3.0
2.0
1.0
0.7
0.5
0.3
0.2
0.1
$V_{CC} = 30$ V
$I_C/I_B = 10$
Q_T
Q_A
10 20 30 50 70 100 200 300 500
I_C, COLLECTOR CURRENT (mA)

FIGURE 5 — TURN-ON TIME

t, TIME (ns)
100
70
50
30
20
10
7.0
5.0
$I_C/I_B = 10$
t_r @ $V_{CC} = 30$ V
t_r @ $V_{CC} = 10$ V
t_d @ $V_{EB(off)} = 2$ V
t_d @ $V_{EB(off)} = 0$
10 20 30 50 70 100 200 300 500
I_C, COLLECTOR CURRENT (mA)

FIGURE 6 — RISE AND FALL TIMES

t, TIME (ns)
100
70
50
30
20
10
7.0
5.0
$V_{CC} = 30$ V
$I_C/I_B = 10$
t_r
t_f
10 20 30 50 70 100 200 300 500
I_C, COLLECTOR CURRENT (mA)

FIGURE 7 — STORAGE TIME

t_s', STORAGE TIME (ns)
300
200
100
70
50
30
$t_s' = t_s - 1/8\,t_f$
$I_{B1} = I_{B2}$
$I_C/I_B = 10$ to 20
10 20 30 50 70 100 200 300 500
I_C, COLLECTOR CURRENT (mA)

FIGURE 8 — FALL TIME

t_f, FALL TIME (ns)
100
70
50
30
20
10
7.0
5.0
$V_{CC} = 30$ V
$I_{B1} = I_{B2}$
$I_C/I_B = 20$
$I_C/I_B = 10$
10 20 30 50 70 100 200 300 500
I_C, COLLECTOR CURRENT (mA)

SMALL-SIGNAL CHARACTERISTICS

NOISE FIGURE

$V_{CE} = 10$ Vdc, $T_A = 25°C$

FIGURE 9 — FREQUENCY EFFECTS

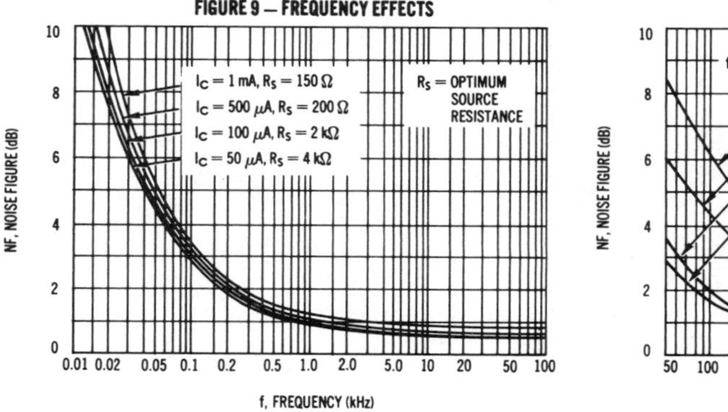

FIGURE 10 — SOURCE RESISTANCE EFFECTS

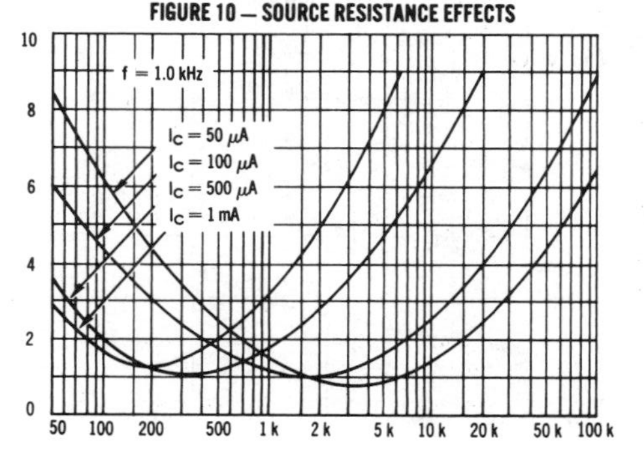

h PARAMETERS

$V_{CE} = 10$ Vdc, f = 1 kHz, $T_A = 25°C$

This group of graphs illustrates the relationship between h_{fe} and other "h" parameters for this series of transistors. To obtain these curves, a high-gain and a low-gain unit were selected from both the 2N4400 and 2N4401 lines, and the same units were used to develop the correspondingly-numbered curves on each graph.

FIGURE 11 — CURRENT GAIN

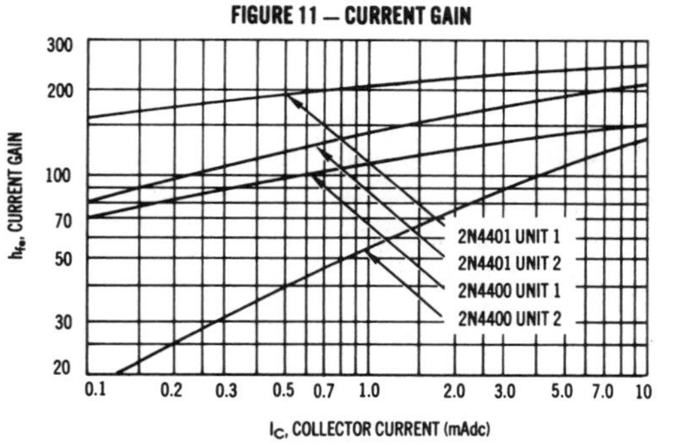

FIGURE 12 — INPUT IMPEDANCE

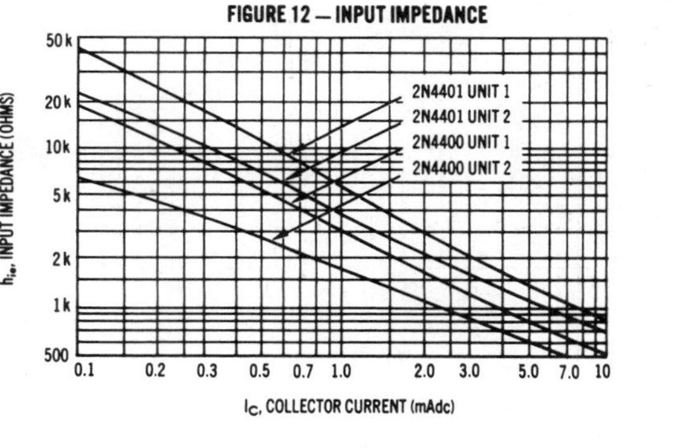

FIGURE 13 — VOLTAGE FEEDBACK RATIO

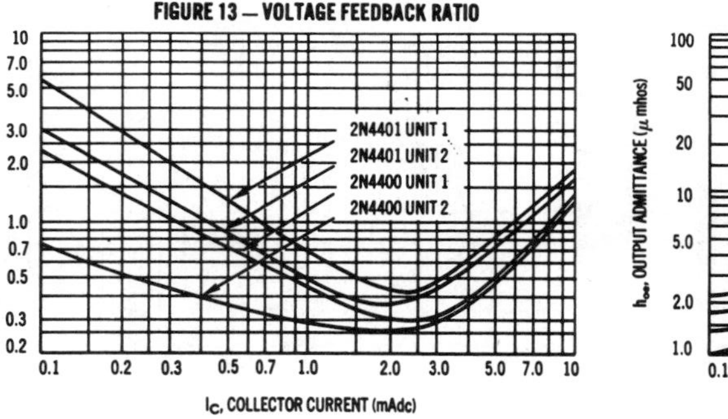

FIGURE 14 — OUTPUT ADMITTANCE

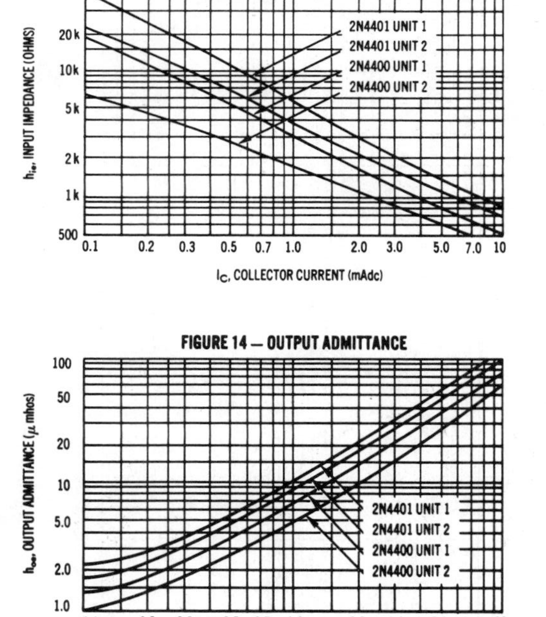

STATIC CHARACTERISTICS

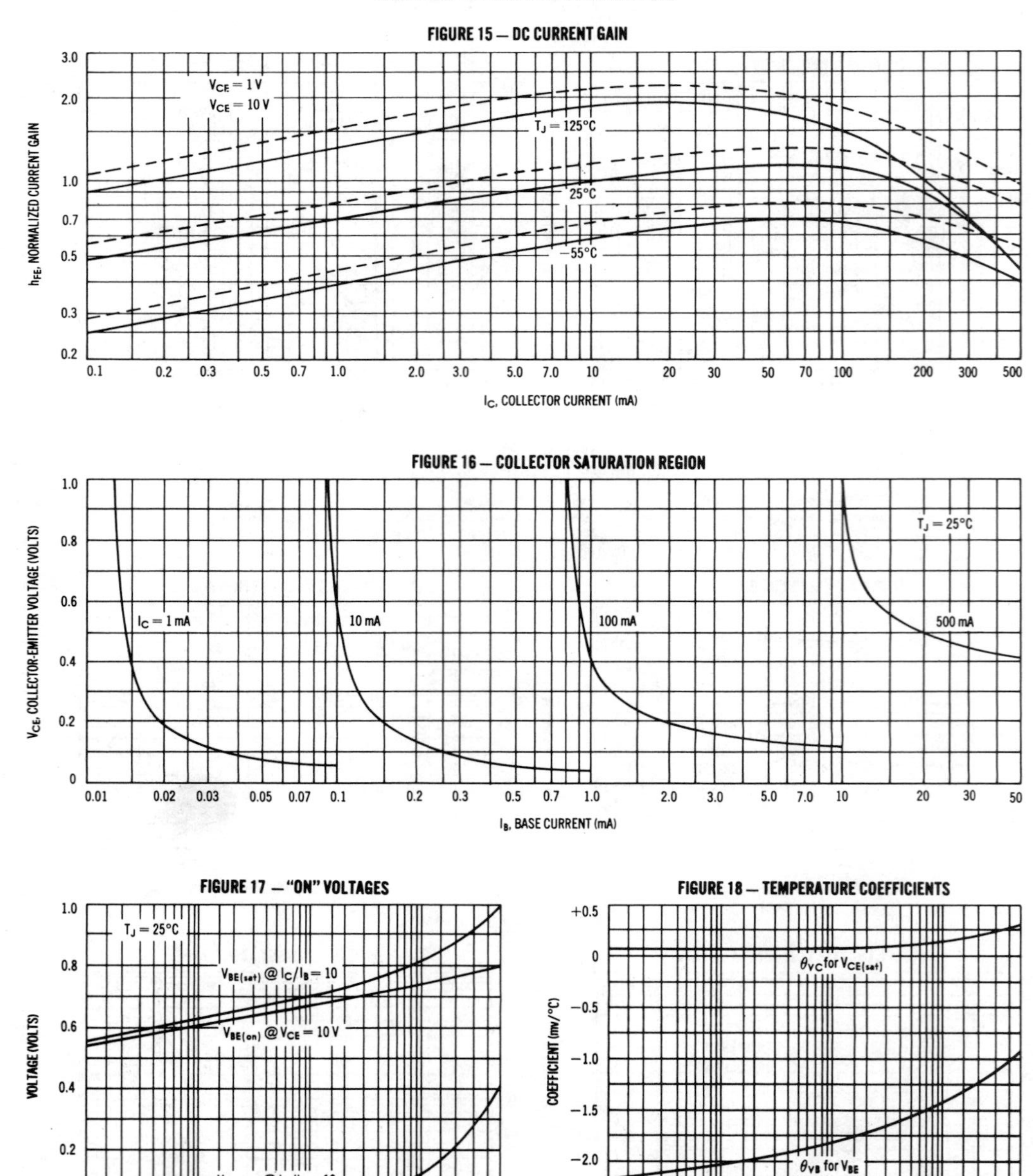

Transistor/Diode Arrays

LM194/LM394 supermatch pair

General Description

The LM194 and LM394 are junction isolated ultra well-matched monolithic NPN transistor pairs with an order of magnitude improvement in matching over conventional transistor pairs. This was accomplished by advanced linear processing and a unique new device structure.

Electrical characteristics of these devices such as drift versus initial offset voltage, noise, and the exponential relationship of base-emitter voltage to collector current closely approach those of a theoretical transistor. Extrinsic emitter and base resistances are much lower than presently available pairs, either monolithic or discrete, giving extremely low noise and theoretical operation over a wide current range. Most parameters are guaranteed over a current range of 1μA to 1 mA and 0V up to 40V collector-base voltage, ensuring superior performance in nearly all applications.

To guarantee long term stability of matching parameters, internal clamp diodes have been added across the emitter-base junction of each transistor. These prevent degradation due to reverse biased emitter current—the most common cause of field failures in matched devices. The parasitic isolation junction formed by the diodes also clamps the substrate region to the most negative emitter to ensure complete isolation between devices.

The LM194 and LM394 will provide a considerable improvement in performance in most applications requiring a closely matched transistor pair. In many cases, trimming can be eliminated entirely, improving reliability and decreasing costs. Additionally, the low noise and high gain make this device attractive even where matching is not critical.

The LM194 and LM394/LM394B/LM394C are available in an isolated header 6-lead TO-5 metal can package. The LM194 is identical to the LM394 except for tighter electrical specifications and wider temperature range.

Features

- Emitter-base voltage matched to 50μV
- Offset voltage drift less than 0.1μV/°C
- Current gain (h_{FE}) matched to 2%
- Common-mode rejection ratio greater than 120 dB
- Parameters guaranteed over 1μA to 1 mA collector current
- Extremely low noise
- Superior logging characteristics compared to conventional pairs
- Plug-in replacement for presently available devices

Typical Applications

Low Cost Accurate Square Root Circuit
$I_{OUT} = 10^{-5} \cdot \sqrt{10\,V_{IN}}$

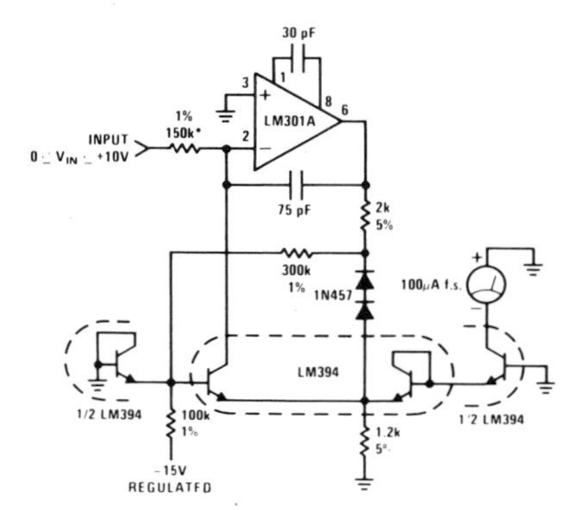

Low Cost Accurate Squaring Circuit
$I_{OUT} = 10^{-6}\,(V_{IN})^2$

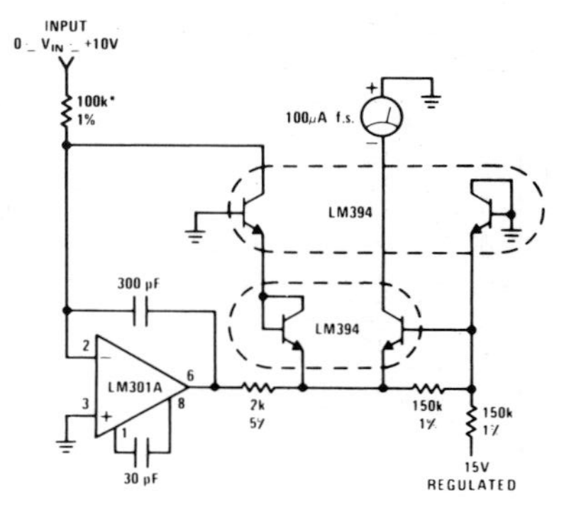

*Trim for full scale accuracy

Absolute Maximum Ratings

Collector Current	20 mA
Collector-Emitter Voltage	V_{MAX}
Collector-Emitter Voltage	40V
LM394C	20V
Collector-Base Voltage	40V
LM394C	20V
Collector-Substrate Voltage	40V
LM394C	20V
Collector-Collector Voltage	40V
LM394C	20V
Base-Emitter Current	±10 mA
Power Dissipation	500 mW
Junction Temperature	
LM194	−55°C to +125°C
LM394/LM394B/LM394C	−25°C to +85°C
Storage Temperature Range	−65°C to +150°C
Lead Temperature (Soldering, 10 seconds)	300°C

Electrical Characteristics (T_J = 25°C)

PARAMETER	CONDITIONS	LM194			LM394			LM394B/LM394C			UNITS
		MIN	TYP	MAX	MIN	TYP	MAX	MIN	TYP	MAX	
Current Gain (h_{FE})	V_{CB} = 0V to V_{MAX} (Note 1)										
	I_C = 1 mA	500	700		300	700		225	500		
	I_C = 100μA	400	550		250	550		200	400		
	I_C = 10μA	300	450		200	450		150	300		
	I_C = 1μA	200	300		150	300		100	200		
Current Gain Match (h_{FE} Match) $= \frac{100\,[\Delta I_B]\,[h_{FE(MIN)}]}{I_C}$	V_{CB} = 0V to V_{MAX}										
	I_C = 10μA to 1 mA		0.5	2		0.5	4		1.0	5	%
	I_C = 1μA		1.0			1.0			2.0		%
Emitter-Base Offset Voltage	V_{CB} = 0, I_C = 1μA to 1 mA		25	50		25	150		50	200	μV
Change in Emitter-Base Offset Voltage vs Collector-Base Voltage (CMRR)	(Note 1) I_C = 1μA to 1 mA, V_{CB} = 0V to V_{MAX}		10	25		10	50		10	100	μV
Change in Emitter-Base Offset Voltage vs Collector Current	V_{CB} = 0V, I_C = 1μA to 0.3 mA		5	25		5	50		5	50	μV
Emitter-Base Offset Voltage Temperature Drift	I_C = 10μA to 1 mA (Note 2)										
	$I_{C1} = I_{C2}$		0.08	0.3		0.08	1.0		0.2	1.5	μV/°C
	V_{OS} Trimmed to 0 at 25°C		0.03	0.1		0.03	0.3		0.03	0.5	μV/°C
Logging Conformity	I_C = 3 nA to 300μA, V_{CB}=0, (Note 3)		150			150			150		μV
Collector-Base Leakage	V_{CB} = V_{MAX}		0.05	0.25		0.05	0.5		0.05	0.5	nA
Collector-Collector Leakage	V_{CC} = V_{MAX}		0.1	2.0		0.1	5.0		0.1	5.0	nA
Input Voltage Noise	I_C = 100μA, V_{CB} = 0V, f = 100 Hz to 100 kHz		1.8			1.8			1.8		nV/√Hz
Collector to Emitter Saturation Voltage	I_C = 1 mA, I_B = 10μA		0.2			0.2			0.2		V
	I_C = 1 mA, I_B = 100μA		0.1			0.1			0.1		V

Note 1: Collector-base voltage is swept from 0 to V_{MAX} at a collector current of 1μA, 10μA, 100μA, and 1 mA.

Note 2: Offset voltage drift with V_{OS} = 0 at T_A = 25°C is valid only when the ratio of I_{C1} to I_{C2} is adjusted to give the initial zero offset. This ratio must be held to within 0.003% over the entire temperature range. Measurements taken at +25°C and temperature extremes.

Note 3: Logging conformity is measured by computing the best fit to a true exponential and expressing the error as a base-emitter voltage deviation.

Typical Applications (Continued)

Fast, Accurate Logging Amplifier, V_{IN} = 10V to 0.1 mV or I_{IN} = 1 mA to 10 nA

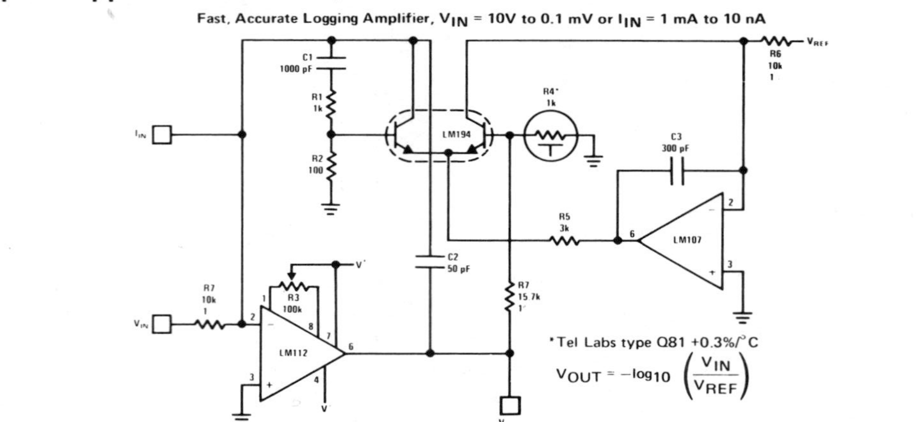

*Tel Labs type Q81 +0.3%/°C

$$V_{OUT} = -\log_{10}\left(\frac{V_{IN}}{V_{REF}}\right)$$

Typical Applications (Continued)

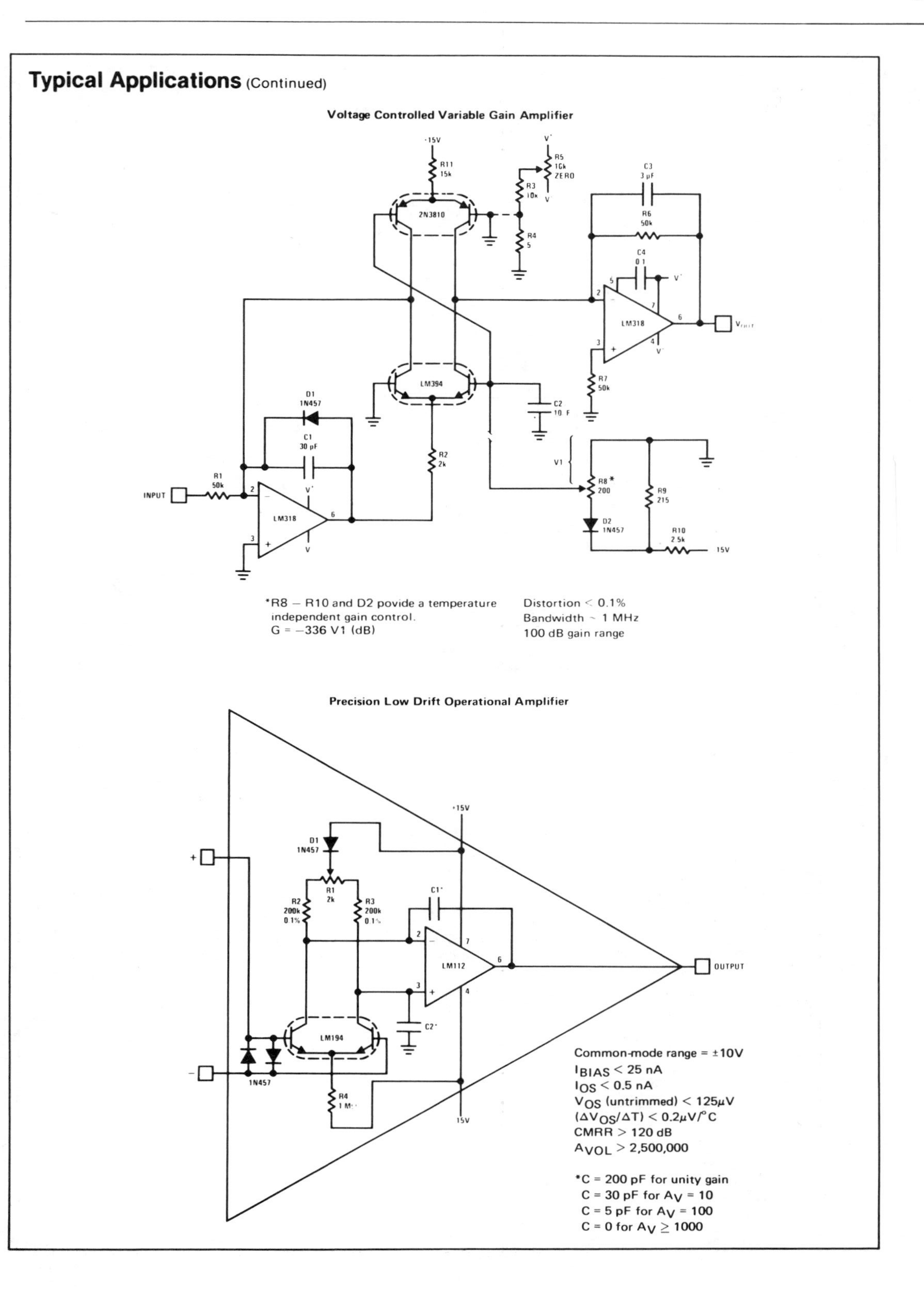

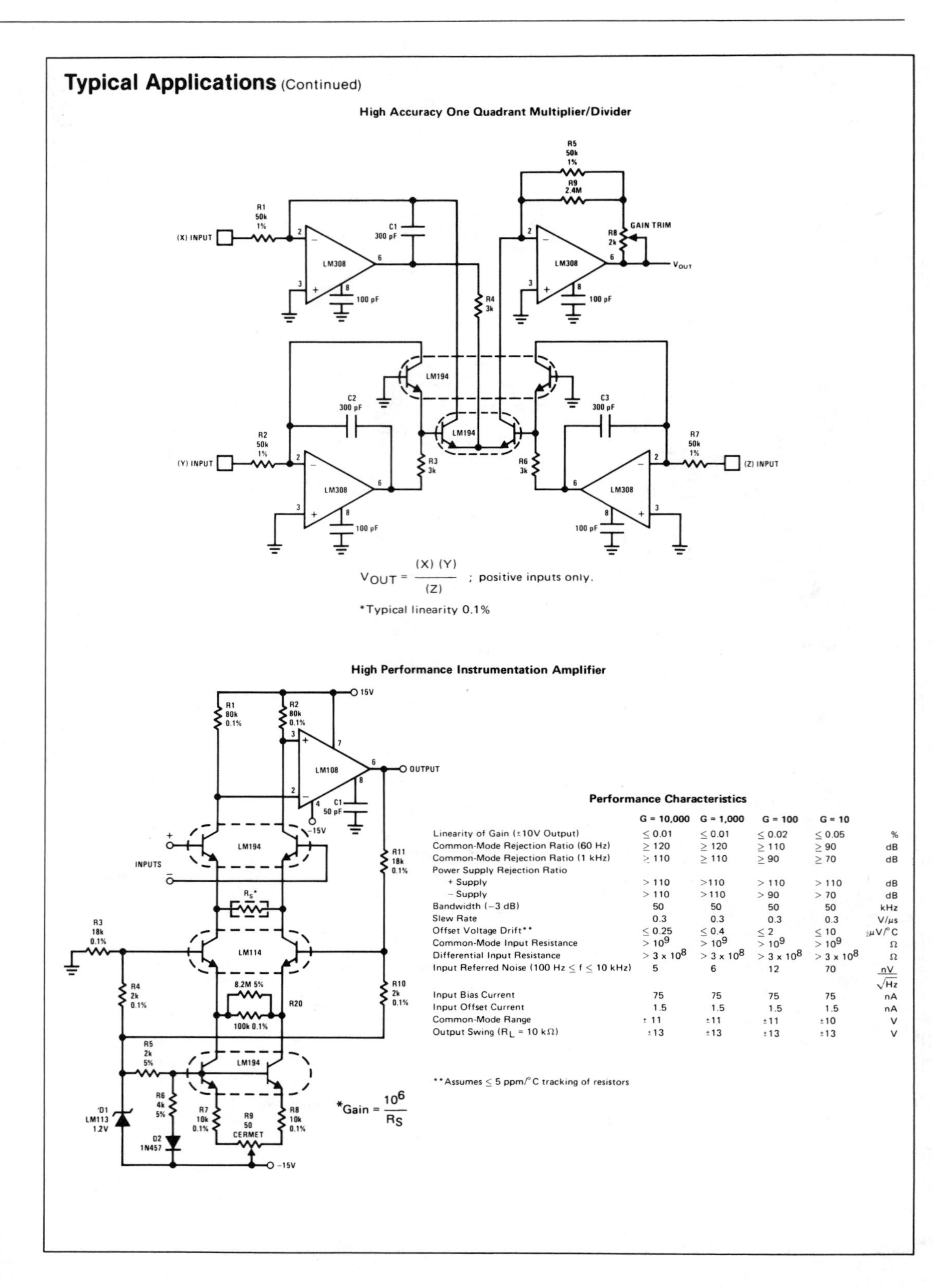

$$V_{OUT} = \frac{(X)\,(Y)}{(Z)} \quad ; \text{ positive inputs only.}$$

*Typical linearity 0.1%

$$^{*}\text{Gain} = \frac{10^6}{R_S}$$

Performance Characteristics

	G = 10,000	G = 1,000	G = 100	G = 10	
Linearity of Gain (±10V Output)	≤ 0.01	≤ 0.01	≤ 0.02	≤ 0.05	%
Common-Mode Rejection Ratio (60 Hz)	≥ 120	≥ 120	≥ 110	≥ 90	dB
Common-Mode Rejection Ratio (1 kHz)	≥ 110	≥ 110	≥ 90	≥ 70	dB
Power Supply Rejection Ratio					
+ Supply	> 110	> 110	> 110	> 110	dB
− Supply	> 110	> 110	> 90	> 70	dB
Bandwidth (−3 dB)	50	50	50	50	kHz
Slew Rate	0.3	0.3	0.3	0.3	V/μs
Offset Voltage Drift**	≤ 0.25	≤ 0.4	≤ 2	≤ 10	μV/°C
Common-Mode Input Resistance	> 10^9	> 10^9	> 10^9	> 10^9	Ω
Differential Input Resistance	> 3 x 10^8	> 3 x 10^8	> 3 x 10^8	> 3 x 10^8	Ω
Input Referred Noise (100 Hz ≤ f ≤ 10 kHz)	5	6	12	70	$\frac{nV}{\sqrt{Hz}}$
Input Bias Current	75	75	75	75	nA
Input Offset Current	1.5	1.5	1.5	1.5	nA
Common-Mode Range	±11	±11	±11	±10	V
Output Swing (R_L = 10 kΩ)	±13	±13	±13	±13	V

**Assumes ≤ 5 ppm/°C tracking of resistors

Typical Performance Characteristics

Small Signal Current Gain vs Collector Current

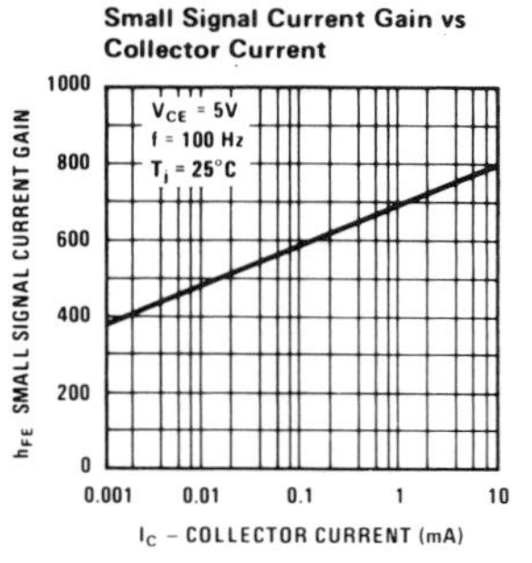

DC Current Gain vs Temperature

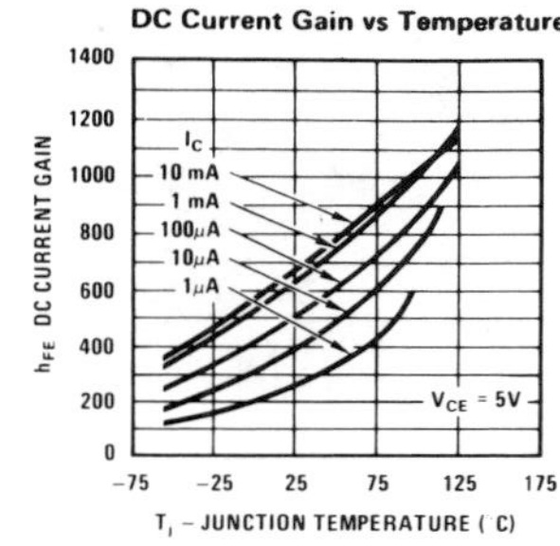
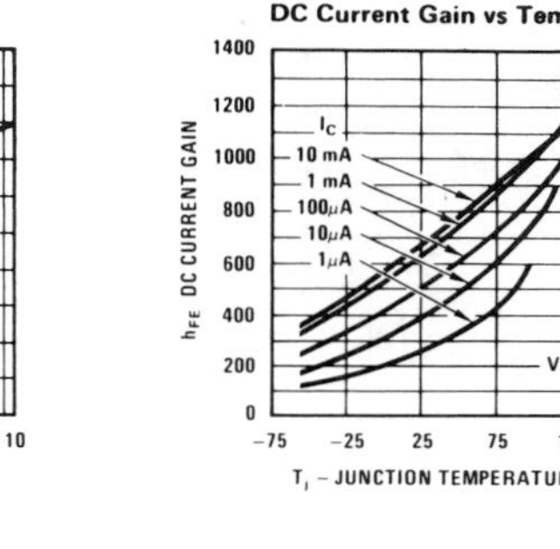

Unity Gain Frequency (f_t) vs Collector Current

f_T (MHz)

I_C – COLLECTOR CURRENT (mA)

Offset Voltage Drift vs Initial Offset Voltage

$\Delta V_{OS}/\Delta T$ (μV/°C)

T_j = 25°C

INITIAL OFFSET VOLTAGE (μV)

Base-Emitter On Voltage vs Collector Current

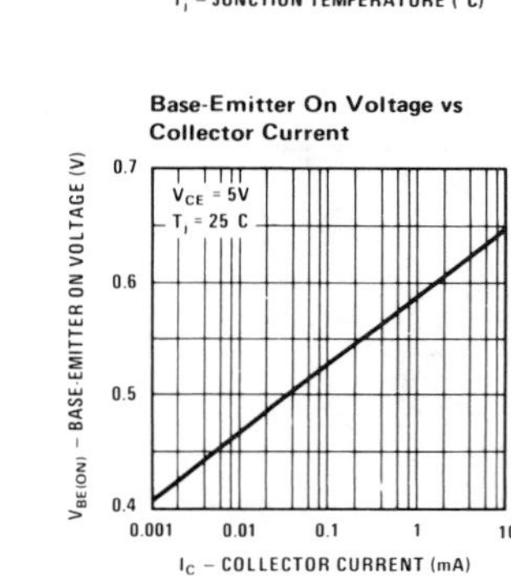

Small Signal Input Resistance (h_{ie}) vs Collector Current

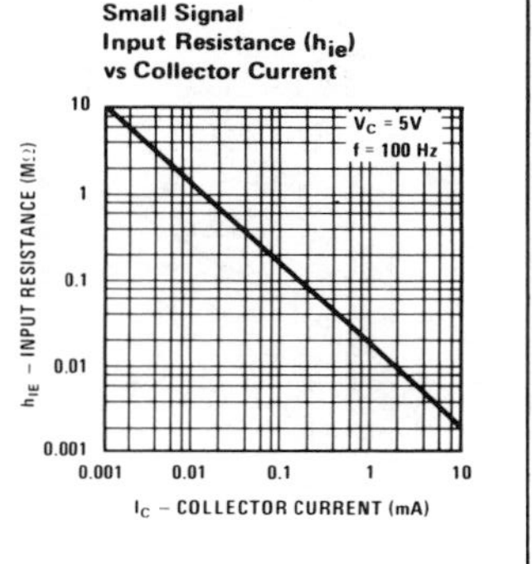

Small Signal Output Conductance vs Collector Current

h_{OE} – OUTPUT CONDUCTANCE (μmhos)

V_{CE} = 5V

I_C – COLLECTOR CURRENT (mA)

Collector-Emitter Saturation Voltage vs Collector Current

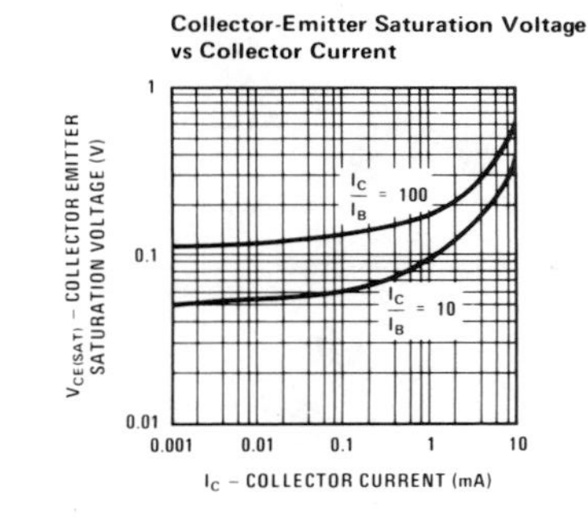

Input Voltage Noise vs Frequency

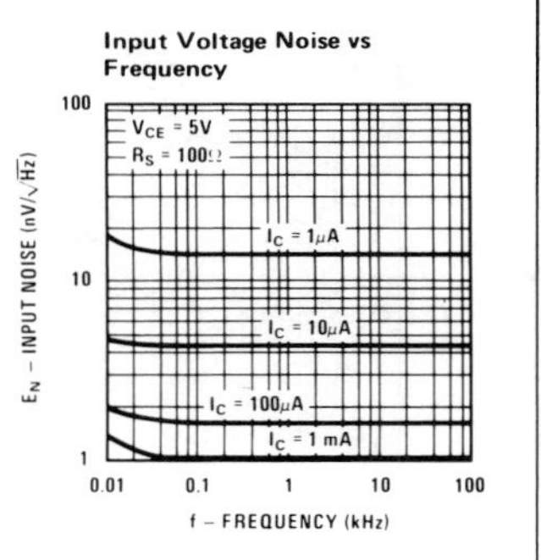

Base Current Noise vs Frequency

I_b – CURRENT NOISE (pA/√Hz)

V_{CE} = 5V, T_j = 25°C

I_C = 10 mA, I_C = 1 mA, I_C = 100μA, I_C = 10μA, I_C = 1μA

f – FREQUENCY (kHz)

Noise Figure vs Collector Current

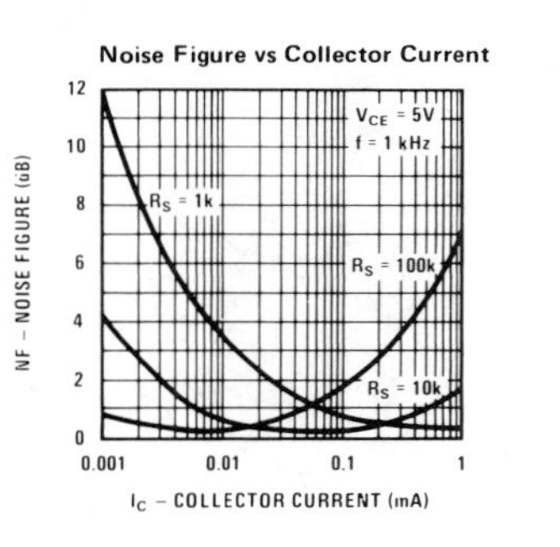

Collector to Collector Capacitance vs Reverse Bias Voltage

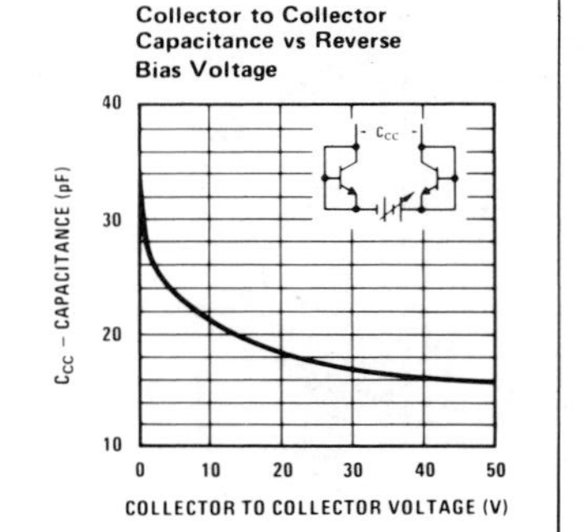

Typical Performance Characteristics (Continued)

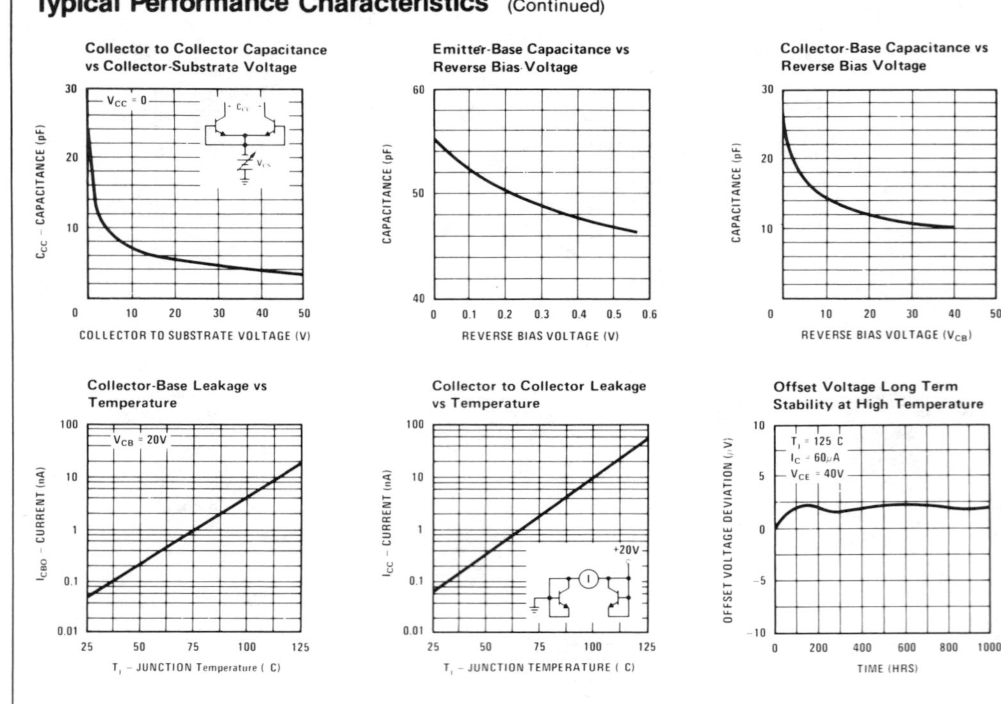

Emitter-Base Log Conformity

V_{CB} = 0V

LOGGING ERROR (mV)

I_C – COLLECTOR CURRENT (A)

Low Frequency Noise of Differential Pair*

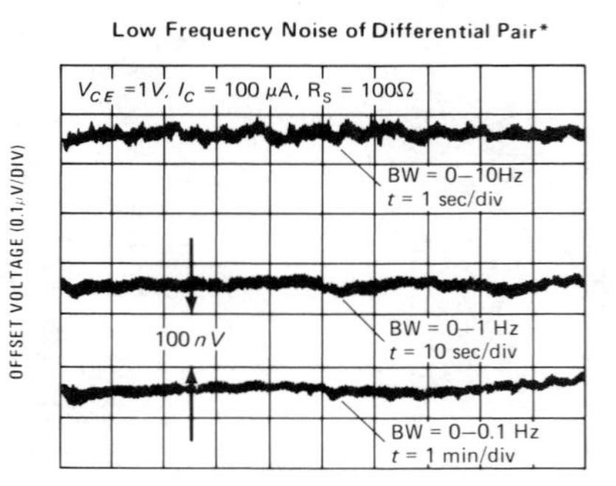

TIME (SEE GRAPH)

*Unit must be in still air environment so that differential lead temperature is held to less than 0.0003°C.

Connection Diagram

Metal Can Package

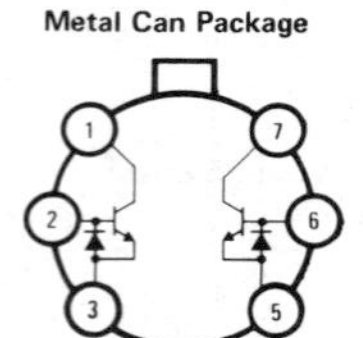

TOP VIEW

Order Number LM194H, LM394H
LM394BH or LM394CH
See NS Package H06C

National Semiconductor

JUNE 1977

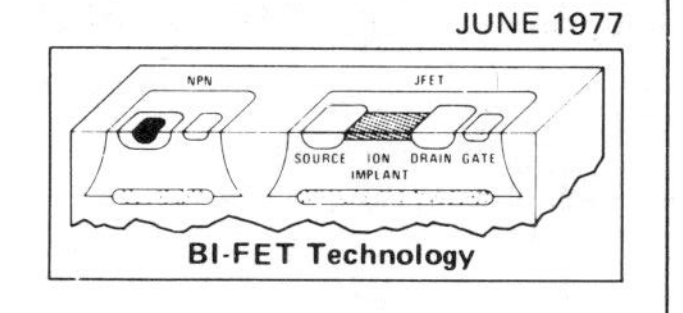

BI-FET Technology

LF155/LF156/LF157 Monolithic JFET Input Operational Amplifiers

LF155, LF155A, LF255, LF355, LF355A, LF355B low supply current
LF156, LF156A, LF256, LF356, LF356A, LF356B wide band
LF157, LF157A, LF257, LF357, LF357A, LF357B wide band decompensated ($A_{V_{MIN}}$ = 5)

General Description

These are the first monolithic JFET input operational amplifiers to incorporate well matched, high voltage JFETs on the same chip with standard bipolar transistors (BI-FET Technology). These amplifiers feature low input bias and offset currents, low offset voltage and offset voltage drift, coupled with offset adjust which does not degrade drift or common-mode rejection. The devices are also designed for high slew rate, wide bandwidth, extremely fast settling time, low voltage and current noise and a low 1/f noise corner.

Advantages

- Replace expensive hybrid and module FET op amps
- Rugged JFETs allow blow-out free handling compared with MOSFET input devices
- Excellent for low noise applications using either high or low source impedance—very low 1/f corner
- Offset adjust does not degrade drift or common-mode rejection as in most monolithic amplifiers
- New output stage allows use of large capacitive loads (10,000 pF) without stability problems
- Internal compensation and large differential input voltage capability

Applications

- Precision high speed integrators
- Fast D/A and A/D converters
- High impedance buffers
- Wideband, low noise, low drift amplifiers
- Logarithmic amplifiers
- Photocell amplifiers
- Sample and Hold circuits

Common Features

(LF155A, LF156A, LF157A)

■ Low input bias current	30 pA
■ Low Input Offset Current	3 pA
■ High input impedance	$10^{12}\Omega$
■ Low input offset voltage	1 mV
■ Low input offset voltage temperature drift	$3\mu V/°C$
■ Low input noise current	$0.01\ pA/\sqrt{Hz}$
■ High common-mode rejection ratio	100 dB
■ Large dc voltage gain	106 dB

Uncommon Features

	LF155A	LF156A	LF157A (A_V = 5)*	UNITS
■ Extremely fast settling time to 0.01%	4	1.5	1.5	μs
■ Fast slew rate	5	12	50	$V/\mu s$
■ Wide gain bandwidth	2.5	5	20	MHz
■ Low input noise voltage	20	12	12	$nV/\sqrt{Hz}$

Simplified Schematic

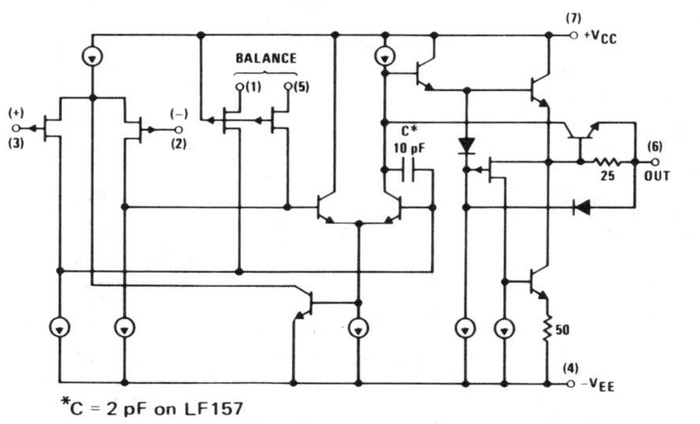

*C = 2 pF on LF157

IM-B20M67/Printed in U.S.A.

Absolute Maximum Ratings

		LF155A/6A/7A	LF155/6/7	LF355B/6B/7B LF255/6/7 LF355B/6B/7B	LF355A/6A/7A LF355/6/7
Supply Voltage		±22V	±22V	±22V	±18V
Power Dissipation (P_d at 25°C) and Thermal Resistance (θ_{jA}) (Note 1)					
T_{jMAX} (H and J Package)		150°C	150°C	115°C	115°C
(N Package)				100°C	100°C
(H Package)	P_d	670 mW	670 mW	570 mW	570 mW
	θ_{jA}	150°C/W	150°C/W	150°C/W	150°C/W
(J Package)	P_d	670 mW	670 mW	570 mW	570 mW
	θ_{jA}	140°C/W	140°C/W	140°C/W	140°C/W
(N Package)	P_d			500 mW	500 mW
	θ_{jA}			155°C/W	155°C/W
Differential Input Voltage		±40V	±40V	±40V	±30V
Input Voltage Range (Note 2)		±20V	±20V	±20V	±16V
Output Short Circuit Duration		Continuous	Continuous	Continuous	Continuous
Storage Temperature Range		−65°C to +150°C	−65°C to +150°C	−65°C to +150°C	−65°C to +150°C
Lead Temperature (Soldering, 10 seconds)		300°C	300°C	300°C	300°C

DC Electrical Characteristics (Note 3)

SYMBOL	PARAMETER	CONDITIONS	LF155A/6A/7A			LF355A/6A/7A			UNITS
			MIN	TYP	MAX	MIN	TYP	MAX	
V_{OS}	Input Offset Voltage	$R_S = 50\Omega$, $T_A = 25°C$		1	2		1	2	mV
		Over Temperature			2.5			2.3	mV
$\Delta V_{OS}/\Delta T$	Average TC of Input Offset Voltage	$R_S = 50\Omega$		3	5		3	5	μV/°C
$\Delta TC/\Delta V_{OS}$	Change in Average TC with V_{OS} Adjust	$R_S = 50\Omega$, (Note 4)		0.5			0.5		μV/°C per mV
I_{OS}	Input Offset Current	$T_j = 25°C$, (Notes 3, 5)		3	10		3	10	pA
		$T_j \leq T_{HIGH}$			10			1	nA
I_B	Input Bias Current	$T_J = 25°C$, (Notes 3, 5)		30	50		30	50	pA
		$T_J \leq T_{HIGH}$			25			5	nA
R_{IN}	Input Resistance	$T_J = 25°C$		10^{12}			10^{12}		Ω
A_{VOL}	Large Signal Voltage Gain	$V_S = \pm15V$, $T_A = 25°C$, $V_O = \pm10V$, $R_L = 2k$	50	200		50	200		V/mV
		Over Temperature	25			25			V/mV
V_O	Output Voltage Swing	$V_S = \pm15V$, $R_L = 10k$	±12	±13		±12	±13		V
		$V_S = \pm15V$, $R_L = 2k$	±10	±12		±10	±12		V
V_{CM}	Input Common-Mode Voltage Range	$V_S = \pm15V$	±11	+15.1		±11	+15.1		V
				−12			−12		V
CMRR	Common-Mode Rejection Ratio		85	100		85	100		dB
PSRR	Supply Voltage Rejection Ratio	(Note 6)	85	100		85	100		dB

AC Electrical Characteristics $T_A = 25°C$, $V_S = \pm15V$

SYMBOL	PARAMETER	CONDITIONS	LF155A/355A			LF156A/356A			LF157A/357A			UNITS
			MIN	TYP	MAX	MIN	TYP	MAX	MIN	TYP	MAX	
SR	Slew Rate	LF155A/6A; $A_V = 1$,	3	5		10	12					V/μs
		LF157A; $A_V = 5$							40	50		V/μs
GBW	Gain Bandwidth Product			2.5		4	4.5		15	20		MHz
t_s	Settling Time to 0.01%	(Note 7)		4			1.5			1.5		μs
e_n	Equivalent Input Noise Voltage	$R_S = 100\Omega$, f = 100 Hz		25			15			15		$nV/\sqrt{Hz}$
		f = 1000 Hz		25			12			12		$nV/\sqrt{Hz}$
i_n	Equivalent Input Noise Current	f = 100 Hz		0.01			0.01			0.01		$pA/\sqrt{Hz}$
		f = 1000 Hz		0.01			0.01			0.01		$pA/\sqrt{Hz}$
C_{IN}	Input Capacitance			3			3			3		pF

DC Electrical Characteristics (Note 3)

SYMBOL	PARAMETER	CONDITIONS	LF155/6/7			LF255/6/7 LF355B/6B/7B			LF355/6/7			UNITS
			MIN	TYP	MAX	MIN	TYP	MAX	MIN	TYP	MAX	
V_{OS}	Input Offset Voltage	$R_S = 50\Omega$, $T_A = 25°C$		3	5		3	5		3	10	mV
		Over Temperature			7			6.5			13	mV
$\Delta V_{OS}/\Delta T$	Average TC of Input Offset Voltage	$R_S = 50\Omega$		5			5			5		$\mu V/°C$
$\Delta TC/\Delta V_{OS}$	Change in Average TC with V_{OS} Adjust	$R_S = 50\Omega$, (Note 4)		0.5			0.5			0.5		$\mu V/°C$ per mV
I_{OS}	Input Offset Current	$T_j = 25°C$, (Notes 3, 5)		3	20		3	20		3	50	pA
		$T_j \leq T_{HIGH}$			20			1			2	nA
I_B	Input Bias Current	$T_J = 25°C$, (Notes 3, 5)		30	100		30	100		30	200	pA
		$T_J \leq T_{HIGH}$			50			5			8	nA
R_{IN}	Input Resistance	$T_J = 25°C$		10^{12}			10^{12}			10^{12}		Ω
A_{VOL}	Large Signal Voltage Gain	$V_S = \pm15V$, $T_A = 25°C$, $V_O = \pm10V$, $R_L = 2k$	50	200		50	200		25	200		V/mV
		Over Temperature	25			25			15			V/mV
V_O	Output Voltage Swing	$V_S = \pm15V$, $R_L = 10k$	±12	±13		±12	±13		±12	±13		V
		$V_S = \pm15V$, $R_L = 2k$	±10	±12		±10	±12		±10	±12		V
V_{CM}	Input Common-Mode Voltage Range	$V_S = \pm15V$	±11	+15.1		±11	+15.1		±10	+15.1		V
				−12			−12			−12		V
CMRR	Common-Mode Rejection Ratio		85	100		85	100		80	100		dB
PSRR	Supply Voltage Rejection Ratio	(Note 6)	85	100		85	100		80	100		dB

DC Electrical Characteristics $T_A = 25°C$, $V_S = \pm15V$

PARAMETER	LF155A/155, LF255, LF355A/355B		LF355		LF156A/156, LF256/356B		LF356A/356		LF157A/157 LF257/357B		LF357A/357		UNITS
	TYP	MAX	TYP	MAX	TYP	MAX	TYP	MAX	TYP	MAX	TYP	MAX	
Supply Current	2	4	2	4	5	7	5	10	5	7	5	10	mA

AC Electrical Characteristics $T_A = 25°C$, $V_S = \pm15V$

SYMBOL	PARAMETER	CONDITIONS	LF155/255/ 355/355B	LF156/256, LF356B	LF156/256/ 356/356B	LF157/257, LF357B	LF157/257/ 357/357B	UNITS
			TYP	MIN	TYP	MIN	TYP	
SR	Slew Rate	LF155/6: $A_V = 1$,	5	7.5	12			$V/\mu s$
		LF157: $A_V = 5$				30	50	$V/\mu s$
GBW	Gain Bandwidth Product		2.5		5		20	MHz
t_s	Settling Time to 0.01%	(Note 7)	4		1.5		1.5	μs
e_n	Equivalent Input Noise Voltage	$R_S = 100\Omega$						
		f = 100 Hz	25		15		15	$nV/\sqrt{Hz}$
		f = 1000 Hz	20		12		12	$nV/\sqrt{Hz}$
i_n	Equivalent Input Current Noise	f = 100 Hz	0.01		0.01		0.01	$pA/\sqrt{Hz}$
		f = 1000 Hz	0.01		0.01		0.01	$pA/\sqrt{Hz}$
C_{IN}	Input Capacitance		3		3		3	pF

Notes for Electrical Characteristics

Note 1: The maximum power dissipation for these devices must be derated at elevated temperatures and is dictated by T_{jMAX}, θ_{jA}, and the ambient temperature, T_A. The maximum available power dissipation at any temperature is $P_d = (T_{jMAX} - T_A)/\theta_{jA}$ or the 25°C P_{dMAX}, whichever is less.

Note 2: Unless otherwise specified the absolute maximum negative input voltage is equal to the negative power supply voltage.

Note 3: Unless otherwise stated, these test conditions apply:

	LF155A/6A/7A LF155/6/7	LF255/6/7	LF355A/6A/7A	LF355B/6B/7B	LF355/6/7
Supply Voltage, V_S	$\pm15V \le V_S \le \pm20V$	$\pm15V \le V_S \le \pm20V$	$\pm15V \le V_S \le \pm18V$	$\pm15V \le V_S$ $\pm20V$	$V_S = \pm15V$
T_A	$-55°C \le T_A \le +125°C$	$-25°C \le T_A \le +85°C$	$0°C \le T_A \le +70°C$	$0°C \le T_A \le +70°C$	$0°C \le T_A \le +70°C$
T_{HIGH}	+125°C	+85°C	+70°C	+70°C	+70°C

and V_{OS}, I_B and I_{OS} are measured at $V_{CM} = 0$.

Note 4: The Temperature Coefficient of the adjusted input offset voltage changes only a small amount (0.5μV/°C typically) for each mV of adjustment from its original unadjusted value. Common-mode rejection and open loop voltage gain are also unaffected by offset adjustment.

Note 5: The input bias currents are junction leakage currents which approximately double for every 10°C increase in the junction temperature, T_J. Due to limited production test time, the input bias currents measured are correlated to junction temperature. In normal operation the junction temperature rises above the ambient temperature as a result of internal power dissipation, Pd. $T_j = T_A + \Theta_{jA}$ Pd where Θ_{jA} is the thermal resistance from junction to ambient. Use of a heat sink is recommended if input bias current is to be kept to a minimum.

Note 6: Supply Voltage Rejection is measured for both supply magnitudes increasing or decreasing simultaneously, in accordance with common practice.

Note 7: Settling time is defined here, for a unity gain inverter connection using 2 kΩ resistors for the LF155/6. It is the time required for the error voltage (the voltage at the inverting input pin on the amplifier) to settle to within 0.01% of its final value from the time a 10V step input is applied to the inverter. For the LF157, $A_V = -5$, the feedback resistor from output to input is 2 kΩ and the output step is 10V (See Settling Time Test Circuit, page 9).

Typical DC Performance Characteristics

Curves are for LF155, LF156 and LF157 unless otherwise specified.

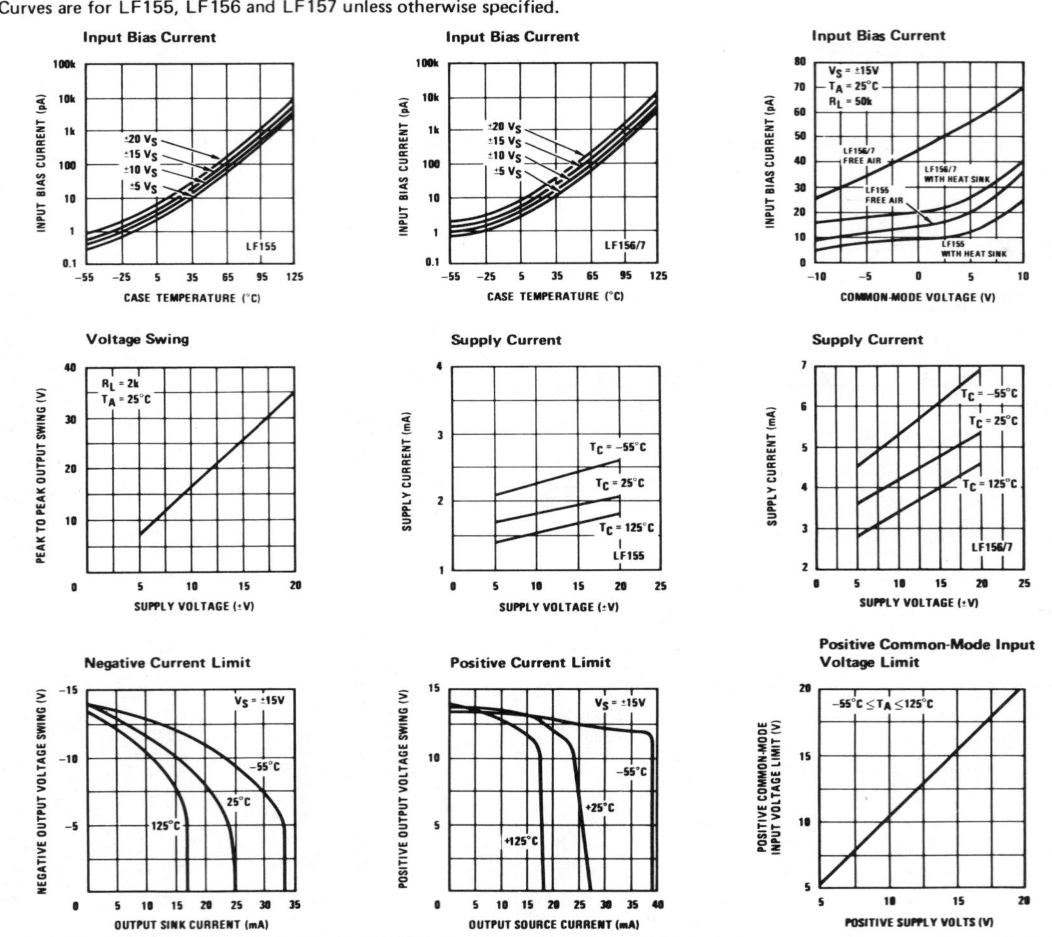

Typical DC Performance Characteristics (Continued)

Negative Common-Mode Input Voltage Limit

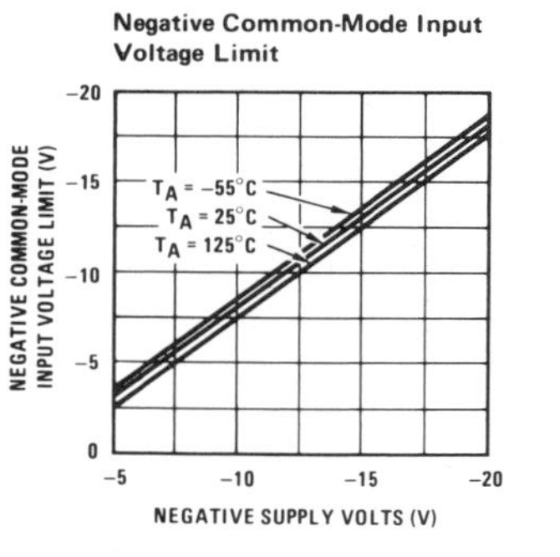

Open Loop Voltage Gain

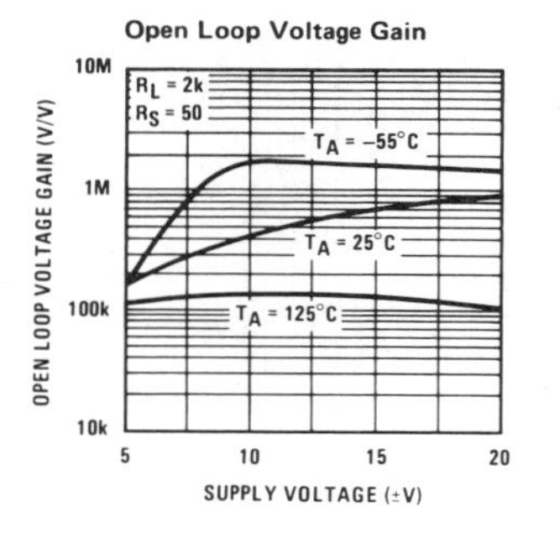

Output Voltage Swing

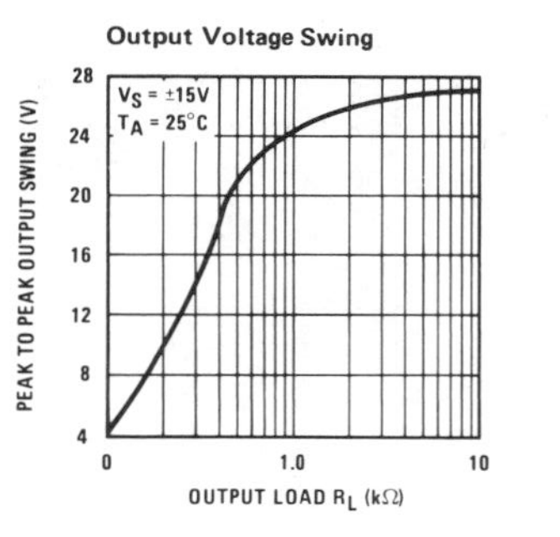

Typical AC Performance Characteristics

Gain Bandwidth

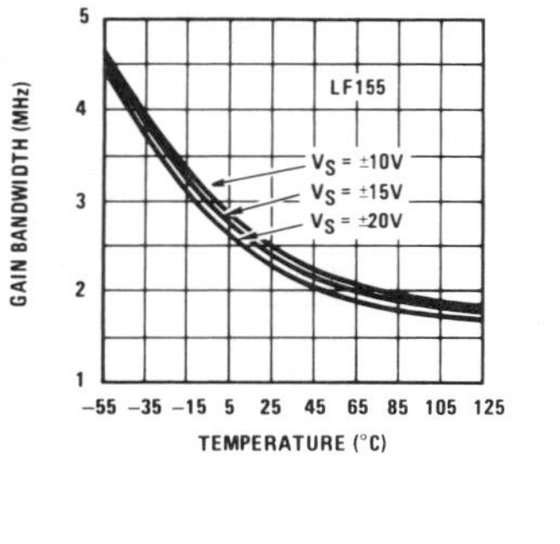

Gain Bandwidth

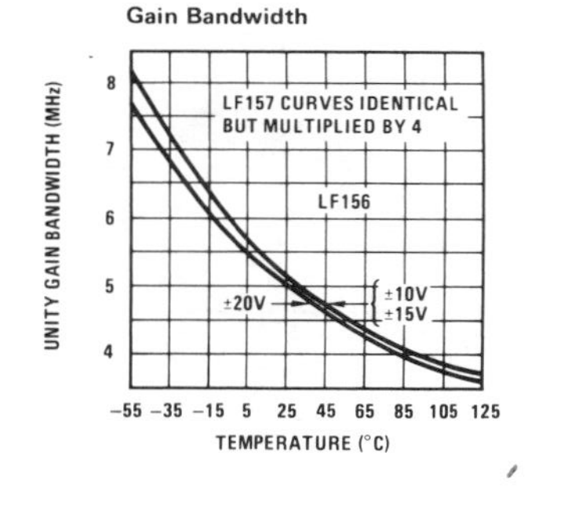

Normalized Slew Rate

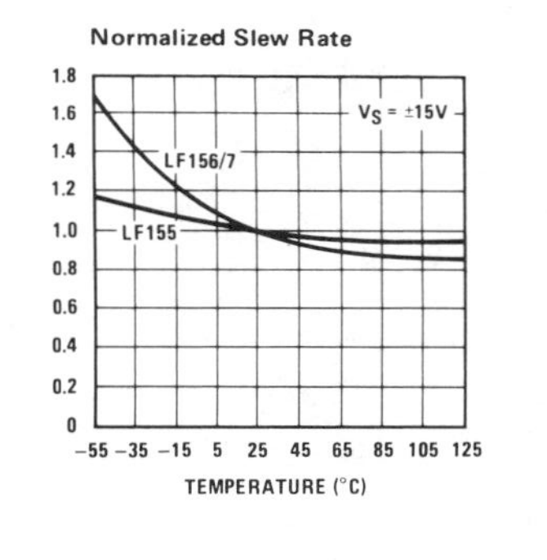

LF155 Small Signal Pulse Response, $A_V = +1$

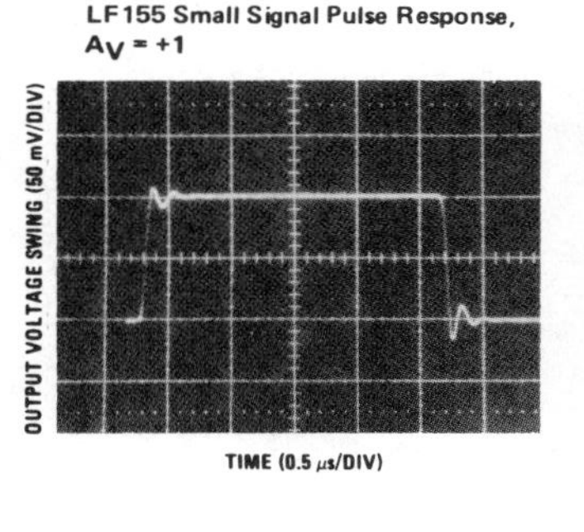

LF156 Small Signal Pulse Response, $A_V = +1$

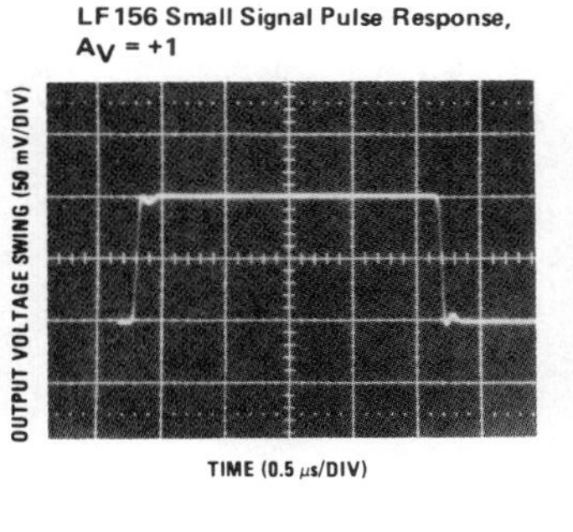

LF157 Small Signal Puls3 Response, $A_V = +5$

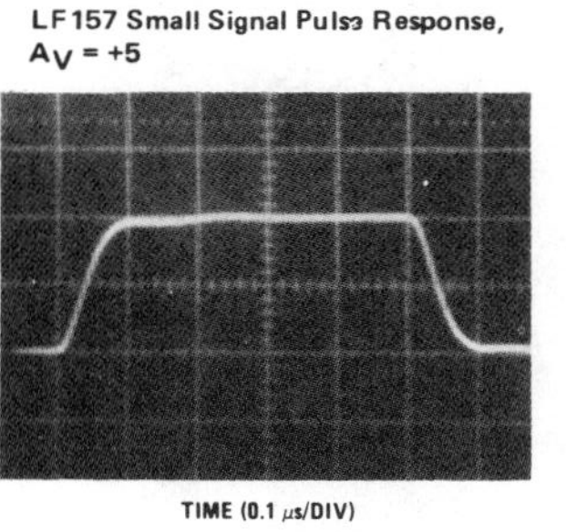

LF155 Large Signal Pulse Response, $A_V = +1$

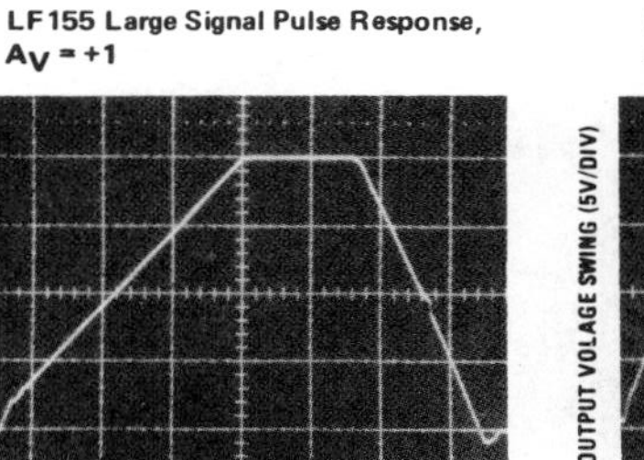

LF156 Large Signal Pulse Response, $A_V = +1$

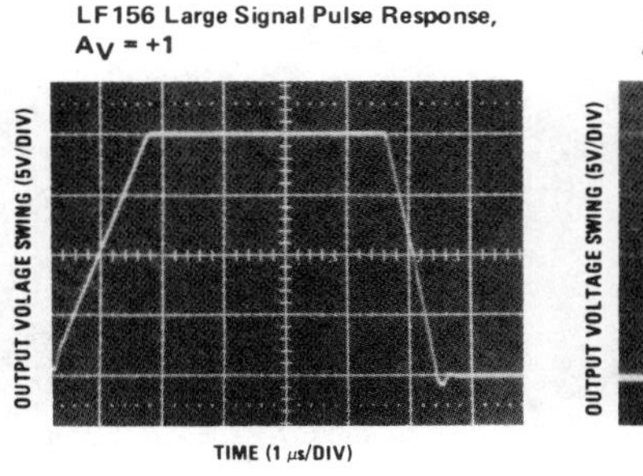

LF157 Large Signal Pulse Response, $A_V = +5$

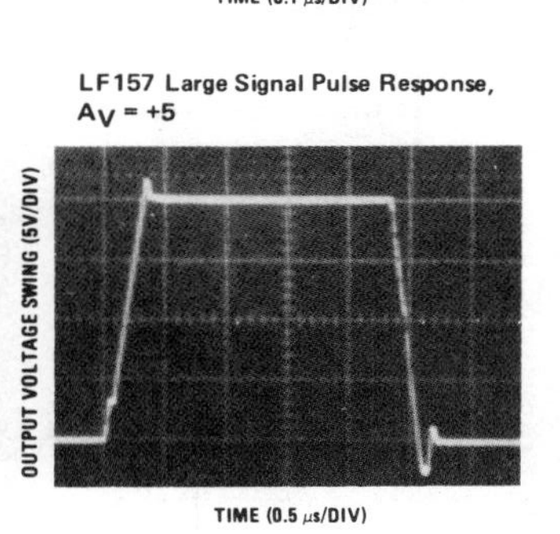

Typical AC Performance Characteristics (Continued)

Inverter Settling Time

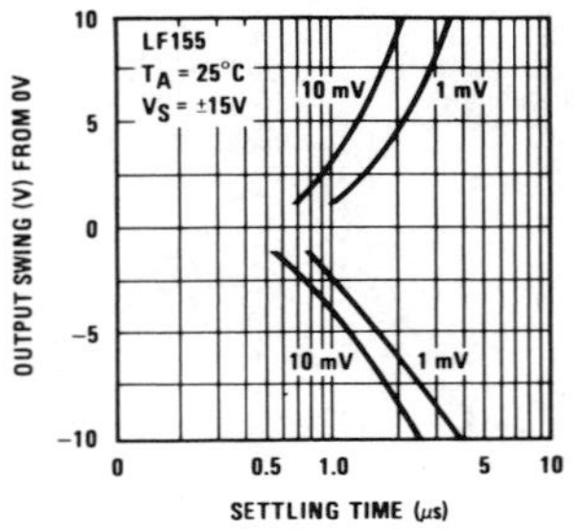

Inverter Settling Time

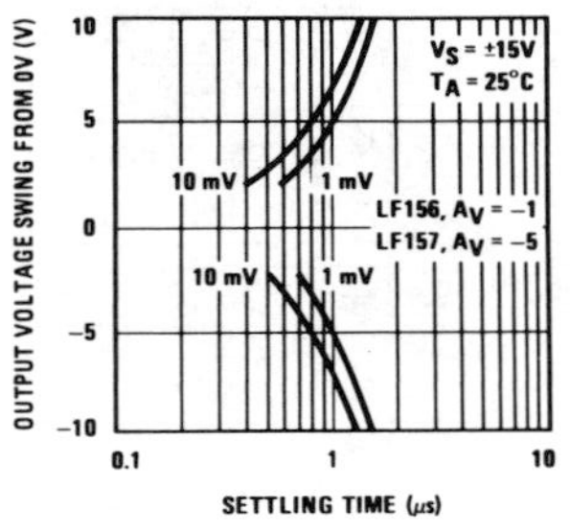

Open Loop Frequency Response

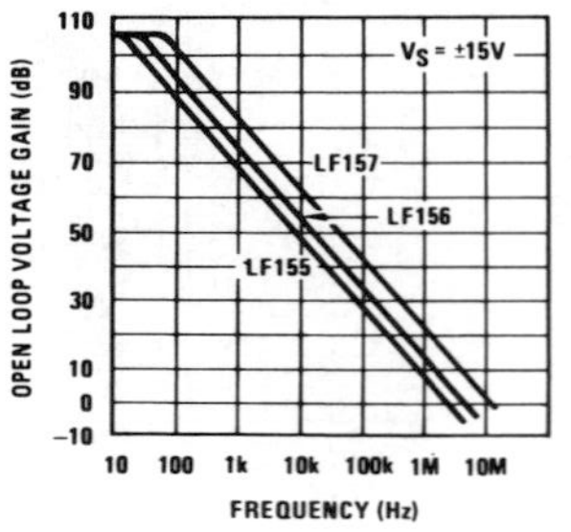

Bode Plot

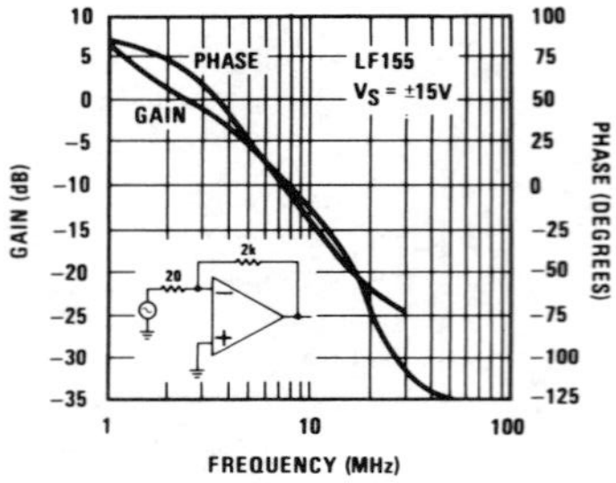

Bode Plot

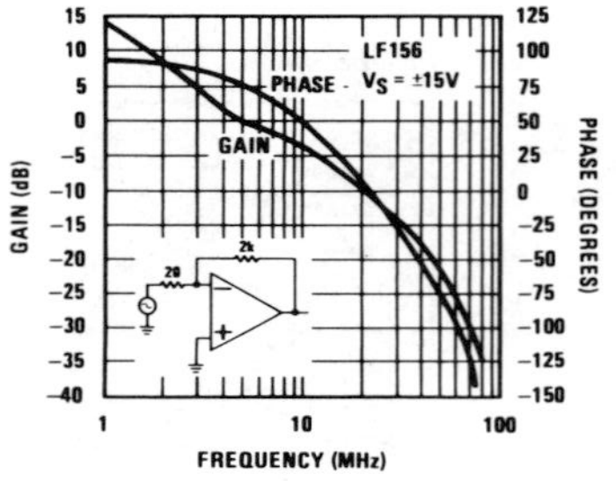

Bode Plot

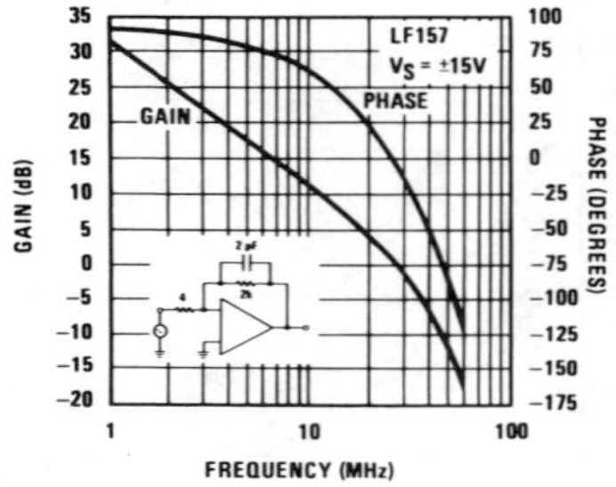

Common-Mode Rejection Ratio

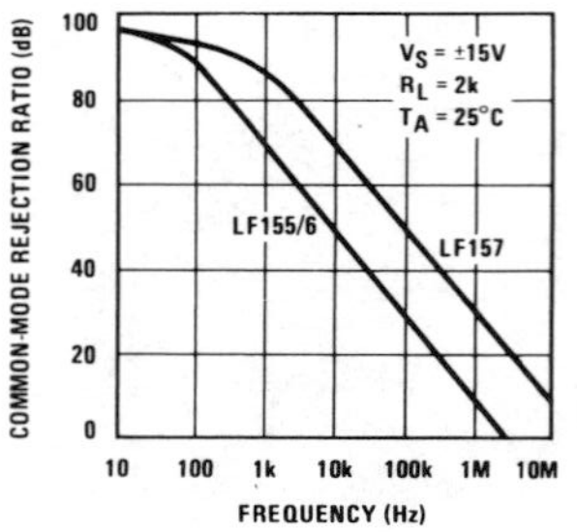

Power Supply Rejection Ratio

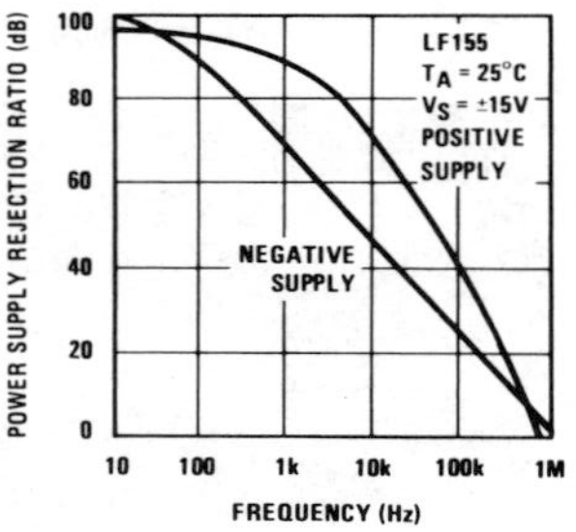

Power Supply Rejection Ratio

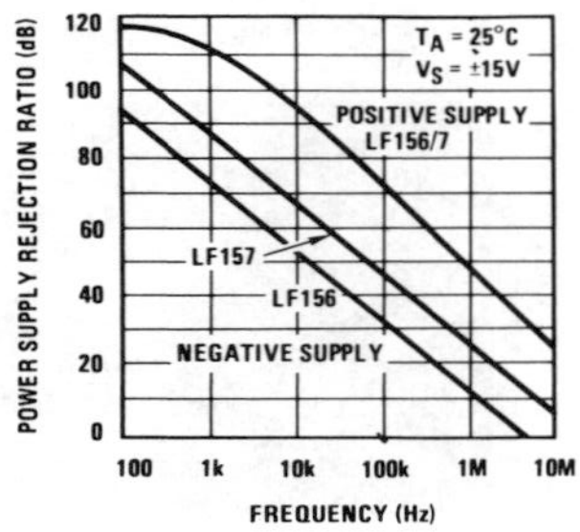

Undistorted Output Voltage Swing

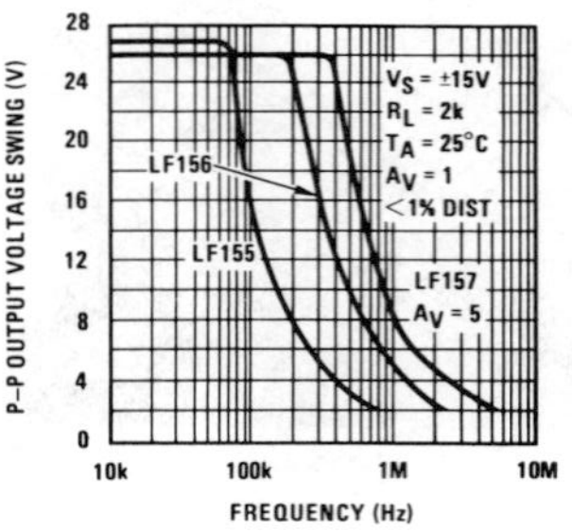

Equivalent Input Noise Voltage

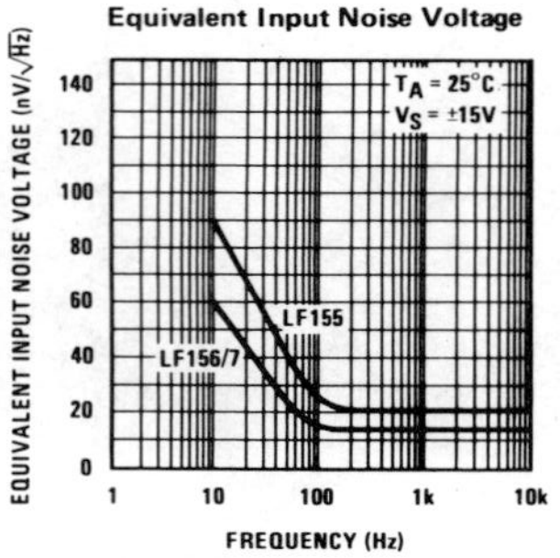

Equivalent Input Noise Voltage (Expanded Scale)

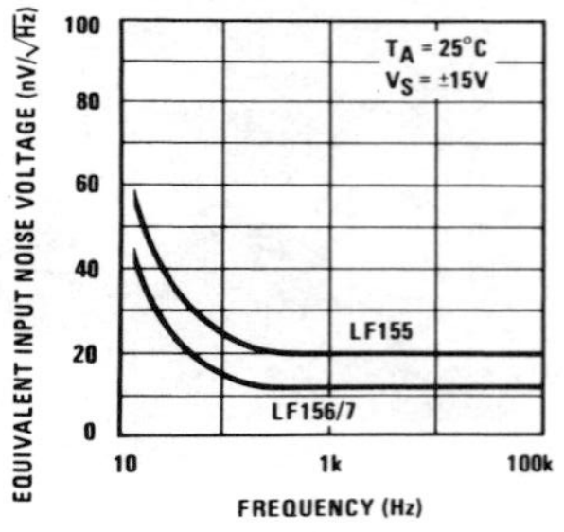

Typical AC Performance Characteristics (Continued)

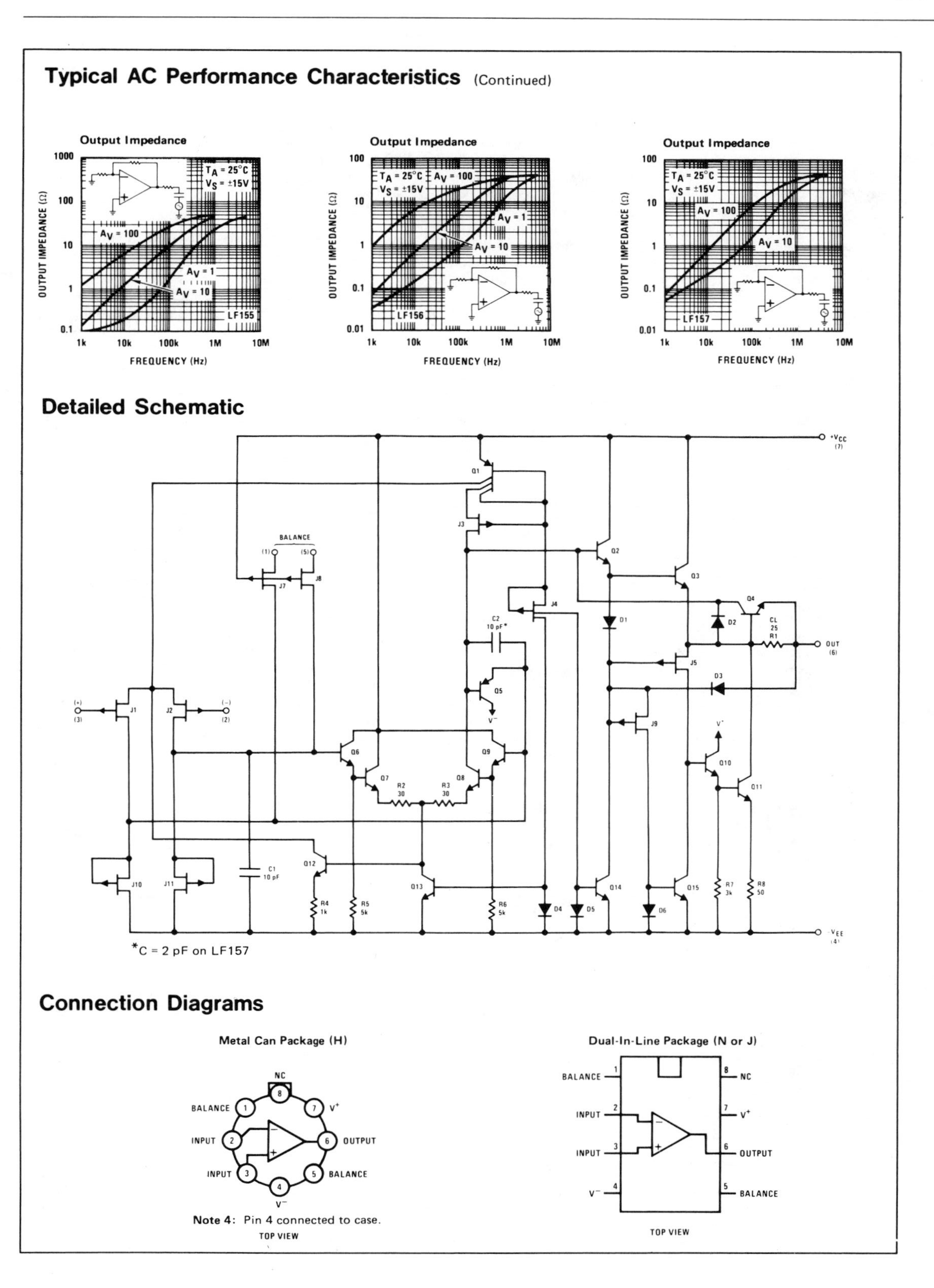

Detailed Schematic

*C = 2 pF on LF157

Connection Diagrams

Note 4: Pin 4 connected to case.

TOP VIEW

TOP VIEW

Application Hints

The LF155/6/7 series are op amps with JFET input devices. These JFETs have large reverse breakdown voltages from gate to source and drain eliminating the need for clamps across the inputs. Therefore large differential input voltages can easily be accomodated without a large increase in input current. The maximum differential input voltage is independent of the supply voltages. However, neither of the input voltages should be allowed to exceed the negative supply as this will cause large currents to flow which can result in a destroyed unit.

Exceeding the negative common-mode limit on either input will cause a reversal of the phase to the output and force the amplifier output to the corresponding high or low state. Exceeding the negative common-mode limit on both inputs will force the amplifier output to a high state. In neither case does a latch occur since raising the input back within the common-mode range again puts the input stage and thus the amplifier in a normal operating mode.

Exceeding the positive common-mode limit on a single input will not change the phase of the output however, if both inputs exceed the limit, the output of the amplifier will be forced to a high state.

These amplifiers will operate with the common-mode input voltage equal to the positive supply. In fact, the common-mode voltage can exceed the positive supply by approximately 100 mV independent of supply voltage and over the full operating temperature range. The positive supply can therefore be used as a reference on an input as, for example, in a supply current monitor and/or limiter.

Precautions should be taken to ensure that the power supply for the integrated circuit never becomes reversed in polarity or that the unit is not inadvertently installed backwards in a socket as an unlimited current surge through the resulting forward diode within the IC could cause fusing of the internal conductors and result in a destroyed unit.

Because these amplifiers are JFET rather than MOSFET input op amps they do not require special handling.

All of the bias currents in these amplifiers are set by FET current sources. The drain currents for the amplifiers are therefore essentially independent of supply voltage.

As with most amplifiers, care should be taken with lead dress, component placement and supply decoupling in order to ensure stability. For example, resistors from the output to an input should be placed with the body close to the input to minimize "pickup" and maximize the frequency of the feedback pole by minimizing the capacitance from the input to ground.

A feedback pole is created when the feedback around any amplifier is resistive. The parallel resistance and capacitance from the input of the device (usually the inverting input) to ac ground set the frequency of the pole. In many instances the frequency of this pole is much greater than the expected 3 dB frequency of the closed loop gain and consequently there is negligible effect on stability margin. However, if the feedback pole is less than approximately six times the expected 3 dB frequency a lead capacitor should be placed from the output to the input of the op amp. The value of the added capacitor should be such that the RC time constant of this capacitor and the resistance it parallels is greater than or equal to the original feedback pole time constant.

Typical Circuit Connections

V_{OS} Adjustment

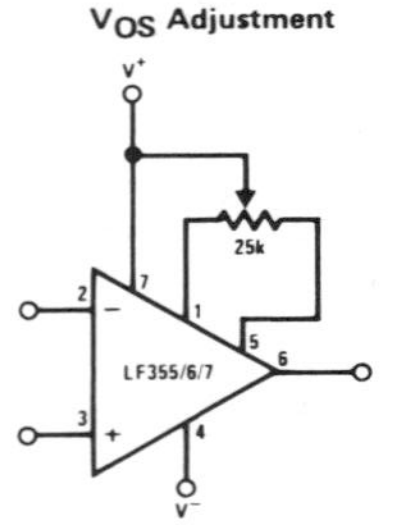

- V_{OS} is adjusted with a 25k potentiometer
- The potentiometer wiper is connected to V^+
- For potentiometers with temperature coefficient of 100 ppm/°C or less the additional drift with adjust is ≈ 0.5 μV/°C/mV of adjustment
- Typical overall drift: 5 μV/°C ±(0.5 μV/°C/mV of adj.)

Driving Capacitive Loads

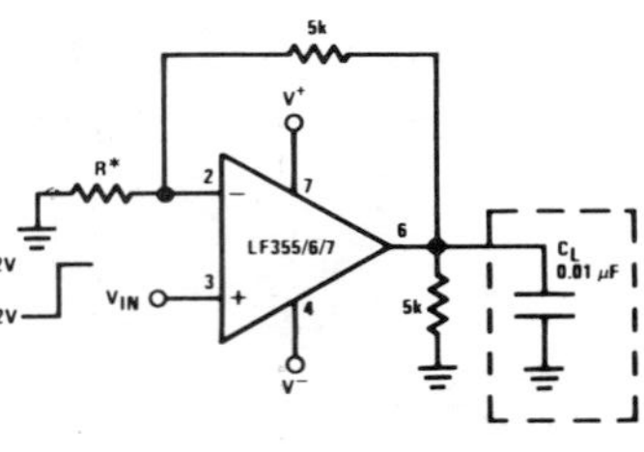

*LF155/6 R = 5k
LF157 R = 1.25k

Due to a unique output stage design, these amplifiers have the ability to drive large capacitive loads and still maintain stability. $C_{L(MAX)} \cong 0.01\ \mu F$.

Overshoot ≤ 20%

Settling time (t_s) ≅ 5 μs

LF157. A Large Power BW Amplifier

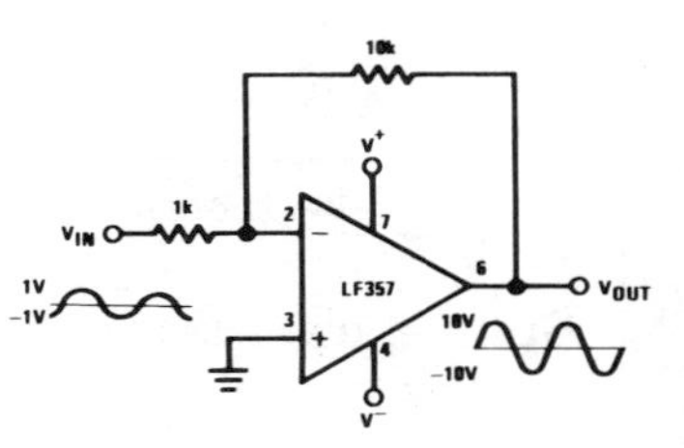

For distortion ≤ 1% and a 20 Vp-p V_{OUT} swing, power bandwidth is: 500 kHz.

Typical Applications

Settling Time Test Circuit

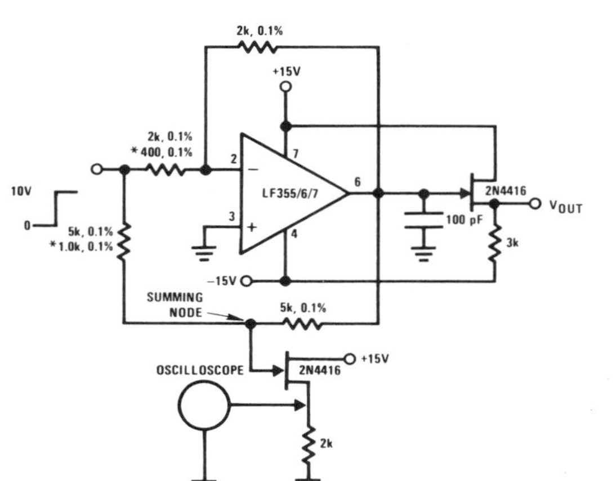

- Settling time is tested with the LF155/6 connected as unity gain inverter and LF157 connected for $A_V = -5$
- FET used to isolate the probe capacitance
- Output = 10V step
- $A_V = -5$ for LF157

Large Signal Inverter Output, V_{OUT} (from Settling Time Circuit)

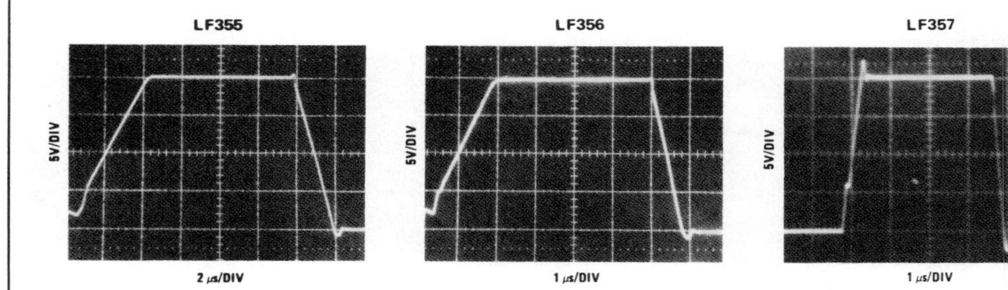

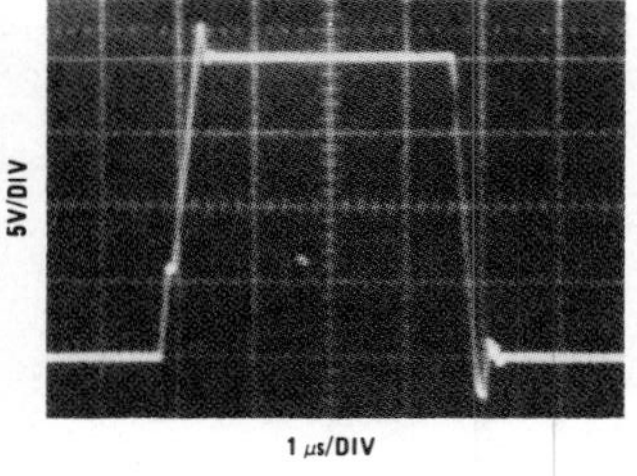

Low Drift Adjustable Voltage Reference

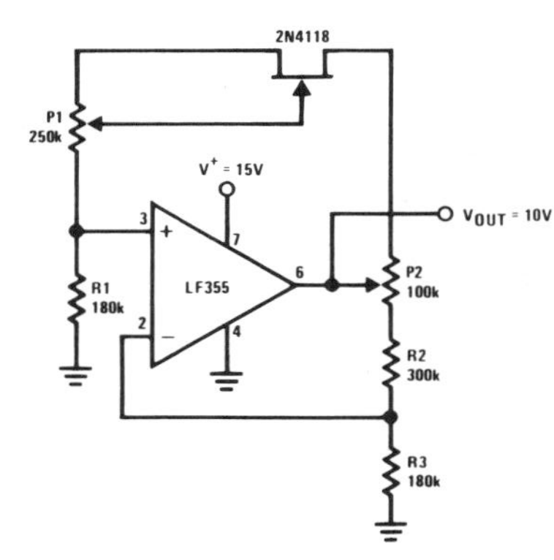

- $\Delta V_{OUT}/\Delta T = \pm 0.002\%/°C$
- All resistors and potentiometers should be wire-wound
- P1: drift adjust
- P2: V_{OUT} adjust
- Use LF155 for
  - ▲ Low I_B
  - ▲ Low drift
  - ▲ Low supply current

Typical Applications (Continued)

Fast Logarithmic Converter

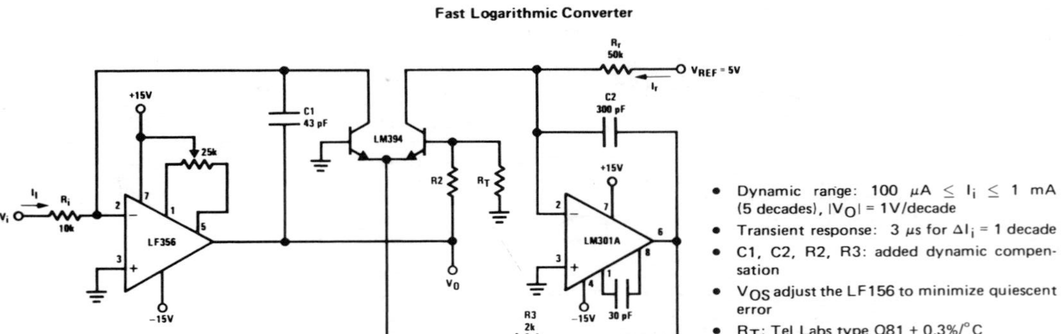

- Dynamic range: 100 $\mu A \leq I_i \leq 1$ mA (5 decades), $|V_O| = 1V$/decade
- Transient response: 3 μs for $\Delta I_i = 1$ decade
- C1, C2, R2, R3: added dynamic compensation
- V_{OS} adjust the LF156 to minimize quiescent error
- R_T: Tel Labs type Q81 + 0.3%/°C

$$|V_{OUT}| = \left[1 + \frac{R2}{R_T}\right] \frac{kT}{q} \ln V_i \left[\frac{R_r}{V_{REF} R_i}\right] = \log V_i \frac{1}{R_i I_r}$$

R2 = 15.7k, R_T = 1k, 0.3%/°C (for temperature compensation)

Precision Current Monitor

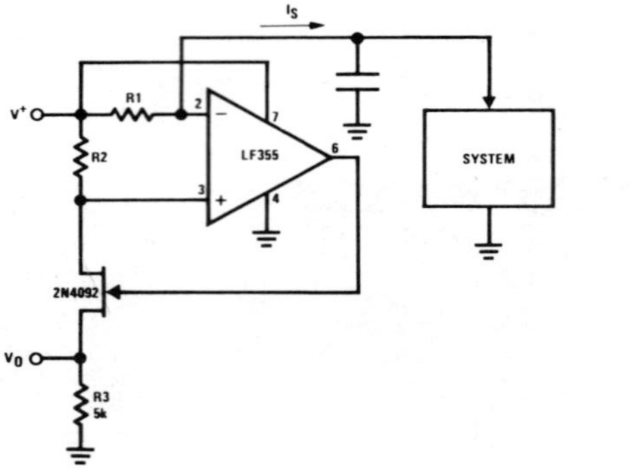

- $V_O = 5$ R1/R2 (V/mA of I_S)
- R1, R2, R3: 0.1% resistors
- Use LF155 for
  - ▲ Common-mode range to supply range
  - ▲ Low I_B
  - ▲ Low V_{OS}
  - ▲ Low supply current

8-Bit D/A Converter with Symmetrical Offset Binary Operation

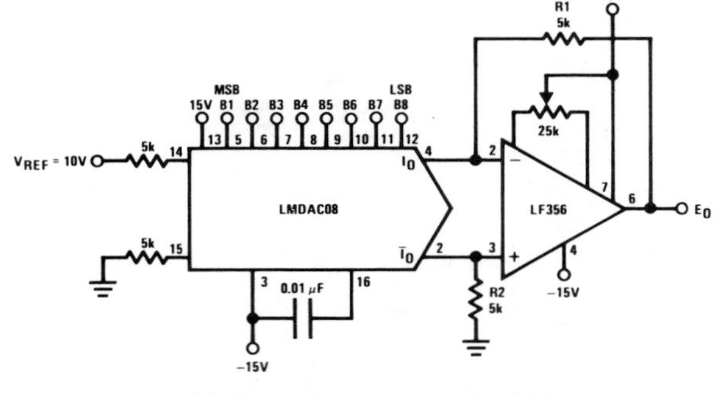

- R1, R2 should be matched within ±0.05%
- Full-scale response time: 3 μs

E_O	B1	B2	B3	B4	B5	B6	B7	B8	COMMENTS
+9.920	1	1	1	1	1	1	1	1	Positive Full-Scale
+0.040	1	0	0	0	0	0	0	0	(+) Zero-Scale
−0.040	0	1	1	1	1	1	1	1	(−) Zero-Scale
−9.920	0	0	0	0	0	0	0	0	Negative Full-Scale

Typical Applications (Continued)

Wide BW Low Noise, Low Drift Amplifier

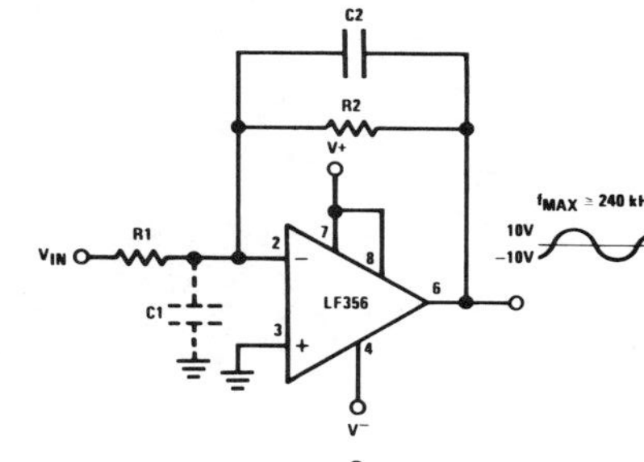

- Power BW: $f_{MAX} = \dfrac{S_r}{2\pi V_P} \cong 240$ kHz
- Parasitic input capacitance C1 ≅ (3 pF for LF155, LF156 and LF157 plus any additional layout capacitance) interacts with feedback elements and creates undesirable high frequency pole. To compensate add C2 such that: R2C2 ≅ R1C1.

Boosting the LF156 with a Current Amplifier

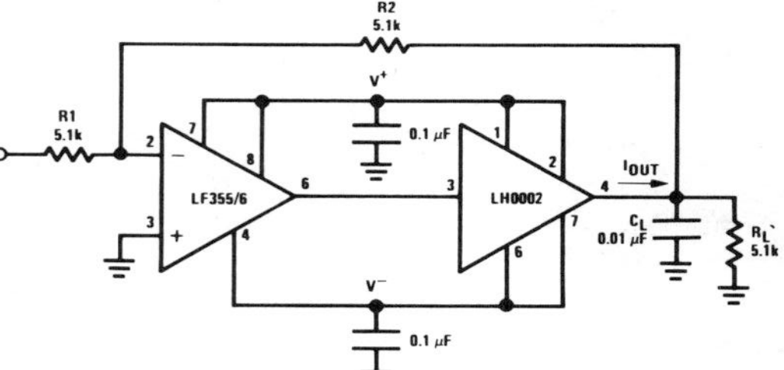

- $I_{OUT(MAX)} \cong 150$ mA (will drive $R_L \geq 100\Omega$)
- $\dfrac{\Delta V_{OUT}}{\Delta T} = \dfrac{0.15}{10^{-2}}$ V/μs (with C_L shown)
- No additional phase shift added by the current amplifier

3 Decades VCO

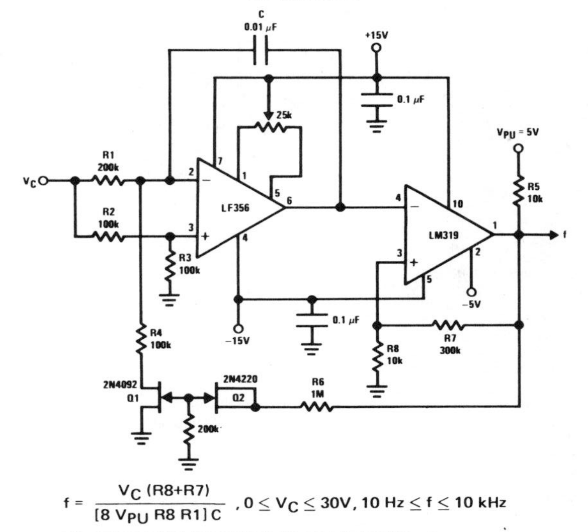

$$f = \frac{V_C\ (R8+R7)}{[8\ V_{PU}\ R8\ R1]\ C},\ 0 \leq V_C \leq 30V,\ 10\ Hz \leq f \leq 10\ kHz$$

R1, R4 matched. Linearity 0.1% over 2 decades.

Isolating Large Capacitive Loads

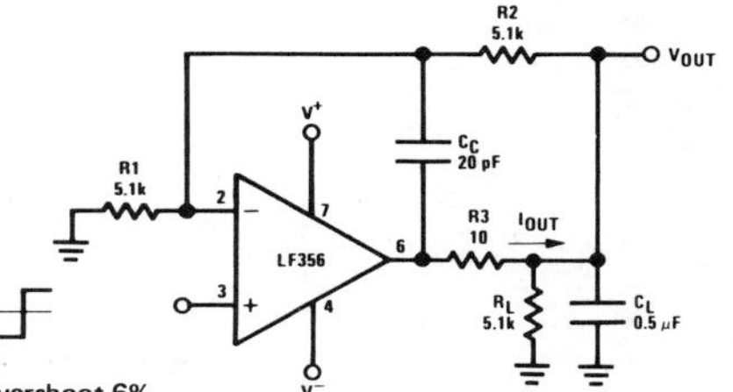

- Overshoot 6%
- t_s 10 μs
- When driving large C_L, the V_{OUT} slew rate determined by C_L and $I_{OUT(MAX)}$:

$$\frac{\Delta V_{OUT}}{\Delta T} = \frac{I_{OUT}}{C_L} \cong \frac{0.02}{0.5}\ \text{V}/\mu\text{s} = 0.04\ \text{V}/\mu\text{s (with } C_L \text{ shown)}$$

Low Drift Peak Detector

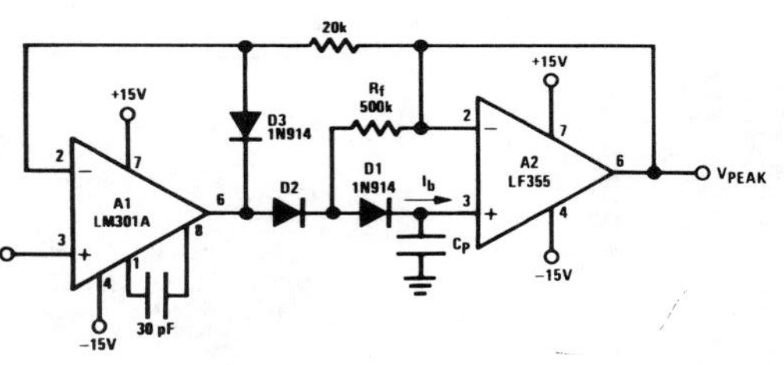

- By adding D1 and R_f, $V_{D1} = 0$ during hold mode. Leakage of D2 provided by feedback path through R_f.
- Leakage of circuit is essentially I_b (LF155, LF156) plus capacitor leakage of C_P.
- Diode D3 clamps V_{OUT} (A1) to $V_{IN}-V_{D3}$ to improve speed and to limit reverse bias of D2.
- Maximum input frequency should be << $1/2\pi R_f C_{D2}$ where C_{D2} is the shunt capacitance of D2.

Non-Inverting Unity Gain Operation for LF157

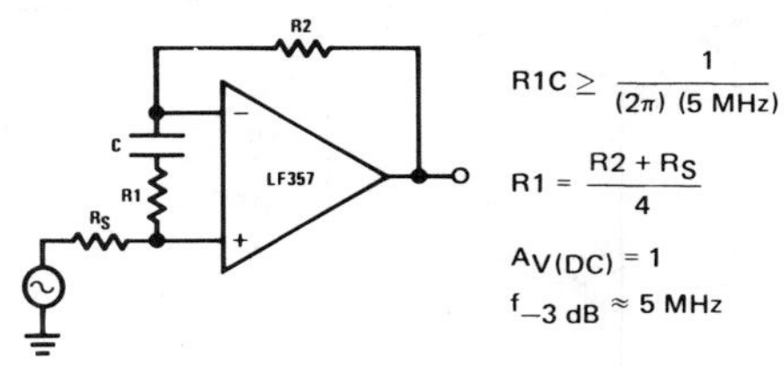

$R1C \geq \dfrac{1}{(2\pi)\ (5\ MHz)}$

$R1 = \dfrac{R2 + R_S}{4}$

$A_{V(DC)} = 1$

$f_{-3\ dB} \approx 5$ MHz

Inverting Unity Gain for LF157

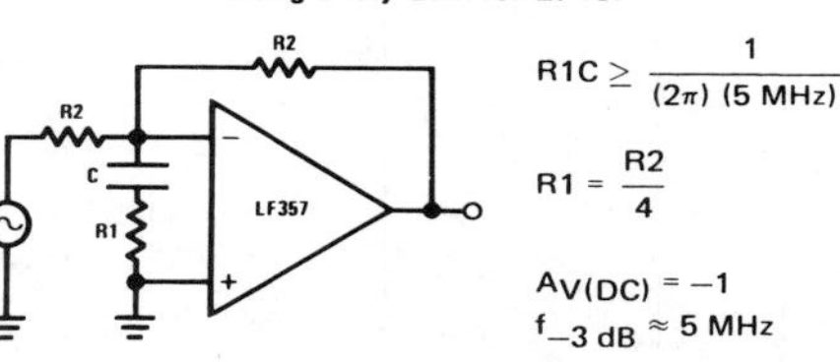

$R1C \geq \dfrac{1}{(2\pi)\ (5\ MHz)}$

$R1 = \dfrac{R2}{4}$

$A_{V(DC)} = -1$

$f_{-3\ dB} \approx 5$ MHz

Typical Applications (Continued)

High Impedance, Low Drift Instrumentation Amplifier

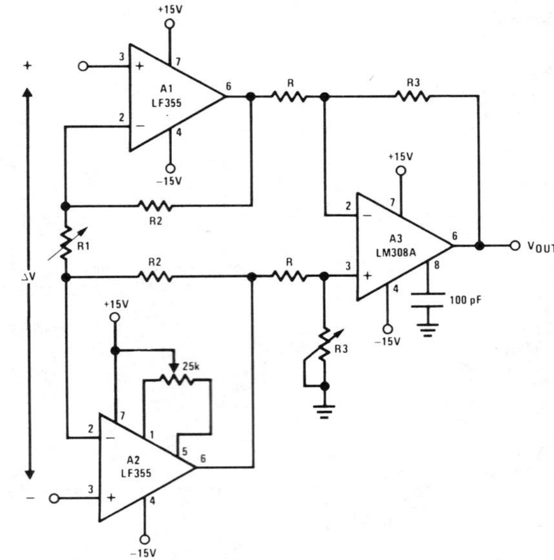

- $V_{OUT} = \frac{R3}{R}\left[\frac{2R2}{R1} + 1\right] \Delta V$, $V^- + 2V \leq V_{IN}$ common-mode $\leq V^+$
- System V_{OS} adjusted via A2 V_{OS} adjust
- Trim R3 to boost up CMRR to 120 dB. Instrumentation amplifier Resistor array RA201 (National Semiconductor) recommended

Fast Sample and Hold

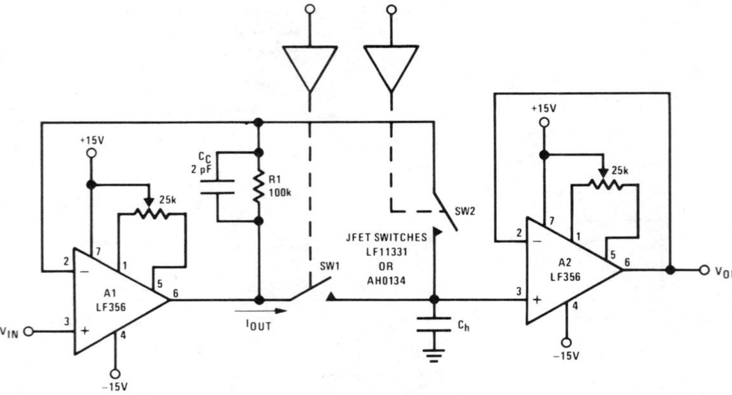

- Both amplifiers (A1, A2) have feedback loops individually closed with stable responses (overshoot negligible)
- Acquisition time T_A, estimated by:

  $T_A \cong \left[\frac{2R_{ON}, V_{IN}, C_h}{S_r}\right]^{1/2}$ provided that:

  $V_{IN} < 2\pi S_r R_{ON} C_h$ and $T_A > \frac{V_{IN} C_h}{I_{OUT(MAX)}}$, R_{ON} is of SW1

  If inequality not satisfied: $T_A \cong \frac{V_{IN} C_h}{20\text{ mA}}$
- LF156 developes full S_r output capability for $V_{IN} \geq 1V$
- Addition of SW2 improves accuracy by putting the voltage drop across SW1 inside the feedback loop
- Overall accuracy of system determined by the accuracy of both amplifiers, A1 and A2

Typical Applications (Continued)

High Accuracy Sample and Hold

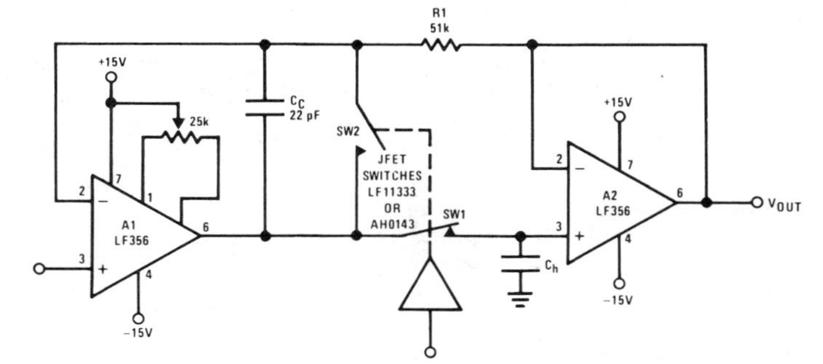

- By closing the loop through A2, the V_{OUT} accuracy will be determined uniquely by A1. No V_{OS} adjust required for A2.
- T_A can be estimated by same considerations as previously but, because of the added propagation delay in the feedback loop (A2) the overshoot is not negligible.
- Overall system slower than fast sample and hold
- R1, C_C: additional compensation
- Use LF156 for
  - ▲ Fast settling time
  - ▲ Low V_{OS}

High Q Band Pass Filter

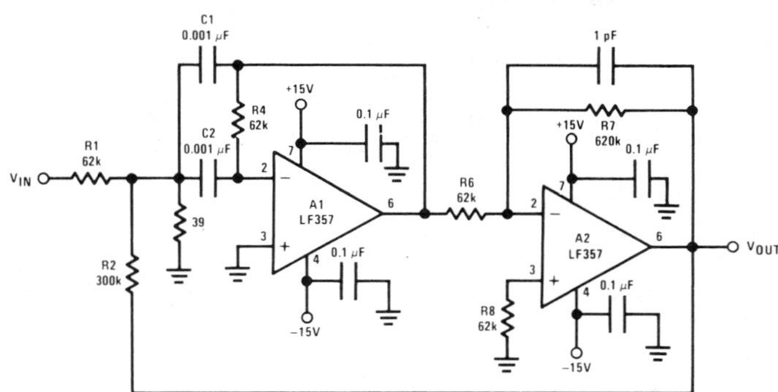

- By adding positive feedback (R2) Q increases to 40
- f_{BP} = 100 kHz

$$\frac{V_{OUT}}{V_{IN}} = 10\sqrt{Q}$$

- Clean layout recommended
- Response to a 1 Vp-p tone burst: 300 μs

High Q Notch Filter

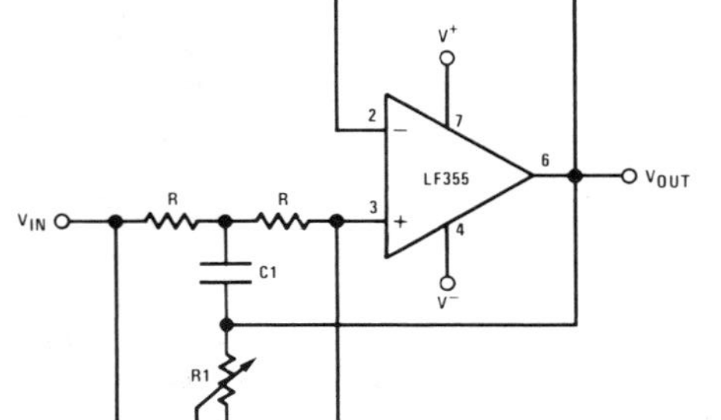

- 2R1 = R = 10 MΩ
  2C = C1 = 300 pF
- Capacitors should be matched to obtain high Q
- f_{NOTCH} = 120 Hz, notch = −55 dB, Q > 100
- Use LF155 for
  - ▲ Low I_B
  - ▲ Low supply current

Definition of Terms

Input Offset Voltage: That voltage which must be applied between the input terminals through two equal resistances to obtain zero output voltage.

Input Offset Current: The difference in the currents into the two input terminals when the output is at zero.

Input Bias Current: The average of the two input currents.

Input Common-Mode Voltage Range: The range of voltages on the input terminals for which the amplifier is operational. Note that the specifications are not guaranteed over the full common-mode voltage range unless specifically stated.

Common-Mode Rejection Ratio: The ratio of the input common-mode voltage range to the peak-to-peak change in input offset voltage over this range.

Input Resistance: The ratio of the change in input voltage to the change in input current on either input with the other grounded.

Supply Current: The current required from the power supply to operate the amplifier with no load and the output midway between the supplies.

Output Voltage Swing: The peak output voltage swing, referred to zero, that can be obtained without clipping.

Large-Signal Voltage Gain: The ratio of the output voltage swing to the change in input voltage required to drive the output from zero to this voltage.

Power Supply Rejection Ratio: The ratio of the change in input offset voltage to the change in power supply voltage producing it. The typical curves in this sheet show values for each supply independently changed. The electrical specification, however, is measured for both supply magnitudes increasing or decreasing simultaneously, in accordance with common practice.

Settling Time: The time required for the error between input and output to settle to within a specified limit after an input is applied to the test circuit shown in typical applications.

Physical Dimensions inches (millimeters)

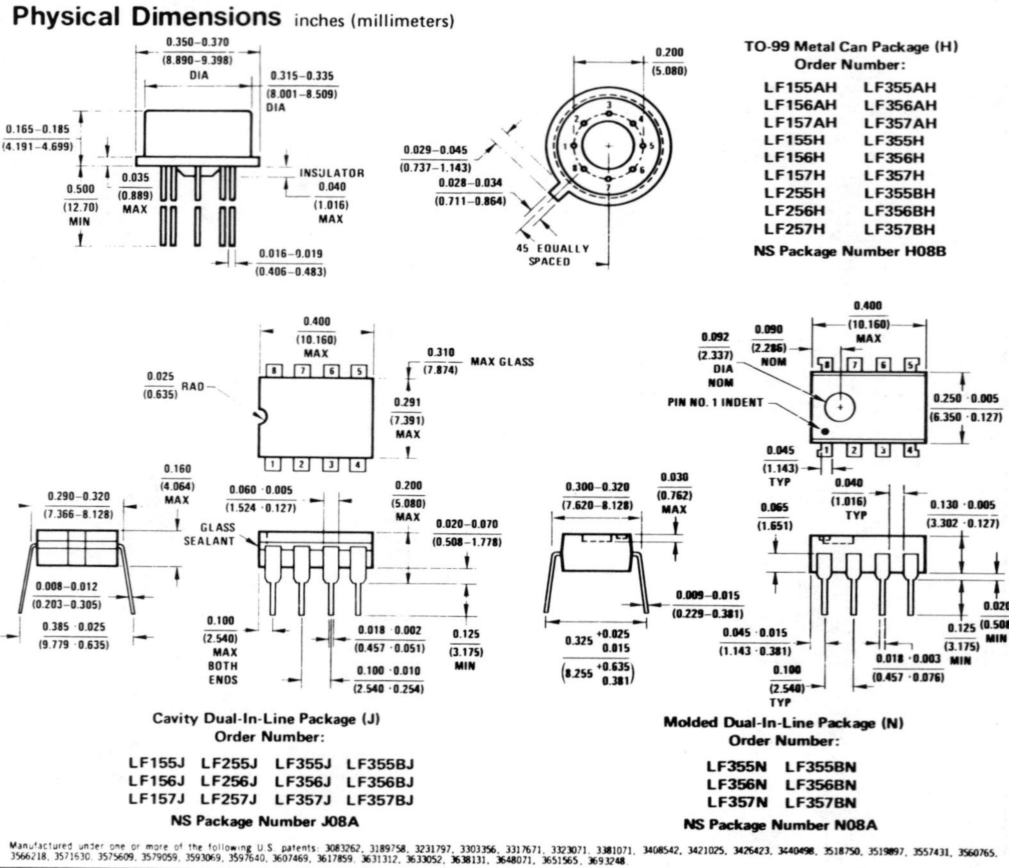

Manufactured under one or more of the following U.S. patents: 3083262, 3189758, 3231797, 3303356, 3317671, 3323071, 3381071, 3408542, 3421025, 3426423, 3440498, 3518750, 3519897, 3557431, 3560765, 3566218, 3571630, 3575609, 3579059, 3593069, 3597640, 3607469, 3617859, 3631312, 3633052, 3638131, 3648071, 3651565, 3693248.

National Semiconductor

Voltage Regulators

LM117/LM217/LM317 3-terminal adjustable regulator

General Description

The LM117/LM217/LM317 are adjustable 3-terminal positive voltage regulators capable of supplying in excess of 1.5A over a 1.2V to 37V output range. They are exceptionally easy to use and require only two external resistors to set the output voltage. Further, both line and load regulation are better than standard fixed regulators. Also, the LM117 is packaged in standard transistor packages which are easily mounted and handled.

In addition to higher performance than fixed regulators, the LM117 series offers full overload protection available only in IC's. Included on the chip are current limit, thermal overload protection and safe area protection. All overload protection circuitry remains fully functional even if the adjustment terminal is disconnected.

Features

- Adjustable output down to 1.2V
- Guaranteed 1.5A output current
- Line regulation typically 0.01%/V
- Load regulation typically 0.1%
- Current limit constant with temperature
- **100% electrical burn-in**
- Eliminates the need to stock many voltages
- Standard 3-lead transistor package
- 80 dB ripple rejection

Normally, no capacitors are needed unless the device is situated far from the input filter capacitors in which case an input bypass is needed. An optional output capacitor can be added to improve transient response. The adjustment terminal can be bypassed to achieve very high ripple rejections ratios which are difficult to achieve with standard 3-terminal regulators.

Besides replacing fixed regulators, the LM117 is useful in a wide variety of other applications. Since the regulator is "floating" and sees only the input-to-output differential voltage, supplies of several hundred volts can be regulated as long as the maximum input to output differential is not exceeded.

Also, it makes an especially simple adjustable switching regulator, a programmable output regulator, or by connecting a fixed resistor between the adjustment and output, the LM117 can be used as a precision current regulator. Supplies with electronic shutdown can be achieved by clamping the adjustment terminal to ground which programs the output to 1.2V where most loads draw little current.

The LM117K, LM217K and LM317K are packaged in standard TO-3 transistor packages while the LM117H, LM217H and LM317H are packaged in a solid Kovar base TO-5 transistor package. The LM117 is rated for operation from −55°C to +150°C, the LM217 from −25°C to +150°C and the LM317 from 0°C to +125°C. The LM317T and LM317MP, rated for operation over a 0°C to +125°C range, are available in a TO-220 plastic package and a TO-202 package, respectively.

For applications requiring greater output current in excess of 3A and 5A, see LM150 series and LM138 series data sheets, respectively. For the negative complement, see LM137 series data sheet.

LM117 Series Packages and Power Capability

DEVICE	PACKAGE	RATED POWER DISSIPATION	DESIGN LOAD CURRENT
LM117	TO-3	20W	1.5A
LM217 LM317	TO-39	2W	0.5A
LM317T	TO-220	15W	1.5A
LM317M	TO-202	7.5W	0.5A

Typical Applications

1.2V–25V Adjustable Regulator

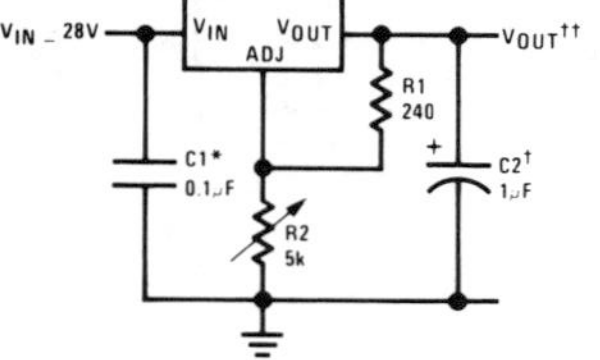

†Optional—improves transient response

*Needed if device is far from filter capacitors

$$\dagger\dagger V_{OUT} = 1.25V \left(1 + \frac{R2}{R1}\right)$$

Digitally Selected Outputs

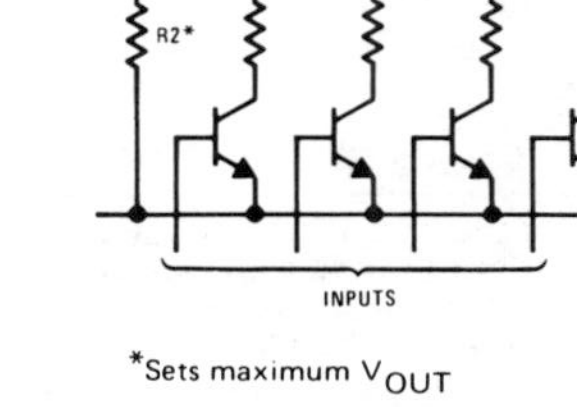

*Sets maximum V_{OUT}

5V Logic Regulator with Electronic Shutdown*

* Min output ≈ 1.2V

absolute maximum ratings

Power Dissipation	Internally limited
Input–Output Voltage Differential	40V
Operating Junction Temperature Range	
LM117	−55°C to +150°C
LM217	−25°C to +150°C
LM317	0°C to +125°C
Storage Temperature	−65°C to +150°C
Lead Temperature (Soldering, 10 seconds)	300°C

electrical characteristics (Note 1)

PARAMETER	CONDITIONS	LM117/217			LM317			UNITS
		MIN	TYP	MAX	MIN	TYP	MAX	
Line Regulation	$T_A = 25°C$, $3V \le V_{IN} - V_{OUT} \le 40V$ (Note 2)		0.01	0.02		0.01	0.04	%/V
Load Regulation	$T_A = 25°C$, $10\ mA \le I_{OUT} \le I_{MAX}$							
	$V_{OUT} \le 5V$, (Note 2)		5	15		5	25	mV
	$V_{OUT} \ge 5V$, (Note 2)		0.1	0.3		0.1	0.5	%
Adjustment Pin Current			50	100		50	100	μA
Adjustment Pin Current Change	$10\ mA \le I_L \le I_{MAX}$ $2.5V \le (V_{IN}-V_{OUT}) \le 40V$		0.2	5		0.2	5	μA
Reference Voltage	$3 \le (V_{IN}-V_{OUT}) \le 40V$, (Note 3) $10\ mA \le I_{OUT} \le I_{MAX}$, $P \le P_{MAX}$	1.20	1.25	1.30	1.20	1.25	1.30	V
Line Regulation	$3V \le V_{IN} - V_{OUT} \le 40V$, (Note 2)		0.02	0.05		0.02	0.07	%/V
Load Regulation	$10\ mA \le I_{OUT} \le I_{MAX}$, (Note 2)							
	$V_{OUT} \le 5V$		20	50		20	70	mV
	$V_{OUT} \ge 5V$		0.3	1		0.3	1.5	%
Temperature Stability	$T_{MIN} \le T_j \le T_{MAX}$		1			1		%
Minimum Load Current	$V_{IN}-V_{OUT} = 40V$		3.5	5		3.5	10	mA
Current Limit	$V_{IN}-V_{OUT} \le 15V$							
	K and T Package	1.5	2.2		1.5	2.2		A
	H and P Package	0.5	0.8		0.5	0.8		A
	$V_{IN}-V_{OUT} = 40V$							
	K and T Package		0.4			0.4		A
	H and P Package		0.07			0.07		A
RMS Output Noise, % of V_{OUT}	$T_A = 25°C$, $10\ Hz \le f \le 10\ kHz$		0.003			0.003		%
Ripple Rejection Ratio	$V_{OUT} = 10V$, $f = 120\ Hz$		65			65		dB
	$C_{ADJ} = 10\mu F$	66	80		66	80		dB
Long-Term Stability	$T_A = 125°C$		0.3	1		0.3	1	%
Thermal Resistance, Junction to Case	H Package		12	15		12	15	°C/W
	K Package		2.3	3		2.3	3	°C/W
	T Package					5		°C/W
	P Package					12		°C/W

Note 1: Unless other wise specified, these specifications apply $-55°C \le T_j \le +150°C$ for the LM117, $-25°C \le T_j \le +150°C$ for the LM217 and $0°C \le T_j \le +125°C$ for the LM317; $V_{IN}-V_{OUT} = 5V$ and $I_{OUT} = 0.1A$ for the TO-5 package and $I_{OUT} = 0.5A$ for the TO-3 package and TO-220 package. Although power dissipation is internally limited, these specifications are applicable for power dissipations of 2W for the TO-5 and 20W for the TO-3 and TO-220. I_{MAX} is 1.5A for the TO-3 and TO-220 package and 0.5A for the TO-5 package.

Note 2: Regulation is measured at constant junction temperature. Changes in output voltage due to heating effects must be taken into account separately. Pulse testing with low duty cycle is used.

Note 3: Selected devices with tightend tolerance reference voltage available.

typical performance characteristics (K and T Packages)

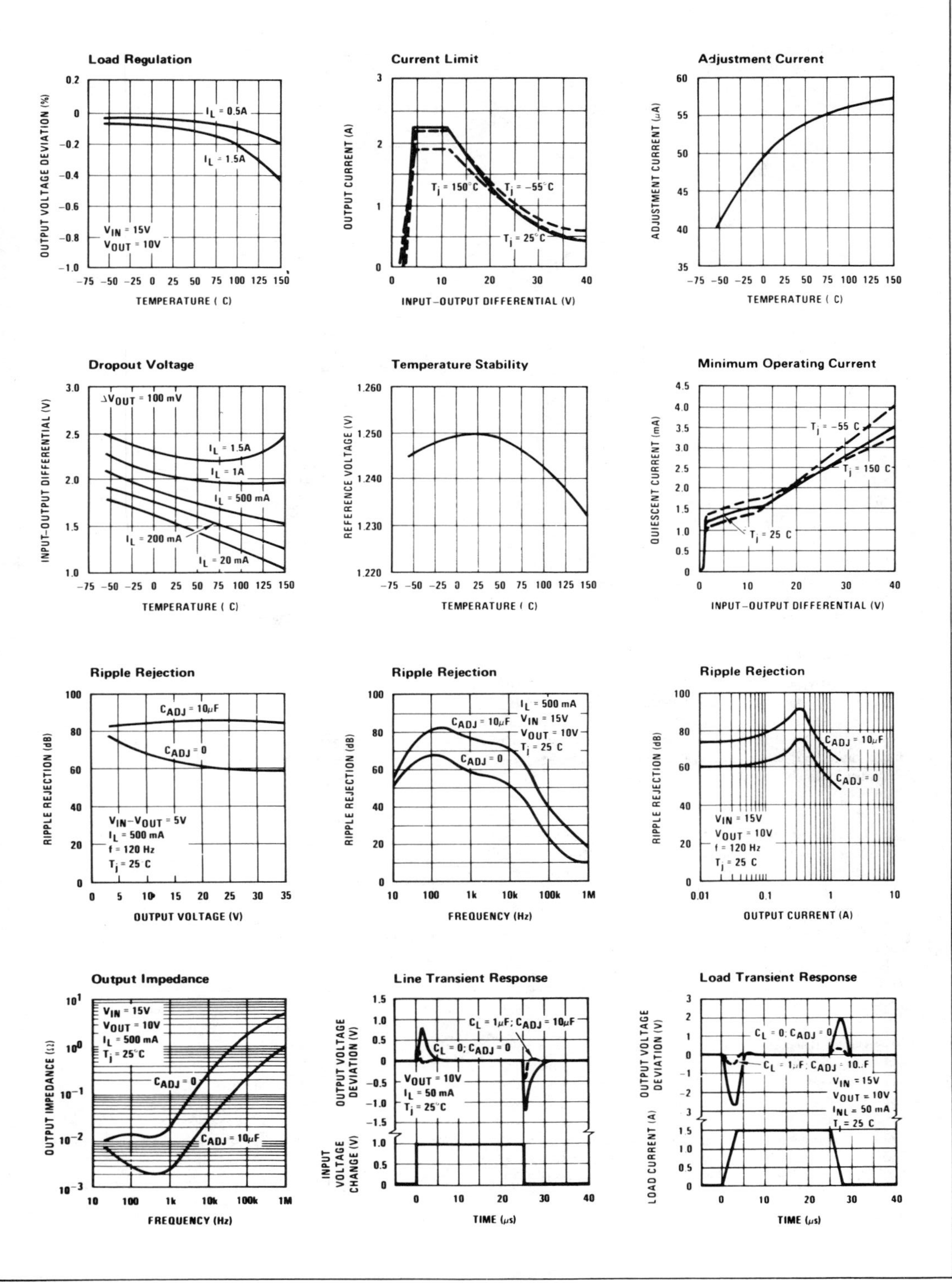

application hints

In operation, the LM117 develops a nominal 1.25V reference voltage, V_{REF}, between the output and adjustment terminal. The reference voltage is impressed across program resistor R1 and, since the voltage is constant, a constant current I_1 then flows through the output set resistor R2, giving an output voltage of

$$V_{OUT} = V_{REF}\left(1 + \frac{R2}{R1}\right) + I_{ADJ}R2$$

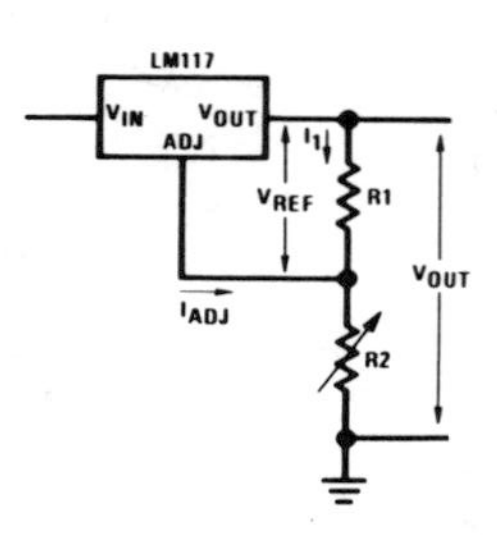

FIGURE 1.

Since the 100µA current from the adjustment terminal represents an error term, the LM117 was designed to minimize I_{ADJ} and make it very constant with line and load changes. To do this, all quiescent operating current is returned to the output establishing a minimum load current requirement. If there is insufficient load on the output, the output will rise.

External Capacitors

An input bypass capacitor is recommended. A 0.1µF disc or 1µF solid tantalum on the input is suitable input bypassing for almost all applications. The device is more sensitive to the absence of input bypassing when adjustment or output capacitors are used but the above values will eliminate the possibility of problems.

The adjustment terminal can be bypassed to ground on the LM117 to improve ripple rejection. This bypass capacitor prevents ripple from being amplified as the output voltage is increased. With a 10µF bypass capacitor 80 dB ripple rejection is obtainable at any output level. Increases over 10µF do not appreciably improve the ripple rejection at frequencies above 120 Hz. If the bypass capacitor is used, it is sometimes necessary to include protection diodes to prevent the capacitor from discharging through internal low current paths and damaging the device.

In general, the best type of capacitors to use are solid tantalum. Solid tantalum capacitors have low impedance even at high frequencies. Depending upon capacitor construction, it takes about 25µF in aluminum electrolytic to equal 1µF solid tantalum at high frequencies. Ceramic capacitors are also good at high frequencies; but some types have a large decrease in capacitance at frequencies around 0.5 MHz. For this reason, 0.01µF disc may seem to work better than a 0.1µF disc as a bypass.

Although the LM117 is stable with no output capacitors, like any feedback circuit, certain values of external capacitance can cause excessive ringing. This occurs with values between 500 pF and 5000 pF. A 1µF solid tantalum (or 25µF aluminum electrolytic) on the output swamps this effect and insures stability.

Load Regulation

The LM117 is capable of providing extremely good load regulation but a few precautions are needed to obtain maximum performance. The current set resistor connected between the adjustment terminal and the output terminal (usually 240Ω) should be tied directly to the output of the regulator rather than near the load. This eliminates line drops from appearing effectvely in series with the reference and degrading regulation. For example, a 15V regulator with 0.05Ω resistance between the regulator and load will have a load regulation due to line resistance of 0.05Ω x I_L. If the set resistor is connected near the load the effective line resistance will be 0.05Ω (1 + R2/R1) or in this case, 11.5 times worse.

Figure 2 shows the effect of resistance between the regulator and 240Ω set resistor.

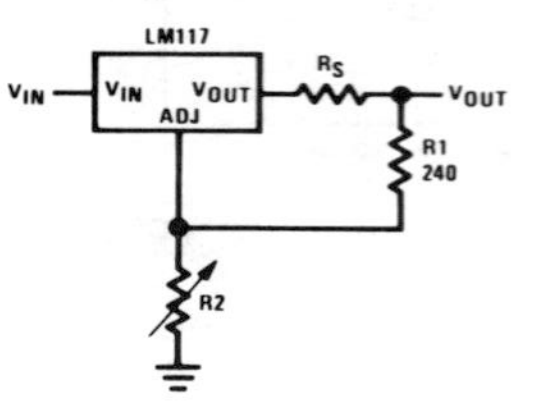

FIGURE 2. Regulator with Line Resistance in Output Lead

With the TO-3 package, it is easy to minimize the resistance from the case to the set resistor, by using two separate leads to the case. However, with the TO-5 package, care should be taken to minimize the wire length of the output lead. The ground of R2 can be returned near the ground of the load to provide remote ground sensing and improve load regulation.

Protection Diodes

When external capacitors are used with *any* IC regulator it is sometimes necessary to add protection diodes to prevent the capacitors from discharging through low current points into the regulator. Most 10µF capacitors have low enough internal series resistance to deliver 20A spikes when shorted. Although the surge is short, there is enough energy to damage parts of the IC.

When an output capacitor is connected to a regulator and the input is shorted, the output capacitor will discharge into the output of the regulator. The discharge current depends on the value of the capacitor, the output voltage of the regulator, and the rate of decrease of V_{IN}. In the LM117, this discharge path is through a large junction that is able to sustain 15A surge with no problem. This is not true of other types of positive regulators. For output capacitors of 25μF or less, there is no need to use diodes.

The bypass capacitor on the adjustment terminal can discharge through a low current junction. Discharge occurs when *either* the input or output is shorted. Internal to the LM117 is a 50Ω resistor which limits the peak discharge current. No protection is needed for output voltages of 25V or less and 10μF capacitance. *Figure 3* shows an LM117 with protection diodes included for use with outputs greater than 25V and high values of output capacitance.

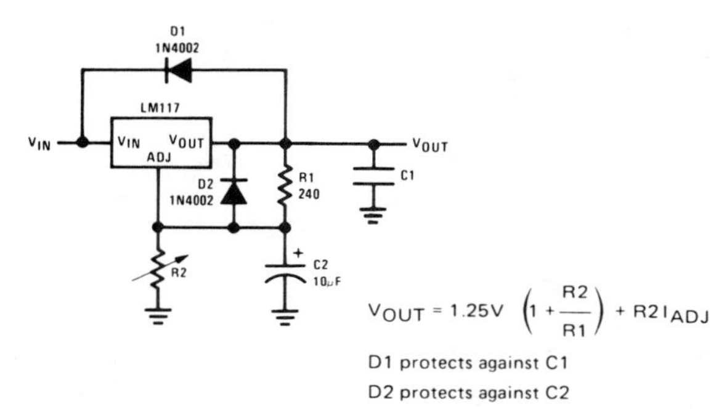

$$V_{OUT} = 1.25V \left(1 + \frac{R2}{R1}\right) + R2 I_{ADJ}$$

D1 protects against C1

D2 protects against C2

FIGURE 3. Regulator with Protection Diodes

schematic diagram

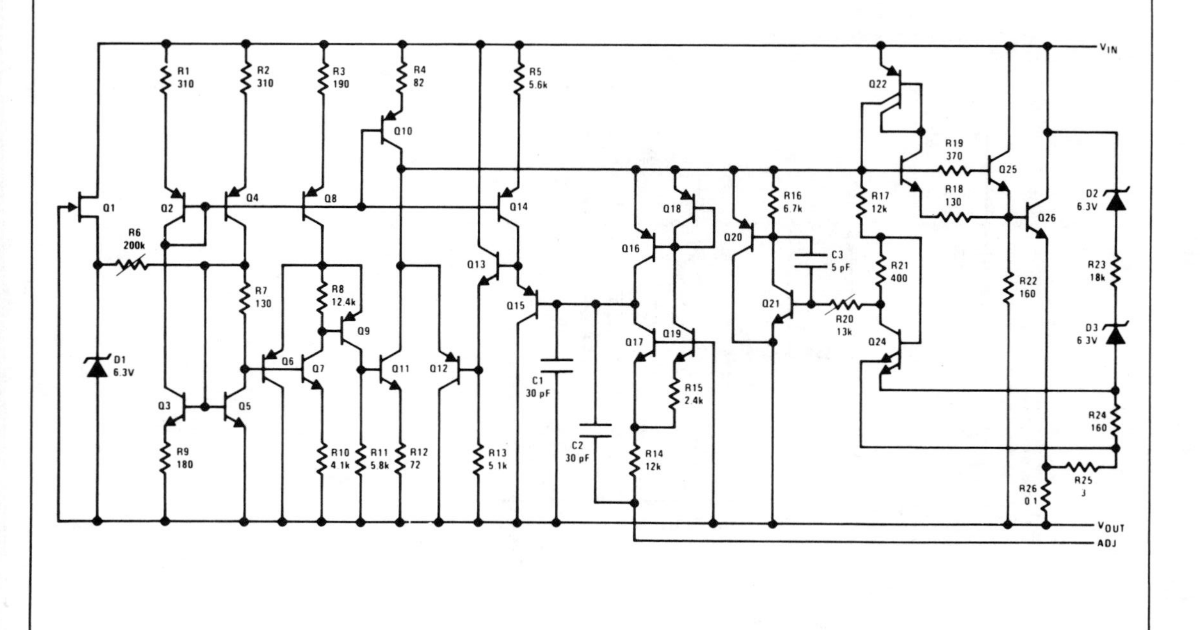

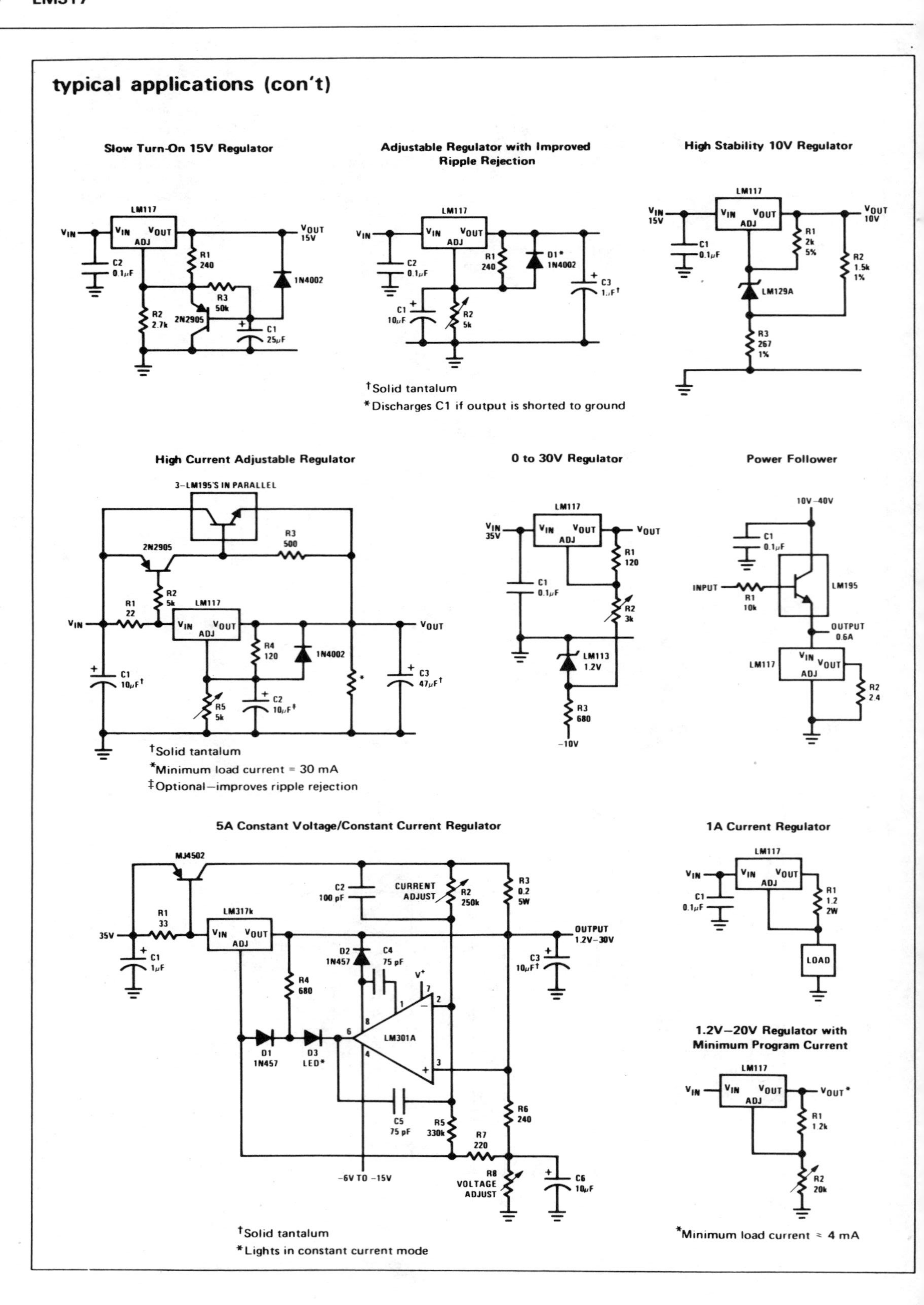
typical applications (con't)
Slow Turn-On 15V Regulator
LM117
VIN
VOUT
ADJ
VOUT 15V
C2 0.1μF
R1 240
R2 2.7k
R3 50k
2N2905
1N4002
C1 25μF
Adjustable Regulator with Improved Ripple Rejection
LM117
R1 240
D1* 1N4002
C2 0.1μF
C1 10μF
R2 5k
C3 1μF†
†Solid tantalum
*Discharges C1 if output is shorted to ground
High Stability 10V Regulator
LM117
VIN 15V
VOUT 10V
C1 0.1μF
R1 2k 5%
R2 1.5k 1%
LM129A
R3 267 1%
High Current Adjustable Regulator
3–LM195'S IN PARALLEL
2N2905
R3 500
R2 5k
R1 22
LM117
R4 120
1N4002
C1 10μF†
R5 5k
C2 10μF‡
C3 47μF†
†Solid tantalum
*Minimum load current = 30 mA
‡Optional—improves ripple rejection
0 to 30V Regulator
LM117
VIN 35V
R1 120
C1 0.1μF
R2 3k
LM113 1.2V
R3 680
−10V
Power Follower
10V–40V
C1 0.1μF
INPUT
R1 10k
LM195
OUTPUT 0.6A
LM117
R2 2.4
5A Constant Voltage/Constant Current Regulator
MJ4502
R1 33
35V
C1 1μF
LM317k
C2 100 pF
CURRENT ADJUST
R2 250k
R3 0.2 5W
OUTPUT 1.2V–30V
C3 10μF†
D2 1N457
C4 75 pF
R4 680
V+
LM301A
D1 1N457
D3 LED*
C5 75 pF
R5 330k
R7 220
R6 240
−6V TO −15V
R8 VOLTAGE ADJUST
C6 10μF
†Solid tantalum
*Lights in constant current mode
1A Current Regulator
LM117
C1 0.1μF
R1 1.2 2W
LOAD
1.2V–20V Regulator with Minimum Program Current
LM117
VOUT*
R1 1.2k
R2 20k
*Minimum load current ≈ 4 mA

typical applications (con't)

High Gain Amplifier

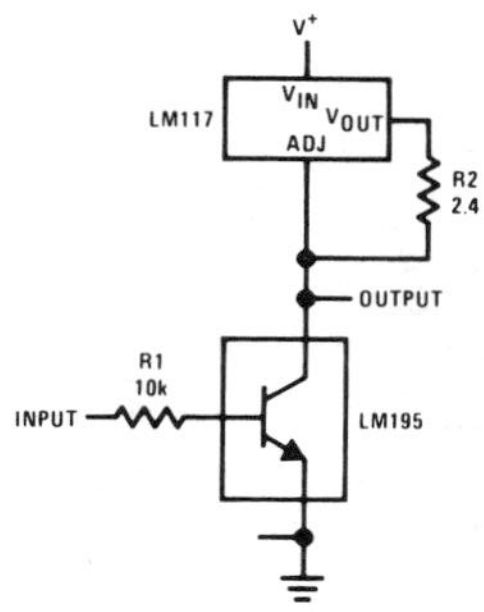

Low Cost 3A Switching Regulator

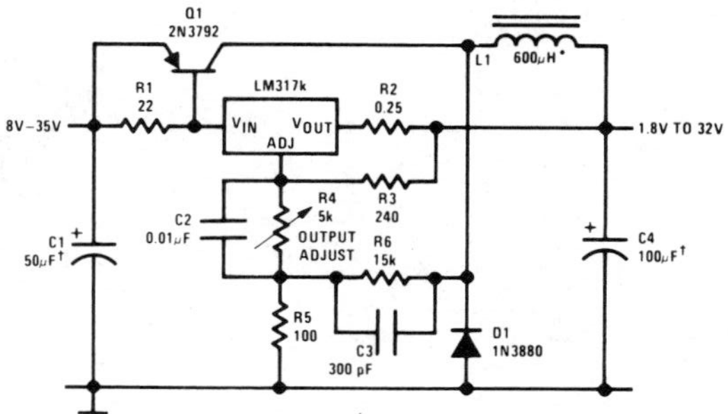

†Solid Tantalum

*Core—Arnold A-254168-2 60 turns

4A Switching Regulator with Overload Protection

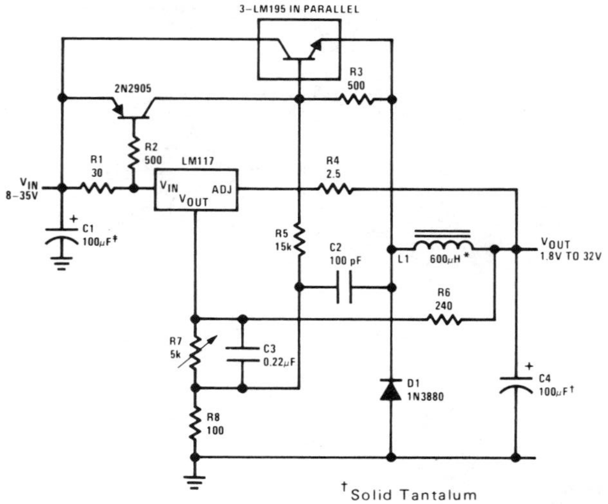

†Solid Tantalum

*Core Arnold A-254168-2 60 turns

Precision Current Limiter

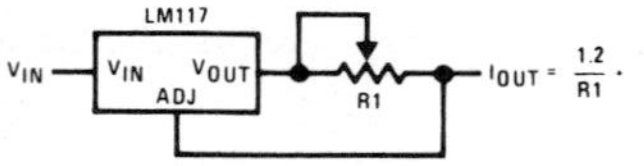

$I_{OUT} = \frac{1.2}{R1}$

$*0.8\Omega \leq R1 \leq 120\Omega$

Tracking Preregulator

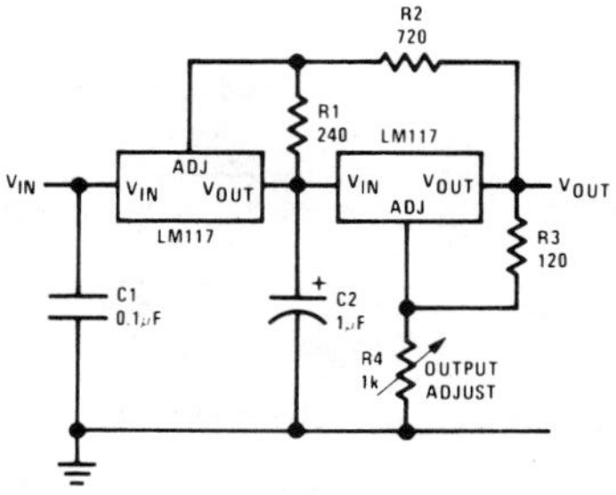

High Voltage Regulator

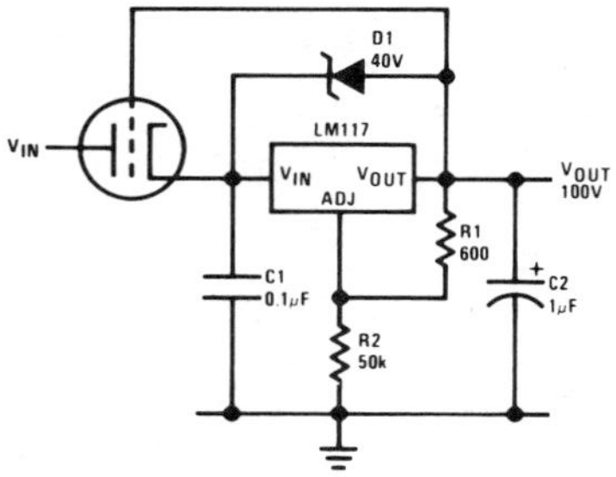

Adjusting Multiple On-Card Regulators with Single Control*

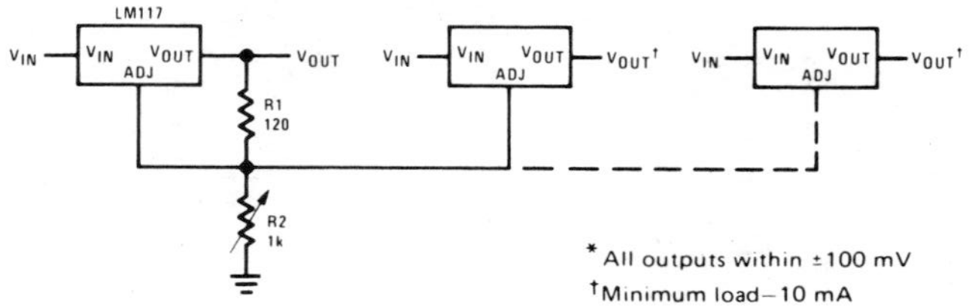

*All outputs within ±100 mV

†Minimum load—10 mA

typical applications (con't)

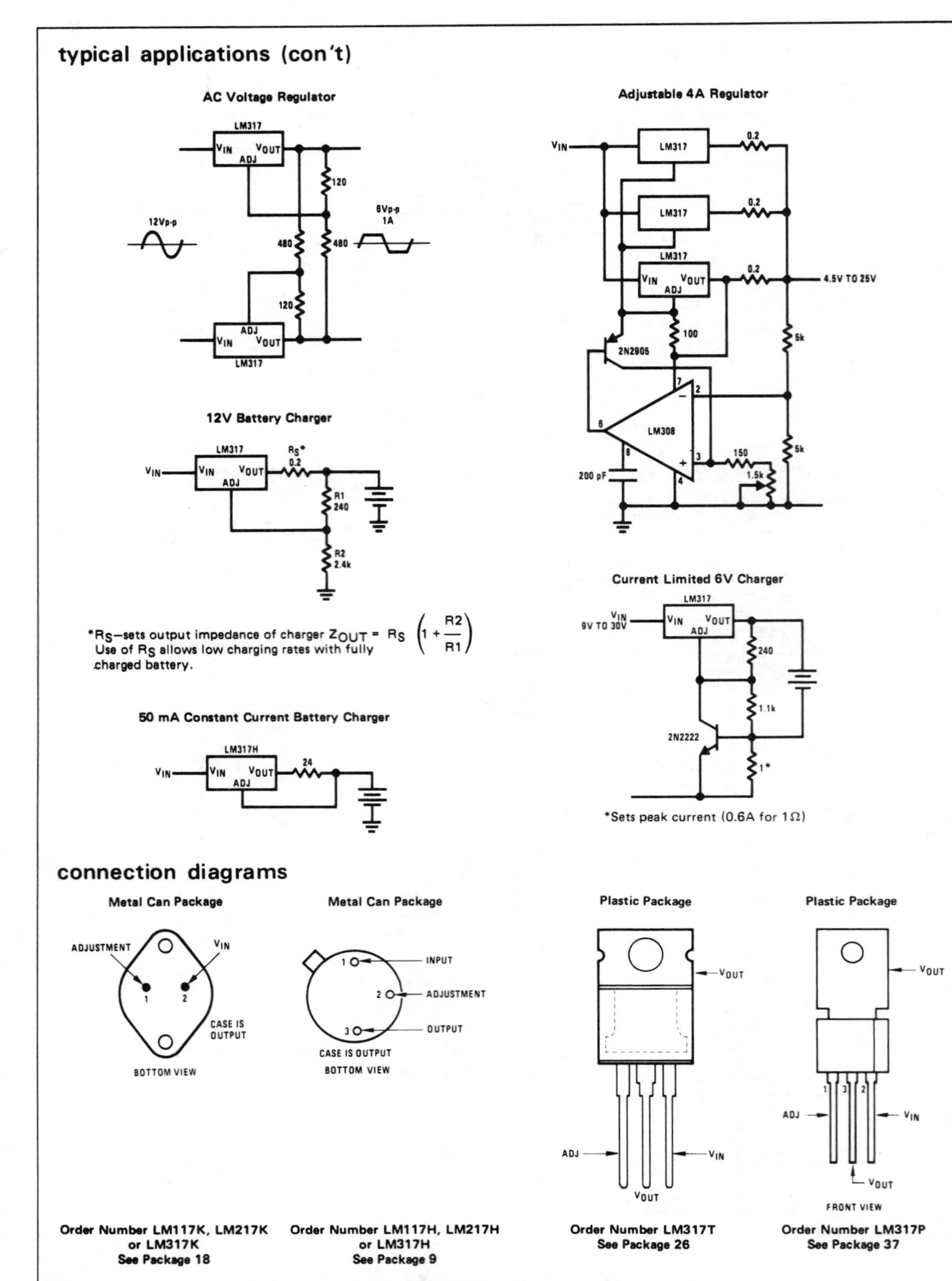

*R_S—sets output impedance of charger $Z_{OUT} = R_S \left(1 + \frac{R2}{R1}\right)$ Use of R_S allows low charging rates with fully charged battery.

*Sets peak current (0.6A for 1Ω)

connection diagrams

Order Number LM117K, LM217K or LM317K See Package 18

Order Number LM117H, LM217H or LM317H See Package 9

Order Number LM317T See Package 26

Order Number LM317P See Package 37

Manufactured under one or more of the following U.S. patents: 3083262, 3189758, 3231797, 3303356, 3317671, 3323071, 3381071, 3408542, 3421025, 3426423, 3440498, 3518750, 3519897, 3557431, 3560765, 3566218, 3571630, 3575609, 3579059, 3593069, 3597640, 3607469, 3617859, 3631312, 3633052, 3638131, 3648071, 3651565, 3693248.

OCTOBER 1978

96LS02
LOW POWER SCHOTTKY DUAL RETRIGGERABLE RESETTABLE MONOSTABLE MULTIVIBRATOR

DESCRIPTION — The 96LS02 is a Dual Retriggerable and Resettable Monostable which uses Low Power Schottky technology to provide wide delay range, stability, prediction accuracy and immunity to noise. The pulse width is set by an external resistor and capacitor. The 96LS02 may utilize timing resistors to 1 MΩ thus reducing required capacitor values. Hysteresis is provided on the inputs for increased noise immunity. The 96LS02 is fully compatible with all TTL families.

- **REQUIRED TIMING CAPACITANCE REDUCED BY FACTORS OF 10 TO 100 OVER CONVENTIONAL DESIGNS**
- **BROAD TIMING RESISTOR RANGE — 1 kΩ to 1 MΩ**
- **OUTPUT PULSE WIDTH IS VARIABLE OVER A 1300:1 RANGE BY RESISTOR CONTROL**
- **PROPAGATION DELAY OF 35 ns**
- **OUTPUT PULSE WIDTH STABILITY OF ±0.5% OVER 0°C to 70°C TEMPERATURE RANGE**
- **OUTPUT PULSE WIDTH STABILITY OF ±0.7% OVER 4.75 V to 5.25 V POWER SUPPLY RANGE**
- **PULSE WIDTH VARIATION OF ±5% FROM UNIT TO UNIT**
- **0.3 V HYSTERESIS ON BOTH TRIGGER INPUTS**
- **OUTPUT PULSE WIDTH INDEPENDENT OF DUTY CYCLE**
- **35 ns TO ∞ OUTPUT PULSE WIDTH RANGE**
- **RESETTABLE IN 20 ns**
- **SAME PINOUT AS 9602, 96L02, 96S02**

LOGIC SYMBOL

V_{CC} = Pin 16
GND = Pin 8

CONNECTION DIAGRAM DIP (TOP VIEW)

*Pins for external timing.

PIN NAMES

		LOADING (Note a) HIGH	LOW
$\overline{I}_0$	Schmitt Trigger Input (Active LOW)	0.5 U.L.	0.625 U.L.
I_1	Schmitt Trigger Input (Active HIGH)	0.5 U.L.	0.625 U.L.
$\overline{C}_D$	Clear Input (Active LOW)	0.5 U.L.	0.625 U.L.
Q	True Pulse Output (Note b)	10 U.L.	5/2.5 U.L.
$\overline{Q}$	Complementary Pulse Output (Note b)	10 U.L.	5/2.5 U.L.

NOTES:
a. 1 TTL Unit Load (U.L.) = 40 μA HIGH/1.6 mA LOW.
b. The Output LOW drive factor is 2.5 U.L. for Military (XM) and 5 U.L. for Commercial (XC) Temperature Ranges.

FUNCTIONAL DESCRIPTION — The 96LS02 Schottky Dual Retriggerable Resettable Monostable Multivibrator has two dc coupled trigger inputs per function, one active LOW ($\overline{I_0}$) and one active HIGH (I_1). Both inputs utilize an internal Schmitt trigger with hysteresis of 0.3 V to provide increased noise immunity. The use of active HIGH and LOW inputs allows either leading or trailing edge-triggering and optional non-retriggerable operation. The inputs are dc coupled making triggering independent of input transition times. When input conditions for triggering are met, the Q output goes HIGH and the external capacitor is rapidly discharged and then allowed to recharge. An input trigger which occurs during the timing cycle will retrigger the 96LS02 and result in Q remaining HIGH. The output pulse may be terminated (Q to the LOW state) at any time by setting the Direct Clear input LOW. Retriggering may be inhibited by tying the $\overline{Q}$ output to $\overline{I_0}$ or the Q output to I_1.

OPERATION RULES

TIMING

1. An external resistor (R_X) and external capacitor (C_X) are required as shown in the Logic Diagram. The value of R_X may vary from 1 kΩ to 1 MΩ.
2. The value of C_X may vary from 0 to any necessary value available. If, however, the capacitor has significant leakage relative to V_{CC}/R_X the timing equations may not represent the pulse width obtained.
3. Polarized capacitors may be used directly. The (+) terminal of a polarized capacitor is connected to pin 2 (14), the (−) terminal to pin 1 (15) and R_X. Pin 2 (14) will remain positive with respect to pin 1 (15) during the timing cycle.
4. The output pulse width t_W for $R_X \geq 10$ kΩ and $C_X \geq 100$ pF is determined as follows:

$$t_W = 0.43\, R_X C_X$$

Where R_X is in kΩ, C_X is in pF, t is in ns OR R_X is in kΩ, C_X is in μF, t is in ms

5. The output pulse width for $R_X < 10$ kΩ or $C_X < 1000$ pF should be determined from pulse width versus C_X or R_X graphs.
6. To obtain variable pulse width by remote trimming, the following circuit is recommended:

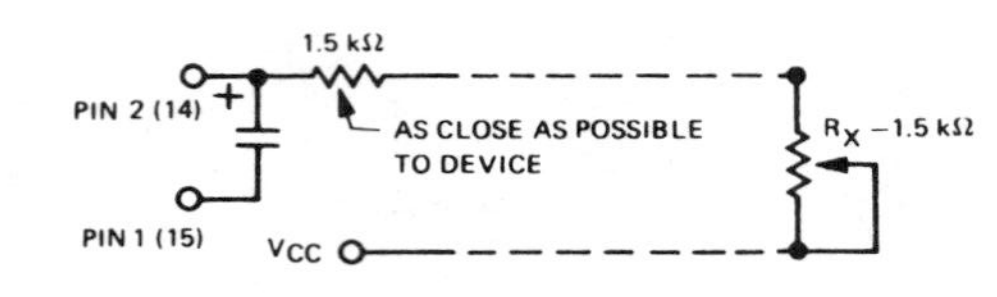

7. Under any operating condition, C_X and R_X (min) must be kept as close to the circuit as possible to minimize stray capacitance and reduce noise pickup.
8. **V_{CC} and ground wiring should conform to good high frequency standards so that switching transients on V_{CC} and ground leads do not cause interaction between one shots. Use of a 0.01 to 0.1 μF bypass capacitor between V_{CC} and ground located near the 96LS02 is recommended.**

TRIGGERING

1. The minimum negative pulse width into I_0 is 15 ns; the minimum positive pulse trigger input I_1 is 30 ns.
2. When non-retriggerable operation is required, i.e., when input triggers are to be ignored during quasi-stable state, input latching is used to inhibit retriggering.

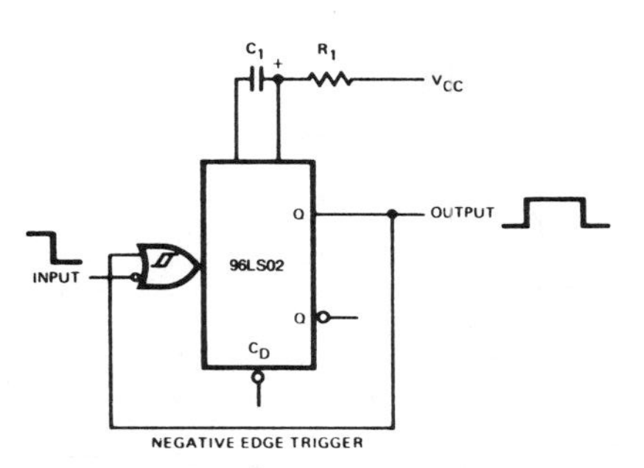

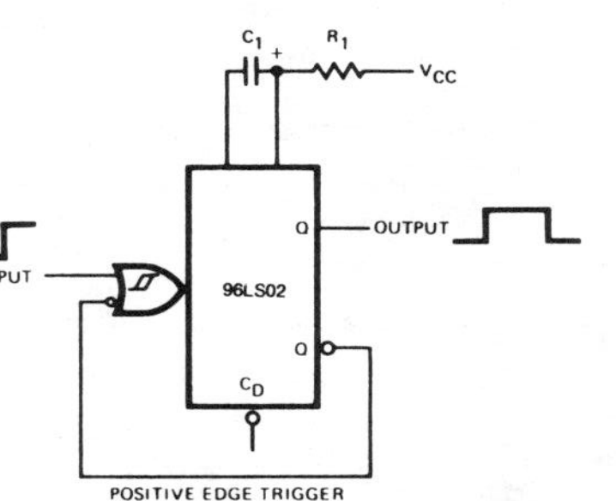

3. An overriding active LOW level direct clear is provided on each multivibrator. By applying a LOW to the clear, any timing cycle can be terminated or any new cycle inhibited until the LOW reset input is removed. Trigger inputs will not produce spikes in the output when the reset is held LOW. A LOW-to-HIGH transition on C_D will *not* trigger the 96LS02. If an input trigger occurs at the same time C_D goes HIGH, the circuit will respond to the trigger.

TRIGGERING TRUTH TABLE

PIN NO'S. 5(11)	4(12)	3(13)	OPERATION
H→L	L	H	Trigger
H	L→H	H	Trigger
X	X	L	Reset

H = HIGH Voltage Level $\geq V_{IH}$
L = LOW Voltage Level $\leq V_{IL}$
X = Immaterial (either H or L)
H→L = HIGH to LOW Voltage Level transition
L→H = LOW to HIGH Voltage Level transition

ABSOLUTE MAXIMUM RATINGS (above which the useful life may be impaired)

Storage Temperature	−65°C to +150°C
Temperature (Ambient) Under Bias	−55°C to +125°C
V_{CC} Pin Potential to Ground Pin	−0.5 V to +7.0 V
*Input Voltage (dc)	−0.5 V to +15 V
*Input Current (dc)	−30 mA to +5.0 mA
Voltage Applied to Outputs (Output HIGH)	−0.5 V to $+V_{CC}$ value
Output Current (dc) (Output LOW)	+16 mA

*Either Input Voltage Limit or Input Current is sufficient to protect the inputs.

RECOMMENDED OPERATING CONDITIONS

PARAMETER	96LS02 MIN	TYP	MAX	UNITS
Supply Voltage V_{CC}	4.75	5.0	5.25	V
Operating Ambient Temperature Range	0	25	75	C

DC CHARACTERISTICS OVER OPERATING TEMPERATURE RANGE (unless otherwise noted)

SYMBOL	PARAMETER		LIMITS MIN	TYP	MAX	UNITS	TEST CONDITIONS (Note 1)
V_{IH}	Input HIGH Voltage Except Pins 4 & 12		2.0			V	Guaranteed Input HIGH Voltage
V_{IL}	Input LOW Voltage Except Pins 4 & 12	XC			0.8	V	Guaranteed Input LOW Voltage
		XM			0.7		
V_{CD}	Input Clamp Diode Voltage				-1.5	V	V_{CC} = MIN, I_{IN} = -18 mA
V_{T+}	Positive-Going Threshold Voltage, Pins 4 & 12				2.0	V	V_{CC} = 5.0 V, T_A = 25° C
V_{T-}	Negative-Going Threshold Voltage, Pins 4 & 12		0.8			V	V_{CC} = 5.0 V, T_A = 25° C
V_{OH}	Output HIGH Voltage	XC	2.7			V	V_{CC} = MIN, I_{OH} = −400 μA, V_{IN} = 0.8 V
		XM	2.5				
V_{OL}	Output LOW Voltage	XC			0.5	V	V_{CC} = MIN, V_{IN} = V_{IH} or V_{IL}; I_{OL} = 8 mA
		XM			0.4		I_{OL} = 4 mA
I_{IH}	Input HIGH Current				20	μA	V_{CC} = MAX, V_{IN} = 2.7 V
					0.1	mA	V_{CC} = MAX, V_{IN} = 10 V
I_{IL}	Input LOW Current				-0.4	mA	V_{CC} = MAX, V_{IN} = 0.4 V
I_{OS}	Output Short Circuit Current (Note 3)		-20		-100	mA	V_{CC} = MAX, V_{OUT} = 0 V
I_{CC}	Quiescent Power Supply Drain				35	mA	Inputs Open, V_{CC} = Max

NOTES:
1. For conditions shown as MIN or MAX, use the appropriate value specified under recommended operating conditions for the applicable device type.
2. Typical limits are at V_{CC} = 5.0 V, T_A = 25°C.
3. Not more than one output should be shorted at a time.

AC CHARACTERISTICS: T_A = 25° C, V_{CC} = 5.0 V, C_L = 15 pF (unless other noted)

SYMBOL	PARAMETER		LIMITS MIN	TYP	MAX	UNITS	CONDITIONS
t_{PLH}	Negative Trigger Input to True Output				55	ns	
t_{PHL}	Negative Trigger Input to Complement Output				50	ns	R_X = 10 kΩ
t_{PLH}	Positive Trigger Input to True Output				60	ns	C_X = 1000 pF
t_{PHL}	Positive Trigger Input to Complement Output				55	ns	
t_{PHL}	Clear Input to True Output				30	ns	V_{CC} = 5 V
t_{PLH}	Clear Input to Complement Output				35	ns	
$t_{W(MIN)}$	Min. Negative Trigger Pulse Width on $\overline{I_0}$				15	ns	C_L = 15 pF
$t_{W(MIN)}$	Min. Positive Trigger Pulse Width on I_1				30	ns	
$t_{W(MIN)}$	Min. Clear Pulse Width				22	ns	
$t_{W(MIN)}$	Min. True Output Pulse Width		25		55	ns	R_X = 1 kΩ, C_X = stray capacity only
t_W	True Output Pulse Width		4.1		4.5	μs	V_{CC} = 5.0 V, R_X = 10 kΩ, C_X = 1000 pF
R_X	Timing Resistor Range		1		1000	kΩ	T_A = 0° C to 70° C, V_{CC} = 4.75 V to 5.25 V
Δt	Max. Change in True Output Pulse Width over Temperature Range	XC			1.0	%	V_{CC} = 5.0 V, R_X = 10 kΩ, C_X = 1000 pF
		XM			3.0		
Δt	Max. Change in True Output Pulse Width over V_{CC} Range	XC			1.5	%	V_{CC} = 4.75 V to 5.25 V; T_A = 25° C, R_X = 10 kΩ, C_X = 1000 pF
		XM			0.8	%	V_{CC} = 4.5 V to 5.5 V

TYPICAL CHARACTERISTICS

OUTPUT t_w vs R_X AND C_X

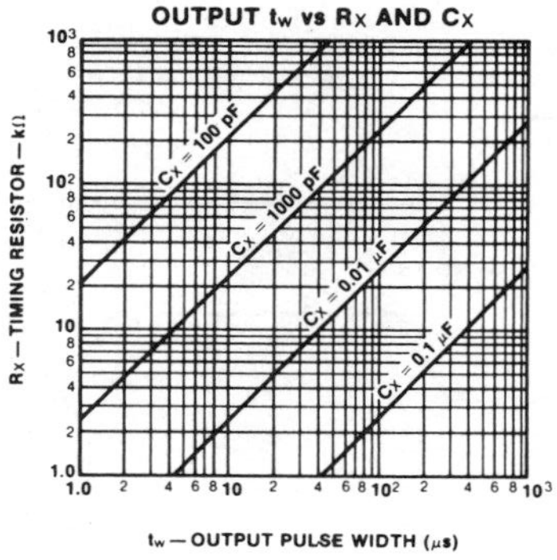

C_X = 100 pF
C_X = 1000 pF
C_X = 0.01 μF
C_X = 0.1 μF

R_X — TIMING RESISTOR — kΩ

t_w — OUTPUT PULSE WIDTH (μs)

I_1 DELAY TIME vs T_A

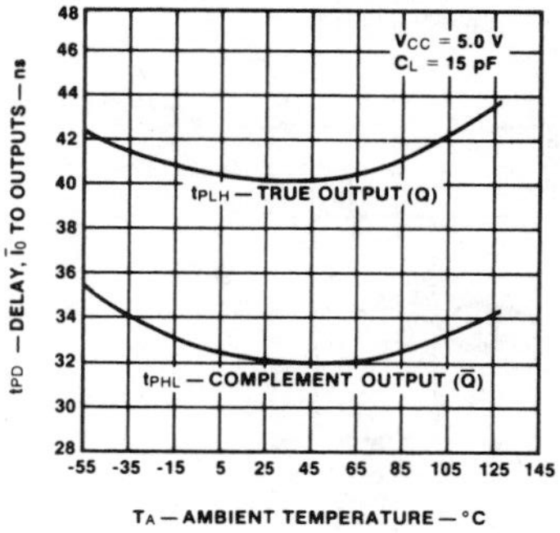

V_{CC} = 5.0 V
C_L = 15 pF

t_{PLH} — TRUE OUTPUT (Q)
t_{PHL} — COMPLEMENT OUTPUT ($\bar{Q}$)

t_{PD} — DELAY, $\bar{I}_0$ TO OUTPUTS — ns

T_A — AMBIENT TEMPERATURE — °C

I_0 DELAY TIME vs T_A

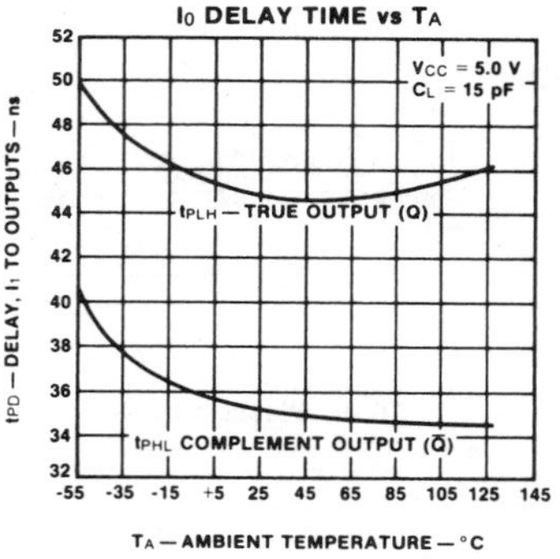

V_{CC} = 5.0 V
C_L = 15 pF

t_{PLH} — TRUE OUTPUT (Q)
t_{PHL} COMPLEMENT OUTPUT ($\bar{Q}$)

t_{PD} — DELAY, I_1 TO OUTPUTS — ns

T_A — AMBIENT TEMPERATURE — °C

PULSE WIDTH VS R_X AND C_X

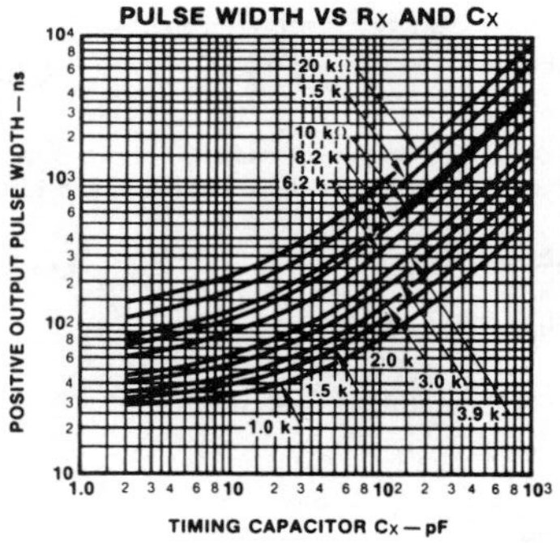

20 kΩ
1.5 k
10 kΩ
8.2 k
6.2 k
2.0 k
3.0 k
1.5 k
3.9 k
1.0 k

POSITIVE OUTPUT PULSE WIDTH — ns

TIMING CAPACITOR C_X — pF

OUTPUT t_w vs T_A

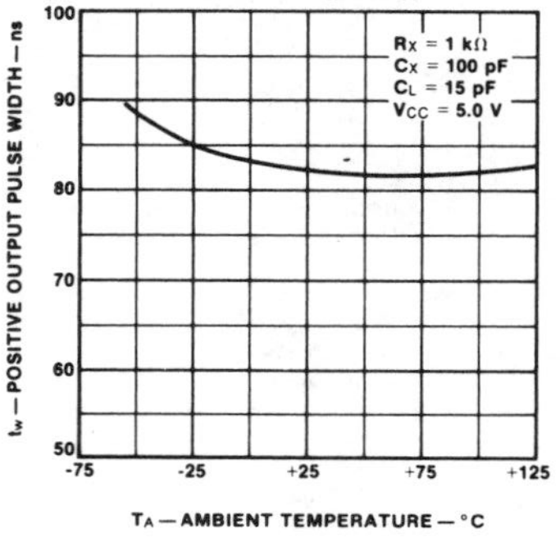

R_X = 1 kΩ
C_X = 100 pF
C_L = 15 pF
V_{CC} = 5.0 V

t_w — POSITIVE OUTPUT PULSE WIDTH — ns

T_A — AMBIENT TEMPERATURE — °C

NORMALIZED Δt_w vs T_A

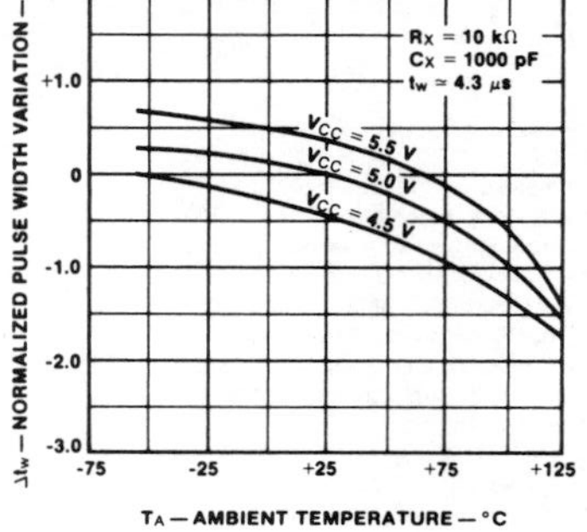

R_X = 10 kΩ
C_X = 1000 pF
t_w = 4.3 μs

V_{CC} = 5.5 V
V_{CC} = 5.0 V
V_{CC} = 4.5 V

Δt_w — NORMALIZED PULSE WIDTH VARIATION — ns

T_A — AMBIENT TEMPERATURE — °C

AC CIRCUITS AND WAVEFORMS

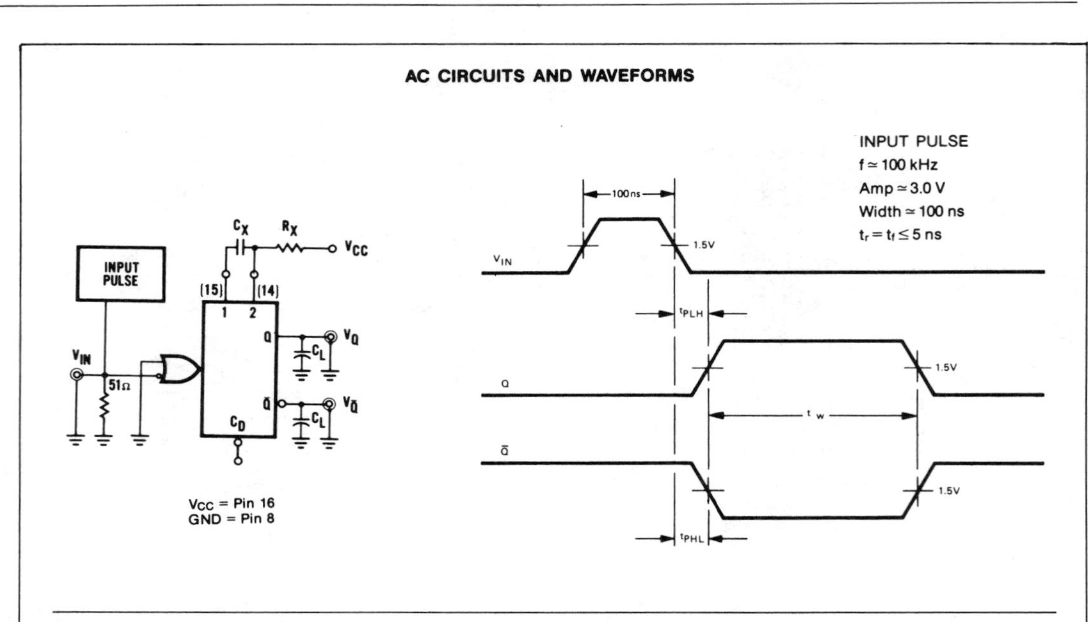

ORDER INFORMATION

Specify 96LS02DC where "D" is for Ceramic Dual In-Line package and "C" is for commercial 0 to 75° C temperature range or 96LS02PC where "P" is for Plastic Dual In-Line package and "C" is for commercial temperature range.

PACKAGE OUTLINE

16-Pin Ceramic Dual In-Line

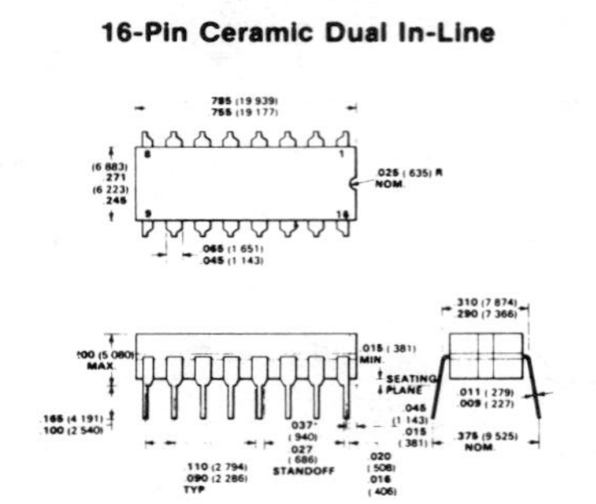

NOTES:
All dimensions in inches (bold) and millimeters (parentheses)
Pins are intended for insertion in hole rows on .300" (7.620) centers
They are purposely shipped with "positive" misalignment to facilitate insertion
Board-drilling dimensions should equal your practice for .020 inch (0.508) diameter pin
Pins are alloy 42
Package weight is 2.0 grams
*The .037/.027 (.940/.686) dimensions does not apply to the corner pins

16-Pin Plastic Dual In-Line

NOTES:
All dimensions in inches (bold) and millimeters (parentheses)
Pins are alloy 42
Pins are intended for insertion in hole rows on .300" (7.620) centers
They are purposely shipped with "positive" misalignment to facilitate insertion
Board-drilling dimensions should equal your practice for .020 inch (0.508) diameter pin
***The .037/.027 (.940/.686) dimension does not apply to the corner pins
Package weight is 0.9 gram

Fairchild cannot assume responsibility for use of any circuitry described other than circuitry embodied in a Fairchild product. No other circuit patent licences are implied.
Manufactured under one of the following U.S. Patents: 2981877, 3015048, 3064167, 3108359, 3117260; other patents pending.
Fairchild reserves the right to make changes in the circuitry or specifications at any time without notice.

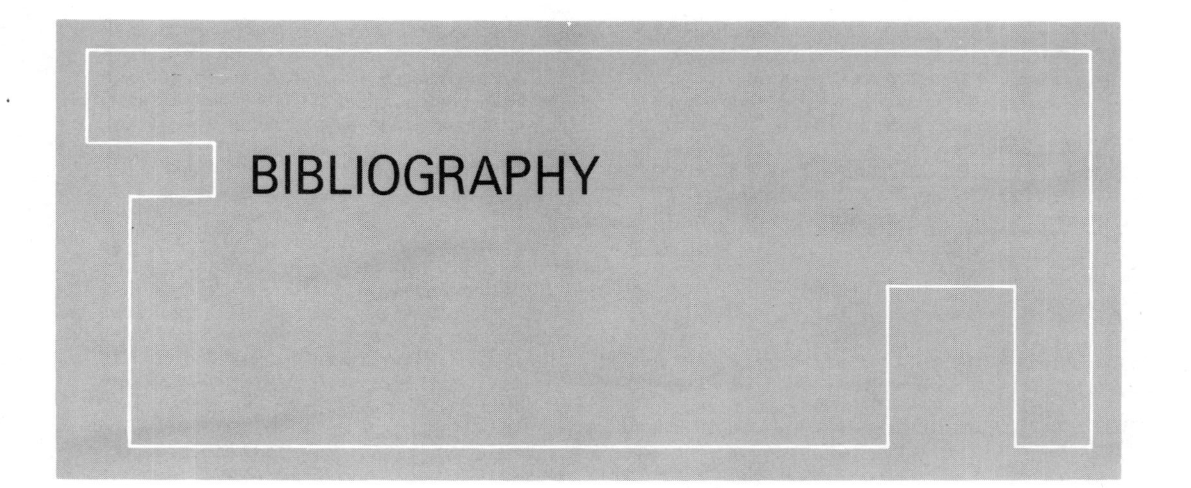

BIBLIOGRAPHY

GENERAL

Handbooks

Fink, D. G., ed. 1975. *Electronic engineers' handbook.* New York: McGraw-Hill. Encyclopedic.

Fink, D. G., and Beaty, H. W., eds. 1978. *Standard handbook for electrical engineers.* New York: McGraw-Hill. Tutorial articles on electrical engineering topics.

Giacoletto, L. J., ed. 1977. *Electronics designers' handbook.* 2nd ed. New York: McGraw-Hill. Excellent tutorials and data.

Reference data for radio engineers. 1968. Indianapolis: Howard W. Sams & Co. General-purpose engineering data.

Master catalogs

EEM: Electronic engineers master catalog. Garden City, N.Y.: United Technical Publications. Thousands of pages of manufacturers' data sheets, plus addresses of companies, their representatives, and local distributors. Extremely useful. Published annually.

Electronic Design's GOLD BOOK. Rochelle Park, N.J.: Hayden Publishing. Similar to the EEM. Published annually.

IC master. Garden City, N.Y.: United Technical Publications. Comprehensive selection guides and thousands of pages of data sheets. Extremely useful. Published annually.

Books

Clayton, G. B. 1975. *Linear IC applications handbook.* Blue Ridge Summit, Pa.: Tab Books.

Eimbinder, J., ed. 1969. *Designing with linear integrated circuits.* New York: Wiley. Good reading for the circuit design enthusiast.

Eimbinder, J., ed. 1970. *Application considerations for linear integrated circuits.* New York: Wiley. More good reading for the circuit design enthusiast.

Hnatek, E. R. 1975. *Applications of linear integrated circuits.* New York: Wiley. Highly readable.

Millman, J. 1979. *Microelectronics—digital and analog circuits and systems.* New York: McGraw-Hill. Highly recommended all-around text and reference.

Senturia, S. D., and Wedlock, B. D. 1975. *Electronic circuits and applications.* New York: Wiley. Good introductory engineering textbook.

Smith, R. J. 1976. *Circuits, devices, and systems.* New York: Wiley. Broad introductory engineering textbook.

CHAPTER 1

Holbrook, J. G. 1966. *Laplace transforms for electronic engineers.* New York: Pergamon Press. Good for learning about the *s*-plane.

Kuo, B. C. 1967. *Linear networks and systems.* New York: McGraw-Hill. Classic network theory.

Purcell, E. M. 1965. *Electricity and magnetism (Berkeley physics course, vol. 2).* New York: McGraw-Hill. Excellent textbook on electromagnetic theory. Relevant sections on electrical conduction and analysis of ac circuits with complex numbers.

Skilling, H. H. 1967. *Electrical engineering circuits.* New York: Wiley. Traditional engineering treatment of the material in Chapter 1.

CHAPTER 2

Camenzind, H. R. 1968. *Circuit design for integrated electronics.* Reading, Mass.: Addison-Wesley. Physics of transistors and integrated circuits; very readable.

Ebers, J. J., and Moll, J. L. 1954. Large-signal behavior of junction transistors. *Proc. I.R.E.* 42:1761–1772. The Ebers-Moll equation is born.

Grove, A. S. 1967. *Physics and technology of semiconductor devices.* New York: Wiley. Principles of fabrication and operation of bipolar and field-effect transistors.

Schilling, D. S., and Belove, C. 1968. *Electronic circuits: discrete and integrated.* New York: McGraw-Hill. Traditional *h*-parameter transistor treatment.

Searle, C. L., Boothroyd, A. R., Angelo, E. J., Jr., Gray, P. E., and Pederson, D. O. 1966. *Elementary circuit properties of transistors (Semiconductor electronics education committee, vol. 3).* New York: Wiley. Physics of transistors.

"Discrete products data books." Soft-cover collections of transistor data sheets are published sporadically under this title by all the transistor manufacturers, in particular Fairchild, GE, Motorola, National, RCA, and TI. Data sheets are essential for circuit design.

CHAPTER 3

Graeme, J. G. 1973. *Applications of operational amplifiers: third generation techniques.* New York: McGraw-Hill. One of the Burr-Brown series.

Graeme, J. G. 1977. *Designing with operational amplifiers.* New York: McGraw-Hill. Applications alternatives.

Jung, W. G. 1975. *IC op-amp cookbook.* Indianapolis: Howard W. Sams & Co. Lots of circuits, with explanations.

Stout, D. F., and Kaufman, M. 1976. *Handbook of Operational Amplifier Circuit Design.* New York: McGraw-Hill. Explicit design procedures.

See also Hnatek, E. R., in the general bibliography.

"Linear databooks." Soft-cover collections of linear IC data sheets and application notes are published approximately every two years under this title by all the linear IC manufacturers, in particular Analog Devices, Fairchild, Motorola, National, RCA, Signetics, and TI. Data sheets are essential for circuit design.

CHAPTER 4

Bruton, L. T. 1980. *RC active circuits: theory and design.* Englewood Cliffs, N.J.: Prentice-Hall.

Clarke, K. K., and Hess, D. T. 1971. *Communications circuits: analysis and design.* Reading, Mass.: Addison-Wesley. Good chapter on oscillators.

Daniels, R. W. 1974. *Approximation methods for electronic filter design.* New York: McGraw-Hill.

Hilburn, J. L., and Johnson, D. E. 1973. *Manual of active filter design.* New York: McGraw-Hill.

Hilburn, J. L., and Johnson, D. E. 1975. *Rapid practical design of active filters.* New York: Wiley. Tables and design procedures.

Johnson, D., Johnson, J., and Moore, H. 1980. *A handbook of active filters.* Englewood Cliffs, N.J.: Prentice-Hall. Up to date.

Lancaster, D. 1979. *Active filter cookbook.* Indianapolis: Howard W. Sams & Co. Explicit design procedure; easy to read.

Sentz, R. E., and Bartkowiak, R. A. 1968. *Feedback amplifiers and oscillators.* New York: Holt, Rinehart, and Winston. Oscillator theory.

Zverev, A. I. 1967. *Handbook of filter synthesis.* New York: Wiley. Extensive tables for passive *LC* and crystal filter design.

See also Graeme, J. G., 1973, for Chapter 3.

CHAPTER 5

Pressman, A. I. 1977. *Switching and linear power supply, power converter design.* Rochelle Park, N.J.: Hayden Book Co.

"Voltage regulator handbook." Soft-cover collections of voltage regulator data sheets and application notes are published sporadically under this title by Fairchild, National, and TI. The "Linear databooks" referenced for Chapter 3 also contain regulator data sheets, which are essential for circuit design.

CHAPTER 6

Penney, W. M., and Lau, L., eds. 1972. *MOS integrated circuits.* New York: Van Nostrand Reinhold. Recommended.

Richman, P. 1973. *MOS field-effect transistors and integrated circuits.* New York: Wiley. Solid-state physics of MOSFETs.

See also Grove, A. S., for Chapter 2.

"FET data book." Soft-cover collections of FET data

sheets and application notes are published every few years under this or similar titles by all the FET manufacturers, in particular General Instruments, Intersil, National, and Siliconix. Data sheets are essential for design.

CHAPTER 7

Morrison, M. 1967. *Grounding and shielding techniques in instrumentation.* New York: Wiley.

Motchenbacher, C. D., and Fitchen, F. C. 1973. *Low-noise electronic design.* New York: Wiley. Recommended for low-noise amplifier design.

Ott, H. W. 1976. *Noise reduction techniques in electronic systems.* New York: Wiley. Shielding and low-noise design.

Sheingold, D. H., ed. 1974. *Nonlinear Circuits Handbook.* Norwood, Mass.: Analog Devices. Highly recommended.

Wong, Y. J., and Ott, W. E. 1976. *Function circuits: design and applications.* New York: McGraw-Hill. Nonlinear circuits and op-amp exotica.

"Data acquisition handbook" or "Linear data book." Soft-cover collections of data sheets and application notes relevant to precision design are published every few years under this or similar titles by many semiconductor manufacturers, in particular Analog Devices, Burr-Brown, Exar, Philbrick, and Precision Monolithics.

CHAPTER 8

Blakeslee, T. R. 1975. *Digital design with standard MSI and LSI.* New York: Wiley. Refreshing approach to practical logic design; includes two chapters of "nasty realities."

Hill, F. J., and Peterson, G. R. 1968. *Introduction to switching theory and logical design.* New York: Wiley. Classic logic design textbook.

Lancaster, D. 1979. *TTL cookbook.* Indianapolis: Howard W. Sams & Co. Practical circuits, good reading.

Lancaster, D. 1977. *CMOS cookbook.* Indianapolis: Howard W. Sams & Co. Good reading, down-to-earth applications. Includes widely used (but rarely mentioned) M^2L (Mickey Mouse logic) technique.

Wickes, W. E. 1968. *Logic design with integrated circuits.* New York: Wiley.

"TTL data book," "CMOS data book." Soft-cover collections of data sheets and application notes are published approximately every two years under these titles by most semiconductor manufacturers, in particular Advanced Micro Devices (AMD), Fairchild, Motorola, National, RCA, Signetics, and TI. These data books are essential.

CHAPTER 9

Gardner, F. M. 1979. *Phaselock techniques.* New York: Wiley. New edition of the classic and indispensable PLL book.

Hnatek, E. R. 1976. *A user's handbook of D/A and A/D converters.* New York: Wiley. Applications.

Jung, W. G. 1978. *IC converter cookbook.* Indianapolis: Howard W. Sams & Co. Using modern converter ICs.

Schmid, H. 1970. *Electronic analog/digital conversions.* New York: Van Nostrand Reinhold. General-purpose conversion techniques.

Sheingold, D. H., ed. 1972. *Analog-digital conversion handbook.* Norwood, Mass.: Analog Devices. A/D and D/A converter fundamentals; emphasis on limits of performance.

Stearns, S. D. 1975. *Digital signal analysis.* Rochelle Park, N.J.: Hayden Book Co. Signal processing and digital filtering.

"Conversion products data book," "Data acquisition data book." Soft-cover collections of data sheets and application notes are published annually under these titles by the semiconductor manufacturers, in particular Analog Devices, Burr-Brown, Datel, Hybrid Systems, and Philbrick.

"Interface data book." Soft-cover collections of data sheets and application notes are published every few years under this title by the semiconductor manufacturers, in particular Fairchild, Motorola, National, and TI. Both Sprague and TI also offer data books of their peripheral driver circuits.

CHAPTER 10

Sloan, M. E. 1980. *Introduction to minicomputers and microcomputers.* Reading, Mass.: Addison-Wesley. Emphasis on computing; software-oriented.

"PDP-11 processor handbook." Booklets under this general title are published by Digital Equipment Corp., Maynard, Mass.

"Programmer's reference manual." Several booklets under this general title are published by Data General Corp., Southboro, Mass.

CHAPTER 11

Barden, W., Jr. 1978. *The Z-80 microcomputer handbook.* Indianapolis: Howard W. Sams & Co. All about the Z80: hardware and software.

Burton, D. P., and Dexter, A. L. 1977. *Microprocessor systems handbook.* Norwood, Mass.: Analog Devices. Unusual applications.

Osborne, A. 1979. *An introduction to microcomputers, vol. 1: Basic concepts.* Berkeley, Calif.: Adam Osborne & Associates.

Discussions of most major microprocessors. See also volume 2 *(Some real microprocessors)* and volume 3 *(Some real support devices).*

Peatman, J. P. 1977. *Microcomputer-based design.* New York: McGraw-Hill. Broad view of applying microprocessors.

Sargent, M., III, and Shoemaker, R. L. 1980. *Interfacing small computers to the real world.* Reading, Mass.: Addison-Wesley. Interfacing microcomputers to laboratory equipment and other instrumentation; includes problems and hands-on laboratory exercises.

Slater, M., and Bronson, B. 1979. *Practical microprocessors.* Santa Clara, Calif.: Hewlett-Packard Corp. Manual to accompany the 5036A microprocessor lab.

Souček, B. 1976. *Microprocessors and microcomputers.* New York: Wiley. Emphasis on microcomputers and programming.

Also, manuals on the 8085 (Intel "MCS-85 user's manual"), 8086 (Intel "MCS-86 user's manual"), 6809 (Motorola), and Z8000 (Zilog) make good reading, when you get down to the nitty-gritty.

CHAPTER 12

Coombs, C. F., Jr., ed. 1979. *Printed circuits handbook.* New York: McGraw-Hill. Design, fabrication, and application of PC boards.

"Technical manual and catalog." Westlake Village, Calif.: Bishop Graphics. Frequently revised product catalog and information for PC layout.

CHAPTER 13

Carlson, R. S. 1975. *High-frequency amplifiers.* New York: Wiley. RF transistor amplifiers.

Klapper, J., and Frankle, J. T. 1972. *Phase-locked and frequency-feedback systems.* New York: Academic Press. Phase-locked loops in radiofrequency communications.

Panter, P. F. 1965. *Modulation, noise, and spectral analysis.* New York: McGraw-Hill. Comprehensive coverage of modulation and detection.

Skolnik, M. I., ed. 1979. *Radar handbook.* New York: McGraw-Hill. Incredible compendium of radar information.

Viterbi, A. J. 1966. *Principles of coherent communication.* New York: McGraw-Hill. A classic; modulation theory.

"The radio amateur's handbook." Newington, Conn.: American Radio Relay League. Published annually, this is the standard handbook for radio amateurs.

"RF transistor data book." Soft-cover collections of data sheets and application notes are published sporadically under this and similar titles by the transistor manufacturers, in particular Motorola, RCA, Siliconix (VMOS), and TRW.

CHAPTER 14

Chappell, A., ed. 1978. *Optoelectronics: theory and practice.* New York: McGraw-Hill. Optical engineering.

Luppold, D. S. 1969. *Precision dc measurements and standards.* Reading, Mass.: Addison-Wesley.

"Electro-optics handbook." Harrison, N.J.: RCA Corporation (revised sporadically). Detectors, lasers, image tubes.

"Temperature measurement handbook." Stanford, Conn.: Omega Engineering Corp. (revised often). Thermocouples, thermistors, resistance thermometers.

Hewlett-Packard application notes: AP52-2 ("Timekeeping and frequency calibration"), AP150 ("Spectrum analyzer basics"), AP200 ("Fundamentals of electronic counters"), and AP200-2 ("Fundamentals of quartz oscillators"). They are available without charge from the Hewlett-Packard Corp., Palo Alto, Calif.

See also the annual product catalogs from the Hewlett-Packard Corp., John Fluke Co., and Princeton Applied Research.

INDEX

Page numbers in italics refer to figures or tables.